Selected Papers on

Optical Techniques for Industrial Inspection

Books in SPIE's Milestone Series
Brian J. Thompson, Series Editor

Acousto-Optics (A. Korpel)
Adaptive Optics and Speckle Imaging (D. Crowe)
Adaptive Optics for Atmospheric Compensation (J.E. Pearson)
Apodization—Coherent Optical Systems (J.P. Mills, B.J. Thompson)
Applications of Polarized Light (B.H. Billings)
Architectural Lighting (M.S. Rea)
Astronomical Optics (D.J. Schroeder)
Atmospheric Optics (V.E. Zuev)
Automatic Object Recognition (H. Nasr)
CO_2 Lasers (J.D. Evans)
Characterization of Optical Coatings (M.R. Jacobson)
Coherence and Fluctuations of Light (E. Wolf, L. Mandel)
Coherence and Radiometry (A.T. Friberg)
Coherent Optical Processing (F.T.S. Yu, S. Yin)
Colorimetry—Fundamentals (D.L. MacAdam)
Computer-Controlled Optical Surfacing (R.A. Jones)
Computer-Generated Holograms and Diffractive Optics (S.H. Lee)
Confocal Microscopy (B.R. Masters)
Coupled-Mode Theory in Guided-Wave Optics (D. Hall)
Cryogenic Optical Systems (G.R. Pruitt)
Deposition of Optical Coatings (M.R. Jacobson)
Design of Optical Coatings (M.R. Jacobson)
Diffraction Gratings (D. Maystre)
Digital Image Processing (M.M. Trivedi)
Digital Image Restoration (M.I. Sezan)
Dye Lasers (F.J. Duarte)
Effects of Aberrations in Optical Imaging (V.N. Mahajan)
Electron Optics (P.W. Hawkes)
Electronic Speckle Pattern Interferometry: Principles and Practice (P. Meinlschmidt, K.D. Hinsch, R.S. Sirohi)
Ellipsometry (R.M.A. Azzam)
Fiber Optic Communications (L.D. Hutcheson, S.C. Mettler)
Fiber Optic Local Area Networks (E. Rawson)
Fiber Optic Gyroscopes (R.B. Smith)
Fiber Optic Sensors (R. Willsch, R.Th. Kersten)
Fourier Optics (M. Abushagur, H.J. Caulfield)
Free-Space Laser Communications (D.L. Begley)
Free-Space Laser Communications II (D.L. Begley)
Fundamentals of Lasers (W.T. Silfvast)
Fundamentals of Optoelectronics (G.R. Little)
Fundamentals of Quantum Optics (G.S. Agarwal)
Gaussian Beam Mode Optics for MM-Wave & Terahertz Systems (J.C.G. Lesurf)
Geometrical Aspects of Scattering (P.L. Marston)
Gradient Index Optics (D. Moore)
High-Power Lasers (J.M. Soures)
Holographic and Diffractive Lenses and Mirrors (T.W. Stone, B.J. Thompson)
Holographic Particle Diagnostics (C.S. Vikram)
Holographic Recording Materials (H.I. Bjelkhagen)
Holographic Storage (G.T. Sincerbox)
Industrial Machine Vision Systems (B.G. Batchelor, P.F. Whelan)
Interference, Interferometry, and Interferometric Metrology (P. Hariharan, D. Malacara)
Image Coding and Compression (M. Rabbani)
Image Tubes (I.P. Csorba)
Infrared Design (R.B. Johnson, W.L. Wolfe)
Infrared Fiber Optics (J.A. Harrington)
Instrumentation in Astronomy (W. Livingston)
Interferometry (P. Hariharan)
Laser Beam Diagnostics (R.N. Hindy, J.H. Hunt)
Laser Damage in Optical Materials (R.M. Wood)
Laser Design (H. Weichel)
Laser Distance Measurements (T. Bosch, M. Lescure)
Laser Doppler Velocimetry (R. Adrian)
Laser Isotope Separation (J.W. Eerkens)
Laser Radar (G.W. Kamerman)
Laser Safety (D.H. Sliney)
Laser Scanning and Recording (L. Beiser)
Light Scattering (M. Kerker)
Linear Optical Composite Materials (A. Lakhtakia)
Liquid Crystals for Optics (S.D. Jacobs)
Microdensitometry (R.E. Swing)
Model-Based Vision (H. Nasr)
Morphological Image Processing (T. Szoplik)
Multiple Scattering in Plane Parallel Atmospheres and Oceans: Methods (G.W. Kattawar)
Natural and Artificial Compound Eye Sensors (J. Sanders)
Natural Optical Activity (A. Lakhtakia)
Nonimaging Optics (R. Winston)
Nonlinear Optics (H.E. Brandt)
Optical Chaos (F.T. Arecchi, R.G. Harrison)
Optical Computing (H.J. Caulfield, G. Gheen)
Optical Correlators (S. Jutamulia)
Optical Fiber Technology (D.B. Keck)
Optical Fibers in Medicine (A. Katzir)
Optical Methods in Surface Metrology (D. Whitehouse)
Optical Microlithography (H. Stover)
Optical Moiré and Applications (G. Indebetouw, R. Czarnek)
Optical Neural Networks (S. Jutamulia)
Optical Remote Sensing Theory and Measurements (J.A. Smith)
Optical Shop Metrology (D. Malacara)
Optical Storage (G.T. Sincerbox, J.M. Zavislan)
Optical Techniques for Industrial Inspection (P. Cielo)
Optical Tolerancing (G.E. Wiese)
Optical Transfer Function: Foundation and Theory (L. Baker)
Optical Transfer Function: Measurement (L. Baker)
Optomechanical Design (D.C. O'Shea)
Optometry (W.M. Rosenblum, W.J. Benjamin)
Performance Evaluation of Signal and Image Processing Systems (F.A. Sadjadi)
Particle Image Velocimetry (I. Grant)
Photochemistry (D. Neckers)
Photoconductivity (N.V. Joshi)
Photodymanic Therapy (D. Kessel)
Photon Statistics and Coherence in Nonlinear Optics (J. Peřina)
Photonic Switching (A. Marrakchi)
Photopolymers: Physics, Chemistry, and Applications (R.A. Lessard, G. Manivannan)
Photorefractive Materials (F.M. Davidson)
Polarization (B.H. Billings)
Precision Stabilization and Tracking Systems for Acquisition, Pointing, and Control Applications (M. Masten, L. Stockum)
Prisms and Mirrors—Design and Fabrication (R. Hradaynath)
Quasielastic Light Scattering (B. Chu)
Radiometry (I.J. Spiro)
Rare-Earth-Doped Fiber Laser Sources and Amplifiers (M.J.F. Digonnet)
Resonant and Collective Phenomena in Quantum Optics (G.S. Agarwal)
Scalar Wave Diffraction (K.E. Oughstun)
Scanning Acoustic Microscopy (B.T. Khuri-Yakub, C.F. Quate)
Scanning Probe Microscopes (Y. Martin)
Scattering in the Atmosphere (C.F. Bohren)
Schlieren Optics (J.R. Meyer-Arendt)
Scientific and Engineering High Speed Photography (D.L. Paisley)
Semiconductor Diode Lasers (J.J. Coleman)
Semiconductor Infrared Detectors (A. Rogalski)
Sensor and Data Fusion (F. Sadjadi)
Silica Integrated Optical Circuits (H.M. Presby)
Single-Mode Optical Fibers (A. Brozeit, K.D. Hinsch, R.S. Sirohi)
Solar Radiation and Solar Thermal Systems (D. Osborn)
Solid State Lasers (R.C. Powell)
Speckle Metrology (R.S. Sirohi)
Surface-enhanced Raman Scattering (M. Kerker)
Temperature Sensing: Optical Methods (R.D. Lucier)
Tissue Optics (V.V. Tuchin)
Turbulence in a Refractive Medium (E.L. Andreas)
Ultrafast Laser Technology (T.R. Gosnell, A.J. Taylor)
Ultraviolet Optics and Technology (R.E. Huffman)
Underwater Optics (F.M. Caimi)
UV, VUV, and X-Ray Lasers (R.W. Waynant, M.N. Ediger)
Visual Communication (T.R. Hsing, A.G. Tescher)
Zone Plates (J. Ojeda-Castañeda, C. Gómez-Reino)
Zoom Lenses (A. Mann)

SPIE Milestone Series
Volume MS 135

Selected Papers on
Optical Techniques for Industrial Inspection

Paolo Cielo, *Editor*
National Research Council Canada
Industrial Materials Institute
Boucherville, Québec

Brian J. Thompson
General Editor, SPIE Milestone Series

SPIE OPTICAL ENGINEERING PRESS

A Publication of SPIE—The International Society for Optical Engineering
Bellingham, Washington USA

Library of Congress Cataloging-in Publication Data

Selected papers on optical techniques for industrial inspection / Paolo G. Cielo, editor.
p. cm. — (SPIE milestone series ; v. MS 135)
Includes bibliographical references and index.
ISBN 0-8194-2531-1 (alk. paper)
1. Engineering inspection. 2. Quality control—Optical methods. I. Cielo, Paolo G. II. Series.
TS156.2.S44 1997 97-3031
670.42'5—dc21 CIP

ISBN 0-8194-2531-1

Published by
SPIE—The International Society for Optical Engineering
P.O. Box 10, Bellingham, Washington 98227-0010 USA
Telephone 360/676-3290 (Pacific Time) • Fax 360/647-1445

This book is a compilation of outstanding papers selected from the world literature on optical and optoelectronic science, engineering, and technology. SPIE Optical Engineering Press acknowledges with appreciation the authors and publishers who have given their permission to reprint material included in this volume. An earnest effort has been made to contact the copyright holders of all papers reproduced herein.

Printed in the United States of America

Introduction to the Series

There is no substitute for reading the original papers on any subject even if that subject is mature enough to be critically written up in a textbook or a monograph. Reading a well-written book only serves as a further stimulus to drive the reader to seek the original publications. The problems are, which papers, and in what order?

As a serious student of a field, do you really have to search through all the material for yourself, and read the good with the not-so-good, the important with the not-so-important, and the milestone papers with the merely pedestrian offerings? The answer to all these questions is usually yes, unless the authors of the textbooks or monographs that you study have been very selective in their choices of references and bibliographic listings. Even in that all-too-rare circumstance, the reader is then faced with finding the original publications, many of which may be in obscure or not widely held journals.

From time to time and in many disparate fields, volumes appear that are collections of reprints that represent the milestone papers in the particular field. Some of these volumes have been produced for specific topics in optical science, but as yet no systematic set of volumes has been produced that covers connected areas of optical science and engineering.

The editors of each individual volume in the series have been chosen for their deep knowledge of the world literature in their fields; hence, the selection of reprints chosen for each book has been made with authority and care.

On behalf of SPIE, I thank the individual editors for their diligence, and we all hope that you, the reader, will find these volumes invaluable additions to your own working library.

Brian J. Thompson
University of Rochester

Selected Papers on

Optical Techniques for Industrial Inspection

Contents

Section Two
Optical Gauging and Profiling

Section Three
Optical Techniques for Nondestructive Testing

Section Four
Optical Techniques for Process Control

Preface

A variety of optical techniques have been developed over the last decades for monitoring industrial processes, including on-line systems for defect detection, on-line or at-line devices for statistical quality control, nondestructive methods for product evaluation, and optical sensors for process control. This is a diversified, rapidly evolving and need-oriented discipline, whose recent growth reflects the accelerating trend of modern industry toward flexible automation, coupled with the availability of increasingly powerful, compact, and rugged optical devices.

In the broadest industrial sense, inspection may be defined as the function of determining the conformance of the product or process to requirements. In the materials processing and manufacturing industry, the main inspection requirements are in the fields of product specification and industrial process monitoring. The technology is evolving from traditional, off-line measurements dominated by inspector skills to continuous monitoring of the relevant production parameters through suitable on-line sensors. Because of the hostile environment often encountered in the materials processing industry and the continuous product motion along the production line, new generations of noncontact sensors, such as optical and laser-based sensors, have been developed. Often, such sensors must be specifically adapted to each particular application and should introduce no perturbation to the industrial process.

Optical techniques are particularly well suited to industrial inspection tasks mainly because of their noncontact nature and sharp imaging capabilities. Optical sensors can be used for remote inspection of high temperature products or otherwise inaccessible parts, provided they are in a line-of-sight relation with the sensor. Optical fibers are being used increasingly to decouple the sensing area from the possibly remotely located signal processing unit. Moreover, the imaging capabilities of optical systems enable the simultaneous collection of large amounts of information and hence the possibility of compensating for variable part position and orientation with respect to the sensor. This is an important requirement in a flexible automation industrial environment.

Optoelectronic devices have gradually become more compact, affordable, rugged and reliable over the last decades. The development of such components is mainly a spin-off of successful technologies such as fiber optic communications and compact disk devices. One example is the laser diode, much more compact and offering a much larger reliability and longer lifetime than the earlier gas lasers, at prices that are rapidly approaching those of inexpensive incoherent light sources. Other important advances in optical technology that are directly relevant to optical inspection requirements are in the fields of optical detectors, compact staring-array devices, and signal processing capabilities.

Because industrial sensors are determined by the industrial needs, optical techniques for industrial inspection are best classified in terms of their function. Accordingly, the five sections of this volume are organized in terms of the kind of industrial requirements to which such techniques are devoted. These are broadly defined classes, often encompassing sensors and systems of widely different configurations. The overall link is the basic optical technology and the reliance on recurring devices. Within each section, the paper chronology reflects the historical perspective into which each subset of optical techniques has been framed.

It is impossible, in such a wide collection of techniques, to include the work of every researcher who contributed significantly to the development of this still rapidly evolving discipline. This volume is thus an attempt to present a summary of the main contributions to the different developments that have led to important optical inspection tools in modern industry. Although an effort has been made in the choice of papers to pinpoint the basic contributions that have made possible the development of such techniques, the emphasis is on methods that have been demonstrated to be appropriate for practical installation on the production line.

Paolo Cielo
National Research Council Canada
Industrial Materials Institute
Boucherville, Québec, Canada
(February 1997)

Introduction

Section One: Surface Inspection

The papers in this section deal with optical techniques for monitoring random surface morphology features. Deterministic shape and size metrology techniques are instead included in the next section. Random surface morphology features, such as surface roughness, glossiness, etc. can be analyzed by surface-integrated techniques, such as monitoring the angular light scattering pattern, or sequential surface scanning methods, such as focused-beam surface profilometry. This first section includes papers covering both of these two approaches. Finally, the last three papers of this section refer to historically significant industrial techniques for on-line detection of surface defects. These last papers may be considered as intermediate between the two previous surface analysis methods, in the sense that they are typically based on the detection of abnormal surface roughness areas on the product surface, and they illustrate the industrial significance of the methods described by the above papers.

Section Two: Optical Gauging and Profiling

This section includes papers on optical techniques for industrial product sizing and shaping. The emphasis is always on techniques of proven industrial relevance. Several approaches are being used to continuously monitor on-line the product shape and dimensions. They range from simple on-off sensors to determine the longitudinal size of a product being driven on a conveyor belt, up to sophisticated three-dimensional imaging techniques for part sorting or robot guiding. The first paper in this section describes an early application of the infrared absorption method, which has become widely used in industry for on-line thickness monitoring of extruded thin films and transparent coatings. The other papers deal with both two-dimensional and three-dimensional imaging techniques. They include methods based on triangulation, moiré topography, and time-of-flight scanning, and several relevant papers relate about practical industrial installations in the production line.

Section Three: Optical Techniques for Nondestructive Testing

Proof testing is often performed on finished products to detect any gross anomaly before release to the user. In many cases, statistical testing may be performed by destructive methods, but for safety-sensitive products such as aeronautical components or nuclear industry parts, which must be inspected for defects after manufacturing or periodically during service, nondestructive techniques are required. A number of optical techniques for nondestructive testing, often complementary to other nonoptical techniques such as ultrasound, x-ray, or electromagnetic methods, have been developed over the last decades. Generally, optical techniques for nondestructive testing are not yet recognized as standard techniques for routine inspection in the industrial environment, but they are receiving increasing consideration, mainly because of their noncontact nature and suitability for rapid scanning of large surfaces. The papers in this section have been chosen in terms of both their historical significance and their industrial relevance and include in-plant holographic and speckle pattern metrology, laser-ultrasound inspection, and thermographic nondestructive testing.

Section Four: Optical Techniques for Process Control

On-line sensors are a crucial component in the process control loop. Simple and rugged sensors are required to measure on the production line process variables such as temperature, pressure, or flow. Optical devices are being increasingly applied to such tasks, mainly because of their noninvasiveness and easy installability at the roots of the process, thus reducing time delays and loop-stability problems in feedback controls. This section contains papers describing sensors for the measurement of process parameters, while optical sensors for either process or quality control based on the direct measurement of product parameters are reviewed in the section on materials characterization. In this section an important number of papers are devoted to infrared temperature sensors, which are widely used in the industrial processes. Papers illustrating both pioneering industrial demonstrations of such sensors and technologically significant new infrared pyrometry techniques have been included. Velocity sensors based on either Doppler anemometry or correlation are described in the following four papers. The other papers focus on industrial applications of fiber optic sensors.

Section Five: Optical Techniques for Materials Analysis

This section contains papers relevant to optical technology for product analysis in the industrial environment, both for statistical quality control and for process control. The interaction of light with matter may provide information on the intrinsic properties of the materials composing the industrial product, such as their chemical compositions or their microstructural states. Usually, on-line product analysis provides a more direct information than the measurement of indirect process variables for product quality assurance. On-line or at-line spectroscopic methods are the object of several papers in this section, from simple nondispersive spectroscopic analyzers to FTIR and Raman spectrometers using fiber optics or other sampling methods for product analysis in the production line. Particular attention is devoted to laser-induced plasma spectroscopy, first demonstrated in

1966, which recently has been finding convincing in-situ applications for remote surface analysis of opaque products. Finally, the last four papers deal with industrial applications of light scattering techniques.

Paolo Cielo
National Research Council Canada
Industrial Materials Institute
Boucherville, Québec, Canada
(February 1997)

Section One
Surface Inspection

Reprinted from *Precision Engineering*, Vol. 3(2), pp. 61-83 (April 1981).

Optical techniques for on-line measurement of surface topography

T.V. Vorburger and E.C. Teague*

Optical techniques have great potential for non-destructive and on-line measurements of surface roughness during manufacturing. This paper reviews the state-of-the-art in a number of optical techniques including specular reflectance, total integrated scatter, diffuseness, angular scattering distributions, speckle, ellipsometry, and interferometry. The distinction is drawn between the more quantitative but slower profiling techniques and less quantitative, parametric techniques, which are faster and hence more useful for high-speed monitoring of surfaces. Overall, no currently available technique combines accuracy and speed and is therefore suitable as an on-line metrological tool for roughness measurement of engineering surfaces. Speckle techniques hold perhaps the greatest potential as accurate, high-speed metrological tools

Current emphases on energy and material conservation, 100% inspection of critical parts, miniaturization, and high accuracy dimensional tolerances[1-4†] have greatly increased the demand for non-destructive, high speed measurements of part surface topography[5-7]. As an example of energy conservation, aerodynamic and hydrodynamic drag and its relationship to skin and hull roughness has become a major factor in the construction of planes and ships. These same pressures for conservation coupled with demands for higher quality and reliability have contributed to the trend towards 100% inspection. In both miniaturization and high accuracy dimensional tolerancing, the dimensions are comparable to surface roughness heights; thus it is no longer possible to isolate roughness specifications from tolerances[4].

In response to such pressures, many new measurement methods are being investigated and there are, in fact, working on-line systems for measuring surface roughness. One example is at an automotive assembly plant in Puerto Rico, where a laser-based system is used to inspect the surface roughness of the shafts of power steering pumps.[7]

This review considers alternative techniques to the conventional stylus instrument, primarily in terms of their potential for non-destructive evaluation and remote, high-speed, 100% on-line surface measurement in the metal-working industry. On-line measurements, ie real-time, high speed, automated measurements that can be performed routinely and without disruption between stages of manufacturing, are distinguished from in-process measurements made while the part is being machined or formed. In-process measurement offers the potential for adaptive control to optimize the finished parts automatically, but it must deal with the environmental problems of part vibration, debris, and surface coolants and lubricants that are generally present during the forming process. Some of the techniques described in this review could be modified or augmented to facilitate in-process measurements, although the changes required will not be discussed.

**National Bureau of Standards, US Department of Commerce, Washington, DC 20234, USA*

†The References at the end of this review are grouped under a number of subject headings. Thus, contrary to normal practice in this journal, they are not numbered consecutively in the order that they are first referenced in the text.

This review concentrates on optical techniques and is directed toward defining the conditions for valid use of these techniques for non-destructive evaluation and on-line measurement of surface topography. Each technique is discussed in terms of a number of characteristics including measurement speed, resolution, and range. Non-optical techniques, including capacitative and inductive techniques, thermal comparators, and pneumatic gauges, also show some promise and are discussed elsewhere[8,9]. Since the emphasis is on engineering rather than optical surfaces, most of the work discussed here is not covered by previous reviews[22-25,141] of the subject.

Stylus instruments

The standard method for surface topography measurement has been the stylus technique[8], which, within the bounds that stylus displacements represent a convolution of the tip geometry and the surface microgeometry, yields directly the surface profile rather than some average property of surface deviations. The output may be digitized and analyzed in a computer[10-12] to yield a variety of statistical parameters and functions[12-16] for characterizing the surface with confidence that these closely represent true profile parameters.

Fine resolution and wide range are also important features of stylus instruments. The ultimate vertical resolution can be as small as 0.1 nm rms[68] and the range can be as much as 100 μm. The horizontal resolution is limited by the stylus tip width[17-18]; it is typically several μm but can be as small as 0.1 μm, more than acceptable for most topographic measurements.

The stylus instrument has a number of shortcomings, however. The transducer and the stylus tips are fragile, so the instrument must be used in a fairly quiet, clean environment. Without proper use, the sharp stylus can damage the surface and, depending on set-up time, a surface profile measurement can take from one to ten minutes, which is slow for most on-line applications. Finally, conventional use of stylus instruments only generates two-dimensional surface profiles rather than three-dimensional surface maps and therefore, the fraction of the surface area examined is quite small. Thus its usefulness is limited for such applications as nondestructive evaluation, inspection of the entire surface of a product, on-line surface measurement, and in-process measurement during manufacture.

Engineering surfaces

Engineering surfaces are produced by a wide variety of processes and their structures may be highly varied and complex. A milled surface, for example, has a strong lay pattern, is highly regular and periodic, whereas a bead blasted surface is isotropic (ie roughness measurements along any direction yield similar results) and its structure is highly random. Most engineering surfaces, however, have a certain degree of anisotropy and a combination of both periodic and random features. To quantify the study of surface topography a number of statistical functions and parameters have been developed. These characterize the two basic aspects of topography: the heights of the asperities (amplitude) and the longitudinal spacings between the asperities (wavelength).

Some of the more fundamental parameters and functions are briefly defined below‡. For simplicity the definitions are expressed in terms of a two dimensional surface profile $y(x)$, but there are equivalent three dimensional expressions. A key question in this review is how the parameters measured by optical techniques can be related to these fundamental surface quantities.

Roughness average (R_a), the internationally standardized nomenclature[19] but also known as centre line average (cla) or arithmetic average (aa), represents the average deviation of a surface profile about its mean line, defined as:

$$R_a = L^{-1} \int_0^L |y(x)|\, dx\,, \tag{1}$$

where x is the distance along the surface, $y(x)$ defines the height of the surface profile about the mean line (Fig 1) and L is the sampling length. R_a is the most widely used surface parameter in the world, partly because it is easy to calculate.

rms roughness (R_q), also known as σ in optics and statistics, is defined by:

$$R_q = \left(L^{-1} \int_0^L y^2(x)\, dx\right)^{1/2}. \tag{2}$$

Since R_q is equal to one standard deviation of the profile about the mean line, a number of theoretical calculations of surface properties yield results which are naturally expressed in terms of R_q rather than R_a. R_q is therefore a more significant parameter in theoretical work. R_q is always greater than or equal to R_a.

Amplitude density function (adf), the probability density distribution of surface heights which characterizes all of the amplitude properties of a surface profile, is found by plotting a histogram of the profile points in the vertical direction (Fig 1). A number of surface amplitude parameters including R_a and R_q can be calculated from the adf[10,16]. If a surface were produced by a truly random process, the adf would be a Gaussian distribution of surface heights[20] given by:

$$\mathrm{adf}(y) = (2\pi R_q^{\,2})^{-1/2} \exp(-y^2/2R_q^{\,2}). \tag{3}$$

In practice, most engineering surfaces yield adfs which are fairly close to Gaussian distributions[14], the exceptions being highly periodic surfaces such as those produced by milling. The adf for a ground surface (Fig 2) has the bell shape typical of adfs for many kinds of surfaces.

Power spectral density function (psd), an important function for characterizing both the asperity amplitudes and spacings, is calculated by Fourier decomposition of the surface profile into its sinusoidal component spatial frequencies f. For a 2-d surface profile:

$$\mathrm{psd}(f) = L^{-1} \left| \int_0^L y(x) \exp(-i2\pi f x)\, dx \right|^2. \tag{4}$$

The inverse of each spatial frequency f is the spatial wavelength Λ of that sinusoidal component. The psd for the ground surface (Fig 2) shows a fairly smooth distribution of surface wavelengths, characteristic of a random surface, with some evidence for a strong periodic component indicated by the spike at small f (large Λ). Note that in all forms of roughness measurement one must take into account the approximate bandwidth of the probe. In the case of stylus instruments the bandwidth is limited at high spatial frequencies by the stylus tip radius and at low spatial frequencies by either a high pass electronic filter with specified cut-off length or ultimately by the trace length. The sharp drop in the psd of Fig 2 below $f \sim 0.005$ μm^{-1} ($\Lambda \sim 200$ μm) was due to the electronic filtering. Since the bandwidth limits of any particular technique influence the measured values of roughness parameters, one must take these limits into account when comparing the results of different techniques. A characteristic spatial frequency ($\bar{f}$) of the surface can be calculated by taking the first moment of the psd:

$$\bar{f} = \int_0^\infty f\, \mathrm{psd}(f)\, df \Big/ \int_0^\infty \mathrm{psd}(f)\, df. \tag{5}$$

A characteristic spatial wavelength can be defined as the inverse of $\bar{f}$.

Autocorrelation function (acf), another function for characterizing both the wavelength and amplitude properties of a surface, is a quantitative measure of the similarity between a laterally shifted and an unshifted version of the profile given by:

$$\mathrm{acf}(\tau) = L^{-1} \int_0^L y(x)\, y(x+\tau)\, dx, \tag{6}$$

where τ is the shift distance. The acf for the ground surface is plotted in Fig 2. It was normalized to have a value of unity at a shift distance of zero. This suppresses any amplitude information in the acf but allows a better comparison of the wavelength information in the various profiles. The autocorrelation length a is derived from the acf and is used for estimating characteristic wavelengths on random surfaces. It is defined as the shift distance at which the value of the acf (or its envelope) drops to a certain fraction (usually 10%) of the zero shift value. Points on the surface profile that are separated by more than a correlation length may be considered as uncorrelated, ie por-

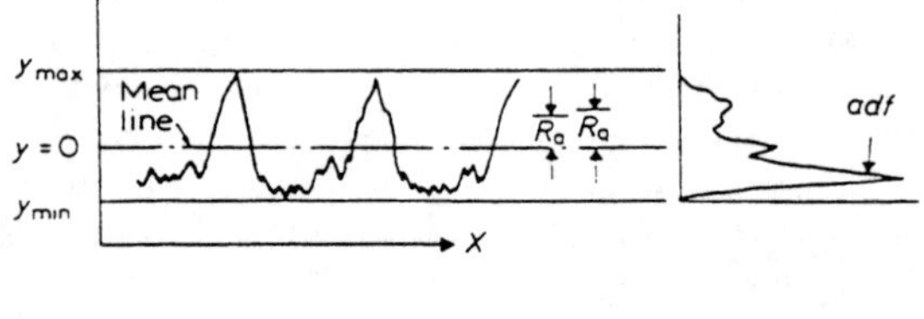

Fig 1 Schematic surface profile y(x) showing two roughness amplitude parameters, R_a and R_q, and the amplitude density function (adf)

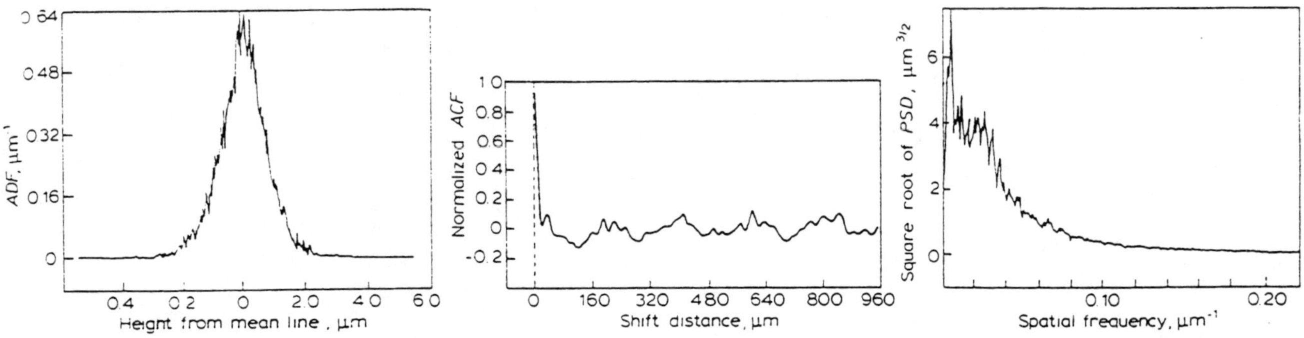

Fig 2 adf, psd, *and* acf *surface functions calculated from stylus profiles of a ground surface. The same set of profile data was used to generate each function*

tions of the surface represented by these points were produced by separate surface forming events. Correlation lengths may range from the infinite correlation length for a perfectly periodic waveform to zero for a completely random waveform. The ground surface of Fig 2 has a rather small correlation length of 15 μm. It should be noted that bandwidth limits affect the shape of both psd and acf.

Average slope (S_a) and rms slope (S_q) are hybrid parameters which combine both the amplitude and wavelength properties of the surface. For a continuous surface profile they can be defined as:

$$S_a = L^{-1} \int_0^L | dy/dx | \, dx$$
$$S_q = (L^{-1} \int_0^L (dy/dx)^2 \, dx)^{1/2} \qquad (7)$$

The profile derivative dy/dx is sensitive to the resolution of the technique used to measure the surface profile. Hence, the measured values of the slope parameters, S_a and S_q, depend on the high spatial frequency cut-off of the measuring instrument[21].

Because they can be produced by any of a number of processes, the regime of engineering surfaces is wide ranging. Typical values for R_a range from 0.1 μm to 10 μm and typical wavelengths may range from 5 μm to 800 μm. In general, surface structures with wavelengths greater than 800 μm are classified as waviness or errors of form rather than roughness.

The engineering surface regime is also to be distinguished from the much smoother regime of optical surfaces, which are produced primarily by polishing, annealing, or diamond turning. Optical surfaces have R_a's that are typically between 1 and 10 nm but that may be as small as several tenths of a nanometer[25,141].

Optical measurements

This discussion of a number of commonly used optical techniques for measuring surface texture gives, for each technique, a general description of the method and its strengths and weaknesses, and comment on its stage of development, ie whether there are commercial instruments or practical working devices. At the outset it is important to distinguish two general classes: namely profiling and parametric techniques.

With a profiling technique, such as interferometry or the stylus method, the topographic information is derived from a point-by-point scan of surface height y as a function of distance x along a straight line on the surface. The resulting profile must be analysed by either analog or digital methods, to derive parameters. A parametric technique, on the other hand, directly produces a measurable parameter that represents some property of the surface topography averaged over the illuminated area. Parametric techniques are fast but the results are hard to quantify in terms of surface topographic parameters such as R_q, because the measured optical parameters depend on the surface topography in a complex fashion.

‡ *Editor's note: For further discussions of parameters see the papers by T.R. Thomas on pp 91–104 of this issue.*

Parametric techniques

When a collimated beam of laser light is reflected by a rough surface, the radiation is scattered into an angular distribution (Fig 3) according to the laws of physical optics. The intensity and the pattern of the scattered radiation depend on the roughness heights, the spatial wavelengths, and the wavelength of the light. In general, small spatial wavelength components diffract the light into large angles relative to the specular direction and long spatial wavelength components diffract the light into small angles. For most surfaces, the psd has a broad spectrum of spatial components, and the light is therefore diffracted over a range of angles.

For very smooth surfaces ($R_q \ll \lambda$), most of the reflected light propagates in the specular direction. As R_q increases, the intensity of the specular beam decreases while the diffracted radiation increases in intensity and becomes more diffuse. In addition, the angular distribution of diffuse radiation consists of a fine grainy structure called speckle, which shows up as intensity contrast between neighbouring points in the scattered field. Finally, the light wave may undergo a change in its polarization state upon reflection from the surface. All of these phenomena, the relative intensity or reflectance in the specular direction, the total intensity of the scattered light, the diffuseness of the angular scattering pattern, the speckle contrast, and the polarization, depend on the surface roughness, and all five have served as the bases for potential surface measuring instruments.

Specular reflectance

The simplest approach is to measure the specular reflectance of the rough surface by measuring the intensity of the specular beam[26–44] (detector 3 in Fig 3). A number of research groups have contributed to this field but much of the pioneering work was done by the group of Bennett *et al*[27–30]. Commercial instruments following this general approach are sometimes called glossmeters[45], and in fact gloss measurement for its own sake has been pursued for more than 60 years. The measurement of roughness relies on the inverse correlation which exists between the specular reflectance ρ and the roughness. If R_q is much smaller than λ and the surface has a Gaussian height distribution, the correlation between ρ and R_q is described by[46]:

$$\rho = \rho_0 \exp\,[-(4\pi R_q \cos\theta_i/\lambda)^2\,], \qquad (8)$$

where θ_i is the angle of incidence measured with respect to the sample normal and ρ_0 is the total reflectance of the rough surface. ρ_0 is found by measuring the total light intensity scattered into all directions including the specular direction. If roughness-induced, light-absorption processes[47] are negligible, ρ_0 is equal to the specular reflectance of a perfectly smooth surface of the same material. It should be added that the values of R_q derived from Eq (8) result from a bandwidth limited sampling of the surface wavelengths, the limitations coming from the sizes of the detector aperture and the illuminating beam and the optical wavelength. This point is discussed more fully later.

For rougher surfaces ($R_q \geqslant \lambda/10$) the true specular beam effectively disappears so ρ is no longer measurable. For example, at normal incidence Eq (8) says that ρ/ρ_0 = exp (-2.5) when $R_q = \lambda/8$, and ρ/ρ_0 = exp (-158) when $R_q = \lambda$. However, even for rougher surfaces there are a number of empirical studies that show an inverse correlation between the light intensity scattered in the specular direction and the surface roughness[33,35,37,38].

The main advantage of this method is speed. If the instrument is properly calibrated, one measurement immediately yields a value for R_q. Therefore, the specular reflectance method is ideal for routine comparisons of similar surfaces. The technique is capable of studying isotropic or anisotropic surfaces and the ultimate resolution of about 1 nm[40] or better is excellent. Although a number of experiments[34–39,42,43] show good correlation between the optical results and either R_a or R_q as measured by the stylus technique, there can be appreciable scatter in the data. Fig 4, taken from an early review by Birkebak[42], shows that the slope of the graph of $R_{q\ \text{optical}}^{**}$ versus $R_{q\ \text{stylus}}$ varied from one observer to the next and that the scatter in the results for each observer was about 10%. These effects have been ascribed to the finite radius of curvature of the stylus and to differences in the stylus radii used by the different observers[42], but it is likely that the inconsistencies of Fig 4 may also stem from the optical model relating ρ to R_q. Eq (8) was derived on the basis of the scalar theory of optical scattering, rather than the more correct vector theory, and assuming a Gaussian distribution of surface heights[46]. Another problem is that of normalizing the measured intensities to ρ_0 for each material under inspection. In addition, Eq (8) describes only the specular component of the scattered light and does not take into account any diffusely scattered light which may happen to be scattered in the specular direction. Hence, it is accurate only if the diffuse component of light scattered into the detector is negligible compared with the true specular component. This means that the detector aperture should be small (~0.01 sr) so as to intercept as little of the diffusely scattered light as possible, but this makes the alignment of the optical system more critical and could make the apparatus sensitive to surface vibrations.

Finally, even for fairly small apertures the diffuse component will be significant unless ρ/ρ_0 is fairly close to unity. This implies that R_q must be much less than the optical wavelength λ, which means that HeNe lasers with λ = 0.6328 μm are suitable only for measuring the smoothest surfaces ($R_q < 0.1$ μm). For most engineering surfaces (0.1 μm $< R_q <$ 2 μm) one must use infrared radiation (1 μm $< \lambda <$ 20μm)[28–30,39,43].

The importance of the normalizing reflectance ρ_0 and the optical modelling problem are well shown by the work of Vashisht and Radhakrishnan[31–32]. Their results (Fig 5) for gloss versus average roughness R_a, measured by a stylus instrument, show large deviations from the

***Determined from measurements of* ρ

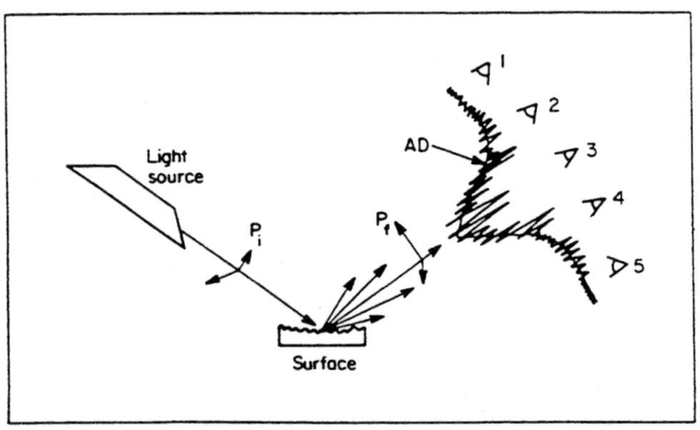

Fig 3 Angular distribution (ad) of light scattered by a rough surface. The gross ad envelope, shown by the smooth line, consists of speckle fine structure shown by the jagged peaks. The polarization state of the light (P_i) is also changed upon scattering (P_f). Detector 3 measures the intensity in the specular direction. Detectors 1, 2, 4,5 measure other components of the ad

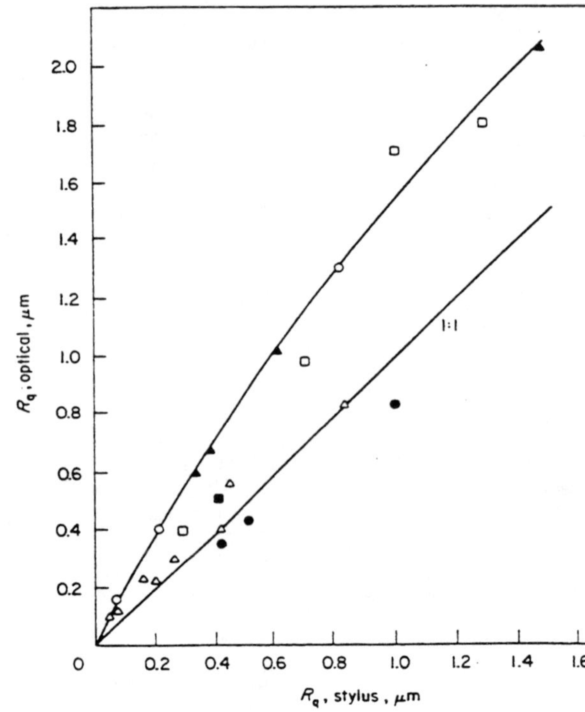

Fig 4 Comparison of rms roughness R_q *as determined by specular reflectance with* R_q *as determined by stylus methods. The figure is a compilation of results for both metal and dielectric surfaces from the work of six groups (as shown by the different symbols) and is taken from a review by Birkebak[42]. The two lines suggest a range of possible fits to the various pieces of data*

monotonic behaviour expected for each material. (Their gloss parameter is a monotonic function of the light intensity scattered in the specular direction). This result indicated either a dependence of ρ_o on the manufacturing process or a total breakdown of the optical model as R_a increases.

Two variations of the specular reflectance method are worthy of note. Hensler[36] and Tanner and Fahoum[33,34] avoided the $R_q \ll \lambda$ requirement and the reflectance normalization problem by measuring ρ for a range of incidence angles θ_i near grazing and fitting their results for ρ versus $\cos^2\theta_i$ to predictions based on models for the surface height distribution. Both experiments show good correlation between the optical and stylus results for roughness. Hensler's results[36] were achieved with the assumption of a Gaussian distribution of surface heights, but Tanner and Fahoum[33,34] required skewed forms of the surface height distribution to fit the data properly, so their analysis was necessarily rather painstaking. Nevertheless, it is conceivable that after further development this method could be adapted to high-speed, on-line measurement with the simultaneous use of several light sources pointing at the surface with different angles of incidence, several detectors, and a microprocessor to compute roughness.

The sensitivity to vibration may be overcome by using optical fibre bundles, which effectively bring the source and detector very close to the surface. Half the bundle transmits incident light to the surface and the other half transmits the reflected light to a detector. This feature was used in early instruments[38,41] and in a more recent one[37]. The fibre bundles, however, give wide-angle source and detector apertures, which works against the accuracy of Eq (8).

A final drawback of this technique is that it is primarily a function of surface amplitude. It is not suitable for measuring surface wavelengths.

In summary, specular reflectance, at its present stage of development, may be used as a fast, on-line technique for measuring R_q provided that $R_q < \lambda$ and the vibration of the specimens can be minimized. In the regime where R_q is approximately equal to λ, most of the previous empirical results suggest that this sort of device may be used as a comparator to distinguish between the roughnesses of similar surfaces. It seems clear that, for all applications, the instrument must be carefully calibrated beforehand for each type of surface.

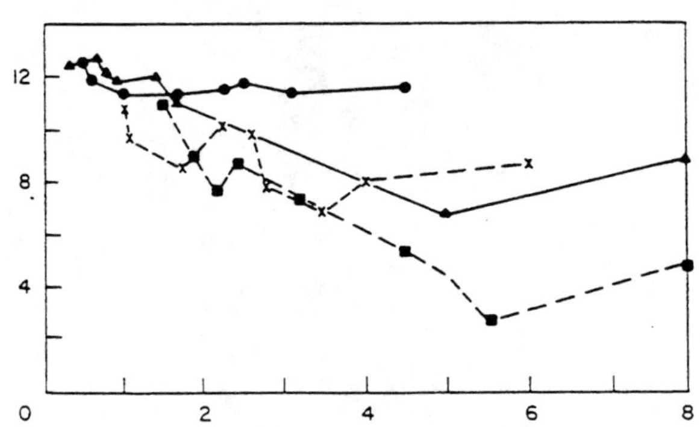

Fig 5 Results for gloss vs. R_a taken from Vashisht and Radhakrishnan[32] ●-Al, ▲-Brass, X-Mild steel, ■-Cast iron. In this work the gloss is a measure of the light intensity scattered in the specular direction

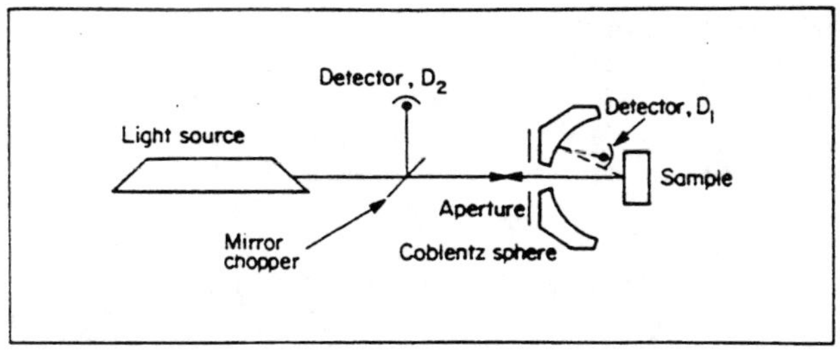

Fig 6 Schematic diagram of tis apparatus. Mirror chopper alternately throws light onto the sample and D_2

Total integrated scatter

The total integrated scatter (tis) method is complementary to specular reflectance. Instead of measuring the intensity of the specularly reflected light, one measures the total intensity of the diffusely scattered light[48–57]. Since tis is the ratio of this quantity to the total intensity of scattered radiation (both specular and diffuse), tis can be related to R_q by normalizing and rearranging Eq (8)[49]:

$$\text{tis} \equiv 1-\rho/\rho_o = 1-\exp[-(4\pi R_q \cos\theta_i/\lambda)^2] \approx (4\pi R_q \cos\theta_i/\lambda)^2. \quad (9)$$

The tis system developed by Bennett *et al*[49,50] (Fig 6) includes an aluminized collecting Coblentz sphere C, which focuses the diffusely scattered light onto a detector D_1 and a mirror chopper which alternately throws the incident light onto the detector D_2 or the specimen S. Although D_2 measures the incident rather than the scattered light intensity, the errors due to transmittance and absorption were small for the materials and surfaces that Bennett *et al* studied. One variation of the technique[58] uses transmitted light and relies on total internal reflection of the specular beam by the surface under test. It is suitable only for transparent materials.

tis has much the same strengths and weaknesses as specular reflectance, *ie*, the technique is fast, is based on an approximate theory, is practical only for surfaces with $R_q \ll \lambda$, and has similar bandwidth limitations. The fundamental difference between the two is that the tis is more directly related to R_q. In this respect, the calculation of R_q from tis via Eq (9) is much less sensitive to errors in ρ_o than is the calculation of R_q from ρ via Eq (8). Indeed, tis is a fundamental quantity for the functioning and testing of optical components[52]. In many optical applications, if the tis can be measured accurately and routinely, the roughness need not be.

Using the technique for practical roughness measurements presents several problems, however. To begin with, the experimental set-up is more elaborate than that required for specular reflectance. The fact that some sort of collecting sphere must be set up close to the surface implies that this technique is more suited to the laboratory than to on-line measurement. Moreover, the bandwidth limitation arises from the fact that the collecting optics normally contain some sort of aperture or stop to prevent the specular beam from being detected along with the scattered radiation. Such an aperture or stop inevitably also blocks a portion of the scattered light and hence limits the surface wavelengths that can be detected[52].

The net result is that the technique does not seem to have been successfully applied to surfaces with R_q greater than about 0.01 μm. Even in the regime of very smooth surfaces, the roughness results obtained from tis do not seem totally reliable. Fig 7 shows a comparison of results for R_q obtained from tis and obtained using feco (fringes of equal chromatic order), an interferometric technique that will be discussed later. The results are a compilation of measurements from several groups[51,53,55]. For glass and fused silica surfaces these data show superb agreement (~ 0.1 nm) between the two techniques but for the Cu surfaces the tis results differ significantly from the feco results. The probable reason that the Cu surface appeared rougher in feco than in tis relates to the different bandwidth limits of the two techniques and to the fact that the Cu surface was produced by diamond turning[57]. This machining technique produces a long wavelength, periodic structure on the surface and it is likely that this structure was measured by feco but missed by tis whose bandwidth range was more limited. Further discussion of this problem is given later.

It appears then that tis is an important technique for rapid measurements of the optical scattering characteristics of very smooth surfaces (say $R_q \leqslant 0.01\ \mu m$) with short wavelengths. Like specular reflectance, tis is not suitable for measuring surface wavelength parameters.

Diffuseness of scattered light

The third approach relies on the observation that, over a large roughness range, the pattern of scattered radiation becomes more diffuse with increasing surface roughness. Hence, the goal here is to measure a parameter which characterizes the diffuseness of the scattered radiation pattern and to relate this parameter to the surface roughness. In principle, this technique avoids the problem of normalizing measured reflectance results for each material studied, but in practice, calibration curves must be generated for each material and surface finishing process. The approach, then, is based on empirical correlations and has been used to study surfaces with moderate-to-high roughness where the proposed theories are largely intractable, rather than optical surfaces.

Two types of instruments will be described. The first measures the ratio of the specular intensity (detector 3 in Fig 3) to the intensity at one off-specular angle (detector 4, say). Since this ratio generally decreases with increasing surface roughness, it could provide a measure of the roughness itself. Peters[59] used this technique with detector 4 held 40° off specular to determine the roughness of cylindrical parts while they were being ground. His results show good correlation between the diffuseness and R_a over a range of R_a up to 0.3 μm. Even under different lubrication conditions (oil, water, dry) the results are well fitted by a single curve.

A similar instrument[60] is being developed by Corey to measure R_a in the range 0.2 μm to 2 μm for high-speed quality control of the surface finish of machined hemispherical parts. Essential features of the instrument are its non-destructive capability and its ability to scan the entire surface of the part. In its current stage of development, the instrument uses the ratio of the intensity measured 15° off-specular to the specular intensity to yield a value for R_a. In order to make meaningful R_a measurements for a particular type of surface, a set of roughness specimens that have been manufactured in a way similar to the test specimens with known R_a values will be required.

Another system, developed by Takeyama *et al*[61], measures the ratio of the specular intensity at the surface-normal to the back-scattered intensity 30° off-normal. This system is designed to be quite insensitive to surface vibration with its use of fibre optics bundles to transmit the incident and reflected light. Indeed, the measured signals from spinning parts are fairly stable with time. Takeyama *et al*'s ratio measurements were performed on machined surfaces with high peak-to-valley roughness ranging from 5 to 80 μm. They found that the experimental curves of intensity ratio against roughness were a function of the tool radius used to machine the surfaces. This dependence on the manufacturing process again implies a need for a set of calibration specimens. Takeyama *et al* claimed that the curves of intensity ratio versus roughness were independent of the surface material, but this claim does not seem to be supported by all of their data.

A distinctly different method, employing a commercial laser scanning analyser (lsa), was used by Clarke and Thomas[62]. In the lsa (Fig 8), originally developed for on-line scanning to detect surface defects, a laser spot is

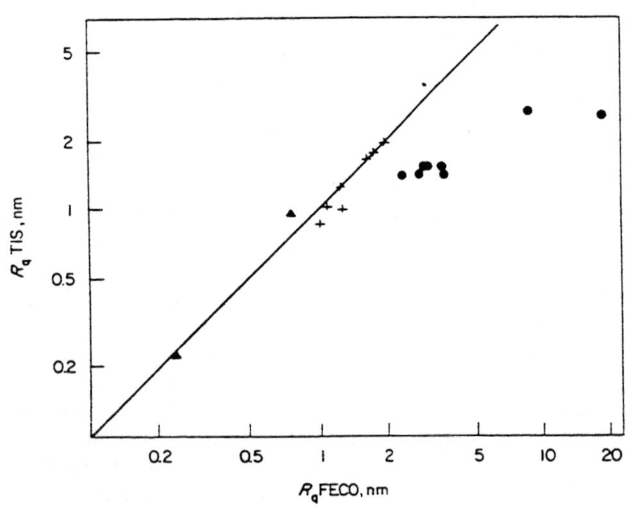

Fig 7 Graph showing a comparison of experimental results for R_q obtained by tis with those obtained by feco. + polished glass surfaces overcoated with silver to improve the reflectivity[51], ▲ polished fused silica overcoated with silver[53] and ● diamond-turned Cu[55]

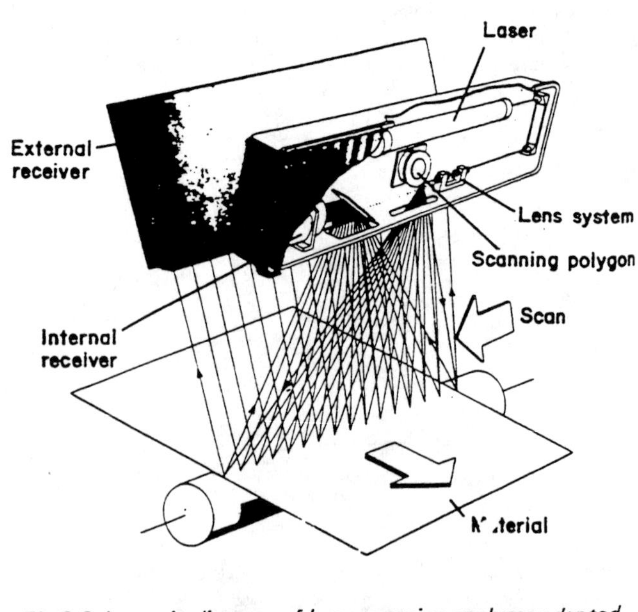

Fig 8 Schematic diagram of laser scanning analyser adapted by Clarke and Thomas to study surface roughness[62]

rapidly and repeatedly scanned over the surface of the material. A portion of the reflected light is collected by a slit detector and the resulting electrical signal is displayed on an oscilloscope and synchronised with the scanning position of the laser spot. The resulting trace contains a peak, which corresponds to the position on the surface where the laser spot is reflected specularly into the detector, and wings on each side of the peak, which fall off as the spot is moved away from the specular direction. Although the instrument is capable of very high speed inspection of large surface areas, the scanning geometry is highly complex. As the laser spot scans across the surface, the angle of incidence, the scattering angle, and the scattering plane all change simultaneously. Nevertheless, Clarke and Thomas attempted to correlate the peak width with R_a as measured by a stylus instrument. They measured the peak width for 22 surface specimens manufactured by several different processes and having R_a between 0.06 μm and 20 μm. Although their results show a general dependence of peak width upon R_a, the data are so scattered as to indicate that this technique is not yet feasible for monitoring R_a. The scatter is unacceptable even among the data taken for each surface forming process. Perhaps further research with this instrument will uncover some techniques or parameters suitable for an unambiguous surface roughness measurement. It is certainly an attractive instrument from the points of view of speed and ease of alignment.

The diffuseness measuring instruments discussed in this subsection are all capable of rapid on-line surface inspection. They were all used to measure rougher surfaces than either the specular reflectance or the tis instruments discussed previously. In this roughness regime ($R_a \gtrsim 1$ μm) any justification that these instruments can be used for accurate measurement of surface roughness is highly empirical and has little basis in optical scattering theory, which is largely intractable in its present forms. The technique is primarily useful as a comparator to distinguish between the roughnesses of similar surfaces rather than as a metrological tool for measuring real surface parameters. The instrument developed by Takeyama *et al*[61] seems to be the most promising, but this type of system has not been tested on surfaces with R_a less than 2 μm.

Angular distributions

In principle, the entire angular distribution of the scattered radiation contains a great deal of information about the surface topography. In addition to rms roughness, measurements of the angular distributions (ad)[33,63–80], can yield other surface parameters such as the average wavelength or the average slope. The angle of incidence is normally held constant and the ad is measured by an array of detectors[70] (see Fig 3) or by a movable detector and is stored as a function $I(\theta_s)$ where θ_s is the angle of scattering. A good example of the resulting ad is shown in Fig 9, taken from a paper by Church *et al*[65]. This group used a nonconventional method of scanning the scattering angle and the incidence angle simultaneously by holding both the source and the detector fixed and rotating the specimen. The curves shown in Fig 9 were taken for a diamond-turned Au specimen illuminated by a HeNe laser. The upper one shows the ad in the plane of incidence and perpendicular to the predominant lay of the surface. It contains an intense specular beam at θ_s=0, a broad scattering distribution due to the random component of the roughness, and a series of

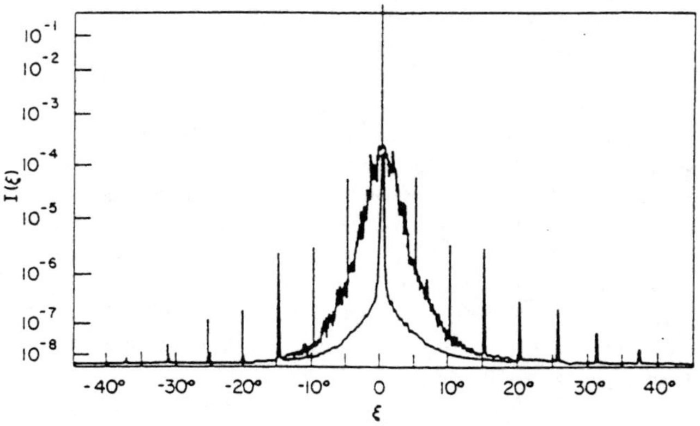

Fig 9 Experimental ad scattering spectrum of a diamond-turned Au surface[65]*. The upper curve shows a scan made perpendicular to the lay direction. The lower curve shows a scan parallel to the lay direction. The angle ξ, equal to θ_s-θ_i, is the angle between specular and scattering directions*

discrete lines due to a periodic component of the roughness caused by the feed rate of the diamond tool. The lower curve is an ad measured parallel to the lay direction and it shows another broad distribution characteristic of the random roughness pattern in this direction. In principle then, one can distinguish between effects due to periodic and random roughness components and can detect the directional properties of surfaces.

The kind of surface information that may be obtained from angular distributions depends on the roughness regime. For $R_q \gg \lambda$ and surface spatial wavelengths $\Lambda \gg \lambda$, one is working in the geometrical optics regime where the scattering may be described as purely specular scattering from a series of facets[67]. The ad is therefore related to the surface slope distribution, and its width is a measure of the characteristic slope of the surface.

As R_q and Λ decrease, the distribution becomes a much more complicated function of both surface slopes and heights and is difficult to interpret. Some work has been done by Leader[71], Smith *et al*[43,79–80] and Chandley[78] for surfaces with $R_q \sim \lambda$. Leader's comparisons between calculated and measured ad's yield interesting, qualitative information about the topography of painted and dielectric surfaces.

The most interesting regime is where $R_q \ll \lambda$ because then it can be shown theoretically that the ad should directly map the power spectral density psd ($\bar{F}$) of the surface topography[63–68,81] where $\bar{F}$ is the two dimensional surface spatial frequency. If the incident light is normal to the surface, the scattering angle θ_s with respect to the normal is related to $\bar{F}$ by:

$$\sin\theta_s = (\lambda/2\pi)\,|\bar{F}| \tag{10}$$

and the angular distribution $I(\theta_s)$ of the scattered radiation is related to psd($\bar{F}$) by:

$$I(\theta_s) = \lambda^{-4}\,F(\theta_s)\,\mathrm{psd}(\bar{F}), \tag{11}$$

where $F(\theta_s)$ is a geometric factor depending on the scattering angles, which is independent of the surface topography[68]. This theory is tractable for both isotropic surfaces and surfaces with lay.

Eq (10) essentially determines the bandwidth of surface wavelengths which can be measured by the ad technique. The bandwidth is limited at the low end by the maximum scattering angle of 90° and at the high end by the minimum scattering angle which can be measured with respect to the specular beam[67]. A practical limit for the minimum angle is about 0.5° [67]. The resulting bandwidths for $\lambda = 0.6328$ μm and 10.6 μm are shown in Fig 10 along with the approximate bandwidth for a stylus instrument which is limited at the low end by the stylus radius and at the high end by the length of the trace. The measurable bandwidth of the optical instruments is much narrower than that of the stylus, an observation to keep in mind when comparing the results of measurements by the two techniques. It should be noted that the lower bound of 0.1 μm represents the state of the art for styli under favarable conditions of roughness height and slope. Some research instruments have stylus radii between 1 and 2.5 μm, although the ultimate step height resolution of such instruments can be significantly smaller than the radius.

A number of experimental systems and one commercial instrument[70] have been developed to study surfaces using ad in the smooth surface regime. However, the experimental studies show that, while the shapes of the measured ads are in reasonable agreement with those calculated from stylus measurements via Eq (11), the magnitudes differ significantly over a range of angles. Early results by Elson and Bennett[68] showed scattering intensities that were a factor of 10^3 lower at large scattering angles than the predictions based on stylus measurements. Edwin[74], on the other hand, observed anomalously high intensities at large scattering angles, presumably coming from scattering by isolated surface defects or particles on the surface. Recent results by Elson, Rahn, and Bennett[69] show fairly good agreement between the measured and predicted angular scattering distributions from smooth, optical surfaces ($R_q \sim 0.7$ nm) with multilayer coatings. Thwaite's experiments[77] on periodic surfaces with R_a between 0.05 and 0.2 μm show good correlation between the positions of discrete spikes in the power spectral density derived from ad and those derived from stylus measurements. The spikes result from the various periodic components of the surface profile. The two techniques, however, yielded significant differences in the relative intensities of the discrete components. One implication of this result is that the ad technique may be better suited to measuring wavelength parameters than roughness amplitude parameters.

The work of Thwaite[77], Stover[63–64] and Church *et al*[65–67] illustrates how the ad technique may be adapted to studying individual periodic components of surface roughness resulting from machining. A detector positioned at a particular scattering angle is sensitive to a narrow range of spatial frequencies. Any strong periodic component of roughness in this range, due to the feed

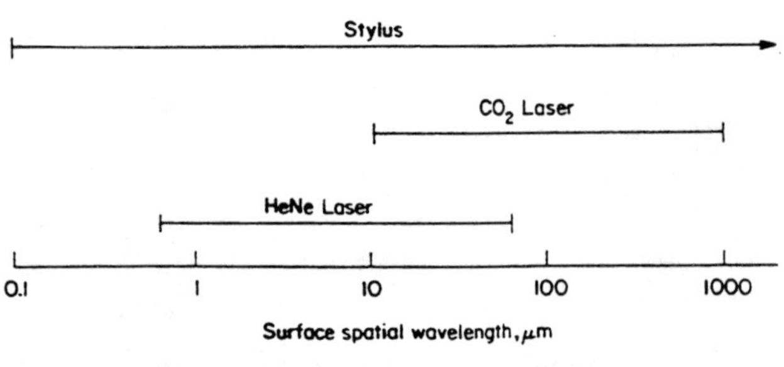

Fig 10 Approximate bandwidths of surface spatial wavelengths spanned by the stylus technique and by light scattering with HeNe and CO_2 lasers[67]

rate of the machine tool, for example, can be detected and measured independently from other sources of roughness. Changes in these discrete components can be monitored and related to parameters of the machining process.

The integral of the power spectral density over a band of spatial frequencies yields R_q for that particular band. Analogously, the integral of the ad curve over angle yields the tis[66] and hence yields an estimate of R_q, the bandwidth being determined by the range of measured angles. This principle has been tested for optical surfaces by Stover[63] and reasonable agreement for R_q has been obtained between the ad method and other methods.

The above principle is also used in a commercial ad instrument developed by Hard and Nilsson[70]. One novel feature of their instrument is a rotating-grating heterodyne technique to achieve high sensitivity and immunity to stray light. The big limitation is that the diode-array detector only spans scattering angles from 0.4 to 2° so that the instrument senses only a very limited range of spatial wavelengths.

The foregoing experiences suggest that ad's can reveal much qualitative detail about the amplitude and wavelength composition of surface topography. The technique is most suitable for smooth, optical-quality surfaces. It can also be adapted for studying periodic components of surface roughness imposed by a machining process. When the ad is measured sequentially by a movable detector, the technique is about as slow as the stylus technique. On the other hand, diode arrays allow for high speed data collection and are incorporated into a commercial system[70]. Measurements of R_q for very smooth surfaces are possible with ad instruments by integrating the contributions from all scattering angles to yield tis.

Scattering methods

In summary, the four scattering methods described so far are generally limited by available theories to studies of surface whose R_q is much less than λ. There are a few studies, mostly empirical, which have pushed beyond this limit. With a HeNe laser as the light source, the above constraint means that these techniques have been used mainly on optical quality surfaces where $R_q < 0.1$ μm. Within that limited regime, they can provide high speed quantitative measurements of the rms roughness of both isotropic surfaces and those with a pronounced lay. With rougher surfaces, ad may be useful as a comparator for monitoring both amplitude and wavelength surface properties. The ultimate vertical resolution is 1 nm or better but the horizontal range is limited to fairly short surface wavelengths. Both the vertical and horizontal ranges can be increased by using long wavelength (infrared) radiation, but there is an accompanying loss of vertical[67] and horizontal resolution.

There are no particular limitations in the part geometries which may be studied by these techniques. Most of the previous applications have been confined to measurements of outside surfaces but two instruments have been developed to monitor the surfaces of cylinder bores[82–83]. Many existing instruments may be sensitive to part vibration because they use fairly narrow apertures to collimate the optical beams. The clear exceptions are the instrument developed by Takeyama *et al*[61], which uses fibre optics, and the laser scanning analyser used by Clarke and Thomas[64]. The latter instrument dynamically scans the surface over a range of incident angles; therefore, the specular direction can be easily located as the peak in its time resolved output trace.

These classes of techniques include a number of practical, working prototypes and several commercial instruments.

Ellipsometry

Ellipsometry is conventionally used to measure the index of refraction of solids and the index of refraction and thickness of surface films[84]. The range of film thickness which can be measured extends from less than 1 nm[85] to hundreds of nanometers[86].

The technique measures the change in the polarization state of a beam of light when it is reflected from a surface. In traditional null ellipsometry the important quantities are the angle of incidence of the beam of light and the rotational positions of the polarizing and analysing elements (P and A in Fig 11) in the light path that produce a null in the detector[84]. These measurements allow one to determine r_p/r_s, the ratio of the complex reflection coefficients for the p- and s-components of the electromagnetic field. The electric field component of the p-wave is parallel to the plane of incidence, whereas the electric field component of the s-wave is perpendicular to the plane of incidence. The ratio r_p/r_s is customarily related to the important ellipsometry parameters, Δ and Ψ, by the following definition:

$$r_p/r_s = \tan \Psi \, e^{i\Delta}, \qquad (12)$$

where $\tan \Psi$ is equal to the ratio of the magnitudes of the p and s reflection coefficients and Δ is equal to the phase shift between them. The angles, Δ and Ψ, in turn may then be used with the Fresnel equations to calculate the optical constants of solids, and the thickness of transparent material through which the light passes[84,87].

Ellipsometry measurements of the index of refraction are sensitive to a number of surface properties including composition[88]; surface structure such as damage[89,90], defects[91], or surface crystal faces[92]; temperature[93]; strain state[94,95]; and surface roughness[96–107,81].

Since the roughening of surfaces can significantly change the results from the ellipsometer, thereby obscuring the other surface changes to be observed, the question arises whether ellipsometry can be used to measure the surface roughness of engineering surfaces directly. The investigations so far have been largely empirical and they are in disagreement.

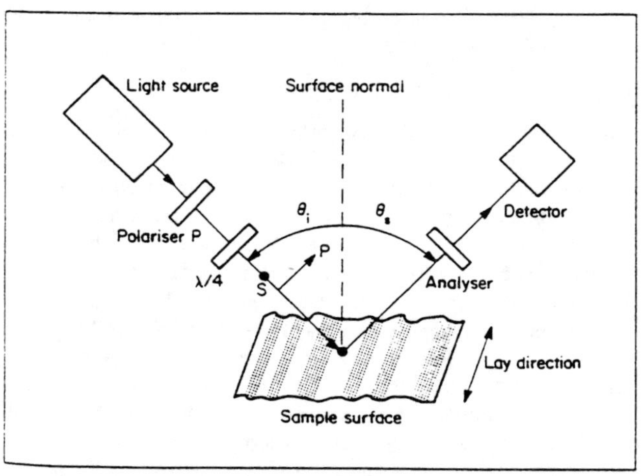

Fig 11 Schematic diagram of an ellipsometer. The surface normal is constructed with respect to the average plane of the surface. The detector and analyser are normally positioned in the specular direction, ie $\theta_s = \theta_i$

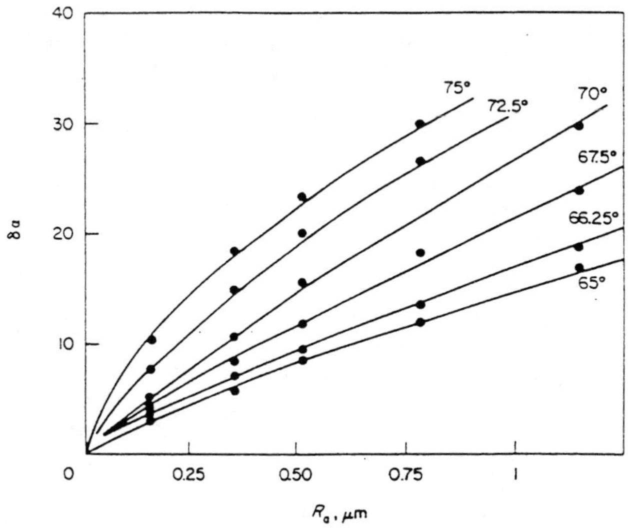

Fig 12 Lonardo et al.'s results for changes in the parameter α vs R_a with different angles of incidence θ_i

Lonardo *et al*[96–98] studied ellipsometry as a potential tool for in-process detection of surface roughness during manufacture. They worked with ground and polished steel surfaces with R_a, measured by stylus techniques, ranging approximately from 0.01 μm to 1.1 μm and found that both Δ and Ψ decreased with increasing R_a. In the light of this result, they formulated and measured a parameter α which combined the effects of Δ and Ψ in order to increase the sensitivity of ellipsometry to R_a. The parameter α, given by:

$$\tan 2\alpha = \tan 2\Psi \cos (\Delta - 180°) \qquad (13)$$

was found to vary almost linearly with R_a (Fig 12).

Smith[99] measured changes in Δ and Ψ for Al surfaces whose roughnesses were varied by acid etching the surfaces for various times. R_q and the rms slope S_q of these surfaces were estimated by the technique of specular reflectance. In one experiment, the ellipsometry angle Ψ was observed to decrease monotonically as the etch time and the estimated R_q increased. In another experiment, the data show that Ψ decreased monotonically with increasing S_q for certain angles of incidence but not for others. The interpretation of the experimental results for Δ were more complicated because the oxide film thickness as well as the roughness was expected to change with etching time and the angle Δ is sensitive to both the oxide film thickness and surface roughness. In spite of this problem, the overall results show that the ellipsometry parameters, Δ and Ψ, cannot be related simply to R_q and S_q.

Vorburger and Ludema's results[101] lead to the same conclusion. They measured the ellipsometry parameters for six lapped and ground surfaces of a comparison standard typical of those used in the machine-tool industry, and they compared the results with surface parameters measured by a stylus instrument.

To isolate the effects due to surface topography from other surface effects, the data were taken under four sets of experimental conditions with varying angles of incidence, optical wavelengths, and surface compositions. Under the first set of experimental conditions, the angle of incidence θ_i was 65°, the optical wavelength λ was 546.1 nm, and the data were taken on the bare surface of the Ni specimen. Under the second set, θ_i was 60° and λ was 632.8 nm. Then, a thin film of Al (~20 nm) was evaporated on the surface and data taken from $\theta_i = 65°$ and $\lambda = 632.8$ nm. Finally a thin film of Au (~ 150 nm) was evaporated on the Al and data were again taken for $\theta_i = 65°$ and $\lambda = 632.8$ nm.

The changes in Δ are plotted in Fig 13 as a function of R_a. For these data the plane of incidence was perpendicular to the lay direction. Since the trends for the four different sets of experimental conditions are quite similar, Vorburger and Ludema concluded that the experiment had isolated a topographic effect in the ellipsometry results.

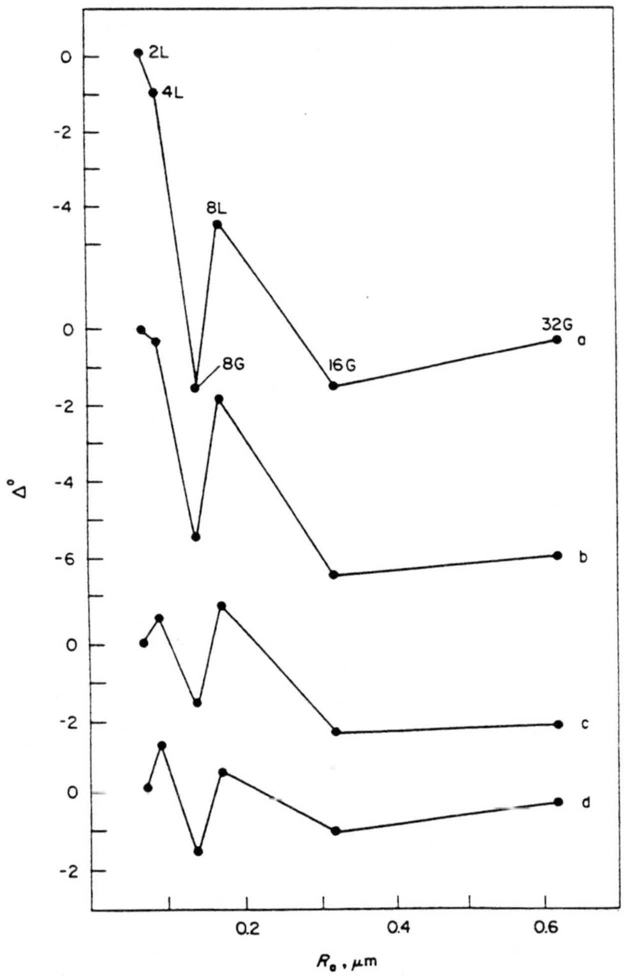

Fig 13 Changes in Δ (degrees) vs. roughness average R_a for six patches of a roughness comparison standard. '2L' means that the patch was a replica of a lapped surface with a nominal R_a of 50.8 nm (2μin), '32G' indicates a replica of a ground surface with a nominal R_a of 812.8 nm (32μin), and so forth

a — $\lambda = 546.1$nm, $\theta_i = 65°$, Ni surface,
b — $\lambda = 632.8$nm, $\theta_i = 60°$, Ni surface,
c — $\lambda = 632.8$nm, $\theta_i = 65°$, 20nm Al evaporated on Ni surface,
d — $\lambda = 632.8$nm, $\theta_i = 65°$, 150nm Au evaporated on top of 20nm Al on Ni surface.
Curves b, c, and d have been shifted arbitrarily for clarity. The data were taken from Ref. 101

The behaviour was quite complicated, however. The results did not vary monotonically with R_a as shown, nor did they vary monotonically with either R_q, S_a, or S_q. Similar behaviour was observed for Ψ and for Lonardo *et al*'s α parameter on these surfaces. In general, it is not at all clear how the parameters of the surface geometry affect the ellipsometry results.

To summarize then, the previous investigations are in sharp disagreement over whether ellipsometric parameters can provide any direct, quantitative measure of the surface roughness R_a or of any other roughness parameter. On the one hand, these results suggest that the use of ellipsometry as a tool to measure surface roughness would be problematical. On the other, it is clear that surface roughness can easily produce errors in the interpretation of ellipsometry measurements of other surface quantities.

If the key problem of quantification could be solved, ellipsometry would have several strengths as a surface measurement technique. The ellipsometer could be automated to produce rapid measurements of Δ and Ψ[108] or α[96–98]. The sensitive R_a range of the technique seems to be approximately 0–1 μm, which is in the range of interest for machined surfaces. Finally, ellipsometry measurements should not be sensitive to fluctuations in the scattered-light intensity due to surface vibration since the technique measures the polarization state rather than the intensity of the scattered light.

Speckle

When a rough surface is illuminated with partially coherent light, the reflected beam consists in part of random patterns of bright and dark regions known as speckle. These patterns can be interpreted in terms of Huygen's principle whereby the intensity at a field point is caused by the interference of wavelets, scattered from different points within the illuminated area, with their phases randomized by height variations of the surface. The spatial pattern and contrast of the speckle depend on the optical system used for observation, the coherence condition of the illumination, and the surface roughness of the scatterer. This section gives a brief summary of these three aspects of speckle patterns and discusses how speckle has been utilized in instruments which appear promising for on-line measurement of surface roughness. More thorough presentations of speckle phenomena and of speckle measurements of surface roughness are given elsewhere[109–113].

Effects of the optical system

The spatial variations classified as speckle should be carefully distinguished from those variations termed the angular distribution, ad. As mentioned earlier, speckle is the local intensity variation between neighbouring points in the overall ad. One means of clarifying the distinction between speckle and the ad is to note that speckle is the intensity noise that is usually averaged out to obtain the ad. An example of this speckle noise superimposed on the structure of the ad may be seen in the ad profile given in Fig 9. In terms of the angular diameter Δ_0 of the detector aperture subtended at a scatterer, ad measurements typically employ minimum Δ_0 of the order of tenths of a degree, while speckle contrast measurements require Δ_0 of the order of hundredths of a degree or less.

Speckle may be observed either in the simple 'Fresnel' configuration (Fig 14(a)) or the 'Fraunhofer' system (Fig 14(b)), the designations being given by analogy with regular diffraction phenomena. In Fig 14(a), the speckle pattern

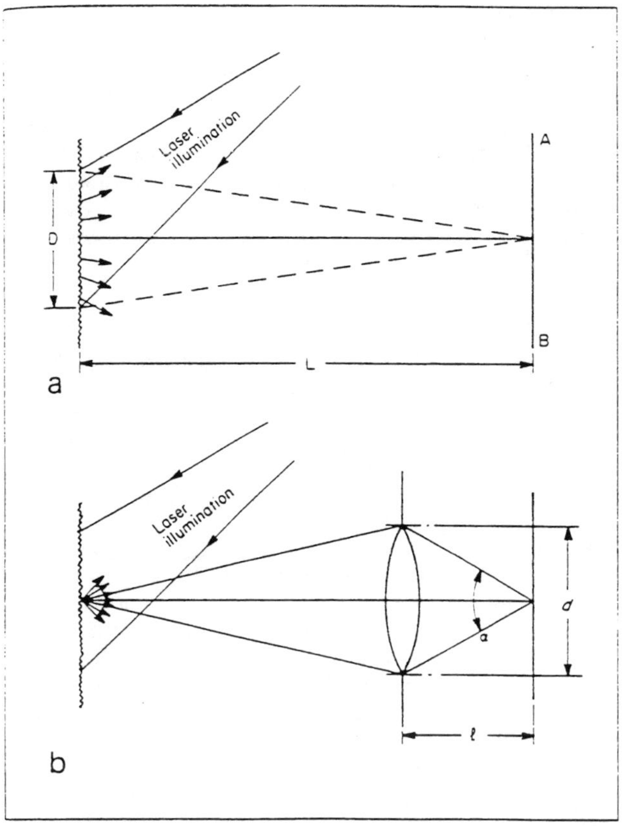

Fig 14 Illumination and recording configurations for observing speckle in coherent light reflected from a rough surface. Note that (b) represents the configuration for observing speckle with the human eye

is found in the whole of the space in front of the optically rough surface and the screen AB has a section of the three-dimensional interference pattern projected onto it[114]. For a circular patch of diameter D, illuminated by light of mean wavelength λ, the average size $<\delta>$ of the speckles is given by the approximate formula:

$$<\delta> = 1.2\,\lambda\, L/D, \qquad (14)$$

where L is the distance from the illuminated surface to the screen. $<\delta>$ is thus approximately equal to the diameter of the Airy disc[111] that would be produced in reflection from a highly polished disc the size of the illuminated region. The power spectrum of the spatial variation, for surfaces producing path differences greater than one wavelength, is nearly a triangular function[115] with a maximum at zero frequency and a cut-off frequency of $D/(\lambda L)$. Fig 15 shows a speckle pattern obtained at a distance of 1 m from a 1 mm diameter area of ground glass illuminated by a HeNe laser. Speckle size is approximately 1 mm in accord with Eq (14).

When the interference pattern is imaged with an optical system such as the one depicted in Fig 14(b), the size $<\delta>$ of the speckle is dependent on the angular aperture α subtended by the exit pupil of the imaging system at the image plane. Eq (14) is replaced by $<\delta> = 1.2\,\lambda/\alpha \approx 1.2\,\lambda\ell/d$, where ℓ and d are as defined in Fig 14(b). Thus for systems with apertures, the area of the object plane contributing to speckle formation is not the entire illuminated area but just the area of the point spread function of the optical system.

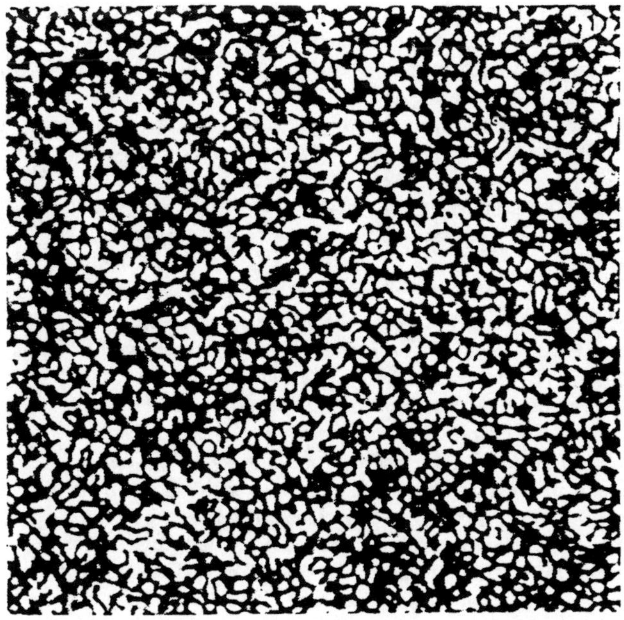

Fig 15 A speckle pattern recorded as shown in Fig 14(a) with light transmitted through a ground glass specimen with surface roughness amplitude much larger than the wavelength of the illumination. (From Dainty[112])

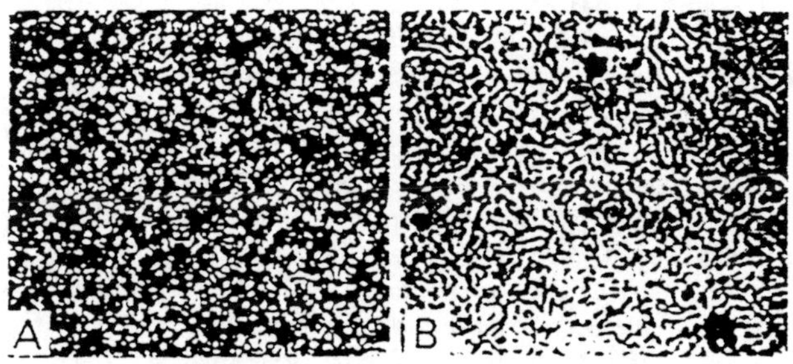

Fig 16 Speckle patterns recorded as shown in Fig 14(b) with light transmitted through ground glass specimens with surface roughnesses R_q, (a) 1.3 μm and (b) 0.2 μm (From Fujii and Asakura[116])

Fig 16 shows two speckle patterns[116] as they would be observed through a microscope at the image plane of an optical system similar to Fig 14(b) with two different ground glasses being illuminated by a HeNe laser. Properties of the optical system used to obtain the speckle patterns in Fig 16 were not specified but subsequent work by Fujii *et al*[117] suggested that α was probably 0.01 to 0.02 radians. Speckle size in typical imaging systems is of the order of 40 μm.

The number of scatterers which contribute to the interference producing a speckle pattern is an important factor that determines the properties of the pattern. As pointed out by Dainty[118], the minimum possible scatterer size (phase decorrelation area of a scattering surface) is approximately λ^2. For an aberration-free lens of numerical aperture, na, the area of the point spread function is λ^2/na^2. Thus the maximum number of scatterers contributing to a particular image point is $1/na^2$. For an illuminated area of diameter D, the maximum number of scatterers making a contribution is approximately D^2/λ^2. In a majority of the applications and instruments discussed below, it is assumed that a large number of scatterers contribute to the pattern. This condition is probably satisfied for most practical situations but will not hold for high resolution optical systems or highly focussed beams. If 100 scatterers is chosen as the lower bound for a sufficiently large number to obtain acceptable statistics, then the properties of speckle patterns should be highly dependent on surface statistics for *na* greater than about 0.1 and illuminating beam diameters less than about 10λ. In these regimes, relationships between surface characteristics and speckle are not well understood but are currently active areas of investigation[109,110,112].

The effect of surface scatterer motion on the properties of speckle patterns is particularly important for understanding the potential of speckle as an on-line measurement technique. The basic concept[111] governing these effects is that a three-dimensional interference pattern fixed relative to the scattering surface exists in the space of the scattered illumination. If a surface is illuminated with a parallel beam of light at an arbitrary angle of incidence, then the speckle pattern follows any translation motion of the surface in its mean plane. The mean plane refers to the ideal flat surface about which the peaks and valleys of the surface topography are distributed and which gives rise to the specularly reflected light beam. The bright and dark regions of the speckle patterns extend over three-dimensional cigar-shaped volumes of length $4\lambda/\alpha^2$. Object-image plane displacements over distances comparable to this dimension should, therefore, cause no first order changes in the speckle patterns.

Having these ideas as a background, the two broad classes of measurement methods for determining the roughness properties of a surface from speckle patterns can be discussed: speckle contrast and speckle pattern decorrelation. In both methods, the roughness properties are obtained from the speckle patterns by empirically relating either the contrast or the degree of pattern correlation to the roughness of the surface under study. The empirical relationship is then interpreted with first order theories of speckle pattern formation, which generally assume that only single scattering of the electromagnetic wave takes place and that the scattering surfaces can be characterized by a Gaussian distribution of heights with a correlation length much less than the dimensions of the scattering region. It is doubtful that these assumptions are fully satisfied in most practical scattering problems. The results given below, however, demonstrate that within these two classes of measurement techniques there are methods to determine R_q of a surface over the large range of 0.01μm to 30μm with a reasonable degree of confidence. One may equally well treat speckle patterns produced by transmitted light from transparent objects with rough surfaces; the majority of speckle experiments have been performed this way. However, because of the application to metal surfaces intended here, most of the examples have been chosen from work in the reflection mode.

Speckle contrast

In speckle contrast measurements the intensity variations are quantified in terms of an average contrast V defined as the normalized standard deviation σ_I of intensity variations at the observation plane. Let $I(r)$ represent the intensity at the observation point r, then V is defined as

$$V = \sigma_I/\langle I(r)\rangle = [\langle I^2(r)\rangle - \langle I(r)\rangle^2]^{1/2}/\langle I(r)\rangle, \quad (15)$$

where the brackets $\langle\,\rangle$ indicate an average over a large number of intensity measurements. The intensity variations are determined by either moving the specimen and thereby the speckle pattern past a fixed detector with an aperture smaller than the speckle size or by moving the detector through the speckle field. High-contrast speckle patterns are produced when all the interfering wavelets have sufficient phase difference ($>2\pi$) to give complete destructive interference at some points in the pattern and the illumination has a high degree of spatial and temporal coherence. Spatial coherence means that the phases of the electromagnetic field at two points spaced across the propagating wavefront are highly correlated and temporal coherence that the phases of the field at two points spaced along the direction of light propagation are highly correlated.

Speckle contrast is unity for fully coherent monochromatic light illuminating a surface whose R_q is much larger than λ so that the wavelet phases are uniformly distributed over the interval from 0 to 2π. Correspondingly, for coherent monochromatic illumination, as the reflecting surface becomes smoother and less complete destructive interference occurs, the contrast decreases toward zero.

For any roughness value a decrease in speckle contrast occurs when the degree of either spatial or temporal coherence of the illumination is reduced. For temporal

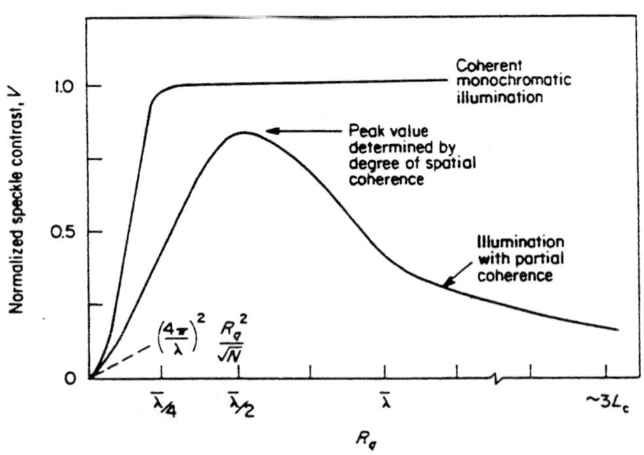

Fig 17 Schematic graph of the variation of speckle contrast with surface roughness and degree of coherence of illumination (distilled from ideas found in Refs. 112, 116 and 120–123)

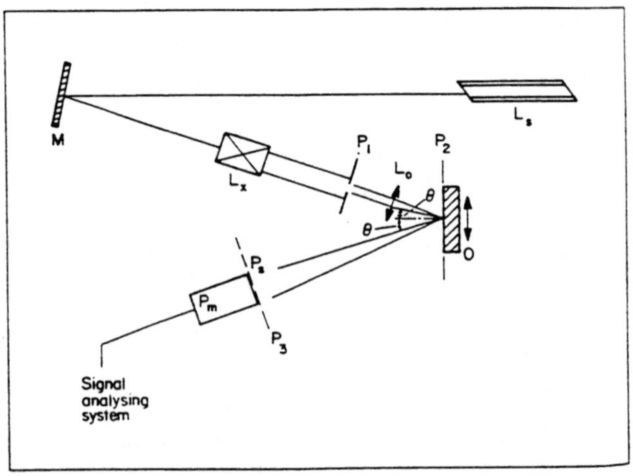

Fig 18 Aparatus used by Fujii and Asakura for measuring surface roughness dependence of speckle contrast. (From Asakara[124])

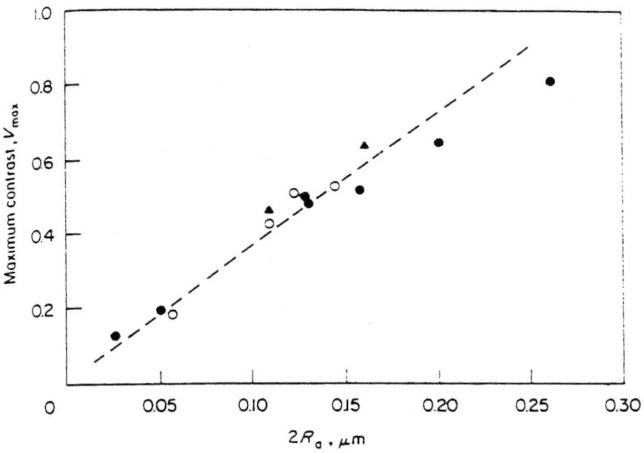

Fig 19 Maximum average contrast V_{max} of speckle intensity variations as a function of R_a roughness of metal surfaces. Lapped alloy steel plates, diamond turned aluminium and brass, and silicon-carbide polished alloy steel surfaces were included in the results shown

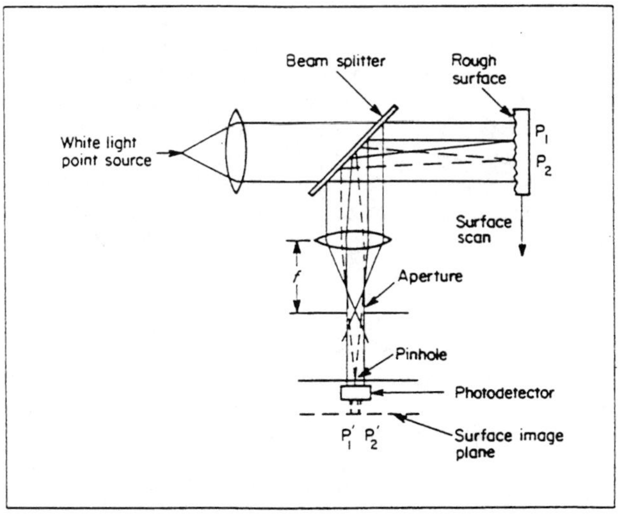

Fig 20 Apparatus used by Sprague for performing white light speckle contrast measurements. (From Ref. 125)

coherence the contrast decreases when the coherence length L_c of the illuminating wave field begins to have dimensions comparable to the roughness heights. In this regime as the roughness becomes larger than the dimensions of the region of coherence, an increasingly smaller number of scatterers give coherent contributions to the scattered field. If $\Delta\nu$ represents the frequency bandwidth of the broadband illumination, then $L_c \approx c/\Delta\nu$ where c is the speed of light; equivalently $L_c = (\lambda^2/\Delta\lambda)$ where λ is the mean wavelength of illumination and $\Delta\lambda$ is the effective width of the wavelength spectrum.

The effects due to decreasing spatial coherence are more difficult to conceptualize. Dainty[119] showed that the speckle pattern produced in an image plane by partially spatially coherent illumination is simply the weighted sum of the intensities of speckle patterns produced by different angles of incidence of coherent illumination. The decrease in contrast for illumination with partial spatial coherence may then be thought of as arising from a washing-out of the interference from this averaging over different angles of effective incidence. Fig 17 summarizes the combined effects of different degrees of coherence and scatterer surface roughness on speckle contrast.

Experiments to relate surface roughness to the contrast of speckle patterns produced by coherent monochromatic illumination were first performed by Fujii and Asakura[124]. In their apparatus, arranged for reflection measurements (Fig 18), light from a HeNe laser L_s was expanded by the lens arrangement L_x so that it uniformly illuminated an aperture P_1. The light transmitted by P_1 was focused by lens L_0 onto the test surface at P_2, which during measurements was translated along the direction shown by 0. The scattered light produced a speckle pattern in the far field region of the illuminated surface whose intensity variations were detected by a photomultiplier P_m apertured by the pinhole P_s whose diameter was made smaller than the average grain size of the speckle. Measurements of V were performed for ten sample objects with various amplitudes and types of surface roughness.

Results of these contrast measurements (Fig 19) show that a strong, linear correlation exists between V and R_a determined by stylus profilometry for R_a values up to 0.13 μm. It is important to note that these data were obtained only after V was maximized by varying the illumination spot size, although this adjustment does not seem to be critical. The roughness range to which the coherent monochromatic speckle contrast is sensitive can be increased by changing the incident angle θ_i so that the vertical projections of roughness heights along the direction of propagation remain less than about 0.1 λ, or by increasing the illumination wavelength. However, the use of a CO_2 laser with a wavelength of 10.6 μm would only extend the upper range to about 1 μm.

The first measurements to relate polychromatic speckle contrast and surface roughness were done by Sprague[125]. The source in his experimental apparatus (Fig 20) was a tungsten-zirconium arc which in combination with the spectral response of his photodetector produced a mean wavelength for the experiment of about 0.5 μm and a coherence length of about 1.5 μm. From intensity variations as a function of specimen position Sprague calculated the arithmetic average of absolute deviations from a mean value rather than the rms V given by Eq (15). Sprague's paper is particularly useful in that it gives specific dimensions for the apparatus. Pertinent dimensions were: distance from aperture stop to pinhole — 125 mm; pinhole diameter — 35 μm; aperture diameter — 3.75 mm; magnification of lens— 11; distance of pinhole from image plane — 288 mm. The detector was placed ahead of the image plane to discriminate against low spatial frequency components of the surface. Results of these measurements (Fig 21) show a good monotonic relation between polychromatic speckle contrast and surface roughness as determined with a stylus instrument for roughnesses from about λ/2 up to about $2L_c$, ie 0.2 μm to 3.2 μm. Again in contrast to the specular reflectance and diffuseness measurements, the correlation of contrast to roughness seems to be relatively independent of the type of finishing process used to form the surface.

The results shown in Fig 21 are in substantial agreement with the subsequent theory of polychromatic speckle[112,121], which for normally incident illumination and viewing at the centre of the reflected light beam gives:

$$V^2 = 1/[1+(4\Delta\nu R_q)^2]^{1/2}. \tag{16}$$

Eq (16) also indicates that use of light with a narrower spectral range than the arc used by Sprague could potentially extend the range of the polychromatic speckle contrast method. Parry[126] has explored this possibility and shown that use of six dominant lines of an argon ion laser ($\Delta\nu \approx 0.04$ μm) extends the upper limit of R_q to about 5 μm.

Nakagawa and Asakura[127] have recently re-examined the contrast dependence of polychromatic image speckle on surface roughness. They studied the variation of speckle contrast for light transmitted through ground glass plates with R_q of 0.33, 0.61, 1.9 and 6.6 μm and various illumination coherence lengths of 1.5, 2.3, 3.2, 4.8, 6.3 and 15μm. For R_q less than about 1 μm they found the same behaviour as did Sprague. However, for $R_q > 1\,\mu$m the contrast was found to increase monotonically with R_q. This increase was attributed to a decrease in the number of scatterers (for larger roughness values) within the point spread function of their optical system. Sprague's experimental configuration was such that for all roughness values measured, a sufficient number of scatterers was included in the illuminated region that no increase in contrast was observed.

The average contrast V of image plane speckle patterns as a function of the coherence length of partially spatially coherent illumination and surface roughness has been studied theoretically and experimentally by Fujii and Asakura[116]. Controlled partial spatial coherence was obtained by changing the diameter of a pinhole at the focal point of a lens used for converging light from a high-pressure mercury-arc lamp. They defined an optical roughness parameter Ψ:

$$\Psi = (2\,\pi/\lambda)^2\, R_q\, (R_q/r_\phi), \tag{17}$$

where r_ϕ is the correlation length of the phase variations of the wavefronts adjacent to the scattering specimen.

Fig 22 shows the combined results of their work. It assumes that r_ϕ is equal to the autocorrelation length a of the surface roughness which they obtained by profilometry. This assumption is probably not valid since Goodman[128] has shown that the ratio of the correlation area of a reflected (or transmitted) wave to that of the surface is a sharply decreasing function of R_q; the ratio of the two correlation areas has dropped to only 0.1 for $R_q \approx 0.25\,\lambda$. Regardless of this cautionary note, however, their results show that the speckle contrast may not be simply related to just the R_q value of the surface roughness; other surface roughness characteristics such as the rms gradient, R_q/r_ϕ, in Eq (17) will also be needed to determine the relationship between surface roughness and speckle contrast.

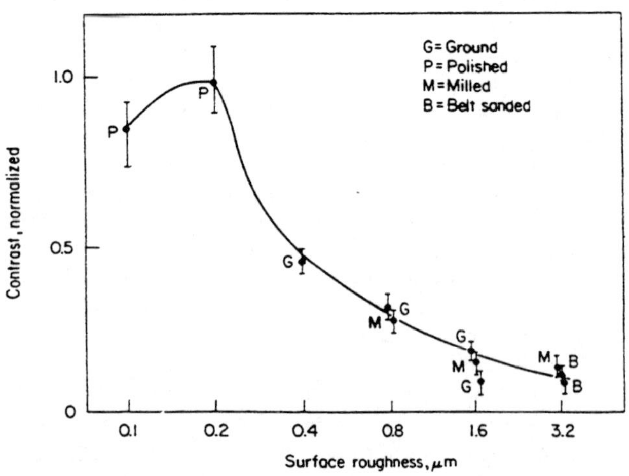

Fig 21 Normalized arithmetic average of absolute deviations of polychromatic speckle intensity as a function of specimen surface roughness as measured with a profilometer. (From Ref. 125)

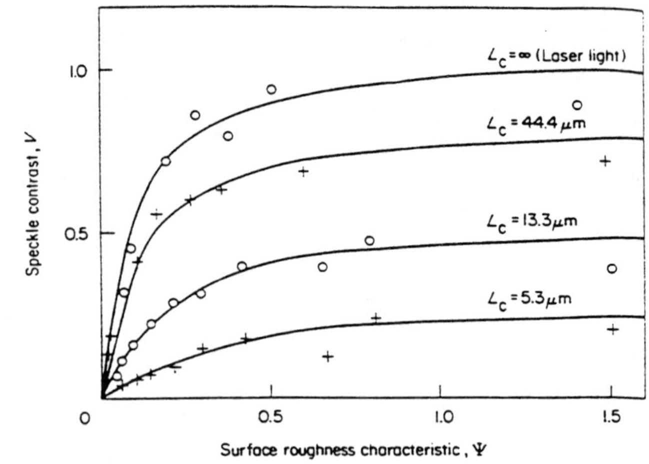

Fig 22 Theoretical calculations (-) and experimental measurements (+'s and O's) of speckle contrast as a function of the surface roughness characteristic Ψ. The contrast was measured with the imaging configuration of Fig 14(b) using transmitted illumination. Estimated coherence lengths L_c of the partially spatially coherent illumination were ∞, 44.4, 13.3 and 5.3 μm

Speckle pattern decorrelation

The second broad class of techniques for relating surface roughness and speckle is speckle pattern decorrelation measurement. Here two speckle patterns are obtained from the test surface by illuminating it with different angles of incidence or different wavelengths of light. Correlation properties of the speckle patterns are then studied by recording the patterns on the same photographic plate by double exposure[129÷131], or by photoelectric detection of the two patterns[132–135]. The primary attribute of this type of speckle measurement is that R_q values as large as 30 to 50 μm can be measured.

Theoretical work by Parry[136] and Goodman[137] form the basis for decorrelation measurements. The basic idea for intensity correlation measurements can be obtained from an equation derived by Parry[136] for the intensity cross-correlation function, γ, which is a measure of the similarity of two speckle patterns. γ is given by:

$$\gamma = \exp[-R_q^{\,2}(k_1-k_2)^2]\ \exp[-\beta\,|k_1 r_1 - k_2 r_2|^2], \tag{18}$$

where $k_i = (2\pi\cos\theta_i)/\lambda_i$, β is a constant proportional to the laser beam spot size on the scatterer, and r_i describes the observation points in a far field plane. θ_i is the angle of incidence of the light of wavelength λ_i. Thus, the value of γ is determined by the two radial scales, k_1 and k_2, of the speckle patterns. If $k_1 > k_2$, the speckle pattern at k_1 is a radially contracted replica of the pattern for k_2. The degree of replication depends on the first exponential, which is controlled by the surface roughness of the test specimen. If $R_q^{\,2}(k_1-k_2)^2 \ll 1$, a high degree of correlation is found and a marked radial structure appears[121]. If $R_q^{\,2}(k_1-k_2)^2 \gg 1$, the two patterns are totally uncorrelated and the radial structure disappears. As regards the second factor, experimental conditions are generally chosen such that $k_1 r_1 = k_2 r_2$ to maximize γ. The large range of R_q for the technique is clearly determined by the experimental capabilities for producing small values of $k_1 - k_2$. As indicated earlier, different values of k are obtained by either fixing θ and varying λ or vice-versa.

Fujii and Lit[132] studied a measure, γ_{rms}, of the correlation related to the γ of Eq (18). γ_{rms} is the rms difference of the two normalized intensities. They found:

$$\gamma_{rms} = 2 \{ 1\text{-}\exp [-\sigma_s^2 (k_1 - k_2)^2] \}^{1/2},$$

where σ_s was assumed to be $2R_q \cos\theta_i$ for reflecting surfaces and R_q ($n \cos\theta_i - \cos\theta_o$) for transmitting surfaces. Their optical system for measuring $I(k_1)$ and $I(k_2)$ is shown in Fig 23. Major components of the system were the multiline argon laser, pinhole P_a and a simple spectrometer. The spectrometer consisted of a lens L_1, a prism P_r, a mirror M and two photomultipliers D_1 and D_2. By translating the object surface O_t or O_r across the illuminating light, varying intensities were produced at D_1 and D_2 as a function of time. $I(k_1)$ and $I(k_2)$ were then analysed with analog circuitry to obtain the mean values necessary to compute γ_{rms}. Integration times were 10 s and corresponded to scanning 8.4 mm of the object surface.

The excellent agreement between theoretical calculations and experimental data obtained by Fujii and Lit is shown in Fig 24. σ_s was computed from the relation given after Eq (19). To calculate R_q they used the relation $R_q = (\pi/2)^{1/2} R_a$, which is correct for surfaces with Gaussian

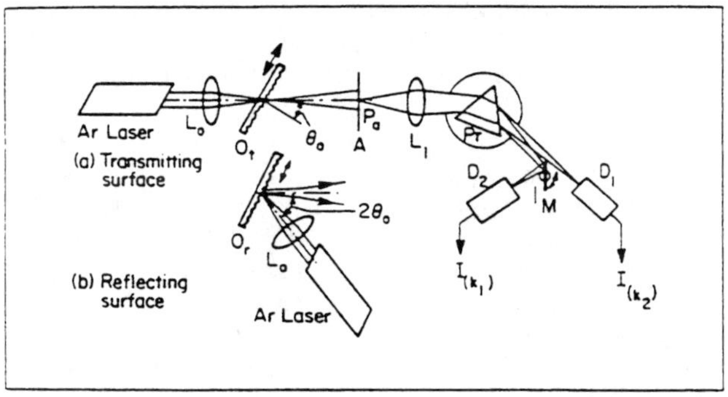

Fig 23 Optical system used by Fujii and Lit for performing speckle wavelength decorrelation measurements of surface roughness. (From Ref. 132)

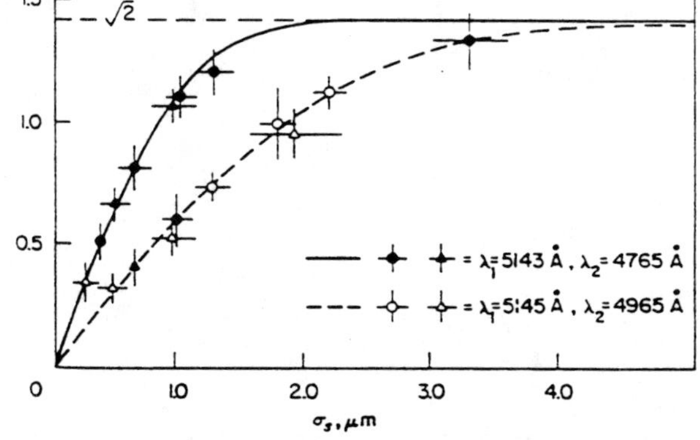

Fig 24 A comparison of experimental results and theoretical calculations for the rms difference between two speckle pattern intensities, γ_{rms}, as a function of surface roughness. The circles are for ground glass surfaces and the triangles are for ground metal surfaces. The bars indicate standard deviations. (From Ref. 132)

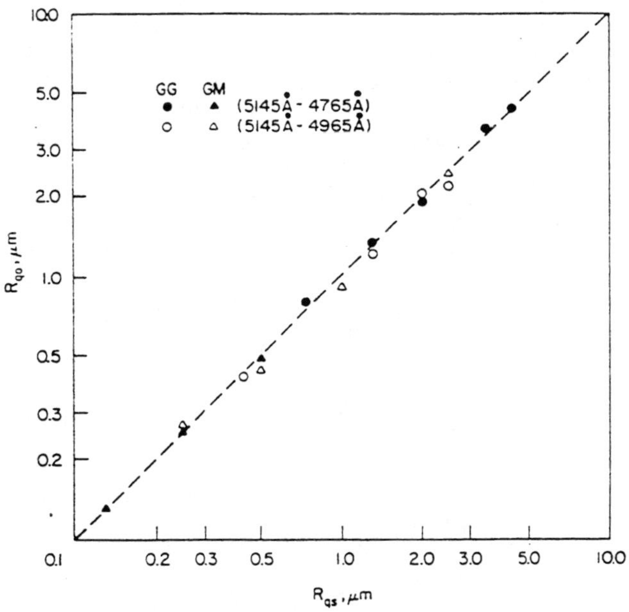

Fig 25 A transformation of the data in Fig 24 to show the correlation between stylus measurements of rms roughness, R_{qs}, and those obtained from speckle decorrelation measurement, R_{qo}. (From Ref. 132)

height distributions. R_a was obtained from stylus measurements of the surface roughness. A comparison between the R_q values from stylus measurements and those computed from Eq (19) with measurements of γ_{rms} are shown in Fig 25. The correlation between the two sets of data is better than 90% over a range of R_q from 0.13 μm for a metal surface to 6 μm for a ground glass plate[132].

Somewhat similar optical system configurations have been employed by Giglio *et al*[133] and Wykes[134]. However, Giglio *et al* processed the electronic signals from the photodetectors with a commercially available cross-correlator to obtain the γ of Eq (18). From plots of γ as a function of $\Delta k = k_1 - k_2$ they deduced R_q values for four ground glass surfaces independently from stylus measurements. They related these to R_a values using the relation $R_q = (\pi/2)^{1/2} R_a$ and compared the results with stylus measurements of R_a. The stylus measurements of the glasses gave R_a as 0.5 μm, 1.5 μm, 2.5 μm, and 6 μm while speckle decorrelation measurements gave 0.5 μm, 1.48 μm, 2.05 μm and 4.8 μm respectively. The assumption of a Gaussian height distribution and its effects on the validity of the theory and on the relationship between R_a and R_q are the most probable explanations for the discrepancies.

Changing the k in Eq (18) by changing the angle of incidence has been studied by Leger and Perrin[135]. Their experimental set-up for superimposing the fields of two speckle patterns is shown in Fig 26. A Michelson interferometer I_1 is used to illuminate the test surface with two plane waves through the focal system of lenses L_1 and L_1'. The angle difference between the two incident beams $\delta\theta_1$ can then be varied by varying the orientation of M_1 or M_1'. The detecting interferometer I_2 is an identical system so that the orientation of M_2 or M_2' can be adjusted to give the proper conditions for speckle pattern interference. In order to measure the visibility of interference fringes automatically, the phase of one of the incident wavefronts is modulated by mounting M_1 on a piezoelectric displacer.

Visibility is then given by the modulation depth, V_F, of the photoelectric signal at *D*.

To overcome the critical adjustments of the system shown in Fig 26, Leger and Perrin incorporated mirrors M_1 and M_2 into a single block so that $\delta\theta_1 = \delta\theta_2$ automatically. Also prisms were incorporated into the two arms of the interferometer I_2 to permit measurements of the modulation depth without any optimization of $\delta\theta_2$ for each $\delta\theta_1$. With these refinements the instrument was configured such that the rotation of the block containing mirrors M_1 and M_2 could be graduated directly as a function of R_q. R_q was obtained from V_F using the equation

$$V_F/V_F{}^{max} = \exp\left[-2(2\pi R_q \sin\theta\ \delta\theta/\lambda)^2\right], \tag{20}$$

where $\delta\theta = \delta\theta_1 = \delta\theta_2$ and $\theta_1 = \theta_2 = \theta$ and $V_F{}^{max}$ is the value of the modulation depth measured for small values of $\delta\theta$. The procedure used by Leger and Perrin was to obtain plots of V_F against $\delta\theta$ and to assume that R_q was inversely proportional to the $\delta\theta$ at which $V_F/V_F{}^{max} = 0.5$; for $\theta = 45°$ an approximate solution to Eq (20) under this condition is $R_q \approx 0.1\lambda/\delta\theta$. Note that except for the factor of two, the argument of the exponential is equal to that of Eq (18). (The trigonometric identity, $\cos(\theta + \delta\theta) = \cos\theta - \sin\theta\delta\theta$ for small $\delta\theta$, leads to the equality.)

Nominally isotropic metal surfaces were used for the measurements. The angle θ was equal to 45°. A comparison of the theoretical and experimental results (Fig 27) shows good agreement for surfaces with R_q ranging from 8 to 32 μm.

Starting from Sprague's first experiments, the field of determining surface roughness from the properties of speckle is only eight years old. As the examples of this subsection have demonstrated, the speckle patterns are rich in information about the microtopography of a test surface. However, because of its relative immaturity, the field has not yet yielded techniques for obtaining characterizations of roughness other than the R_q value. While the second-order statistical properties of speckle patterns, ie decorrelation techniques, have been used to obtain R_q values, attempts to obtain higher order statistical properties of the surface roughness such as the autocorrelation length will require further research.

We are not aware of any commercial instrument to measure roughness from speckle. Yet the range of R_q values measurable with speckle techniques and their apparent insensitivity to the type of material and type of surface forming process indicate that these techniques have a high potential for on-line measurements. Within the limits which have been discussed for the theoretical models, some of the speckle techniques offer the possibility of determining 'absolute' roughness values over a wide range. However, in evaluating theories of speckle one should not let the elegant and sophisticated statistical models overshadow the limiting assumptions of Gaussian height distribution, single scattering, etc, for the same physical processes which govern speckle and all light scattering.

Profiling techniques

Interferometry

Interferometry is an important technique for measuring the roughness and figure of high-quality, optical surfaces[138–143]. One important type is the Michelson two-beam interferometer[140] (Fig 28). An optical wavefront is split into two coherent beams by a partially transmitting mirror M. One beam is reflected from a smooth flat reference surface R while the other is reflected from the surface under test S. The reflected beams are then recombined at M. Under conditions of perfect alignment a circular interference pattern of light and dark fringes is observed at 0. With a small amount of tilt between either S or R, the interferometer will produce a set of generally parallel fringes. The undulations of each fringe reveal the peaks and valleys of surface S in a manner similar to that of a roughness profile obtained from a stylus instrument (Fig 29). Interferometry enables one to measure surface heights directly in terms of optical wavelengths. Its ultimate resolution using multiple beams can be as small as 0.05 nm peak-valley[144] but is generally about 1 nm rms[141].

There are, however, undesirable features which limit the applicability of interferometry for on-line measurement of a wide range of surfaces. Firstly, since the interferometer normally produces an optical fringe, added instrumentation and data processing are required to convert the fringe into profile data. Secondly, the vertical range is limited. If the surface roughness is greater

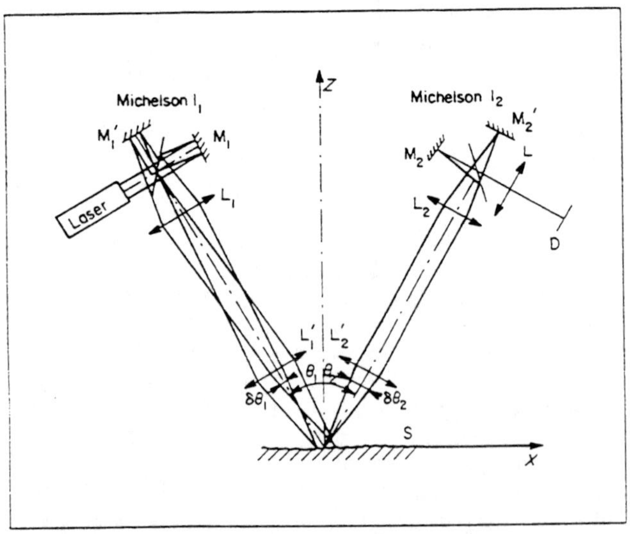

Fig 26 Experimental set-up used by Leger and Perrin for performing speckle angle decorrelation measurement of surface roughness. (From Ref. 135)

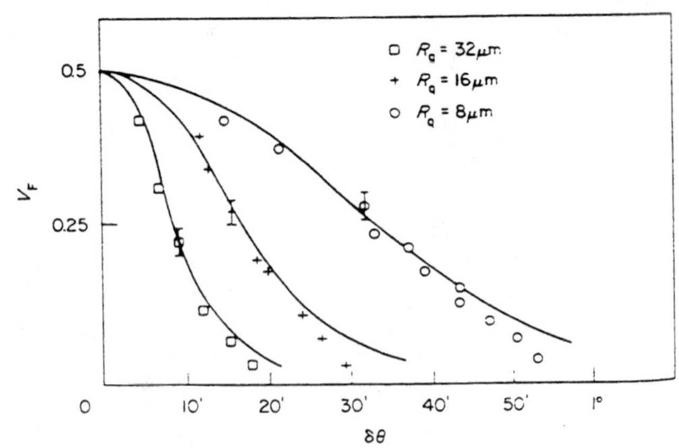

Fig 27 Experimental results and theoretical curves for the decrease of the visibility, V_f, of fringes by increasing the difference between the angles of incidence for different values of R_q. The theoretical curves are calculated from Eq (20) with $V_f{}^{max} = 0.5$

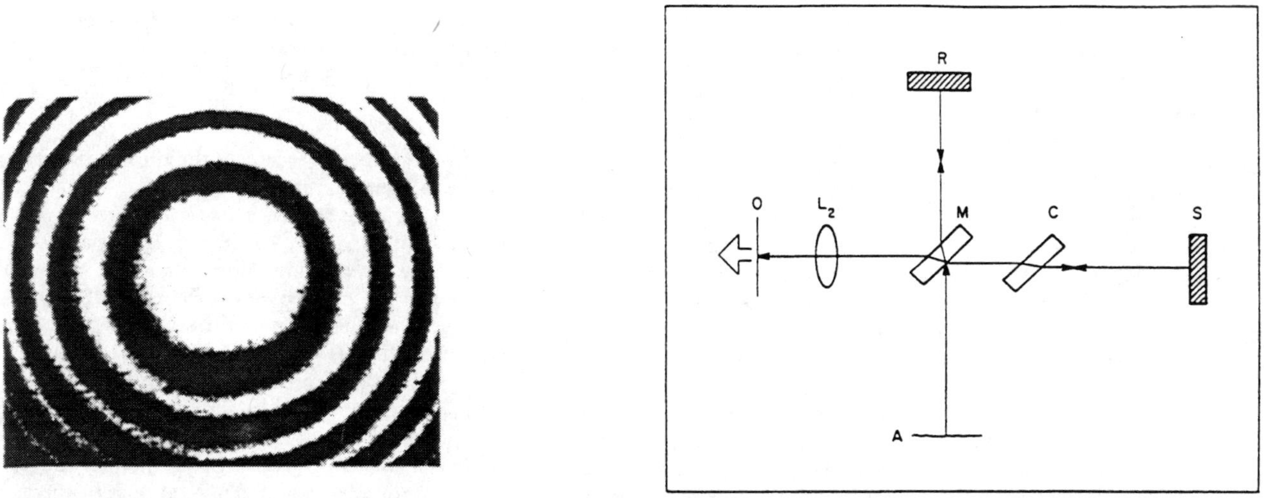

Fig 28 Michelson two-beam interferometer. A is an extended source of monochromatic light. The fringes are observed at O. C is a compensator positioned to keep the optical path lengths of the two beams the same. The photo was taken from Ref. 138. It is actually a pattern of Newton's rings but the form is similar to what is seen in the interferometer

than λ/2, the interference fringes tend to merge into one another making it impossible with conventional interferometry to track one fringe at a time. Thirdly, like stylus instruments, interferometers produce a series of 2-d profiles rather than 3-d surface maps.

A feco scanning interferometer has been developed by Bennett[141]. It uses a television camera to scan the image of a single interference fringe and the resulting electrical signals are then fed to a signal averager and stored in a minicomputer. In spite of this automated data handling, the feco interferometer is essentially a profiling instrument and hence acquires data slowly. The feco technique is a form of multiple beam[138] rather than two-beam interferometry. Bennett's feco system has a vertical resolution of about 0.8 nm rms and a horizontal resolution of 2 μm.

Surface roughness greater than λ/2 has been measured successfully by the heterodyning interferometer of Pettigrew and Hancock[145]. The instrument is operated at two wavelengths, λ_1 and λ_2, in such a way that the effective wavelength of the interferometer λ_{eff} is given by

$$\lambda_{eff} = \lambda_1 \lambda_2/(\lambda_2-\lambda_1). \qquad (21)$$

By choosing wavelengths λ_1 and λ_2 that are approximately equal, λ_{eff} can be made quite large so that fairly rough surfaces may be studied. This instrument has a proven roughness range of about 10 μm peak-to-valley, and its horizontal resolution is approximately 2.5 μm. It is similar to a stylus instrument in that the output profile is not an optical fringe but rather a time sequence of surface heights measured as the surface is scanned under the instrument.

Another heterodyning interferometer which produces a profile has been developed recently by Sommargren[146]. It has a height sensitivity of 0.1 nm, but its uniqueness lies in the fact that the surface scan and the resulting surface profiles are circular rather than straight.

Like stylus instruments interferometers are not ideally suited for rapid, on-line surface measurement. Interferometry is also highly sensitive to surface vibrations and it requires added instrumentation to convert the optical fringes into profile data. The range of the technique is fairly limited and hence unsuitable for studying highly curved surfaces, such as the inside bores of parts. Finally,

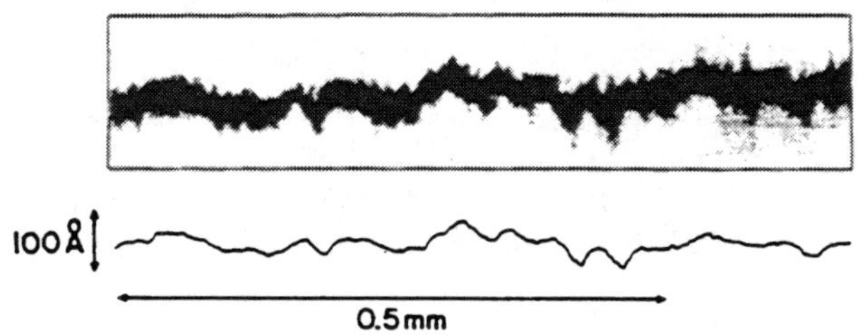

Fig 29 feco interference fringe profile of a Cu surface machined by single-point diamond turning[141]

interferometry generally requires surfaces that have a high reflectivity so that the contrast of the fringes will be high.

On the positive side, the vertical resolution of interferometry, within the horizontal bandwidth of the system, can be about 1 nm or better and the technique is suitable for studying both isotropic and anisotropic surfaces. For these reasons interferometry has been used mainly to study the fine microroughness of highly reflecting smooth surfaces.

Motycka[147], Ribbens[148–150] and Kwon *et al*[151] have proposed variations of the interferometric technique which rely on the fringe visibility to determine surface roughness. These are parametric rather than profiling techniques and therefore, they should be more adaptable for fast, on-line measurements. The CO_2 interferometer developed by Kwon *et al*[151] has been used to measure rough optical surfaces with R_q as great as 2.2μm and it therefore shows promise. However, the methods of Motycka[147] and Ribbens[148–150] are still in the proposal or early experimental stages and there is no telling whether they can be used with engineering surfaces.

Other profiling techniques

In addition to the major classes previously mentioned, a number of other optical techniques have been proposed and developed for surface texture measurement. Four profiling techniques for which working prototypes have been developed and briefly described here. The first two rely on principles of geometrical optics for profile measurement.

In a microscope objective lens, the position of the image plane depends sensitively on the position of the source plane. Accordingly, the optical profiling instrument developed by Sawatari and Zipin[152] detects the height of a point on a surface by measuring the position of its image plane formed in a research microscope. The instrument detects the image plane focus position by photodetection techniques similar to those developed by Simon[153] and Simpson *et al*[154–155]. It has a horizontal resolution of ~ 10 μm and a vertical range of ~ 12.4 μm. Because it has a slit aperture rather than a circular aperture, the optical profile transducer is designed to measure surfaces with a one-dimensional lay; however, the basic principles of the Simon-Simpson technique do not preclude the measurement of isotropic surfaces. A similar instrument has been developed by Arecchi *et al*[156] who claim an ultimate vertical resolution of ~0.1μm and no limitation on the isotropy of the surface.

Sato *et al*[157-159] have constructed two instruments for measuring surface profiles. Both use the method of light sectioning whereby the height of the surface is determined by the lateral position of the spot cast on the surface by an obliquely incident light beam. One of these instruments has been used on-line to measure the roughness of spinning parts during machining. The resolution limit is about 1 μm, however.

An instrument developed by Whitefield[160] uses optical Fourier transform techniques to measure surface profiles. Accurate profiles of isolated pits and scratches have been obtained at high resolution but the usefulness of this instrument for measuring random and periodic roughness texture has yet to be demonstrated.

Surface profiles have also been obtained by Goedgebuer and Vienot[161–163] with a new technique known as temporal holography, which combines the methods of two beam interferometry, spectroscopy, and holography to produce surface roughness profiles. This work is still in the experimental stage and more work is required before an adequate evaluation can be made.

Summary

The optical techniques emphasized in the previous sections are summarized in Table 1. Many of the entries are derived by assuming that the light source is a HeNe laser with λ = 0.6328 μm which illuminates a 1 mm spot on the surface. The distinction between profiling and parametric techniques determines many of the other properties in the table. Parametric techniques are faster than profiling techniques so they are more suited for on-line measurements in real time. Their results, however, are harder to quantify in terms of surface topographic parameters. In the present stage of development, parametric instruments are useful only as comparators in the engineering surface regime (0.1 μm $< R_a <$ 10 μm), ie roughness measurements can be made provided that careful calibrations have been done with known samples of the same material that have been formed by the same processes as the unknown surface. Considerable development remains to be done before any of the parametric techniques will be adequate as metrological tools like the stylus, whose measurements are understood from first principles and which is generally useful for measuring geometric parameters of a wide range of surfaces.

Parametric techniques are inherently three-dimensional. The sampling area is limited by the size of the illumination spot which is typically about 1 mm. The spot size may be increased or decreased by proper focusing conditions but it should be large enough so that the sampled area includes a statistically large number of surface peaks and valleys. It is important to note that although their area resolution is limited by the spot size, the parametric techniques are sensitive to surface wavelengths as small as the optical wavelength.

Profiling techniques are much slower since they gather data sequentially one element at a time, but they

Table 1 Summary of optical techniques for surface roughness measurement

Technique	Quantitative? Capable of absolute measurements?	Can measure amplitude(A) & wavelength(W) parameters?	Speed (number of detector readings)	Horizontal resolution	Approximate vertical resolution, μm	Vertical range of R_a, μm	Surface wavelength range, μm	Limits of vibration⁺	Can study isotropic or anistropic surfaces?	Can study inner surfaces of parts?	Working devices existing?
Parametric											
Specular Reflectance	For $R_q < \lambda$	A	2	~ 1mm	$\lesssim 0.001$	~ 1	0.6–100	0.1D*	Yes	Yes	Yes
tis	For $R_q << \lambda$	A	2	~ 1mm	$\lesssim 0.001$	0.1	0.6–100	0.1D*	Yes	No	Yes
Diffuseness	Not yet	A	2	~ 1mm	~ 0.01	10	0.6–100	0.1D*	Yes	Yes	Yes
ad	For $R_q << \lambda$	AW	10–100	~ 1mm	$\lesssim 0.001$	> 1	0.6–100	0.1D*	Yes	Yes	Yes
Laser Speckle Contrast	$R_q < 0.1\ \lambda$	A	100	~ 10 λ	~ 10% of range	0.1	0.6–100	$4\lambda/\alpha^2$	Yes	‡ Q	No
Polychromatic Speckle Contrast	$\lambda/2 < R_q < 2L_c$	A	100	~ 1mm	~10% of range	2–5	0.6–100	$4\lambda/\alpha^2$	Yes	Q	No
Speckle Correlation	$\lambda < R_q < 10\lambda$	A	10	~ 1mm	~10% of range	1–30	0.6–100	$4\lambda/\alpha^2$	Yes	Q	No
Ellipsometry	Not yet	–	2	~ 1mm	<0.001	~ 1	?	?	Yes	Q	No
Profiling											
Interferometry	Yes	AW	1000	2 μm	$\lesssim 0.001$	< 1	1–500	Limited by the desired resolution	Yes	No	Yes

⁺ D* is the diameter of the detector aperture

‡ Q — questionable

lend themselves to quantification. The horizontal resolution is very high; in Bennett's feco system[141] it is ~ 2 μm. Horizontal resolution plays a different role for profiling techniques insofar as it determines both the minimum size of the sampled area and the lower limit of surface wavelengths which the profiling technique can sense.

In the table, the speed of each technique was estimated by estimating the number of individual detector readings that would be required for a measurement of R_q. In the case of specular reflectance, one could get by with a single measurement of ρ and one of ρ_0, whereas in the case of interferometry with a digitized instrument, 1000 elemental height measurements might be needed to obtain a precise value for R_q.

The column headed 'limits of vibration' gives an order-of-magnitude estimate of the peak-to-valley surface vibration that would significantly affect the accuracy of each technique. In most cases it is determined by surface motion that would cause significant motion of the reflected light out of the detector aperture D^*. Ten percent of the detector aperture is assigned as the upper limit for this motion. The requirement in interferometry is much more stringent because any motion of the surface directly carries over as an effective change in surface height. Hence, the vertical motions in interferometry must be less than the desired resolution of the roughness height measurements.

'Working devices' refers to commercial and other instruments that have been developed beyond the research and development stage and are being relied upon to make some sort of practical roughness measurement. It is felt that only ellipsometry and the speckle techniques are primarily in the experimental stages.

Overall there is no currently available technique which is adequate as an on-line metrological tool for roughness measurements of engineering surfaces. Specular reflectance, diffuseness, and angular distribution measurements show some promise. At the present state of the art these techniques can be used in a production line to sense differences between good and bad surfaces from a set which has been well characterized already.

One problem with ellipsometry is that it is sensitive to many surface properties in addition to topography. Interferometry is a quantitative technique but, at present, it is rather slow for on-line measurements. The problem can be solved perhaps with higher-speed computer scanning and processing of the optical fringes to obtain topographic parameters.

Speckle techniques perhaps hold the most promise as future on-line tools for surface roughness measurement. They can easily be adapted to the measurement of moving surfaces and early results suggest that roughness measurements can be quantified over the entire engineering surface regime.

Acknowledgements

The authors are grateful to R.D. Young for his support and encouragement during the course of this work and to E. Wolf, J.M. Bennett, E.L. Church, and J.C. Dainty from whom we learned a great deal. Thanks also go to F. Scire for assistance in searching the literature, to T. Lettieri, D.R. Flynn, G. White, and D. Nyyssonen for their careful reading of the manuscript, and to C. Stull and M. Johnson for assistance in preparing the manuscript. This work was supported in part by the Nondestructive Evaluation Program at the National Bureau of Standards and by NASA contract L4718B.

References

General

1. **McKeown P.A.** Why Precision? *Prec. Eng., 1979,* **1**, *59*

2. **Treywin E.T.** Trends in Measurement. *Prec. Eng., 1979,* **1**, *143*

3. **Young R.D.** Frontiers in Microtechnology. *Prec. Eng., 1980,* **2**, *3*

4. **Whitehouse D.J.** Micromechanics: A Cinderella Subject. *Prec. Eng., 1980,* **2**, *55*

5. **Merchant M.E.** Delphi-Type Forecast of the Future of Production Engineering. *CIRP Annals, 1971,* **20**, *213*

6. **Young R.D., Vorburger T.V. and Teague E.C.** In-Process and On-line Measurement of Surface Finish. *CIRP Annals, 1980,* **29**, *435*

7. **Schneider E.J.** Recent American and International Developments in the Assessment of Surface Quality and Their Effect on the Future. *Wear, 1979,* **59**, *17*

8. **Thomas T.R.** Surface Roughness Measurement: Alternatives to the Stylus. *No. 4.2, NELEX 78, (National Engineering Laboratory, East Kilbride, Glasgow, 1978)*

9. **Radhakrishnan V.** Application of Inductive Heads for Non-Contact Measurement of Surface Finish. *in Proc. Int. Conf. on Prod. Eng. 1977 (Inst. of Eng., Calcutta, 1977) p. 80*

10. **Teague E.C.** Evaluation, Revision, and Application of the NBS Stylus/Computer System for the Measurement of Surface Roughness. *NBS Tech. Note 902, (US Dept. of Commerce, Wash., DC, 1976).*

11. **Vorburger T.V., Teague E.C. and Scire F.E.** FAST Facility Available for Engineering Needs. *Dimensions/NBS* **62**, *No. 11, 18 (Nov. 1978)*

12. **Whitehouse D.J.** Stylus techniques. *in Characterization of Solid Surfaces, ed. P.E. Kane and G.R. Larrabee (Plenum, New York, 1974) Chap. 3*

13. **Peklenik J.** New Developments in Surface Characterization and Measurements by Means of Random Process Analysis. *Proc. I. Mech. E., 1967-68,* **182** *Pt 3K, 108*

14. **Sayles R.S. and Thomas T.R.** Measurements of the Statistical Microgeometry of Engineering Surfaces. *Trans. ASME, F: J. Lub. Tech., 1979,* **101**, *409*

15. **Whitehouse D.J.** Surfaces — a Link Between Manufacture and Function. *Proc.I.Mech.E., 1978,* **192**, *179*

16. American National Standards Institute B46.1-1978, Surface Texture *(ASME, New York, 1978)*

17. **Jungles J. and Whitehouse D.J.** An Investigation of the Shape and Dimensions of Some Diamond Styli. *J. Phys. E., 1970,* **3**, *437*

18. **Vorburger T.V., Teague E.C., Scire F.E. and Rosberry F.W.** Measurements of Stylus Radii. *Wear, 1979,* **57**, *39*

19. For discussions of surface texture parameters see Ref. 10-15 and Ref. 16, Appendix C.3.

20. **Williamson J.B.P., Pullen J. and Hunt R.T.** The Shape of Solid Surfaces. *in Surface Mechanics, ed. F.F. Ling (ASME, New York, 1969) p.24*

21. **Thomas T.R. and Sayles R.S.** Some Applications of Statistical Surface Measurements to Engineering Problems. *Mecanique Materiaux Electricite, Jan 1978, No. 337*

Previous reviews of optical techniques for studying surface texture

22. **Bennett H.E. and Bennett J.M.** Precision Measurement in Thin Film Optics. *in Physics of Thin Films, Vol 4 ed. G. Hass (Academic Press, New York, 1967) p. 1*

23. **Young R.D.** Eight Techniques for the Optical Measurement of Surface Roughness. *NBSIR 73-219 (US Dept. of Commerce, Wash., DC, 1973)*

24. **Welford W.T.** Optical Estimation of Statistics of Surface Roughness from Light Scattering Measurements. *Opt. Quant. Elect., 1977,* **9**, *269*

25. **Elson J.M., Bennett H.E. and Bennett J.M.** Scattering from Optical Surfaces. *in Applied Optics and Optical Engineering, Vol. 7 (Academic Press, New York, 1979) chap. 7*

Scattering

26. **Chinmayanandam T.K.** On the Specular Reflection from Rough Surfaces. *Phys. Rev., 1919,* **13**, *96*

27. **Bennett H.E. and Porteus J.O.** Relation between Surface Roughness and Specular Reflectance at Normal Incidence. *J. Opt. Soc. Am., 1961,* **51**, *123*

28. **Bennett H.E.** Specular Reflectance of Aluminized Ground Glass and the Height Distribution of Surface Irregularities. *J. Opt. Soc. Am., 1963,* **53**, *1389*

29. **Porteus J.O.** Relation between the Height Distribution of a Rough Surface and Reflectance at Normal Incidence. *J. Opt. Soc. Am., 1963,* **53**, *1394*

30. **Bennett H.E.** Measurement of Smooth Surface Finishes. *Ind. Qual. Control, 1964,* **20**, *1*

31. **Vashisht S.K. and Radhakrishnan V.** Gloss Meter for Surface Evaluation. *J.Inst.Eng. (India), 1974,* **55**, *PtME3, 144*

32. **Vashisht S.K. and Radhakrishnan V.** Surface Studies with a Gloss Meter. *Tribology Int., 1974,* **7**, *70*

33. **Tanner L.H. and Fahoum M.** A Study of the Surface Parameters of Ground and Lapped Metal Surfaces, Using Specular and Diffuse Reflection of Laser Light. *Wear, 1976,* **36**, *299*

34. **Tanner L.H.** The Use of Laser Light in the Study of Metal Surfaces. *Opt. and Laser Tech. 1976,* **8**, *113 and* A Comparison of Talysurf 10 and Optical Measurements of Roughness and Surface Slope. *Wear, 1979,* **57**, *81*

35. **Murray H.** Exploratory Investigation of Laser Methods for Grinding Research. *CIRP Annals, 1973,* **22**, *137*

36. **Hensler D.H.** Light Scattering from Fused Polycrystalline Aluminum Oxide Surfaces. *Appl. Opt., 1972,* **11**, *2522*

37. **Lin G.C.I., Shea T. and Hoang K.** Measurement of Surface Roughness with a Laser Beam. *in Aust. Conf. on Manuf. Eng. (Instn. Engrs. Aust., Barton, 1977) p. 132*

38. **Spurgeon D. and Slater R.A.C.** In-Process Indication of Surface Roughness Using a Fibre-Optics Transducer. *Proc. 15th Int. Machine Tool Design and Research Conf. ed. by S.A. Tobias and F. Koenigsberger (Macmillan, London, 1975) p. 339*

39. **Depew C.A. and Weir R.D.** Surface Roughness Determination by the Measurement of Reflectance. *Appl. Opt., 1971,* **10**, 969

40. **Cunningham L.J. and Braundmeier Jr A.J.** Measurement of the Correlation between the Specular Reflectance and Surface Roughness of Ag Films. *Phys. Rev. B, 1976,* **14**, *479*

41. **Adkins H.** A Look at Surface Finish. *Am. Mach., 8 Sept. 1969,* **113***(18), 111*

42. **Birkebak R.C.** Optical and Mechanical RMS Surface Roughness Comparison. *Appl. Opt., 1971,* **10**, *1970*

43. **Smith T.F. and Hering R.G.** Comparison of the Beckmann Model with Bidirectional Reflectance Measurements. *presented at the ASME-AICHE Heat Transfer Conf., Atlanta, 1973, ASME publ. 73-HT-11*

44. **Chandley P.J.** Surface Roughness Measurements for Coherent Light Scattering. *Opt. Quant. Elect., 1976,* **8**, *323*

45. **Christie J.S.** Review of Geometric Attributes of Appearance. *J. Coatings Tech., 1979,* **51**, *64*

46. **Beckmann P. and Spizzichino A.** The Scattering of Electromagnetic Waves from Rough Surfaces. *(Pergamon, New York, 1963) Chap. 5*

47. **Endriz J.G. and Spicer W.E.** Study of Aluminium Films. I. Optical Studies of Reflectance Drops and Surface Oscillations on Controlled-Roughness Films. *Phy. Rev. 1971,* **B4**, *4144*

48. **Decker D.L., Bennett J.M., Soileau M.J., Porteus J.O. and Bennett H.E.** Surface and Optical Studies of Diamond Turned and Other Metal Mirrors. *Opt. Eng., 1978,* **17**, *160*

49. **Bennett H.E.** Scattering Characteristics of Optical Materials. *Opt. Eng., 1978,* **17**, *480*

50. **Archibald P.C. and Bennett H.E.** Scattering from Infrared Missile Domes. *Opt. Eng., 1978,* **17**, *647*

51. **Bennett H.E. and Stanford J.L.** Structure-Related Optical Characteristics of Thin Metallic Films in the Visible and Ultraviolet. *J. Res. NBS, 1976,* **80A**, *643*

52. **Bennett J.M., Burge D.K., Rahn J.P. and Bennett H.E.** Standards for Optical Surface Quality Using Total Integrated Scattering. *in Contemporary Optical Systems and Components Specifications, Proc. SPIE, 1979,* **181**, *124*

53. **Hodgkinson I.J.** A Simple Scatter Method for Optical Surface Roughness and Slope Measurements of Polished Fused Silica. *J. Phys E, 1970,* **3**, *341*

54. **Arnold J.B., Steger P.J. and Saito T.T.** Tool Feed Influence on the Machinability of CO_2 Laser Optics. *Appl. Opt., 1975,* **14**, *1777*

55. **Saito T.T. and Kurdock J.R.** Diamond Turning and Polishing of Infrared Optical Components. *Appl. Opt., 1976,* **15**, *27*

56. **Thomas N.L.** Low-Scatter, Low-Loss Mirrors for Laser Gyros. *in Laser Inertial Rotation Sensors. Proc. SPIE, 1978,* **157**, *41*

57. **Bennett J.M., Archibald P.C. and Klugman A.** Scattering and Surface Finish of Unconventional Mirror Materials. *presented at the Workshop on Optical Fabrication and Testing, OSA Spring Conf. on Applied Optics (Mills College, Oakland, CA, June 1980)*

58. **Hausler G. and Hohberg G.** Streulicht Messungen an Polierten Glasoberflachen. *Optik, 1970,* **30**, *437*

59. **Peters J.** Messung des Mitterauhwertes Zylindrischer Teile Während des Schleifens, *VDI - Berichte, 1965,* **90**, *27*

60. **Corey H.S.** Surface Finish from Reflected Laser Light. *in Advances in Optical Metrology, Proc. SPIE, 1978,* **153**, *27*

61. **Takeyama H., Sekiguchi H. and Murata R.** In-Process Detection of Surface Roughness in Machining. *CIRP Annals, 1976,* **25**, *467*

62. **Clarke G.M. and Thomas T.R.** Roughness Measurement with a Laser Scanning Analyzer. *Wear, 1979,* **57**, *107*

63. **Stover J.C.** Roughness Measurement by Light Scattering. *in Laser Induced Damage in Optical Materials: 1974, NBS Special Publication 414 (Dept. of Comm., Wash., DC, 1974) p. 163*

64. **Stover J.C.** Roughness Characterization of Smooth Machined Surfaces by Light Scattering. *Appl. Opt., 1975,* **14**, *1796 and* Spectral Density Function Gives Surface Roughness. *Laser Focus, Feb. 1976,* **12***(2), 83*

65. **Church E.L., Jenkinson H.A. and Zavada J.M.** Measurement of the Finish of Diamond-Turned Metal Surfaces by Differential Light Scattering. *Opt. Eng., 1977,* **16**, *360*

66. **Church E.L., Jenkinson H.A. and Zavada J.M.** Relationship between Surface Scattering and Micro-topographic Features. *Opt. Eng., 1979,* **18**, *125*

67. **Church E.L.** The Measurement of Surface Texture and Topography by Differential Light Scattering. *Wear, 1979,* **57**, *93*

68. **Elson J.M. and Bennett J.M.** Relation between the Angular Dependence of Scattering and the Statistical Properties of Optical Surfaces. *J. Opt. Soc. Am., 1979,* **69**, *31*

69. **Elson J.M., Rahn J.P. and Bennett J.M.** Light Scattering from Multilayer Optics: Comparison of Theory and Experiment. *Appl. Opt., 1980,* **19**, *669*

70. **Hard S. and Nilsson O.** Laser Heterodyne Apparatus for Roughness Measurements of Polished Surfaces. *Appl. Opt., 1978,* **17**, *3827*

71. **Leader J.C.** Analysis and Prediction of Laser Scattering from Rough-Surface Materials. *J. Opt. Soc. Am., 1979,* **69**, *610*

72. **Hildebrand B.P., Gordon R.L. and Allen E.V.** Instrument for Measuring the Roughness of Supersmooth Surfaces. *Appl. Opt., 1974,* **13**, *177*

73. **Heitmann D. and Permien V.** Determination of Surface Roughness From Light Scattering Measurements. *Opt. Comm., 1977,* **23**, *131*

74. **Edwin R.P.** Light Scattering as a Technique for Measuring the Roughness of Optical Surfaces. *J. Phys. E., 1973,* **6**, *55*

75. **Birkebak R.C. and Eckert E.R.G.** Effects of Roughness of Metal Surfaces on Angular Distribution of Monochromatic Reflected Radiation. *Trans. ASME, J. Heat Transf., 1965,* **86**, *85*

76. **Semplak R.A.** 0.64 μm Scatter Measurements from Teflon and Various Metallic Surfaces. *Bell Sys. Tech. J., 1965,* **44**, *1659*

77. **Thwaite E.G.** The Direct Measurement of the Power Spectrum of Rough Surfaces by Optical Fourier Transformation. *Wear, 1979,* **57**, *71 and* Power Spectra of Rough Surfaces Obtained by Optical Fourier Transformation. *CIRP Annals, 1980,* **29**, *419*

78. **Chandley P.J.** Determination of the Autocorrelation Function of Height on a Rough Surface from Coherent Light Scattering. *Opt. Quant. Elect., 1976,* **8**, *329*

79. **Smith T.F. and Hering R.G.** Bidirectional Reflectance of a Randomly Rough Surface. *Prog. in Astronautics and Aeronautics, 1972,* **29**, *69 and* Surface Roughness Effects on Bidirectional Reflectance. *Prog. in Astronautics and Aeronautics, 1974,* **35**, *145*

80. **Smith T.F., Suiter R.L. and Kanayama K.** Bidirectional Reflectance Measurements for One-Dimensional Randomly Rough Surfaces. *Proc. 14th Thermophysics Conf. (AIAA, New York, 1979)*

81. **Church E.L. and Zavada J.M.** Residual Surface Roughness of Diamond-Turned Optics. *Appl. Opt., 1975,* **14**, *1788*

82. **West R.N. and Stocker W.J.** Automatic Inspection of Cylinder Bores. *Metrology and Inspection, July 1977,* 9*(4), 9*

83. **Bailey W.** Optical Inspection of Cylinder Bores. *Tribology Int., 1977,* **10**, *319*

Ellipsometry

84. For an extensive review of the subject, see **Azzam R.M.A. and Bashara N.M.** *Ellipsometry and Polarized Light* (North Holland, Amsterdam, 1977)

85. **McCrackin F.L., Passaglia E., Stromberg R.R. and Steinberg H.L.** Measurement of the Thickness and Refractive Index of Very Thin Films and the Optical Properties of Surfaces by Ellipsometry. *J. Res. NBS, 1963,* **67A**, *363*

86. **Archer R.J.** Determination of the Properties of Films on Silicon by the Method of Ellipsometry. *J. Opt. Soc. Am., 1962,* **52**, *970*

87. **Kruger J. and Hayfield P.C.S.** Ellipsometry in Corrosion Testing. *in Handbook on Corrosion Testing and Evaluation, W.H. Ailor, ed. (John Wiley and Sons, New York, 1971) p. 783*

88. Ref. 84, p. 419

89. **Vedam K. and So S.S.** Characterization of Real Surfaces by Ellipsometry. *Surface Sci., 1972,* **29**, *379*

90. **Ibrahim M.M. and Bashara N.M.** Ellipsometric Study of 40eV Ion Damage in Silicon. *Surface Sci., 1972,* **30**, *632*

91. **Vedam K.** Characterization of Defects in Real Surfaces by Ellipsometry. *Surface Sci., 1976,* **56**, *221*

92. **Meyer F., Kluizenaar E.E. and den Engelsen D.** Ellipsometry Determination of Optical Anisotropy of Gallium Arsenide. *J. Opt. Soc. Am., 1973,* **63**, *529*

93. **Van der Meulen Y.J. and Hien N.C.** Design and Operation of an Automated High Temperature Ellipsometer. *J. Opt Soc. Am., 1974,* **64**, *804*

94. **Yokota H., Sakata H., Nishibori M. and Kinosita K.** Ellipsometric Study of Polished Glass Surfaces. *Surface Sci., 1969,* **16**, *265*

95. **Primak W.** Determination of Small Dilatations and Surface Stress by Birefringence Measurements. *Surface Sci., 1969,* **16**, *398*

96. **Lonardo P.M.** Measurement of Smooth Surface Roughness by Means of a Photometric Method. *CIRP Annals, 1974,* **23**, *189*

97. **Chiesorin P. and Lonardo P.M.** A New Sensor of Surface Roughness for Process Control System. *SME Tech. Paper MS77-216, 1977,*

98. **Lonardo P.M.** Testing a New Optical Sensor for In-Process Detection of Surface Roughness. *CIRP Annals, 1978,* **27**, *531*

99. **Smith T.** Effect of Surface Roughness on Ellipsometry of Aluminum. *Surface Sci., 1976,* **56**, *252*

100. **Smith T. and Lindberg G.** Interpretation of Ellipsometric Data for Transparent Films on Rough Metal Surfaces. *Surface Tech., 1979,* **8**, *1*

101. **Vorburger T.V. and Ludema K.C.** Ellipsometry of Rough Surfaces. *Appl. Opt., 1980,* **19**, *561*

102. **Fenstermaker C.A. and McCrackin F.L.** Errors Arising from Surface Roughness in Ellipsometric Measurement of the Refractive Index of a Surface. *Surface Sci., 1969,* **16**, *85*

103. **Ohlidal I. and Lukes F.** Ellipsometric Parameters of Rough Surfaces and of a System Substrate-Thin Film with Rough Boundaries. *Opt. Acta, 1972,* **19**, *817*

104. **Ohlidal I., Lukes F. and Navratil K.** Rough Silicon Surfaces Studied by Optical Methods. *Surface Sci., 1974,* **45**, *91*

105. **Church E.L. and Zavada J.M.** Effects of Surface Microroughness in Ellipsometry. *J. Opt. Soc. Am., 1976,* **66**, *1136A*

106. **Azzam R.M.A. and Bashara N.M.** Polarization Characteristics of Scattered Radiation From a Diffraction Grating by Ellipsometry with Application to Surface Roughness. *Phys. Rev., 1972,* **B5**, *4721*

107. **Aspnes D.E., Theeten J.B. and Hottier F.** Investigation of Effective Medium Models of Microscopic Surface Roughness by Spectroscopic Ellipsometry. *Phys. Rev, 1979,* **B20**, *3292*

108. See for example **Muller R.H.** Present Status of Automatic Ellipsometers. *Surface Sci., 1976,* **56**, *19* and other articles in the same issue

Speckle

109. Laser Speckle and Related Phenomena. *ed. J.C. Dainty (Springer-Verlag, Berlin, 1975)*

110. Speckle Metrology. *ed. R.K. Erf (Academic Press, New York, 1978)*

111. Laser Speckle and Applications in Optics *M. Francon, Translated by H.H. Arsenault (Academic Press, New York, 1979)*

112. **Dainty J.C.** The Statistics of Speckle Patterns. *in Progress in Optics, Vol. 14, ed. E. Wolf (North Holland, Amsterdam, 1976)*

113. Speckle in Optics. *J. Opt Soc. Am., 1976,* **66***(11)*

114. **Ennos A.E.** Speckle Interferometry. *in Progress in Optics, Vol. 16, ed E. Wolf (North Holland, Amsterdam, 1978) p. 236*

115. **McKechnie T.S.** Measurement of Some Second-Order Statistical Properties of Speckle. *Optik, 1974,* **39**, *258*

116. **Fujii H. and Asakura T.** A Contrast Variation of Image Speckle Intensity under Illumination of Partially Coherent Light. *Opt Comm., 1974,* **11**, *32*

117. **Fujii H., Asakura T. and Shindo Y.** Measurement of Surface Roughness Properties by Using Image Speckle Contrast. *J. Opt Soc., 1976,* **66**, *1217*

118. Page 8 in Ref. 112.

119. Page 22 in Ref. 112.

120. **Goodman J.W.** Statistical Properties of Laser Speckle Patterns. *in Ref. 109*

121. **Parry G.** Speckle Patterns in Partially Coherent Light. *in Ref. 109.*

122. **George N.** and **Jain A.** Space and Wavelength Dependence of Speckle Intensity. *Appl. Phys., 1974,* **4**, *201*

123. **Pedersen H.** Theory of Speckle Dependence on Surface Roughness. *J. Opt. Soc. Am., 1976,* **66**, *1204*

124. **Asakura T.** Surface Roughness Measurement. *in Ref. 110*

125. **Sprague R.A.** Surface Roughness Measurement Using White Light Speckle. *Appl. Opt., 1972,* **11**, *2811*

126. **Parry G.** The Scattering of Polychromatic Light from Rough Surfaces-First Order Statistics. *Opt. Quant. Elect., 1975,* **7**, *311*

127. **Nakagawa K. and Asakura T.** Contrast Dependence of White Light Image Speckles on Surface Roughness. *Opt. Comm., 1978,* **27**, *207*

128. Page 63 in Ref. 120.

129. **Mendez J.A. and Roblin M.L.** Utilization des franzes d'interference en lumière diffuse pour l'étude de l'état de surface d'un diffuseur. *Opt. Comm., 1975,* **13**, *142*

130. **Tribillon G.** Correlation entre deux speckles obtenus avec deux longueurs de l'onde — application à la mesure de la rugosite moyenne. *Opt. Comm., 1974,* **11**, *172*

131. **Leger D., Mathieu E. and Perrin J.C.** Optical Surface Roughness Determination Using Speckle Correlation Techniques. *Appl. Opt., 1975,* **14**, *872*

132. **Fujii H. and Lit J.W.Y.** Surface Roughness Measurement Using Dichromatic Speckle Pattern: An Experimental Study. *Appl. Opt., 1978,* **17**, *2690*

133. **Giglio M., Musazzi S. and Perini U.** Surface Roughness by Means of Speckle Wavelength Decorrelation. *Opt. Comm., 1979,* **28**, *166*

134. **Wykes C.** De-Correlation Effects in Speckle-Pattern Interferometry. *Opt. Acta, 1977,* **24**, *517*

135. **Leger D. and Perrin J.C.** Real-Time Measurement of Surface Roughness by Correlation of Speckle Patterns. *J. Opt. Soc. Am., 1976,* **66**, *1210*

136. **Parry G.** Some Effects of Surface Roughness on the Appearance of Speckle in Polychromatic Light. *Opt. Comm., 1974,* **12**, *75*

137. **Goodman J.W.** *Stanford Electronics Lab. Tech Report No. 2302-1 (1963)*

Interferometry

138. **Tolansky S.** *Surface Microtopography. (Interscience, New York, 1960).*

139. **Koehler W.F. and White W.C.** Multiple-Beam Fringes of Equal Chromatic Order Part VI. Methods of Measuring Roughness. *J. Opt. Soc. Am., 1955,* **45**, *1011*

140. **Baird K.M. and Hanes G.R.** Interferometers. *in Applied Optics and Optical Engineering, Vol. IV, ed. R. Kingslake (Academic Press, New York, 1967) Chap. 9*

141. **Bennett J.M.** Measurement of the RMS Roughness, Autocovariance Function and Other Statistical Properties of Optical Surfaces Using a FECO Scanning Interferometer. *Appl. Opt., 1976,* **15**, *2705*

142. **Dietz R.W. and Bennett J.M.** Smoothness and Thermal Stability of Cer-Vit Optical Material. *Appl. Opt., 1967,* **6**, *1275*

143. **Moody J.C.** Measurement of Ultrafine Surface Finishes. *ISA Trans., 1968,* **7**, *67*

144. **Lang J.E. and Scott G.D.** Resolution Limits in Multiple-Beam Interferometry. *J. Opt. Soc. Am., 1968,* **58**, *81*

145. **Pettigrew R.M. and Hancock F.J.** An Optical Profilometer. *Prec. Eng., 1979,* **1**, *133*

146. **Sommargren G.E.** Heterodyne Profilometer for Measuring Surface Microstructure. *J. Opt. Soc. Am., 1979,* **69**, *1404A*

147. **Motycka J.** Proposed Interferometric Method of Measurement of Roughness and Autocorrelation Function in Smooth-Finished Surfaces. *Appl. Opt., 1969,* **8**, *1435*

148. **Ribbens W.B.** Interferometric Surface Roughness Measurements. *Appl. Opt., 1969,* **8**, *2173*

149. **Ribbens W.B.** Surface Roughness Measurement by Holographic Interferometry. *Appl. Opt., 1972,* **11**, *807*

150. **Ribbens W.B.** Surface Roughness Measurement by Two Wavelength Holographic Interferometry. *Appl. Opt., 1974,* **13**, *1085*

151. **Kwon O., Wyant J.C. and Hayslett C.R.** Rough Surface Interferometry at 10.6 μm. *Appl. Opt., 1980,* **19**, *1862*

Profiling and other optical methods

152. **Sawatari T. and Zipin R.B.** Optical Profile Transducer. *Opt. Eng., 1979,* **18**, *222*

153. **Simon J.** New Noncontact Devices for Measuring Small Microdisplacements. *Appl. Opt., 1970,* **9**, *2337*

154. **Simpson J.A.** Use of a Microscope as a Non-contacting Microdisplacement Device. *Rev. Sci. Instrum., 1971,* **42**, *1378*

155. **Hartman A.W., Rosberry F.W. and Simpson J.A.** A Non-Contacting Length Comparator with 10 Nanometer Precision. *Opt. Eng., 1973,* **12**, *95*

156. **Arecchi F.T., Bertani D. and Ciliberto S.** A Fast Versatile Optical Profilometer. *Opt. Comm., 1979,* **31**, *263*

157. **Sato K. and Sato H.** Development of an In-Process Sensor for Surface Roughness by a Laser Beam. *Proc. 16th International Machine Tool Design and Research Conference, ed. F. Koenigsberger and S.A. Tobias (Macmillan, London, 1976) p. 171*

158. **Mitsui K. and Sato H.** Frequency Characteristic of Cutting Process Identified by an In-Process Measurement of Surface Roughness. *CIRP Annals, 1978,* **27**, 67

159. **Uchida S., Sato H. and O-hori M.** Two Dimensional Measurement of Surface Roughness by the Light Sectioning Method. *CIRP Annals, 1979,* **28**, *419*

160. **Whitefield R.J.** Noncontact Optical Profilometer. *Appl. Opt., 1975,* **14**, *2480*

161. **Goedgebuer J.P., Lacourt A. and Vienot J.C.** Holographic Behaviour of Channelled Spectra in Space and Time Optics. *Opt. Comm., 1976,* **16**, *99*

162. **Vienot J.C., Goedgebuer J.P. and Lacourt A.** Space and Time Variables in Optics and Holography: Recent Experimental Aspects. *Appl. Opt., 1977,* **16**, *454*

163. **Goedgebuer J.P. and Vienot J.C.** Space-Time Optics in Shape and Surface Metrology. *1st European Congress on Optics Applied to Metrology. Proc. SPIE, 1977,* **136**, *3*

Additional papers on optical techniques

Ribbens W.B. and Lazik G.L. Use of Optical Data Processing Techniques for Surface Roughness Studies. *Proc. IEEE (Letters), 1968,* **56**, *1637*

Felstead E.B. Additional Optical Data Processing Techniques for Surface Roughness Studies. *Proc. IEEE (Letters), 1969,* **57**, *343*

Kopf U. Die kohärenz-optische Messung der Rauhtiefe vor metallischen Oberflachen. *Z. Angew. Phys., 1971,* **31**, *156*

Aas J.A. Reconstruction of Surface Profiles from Their Diffraction Spectra. *Appl. Opt., 1972,* **11**, *1579*

Whitehouse D.J. Modern Methods of Assessing the Quality and Function of Surface Texture. *Soc. Manuf. Engrs., 1972, 1Q72-206*

Gesell T.F., Arakawa E.T., Williams M.W. and Hamm R.N. Optical and Photoelectric Properties of Mg in the Vacuum Ultraviolet. *Phys. Rev., 1973,* **B7**, *5141*

Bodesheim J. and Otto A. On the Quantitative Measurement of the Roughness Spectrum of Silver Films. *Surface Sci., 1974,* **45**, *441*

Levin A.I. Determination of the Surface-Finish Parameters by Means of the Total Internal Reflection Method. *Izmer. Tekh., 1974, No. 4, 40*

Saito T.T. and Simmons L.B. Performance Characteristics of Single Point Diamond Machined Metal Mirrors for Infrared Laser Applications. *Appl. Opt., 1974,* **13**, *2647*

Obradovich K.A. and Solodukho F.M. Reflectometric Methods of Surface Roughness Measurement. *Izmer. Tech., 1975, No. 1., 36*

Braundmeier Jr A.J. and Hall D.G. Some Measurements of the Angular Distribution and Polarization of Surface Plasmon Radiation from Ag Films. *Surface Sci., 1975,* **49**, *376*

Kuchin A.A. Optical Instruments for Surface Roughness Measurements. *Izmer. Tekh., 1975, No. 1, 33*

Lonardo P.M. Theoretical Determination of the Reflection Curves and Introduction of a New Method for Measuring Surface Gloss. *CIRP Annals, 1975,* **24**, *491*

Hornauer D.L. Light Scattering Experiments on Silver Films of Different Roughness Using Surface Plasmon Excitation. *Opt. Comm., 1976,* **16**, *76*

Kapitza H. Influence of Surface Roughness on the Reflection of Gold Films in the Region of Surface Plasmon Excitation *Opt. Comm., 1976,* **16**, *73*

Karpinski T. and Tukianowicz Cz. Surface Roughness Measurements by Photometric Method. *CIRP Annals, 1977,* 25, *371*

Kessler E.G. Initial Phase in the Development of an Automatic, Optical Scatter Inspection Station. *Tech. Rep. ARLCD-TR-77033 (USARRADCOM, Dover, New Jersey, 1977)*

Medvedeva M.G. A Laser Method of Determining the Surface Finish. *Soviet Journal of Non Destructive Testing, 1979,* 15, *77*

Ushioda S., Aziza A., Valdez J.B. and Mattel G. Effects of Surface Roughness on Surface Polaritons. *Phys. Rev., 1979,* B19, *4012*

Scott W. and Baul R.M. Preliminary Evaluation of a Proximity Transducer for the Measurement of Surface Texture. *Wear, 1979,* 57, *33*

deKorte P.A.J. and Laine R. Assessment of Surface Roughness by X-Ray Scattering and Differential Interference Contrast Microscopy. *Appl. Opt., 1979,* 18, *236*

Church E.L. The Role of Surface Topography in X-Ray Scattering. *Proc. SPIE, 1979,* 184, *196*

Noll R.J. Effects of Scattering on X-Ray Imaging. *Opt. Eng., 1980,* 19, *249*

Pernick B.J. Surface Roughness Measurement with an Optical Fourier Spectrum Analyzer. *Appl. Opt., 1979,* 18, *796*

Representative theoretical papers and books

Rice S.O. Reflection of Electromagnetic Waves from Slightly Rough Surfaces. *Comm. Pure Appl. Math., 1951,* 4, *351*

Davies H. The Reflection of Electromagnetic Waves from a Rough Surface. *Proc. IEE, 1954,* 101, *209*

Barrick D.E. Rough Surface Scattering Based on the Specular Point Theory. *IEEE Trans. Ant. and Prop., 1968,* AP-16, *449*

Houchens A.F. and Hering R.G. Bidirectional Reflectance of Rough Metal Surfaces. *AIAA Progress in Astronautics and Aeronautics, 1968,* 20, *65*

Valenzuela G.R. The Effective Reflection Coefficients in Forward Scatter From a Dielectric Slightly Rough Surface. *Proc. IEEE, 1970,* 58, *1279*

Kroger E. and Kretschman E. Scattering of Light by Slightly Rough Surfaces or Thin Films Including Plasma Resonance Emission. *Z. Phys., 1970,* 237, *1*

Marathay A.S., Heiko L. and Zuckerman J.L. Study of Rough Surfaces by Light Scattering. *Appl. Opt., 1970,* 9, *2470*

Chandley P.J. and Welford W.T. A Reformulation of Some Results of P. Beckmann for Scattering from Rough Surfaces. *Opt. Quant. Elec., 1975,* 7, *393*

Croce P. and Prod'homme L. Etude par Diffusion Lumineuse de la Nature des Surfaces de Verre Poli. *Nouv. Rev. Optique, 1976,* 7, *121*

Agarwal G.S. Interaction of Electromagnetic Waves at Rough Dielectric Surfaces. *Phys. Rev., 1977,* B15, *2371*

Toigo F., Marvin A., Celli V. and Hill N.R. Optical Properties of Rough Surfaces: General Theory and the Small Roughness Limit. *Phys. Rev., 1977,* B15, *5618*

Leader J.C. Incoherent Backscatter from Rough Surfaces: The Two-Scale Model Reexamined. *Rad. Sci., 1978,* 13, *441*

Sung C.C. and Eberhardt W.D. Explanation of the Experimental Results of Light Backscattered from a Very Rough Surface. *J. Opt. Soc. Am., 1978,* 68, *323*

Carniglia C.K. Scalar Scattering Theory for Multilayer Optical Coatings. *Opt. Eng., 1979,* 18, *104*

Elson J.M. and Bennett J.M. Vector Scattering Theory. *Opt. Eng., 1979,* 18, *116*

Bass F.G. and Fuks I.M. *Wave Scattering from Statistically Rough Surfaces. trans. by C.B. Vesecky and J.F. Vesecky, (Pergamon, New York, 1979)*

Ohlidal I. Expression for the Reflectance of Randomly Rough Surfaces Derived with the Fresnel Approximation. *Appl. Opt., 1980,* 19, *1804*

Shen J. and Maradudin A.A. Multiple Scattering of Waves from Random Rough Surfaces. *Phys. Rev., 1980,* B22, *4234*

DeSanto J.A. Scattering from a Rough Interface. *Rad. Sci. (to be published)*

Reprinted from *Optical Engineering*, Vol. 24(3), pp. 408-413 (May/June 1985).

Optical roughness measuring instrument for fine-machined surfaces

Rainer Brodmann
Oskar Gerstorfer
Optische Werke G. Rodenstock
Munich, Federal Republic of Germany

Gerd Thurn
Technical University, Berlin
Institute of Measurement and Control
Berlin, Federal Republic of Germany

Abstract. The roughness measuring instrument described is based on light scattering and is suitable in a wide range of applications, especially in micromachining. The most important properties are the sensitivity in the measuring range from below 0.005 μm up to 2 μm (R_a value), the independence of the reflection coefficient due to normalization, and the larger tolerance of measuring distance of ±2 mm.

Subject terms: surface metrology; surface roughness; nondestructive testing; light scattering.

Optical Engineering 24(3), 408-413 (May/June 1985).

CONTENTS

1. INTRODUCTION

The measurement of the surface roughness of engineering surfaces is becoming increasingly important in industry. The main reasons for this can be found in the more stringent demands on quality, the miniaturization in precision mechanics, and the greater demands on the function of the surface.[1-3] A typical example is the manufacture of video recorders, which contain a great number of small, high precision parts. The surface of the diamond-turned head wheel and the polished parts of the video head have a direct influence on the wear behavior of the videotape and on the quality of the picture reproduction.[4] A further example of the importance of surface roughness is diamond-turned aluminum disks that are used as base material for magnetic data memories. The required roughness depth here is in the range of $R_a \approx 0.01$ μm (the roughness average R_a is the average absolute deviation of the surface profile about the mean line).

Even our most modern, state-of-the-art production technology is not in a position to guarantee a specified surface quality from the setting of certain machine parameters alone. Primarily responsible for this are the facts that tool wear cannot be calculated and that chance disruptions in the production process can occur at any time. For this reason, it is desirable to have in the actual production area a quick control of the surface quality of all parts produced.

Invited Paper ME-106 received Sept. 20, 1984; revised manuscript received Dec. 3, 1984; accepted for publication Dec. 3, 1984; received by Managing Editor Feb. 1, 1985.

Until now, the usual method to determine the surface roughness was the stylus method.[5,6] However, particularly in the area of fine machining, this process has fairly large disadvantages with regard to the desired fast quality control: (a) The stylus exerts a pressure on the surface to be tested so that the surface is damaged and the stylus wears. (b) Depending on the setup time, a profile measurement can take from several seconds to several minutes. (c) The three-dimensional topography is always reproduced as a two-dimensional profile that in many cases gives only an imperfect description of the surface.

Recent investigations made at the Physikalisch Technische Bundesanstalt in Braunschweig, West Germany, have shown that profile measurements with conventional stylus instruments (sylus radius 2 to 3 μm) in the range $R_a < 0.1$ μm are affected by large systematic errors.[6,7]

A suitable way to avoid the shortcomings of the stylus method is the use of optical measuring processes whose main advantages are their noncontact nature and their faster measuring rate. Some measuring principles also allow the further advantage of an area-sampling technique. References 8 through 10 show an overview of the present state-of-the-art techniques in optical roughness measurement.

The present paper describes an optical instrument for measuring surface roughness that works on the basis of the light scattering method. The main advantages of this instrument over other optical methods can be found in its simple design and in its good sensitivity for studying finely machined surfaces. The instrument is available commercially and has been used in many areas of application.[11,12]

2. THEORETICAL PRINCIPLES

When a bundle of light A falls on a surface, it is scattered by the microscopic irregularities of the surface (Fig. 1). In addition to the directly reflected beam A′, there also occurs scattered light at the scattering angles $\varphi_1, \varphi_2, \ldots, \varphi_n$. A single detector can register the intensity distribution $I(\varphi)$, which depends on the scattering angle. The distribution of the scattered light in a plane perpendicular to the plane of incidence is shown in Fig. 2. Different machining processes

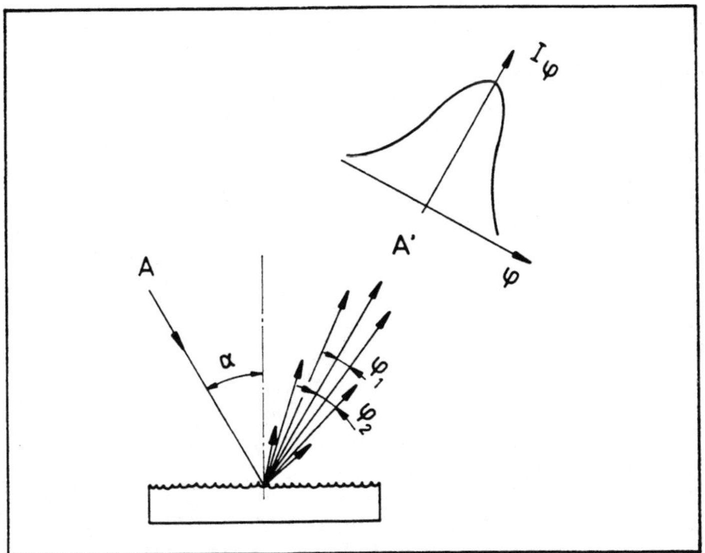

Fig. 1. Light scattering due to surface roughness (φ = scattering angle, I_φ = intensity signal).

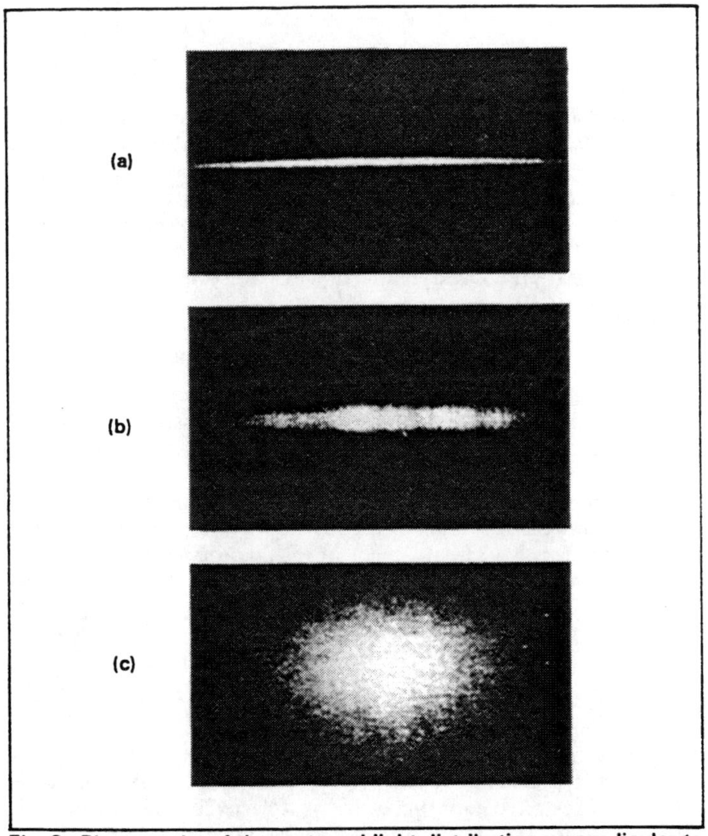

Fig. 2. Photographs of the scattered light distribution perpendicular to the scattering plane of (a) ground surface, (b) turned surface, and (c) cold-rolled surface with $R_a = 0.1\ \mu m$. The photographs span an angle of about 25° from left to right.

can lead to greatly differing scattered light distributions even though the roughness structures show equal R_a values.

Theoretical relationships can be shown to exist between the scattered light distribution and the roughness parameters, with certain restrictions.

Figure 3 offers an overview of different models. The methods used to describe the scattered reflection are based on either a diffraction model or a mirror facet model. A description of these two models can

Description of Scattering	Model		Roughness Parameter	Restrictions
TIS Method	Diffraction Model	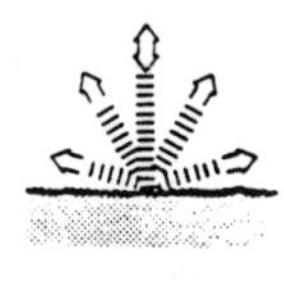	R_q	$R_q < 0{,}01\,\mu m$ Gaussian
Fourier Method			$P(s), R_q$	$R_q < 0{,}1\,\mu m$
Beckmann Method			R_q, T	Gaussian
Profile Angle Method	Facet Model	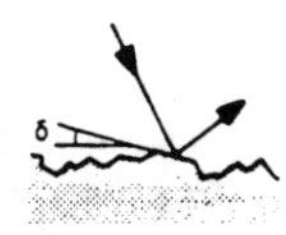	$p(\delta), R_q$	no Diffraction

Fig. 3. Overview of the methods used to describe the scattering reflection. R_q = rms roughness value, P(s) = power spectral density, s = spatial frequency, T = correlation length, p(δ) = probability density, δ = profile angle.

be found in Refs. 11 and 13 through 17. The characteristic roughness values and the characteristic surface functions that can be determined with these models are given in the roughness parameter column of Fig. 3. In this column R_q equals the root mean square (rms) roughness value, P(s) is the power spectral density of the surface texture with the spatial frequency s, T is the correlation length, and p(δ) is the probability density of the profile angle δ of the surface.

The validity range of the described methods is limited by different surface conditions. The most important are the roughness range, the presence of a certain roughness distribution (e.g., Gaussian distribution G), and, in the case of the profile angle method, the neglect of diffraction. In addition to roughness, the scattered light distribution is also influenced by the Fresnel reflection coefficients and the light scattering properties of the material. Of particular importance are the geometrical features such as form and waviness of the workpiece. For the practical application of scattered light measurement, the entire scattering process can be seen as a complex superposition of these partial processes.

A direct calculation of standardized characteristic roughness values such as R_a, R_q from the scattered light distribution is problematic due to the precise knowledge of the electromagnetic boundary conditions that would be required. Such a calculation would restrict the practical application of scattered light measurement to only a few special cases. For this reason, characteristic values of the scattered light distribution are used to evaluate the surface quality, and these are correlated with standardized stylus parameters. The determination of the characteristic values is based on mathematical-statistical methods.

The scattered light intensity $I(\varphi)$ is measured by a sensor system and is normalized for further processing:

$$p(\varphi) = \frac{I(\varphi)}{\int_{-\pi/2}^{\pi/2} I(\varphi)\,d\varphi} \ . \tag{1}$$

The normalized function $p(\varphi)$ is taken as a density function. The variance S and the mean value B are calculated as parameters of this function:

$$S = \int_{-\pi/2}^{\pi/2} p(\varphi)\,(\varphi - \overline{\varphi})^2\,d\varphi \ , \tag{2}$$

$$B = \int_{-\pi/2}^{\pi/2} p(\varphi)\,|\varphi - \bar{\varphi}|\,d\varphi \;, \tag{3}$$

where

$$\bar{\varphi} = \int_{-\pi/2}^{\pi/2} p(\varphi)\varphi d\varphi \tag{4}$$

is the mean value of the distribution.

The integration limits $-\pi/2$ and $\pi/2$ are theoretical limits; in practice they are determined by the aperture angle of the sensor. The parameters S and B describe the scattering of directional light from surfaces and are termed "optical scattering characteristic values." They can be used directly as measures of the optical quality of a surface.

A comparison of these values with known characteristic values such as the half-width shows that the precision of measurement of the parameter S is greater due to the averaging over all scattering angles. Furthermore, determination of this type of parameter will still produce easily interpretable results in practice, even for complicated curve forms that, for example, possess several maxima.

In the technical configuration of the measuring system, a discrete measurement process occurs due to the finite number of detectors in the diode array and to the digital processing of the measured data. The corresponding equations with discrete values are

$$S_N = K^2 \sum_{i=1}^{n} (i - \bar{i})^2 p_i \;, \tag{5}$$

$$B_N = K \sum_{i=1}^{n} |i - \bar{i}|\, p_i \;, \tag{6}$$

where

$$\bar{i} = \sum_{i=1}^{n} i\, p_i \;, \tag{7}$$

$$p_i = \frac{I_i g_i}{\sum_{i=1}^{n} I_i g_i} \;. \tag{8}$$

In these equations, i equals the diode number of the linear photodiode array, p_i is the normalized intensity value at the *ith* diode, I_i is the signal of the *ith* diode, $\bar{i}$ is the mean value of the diode number ascertained using the intensity values, g_i is a weight factor, and K is a constant that is given by the geometrical-optical data of the sensor.

The resolution limits for the parameters S and B, and also the measuring time, are influenced by the number of diodes used and the precision with which the detector signals are processed. A technically justifiable compromise is achieved by using a diode array consisting of 20 individual elements and 12-bit signal processing. The time for one measurement will then be approximately 40 ms.

In Eq. (8) the weight factors g_i occur as coefficients. Therefore, they can be used in a predetermined analysis of the scattered light distribution. For example, they can be used to highlight or subdue certain portions of the light curve. If, for example, the factors g_i are made equal to zero for a certain range of the scattering angle, then either the scattered light or the normally reflected light can be evaluated separately [Fig. 4(a)]. The g_i factors can also be used to extend the dynamic range of the sensor system. However, this process is limited to a certain form of the scattered light curve that must be known, at least approximately, beforehand.

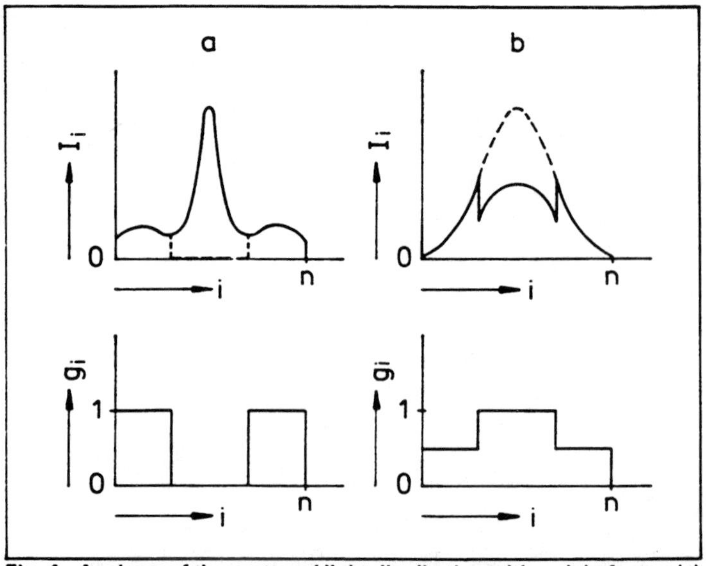

Fig. 4. Analyses of the scattered light distribution with weight factor: (a) turning off the middle diodes and (b) attenuation of the middle diodes. I_i = intensity signal of one single diode, g_i = weight factor.

In Fig. 4(b) the signals of the central detectors are diminished by a constant factor. This can be done, for example, electronically or by using an optical process. The whole gain of the measuring loop can then be increased by this factor. In the digital processing of the measured value, this is taken into consideration by corresponding g_i factors so that the actual light distribution is reconstructed. In practice, this means that the dynamic range of the system can be increased by a factor of 10.

3. TECHNICAL SETUP

The measuring system consists of an optoelectronic sensor, an analog preprocessor, and a microcomputer to control the system and to calculate the parameters. Figure 5 shows the arrangement of the optoelectronic parts in the sensor.

The luminous area of an IR LED, L, (wavelength $\lambda = 0.8\ \mu m$) is imaged onto the specimen by means of the collimator lens K and the measuring lens M. The diameter of the illuminated area,

$$y' = \frac{f_M}{f_K} y \;, \tag{9}$$

can be predetermined within certain limits and in the standard version of the instrument is equal to 1.8 mm.

The bundle of rays strikes the surface at an angle of $\alpha = 8.4°$ with the standard measuring lens (f'_M = 34 mm, acceptance angle φ = 15°). The scattered light that occurs is collected by the measuring lens and with the same lens is directed to the diode array A.

This asymmetric arrangement of the optical components provides a number of advantages. Not using a beam splitter results in a fourfold intensity gain and also allows the diode array to be positioned in the Fourier plane of the measuring lens. This means that all rays coming from the surface at an angle φ will be focused to one point of the back focal plane. In this way, a well-defined mathematical relation results between image height y″ and the ray inclination φ on the lens side, given by

$$y'' = f'_M \tan\varphi \;. \tag{10}$$

Thus, each diode shows the intensity of a defined scattering angle area.

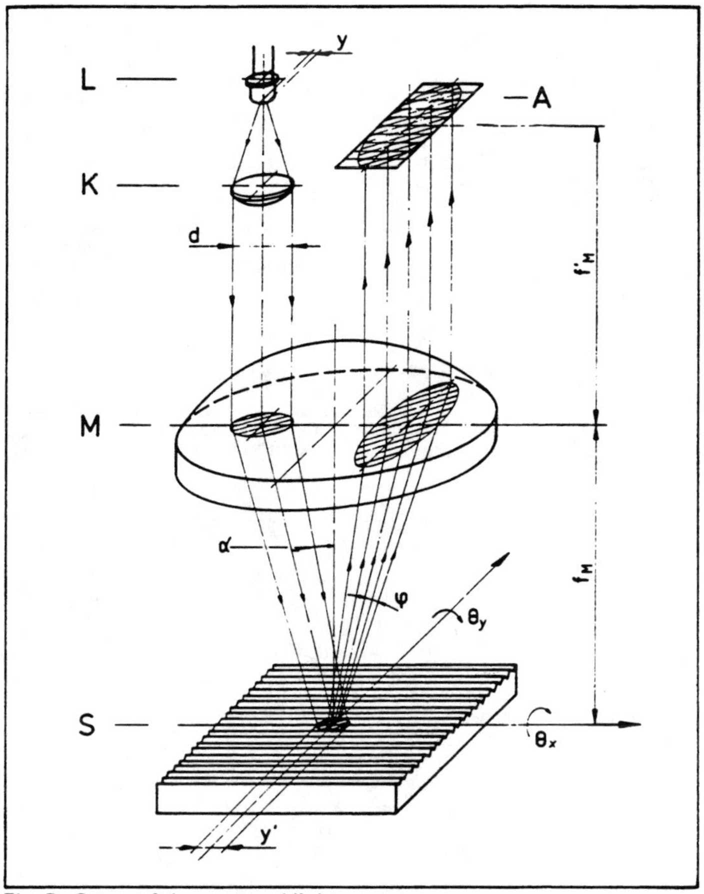

Fig. 5. Setup of the scattered light sensor.

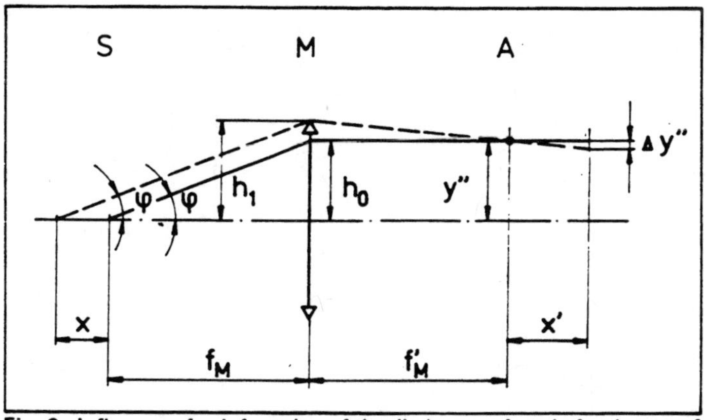

Fig. 6. Influence of a defocusing of the diode array A and of a change of the specimen distance on the image height y" (M = measuring lens).

The correlation between scattering angle φ and diode No. i of the array is calculated to be

$$\tan\varphi = \frac{\ell}{n f'_M} i \,, \tag{11}$$

where ℓ equals the array length, n is the number of diodes, and f'_M is the focal length of the measuring lens.

The factor K in Eqs. (5) and (6) then corresponds to

$$K = \frac{\ell}{n f'_M} \,. \tag{12}$$

Since the specimen is in the front focal plane of the measuring lens, the illuminated area has the optical effect of a telecentric diaphragm for the scattered rays. In this way, all rays become axis-parallel on the image side. The advantage of this system is a certain insensitivity of the axial position of the array with respect to the measuring lens.

The effect of a defocusing x' of the diode array A on the image height y" for a simultaneous specimen displacement x is shown in Fig. 6. By using trigonometric relations, the following expression can be deduced for the change in the image height:

$$\Delta y'' = \frac{x x'}{f'_M} \tan\varphi \,. \tag{13}$$

From Eq. (13) we can deduce that $\Delta y'' = 0$ when $x' = 0$, i.e., the diode array is exactly in the focal plane. In this case, the specimen can be displaced axially over a large range since, in accordance with Eq. (10), the image height y" is dependent only on the scattering angle φ. In practice, the possible displacement x is restricted by the vignetting caused by the measuring lens. Therefore, x is limited to $\pm 0.1 f'_M$.

A simultaneous displacement of specimen and array ($x \neq 0$; $x' \neq 0$) results in a change in the image height. With the standard optical system ($f'_M = 34$ mm, $\varphi = 15°$), for example, a displacement of $x = x' = 2$ mm produces a relative change of the image height of 0.3%, which is negligible for practical applications.

The relations shown above apply in practice only with a good, angle-corrected lens, taking into account the requirements for Fourier optics.

4. RESULTS

4.1. System properties

Figure 7 shows a comparison of different distribution curves. An ideal surface [optically polished Si, Fig. 7(a)] reflects the incident rays specularly, and the intensity distribution of the collimator bundle hits the array with the diameter d (see Fig. 5).

A reflection grating (Al evaporated) produces scattered light perpendicular to the groove direction as a result of diffraction. For almost vertical incidence the relationship

$$\sin\varphi = \frac{\lambda}{a} \tag{14}$$

applies for the first order of diffraction, where λ is the light wavelength and a is the groove spacing.

When $\lambda = 0.8$ μm and $a = 3.3$ μm, then $\varphi \approx \pm 14°$. The fact that the intensity in the positive order at +14° is greater than that in the negative order is a result of the blaze of the grating ($\lambda_{blaze} = 2$ μm). The maximum at the center is shown by the share of the zeroth order [Fig. 7(b)].

The scattered light in the zeroth order, measurement in the groove direction [Fig. 7(c)], should be low for a blazed grating if the grooves have no substructure. The measured intensity distribution in Fig. 7(c) can be explained only by inexact grooves. Figure 8 shows a SEM picture of the measured grating. Some microstructures can clearly be seen on the top of the profile. These may cause the scattered light in the zeroth order [Fig. 7(b)]. This scattered light appears also when measuring along the groove direction [Fig. 7(c)].

With rougher textures, the diffraction can be neglected as to its importance to the scattered light, and the simplified mirror facet model applies. As an example, we can again take a sawtooth profile. The surface is turned aluminum. Figure 9 shows the scattered light distribution together with the profile measured. The scattered light maximum can be deduced directly from the geometrical light reflection. It should be remembered here that in reflection, twice the profile angle is measured.

The most important system properties of the scattered light sensor have been collected together in Fig. 10. Here, the relative change of the S_N value for a change in the distance x of the specimen from the sensor [Fig. 10(a)] was measured, as was the relative change for a tilting of the specimen by the angles θ_x and θ_y [Figs. 10(b) and 10(c)].

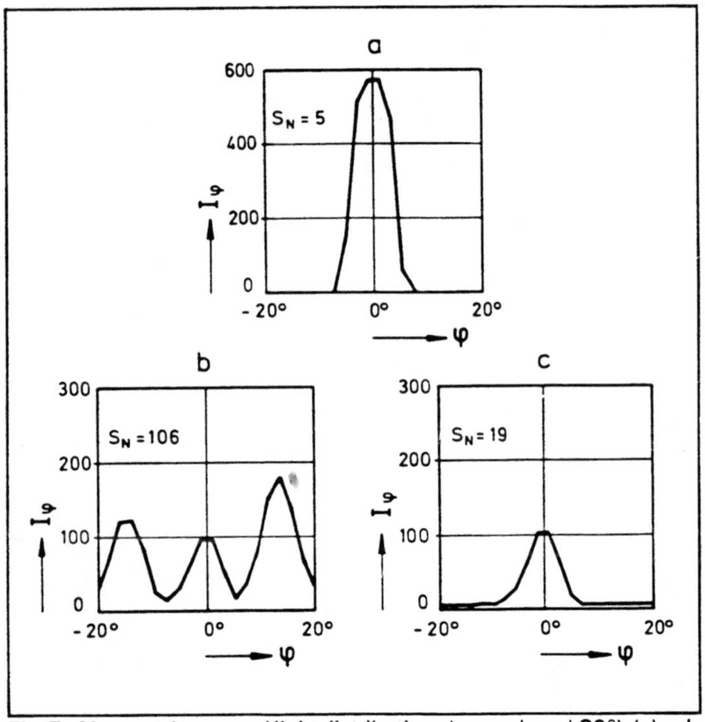

Fig. 7. Measured scattered light distributions (sensor lens ±20°): (a) polished, flat silicon disk $R_a < 0.003\ \mu m$; (b) blazed aluminum grating (300 lines/mm, a = 3.3 μm, h = 1 μm), array normal to groove direction; (c) same as (b) except array in groove direction.

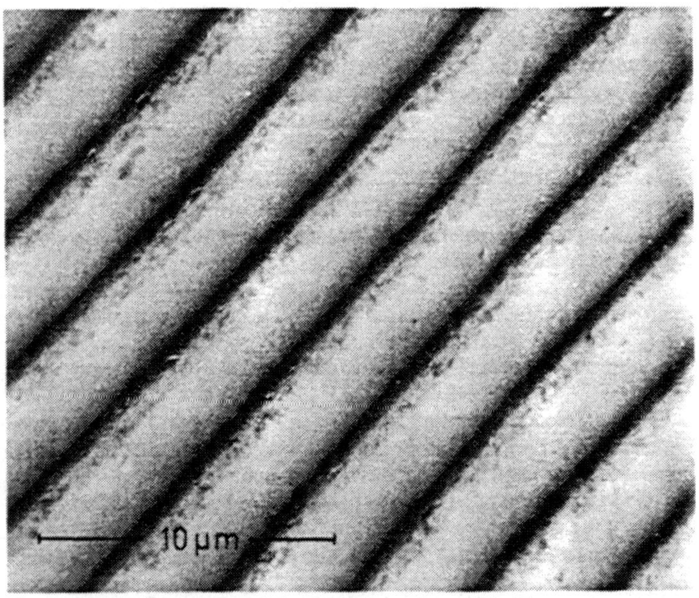

Fig. 8. SEM photograph of the blazed grating of Fig. 7.

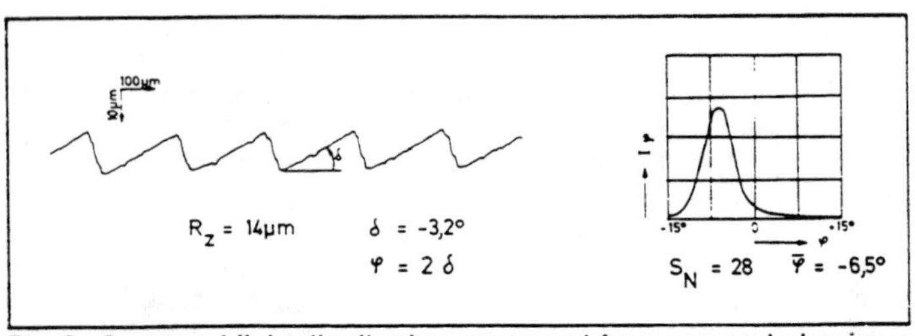

Fig. 9. Scattered light distribution, measured from a turned-aluminum surface (a piston for a combustion engine).

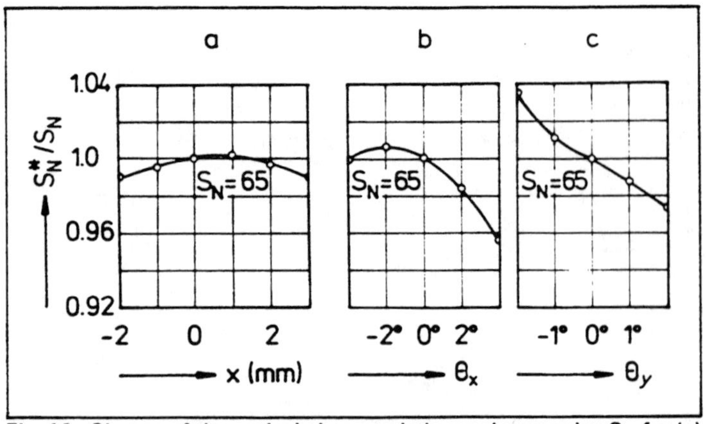

Fig. 10. Change of the optical characteristic roughness value S_N for (a) distance change x, (b) tilting by θ_x, and (c) tilting by θ_y.

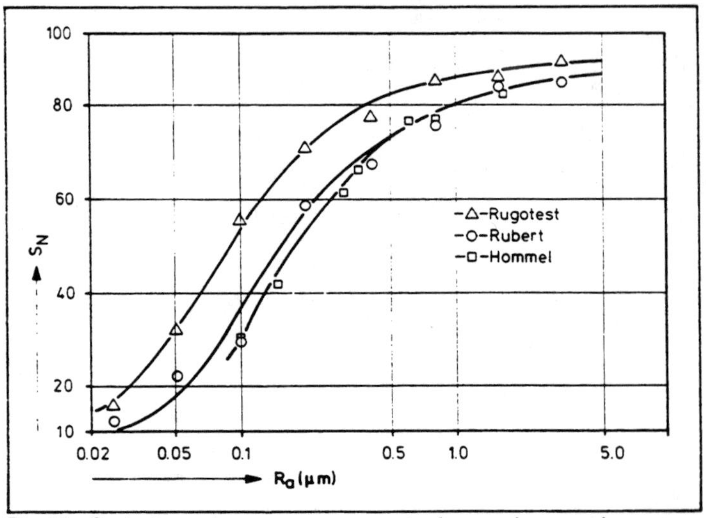

Fig. 11. Optical roughness measurement of ground comparison standards (sensor lens ±15°).

The value x = 0 and $\theta_x = \theta_y = 0°$ refers to the untilted specimen at an optimal distance.

The measurements were carried out on a flat-ground specimen of average roughness ($S_N = 65$, $R_a = 0.3\ \mu m$). The asymmetric shape of the curve for tilting of the specimen can be attributed to a slight asymmetry of the intensity distribution [Fig. 2(b)] and the tolerances in the alignment of the optical components. If a measurement uncertainty factor of 2% is permitted, a flat specimen can be deviated by ±2 mm from the optical measuring distance or tilted by ±2° in the θ_x direction or ±1° in the θ_y direction. The angle insensitivity in the θ_x direction can be explained by the property of the characteristic value S_N according to Eqs. (2) and (5).

4.2. Measuring specimens in the fine roughness range

The correlation of the optical characteristic value S_N with roughness parameters from a mechanical stylus instrument is an important requirement for practical use. It has already been mentioned (in Sec. 2) that there is no direct connection between the S_N value and the stylus characteristic value. A correlation can only be found by making a comparison measurement.

Figure 11 shows the results of an investigation using plane-ground comparison standards made by the Rugotest, Hommel, and Rubert companies. The finest specimen has the value $R_a = 0.025\ \mu m$; the roughest specimen has the value $R_a = 3.2\ \mu m$. Up to $R_a = 1$ to 2 μm the correlation obtained is good, and the sensitivity is especially

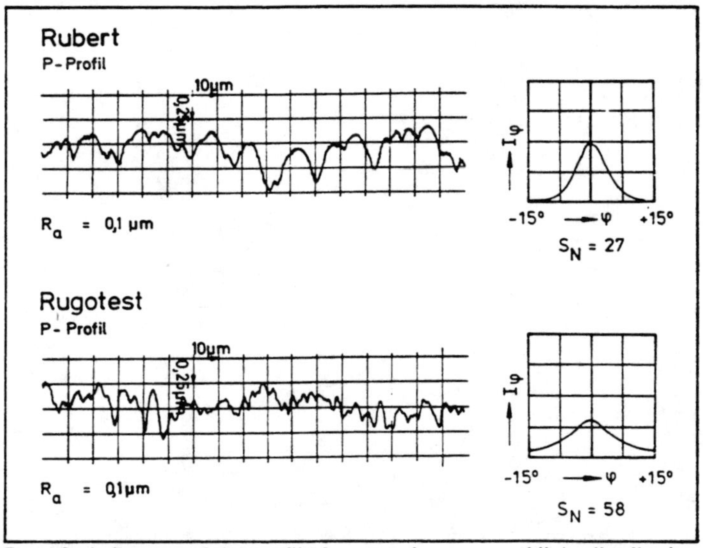

Fig. 12. Influence of the profile form on the scattered light distribution for two ground samples.

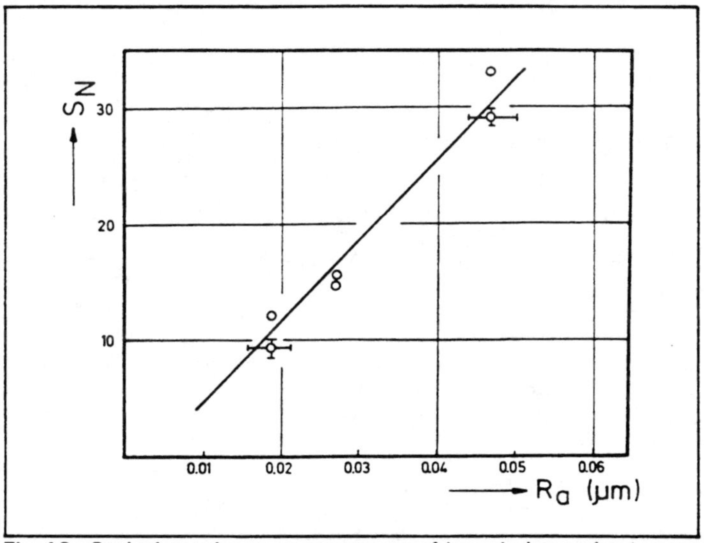

Fig. 13. Optical roughness measurement of honed piston pins (sensor lens ±15°).

high in the range $R_a = 0.02$ to $0.5\ \mu m$. The sensitivity decreases with larger roughness values, and the curve moves toward saturation. In this range, the intensity distribution on the array approaches a rectangular curve.

The most striking feature is the different position of the characteristic curves of the Rubert specimen and the Rugotest specimen. An explanation for this behavior can be found in the example showing specimens with $R_a = 0.1\ \mu m$ in Fig. 12. The left sides show profile measurements taken with a stylus instrument; on the right the scattered light distributions and the S_N values are shown. The profile of the Rubert specimen, with its larger distances within the roughness textures, produces a considerably smaller S_N value than that of the Rugotest specimen, with its larger proportion of high spatial frequencies. It is therefore important to remember when setting up an S_N-R_a characteristic curve that the optical scattered light method is sensitive to both the vertical and the horizontal roughness parameters.

An investigation of honed piston pins (Fig. 13) shows that, in practice, no problems occur with fine-machined surfaces of different roughness texture due to the predetermined production process. Further examples can be found in Ref. 12.

TABLE I. Optical and Mechanical Measurement of the Surface Roughness of Polished Plane Silicon Disks

Specimen	Polishing time (min)	Mechanical method $\bar{R}_a \pm s$ (µm)	Optical method $\bar{S}_N \pm s$
S 1	30	0.061 ±0.04	7.50 ±0.5
S 2	60	0.017 ±0.008	5.73 ±0.09
S 3	180	0.006 ±0.004	5.10 ±0.02
S 5	240	0.003 ±0.001	5.02 ±0.01
S 11	360	0.002 ±0.001	4.99 ±0.02

The results of measurements on optically polished silicon disks are summarized in Table I. These investigations show the lower limit of the measuring range and are carried out with the g_i factors [described in Eq. (8)] chosen to increase the measuring dynamics. In the evaluation, the specular portion of the backscattered light has been suppressed by a factor of eight. The optical roughness parameter S_N was determined from 24 independent, individual measurements on different parts of the specimen. The R_a is the mean of five individual measurements. With the stylus measurement, the profiling measuring method makes its disruptive influence clearly noticeable since the stylus can only offer a purely random evaluation of the holes in the surface with its contact length of 1.5 mm. This leads to discrepancies in the R_a value.

On these extremely fine surfaces, the light scattering method shows its advantage not in the higher sensitivity but in faster readings and the nondestructive testing capabilities.

5. SUMMARY

The measurement of light scattering to determine the roughness of a fine machined surface offers many advantages compared to the mechanical stylus method. Among them are nondestructive testing, higher measuring rate, and the inspection of an area instead of a line. The optical roughness parameter S_N, introduced in this paper, is based on the statistical evaluation of the intensity signal in a certain scattering angle range. For similar machining processes a good correlation to the well-known roughness parameter R_a exists. The application of the instrument is in the field of automatic testing, direct production control, and on-line measurement.

6. REFERENCES

1. R. D. Young, Prec. Eng. 1, 143 (1979).
2. D. J. Whitehouse, Proc. Inst. Mech. Eng. 192(19), 179 (1978).
3. M. F. DeVries, M. Field, and J. F. Kahles, Ann. CIRP 25(2), 569 (1976).
4. K. H. Kotter, Feinwerktechnik und Meßtechnik 91(5,6), 212 (1983).
5. T. R. Thomas, *Rough Surfaces*, Chap. 2, Longman, New York (1982).
6. W. Hillmann, Technisches Messen 47, pp. 169, 209, 273 (1980).
7. K. Eckolt, Physikalisch Technische Bundesanstalt (Braunschweig, West Germany), Report ME - 44 (1983).
8. T. V. Vorburger and E. C. Teague, Prec. Eng. 3, 61 (1981).
9. E. C. Teague, T. V. Vorburger, and D. Maystre, Ann. CIRP 30, 563 (1981).
10. H. J. Warneke and R. J. Ahlers, Proc. IMEKO - Symposium on Measurement and Estimation, Bressone, Italy (May 1984).
11. R. Brodmann, G. Thurn, and Th. Gast, Ann. CIRP 33(1), 403 (1984).
12. R. Brodmann, H. Paisdzior, and O. Gerstorfer, Werkstatt und Betrieb 11, 701 (1984).
13. P. Beckmann and A. Spizzichino, *The Scattering of Electromagnetic Waves from Rough Surfaces*, Macmillan, New York (1963).
14. J. M. Elson, H. E. Bennett, and J. M. Bennett, in *Applied Optics and Optical Engineering*, Vol. 7, R. R. Shannon and J. C. Wyant, eds., p. 191, Academic Press, New York (1979).
15. E. L. Church, Wear 57, 93 (1979).
16. E. G. Thwaite, Ann. CIRP 29(1), 419 (1980).
17. G. Thurn, "Automatisierte Oberflächenprüfung durch rechnergestütztes Messen der Streulichtverteilung," Ph.D. Thesis, Technical University of Berlin, West Germany (1984).

Reprinted from *Optical Techniques for Industrial Inspection,* P.G. Cielo, editor, Proc. SPIE Vol. 665, pp. 25-31 (1986). ©1986 SPIE.

On-line glossmeter for stainless steel sheets

Yoshiaki Shibata, Yuichiro Asano, Kunio Kurita

Instrumentation & Control Research Dept.
Technical Research Division, Kawasaki Steel Corp.
1, Kawasaki-cho, Chiba City, 260 Japan

Motoji Shiozumi

Equipment Maintenance Dept.
Hanshin Works, Kawasaki Steel Corp.
2-88, Wakinohamakaigan-dori, Chuo-ku, Kobe, 651 Japan

Abstract

The determining factors of the glossiness, the relation between surface roughness parameters and glossiness, and the method of discrimination of the grade of glossiness by detecting light reflection intensities were investigated. A new glossmeter was developed as results of the investigation. The grade of the surface glossiness of stainless steel sheets determined by this method coincided with the results by visual inspection with the accuracy of 90%.

Introduction

The control of surface glossiness is one of the important factors for the production of stainless steel sheets. Glossiness means in this paper the sensory glossiness, which can be sensed psychologically.

The glossiness is usually determined by light reflection, and a number of methods have been proposed. They are (1) to measure specular surface glossiness, (2) to measure contrasting glossiness, or (3) to measure sharp glossiness. The glossiness by the first method, however, does not coincide well with the sensory glossiness for low surface roughness. The glossiness by the second method shows low accuracy when the objects have specular surfaces. The glossiness by the third method coincides well with the sensory glossiness, but owing to human factors, quantitative indication is impossible, and therefore its application to on-line measurement is difficult. With these disadvantages, the glossiness is in most cases determined by visual inspection and the development of an on-line measurement equipment has long been desired.

In this paper, factors which affect the glossiness, and relations between glossiness and laser beam reflection intensity were analyzed. And a method to determine glossiness from laser beam reflection intensities was investigated, the result of which has led to the development of an on-line glossiness measurement equipment for a stainless steel sheet production line.

Principle

Relation between glossiness and roughness

It was expected that the surface glossiness of steel sheets would be determined by the characteristics of the spectrum of the light reflection and by the characteristics of surface roughness. It turned out, however, that the difference in the spectrum of the reflected light was very small among the samples with various grades of glossiness determined by visual inspection. It was concluded, therefore, that the effect of the surface physical properties on the glossiness was very small. In turn, the relation between the glossiness and the surface roughness was investigated.

In general, a surface with an irregular geometrical profile is characterized by such parameters as σ, Ra, T, Sm, and Ra/Sm, where σ is the standard deviation in the height distribution, Ra is the averaged roughness, T is the correlation distance of the surface profile, Sm is the averaged spatial period of the surface profile, and Ra/Sm means the steepness of surface profile, all of which can be measured by a stylus roughnessmeter.

The relation between Ra and the sensory glossiness is shown in Fig.1. Although the measured values of Ra vary in some degree for each grade of the sensory glossiness, it is observed that the averaged values of Ra for each grade, shown as | in the figure, correspond well with the grades of the sensory glossiness. The relation between Ra/Sm and the sensory glossiness is shown in Fig.2. Some degree of correlation is also observed between them. It is resulted from these facts that the surface glossiness is related to the surface roughness, and the glossiness decreases with the increase of Ra and Ra/Sm.

Relation between roughness and light reflection characteristics

There are several papers which describe theoretical investigations concerning the light reflection characteristics of rough surfaces. Among them, Beckmann's theory[1] may be the most general one from the viewpoint of statistics. Based on this theory, an on-line roughnessmeter[2] for steel sheets was developed by the authors. The principle of the optical measurement system of surface roughness is summarized as follows.

The surface is assumed to be perfectly conductive and isotropic. It is also assumed that the height profile of the rough surface is normally distributed and that its autocorrelation function is represented as Gaussian process. Suppose an incident light beam with the wavelength λ is projected at the incident angle $\theta 1$ onto the rough surface which is characterized by σ and T as shown in Fig.3. Then the specular reflection intensity Is whose reflection angle $\theta 2$ is equal to the incident angle $\theta 1$ can be related to σ and T as follows.

$$Is = \exp(-g) \quad \text{if } g << 1, \tag{1}$$
$$Is = f(\sigma/T) \quad \text{if } g >> 1, \tag{2}$$

where

$$g = (2\pi\sigma/\lambda)^2(\cos\theta 1 + \cos\theta 2)^2, \tag{3}$$
$$\sigma = 1.25Ra. \tag{4}$$

As the value of T is proportional to Sm, the values of Is in eqs. (1) and (2) are functions of Ra and Ra/Sm, respectively. To satisfy the conditions of $g << 1$ and $g>>1$, two incident angles or two wavelengths are necessary.

Outline of on-line surface roughnessmeter

The on-line surface roughnessmeter was developed according to the above-mentioned principle. It was designed for bright finish steel sheets and installed in the temper processing line at Chiba Works.

The value of Ra of bright finish steel sheets ranges between 0.1 μm and 0.5 μm. When two He-Ne laser beams (λ = 633nm) are projected at the incident angles of 75° and 10°, respectively, the conditions described in eqs. (1) and (2) are approximately satisfied. The reflected light distribution of each beam is measured by a rotating silicon photodiode on a circular arc. Because the detector is scanned, the specular reflection intensity Is and the total reflection intensity are always measured exactly in spite of strip vibrations. The values of the total reflection intensities are used for correcting errors of Is caused by the change of total reflectivity. The measuring cycle is 2 sec. The measurement accuracy was found to be ± 10% for Ra and ± 20% for T, by comparing the values of Ra and T measured by this on-line equipment and those measured by a stylus roughnessmeter.

Relation between glossiness and light reflection characteristics

As mentioned before, the glossiness correlates with both Ra and Ra/Sm. Accordingly, it is expected that the glossiness correlates with the specular reflection intensities Is in eqs. (1) and (2), and can be measured by applying the optical measurement method of the surface roughness. In order to confirm this expectation, the relation was investigated between the glossiness and the specular reflection intensities under each condition of the parameter g.

The values of Ra of stainless steel sheet surfaces range between 0.01 μm and 0.1 μm. In order to satisfy the two conditions of $g >> 1$ and $g << 1$ in this range of Ra with two wavelengths and one incident angle, one of the two wavelengths falls in the invisible range. Because the measured glossiness must be close to the result by visual inspection, using a wavelength in the invisible range may provide irrelevant information concerning the surface glossiness. In that sense, the method to measure the surface glossiness should use one common visible wavelength with two incident angles.

When the specular reflection intensity at the incident angles θ is represented as Is(θ), the actual measurement conditions result in

$$g < 4 \ : \ Is(70) = f1(Ra), \tag{5}$$
$$g > 10 \ : \ Is(10) = f2(Ra/Sm), \tag{6}$$

where λ = 488nm, and f1 and f2 mean functions.

According to above-mentioned conditions, Ra and Ra/Sm are obtained by measuring the specular reflection intensities, Is(70) and Is(10), and the glossiness is obtained utilizing the results of Fig.1 and Fig.2 effectively. The result on the relation between the glossiness and Is(10) is shown in Fig.4. It is observed that the averaged value of Is(10) increases with the increase of the glossiness.

Method to discriminate the grade of glossiness

Although the glossiness correlates with both Is(10) and Is(70), it is inadequate to determine the glossiness from either Is(10) or Is(70) alone, because the measured values of Is(10) (or Is(70)) for each grade of glossiness overlap those of the adjacent grade. An effective method is necessary which determines the glossiness accurately in spite of overlapping. It tured out that it is effective to locate each pair of measured values of Is(10) and Is(70) on the two-dimensional plane whose axes are Is(10) and Is(70), and to utilize the discriminant function which draws the boundary lines so that the degree of overlapping on that plane becomes minimum.

Fig.5 shows the result in which each boundary line was obtained with the samples whose glossiness was determined by off-line measurement while the samples were held firmly. By using these boundary lines, the optically measured grade of glossiness coincided with the grade of sensory glossiness by visual inspection at the rate of 90%. The glossiness of surfaces of stainless steel sheets can be obtained by measuring Is(10) and Is(70) and examining the area in which the point (Is(10), Is(70)) exists on the two-dimensional plane shown in Fig.5.

Outline of the equipment

An outline of the on-line glossmeter is shown in Fig.6. An Ar laser is used as a light source, whose wavelength is 457nm and the output power is 2W as multilines. The laser beams are guided by two optical fibers to the detecting head, which is located about 30m distant from the Ar laser, and projected onto the surface of stainless steel sheet at the incident angles of 10° and 70° respectively. The laser beam power at the exit of the optical fiber is about 20mW. The intensities of the reflected light and the reference light are measured by silicon photodiodes D1~D4. The silicon photodiodes for the measurement of the reflected light are scanned along an arc which has its center at the laser spot on the stainless steel sheet surface, searching the peak of each reflected light intensity. The discriminant function in the computer memory decides the grade of glossiness.

It is necessary to change the discriminant function for different types of steel sheet, because the light reflection characteristics of the surface and the basis of the discrimination are different. The computer system can handle such cases. Fig.7 shows a block diagram of this system. This equipment has been installed in No.2 skinpass line at Hanshin Works.

Result of on-line experiment and discussion

The glossiness determined by this equipment in the skinpass line for stainless steel sheets and the result by visual inspection were compared in order to investigate the distribution of each grade of glossiness and the effectiveness of the boundary lines. The stainless steel sheets of type 430 with No.2B finish, which have three grades, were tested by the on-line glossmeter. The result is shown in Fig.8. It is indicated that the boundary lines derived from the data of sample sheets are not applicable, although the data are indeed classified into three grades. In other words, for the same grade of glossiness, the on-line measurement provided larger values of Is(10) and lower values of Is(70) than the off-line measurement with sample sheets provided.

The difference probably results from the difference of the measuring conditions between the on-line measurement and the off-line measurement: the difference of the illuminating light, of the field of view, and of the speed of sheets. It is important after all that when deriving the boundary lines, the results of visual inspection by operators should be taken into account as well as the measurements by this equipment, and a large amount of data is required to obtain the optimum boundary lines for the most accurate discrimination.

In addition, two questions remain unsolved: how to correlate the representative grade of glossiness by visual inspection to the glossiness measured by this equipment, when the measured values of Is(10) and Is(70) vary along the longitudinal direction of a steel strip, and how to classify the glossiness whose grade lies near the boundary line.

The characteristics of specular surfaces and of diffuse surfaces are discussed as follows.

Stainless steel is classified into several types. Type 304 or 430, for example, are based on content, and No.2B finish or bright annealed finish are based on the finishing processes. Boundary lines must be determined for each type of stainless steel and for each finish, because the standard grade of the glossiness is arranged for each type of stainless steel sheet.

Fig.9 shows the discrimination of the glossiness for the bright annealed finish surfaces. By comparing this figure with Fig.5, the inclination angle of each boundary line is steeper, causing the discrimination of the grade of glossiness for bright annealed finish surfaces to depend largely on the value of Is(10).

As shown in eqs.(5) and (6), Is(70) corresponds to Ra, and Is(10) corresponds to Ra/Sm. That is, also as shown in Figs.10 and 11, the glossiness of bright annealed finish surfaces has less correlation with Ra (i.e. Is(70)) than the glossiness of No.2B finish surfaces. This fact results from the characteristics of the surface, that is, whether it is specular or diffuse.

As shown in eqs.(1) and (2), the condition $g \ll 1$ corresponds to specular reflection, and the condition $g \gg 1$ corresponds to diffuse reflection. Is(70) represents the degree of specular reflection, and Is(10) represents the degree of diffuse reflection.

On the other hand, the glossiness by visual inspection depends on whether the surface is specular or diffuse. Inspectors observe the ratio of specular reflection to diffuse reflection when they try to determine the grade of surface glossiness. Therefore, in case of specular surfaces, inspectors judge the grade of glossiness using slight differences of diffuse reflection, and in case of diffuse surfaces, they judge the grade using slight differences of specular reflection.

This analysis explains the difference in the reflection characteristics between No.2B finish surfaces and bright annealed finish surfaces. That is, for bright annealed finish surfaces with strong specular reflection, Is(10) correlates well with the glossiness, and for No.2B finish surfaces with strong diffuse reflection, Is(70) correlates well with the glossiness.

Because commercially available glossmeters detect mainly characteristics of the specular reflection in principle, it is difficult to determine precisely the grade of glossiness, when the major factor determining the sensory glossiness is a slight difference of diffuse reflection. Therefore, these have limitations for their applications. On the contrary, the glossmeter mentioned in this paper has an advantage in that the method is similar to visual inspection, as it detects both the specular condition and the diffuse condition simultaneously.

Conclusion

A method was investigated whose classification of the grade of surface glossiness would coincide well with the result by visual inspection. The following results were obtained.

(1) The roughness parameters Ra and Sm are main factors which affect the glossiness of surfaces.

(2) The grade of glossiness is discriminated by measuring the specular reflection intensities, as the values of Ra and Sm are obtained from the specular reflection intensities at two incident angles.

(3) Owing to a discriminant function, the grade of glossiness of stainless steel sheets obtained by the above-mentioned method coincide well with the grade evaluated by visual inspection.

Based on these results, an on-line glossmeter has been developed, which is expected to take the place of conventional visual inspection. Further accumulation of data will improve the precision of discrimination.

References

1. P.Beckmann et.al., "The Scattering of Electromagnetic Waves from Rough Surfaces," Pergamon Press (1963)
2. Y.Asano et.al., "On-line Surface Roughness Measurement of Cold-Rolled Steel Sheets by Light Reflection Method," Proc.IMECO X-th World Congress (1985) Vol.4, p.151

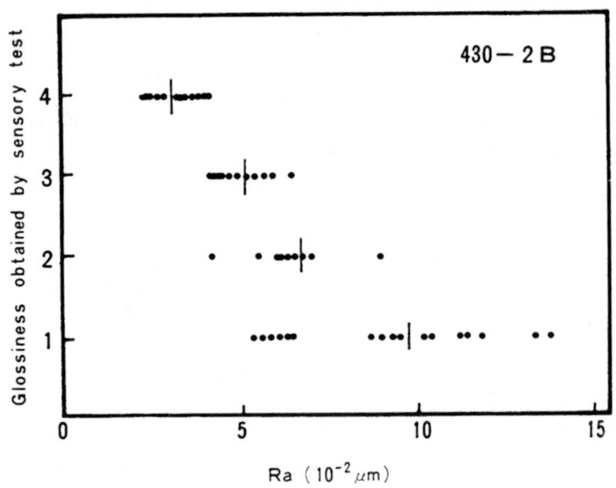

Fig.1 Experimental result showing the correlation between the averaged roughness Ra measured by a stylus roughnessmeter and the grade of glossiness determined by visual inspection. Samples are stainless steel sheets,type 430,No.2B finish.

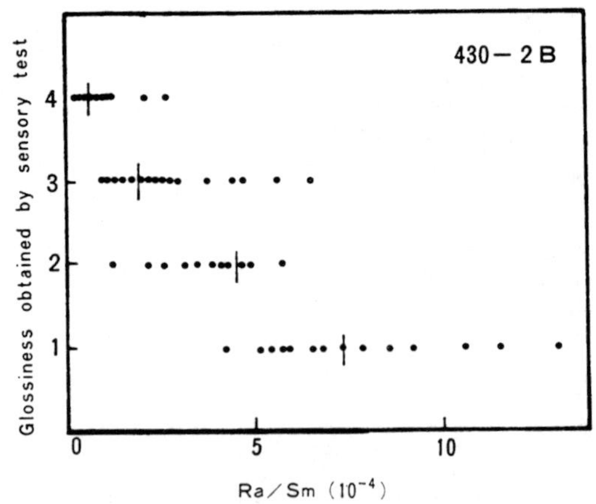

Fig.2 Experimental result showing the correlation between Ra/Sm measured by a stylus roughnessmeter and the grade of glossiness determined by visual inspection. Samples are stainless steel sheets, type 430, No.2B finish.

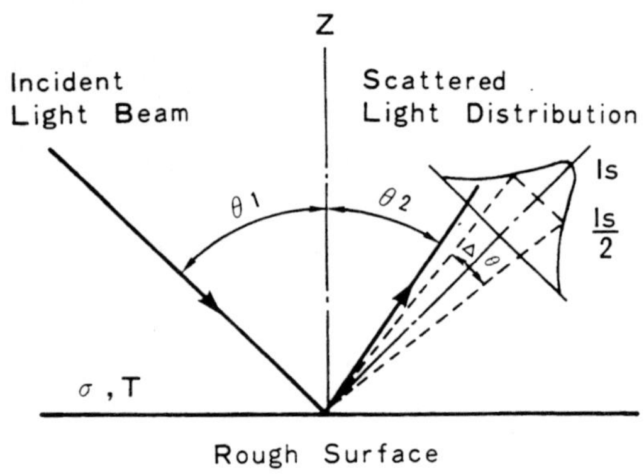

Fig.3 Schema of light scattering from a rough surface.

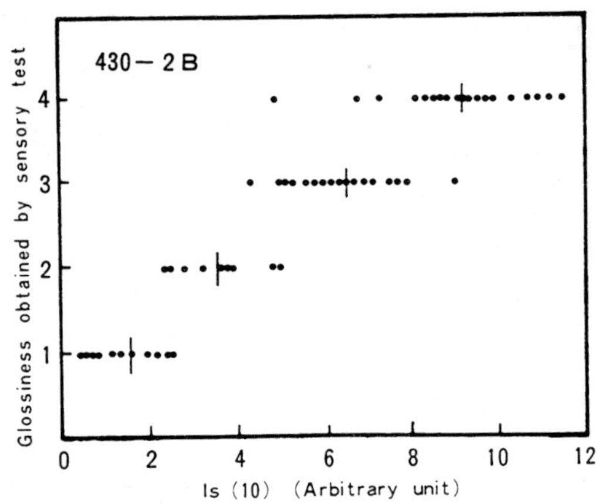

Fig.4 Experimental result showing the correlation between the specular reflection intensity Is(10) at the incident angle of 10° and the grade of glossiness determined by visual inspection. Samples are stainless steel sheets, type 430, No.2B finish.

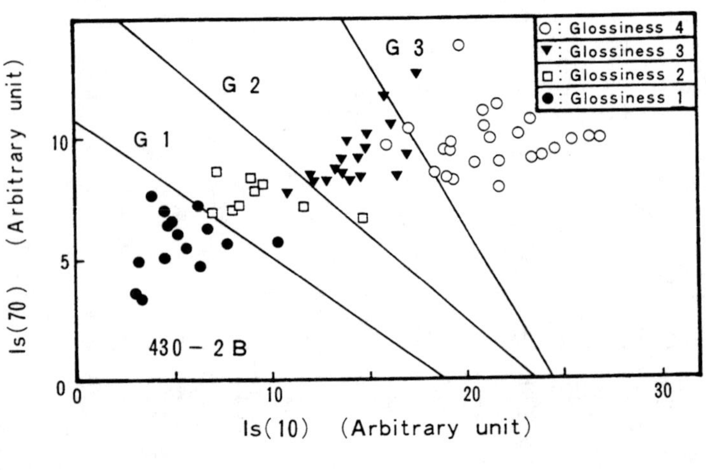

Fig.5 Experimental result showing the classification of glossiness of stainless steel sheets, type 430, No.2B finish based on the discriminant function. Each boundary line is determined so that the degree of overlapping between adjacent grades on the two-dimensional plane becomes minimum.

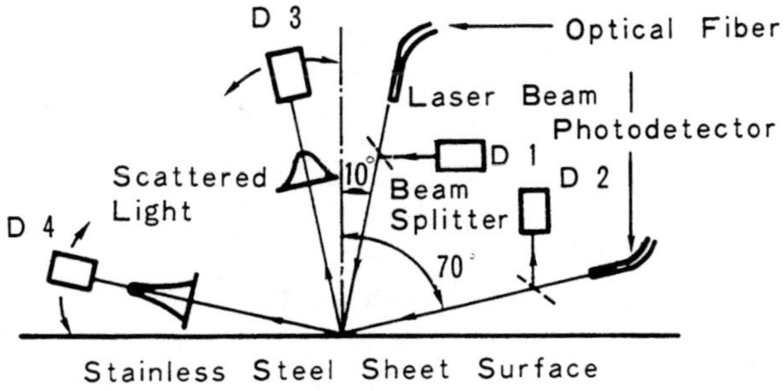

Fig.6 Schema of the detecting head of the on-line glossmeter.

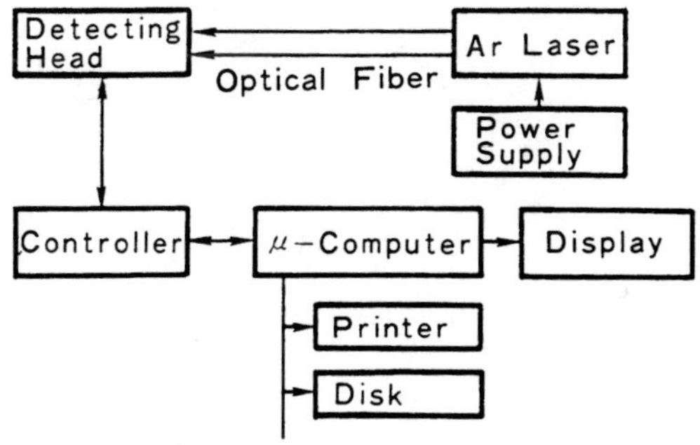

Fig.7 Block diagram of the system.

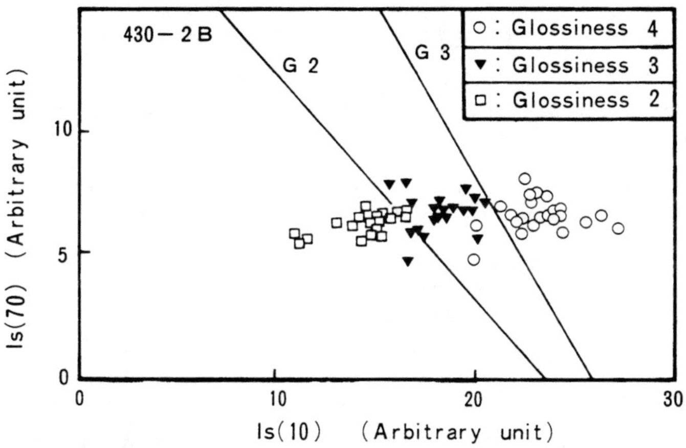

Fig.8 Experimental result showing the classification of glossiness of stainless steel sheets, type 430, No.2B finish, which were tested by the on-line glossmeter.

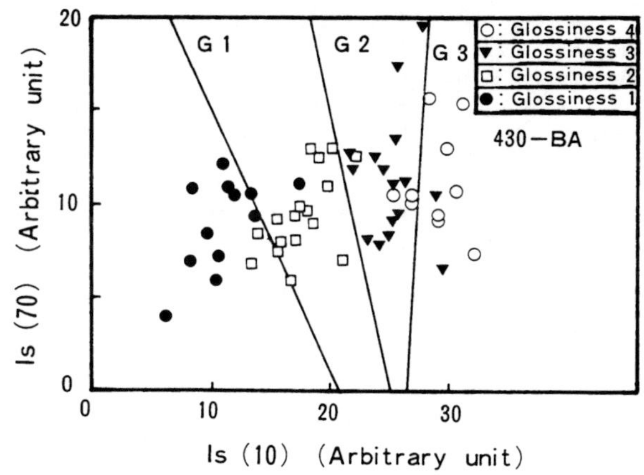

Fig.9 Experimental result showing the classification of glossiness of stainless steel sheets, type 430, bright annealed finish.

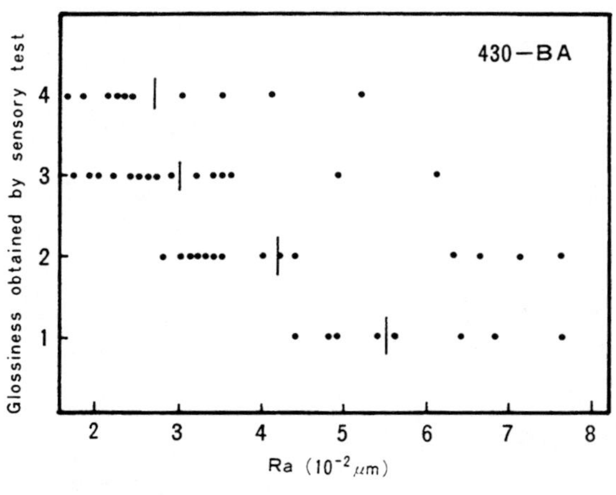

Fig.10 Experimental result showing the correlation between Ra measured by a stylus roughnessmeter and the grade of glossiness determined by visual inspection. Samples are stainless steel sheets, type 430, bright annealed finish.

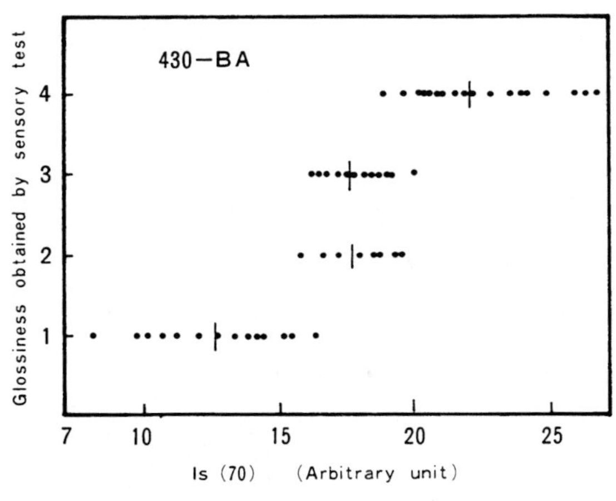

Fig.11 Experimental result showing the correlation between the specular reflection intensity Is(70) at the incident angle of 70° and the grade of glossiness determined by visual inspection. Samples are stainless steel sheets, type 430, bright annealed finish.

Reprinted from *Automated Inspection and High-Speed Vision Architectures II,* M.J.W. Chen, editor, Proc. SPIE Vol. 1004, pp. 10-16 (1988). ©1988 SPIE.

In process optical measurement of micro profile on cold rolled steel plates

H.Schmalfuß, K. Weber, W. Leyer

ERWIN SICK GmbH, D-8000 München, Fed. Rep. of Germany

ABSTRACT

The knowledge of surface roughness profiles or roughness parameters are necessary for a great variety of industrial production processes. In most cases at this time a in-process measurement is not possible due to different boundary conditions given by these processes.

It is very clear that the classic mechanical method does not allow in-line measurements. Therefore optical methods were investigated by a large couple of R&D groups.

The state of the art is that some optical roughness measurement systems you can buy commercially. But there is a general disadvantage of these systems because they do not cover all the possible applications as well as necessary.

In our paper will give first a short overview on the field of optical roughness measurement methods. Second we will introduce you into a special application, the in-process measurement of cold rolled steel. Third we will explain the basic principles and limitations of our new method and we will show you the results of our first industrial field tests.

1 INTRODUCTION

The in process measurement of micro profiles and/or texture (roughness) parameters is impossible by using the classical stylus instruments.

Therefore for a large couple of years optical methods were investigated to overcome the limitations of tactile systems.

The state of the art is that the following types of optical profilometers are more or less commercially available:

* Laser light stylus (Rodenstock, Perthen)/1/
* Light scattering technique (Rodenstock/sn-value/)/2/
* White light speckle method
* Structured light illumination technique
* Shadow method
* Far IR illumination (e.g. 10 m)/3/

But all these different methods do not cover all the real needs of industry, such as:

* Comparable results to stylus instruments
* fast
* non-contact
* measurement of "rough" surfaces (Ra»2 μm)
* insensitive against height variations and surface tilts
* easy maintenance and simple to use

Therefore we decided to develop an instrument which meets all the above mentioned conditions for a specific application, the in line measurement of cold rolled steel:

* indirect method, must be insensitive against vibrations
* working distance at least 50 mm
* range of profile heights up to 25 µm minimum, (50 µm desirable)
* range of material speeds up to 30 m/sec.
* generation of the micro profile line as base for the deduction of different roughness parameters

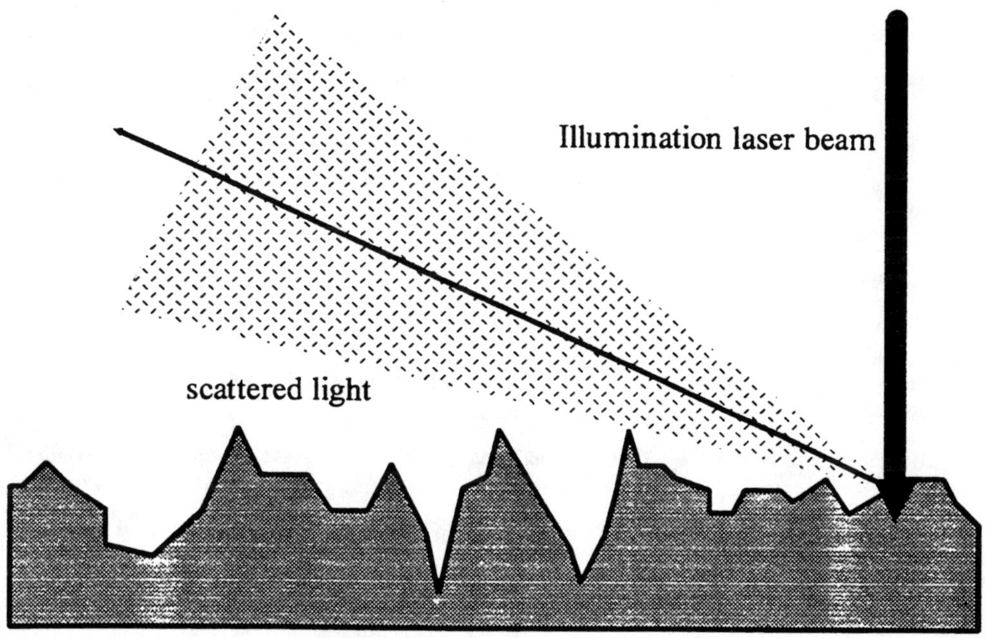

Figure 1: Basic Idea

2 Basic idea

The basic idea of our system is not to measure the scattered light distribution or to measure the height profile directly. Our system measures the first derivative of the height profile. In this case it is necessary that the surface can be described as an assemble of small plane mirrors, each mirror with different tilt angle in space. A metal surface can be considered to consist of a great number of such elementary mirrors. Figure 1 shows the principle.

If a laser spot of a diameter corresponding to the element size is projected on a specific elementary mirror, the incident ray will be reflected by an angle depending on the inclination of this element. If now the metal surface is moving and the angle of reflection is measured in this direction at constant distances, which correspond to the size of these surface elements, a sequence of angles of inclination of surface elements in this direction is obtained.

From these data, by summation and some known filtering operations, the micro profile in this direction is obtained. On base of this micro profile the different roughness parameters can be calculated in the usual manner (e.g. DIN rules).

Hence, we had to develop an angle measuring system, which is able to average in a very short time the scattered intensity to a mean angle αi per position i.

3 Instrumental realisation

The scheme of the sensor head is illustrated in Figure 2. The collimator lens system of a laser diode focuses a fine laser spot on the surface of the moving steel plate. The angle of incidence is about 10 degrees, the spot diameter 15 m. Within the angle of regular reflection the central collecting lens L0 and the central receiver E0 are arranged on a semicircle, which supports all collecting lenses for angular measurement of scattering reflected light. The semicircle and with it the plane of angular measurement is arranged in the direction of movement, see front view of fig. 2 and perpendicular to the plate surface, see side view of Fig. 2

Each of the collecting lenses images the laser spot on the corresponding PIN-diode. The receiving diodes are connected with resistors and hereby an angle sensitive diode array (ASD) is realised. This is an analogy to the well- known position sensitive diode (PSD).

Analogous to PSD, the described ASD gives an average angle information even if the incident radiation is distributed irregularly. The averaging process can be made fast enough to allow measuring sequences of more than 2 MHz.

The local averaged scattering angles are represented by the two currents Ia and Ib of the ASD.

Here some further specifications of the first realisation of the sensor head:

- angular receiving range of ASD : 70 degrees
- diameter of the semicircle of lenses: 150 mm
- number of lenses and receivers: 21
- free working distance: 50 mm

The schematic diagram of figure 3 shows, how the two currents Ia and Ib of the ASD are processed in the subsequent electronic assemblies.

First, they are amplified and converted by fast operational amplifiers and the resulting voltages Ua and Ub then are digitised.

To obtain the constant distance x = 10 μm of the measuring points the material speed controls the A/D conversion frequency.

The subsequent arithmetic logic unit calculates the local angle of scattered reflected light i at the position i by using the equation, stored in a LUT:

$$\alpha i = 90 \frac{Ua - Ub}{Ua + Ub} \text{(degrees)}$$

and finally the height S of the surface element at the position i is determined:

$$Si = tg\ (0.5 * i)$$

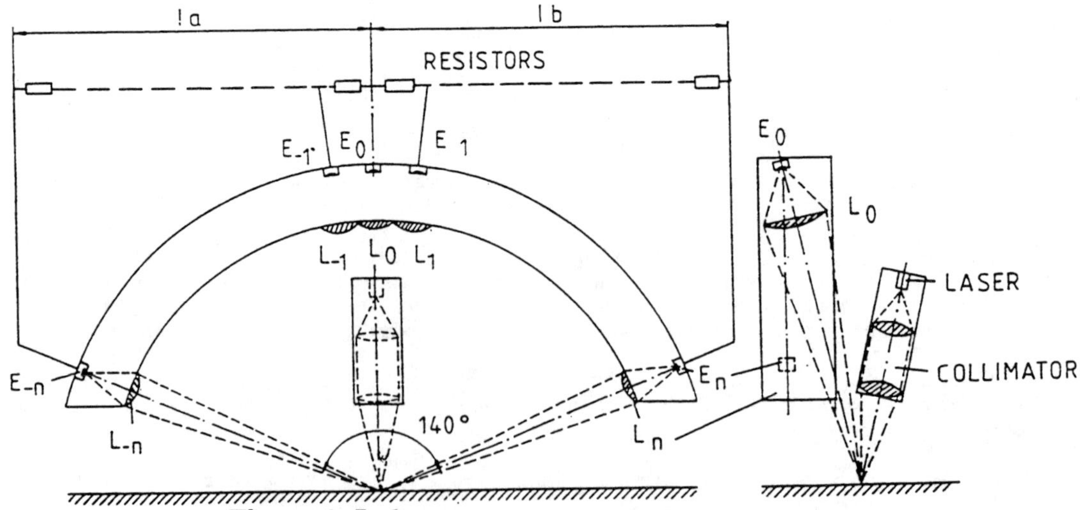

Figure 2: Lab system

front view , side view

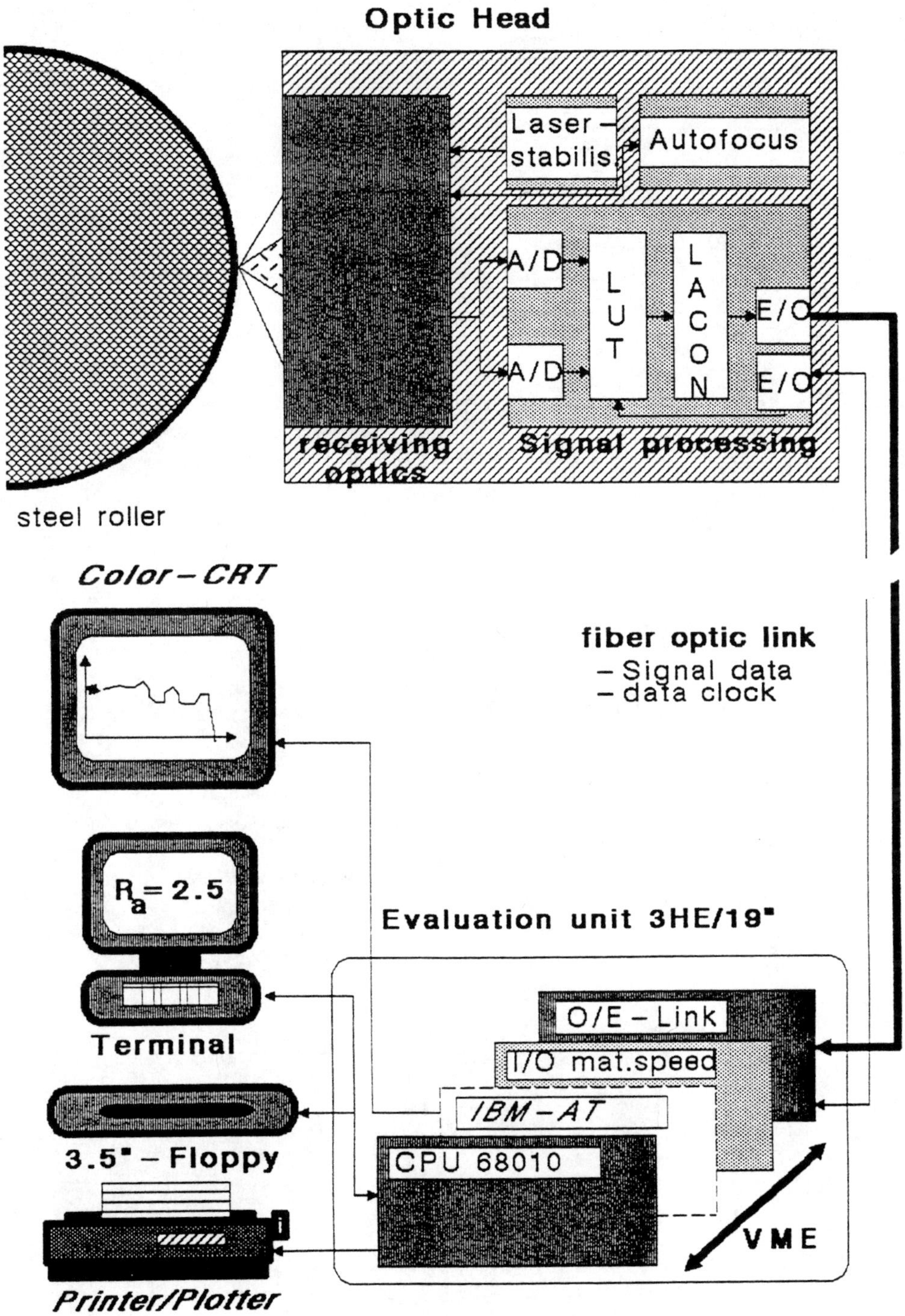

Figure 3: SORM System , new concept

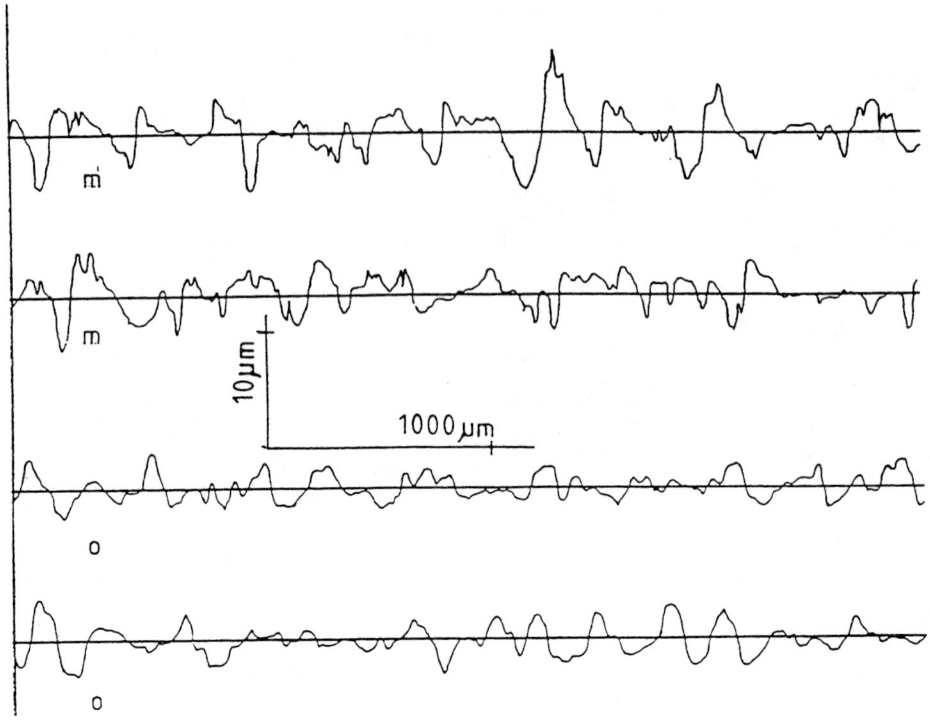

Figure 4: Comparisson: mechanical (m) / optical (o) profiles

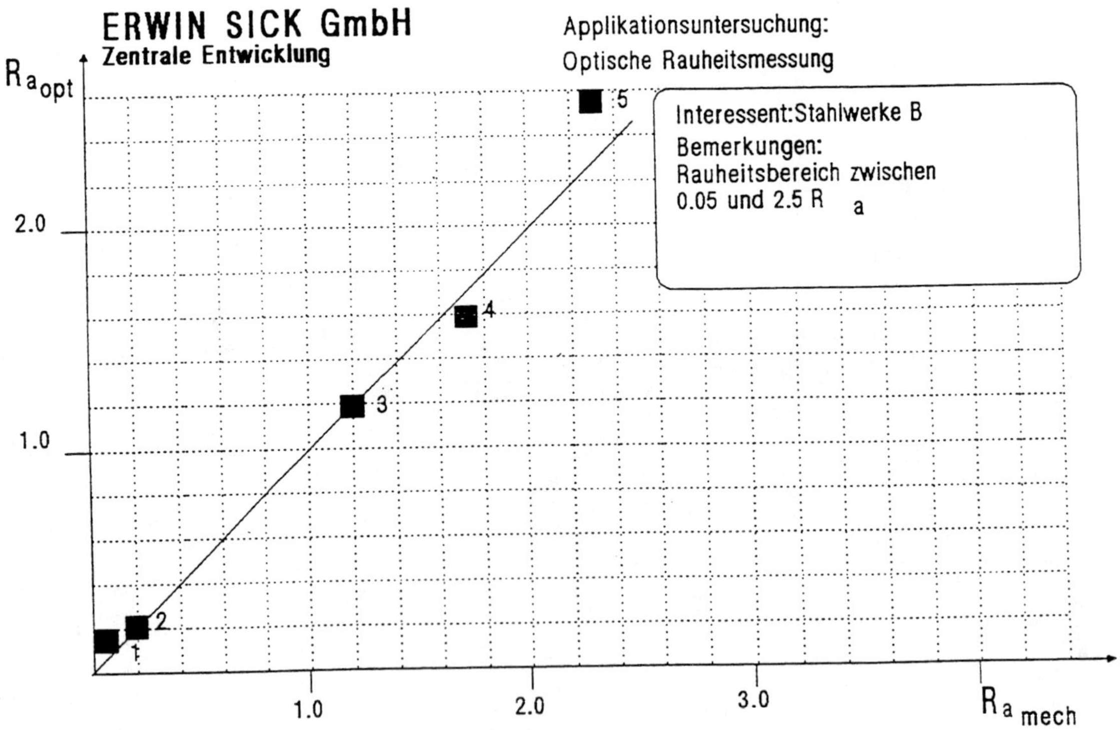

Figure 5: Correlation Ra(mech)/Ra(opt)

The height profile data then are stored in a RAM.

After storing the height profile data, the same operations as in stylus instruments are used to filter out the skewness of the plate and the waviness of the surface to produce the corrected micro profile coordinates.

These calculations are now done by software. The next generation system will have a special DSP chip to provide real time filter capability.

From these data, any roughness parameters , such as Ra and Rz can be calculated in the standardised way.

4 Results

With a lab type sensor head optical roughness measurements were performed and compared with results of mechanical stylus measurements on a great number of samples of cold rolled steel plates.

Hereby the samples were fixed on a rotating disc and were moved with up to 20 m/s .

For demonstration, Fig. 4 shows typical results of a series of mechanical and corrected optical measurements on cold rolled steel samples.

One test was performed in a cold rolling mill K, where the steel was annealed in coils before cold rolling. The results of optical measurements are shown in Fig. 5. On the abszissa the scale shows the number of measurements, counting from no. 1 to no. 330. The black columns in direction of the ordinate give the optical Ra values of these measurements.

An interesting effect is the higher roughness in the middle of each coil compared with both ends. This may be a consequence of differences in the variation of temperature versus time during the annealing process or may be caused by the discontinuous temper process.

At the end of every coil, a sample was cut for stylus measurements and Figure 6 shows the correlation between optical and mechanical measurements of Ra.

In another plant H with an continuous annealing process the same test set up was used. Figure 7 shows the results. Hereby the Ra-values are rather constant in time.

5 Conclusion

Due to the positive results of the two field tests we decided to develop a series of such a measurement system. The following improvements will be realised:

* addition of an automated focusing system to work in the best focus even with changing thicknesses of the metal plates;
* faster electronic equipment and larger fast memory to realise a measuring length of 320 mm and a shorter pause of less than 10 sec. between two measuring positions;
* mass memory provided in the microprocessor unit
* fiber optics connection between sensor head and PC to allow fast data rates over long distances.

6 Acknowledgement

The authors wishes to thank the R&D group for their fruitful co-operation and our industrial partners for given us the help and assistance in carrying out the field tests.

7 References

/1/ Ch. Gorecki, G. Tribillon and J. Mignot, Profilom tre optique en lumi re blanche, J. Optics (Paris) 14 (1). 19-23 (1983)

/2/. R. Brodmann, O. Gerstorfer and G. Thurn, Optical roughness measuring instrument for fine-machined surfaces, Opt. Eng. 24 (3) 408-413 (1985).

/3/ G.A. Monfort and A.A. Bragard, Proc d de mesure en continu de la rugosit de la surface d'un produit lamin froid, Demande de brevet europ en 0 229 603, publ. 22.7.87.

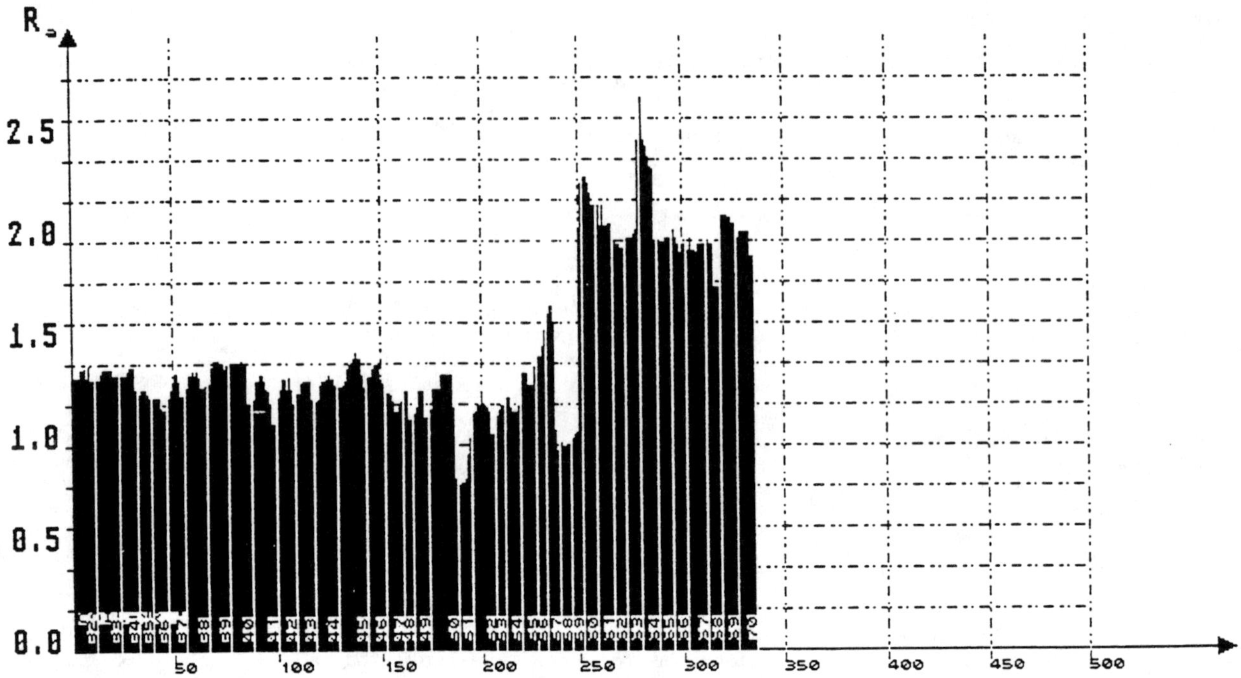

Figure 6: Field test results "K"

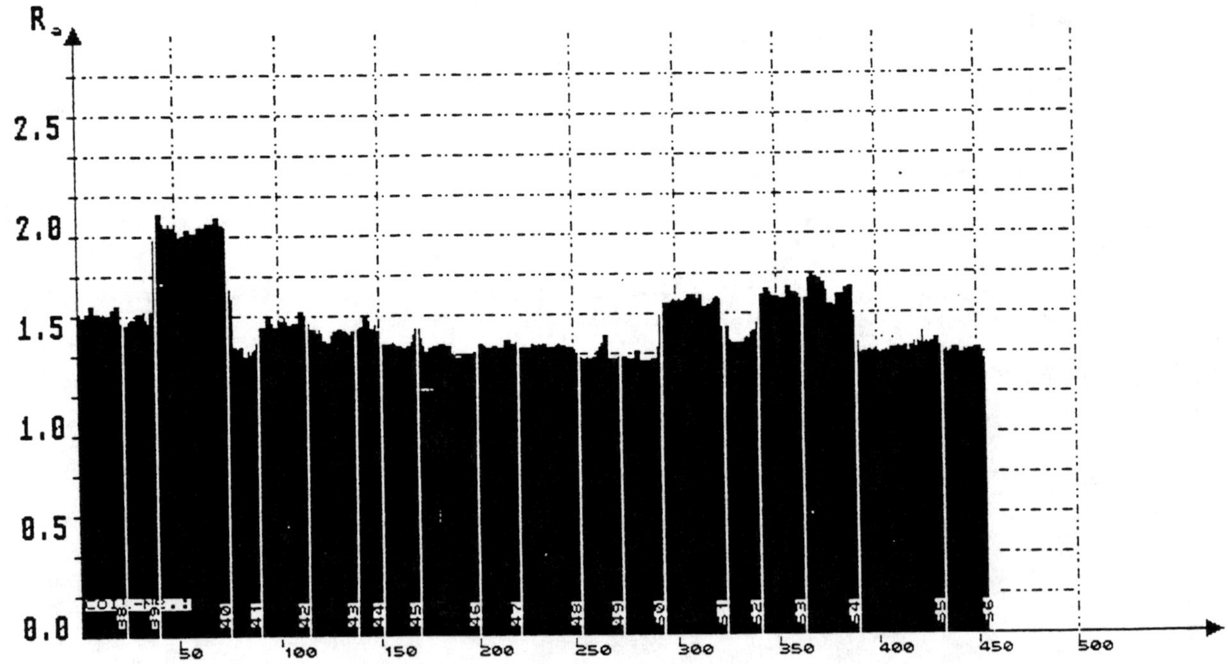

Figure 7: Field test results "H"

Reprinted from *Optical Engineering*, Vol. 24(3), pp. 414-418 (May/June 1985).

Specification and control of surface finish: empiricism versus dogmatism

K. J. Stout
Coventry (Lanchester) Polytechnic
Department of Production Engineering
Priory Street
Coventry, CV1 5FB, England

C. Obray
Coventry (Lanchester) Polytechnic
Department of Statistics
Priory Street
Coventry, CV1 5FB, England

J. Jungles
JAR MicroEngineering Ltd.
63 Alexandra Street
Thurmaston, Leicester, LE4 8FE, England

Abstract. This paper reviews the development of the analysis of surface finish from the early attempts in the 1930s to the present day. The development of parameters used in surface analysis is shown in context with the instrumental techniques available at the time, and it is argued that characterization based on graphical and experimental convenience has influenced industrial practice. As the requirements of manufacture and functional performance have been stretched by advancing technology, many industrialists have been forced to accept the fact that existing specification practices are limited and have sought alternative descriptions based on well-established techniques; but these techniques themselves are limited, their suitability to in-process measurement being practically nonexistent. It is shown that attempts have been made recently to develop optical methods of assessing surface finish using traditional parameters such as R_a. This paper suggests that it may be time to look toward a new form of specification that is more suited to assessment by optical transducers, and some methods of assessment are proposed. To support this view, a simple low-cost device is discussed that can be calibrated to give R_a but that also presents information in a more relevant empirical way that may be more valid than the existing parameter specification.

Subject terms: surface metrology; surface finish; optical technique; stylus technique; roughness average; plunge grinding.

Optical Engineering 24(3), 414-418 (May/June 1985).

CONTENTS

1. INTRODUCTION

"Emperici" and "Dogmati" were words coined by the Greeks to denote different types of physicians. Emperici were experimenters who generally scorned theory. The successes of the Emperici were in bone setting, surgery, and in fields that demanded fast and unique solutions. Emperici were often branded as quacks because of the lack of theory surrounding their craft. Dogmati, on the other hand, could produce elegant theories for the use of leaches and similar practices, and evolved a professionalism based upon decorum of work and action. Dogmati could seldom be accused of harming a patient, while the actions of the Emperici might easily be related to the reactions of the patient. Medicine has come a long way since the early Greeks, and this has been largely due to the foregoing two-pronged attack upon illness and disease.

Invited Paper ME-107 received Aug. 2, 1984; revised manuscript received Feb. 27, 1985; accepted for publication Mar. 5, 1985; received by Managing Editor Mar. 7, 1985. This paper is a revision of Paper 525-04 which was presented at the SPIE conference on Measurement and Effects of Surface Defects and Quality of Polish, Jan. 21-22, 1985, Los Angeles, CA. The paper presented there appears (unrefereed) in SPIE Proceedings Vol. 525.

The situation in the surface finish world in many ways mirrors the early days of medicine. We have our dogmatists who would tell us that everything we need to know about the texture of machined surfaces can be found by feeling the surface with a diamond and applying certain statistics to the data so derived. We also have our empericists who proclaim success by uncommon means of production and measurement. This paper outlines the surface finish characterization techniques that are dogmatic in character, and contrasts these to some empirical methods that are struggling to gain acceptance.

2. DEVELOPMENT OF SURFACE CHARACTERIZATION—COLLECTION OF SERIAL DATA BY STYLUS INSTRUMENTS

The first steps in surface assessment by direct measurement were proposed by the German engineer Gustav Schmalz approximately 55 years ago.[1] Initially, the dominant method of measuring surface irregularities was by mechanically magnifying the vertical displacement of a stylus as it passed over a surface. The magnified surface was recorded on a smoked glass plate. It was not surprising that the early measures of surface roughness were directed at establishing the peak-to-valley roughness. This initial peak-to-valley parameter, which was originally called R_T but has recently been redefined as R_y,[2] is shown in Fig. 1(a).

It was soon realized that the peak-to-valley roughness of a surface could be very misleading, as its magnitude was dependent on the highest asperity (peak) to deepest scratch (lowest valley). A single unrepresentative feature was capable of yielding a value that could exceed by three to five times the more typical peak-to-valley roughness heights. R_y is therefore an extreme value parameter. In an

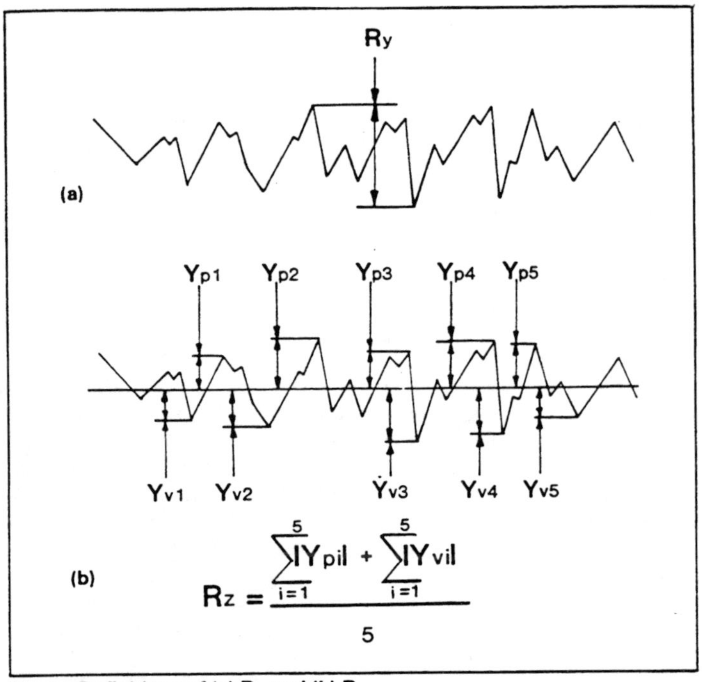

Fig. 1. Definitions of (a) R_y and (b) R_z.

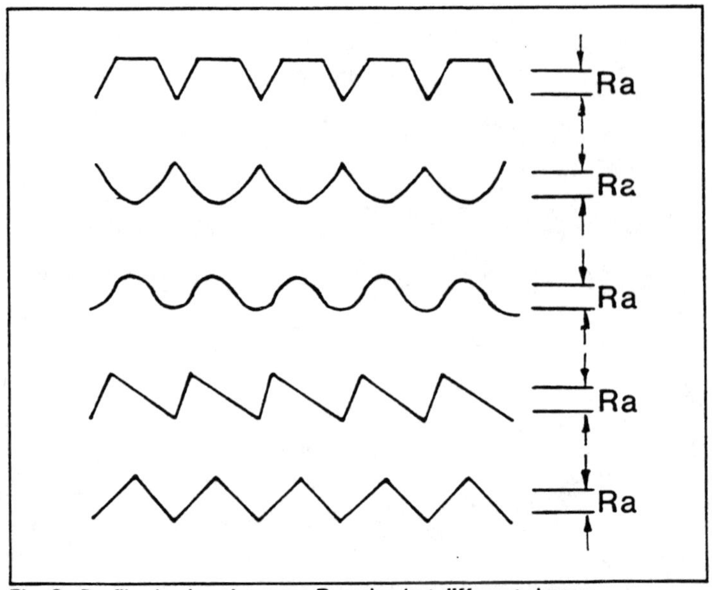

Fig. 2. Profiles having the same R_a value but different shapes.

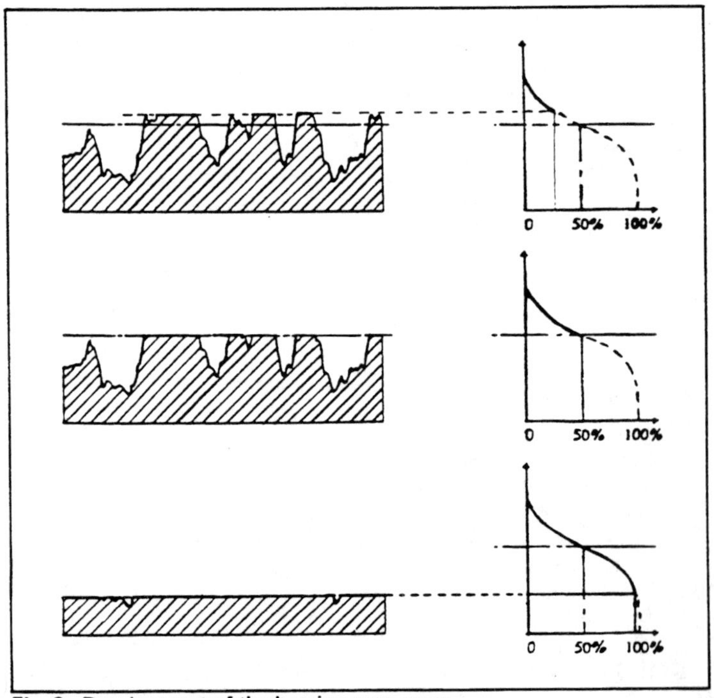

Fig. 3. Development of the bearing area curve.

attempt to overcome this disadvantage, a new parameter that takes the average of the five highest peaks and the average of the five lowest valleys was devised. This parameter was termed the ten-point height R_z [Fig. 1(b)]. A variety of other peak-to-valley roughness parameters were also proposed, many of which were slight variations on a theme and all of which had severe limitations.

As instrumental techniques developed, it was soon recognized that analog techniques could be used to determine the average roughness of the profile. This average roughness value was originally called center line average (CLA) and is now termed R_a. The average roughness value, introduced because of its instrumental simplicity, soon became widely used. Those more knowledgeable recognized that this parameter had to be specified along with details of the machining process by which it was to be achieved.[1]

From this logical development, which was limited by the available instrumental techniques, the early American, British, and Russian standards were developed. These national standards, finally leading to the ISO standard,[2] provided a basis for the communication of the quantitative assessment of the roughness of engineering components. As more industries tried to specify roughness in attempts to control the quality of manufacture and to relate the surfaces produced to functional attributes, it became evident that these parameters had severe limitations. This was particularly true for surfaces employed for tribological functions, which cover a wide range of manufactured components. These surfaces have regularly been the subject of surface specification. It was noted that descriptions of the scale of a surface were of extremely limited value and that parameters that describe the surface shape were often equally or more applicable. The limitation of the scale parameter R_a is shown in Fig. 2, because R_a does not uniquely specify the surfaces there.

A further parameter that attempted to describe the shape of the surface was defined during this early period. This was known as the bearing area curve,[3] which essentially is the cumulative height distribution, as shown in Fig. 3.

During the 1970s the wide availability of digital computers enabled metrologists to experiment with an even greater range of possibilities; thus parameters such as peak curvatures, valley curvatures, and peak spacing were developed in an attempt to enhance the understanding of functional performance, but all of these had limited value. It was during this period that descriptions suited to specifying surface shape were introduced, these being the hitherto well-defined statistical parameters of skewness and kurtosis. Skewness describes the asymmetry of a profile, while kurtosis describes the peakedness or spikiness of a profile. These parameters are shown diagramatically in Figs. 4(a) and 4(b). The additional descriptors still do not provide a complete surface characterization, and hence specification is still incomplete.

3. OPTICAL METHODS OF SURFACE ASSESSMENT

An indirect method of surface assessment is achieved by projecting an optical beam at the surface and examining the scattered energy field reflected from the surface.

Optical methods of surface inspection and surface texture characterization are attractive from a number of viewpoints: they are noncontacting and so do not degrade the surface, and they are able to achieve 100% coverage of the surface material in real time.

In the late 1960s optical scanners were devised, frequently using

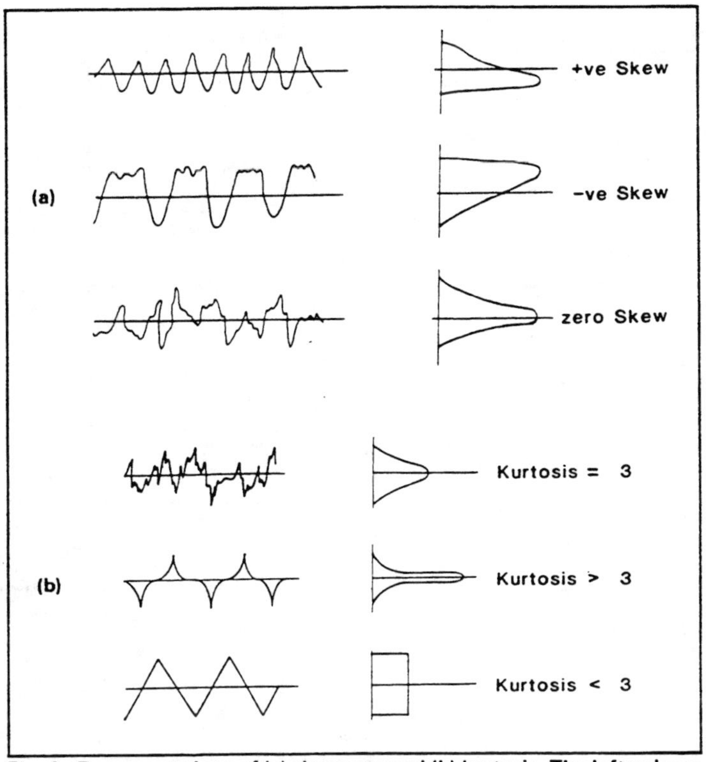

Fig. 4. Representations of (a) skewness and (b) kurtosis. The left column shows various types of surface profiles, and on the right column are the corresponding height distributions.

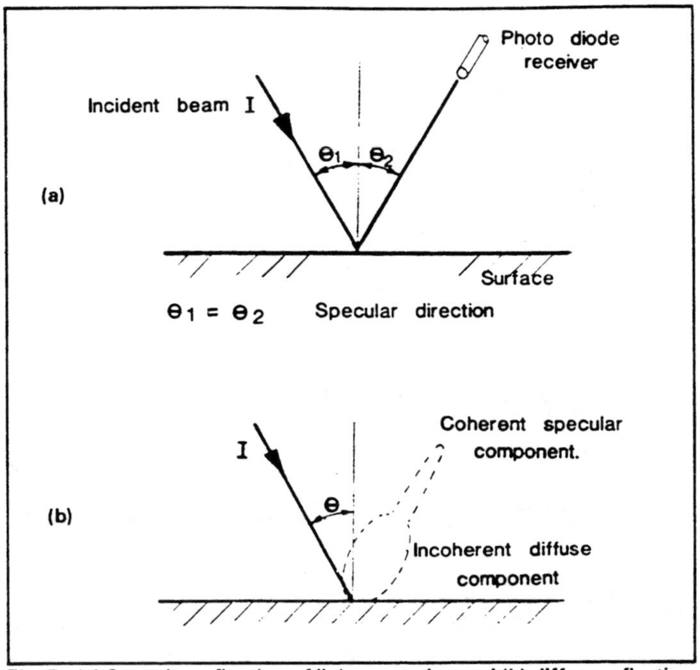

Fig. 5. (a) Specular reflection of light scattering and (b) diffuse reflection components of light scattering.

finely controlled parallel laser beams for detecting surface blemishes. The empiricists tuned the system so that variations in the specular and/or off-specular response(s) could be interpreted as signaling the presence of a surface defect (see Ref. 4, for example). Soon after, the science had progressed so that certain defects could not only be detected but could also be classified with reasonable accuracy.[5]

In order to assess the performances of such detector/classifier systems as a function of surface roughness and to attempt some degree of optimization, the physics of the interaction of a laser beam with a rough surface had to be taken into account; i.e., the empiricist needed the dogmatist to understand the measurement system. Accordingly, during approximately the same period, other workers were developing scattering theories on the basis of various model surfaces. Twersky's[6] concept of a rough surface was that of a regular or irregular distribution of protuberances on a perfectly conducting plane, Ament's[7] model surface consisted of randomly spaced half-planes, while Beckmann[8] described a half-plane facet model by a Markov chain. The model surface of Du Castel et al.[9] had a number of brilliant points per unit area from which orginated the scattered field. The ideas elucidated in these early papers are relevant both to a geometrical facet view of scattering and to a wave theory physical optics approach, although the model surfaces in most cases are unrealistic.

Beckmann[10] developed a theory appropriate to laser light for both strong and weak scatterers. His theory provided a theoretical basis for optical methods of estimating surface roughness from measurements of the far-field reflected light. In brief, his theory yielded the average intensity scattered in and away from the specular direction, and in some cases the complete distribution about those mean values was specified. The scattered field consists of (1) a specular coherent component, which vanishes everywhere except in the precise neighborhood of the specular direction [Fig. 5(a)], and (2) a diffusely scattered incoherent component [Fig. 5(b)]. Broadly speaking, the coherent component depends on the variance of surface height, while that scattered away from the specular direction depends additionally on the autocorrelation function of surface heights. Several workers have estimated these first- and second-order statistics of surface heights from light scatter data based on Beckmann's theory, cf. Porter and Mundy,[11] Obray,[12] and Chandley.[13,14]

For certain scales of roughness the diffuse scatter depends critically on surface slopes. For normal incidence light, $\lambda = 0.914$ μm and the rms deviation $R_q = 0.23$ μm, the full solution approaches that given by geometrical optics,[12] which suggests that for all but the smoothest of ground surfaces, the diffuse component is dominated by the distribution of surface slopes. Tanner[15] quantified the reduction of the angular spread of the diffuse light distribution relative to that of the surface slopes. The roughness experienced in a wide range of engineering surfaces may be considered as being "optically rough"; nevertheless, these surfaces fall within the range of the techniques mentioned above.

Although measurement by instruments based on the physical principles outlined above are feasible, the data obtained fall short of a complete surface characterization. This limitation is therefore shared with the direct methods of measurement previously discussed.

4. SURFACE CHARACTERIZATION: DOGMATIC APPROACH

We have discussed the two classes of surface assessment most commonly used today. Assessment by tactile instrument is limited by the slowness of serial data capture. In addition, current tactile instruments are primarily suited for use in the metrology laboratory, not in the manufacturing environment. A further major drawback of assessment by serially captured data is that the characterization achieved only relates to a single line on the surface. In normal interaction, surfaces come into contact over an area.

Optical assessment of surfaces has the advantages of speed and the fact that an area, rather than a line, is surveyed. One limitation is that the types of parameters that can be evaluated by these instruments, such as R_q, correlation length, and profile slope, are also within the scope of a stylus instrument.

We now take the dogmatic approach and attempt a comprehensive characterization of a surface profile. We must realize that no assessment will yield a full surface characterization with one parameter. To achieve an adequate statistical characterization of a surface it is necessary to define five features: (a) a measure of scale of roughness (R_a or R_q), (b) a measure of the shape of roughness (skewness), (c) a measure of the peakedness of roughness (kurtosis), (d), a measure of

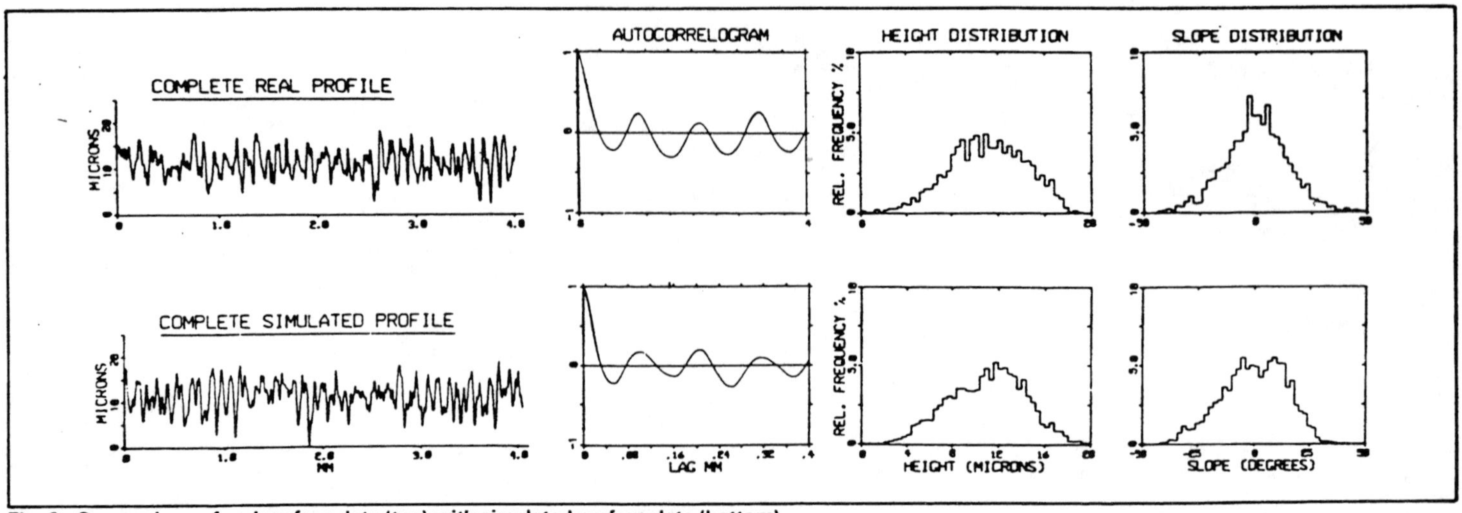

Fig. 6. Comparison of real surface data (top) with simulated surface data (bottom).

the random nature of the surface, and (e) a measure of the periodicity of the surface. Elements (d) and (e) may be analyzed, for example, by the autocorrelation function, which is then characterized in terms of its random and periodic components. The periodic component can be modeled using harmonic regression; the random component can be modeled using autoregressive moving average processes.[16] It should be emphasized that data collected in these five ways adequately define a profile class but do not necessarily describe the characteristics of an area of a surface.

To demonstrate the adequacy of this characterization, Fig. 6(a) shows a surface measured by a stylus instrument and characterized by the five features listed above; Fig. 6(b) is a simulation of the same surface using the numerical values obtained during characterization to generate the modeled trace. Clearly, the simulation could be a random sample taken from the original surface.[16]

5. SURFACE CHARACTERIZATION: EMPIRICAL APPROACH

In practice, provided that the machining process remains unchanged, it is possible to monitor surfaces by examining one controlling parameter that has significance and that is fundamental to the general character of the surface. Such parameters include R_q or R_a, skewness, or other parameters such as rms surface slope that are combinations of the fundamental features discussed above. It is therefore possible to use either tactile methods or optical methods in a comparative assessment. However, if the machining process is changed, or if two surfaces that have been produced by different processes are to be compared, a full profile characterization is necessary.

Since much industrial measurement is done by comparing components against an acceptable standard, either tactile or optical techniques are appropriate, as long as the parameter being measured is related to either R_q, skewness, rms slope, or a combination of these. Optical measurement by either specular reflection or diffuse reflection is appropriate. The former phenomenon can be related to R_q or a good approximation of R_a, whereas the latter can be related to a combination of R_q (R_a) and rms surface slope.[17]

To demonstrate this point a simple low-cost instrument called the Compari-surf, which measures scattered light intensity, was calibrated for R_a against a set of surface roughness blocks produced by plunge grinding. In plunge grinding, the feed of the wheel is directly into the surface rather than across it. The instrument gives a reading between 0 and 100, depending on the proportion of diffuse light reflected from the surface into and around the specular direction, with a reading of 100 representing 100% reflection. A segment of the curve produced is shown in Fig. 7, and similar curves have been published previously.[18] After calibration, the instrument was used as

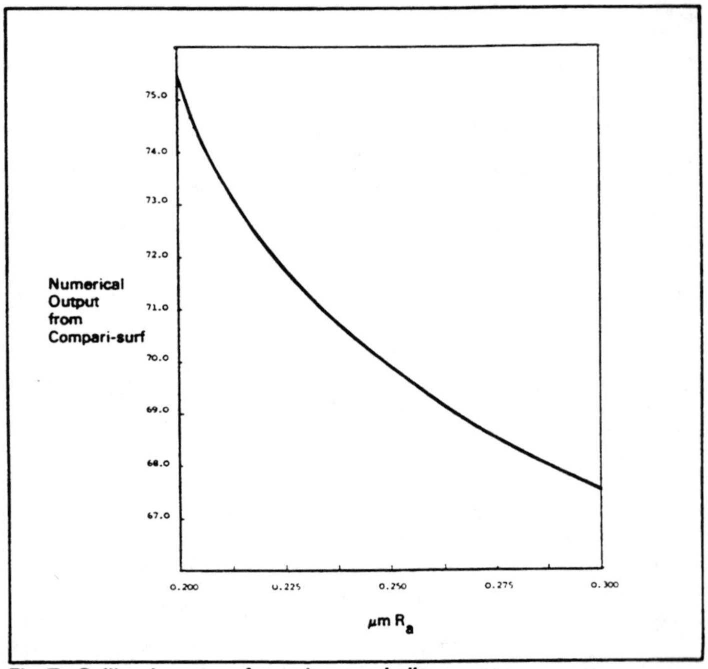

Fig. 7. Calibration curve for a plunge-grinding process.

a comparator to assess a number of plunge-ground surfaces of differing roughness. These surfaces were later measured using a stylus instrument to obtain an assessment of R_a by traditional means. Tables I and II compare the two methods of assessment for one of these surfaces. The results show that the values obtained for R_a by the two methods are similar and are well within the variations expected when taking small samples of data from large populations. The ranges of values obtained from 20 samples taken from the surface by the two methods of assessment were of a similar order of magnitude, and these are also given in the tables. The statistics for the other plunge-ground surfaces are comparable to those given in Tables I and II. Similar statistics for centerless ground surfaces are given in Ref. 18.

6. CONCLUSIONS

It is suggested that when a comparative assessment of surfaces is undertaken, only one parameter that is significant in a full dogmatic surface characterization needs to be monitored. This parameter may be R_q, R_a, skewness, or rms slope.

Although much of the research currently being conducted on

TABLE I. R_a Values Obtained for a Plunge-Ground Surface Using the Talysurf Stylus Instrument

Sample	R_a
1	0.267
2	0.241
3	0.253
4	0.260
5	0.241
6	0.258
7	0.246
8	0.259
9	0.271
10	0.258
11	0.237
12	0.260
13	0.246
14	0.270
15	0.232
16	0.256
17	0.260
18	0.274
19	0.259
20	0.268
Mean	0.256
Standard deviation	0.012
Maximum value	0.274
Minimum value	0.232
Range	0.042
Maximum percentage variation from mean	+ 7.03 −10.34

TABLE II. Values of R_a Obtained for the Same Plunge-Ground Surface (as Table I) by Use of the Compari-surf and Calibration Curve

Sample	Numerical output from Compari-surf	R_a
1	67.6	0.290
2	69.8	0.253
3	69.5	0.263
4	70.0	0.248
5	70.3	0.240
6	69.9	0.251
7	68.2	0.288
8	72.7	0.215
9	68.2	0.288
10	68.2	0.287
11	71.5	0.230
12	68.7	0.267
13	68.2	0.286
14	69.7	0.257
15	69.1	0.262
16	68.9	0.267
17	68.7	0.268
18	70.1	0.246
19	69.1	0.264
20	70.0	0.248
Mean		0.261
Standard deviation		0.021
Maximum value		0.290
Minimum value		0.215
Range		0.075
Maximum percentage variation from mean		+11.1 −12.4

optical techniques attempts to describe traditional parameters, the work is unlikely to become very significant unless the five features of a full surface characterization can be achieved. Current optical analysis provides only two of the five parameters necessary, and hence the incomplete analysis indicates that, realistically, optical techniques are limited to comparative assessment.

7. REFERENCES

1. R. E. Reason, Wear 57, 1 (1979).
2. ISO Standard No. ISO/DIS 4287/1.2, "Surface Roughness—Terminology Part 1: Surface and Its Parameters," (1984).
3. D. J. Whitehouse, T. G. King, and K. J. Stout, Surf. Technol. 6, 259 (1978).
4. R. Brook, Metron. 3(8), 219 (1971).
5. W. J. Hill, "Defect Recognition in Automated Surface Inspection," Ph.D. thesis, The City University, London, Department of Systems Science (1971).
6. V. Twersky, J. Acoust. Soc. Am. 29(2), 209 (1957).
7. W. S. Ament, Trans. IRE AP-4, 369 (1956).
8. P. Beckmann, Acta Technica 2(4), 311 (1975).
9. F. Du Castel, A. Spizzichino, et al., J. Res. Nat. Bur. Stand. 66(3), 273 (1962).
10. P. Beckmann, *The Scattering of Electromagnetic Waves from Rough Surfaces*, Pergamon, London (1963).
11. G. B. Porter and J. L. Mundy, in *Techniques and Applications of Image Understanding*, J. J. Pearson, ed., Proc. SPIE 281, 176 (1981).
12. C. D. Obray, "Application of Scattering Models to Optical Surface Inspection," Ph.D. thesis, The City University Department of Systems Science (1983).
13. P. J. Chandley, Opt. & Quant. Electron. 8, 323 (1976).
14. P. J. Chandley, Opt. & Quant. Electron. 8, 329 (1976).
15. L. Tanner, Opt. & Laser Technol., p. 113 (June 1976).
16. T. A. Spedding, W. Watson, T. G. King, and K. J. Stout, "Sampling Variation in Surface Profile Skewness and Kurtosis Measurements," Proc. 8th Leeds/Lyon Symposium on Tribology, Inst. of Mechanical Engrs. publ., p. 49 (1981).
17. L. H. Tanner and M. Fahoum, Wear 36, 299 (1976).
18. K. J. Stout, Prec. Eng. 6, 35 (1984).

Reprinted from *Optical Engineering*, Vol. 18(2), pp. 222-225 (March-April 1979).

Optical Profile Transducer

Takeo Sawatari
Bendix Research Laboratories
20800 Civic Center Drive
Southfield, Michigan 48076

Richard B. Zipin
Bendix Automation & Measurement Division
721 Springfield Street
Dayton, Ohio 45401

Abstract

A noncontacting optical transducer, which measures the profile of machined surfaces, was constructed and tested. The profile transducer consists of a microscope objective, a point-source illuminator, and two photodetectors. The dynamic range of the instrument is typically 500 microinches with an accuracy of better than 10 microinches. The output of this transducer is a profile signal along a line on the surface under examination. The profile of a machined surface obtained by the optical-transducer correlates very well with that obtained by a mechanical Proficorder® which uses a diamond stylus. Principles of operation, results of various performance tests, and potential applications of the optical-transducer are discussed.

1. Introduction

The measurement of the profile of surfaces is important to many applications. The mechanical profiling instrument, which uses a diamond stylus, is a standard instrument for this purpose. Stylus instruments require the stylus tip to contact the test surface. The contact measurement can cause wear of the stylus tip or damage of the test surface.

On the other hand, optical techniques are, in general, noncontacting methods. Many optical techniques have been proposed to measure the surface profile of various materials. Optical interferometers of various types, such as two-beam, multibeam, scanning, holographic, and moire interferometers provide fine interference fringes presenting the surface profile of test samples. Optical imaging devices, such as a section microscope, can also measure the surface profile. A brief review paper[1] (with extensive references) on the above techniques is given by J. M. Bennett.

However, the problems of such conventional optical instruments are (a) the output of the instruments is an interference fringe which is not like a profile trace obtained by the stylus instrument (which is the standard instrument in the U. S.) and the interpretation of the output (interference fringes) is sometimes rather cumbersome; and (b) the region of the surface to be measured is often restricted by the field of view of the optical system used.

We describe an optical instrument which is noncontacting in nature and which provides output similar to that of the stylus instrument. The instrument was designed to be applicable only for a one-dimensional lay of surface texture. A prototype of the instrument was constructed and tested.

The principle and characteristics of the prototype instrument are discussed in Section 2 and an experimental result of the instrument along with the result of a comparison test are shown in Section 3. In a latter part of Section 3, experimental results of a rotary machine part are also described.

2. Principle and Characteristics

A schematic of the system developed is shown in Figure 1. An image of a pinhole source (S) is projected on a surface of a test sample (O) through the first beamsplitter (BS_1) and the microscope objective (L). The light reflected from the sample surface passes through the objective lens again and forms images at positions P_1 and P_2 through the second beamsplitter (BS_2). A large area photodetector measures the total intensity (I_1) arrived at the image position P_1. Another detector with a slit in front of it is placed at a front defocused plane (a distance (d) from the image position P_2)

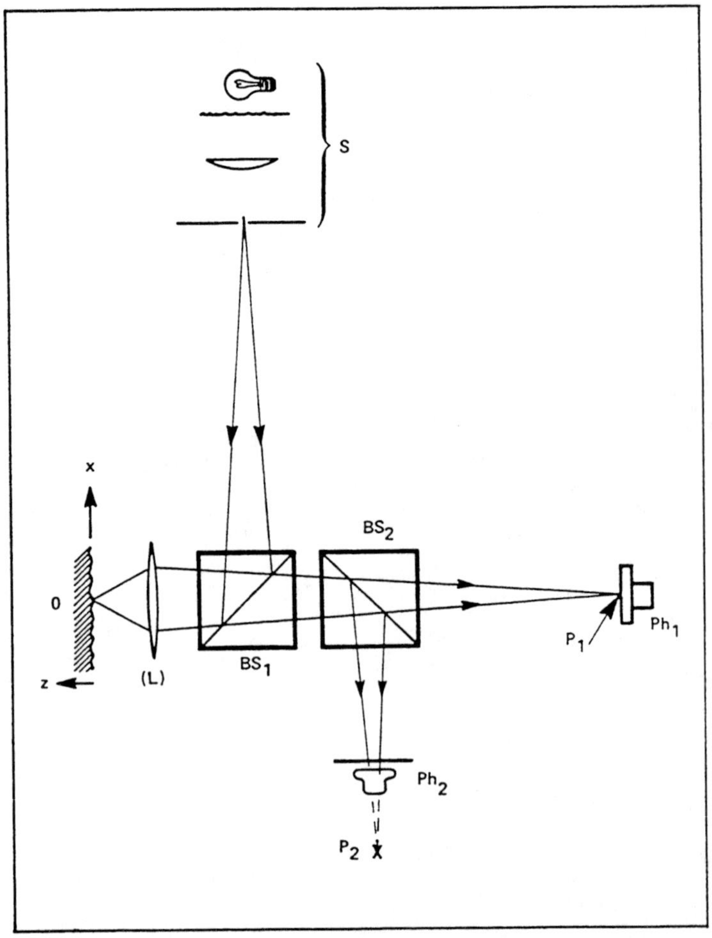

Figure 1. Schematic showing the principle of the instrument where S is the point source; L, microscope objective; O, test sample; BS_1 and BS_2, beamsplitters; P_1 and P_2, image locations; and Ph_1, large area photodetector; and Ph_2, photodetector with a slit.

1536 received October 2, 1978. This paper was presented at the SPIE seminar on Optical Metrology, August 28–29, 1978, San Diego, and appears in SPIE Proceedings Volume 153.

of the image P_2 and measures the intensity (I_2) of a part of the light which passes through the slit.

The intensity I_1 is a measure of the reflectivity of the sample surface and the intensity I_2 is a measure of the longitudinal displacement (i.e., the displacement along the optical axis) of the sample surface O from a certain reference plane (say, the mean-surface location of the object sample). The output of the system is the normalized signal (I_2/I_1) which is displayed as a function of the lateral sample location (x). The lateral location (x) is mechanically scanned by moving either the optical system or the test sample.

Now let's examine why the output (I_2/I_1) is proportional to the longitudinal displacement (z) of the sample surface (O). If the sample surface is displaced a small amount (z) longitudinally, the position of the image (P_2) shifts by the amount of z' ($=M^2z$), where M is the lateral magnification of the imaging system. Using a simple geometrical optics approximation for $z \ll d/M^2$, the intensity fluctuation as a function of a displacement (z), which is measured from the reference position ($z = 0$), is given as

$$I_2(z) = \frac{2\Delta s I_1(b-d)}{\pi d^2 r_0} M^2 z + I_2(0) \tag{1}$$

where $I_2(z)$ and $I_2(0)$ are the intensities of the slit detector at the longitudinal surface positions z and 0, respectively, Δs is the width of the slit, r_0 is the radius of the pupil of the objective lens, and b is the distance from the pupil plane of the microscope objective to the initial image point P_2 (for $z = 0$). Defining the output of the system as $I_2(z)/I_1$, we find that the output is proportional only to the displacement (z). Note that z is a function of x and the function z(x) is the profile of the surface. In order to obtain the output from the two detector signals, an analog electronic divider (AD 532) was used along with low noise amplifiers (AP 504).

The characteristics of our prototype, a photograph of which is shown in Figure 2, are summarized in Table 1. Results for a 21× microscope objective are given. The lateral resolution is determined by the pinhole size (300 μm in diameter) and the magnification ($M = 30$) of the system. The standoff distance is fixed by the design of the lens. The dynamic range, linearity, and errors in the table are estimated from the curve shown in Figure 3, where $I_2(z)/I_1$ is plotted using a 21× microscope objective. To obtain this curve, a flat surface was placed at the plane ($z = 0$) and the output of the system was recorded as a function of a longitudinal

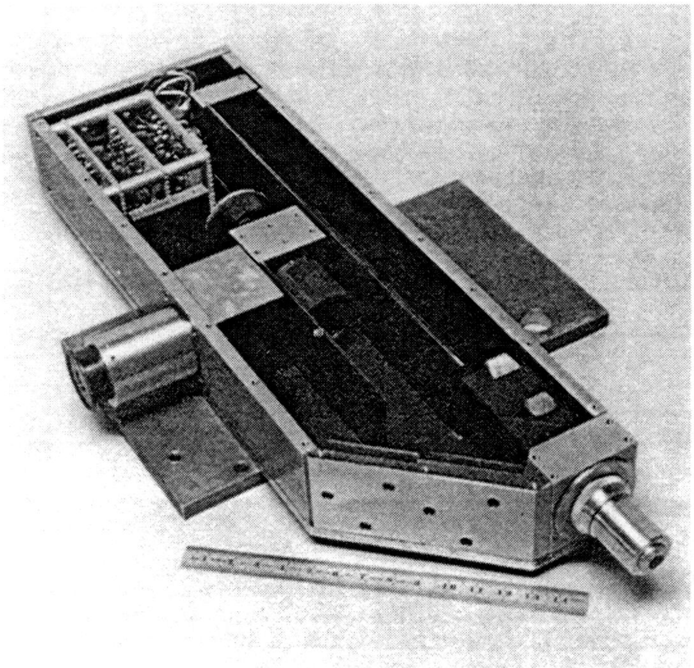

Figure 2. Photograph of the prototype (uncovered).

Table 1. Characteristics of Optical Profile Transducer.

Characteristic of the Model	Values
Standoff distance	54.5 mil
Linear range	0.5 mil
Linearity	±1.0%
Error	0.01 mil (maximum)
Minimum spot size (lateral resolution)	0.4 mil in diameter of geometrical sense
Scanning velocity	90 mil/sec (110 Hz cut-off frequency) of electronic processor

replacement (z) of the surface.

This instrument, however, provides an accurate profile (given in Eq. (1)) of the surface lay only when a surface having one-dimensional lay is scanned with the slit direction perpendicular to the lay direction. The characteristics of the instrument shown in Table 1 is, therefore, applicable only when a one-dimensional lay surface is tested. The reason for this limitation of the instrument is

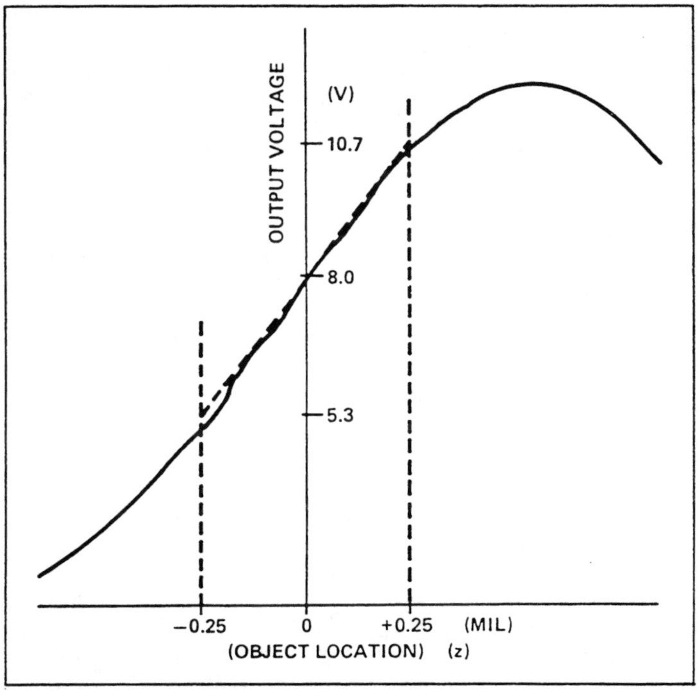

Figure 3. The output of the instrument as a function of longitudinal displacement (z) of the flat sample surface.

due to the use of the slit detector which is one-dimensional in nature. Simon[2] had proposed a similar optical system using pinholes instead of slits. His system uses two detectors (with a pinhole in front of them); one is at a front defocused plane and the other is at a back defocused plane. The output of his system is $(I_1 - I_2)/(I_1 + I_2)$, where I_1 and I_2 are the intensities measured by each detector. With his method, however, the profile of a machined surface showed a gross error[3] (10 times larger fluctuation than the expected value). This is due to the difference in magnification between the two defocused images.

3. Test Results

3.1 Standard Sample Test

In order to examine the performance of the prototype transducer, we first used a surface roughness standard as a test sample. The test sample consists of one-dimensional triangular grooves; the peak-to-valley height is 0.0005 in. and the peak-to-peak distance is 0.0037 in. Figure 4 shows an example of the trace obtained by a 21× objective system, demonstrating high fidelity.

In order to compare the machined surface profile traces obtained by the optical transducer and the stylus instrument, we used a standard machined surface with a surface roughness arithmetic average (AA) of 63 μ inches. The test sample has one-dimensional surface grooves. Ten different portions were scanned along the direction perpendicular to the direction of the surface grooves. (Note the slit is oriented parallel to the scanning direction.) The interval of each scan line was 0.002 in. The same portion of the sample was then scanned by a stylus instrument (0.0005-in. radius diamond stylus) several times; each time the

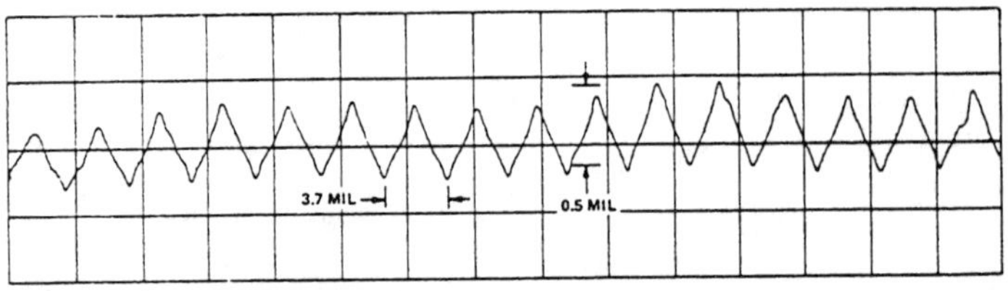

Figure 4. Profile trace of a calibration block with one-dimensional triangular grooves (P-V, 0.0005 in.; p-p, 0.0037 in.).

scan was made along a slightly shifted line. The two sets of data (one obtained by the optical transducer and the other by the stylus instrument) were thoroughly examined. Figures 5(a) and 5(b) are the optical and the mechanical traces, respectively. Although the lateral coordinates are not identical in the traces, the similarity of the traces (in detailed structure) is obviously seen, especially in each point numbered in both traces. The heights of the peak points of both traces were also compared. Although there was not an exact one-to-one correspondence, the difference of heights in the two traces is equivalent to the difference of the heights obtained by two stylus instruments. These results show that the optical transducer provides a machined surface profile which is almost identical to that obtained by a stylus instrument, as long as the surface lay is one dimensional.

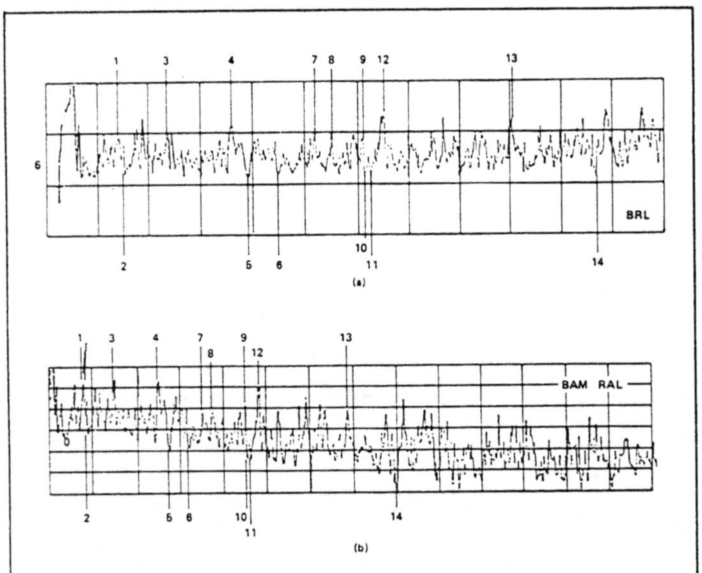

Figure 5. Comparison of trace of a standard machined surface (63 μin. AA), (a) by the optical transducer and (b) by stylus instrument.

3.2 Rotary Profiling Test

For an example of the application of the prototype, we investigated the applicability of the instrument for inspection of roundness of a cylindrical machined part.

A six-lobed part was examined on a rotary spindle machine and polar charts of its radius variations (~70 μin) were drawn. Figure 6 shows the profile chart obtained when the instrument filter response was set at 15, 50, 150, and 500 undulations/revolution. The chart of Figure 6 was made when the part was traced using a 0.0005-in. radius diamond stylus on the gage head.

The workpiece was not moved on the spindle when the filter settings were changed. The phase shifts exhibited by the chart are due to the response time of the filters. It should be noted that the spindle and the polar chart table are both rotated by the same drive mechanisms and move counterclockwise.

Figures 7(a), 7(b), 7(c) and 7(d) are polar charts of the same six-lobed part made using the optical transducer fitted with a 21× objective. For these charts, the response settings were again 15, 50, 150 and 500 undulations/revolution, respectively. We see the

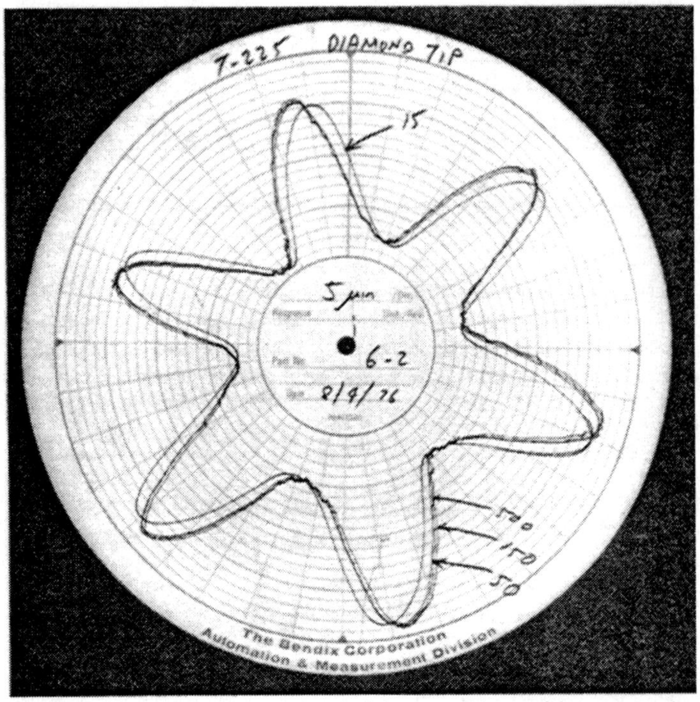

Figure 6. Rotary profile of the six-lobe part taken with a contacting tracer having a 0.0005-in. radius diamond stylus. The signal is filtered to pass up to 15, 50, 150 and 500 undulations per revolution.

same charts in their broad aspects. However, they differ from the charts made with the mechanically contacting gage head in the degree and type of detail exhibited. The detail seen on the charts is not noise. Often, we allowed a chart to be written for several turns of the spindle, and the repeatability of the trace was very good.

The six-lobed part was generated by lapping, and the predominant lay of its fine surface scratches is circumferential. The charts of Figure 7 were therefore made with the slit parallel to the axis of rotation of the spindle, i.e., normal to the circumferential direction of the workpiece surface. However, the high frequency component seen in Figure 7(c) and (d) is due to the surface lay in the axial direction of the cylinder. If the instrument is oriented 90° on the optical axis to orient the slit direction parallel to the surface structure, a large amount of high frequency component appears in the trace.

The above results show that the instrument works as expected and that it can be used for profiling of a cylindrical object even though the surface of the test sample has a certain amount of axial direction scratches.

4. Summary

A prototype of an optical profile-transducer, which consists of a microscope objective, two beamsplitters, a point source and two detectors (one with a slit), was constructed. The instrument was designed to provide an output similar to that of a stylus instrument, but can be used only for surfaces with a one-dimensional lay. The experimental results show high fidelity and good correla-

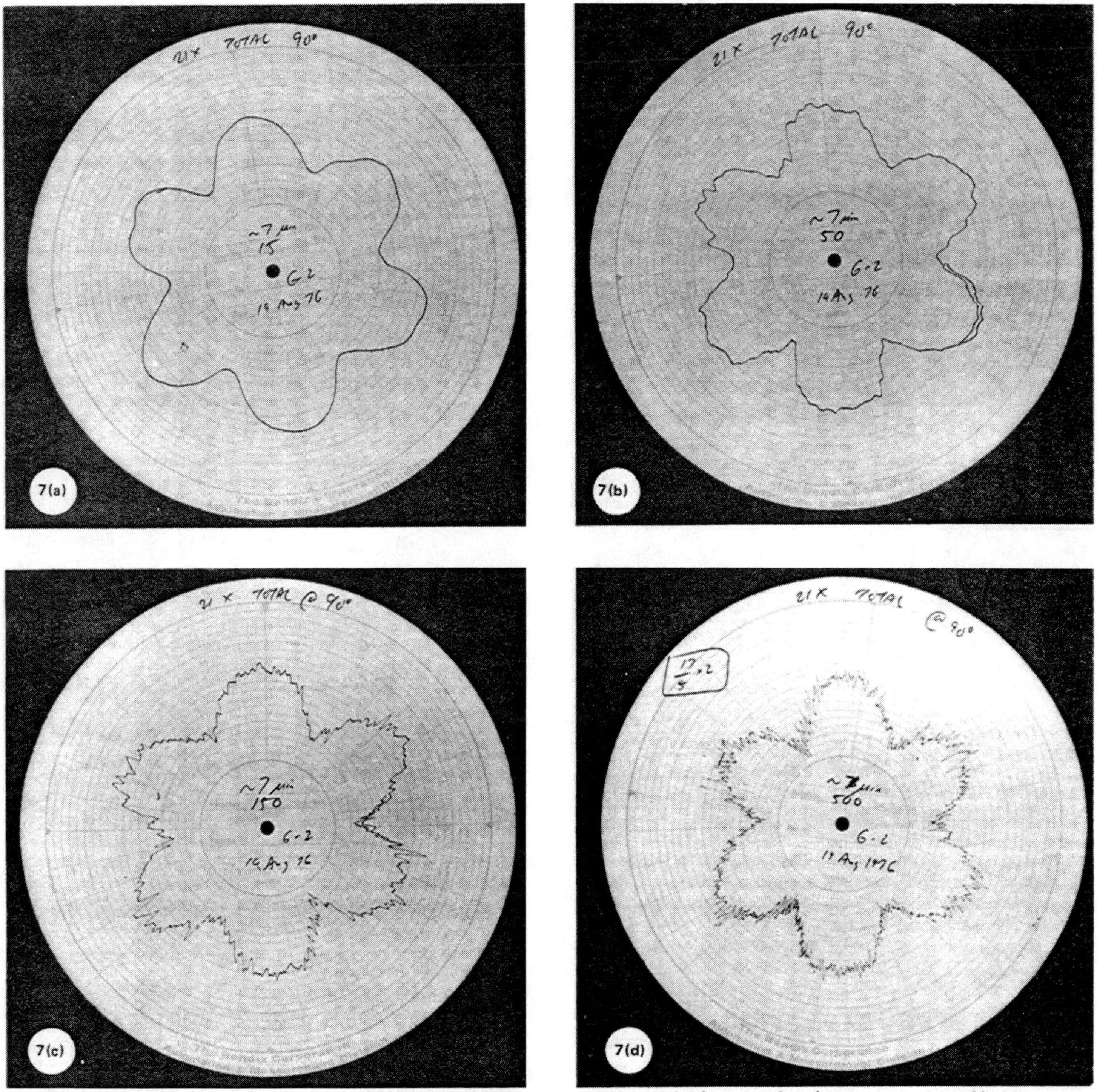

Figure 7. (a) Rotary profile of the six-lobe part taken with the optical transducer. The slit detector is parallel to the axis of rotation and the signal is filtered to pass up to 15 undulations per revolution. (b) Same as Figure 7(a) with filter setting of 50 undulations per revolution. (c) Same as Figure 7(a) with filter setting of 150 undulations per revolution. (d) Same as Figure 7(a) with filter setting of 500 undulations per revolution.

tion with the output of a good stylus instrument. The rotary profile test shows that the instrument can even be used as a noncontacting gage probe for clean and rather smooth surfaces. If the predominant lay of the fine surface structure of the test sample is circumferential.

The applications of the instrument will be (a) high-speed inspection applications where stylus instruments require frequent and costly changes of the stylus tip because of stylus-tip wear; (b) inspection of soft materials, where contacting profile instruments can be in error due to distortion of the surface; and (c) gaging of parts with highly polished surfaces, where again noncontacting is essential.

Acknowledgments

The authors would like to thank P. N. Keating for his valuable suggestions and R. F. Koppelmann for his assistance in electronic design. Thanks are also extended to M. W. Beaubien and E. J. Van Hoey for their technical assistance.

References

1. J. M. Bennett, Appl. Opt., *15*, 11, 2705 (1976.)
2. J. Simon, Appl. Opt., *9*, 10, 2337 (1970).
3. Private communication with G. G. Goetz.

Reprinted with permission from *Applied Optics*, Vol. 21(17), pp. 3200-3208
(September 1, 1982). ©1982 Optical Society of America.

Optical profilometer: a new method for high sensitivity and wide dynamic range

Yeshaiahu Fainman, Ehud Lenz, and Joseph Shamir

The noncontact optical measuring method is based on the focusing properties of simple optical elements. High sensitivity is achieved by the utilization of differential detection and electronic feedback systems. Preliminary measurements indicate that height sensitivities of the order of 0.1 μm with spatial resolution of better than 2 μm are easily obtained, while measurements in the millimeter range are also feasible. For optimization purposes, a mathematical analysis has been performed using operator algebra and the characteristics of the Wigner distribution function.

I. Introduction

The measurement of 3-D profiles is an important requirement in many engineering fields. The most popular profiling instrument is a mechanical stylus scanner. The main drawback of a stylus instrument is the mechanical contact required for performing the measurement. The mechanical contact with a loaded stylus will cause wear of the stylus itself and damage the measured surface. Also, the spatial resolution of these contact measurements is restricted by the size of the stylus point that degrades during operation due to wear and contamination. A detailed analysis of stylus measurements is given in Ref. 1.

To overcome the difficulties induced by the use of contacting measuring systems, a number of noncontacting optical methods were devised.[2] Many of these techniques use the properties of imaging systems,[3–7] augmented sometimes by automatic procedures[4] or more sophisticated interferometric methods.[8,9]

In the present work we describe a new optical method which is easier to apply than interferometric methods and does not require polished surfaces. Nevertheless its sensitivity may approach that of medium-accuracy interferometry. The basic physical principles and experimental results are described in Secs. II–IV, while Sec. V is devoted to a rigorous mathematical analysis useful for system optimization.

Joseph Shamir is with Texas Tech University, Department of Electrical Engineering, Lubbock, Texas 79409; the other authors are with Technion-Israel Institute of Technology, Department of Mechanical Engineering, Haifa, Israel.

Received 5 February 1982.
0003-6935/82/173200-09$01.00/0.

II. Measuring Principles

The measuring principles are illustrated in Fig. 1. A light beam traversing a beam splitter, B_1, is focused onto the surface of the investigated object, O, by lens L_1. When the surface is exactly at the focal plane of the lens, a large fraction of light is backscattered and collimated by L_1. The backward-traveling collimated beam is partially reflected by B_1 into lens L_2 (having an aperture function P) to be focused at f_2. Beam splitter B_2 is inserted in the beam to produce a second focal plane at f_2' which is spatially separated from f_2.

Following the results of the mathematical analysis given in Sec. V, two identical spatial filters were prepared to match the light-intensity distribution at the focal plane of L_2 in such a way that they pass all the power only if positioned exactly at the focal plane. Any displacement from the focal plane will reduce the power detected by the two large detectors D_1 and D_2. If filter F is placed behind the focal plane, a symmetric position may be found in front of the focal plane for F' so that the detected power in the two detectors is identical. The effect of a small longitudinal displacement of the object (direction x in Fig. 1) will be an effective shift of the focal planes f_2 and f_2' approaching one spatial filter and retreating from the other. As a result, the power detected by one detector will increase, while the second detector will show an approximately identical decrease of power. A differential amplification of the two detector signals (by differential amplifier DA) will give a very good indication of the amount and sign of the original displacement. To retain linearity and high accuracy, the system is completed by feeding back the amplifier output into a translation stage that restores the original position of the object surface.

Simple physical arguments will be used in Sec. III for estimating the basic characteristics of the system with

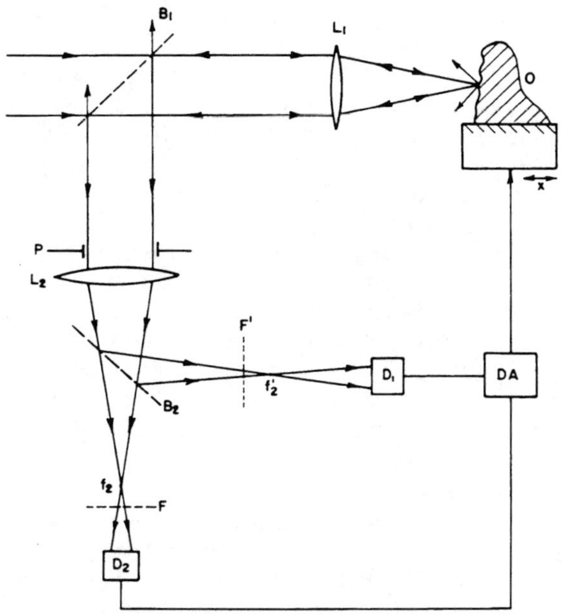

Fig. 1. Schematic configuration of measuring system. The object, 0, mounted on an xy translation stage is illuminated by a collimated beam transmitted through beam splitter B_1 and focused by lens L_1. The filtered Fourier transform of the aperture, P, by filters, F, is analyzed by detectors, D, connected to differential amplifier, DA.

the experimental verification described in Sec. IV. The exact function of the filters can only be understood by invoking a more elaborate mathematical analysis that will be deferred to Sec. V.

III. System Characteristics Estimation

In principle the present method may be used with either a conventional light source or a laser. However, all experiments were performed with a He–Ne laser, which also lends itself to direct mathematical analysis. Thus, we shall assume that the light source will always be a collimated coherent beam characterized by a fundamental Gaussian mode.

Observation of Fig. 1 reveals that the ultimate limit for the resolution of the system (longitudinal as well as lateral) is strongly dependent on the focusing properties of lens L_1. Specifically, the height–contour sensitivity (or surface axial-displacement sensitivity) depends on the focal depth of the lens, while the lateral (or spatial resolution) is determined by the illuminated spot size on the object surface. Assuming a well-collimated laser beam with its waist at the lens surface, the waist at the focal plane w_{of} (spot size) will be given by[10–12]

$$w_{of} = \frac{f\lambda}{\pi w_{oi}} [1 + (f\lambda/\pi w_{oi}^2)]^{-1/2}, \tag{1}$$

where f is the focal length of the lens, λ the wavelength of the illuminating light, the refractive index of the medium is assumed to be 1, and the beam waist at the lens is w_{oi}. If the beam is matched to the f/No. of the lens so that w_{oi} is approximately equal to the lens aperture, one may put

$$f/\text{No.} \cong f/2w_{oi}, \tag{2}$$

With an f/No. of order unity (microscope objective), Eq. (1) may be approximated by

$$w_{of} \cong \frac{2\lambda f/\text{No.}}{\pi}, \tag{3}$$

which gives us an estimate for the lateral resolution. For example, using a microscope objective with numerical aperture (N.A.) of 0.85 at a wavelength of 632.8 nm, a rough estimate for the spatial resolution will be

$$2w_{of} = 2\frac{2\lambda}{2\pi(\text{N.A.})} \cong 500 \text{ nm}. \tag{4}$$

Using the relation for the beamwidth $w(z)$ as a function of z, the distance from the focal plane,

$$w^2(z) = w_{of}^2 [1 + (z/b)^2], \tag{5}$$

where the confocal beam parameter is given by

$$b = \frac{\pi}{\lambda} w_{of}^2, \tag{6}$$

one may conclude that the beam does not change appreciably up to a distance $z = b$, where it also attains its maximum phase curvature—the minimum value of

$$R(z) = z[1 + (b/z)^2]. \tag{7}$$

Thus, working with a high spatial resolution at the focal plane as given by Eq. (4), one will obtain a longitudinal resolution of the order b, which in our present example is ~1.25 μm. However, if one is ready to give up some of the spatial resolution and work outside the waist region ($z \gtrsim b$) where the wave front resembles a spherical wave, the longitudinal sensitivity may be appreciably improved. At a distance $z = b$ the spot diameter is $2w_{of}\sqrt{2}$ (= 1.4 μm in the example). Taking into account beam distortions and aberrations, one may safely estimate a lateral resolution of the order of 2 μm with the longitudinal (axial) resolution determined essentially by the detection procedure to be discussed later.

For any specific application, the relation between lateral and longitudinal resolutions can be optimized within the limitations given above. In our experiments with the objective parameters mentioned in the example, good measuring capabilities were observed, with a misfocusing of ~1 μm yielding a lateral resolution of the order of 2 μm. The longitudinal sensitivity was evaluated experimentally by using as the object a mirror mounted on an interferometrically calibrated piezoelectric transducer. Figure 2(a) represents the voltage output of the differential amplifier (DA in Fig. 1) in an open loop as a function of mirror displacement. The extremely high sensitivity indicated by this plot is limited in practical measurements by electronic and environmental noise as shown in the time recording of Fig. 2(b), by translator quality, and object-surface roughness. Nevertheless, the experiments described in Sec. IV show that an estimate of 0.1-μm resolution based on Figs. 2(a) and 2(b) is quite realistic, with indications for at least one order of magnitude improvement possibilities.

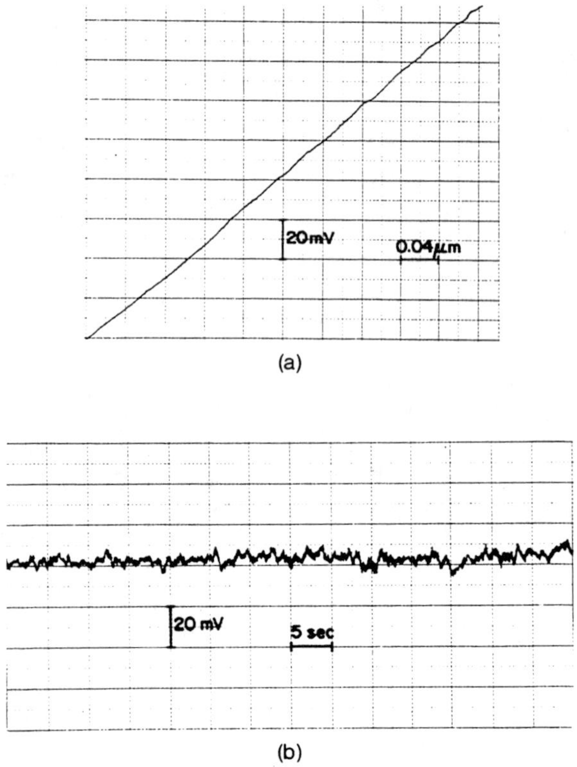

Fig. 2. Sensitivity test of optical system: (a) Differential amplifier output with feedback disconnected as a function of mirror axial displacement; (b) Noise output of stationary system.

IV. Experimental Setup and Measurements

One of the requirements on the measuring system as derived mathematically in Sec. V is $f_1/f_2 \ll 1$, where f_1 and f_2 are the respective focal lengths of lenses L_1 and L_2 in Fig. 1. This condition is easily met if L_1 is a microscope objective (with N.A. = 0.85 in our system) and L_2 is a simple thin lens (in our system f_2 = 220 mm). For the filters F and F', one may use two pinholes, however, better performance is attained by optimizing them according to the derivation of Sec. V. The detectors were simple solar cells, and their differentially amplified outputs were fed into a linear positioning table driven by a step motor to retain the initial position of the object surface. The movement of the stage in the x (longitudinal) direction was guided by a prestressed linear ball bearing to ensure zero play and backlash. The resolution of the stage was 1 μm, and maximum speed was 2 mm/sec. A smooth translation of the object in the y (lateral) direction was achieved by a pair of granite gauge blocks sliding one upon the other. The translation was driven by a dc motor, and the position detected by a precision potentiometer. The lateral scanning speed was limited by the slope of the object and its effect on the longitudinal motion of the step motor. In our experiments scanning speeds were in the range of 0.2–1 mm/sec.

In all the measurements to be described the plots were obtained by using a DAC to indicate the analog position of the step motor, which is a measure of surface height against the reading of the potentiometer that marks the lateral position of the object. Whenever possible the measurements were verified using a stylus instrument with two operation modes: a stylus of 25-μm point radius with 2.5-g loading and a stylus of 2.5-μm point radius with 80-mg loading.

The first measurement was performed on a standard block with 1-D triangular grooves. Figure 3(a) is the optical measurement and Fig. 3(b) is the measurement performed with the 2.5-μm stylus. Both measurements are in exact agreement with the data given for the block, with an appreciable deviation for the larger stylus [Fig. 3(c)]. These stylus measurements are analyzed in Appendix A, with the conclusion that our optical measuring method has at least the resolution (longitudinal and lateral) of a 2.5-μm point stylus.

The second measurement was made on a milled-steel surface. Figure 4 is a result of this measurement. A number of scans were performed in the same region to obtain a correlated set for the three measurements [(a) optical, (b) small stylus and (c) large stylus]. The profiles are the same except for the fine structure that

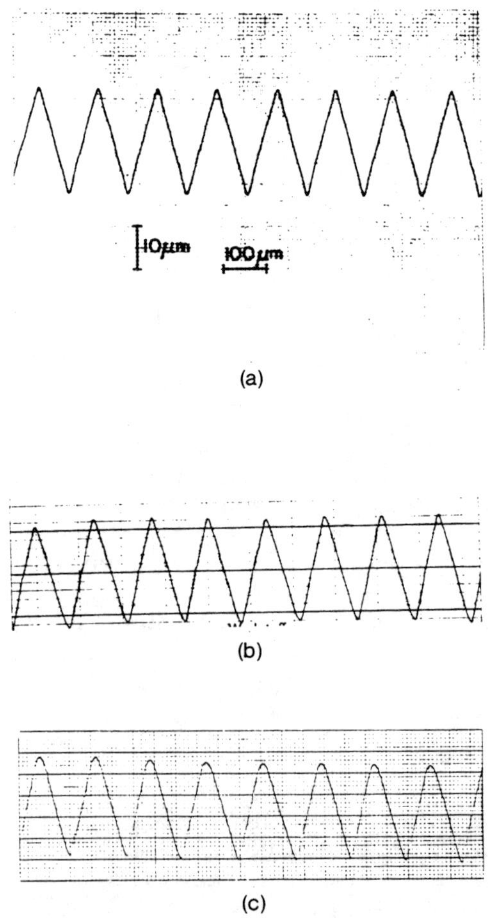

Fig. 3. Measurements on triangular groove standard block: (a) Optical profilometer; (b) Stylus gauge with a 2.5 μm point; (c) Stylus with a 25-μm point.

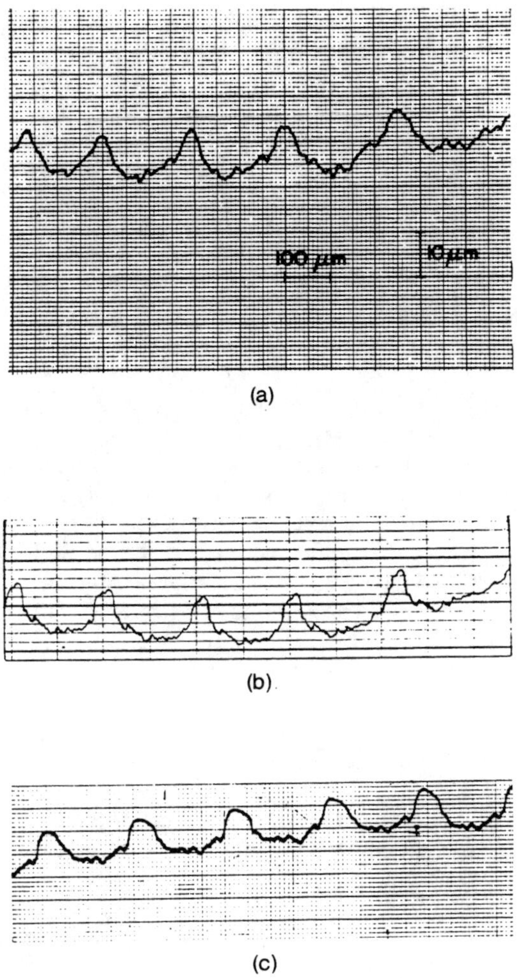

Fig. 4. Same as Fig. 3 but for a milled-steel surface.

cannot be resolved by the step motor used (1 μm) nor by the 25-μm stylus. The rounded peaks of Fig. 4(c) were due to mashing down material by the heavily loaded stylus—this could be easily observed under a microscope.

The above two examples of measurements were in the range where stylus instruments are applicable. As indicated by the discussion related to Fig. 2, much higher resolutions are possible, at least for polished surfaces. Here, however, we demonstrate the capabilities of the method for wider-range measurements where stylus instruments are inapplicable. Unfortunately, wide-range measurements involve large working distances requiring microscope objectives of lower N.A. that reduce the resolution, both lateral and longitudinal. Figure 5 is the profile recording of a rough emery paper performed with an objective of N.A. = 0.65. Naturally, such a sample could not be measured by a stylus for comparison, but the average profile corresponds quite well to the manufacturer number for average grain size.

The final experiment was performed with an objective of N.A. 0.33. Figure 6 is the profile of the cutting edge of a drill. the 120° edge angle (taking into account the different scales along the two coordinates) corresponds to the required shape of the drill.

V. Mathematical Analysis

The physical principles of the method are intuitively clear. Nevertheless, a mathematical analysis is very useful for system optimization and error estimation. The analysis is greatly simplified by the use of a recently developed Operator Algebra (OA),[13] together with a novel application of the Wigner distribution function (WDF).[14,15]

The practical purpose of this section is to evaluate the structure of a good spatial filter to be placed in front of the detectors and to estimate the response of the measuring system.

Starting from the object surface, we describe its amplitude reflectively by two complex functions: a deterministic complex function $\gamma(\rho)$ describing the reflectance of the surface (amplitude) and profile (phase) and a random function $\phi(\rho)$ due to the roughness of the surface, which usually yields the speckle

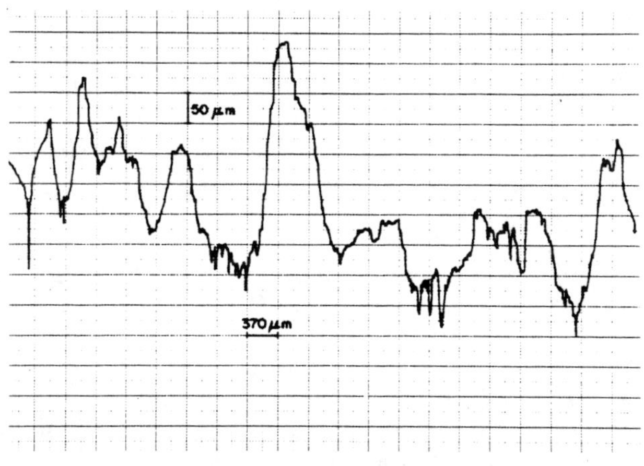

Fig. 5. Optical measurement of rough emery-paper surface.

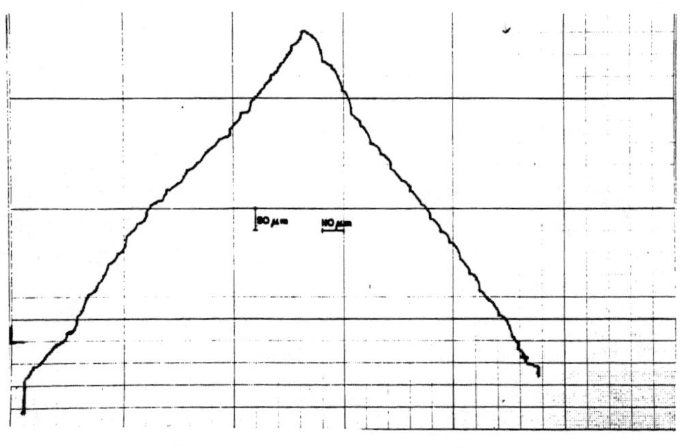

Fig. 6. Optical measurement of a drill cutting edge. (Note the difference in the x and y scales.)

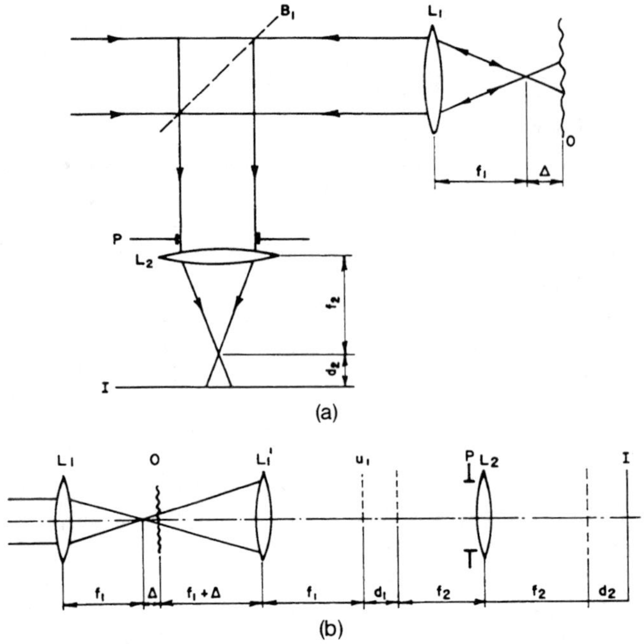

Fig. 7. One arm of the optical system of Fig. 1: (a) The actual system without the beam splitter B_2; (b) The unfolded equivalent form of (a).

pattern.[16] A plane wave reflected from the object surface will have a complex amplitude

$$\gamma(\rho)\exp[\phi(\rho)]. \tag{8}$$

For convenience, we consider one arm of the optical system [Fig. 7(a)] and analyze it in an unfolded form [Fig. 7(b)].

In the unfolded optical system the object surface is represented by a transparency, 0, having the transfer function given in expression (8) placed at a small distance, Δ, from the focal plane of lens L_1. The reflected field that travels back to L_1 is represented by the transmitted wave traveling a distance $f_1 + \Delta$ to the image of L_1, L_1', where f_1 is the focal length of L_1. Our immediate aim here is to derive the optical field distribution at the observation plane I as a function of system parameters and, particularly, its dependence on the misfocusing Δ.

We start the calculation at the focal plane of L_1 and represent the field distribution there by its complex amplitude, $a(\rho)$, assumed to be known. The argument, ρ, is the transversal coordinate, and we use the notation and calculation rules of the OA to derive the field propagation through the system. The propagation from the focal plane to the object can be evaluated by operating on the distribution $a(\rho)$ with the Fresnel free-space propagation operator (FPO), $\mathcal{R}[\Delta]$, that represents the Kirchhoff diffraction integral. Thus, the object illumination will have the complex amplitude

$$\mathcal{R}[\Delta]a(\rho).$$

Using Eq. (8) for the object-transfer function, the complex amplitude immediately after reflection by the object [transmission by 0 in Fig. 7(b)] will be

$$u_r = \gamma(\rho)\exp[i\phi(\rho)]\mathcal{R}[\Delta]a(\rho). \tag{9}$$

The reflected field is propagated again a distance Δ and then Fourier transformed by lens L_1 at its front focal plane, which is the back focal plane of L_1' in the unfolded system of Fig. 7(b). Continuing reference to the unfolded equivalent system, the field distribution u_1 may be represented in the form

$$u_1 = \mathcal{V}\left[\frac{1}{\lambda f_1}\right]\mathcal{F}\mathcal{R}[\Delta]\gamma(\rho)\exp[i\phi(\rho)]\mathcal{R}[\Delta]a(\rho), \tag{10}$$

where $\mathcal{V}[1/\lambda f_1]$ is a scaling operator that induces a change in the argument scaling of any function written on its right. The Fourier transform (FT) operator $\mathcal{F}$ performs a Fourier transformation on the function, that may be a whole expression written to its right. Further propagation of the field through a distance $d_1 + f_2$, with f_2 the focal length of L_2, is again represented by an FPO, and the resulting expression is multiplied (from the left) by the lens-aperture transfer function, $\mathcal{P}$. Transmission through the thin lens L_2 adds a quadratic-phase factor,

$$\mathcal{Q}[-1/f_2] \equiv \exp\left[\frac{-jk}{2f_2}(x^2+y^2)\right],$$

where $k = 2\pi/\lambda$. A further propagation of a distance $f_2 + d_2$ to the observation plane I completes the derivation, yielding

$$u_i = \mathcal{R}[d_2+f_2]\mathcal{Q}\left[-\frac{1}{f_2}\right]\mathcal{P}(\rho)\mathcal{R}[d_1+f_2]\mathcal{V}\left[\frac{1}{\lambda f_1}\right]\mathcal{F}\mathcal{R}(\Delta)\gamma(\rho) \times \exp[i\phi(\rho)\mathcal{R}[\Delta]a(\rho). \tag{11}$$

This general relation may be used to derive the field distribution for any surface and coherent source. However, in the present case calculations can be greatly simplified by some assumptions relating to a practical system:

First, we note that no speckle effect was observed during the measurements. The analytic reasoning behind this observation is as follows: we recall that the misfocusing, Δ, is very small and is maintained by the feedback system around the value b [Eq. (6)]. At this point, the radius of the illuminated spot is $w_{of}\sqrt{2}$, with w_{of} given by Eq. (4). The minimum possible phase-deconvolution area is of the order of λ^2, thus, the maximum number of independent scatterers in the illuminated area is of the order of $\pi(w_{of}\sqrt{2})^2/\lambda^2 \cong 2/\pi(\mathrm{N.A.})^2$ (see also Ref. 17). With our high-resolution system (N.A. = 0.85) this limits the number of scatterers to one to three even for slightly larger defocusing and nonideal optical systems. For such a small number of scatterers, it was shown by Uozumi[18] that for a weak random-phase standard deviation, the intensity-probability density-function is concentrated around the intensity distribution of a single scattering cell which has a δ-probability density function. Experiment indicates that one may assume that our surfaces have a height (or phase) correlation length large compared to the minimum size of the point spread function[17–21] and the speckle effect may be completely ignored. This may not be the case for measurements with a small numerical aperture, however, in that system the resolution is adequately reduced to make the effect of surface roughness undetectable.

The conclusion from the above considerations is that we may ignore the random-phase variation across the illuminated region. Furthermore, the deterministic part of the surface function may also be assumed not to vary much throughout this area; thus, the whole surface function (8) may be approximated by a constant. Being interested in relative intensities only, this constant, together with all other constants, may be ignored. Finally, we assume that the focal-point illumination, $a(\rho)$, is so small that it may be approximated by a δ function.

Substituting these assumptions into Eq. (11), combining the two FPOs according to OA rules after the extraction of the surface function (8) and interchanging the order of the two scalar functions, $\mathcal{P}$ and $\mathcal{Q}$, leads to the simplified form

$$u_i \propto \mathcal{R}[d_2 + f_2]\mathcal{P}(\rho)\mathcal{Q}\left[-\frac{1}{f_2}\right]\mathcal{R}[d_1 + f_2]\mathcal{V}\left[\frac{1}{\lambda f_1}\right]\mathcal{F}\mathcal{R}[2\Delta]\delta(\rho). \tag{12}$$

Using the representation[13]

$$\mathcal{R}[d] \propto \mathcal{F}^{-1}\mathcal{Q}[-\lambda^2 d]\mathcal{F} \tag{13}$$

for one FPO we obtain

$$\begin{aligned} u_i &\propto \mathcal{R}[d_2 + f_2]\mathcal{P}(\rho)\mathcal{Q}\left[-\frac{1}{f_2}\right]\mathcal{R}[d_1 + f_2]\mathcal{V}\left[\frac{1}{\lambda f_1}\right] \\ &\quad \times \mathcal{F}\mathcal{F}^{-1}\mathcal{Q}[-\lambda^2 2\Delta]\mathcal{F}\delta(\rho) \\ &= \mathcal{R}[d_2 + f_2]\mathcal{P}(\rho)\mathcal{Q}\left[-\frac{1}{f_2}\right]\mathcal{R}[d_1 + f_2]\mathcal{Q}\left[\frac{-2\Delta}{f_1^2}\right], \end{aligned} \tag{14}$$

where we used the fact that the Fourier transform of a delta function is a constant, cancelled the FT operator with its inverse, and scaled the remaining $\mathcal{Q}$ factor as indicated by the $\mathcal{V}$ operator. Relation (14) is already a quite simple form, but it can be further simplified by considering the parameters of our specific optical system: Under optimal adjustment conditions the output plane should be the image of the point source projected onto the object surface. The ideal way of meeting this condition is $\Delta = d_1 = d_2 = 0$.

However, resolution considerations discussed earlier require a small initial misfocusing ($\Delta \neq 0$), to which another displacement, Δ', is added during measurement. Since the magnitude of the feedback required to cancel Δ' is our measurement, we must determine the effect of the displacement on the detection system. Since the components of the optical system may be adjusted at will (their relative spacings are not affected by the feedback system), one may put $d_1 = 0$ and evaluate d_2 that retains imaging for a small object displacement, Δ. Using the second representation for the remaining FPOs,

$$\mathcal{R}(d) \propto \mathcal{Q}\left(\frac{1}{d}\right)\mathcal{V}\left(\frac{1}{\lambda d}\right)\mathcal{F}\mathcal{Q}\left(\frac{1}{d}\right), \tag{15}$$

and putting $d_1 = 0$ one obtains from relation (14),

$$\begin{aligned} u_1 \propto \mathcal{Q}\left(\frac{1}{d_2 + f_2}\right)\mathcal{V}\left[\frac{1}{\lambda(d_2 + f_2)}\right]\mathcal{F}\mathcal{Q}\left(\frac{1}{d_2 + f_2}\right)\mathcal{P}(\rho)\mathcal{Q}\left(-\frac{1}{f_2}\right)\mathcal{Q}\left(\frac{1}{f_2}\right) \\ \times \mathcal{V}\left(\frac{1}{\lambda f_2}\right)\mathcal{F}\mathcal{Q}\left(\frac{1}{f_2}\right)\mathcal{Q}\left(\frac{-2\Delta}{f_1^2}\right). \end{aligned}$$

Performing the rightmost FT operation, commuting the scalar $\mathcal{P}$ with its left-hand $\mathcal{Q}$ factor, and combining all adjacent $\mathcal{Q}$ factors, we have

$$\begin{aligned} u_i \propto \mathcal{Q}\left(\frac{1}{d_2 + f_2}\right)\mathcal{V}\left[\frac{1}{\lambda(d_2 + f_2)}\right]\mathcal{F}\mathcal{P}(\rho) \\ \times \mathcal{Q}\left[\frac{1}{d_2 + f_2} - \frac{1}{f_2 - 2\Delta f_2^2/f_1^2}\right]. \end{aligned} \tag{16}$$

Imaging is obtained, together with the FT of the aperture function $\mathcal{P}$, by eliminating the right-hand $\mathcal{Q}$ factor

$$\frac{1}{d_2 + f_2} - \frac{1}{f_2 - 2\Delta f_2^2/f_1^2} = 0,$$

or

$$d_2 = -2\Delta f_2^2/f_1^2. \tag{17}$$

The satisfaction of Eq. (17) reduces Eq. (16) to the simple form

$$u_i \propto \mathcal{Q}\left(\frac{1}{d_2 + f_2}\right)\mathcal{V}\left[\frac{1}{\lambda(f_2 + d_2)}\right]\mathcal{F}\mathcal{P}(\rho), \tag{18}$$

which is a scaled Fourier transform of the aperture function $\mathcal{P}$ multiplied by a quadratic-phase factor that is not detectable by a power detector. Equation (17) indicates that d_2 is a linear amplification of Δ by a factor $(f_2/f_1)^2$. Thus, to obtain a large amplification factor, one should take

$$f_2/f_1 \gg 1. \tag{19}$$

Due to the linear dependence of d_2 on Δ [Eq. (17)], any additional displacement Δ' introduced by the surface will be amplified by the same factor, resulting in an additional shift of the image plane by an amount

$$d_2' = -2\Delta' f_2^2/f_1^2. \tag{17a}$$

Considering only small displacements, we shall still have for the overall shift

$$d_2 \ll f_2, \tag{20}$$

resulting in an undetectable change of the image scaling [label of the scaling operator in Eq. (18)]. Thus, a simple, straightforward detection method will be insufficient for our purpose unless some modification of the system is made. A number of such modifications may be possible, and here we describe one which was found very useful in obtaining high sensitivities.

In our system we use a large detector that integrates all the power reaching the output plane. In front of this detector we place a matched filter that transmits the power spectrum of the aperture function $\mathcal{P}$. Such a filter can be obtained by a simple photograph of the power spectrum under optimum adjustment of the

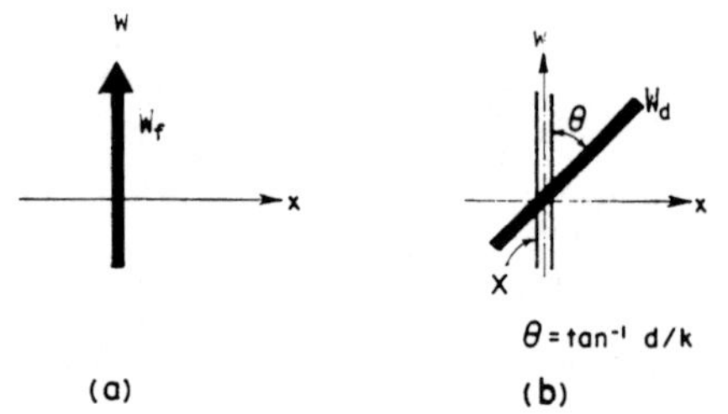

Fig. 8. The Wigner distribution function of a 1-D infinite aperture. The x-w plane has only meaning in an abstract mathematical space, while the actual field distribution is a delta function—a single point on the x axis. (a) Focused image, (b) misfocused image.

system. Intuitively, it is clear that any defocusing will result in a decrease of the detected power, however, an exact calculation cannot be done for a general case. Nevertheless, a fair estimate is possible by utilizing some of the properties of the Wigner distribution function.[14,15]

Restricting ourselves to 1-D functions for clarity, the WDF $W(x,w)$ of a function $u(x)$ may be evaluated in two ways:

$$W(x,w) = \int_{-\infty}^{\infty} u(x + \tfrac{1}{2}x')u^*(x - \tfrac{1}{2}x') \exp(-iwx')dx' \quad (21)$$

or

$$W(x,w) = \int_{-\infty}^{\infty} U(w + \tfrac{1}{2}w')U^*(w - \tfrac{1}{2}w') \exp(-ixw')dw', \quad (22)$$

where $U(w)$ is the Fourier transform of $u(x)$.

The energy distribution (or intensity) can be evaluated either in the physical plane or the frequency plane:

$$|u(x)|^2 = \frac{1}{2\pi} \int W(x,w)dw \quad (23a)$$

$$|U(w)|^2 = \int W(x,w)dx. \quad (23b)$$

For our present purpose an important characteristic of the WDF is that if a field propagates through a linear system, its WDF remains unchanged except for a linear transformation of its coordinates. In particular, propagation through free space may be represented as a shear in the xw space.

Denoting the WDF at the focal plane of our measuring system by $W_f(x,w)$ and the field propagated a distance d from that plane by $W_d(x,w)$, one can show[14] that in the Fresnel approximation the two functions will be related by

$$W_d(x,w) = W_f\left(x - \frac{d}{k}w,w\right) = W_f(x,w)*\delta\left(x - \frac{d}{k}w\right), \quad (24)$$

where the delta function is the WDF of a quadratic-phase factor.

In our filtering procedure we assume a spatial filter with *intensity* transfer function $H(x)$, resulting in an intensity distribution of the incident field at the detector immediately after the filter,

$$I(x) = H(x)|u_d(x)|^2 = \frac{1}{2\pi} H(x) \int_{-\infty}^{\infty} W_d(x,w)dw$$

$$= \frac{1}{2\pi} \int H(x)W_d(x,w)dw. \quad (25)$$

The transfer function $H(x)$ may be included in the integration as it is independent of the integration variable w. The total detected energy will be the integral of this quantity over all space:

$$E(d) = \int_{\infty}^{\infty} I(x)dx = \frac{1}{2\pi} \iint_{\infty}^{\infty} H(x)W_d(x,w)dxdw. \quad (26)$$

If $H(x)$ is prepared in such a way that it transmits all the energy for $d = 0$ for which W_d is replaced by W_f in Eq. (26), we shall always have

$$E(d) \leq E(0). \quad (27)$$

One can make such a filter by recording the intensity distribution at the focal plane on a high-contrast film to obtain the transfer function

$$H(x) = \begin{cases} 1; & |u_f(x)|^2 \neq 0 \\ 0; & |u_f(x)|^2 = 0. \end{cases} \quad (28)$$

For this case, Eq. (26) can be written in an alternative form:

$$E(d) = \int_{-\infty}^{\infty} \left[\int_X W_u(x,w)dx\right] dw, \quad (26a)$$

where X denotes the collection of all space points for which $H(x) \neq 0$. Inequality (27) follows automatically since the total energy in the unfiltered wave is independent of d.

Using these relations one may estimate the sensitivity for some specific aperture functions $\mathcal{P}$. The ideal aperture is an infinite one, having a delta function as its Fourier transform. The WDF for this case will be

$$W_f(x,w) = \int_{\infty}^{\infty} \delta(x + \tfrac{1}{2}x')\delta^*(x - \tfrac{1}{2}x') \exp(-iwx')dx' = \delta(x), \quad (29)$$

which is the w axis in the xw plane [Fig. 8(a)]. This distribution can be made into a filter that restricts transmission to a line in the xw plane. (It should be noted at this point that this xw plane has only an abstract mathematical meaning and cannot be observed.)

In the ideal case the width of X is negligible, and the energy transmitted from a misfocused distribution having the inclined WDF [from Eq. (24)]

$$W_d(x,w) = \delta\left(x - \frac{d}{k}w\right) \quad (30)$$

and shown in Fig. 8(b), will drop rapidly to zero for any finite value of the angle θ. In a physical situation, however, the aperture will be finite, and, as long as it is large, the distribution at the focal plane may be approximated by the central lobe of a sinc function for a 1-D slit aperture, $\mathcal{P}$, or the central lobe of the Bessel function for a circular aperture in 2-D. Assuming, for simplicity, the 1-D slit of finite width, the clipped WDF at the FT plane will be as shown in Fig. 9(a). Any displacement by an amount d from the FT plane will rotate the distribution in the xw plane [Figs. 9(b) and 9(c)] according to Eq. (24). It is evident that if the distribution at the FT plane is made into a filter, the integral over the area X (shaded in the figure) representing the

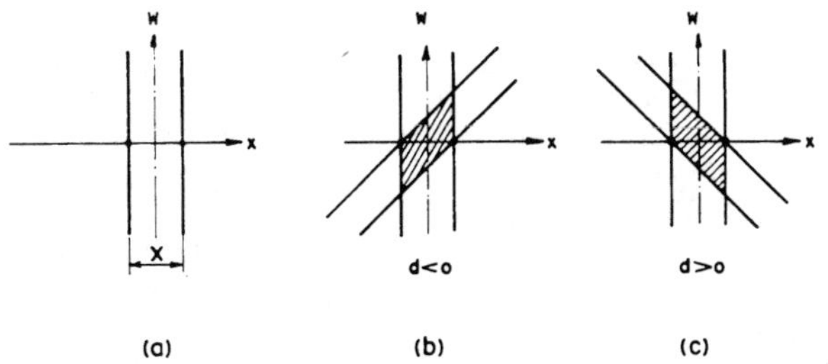

Fig. 9. As in Fig. 8 but for a 1-D slit of infinite width. S represents the size of the central lobe of the sinc function. (b) and (c) are misfocusing in opposite directions.

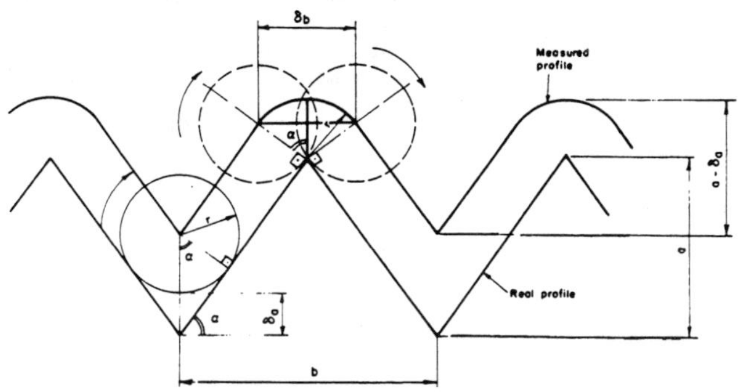

Fig. 10. Profiling a triangular grid by a stylus point of radius r. The measure profile corresponds to the trajectory of the center of the curvature of the stylus tip. δ_a and δ_b are the measure for profile distortions.

transmitted power will decrease with the increase of inclination dependent on d.

The measuring procedure can now be explained with reference to Fig. 9: With the system properly aligned, two identical filters are prepared photographically at the FT plane of the aperture function. Naturally, requirement (28) cannot be fulfilled exactly, but a good approximation is possible. Replacing the filter at a distance d from the FT plane transmits a fraction of the total power depending on the magnitude of d but not on its sign. Referring back to Fig. 1, one of the filters, F, is placed beyond the FT plane, while the other one, F', is placed in front of the FT plane at a symmetrical distance in such a way that there will be no output from the DA at the chosen zero position of the object surface. Any surface displacement, Δ', will induce a change in the two detected powers approximately identical in magnitude and opposite in sign, resulting in an amplified output from the DA proportional to the surface displacement.

VI. Conclusions

A noncontact optical method for profile measurement has been described. The utilization of a differential-detection method, together with a nulling-feedback system facilitates high-longitudinal sensitivity. The lateral resolution and ease of use is comparable with that of a stylus gauge, but it has a wider range of applicability.

Naturally, the main advantage of the method is its noncontact measuring procedure, with a sensitivity approaching interferometric techniques. With appropriate modifications this method may be applied also for the measurement of thickness and refractive-index variations in transparent objects and for displacement measurements.

In our experiments a displacement sensitivity better than 50 nm was demonstrated, limited only by system noise of the same order of magnitude. However, the actual profile measurements were limited by the translation stage driven with a step motor of 1-μm steps. For this reason there was no interest, at this point, in improving the optical detectability which should be achievable by the application of the theory. Apart from the translation-stage limitations, the ultimate sensitivity of the system depends on the quality of the optical components and on the shape of the aperture and filters and their exact positioning. The influence of the aperture and filter functions predicted by theory are open to further investigation, but, most probably, this will only be important for systems with all other imperfections brought to a limit.

From the theoretical point of view, the application of a recently developed operator algebra and the Wigner distribution function demonstrate their usefulness in analyzing practical optical systems.

This publication constitutes a part of the D.Sc. thesis of Y. Fainman.

Joseph Shamir is on leave from the Department of Electrical Engineering, Technion-Israel Institute of Technology, Haifa, Israel.

We appreciate the constructive comments of two unidentified reviewers that were of great help in improving the presentation of this work.

Appendix A Stylus Profiling

Figure 10 shows the trajectory of the center of curvature for a stylus point with spherical shape scanning the triangular profile. It is clearly evident that the measured peak-to-peak value of the profile will decrease with increasing point radius, together with an enhanced rounding of the tops. The relevant parameters defined in the figure are easily derived from trigonometric considerations to be

$$\delta a = r/\cos\alpha - r, \tag{A1}$$

with

$$\tan\alpha = 2a/b, \tag{A2}$$

$$\delta_b = 2r \sin\alpha. \tag{A3}$$

In the standard surface analyzed (Fig. 3) the given parameters were $a = 24\ \mu$m and $b = 140\ \mu$m. These values yield for the large stylus $r = 25\ \mu$m, $\delta_a \cong 1.4\ \mu$m, and $\delta_b \cong 16\ \mu$m, in good agreement with results in Fig. 3(c). For the smaller stylus, $r = 2.5\ \mu$m, the deviation from the true profile is negligible on the scale of Fig. 3(b), thus, one may take this as the true profile to be compared with the optical profilometer plot of Fig. 3(a).

References

1. J. M. Bennett and J. H. Dancy, Appl. Opt. **20,** 1785 (1981).
2. J. M. Bennett, Appl. Opt. **15,** 2705 (1976).
3. T. Sawatari, and R. B. Zipin, Proc. Soc. Photo-Opt. Eng. **153,** 8 (1978).
4. F. T. Arecchi, D. Bertani, and S. Ciliberto, Opt. Commun. **31,** 263 (1979).
5. S. Uchida, H. Sato, and M. O-hori, Ann. CIRP **28,** 419 (1979).
6. M. Shiraishi, Bull. Japan Soc. Prec. Engg. **13,** 133 (1979).
7. L. Parker, Laser Focus, 40–46, (1980).
8. J. Shamir, Opt. Eng. **19,** 801 (1980).
9. Y. Fainman, E. Lenz, and J. Shamir, Appl. Opt. **20,** 158 (1981).
10. H. Kogelnik and T. Li, Proc. IEEE **54,** 1312 (1966).
11. A. Yariv, *Introduction to Optical Electronics* (Holt, Reinhart & Winston, New York, 1971).
12. H. Kogelnik, "Propagation of Laser Beams," in *Applied Optics and Optical Engineering*, R. R. Shannon and J. C. Wyant, Eds. (Academic, New York, 1979).
13. M. Nazarathy and J. Shamir, J. Opt. Soc. Am. **70,** 150 (1980).
14. M. J. Bastiaans, "The Wigner Distribution Function and Its Applications to Optics," *ICO Mexico 1980 Conference on Optics in Four Dimensions* (Pergamon, New York, 1980).
15. H. O. Bartelt, H. H. Brenner, and A. W. Lohnmann: Opt. Commun. **32,** 32 (1980).
16. Y. Fainman, J. Shamir, and E. Lenz, Appl. Opt. **20,** 3526 (1981).
17. J. C. Dainty, Prog. Opt. **14,** 3 (1976).
18. J. Uozumi and T. Asakura, Opt. Acta **27,** 1345 (1980).
19. J. W. Goodman, "Statistical Properties of Laser Speckle Patterns," in *Laser Speckle and Related Phenomena*, J. C. Dainty, Ed. (Springer, Heidelberg, 1975).
20. S. Lowenthal and H. H. Arsenault, J. Opt. Soc. Am. **60,** 1478 (1970).
21. J. Uozumi, H. Fujii, and T. Asakura, J. Opt. Soc. Am. **67,** 808 (1977).

Reprinted from *Measurement and Effects of Surface Defects and Quality of Polish*,
L.R. Baker, H.E. Bennett, eds., Proc. SPIE 525, pp. 174-180 (1985). ©1985 SPIE.

Optical profilers for surface roughness

James C. Wyant

Optical Sciences Center, University of Arizona
Tucson, Arizona 85721

Abstract

Different possible interferometers for surface roughness measurements are described. Advantages and disadvantages of optical techniques compared to stylus instruments are discussed. Important features of phase-shifting techniques for surface roughness measurements are given. The WYKO NCP-1000M optical profiler is described along with techniques for measuring surfaces having an rms roughness on the order of 0.1 nm.

Introduction

The measurement of surface roughness is important in areas of optics such as mirrors used in laser gyro systems, diamond-turned optics, contact lenses, optical disks, soft lenses, contact lenses, polygon scanners, master molds, and optical fibers. Surface roughness characterization is also of much interest in the magnetic storage area where the roughness of magnetic tape, floppy disks, hard disks, and magnetic heads is very important.

A tool commonly used to measure surface roughness is a stylus instrument that amplifies and records the vertical motion of a stylus as it is moved at a constant velocity across the surface to be measured. Stylus instruments are popular because they are readily available and they can have a high lateral resolution of a few tenths of a micrometer with a height resolution of up to 0.1 to 0.2 nm.[1] A problem with the stylus instrument is that it involves mechanical contact at the surface at extremely high pressure. That contact can easily damage soft surfaces, and may even damage hard surfaces or the stylus. A non-contact method of measuring surface roughness is preferred to a contact method.

Optical interferometry methods have long been used to measure surface roughness. A wide variety of interferometry attachments can be added to a microscope to measure surface roughness. With the standard interferometric methods that use visual evaluation of fringe patterns produced in an interferometer system, both the accuracy and the measurement time are unacceptable for many situations. However, if heterodyne interferometry or phase-shifting interferometry is used with almost any conventional interference microscope attachment, the resulting system can give excellent accuracy and measurement speed.

Interference microscope attachments

Interferometers that might be used with a microscope can be broken into two major categories: 1) interferometers that measure the surface height variations directly and 2) interferometers that measure the differential (slope) of the surface height variations. Both types of interferometers have their advantages and disadvantages. Differential interferometers are generally simpler and are less affected by vibration. However, as shown below, they are inherently less accurate for measuring surface height variations than are interferometers that measure the height variations directly, and since they measure surface slope errors in only one direction, sample orientation is very important.

Surface measuring interferometers

A large number of interferometers might be used with a microscope to measure surface height variations directly.[2-6] We will mention only a few of the most popular ones below.

Figure 1a shows how a Fizeau interferometer might be used to measure surface roughness. A problem with the Fizeau is that the reference surface is very close to the sample, and contact (and hence damage) is probable. If the surface being tested has a high reflectivity, the reference surface in the Fizeau can be coated to give high finesse fringes. The high finesse fringes are useful if visual interpretation of the fringes is being used. However, if heterodyne or the phase shifting interferometry techniques described below are used to improve the accuracy and speed of the measurement, the high finesse fringes are actually a problem rather than an advantage.

Figure 1b is a diagram of a Linnik interferometer. The Linnik has the advantage of a large working distance and a wide variety of magnifications from at least 1X to 60X. The major disadvantage is that a matched pair of high quality microscope objectives must be used.

Figure 1c shows a diagram of a Michelson interferometer attachment. The Michelson has the advantage over the Linnik that only a single objective is required, however, the objective must be corrected for the thickness of the glass in the cube beamsplitter, the beamsplitter must be very high quality, and due to the large piece of glass that is put in the system, the working distance of the interferometer is short. In either the Linnik or the Michelson, the reference surface can be replaced with a surface having different reflectivities. Curved reference surfaces can be used if the sag of the surface is not larger than the depth of focus of the objective.

Figure 1d is a diagram of a Mirau interferometer attachment. The Mirau typically has a working distance between that of the Linnik and the Michelson. Since the size of the reference spot is equal to the field of view, if normal microscope objectives are used, the maximum field is on the order of 1 mm; while the Michelson or Linnik can have a field of several millimeters. Different reflectivity reference surfaces can be used. The major advantages of the Mirau are its simplicity, relative low cost, and the ease with which excellent quality interferometers can be fabricated.

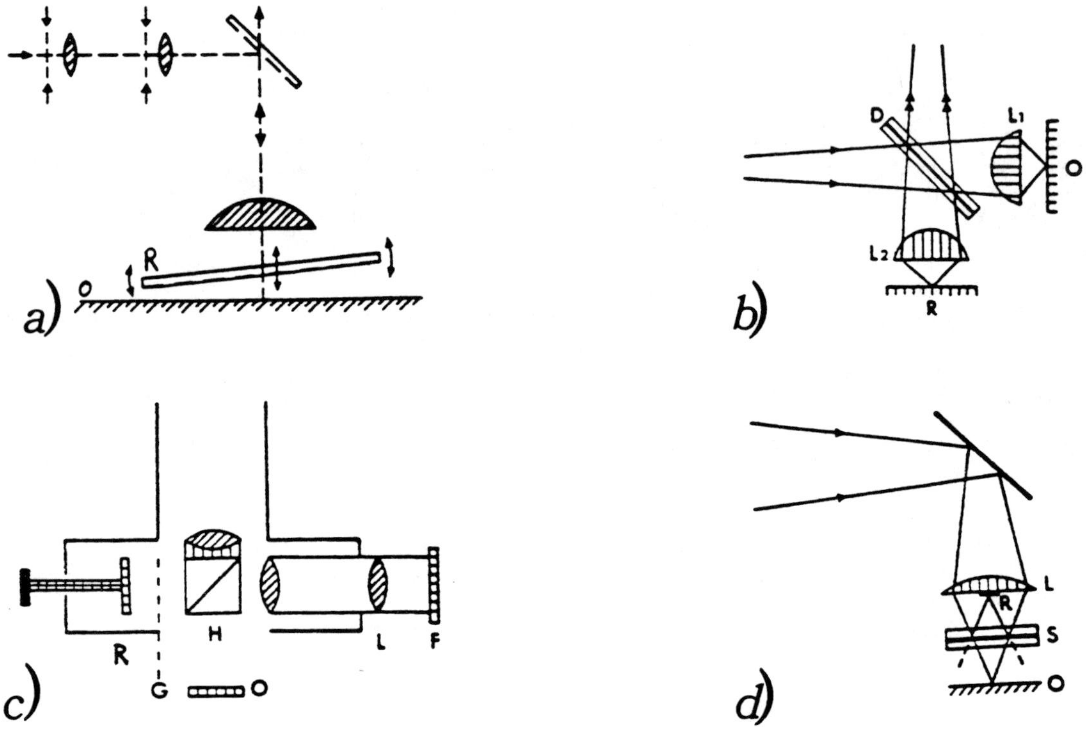

Fig. 1. Several surface measuring interferometers: a) Fizeau; b) Linnik, c) Michelson; and d) Mirau. O is the object surface and R is the reference surface.

Differential interferometers

The Normaski or differential interference contrast (DIF) interferometer is commonly used in optical inspection because it has high sensitivity, is easy to operate, and is insensitive to vibration. A typical differential interferometer is shown in Fig. 2. The optical phase difference between the two interfering beams is proportional to the height difference between points separated by the shear introduced in the interferometer. The major problem with the interferometer is that the interferometer measures the slope of the surface height variations, and if the item of interest is the surface height variations themselves, the slope data must be integrated to get the height variations. It can be shown that the measurement error of the profile height goes as

$$u_K = u_0\sqrt{K}\left(1 - \frac{K}{N}\right)^{1/2} \quad (K = 1, 2, 3 \ldots N) \tag{1}$$

where u_0 is the height error of a single height measurement. Curves showing how the error distribution varies along a profile are given in Fig. 3. The maximum error occurs in the center of the profile measurement and is given by

$$u_K = \frac{1}{2}u_0\sqrt{N} \tag{2}$$

where again u_0 is the height error of a single measurement and N is the number of measurements.[6] In an interferometer that measures the surface directly, the error would be only u_0, independent of the number of measurements and the length of the profile.

One situation where the differential interferometer is useful for measuring actual surface roughness occurs when a perfectly random surface is measured, the only item of interest is the rms roughness, and the shear is greater than the correlation length. For this situation, the rms of the surface height is given by

$$rms_{actual} = (rms_{measured})/\sqrt{2} \tag{3}$$

where $rms_{measured}$ is the rms coming directly from the differential interferometer measurement.[7]

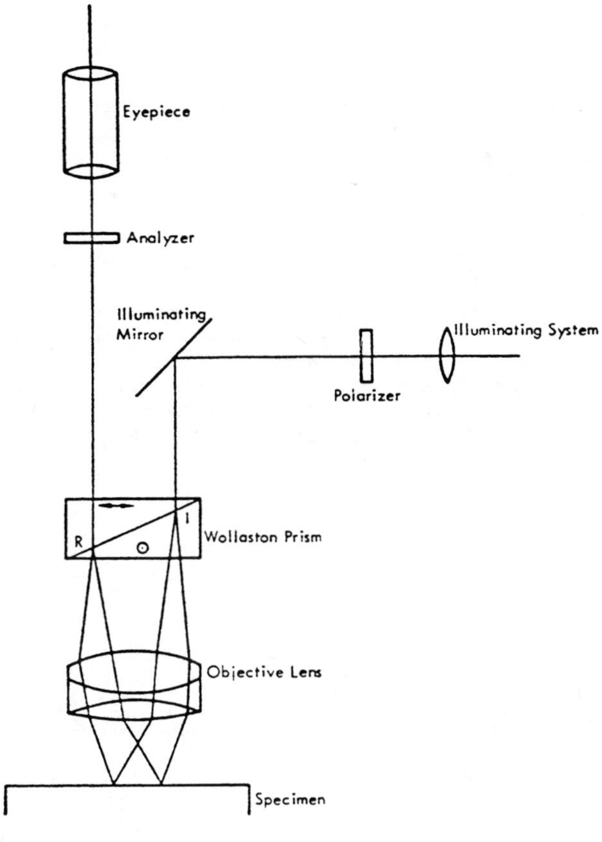

Fig. 2. Differential interferometer.

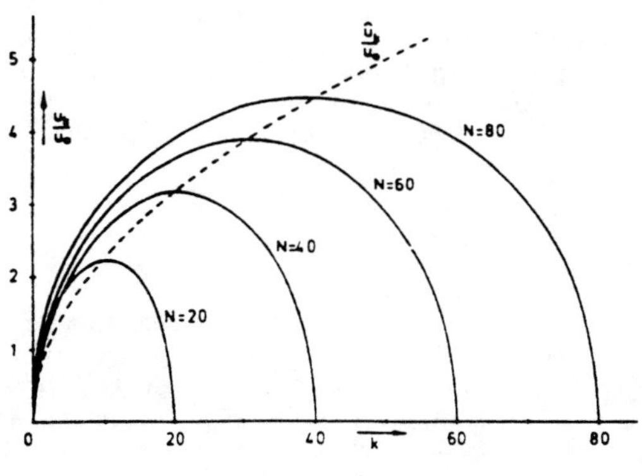

Fig. 3. Error distribution along the profile (N-total number of measurement intervals).[6]

Phase shifting interferometry

While phase shifting interferometry (PSI) techniques are not new (they date back at least to the 1950's[8] and 1960's[9]), it is only recently that they have been used in surface roughness measurement interferometers. With the introduction of solid state detector arrays and microprocessors, PSI techniques are becoming commonly used.[10-12]

PSI can be used with almost any two-beam interferometer. The basic idea is that if the phase difference between the two interfering beams is made to vary in some known manner (such as changing in discrete steps or changing linearly with time) then, as described below, three or more measurements of the intensity distribution across the pupil for different phase differences between the interfering beams can be used to determine the phase distribution across the pupil. The most common way to vary the phase difference between the two interfering beams is to mount the reference mirror on a piezoelectric (PZT) transducer and vary the voltage to the PZT. Other common techniques include polarization techniques such as the use of a rotating half-wave plate in a circularly polarized beam of light,[13] translating a diffraction grating,[14] or using an acousto-optic Bragg cell.[15]

The algorithm for determining the phase can be derived as follows: Let the intensity of the interference pattern, I(x,y) be given by

$$I(x,y) = I_A + I_B \cos[\phi(x,y) + \alpha] \tag{4}$$

where $\phi(x,y)$ is the phase distribution that is being measured and α is the phase difference between the two interfering beams that is being varied. It is well known from synchronous detection techniques in communication theory that if $I_n(x,y)$ is the intensity of the interference pattern obtained by stepping the phase an amount $n(2\pi/N)$, then

$$\tan\phi(x,y) = -\frac{\sum_{n=0}^{N-1} I_n(x,y)\sin[n2\pi/N]}{\sum_{n=0}^{N-1} I_n(x,y)\cos[n2\pi/N]} \quad . \tag{5}$$

If N = 3, then

$$\tan\phi(x,y) = \sqrt{3}\,\frac{(I_2 - I_1)}{2I_0 - (I_1 + I_2)} \quad . \tag{6}$$

Likewise, if n = 4, then

$$\tan\phi(x,y) = \frac{I_3 - I_1}{I_0 - I_2} \ . \tag{7}$$

It follows from Eq. (4) that since the equation for the intensity of the interference pattern contains only three unknowns, I_A, I_B, and $\phi(x,y)$, regardless of the value of α, only three measurements of the intensity distribution need to be made to determine $\phi(x,y)$. If for these three measurements the controllable portion of the phase difference is $-\alpha$, 0, and α, $\phi(x,y)$ is given by

$$\tan\phi(x,y) = \frac{1 - \cos\alpha}{\sin\alpha} \ \frac{I_1 - I_3}{2I_2 - (I_1 + I_3)} \ . \tag{8}$$

In practice, it is generally easier to take the data while the phase difference varies linearly with time rather than in discrete steps because if discrete steps are used, there must be a sufficient delay between changing the phase and taking data for any transients to damp out. It can be shown that the same equations derived above for discrete steps are valid for the linear varying phase situation.[16] The only difference between the two situations is a constant phase term across the pupil, and the contrast or modulation of the interference pattern given in Eq. (1) is reduced. The constant phase term is of no interest and the reduced modulation drops out when we do the division in the phase calculation.

If the sign of the numerator and denominator of Eq. (5) is checked, the phase $\phi(x,y)$ is measured modulo 2π. If the phase difference between adjacent detector elements is less than π, then the 2π discontinuities in the measured phase can be removed. The procedure is to check the phase between adjacent pixels, and if the phase difference is greater than π, then a multiple of 2π is either added or subtracted to make the difference less than π.

The major problem with PSI is obtaining the proper shift between data frames. If the phase shift is incorrect, a nearly sinusoidal phase error of twice the spatial frequency of the interference fringes is obtained as illustrated in Fig. 4 for the case where the phase shift between frames is supposed to be 90°, but is actually 88°. A convenient way to reduce this error is to make two sets of measurements, where the initial phase difference between the two sets differ by approximately 90°, and then average the two sets of measurements.

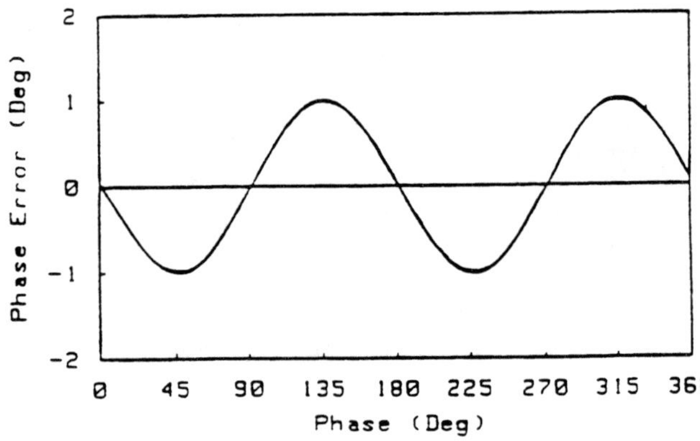

Fig. 4. Phase error versus phase for a phase step of 88°, rather than 90°.

The real power of the phase shifting algorithm is that because of the subtraction process, fixed pattern noise across the detector is subtracted out, and, through the division process, gain or intensity variation across the detector drops out. For these reasons, the phase shifting technique is very useful with solid state detector arrays. The use of solid state detector arrays offers large density of detector elements in a small package, integration of high frequency irradiance noise, and a parallel-in/serial-out philosophy that reduces electronic processing to an absolute minimum.

WYKO NCP-1000M optical profiler

A non-contact optical profiler we have been working on for several years, first at the University of Arizona, and then at the Wyko Corporation, is shown schematically in Fig. 5 and an actual photograph of the instrument is shown in Fig. 6. The Mirau interferometer is used in the 10X and 20X magnification version of the instrument and a Michelson interferometer is used for the 2.5-X version. Figure 5 shows the Mirau version. Table I gives optical specifications of the instrument. The integrating three-bucket version of phase shifting interferometry described above is used to obtain data. A PZT is used to move the reference surface during the taking of the data. A Fairchild Model CCD 1300R 1024-element CCD camera is used as the detector. The detector output is digitized to 10 bits. A very good feature of the instrument is that the operator can look through the microscope to see where the data are being taken and when he sees where he wants to take data, he can simply press a button and the data are taken in a matter of milliseconds.

The surface height errors are calculated and analyzed using an HP 9816 desktop computer. Tilt and curvature can be subtracted from the profile data if desired. Typical output plots are shown in Fig. 7. If only a portion of the profile is of interest, the portion of interest can be selected using a cursor and the knob on the 9816 keyboard.

Data acquisition time is less than 50 ms. About 3 s are required to obtain the profile plots. For a single frame, the rms repeatability of the data for smooth surfaces is on the order of 0.4 nm. By averaging several data frames, the noise can be reduced by the square root of the number of data frames averaged. Typically four data frames are averaged to give an rms repeatability of about 0.2 nm. If desired, 16 data frames can be averaged to give an rms repeatability on smooth surfaces of less than 0.1 nm.

The roughness of the reference surfaces used in the Mirau range from 0.4 to 0.7 nm. There are two techniques that can be used to measure surfaces smoother than the reference surface. First, if a super smooth sample is available, it can be tested and the resulting data can be stored on disk to serve as a reference data set that can be subtracted from the measured data.

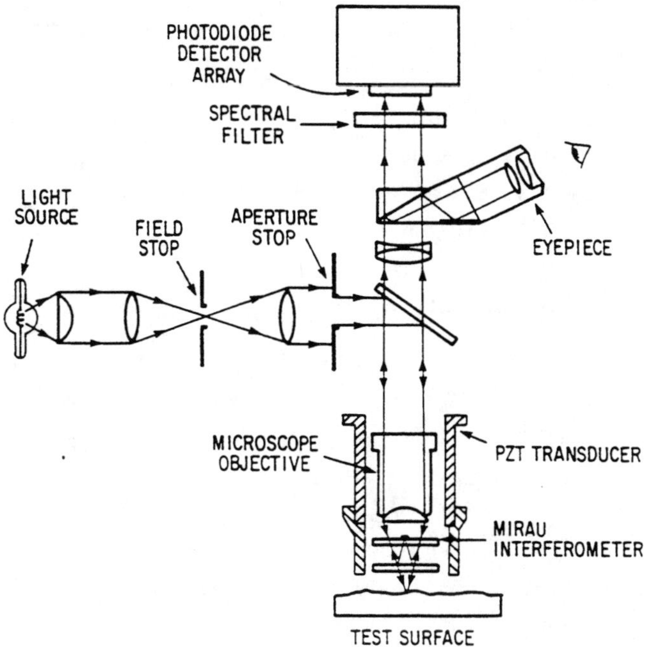

Fig. 5. Optical schematic of WYKO NCP-1000M optical profiler.

If a super smooth reference surface is not available, a reference data set can be generated by measuring a reasonably smooth sample several times (four to eight times is generally sufficient), where between each measurement the sample is moved a distance larger than the correlation length of the surface roughness of the sample. All the sample measurements should be stored on disk. The computer can then take the average of these measurements and this average data set can serve as the reference data. This technique works because the errors arising from the reference surface are in all measurements, and, by moving the sample a distance large compared to the correlation length between measurements, errors in the sample tend to average out. For strictly random errors, the errors in the sample will be reduced by the square root of the number of measurements made.

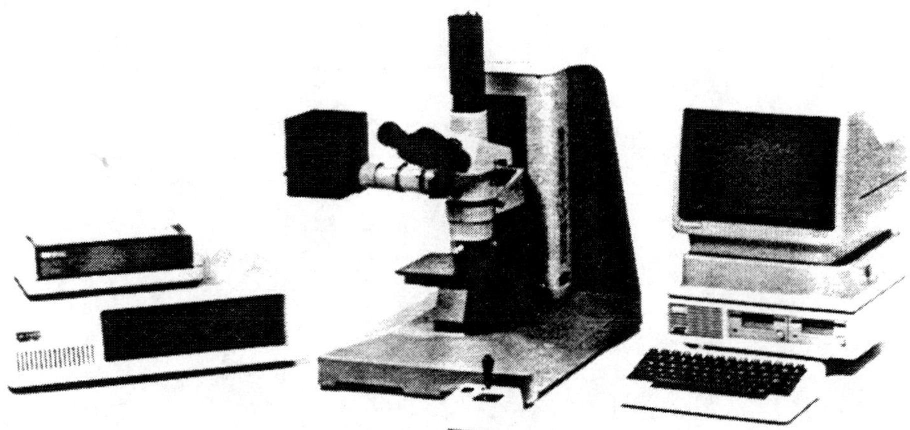

Fig. 6. WYKO NCP-1000M optical profiler.

Table I. Optical Specifications of WYKO NCP-1000M Profiler

Magnification	2.5	10	20
Profile Length (mm)	5.3	1.3	0.665
Spatial Sampling Interval (μm)	5.2	1.3	0.65
Working Distance (mm)	12	4.5	2.5
Maximum Surface Heights (μm)	6.5	5.2	2.0
Maximum Surface Slope (Degrees)	1.6	6.4	12.4

If the only item of interest is the rms roughness, there is an easier method for removing errors on the reference surface. The procedure is to first take profile data and store it on disk. Move the sample more than the surface roughness correlation distance and take a second profile data set. If the two data sets are subtracted, errors on the reference surface will subtract out since they are the same for the two data sets, and the rms surface height calculated will by $\sqrt{2}$ times the actual value. By using either this method or the method described above, surfaces having an rms roughness of less than 0.1 nm can be measured in a matter of minutes.

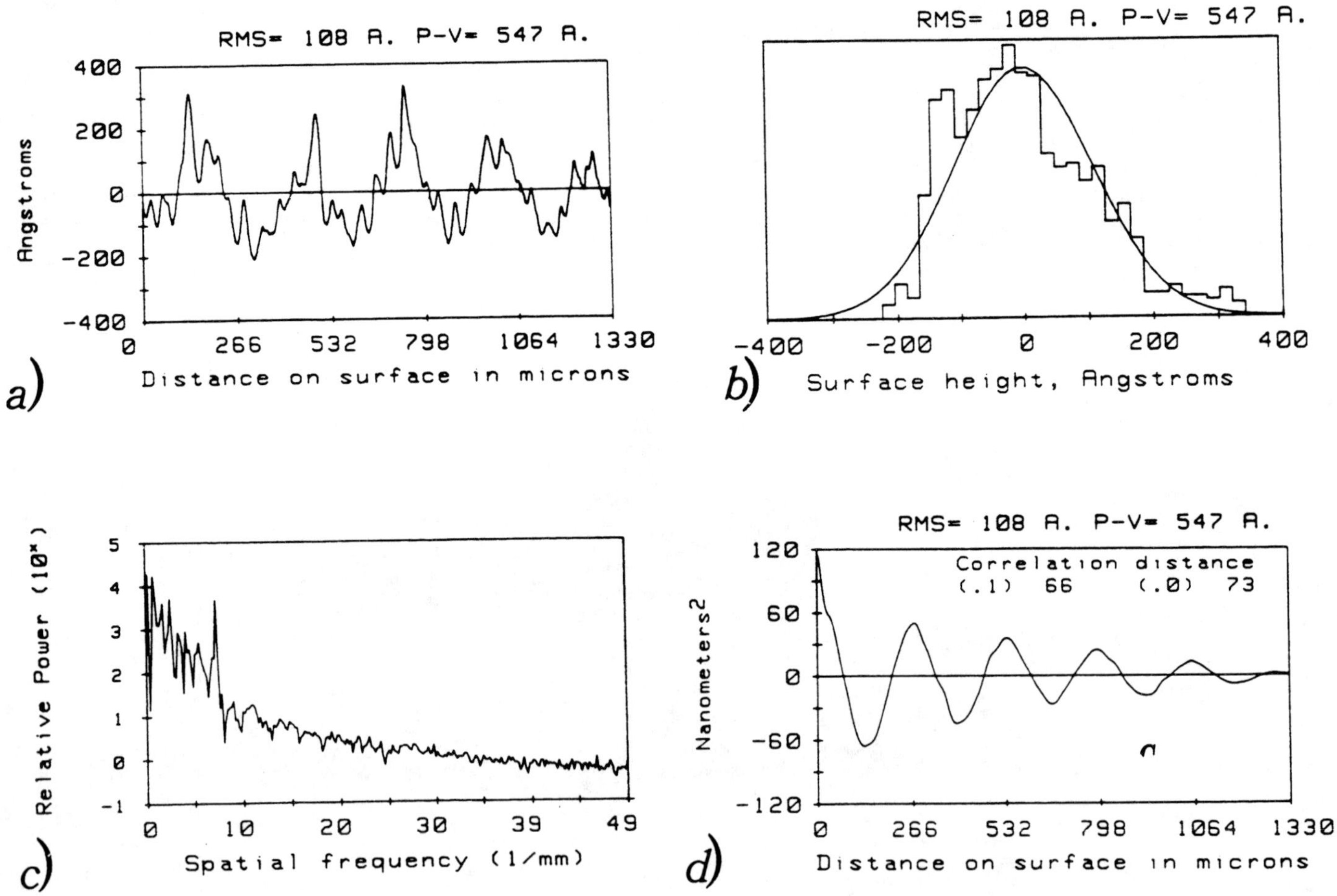

Fig. 7. Typical output plots. a) Profile, b) histogram, c) power spectrum, and d) autocovariance.

Summary and conclusions

Several different interferometers can be combined with an optical microscope to measure surface roughness. For the most precise measurements, interferometers that measure surface height errors directly are preferred to interferometers that measure the slope of the height errors. By using phase shifting techniques, solid state detector arrays, and microcomputers, instruments have been made that can measure surface heights to within 0.1 nm or less in a matter of a few seconds to a minute. The major disadvantage of optical techniques to mechanical stylus techniques is that an optical system using visible light has a lateral resolution on the order of 1 μm, while it is possible for stylus systems to have a lateral resolution of 0.1 to 0.2 μm.

References

1. J. M. Bennett and J. H. Dancy, "Stylus Profiling for Measuring Statistical Properties of Smooth Optical Surfaces," Appl. Opt. Vol. 20, p. 1785. 1981.
2. M. J. Lavan et al., "Digital Wavefront Measuring Interferometer for Testing Optical Surfaces and Lenses," Appl. Opt. Vol. 13, p. 2693. 1974.
3. G. E. Sommargren, "Optical Heterodyne Profilometry," Appl. Opt. Vol. 20, p. 610. 1981.
4. C. W. Chen and P. W. Mayman, "Use of Phase Measuring Interferometry for Surface Characterization," Proc. SPIE Vol. 316, p. 9. 1981.
5. J. C. Wyant, C. L. Koliopoulos, and B. Bhushan, "An Optical Profilometer for Surface Characterization of Magnetic Media," ASLE Reprint 83-AM-6A-1. 1983.
6. G. Makosch and B. Drollinger, "Surface Profile Measurement with a Scanning Differential AC Interferometer," Appl. Opt. Vol. 24, p. 4544. 1984.
7. K. L. Underwood and J. C. Wyant, "Quantitative Surface-Roughness Measurement Using a Shearing Interferometer," J. Opt. Soc. Am. Vol. 72, p. 1771. 1982.
8. J. Terrien, "Observations Photoelectriques a L'interferometre de Michelson," J. Phys. Radium, Paris Vol. 19, p. 390. 1958.
9. P. Carre´ "Installation et Utilisation due Comparateur Photoelectronique et Interferentiel du Buneau International des Poids et Measures," Metrologia Vol. 2, p. 13. 1966.
10. J. H. Bruning et al., "Digital Wavefront Measuring Interferometer for Testing Optical Surfaces and Lenses," Appl. Opt. Vol. 13, p. 2693. 1974.
11. J. C. Wyant, "Use of an AC Heterodyne Lateral Shear Interferometer with Real Time Wavefront Correction Systems," Appl. Opt. Vol. 14, p. 2622. 1975.
12. J. C. Wyant and C. L. Koliopoulos, "Phase Measurement System for Adaptive Optics," Agard Conf. Proc. No. 300. 1981.
13. G. E. Sommargren, "Up/Down Frequency Shifter for Optical Heterodyne Interferometry," J. Opt. Soc. Am. Vol. 65, p. 960. 1975.
14. J. C. Wyant, "Double Frequency Grating Lateral Shear Interferometer," Appl. Opt. Vol. 12, p. 2057. 1973.
15. J. F. Ebersole and J. C. Wyant, "Collimated Light Acousto-optic Lateral Shearing Interferometer," Appl. Opt. Vol. 13, p. 1004. 1974.
16. J. Greivenkamp, "Generalized Data Reduction for Heterodyne Interferometry," Opt. Eng. Vol 23., p. 350. 1984.

Reprinted with permission from *Optica Acta*, Vol. 25(12), pp. 1187-1196 (1978).

A design study of an automatic system for on-line detection and classification of surface defects on cold-rolled steel strip

A. J. BARKER and R. A. BROOK
Sira Institute Ltd, South Hill, Chislehurst, Kent, England

(*Received* 1 *October* 1978)

Abstract. In collaboration with the British Steel Corporation, Sira Institute Ltd is currently pursuing the development of sophisticated equipment for the automatic inspection of cold rolled steel strip. This paper will examine the special difficulties associated with the application of optical inspection techniques to a dull metallic surface at line speeds of up to 50 m s^{-1}.

The development work has centred on the use of high-speed scanned laser systems, unique in their ability to provide diffraction-limited resolution with intrinsically high optical throughput. In order to detect and delineate all significant defects down to the most subtle imperfections, the polar scattering characteristics of the surface have been exploited and optimum signal processing techniques are being identified. The paper will discuss some of the optical, mechanical and electronic constraints imposed by the inspection requirement.

1. Introduction

In collaboration with the British Steel Corporation and The City University, London, Sira Institute Ltd is currently developing an advanced automatic surface inspection system for cold-rolled steel strip based on measuring the intensity of reflected laser light. Although capable of application to a range of strip materials, the techniques have been optimized for the inspection of cold-rolled strip, whose dull, optically rough surface poses a particularly demanding inspection task. This paper presents an outline of the development programme, examining firstly the constraints and performance criteria dictated by the inspection requirement, then discussing the reasoning behind the choice of a laser scanning technique and finally describing the optimization of its performance over the full range of defects to be detected.

2. Inspection requirement

The inspection requirement is determined by a list of surface defects which it is important firstly to detect and then as far as possible to name so as to identify their cause and origin. Conditions on-line at the point where the equipment is to be installed also play an important part in determining system design. All aspects of the requirement have to be satisfied for the equipment to be successful. Thus, for example, the ability of a system to detect all significant defects at a low throughput speed under controlled laboratory conditions would be of solely academic interest unless a robust, high-speed industrial version were an engineering possibility. Furthermore, insensitivity to only one of the significant defects would drastically reduce the utility of the system, however good its performance on the remainder.

Table 1. Summary of the most significant defects occurring on cold rolled strip.

Description of defect	Optimum viewing direction 1. Specular (bright ground) 2. Just off-specular 3. Well off-specular (dark ground)	Typical defect contrast
1. Longitudinal defects producing elongated broken line in rolling direction :		
skin lamination, seam, sliver, scratch.	1	high
2. Transverse defects producing line or crease across the moving strip :		
sticker wrench	2, 3	low
chatter marks	1	low/medium
coil break	3	low
edge strain	1	low
3. Localized mechanical indentations :		
steel pits and digs	1	high
scrap marks	1	low
roll marks	1, 3	low
pick-up	1, 2	low/medium
coil digs	1	high
indents	1, 3	low
4. Fleck pattern defects affecting extended area of strip :		
fleck scale, jet scale, scale pits, sand pits, rust spots	1	medium
5. Other large area defects :		
carbon pick-up	1	low/medium
feather marks	3	low
water strain	1	low/medium
6. Gross defects :		
holes, serrated edge, oxidized edge	1	high

The table summarizes the range of surface defects on cold-rolled steel strip which are considered to be of significance. There are three noteworthy features of the defect heirarchy.

(1) The wide variation of optical contrast between gross surface defects such as holes or pits and subtle creases or wrinkles such as sticker wrench.

(2) The dependence of defect contrast on viewing angle, relative to the direction of incident illumination : an optimum angle or range of angles exists for many of the significant defects. Most frequently, the direction of specular reflection is the optimum viewing angle, but some of the more subtle surface defects yield maximum contrast at other angles and may even be invisible in the specular direction.

(3) The importance of defect shape on the strip in determining the identity (i.e. the name of the defect type).

When installed in a steelworks, the equipment is required to operate on strip up to 1500 mm wide moving at 1000 m min^{-1}. The equipment would normally be located over a roller, thereby reducing tilt and flap of the surface to a minimum. Ambient illumination levels are moderately high and vary according to the level of sunlight entering through the roof, the level of internal lighting and the passage of overhead cranes.

In addition, due to the presence of heavy machinery and the nature of the steel production process, the installation must be designed to operate satisfactorily in the presence of heavy mechanical vibration, inductive electrical interference and an atmosphere frequently bearing particles of metal and lubricant which could rapidly contaminate optical surfaces.

3. Choice of scanning system

For the purpose of this discussion, the various possible scanning system configurations which might meet the inspection requirement can be classified into two main types according to whether system resolution is defined by the illumination projection system or the reflected light collection system. If the incident illumination is scanned (flying-spot system), then the optical resolution of the inspection system is defined by the cross-section of the illuminating beam projected onto the inspected surface. If the incident illumination is static, then some form of image scanner (either serial or parallel) is required to maintain adequate sensitivity to the smallest defects of interest. The image scanner could comprise an array of individual sensors rendered conjugate to the surface by a static imaging system ; alternatively, a single detecting element in the image plane of a scanning optical system could provide sequential signals at a fast enough rate to provide complete surface coverage.

In both cases total coverage along the length of the strip is provided by the movement of the strip past a stationary inspection ' window ' beneath the scanner spanning the full width of the product.

We can show that the constraints imposed by the inspection requirement favour a coherent-illumination flying-spot system as the preferred option for development. Laboratory experiments have shown that monitoring the angular distribution of reflected light from at least three distinct directions (see figure 1) is necessary to provide adequate sensitivity to the full range of important surface defects on cold-rolled strip. This finding alone puts the scanned collector option at a disadvantage because of the need to triplicate a high-resolution scanning system. Furthermore, maintaining mutual intersection and synchronization of the three independent viewing points at the surface of a possibly tilting, vibrating strip would be quite difficult. With a scanned projection system the optical resolution is defined by a single incident beam ; each collection

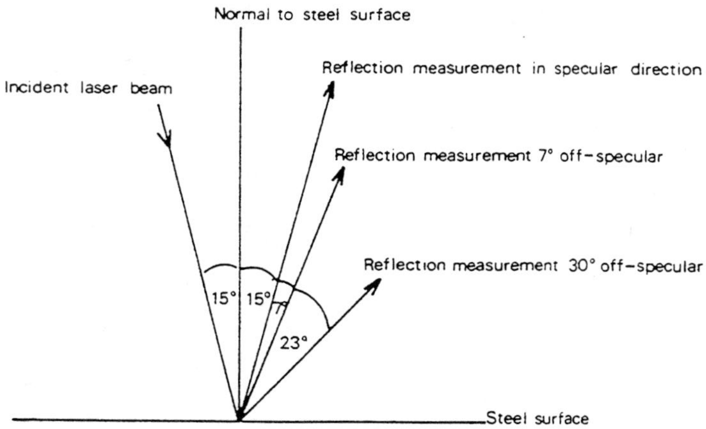

Figure 1. Minimum set of viewing directions in the plane of incidence for detecting all significant surface defects on cold-rolled steel strip.

channel could comprise nothing more than a detector sited at the correct viewing angle, triplication of which is neither complex nor expensive. Spatial synchronization in the three viewing channels occurs automatically and allows real-time signal processing to be carried out efficiently on any combination of channels simultaneously.

Hence a flying-spot scanner has significant advantages over scanned collection systems in terms of overall simplicity and ease of multichannel synchronization. In order to decide whether a laser or a conventional source should be used to generate the scanned illumination, it is necessary first of all to determine the available light levels. A flying-spot laser system will always out-perform the equivalent configuration based on a conventional source, at least in terms of energy throughput for a given spot size and depth of focus, simply because the intrinsic brightness of laser sources exceeds that of the brightest practical conventional sources by two to three orders of magnitude. However, laser speckle and other possible side-effects due to the coherence of laser radiation can add unwanted noise to a reflection measurement. Hence incoherent illumination is often preferable for measurement of subtle changes in reflectivity, provided that shortage of light is not a problem at the required speed of inspection.

Figure 2 shows the angular variation of the intensity of light reflected from a typical fairly close-textured steel surface. The solid line was obtained using He–Ne laser illumination and the dashed line using tungsten illumination. There are two notable features of figure 2.

(1) The difference between the coherent and incoherent curves is not very marked, although there is some indication of secondary structure on the laser curve.

(2) The polar distribution is such that moving away from the specular direction in the plane of incidence, the measured intensity is reduced by one-half at $\sim 15°$ off-specular and by 2 orders of magnitude at $\sim 40°$ off-specular (the precise angles depend on the surface finish).

Resolution of the smallest defects noted in the table at a line speed of 1000 m min^{-1} requires a spot width of the order of 1 mm and an electronic system bandwidth approaching 25 MHz. In order to maintain an adequate depth of

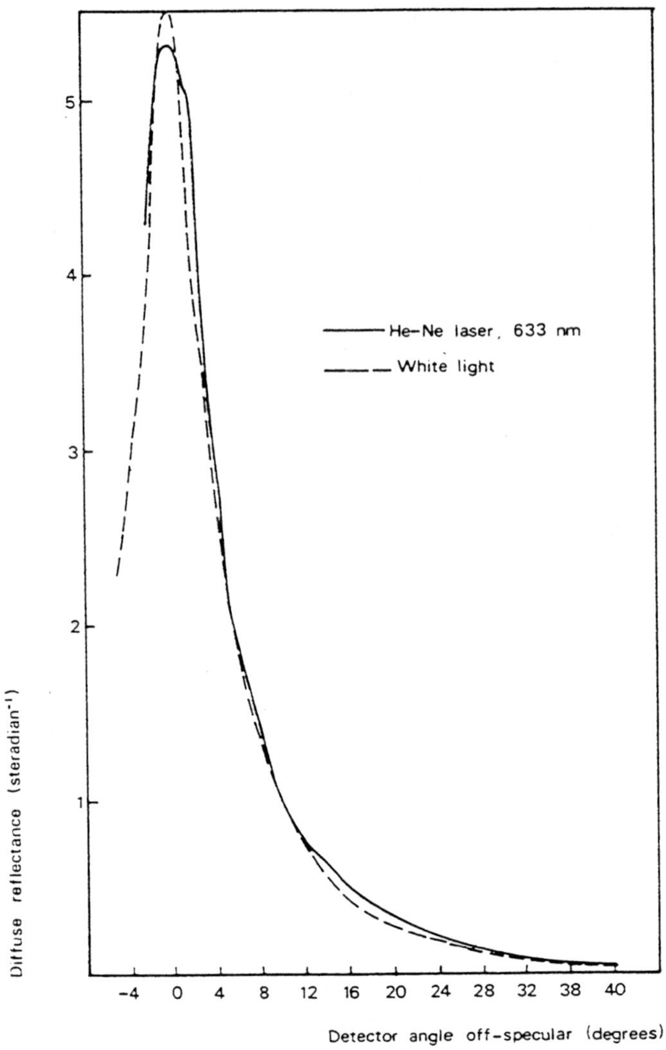

Figure 2. Scatter from a fine matt steel surface.

focus and also to keep the necessary size of the optical scanning element within reasonable limits, the maximum practical convergence angle for the scanned incident beam is about $\frac{1}{2}°$ total, corresponding to a solid angle of 6×10^{-5} sr. A typical scanner would be designed to operate with a beam convergence angle somewhere between this maximum of $\frac{1}{2}°$ and a minimum value of $\sim \frac{1}{10}°$ which is necessary to retain a 1 mm spot in the presence of diffraction.

Assuming that the source of illumination is a 5 mW laser, the beam power incident on the steel surface is likely to be in the region of 3 mW after passing through the spot projection optics. The largest readily available standard photodetector having a response consistent with a 25 MHz system bandwidth is a 75 mm photomultiplier ; to accommodate coverage of the full strip width the detector has to be situated approximately 2 m from the surface of the strip, subtending a solid angle of $1{\cdot}1 \times 10^{-3}$ sr. A full-width single-scanner system is preferred to a bank of scanners stacked side-by-side across the strip since, although the stacking arrangement eases the requirement on resolution and bandwidth, it is more expensive and again presents difficulties with synchronizing and matching the signals from the separately inspected zones. From figure 2, it can be seen that the available optical power at the preferred collection points in the specular direction, at about 7° off-specular and 30° off-specular, will be 15, 6 and 0·5 μW respectively. With these power levels, system operation at 25 MHz would be subject to limiting signal/peak–peak random noise ratios (due to photon statistics) of 50 : 1, 30 : 1 and 10 : 1 (order of magnitude figures) along the three viewing axes.

The noise levels in the specular and 7° off-specular channels (2 and 3 per cent peak-to-peak respectively) would be acceptable. The 10 per cent peak-to-peak noise level in the 30° off-specular channel is rather high, but could be reduced by restricting the inspection bandwidth in this channel. This turns out to be practicable since the majority of defects for which 30° off-specular collection is required extend along at least 10 mm of the scan line or produce signals of very large amplitude which can be detected even in a reduced bandwidth.

These S/N ratios could be improved if necessary by increasing the laser power employed and/or changing the laser wavelength from red to blue/green to match the optimum photomultiplier response.

In contrast, the best available S/N ratios using a conventional source with a $\frac{1}{2}$° incident beam convergence would be five to ten times lower, with no possibility of improvement by increasing the input power. Hence the use of a laser source is highly desirable.

4. Optical system design

Other aspects of the reflection characteristics of cold-rolled steel have a direct bearing on the design of an optimum flying-spot inspection system. In what follows, ' surface ' noise is taken to mean the signal variations caused by changes in the collected light intensity as the incident beam moves over the relatively rough non-defective steel surface.

Coherence of the laser source results in surface noise levels approximately twice as high as those obtained using an incoherent source. The higher noise level is largely attributable to speckle which adds noise to the polar distribution of light reflected from different areas of the steel surface. Since, for reasons discussed above, the use of a conventional source is not practical, the speckle noise can be reduced only by increasing the effective solid angle Ω_c of collection (mean signal/surface noise $\propto \Omega_c$, approximately).

The ratio of mean signal to surface noise measured at any fixed angle off-specular does not depend on the angle of incidence employed, except in a manner predictable from the increased area of surface illuminated by an incident beam of fixed cross-section (mean signal/surface noise $\propto(\cos \theta_{\text{incident}})^{-1/2}$, approximately). Hence, it is not essential to maintain a constant angle of incidence in the spot projection optics in this respect.

The ratio of mean signal to surface noise varies markedly with off-specular collection angle. Close to the specular direction the ratio changes typically by 5 per cent per degree. The optical system should therefore provide a constant off-specular viewing angle over the full inspection width, not only to maintain the identity of the three collection axes shown in figure 1, but also to reduce the degree of adaptation to varying noise levels required of the signal processing system.

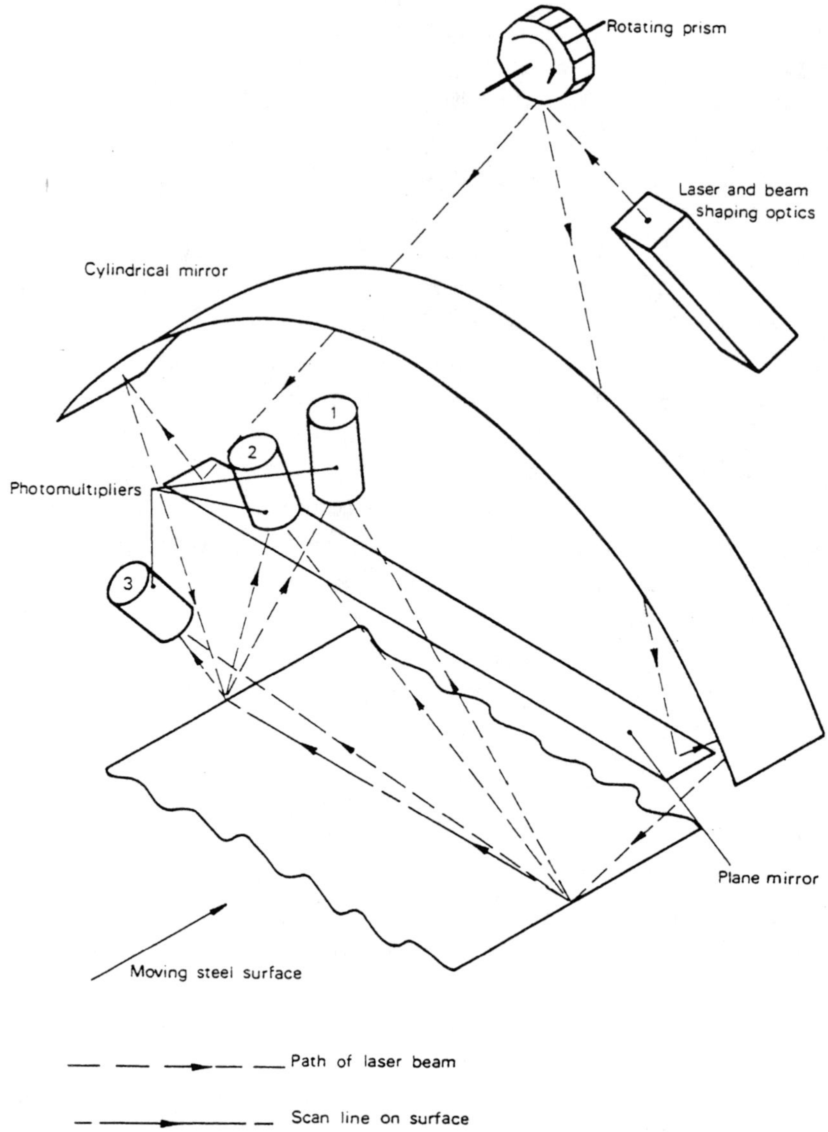

Figure 3. A high-speed scanner.

Figure 3 shows one possible form of high-speed scanner which satisfies the above criteria. The adoption of converging incidence eliminates the need for any collection optics in front of each detector, thereby minimizing system complexity. The scanned beam is generated by reflection from a 12-facet prism rotating on an air bearing.

Flying-spot systems tend to be prone to ambient light pick-up since the field of view of the detector spans the full inspection width. Ambient light rejection from within the field of view of the detector can be improved by optical filtering. In principle, the use of a laser source should allow extremely high ambient rejection by employing a narrow-band interference filter. Unfortunately, these filters are unsuitable because of their angular sensitivity and limited size. Nevertheless, useful rejection can be achieved by employing simple gelatine filters—a Kodak Wratten 25, for example, is well suited to the He–Ne laser wavelength when used in conjunction with an S20 photocathode. The 30° off-specular channel is most susceptible to interference from ambient illumination, since the collected laser power is relatively low. It is essential to ensure that no bright source in the vicinity of the inspection equipment can be reflected specularly from the steel surface into the 30° off-specular channel.

Finally, on-line operation under mill conditions calls for a mechanically robust housing for resistance to impact, and a pressurized enclosure surrounding the optics with clean air to minimize build-up of dust and dirt.

5. Electronic signal processing system design

5.1. *Defect detection*

The majority of the signals produced by the detectors comprise combinations of amplitude and frequency which are of no significance. The first task for the signal processing system is to separate the relatively infrequent signals caused by defects on the surface from the background noise. The main components of this background noise originate from variations of the laser speckle pattern at the detector, small local changes in contrast caused by the rough surface structure, and random photon and detector shot noise. The overall surface reflectivity commonly varies slowly from side to side and along the length of the strip. Local contrast of any significance is detected by filtering and controlled thresholding, which yields signals indicating the presence of defects and other surface marking, along with a few 'false alarms' generated by noise. This process immediately reduces the data rate very significantly.

The next step is to distinguish inconsequential surface marking, small negligible defects, and false alarms from significant defects. This requires discrimination on the basis of the amplitude and width of the signal, its spatial relationship to other detected signals, and possibly the use of other measured parameters which are subsequently employed in attempting to identify the defects by name. The processing needed to do this may be quite time-consuming and it is therefore crucial that the threshold levels set to detect defect signals do not result in a large number of false alarms. Since the surface noise varies considerably according to surface finish the thresholds must be set optimally for each finish to avoid insensitivity to noticeable defects on the less noisy surfaces. Surface finish may vary from one coil to another, and since it is not practicable to resort to manual adjustment of thresholds in each of the several analogue signal channels, automatic methods have to be used to set the thresholds in relation to a measured value of background noise level. This is quite tricky since some defects, such as fleck scale, comprise a global distribution of small black marks across large areas of the strip. It is important to avoid interpreting the onset of such relatively dense small signals as an increase in noise level since this would produce an erroneous alteration in threshold levels.

Some low-contrast defects are particularly difficult to see and are not adequately detected by simple 'instantaneous' filtering and thresholding of line-scans. The contrast and signals can be enhanced in several ways, including combining signals from the different detector channels and by applying two-dimensional filtering to the 'array' of signals produced by the raster of sequential line scans.

Inconsequential contrast, which is of significant amplitude and sufficiently localized to be confused with defects, is usually associated under on-line conditions with the presence of a lubricant on the surface or minor local distortion of the strip. Such features will be detected and therefore must be classified correctly and ignored.

5.2. *Identification of defects by type*

5.2.1. *Introduction*

Defects are identified by inspectors according to shape, size and contrast, variation of appearance with angle of view and illumination, and by using contextual information and experience concerning the frequency with which defects are likely to occur. The automatic system is capable of acquiring some of this information, its main limitations being in picture resolution and the fixed angles of illumination and viewing. The automatic system also lacks ' long-term experience ', but some data describing the relative likelihood of occurrence of the different types of defect can be incorporated into a classifier. However, under on-line, high-speed production conditions the human inspector is subject to even greater limitations arising from the restricted response of the human eye and brain system.

Present-day developments are seeking to raise the level of sophistication in the signal processing system to the point where the speed, classification accuracy and objectivity of automatic equipment are sufficient for production purposes. This means that on high-speed lines the automatic system has to be capable of equalling the performance of a good inspector watching strip which is moving, but at a lower speed, allowing him to see and make judgements on the majority of the significant defects.

Determination of the extent to which the limited resolution, viewing angles and ' experience ' which can be realized in a practical inspection machine will allow such a target performance to be reached in practice is the subject of the present project.

5.2.2. *Method*

In an attempt to classify defects, the first stage must be to remove false alarms. These will be detected signals which, by virtue of duration, amplitude and distance from other features are of no significance and would be passed by an inspector. This process eliminates signals from tiny and isolated flecks, rust spots and small anomalies in surface structure, and removes triggers generated by photon noise variations which exceed the detection threshold. It is important to note that in this respect it is obvious that some degree of classification is needed to produce detection performance with a sufficiently small number of false alarms.

Classification can then proceed based on an analysis of measurements, made firstly on isolated contiguous features detected on the surface, and then on areas within which like features have been associated.

A number of possible pre-processing and classification schemes are being evaluated by large-scale computer simulation of alternative options, using data acquired from a fully engineered scanning system, but under laboratory conditions. Naturally, considerable differences exist between laboratory and on-line conditions ; for example, samples handled off-line are liable to spurious marking and do not show changes in surface condition which may occur over a large area (the sample size is typically 30 cm square). Nevertheless, the difficulties of conducting intensive on-line work under production conditions and of retrieving data from the scanner and the material in question would not permit this investigation to be carried out on-line.

6. Conclusions to date

Results so far obtained show that

(1) many of the significant defects are easy to detect, but some of the classes of similar generic type are likely to be difficult to distinguish automatically,

(2) signal-enhancement techniques are quite effective on low contrast defects which are difficult to see, but in most cases need to be followed up by analysis of spatial parameters to remove false alarms, and

(3) relatively sophisticated adaptive (' intelligent ') pre-processing is needed to detect defects adequately and minimize false alarms.

None of this is particularly surprising, in fact these general results were predicted at the outset on the basis of previous experience. The aim of the project presently in progress is to produce a specification for the construction of an inspection system which meets the target performance requirement as closely as possible, using realizable technology and at reasonable cost. An important part of this outcome is a statement of performance, in quantitative terms, based on this detailed off-line evaluation.

ACKNOWLEDGMENT

This paper is published with the permission of the British Steel Corporation.

Reprinted from *Optical Techniques for Industrial Inspection,* P.G. Cielo, editor, Proc. SPIE Vol. 665, pp. 17-24 (1986). ©1986 SPIE.

Hot Slab Surface Inspection By Laser Scanning Method

Toshiro Matsubara

R & D Laboratories-I, Nippon Steel Corporation
1618 Ida, Nakahara-ku, Kawasaki, Japan

Toshio Toyota

R & D Laboratories-I, Nippon Steel Corporation
1-1-1 Edamitsu, Yahatahigashi-ku, Kitakyushu-City, Japan

Akihiro Fujiyama

Equipment Dept. Yawata Works, Nippon Steel Corporation
1-1-1 Edamitsu, Yahatahigashi-ku, Kitakyushu-City, Japan

Abstract

An optical flaw detector with laser as the external light source, which is called LST (laser scanning tester), has been developed. This equipment automatically inspects the entire surface of hot slabs. The results are used to examine the suitability of those slabs for hot charge rolling. The characteristics of LST are its high optical resolving power and the signal processing method with which two-dimensional information on the type of the flaw is processed. For the opening width of 0.4mm and over, the detection ratio is nearly 100%. This equipment started commercial operation in January 1983 in Nippon Steel's Yawata Works and its application has increased the hot charge rolling ratio.

Introduction

In recent years, iron and steel making processes have been directly linked and made continuous with the aim of saving the energy and rationalizing the processing steps, and the hot charge rolling (HCR) of slabs has firmly established itself in Japanese steel industry. For the HCR process, hot slab quality inspection technique is important as well as the technique of making slabs with no defect. In particular, the technique of detecting surface flaws of hot slabs have been a subject of intensive study of late.[1] Optical[2,3], eddy-current[4], induction heating[5] and various other testing methods have been proposed, and some of them are already in practical application.

Nippon Steel's Yawata Works has developed and installed a unique flaw inspection system[6], which is the combination of laser scanning tester[7,8], and eddy-current tester[6]. This system now inspects the surface of the continuously casted hot slabs for sheets in the HCR process. This paper introduces the entire system briefly and the laser scanning tester in detail.

Requirements for the tester

When deciding the method to inspect hot slabs, four important factors, such as, (1) sorts of flaws and their features (2) surface temperature (3) surface condition (4) size of slabs, must be considered. The major types of surface flaws in the continuously casted slabs are shown in Table 1. The longitudinal surface crack is the predominant type of the surface flaw, and must be detected without failure.

Other factors such as surface temperature, surface condition, and size of slabs are also shown in Table 1. Optical testing method has some strong points and some weak points as shown in Table 2.

Yawata Works regarded flaw inspection as "automatic inspection of all slabs to increase the HCR ratio of slabs for cold-rolled sheets and tinplates" and studied the optimum system under the constraints of operating and equipment conditions. As a result, it decided to adopt the system that employs the optical testing method mainly and the eddy-current testing method partly. The way of thinking is as follows:

(1) Cold-rolled sheets and tinplates are rolled to the thickness of several hundredths of that of the slabs and minute slab surface flaws are harmless except in the case of the products of certain quality. Therefore, the optical testing method can be applied to a considerable percentage of these products.
(2) The surface temperature of a hot slab stretches across the Curie temperature and therefore the hot slab cannot be inspected on all the surface by the eddy-current

testing method in that condition. An impact cooler or similar equipment which cools the slab surfaces below the Curie temperature is impossible to install because of the mill layout and is contrary to the purpose of energy savings.

(3) The optical testing method is suitable for general inspection because of its strong points shown in Table 2.

(4) The edges of the slabs should be inspected by the eddy-current testing method because they usually make the insensitive zones of optical method, their temperature is lower than the Curie temperature and they often develop almost closed cracks.

The following are the requirements to be satisfied when applying the optical testing method to the detection of hot slab surface flaws:

(1) To increase the resolving power to such a degree that the sufficient signal levels can be obtained from very slightly open cracks.

(2) To remove the effect of the light spontaneously emitted from the slab and the external light as much as possible.

(3) To perform signal processing to eliminate disturbance such as surface irregularities and color non-uniformity as much as possible and extract flaw signals effectively.

(4) To actualize the high-speed signal processor which can deal with the enormous information, which increases as the resolving power rises.

(5) To take some measures to make the slab surface clean to some degree before the inspection.

LST is an optical tester which satisfies the requirements described above.

Table 1. Important factors for hot slab surface inspection

1. Sorts of flaws and their features	a) Longitudinal surface cracks b) Transverse surface cracks	Length: Several tens - hundreds mm Width: Usually less than 1 mm Depth: Several mm
	c) Edge surface cracks	Length: Several tens mm Width: Very narrow Depth: Several mm
	d) Pinholes	Open or closed 1 - several mm in diameter
	e) Slag inclusions	Filled with slag
2. Surface temperature	a) High temperature (600 - 1,000°C) b) Not spatially uniform within a piece of slab c) Varies with conditions of operation d) Usually stretches across the Curie temperature (about 750°C)	
3. Surface condition	a) Not smooth (scale, oscillation mark, etc.) b) Not clean c) Not uniform (optically, electromagnetically, etc.)	
4. Size of slabs	Large	Length: 5 - 15m Width: 1 - 2m Thickness: 200 - 300mm

Table 2. Strong points and weak points of optical surface inspection

Strong points	Weak Points
1. Mechanical reliability is high because of noncontact.	1. Closed flaws can not be detected.
2. It is easy to inspect wide area at high speed.	2. Depth of flaws can not be directly determined.
3. Temperature constraint on inspection is mild. (Curie temperature does not matter.)	3. Inspection is affected by surface nature of slabs.

Optical head

The basic principle of LST is shown in Figure 1. The laser beam is very narrowly focused and projected to the surface nearly vertically. When the reflected light is received aslant, the scattered light is received where there is no flaw, but where there is an open flaw like a crack, the scattered light is intercepted by the sidewalls of the flaw and the amount of received light is sharply reduced. The change in the amount of light is converted into the electrical signal by the light sensor and is processed. This means is a kind of flying-spot method, and the thinner the focused beam is, the higher the resolving power becomes.

Figure 2 shows the basic configuration of the projecting and receiving section of LST. The light from the argon laser is converted into a thick parallel beam of light by the collimator, and by the combination of the revolving polygonal mirror and parabolic mirror it is made to scan the slab surface in the transverse direction with very sharp focus. This optical system is called the $f\theta$ mirror system and has such features that, in the transverse direction of the slab, (1) the laser beam can be focused into an almost uniform spot diameter, (2) the scanning speed is constant, (3) and the laser beam can be projected nearly vertically.

The beam diameter on the focal plane is given by the following equation.

$$d_O = \frac{4}{\pi} \frac{f \cdot \lambda}{D}$$

where
f : focal length of the parabolic mirror
D : incident beam diameter of the parabolic mirror
λ : wavelength of the laser light

Using the values of f, D, λ given in Table 3, theoretical value of d_O is approximately 0.1mm. Actual value of d_O is more than 0.1mm under the influence of aberration, misalignment of optics etc., but it is less than 0.2mm. Accordingly, the optical resolving power of LST is better than 0.2mm.

Four light sensors (photomultiplier tubes) are installed, upstream, downstream, right and left of the slab respectively, so that signals can be added to minimize signal level variations in the transverse direction.

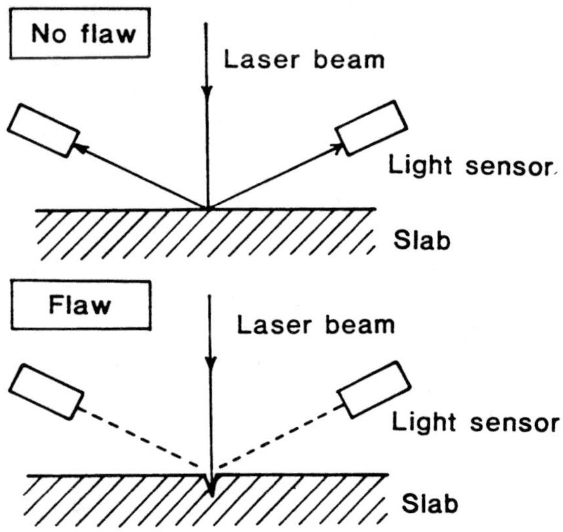

Figure 1. Basic Principle of LST

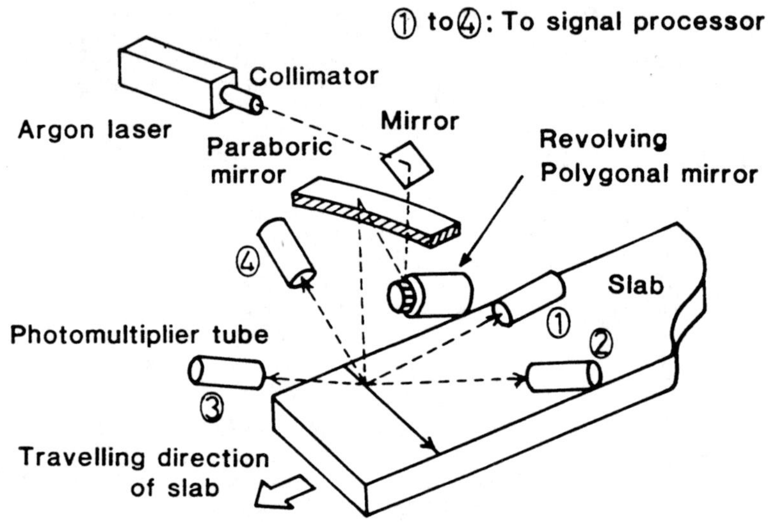

Figure 2. Basic configuration of the projecting and receiving section

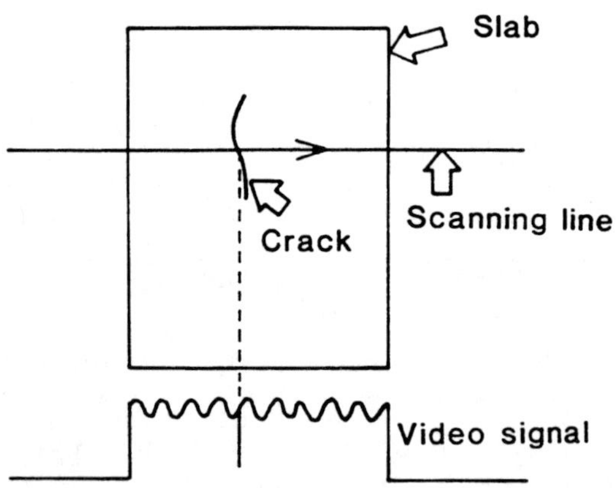

Figure 3. Waveform of received light signal

Table 3. Optical specifications of LST

Laser	
Type	Ar ion laser
Wavelength	0.4880 μm
Power	1.2W (Maximum) 0.3-0.9W (In operation)
Transverse mode	TEMoo
Polarization	Linear polarization
Polygonal mirror	
Incident beam diameter	About 10mm
Number of surfaces	12
Rotating speed	3,600rpm (Constant)
Parabolic mirror	
Focal length	1,700mm
Scanning rate	720Hz
Scanning width	About 1,200mm
Focused beam diameter	0.1-0.2mm
Light receiving section	
Photosensitive device	Photomultiplier tube
Optical filter	Interference filter

The important thing with the light reception is to remove the effect of the light spontaneously emitted from the slab surface and the external light as much as possible. It is important to set an enclosure around the projecting and receiving section, to hold the field of view of the light receiver to the minimum necessary extent, and to select the wavelength of the laser properly. The argon laser has intense oscillation lines at wavelengths of 0.5145 and 0.4880 μm, and the output at 0.5145 μm is about 20% higher than the other. According to Planck's radiation formula, the light spontaneously emitted from the slab is about three times as intense in the neighborhood of 0.5145 μm as in the neighborhood of 0.4880μm when the surface temperature is 1,100K (827°C). The wavelength of 0.4880μm is better suited to eliminate the adverse effect of the emitted light.

A set of light projecter and light receivers of LST covers the width of about 1,200mm. In the case of ITV method, for example, four or five high-resolution television cameras would be necessary and the illumination would be difficult if we wish to actualize the resolving power and the inspection width equivalent to those of an LST's optical head.

The received light signal for one scanning line is schematically shown in Figure 3. Table 3 gives the optical specifications of LST.

Signal Processing

Conventional optical flaw detectors process signals per scanning line or one-dimensionally. Two-dimensional processing has been recently introduced for the effective utilization of flaw pattern information, but it is not very widely used yet.

The two-dimensional signal processing technique of LST has made it possible to improve the S/N ratio and also to detect the flaws which are like a specific standard flaw pattern selectively. This process mainly consists of two circuits, one is a "real-time two-dimensional correlation circuit" which deals with gray level data, and the other is a "continuity check circuit" which for on-off data. Figure 4 is the signal processing block diagram of LST.

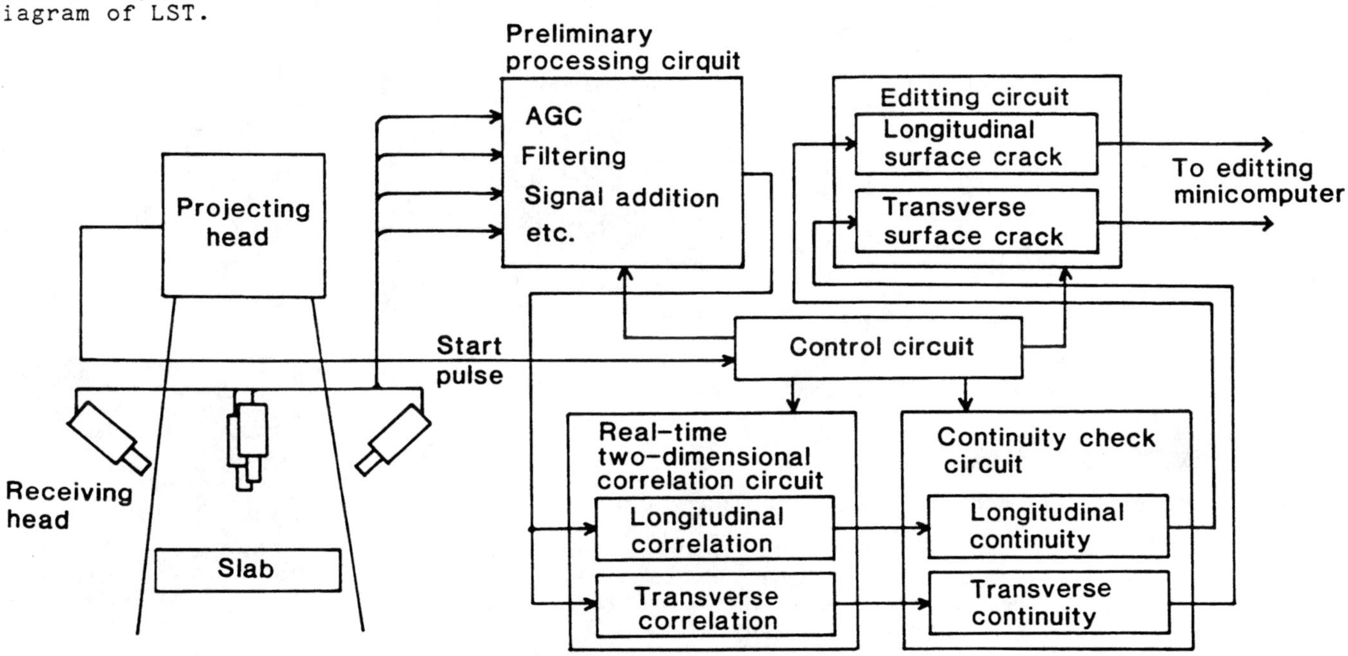

Figure 4. Signal processing block diagram

Real-time two-dimensional correlation circuit

Figure 5 shows the concept of the two-dimensional correlation and the basic configuration of the real-time two-dimensional correlation circuit. Video signals for the entire width of the slab are stored as the two-dimensional digital data for the multiple scanning lines. The contents of the memory are rewritten sequentially as the scanning process proceeds. At the same time, data on a certain small portion are converted from digital to analogue data and are delivered as a parallel analogue image. Correlation is performed (achieved by weighted addition) between the analogue image and the prepared standard pattern, and the results are turned into the binary form by the comparator.

This process is the same as the filtering operation in the image processing field. LST's circuit performs the operation in real time, because the processing is done only by the fast hardware. Figure 6 shows the typical effects of two-dimensional correlation. Using the apparatus whose configuration is as shown in Figure 7, an input image consisting of vertical, horizontal and oblique lines was taken with the television camera and calculated for correlations with the respective standard patterns and the results (before the binary conversion) were displayed on the monitor. With the standard pattern of the vertical line, for instance, the vertical lines in the input image clearly remain, the horizontal lines completely disappear and the oblique lines lightly remain.

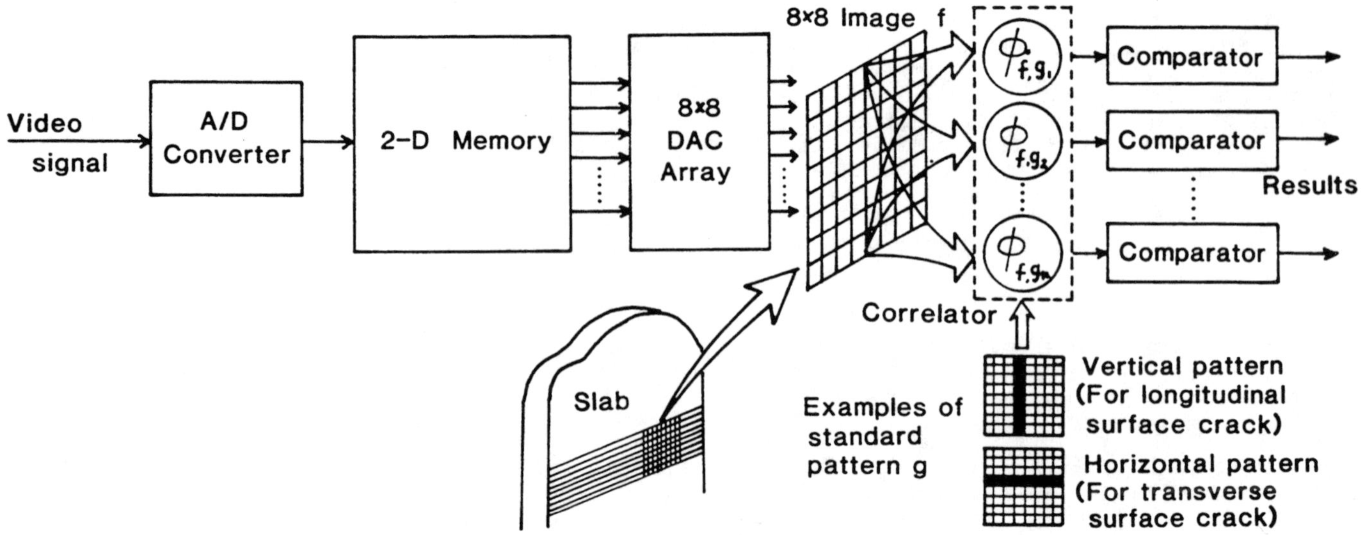

Figure 5. Concept of real-time two-dimensional correlation and basic configuration of circuit

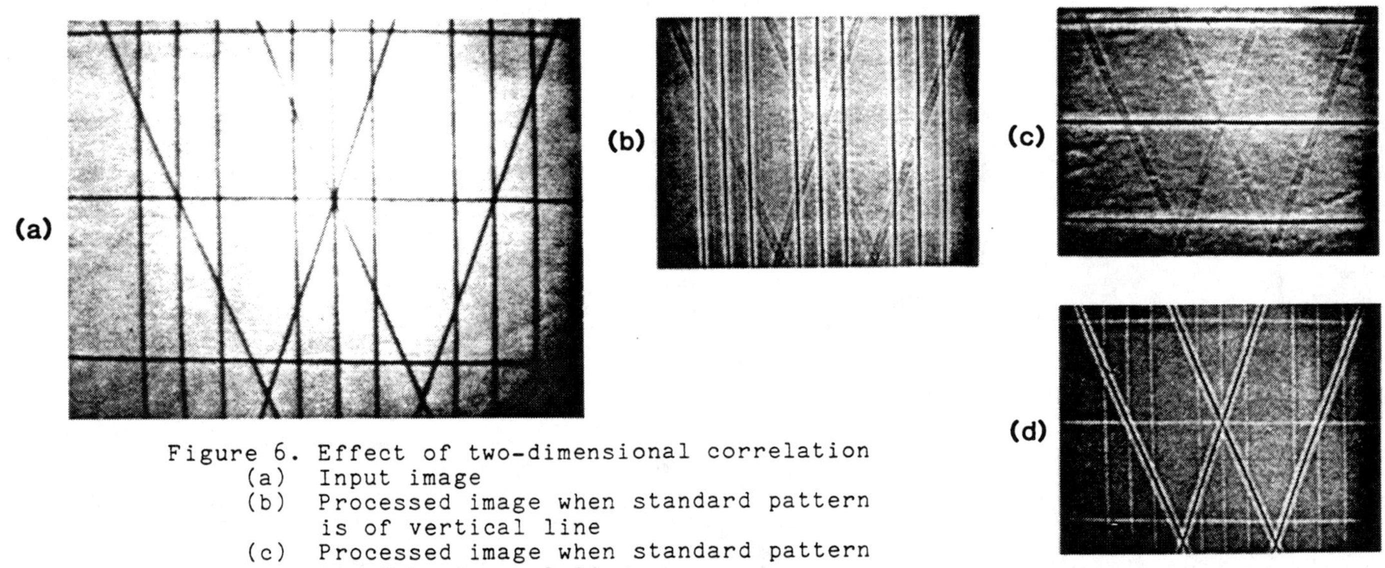

Figure 6. Effect of two-dimensional correlation
(a) Input image
(b) Processed image when standard pattern is of vertical line
(c) Processed image when standard pattern is of horizontal line
(d) Processed image when standard pattern is of oblique lines (X shaped)

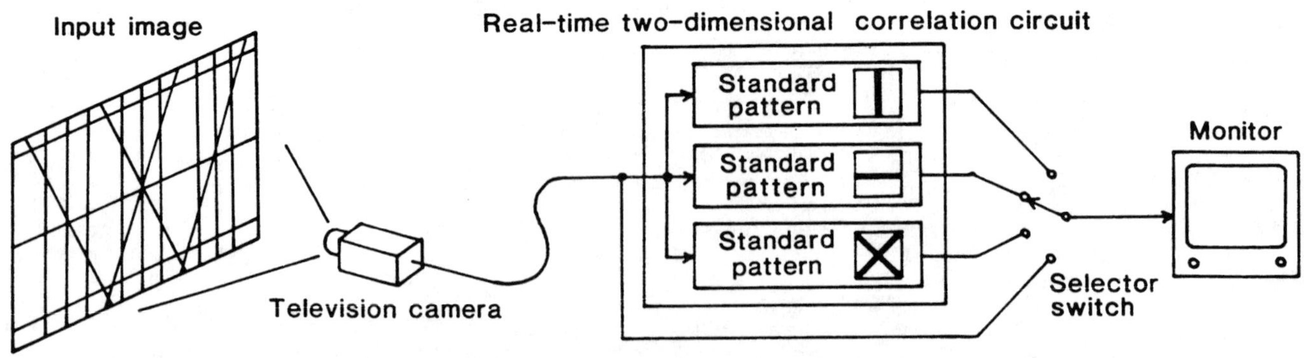

Figure 7. Configuration of apparatus for confirming effect of real-time two-dimensional correlation circuit

Configuration of the whole LST system

LST is installed and operating on No.1 and No.2 continuous slab casters at Yawata Works. The configuration of the whole LST system is shown in Figure 8 and its main specifications are given in Table 4. LST inspects the top and bottom surfaces of the slab. A set of light projecting and receiving heads can inspect the slab width of about 1,200mm. No.1 caster, which produces slabs with the maximum width of 2,100mm, has two sets of light projecting and receiving heads arranged in the transverse direction, while No.2 caster, which casts 950mm wide slabs, covers the entire width of the slab by one set of heads.

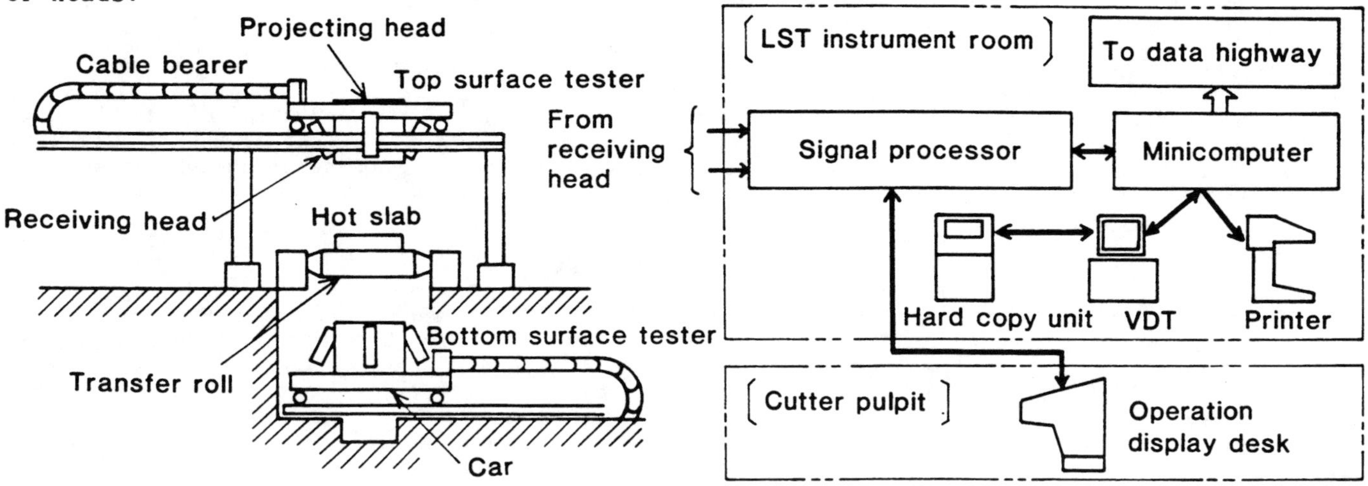

Figure 8. Configuration of LST

Table 4. Main specifications of LST

Line conditions	
Slab width	800 to 2,100mm (No.1 caster) 950mm (No.2 caster)
Slab thickness	250mm
Slab surface temperatur	900°C or less
Specifications of LST	
Number of heads	2 sets each for top and bottom surface (No.1 caster) 1 set each for top and bottom surface (No.2 caster)
Slab-head distance	About 800mm
Scanning line spacing	0.35mm (Depends on inspection speed)
Insensitive zone at edge	About 15mm
Output	Display for the operators and transmission to the host computer
Other functions	Detection of various abnormal conditions and faults

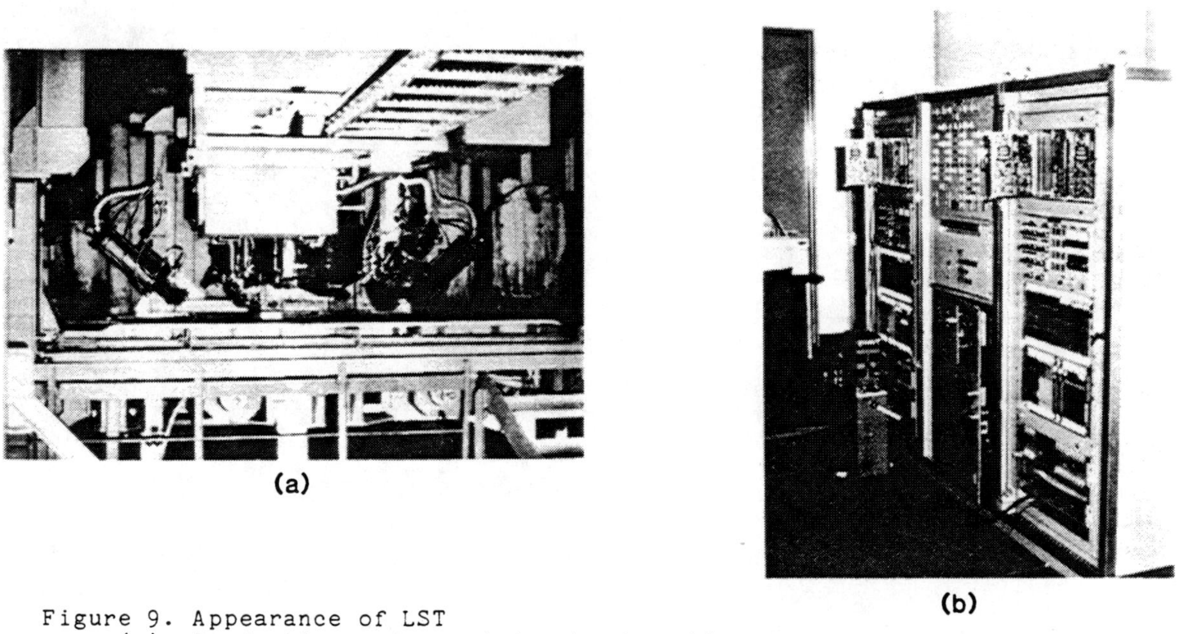

Figure 9. Appearance of LST
(a) Projecting and receiving head section
(b) Signal processor panel

Flaw present or absent signals, the output from the signal processor, are edited by minicomputer per type of flaw and slab surface, and sent as crack length (block length) information through the data highway to the host computer.

Figure 9 shows the appearance of the light projecting and receiving heads (for inspecting the bottom surface of No.2 caster) and the signal processor (for No.2 caster).

Hot slabs are covered with loose scale and impurities from cooling water, and as a result, the received light signal waveforms change to a considerable extent. This sharp change leads to incorrect detection in the extreme case, but this problem is difficult to solve by the signal processing technique alone. Some means is thus required for cleaning the slab surfaces. Yawata Works has rotating wire brushes or polishers installed as low-cost, compact units. The slab surfaces are effectively cleaned by these units just before they are inspected by LST.

Detecting Capability

Figure 10 shows the results of the test on the capability of LST to detect longitudinal surface cracks. Crack length is represented on the horizontal axis, open crack width is plotted on the vertical axis and whether or not the respective flaws were detected by LST is indicated. As is evident from the diagram, almost all longitudinal surface cracks open to 0.3mm or wider can be detected by LST. This means that all the flaws, over 0.4mm in opening width and over 100mm in length and considered as harmful flaws, can be detected by LST. The incorrect flaw judgment rate (percentage of sound slabs being wrongly judged as defective) is a few percent.

Transverse surface cracks rarely occur and not actually verified, but they are thought detectable by LST to the same degree as longitudinal surface cracks. When the flaw is oriented parallel to the laser scanning lines, the laser scanning lines may not cross the flaw and the detecting capability may lower. To avoid this possibility, the scanning lines are inclined about 20° to the transverse line. The scanning line spacing is 0.35mm. As the slab surface is scanned aslant in this way, the scanning lines can cross transverse cracks sufficient number of times.

The maximum surface temperature at which the hot slab can be inspected by LST is about 900°C. When the surface temperature rises above this level, the S/N ratio may drop under the influence of the spontaneously emitted light.

Signals suddenly change at slab edges and are often misjudged as flaw signals. These incorrect signals must be eliminated by the masking. The insensitive zone width created at the edge as a result is about 15mm. In Yawata Works, these zones are inspected by ECT[6].

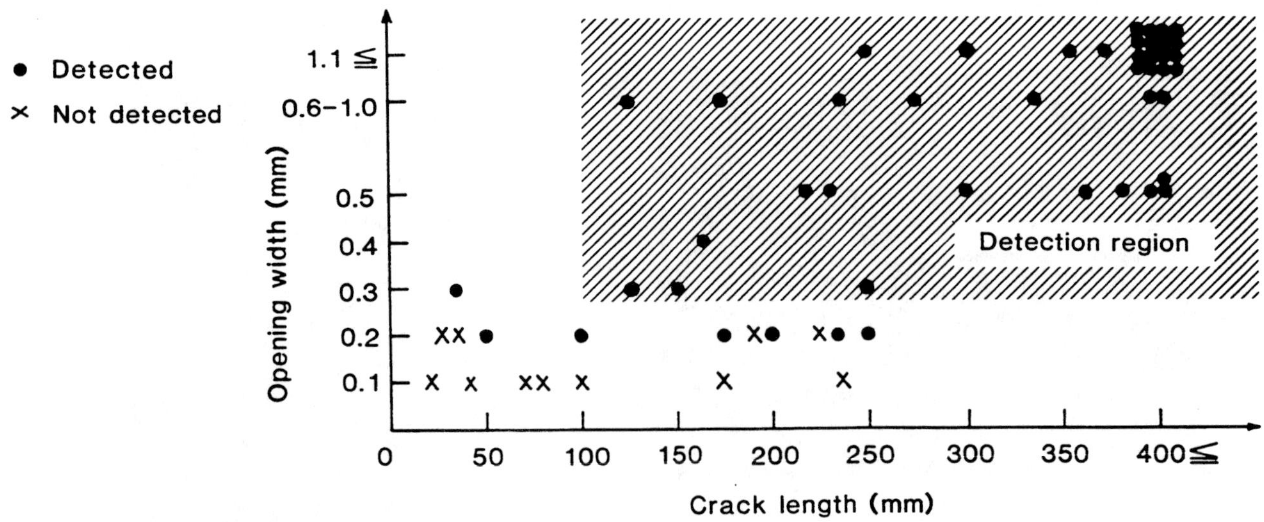

Figure 10. Results of test on detection of longitudinal surface cracks

Total System

Figure 11 shows the total slab flaw inspection and slab flow control system. LST is installed after the cutter and inspects the top and bottom surfaces of each slab. ECT is installed before the cutter and its inspection results are correlated with each slab by the continuous casting process computer. The results of inspection by LST and ECT are transmitted to the central computer which judges whether each slab is suited for HCR or not, determines the appropriate slab route and controls the finishing line accordingly.

Some pinholes and slag inclusions have no opening and cannot be detected by LST. They are to be taken care of by the operational control.

HCR is formerly performed mainly for cold-rolled sheets. Since its commercial startup (for No.1 caster in January 1983 and for No.2 caster in March 1983) the hot slab surface inspection system has expanded HCR to include tinplates and has markedly increased the HCR ratio from about 30% to about 70%, accomplishing large energy savings and processing step savings.

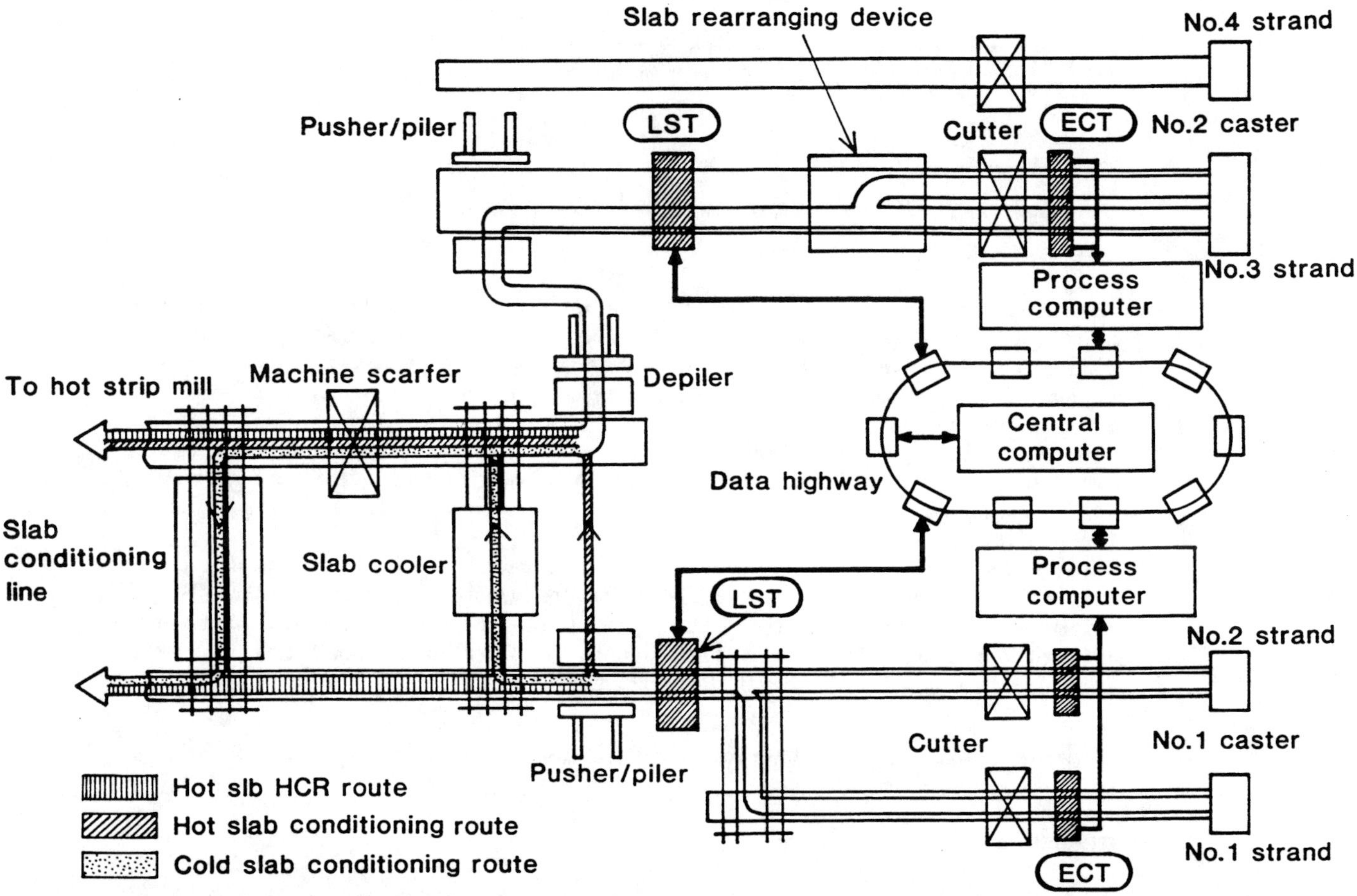

Figure 11. General diagram of slab flaw inspection and slab flow control system

Conclusions

LST, an optical flaw detector, has been developed. The hot charge rolling ratio of slabs has been remarkably increased by the effective utilization of it and ECT.

The signal processing technique incorporated in LST has the possibility of further advance in line with recent rapid progress of semiconductor technology and is expected to find use in various flaw detectors.

References

1 Nakakura, M. : Technical Exchange Session, Brussels, May 1983, IISI.
2 Iwasaki, M. et al. : TETSU-TO-HAGANE, 70(9), 1194(1984)
3 Uesugi,M. et al. : TETSU-TO-HAGANE, 70(9), 1188(1984)
4 Hiroshima, T. et al. : TETSU-TO-HAGANE, 70(9), 1202(1984)
5 Iwai, K. et al. : TETSU-TO-HAGANE, 70(9), 1181(1984)
6 Moritama, N. et al. : TETSU-TO-HAGANE, 70(5), S408(1984)
7 Hirakawa, N. et al. : TETSU-TO-HAGANE, 67(4), S137(1981)
8 Moritama, N. et al. : TETSU-TO-HAGANE, 70(5), S409(1984)

Reprinted with permission from Elsevier Science Publishers B.V., from *Industrial Metrology*, Vol. 1, pp. 155-164 (1990). ©1990 SPIE.

Laser scanner versus CCD camera: a comparison*

Harald J. Schmalfuß

Erwin Sick GmbH, GB Surface Inspection, 8000 München, F.R.G.

Abstract. A big challenge for automated industrial inspection systems is the hundred percent in-line surface inspection. Because of material speeds up to 10 m/s and material widths of several meters the amount of data received by the sensor systems is very large. In most cases the material flow is endless and therefore it is necessary that the complete inspection system works in real-time.

Nearly 25 years ago the first systems based on laser scanning technology were introduced in industry. At the moment more than one thousand laser scanning systems are working worldwide. During the last five years a competitive sensor system, the CCD camera, became commercially available.

In the present paper we will give an overview and, if necessary, a closer look of the different items of both technologies. Advantages and limitations of both systems are discussed. This discussion is not based on pure theory but mostly on real experiences. Therefore a couple of industrial applications are shown. In conclusion we will give an outlook on future developments in sensor technology and evaluation techniques.

Keywords. Surface inspection, laser scanning, CCD-camera, 2-D signal processing.

1. Introduction

During the last five years a new trend has rapidly increased: automatic in-process inspection. The reason for this trend is very clear because the production process itself has been automated to a very large extent, but the quality assurance is done mostly by human beings.

Especially the European industry is forced to stabilize their quality standards on a high level and to reduce their production costs at the same time. Therefore it is clear that product inspection has to be automated. This knowledge is not new. Exactly 25 years ago the first laser scanning systems were installed in industry.

* Reprinted with permission from the Proceedings of the International Congress on Optical Science and Engineering, Hamburg, F.R.G., September 1988, *Proc SPIE*, Vol. 1010 (Industrial Inspection), pp. 140–147.

But, as we will discuss later, a good sensor system is only one half of an inspection system.

During the last five years another rapidly increasing technique has entered the scene: electronic image processing systems. Thus we have the opportunity to combine high level sensors and fast electronic components to get vision systems which are able to satisfy the strong industrial boundary conditions for automated in-process quality control systems.

At the moment there exist two competitive sensors: the well-known CCD cameras (CCD) and the established laser scanners (LSC). In the present paper we will focus our attention to real-time surface inspection techniques and discuss the advantages and limitations of CCDs and LSCs with respect to these applications.

Due to the boundary conditions surface inspection is a very delicate application for automated inspection systems:

– *Web length*: up to "endless".
– *Web width*: up to several meters.
– *Web speed*: up to several hundred meters per second.
– *Defect size*: 1/4000 or less of web width.
– *Material*: steel or other types of metal; wood; rubber; woven or non-woven fabrics; paper; foils (different types); coated foils (e.g. light sensitive film); ceramics.
– *Surface structure*: with or without texture; reflective or diffuse; transparent, translucent or opaque.
– *Environmental conditions*: water spray; oily; dusty; vibrations; electronic and magnetic influences; stray light.

2.1. Sensor definition

In our sense a sensor is more than a piece of silicon which converts photons into electrons. In our sense a sensor consists of the following components:

– light source(s);
– illuminations optics;
– imaging optics;
– opto/electronic converters.

For our discussion we will introduce some parameters:

Harald J. Schmalfuß was born in Hof/Saale, Fed. Rep. Germany in 1949. He received his MS in Physics in 1975 and his PhD in 1979. From 1979 to 1986 he worked at the Battelle Institut, Frankfurt as Section Manager "Inspection Systems". Since 1986 he has been with Erwin Sick GmbH, first as R&D Manager at the Central Development Laboratory, and since 1989 as Manager "Electronic Systems".

Geometrical properties
–angle of incidence;
–illuminating aperture;
–angular distribution of reflectance: specular and diffuse;
–main angle of reflectance;
–receiver aperture;
–angle of observation; bright field (transmission and reflection); near dark field (transmission and reflection); far dark field (transmission and reflection).

Light source properties
–spectral light distribution;
–temporal and spatial coherence;
–polarisation;
–pulsed or continuous wave.

Illumination/imaging optics
–"Köhler" type;
–critical type;
–intermediate type or none of them.

Receiver properties
–detector size;
–speed;
–spectral sensitivity;
–signal/noise ratio;
–single element detector (photo multiplier tube, Si–PIN–diode, avalanche diode);
–position sensitive detector (PSD);
–array detector (CCD, CID, etc.), 1 D and 2 D.

When designing a CCD or LSC sensor you have to handle all these parameters in relation to the problem you have to solve.

2.2. The laser scanner

A laser scanner (Fig. 1) consists of two parts: an illumination part and an "imaging" part. The *illumination part* consists of a light source and "illumination" optics (beam shaping optics, beam deflection system, beam guiding optics). The *imaging part* consists of imaging optics and an O/E-converter.

2.2.1. Light sources

There are different types of lasers in practical use:
–HeNe-laser (5–15 mW cw power);
–Near IR semiconductor laser (up to 30 mW, 780–830 nm);
–Ar^+-ion laser (air cooled; 514, 488 nm);
–HeCd-laser;
–Nd:YAG (diode pumped, frequency-doubled: 532 nm).

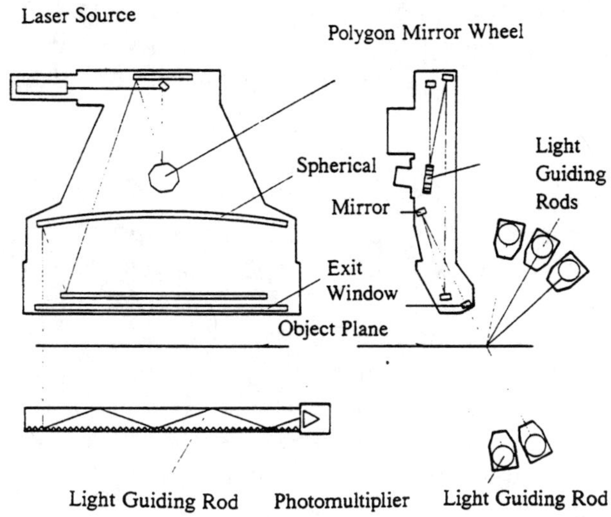

Fig. 1. Schematic outline of a laser scanner. Front view: scanner and receiver (light guiding rod). Side view: Arrangement of receivers.

The trend is towards gas lasers being more and more replaced by semiconductor or solid state lasers. At the moment the HeNe-laser is the first "victim". Very interesting candidates for the future are the Nd:YAG-laser and other frequency-doubled semiconductor lasers.

For industrial applications it is necessary that life times of more than 20000 hours are realistic. Semiconductor lasers will reach life times of more than 40000 hours (5 years continuous operation).

The Near-IR-lasers are somewhat critical because of the necessary safety properties.

2.2.2. Light deflection systems

Four types of deflection systems are in use:
- polygon mirror wheels;
- galvanometer mirrors;
- acousto-optic cells;
- holographic optical elements (HOEs).

The galvanometer mirrors offer the largest variety of components, but they are limited in different parameters: A large scan angle and/or a reasonable beam diameter causes slow scan rates (some few hundreds Hz). A high scan rate (up to 20 kHz) causes small mirror sizes and only sinusoidal motion functions. An advantage is that the center of deflection coincides with the mirror axis. This fact simplifies the optical corrections.

The acousto-optic cell is a very fast and flexible deflection system, but it is very limited in its scan angle (only a few degrees).

In theory the HOE's combines a couple of advantages. But at this time practical limitations reduce the variety of applications.

For industrial applications that require high scan rates, large scan angles and high resolution, the diamond machined polygon mirror wheel is a very powerful solution. The surface quality and the optical properties are excellent. The scan rate exceeds 6 kHz.

2.2.3. Beam guiding optics

The simplest and cheapest optical system is no optical system. This is widely used in so-called "V"-scanners, which have no optical system behind the deflection system. The disadvantages are:

–the "plane" of focus is part of a circle;
–the speed of the light spot is not constant;
–the angle of incidence is not constant.

A better solution from the optical point of view are the divergent f–θ scanners. They have a specially designed optical system which reduces the focusing error and the speed variation.

The best, but most expensive solution is the telecentric scanner which avoids the above mentioned disadvantages by using a combination of mirror and glass optics. The largest disadvantage is that you are forced to have at least one optical component (i.e. a spherical mirror) which is as large as the wanted length of the scan line.

In conclusion, a laser scanner is an illumination system which provides a lot of powerful system parameters:

–illumination aperture: zero;
–illumination angle: fixed;
–light intensity: variable with semiconductor lasers;
–high scan frequency: up to 12 kHz;
–high resolution: 1/4000–1/20 000 of scan length;
–scan length: 40–2000 mm telecentric, fixed.

2.2.4. Receivers

The interaction of light with the material surface can influence the intensity and/or the spatial distribution. By using a laser scanner you can use more than one receiver, as shown in Fig. 1.

A very convenient type of receiver for a telecentric scanner is the light guiding rod. Hereby the reflected or transmitted light from the scan line is imaged by a long cylindrical lens system onto the rear surface of a light rod. The diameter of this light rod is approximately 50 mm (2 inch). Then the light is deflected by diamond machined triangle grooves into the axis of the rod. By total internal reflection the light reaches the end of the rod. At this end the receiver is mounted. The efficiency of such a system is approximately 90%. This type of receiver is very compact, so it is very simple to combine two or three receivers close together.

The receiver aperture is approximately 15 degrees. If a better angular resolution is necessary, a mirror type receiver will be used. This receiver is nearly a copy of the illuminating system. The most important optical component is a large parabolic mirror which focuses the incoming light onto the receiver. The angular resolution is a few degrees.

For "V"-scanners in principle the same types of receivers can be used.

The advantage is that single element detectors are used, e.g. PMT's. In comparison with Si–PIN-diodes the PMT has a higher dynamic range (nearly 1000 times better), a better sensitivity and a larger photosensitive area. But it is rather expensive and, due to the large sensitive area, slow (15 MHz max.).

An avalanche diode has nearly the same sensitivity but only a very small sensitive area (1.5 mm radius max.).

The use of Si–PIN-diodes is only possible in bright-field applications. For dark-field applications the sensitivity is too low.

2.3. The CCD camera system

The *illumination part* of a CCD camera system (Fig. 2) consists of illumination source(s) and illumination optics.

The *imaging part* consists of imaging optics, receiver array, and pre-processing electronics.

In principle you can use the same optical systems as with the LSC's. You have only to replace the laser by a standard light source and the PMT by a camera. Table 1 gives an overview of the different light sources.

2.3.1. Light sources

The most severe problem in using CCD arrays is the illumination problem. The most popular light sources are:

–the fluorescent tube;

–the tungsten filament lamp.

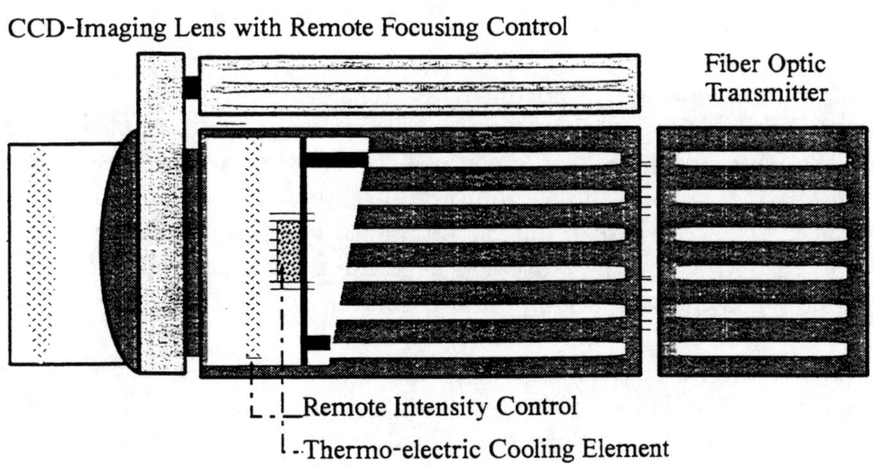

Fig. 2. Schematic outline of our CCD camera.

Table 1
Characteristics of light sources

Light source	Electrical power P(W)	Radiant flux Φ_e(W)	Efficiency (%)	Luminous flux Φ (lm)	Luminous efficiency η (lm/W)
Tungsten lamp	100	9	9	1 380	14
Halogen lamp	100	13	13	2 000	20
Fluorescent tube	40	9	23	3 200	80
Hg high pressure	400	52	13	21 000	52
Halogen metal vapor	385	77	20	28 000	73
Na high pressure	400	120	30	40 000	100

Source: H.-H. Marguerre, *Elektronik* **14** (1988) 87–97.

The advantages of a fluorescent tube are:
–simple high-frequency driving unit (35 kHz);
–long (2500 mm, 100″) homogeneous light band;
–cheap;
–simple use, easy maintenance.
The big draw back is the low light flux of only 2 stilb. Because the CCD array is essentially a silicon type detector with a very small sensitive area (13×13 μm or 5×7 μm), a high light intensity is needed to drive the sensor with a high line scan frequency.

In practice the fluorescent light tube is only usable in pure transmission applications, such as foil inspection or shadow imaging.

For the reflection case the tungsten filament lamp is often used. To avoid frequency mixing problems between the scan rate and the standard 50 or 60 Hz AC frequency you have to drive these lamps with DC. The efficiency of a tungsten lamp is only 3%. This means that 97% of the electrical power is converted into heat, and therefore a lot of electrical power, e.g. 4–8 kW, is needed to get enough light.

Another disadvantage is that the light intensity distribution is not very homogeneous. This is a problem when this device is used in a direct reflection case.

For the illumination of traffic scenes a very interesting light source exists: the sodium lamp, especially the sodium high-pressure lamp. The light efficiency and the total flux is very high. This gas discharge lamp is driven by a standard AC unit and therefore a fully modulated light intensity is obtained. For the application in conjunction with a CCD it is necessary to have a high-frequency driving unit.

Commercially no sufficient driving unit was available. Therefore we developed such a device. Now we are able to drive a sodium or another discharge lamp with 20–30 kHz. A combination of a cylindrical mirror and a cylindrical lens is used to image the bulb onto the object and to produce a very homogeneous light intensity distribution in the object plane.

In this area a lot of further R&D-efforts are necessary to improve the illumination situation.

If "Köhler-type" illuminations systems can be used, it is much easier to build up high-quality illumination systems, because use can be made of the complete know-how of the microscope research field, for instance.

2.3.2. The CCD array

The use of CCD-based cameras is very popular because no mechanical system is included. The cameras are very small in size, conventional photographic lenses can be used, they are widely insensitive against shock and vibration and the data sheets offer very astonishing technical data. However, be careful with these data sheets. There exist manufacturers which offer a read-out clock rate of 40 MHz, but in reality you have to multiplex 4 parallel signals each with 10 MHz. Normally they offer a clock rate of 20 MHz. If the arrays used are shorter than 2048 pixels this rate is correct, but by using 2048 or longer arrays the clock rate may drop down to 8–12 MHz due to sample-and-hold problems between the odd and even channel. The manufacturers offer a dynamic range of 1:5000 or more. However, due to boundary conditions, such as temperature and exposure time, you will be glad if you reach a dynamic range of 1:256.

For the specific use in surface inspection problems we have developed our own camera system. This camera incorporates the following features:

- remote focus control;
- remote intensity control;
- temperature stabilisation;
- fiber optic transmitter module;
- adaptable to different sensor families (Fairchild, Reticon, etc.).

3. Comparison

CCD or LSC? That's the question!—Well formulated, but wrong! If you have an inspection problem you need a partner who has enough mechanical, optical and electronical know-how to do a very careful investigation for you. At the end of this investigation the decision between LSC and CCD has to be a pragmatic decision. The best sensor type has to be combined with the best evaluation technique.

In my opinion it is not the time for a "religion" war between the two sensor types.

Figure 3 shows the different behaviour of a laser scanner and a CCD array. The laser scanner convolves the object with the light spot distribution. The CCD camera multiplies the object with a sampling comb convolved with a rectangular aperture.

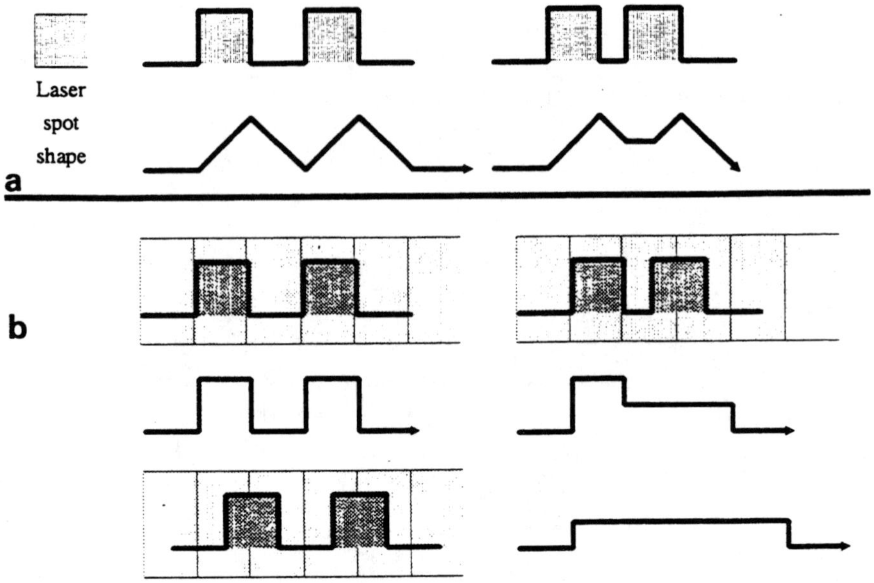

Fig. 3. Influence of the different "scanning" techniques. (a) Laser scanner: original signals and resulting signals. (b) CCD camera: influence of defect size and location with respect to size and location of the CCD pixels.

If you approach the sampling frequency you must take into account Shannon's sampling theorem.

In general the laser scanner offers a possible optical data rate of 80 MHz or more with 10—12 bit grey scale resolution.

This amount of data is far beyond all capabilities of existing receiver systems and image processing techniques. To make use of such a data rate it is necessary to have better receivers (e.g. large area or arrayed avalanche diodes) and much faster digital signal processors (ECL-type). Figure 4 shows the two principle ways to evaluate the data stream.

At the moment the LSC and the CCD are very comparable in spatial resolution and speed, because two or more CCD-cameras can be very easily combined to longer imaging systems.

The disadvantage of the CCD is that the illumination problems are not completely solved yet. It is necessary to improve the sensitivity of the CCD (avalanche arrays) and the power of the light sources. But I do not believe that CCDs will reach the power of LSCs.

On the other hand the use of a CCD is very much simpler than that of a LSC. Therefore for a large variety of applications the CCD will have the better cost/efficiency ratio. This is especially correct if CCD and telecentric LSC are compared. If very broad material is to be inspected, again the CCD is better, because it is much easier to combine CCDs than LSCs into longer systems.

Concluding this paper, Table 2 gives a short overview of the pro's and con's.

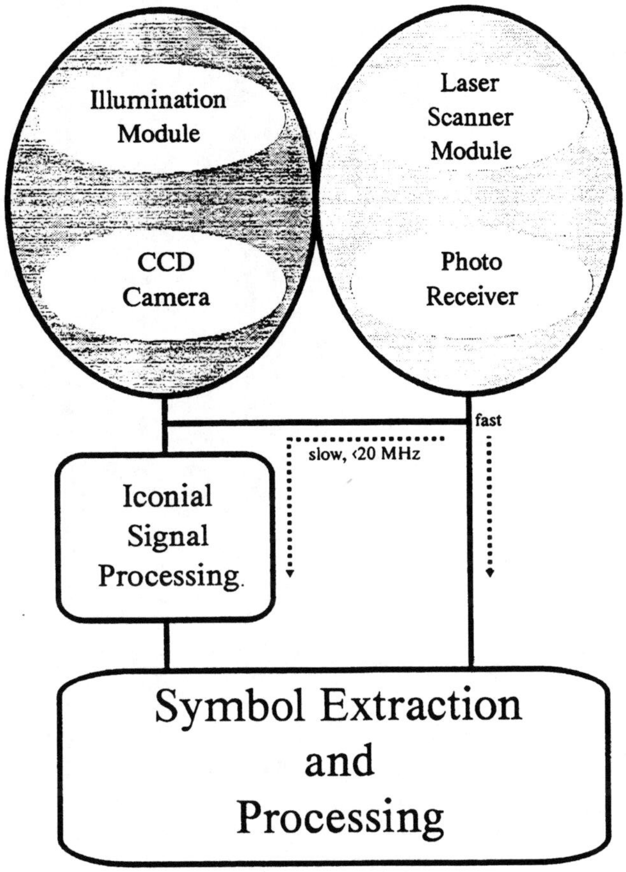

Fig. 4. Principal data paths.

Table 2
Comparison of laser scanner and CCD camera

	LSC	CCD
Spatial resolution	++	+
Scan frequency	++	+
Scan length	+	++ (multiple cameras)
Color detection	0	+
Interference problems	yes	no
Energy density (J/m^2)	low	high
Illumination aperture	0	variable
Receiver aperture	variable	variable
Receiver sensitivity	high	low
Mechanical complexity	high	low
Optical complexity	high	low–high
Maintenance	0	+

Section Two
Optical Gauging and Profiling

Reprinted from *Industrial and Civil Applications of Infrared Technology*,
W.D. Lawson, editor, Proc. SPIE Vol. 110, pp. 30-34 (1977).

USES OF THE INFRARED ABSORPTION GAUGE FOR ON-LINE GAUGING OF COATINGS AND FILM THICKNESS

Roger F. Edgar
Infrared Engineering Limited,
Maldon, Essex, Great Britain

Abstract

The operating principle and performance of the infrared absorption gauge is examined. Its application to both measurement of coatings and the thickness of plastic films is considered.

Introduction

Coated papers and plastic films are commonly manufactured in sheets or webs which may be up to 5 metres wide and which are produced at speeds as high as several metres per second. There is a recognised need for instrumentation to monitor the quality of these materials continuously during manufacture, since considerable quantities and significant costs accrue over the time taken to carry out a laboratory check of coating weight or film thickness. In any event, even when a competent quality control laboratory is available on plant, there may be serious risks imposed by the need to scale up a million times from the few grammes measured by the laboratory to the few tonnes produced by the plant. The requirement for on line measurement, which sould preferably be automatic, continuous and non destructive is quite clear and need not be further justified.

The on line instrument most commonly used hitherto has been the beta absorption gauge. A single transmitter and receiver combination is used for film thickness, whereas a dual, differential form, as shown diagramatically in Figure 1 is required when a coating weight is to be measured. In order to observe the

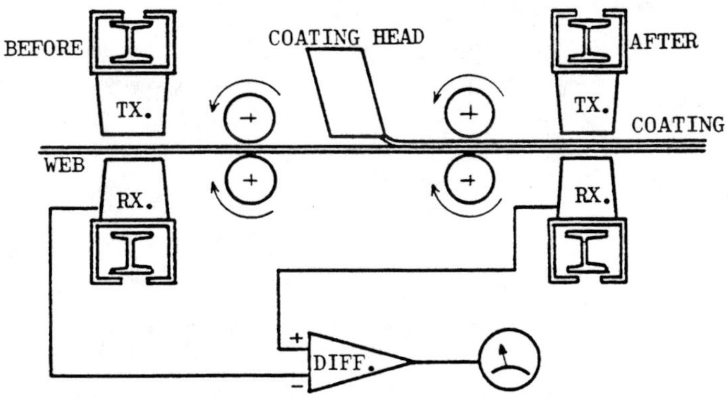

Figure 1. Beta Absorption Gauge.

whole width of the sheet, it is customary to use a scanning frame on which the sensing transmitter and receiver are mounted so that they may be traversed back and forth across the width of the sheet. In the case of the coating gauge, problems of synchronisation of the movement of the heads measuring before and after coating may arise. However, the major disadvantage of the differential beta gauge is that the coating weight is frequently a relatively small percentage of the weight of the coated and uncoated substrate so that fractional errors in these weights are much magnified when their difference is taken. The corresponding infrared absorption gauge, operating by backscatter on the coated side of the web is vastly simpler as a comparison of Figures 1 and 2 will confirm.

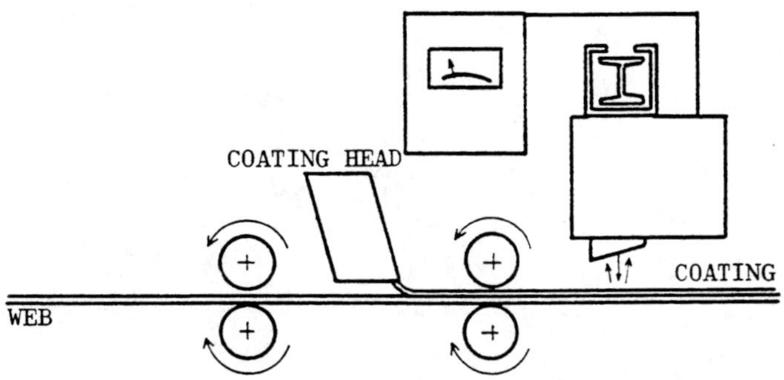

Figure 2. Infrared Absorption Gauge.

Coating Weight Gauges

Operating Principle

The infrared gauge used for coating weight measurement is essentially the same as that which is used for backscatter moisture measurement. Figure 3 shows the significant components of the sensing head of such a gauge, which operates as follows. Near infrared radiation from a source is focussed by suitable optical components onto the applied coating that is to be measured. (The infrared source is normally a compact filament lamp, underrun so as to extend its operating life.) The focussed beam of radiation passes through an optical filter system and falls onto the coating. The purpose of the filter system is to isolate from the broad spectrum of infrared radiation produced by the source, pulses of energy at two or more selected infrared wavelengths. Conventionally this is done by mounting optical interference filters into a wheel which is spun at a convenient speed in the range 1000 to 3000 rpm. Thus pulses of infrared energy at the chosen wavelengths are focussed onto the coated material, the area over which they fall defining the sampling area. These pulses interact with the coating, are partially absorbed and scattered by both the coating and the substrate and a part of the scattered radiation falls on the collecting optics whence it is fed to the infrared detector.

The transmission bands of the optical interference filters are selected with reference to the infrared spectrum of the coating material that is to be measured. One filter is chosen so as to pass radiation at a wavelength which the coating absorbs fairly strongly, whilst the other is at an adjacent so called reference wavelength at which optical absorption by the coating is much weaker.

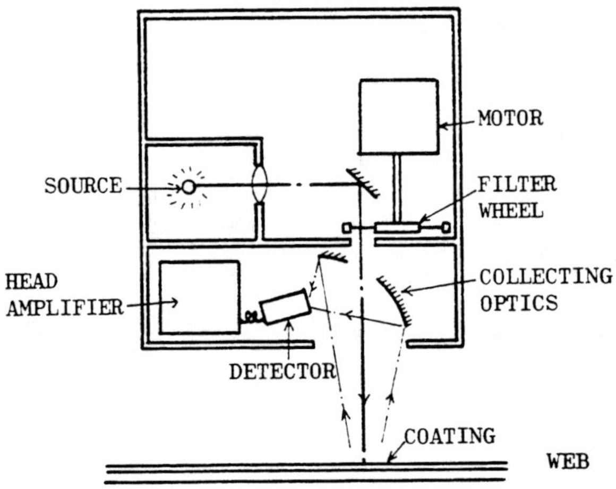

Figure 3. Backscatter Sensing Head.

Referring to Figure 4, which shows transmission spectra of thin layers of polythene and water, we see that if a polythene coating was to be gauged, an absorption wavelength of 2.35 microns, combined with a reference wavelength of 2.15 microns would be appropriate.

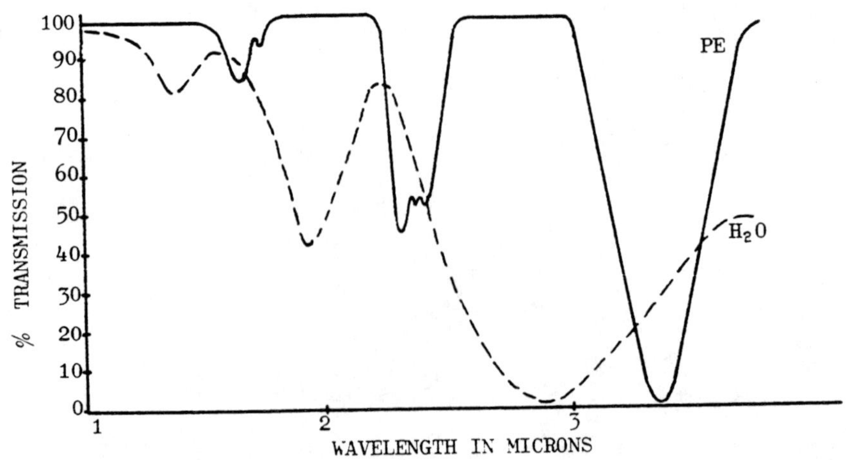

Figure 4. Transmission IR Spectra of Polythene and Water.

If the paper substrate were uncoated, the pulses received by the detector would be of similar amplitude for both absorption and reference wavelengths. If a layer of clear polythene is now deposited on the paper, the pulse at the absorption wavelength will be somewhat attenuated before it reaches the paper surface from which it is scattered and it will be further attenuated on its way out of the polythene, whereas the pulse at the reference wavelength will be substantially unaffected. The train of pulses from the detector is transmitted along with appropriate synchronising information to an electronic unit in which it is processed. By synchronous gating of the pulse train followed by rectification it is possible to derive two dc signals representing the backscattering efficiency of the coated surface at the two wavelengths. These signals are manipulated so as to yield an output proportional to the applied coating weight. For example, in the Infrared Engineering instrument, an algorithm is used which generates an output signal of the form

$$OP = a + b \log \frac{(S\lambda \text{ reference})}{(S\lambda \text{ absorption})} \qquad (1)$$

where $S\lambda$ reference and $S\lambda$ absorption are the received signals at the two wavelengths, and a and b are constants which an instrument engineer would recognise as zero shift (offset) and sensitivity (gain). (Implicit in this formula is the assumption that the interaction of the radiation with the coating can be described reasonably well by the Beer Lambert absorption law.)

The formula above demonstrates an important practical advantage of the technique: it is based on a ratio. If the source brightness changes, or the distance from the sensing head to the coated material changes, or the viewing window becomes dirty, to a good approximation $S\lambda$ reference and $S\lambda$ absorption will be equally affected and the measurement will not change.

Scope of Measurement

Having demonstrated the the infrared absorption technique is capable of measuring the weight of a range of coatings that are applied to paper and other substrates, we may now consider what types of coating can be accommodated by the technique, the ranges of coat weight and the accuracies that can be achieved. Coatings may be measured in either wet or dry state. The gauging of dry coatings is usually restricted by considerations of the infrared spectra of the coating materials, few inorganic materials exhibiting suitable absorption bands. Wet coating measurement is more flexible in that almost all organic solvent or water based coating solutions exhibit suitable infrared absorption features that can be utilised for gauging.

For those few cases where there is a choice of wet or dry gauging, it is instructive to compare the merits and demerits of the two methods. For materials with a satisfactory infrared spectrum, which includes most extruded plastics and rubber based or acrylic adhesives, for coat weights up to $100 g.m^{-2}$ or higher, dry coat gauging can offer accuracies of $\pm 0.5 g.m^{-2}$ or better. This measurement is on the final coated product after setting or drying. Dry coat weight has the merit of being unaffected by the concentration of the coating solution, but by the time the measurement is made the material will have passed through a drier, and the inevitable time lag may reduce the value of the measurement as an aid to coater control. Wet coating measurements on the other hand do not suffer from this time lag limitation. The main disadvantage is that the precision of the measurement is generally set and limited by the precision with which the solids ratio is maintained. This is because it is in effect the solvent that is gauged and the solids component inferred from it. There are of course ways in which solids ratio can be continuously monitored and a correction signal generated. However, this significantly complicates what is otherwise an elegant and cost effective monitoring technique.

Dry Coat Gauging

As indicated above, the infrared gauge is suitable for gauging of polythene and other extruded coatings and a wide range of adhesives. One restriction that should be noted is that the presence of a pigment or filler in the coating limits the weight range over which the measurement can be made, since the pigment prevents the infrared radiation from reaching the interface between coating and substrate from which the scattering takes place.

Solvent Coating Gauging

Many commonly used industrial solvents, such as toluene, methylethyl ketone, acetone, etcetera have suitable near infrared absorption bands, and appropriate filters may be fitted to an infrared absorption gauge so as to allow the gauging of, for example, silicone release coatings. Here, because the proportion release material to solvent is low, the accuracy that can be obtained referred to the release material is very high, $\pm 0.05 g.m^{-2}$ being readily achievable. The other main use of infrared gauges in solvent coating is to monitor the weight of adhesive coats. Here, accuracies of $\pm 0.5 g.m^{-2}$, referred to the final product may be achieved.

Because the solvents are highly volatile, the sensing head must be mounted very close to the coating head and is therefore in what is a hazardous area. It is normal to use an air purge protection system in this location, with an air pressure switch to give an automatic shutdown in the event of purge air failure and of course, a clearance delay on the restoration of air supply.

Aqueous Coating Gauging

The use of an infragauge in which the filter wavelengths have been selected to give maximum sensitivity to water is a useful way to monitor coatings of essentially inorganic material such as clays and mineral pigments. Such coatings are widely used to improve the surface properties of paper. In this case the range of the measurement is in principle limited by the 'hiding' of the substrate by the pigment particles as the coat weight is increased. However, on a range of trials with clay coatings with dry weights up to $50g.m^{-2}$, the loss of sensitivity that might be caused by this phenomenon has not been observed. This can be ascribed to two reasons. Firstly the clay particles are probably ground and sieved to a size distribution at which their light scattering power is most effective for visible light. At the longer infrared wavelenghts, they are much less efficient scatterers so that a significant part of the infrared radiation penetrates to the substrate. Secondly, the refractive index mismatch between clay and water, on which the scattering depends is far less for the wet coating than it is for the clay/air boundaries in the dry coating. The combined effect of these two features is that the weight range over which the method operates usefully exceeds the weight range over which coatings are normally made.

Film Thickness Gauges

Operating Principle

Figure 5 shows schematically the optical arrangement required for a gauge to measure the thickness of an organic film, such as nylon or polythene. Save for the fact that the detector and head amplifier are located in a separate receiver unit mounted on the opposite side of the web from the unit containing the source and filter wheel, the operation is equivalent to that of the backscatter gauge which we have already described in detail.

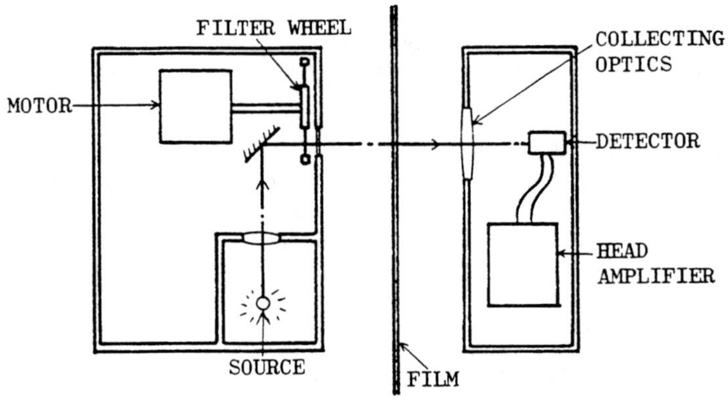

Figure 5. Transmission Infrared Gauge.

Referring to the polythene spectrum in Figure 4 we see that three absorption bands are evident, at wavelengths of about 1.7, 2.35 and 3.4 microns in order of increasing strength. A suitable choice of absorption band allows a very great range of film thickness to be measured. For example, at 3.4 microns, a measurement range of 0 to 15 microns thickness is possible with an accuracy better than ± 0.05 microns. At the other extreme, the band at 1.7 microns allows a range 0 to 20 millimetres, with an accuracy better than ± 10 microns. This ability to select an absorption band to match the thickness range of interest is a very useful feature of the infrared absorption technique.

Scope of Measurement

The infrared transmission gauge is effective on unpigmented or lightly pigmented films of polythene, polypropylene and similar plastic films and sheets. As we have noted, it is frequently possible to select the infrared filter combination so as to optimise the measurement performance over the range of particular interest. As well as basic choice of filter wavelengths, filter bandwidth may be used as a variable. For example, the range of measurement at a strong absorption band can be extended by increasing the bandwidth of the filters. (Filters with fractional bandwidths between 10% and 0.5% are fairly easily obtainable and regularly used.) A further reason for using a broad filter bandwidth is to minimise the effect of optical interference fringes which may occur if the surfaces of the film are optically flat.

An interesting specialised application is the use of a multiwavelength gauge to measure simultaneously the two component thicknesses in a co-extruded laminate of nylon and polythene. Here one wavelength is chosen at which both nylon and polythene exhibit a CH absorption and a second at which there is significant absorption by the NH band which occurs only in the nylon. In principle a common reference wavelength could be used, although in practise it is preferable to use two, one adjacent to each of the absorption wavelengths. Such a gauge provides an elegant solution to an extremely difficult measurement problem, for which it is very hard to imagin any possible alternative which is non destructive.

It is difficult to generalise about the effectiveness of the measurement when the material is pigmented, but a measurement is frequently possible, although the calibration will obviously be effected by the pigment, which modifies the optical properties of the material in the infrared as well as the visible. With thinner, lightly pigmented films, it is on occasions possible to derive a satisfactory thickness measurement using a backscatter gauge. This is particularly convenient for measurements on 'bubbles' of film extruded as a cylinder whose internal space is inaccessible.

Conclusions

The infrared absorption gauge affords an effective solution to a wide range of measurement problems encountered in industries such as paper, plastics, textiles where materials are produced and processed in sheet form. Because it derives a measurement based on a ratio, its setting up and use is relatively simple and the measurement affords a high degree of stability. This makes it suitable for use on plant as well as in the laboratory.

Reprinted with permission from *Iron and Steel Engineer,* Vol. 55(7), pp. 43-50 (July 1978).

Mill applications using self scanning diode arrays

David L. Burk, Director, Manufacturing Methods, and **Nicholas F. Simcic**, Senior Special Projects Engineer, Flat Rolled Products Div., Allegheny Ludlum Steel Corp., Brackenridge, Pa.

A number of applications of a self scanning diode array are well suited for steel mill service. Some of these applications are width and edge checks on strip annealing lines, hole and edge checks on a coil prep line, hole detection on a finishing line and width determination during hot rolling.

A RELATIVELY new sensor is described as a photosensitive diode array. The basic diode is a silicon photosensitive cell, therefore, its spectral sensitivity is in the red end of the visible spectrum and in the near infrared. The silicon diode is familiar in steel mill technology, as it is the basis for several commercially available automatic brightness pyrometers. The latter are commonly used as temperature measuring devices on hot strip mills. The most familiar use of silicon cells, of course, is in space hardware for electrical generation from solar radiation.

The device used in Allegheny Ludlum's applications consists of a large number of silicon cells side-by-side in a linear array. The manufacturer produces the array in miniaturized form through microcircuit technology. The individual cell in one of the units used is 1 x 6 mils and there are 512 cells in the array, so it is slightly more than ½ in. long.

The chip containing the array is mounted in the image plane of a lens so that the array cuts across the image of the object to be measured. Fig. 1 is a general schematic of an application set up. The situation depicted is that of an incandescent object, eg, hot strip. The enlarged view depicts its image on the cells of the array. Each cell with a high level of light falling on it generates a current which adjusts the charge on the capacitor associated with that cell. The capacitor is an integral part of the microcircuit chip containing the array. At a repeat interval, appropriate for the application, the array is electronically scanned from one end to the other to determine the charge on each capacitor. The charge level becomes represented in the camera output as a voltage level associated with each of the individual cells of the array. The number of highly illuminated diodes is a measure of the width of the object.

As seen by this description, the array's primary function is as a scale, ie, a dimension measurer. However, by providing some simple, selectable logic in the control unit, the usefulness of the sensor is greatly expanded.

Fig. 2 is a photograph of the sensing head with lens removed. The head of this particular unit is about the size of an 8-mm movie camera, ie, 2.8 x 4 x 6 in. The chip with its linear array is visible. The dark line is the array of 512 cells which, at this magnification, cannot be individually resolved. The chip is mounted on a circuit board in the head, which also contains other boards with scan timing controls as well as the other electronics. Fortunately, for mill applications one does not have to get involved in those details, aside from selection of the scan repeat rate.

The complete unit, as used in the mill for many of the applications (Fig. 3) shows the sensing head and control box. The latter has the power supplies, the logic selectors, a digital display and an analog output.

Fig. 1 — Linear array application.

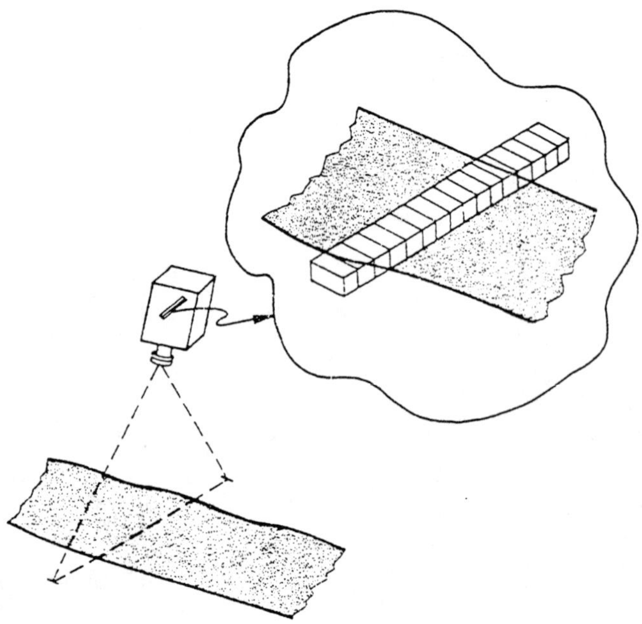

Fig. 2 — Sensing head without lens showing diode array microcircuit chip.

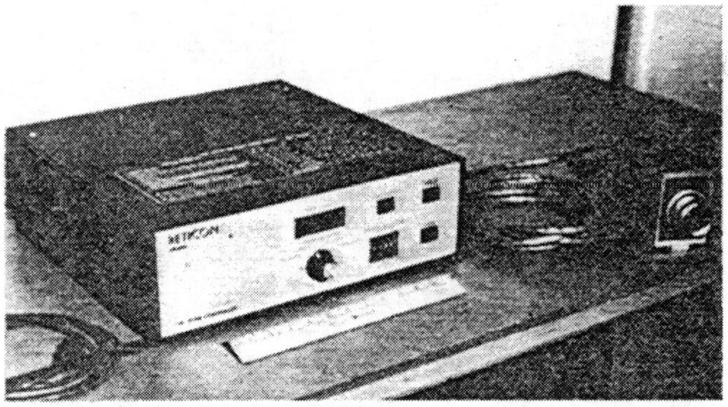

Fig. 3 — Control unit and camera.

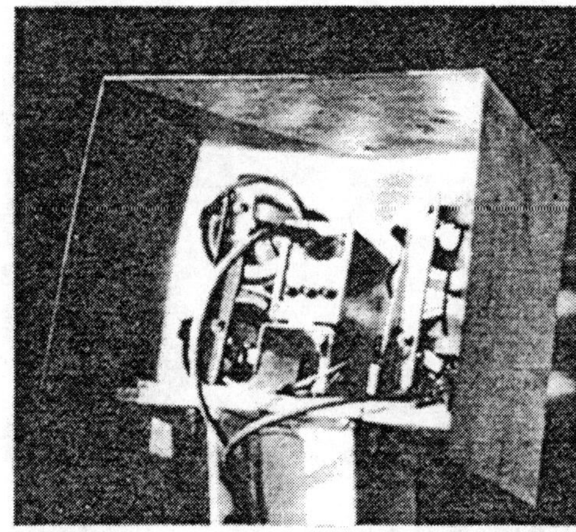

Fig. 5 — Rear view of the camera and high-intensity spotlights.

Mill applications

There is an interest in exploiting this new technology because it has provided a means to perform some difficult inspection and process performance measuring jobs in a reliable and relatively low cost manner. In fact, the original unit has been used as a prototype for a growing number of possible applications; thus, it has not really been used for the purpose originally envisioned.

Strip width measurement — The most straightforward use of the linear array is simply to continuously measure the width of strip as it moves over a processing line. Involved in the following examples was the result of a need for an inspection aid. The inspector was required to judge whether the hot band could be processed into the planned trim width. If not, he was to specify what the cut width should be. These judgments had to take into account edge checks as well as the minimum width of the coil. Even at a relatively slow line speed, under 100 fpm, this could not be satisfactorily handled by one inspector on this 2-strand line. A standard Reticon unit was installed on each strand with the width displayed on a recorder at the inspection station. Fig. 4 shows the camera installation. The control units are in the enclosure at the far right. The recorders are at the inspection station at the far end of the line. The strip is viewed by the camera just after it leaves the roll which carries it off the upper deck. The strip at this point on the processing line is not incandescent, so the strip has to be backlighted. The strip is still very hot, though (about 600 F) but a maintenance-free back-lighting arrangement has been achieved by placing high-intensity spotlights at the camera (Fig. 5) and putting water-cooled reflectors behind the strip edges (Fig. 6). Since the spots illuminating the left side of the strip are on the left side of the camera, the shadow is behind the strip and not seen by the measuring camera. The plates are given an extremely high reflectivity by covering them with self adhesive Scotchlite, the material used on highway signs. It has the unique property of having its highest reflectivity in the direction of incidence of the light source. So, in effect, the camera thinks it is looking directly into the spotlights, but they are mounted right at the camera where it is cool. The 12-v spots are operated at 11 v, and none of the eight has burned out in two years of operation. The only maintenance on this illumination system was during the winter when the tubes on the plates froze and split when the line was down. At the same time, the Scotchlite was replaced.

The spots also illuminate the strip, but because of the oxide, its reflectivity is so diffuse and low compared to the extremely bright plates beyond the strip that even at the large lens aperture used (f2.8) it is impossible to get a signal from the strip itself.

To the diode array, the field of view appears to be fully illuminated except where there is strip. Thus, for this direct width measuring setup the logic which is selected by control panel thumbwheels is to simply read out the number of dark diodes. Fig. 7 shows the control units. The digital display is the number of dark diodes. A D/A converter is used to provide a signal which is directly proportional to the width of the strip, ie, the number of dark diodes. The recorder sensitivity is set to show this as a trace on the chart which reads directly in inches of width. Fig. 8 shows the recorder rack at the inspection station which is about 500 ft from the camera installation. Each recorder has a second channel which indicates the strip thickness as seen by the x-ray gager mounted near the width measuring installation.

Fig. 9 shows a recorder trace. Coils No. 2 and 1 were individually rolled on the hot strip mill and built up into a double coil by butt welding at a coil prep line prior to this hot rolled band normalizer operation. The butt weld is indicated. The

Fig. 4 — Camera installation at width measurement station.

Fig. 6 — Water-cooled Scotchlite covered reflectors behind the strip.

Fig. 7 — Control units in industrial enclosure.

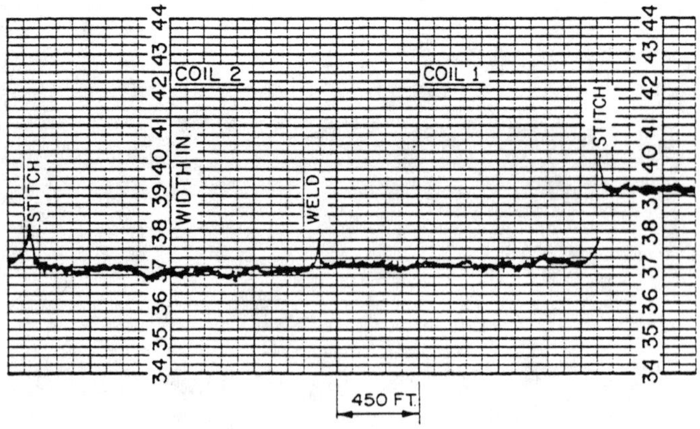

Fig. 9 — Example of a recorder trace.

stitches are the lap joints put in to get the double coil across this continuous normalizer. The overlapping area often is offset so that the short area is wider, giving spikes on the trace at the stitch. These are two good coils, ie, the widths are quite uniform within each coil, and they are both within a $1/4$ to $3/8$ in. of the ordered width. The coil at the extreme right was ordered 2 in. wider.

The width resolution did not need to be extremely high on this installation, ie, judgments as to making trim width or not are $1/4$ to $1/2$ in. judgments. A single camera installation using a 512-diode array was selected which, when used with a 25-mm lens, gives a 54-in. field of view and locates the camera 9 ft back from the strip. The unit resolution, ie, the width represented by one diode, becomes 0.105 in. (54/512). Thus, the maximum resolution for this particular setup can be stated as about $\pm 1/8$ in. Areas where one diode goes off and on can be seen on this trace. For example, just to the right of the weld the thinnest part of the trace is the width of the felt tip pen, and the next widest is the change of one diode which takes up about $1/8$ in., ie, one half of the smallest ruled division, which is $1/4$ in.

In using these traces the inspector does several things:

1) Puts the heat and coil identity directly on the recorder chart so that it can be used off-line for detailed studies.
2) Records two process performance numbers on the document which goes with the coil—these are the minimum width and the width which best represents the bulk of the coil. These numbers are put together weekly by the data analysis section as histograms which are reported as the process performance by the hot strip mill for the width of this product.
3) Indicates either OK for planned trim width or what the cutdown width should be.

Fig. 10 shows coils with edge checks 1, 2 and 4 in. deep. Edge checks show as a pen line with about half the amplitude of the actual depth of the check. Even at this relatively slow line speed the checks produce a narrow width reading of only very short duration. The diode length and scan rate, 1 mil and 5 ms, respectively, have been selected to give maximum resolution along the length of the strip to optimize the apparent check depth. The recorders are very fast response time units, about 0.3 s full scale, but even then the apparent pulse amplitude is down by a factor of about 2. As can be seen, this is still quite adequate for recording the presence of edge checks so that the inspector can evaluate the coil.

Fig. 8 — Recorder rack at inspection station.

Fig. 10 — Example of coils with 1, 2 and 4-in. deep edge checks.

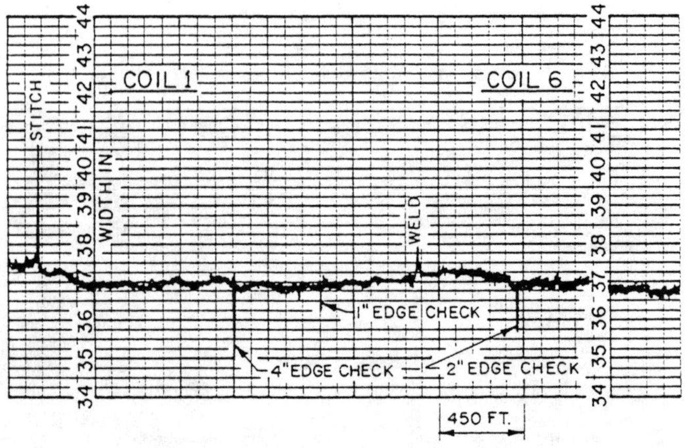

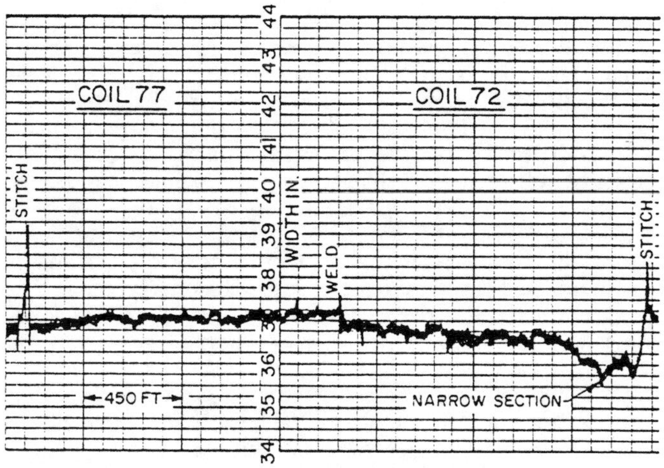

Fig. 11 — Example of a narrow coil situation.

Fig. 13 — An incandescent heating element mounted under the table provides back-lighting.

Fig. 11 shows a narrow coil situation which can sometimes occur. The inspector would note 35⅝ in. as the narrowest part of coil No. 72, probably 36¾ in. as the representative width, and call for a cutdown width 1 in. under the planned trim width. From these recorder representations of the width and edge check condition of the entire coil, the trim width decision is easily made by the time the coil is off the line, even for the 2-strand operation. In fact, because the data is being continuously recorded, the inspector can leave the station temporarily to perform other duties, ie, cut and identify samples, double check identity, or even go down the line to visually inspect when an edge check pulse has occurred. Five minutes elapse from the time the defect passes the camera until the strip is coiled.

Fig. 12 — Overall view of coil prep line inspection station.

Coil prep line—hole and edge check detection — The coil prep line has the usual conflict between wanting to run at high line speed (600 fpm) for productivity reasons, and needing to visually detect defects, such as holes, scabs and edge checks so that the line can be stopped and defects replaced by repair welds. The visual inspection is further complicated by the fact that the line is very compact, so there is no opportunity to see a long run of strip.

The linear diode array equipment has been installed on this line as an aid for the high-speed inspection job. The setup is designed to detect holes and edge checks. Visual inspection for scabs is still necessary, but most major scabs have holes associated with them, so this relatively low cost (less than $9000) simple, no-moving-parts optical transmission installation gets most of the job done without the complexity or cost of a full surface scanning inspection system.

Fig. 12 is an overall view of the installation at the coil prep line. The enclosure with the control box is seen behind the line on the left of the picture. The camera is mounted 5 ft above the table and uses a 16-mm lens. The shiny square unit in the middle is the readout device—a memoscope—which is located where the operator-inspector can see it while visually observing the strip.

Hole detection was a requirement of the system, so the strip had to be back-lighted across its entire width. This is accomplished by a 65-in. long x ¾-in. wide high-temperature heating element. Fig. 13 is a view of the table without any strip, with the incandescent heater visible. It operates on 3600 w and attains a surface temperature of about 1600 F. An air jet system is used by the operator to occasionally blow loose scale

Fig. 14 — Hole and edge check detector output display—memoscope.

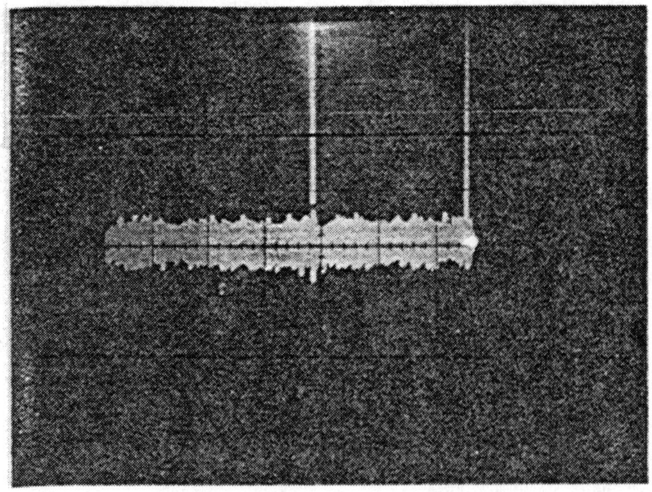

Fig. 15 — Memoscope display—hole and edge check detector indicating holes.

from the heater surface. The heating element is replaced about every three to four months.

The readout logic selected for this application is different than the all-dark-diodes mode described earlier. Here, it is advantageous to enhance the effect on the output when a hole exists. To do this, the logic thumb-wheels are set to give the number of diodes in the first dark. In the absence of a hole, the camera output is still proportional to the strip width, but a hole looks like an abrupt, large change in width. The other defect of interest, the edge check, is that already, ie, an abrupt change in width. To further enhance the change in output due to these defects, this apparent width signal is differentiated, so the final signal consists of strong pulses when the defects pass through the field of view of the camera.

The signal is displayed for the operator of the line on a memoscope (Fig. 14). It continually displays all signals until erased. The slowest horizontal sweep rate was decreased to give 1 cm/min, so the spot slowly traverses as the coil is being run. A noise level of about $\pm 1/2$ cm on the scope display is always present with spikes superimposed when a defect occurs. Fig. 15 shows the display when holes exist. The weld hole is drilled near the butt weld for weld detection at final gage. It is $1/4$ in. dia. Its presence in all butt welded coils provides a continuing built-in audit of the detection system, ie, the operator can readily tell if the system is alive. Fig. 16 is the scope display from a coil with an edge checked condition with checks up to about 2 in. deep. Fig. 17 shows the display from a test coil with simulated defects. It can be seen that checks produce both positive and negative pulses, and that their amplitude is proportional to the check depth. Holes give only large, positive pulses. In operation, an acceptance line can be placed on the screen so that any time a pulse occurs above this line the operation must be stopped and the defect examined. Originally, the intention was to automatically stop the line for such pulses, but with the memoscope the operator is able to keep running to assure that there is not another defect coming up, then reverse, return to the defect and make a useful repair weld.

The console from which the line is controlled has had pushbuttons added which provide for reset/erase of the memoscope, putting power on the strip heater and activating the air jets to clean the heater.

The system has been in use on this coil prep line where there is a tremendous amount of mechanical thumping and banging, varying voltages, etc., for 15 months with only one minor power supply circuit board failure. Replacement with a spare board resulted in negligible downtime.

Hot width — A major area of possible application for the measuring capability of the self scanning linear diode arrays is in hot working operations. Here, the visible radiation from the material can be imaged directly without auxiliary illumination. The systems engineering for a successful application is somewhat more complicated and is handled by the manufacturer through the use of a fast microprocessor which can be programmed to provide a broader range of logic than is available in the standard control box.

Allegheny Ludlum operates a Reticon width gage designed specifically for use in the hot mill. It is being used to view the width of the material as it exits the reversing rougher. It will also be tried at the exit of the bloomer to measure slab width and at the exit of the hot tandem mill to measure the hot rolled band width. The latter site has a width gage but the other two sites have not previously been instrumented.

Fig. 18 is an overall view of the installation. The rougher takes slabs of different widths, 18 to 50 in., and reduces their thickness from about $8^1/_2$ in. down to about 1 in. The hot tandem mill is to the left beyond the field of view of the picture. The single camera is placed on top of the mill, looking downstream, at the location where the side guides widen out. The camera looks the same as the others previously shown, but in a concession to the more hostile environment, it is mounted in a cylindrical, sealed, water-cooled enclosure. Fig.

Fig. 16 — Memoscope display—hole and edge check detector indicating edge checks.

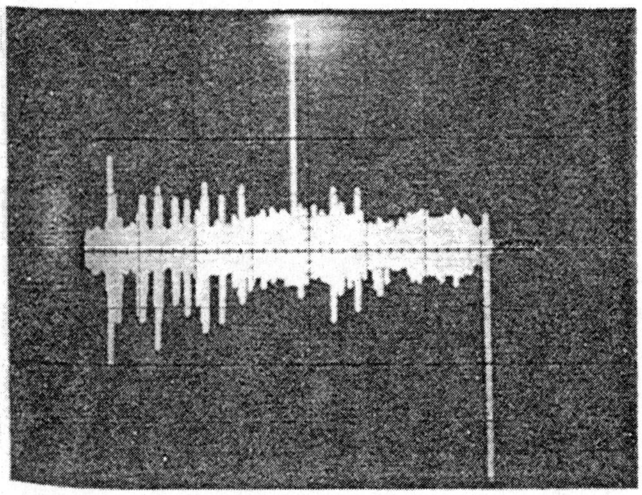

Fig. 17 — Memoscope display—hole and edge check detector. Test coil with drilled holes and simulated edge checks.

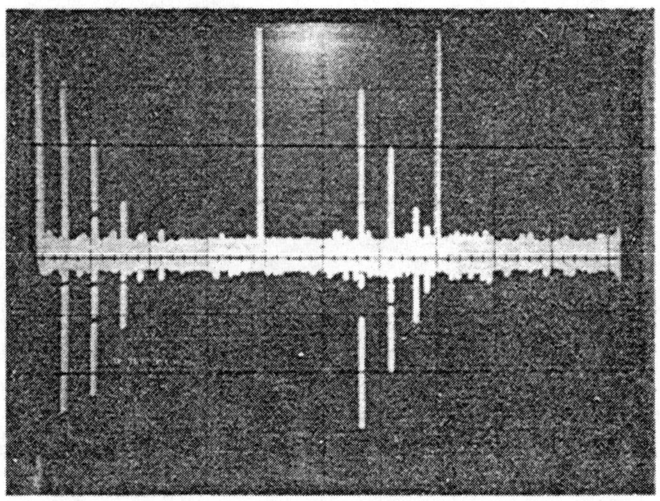

Fig. 18 — Overall view of hot width installation at the reversing rougher.

Fig. 19 — Camera enclosure is mounted on top of the mill.

19 is a closer view of the camera enclosure and mounting location. Because of the larger object distance (25 ft) a 135-mm lens is used giving a 60-in. field of view. The array chosen for this application contains 1024 diodes, each 1 x 6 mils long. The control box is placed in the motor room behind the mill. Fig. 20 shows the unit. It is packaged directly in an industrial enclosure. There is a digital display (the rectangle in the upper left of the picture) which can be used as a remote readout in the operator's pulpit. The update rate for this display is selectable by thumbwheels in the control cabinet. Also, the previous reading is held unless the width changes by more than 1/8 in. although internally the unit resolution is much higher.

A major complication in hot working applications is the wide range of brightness which generally must be accommodated. Fig. 21 shows a bar which has a surface temperature of about 1900 F. Fig. 22 shows a bar which has a surface temperature of over 2300 F.

In the previous applications, part of the initial setup was the selection of the lens aperture to bring the brightness of the light source into the range which could be handled by the diodes. Since the source was of constant brightness, this was a one-time adjustment. Usually in a hot working operation it will be necessary to measure materials at a given installation which are being worked at widely different temperatures. It is not practical to be manipulating the lens aperture according to expected temperature. Therefore, in a self scanning diode array, the agent used to do the automatic exposure control is the adjustment of the scan repeat time. During the time between scans, the charge on the diode's capacitor is being adjusted at a rate proportional to the level of illumination. So, to keep within the dynamic range of the diodes, this time is electronically adjusted inversely proportional to the brightness, ie, short times (high-frequency scanning) for the brightest objects. The exposure algorithm to be programmed will probably need to be tailored to the application. It will take into account the fact that the big brightness changes occur from one piece to the next with the brightness changing only slowly during the working of a piece. Allegheny Ludlum set-

Fig. 20 — Control unit is mounted in motor room.

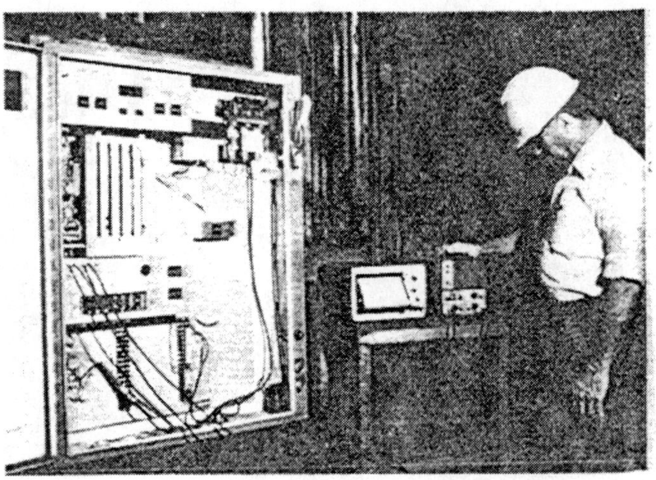

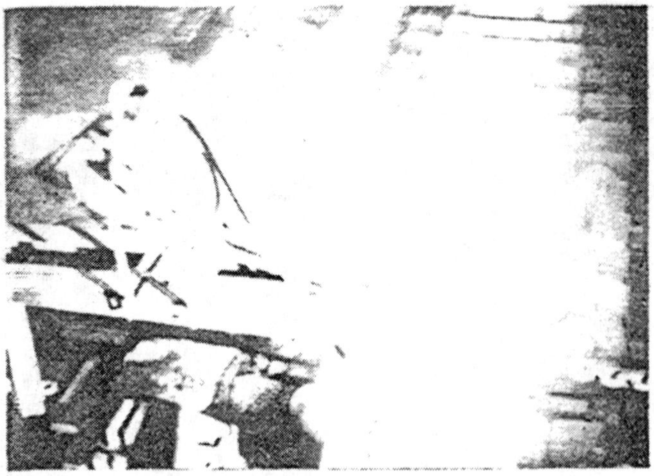

Fig. 21 — Low-temperature bar in camera's field of view.

tled on an algorithm like the following: determine that a hot object exists in the field of view (voltage present on any diode of the array), go from the fastest scan rate toward the slowest until the hottest diodes reach a reference level just below saturation and update infrequently (rate is thumbwheel selectable). This procedure provides the automatic exposure control, and the initial setup is accomplished in about five scans.

The other major complication in the hot working environment has to do with scabs, scale, water pools, smoke, etc, usually present in varying degrees which can interfere with the measurements. For example, a straightforward, all-bright-diodes logic would give a very noisy representation of the width in the presence of this condition. At the very least, an electronic peak-picker would be needed. To minimize these interferences, a width algorithm is used which on each scan determines the first diode above a fixed (but manually selectable at setup) threshold voltage level and the last diode above that level. The numerical difference is taken as the number of diodes representing the width. Thus, it is a form of edge detector and ignores the conditions on the surface of the hot object.

The output of the unit can be further engineered to provide process controls of various types, or it can be used directly as a hot caliper to indicate width. The latter use is illustrated in Fig. 23. The last pass is the final bar width being sent on to the hot tandem mill. It is apparent that the bar width is being indicated to 1/8 in., which is adequate since direct data at this

Fig. 22 — High-temperature bar in camera's field of view.

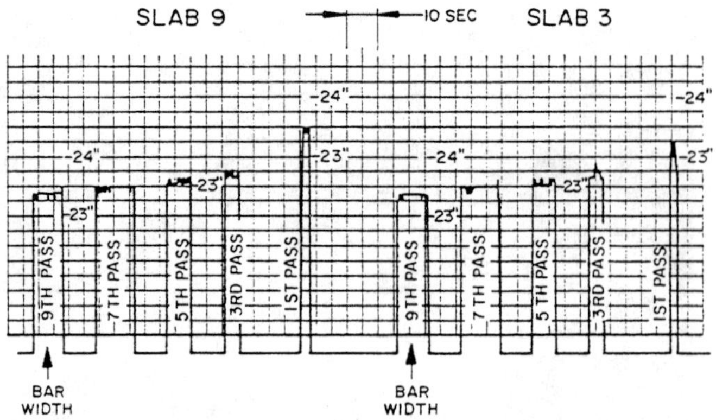

Fig. 23 — Recorder trace—hot width at exit reversing rougher.

site was not available before. The uncertainty is determined more by the foreign material than the instrument.

The first pass trace shown on the figures gives some indication of the speed with which the algorithms are carried out. Because the slab is still only about 19 ft long, it barely reaches to the camera's field of view, so it is only available for 1 to 3 s. Yet, even that short trace shows more than one readout update which has the slowest repeat rate of all, being set for about 1/2 s. Examination of the camera output with an oscilloscope shows that for the highest temperature material, the rate at which the linear array is scanned is about 1 scan every 10 ms, while for the other material which is about 400 F cooler, the array is scanned once very 30 ms. Each of the scans puts out a width determination so it can be seen that the bar is being examined rather closely.

Although the primary interest is in the bar width, spreading data during the roughing operations is available if a width correction is made for the progressive change in thickness as the material goes from slab to bar. The correction is necessary because the surface-to-camera distance is less for the slab than it is for the bar. This correction is shown on these figures as the changing position of the superimposed width scale as the pass schedule is carried out. It can be seen that the bars of Fig. 23 spread only about 1/4 in. from the first to the last pass.

Other applications

There are many other applications where this technology can be utilized. For example, Fig. 24 shows a utility unit—a standard camera and control with a high-intensity fluorescent light used as a back light. This unit has been used on several processing lines for width recording of special lots of material, for hole detecting and alarming, and for prototype investigations.

Fig. 24 — Utility unit with high-intensity fluorescent light. Used at various locations for width measurement and/or hole detection.

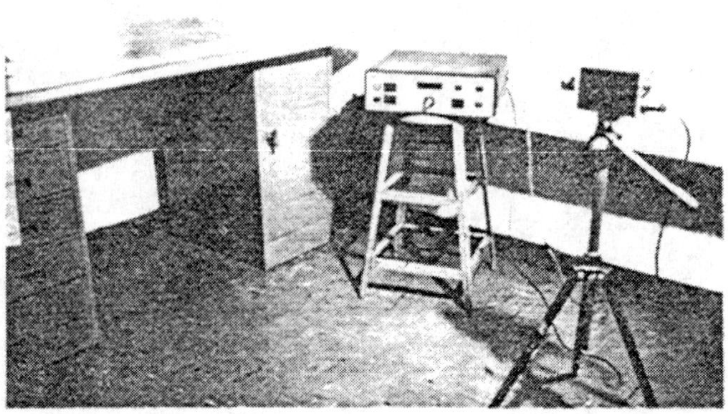

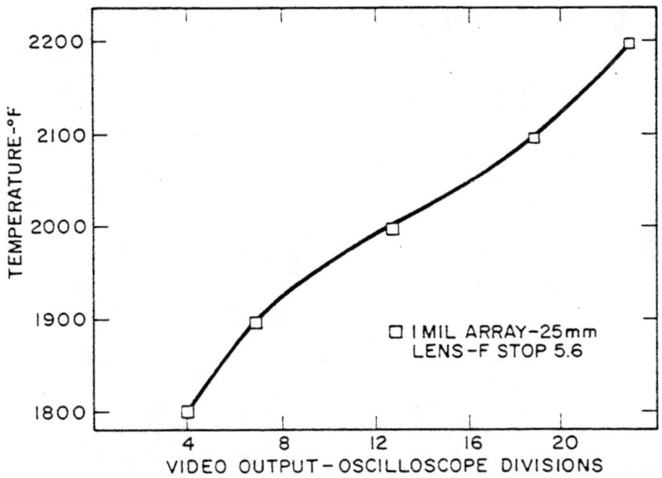

Fig. 25 — Calibration curve indicating possible application of array as a small-target brightness pyrometer.

One interesting use is to go back to thinking of the diode as a brightness pyrometer. The standard unit is appropriate for such an application because the scan rate will be constant. Fig. 25 shows the output of diodes when looking at the tungsten ribbon of a pyrometer calibration lamp. The temperature senstivity is reasonable and not far from linear. It looks like a rather useful curve. If one thinks of the array as a line of very small target pyrometers, several applications involving temperature distributions over small areas can be imagined.

Summary

The major parameters to be considered in any application of this diode array technology are 1) what measurement resolution is desired, ie, how many diodes; 2) how long does the diode have to be to give a usable signal in the bright area; 3) this length, as projected on the field of view, coupled with the speed of the object, sets a practical upper limit on the dynamic resolution, eg, a change in the parameter like width cannot be perceived any more quickly than the time it takes a point on the object to travel the length of the diode. This sets an upper limit for the particular application on the scanning repeat rate which will be useful.

Major benefits of the linear diode array technology are: 1) a wide range of measurements are available within the engineering constraints; 2) in many applications this new technology provides a price breakthrough compared to more conventional systems and 3) the simplification due to the all-electronic mode of operation produces the high reliability which is so important in mill applications.

Reprinted from *Solving Quality Control and Reliability Problems with Optics,* J.J. Amodei, H.N. Lowell, eds., Proc. SPIE 60, pp. 52-56 (1975).

APPLIED OPTICS IN THE EUROPEAN COAL AND STEEL COMMUNITY - ITS APPLICATION TO QUALITY CONTROL

R.V. Williams
British Steel Corporation, Corporate Engineering Laboratory
140 Battersea Park Road, London SW11, England

Introduction

Under the treaty of Paris the European Economic Community supports research and development related to the manufacture and processing of steel products. Optical methods have played an honourable part in this programme. They lend themselves especially to quality control during the manufacture of semi finished products and to inspection of finished material. Parallel with co-operative work sponsored by the Community, Steel Companies in the member countries of the EEC have carried out optical developments related to their own needs and instrument firms have developed equipment in response to meet market forces. This paper discusses optical developments in the European Steel Industry originating in each of these areas.

It is useful to consider instruments by application area within a works rather than by technique; while there is a concentration on steelworks applications, most of the instruments have quality control uses in other industries. Following the production path through a typical works, the first application we meet is to check dimension accuracy in a heavy mill.

Slab Thickness

Before entering a hot rolling mill, steel is reduced from ingot dimensions to slabs, which are the order of 100-300 mm thick. Mill Control schemes under consideration at present call for a monitor of slab thickness at several positions across the slab width. The Centre de Recherches Metallurgiques of Liege have developed a laser based gauge suitable for this purpose.[1] Figure 1 illustrates its operation. Gas lasers are mounted

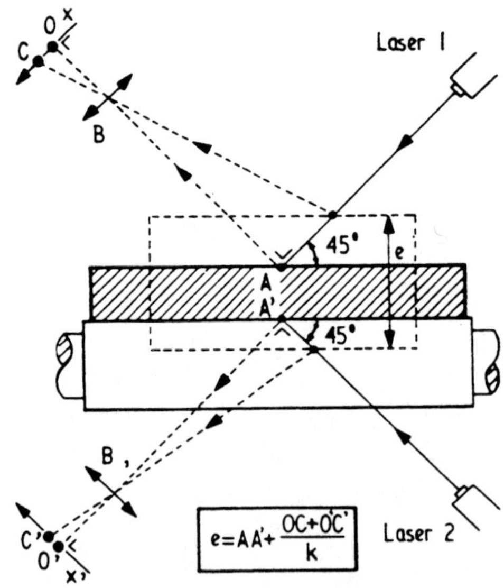

FIG. 1 MEASURING SLAB THICKNESS BY TRIANGULATION

a. Optical diagram

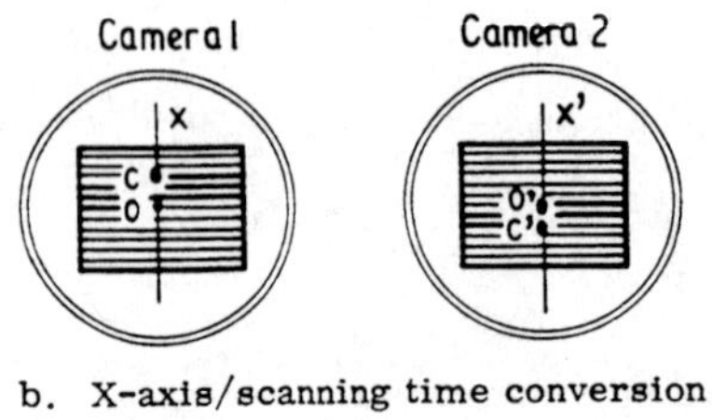

b. X-axis/scanning time conversion

above and below the material (which can be at temperatures of up to 1100 °C) so that beams fall onto the slab at 45° to its surface. Television cameras are mounted to view the points of contact between the beam with the slab surfaces. It can be easily seen from the diagram that the thickness variation is translated by optics into a line location in a television raster. Many tests in the works situation showed that systems accuracy is of the order of ± 2.5 mm. Errors are due to the sensitivity of the apparatus to surface defects, which are severe in slabs. For products of higher surface quality the gauge accuracy is better than ± 1 mm. This instrument has worked well in mill applications although higher precision would be desirable to meet all computer control requirements.

Velocity Measurement

There are many places in a steel plant where the velocity of stock, (hot or cold) has to be measured. A common example is the cut-up line, where billets are cut to length to satisfy customers requirements. A billet is fed into the hot saw until the required length can be cut off. If the billet velocity is measured during positioning, then simple integration gives the movement with respect to a hot metal detector. In this way a combination of an IR detector and a velocity sensor can be used to cut accurately to length.

To meet these needs workers in both France and Germany have separately developed velocity measuring devices under EEC funding. The German equipment(2) is more advanced and is now made commercially. In Dusseldorf, they have developed a laser system in which a coherent beam is reflected off the moving material back to a detector, as shown in Figure 2. The beam is reflected off the object in a diffused form and interference

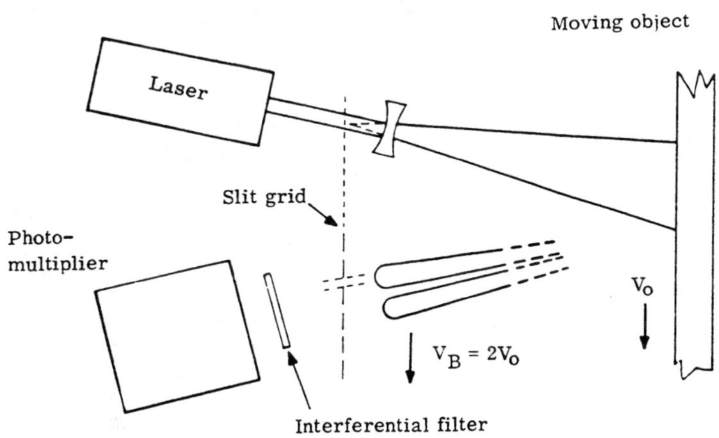

Fig.2 Measuring Rolling Speed by Lasers

occurs in the equipment producing a speckle-like distribution. When the object is moving, this pattern moves past the optical lattice and activates a photo-multiplier. The maximum amplitude of the frequency spectrum of the signal is uniquely determined by the stock velocity. Any variation in the speed of the object brings this frequency to a new level. Over a dynamic range of 0.5-60 m/sec, accuracy is better than ± ½%. This method is applicable to situations where the stock moves along a well defined path without too much flutter. If flutter is excessive the accuracy is considerably reduced. The French development(3) although not quite as well developed, overcomes this difficulty in an elegant manner. A simple pair of lenses are placed about 0.5 m apart, about ½ m above the moving material. Photo detectors pick up fluctuations in radiation from the hot stock. A cross-correlator conditions the fluctuating photo diode signals into a velocity vector. This equipment has not been as rigorously tested as that in Germany but has the great advantage of being insensitive to pass line variations. Figure 3 illustrates this device, which is capable of ± ½% accuracy over the range 0.5 to 60 m/s.

Measurements on Round Bars and on Wire

It is useful to mention, for the record, instruments now well established which measure the diameter round material, hot or cold. These have been developed primarily to measure material in the diameter range 5 mm to about 100 mm. Out of a large number of proposals and prototype instruments, two main contenders emerged, and are being marketed. One is based on the use of a laser; and other couples a sophisticated lens arrangement to a conventional light source. Both these instruments perform well in works situations.

Hot Rolled Steel Strip Velocity Measurement

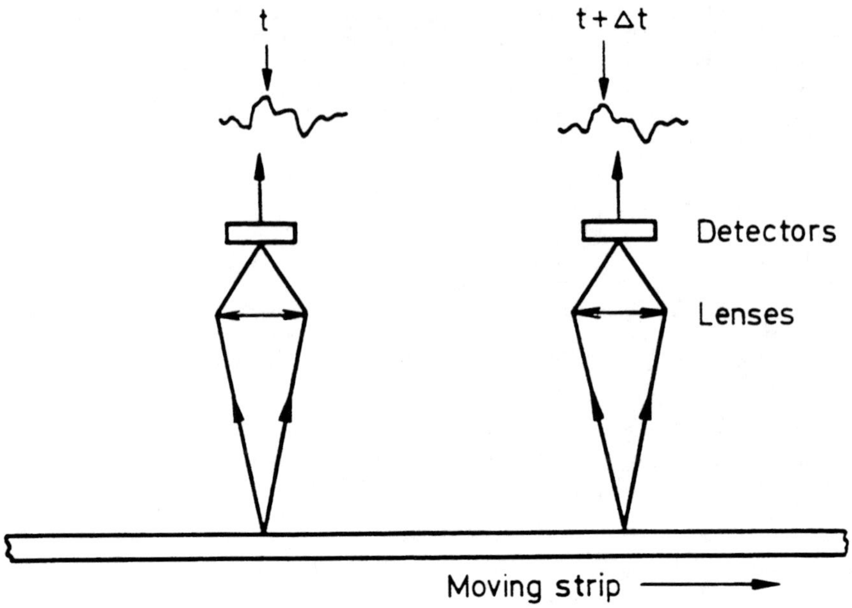

The laser instrument is manufactured by the Davy United Instrument Company and that based on the conventional light source by the Decca Group. The basic concept is similar in both cases, in that a beam is scanned across the stock and the lateral extent of the shadow is transferred into the time domain by a rotating mirror. The laser gauge is designed especially for large diameter (50-100 mm) material. In the Decca gauge parallel light casts a shadow in the plane of a narrow slit. A rotating mirror transforms this image into a time dependent signal picked up by a photo tube. Measurement of the time taken for the shadow to pass across the slit coupled with the rotational speed of the rotating mirror gives a measure of the bar diameter. Great precautions have to be taken to ensure that the optical parts of this type of equipment are impervious to the environment, especially to water splashes and steam. These problems have been successfully overcome in both cases and both have now given at least two years of satisfactory service application.

A more difficult problem is the measurement of the diameter of wire. In Germany an approach has been made to this problem by rotating a lens assembly, right around the wire path-line. In this way, more than one dimension can be measured and it is claimed that a complete diameter profile is built up. This gauge is still at an early stage of application, and it remains to be seen how reliable it will prove to be in practice. The laser has made the measurement of a cross-sectional profile of round material, (or any diameter), possible. If this is carried out with sufficient precision then surface defects are also detected in a way not possible by electro-magnetic methods. For example, eddy current devices will not readily pick up surface defects parallel to the direction of travel of the stock, whereas a precise measurement of profile will reveal all surface defects; within the range of resolution of the equipment. The BFI in Germany are developing under ECSC funding an instrument for this purpose. Still at an early stage, a series of lines are developed in an interferometer and are projected onto the stock. Viewing at right angles to the projection direction, gives a measure of deviation from a circular cross-section. This interesting development is still at the laboratory stage, but progress will be closely followed, as soon as it enters works service.

Surface Inspection

The Steel industry is becoming increasingly aware of the need to improve methods of inspection of flat material. It is important, for example, that cold reduced steel and tin plate be sold free from surface blemishes, which would be revealed as a defect in motor car bodies or in cans used for selling food. For many years now manual inspections have been carried out on tin plate and other products sold to light surface quality specifications. Higher line speeds and increasing customer's demands, coupled with the increasing difficulty of obtaining labour prepared to undertake repetitive inspection tasks have forced steel companies to the use of mechanical aids to inspection. Fully automatic inspection is some way in the future. In the meantime an inspector is given warning that defect material is on its way through the line. He can then make certain that his attention is fully given to the line when defect is due to arrive in his field of vision. It is also possible to supplement the inspector by allowing machine inspection to be used for the reverse side of material.

One such aid to inspection, developed by the SIRA Institute, is illustrated in Figure 4. A laser beam is scanned across the width of the material to be inspected; reflected

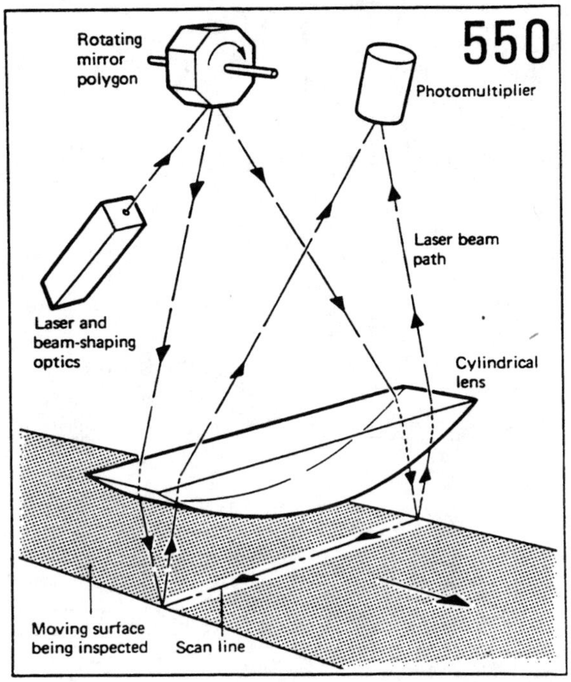

Fig.4 SIRA Surface Inspection Equipment

or scattered light is collected in a photo-multiplier. Other versions using more sophisticated optics are also being developed as well as a transmission version. In all these cases, a video-print is given to aid an inspector. There is also a facility for recording all the defects of a coil so that an overall assessment of coil quality is available for recording or inspection purposes. This equipment is complementary to an earlier device in which the field was scanned rather than the source of light. SIRA Institute have produced a plot of defect size against rate of inspection, (Figure 5) which indicates which type of equipment is most suitable for particular inspection application.

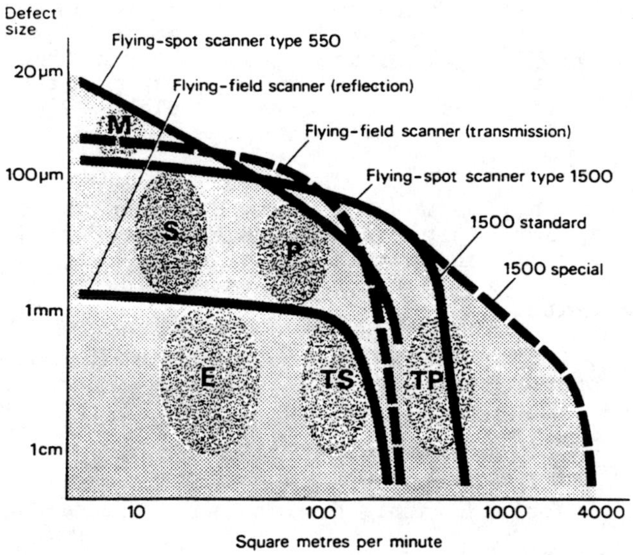

Fig.5 Application of Surface Inspection Equipment

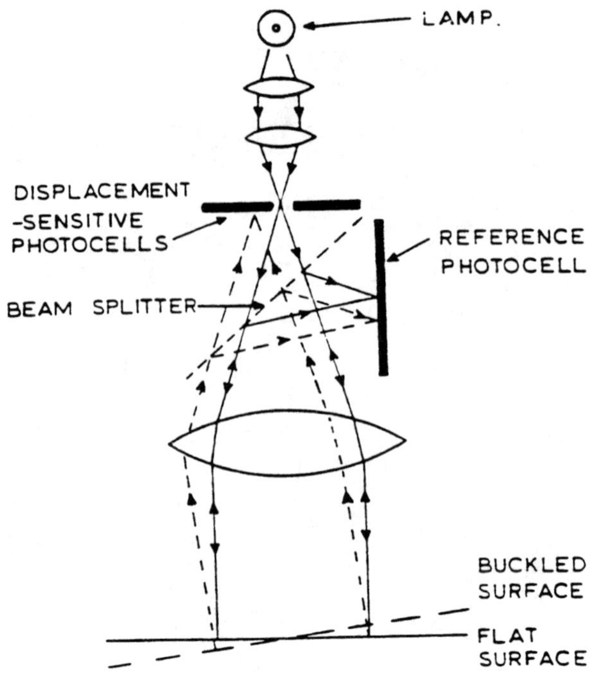

Fig.6 Optical Arrangement of BNF Manifest Shapemeter

The Ferranti Company have also produced similar equipment using gas laser beam scanned across the material by a rotating mirror. Both instruments have the capability of inspecting at specular angles or to receive scattered light. It is also possible, using external receiver units, to inspect at more than one angle i.e. specular and scattered at the same time or at two scattered angles. In this way the inspection equipment can be tailored to the type of defects.

Product Flatness

The most important quality parameter of steel sheet for many applications is its flatness. This is especially true of electrical steels where wrinkled transformer laminations give rise to excessive electrical noise. High speed can making machinery also calls for very flat feedstock. Optical methods are being used to monitor the shape of flat rolled steel when unstressed. The most advanced instrument at present is Japanese in origin, but the BNF Metals Technology Centre in the UK is developing a simple direct reading instrument[4] (Figure 6). A lens beam light from a conventional source down onto the stock. Displacement sensitive detectors mounted either side of a focal position are combined with a reference photocell to detect bukkling of the material.

Calibration is independent of line speed and of a variety of finishes on non-ferrous and ferrous metals. In tests on Zinc cut-up line, a measurement accuracy of ± 0.01% was demonstrated. A higher sensitive version of the instrument is under construction for stretched and levelled products. In this the optical design is being modified so that a single calibration will suffice for a wider range of surface finishes.

What of the Future?

The instruments described above are all in active use and at an advanced stage of development. They have demonstrated that optics can be used in rugged industrial environments and have given hope for future developments and extensions of optical technology; for example, the surface defect monitor should lead to automatic inspection when pattern recognition technology has sufficiently advanced.

A more difficult challenge lies in extension of dimensional measurement to such problems as checking the cross-section of structural beams and monitoring the extent of 'crown' on thick plates and slabs. There is little doubt that optics will rise to the occasion.

References

1. Pirlet, R., Camber of Sections, applications of Lasers in Iron and Steelmaking, eur 5094, EEC Brussels.
2. Schulz, U., Private Communication, VDEh, Dusseldorf.
3. Petit, P., Versux, P., and Barollier, J.P., Hot Rolled Steel Strip Velocity Measurement, Proc. SIMAC 74, Institute of Measurement and Control, London.
4. Reynolds, P.M. Proc. Metals Soc. Conference, Autumn 1974, Rolling Technology.

Reprinted from *Optical Engineering*, Vol. 20(6), pp. 953-961
(November-December 1981). ©1981 SPIE.

Laser electro-optic system for rapid three-dimensional (3-D) topographic mapping of surfaces

Martin D. Altschuler
Hospital of the University of Pennsylvania
Department of Radiation Therapy
3400 Spruce Street (Mail Stop 522)
Philadelphia, Pennsylvania 19104

Bruce R. Altschuler
United States Air Force School of Aerospace Medicine
Dental Investigation Service
Brooks AFB, Texas 78235

J. Taboada
United States Air Force School of Aerospace Medicine
Laser Effects Branch
Brooks AFB, Texas 78235

Abstract. A method is described for high-resolution remote three-dimensional mapping of an unknown and arbitrarily complex surface by rapidly determining the three-dimensional locations of M × N sample points on that surface. Digital three-dimensional (3-D) locations defining a surface are acquired by (1) optically transforming a single laser beam into an (expanded) array of M × N individual laser beams, (2) illuminating the surface of interest with this array of M × N (simultaneous) laser beams, (3) using a programmable electro-optic modulator to very rapidly switch on and off specified subsets of laser beams, thereby illuminating the surface of interest with a rapid sequence of mathematical patterns (space code), (4) image recording each of the mathematical patterns as they reflect off the surface using (a) a wavelength-specific optically filtered video camera positioned at a suitable perspective angulation and (b) appropriate image memory devices, (5) analyzing the stored images to obtain the 3-D locations of each of the M × N illuminated points on the surface which are visible to the camera or imaging device, and (6) determining which of the laser beams in the array do not provide reflections visible to the imaging device. Space coding of the light beams allows automatic correlation of the camera image (of the reflected spot pattern from the surface) with the projected laser beam array, thus enabling triangulation of each illuminated surface point. Whereas ordinary laser rangefinders aim and project one laser beam at a time and expect to receive one laser beam reflection (bright dot image) at a time, the present system is optical (nonmechanical and vibration-free) and can collect all the data needed for high-resolution 3-D topographic mapping (of an M × N sample of surface points) with the projection of as few as 1 + $\log_2 N$ light patterns. In some applications involving a rapidly changing time-dependent environment, these 1 + $\log_2 N$ patterns can be projected simultaneously in different wavelengths to allow virtually instantaneous data collection for a surface topography. The hardware and software used in determining the (x,y,z) location of each surface dot can be made highly parallel and can handle noise as well as multiple-grazing reflections of laser beams. In common with many other active rangefinder devices, the proposed method is unambiguous in determining the topography of all nonspecular, illuminated, and visible surfaces within its operating (stereo) range, is simple to set up and calibrate, requires no *a priori* knowledge of the object to be inspected, has a high signal-to-noise ratio, and is largely insensitive to material textures, paint schemes, or epidermal properties that mask surface features to inspection by passive topographic devices.

Keywords: three-dimensional (3-D) imaging; automated replication; robot vision; topographic mapping; pattern recognition; artificial intelligence; photogrammetry; electro-optics; laser; imaging.

Optical Engineering 20(6), 953-961 (November/December 1981).

CONTENTS

1. INTRODUCTION

Interest in robot vision has greatly increased recently because providing a vision capability to an industrial robot would enhance its versatility and generic utility in a factory/assembly environment. For example, computer-aided manufacturing may be improved if a real-time fully-three-dimensional (3-D) computer-aided inspection system were on-line.

A standard videocamera for robot vision provides a two-dimensional image which usually contains insufficient information for a detailed three-dimensional reconstruction of an object. (This is not always a problem, however, if the objects of interest in the robot/inspection environment can be mathematically defined and/or labeled in advance.)

To obtain the additional information needed for three-dimensional mapping of objects with complex surface shapes, a scene can be analyzed passively by stereo photogrammetry or actively with rangefinders and coded illumination. Passive stereophotogrammetry generally requires a human operator to determine corresponding scene positions in different photographs,

Invited Paper 5058 received Apr. 28, 1980; revised manuscript received Feb. 11, 1981; accepted for publication Feb. 16, 1981; received by Managing Editor Mar. 12, 1981. This paper is a revision of Papers 182-24 and 182-25 which were presented at the SPIE seminar on Imaging Applications for Automated Industrial Inspection & Assembly, Apr. 19-20, 1979, Washington, D. C. The papers presented there appear (unrefereed) in SPIE Proceedings Vol. 182.

and is therefore too slow for real-time applications. Automated passive stereophotogrammetry requires considerable analysis.[1-4] Methods of actively interrogating a scene (by applying various kinds of light to the scene) have been used in recent years. Laser rangefinders[5,6] project one beam onto the scene at any instant; thus there is no difficulty in correlating the illuminated position in the scene with its images in each of two cameras. Although this method requires as many "images" as there are sample points in the scene, very rapid sequential laser rangefinders may soon be possible.[7] Holography requires the interference of phase coherent light beams (one beam scattered off the scene and one reference beam), but the *scene* must be free of vibrations, and to extract numerical data is often difficult. Three-dimensional information has also been obtained by illuminating the scene from different directions[8-11] and by applying light grids[12,13] and light stripes.[14-17] The light stripe method appears to have been adapted recently for commercial use to create 3-D busts and sculpture.[18]

The system described in this paper analyzes a sequence of laser dot patterns which are rapidly projected onto a surface and viewed from one or more suitable perspectives. An example of a system consists of

(1) a laser beam array generator:
  (a) a single laser,
  (b) a lens and shearing plate assembly that expands and partitions the primary laser beam into a two-dimensional (usually rectangular) array of, say, $M \times N$ separate laser beams (where M and N are typically about 128).[19,20]
  (c) a spatially programmable electro-optic light-modulating device to sequence the $M \times N$ beams through several (for example) binary-encoded patterns;
(2) an optical image recorder:
  (a) one or more video cameras (with wavelength-specific optical filters), each of which captures in digital form (or transmits in analog form to an A/D converter) the image of each coded pattern as reflected from the surface and seen from the particular camera perspective,
  (b) a device to synchronize all the TV cameras with the patterns generated by the electro-optic device,
  (c) a buffer storage device to hold a sequence of images, and/or a device to rapidly transfer image data to a computer;
(3) software:
  (a) software which rapidly decodes the sequence of TV images and calculates the position (x,y,z) of the surface at each visible dot,
  (b) a software warning capability which can automatically detect inconsistent or incomplete data (e.g., from incorrectly pointed TV cameras) and can suggest corrections to the operator,
  (c) software for image processing and error detection (possibly with an extra parity bit image),
  (d) software for interpolation between surface points to obtain a continuous surface,
  (e) software for fully-three-dimensional pattern recognition, motion detection, etc., depending on application.

An array of laser beams, subsets of which can be turned on and off by an electro-optic shutter under computer control, can be perceived as an "active camera." The present paper then discusses a modified method of stereophotogrammetry using one active camera and at least one "passive camera" rather than two passive cameras as in conventional stereophotogrammetry. (See also the preliminary accounts of this system.[21,22]) More than one passive camera may be used to view the projected patterns from more than one viewing angle if the surface of interest is rough or convoluted. Systems using several active cameras (projectors of laser beam arrays each at a selected wavelength) and several passive cameras can also be used for those applications requiring both low-resolution global mapping of large surfaces and simultaneous high-resolution mapping of selected areas, or where simultaneous viewing of multiple surfaces is desired, or if the surfaces of interest are changing in time.

The active-passive camera system has several advantages over strictly passive camera systems. Industrial parts of varied composition, material reflectivity, material finish, and textural or paint combinations, can produce artifacts or ambiguities when straightforward passive video imaging is used; interpretation difficulties and boundary definition for pattern recognition may become difficult especially when convolutions occur in complex shapes. The projection of discrete beams creates unambiguous reflective peaks (bright dots) on the object surface that are largely independent of surface characteristics, and are detectable despite peak intensity variation between dots on mixed textural surfaces. Natural protective coloration of biological specimens in their natural habitat could mask passive analysis but are clearly measurable using active sensing.

The projection of a laser beam array onto a surface produces a dot pattern image in a viewing camera. With highly convoluted surfaces, the dot images may appear much less ordered than the original beam array. Space coding, however, tags each column in the laser beam array so that the beam array column of each dot seen on the surface is uniquely identified no matter how randomized the dot images may become. Thus the reflected dot pattern (passive image) can be correlated automatically with the original beam array projection pattern (active image) to permit point-by-point triangulation of the sample points on the 3-D surface.

Beam reflections (bright dots) hidden from the passive camera sensor can be "detected" as missing by the software. This is done by taking attendance, that is, by matching the $M \times N$ discrete beams projected with those beams whose reflections are imaged. Knowledge of which beams are not imaged provides the feedback information needed to reposition a (passive or active) camera during automatic topographic scanning of a 3-D object.

Various stripe projection methods can also be space coded but require somewhat more analysis to provide information than does the discrete dot (beam array) method described here.

A space code for an array of beams arranged in M rows and N columns reduces the number of images, I, necessary for correlating all light spots seen on the surface to $I = 1 + \log_2 N$ (compared with $I = M \times N$ for a laser scanner), where N is also the number of columns of the electro-optic shutter which can be individually switched. For convenience, the value of N is usually chosen to be a power of two.

2. MATHEMATICAL METHOD

When an array of laser beams illuminates a surface, at least some of the illuminated surface positions can be imaged by an image recording device (e.g., a video camera) at a suitable perspective. The passive image plane then contains a large number of bright dots each caused by a light ray connecting an illuminated position on the surface with the passive camera focus. The information in the passive image plane (that is, the projection of some of the 3-D surface spots onto the 2-D image plane) is by itself insufficient to determine the 3-D positions of the illuminated spots on the surface.

Suppose an array of laser beams diverges from some focal point or laser source and passes through a transparent active image (shutter) plane. If (1) a particular laser beam (identified by its intersection with the transparent active image plane) illuminates a spot on the surface of interest, and (2) the image of that spot can be located in the passive camera plane, then the 3-D spatial position of the surface spot can be determined (just as in stereophotogrammetry) provided the homogeneous 4×4 transformation matrices (containing parameters of orientation, displacement, perspectivity, scaling, etc.) are known for the active camera (laser source and shutter) and the passive camera.

The passive camera image (x^*,y^*) of a point (x,y,z) in the scene is given by the perspective transformation[22-24]

$$(T_{11}-T_{14}x^*)x + (T_{21}-T_{24}x^*)y + (T_{31}-T_{34}x^*)z + (T_{41}-T_{44}x^*) = 0 \tag{1}$$

$$(T_{12}-T_{14}y^*)x + (T_{22}-T_{24}y^*)y + (T_{32}-T_{34}y^*)z + (T_{42}-T_{44}y^*) = 0 . \quad (2)$$

If the scene-to-image transformation matrix T is known, we have for each laser dot visible to the TV camera the known quantities T_{ij}, x^*, y^* and two equations for the three unknowns x, y, z. We need one more equation. Suppose our laser beam array passes through an electro-optic shutter, so that the intersection of a beam with the shutter plane has a unique position (u,w) in that plane. Then the laser beam array can also be described in terms of a perspective transformation

$$(L_{11}-L_{14}u)x + (L_{21}-L_{24}u)y + (L_{31}-L_{34}u)z + (L_{41}-L_{44}u) = 0 \quad (3)$$

$$(L_{12}-L_{14}w)x + (L_{22}-L_{24}w)y + (L_{32}-L_{34}w)z + (L_{42}-L_{44}w) = 0 , \quad (4)$$

where L is the scene-to-laser transformation matrix, (u,w) identifies the particular beam in the shutter plane, and (x,y,z) is (as before) the (unknown) position on the surface (in the scene) that the laser beam hits.

We apply space coding to associate with each image point (x^*,y^*) the value u of the corresponding laser beam. We then have the given quantities T_{ij}, L_{ij}, x^*, y^*, u and solve Eqs. (1), (2), (3) above for the three unknowns x, y, z provided the equations are nonsingular. Thus for each image point (x^*,y^*) we can obtain the corresponding surface position (x,y,z).

The camera equations by themselves give two equations for three unknowns and thus determine only the ray from the scene to the camera. The first equation for the laser perspective transformation (with u given by the space code) provides the plane which intersects the ray to the camera. Clearly, a well-conditioned solution for x, y, z requires that the laser and camera parameters (in particular, the laser and camera positions and the space coding planes u = const) are such that the solution rays from the camera are not nearly parallel to the mathematical planes determined by L_{ij} and u = constant. Well-conditioned solutions (accurately determined positions) should be obtainable as long as the points in the scene are not extremely distant from the camera laser system (where all distances are measured relative to the camera-laser separation distance). Once we find x, y, z, we can calculate w from the last equation so that we can later determine which laser beams (u,w) have been imaged.

3. SPACE CODING

We now describe the space coding technique. Suppose we have an $M \times N$ array of laser beams (M rows, N columns) which pass through an electro-optic shutter plane u,w. Let the centroid of beam (n,m), where $1 \le m \le M$ and $1 \le n \le N$, intersect the shutter plane at some position (u_n,w_m), such that $n-1 \le u_n/a < n$ and $m-1 \le w_m/b < m$, where a is the distance between the midlines of the adjacent columns of the laser beam array and b is the distance between the midlines of the adjacent rows of the laser beam array. With this definition, that beam of the laser beam array which passes through the shutter plane at position (u,w) is identified with the unique integer pair

$$(n,m) = (1 + \mathrm{flr}(u/a), 1 + \mathrm{flr}(w/b)) , \quad (5)$$

where flr(x) = largest integer contained in real number x.

We design the electro-optic shutter to have N separately controllable columns (one for each column of the laser beam array) so that if we apply an electric signal to one of N input wires, say wire n, the domain $\{(u,w): (n-1)a \le u < na\}$ of the laser shutter will become opaque. With such a shutter we can control which beams of the laser beam array are transmitted and which are blocked, and in this way encode patterns in the array of transmitted beams. By projecting a sequence of $1 + \log_2 N$ binary laser beam patterns, we can determine uniquely, for any dot image seen in the passive image plane, the address n of the shutter column of the corresponding laser beam.

As an example, suppose we have a 200×16 laser beam array passing through an electro-optic shutter with 16 controllable columns labeled by 1 + flr(u/a). We then sequentially project (and image) the following patterns: (1) the entire laser beam array, (2) the higher-numbered half of the array (columns 16 through 9 transparent; columns 8 through 1 opaque), (3) alternate quarters of the array (columns 16 through 13 and 8 through 5 transparent; all others opaque), (4) alternate eighths of the array (columns 16, 15, 12, 11, 8, 7, 4, 3 transparent; all others opaque), (5) alternate columns of the array (even columns transparent; odd columns opaque). Altogether $1 + \log_2 N$ patterns must be projected and imaged for a binary space code. In the case just illustrated, N = 16, so that $1 + \log_2(16) = 5$. Then for any spot imaged by the camera during the first pattern (all the beams), the value of flr(u/a) of the corresponding beam in the laser beam array is given by the binary number obtained by the appearance or nonappearance of the spot in the images of subsequent patterns. For example, suppose a surface spot appears in the passive camera images of patterns 1, 2, and 4 but not in patterns 3 and 5. Then

$$\mathrm{flr}(u/a) = 1010_2 = 10$$

(corresponding to its appearance or nonappearance in patterns 2, 3, 4, 5) so that

$$n = 1 + \mathrm{flr}(u/a) = 11$$

$$\text{shutter domain} = \{(u,w): 10\,a \le u \le 11\,a\} .$$

The apparatus can be set up and calibrated so that a more precise position value u_n on the shutter plane is known for each column n of the laser beam array. Thus once the shutter column n = 1 + flr(u/a) is determined, the value of u_n is known for the laser beam.

If the transformation parameters T_{ij} and L_{ij} are known, we can obtain (x,y,z) for each (x^*,y^*,u). Once (x,y,z) is known, w can be found with the last equation. Any beam (u,w) not visible to the camera can be found by listing all the projected beams and then eliminating from the list those that *are* visible.

In Fig. 1 we illustrate the space coding for a 6×8 array. Only four spots on the scene are visible to the camera in this example. The laser shutter column corresponding to each image point is found in Table I from the data of Fig. 1.

TABLE I.

Image point	Binary number	flr(u/a)	n
A	000	0	1
B	011	3	4
C	110	6	7
D	010	2	3

The space code method, which allows us to use one passive camera and one active camera, should be compared with the case of stereophotogrammetry in which one must use two cameras and know the corresponding transformation matrices T^a_{ij}, T^b_{ij} and the corresponding image points (x_a^*,y_a^*), (x_b^*,y_b^*) for the same unknown point (x,y,z) of the scene; hence four equations are solved for the three unknowns x, y, z with inconsistencies treated by least squares.[23,24]

4. OBTAINING THE TRANSFORMATION PARAMETERS

In a laboratory the camera and laser beam projector can be fixed and the parameters T_{ij} and L_{ij} measured directly.

In portable operation, as for example a medical or industrial ap-

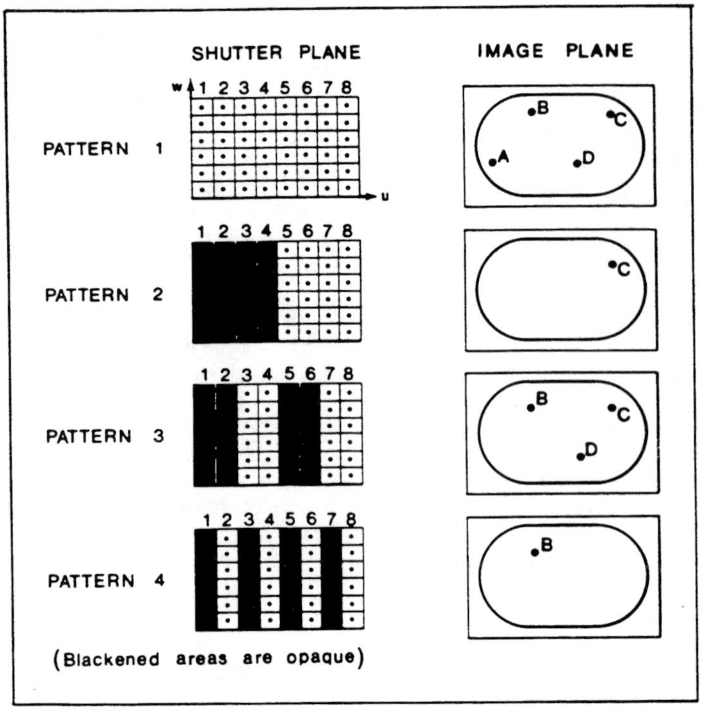

Fig. 1. The array of laser beams is space coded by the sequence of patterns shown at the left, which is obtained by controlling the opaqueness of the columns of the shutter plane. The intersections of the laser beams with the scene provide bright spots on the scene which can be imaged from a suitable perspective by (say) a TV camera (provided there are no obstructions in the line of sight between the scene and the camera). The TV camera is synchronized with the shutter so that for each projected pattern a TV image is obtained. In the illustration, the TV image sees four spots of the scene corresponding to four laser beams. The columns of the shutter plane through which these beams passed can be determined by the space coding. Point B is seen in pattern 3 and pattern 4 but not in pattern 2 and so correlates with shutter column $1 + 011_2 = 1 + 3 = 4$.

plication, it may be necessary to set up and disassemble the equipment periodically or per object to be mapped, so a rapid calibration scheme is needed which a nonexpert can immediately use.

One simple calibration method is to insert into the scene a prism whose faces are known planes. To find the T transformation matrix, we locate six vertices of the prism in the passive image plane, so that we know the values (x_k^*, y_k^*) and (x_k, y_k, z_k) for i = 1, . . . ,6. We can then set $T_{44} = 1$ and solve eleven equations for the eleven unknowns T_{ij} of Eqs. (1) and (2).[23,24] To obtain the L transformation matrix, we project the coded laser beam array and image the prism. If, say, $N \geq 7$ suitable points are imaged, we can obtain coefficients L_{11}, L_{21}, L_{31}, L_{41}, L_{14}, L_{24}, L_{34} of Eq. (3) by setting $L_{44} = 1$ and minimizing

$$F = \sum_{k=1}^{N} \Big[(L_{11}-L_{14}u_k)x_k + (L_{21}-L_{24}u_k)y_k + (L_{31}-L_{34}u_k)z_k + (L_{41}-u_k) \Big]^2 , \quad (6)$$

where the subscript k refers to one of the N points imaged by the passive camera. The values (x_k, y_k, z_k, u_k) can be found for each imaged point k because we know the T_{ij} (from before), the image position (x_k^*, y_k^*), the space code, and the planes of the prism. The least-squares solution is obtained by solving the set of seven linear equations

$$\partial F / \partial L_{ij} = 0 , \quad (7)$$

one for each of the seven unknown L_{ij} coefficients. A solution exists if the 7×7 matrix is nonsingular. In most cases, a nonsingular matrix occurs if the prism is within the active-passive stereo range and the N points appear on more than one face of the prism. In this method, the accuracies of the L_{ij} values depend on the accuracies of the T_{ij} values. This automatic calibration procedure does not provide the L_{ij} coefficients of Eq. (4), so that solving for the shutter row w after x,y,z has been obtained (that is, attendance-taking information) would not be possible. To obtain the L_{ij} of Eq. (4), a second automatic calibration procedure involving a row-coded electro-optic shutter can be easily appended.

The easiest situation of all would be to do all calculations relative to a coordinate system fixed to the laser projector or the main body of the robot and to sense independently the camera orientation and other parameters with respect to this reference frame (say by means of separate optical, mechanical, or inertial guidance systems). With such sensing capability, the projector and the sensor can be independently manipulated and need not be limited to predetermined, precalibrated viewing positions.

No matter what calibration is used, the cameras, active or passive, need not be at the same relative distance from the object of interest but can be placed at any convenient locations within the stereo range. The only restriction in positioning the active and passive cameras is that the solution of Eqs. (1), (2), (3) must exist and be well conditioned.

5. OPTICS FOR LASER BEAM ARRAY GENERATION

Laser light can be expanded with a lens and convolved into rectangular grid patterns by a parallel arrangement of fringe-generating shearing plates. However, for spatial encoding and convenient light image capture by videocamera, an array of lighted, discrete (Gaussian isolable) points is desirable. We now describe a series arrangement of shearing plates to generate the discrete (bright) beam array pattern required.

The principle used in creating a fringe array is outlined in Fig. 2.[19,20] Interference fringes are generated by front and back reflection from a shearing plate. The condition for the creation of N interference fringe lines is[19]

$$N = (2d/\lambda)[\tan^{-1}(W_0/2f_1)][\sin 2\alpha]/(n^2 - \sin^2\alpha)^{1/2} , \quad (8)$$

where d is the shearing plate thickness, n is the index of refraction of the plate, W_0 is the diameter of the incident laser beam, f_1 is the focal length of the focusing lens, λ is the wavelength of the laser light, and α is the incident angle of the principal ray.

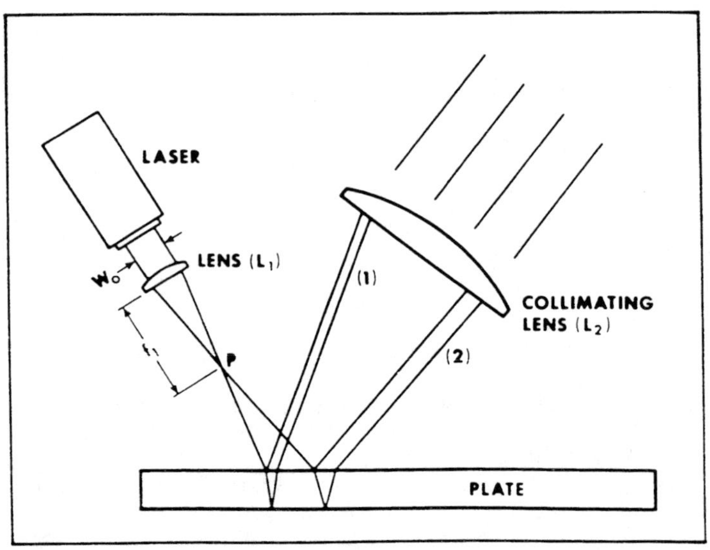

Fig. 2. Generation of parallel fringes.

If fringe lines produced by one plate are incident on a second partially silvered plate in such a way that the plane of incidence of the second plate is perpendicular to the first, the fringe lines will again be broken by interference minima, and the light reflected from the second plate will constitute an expanding rectangular array of lighted beams emanating from a common point. This light array can be recollimated and spatially filtered to produce a sharp discrete array of laser light beams. The final array may then be caused to converge for high resolution pattern recognition onto microscopic objects, or to expand at a convenient angular divergence to map larger objects. A suitable combination of lenses is used. The relative size of the beams with respect to the distances between beam centers (measured in a plane perpendicular to the array path, i.e., parallel to the shutter plane) remains constant as the total array size and individual beam diameters increase over distance in a diverging pattern.

A decided advantage of this method to generate beams in an ordered array is that the discrete beams that form the pattern are the product of diffraction, the beams behaving as individual lasers. Patterns will not deteriorate at long range (in a vacuum), and can (theoretically) be converged to wavelength separation between beams. This is in sharp contrast to forming a beam pattern by conventional masking apertures. Masks result in diffraction effects that seriously limit the useful range of the beam array and its degree of useful convergence.

A laboratory demonstration prototype of the basic hardware concept is illustrated in Fig. 3. The system generates and projects a programmed sequence of coded arrays of discrete bright laser beams onto an object or specimen. A 2''-diameter plaster stone model of a human premolar tooth (object) is seen at the lower right mounted on a lens holder. The bright beams are generated optically

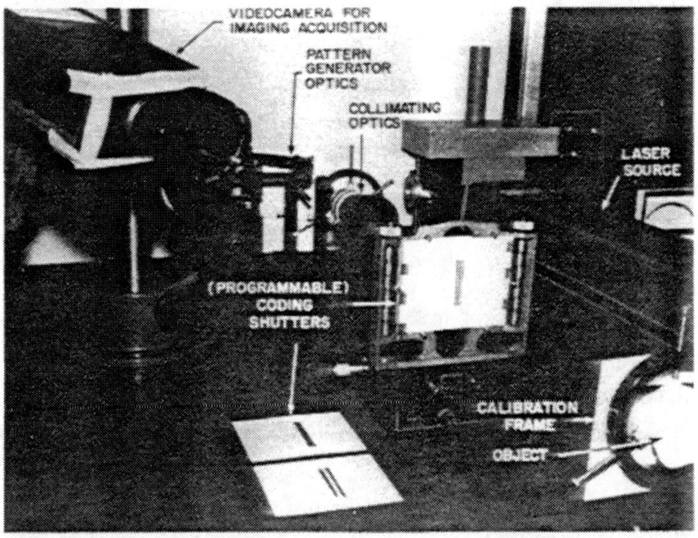

Fig. 3. Laboratory demonstration prototype of Topographic Mapping System using manually placed cards to illustrate pattern changing for space coding.

from a single linearly polarized laser source and an optical arrangement which includes beam expansion, two optical shearing plates, and beam steering devices (''pattern generating optics'' of Fig. 3). Any suitable laser wavelength or set of wavelengths can be used. For this demonstration, the 514.5 nm line of a model 165 Spectra-Physics argon-ion cw laser was chosen. The optical pattern may be intermittent or continuously generated depending on the laser source, configuration, and application requirement. Multiple-source projection is possible provided that either synchronization or selective wavelength filtering is used with the scene capture imaging device; for *simultaneous* multiple source projection, the imaging device(s) must have multispectral discrimination capability.

6. HARDWARE FOR BEAM ARRAY CODING

The beam array once generated is recollimated as required by collimating optics (Fig. 3) and then passed through a programmable set of rectangular (columnar) shutters. The shutter assembly can be constructed from mechanically actuated slits, rotating assemblies of rulings, liquid crystal gates, or electro-optic or acousto-optic crystal gates. The function of the shutter is to eliminate selected columns (or rows) of beams from the laser beam array. For an electro-optic shutter, each individual shutter column (or row) becomes transmissive or opaque to the projected light beams depending on the particular electrical signal it receives. The collective effect of blocking individual columns (or rows) dictates the particular pattern being transmitted. The shutter columns (or rows) are controlled by a computer or microprocessor which drives the shutter to transmit a programmed sequence of two-dimensional array patterns onto a surface of interest. The reflected image of each new beam array pattern from the surface is recorded by one or more image acquisition devices (such as a videocamera). The shutter and the imaging devices are synchronized by a microprocessor.

The manually positioned (programmable) coding shutters shown in Fig. 3 are presented simply to illustrate the basic concept of shutter pattern reconfigurations required for space coding. A larger (128 × 128) beam array was masked by a white card set in a holographic plate holder to create an (illustrative) eight-column-wide pattern. The space code sequence begins when an unshuttered (uncoded) beam array is first projected on the surface of interest and is recorded by a suitable videocamera and storage electronics. In this demonstration a standard closed circuit (RS-170 compatible) General Electric model TE-26 videocamera was used (upper left corner of Fig. 3) and focused on the tooth model (seen in right foreground of Fig. 3).

After this first view is recorded the shutter is changed or reconfigured. For illustrative purposes, the second shutter configuration may be represented by the top white card lying on the optical table in Fig. 3. This shutter, when placed in the shutter holder, eliminates four consecutive columns of beams while allowing four consecutive columns of beams to pass unimpeded. The scene is again recorded to obtain the second view. The third shutter, (bottom white card on optical table, Fig. 3) replaces the previous shutter to produce a new pattern code. Figure 4 is a photograph of the video image recording of the third shutter pattern as projected on a tooth model. (Compare this camera view with pattern 3 of the shutter plane in Fig. 1.) The fourth shutter (seen in the beam array path, Fig. 3) is then placed and the scene recorded to produce the fourth view.

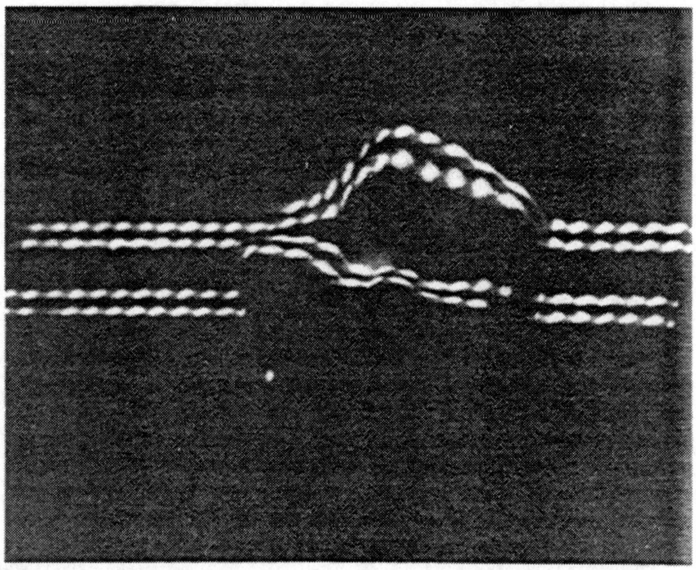

Fig. 4. An example of one shutter coded pattern as projected on a tooth model using an 8 columnar (programmable) shutter.

Manually placed shutters as illustrated (Fig. 3) in the laboratory demonstration (even if engineered into a filmstrip mechanical device) would be cumbersome and prone to failure. The optimum method for blocking selected columns (or rows) is to use a fixed, vibrationless, programmable spatial light modulator (PSLM) with exclusively electro-optic components and properties.

A laboratory demonstration model using a 128 (vertical) column electro-optic PSLM is seen in Fig. 5. The PSLM is visible in the center of the figure, just below the videocamera lens and above the top of the laser (Fig. 5). The tooth model and support is to the extreme right in the figure. In this photo, the PSLM has been programmed with four sets of "ON" (transmisive) blocks of columns alternating with four sets of "OFF" (opaque) blocks. Each set has sixteen successive columns "ON" or "OFF." A polarizing plate mounted in a metal frame appears just in front of the shutter. Not visible (hidden by the videocamera) is an interface and microprocessor which controls the PSLM.

Since space coding image conservation made a log-base-2 width system desirable and since standard U.S. videocameras of 480

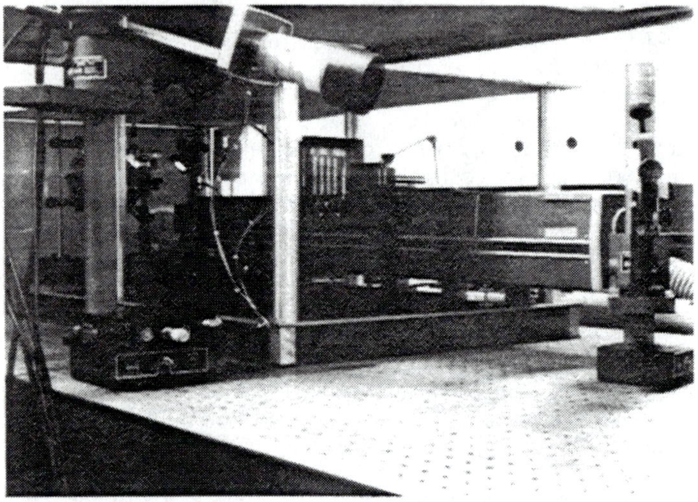

Fig. 5. Laboratory demonstration breadboard in which an electro-optic programmable spatial light modulator (PSLM) is substituted for manual cards (illustrated in Fig. 3) to provide space coding. The PSLM blocked selected (preprogrammed) columns (or rows depending on orientation) in the projecting beam array path. The PSLM (center of photo below videocamera lens and above laser) is shown with 16 consecutive rows on, 16 off, etc., or 128 columns in which eight sets of 16 columns are alternately on (transmissive) or off (opaque or black). A microprocessor (hidden by videocamera lens) controls the PSLM pattern. Any combination of the 128 individual columns in the PSLM may be selected for transmission or opaqueness. Tooth model is at right center fixed to a magnetic mount.

useful scanned lines were available in the laboratory, a total column number of 128 was chosen for this demonstration. The PSLM module was constructed by Applied Optics Corporation (a division of ITEK Corporation) and mounted on a glass plate with a total optical (light modulating) area of 2×2 inches square. The module contained integral (embedded) peripheral electronics to assist in loading the pattern. Each column of electro-optic material in the shutter is individually programmable. When an appropriate voltage signal is applied to this material it will exhibit half-wave plate behavior. Since the beams in the array have definite polarization derived from the laser source, any beam passing through an energized column would be eliminated by an appropriately oriented analyzing polarizer placed beyond the shutter.

An interface was devised by Capt. Michael Tworek, USAF, so that any suitable microprocessor could be used with the PSLM. Moreoever, the requirement to load the PSLM with 128 bits of clocked serial data from the microprocessor is avoided. Instead, a 16-bit parallel-load-to-serial-data-out capability was created where eight 16-bit words are loaded into the interface to properly code the 128 columns of a selected configuration. The interface matched an available Intel SBC 80/20 single-board computer to the PSLM liquid crystal materials rise time of 500 ms. Were a faster PSLM to be used, the existing interface could be redesigned with a faster clock to load a pattern in 1 ms. Two PSLM modules and interfaces were assembled in the optical path so that either vertical columns or horizontal rows, or a combination, could be space coded. Since the present PSLM module is too slow for switching between video raster scans, the preprogrammed patterns were loaded into the microprocessor and a manual binary switching method was used to reconfigure the shutter. This was sufficient to demonstrate space coding by a PSLM in an active camera. A second generation PSLM can be constructed to permit reconfiguring the PSLM columns within (and synchronized to) the vertical blanking time of a standard video raster scan, or during the erase/recovery cycle of a solid-state imaging camera. A 100 μs reconfiguration rate could be achieved with little difficulty (well within the 1.4 ms of the vertical blanking time) using PLZT material, and even 1 μs reconfiguration rates are theoretically possible for some candidate methods.[25-28] The limiting factors would then be the videocamera image acquisition and throughput/erase/recovery cycle, and the ability of a suitable device to store or process incoming data.

7. BEAM ARRAY PROJECTION ONTO THE SCENE

The choice of optics required to adjust the total area illuminated by the beam array, as patterned by the coding subsystem, is determined by the relative size and distance of the object from the beam array projector, and the resolution required for mapping the scene or surface of interest. Either a nondiverging, converging, or diverging beam array may be used to suit a particular application. A converging optical arrangement may be used to project a higher spatial frequency array onto a very small object. For example, a 10 dot/mm spatial frequency array (100 μm dot separation) has been projected onto a surface and viewed through a microscope objective. A zoom lens may quickly and conveniently alter the spatial frequency and size of the beam array projected field for multipurpose uses, or for changing from wide area to detailed scanning. Suitable rotatable and gimbal-mounted optical devices may be used to accurately control the direction of beam array projection. The laser beams in the array are focused at infinity by the beam array generating optics. The laser beam array will project through space for a considerable distance until impeded by an object or surface.

8. IMAGE ACQUISITION

A videocamera records from a suitable perspective the images of the sequence of coded beam array patterns reflected off the surface of interest. In Fig. 6, a full (128×128) pattern is intercepted by a 2"-diameter plaster premolar model which was bordered by black rubber. A General Electric TE26 vidicon camera with a GE f/2.5 to 22 and 20 to 80 mm tele-zoom was used to record the pattern. Illumination of the object was approximately 10 $\mu W/cm^2$ by 514.5 nm laser light. Ambient daylight fluorescent lighting was present. A $\pm$ 10λ selective filter was used in front of the telephoto lens. The image was recorded on an IVC 1" reel-to-reel video recorder and played back on a black and white 19" monitor. A 35 mm still camera recorded this photograph (Fig. 6) during playback. A dot in the lower left corner (beyond the model) is an artifact in the monitor. Approximately three thousand sample points are visible.

Figure 7 is photographed in real time in the laboratory. Note ambient lighting levels seen in Figs. 5 and 7. A two "ON," two "OFF" alternating pattern was being projected.

Figure 8 is another image of the object shown in Figs. 6 and 7. In this image the pattern was configured with alternating eight columns on and eight columns off. Variations in dot brightness (intensity) and adjacent dot to dot spacing are caused by surface curvatures of the object.

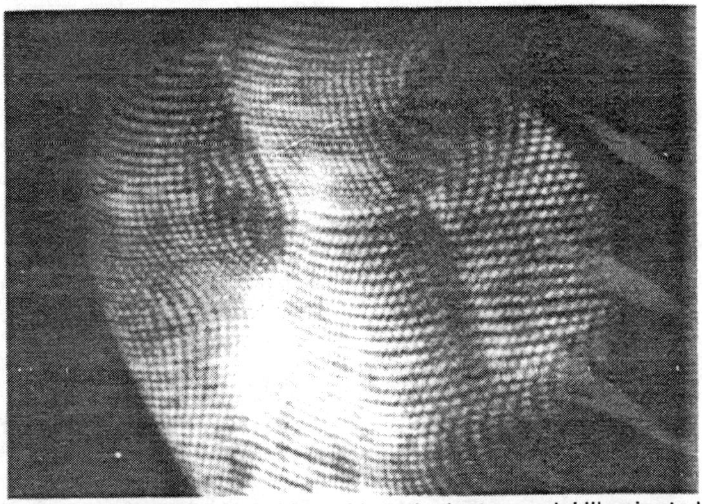

Fig. 6. Two-inch-diameter bicuspid tooth plaster model illuminated by full beam array pattern (128 × 128). Approximately 3000 dot reflections are visible. Photo was taken of video monitor while videorecording playback. Object was illuminated with a 514.5 nm argon laser in a daylight-fluorescent lighted laboratory. Videocamera had a selective-wavelength filter in front of its lens to remove ambient lighting (black cylinder housing on videocamera lens is seen in top left of Fig. 5).

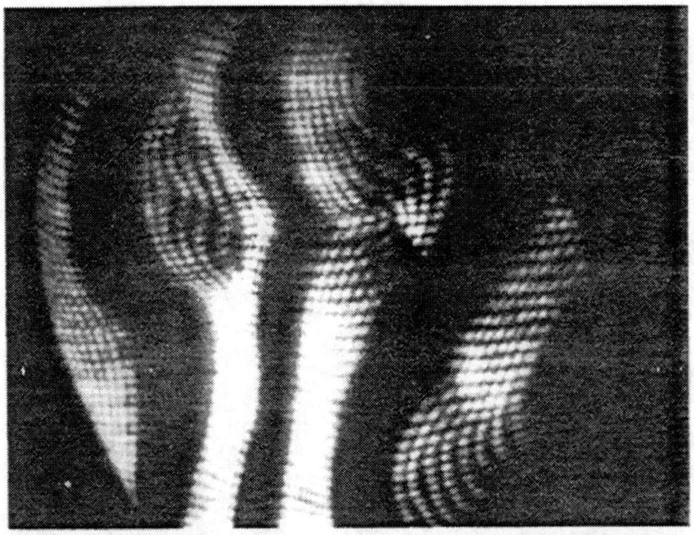

Fig. 8. Photograph of same object as in Fig. 6 but with the beam array now coded by the PSLM into one of the binary patterns. A total of eight different binary patterns are sequentially projected to completely code a 128 × 128 beam array and tag each array column for identification in the image.

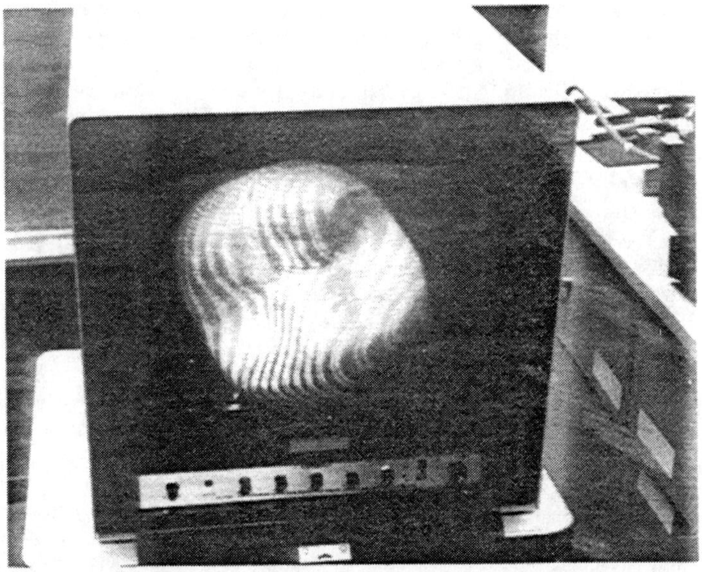

Fig. 7. Photograph of videomonitor viewing in real time a two-on-two-off consecutive pattern projected onto the tooth model. Ambient lighting in laboratory is used for this exposure.

Fig. 9. White-light digitized image of prism block displayed on a computer graphics terminal after playback from digital (computer) tape.

Figures 9 and 10 show digitized images of a prism played back from computer tape onto a Comtal display unit. Figure 10 shows the passive image of a laser beam array pattern coded by a PSLM and reflected off the prism.

For a standard composite video signal, a raster scan of each image is completed in 1/30th second. During the interval between the end of the preceding raster scan and the beginning of a new scan, the electro-optic shutter reconfigures. A microprocessor, minicomputer, or clock synchronizes the videocamera, beam array shutter, and video-recording device. A video recorder or video disk system may be used to initially record the images. An alternative method would be solid-state storage of individual video images on charge-coupled device (CCD) or random access memory (RAM) chips with high speed sequential or parallel analog to digital (A/D) conversion. A 512 × 512 × 8 pixel array from a 1/30th second raster scan may be stored in digital form (using commercially available devices) with each point mutually addressable.

For a 16,000 point topographic spatial map, a total of eight coded images is required. Total storage time is thus 8/30th second (= 267 ms), and eight images must be stored for processing. While adequate for many applications, faster speeds are often desirable. If charge-injection device (CID) or CCD direct imaging technologies are used, then the speed of total scene capture (that is, eight views for a 16,000 point 3-D) might be extremely rapid. A two millisecond time cycle is reasonable for acquisition of all data need-

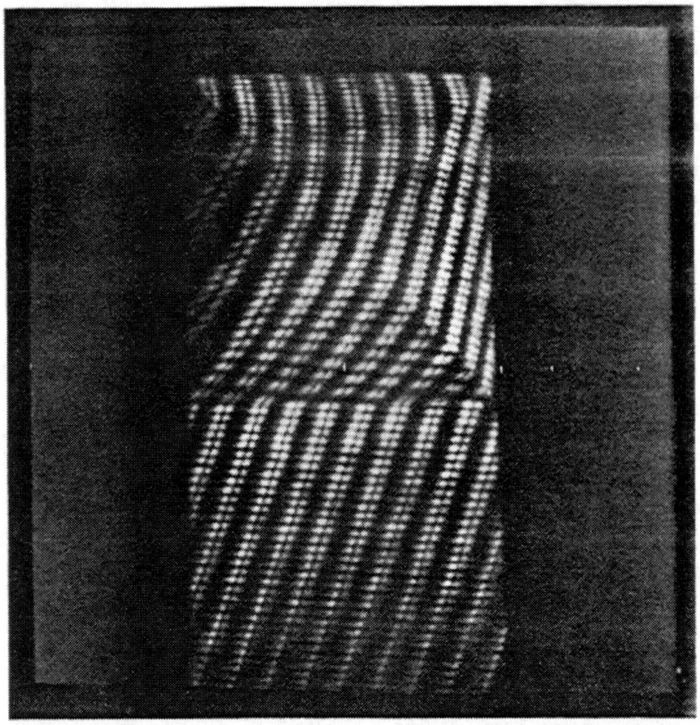

Fig. 10. Digitized image of the same prism block (displayed on a computer graphics terminal) with the 128 × 128 laser beam array shown coded by a pattern of the 128-column PSLM.

ed to define the three-dimensional coordinates (x,y,z) of a beam array projection such as the one on the bicuspid tooth model shown in Fig. 6. A 512 × 512 CID or CCD array matrix would have to be developed for rapid acquisition of data, in which the output would entail simultaneous (parallel) output of rows or blocks of rows of the array (rather than using the RS-170 output rate serial-scan methods presently used to emulate the vidicon camera raster scan). Present commercial devices have designed the imaging array output serially only, unfortunately. A parallel-processing computer, or a hierarchy of serial-processing computers would be needed to absorb data rapidly.

In certain applications pertaining to a rapidly changing environment (e.g., assembly lines, human motion, vision aids for the blind,[29] etc.), all the necessary coded images could be obtained by using a cluster of $1 + \log_2 N$ separate active cameras, each projecting a single beam array pattern at a different laser wavelength and a cluster of $1 + \log_2 N$ separate selectively filtered passive cameras, each receiving the corresponding pattern reflections in the corresponding wavelength. Such a $(1 + \log_2 N)$-fold parallel active-passive camera system would accomplish data collection for an entire topography in one cycle time. The calibration for the cluster cameras would be only slightly more complicated than for the sequential cameras.

Parallel processing of data could result in "almost instantaneous" scene capture and analysis; an updated 3-D map could possibly be calculated every second, or faster. A two millisecond scene capture with one to ten seconds minicomputer processing would be adequate for many medical and industrial applications. A high-speed synchronized cinematic process could be used as an alternative for image acquisition, but it would require a possibly unacceptable postprocessing film development delay.

One or more scene capture devices (e.g., videocameras) may be placed around the scene which is being illuminated by the space coded beam arrays. Gimbal mounts may be used for these cameras, as well as various robot arm mountings, and various electro-optic and/or acoustic aimers can be used to keep the cameras focused and centered on the scene, even if the camera(s) and beam array projector are mobile. Calibration and positioning requirements have already been discussed. Camera-projector angulations can be calculated automatically by the minicomputer dedicated to the 3-D analysis. Suitable lens systems, determined by the user for each application, are mounted on the imaging cameras.

User-defined narrow-bandpass optical filters for each relevant laser wavelength are chosen for viewing in ambient lighting conditions. If eye safety constraints require low levels of laser light for measuring human subjects,[29] image intensification may be required. A beam array pattern illuminating at 1 $\mu W/cm^2$ at 514.5 nm has been detected reflecting off a coarse black cardboard when appropriately selected videocameras or externally mounted image intensifiers were used.

9. FURTHER DATA PROCESSING AND OUTPUT

Data processing is required for (1) hardware control to synchronize the programmable laser-shutter encoder with the image acquisition system, and (2) the calculation of the three-dimensional number maps for the surface topography. The data processor may also verify the accuracy of the camera orientation vis-â-vis the laser surface target and warn of reflected points hidden to the camera by convolutions in the surface topography. Data output may be in the form of digitally stored numerical tables. Numerous applications and postprocessing methods are possible once a numerical table of the 3-D coordinates of sufficient sample points of a surface is available. For example, graphic display output and numerical control of machines are possible. Serial maps may be rapidly updated to discern movements and distortions occurring on a surface being scanned by the topographic system. Straightforward image subtraction techniques using sequentially generated output maps could lead to notification and 3-D vector analysis, in real time, of objects entering or moving within even the most cluttered scene area.

10. DISCUSSION

The objectives of this effort were to define and conceptually create the technological capability: (1) to obtain unambiguously and from a distance the three-dimensional coordinates of a large number of sample points of any open uncovered surface of any object (of any size) and thereby to determine the surface topography of the object, (2) to map a surface rapidly, remotely, nondestructively, and automatically, (3) to obtain, where possible, spatial data from as few as one scene-viewing direction, (4) to generate the topographic data in real time, (5) to design a simple-to-operate, self-contained, portable system whose output requires no human processing and whose setup requires no tedious calibration or excessive expertise, (6) to design a device so that its software determines insofar as possible the geometric transformations among the sensor, emitter, and scene (to enable decoding of the image data), (7) to design a system for maximum versatility and ruggedness for use in hazardous industrial, clean room clinical and surgical, and military environmental conditions, (8) to provide human or machine interactive capability for design fabrication, pattern recognition and scene analysis in three dimensions, and servo-controlled robot-vision-coupled applications.

The specific concept described and demonstrated by laboratory breadboarded means, and outlined mathematically, has certain uniquely interesting features that lend themselves to industrial and medical applications, were the concept developed further.

(1) The same apparatus could map microscopic or large-size objects merely by choosing suitable end objective lenses. The beam array is composed of discrete beams at infinite focus, so that an object intercepting a beam need not be at a predetermined distance from the origin of the array. Pattern array generation is the product of diffraction, so that the array may be enlarged or reduced at will or projected a long distance, avoiding the diffraction-limiting effects that occur when "grids" are produced with limiting apertures or masking slits. No moving parts or mechanical devices are used for array generation or for space coding, and the electro-optic can-

didate materials appear to have long duty cycles, possibly permitting the active camera optics to be sealed for maintenance-free use. With reasonable selection of optics, the array generation would work with most laser wavelengths or power outputs.

(2) The active (projector) and passive camera(s) can be positioned (moved) independently at arbitrary axes (e.g., on independent robot arms) for observation of the scene from different perspective directions after each set of data is acquired. Thus a fixed rotation axis for the camera system and/or the part to be inspected is not necessary.

(3) The active illumination of medical or industrial parts eliminates ambiguities obtained with passive systems due to (a) natural protective coloration, and (b) variations in reflective properties as a result of mixed textures, machine surface finishings, differing materials, coatings, and clothings. Active beam illumination creates discrete peak reflections (even if of variable amplitude) on the object surface, and a known finite array of ordered beams can be accounted for as either present or missing.

(4) Any suitable recording sensor(s) could be used for the passive camera(s). A videocamera output by whatever means could be rapidly converted to a standard TV broadcast system for remote transmission of image data in very near real time, permitting remote monitoring of automated system 3-D measurement. With lasers of different wavelengths and correspondingly filtered videocameras, more than one beam array pattern may be projected and simultaneously and independently viewed. Thus simultaneous coarse/fine scanning and global/regional scanning of a large object or parts (for positioning or comparative dimensional analysis) is possible.

(5) Gross deformations may be comparatively and accurately measured against a data base. Fully-3-D deformation analysis in gross increments is especially useful for comparison of rough castings or forgings, measurements of bubble expansion, and choosing which of a set of similar large industrial assemblies should be mated to another with the least hand fitting.

11. ACKNOWLEDGMENTS

We would like to thank our colleagues Drs. J. Posdamer and G. Frieder of the Department of Computer Science, SUNY Buffalo, for their comments and suggestions.

The work of Martin D. Altschuler was supported in part by NIH grants HL18968, HL4664, and GM-381. The other authors were supported in part by the United States Air Force School of Aerospace Medicine, Aerospace Medical Division (AFSC), Brooks AFB, Texas; Defense Advanced Research Projects Agency; Integrated Computer Aided Manufacturing (AFWAL) (AFSC); and NASA, Johnson Space Center.

We would further like to thank Capt. M. Tworek, USAF, for programming and constructing an interface for the electro-optic shutter, Col. (Dr.) J. M. Young, and Col. (Dr.) R. L. DeHart of USAFSAM, for tireless administrative encouragement.

12. REFERENCES

1. Sobel, I., 3rd Int. Joint Conf. on Artificial Intelligence, 648 (1973).
2. Shapira, R., 5th Int. Joint Conf. on Artificial Intelligence, 628 (1977).
3. Gennery, D. B., 5th Int. Joint Conf. on Artificial Intelligence, 576 (1977).
4. Moravec, H. P., CMU-R1-TR-3 (Carnegie-Mellon Robotics Institute), (1980).
5. Dreyfus, M. G., Dreyfus-Pellman Corp., Contourmate 7000 System (1979).
6. Lewis, R. A., and Johnston, A. R., 5th Int. Joint Conf. on Artificial Intelligence, 762 (1977).
7. Kanede, T., Proc. SPIE 283, 48 (1981).
8. Horn, B. K. P., MAC TR-79, MIT Project MAC (Nov. 1970).
9. Horn, B. K. P., Obtaining Shape from Shading Information, in *The Psychology of Computer Vision* (P. H. Winston, ed.), McGraw-Hill, New York (1975).
10. Shirai, Y., and Tsuji, S., 2nd Int. Joint Conf. on Artificial Intelligence, 71 (1971).
11. Woodham, R. J., 5th Int. Joint Conf. on Artificial Intelligence, 635 (1977).
12. Will, P. M., and Pennington, K. S., 2nd Int. Joint Conf. on Artificial Intelligence, 66 (1971).
13. Rocker, F., and Kiessling, A., 4th Int. Joint Conf. on Artificial Intelligence, 669 (1975).
14. Sirai, Y., and Suwa, M., 2nd Int. Joint Conf. on Artificial Intelligence, 80 (1971).
15. Agin, G. J., and Binford, T. O., 3rd Int. Joint Conf. on Artificial Intelligence, 629 (1973).
16. Popplestone, R. J., Brown, C. M., Ambler, A. P., and Crawford, G. F., 4th Int. Joint Conf. on Artificial Intelligence, 664 (1975).
17. Lovesay, E. J., Proc. Symp. Comm. V, ISP Biostereometrics, 74-311, 147 (1974).
18. Solid Photography advertisement with photograph of subject illuminated by light stripes.
19. Taboada, J., and Altschuler, B. R., Appl. Optics, 15, 597 (1976).
20. Altschuler, B. R., Taboada, J., and Segreto, V. A., Proc. SPIE 70, 347 (1975).
21. Altschuler, B. R., Taboada, J., and Altschuler, M. D., Proc. SPIE 182, 192 (1979).
22. Altschuler, M. D., Altschuler, B. R., and Taboada, J., Proc. SPIE 182, 187 (1979).
23. Sutherland, I. E., Proc. IEEE, 62 (2), 453 (1974).
24. Rogers, D. F., and Adams, J. A., *Mathematical Elements for Computer Graphics*, McGraw-Hill, New York (1976).
25. Land, C. E., Opt. Eng. 17 (4), 317 (1978).
26. Oliver, D. S., Opt. Eng. 17 (3), 288 (1978).
27. Blaha, W. P., Lipton, L. T., Wiener-Avnear, E., Grinberg, J., Reif, P. G., Casasent, D., Brown, H. B., and Markevitch, B. V., Opt. Eng. 17 (4), 371 (1978).
28. Cutchen, J. T., Harris, Jr., J. O., and Laguna, G. T., Sandia Laboratories Tech. Report SLA-73-0777 (1973).
29. Altschuler, M. D., Posdamer, J. L., Frieder, G., Manthey, M. J., Altschuler, B. R., and Taboada, J., Proc. Int. Conf. on Cybernetics and Society, ISSN 0360-8913, Cambridge, Massachusetts, p. 1000, Oct. 1980.

Reprinted from *Optical Techniques for Industrial Inspection,* P.G. Cielo, editor, Proc. SPIE Vol. 665, pp. 61-67 (1986). ©1986 SPIE.

Optical systems for dimensional control in hostile environments(*)

R. Pirlet, J. Boelens and J. Salmon

Centre de Recherches Métallurgiques, 11, Rue Ernest Solvay, B-4000, Liège, Belgium

P. Balthasart and J. Lückers

CEREMCO, 24, Chaussée Churchill, B-4320 Saint-Nicolas, Belgium

Abstract

For many years, C.R.M. has been involved in the application of optoelectronic noncontact methods for inspecting steel products especially during hot rolling. Among the first prototypes, the OPTIMAC gauge measures the slab thickness on roughing mills according to an optical triangulation method.

A measuring system, called ROMETER-5 has been developed for the objective assessment of the flatness of hot steel strips at the runout table of the finishing mill, i.e. the wavy length of 5 strip ´fibers´. Relative differences in fiber lengths as small as 10^{-6}, undetectable by visual inspection, can be determined.

On a wide flange beam mill, a new system measures the flanges width of hot H-beams whose width x height dimensions are ranging from 100 x 100 up to 500 x 1100 (mm^2). This automatic system, called DIMPRO, provides measurements with a $\pm$ 1 mm-accuracy in the whole range without requiring any additional sliding guides of the beams in the reversing mill.

1. Introduction

In many processes in the steel industry, the dimensions of the products must be controlled during and after manufacturing in order to assess the quality and to avoid breakdowns in the production lines. Contactless measurements are often required, especially on hot moving products. Moreover, the continuous improvements in steel quality and productivity conduce to the development of additional instrumentation. With this respect, the modern optoelectronic techniques are widely applied to new gauges for measurement in the harsh environment of the steel industry.

2. Optical triangulation-based measurement

A well-known method to measure dimensions is based on the monostatic triangulation operating on the principle of image shift. It consists in solving a triangle, one dimension of which being the variable to be measured. As an example in Figure 1, an object whose thickness has to be measured is laying on a reference plane. A light beam, issued from a source S, is vertically impinging onto this object. At the impact point M, the optical energy is partially diffused in the direction of the receiver whose lens focuses it on the sensitive surface of the detector at point I. Any thickness change (e to e´) will shift the image point from I to I´. For a given layout, the thickness e is a well-defined function of the I location.

Due to the important advancement in optoelectronics during the last twenty years, new analog and digital receivers have been manufactured. At the same time, the gauges based on triangulation methods widely used infrared-emitting diodes (leds) and laser sources. The infrared-emitting diodes are semiconductors having a strong diverging radiation usually in the near infrared spectrum and a quite narrow optical bandwidth (20 to 50 nm). The He-Ne lasers and the lasers diodes are mostly used in triangulation-based systems : the most popular He-Ne lasers emit a monochromatic red beam (633 nm), having narrow divergence and low power (1-10 mW), the lasers diodes have similar power caracteristics but radiate in the near infrared band with a large divergence. These sources have long life times (more than 15000 hours) ; their narrow spectrum facilitates the filtration of disturbing radiation such as the radiation emitted by hot products.

At the receiver side, the one-cell detectors need mechanical scanning to cover the whole field of view ; they are often replaced by photodetectors allowing a direct detection of the spot location. Besides the well-known TV camera tubes (Vidicons,...), the following detectors, using Silicon as sensitive material, are available :

- the analog sensitive photodiodes (Figure 2a) operate, in a first approach, as photopotentiometers (B´). At the impact point, the light is converted into a proportional electric current dividing itself between the two branches of the potentiometer. By measuring the currents at both ends, the abscissa of the beam impact point can be calculated ;
- the linear multi-elements array detectors combine a reduced number of photocells on a same substrate. By processing the signals from each cell, the spot position can be deduced. The advent of the charge transfer technology (CCD) permits the production of linear arrays fitted with several thousands detectors (Figure 2b). The radiated energy is directly converted into electrical signals available as a succession of pulses at the output pin of the circuit.

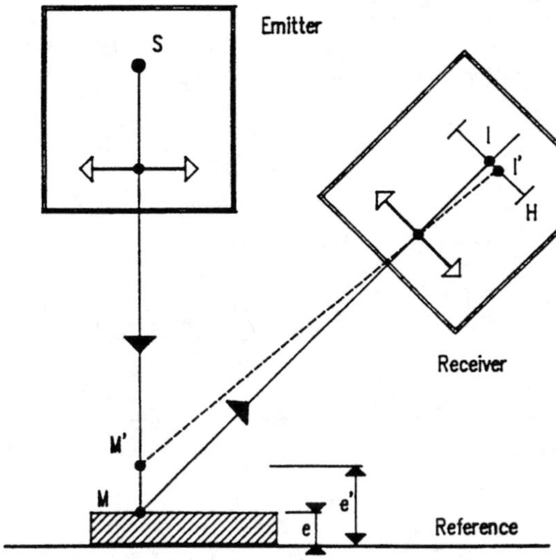

Fig. 1 - Principle of the distance measurement.

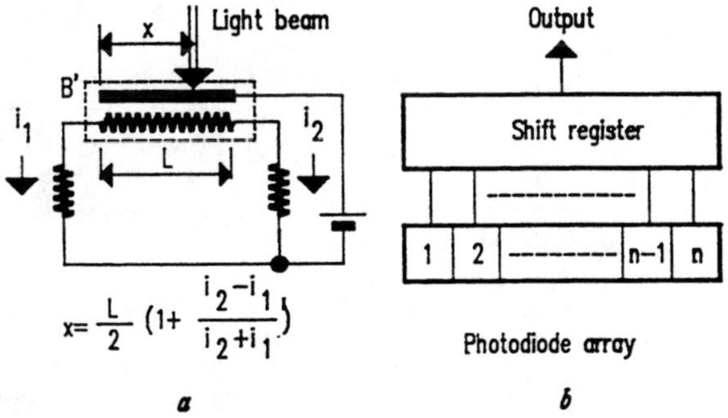

Fig. 2 - Position detectors.

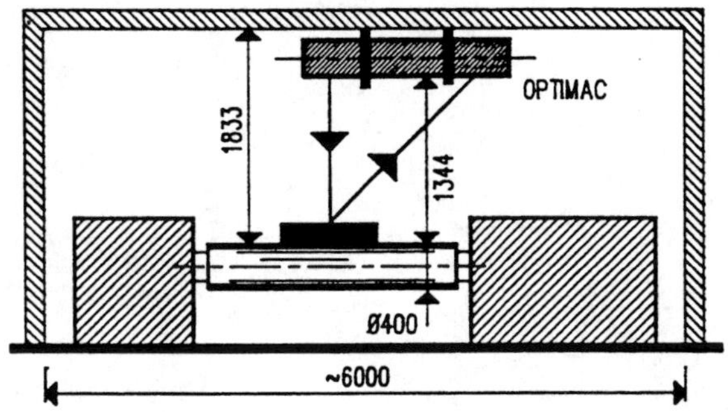

Fig. 3 - Thickness measurement with the OPTIMAC gauge.

Following examples show how the optical triangulation technique has been used in CRM to solve measurement problems in the steel industry.

3. Slab thickness measurement at the roughing mill

The OPTIMAC gauge has been developed to measure the thickness of semi-products (slabs) at the roughing mill. The gauge is usually installed in front of the roughing stand and its measurement data are compared with the position of the screws, i.e. with the gap between the horizontal rolls. In case of discrepancy, the roller table is switched off. As a matter of fact, when the slab thickness is slightly higher than the value allowed by the roll gap, the bite occurs but the slab cannot be rolled due to the excessive generated rolling pressure ; breakdowns in rolls, shafts and motors can happen.

Both emitter and receiver are fastened to a rigid steel section and located in a water-cooled housing fitted with two windows (Figure 3). This housing is hung under a rigid gantry anchored to the floor on both sides of the roller table. The light issued from a halogen lamp is modulated by a chopper and focused onto the target. The detector is an analog PIN sensitive diode.

For a thickness measuring range from 90 to 310 mm, the accuracy on products is ± 2 mm, which is adequate to identify the product to be rolled. This gauge has been continuously used for more than ten years in a Belgian steelwork. For the following applications requiring a higher accuracy, digital detectors were preferred Moreover, their better resolution, linearity and insensitivity to optical intensity changes were appreciated.

4. Measuring hot strip flatness

Hot slabs are rolled into strips (usual range : 1 to 2 m in width, 2 to 10 mm in thickness, 10 to 20 t in weight) by a line of stands (8 to 10) equipped with 4 horizontal rolls (or 6 in the most advanced mills). Downwards, the strips are cooled down on a runout table and finally coiled. The strips can be sold as such or rolled in a cold mill, annealed, etc... In the worldwide competition, the steelmakers have to control the strip flatness during hot rolling.

4.1. Principle of the flatness measurement

The C.R.M. gauge, so-called ROMETER-5, determines, on the one side, differences between the wavy lengths-i.e. the real strip length along the defects - on 5 selected points accross the strip width and, on the other side, RO-indexes which define flatness, respectively on the edges with respect to the middle (RO135 - Figure 4) and between both edges (RO15). Similar RO-indexes are calculated between L2, L3 and L4. As an example, a sinusoidal defect with A mm in the magnitude and a wavelength λ can be defined with a RO-value according to Figure 4.

The operating principle is illustrated in Figure 5. The longitudinal section of the strip L has to be calculated in real-time. For this, the strip levels above a reference plane (the runout table) Y0, Y1, Y2... Yn are measured periodically at the moments t0, t1, t2 ... tn respectively (for instance at 2.000 Hz) and the portion L can be obtained according to Pythagora's relationship where the rolling speed is known at every time.

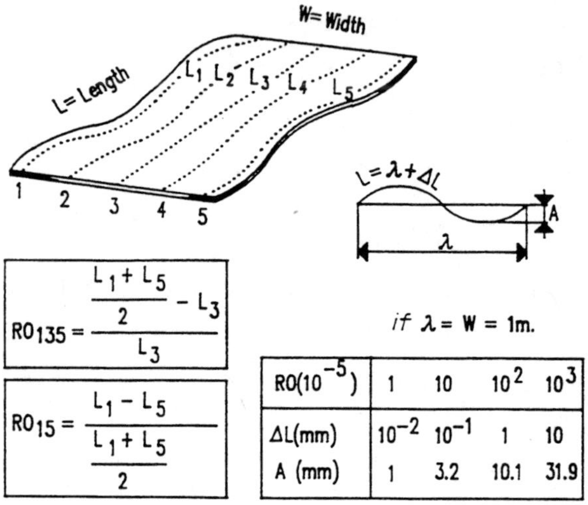

Fig. 4 - Definition and meaning of RO-indexes.

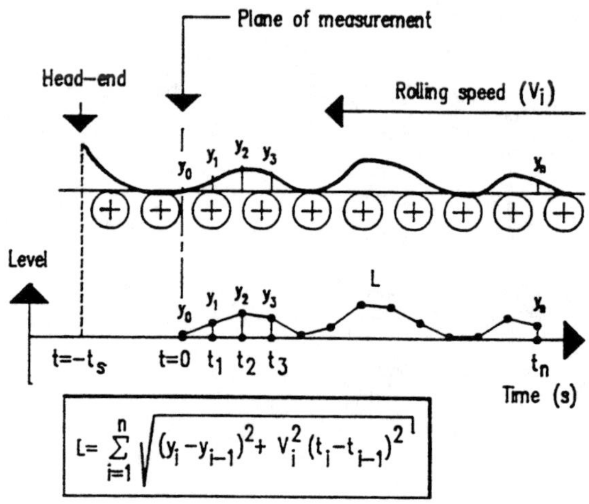

Fig. 5 - Principle of flatness measurement.

The detection of wavy edges for every width in the range from 600 up to 2.000 mm and the selected intermediate fibers between center and edges requires presetting two laser beams in order to illuminate the strip at constant distance from the edges and two other ones onto the fibers located for instance at equal distance respectively between center and edges. The most reliable solution consisted in directing each laser beam by a rotating mirror ; on the receiver side, the optical axis of the lens is oriented in the same way (Figure 6). Five telemeters have been set up for covering the whole width range ; they also permit a fast tracking of the edge location when the strip is shifting from one side to the other during rolling (Figure 7). These facilities are also valuable for measurements on heavy plates and for 2 or 3D-sizing.

4.2. Description of ROMETER-5

Each telemeter consists in an He-Ne Laser (5-10 mW) and a line scan camera fitted with 1024 photo-diodes. The sensitivity of the level measurement amounts typically to 0.05 mm in the field of view and the flatness index amounts to 10^{-6} by processing the level data. The rotating mirrors are fixed on a table powered by stepping motors ; an optical encoder locates the shaft position.

A microprocessor-based unit continuously controls the presetting sequence and handles the data delivered by the camera for determining the required strip lengths. The 5 telemeters are linked to a central minicomputer which collects useful data from the mill instrumentation (rolling speed, strip width,...), presets the motors and delivers results to the gauge peripherals and to the mill computer.

The measuring components were mounted into two rigid water-cooled boxes fastened on both sides of the runout table at a suitable height for preventing any breakdown by vertical motion of uncontrolled strips. These measuring boxes can be conveniently fastened below a heavy gantry or into an air-conditioned cabin.

Numerous investigations were achieved for operating free of disturbances such as heat radiation, dust, vapor and vibrations.

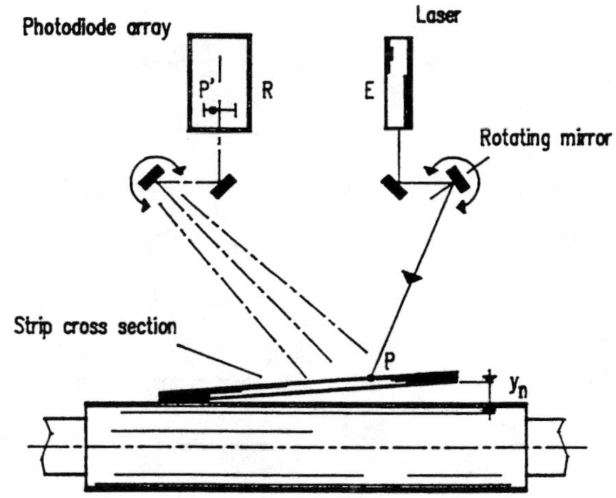

Fig. 6 - Principle the strip level measurement.

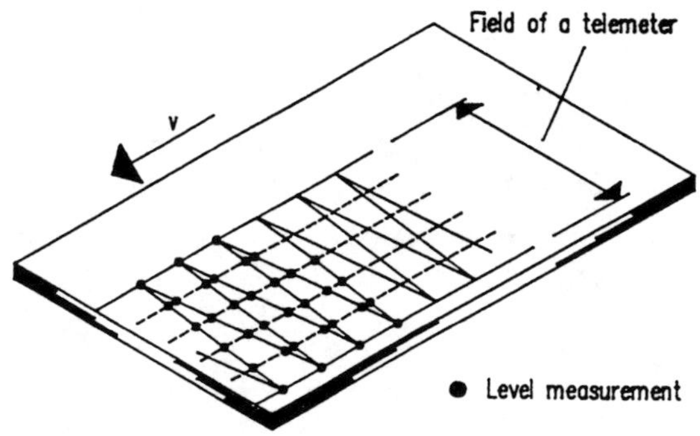

Fig. 7 - Scanning a sheet.

4.3. Industrial results

The ROMETER delivers listings of the RO-evolution along the products. For special purposes, the analog outputs of the five telemeters can be recorded. As an example, Figure 8 shows records of 140 Hz measurements along the whole strip length. Unequal wavy edges can be noted. Difficulties in the reliable visual discrimination between flatness defects and strip vertical oscillations are evident, mainly before coiling. An overall examination of Nr. 3580 strip reveals wavy (RO135 > 0) and unequal (RO15 = 0) edges. The defects are rather big at the head-end and remain detected when the strip is being tensioned by the coiler ; the flatness becomes better afterwards and finally decreases at the tail-end. These conclusions are deduced from the RO135-values and confirmed by the RO234-evolution (Figure 9).

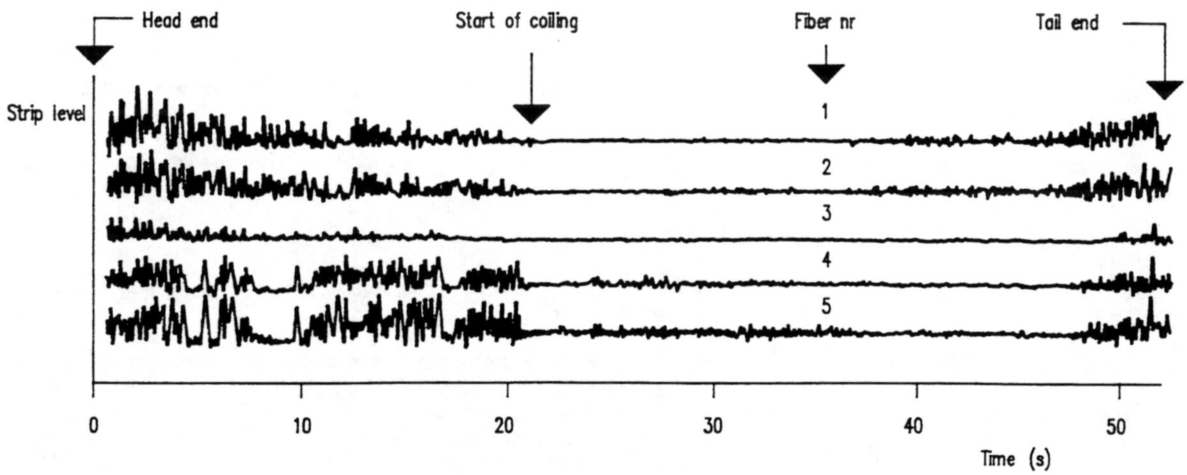

Fig. 8 - Measurement data before and during coiling.

The ROMETER permits adjustments of the levelling of the last stand, i.e. screwing the upper rolls down on one side and up on the other in order to get strip edges in the same length. Figure 10 shows, as examples, manual levelling adjustments and their favourable influence on the measured RO15 a few seconds later on. An automatic feedback on the mill control is also being applied as a major advantage of this shapemeter.

Equipments were installed in the output of three European finishing mills.

Coil 3580

RO (.10^{-5}) w = 1113mm

R0135	R0234	R015	R024
153	128	117	138
146	95	94	85
107	124	72	67
19	25	13	13
0	0	0	0
2	4	2	3
0	2	1	2
2	3	2	3
3	1	1	1
14	21	6	8

Coil 3702

RO (.10^{-5}) w = 1113mm

R0135	R0234	R015	R024
1	0	0	−1
0	−1	0	0
0	0	0	0
1	1	0	0
0	0	0	0
0	0	0	0
0	0	0	0
0	0	0	0
0	0	0	0
0	0	0	0
0	0	0	0
0	2	0	0
0	0	0	0
0	0	0	0
0	−1	−5	−3

Fig. 9 - Measurement data on two coils.

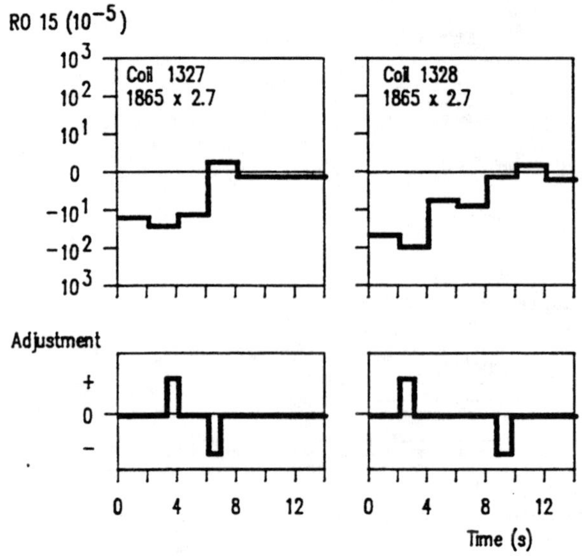

Fig. 10 - Rolls levelling adjustments.

5. Beams width measurement

The steel beams are usually issued from cast semi-products which take gradually the wanted H-shape during rolling by several groups of stands : the web is laminated by horizontal rolls, the two flanges are made thinner by vertical stands. At each pass, the rolls must be shifted by the screws according to the rolling sequence which leads to the final dimensions. It is nowadays important to collect continuous measurements during rolling in order to optimize the sequence at each pass and to improve the dimensions adjustments. These advantages are mostly important in rolling products with special dimensions requested by the customers. The C.R.M. gauge, so-called DIMPRO, is an industrial solution to on-line measurements of the two flange widths.

5.1. Measurement principle

The major difficulties of the beam measurement are bound to the nature of the product itself and to the rolling process i.e. :

- the width is ranging from 100 to 500 mm ; the height : from 100 to 1100 mm ; the required accuracy amounts to ± 1 mm in the whole range ;

- the surface temperature of the beams amounts to 1100°C ;

- during rolling, the beams are not only moving longitudinally by also vertically above the rollers and transversely between the lateral guides of the runout table.

These requirements forbid any direct contact measurements and move the choice towards optical methods. The selected method is based on the measurement of the coordinates of the 4 edges of the beam section. The Figure 11 shows how such a point P is located. This edge is at the intersection between a vertical V line, at distance D from a reference fixed point, and an oblique line which separates the light and the shady parts of the source. The D distance is measured by a commercial contactless telemeter. The angular position of the oblique line is measured by the abscissa of the photodiode where the transition takes place on the line scan array of the receiving camera. For improving the optical contrast on the detector and to be free of the variations of the steel luminosity, a high emitting source has been located behind the beam.

The same principle has been applied at the 4 points of the section ; the width of the flanges are computed by difference (Figure 12). Reflecting mirrors are set up in the optical lines of sight for reducing the outer dimensions of the receiving box. The range of flange width is so wide that the measuring layout for the upper edges is arranged so that the effective field of view remains free of geometrical distortions. For this aim, a tilting mirror system has been mounted in the optical path : the upper line of sight is used for the higher widths ; the lower one for narrow flanges. Thanks to this device, the gauge owns 6 lines of sight for 4 cameras. The layout is designed to obtain the same image/object ratio for the 6 lines of sight (i.e. the same triangulation magnification). Consequently, many errors are neutralized. Moreover for getting a sharp transition between the illuminated and the dark segments of the photodiode array, the lens must be focused on the sighted flange edge. Rather than using an autofocus system introducing unacceptable tranverse misaligments of the lens,it has been decided to translate the receivers for keeping the lens/object distance constant independently of the H-height.

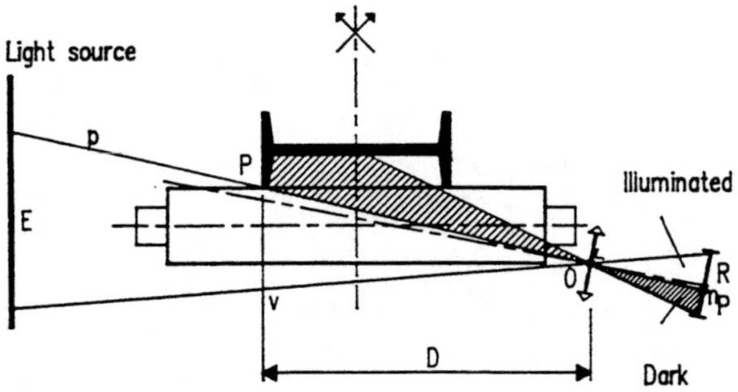

Fig. 11 - Principle of the position measurement of a wing edge.

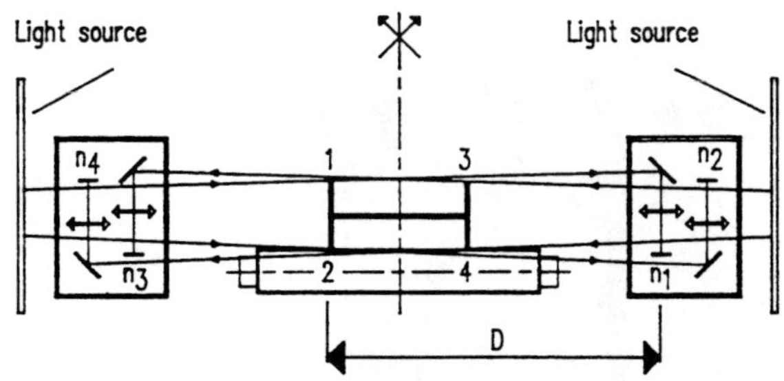

Fig. 12 - Optical layout of the width gauge.

5.2. Description of the DIMPRO gauge

The source boxes, located at each side of the rollers table contain 35 iodine lamps which are supplied with reduced voltage for increasing their life time. The lenses are focused on the beam flanges and therefore the lamps, out of the field of focus, appear as a continuous line of light. The support of the 2 receivers sighting the same flange may be shifted by a stainless steel screw powered by a stepping motor.

The cameras measuring the upper and the lower edges are fitted respectively with 1024 and 512 photodiode arrays. In the first system developed for a C.R.M.-affiliated company, the upper cameras cover a vertical field of view amounting to 407 mm and the lower cameras cover only 118 mm which is adequate for taking into account the slight vertical motion and the straightness defects of the beams during hot rolling. The transverse location of the flanges on the rollers are measured with two infrared telemeters. The water-cooled boxes containing respectively the cameras and the lamps are made with stainless steel (Figure 13). A cooled electronic cabinet contains the main units of the equipment i.e. the units driving the cameras, the motors and supplying the lamps. The temperature control unit, the acquisition and data processing units and peripheral equipments were also located in this cabinet.

The data are processed by a Data General Nova computer having 32 Kbytes central memory and a multiplication/division unit. It is fitted with the required interfaces to the peripherals. The major function of the computer is to handle in real-time the measurements and to determine instantaneous and average values of the flange width (for instance by averaging 10 instanteneous measurements at 125 Hz). All these data are displayed in front of the operators, printed out and logged in the process computer.

5.3. First results

The accuracy of the gauge has been checked periodically. The first controls of the gauge performances were carried out in C.R.M. laboratory with the use of reference plates simulating the beams of all the range of dimensions (100 x 100 to 500 x 1100 mm) and locations (± 100 mm transversely ; +50, -10 mm vertically) of the beams on the runout table. They have proved an accuracy of ± 1 mm even in the worse conditions. This result has been corroborated by calibration tests in the mill. The measurements proved also to be free of disturbances such as temperature and rolling speed of the products. The reproducibility has been determined over several months ; typical results are displayed in Figure 14.

A few results showing the evolution of the flange widths of 9 beams at three passes downwards the mill (odd passes) are shown in Figure 15. The gradual reduction of the width can be observed. The final dimensions of the beam will be obtained after a last pass in this mill, a final pass in the finishig mill and the cooling down to the ambiant temperature.

6. Conclusions

The gauges described in this paper are examples of C.R.M. application of optical triangulation-based methods for carrying out intricate measurements in hostile environments. The field of application of these methods is being extended in order to replace contact measuring systems. These gauges can also be used for testing other non-metallic products and for fulfilling many specifications.

Fig. 13 - Emitter and receiver boxes.

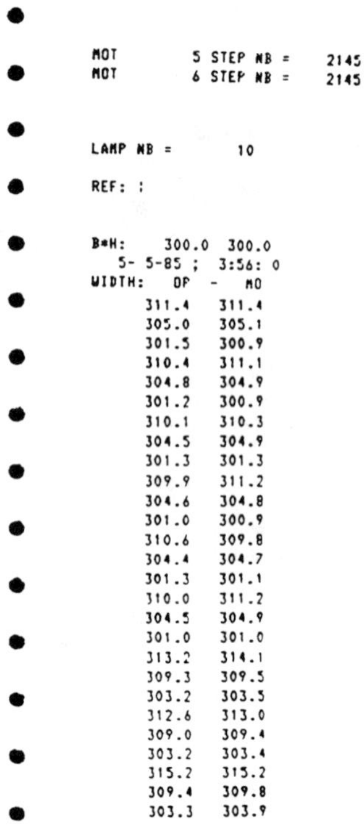

```
MOT          5 STEP NB =    2145
MOT          6 STEP NB =    2145

LAMP NB =         10

REF: :

B*H:     300.0  300.0
   5- 5-85 ;  3:56: 0
WIDTH:    DP   -   MO
        311.4   311.4
        305.0   305.1
        301.5   300.9
        310.4   311.1
        304.8   304.9
        301.2   300.9
        310.1   310.3
        304.5   304.9
        301.3   301.3
        309.9   311.2
        304.6   304.8
        301.0   300.9
        310.6   309.8
        304.4   304.7
        301.3   301.1
        310.0   311.2
        304.5   304.9
        301.0   301.0
        313.2   314.1
        309.3   309.5
        303.2   303.5
        312.6   313.0
        309.0   309.4
        303.2   303.4
        315.2   315.2
        309.4   309.8
        303.3   303.9
```

Fig. 15 - Print-out of measurement data.

Nominal dimensions		Drive side		Cabin side	
Width x height (mm)		84/08/24	85/06/05	84/08/24	85/06/05
389,1	650	389	388,9	388,9	388,6
287	400	287	287	286,9	285,9
500	1100	499,8		500	
500	1000		500,8		500,4

Fig. 14 - Long-term reproducibility tests.

7. References

1. Pirlet R., Franssen R., Laforge V., Surleau F., Distances and dimensions gaging in iron and steelmaking, Acta Imeko pp. 389-394, 1976.

2. Pirlet R., Mulder J., Adriaensen D., Boelens J., Lochen Cl., A noncontact system for measuring hot strips, Iron and Steel Engineer, pp. 45-50, July 1983.

3. Salmon J., Boelens J., Pirlet R., Haler G., Reimen V., Mesure de la largeur des poutrelles en cours de laminage, jauge DIMPRO, C.R.M. report S 3/86.

(*) The researches were carried out under the joint sponsorship of the European Coal and Steel Community and the Institut pour l´Encouragement de la Recherche Scientifique dans l´Industrie et l´Agriculture (Belgium).

Reprinted from *Optics, Illumination, and Image Sensing for Machine Vision,*
D.J. Svetkoff, editor, Proc. SPIE Vol. 728, pp. 152-159 (1986).

A Novel, High Range-to-Resolution Ratio, Optical Sensing Technique for High Speed Surface Geometry Measurements

Robert D. Lorenz

Department of Mechanical Engineering, University of Wisconsin-Madison
1513 University Avenue, Madison, WI 53706

Abstract

A novel combination of image formation theory, along with optimal digital estimation theory was employed in development of a new optical probe suitable for a variety of high speed sensing applications, such as in-process manufactured part inspection. Experimental results demonstrate a robust, 40,000:1 range-to-resolution ratio. The sensing strategy also utilizes multiple, simultaneous illumination beams such that it is possible to accurately estimate surface orientation for robotic guidance applications. A significant element in this research was the development of a technique which is least influenced by surface finish roughness as commonly found on manufactured surfaces. Experimental results are provided to demonstrate the validity of the theoretical models.

Introduction

A large number of industrial automation problems require rapid measurement of surface geometry, such as automated in-process inspection or in-process guidance of robotic work heads. The solutions to such problems are largely dependent on development of sensing technologies which possess the following attributes:

- high resolution, of at least ±0.0001 inches,
- sufficient range that the servo system spatial dynamic requirements will diminish to acceptable levels for high speed motion,
- minimum influence of surface finish on the accuracy of the measurements, and
- orientation measurement sufficient to allow intelligent robotic guidance.

A number of commercial optical probes have been developed based on ray optics triangulation techniques.[1,2,3] A general optical triangulation configuration is shown in Figure 1 where the primary output consists of an image location measurement.

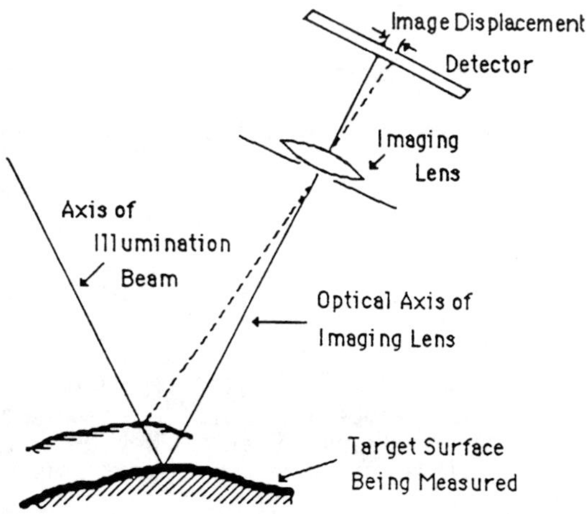

Fig. 1 Optical Triangulation Probe

The currently available probes generally do not provide substantial advantages over classical contacting style probes[4] because they possess a narrow sensing range and thus the probe or part must be precisely servoed at slow speeds to follow the part contour. The potential speed of optical triangulation sensing is not realized by such a requirement. In addition, such probes do not measure surface normal. This measurement or some surrogate is required for robotic guidance along surfaces.[5]

University researchers have sought to develop techniques which provide some combination of high speed,[6] wide sensing range,[7] and surface normal measurement.[8] The general approaches taken by both the commercial and the research probe investigators have not included either image formation theory or optimal estimation theory. As a result, the goals outlined above do not appear to have been achieved in prior work.

This paper begins by presenting a discussion of the limitations of the existing optical triangulation technologies. This discussion leads into a formulation of an image formation and estimation theoretic approach which will allow the range, resolution and surface finish dependency requirements to be achieved.[10] This theoretic discussion is followed by a formulation of techniques for high speed signal processing and continuous surface orientation measurement.[10,11] The experimental verification of the formulation is presented in the final section.

Limitations of current optical triangulation techniques

All of the optical triangulation approaches are based on measurement of the location of the image at the detector. Thus, a key issue in determining measurement performance is the ability to accurately resolve image location. The resolution in this measurement, in addition to the length of the detector provide a bound for the range-to-resolution ratio of a given design. The major limitations of current optical triangulation techniques are then based on:
detector type, image location analysis approach, illumination shape, and illumination orientation constraints.

Detector limitations

There are two classes of detectors which are commonly used in optical triangulation probes: lateral effect photodiodes, and linear photodiode arrays.

Lateral effect photodiodes are single element, analog devices which may be viewed as possessing a one-dimensional sensing capability. A simplified 1-D model is shown in Figure 2.

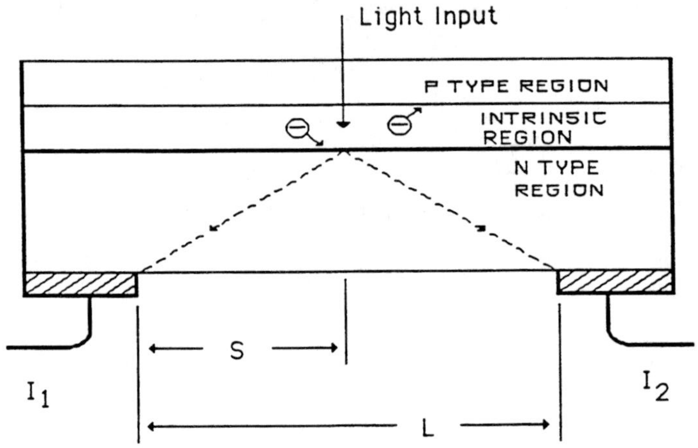

Figure 2 Simplified Model of a Lateral Effect Photodiode

They may be modeled as dividing the total photon-induced current between the two output terminals depending on the distance between the incident radiant flux and the two output pins. This relationship may be modeled in one dimensional space as

$$\frac{I_1}{I_1 + I_2} = \int_0^L P_A(S) \cdot R_\lambda \cdot \frac{S}{L}\, dS \tag{1}$$

where the incident flux distribution $P_A(s)$ is taken as uniform in the transverse plane. This output relationship is seen to be a first moment or centroid calculation. Thus, independent of image shape, a centroid algorithm is inherently used by this device. Image shape information is irretrievably lost. Such devices are limited in accuracy by the uniformity of the bulk properties such as bulk resistivity. Typical range-to-useable resolution ratios are 1000:1.

A second class of detector, the linear multi-element photodiode array, consists of a series of discrete, but monolithically constructed photodiode elements as shown schematically in Figure 3.

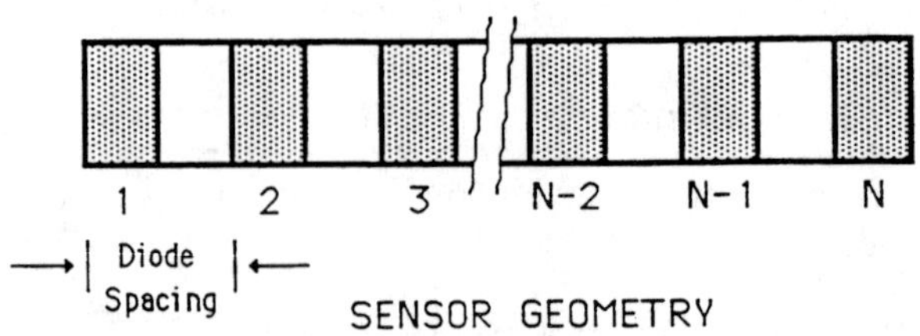

Figure 3 Linear Photo-Diode Array Geometry

Each of these photodiodes (pixels) produce photon-induced current resulting from the spatially integrated incident flux on their respective aperture area. Due to the low levels of currents produced by such elements, an equivalent capacitor is used to integrate the photon currents to measurable voltages (or charges). Thus, the photodiode array element output may be modelled as a spatial and a temporal integration as follows:

$$e_{diode} = \int_0^T \frac{1}{C} \int_{x_i}^{x_{i+1}} P_A(x) \cdot R_\lambda \cdot dx\, dt \qquad (2)$$

The image produced by such devices appears as a spatial, sample & hold, (or boxcar) shape as shown in Figure 4.

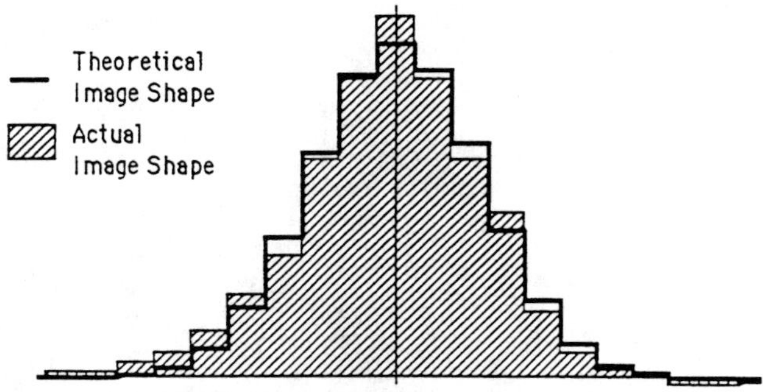

Figure 4 Spatial Output of a Linear, Photodiode Array

Unlike the lateral effect photodiode, linear diode arrays provide image shape data, albeit of a sampled form. The relative size of the image to the pixel will determine the ability to reconstruct the image shape. Because of this, the signal processing algorithm and image size/shape determine the limits of the spatial image location estimation for linear photodiode arrays. Thus, linear diode arrays of ≥1000 pixels will provide at least as much image location range-to-resolution as lateral effect photodiodes.

Image location analysis limitations

The lateral effect photodiode is inherently limited to perform the centroid image location algorithm. However, the linear diode array will allow any form of processing, dependent only on the compatability with the sampled form of the output. The current technology of triangulation probes provides centroid-type of algorithms. This is largely a result of the uncontrolled nature of the image shape in the existing probe designs. If image shapes and size is allowed to vary over the range of the probe, but a generally symmetrical shape is produced, then a centroid algorithm is an unbiased estimator. Unfortunately the centroid algorithm is unfavorably affected by image size (width). A larger image size will produce a larger image location variance for a given level of image noise. In the limit, this effect would force image size to be one pixel. Such a result would imply range-to-resolution ratios equal to the pixel length of the detector.

Illumination shape limitations

For any given illumination wavelength, the image shape in the general triangulation probe of Figure 1 is dependent on the illumination shape, the surface spatial model, and the impulse response of the lens system. Convolution of these terms will allow prediction of the image shape. Of these terms, the surface spatial model is generally unknown. However, the illumination shape and the impulse response of the lens system are probe design optimization issues.[11] For most of the current technology triangulation probes, the illumination beam is not formed to produce a desired shape in the detector image. Thus, more optimal, shape-based location estimators are not easily applied to the existing design of triangulation probes .

Illumination orientation limitations

A fundamental consideration in any optical instrument is the light intensity at the detector. Orientation of the illumination beam and detector to the surface is a primary variable in such a design decision. For triangulation probes viewing diffuse surfaces, the orientation of the illumination beam and detector is a very insensitive term. However, for more specular surfaces the instrument operation will be enhanced by near specular alignment. Most of the current technology triangulation probes assume illumination normal to the surface and thus are configured for diffuse surface measurement. Near specular surfaces are not easily measured with such configurations. Despite such limitations, most of the current triangulation probes do not provide a surface normal measurement so that alignment may be sensed and orientation controlled.

A new image formation and estimation theory approach

To make a fundamental improvement in the resolution and range-to-resolution ratio of optical triangulation probes it is imperative that a significant improvement in the image location estimation algorithm be achieved. To do this requires contributions from two domains: image formation theory and estimation theory. The most fundamental is from estimation theory, where sub-pixel resolution must be demonstrated. It will be shown that under even substantially noisy circumstances, that such resolution is achievable.

An estimation theory approach for attaining sub-pixel resolution

It is well understood that optimal estimators contain all possible information about the expected characteristics of a signal, that is to say, only unknown noise is left unexplained. Thus, if both the theoretical image shape and the lens system impulse response is known, then an optimal estimator must use this information. Failure to do so will produce higher uncertainty than is optimum. If the theoretical image shape function is known, then it is possible to employ a common signal processing technique for extracting an estimate of its location. The technique, digital image cross correlation is applied by shifting the theoretical image shape until the product of it and the actual image is maximized. The spatial location estimate then becomes:

$$X_c = S \Big|_{R_{I_A, I_d}(S) = \text{Max}} \tag{3}$$

where

$$R_{I_A, I_d}(S) = \frac{1}{L} \int_0^L I_A(x_i) \cdot I_d(x_i - s)\, dx \tag{4}$$

and $R_{Ia,Id}$ is the cross correlation estimate of the theoretical image with the actual image,
I_a is the actual image amplitude as a function of displacement on the detector,
I_d is the desired, theoretical image amplitude as a function of displacement on the detector,
L is the detector length, and
X_i, S are distances along the detector.

This cross correlation function (R_{xy}) has a smooth, unimodal behavior even with severly distorted image shapes, as shown in Figure 5.

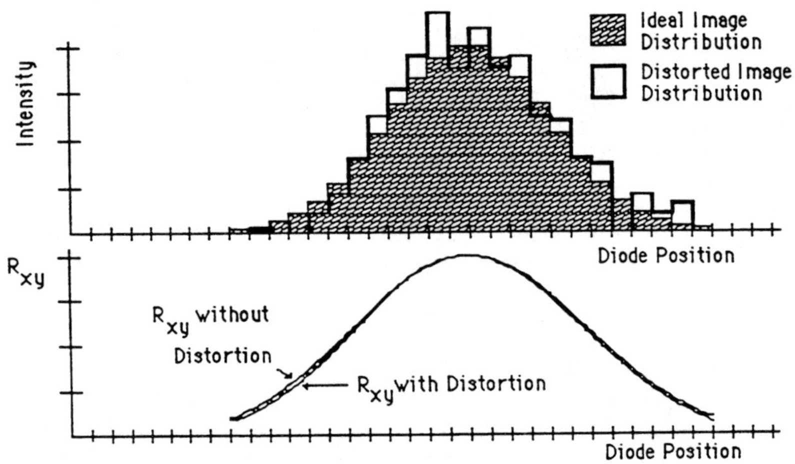

Figure 5 Unimodal Behavior of Cross Correlation Algorithm under Conditions of Signal Noise

The image location estimate produced by this algorithm is very robust and can be shown to be unbiased and efficient. To further optimize the image location estimate, however, it is possible to vary the shape and width of the image.

An image formation approach to optimize both the image location estimate and the signal processing task

Image formation design variables permit at least two additional degrees of freedom in improving the noise rejection properties of the Max[R_{xy}] location estimator and in minimizing the signal processing overhead. The two, independent, manipulated design variables include the image width and the image shape. If various shapes are evaluated then the optimum combination of shape and width may be selected. With extensive numerical optimization which includes detector saturation effects, it can be shown that width of approximately 10-15 diodes is optimum for shapes which have smooth tails and are heavily corrupted by image noise due to surface finish on the measured part surface.[9]

Surface orientation measurement

To achieve a measure of the surface normal as needed for automated robotic guidance on surfaces[5], it is possible to apply the beam splitting to effectively generate multiple simultaneous probing points. Figure 6 shows a two beam implementation which can be used for surface slope measurement in the triangulation plane.

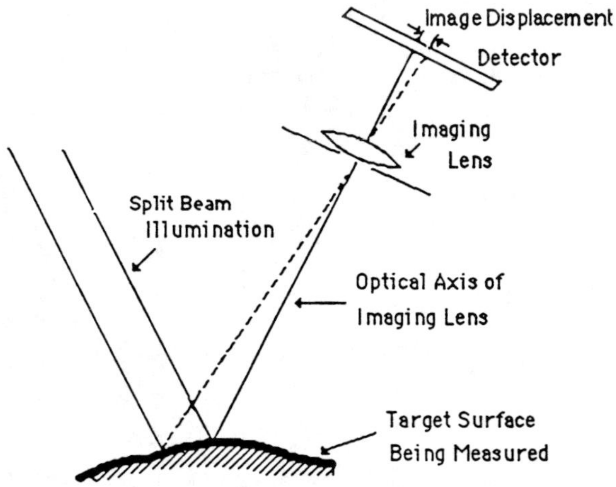

Figure 6 Beam Splitting Configuration for Surface Normal Measurements.

From three spatial points which are not co-linear it is possible to find the three coefficients of a plane through the points. Such plane-fitting routines are common with contacting probe measurements but are usually obtained via sequential measurements. The parallel acting nature of optical probes will allow continous determination of the path rate of change needed for robotic guidance. For robotic guidance purposes the precision of the slope measurement can be within a relatively broad tolerance (the possible servo following error).

Experimental results

A two-beam probe was constructed to investigate the performance which may be achieved using the image formation, optimal estimation principles, and 2-D surface normal principles described above. The configuration of the experimental probe is shown in Figure 7.

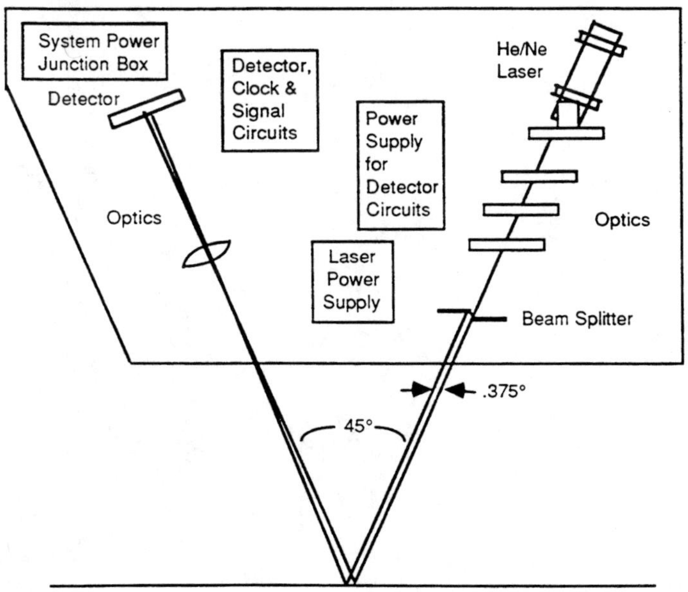

Figure 7 Experimental Probe Design

The probe was installed on the Z-axis of a modern CNC machining center. The real time computations for the initial tests were performed on a DEC PDP 11/34. Currently the real time software runs on an IBM PC AT. A variety of tests and over two years of use have demonstrated the following attributes.

Resolution

To evaluate resolution the probe Z-axis was stepped in 0.0001 in. increments. The probe output was compared to the reading of the machine's axis resolver. Results with target surfaces of both machined cast iron and painted steel are shown in Figure 8.

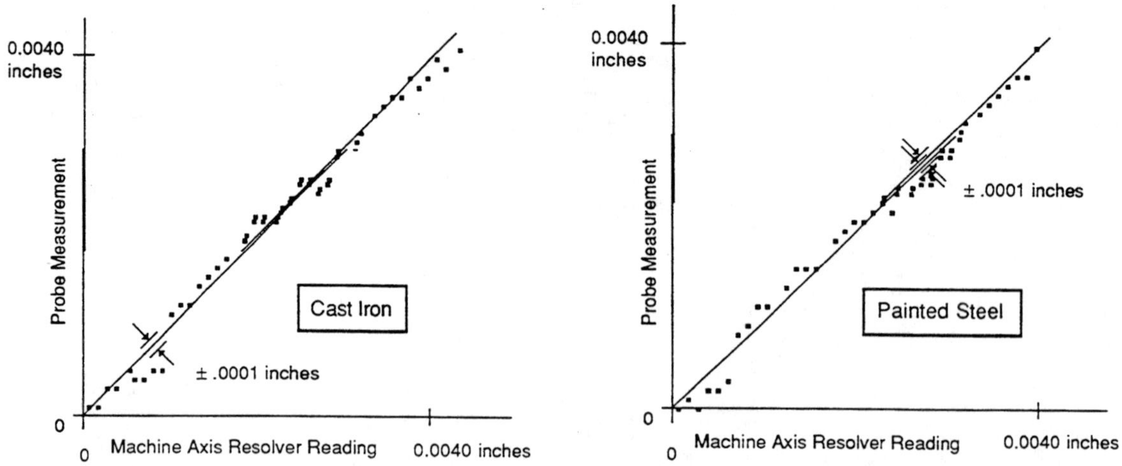

Figure 8 Incremental Resolution of the Experimental Probe

The results are very repeatable given a repeated starting position of the motion. From the results it can be concluded that the probe resolution is comparable to that of the machine (± 0.000050 inches) motion.

Linearity and range-to-resolution ratio

The linearity of the probe measurement (after inversion of the triangulation equations) was evaluated over a 4 inch range. The probe's measurment was compared to the machine's feedback resolver. Figure 9 shows the linearity and enlarged deviations plots for the same machined cast iron and painted steel targets used in investigating resolution.

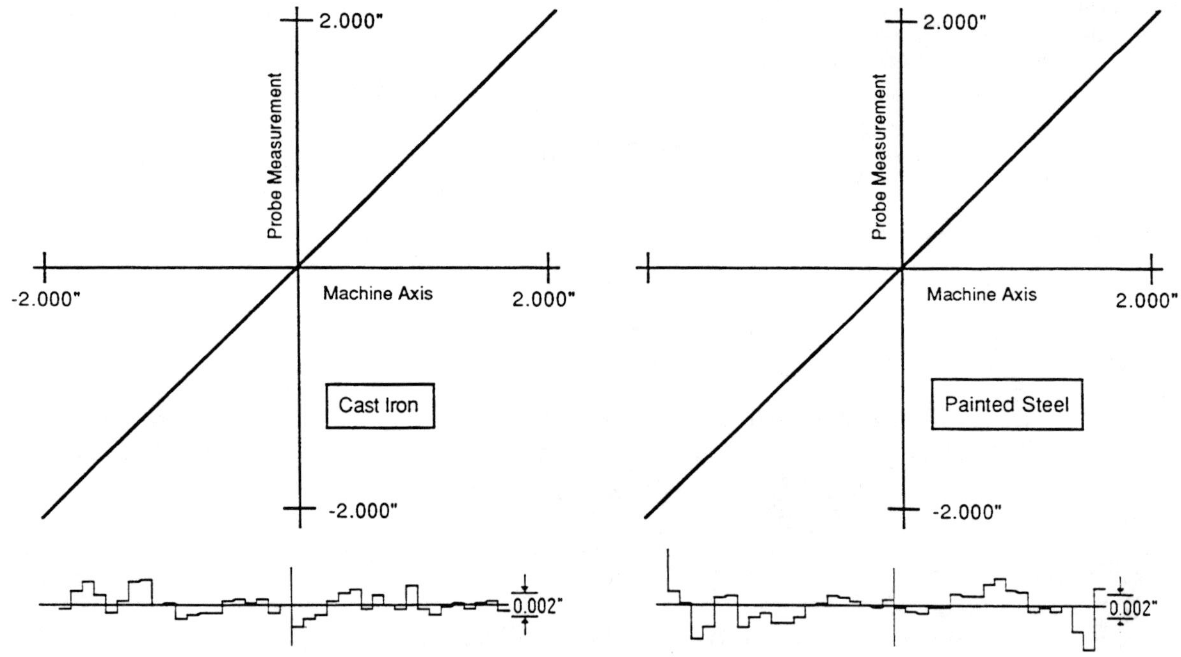

Figure 9 Linearity Evaluation of the Experimental Probe

In both cases, the linearity deviations on the 4 inch scale fell within the pixel resolution of the plotting system.

Surface normal

The linearity of the probe's surface normal measurement was evaluated over a ±30 degree range by using a servo driven indexing head of 0.0587 degrees resolution. The probe's measurment was compared to the indexing head's encoder. Figure 10 shows the linearity and enlarged deviations plots for a machined steel and a very diffuse cardboard target.

The linearity deviations on both the cardboard and the machined steel fell within the a 1 degree resolution for surface normal measurement. The range was ± 30 degrees for a very diffuse surface and ± 6 degrees for a more specular surface.

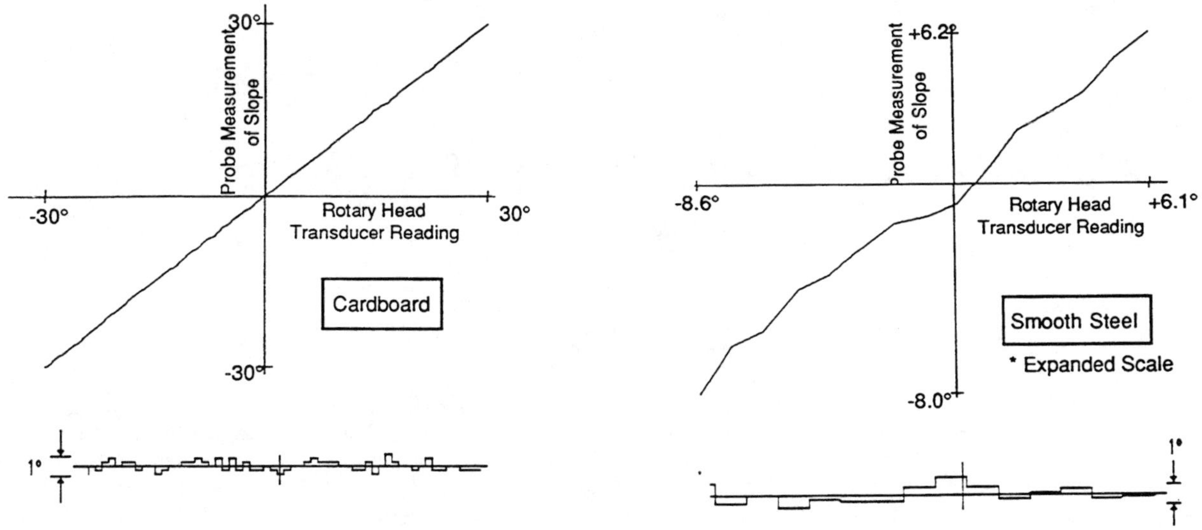

Figure 10 Linearity Evaluation of the Experimental Probe for Surface Normal Measurements

Conclusions

A unified, rigorous theory for optical triangulation probe design has been presented. It is based on optimal estimation theory and image formation theory. The key theoretical contributions include

- development of a means of attaining sub-pixel resolution by applying digital image correlation and by controlling the image formation process
- optimization of the image formation process to reduce sensitivity to target surface finish
- development of a means of directly measuring surface normal as required for robotic guidance problems.

The key experimental contributions achieved by this research and presented in this paper include development of a probe which:

- demonstates a useable resolution of 0.0001 inches,
- demonstrates acrange of 4.0000 inches, thus providing a 40,000:1 range to resolution ratio which far exceeds the 1000:1 ratio commonly found on currently available sensors,
- demonstrates ability to achieve the above range-to-resolution ratio performance on surface finishes such as cast iron, painted steel, milled steel, formed steel, and opaque plastic surfaces, and
- demonstrates surface normal measurement range of up to ± 30 degrees.

Acknowledgement

A patent application on the probe design approach described in this paper has been filed by Rexnord Corporation on behalf of Dr. Lorenz.

References

1. Renishaw Inc., "OP2 Laser Scanning Probe User's Guide," Provisional Issue 2, Schaumberg, IL, January 9, 1986.
2. Diffracto, Ltd., "Laser Probe Digital Ranging Sensor," Product literature, Windsor, Canada, June,1981.
3. Selcom AG., "Gauge Probe," Selective Electronic Co. A. G., Sweden, Jan. 15, 1983.
4. Renishaw Metrology Limited., "3 Dimensional Probes for Co-ordinate Measuring Machines," Product Literature, Gloucestershire, U. K., 1984.
5. Lau, K., Development of an Automated Guidance and Coordinate Measuring System for Coordinate Measuring Machines and Robotic Devices, Ph.D. Dissertation, University of Wisconsin-Madison, 1982.
6. Berarducci, T., A Microprocessor-Based Electro-Optical, Non-Contacting Gear Inspection System, M.S.E.E. Thesis, Case Wester Reserve University, Cleveland, May, 1982.
7. Nimrod, N., Margalith, A., and Mergler, H.W., "A Laser-Based Scanning Range Finder for Robotic Applications," Proceedings of the 2nd International Conference on Robot Vision and Sensory Control, Nov. 2-4, 1980, Stuttgart, West Germany.
8. Kanade, T., Sommer, T., An Optical Proximity Sensor for Measuring Surface Position and Orientation for Robot Manipulation, Tech Report TR-83-15, Carnegie Mellon University Robotics Institute, 1983.
9. Lorenz, R. D., The Theory and Design of Optical/Electronic Probes for High Performance Measurement of Parts, Ph.D. Dissertation, University of Wisconsin-Madison, 1984.
10. Lorenz, R. D. and Schade, D. K., "Synthesis of a High Speed Signal Processing System for a Laser-Based, Optical/Electronic Measurement Sensor," Proceedings of the 1986, IEEE Industrial Electronics Conference, Oct 1-3, 1986.
11. Lorenz, R. D., "Microcomputer-Aided Design of an Optical Triangulation Probe," Journal of Computers in Mechanical Engineering, Vol.2 No. 5, March, 1984.

Reprinted with permission from *Applied Optics,* Vol. 23(21), pp. 3837-3844 (November 1, 1984).

Laser range finder based on synchronized scanners

Marc Rioux

A new geometrical arrangement is proposed to improve performances of optical triangulation. Two scanners in synchronization allow a linear position sensor to be used for surface topography measurement. Besides a large increase in speed of measurement (approximately megahertz), the new geometry allows considerable reduction of the optical head size compared with usual geometries, so the shadow effects are reduced proportionally. It also provides a means to obtain a very large field of view without compromising on resolution. Geometrical analysis and experimental results are presented.

I. Introduction

For many years, much of the work in 3-D measurements involved passive techniques. One field of applications is earth topographic measurements where active techniques are impractical. Another one is scene analysis for miltiary applications, where for obvious reasons of security, passive techniques are highly desirable. More recently the importance of 3-D vision in robotics was recognized, and research activities in this field are growing. The industrial environment provides new constraints and limitations to the applicability of usual techniques such as difficult environment, cost, and compactness, but, on the other hand, the proximity of objects allows active methods to be used to get 3-D data much more easily than passive techniques. Also, objects to be identified and manipulated are usually geometrically quite simple. Furthermore, objects of the same class are quite similar to each other (in contrast to natural objects in natural scenes) so that discrimination between classes should be easier than in natural scenes, especially with the use of 3-D measurements of their surface profile and some sort of compact geometrical description.

Active methods where a beam of light is superimposed to the naturally lightened scene greatly simplify the signal processing to be done to recover distance information. Among the various techniques described in the literature,[1-15] we selected active triangulation as an attractive approach that has the potential to evolve toward a low-cost 3-D camera. Basic elements of such a system consist of a light source that is usually a laser, a scanning mechanism to project the light spot onto the object surface, and a position sensor with a collecting lens looking off axis for the light spot. The distance measurement is done by trigonometric algebra applied to the projection direction (scanner angular position) and the detection direction determined by the light spot position on the sensor and the principal point of the collecting lens.

A large angular separation has to be used with usual triangulation technique to get a good compromise between resolution along range and field of view. The limitations imposed by a large angular separation are twofold; first, continuous profile measurement is prevented by severe shadow effects, second, the optical head consisting of a scanner plus a camera is bulky. What we propose here is a new scanning mechanism that allows a compact camera head to be designed without compromising resolution and field of view and also limits shadow effects to a minimum. The basic idea is to synchronize the projection of a light spot with its detection. This is realized using two scanners instead of one. The first one projects a beam on the surface to be measured, and the second one is coupled to the receiving sensor to follow the spot.

II. Geometrical Analysis

A. Usual Triangulation Geometry

Let us start the analysis with the usual triangulation geometry. In Fig. 1, we have a beam of light originating from position d along the X axis. The light beam projecting at an angle θ_0 defines a reference point $(d/2,l)$ that will be used for calibration purposes. At the origin there is a lens (not shown) of focal length f used to focus the light on a position sensor aligned parallel to the X axis and in focus at $-fl/(l-f)$ along the Z axis. We assume here that d, l, and f are known and are, respectively, the distance between the scanner axis of rotation and the principal point of the lens, the distance between

The author is with National Research Council of Canada, Division of Electrical Engineering, Ottawa K1A 0R8.

Received 28 April 1984.
0003-6935/84/213837-08$02.00/0.

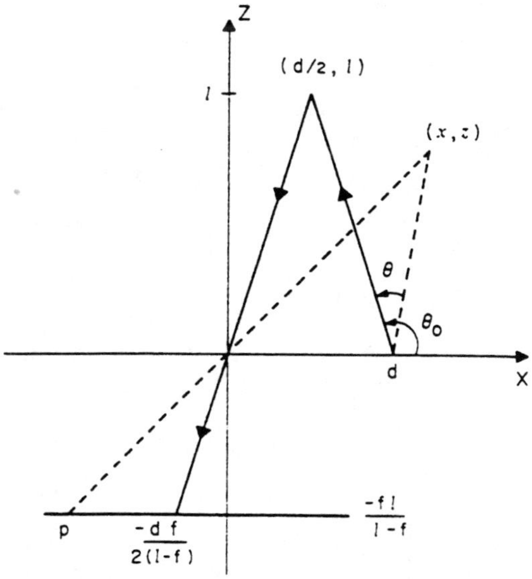

Fig. 1. Conventional triangulation geometry.

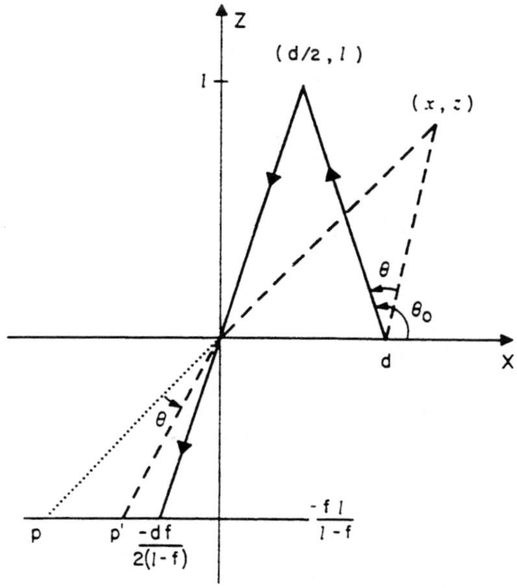

Fig. 2. Synchronized scanning geometry.

the common axis of projection and detection $(0,0-d,0)$ and the reference point, and the focal length of the lens used to collect light from the scene. Under rotation of the scanner, the light beam rotates to another angular position $\theta_0+\theta$ (θ is negative in Fig. 1). The spot of light on the position sensor moves from $df/2(l-f)$ to a new location p due to the intersection of the projected light beam with the object surface at x,z. By trigonometry we find the relation between the coordinates (x,z) and the parameters of the geometry as

$$x = d\cdot p\left[p+\frac{f\cdot l(2l\cdot\tan\theta+d)}{(l-f)(d\cdot\tan\theta-2l)}\right]^{-1}, \tag{1}$$

$$z = -d\left[\frac{p(l-f)}{f\cdot l}+\frac{2l\cdot\tan\theta+d}{d\cdot\tan\theta-2l}\right]^{-1}. \tag{2}$$

B. Synchronized Scanning Geometry

Consider now a synchronized scanning geometry as shown in Fig. 2. The only difference here is the addition of a scanner which has its axis of rotation set at (0,0). The scanner moves synchronously with the one at point $(d,0)$ (the projection scanner) and has the effect of canceling its angular movement. The net result is to bring the position of the spot on the sensor closer to the reference point $-df/2(l-f)$. That is illustrated in Fig. 2 as a trajectory from point x,z to point p' and point p. Position p is the position that would be measured without synchronization. Therefore, to find coordinates (x,z) with synchronization, we relate p to p', which is

$$p = p'+\frac{f\cdot l\cdot\tan\theta}{l-f}\left[1-\frac{p'(l-f)\cdot\tan\theta}{f\cdot l}\right]^{-1}, \tag{3}$$

and we use Eqs. (1) and (2). It can be seen from Fig. 2 that a change in the position of the light spot along the Z axis produces an equivalent angular shift for both geometries (usual and synchronized), but a change along the X axis produces a much smaller angular shift in the case of the synchronized geometry. In other terms, the position sensor is mainly used to measure range with synchronization, while in usual geometries a large portion of the position sensor area is also used to measure x coordinates. That is interesting because the scanner mirror position can be precisely obtained. The net gain with synchronization is that, with the same position sensor used for usual geometries, we can increase the focal length of the lens used to collect the light and obtain an increased resolution in range without reduction of the field of view. A more detailed description of the effect of synchronization is described in Sec. II.C.

C. Trajectory of the Intersection of Axes

In contrast to the usual triangulation geometry, a single point on the position sensor defines a surface in the object space. As an example, consider in reference to Fig. 2 that

$$p = -\frac{df}{2(l-f)} \tag{4}$$

for all angular rotation θ. This represents the trajectory of the intersection of the previously defined axis of detection [axis (1) in Fig. 3] and the projected light beam (2). As both scanners are synchronized, both axes rotate at the same angular speed (see Fig. 3). Axis (1) can be written as

$$z_1 = x\tan(\theta-\theta_0), \tag{5}$$

while axis (2) is

$$z_2 = (x-d)\tan(\theta_0+\theta), \tag{6}$$

we know that

$$\tan\theta_0 = \frac{-2l}{d}. \tag{7}$$

At the intersection of both axes,

$$z_1 = z_2. \tag{8}$$

This gives the trajectory of the intersection of the two axes as

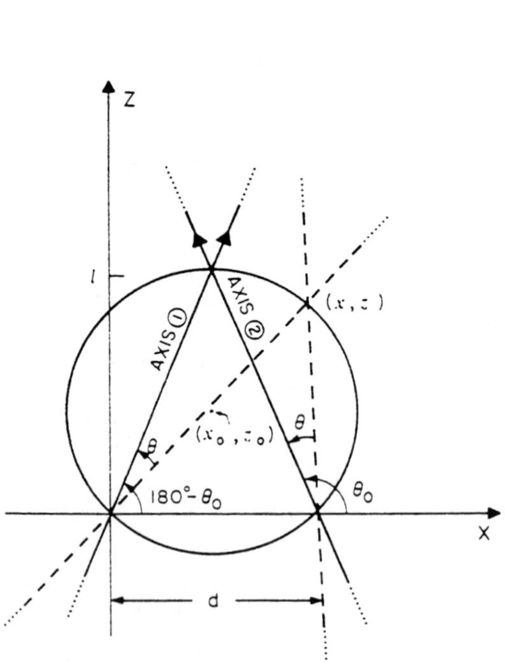

Fig. 3. Trajectory of the intersection of axis.

$$x^2 - dx - \frac{l}{d} - \frac{d}{4l} + dz + z^2 = 0. \tag{9}$$

This can be rewritten as

$$(x - d/2)^2 + \left(z - \frac{l^2 - d^2/4}{2l}\right)^2 = \left(\frac{l^2 + d^2/4}{2l}\right)^2, \tag{10}$$

which is a circle of radius R;

$$R = \frac{l^2 + d^2/4}{2l} \tag{11}$$

centered at the coordinates

$$x_0 = d/2, \tag{12}$$

$$z_0 = \frac{l^2 - d^2/4}{2l}. \tag{13}$$

Noticed that the trajectory of the intersection of the axes is a circle inscribing the reference point and both scanners' axes of rotation.

For small angular separations between the projection and detection axes (which reduce shadow effects), the diameter of the circle is equal to $\sim l$.

The fact that the trajectory of the intersection of axes is on a circle can be used advantageously for a class of inspection applications where profile measurements of spherically or near spherically shaped objects has to be made (rf antennas as an example). The advantage here is that low-resolution high-speed position sensors can be used to get very high-resolution profiles. This is because we know the reference surface, and we measure only the difference between the object's shape and the trajectory.

D. Bidimensional Scanning

Consider now a dual-axis synchronized scanner as shown in Fig. 4. The geometry is similar to the one described in Fig. 2, but with the addition of a deflecting mirror which has its rotation axis parallel to the X axis and is located at a distance h along the Z axis from the synchronized line scanners. Angular rotation of this deflecting mirror is assumed to be known. Then the 3-D coordinates of a surface measured by the system are

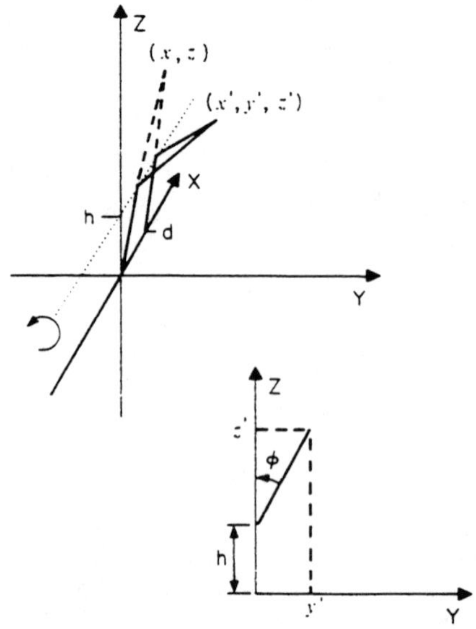

Fig. 4. Dual-axis synchronized scanning geometry.

$$x' = d \cdot p \left[p + \frac{fl}{(l - f)\alpha}\right]^{-1}, \tag{14}$$

$$y' = [h - \alpha(x' - d)] \sin\phi, \tag{15}$$

$$z' = h + \{[-h + \alpha(x' - d)] \cos\phi\}, \tag{16}$$

$$\alpha = \frac{d \cdot \tan\theta - 2l}{2l \cdot \tan\theta + d}, \tag{17}$$

where ϕ is twice the rotation angle of the deflecting mirror. Notice that ϕ is negative in Fig. 4. Then the reference surface for the trajectory of the intersection of the axes will be spherical if h equals z_0 [see Eq. (13)]. For other values of h the reference surface is toroidally shaped with a radius of curvature given by Eq. (11) along the x axis and a radius of $l - h$ along the y axis.

III. Autosynchronized Scanners

A. Scanner Geometry

Two separate scanning mechanisms can be used to provide the rotation of projection and detection axes, but it requires critical adjustments to get stable synchronization. A perfect synchronization can be obtained using polygonal or pyramidal mirrors or a plane mirror coated on both sides. The idea is to use one facet of the scanner to project the light beam and one other facet of the same scanner for detection. Among the numerous geometrical arrangements, the one depicted in Fig. 5 has been implemented in our laboratory. It uses a pyramidal mirror consisting of six facets. A light beam from source S is deflected by a facet of the scanner to fixed mirror M_1. A second deflection occurs on deflecting mirror M_3. After scattering of the light onto the objects surface, the beam is deflected again on M_3 to another fixed mirror M_2. Then we use the opposite facet of the pyramidal mirror to deflect the detection

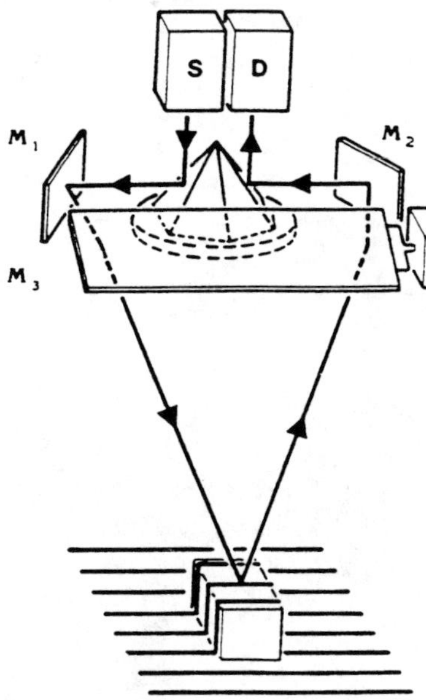

Fig. 5. Autosynchronized geometry using multifacet pyramidal mirror.

beam to position sensor D. Using a multifaceted mirror eliminates the need of trying to synchronize two scanners. The reference point discussed in Sec. II can easily be set to any desired value by angular adjustments of fixed mirror M_1 and M_2. Advantages of a pyramidal mirror for this application are (1) it allows a large collecting area for small inertia (compared to a polygon), and (2) angular coverage is a factor of 2 smaller than a polygon (with the same number of facets), which doubles the duty cycle. Deflecting mirror M_3 is scanning at a very low speed, typically 2 orders of magnitude slower than the pyramidal mirror.

B. Position Sensor

Usual triangulation methods require a dual-axis position sensor to do surface profile measurements. One very interesting property of a synchronized approach is that it requires a linear position sensor only, which increases the bandwidth from ~10 kHz to a megahertz. Also, the physical size of the position sensor can be much smaller due to subtraction of scan angles. Then to take advantage of the speed, we chose the smallest lateral effect photodiode available on the market. These are made of a photodiode deposited on top of a resistive layer. The electrical equivalent circuit of the device is shown in Fig. 6. The current produced by the photodiode is split by the resistive layer so that the position information can be simply obtained by dividing the difference of the currents by their sum. On the left-hand side of the figure, there is a physical description of the sensor along with the position signal characteristic as a light spot scans its surface. Interesting features of the lateral effect photodiode are insenitivity to large amounts of defocusing, a response time of 500 nsec, and an output position signal that gives the centroid of the incident light spot. Another property is that we can get a reflectance map of the scene under scan that is in the perfect registration with the 3-D image. Indeed, the sum of the current is the amount of light diffused by the surface in the direction of the camera which gives a picture equivalent to what a TV camera provides. Interest of combined intensity and 3-D images have been discussed by Nitzan *et al.*[16]

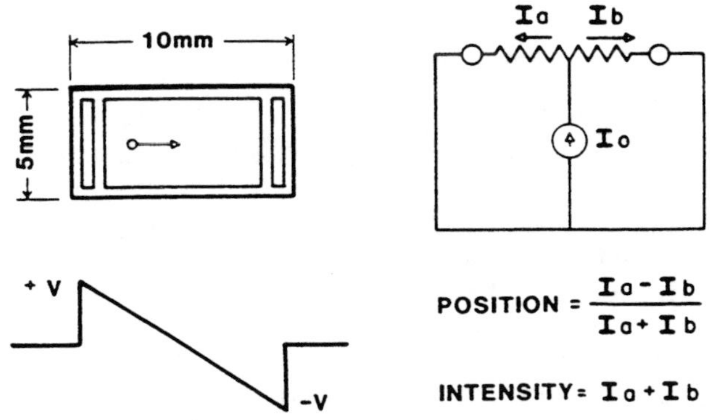

Fig. 6. Lateral effect photodiode.

Although CCD linear arrays are much slower (~10 kHz), they provide higher resolution than lateral effect photodiodes at extra cost. The output signal from a CCD does not provide a direct measurement of the light spot position, and to take advantage of the high geometrical precision, it is necessary to use some kind of peak detection circuitry. They also provide high sensitivity allowing measurements to reach larger distances. As a comparison, the use of a lateral effect photodiode as a position sensor is limited to object–camera distances of the order of a meter using low-power lasers such as He–Ne and cw diode lasers. On the other hand, the use of CCD allows us to increase that distance to ~10 m. Another comparison between the two sensors is to use the resolution-bandwidth product, which is, at a distance of 1 m, of the order of 100 MHz for lateral effect photodiodes (100 elements of resolution at 1 MHz) and 40 MHz for CCD using a peak detection algorithm having half a pixel of resolution.

A third type of position sensor is the dual-cells photodiode. Useless with usual triangulation geometries, this type of sensor can be used advantageously with synchronized approaches. It is made of two small photodiodes located side by side. As for lateral effect photodiodes, the position signal is provided by the division of the difference of the currents generated by each cell by the sum of those currents. Its main advantage is a bandwidth of more than 30 MHz (potential resolution-bandwidth product of ~1 GHz). Signal processing circuitry required to cope with such a speed is the limiting factor of dual cells. It is also sensitive to geometrical distortion of the spot and has a very small displacement range. Nevertheless, its low cost makes it attractive for future development such as video rate 3-D cameras.

Beside these well-known position sensors, there are simple geometrical arrangements that were suggested in the literature[8,9] that provide high-speed high-resolution position sensing using a single photodetector. Such arrangements can also be used with synchronized geometries with the advantage of increased dynamic range along the Z axis.

IV. Experimental Results

An experimental phototype based on the schematic shown in Fig. 5 has been realized. A photograph of it is shown in Fig. 7. Although the design was made to use a diode laser, we got our first experimental results with a He–Ne laser of 15 mW (not shown). At this time we are installing a RCA CDH-LOC diode laser. Exceptional beam quality and high cw power available from it (20 mW) motivates our choice. The pyramidal scanner consists of six facets of 5 cm^2 each. The angular speed adjustment ranges from 60 to 1800 rad/sec. This scanner provides X axis scanning, while a stepper motor driven flat mirror provides Y axis scanning. A collecting lens used to focus scattered light from the objects surface is shown on the schematic, fixed to the front end of the optical head. The lens has a 10-cm focal length, and its output goes to a position sensor located on top in the rectangular black box. This box contains the lateral effect photodiode plus two stages of amplification on both sensor outputs. Those signals are sampled at 600 kHz, converted to digital using a 12-bit ADC, processed to obtain the position information, and sent to a memory. We can see on the prototype table a shoe last that has been scanned. The 3-D data in memory are displayed on a TV monitor as an isometric view. Image size is 256 pixels/line, 128 lines. For clarity, we display only half of the lines on an isometric view. Scanning time is <1 sec, and numerical computation (difference divided by sum) is done in real time. The pyramidal scanner rotation speed is set to 1000 rad/sec. Our field of view is ~20°, so our spatial resolutions along the X, Y, and Z axes are, respectively, 1, 2, and 0.4 mm for a field of view of 250 × 250 × 100 mm. A more detailed description of the camera output can be made using the photograph of Fig. 8. As mentioned earlier in the text, lateral effect photodiodes provide simultaneously an intensity image similar to what a TV camera would produce plus a tridimensional image. The first two pictures shown at the top of Fig. 8 are intensity images obtained by summing the two position sensor outputs for a shoe last and a human hand. The two pictures at center show the 3-D signals as computed from the division of the difference by the sum of the position sensor outputs. Each pixel (voxel or surfel . . .) displays range information as grey levels, white being the shortest distance and black the farthest. Our display is made on thirty-two levels only (5 bits), although the 3-D signal has a resolution of 8 bits. A more detailed view can be obtained using an isometric projection as shown at the lower part of the picture. The only inconvenience with this type of representation is that we have to hide a part of the available measurements to improve the quality of 3-D perception.

V. Discussion

The usual triangulation technique shows severe shadow effect problems associated with the requirement of a large angular separation between projection and

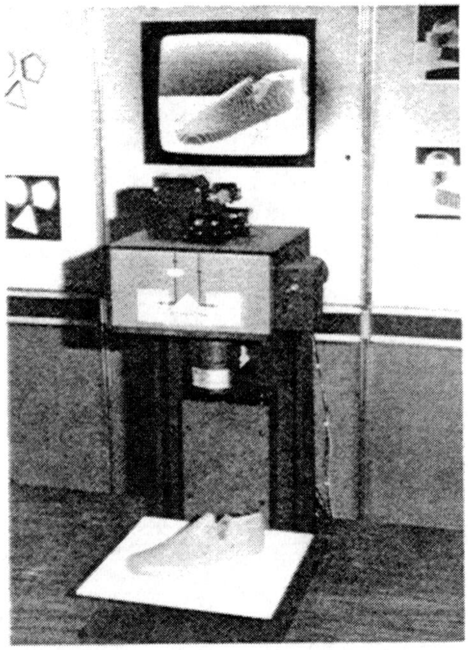

Fig. 7. Experimental prototype photograph showing the optical head, the object (shoe last) that has been scanned, and a display of the resulting measurements.

detection axis in order to get a reasonable range resolution (Z axis). Shadow effects are of two types (see Fig. 9):

(a) the points on the surface that the projection beam cannot reach (points under the dashed zone on the left side in Fig. 9);

(b) the points on the surface that the position sensor cannot read (points under the dashed zone on the right side in Fig. 9).

Another way to get a higher resolution is to increase the focal length of the collecting lens, but then the field of view (along the X and Y axes) is proportionally reduced. Practically the usual triangulation geometry will use an angular separation of ~30°, which results in a very bulky optical arrangement. In contrast, synchronization of scannings removes the tight relationship between X, Y, and Z parameters. Scanner characteristics define the field of view, while the position sensor is mostly used to monitor range. In those conditions, we can bring the projection and the detection axis as close as we can without any reduction of the field of view. Then the resolution along Z can be set by the focal length of the collecting lens without any limitations on the other two variables. Our experimental prototype has an angular separation of 10°, which is limited by the size of the pyramidal mirror. Smaller angular separations can be obtained using polygonal scanners. Looking back at Fig. 8, we see that the shadow effect represents a very small proportion of the image, although the surface under measurement shows a very broad range of surface orientations. In an attempt to reduce the shadow effect, we have to keep in mind that the removing of all shadows is a very difficult task. Points that are not illuminated by the projecting

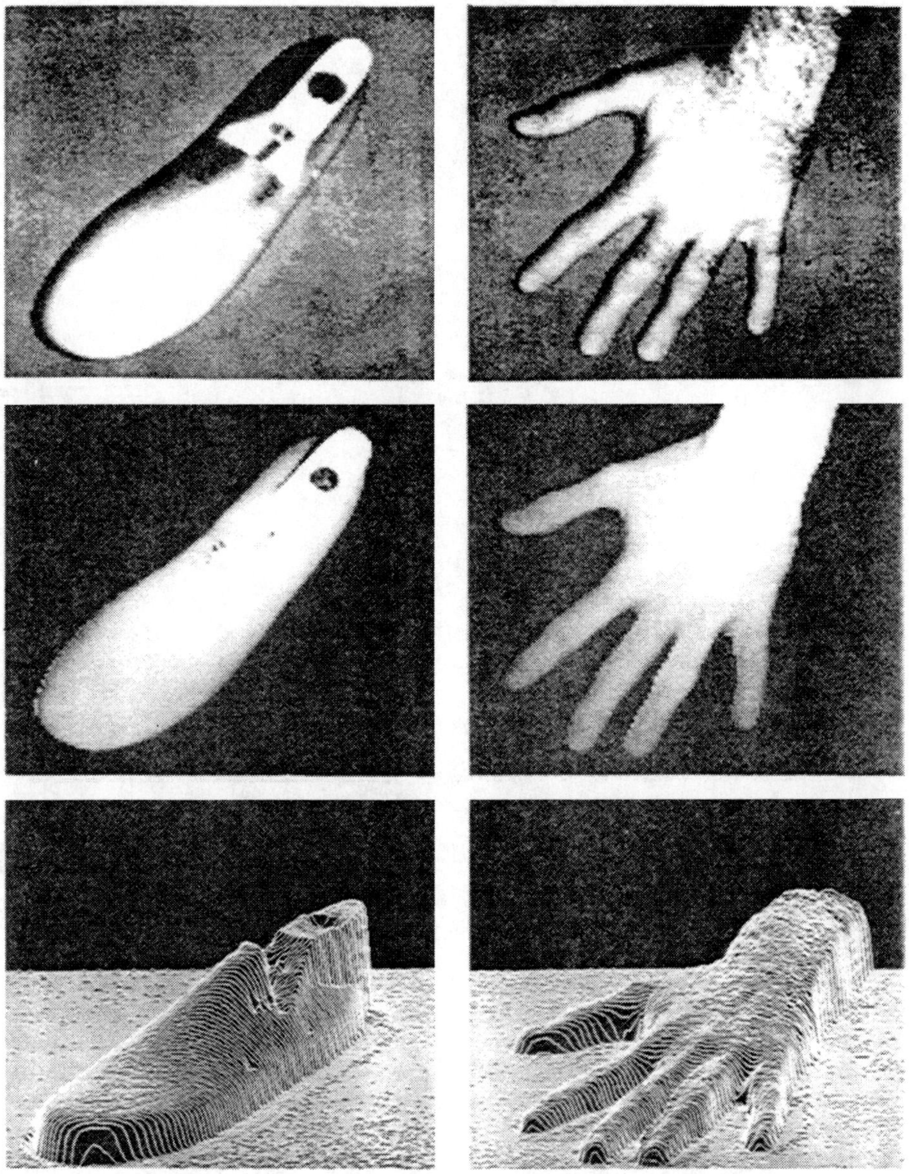

Fig. 8. The 3-D camera outputs for two different inputs. Top photographs show intensity output, while the four others show the 3-D data.

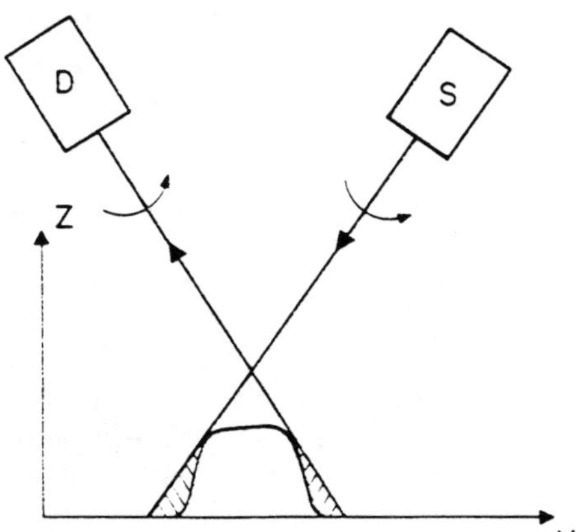

Fig. 9. Shadow effects.

beam cannot be read in any geometry. The only way to get around this difficulty is to either rotate the object or translate the camera head. If not practical, a more complex geoemtry as shown in Fig. 10 can be used to get measurements everywhere the projecting beam can reach (except for small deep holes). The arrangement is based on the use of two sensors looking on either side of the projecting beam while keeping synchronization. Notice that there is a 90° rotation of the axis at the pyramidal mirror surface (top view) on Fig. 10. For rotating objects, a geometry is suggested in Fig. 11. Here the reference point is set to be on the rotation axis of the object. Because of the rotation of the object, there is no need for a deflecting mirror, so the optical arrangement can be very simple and compact using polygonal scanner. The camera head can also be rotated as shown

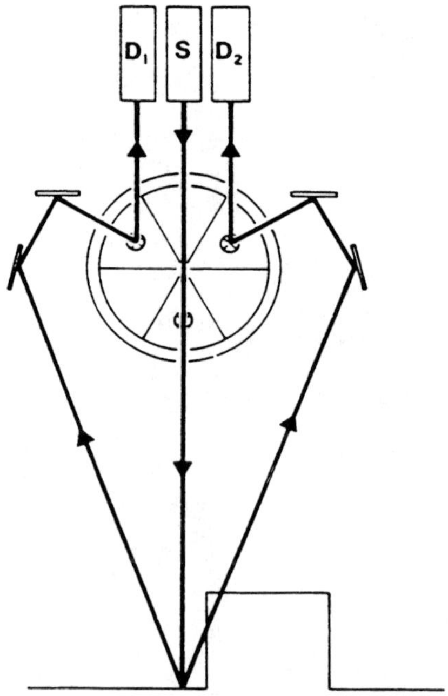

Fig. 10. Shadow effects reduction using two sensors.

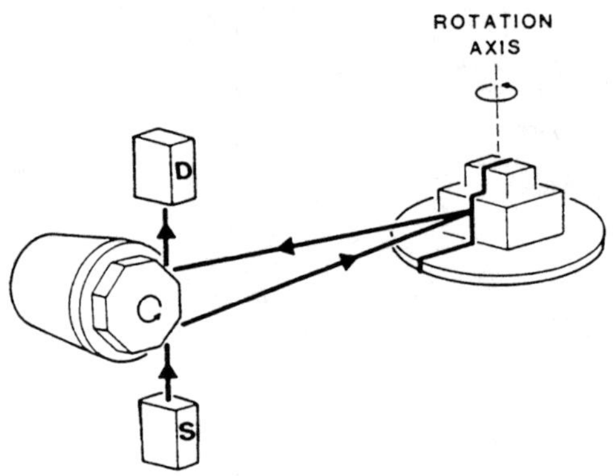

Fig. 11. Measurements on a rotating object.

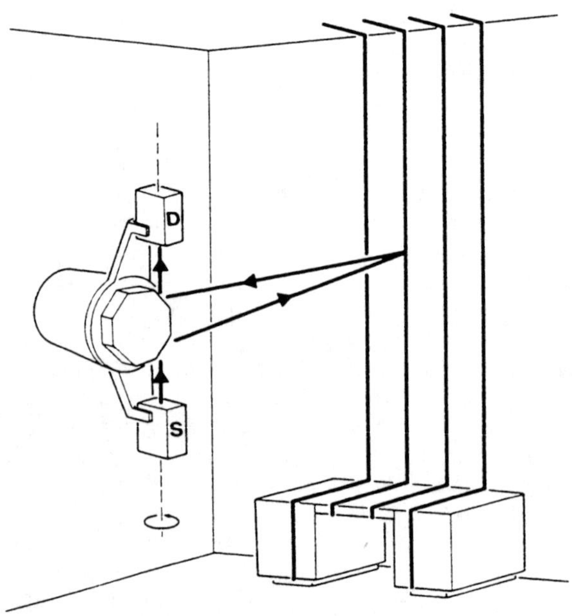

Fig. 12. A 360° range camera.

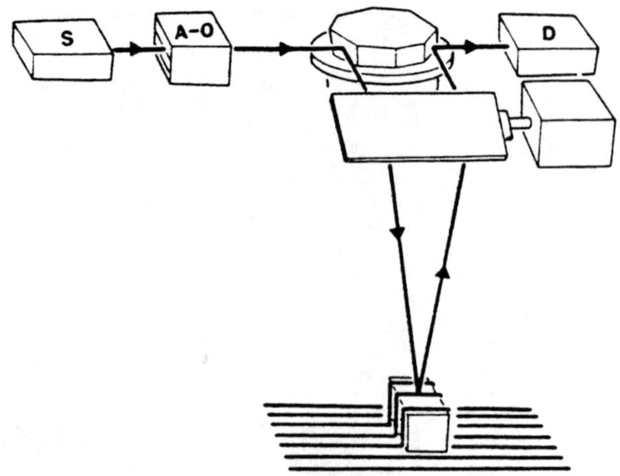

Fig. 13. Use of an acoustooptic deflector to produce a reference surface of any shape.

in Fig. 12 to map the interior of a room. Such an arrangement can be useful in robot guidance, where 360° scanning is attractive. One of the objectives of this project is related to robotic research, in particular, automatic assembly. Having the possibility to grab 3-D data of a scene in a fraction of a second has many practical advantages, especially in an industrial environment. Among them is a reduction in interference caused by mechanical vibrations. Furthermore, moving objects can be measured. Coupled to a robot arm, a 3-D camera can be used in conjunction with an appropriate vision processor to identify parts in the scene. On the other hand, we have to remember that a 3-D camera provides only the raw data (with noise), and there is still a lot of fundamental research to be done to use the 3-D data efficiently for object recognition. At least, one advantage of a 3-D camera is that the measurement of the object geometry is not a projection (as 2-D imagery). Indeed, we are actually measuring without ambiguity the surface profile of the objects under examination.

In terms of signal characteristics, the output from a 3-D camera as described here is very similar to a TV signal. In a 3-D camera the only difference is that the amplitude of the signal is related to geometrical characteristics of the object. In contrast, 2-D cameras (like ordinary TV cameras) provide an output signal amplitude that is not geometrically related to the object, instead it represents the surface reflectance properties of the object combined with the ambient light conditions of orientation, intensity, and spectral characteristics. In addition, the net result is affected by the orientation of the object and by the proximity of other objects. This is why 3-D feature extraction from a 2-D image of a scene is so difficult to realize. A 3-D camera using an active approach also provides good immunity to background illumination. This is done by using an interference filter that allows only the laser light to reach the sensor.

Resolution needed for object recognition is expected to be low (~1% in each axis). In contrast, inspection of manufactured objects, which is a big potential market for 3-D cameras, requires usually much higher resolutions. This can be met by CCD linear arrays, when high-precision measurements are needed, or by an arrangement such as shown on Fig. 13, when comparison of an object has to be made with a reference (inspection of PC Boards for missing parts, for example). This optical arrangement uses an acoustooptic deflector to change the angular orientation of the projecting beam prior to synchronization, so we can program a reference surface of any shape for inspection. It means that, in normal operation where the inspected object is similar to the reference, the spot position on the sensor is stable. Now, if any part of the object is missing or distorted, a change in the position of the spot occurs. The departure of the new position from the stable one is proportional to the error and can be used to identify the missing part or the nature of the distortion.

VI. Conclusions

A laser range finder based on triangulation has been described. New geometries are suggested using synchronized scanners. Experimental results show a response time shorter than a microsecond using a lateral effect photodiode as a position sensor. This sensor also provides an intensity image which is in perfect registration with the 3-D one. Synchronized scanning allows very large fields of view to be obtained without compromising resolution along range axis. With the increasing development of diode lasers and holographic scanners, the new geometries proposed should lead to the realization of compact low-cost high-speed 3-D cameras. We see besides robotic applications, inspection of manufactured parts, as well as medical application such as prostheses fabrication and diagnostic instrumentation.

References

1. V. Bodlaj and E. Klement, "Remote Measurement of Distance and Thickness Using a Deflected Laser," Appl. Opt. **15,** 432 (1976).
2. J. D. Boissonnat and F. Germain, "A New Approach to the Problem of Acquiring Randomly Oriented Workpieces of a Bin," in *Proceedings, Seventh International Joint Conference on Artificial Intelligence, 1981*, pp. 796–802.
3. M. D. Altschuler *et al.*, "Laser Electro-Optic System for Rapid Three-Dimensional (3D) Topographic Mapping of Surfaces," Opt. Eng. **20,** 953 (1981).
4. M. Bernasconi *et al.*, "Accuracy of Measurement Through Stereo Images," in *Proceedings, Fifth International Conference on Automated Inspection and Product Control, 24–26 June*, (1980), p. 309.
5. F. Bien *et al.*, "Absolute Distance Measurements by Variable Wavelength Interferometry," Appl. Opt. **20,** 400 (1981).
6. J. M. Burry, "Contouring in Real-Time with Moiré Interference," Opt. Commun. **41,** 243 (1982).
7. Z. Fuzessy and N. Abramson, "Measurements of 3-D Displacement: Sandwich Holography and Regulated Path Length Interferometry," Appl. Opt. **21,** 260 (1982).
8. D. P. Himmel, "A Laser Measuring System for Automatic Industrial Inspection," in *Proceedings, Fourth International Joint Conference on Pattern Recognition, 1978* (IEEE, New York), pp. 952–954.
9. G. Indebetouw, "A Simple Optical Non-Contact Profilometer," Opt. Eng. **18,** 63 (1979).
10. K. E. Morander, "The Optocator. A High Precision, Non-Contacting System for Dimension and Surface Measurement and Control," in *Proceedings, Fifth International Conference on Automated Inspection and Product Control* (1980), pp. 393–396.
11. C. J. Page and H. Hassan, "Non-Contact Inspection of Complex Components Using a Rangefinder Vision System," in *Proceedings, First International Conference on Robot Vision and Sensory Controls* (1981), pp. 245–254.
12. F. Pipitone, "A Ranging Camera for 3-D Object Recognition," *Midwest Symposium on Circuits and Systems* (1979), pp. 339–343.
13. T. Sawatari and R. B. Zipin, "Optical Profile Transducer," Opt. Eng. **18,** 222 (1979).
14. J. Taboada, "Coherent Optical Methods for Applications in Robot Visual Sensing," Proc. Soc. Photo-Opt. Instrum. Eng. **283,** 25 (1981).
15. T. Kanade and H. Asada, "Non-Contact Visual Three-Dimensional Ranging Devices," Proc. Soc. Photo-Opt. Instrum. Eng. **283,** 48 (1981).
16. D. Nitzan *et al.*, "The Measurement and Use of Registered Reflectance and Range Data in Scene Analysis," Proc. IEEE **65,** 206 (1977).

The author wishes to thank P. Chartier and J. Domey for many discussions and help during the implementation of the 3-D camera system. I would also like to thank L. Cournoyer and R. Misner for their excellent technical support in realizing the programming and constructing the experimental setup.

Reprinted from *Optics, Illumination, and Image Sensing for Machine Vision VII*, D.J. Svetkoff, editor, Proc. SPIE Vol. 1822, pp. 150-158 (1992). ©1992 SPIE.

Physical limits of 3D-sensing

Gerd Häusler, Jürgen M. Herrmann

University of Erlangen, Institute of Physics
Staudtstr. 7, D 8520 Erlangen, Germany

ABSTRACT

A fundamental limit for the distance uncertainty of coherent 3D-sensors is presented. The mimimum distance uncertainty is given by $\delta z = 1/2\pi \cdot \lambda/\sin^2 u$, with the aperture of observation $\sin u$ and wavelength λ. This distance uncertainty can be derived via speckle statistics for different sensing principles, and surprisingly the same result can be obtained directly from Heisenberg's uncertainty principle for a single photon. Because speckles are the main reason for distance uncertainty, possibilities to overcome the speckle problem are discussed. This leads to an uncertainty principle between lateral resolution and longitudinal distance uncertainty. A way to improve the distance uncertainty without sacrificing lateral resolution is the use of temporally incoherent light.

1. INTRODUCTION

3D-sensors supply information about the geometrical shape of an object. This geometrical information is, generally, better suited for solving industrial inspection problems, than the intensity information supplied by TV-cameras: most frequently, the question about the geometry is the primary question in industrial inspection. The geometrical shape is independent of the illumination, and independent of soiling, scratches, etc. Shape data can have a much larger dynamical range than intensity data; post processing and evaluation of 3D-data is much simpler than that of TV-pictures.

If 3D-data have such obvious advantages, why don't we find many 3D-sensors in the factories, instead of TV-cameras? What are the limits of 3D-sensors? The answer is that it is not easy to build accurate and fast 3D-sensors, with a large measuring range. And this is not only a question of technology. Instead, there are physical limitations introduced by the wave equation and by Heisenberg's uncertainty principle that determine the uncertainty of the measured shape.

In this paper we give a comprehensive description of those physical limits and of their origin. The understanding of the origin of distance uncertainty leads to methods to overcome those limits. We restrict ourselves in this paper to essentially coherent sensors, i.e., to sensors that work with spatially and temporally coherent light.

For our purpose, 3D-sensors can be separated into two groups: There are sensors that use more or less incoherent illumination, such as phase measuring Moire[1,2]. A different sensor is based on the projection of fringes onto the object and the evaluation of the fringe contrast in the image of the illuminated object[3]. We will deal with those 'incoherent' sensors in another paper.

Here we will discuss 3D-sensors that are based on more or less coherent illumination. We consider the use of these sensors for the profilometry of optically rough surfaces. A very common sensor principle is to project a light spot onto the object and extract the distance information from the reflected or scattered light. Focus sensing[4], laser triangulation[5,6,7] and shearing interferometry[8] are examples of these principles. It turns out that the ultimate distance uncertainty is independent of the method by which the scattered wave is evaluated. Whether we measure the size of the defocused spot image, or its lateral shift, or the radius of the scattered wave - we observe an ultimate distance uncertainty that cannot be overcome by 'better technology'. What is the origin of this fundamental uncertainty? Can we overcome the limitations by physical reasoning? These questions will be the subject of this paper.

2. THE UNCERTAINTY OF MEASUREMENT IN TRIANGULATION SENSORS

To understand the origin of the uncertainty, we first discuss laser triangulation as the most common principle of commercial sensors (fig. 1).

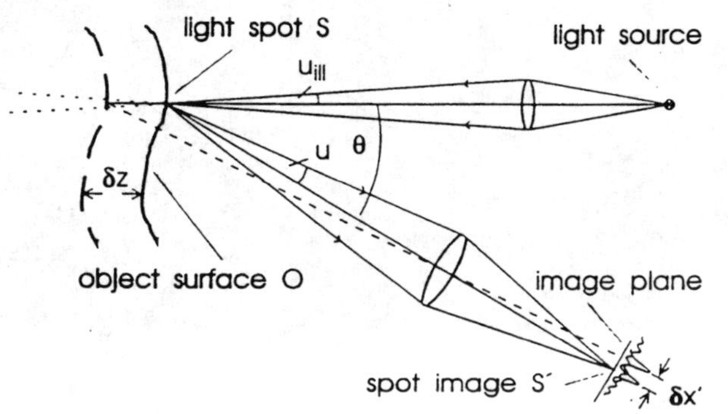

Fig. 1: Principle of laser triangulation. The position of the spot image S' is uncertain within a distance δx'. This corresoponds to a distance uncertainty δz

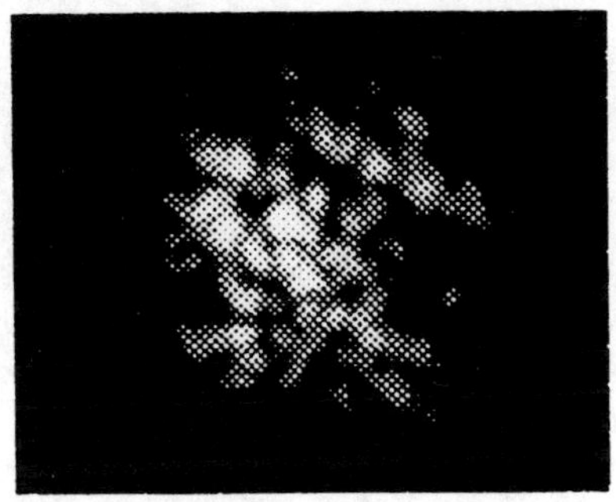

Fig. 2: Typical image S' of the light spot.

A light spot S is projected onto the object under test. In the image plane we get an image S' of this spot. The typical spot image is shown in Fig. 2. Its center of gravity (cog), as a measure of its location, depends on the microtopology of the rough surface, which varies statistically. As a consequence, the spot location can be determined only with a statistical uncertainty.

This is illustrated in fig. 3: We project S again onto a fixed position, marked by the reticle in the object plane. We observe the apparent position of S' in the image plane, while we shift the object in lateral x-direction. Dependent on the statistical local microtopology, the apparent spot position is 'dancing' around its true position at the reticle. We are mislead about its true position.

The statistical error δx of the spot location is transferred via the geometry of the sensor to a statistical distance error δz:

$$\delta z = \delta x'/(\beta \sin\Theta) = \delta x/\sin\Theta \tag{1}$$

where β is the magnification of the imaging system. In the following, we will project all variables from the image plane (primed) into the object space (unprimed).

Experiments have been performed to measure the standard deviation δx under various conditions (spot size, observation aperture, coherence), as well as theoretical calculations based on speckle statistics[9]. The theoretical and experimental investigations both yield the same results.

$$\delta x = C\cdot\lambda/(2\pi \sin u) \quad (2), \quad \text{and} \quad \delta z = C\cdot\lambda/(2\pi\ \sin\Theta\cdot\sin u) \quad (2a)$$

with the speckle contrast C, the wavelength of the illumination λ, and the aperture of observation sin$u$, and the angle of triangulation Θ (see. fig. 1). Baribeau and Rioux[10] derived a similar Equation as (2a) for their triangulation system, although they use for the location of the spot image instead of the center of gravity a different algorithm. They investigated the fully coherent case, therefore they did not take into account speckle contrast.

Fig. 3: Although the spot S is always imaged onto the reticle, the locally varying microtopology causes a statistical shift of the position of its image S′. Its true position is concealed to the observer.

Figure 4 confirms the dependence of δx from the aperture of observation. Furthermore, it shows that for spots larger than the resolution of the observation, δx is independent from the spot size.

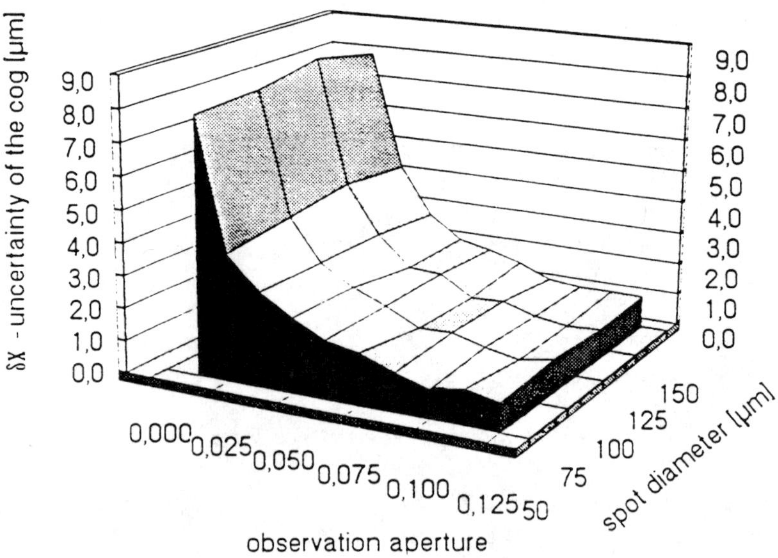

Fig. 4: The uncertainty δx of the spot location decreases with larger aperture of observation and is independent of the spot size. (object: ground glass, laser illumination)

2.1 How to build a good laser triangulation sensor?

Eq. 2b supplies some hints how to build a good laser triangulation sensor: a) use a large aperture of observation, b) use a large angle of triangulation, c) use low speckle contrast.

To a) and b): For macroscopic applications, with free working distances in the range of z=100mm, a practical aperture cannot be much larger than sin$u$ =0.1. The angle of triangulation has to be kept as small as possible, to avoid shading. Let us select Θ=15°. With laser illumination (λ=0.8µm, C=1) we get from Eq. (2b): δz=5µm, as the standard deviation of our sensor.

It has to be mentioned that experimentally we get these results only with cooperative objects. These are objects that satisfy the conditions for the application of speckle theory: the microstructure has to be much smaller than the resolution of the observing lens. Examples are objects with fine ground surfaces, smooth plastics, etc. In practice, however, we fre quently have milled or turned surfaces, with no stationary statistics within the spot size, but with macroscopic structures. The corresponding spot image then is usually deformed, because of a variing object reflectivity or speculiar reflection. Its intensity distribution can strongly deviate from that shown in Fig. 2. This results in a much larger 'deterministic' distance uncertainty than given by Eq. (2b). So Eq. (2b) defines a lower limit of δz. How to overcome the deterministic uncertainty, to a certain extent, will be dealt with in a further paper. Here we will restrict ourselves to the fundamental statistical uncertainty.

We will now turn to the condition c): use a low speckle contrast. Figure 5 illustrates the effect of different light sources upon the lateral uncertainty δx.

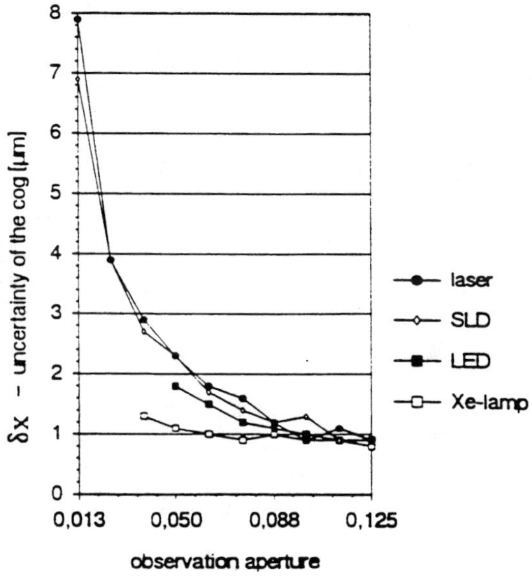

Fig. 5: Uncertainty δx, for different light sources. The uncertainty diminishes with less coherent illumination. The object is a ground glass, with roughness σ_h=3µm.

The effect is mainly due to the different *temporal coherence.* According to Pedersen[11], the contrast in a subjective speckle field (speckle pattern observed in reflection; speculiar direction) is approximately given by

$$C^2 = 1/(1 + (4\sigma_h/l_c)^2)^{0.5}, \tag{3}$$

where l_c is the coherence length of the illumination. Eq. (3) explains that a significant reduction of speckle contrast is only possible, for a roughness σ_h larger than the coherence length. The significance of the speckle contrast in Eq. (2a) is confirmed by the table of Fig.6.

sin*u*	C	δx_{exp}	δx_{theo}	$\delta x_{exp}/\delta x_{theo}$	(S=0.1mm)
0.05	1.0	2.2	2.0	1.1	
0.05	0.7	2.0	1.9	1.0	
0.05	0.5	1.7	1.4	1.2	
0.05	0.3	0.9	0.8	1.1	

$\sin u$	C	δx_{exp}	δx_{theo}	$\delta x_{exp}/\delta x_{theo}$
0.1	1.0	1.1	1.0	1.1
0.1	0.7	1.0	0.9	1.1
0.1	0.6	1.0	0.8	1.2
0.1	0.5	1.0	0.7	1.4

(S=0.1mm)

Fig. 6: Lateral uncertainty δx, for different speckle contrast, and different apertures. δx_{exp} is the experimental value, and $\delta x_{exp}/\delta x_{theo}$ is the ratio of the experimental and the theoretical value, according to Eq. (2).

The effect of low temporal coherence on the depth uncertainty of a triangulation sensor is illustrated in Fig. 7.

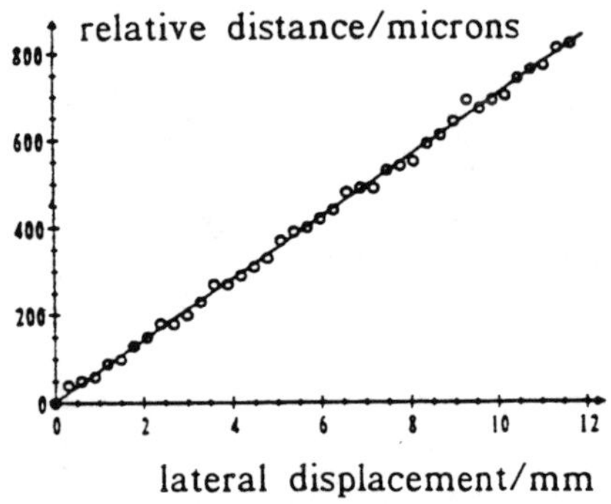

Fig.7a

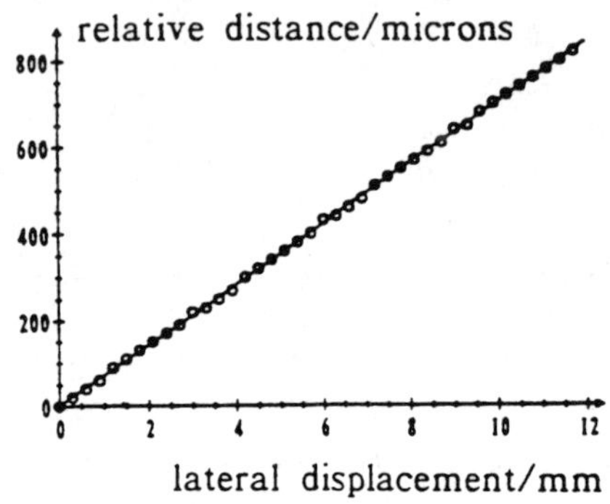

Fig.7b

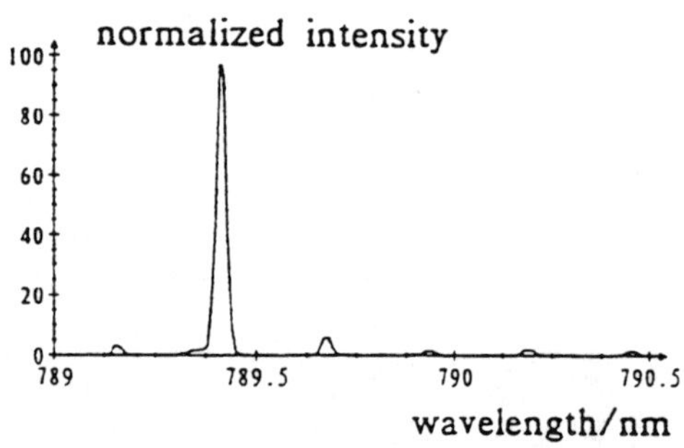

Fig.7c

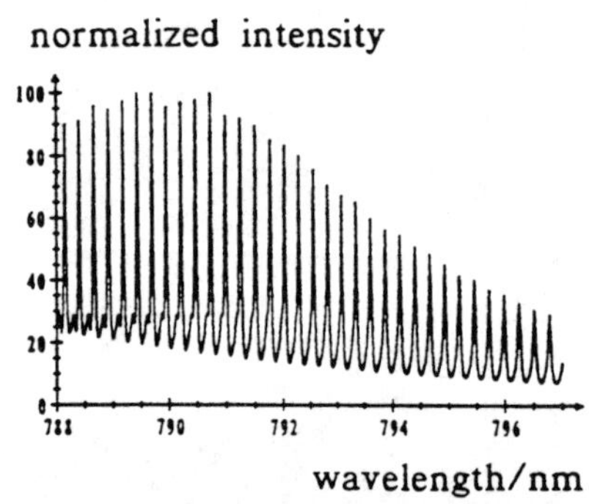

Fig.7d

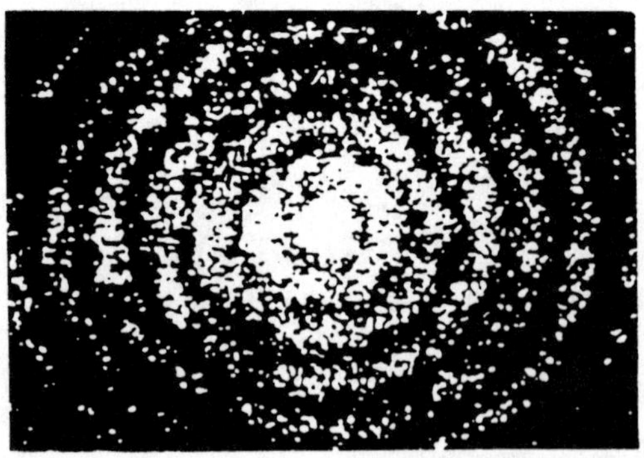

Fig.7e

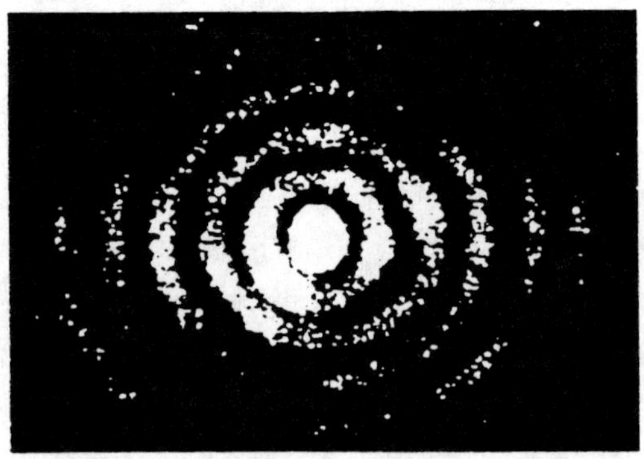

Fig.7f

Fig. 7: The distance uncertainty, with temporally coherent illumination (a) is worse (10.5µm) than for less temporally coherent illumination (b), (3.5µm). The coherence was reduced by driving the semiconductor laser beyond the threshold. The spectra in the coherent and noncoherent case are shown in fig. 7c,d. The corresponding spot images are shown in fig.7 e,f.

A different, and even more effective measure to reduce the speckle contrast is **low spatial coherence.** This is equivalent to a large spot S on the surface, since the coherence area in the spot is approximately equal to the center of the Airy disc. For example, for averaging over n^2 independent coherence areas the spot area has to be n^2 times the area of the Airy disc. The speckle contrast is then reduced by a factor $1/n$, and the distance uncertainty is improved by the same factor. Obviously, there exists an uncertainty principle connecting lateral resolution and longitudinal distance uncertainty[12] in the presence of speckle noise. An improvement of the distance uncertainty δz can only be achieved by sacrificing lateral resolution $1/\Delta x$:

$$\delta z \cdot \Delta x = \text{const.} \tag{4}$$

The same uncertainty relation is valid in the fully coherent case, if the speckle are reduced by scanning a small light spot across the object surface and averaging the distance. It is interesting, that Inglestam[13] derived a relation equivalent to Eq.(4) 40 years ago, for interferometry at non rough surfaces.

3. EXAMPLES OF OTHER SENSORS AFFLICTED WITH A FUNDAMENTAL MEASUREMENT UNCERTAINTY

The relations discussed above are not restricted to triangulation systems. The same principles are valid also for other coherent 3D-sensors. Two additional examples will be given.

The first sensor uses focus sensing for the measurement of the distance z. It is a point sensor that was introduced by Shamir[14] and others. Our modification[4] is adapted for macroscopic objects and can handle mirrorlike surfaces and rough surfaces. The sensor works as follows: a light spot is projected onto the object surface. Via a lens and a beamsplitter an image of this light spot is created. Near the image plane a pinhole is located on the optical axis. The flux through the pinhole is measured, it encodes the distance information. Because the flux depends on the reflectivity of the object, the measured distance depends also on the reflectivity. A possibility to avoid this is to create two images and to use two pinholes, one in front and one behind the image plane. The ratio of the two measured flux values gives the distance to the object, independent of the object reflectivity.

From experiments, it appears that the measured distance is, like in triangulation sensors, afflicted with an uncertainty. The reason for the uncertainty is again speckle noise. The speckles lead to a statistical change of the measured flux Φ through the pinhole. The uncertainty $\delta\Phi$ of the flux depends on the number n of speckles in the pinhole and the average flux $\langle\Phi\rangle$:

$$\delta\Phi = \langle\Phi\rangle / n^{1/2} \tag{5}$$

δz can be derived from the calculation of $z(\Phi)$ and $\delta\Phi$. The calculation is not shown in detail here.

$$\delta z = \partial z(\Phi)/\partial\Phi \cdot \delta\Phi = \ldots \approx 1/4 \cdot \lambda/\sin^2 u \tag{6}$$

One should notice that this is nearly the same result as that derived for triangulation, (compare with Eq. (2b) for $u=\Theta$). Figure 8 shows the measurement uncertainty for two light sources, laser and halogen lamp. The measurement uncertainty is, with laser 99µm (fig.8a), with the white light source 17µm (fig.8b). For both experiments, the aperture of observation was 0.12. A 6 fold improvement of the distance uncertainty was possible, due to speckle reduction.

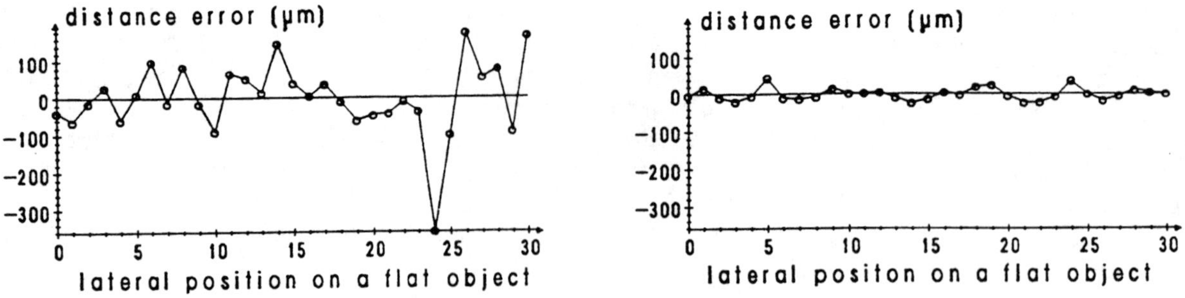

Fig. 8a,b Distance uncertainty for measurements with the confocal optical 'macroscope' with laser illumination, $\delta z = 99\mu m$ and b) illumination with a white light source, $\delta z = 17\mu m$.

Another sensor which shows similar properties is based on shearing interferometry[15]. From a small light spot, projected onto the object, a scattered spherical wave emerges. The radius of this wave is the desired distance z. Determination of the radius is possible by shearing interferometry. Interference fringes are created with a period proportional to the distance z. The sensor achieves a typical experimental distance uncertainty of $\delta z = 0.35$mm at a distance of z=250mm, with an observation aperture sin$u$=0.02. In a former paper[15] we investigated theoretically the influence of speckle on the distance uncertainty. This distance uncertainty is a result of phase errors in the interferogram introduced by speckle. The phase errors can be calculated with the probability density function $p(\gamma)$ for phase errors γ between two points in a fully developed speckle field given by Goodman[16]. With the known phase error in the interferogram the distance uncertainty was calculated. The result is:

$$\delta z = 1/4 \cdot \lambda/\sin^2 u \tag{7}$$

This is again the same result as for focus sensing, and approximately for trianglulation.

Again speckle reduction with less temporally coherent illumination is possible. With a high-pressure mercury lamp a distance uncertainty of 63µm was achieved (aperture 0.013). Compared with Eq.(7), valid for coherent illumination, a 14 fold improvement.

4. DISTANCE UNCERTAINTY AND HEISENBERG 'S UNCERTAINTY PRINCIPLE

The section above shows that a distance uncertainty according to Eq. (7) is not restricted to one special sensing principle, it is of a more fundamental nature. In fact, a distance uncertainty such as in Eq. (6) and Eq. (7), can be derived not only from speckle theory, but also directly from Heisenberg's uncertainty principle, for a single photon. We consider an experiment, where we try to localize one single photon from its location in the image plane of a lens with aperture sin$u$. Heisenberg's principle connects the uncertainty of the z-component of the momentum, δp_z, with the uncertainty of the distance, δz:

$$\delta p_z \cdot \delta z \geq h/4\pi , \tag{8}$$

with h=Planck's constant. Eq. (7) implies that a small desired distance uncertainty, requires a large uncertainty of the z-component of the photon momentum.

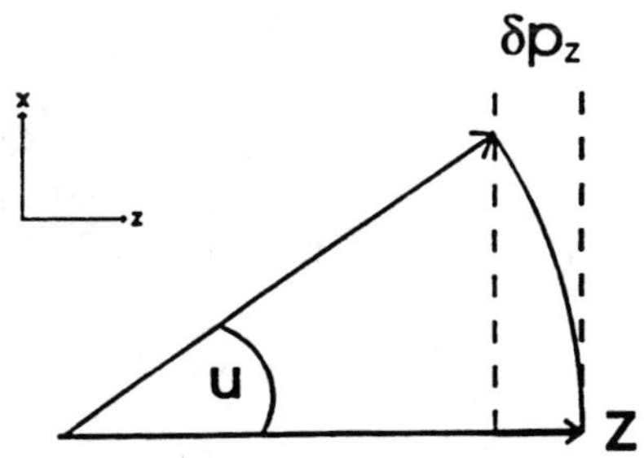

Fig. 9 Uncertainty δp_z of the z-component of the photon momentum, for a given aperture sin*u*.

Fig. 9 shows the uncertainty δp_z of the photon momentum in z-direction for a given aperture sin*u*. Short calculation gives the final result for the distance uncertainty,

$$\delta z \geq 1/2\pi \cdot \lambda/\sin^2 u \,, \tag{9}$$

which is the same as for the speckle experiment with many photons. The reason for this surprising fact is, that the single photon experiment without speckle has the same signal-to-noise ratio as the speckle experiment with many photons - although we would intuitively guess that the depth uncertainty with n photons should be better by a factor $1/n^{1/2}$ than that for a single photon. However this is not true for rough objects. Speckle limits the uncertainty to the large value of the single photon experiment.

It should be mentioned that similar considerations as above, concerning the uncertainty δp_x of the photon momentum in x-direction, result in a uncertainty δx of such as in Eq. (2) (without C) which is important for triangulation.

What are the consequences? Is it possible to measure beyond the limit $\delta z = 1/2\pi \cdot \lambda/\sin^2 u$ without sacrificing lateral resolution? The answer is yes. A high uncertainty in p_z, which is necessary for a small distance uncertainty, cannot only be achieved by large apertures, but also by 'broad band photons'. Then, even with zero aperture, we have a large δp_z, which is related to the coherence length l_c by:

$$\delta p_z = h/l_c \tag{10}$$

We already discussed the use of temporally incoherent light sources as a method to improve the distance uncertainty. This concept is especially successful in our coherence radar[17], combining a depth uncertainty of about 2μm and a small aperture (sin*u*=0.025) at the same time.

5. CONCLUSIONS

Given an observation aperture sin*u*, the minimum distance uncertainty for coherent 3D-sensors is $C/2\pi \cdot \lambda/\sin^2 u$. The uncertainty can only be diminished either by sacrificing lateral resolution and/or by the use of temporally incoherent light. Then the speckle contrast C is lower than 1.

6. ACKNOWLEDGEMENTS

We are grateful to R. Dorsch, for supplying fig. 4, 5, 6, and to S. Will, for supplying fig. 7.

7. REFERENCES

1. M. Halioua, H. Liu, "Optical sensing techniques for 3D-machine vision," Optical Techniques for Industrial Inspection, P.G. Cielo, Editor, Proc. SPIE 665, Quebec, Canada, 1986
2. M. Gruber, G. Häusler, "Simple, robust and accurate phase-measuring triangulation," Optik, vol. 89, no. 3, pp. 118-122, 1992
3. K. Engelhardt, G. Häusler, "Acquisition of 3D-data by Focus Sensing," Appl. Opt., vol. 24, no. 22, pp. 4684-4689, 1988
4. G. Häusler, J. M. Herrmann, "3D-sensing with a confocal optical 'macroscope' ," Optics in Complex Systems, F. Lanzl, H.-J. Preuss, G. Weigelt, Editors, Proc. SPIE 1319, p. 359, 1990
5. M. Rioux, "Laser range finder based on synchronized scanners," Appl. Opt., vol. 23, no. 21, pp. 3837-3836, 1984
6. G. Bickel, G. Häusler and M. Maul, "Triangulation with expanded range of depth," Opt. Eng. vol. 24, no. 6, pp. 975-977, 1985
7. K.G. Seitz, H.J. Tiziani, R. Litschel, "3-D Koordinatenmessung durch optische Triangulation," Zeitschrift Feinwerktechnik und Meßtechnik, vol. 94, no. 7, pp. 423-425, 1986
8. G. Häusler, J. Hutfless, M. Maul, H. Weißmann, "Range Sensing based on Shearing Interferometry," Appl. Opt., vol. 27, no. 22, pp. 4638-4644, 1988
9. R. Dorsch, G. Häusler, J. M. Herrmann, "Laser triangulation: Fundamental uncertainty of distance measurement," subm. to Appl. Opt., Oct., 1992
10. R. Baribeau, M. Rioux, "Influcence of speckle on laser range finders," Appl. Opt., vol. 30, no. 20, pp. 2873-2878, 1991
11. H. M. Pedersen, "On the contrast of polychromatic speckle patterns and its dependence on the roughness," Opt. Acta, vol. 22, no. 1, pp. 15-24, 1975
12. G. Häusler, "About fundamental limits of three-dimensional sensing or nature makes no presents," in Optics in Complex Systems, F. Lanzl, H.-J. Preuss, G. Weigelt, Editors, Proc. SPIE 1319, pp. 352-353, 1990
13. E. Ingelstam, "An optical uncertainty principle and its application to the amount of information obtainable from multiple-beam interferences," Arkiv för Fysik vol. 7, no. 24, pp. 309-322, 1953,
14. J. Shamir, A. Brunfeld and G. Toker, "Non-contact optical profilometer with linearized response and high sensitivity," NDT International, vol. 21, no. 6, pp. 430-434, 1988
15. G. Häusler, J. M. Herrmann, "Range sensing by shearing interferometry: Influence of speckle," Appl. Opt., vol. 27, no. 22, pp. 4631-4637, 1988
16. J. W. Goodman, "Statistical Properties of Laser Speckle Patterns," Laser Speckle and Related Phenomena, J.C. Dainty, Editor, p. 46 Springer-Verlang, Berlin, 1984
17. T. Dresel, G. Häusler, H. Venzke, "3D-sensing on rough surfaces by coherence radar," Appl. Opt., vol. 31, no. 7, pp. 919-925, 1992

Reprinted from *Optical Techniques for Industrial Inspection,* P.G. Cielo, editor,
Proc. SPIE Vol. 665, pp. 150-161 (1986). ©1986 SPIE.

OPTICAL SENSING TECHNIQUES FOR 3-D MACHINE VISION

Maurice Halioua and Hsin-Chu Liu

Center for Optics and NYCOM
New York Institute of Technology
Old Westbury, New York 11568

Abstract

Active optical vision systems rely on the analysis of images on an object illuminated by known structured optical patterns. Approaches differ by the choice of projected pattern (dots, stripes, grids, etc.) and the type of analysis used to extract shape information. A novel approach is reviewed, whereby a sinusoidal grating structure is projected onto an object, and the resulting deformed grating images are detected by a solid state array camera and analyzed by a microcomputer using digital interferometric phase-measuring algorithms. Large arrays of data points can be acquired and processed in parallel with very high accuracy and speed. Two systems are described, one that yields range measurement from a single point of view over a two-dimensional array of points, and a second that permits full 360 degree measurement. The case of parallel laser light illumination, as well as that of general white light diverging illumination are described. Experimental results and performance analysis are presented.

Introduction and review

Optical sensing techniques for 3-D shape measurement have been extensively studied because of their inherent non contact nature and are now being applied to numerous metrological problems including industrial inspection, robotic vision, solid modelling for computer graphics and biostereometrics. In contrast to passive monocular or binocular sensing techniques using image intensity information, active sensing techniques that make use of contrived or structured light illumination are emerging as preferred embodiment in the industrial environment. They rely on the analysis of images of an object illuminated by a known structured optical pattern such as dots, stripes, grids, etc. Their operation, based on the principle of triangulation, may be simply understood with the aid of Figure 1. An incident ray of light hits the object and a reference plane respectively at points D and A, so that A appears displaced at C from an offset observation direction. It is this displacement X that carries information on the height Z = h(X,Y) from the reference plane, or range L-h(X,Y) from the camera I. Various approaches differ essentially by the choice of projected pattern, and the type of measurement and analysis to extract shape information.

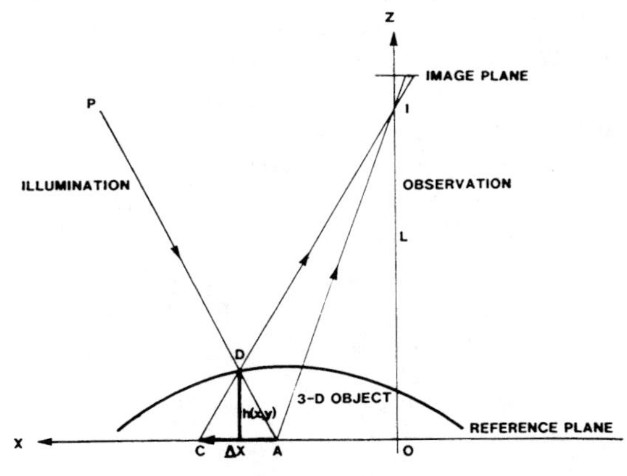

Figure 1. General geometry for optical triangulation

Among the methods that measure directly the geometric light pattern displacement or deformation, single laser point[1-2] or single laser stripe[3-6] scanning are characterized by no measuring ambiguities. However they require scanning mechanisms, which results in slow data acquisition and low stability. Also, the use of direct geometric measurement results in inherently low accuracy and high sensitivity to noise. The use of laser beam arrays of points[7-10] or multi-stripes[11-13], while not requiring mechanical scanning and having the potential for parallel processing, involves much digital processing and suffers from the same inherent low accuracy of direct measurement.

In the case of grating projection, moire techniques[14-16] make use of a second grating (real or virtual) in the image plane to optically demodulate the deformed grating image and produce contour maps of the object height. While moire contour maps are very attractive for visual observation, they are difficult to analyze and quantify: ambiguity may exist with regard to fringe order and sign, noise patterns are generated by the master grating and superimposed onto the useful information, sampling may be highly non uniform, system geometry must be very specific or true contours are not produced[16], and large digital image processing capability is required if some degree of automation is to be achieved. Improvement in the moire method aimed at increasing the accuracy and automating the measurement are based on the use of phase modulation techniques[17-21] which have been essentially pointwise. An alternative to the moire method that has been demonstrated to be more suitable for automated operation, is based on the use of a fast Fourier transform (FFT) analysis to digitally demodulate deformed grating images[22-23]. Measurement of steep object slopes, step discontinuities, the need for very high resolution imaging devices and large computing capability are some of the limitations of the FFT method.

The 3-D sensing methods[24-26] described here are based on the use of digital phase measuring techniques developed for classical interferometry and surmount many of the difficulties encountered in the previously described methods. Thanks to their inherent simplicity and parallel processing capability, they permit acquisition and processing of large amount of data points with high accuracy and speed, with low-cost optical, electronic and computer hardware.

Principle of digital phase measurement

Referring to Figure 1, when a sinusoidal intensity distribution is projected onto a 3-D diffuse object, the mathematical representation of the deformed grating image intensity distribution is similar to that encountered in classical optical interferometry and may be expressed in the general form as

$$I(X,Y) = R(X,Y)[A(X,Y) + B(X,Y)\cos\phi(X,Y)] \tag{1}$$

where $R(X,Y)$ is the object reflectivity function characterizing the surface nature, $A(X,Y)$ the background intensity, $B(X,Y)/A(X,Y)$ the fringe contrast, and $\phi(X,Y)$ the phase that is a function of the object shape $h(X,Y)$. The latter function is dependent on the general system geometry and the nature of the illumination. The surface shape is thus converted into a phase distribution and the methods of digital interferometric phase measurement , well known for their accuracy, can be used for the analysis.

The method using the phase-shifting algorithm[28-29] is most suitable for parallel acquisition and processing with a solid state detector array and a microcomputer. Translating the original projection grating by an amount equal to a fraction p/N of its period, the phase of the 'interferogram' represented by Equation 1 is shifted by $2\pi/N$, yielding a new intensity function I_N. With three or more interferograms corresponding to different phase shifting values, the phase function ϕ can be retrieved independently from the other parameters in Equation 1. For example, in the three-phase algorithm, three interferograms are detected by shifting the phase by $2\pi/3$ increments:

$$I_1 = R\,[\,A + B\cos(\,\phi\,)\,] \tag{2}$$

$$I_2 = R\,[\,A + B\cos(\,\phi + 2\pi/3)\,] \tag{3}$$

$$I_3 = R\,[\,A + B\cos(\,\phi + 4\pi/3)\,] \tag{4}$$

From these three equations, the phase function may be readily computed and is given by

$$\phi = \tan^{-1} \sqrt{3}\,[I_3 - I_2] \,/\, [2I_1 - (I_2 + I_3)] \tag{5}$$

and for the general case of the N-shift algorithm

$$\phi = \tan^{-1} \frac{\sum_{n=1}^{N} [I_n \sin(2n\pi/N)]}{\sum_{n=1}^{N} [I_n \cos(2n\pi/N)]} \tag{6}$$

The phase function is measured modulo 2π, but the absolute phase can be determined by monitoring the computed phases between two adjacent detectors and identifying sharp discontinuities which result from the two transitions.

A good feature of the phase shifting technique is that a high degree of precision can be achieved thanks to the use of multiple frame imaging. Phase values can be measured with a precision attaining 1/1000 of a fringe period, in contrast to about 1/30 for direct single frame measurement. Thus, with a projected grating with pitch in the millimeter range, profile measurements with better than 10 micron accuracy are possible. The technique can be used with linear or area solid state detectors such as photodiode, CCD or CID arrays. Also, since the method requires only the computation of an arc tangent function involving three or more intensity values at each point, a microcomputer system is adequate for control, storage and computation of surface shapes over a large array of data points.

Another similar phase measuring technique, the N-bucket technique[30] which requires continuous linear phase shifting, may also be used in the present application. Other techniques, such as heterodyne detection[31] and phase-lock detection[32] are essentially pointwise methods and are not suitable for parallel processing with large matrix detector arrays.

Phase-measuring profilometry: ranging of a 2-D single view

Projection with parallel coherent light illumination[24]

Principle of measurement. We consider the system geometry shown in Figure 2, where a collimated sinusoidal intensity distribution of period p is projected onto a 3-D object, with an offset angle θ from the observation direction, and where the object shape $h(X,Y)$ is measured from the reference plane. The projected sinusoidal pattern as seen on the reference plane has a period p_o and the intensity produced at a point such as C is given by

$$I_C(X,Y) = A(X,Y) + B(X,Y) \cos(2\pi \, OC/p_o) \quad (7)$$

Point O is the intersection of the imaging optical axis with the reference plane and is assumed, for convenience, to coincide with an intensity maxima of the projected pattern. The argument $\phi_R = 2\pi \, OC/p_o$ of the cosine function is defined as the phase at C and it effectively measures the geometric distance OC. D_c is one detector of the array in the image plane and is used to measure the intensity at C on the reference plane and at D on the object. The intensity observed at D is the same as that which would have been observed at A on the reference plane, modified by the object reflectivity $R(X,Y)$:

$$I_D(X,Y) = R(X,Y)[A(X,Y) + B(X,Y) \cos(2\pi \, OA/p_o)] \quad (8)$$

The difference ϕ_{CD} in phase values for the points C and D, observed by the same detector D_c, can be related to the geometric distance AC:

$$AC = (p_o / 2\pi) \phi_{CD} \quad (9)$$

In turn, AC is related to the surface height h = BD :

$$h = BD = AC \tan \theta_o / (1 + \tan \theta_o / \tan \theta_c) \quad (10)$$

where the angles θ_o and θ_c are as shown in Figure 2. Assuming that θ_c is nearly 90° as is the case for any practical system with large demagnification, the relation can be simplified:

$$h = BD = AC \tan \theta_o \quad (11)$$

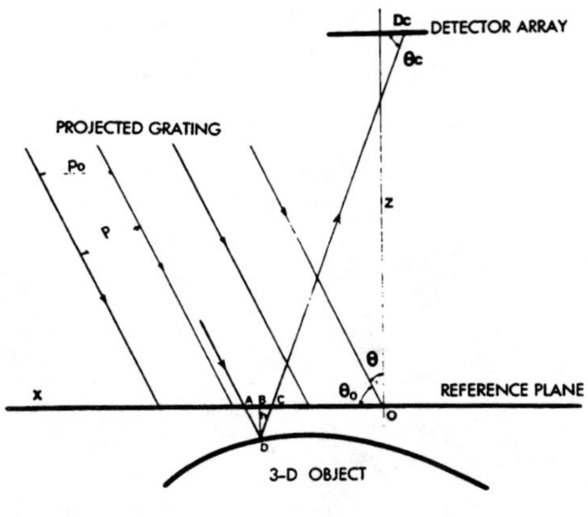

Figure 2. Optical geometry for parallel laser light illumination.

From Equations (9) and (11), the effective wavelength of the system is defined as

$$\lambda_e = p_o \tan \theta_o = p_o / \tan \theta \qquad (12)$$

To measure the phase of the intensity variation in Equations (7) and (8), N frames of intensity data with $2\pi/N$ incremental phase shifts are detected by the array and stored in memory, both for the reference plane and the object surface. Then, the phase shifting algorithm described by Equation (6) is applied and the surface shape is computed using Equation (11) where the geometric parameters are determined by a calibration experiment.

Experimental system and results. To produce a linear sinusoidal intensity distribution, a polarization shearing interferometer is used with a Helium-Neon laser source as shown in Figure 3. The polarized output beam of the laser is spatially filtered and sheared by a Wollaston prism W, producing straight sinusoidal fringes after passage through a suitably oriented polarizer P. The fringe period can be readily changed by axial translation of the Wollaston or by using a set of prisms with different angles. In Figure 3a, an electro-mechanical Senarmont modulator[24] is used as the phase shifter. It consists of a quarter wave plate Q and a rotatable polarizer P, so that a rotation of 180° of the polarizer corresponds to a 2π phase shift. In Figure 3b, a transverse effect electro-optical modulator is used, permitting very fast phase shifting with no moving parts[33]. In contrast to other interferometers, the polarization interferometer is compact, stable and insensitive to external disturbances.

A schematic of the complete system is shown in Figure 4. The results presented here were obtained by using a 128x128 photodiode array, a microprocessor based A/D converter and buffer, and an LSI 11/23 microcomputer. A metallic half cylinder, with two sections having different radii was used as a test object and placed on a reference plane. Figure 5a shows the three interferograms, with $2\pi/3$ incremental phase shift, as detected by the array camera, and Figure 5b shows the computed surface profile plot. The effective system wavelength was 5.67 mm and quantitative analysis of the results was performed by first calibrating the system with the aid of a known step function. Comparing with measurements using a mechanical contact profilometer, the measuring accuracy was found to be about 25 microns or $\lambda_e/200$.

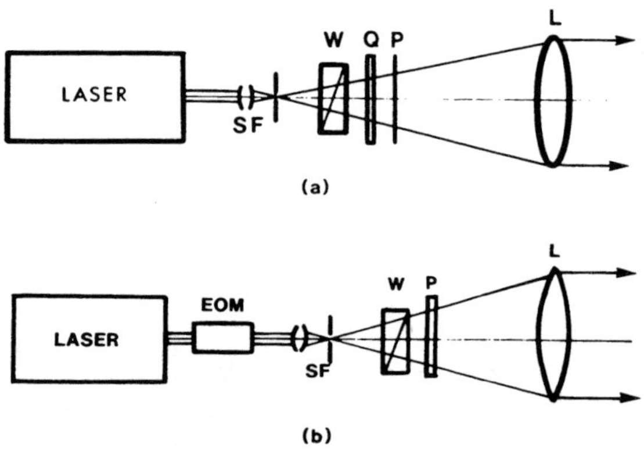

Figure 3. Polarization shearing interferometer configurations for laser light projection.

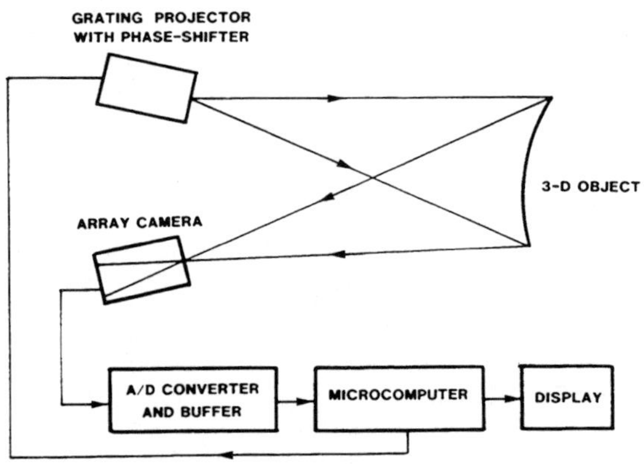

Figure 4. Block diagram of the complete system.

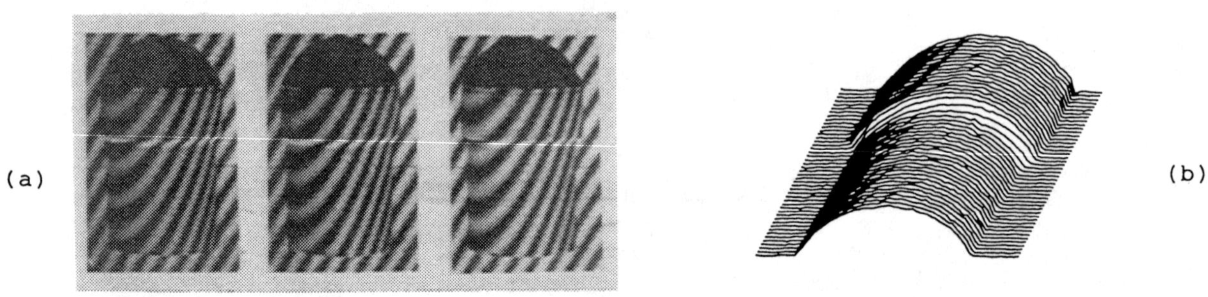

Figure 5. Interferograms (a) and computer plot (b) of a test object.

Projection with diverging white light illumination[25]

Principle of measurement. To make the system more practical in handling large objects, a diverging illumination is used by projecting a white light illuminated sinusoidal transmission grating. Because of the diverging nature of the illumination and because the optical axes of the projection and observation systems are not necessarily parallel, the analysis is more complicated than the previous case and requires a different approach. To deal with this generalized geometry, a new 'phase mapping' algorithm was developed.

The optical geometry of the projection and imaging system is shown in Figure 6. P_1 and P_2 are the centers of the entrance and exit pupils of the projection optics. I_1 and I_2 are the centers of the exit and entrance pupils of the imaging optics. G is a grating with pitch p_o and a sinusoidal intensity transmission. D_c is a photodiode element in the image sensing array. The intensity variation along X on the reference plane can be described by the equation

$$I_R (X,Y) = A(X,Y) + B(X,Y) \cos \phi(X), \qquad (13)$$

where the phase ϕ is a nonlinear function of X because of the divergent nature of the illumination. With respect to a reference point such as O, every point on the reference plane is characterized by a unique phase value. For example, the point C, observed by the detector D_c of the array, has the phase value ϕ_C. The detector array samples the reference plane, as well as the object, and is used to measure the phase at the sampling points by the phase shifting technique. N frames of intensity data, with $N \geq 3$, are recorded and after each frame the grating G is translated by a distance p_o/N . If $I_1, I_2, \ldots, I_N$ are the intensity measurements for a point such as C, then the phase at C is computed with the aid of Equation (6).

A particular detector such as D_c measures the phase ϕ_C at a point C on the reference plane as well as the phase ϕ_D on the point D of the object. A mapping algorithm can then determine a point A on the reference plane such that $\phi_A = \phi_D$. This enables a computation of the geometric displacement AC. C is a known detector location and the position of A, which in general would lie between two sampling points, can be located by linear interpolation using known phase values. From similar triangles P_2DI_2 and ADC, the object height is equal to

$$z = h(X,Y) = (AC/d)\, L\, (1 + AC/d)^{-1} \qquad (14)$$

where d and L are distances as shown in Figure 6. Since in most practical situations d >> AC, because of the large magnifications involved, Equation (14) can be simplified:

$$z = h(X,Y) = (AC/d)\, L \qquad (15)$$

However h(X,Y) is the object height at the X coordinate corresponding to B and not C. From known system geometric parameters, the distance BC can be calculated and determines the exact X coordinate corresponding to the point B.

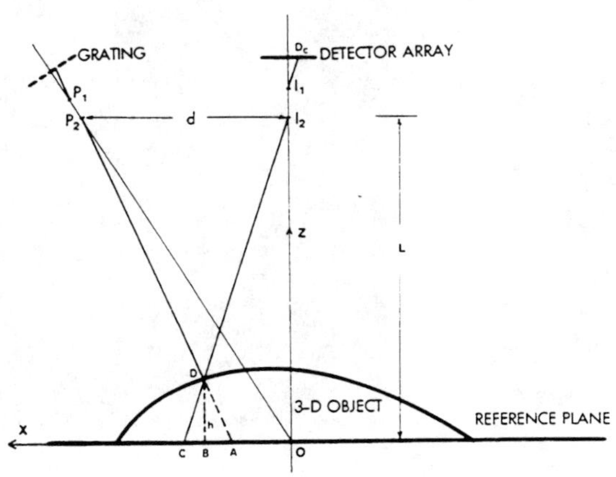

Figure 6. Optical geometry for general diverging white-light illumination.

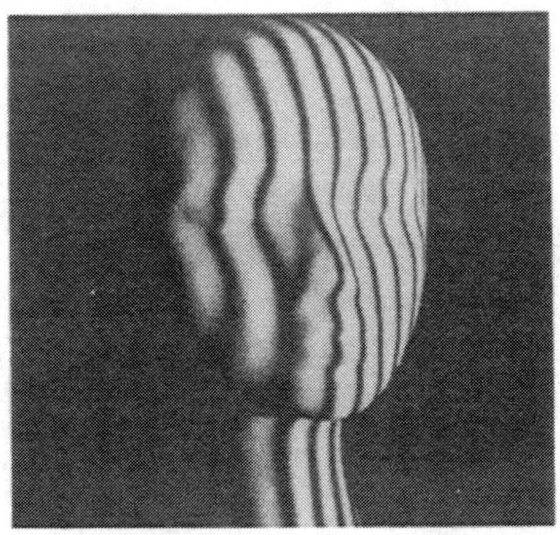

Figure 7. Typical interferogram of mannequin head.

Experimental system and results. A conventional slide projector was modified in order to operate with a grating slide, mounted on a stepper motor driven translation stage. Sinusoidal gratings were recorded by suitably photographying out of focus square wave gratings. The period of the projected grating measured on the reference plane, close to the optical axis of projection, was 15 mm. As previously, deformed grating images were recorded on a 128 X 128 photodiode array detector interfaced to an LSI 11/23 computer for data acquisition and processing. Phase measurement was carried out by the three discrete phase shift implementation given by Equation (5).

A general type of diffuse object--a mannequin face--was measured. Figure 7 shows a deformed grating image produced by this object. Figure 8 is a computer plot representation of the result for the front part of the mannequin face. This object was approximately 150 mm wide and the depth at the tip of the nose measured from a reference plane was about 100 mm. In order to obtain a complete shadow free illumination of such objects and obtain a large angular field of view, two or more projected gratings may be used.

The results of a calibration experiment with a metallic cylindrical test object, which was also measured by a mechanical contact profilometer, are shown in Figure 9. The line profile was generated by the optical method and the 'X' marks indicate measurements made by a manual contact profilometer. Almost perfect agreement within better than 1% was observed, except in the regions with steep slopes where mechanical contact methods are not very reliable. The measuring accuracy was estimated to be better that 0.1 mm or about $\lambda_e/200$.

Once shape data has been computed and stored, computer graphics techniques can be used to manipulate the data. Figures 10a and 10b show differently rotated images of the mannequin head, and Figure 10c shows a solid model rendering.

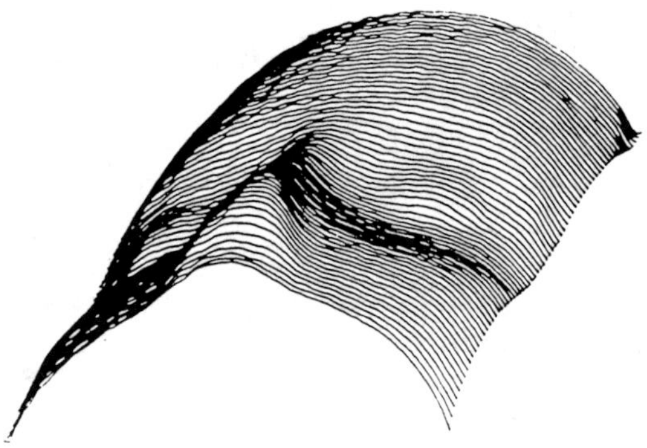

Figure 8. Computer plot of mannequin face.

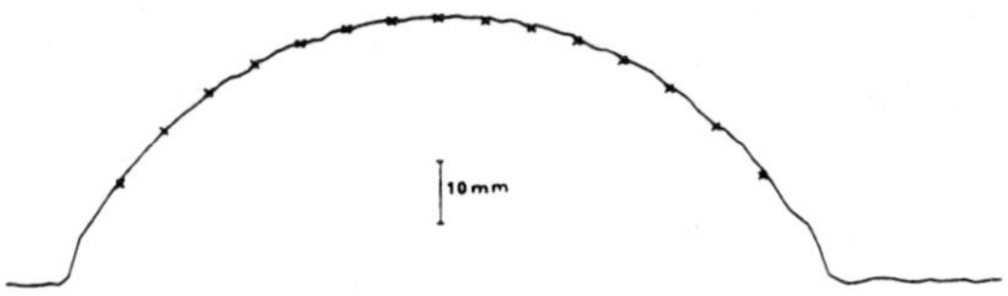

Figure 9. Computer plot of test object and comparison with contact (X) measurement.

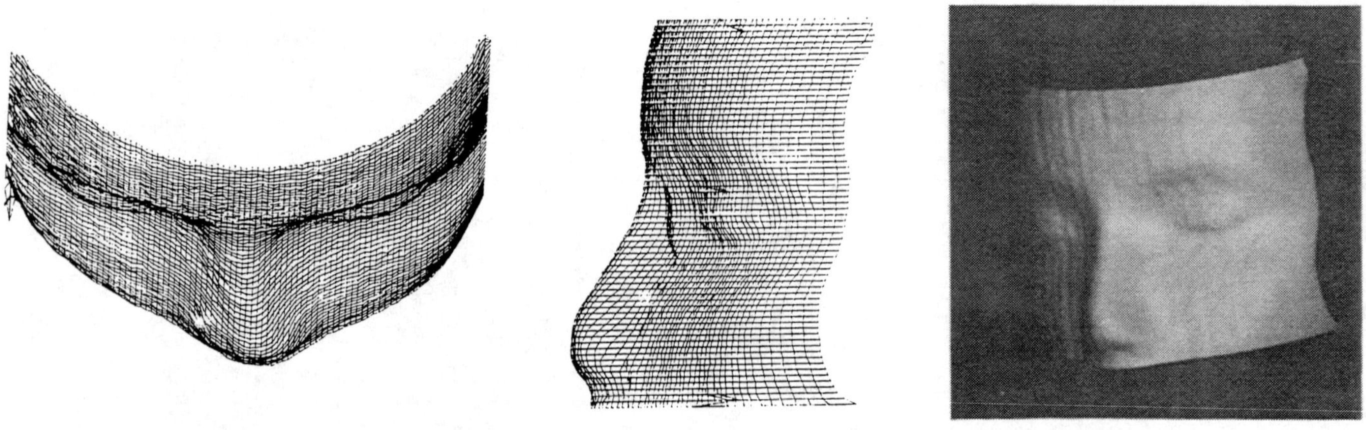

(a) (b) (c)

Figure 10. Computer graphics generated imagery: a. and b. rotated images, c. solid model rendition.

Phase-measuring profilometry: 360 ° measurement[26]

Principle of measurement. To obtain full 360° measurement of a general 3-D object, one approach which has been the subject of intense activity, especially in computer vision research, is the reconstruction of 3-D shapes from several views of the object as seen from several points of observation[34]. It involves the use of elaborate algorithms to match these views, is computation intensive and generally offers limited accuracy. A less frequent approach involves the use of the shadow moire method in conjunction with the 'periphery camera'[35]. It is mechanically cumbersome and requires tedious processing of the fringe pattern recorded on film.

We have extended the use of the phase-measuring profilometry described above to the 360° measurement problem. Here the object is placed on a stepper motor rotation stage and successive line images of the deformed projection grating are recorded as the object is rotated by discrete angles between recordings. The corresponding profiles of the object are then calculated using the phase shifting algorithm and are combined together to form the 360° reconstruction of the 3-D object.

The optical arrangement is schematically shown in Figure 11a for the general case of divergent illumination and non-parallel axes. A sinusoidal grating with horizontal lines is projected onto the object and the resulting deformed grating is reimaged onto a linear photodiode detector array. The phase measuring profilometry method permits the determination of the distance h of an object point D from the reference line, and therefore gives the object line profile in the Y-Z plane. As the object is rotated around the Y-axis in incremental steps, every h-value obtained corresponds to the radial distance Δr of a general point P from a reference virtual cylinder of radius R, as shown in Figure 11b. The complete 3-D object data is obtained through a 360° rotation and is expressed in terms of $r(Y,\theta)$ in cylindrical coordinates. A general object point P can be represented in relation to a global system of coordinates X, Y, Z using the transformation

$$X = R \sin\theta \; [1 - \Delta r/R] \tag{16}$$

$$Z = R \cos\theta \; [1 - \Delta r/R] \tag{17}$$

The values of Δr, for a given profile corresponding to a θ value, are calculated by the phase-shifting technique and phase mapping algorithm as described, then subsitituted into Equations (16) and (17) to obtain the governing equations for a general point description:

$$X = \sin\theta \; (R - AC\; L/d) \tag{18}$$

$$Z = \cos\theta \; (R - AC\; L/d) \tag{19}$$

These expressions represent the "wrapping" of the data to obtain the final 3-D shape with reference to the global X, Y and Z coordinate system.

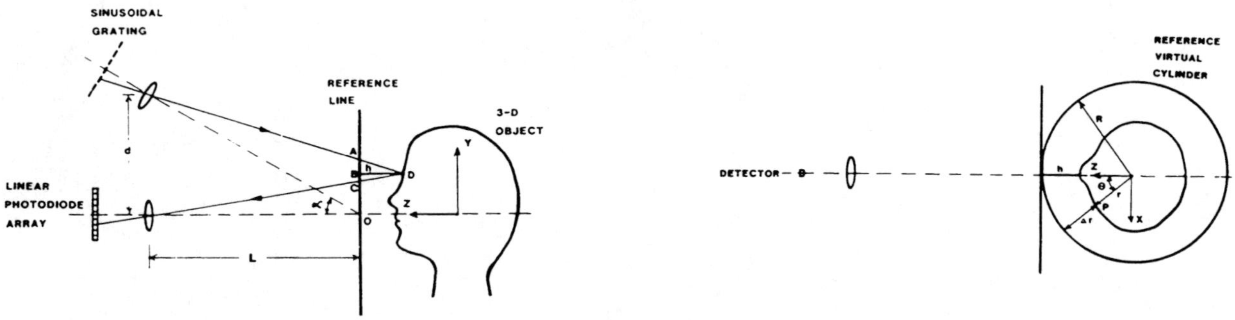

(a) (b)

Figure 11. Optical geometry for 360° measurement with diverging white light illumination.

Experimental system and results

A block diagram of the complete system is similar to that shown in Figure 4, with the object mounted on a stepper motor driven rotation stage. The hardware consists of an ordinary white light slide projector with a translatable sinusoidal grating and an image sensor consisting of a solid state linear detector array configured with a microcomputer. In this preliminary work, we have used a 128X128 photodiode array from which a single column data is read out and then processed on an LSI 11/23 computer. The computer is used to control the phase setting of the projection grating, the object rotation stage, and the array camera, resulting in a fully automated data acquisition system.

The same mannequin head as above was used as a test object. The projected grating pitch on the object was 15 mm and the angle between the projection and imaging axes was 20°, resulting in an equivalent system wavelength of 45 mm. The object was rotated through 360° in steps of 2°, resulting in a total of 180 line profiles and 23,040 data points. Figure 12a shows the result for the mannequin head in the form of a development in the θ direction, directly obtained from the raw data. The "wrapping", obtained by the transformation indicated in Equations (18) and (19), yields the reconstruction of the actual 3-D shape in the global rectangular coordinates. Figures 12b and 12c show the front and side views, Figures 12d and 12e show a vertical and a horizontal cross section, and Figure 12f shows a complete see-through horizontal section representation. With the available 3D data it is also possible to fit mathematical surfaces for the entire 360° object, and a solid model representation can be obtained using computer graphics techniques.

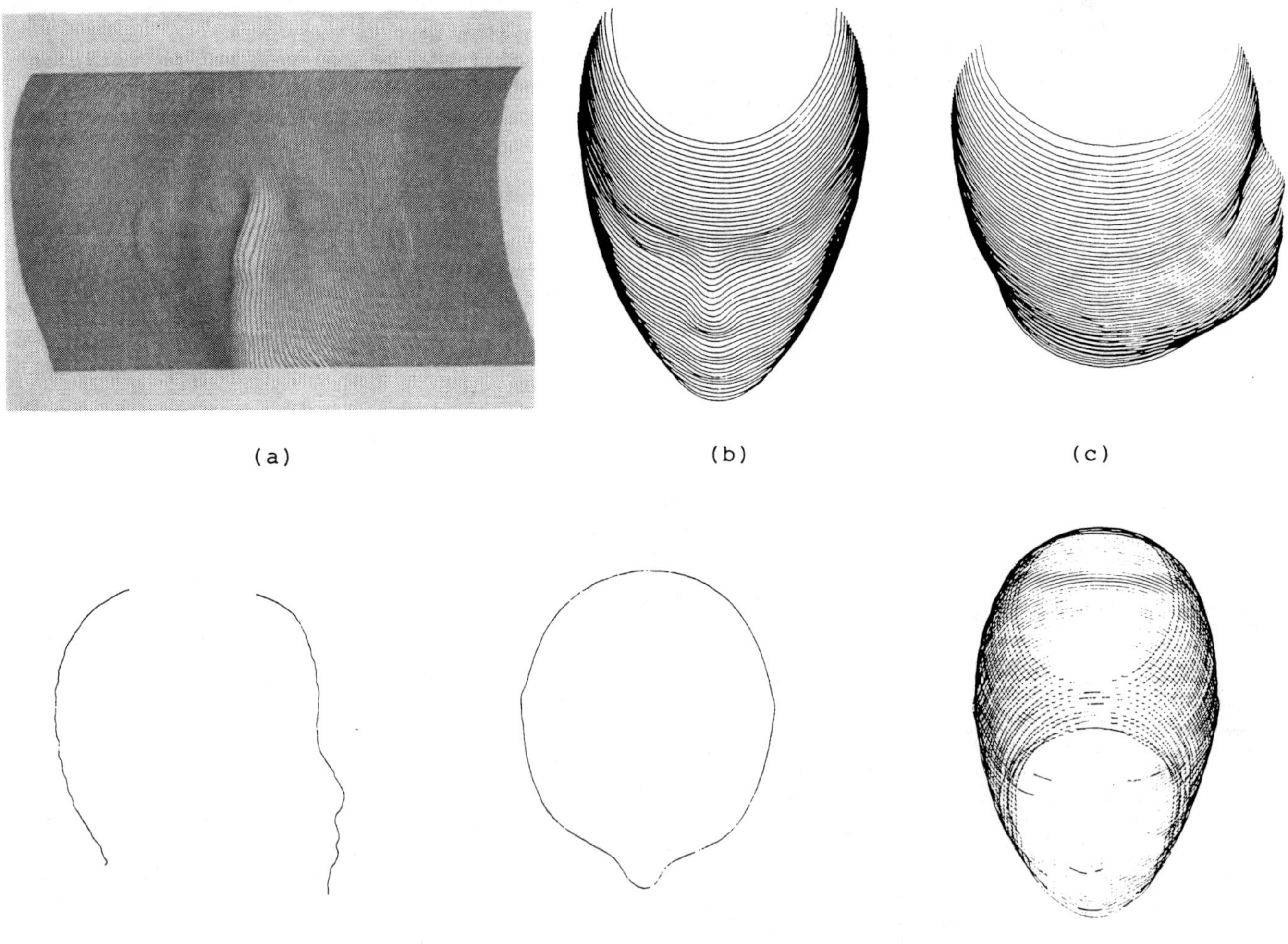

(a) (b) (c)

(d) (e) (f)

Figure 12. Results of 360° measurement of mannequin head: a. unwrapped output data, b. front view, c. side view, d. vertical section, e. horizontal section, g. see-through plot (horizontal Sections).

Performance analysis

Among the various sources of error that can affect 3-D measurement based on triangulation, we will analyze those that particularly affect the phase measuring implementation. In each case the errors will be measured in degrees or fraction of effective wavelength.

Errors due to imperfect sinusoidal fringe profile

A prerequisite for applying phase measuring techniques is that the projected fringe profile be ideally sinusoidal. Analyzing the effect of imperfect sinusoidal fringe profile is particularly relevant for the case of diverging illumination when a grating transparency is used, as it is difficult to produce sinusoidal gratings photographically.

A series of gratings were fabricated and further variation of their profile was obtained by defocusing the slide projector. Figure 13 shows a series of intensity distributions with various degree of deviation from an ideal sinusoidal function. They were obtained by detecting the intensity at a point on the reference plane, with linear translation of the grating. In the Figure, from top to bottom, the deviation from a sinusoid decreases until set number 9, for which the RMS deviation is 1.13%, and the deviation for set numbers 10-14 increases in the opposite direction. In each case, the RMS intensity deviation ΔI is measured and the corresponding RMS phase error $\Delta\phi$ is computed by using the three-phase algorithm (RMS deviation from linear phase function). The results are shown in Table 1. Figure 14 shows the actual phase deviation for selected cases. Thus it is seen that the phase error resulting from imperfect sinusoidal grating can be kept as low as 1.15° or $\lambda_e/313$.

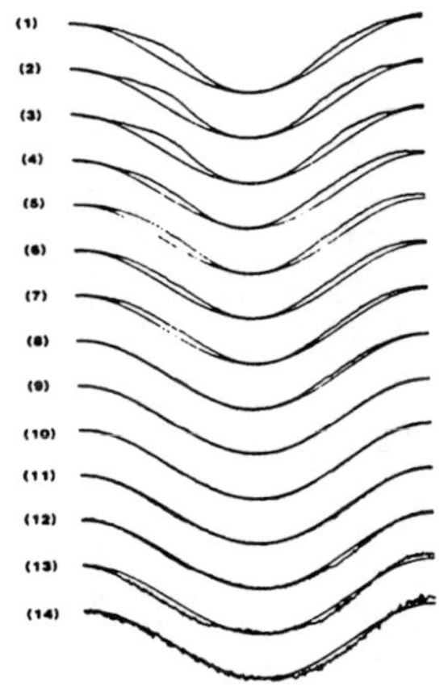

Figure 13. Intensity distribution of imperfect sinusoidal gratings.

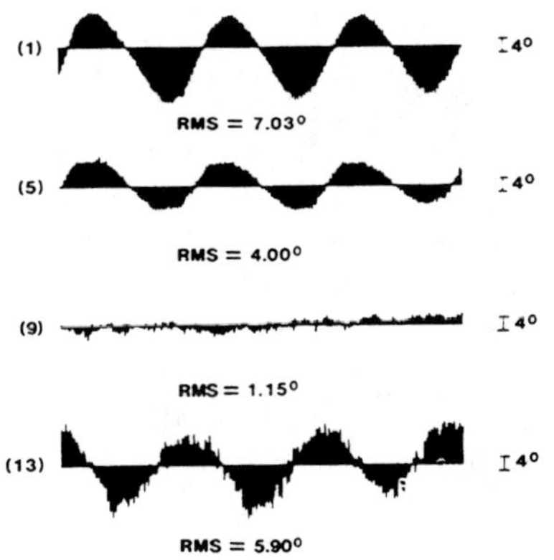

Figure 14. Phase measurement error due to imperfect sinusoidal intensity distribution.

Table 1

Grating #	ΔI RMS (%)	$\Delta\phi$ RMS ($^{\circ}$)
1	9.64	7.035
2	9.77	7.373
3	9.72	7.317
4	8.89	4.905
5	8.69	4.000
6	7.12	3.253
7	6.11	2.719
8	3.11	1.815
9	1.13	1.155
10	1.27	1.323
11	2.26	2.251
12	3.61	3.840
13	6.40	5.903
14	5.52	5.381

Errors due to imperfect phase shift settings

Using the N-phase shifting algorithm assumes that the incremental phase shift is ideally equal to $2\pi/N$ in Equation (6). Deviation from ideal values may be due to the discrete nature of the stepper motor used for grating translation. First, the effect of phase setting error was mathematically simulated by introducing an error ε in the three-phase algorithm, and the results are shown in Figure 15. It is seen that the effect is a spurious periodic variation with twice the original frequency and with amplitude approximatively equal to the phase setting error. Then, phase setting errors were experimentally introduced by modifying the stepper motor setting and the correponding phase errors were computed as before. Results are shown in Table 2 and selected phase error profiles shown in Figure 16. Thus, in the present system, it is shown that the RMS error can be reduced to 1.15° or $\lambda_e/313$. It should be noted that this value mostly represents the electronic system noise since the phase can be set well within a fraction of a degree.

Table 2.

a. Simulation

ε	RMS error (°) *	RMS error (°) **
0	0	0
1	0.701	0.411
2	1.403	0.827
3	2.104	1.247
4	2.805	1.672
5	3.506	2.101
6	4.207	2.536
7	4.908	2.976
8	5.609	3.421
9	6.310	3.873
10	7.011	4.330

* $I_1 = a + b\cos\phi$
$I_2 = a + b\cos(\phi + 2\pi/3 + \varepsilon)$
$I_3 = a + b\cos(\phi + 4\pi/3 + \varepsilon)$

** $I_1 = a + b\cos\phi$
$I_2 = a + b\cos(\phi + 2\pi/3 + \varepsilon)$
$I_3 = a + b\cos(\phi + 4\pi/3 - \varepsilon)$

b. Experimental

ε	RMS error (°)
-12.6	8.784
-11.34	7.903
-10.08	7.028
-8.82	6.153
-7.56	5.283
-6.3	4.424
-5.04	3.576
-3.78	2.747
-2.52	1.981
-1.26	1.345
0	1.155
1.26	1.538
2.52	2.246
3.78	3.040
5.04	3.882
7.54	5.606
8.82	6.474
10.08	7.354
11.34	8.236
12.6	9.116

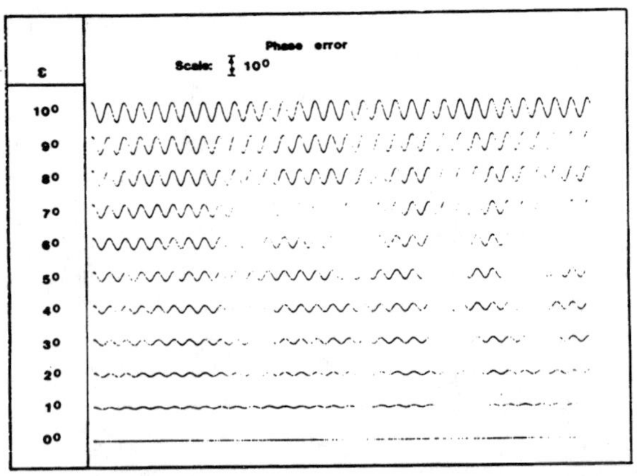

Figure 15. Phase measurement error due to imperfect setting (ε) of the phase shifter (simulation).

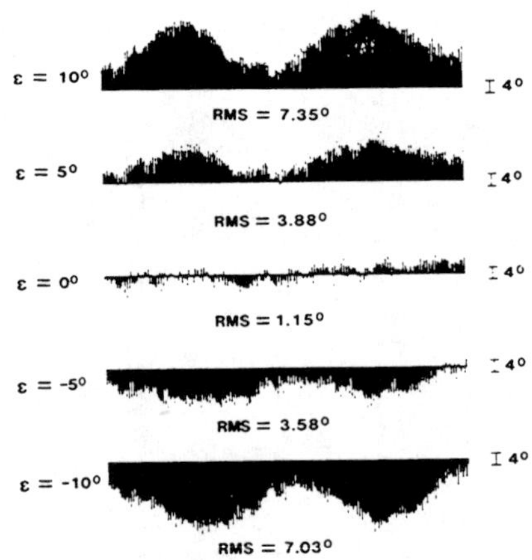

Figure 16. Phase measurement error due to imperfect setting (ε) of the phase shifter (experimental).

Errors due to quantization effects

A primary source of error in phase measurement is the quantization noise due to the finite number of digitized levels present in the detected intensities[29]. Assuming Q levels are present over the range 2B (Equation 1) and assuming the worst case corresponding to 0° phase, the quantization results in an uncertainty $\pm$ B/Q for each intensity measurement. For the case of the three-phase algorithm, one has:

$$I_1 = A + B \pm B/Q \qquad (20)$$

$$I_2 = A - (B/2) \pm B/Q \qquad (21)$$

$$I_3 = A - (B/2) \pm B/Q \qquad (22)$$

The RMS deviation error is

$$\Delta\phi \ \text{RMS} = 1 / 1.732\ Q \qquad (23)$$

and the maximum error is

$$\Delta\phi_{MAX} = 1 / 0.866\ Q \qquad (24)$$

In the present system, for Q = 50, the RMS phase error is 0.7^o or $\lambda_e/514$ and the maximum phase error 1.3^o or $\lambda_e/276$.

When all the above errors are combined, the resulting measuring accuracy may be estimated to be of the order of $\lambda_e/200$.

Conclusion

Novel methods of 3-D shape measurement, based on the projection of a sinusoidal intensity distribution and digital phase measuring techniques, have been reviewed. Systems using parallel laser light illumination or general diverging white light illumination have been described, which can yield single view measurement or full 360° measurement of a diffuse 3-D object. Theoretical analysis and experimental verifications have shown that high measuring accuracy can be attained and is conveniently expressed as a fixed fraction of the system effective wavelength $p/\tan\theta$. The accuracy can thus be set a priori by tuning the period of the projection grating and the angle offset between the projection and the observation optics. In the present system a combined accuracy of about $\lambda_e/200$ has been achieved, so that resolution better than 10 microns can be obtained with a 1 mm projected grating period.

A complete system simply consists of a white light projector with translatable grating (or a compact polarization shearing interferometer with a phase modulator) and a solid state detector array configured with a microcomputer. The system described above used a 128x128 photodiode array, a single-frame buffer and an LSI-11/23 microcomputer. A new system has just been completed, which uses a 380x490 CCD camera, an addressable multiframe buffer card and an IBM AT microcomputer, with current efforts concentrated on improving resolution and speed.

Acknowledgments

This work was supported by NYIT and NYCOM research funds. We wish to thank Alexander Schure, Philip Fleisher, Herbert Fox, Melvin Drossman and Louis Schure for their continued support. We also thank Georges Nomarski and Lance Williams for fruitful discussions; V. Srinivasan, Ravi Krishnamurthy and Fu-Pen Chiang for their contribution to the work presented here.

References

1. T. Kanade and H. Asada, "Non-Contact Visual Three-Dimensional Ranging Devices", S.P.I.E. 283 (3-D Machine Perception), pp. 48-53. 1981.
2. S. Parthasarathy, J. Birk and J. Dessimoz, "Laser Rangefinder for Robot Control and Inspection", S.P.I.E. 336 (Robot Vision), pp. 2-11. 1982.
3. G.J. Agin and T.O. Binford, "Computer Description of Curved Surfaces", Third Inter. Joint Conf. on Artificial Intell., pp. 629-640, Stanford University, 1973.
4. G.B. Porter III and J.L. Mundy, "Noncontact Profile Sensing System for Visual Inspection", S.P.I.E. 336 (Robot Vision), pp. 67-76. 1982.

5. G.J. Agin and P.T. Highnam, "Movable Light-Stripe Sensor for Obtaining Three-Dimension Coordinate Measurements", S.P.I.E. 360 (Robotics and Industrial Inspection), pp. 326-333. 1982.
6. Y. Shirai and M. Suwa, "Recognition of Polyhedrons with a Range Finder", Proc. 2nd Intern. Joint Conf. on Artificial Intell., pp. 80-87, London 1971.
7. M.D. Altschuler, B.R. Altschuler and J. Taboada, "Measuring Surfaces Space-Coded by a Laser-Projected Dot Matrix", S.P.I.E. 182 (Imaging Applications for Automated Industrial Inspection and Assembly), pp. 187-191. 1979.
8. M.D. Altschuler, B.R. Altschuler, and J. Taboada, "Laser Electro-Optic System for Rapid Three-Dimensional (3-D) Topographic Mapping of Surfaces", Opt. Eng. 20, pp. 953-961. 1981.
9. J.L. Posdamer and M.D. Altschuler, "Surface Measurement by Space-Encoded Projected Beam Systems," Computer Graphics and Image Processing 18, pp. 1-17. 1982.
10. B.F. Alexander and K.C. Ng, "A Liquid Crystal Projection System for 3D Imaging," in Proc. 5th Int. Conf. Robot Vision and Sensory Controls, pp. 211-222. Amsterdam 1985.
11. P.M. Will and K.S. Pennington, "Grid Coding: A Novel Technique for Image Processing", IEEE Proc. 60, pp. 669-680. 1972.
12. P. Benoit, E. Mathieu, J. Hormiere and A. Thomas, "Characterization and Control of Three Dimensional Objects Using Fringe Projection Techniques", Nouv. Rev. Optique, t. 6, pp. 67-86. 1975.
13. J.A. Jalkio, R.C. Kim, and S.K. Case, "Three Dimensional Inspection Using Multistripe Structured Light", Opt. Eng. 24, pp. 966-974. 1985.
14. D.M. Meadows, W.O. Johnson, and J.B. Allen, "Generation of Surface Contours by Moire Patterns", Appl. Opt. 9, pp. 942-947. 1970.
15. H. Takasaki, "Moire Topography", Appl. Opt. 9, pp. 1457-1472. 1970.
16. M. Idesawa, and T. Yatagai, "General Theory of Projection-Type Moire Topography", Sci. Papers I.P.C.R. 71, pp. 57-70. 1977.
17. J.C. Perrin and A. Thomas, "Electronic Processing of Moire Fringes. Application to Moire Topography and Elements of Comparison with Photogrammetry", SPIE 136, (1st European Congress on Optics Applied to Metrology), pp. 311-317. 1977.
18. G. Indebetouw, "Profile Measurements Using Projection of Running Fringes", Appl. Opt. 17, pp. 2930-2933. 1978.
19. R.N. Shagam, "Heterodyne Interferometric Method for Profiling Recorded Moire Interferograms", Opt. Eng. 19, pp. 806-809. 1980.
20. D.T. Moore and B.E. Truax, "Phase-Lock Moire Fringe Analysis for Automated Contouring of Diffuse Surfaces", Appl. Opt. 18, pp. 91-96. 1979.
21. T. Yoshizawa, H. Tashiro and S. Nakagawa, "Interpolation of Contour Intervals in Moire Topography by Phase Modulation", Opt. Comm. 42, pp. 399-402. 1982.
22. M. Takeda, and K. Mutoh, "Fourier Transform Profilometry for the Automatic Measurement of 3-D Object Shapes", Appl. Opt. 22, pp. 3977-3982. 1983.
23. M. Takeda, H. Ina and S. Kobayashi, "Fourier-Transform Method of Fringe-Pattern Analysis for Computer-Based Topography and Interferometry", J. Opt. Soc. Am. 72, pp. 156-160. 1982.
24. V. Srinivasan, H.C. Liu, and M. Halioua, "Automated Phase-Measuring Profilometry of 3-D Diffuse Objects," Appl. Opt. 23, pp. 3105-3108. 1984.
25. V. Srinivasan, H.C. Liu, and M. Halioua, "Automated Phase-Measuring Profilometry: A Phase Mapping Approach," Appl. Opt. 24, pp. 185-188. 1985.
26. M. Halioua, R.S. Krishnamurthy, H.C. Liu, and F.P. Chiang, "Automated 360° Profilometry of 3-D Diffuse Objects," Appl. Opt. 24, pp. 2193-2196. 1985.
27. J. Wyant and C. Koliopoulos, "Phase Measurement Systems for Adaptive Optics", pp. 48.1-48.12 in NATO-AGARD Conf. Proc. No.300, 28th Meet. E.M. Wave Propag. Panel, Monterey, CA, 1981.
28. J.H. Bruning, J.E. Gallagher, D.P. Rosenfeld, A.D. White, D.J. Brangaccio, and D.R. Herriott, "Digital Wave Front Measuring Interferometer for Testing Optical Surfaces and Lenses', Appl. Opt. 13, pp. 2693-2703. 1974.
29. C.L. Koliopoulos, "Interferometric Optical Phase Measurement Techniques", Ph.D Thesis, Univ. of Arizona, 1981.
30. J.C. Wyant, "Use of an ac Heterodyne Lateral Shear Interferometer with Real-Time Wave Front Correction Systems", Appl. Opt. 14, pp. 2622-2626. 1975.
31. R.C. Crane, "Interference Phase Measurement", Appl. Opt. 8, pp. 538-542. 1969.
32. R.C. Moore, "Automatic Method of Real-Time Wavefront Analysis", Opt. Eng. 18, pp. 461-463. 1979.
33. M. Halioua, T. Bowins and H.C. Liu, "Holographic Interferometry with Fringe Control and Digital Phase Measurement", to be pulished, Appl. Opt., 1986.
34. B. Bhanu, "Representation and Shape Matching of 3-D Objects", IEEE Trans. Pattern Anal. Machine Intell. PAMI-6, pp. 340-350. 1984.
35. C. Saunders, "Replication from 360 Degree Moire Sensing", in Moire Fringe Topography and Spinal Deformity, pp. 76-80. M.S. Moreland, M.H. Pope, G.W.D. Armstrong, Eds, Pergamon Press 1981.

[Editor's note: A revised and expanded version of this paper has since been published: "Optical three-dimensional sensing by phase measuring profilometry," Maurice Halioua and Hsin-Chu Liu, *Optics and Lasers in Engineering* **11**, 185-215 (1986).]

Reprinted with the permission of the Controller of Her Majesty's Stationery Office
from *Optical Techniques for Industrial Inspection,* P.G. Cielo, editor,
Proc. SPIE Vol. 665, pp. 162-167 (1986).

AUTOMATIC INSPECTION OF QUASI-CYLINDRICAL OBJECTS BY PHASE MEASURING MOIRE TOPOGRAPHY

G. T. Reid and R. C. Rixon
National Engineering Laboratory, East Kilbride, Glasgow G75 0QU, UK

Abstract

A three dimensional machine vision system, based on a phase measuring version of moire topography, is described. By rotating an object in front of the system, surface co-ordinates can be acquired from many thousands of points around the periphery of that object. A computational method for comparing the dimensions of an object with its design specification is described. The application of this method is restricted to the subset of objects which can be described in terms of simple primitive shapes.

Introduction

Recent work in optics and image processing has led to the development of a number of new machine vision techniques[(1-3)] for three dimensional measurement and inspection. Phase measuring moire topography, for example, is a machine vision method which provides automatic measurements of three dimensional surface relief by optically generating, and digitally analysing, the contour maps of surfaces. In common with most other machine vision techniques, phase measuring moire topography is most readily applied to inspection tasks which can be accomplished from a single viewpoint. Since most artefacts have three dimensional solid geometry, however, they require inspection from multiple viewpoints. Consequently, some form of mechanical handling and/or a multi-camera vision system is usually required to provide comprehensive inspection.

This paper describes a version of phase measuring moire topography which retains a single camera but, by rotating the object in front of the vision system, can rapidly acquire the three dimensional co-ordinates of many thousands of points around the surface of a solid object. Having explained the data acquisition process, the paper will discuss some of the techniques which can be used to automatically "understand" the data and to compare the object with the dimensions which exist within the computer aided design model of the object.

Phase measuring moire topography

Detailed descriptions of phase measuring moire topography have been given in previous papers[(4,5)] so the technique will merely be summarised here for completeness.

As shown in Fig. 1, a grating is projected onto the surface of the object while an image of the object is formed in the plane of a second, reference, grating. Interference between the image and the reference grating causes a number of moire fringes to appear superimposed upon the image of the surface of the object. By careful choice of the optical geometry[(6)], we can ensure that the moire fringes describe contours of equal surface height. Defining 'a' and 'b' as geometric constants of the optical system, three dimensional surface co-ordinates (X,Y,Z) can be calculated from the image plane co-ordinates (x,y) and the fringe number, n, through the equations: $Z = a/n$; $X = Zx/b$; $Y = Zy/b$.

Fig. 2 shows a representative moire fringe contour map. In order to unambiguously analyse the image we must establish a means of discriminating between convex and concave surface features and we must isolate the contour pattern from spurious variations in image intensity. Unambiguous image analysis can be achieved after measuring the phase, rather than the intensity, of the moire fringes.

Phase values are calculated after digitizing a series of moire fringe contour patterns. In the simplest case, the first contour pattern is digitized and stored in computer memory before the projection grating is moved, in its own plane, through 1/3 of the grating pitch (ie the phase of the projection grating is increased by $2\pi/3$). A second contour pattern is then digitized and stored in computer memory and the projection grating is moved through a further 1/3 of its pitch before a third contour pattern is digitized and stored. Each time the grating is moved through a fraction of its pitch, the moire fringes move through the same fraction of the contour interval. If the moire fringes have sinusoidal intensity profiles then the intensity of the image at a given pixel location will vary sinusoidally with the position of the projection grating. Having digitized and stored three contour patterns, therefore, we have stored intensity levels I_1, I_2 and I_3 at each pixel location. The phase, $\emptyset$, of the contour pattern at a given pixel location is given by equation (1):

$$\emptyset = \arctan \frac{\sqrt{3}\ (I_1 - I_2)}{2I_3 - I_2 - I_1} \qquad (1)$$

By evaluating equation (1) at every pixel location, we transform the three contour patterns into a phase distribution which is directly related to the topography of the surface while being insensitive to spurious variations in image intensity. Since phase values increase over positively sloping regions, of the surface but decrease over negatively sloping regions, concave and convex surface features are unambiguously

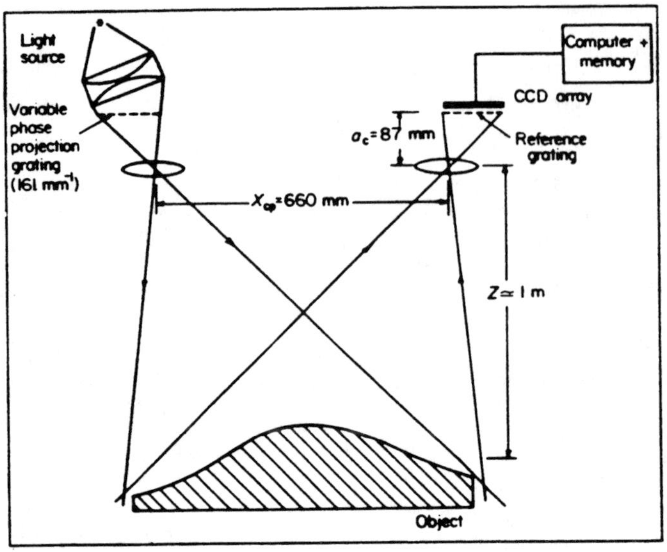

Figure 1. Schematic diagram of phase measuring moire topography.

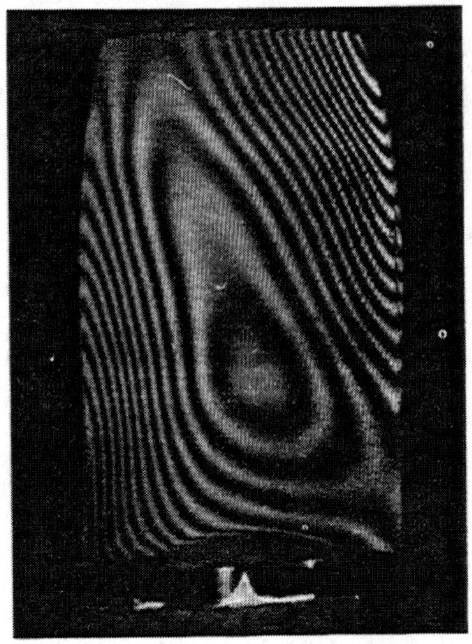

Figure 2. Moire fringe contour map of a fan blade; the contour interval is 2 mm.

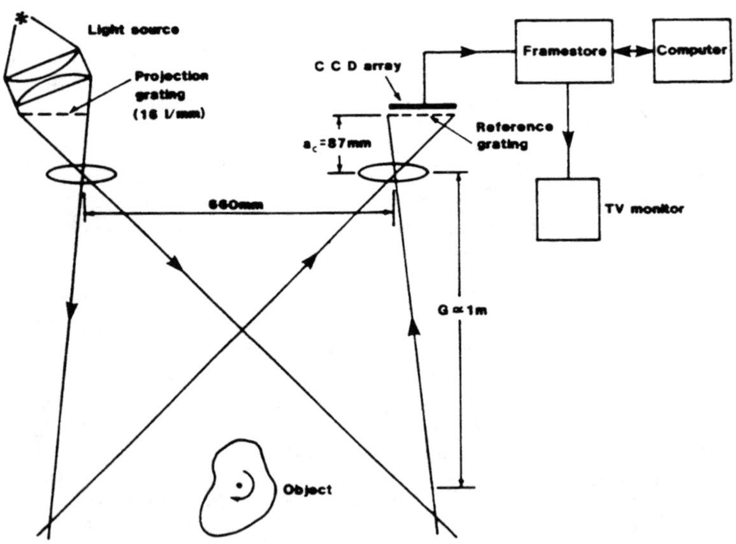

Figure 3. Schematic diagram of a peripheral moire contouring system.

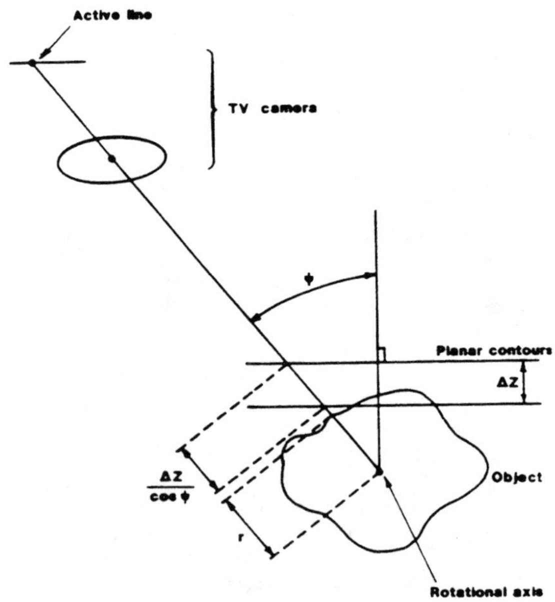

Figure 4. Geometry of peripheral contour formation.

described. Furthermore, since values of phase can usually be calculated to an accuracy of better than $2\pi/50$, the Z co-ordinate of the surface can be determined to within 1/50th of the contour interval.

Having evaluated equation (1) at every pixel location, values of the fringe number, n, are given by $\emptyset/2$ and three dimensional surface co-ordinates (X,Y,Z) can be calculated at each pixel in the image.

Measurements of quasi-cylindrical objects

Phase measuring moire topography is a powerful means of inspecting three dimensional surface relief but it is only after adapting it to the specific requirements of an application that its full potential can be exploited.

When inspecting quasi-cylindrical objects, for example, modification of the apparatus of Fig. 1 allows us to generate peripheral contour patterns which show cylindrically developed images of workpieces.

As shown in Fig. 3, planar contours are generated by the projection moire method of Fig. 1 and are detected by a television camera. The object rotates about an axis which is parallel to the TV scanning lines. One line from each of a number of TV images contributes to the peripheral contour pattern. This 'active' line can be isolated from the rest of the planar contour pattern by capturing a TV image in a frame-store and reading only the active line from the frame-store into computer memory. If, as shown in Fig. 4, the active line lies in the plane defined by the rotational axis and the perspective centre of the camera lens then the planar contours can be interpreted as contours of equal radius where they intersect the active line.

Acquisition of the peripheral contour pattern begins with the digitization of the TV image of a planar contour pattern. The digitized image is held in a frame-store while the active line is read into computer memory. This active line becomes the first line of the peripheral contour pattern. The frame-store memory is then erased. During the above operations, the object rotates through 360 degrees/N (where N represents the number of lines which will eventually be present in the TV image of the peripheral contour pattern). A second TV image is digitized and the active line is read into computer memory to become the second line in the peripheral contour pattern. This process is repeated until the active lines from N TV images, each showing a different part of the surface, have been read into computer memory and the frame-store has been erased for the last time. The peripheral contour pattern is then transferred from the computer to the frame-store whereupon it can be displayed on a TV monitor. Since each TV image contributes only one line to the peripheral contour pattern, it may seem that a linear photodiode array should be used as an image sensor. The use of a TV camera, however, has two important advantages. Firstly, the apparatus can be constructed using widely available digital image processing equipment. Secondly, the apparatus can be used for conventional projection moire work without modification of the hardware.

To a first approximation, the peripheral contours can be regarded as contours of equal radius on a cylindrical development of the surface. We note, however, that if the planar contour interval is Z then, as shown in Fig. 4, the peripheral contour interval is $Z/\cos\psi$. This interpretation of the image is usually adequate for qualitative assessment of the surface relief but a more rigorous analysis[5] is necessary when accurate, quantitative, information is required.

Peripheral contour patterns can be analysed using the phase measuring technique which has already been described. After acquiring three peripheral contour patterns at different positions of the projection grating, phase values are calculated from equation (1), the fringe number n is expressed as $\emptyset/2\pi$ and cylindrical polar co-ordinates of the surface are calculated.

Results

As part of its research program on advanced manufacturing technology, the National Engineering Laboratory has started to investigate methods of improving the interface between machine vision-based inspection equipment and computer aided design (CAD) systems. Ultimately, this work aims to demonstrate that a wide variety of artefacts can be geometrically inspected by comparing their critical dimensions with specifications which have been derived from three dimensional CAD models. The early results of this work illustrate some of the capabilities of phase measuring moire topography while indicating one of the ways in which vision-based inspection equipment might function in a factory of the future.

Fig. 5 shows the solid body CAD model of a bi-conical mandrel, the geometry of which has been defined in terms of conical and cylindrical component shapes or 'primitives'. Data from the CAD model was used to program a multi-axis, numerically controlled machine tool and a mandrel was manufactured. The mandrel, which nominally is 190 mm high and has included cone angles of 40 degrees, was then mounted on a computer controlled rotary table so that the peripheral contour map of Fig. 6 could be generated. Using the phase measuring technique, the moire fringe contours were analysed and cylindrical polar co-ordinates were evaluated at a dense matrix of points on the surface of the mandrel. By plotting every 6th measured point along the rotational axis of the object and every 3rd point along the vertical axis, the wire mesh diagrams of the mandrel shown in Fig. 7 were generated.

Having captured a large number of surface co-ordinates, this data may be further processed with a view to comparing the geometry of the object with the design specification. In many respects, this comparison is at the heart of the inspection cycle.

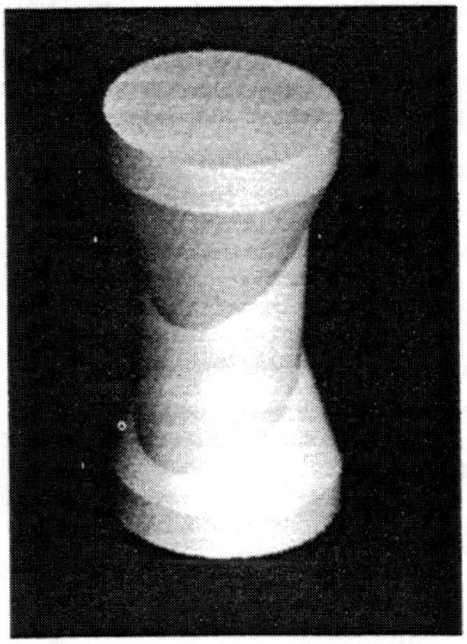

Figure 5. Shaded view of the computer aided design model of the bi-conical mandrel. The model was constructed on the Romulus solid body modeller by Shape Data Ltd.

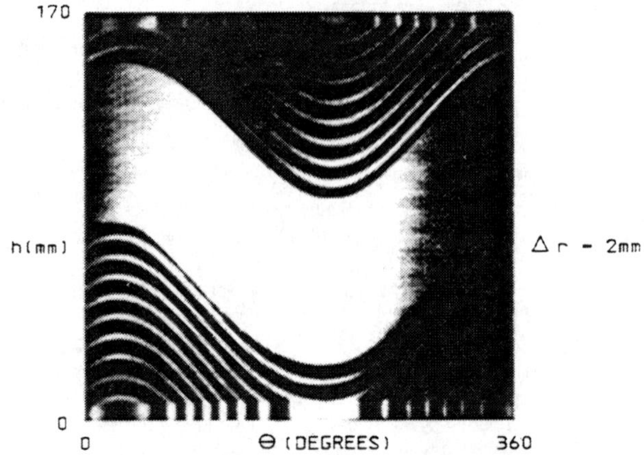

Figure 6. Peripheral contour map of the bi-conical mandrel; the contour interval is 2 mm.

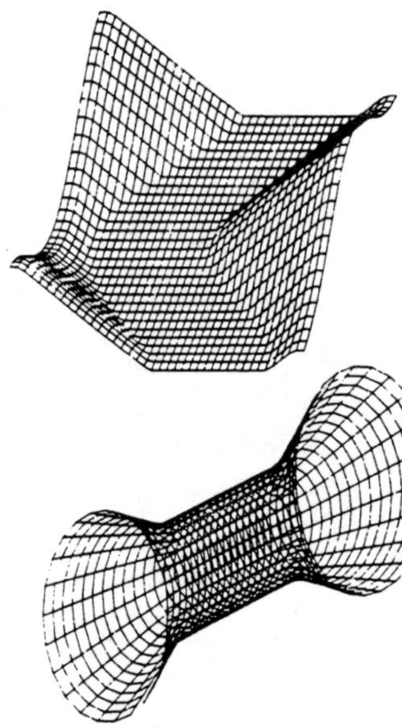

Figure 7. Wire mesh diagrams showing the surface of the bi-conical mandrel. The upper diagram shows the cylindrically developed data while the lower diagram shows the data in undeveloped form.

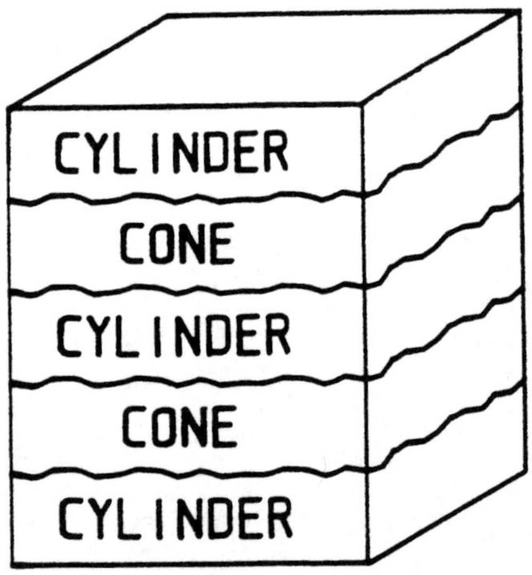

Figure 8. Relative positions of primitive shapes within the bi-conical mandrel.

Before attempting to compare an object with its specified dimensions, we must explicitly determine the orientation and position of the object with respect to the axes of the vision system and must explicitly determine the loci of the boundaries between the primitive shapes which constitute the object. We start this process by providing the vision system with descriptions of the primitives and their relative positions. In the case of the mandrel, we need only specify that, if the object is lying with its axis in the vertical direction, then the primitives will be described and distributed as shown in Fig. 8. At this stage, the vision system has no information about the specified dimensions of the workpiece but, by referring to previously programmed definitions of cylinders and cones, it does know that the workpiece can be described in terms of cylinder diameters, cone angles, axial lengths and axial positions. The vision system is also aware that adjacent primitives are separated by a boundary line and it is aware of the type of measurement which is required within each primitive.

The approximate loci of the boundary lines can now be determined by assuming that the gradient of the data is discontinuous at, and only at, the boundaries between primitives. In most cases, this assumption is valid regardless of direction in which the gradient is calculated. The gradient of the co-ordinate data is then determined, discontinuities are identified and, as shown in Fig. 9, data points are erased from the immediate vicinity of the boundary lines to avoid the risk of localised undersampling of the surface co-ordinates. From the information given in Fig. 8, the vision system can now identify the shape of each primitive region . Drawing on this knowledge, radii, cone angles and relative axial positions and orientations are then fitted to the various primitives by the least squares method. The boundary lines are then reconstructed by calculating the loci of the intersections of adjacent primitives. The vision system is now equipped with an 'understanding' of the data.

Specified dimensions can then be derived from the co-ordinate data and compared with corresponding dimensions in the CAD model of the object. The rigour of the inspection can be increased simply by increasing the number of object dimensions which are compared with the design model.

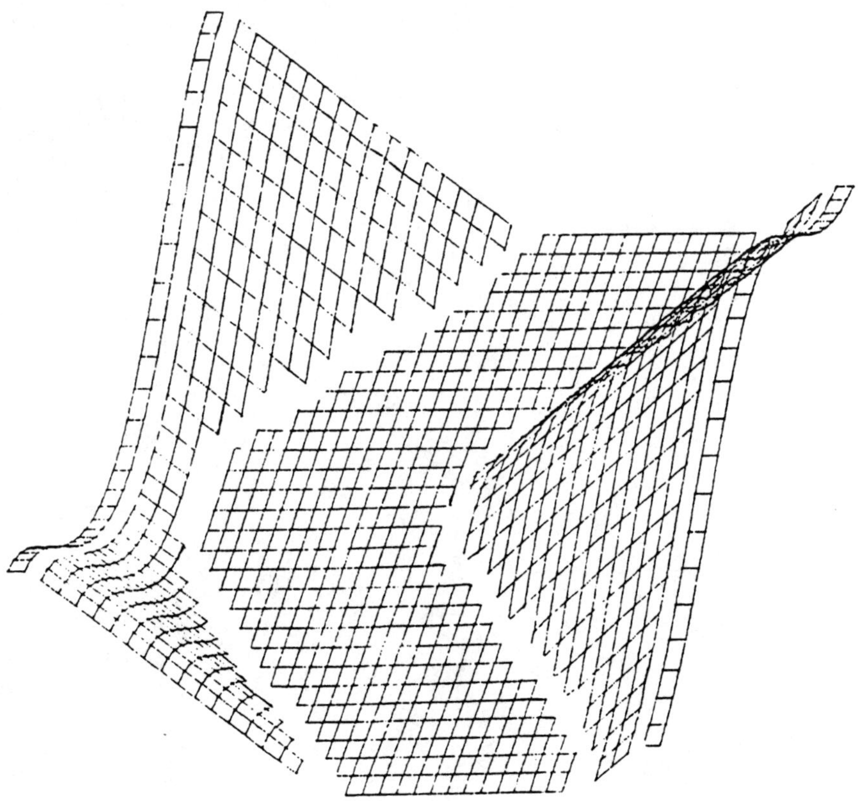

Figure 9. Cylindrically developed co-ordinate data showing the effect of removing points in the vicinity of boundary lines.

If a series of nominally indentical workpieces require inspection then, by ensuring that each workpiece lies in the same orientation as the first, we can make direct measurements of specified dimensions by using a subset of the surface co-ordinates. In this case, the diameter of a cylinder might be calculated from points lying on a cross-section while the angle of a cone might be calculated from points lying on a slant line.

Conclusions

We have described a vision system which exploits moire fringe contouring in order to acquire large numbers of three dimensional co-ordinates from the surfaces of workpieces. We have then described some of the computational methods which can be used to provide the vision system with the ability to compare the dimensions of an object with its design specification. At the present stage of development, these computational methods can only be used for the inspection of objects which can be described in terms as a combination of relatively simple primitive shapes. In principle, however, a similar approach could be followed when inspecting complicated solids of revolution and other, more complex shapes.

Acknowledgements

This paper is published by permission of the Director, National Engineering Laboratory, UK, Department of Trade and Industry. It is British Crown Copyright. This work was funded by the Metrology and Standards Requirements Board of the UK Department of Trade and Industry.

References

1 Strand, T. C., "Optical three-dimensional sensing for machine vision", Opt Engng, Vol 24, pp 33-40, 1985.

2 Rioux, M., "Laser range finder based on synchronized scanners", Appl Opt, Vol 23, pp 3837-3844, 1984.

3 Kanade, T. and Haruhike, A., "Noncontact visual three-dimensional ranging devices", Proc SPIE, Vol 283, pp 48-53, 1981.

4 Reid, G. T., Rixon, R. C. and Messer, H. I., "Absolute and comparative measurements of three dimensional shape by phase measuring moire topography", Opt and Laser Technol, Vol 16, pp 315-319, 1984.

5 Reid, G. T., Rixon, R. C. and Marshall, S. J., "3-D machine vision for automatic measurements of complex shapes", Proc 7th Int Conf on Automated Inspection and Product Control, Birmingham, UK, March 1985, pp 129-138.

6 Reid, G. T., "Phase measuring moire topography", in Optical Metrology, O.D.D. Soares (Ed), NATO ASI Series, Martinus Nijhoff Publishers, The Hague, 1986.

Reprinted from *Optical Techniques for Industrial Inspection,* P.G. Cielo, editor,
Proc. SPIE Vol. 665, pp. 168-173 (1986). ©1986 SPIE.

Recent results of the performance analysis of a 3-D sensor based on time-of-flight measurements

T. Heikkinen *, R. Ahola **, M. Manninen ***, R. Myllylä **

* Technical Research Centre of Finland, Electronics laboratory P.O Box 181 SF-90101 Oulu, Finland

** University of Oulu, Department of Electrical Engineering SF-90570 Oulu, Finland

*** Prometrics Oy, Teknologiantie 4, SF-90570 Oulu, Finland

Abstract

This report presents a constructed 3-D image acquisition system based on time-of-flight measurements. The basic structure of the system is described and the main factors affecting the performance of the system are introduced. Some recent experimental results of resolution tests are presented. The current image resolution is ca. 1 cm at a working range of 2.5 m to 4 m.

Introduction

The importance of 3-D sensors in robotics and in computer aided inspection has increased rapidly during the past few years. Both passive and active methods have been introduced [1]. Passive methods such as stereo utilize standard camera techniques. Such systems require sophisticated algorithms and much computational capacity to solve the correspondence problem. The active methods usually utilize less 'vision oriented' technolgy. The main advantage of these methods is the reduced computational effort.

In active 3-D sensing there are two basic methods for obtaining the required range information: triangulation and time-of-flight measurement [2]. Applications of both of these methods have been reported and the results are encouraging [3,4].

In the present study we are investigating the use of time-of-flight measurements for range-image acquisition [5]. The system constructed utilizes a straightforward time-of-flight principle. Short light pulses are emitted with a high repetition rate towards an object. A small fraction of the light is reflected from the surface of the object and received by the measurement system. The distance is uniquely obtained from the elapsed time-interval between the transmitted and the received light pulses. A range-image of the scene is created by scanning the measurement beam in two directions using a raster scan pattern.

The applications of the 3-D sensor developed in this study are in the field of robotics and computer aided inspection of large objects. The system is designed to work at distances from 2.5 m to 4.0 m. The maximum size of the imaged area at the maximum distance is 2.5 m x 2.5 m.

In this paper we present the basic structure of the 3-D image acqusition system and some recent results of the performance of the system.

3-D image formation

A 3-D image is a set of coordinates of points on an object surface. The system described measures these coordinates in a modified spherical coordinate system. The range coordinate is measured with the time-of-flight rangefinder, and the angular coordinates are measured with the scanning system which directs the measuring beam of the rangefinder. The measured 3-D image is transformed into a cartesian coordinate system. The final image is stored as an orthographic view of the 3-D scene. This image forms a source data-base for a higher level image analysis system [6].

When the image is represented in this way the sampling must be done at equal intervals.In x-direction the sampling is done at equal intervals by the scanner. In y-direction the intervals of the samples are adjusted to equal length by post-processing.

3-D image acquisition system

The 3-D imaging system consists of two main physical units: the sensor and the computer. The sensor unit contains the scanner and the laser rangefinder. The structure of the system is described in figure 1 and the principal functional parameters are shown in table 1.

PARAMETER	DESIGN VALUE
Range	2.5 m - 4 m
Maximum field of view	35 deg. for x and y directions
Maximum image size	512 x 512
Scan rate	1-20 lines/s
Pulse repetition rate	>1 MHz
Number of averaged range measurements/ pixel	>100
Spot size within the range (1/e diam.)	5 mm - 8 mm

Table 1: Functional parameters of the system

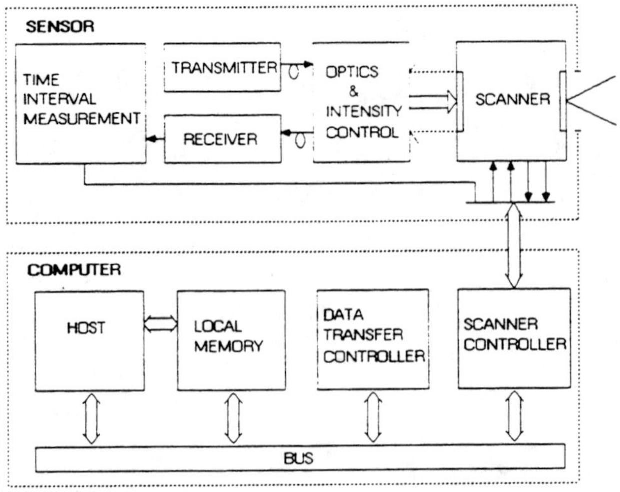

Figure 1: 3-D image acquisition system based on the time-of-flight measurements

Sensor. The time-of-flight rangefinder consists of a transmitter, a receiver and a time interval measurement unit. The transmitter utilizes a pulsed DH-type laser diode. The system emits about 5ns (FWHM) long light pulses with a pulse repetition rate of more than 1 MHz. The peak output power of the pigtailed (100 um) laser diode is about 500 mW. The light pulse is focused to give the smallest spot size (4.9 mm) at the middle operating distance (3m) of the sensor. The START-pulse for the time-interval measurement unit is taken from the transmitted light pulse.

The transmitter and receiver have coaxial optics. The received light pulse is detected and the STOP-pulse for the time-interval measurement unit is generated. Both the final START and STOP timing pulses are generated by a so-called constant fraction timing discriminator (CF). The time-interval between the timing pulses is converted to a continuous analog voltage by a time-to-amplitude converter (TAC). The continuous analog voltage is amplified, filtered and an offset voltage is subtracted to produce the final range information. The bandwidth of the range voltage is from dc to 10 kHz.

In the 3-D imaging system constructed the amplitude control is carried out in three stages. In the first stage an optical attenuator driven by a step motor is used. The attenuator varies the intensity of the received light pulses in a range of 1:100. In the second stage an AGC-amplifier (automatic gain control) is used to control the amplitude of the STOP signals from the preamplifier. The AGC-amplifier can control the amplitude in the range of 1:5. Finally the constant fraction timing discriminator eliminates the effect of the fastest amplitude variations on the timing accuracy. The discriminator can tolerate amplitude variations of 1:2 with an accuracy range of 1 mm.

The scanner consists of two mirrors which are driven by moving iron galvanometers. The system is controlled by servo amplifiers. Both positional and velocity feedback are used. The maximum scan rate is 20 lines per second.

The direction of the transmitted light pulse is determined by measuring the angular positions of the mirrors with the internal capacitive sensors of the galvanometers. To remove the temperature drifts of the sensors and to improve the linearity an internal calibration system is constructed. The system utilizes auxiliary beams which are launched to the reverse sides of the scanning mirrors. The reflected beams are deflected to detectors which indicate the four fixed angular positions of the scanning mirrors. Information of these known angles is used to correct the angular information produced by capacitive sensors. A calibration procedure is executed before each image frame.

Computer. The computer includes a scanner-controller, a host computer and an image memory. The scanner-controller generates the required signals to servo amplifiers to produce the raster scan pattern. To form the 3-D image the controller samples the angle and range signals and transforms the measured coordinates into a cartesian coordinate system. The samples in x-direction are taken at equal intervals, which is why only y- and z-coordinates are stored. Coordinates are transferred into the image memory by a data-transfer controller.

The user controls the image acquisition system through the host. The host computer also compresses the measured 3-D image into the final format by equalizing the sampling intervals in y-direction. In the final range image only the z-coordinates are stored.

Performance factors

The primary quality criteria for a 3-D image are spatial resolution and accuracy. The resolution can be defined as corresponding to the minimum distance between two features which can be resolved. This property typically depends on e.g. the scan speed.

The accuracy in this context can be interpreted as a measure for all aberrations of the 3-D image. This property indicates how accurately the resolvable dimensions and positions can be measured. Where accuracy is concerned we can distinguish two different kinds of errors: offset errors and scale errors. Offset type errors affect only the absolute accuracy of the position measurement. Scale errors also affect the accuracy of dimension and shape measurements. The principal contributors to the resolution and accuracy are shown in figures 2 and 3.

Range resolution. The main factors which affect the resolution of the range measurement are noise, averaging, and amplitude variations of received pulses. The noise of the range signal is mainly caused by the timing jitter of the CF and electrical interference (crosstalk etc.). The jitter depends on the S/N-ratio of the received pulses, the rise time of the received pulses and the residual jitter of the electronics. The noise limited resolution with a single pulse is typically about 10 mm.

Noise limited range-resolution can be improved by making the range measurements with a high rate of repetition and averaging a certain amount of single results. On the other hand this reduces the signal bandwidth and smoothens the edges of the image in the direction of the scan. The amount of reduction of the resolution in direction of the scan depends on the scan speed. The range signal bandwidth is defined by the averaging and the pulse repetition rate. The pulse repetition rate is limited by the minimum pulse width and the maximum allowed duty cycle of the laser diode.

If the amplitude variation of the received light pulse is wide and so rapid that the optical attenuator and the AGC amplifier cannot correct it, an error is generated in the range information due to unideal operation of the constant fraction discriminator. The error is caused by the so-called walk error present in the timing discriminator. With rapid intesity variations the effect is seen as additional noise in the range signal.

Direction resolution. The primary factors which affect the measurement of direction are noise, bandwidth, and spot size. Noise is mainly generated by the angle sensors and electrical interference.

The signal bandwidth affects the direction resolution in the similar manner as it affects the range signal. With a wide bandwidth the noise limits the resolution and with a narrow bandwidth the rise-time of the signal limits the resolution. In this case the errors in angle signals cause misplacement of the pixels in the final image.

The finite size of the laser spot on the surface of an object reduces directional resolution. Range information is obtained somewhere inside the spot. The exact direction depends on the geometry and the reflectance of the surface and the intensity distribution of the transmitted beam. A safe approximation of the spot size limitted directional resolution is the angular size of the spot.

Range accuracy. The main factors in the accuracy of range measurements are nonlinearity, amplitude variation, and drifts. Nonlinearities are mainly caused by the time-interval measurement electronics. Typical nonlinearity values for best analog TAC:s in a few metres' range are from about 1 to 10 mm.

The effect of amplitude variations of the reflected light pulse depend on the amplitude control system and the timing discriminator. The primary mechanisms of errors are the same as in the case of resolution.

Changes in temperature induce drifts to the range finder electronics. Both scale and offset errors are generated. The most crucial effects are temperature-induced changes in the propagation delays of signals in the receiver and the time-interval measurement electronics.

Direction accuracy. In direction measurement nonlinearites and drifts cause inaccuracies. The nonlinearites can be caused both by angular sensors and misalignment of the scanner mirrors. Typical nonlinearity of the capacitive sensors used in this system is 0.15 %.

Drifts are caused by angular sensors (gain, offset), the internal calibration system and the alignment of the mirrors. In both electronics and mechanics drifts are mainly induced by temperature changes. Mechanical vibrations can also cause drifts in the alignment of the calibration system and scan mirrors.

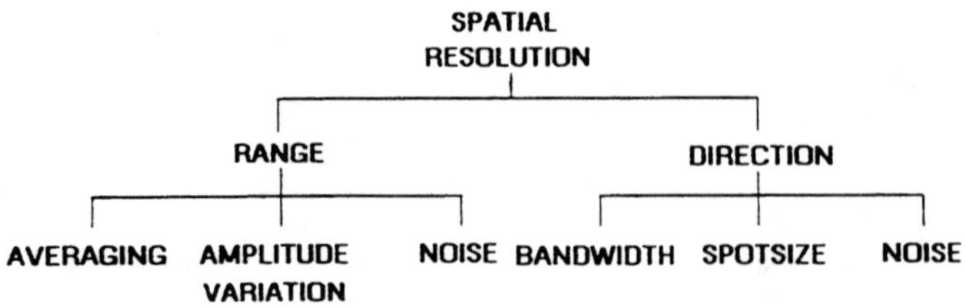

Figure 2: Formation of the spatial resolution of the image

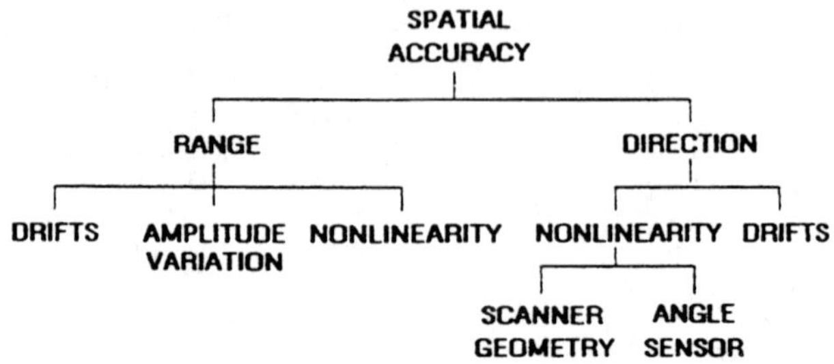

Figure 3: Formation of the spatial accuracy of the image

Experimental results

In recent experiments the spatial resolution of the system has been tested. In these tests the amplitude of the input pulses of the CF-discriminator are controlled only by AGC-stage. Also the final calibrations of the angle and range measurements are not yet completed. Noise limited range resolution is ca. 5 mm in the 10 kHz bandwith. The resolution of angular measurements is ca. 1.5 mrad in 100 kHz bandwidth.

Two examples of the resolution of a line scan are presented in figure 4. The results represent the step response of the system. The target was made of polystyrene foam. These measurements indicate a spatial resolution of about 1 cm with a scan rate of 1 line/s. With a higher scan rate the edges are roundend. With the scan rate of 20 lines/s the resolution of edge detection in transverse direction is about 2 cm. The measuring distance was about 3.5 m.

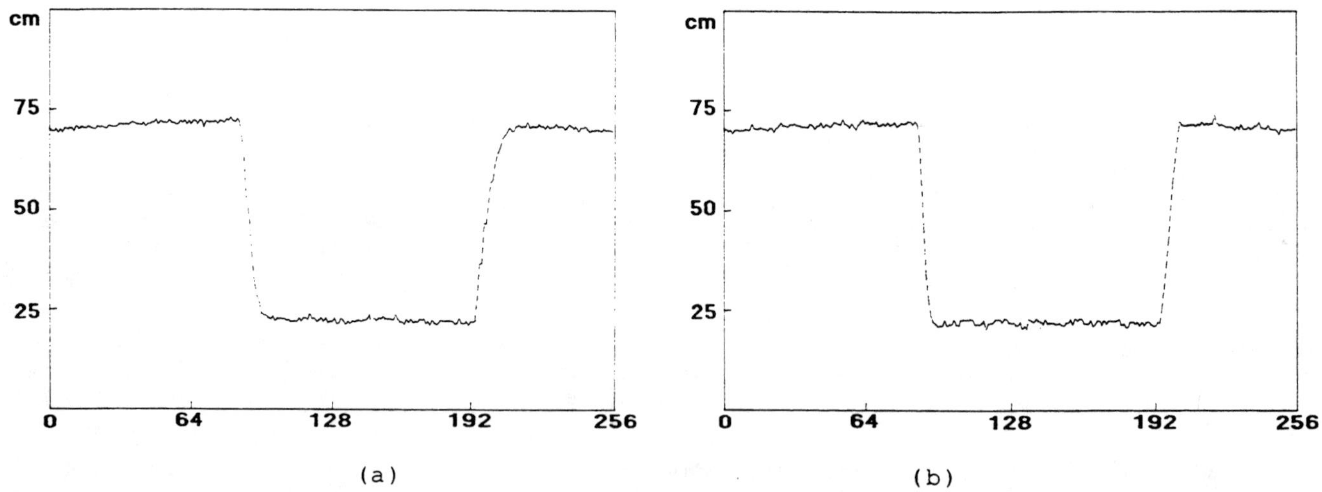

Figure 4: Resolution test of a line scan. The height of the polystyrene target is 49.5 cm and the width 24.5 cm. (a) Scan speed 20 lines/s, (b) scan speed 1 lines/s.

Figures 5 and 6 show two examples of scenes imaged by the constructed system. The visualization is carried out by converting the range values into 8 bit format and displaying these values in the gray levels of a video monitor. The darkest points are nearest and the lightest points are farthest.

Figure 5 represents a 256 x 256 3-D image of a log. The cylindrical surface of the log is covered with bark and the cut surface is plain wood. The picture reveals that the edge between the cut surface and the cylindrical surface is aberrated. A large part of the error can be assumed to be induced by intensity variation.

(a)

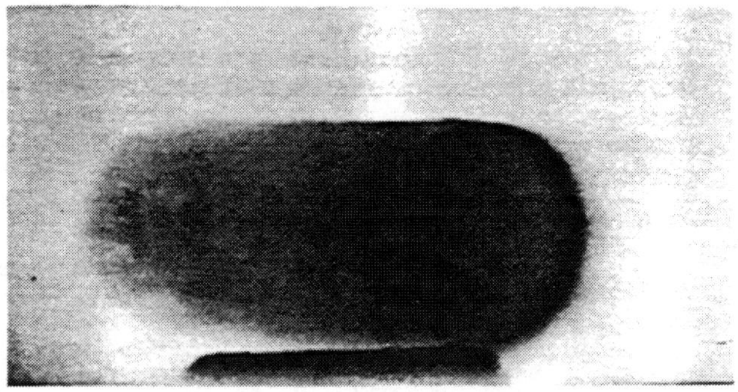

(b)

Figure 5: (a) Photograph of a log. The length of the log is 49 cm and the diameter 26.5 cm. (b) A corresponding 256 x 256 3-D image of the log.

Figure 6 represents a 256 x 256 3-D image of a chair made of woollen material. The reflectance of the surface of the object does not vary very much and the object does not include sharp edges. In this case most of the intensity varation is due to distance variation. In this picture the edges of the object are repeated correctly. Figure 7 presents the corresponding profile of horizontal line n:o 187.

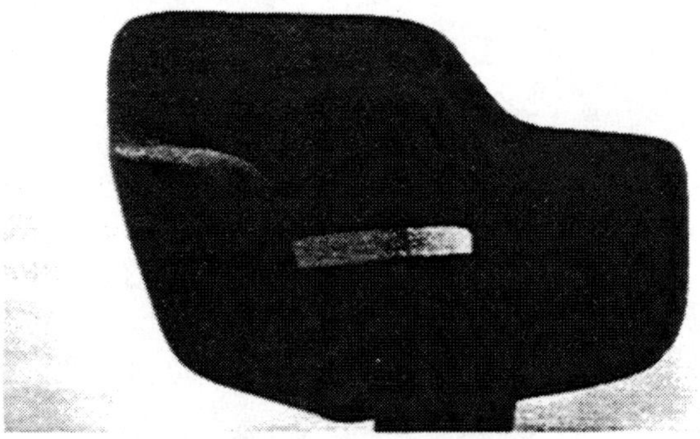

(a)

(b)

Figure 6: (a) Photograph of a woollen chair. The height of the woollen part is ca. 45 cm, the width ca. 65 cm and the depth 58 cm. (b) Corresponding 256 x 256 3-D image

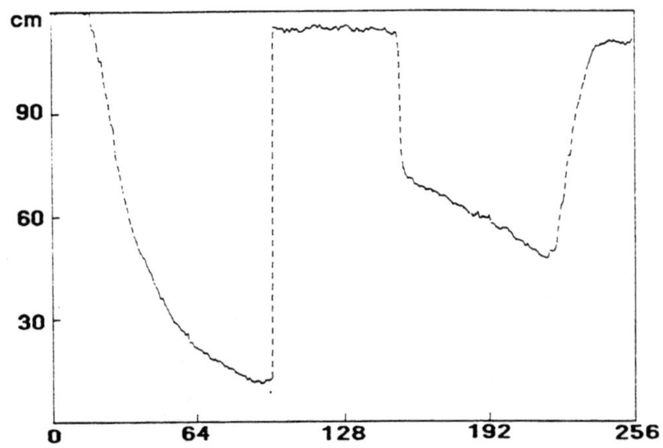

Figure 7: Profile of the horizontal line n:o 187 of the 3-D image presented in figure 6 (b).

Figure 8: 3-D image of the top of a gas bottle. The system cannot get enough signal from the white areas.

Surfaces with steep angles and shiny coatings create problems for active methods. Figure 8 presents a 3-D image of the top of a gas bottle. White areas represent those parts of the surface where the system cannot get enough signal. In this case the problems resemble the illumination problems of passive imaging systems.

Discussion

Recent results of the performance analysis of the constructed 3-D sensor are presented. The measurements are carried out with the sensor which is not at the moment in its final stage. The current spatial resolution is better than 1 cm at a working range of 2.5 m to 4 m. Much of the noise is generated by electrical interference. By improving the electronics the resolution in the final stage is expected to be significantly better approaching the level of 1 mm.

Measurements and tests completed show that the idea of using the scanning time-of-flight distance sensor for 3-D image acquisition is feasible. The amplitude control of the received pulses is crucial to the performance of the system. In such 3-D imaging applications where relative distances, the object surface orientations and the surface reflectance may greatly vary this demand is even more emphasized. The 3-D acquisition system described is designed to be capable of handling received pulses with amplitude variations in the range of 1:1000. Numerous industrial applications can be covered with the three decades dynamic range.

References:

1. Paul J. Besl and Ramesh C.Jain: Three-Dimensional Object Recognition, Computing Surveys, Vol 17, No. 1, March 1985.

2. R. A. Jarvis, A Perspective on Range Finding Techniques for Computer Vision, IEEE Transactions on pattern analysis and machine intelligence, vol. PAMI-5, No 2, March 1983

3. R.E. Sampson, C. W. Swonger, P.W. VanAtta, Real time 3 dimensional image processing for robot applications, RI/SME Conference, June 1985

4. Marc Rioux, Laser range finder based on synchronized scanners, Applied optics, Vol. 23, No. 21, November 1984

5. R.Ahola, T.Heikkinen, M.Manninen, 3D-Image acquisition by scanning time of flight measurements. International conference on advances in image processing and pattern recognition, Pisa Dec. 10-12, 1985

6. H. Ailisto, H.Hakalahti, M.Manninen, Dynamic split and merge method for three dimensional scene analysis, Proc. of 4th SPIE conference on intelligent robots and computer vision, vol 579, September 16-20.1985

Reprinted from *Optics, Illumination, and Image Sensing for Machine Vision II*, D.J. Svetkoff, editor, Proc. SPIE Vol. 850, pp. 210-217 (1987). ©1987 SPIE.

Phase detected triangulation: a new twist on an old technology

Leonard H. Bieman
Kevin G. Harding
Mark Michniewicz
Robert Tait

Industrial Technology Institute
P. O. Box 1485
Ann Arbor, MI 48106

ABSTRACT

A new approach to measuring range using triangulation is presented. In this approach, the triangulation angle is determined by spatially modulating the intensity of light in front of a single channel photodetector and then measuring the phase of the detector output signal. Using a single detector instead of the commonly used linear array or lateral effect photodivide, allows an increased signal to noise ratio and the possibility of faster measurement rates. Two possible application areas discussed are long range measurement (greater than one meter) and measurement of mirror like surfaces.

1. INTRODUCTION

Ranging techniques have become important for a wide variety of applications including the coupling of such systems with scanning devices to produce three dimensional images. Most of the ranging techniques are based on some form of time of flight measurement[1,2] or triangulation[3,4]. The phase detected triangulation technique presented here offers advantages over traditional triangulation approaches and time of flight measurements.

Point triangulation instruments, to make non-contact range measurements, have become important tools in the industrial arena. As seen in Figure 1, the triangulation effect is created by illuminating the point on the surface to be measured with a beam of light incident at some angle to the surface. The point of light is then viewed at some different angle than the illumination beam and imaged back to a detector. As the distance to the surface changes, the position of the image changes as dictated by the trigonometric relationship. The position of the image is traditionally detected with either a linear array or a lateral effect diode. The performance of these detectors limits the performance of the triangulation sensor in two ways. First is the limited signal to noise ratio which requires using an intense beam of illuminating light or measuring only close surfaces or surfaces with a large component of diffuse reflection. Second is the rate of measurement which is generally limited to 1000 Hertz.

Phase detected triangulation is a point triangulation technique which eliminates these shortcomings by using a single channel photo detector. The technique detects the position of the image by spatially modulating the intensity of light at or near the image plane and then re-imaging onto a single channel detector. An example of the technique is illustrated in Figure 2, where the modulation is created by moving a grating at the image plane of the system. Since this modulation is caused by a traveling wave, the distance to

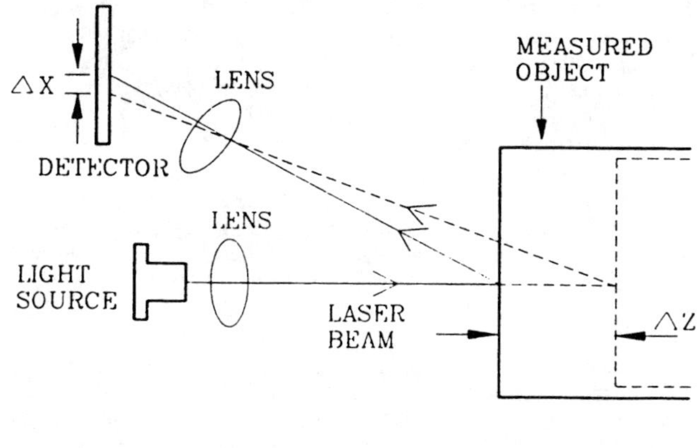

Figure 1. Traditional Point Triangulation

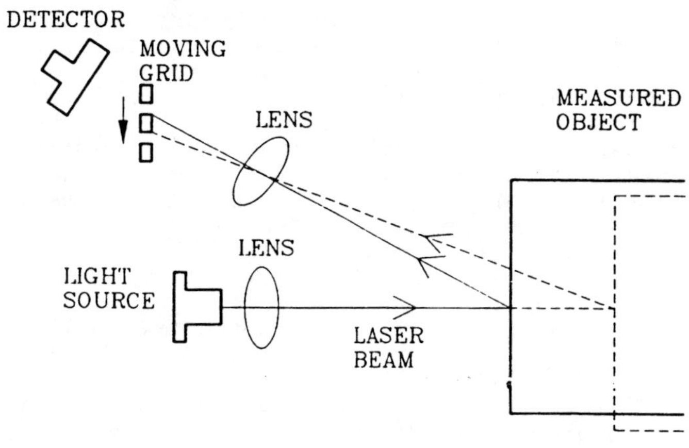

Figure 2. Phase Detected Triangulation

the target is represented by the phase of the modulated signal received at the detector. Hence, position has been converted to phase, making this triangulation method amenable to powerful phase detection schemes. The modulation can be obtained by using a radially-grated disk, which means that the phase detected triangulation sensor is really just a modification of the optical encoder.

To test out the concept, a chopper wheel was used for the intensity modulator, and a lock-in amplifier was used to detect the phase. The two initial tests which produced encouraging results were aimed at two possible application areas for the device. The first test measured an object at a distance of 1350mm using a tight triangulation angle of 11 degrees. The second test used a 45 degree triangulation angle to measure the position of the surface of the mirror. This test demonstrated the ability to measure an object with a small component of diffuse reflection. A third area to be explored is the use of phase detected triangulation to measure with a small triangulation angle (1 or 2 degrees) at longer standoffs (such as 10 or 20 meters). This is another advantage of phase detected triangulation, that it is well-suited for small triangulation angles.

What makes this device exciting is that the set-up tested could be adapted for multiple production and produced economically because the components needed are readily available with high quality at a reasonable cost. That is, the chopping system can be based on optical encoding technology and the emitter and receiver can be diodes developed for fiber optic communications. Thus, the development of phase detected triangulation can ride on the coat tails of large scale development in other areas. Also, because of these cost factors, it is possible to have an improved resolution system using two detectors: one measuring depth coarsely, the other with fine resolution.

The advantages of phase detected triangulation can be summarized as follows:

- Better signal to noise ratio because low-noise, single detectors can be used (communication quality photodiodes, avalanche detectors, and photomultiplier tubes).
- Faster Measurement Rates because the modulation frequencies of up to 500 kHz can easily be achieved.
- Potentially lower cost because the needed components can be obtained from mass volume, high technology areas (optical encoders and lasers and detectors developed for fiber optic communication).
- More control over resolution because the grating spacing can be adjusted as well as the imaging optics.
- Well-suited for small triangulation angles.
- Potential for increased resolution using two detectors (coarse and fine).

This paper includes a description of the measurements made to validate the concept. Then measurement capabilities are established for the technique. The last section is a comparison of phase detected triangulation with other ranging techniques.

2. EXPERIMENTAL RESULTS

To establish the viability of the phase detected triangulation concept, a chopper wheel was used to modulate the light and a lock-in amplifier was used to pick up the phase of the detected signal. The basic experimental set-up is shown in Figure 3. A helium-neon laser light beam is focused on a small spot on the measured object. This spot is imaged onto the chopper wheel and then reimaged onto the silicon detector. Before the light enters the detector, it is filtered with a interference filter to eliminate most of the background light. The lock-in amplifier measures the phase of the detected signal by comparison with a reference square wave signal from the chopper wheel. As the measured object is moved forward and back with a translating table, the phase (measured by the lock-in) changes due to a shift in the position of the image on the chopper wheel. For the measurements, the light passed through a 30 slit chopper wheel at a distance of 25mm from the center of rotation. At this position, the slit width (W) was 2.5mm. The chopping frequency used for this initial testing was 500 Hz.

To try out the phase detected triangulation at a moderately long distance, the geometry shown in Figure 3 was used with $O = 1350$ mm, $I = 140$ mm, and the triangulation angle $\theta = 11$ degrees. The data obtained is shown as squares in Figure 4a. The linear least squares fit is the solid line through the data points. From the fit, it is determined that a one-degree phase change represents a depth change of 0.77 mm. Thus a full 360 degree phase change, which is the ambiguity interval, translates into a measurement depth of 280

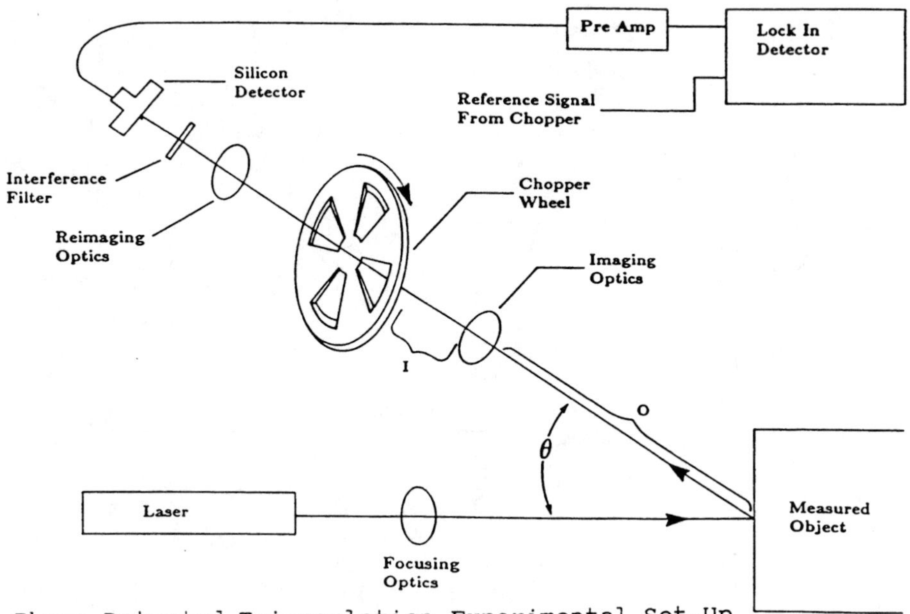

Figure 3. Phase Detected Triangulation Experimental Set-Up

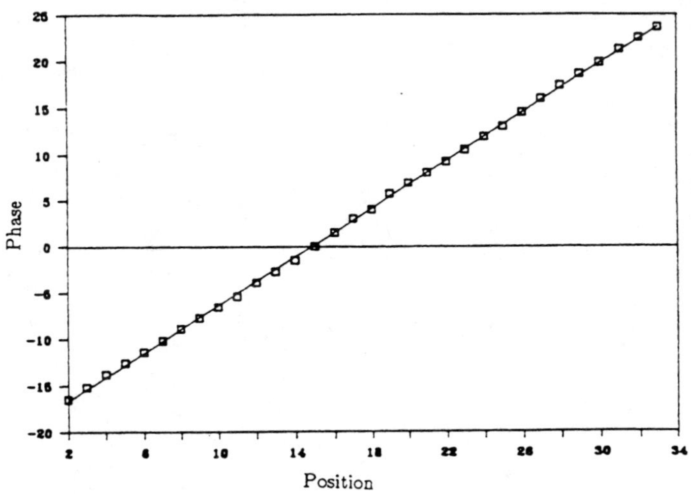

Figure 4a.

Phase (degrees) versus Position (mm)

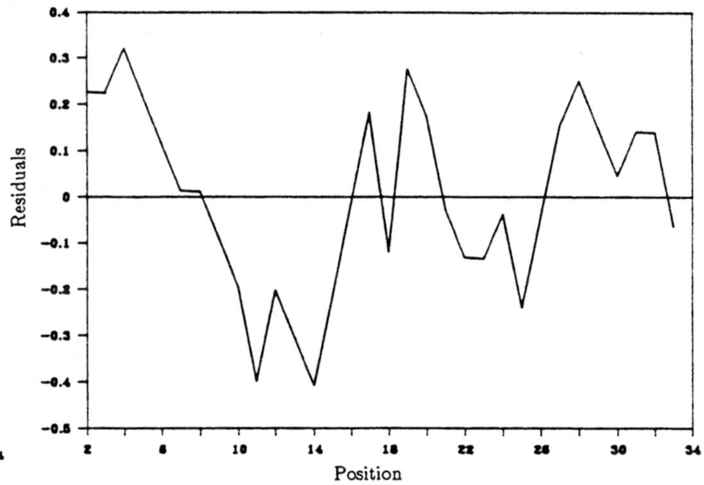

Figure 4b.

Residuals of Fit (degrees) versus Position (mm)

mm. This type of range depth was not shown in Figure 4a because only a linear responding region was characterized.

The resolution of the system can be determined from the scatter of the data. The difference between the data and the fit is found in Figure 4b. The standard deviation calculated from these residuals is 0.20 degrees or 0.15mm in terms of distance. Since the maximum phase change without ambiguity is 360 degrees, the resolution is better than one part in 1000.

The second configuration tried was to determine the displacement of a front surface mirror. The mirror surface was chosen to see how well phase detected triangulation could do with very little diffuse reflection. The geometry was the same as shown in Figure 3, with O = 50mm, I = 230mm, and θ = 45 degrees. The data and fit to the data are shown in Figure 5a. A one-degree phase change represents a 4.3 micron depth change. For the full 360 degree phase change, the ambiguity depth range is 1.5mm. As was the case for the other configuration, the full depth range was not shown in Figure 5a because only the linear region was characterized.

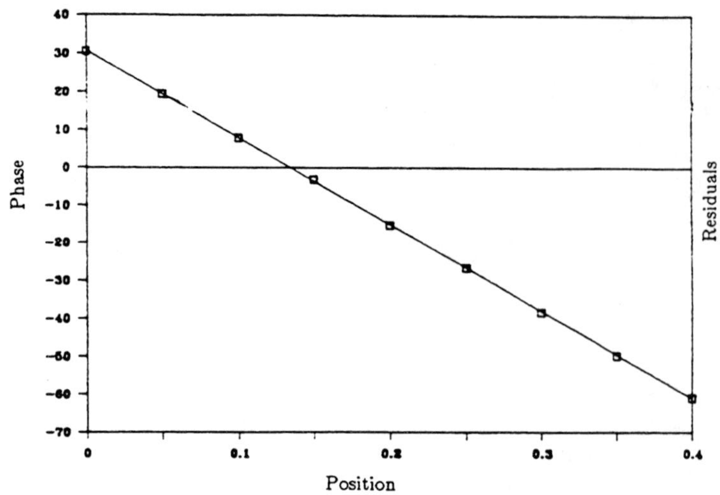

Figure 5a.

Phase (degrees) versus Position (mm)

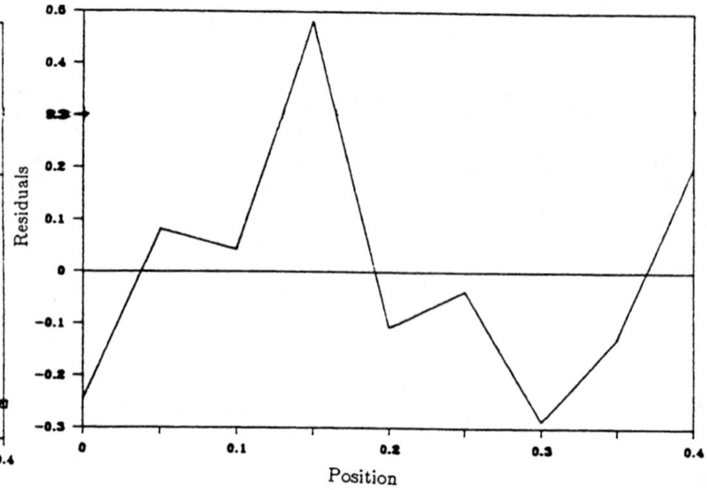

Figure 5b.

Residuals of Fit (degrees) versus Position (mm)

The scatter of the data for this second configuration is found in Figure 5b. The standard deviation calculated from the residuals is 0.26 degrees or 1.1 microns in terms of distance. As with the first case, the resolution for the phase detected triangulation is better than one part in 1000. This is excellent for a measurement using diffuse reflection off a mirror surface.

The measurements just described were made in the spirit of a proof of concept test and are not intended to represent an optimized phase detected triangulation system. Deficiencies that detract from the performance of the system were many. In follow-up work, these deficiencies will be reduced or eliminated. The deficiencies include:

- Laser beam wander which will be eliminated in the future by either spatial filtering or the use of a laser diode.
- Vibrations in the set-up which are inherent in a breadboard set-up, but which can be significantly reduced for a well-designed unit.
- Jitter in the chopping system which can be greatly reduced by a tuned system that runs at one fixed frequency.
- Large chopper slit openings which contribute to non-linearity for large phase shifts (approaching 360 degrees).
- Optics not designed to optimize light collection over the entire measurement range.

The next section will explore how phase detected triangulation can be tailored to the requirements of a specific application.

3. MEASUREMENT CAPABILITIES

In this section we will explore the capabilities of Phase Detected Triangulation in two different modes of operation. First, the equations defining the operating performance will be examined. Then we will look at its performance using a small triangulation angle to measure range at a long standoff. Finally we will explore the use of the system to make high precision measurements at moderate standoff distances.

Before starting to look at configurations, it is useful to establish power requirements needed to achieve a desired depth resolution. From signal processing theory, the received detector power P_r required to achieve a desired resolution can be conservatively approximated by:

$$P_r \sim NEP \cdot 2^n \cdot (f_m)^{1/2} \quad (1)$$

Where NEP is the noise equivalent power of the photodetector, n is the number of bits of depth resolution, and f_m is the rate at which data is acquired. For a low-noise PIN silicon photodetector preamp combination such as EG&G HUV-1100BQ, the NEP is 3 X 10^{-14} W/Hz$^{1/2}$. For even greater sensitivity, a photomultiplier tube might be used with a typical NEP of 1 X 10^{-15} W/Hz$^{1/2}$.

If one desires to obtain a three-dimensional image using phase detected triangulation, the system might be configured as a flying spot laser scanning device. The system can be both time and light efficient because all the laser power is focused on one spot on the object and the image of the spot is refocused onto a single sensitive detector. The scanning might be accomplished by acousto-optic cells, polygon mirrors, resonant scanners, or holographic devices[5,6]. To establish the detector power requirements for such a three-dimensional system, it will be assumed that a 256 by 256 image will be acquired in one second with a depth resolution of 8 bits or one part in 256. Thus, for calculating the P_r, n is 8, and f_m is approximately 50 kHz. Thus, P_r equals 1.7 X 10^{-9} W for the PIN detector system and 5.7 X 10^{-11} W for a photomultiplier tube.

Achieving the 50 kHz measurement rate is reasonable with a phase detected triangulation system. Allowing 10 cycles per measurement point, it requires operation at 500 kHz, which can be obtained without pushing the state-of-the-art. For example a radially-grated disk with 5000 line pairs can easily be rotated at 100 revolutions per second(6000 rpm).

To complete the power requirements analysis, the power received by the detector can be determined if the geometry, optical efficiency and the reflectivity of the object measured are known. The following expression is obtained assuming a lambertian object reflecting surface:[7]

$$P_r \sim (T_e T_r \rho_d r^2 \cos\theta / S^2) \cdot P_e \quad (2)$$

Where, T_e and T_r are the optical transmission of the emitting optics and receiving optics, respectively, ρ_d is the diffuse reflectivity of the surface measured, r is the radius of the effective aperture of the receiving optics, θ is surface tilt angle of the object surface with respect to the line of sight to the receiving optical system, S is the distance between the object and the receiving optics (stand off), and P_e is the power emitted by the optical source which will most likely be a laser diode or helium neon laser.

Using Equations 1 and 2 and conservative estimates for the parameters plugged into these equations, estimates of the standoff range can be made. The same assumptions as described above for Equation 1 will be used and the following assumptions for Equation 2: A system light efficiency (T_e T_r) of 75%; very little diffuse reflection (ρ_d = 1%; a 40mm receiving aperture r; a surface up to 75 degrees away from normal, which means for random surface orientation that over 80% of the surface can be measured cos θ = .25; and laser power P_e of 5mWatts. Plugging these values in yields a range between measurement system and measured object of 3 meters using a PIN silicon detector, and 16 meters using a photomultipler tube. Of course, if some of the stringent measurement requirements were relaxed, the range could be significantly increased. On the other hand, if the range was much smaller, surfaces with hardly any diffuse reflection, such as polished parts, could be measured. For example, if the standoff ranges were 100 mm, the receiving aperture (r) was 10 mm, a PIN silicon detector was used, and all other parameters were left the same, the diffuse reflectivity of the measured surface (ρ_d) need only be 0.02%, which is a very good quality mirror.

Now, two possible configurations for phase detected triangulation will be reviewed. In the first configuration, the triangulation angle will be small and the standoff distance will be large. Such a configuration is shown in Figure 6. From the geometry of the figure, the relationship between the radial grading slit width (W) and the ambiguity depth R can be established. (It should be noted that the technology should be inexpensive enough to have two detection systems: a course resolution and a fine resolution which would eliminate the ambiguity problem.) Assuming that S is much larger than both X and R, then:

$$R \sim 2 \cdot W \cdot S^2 / (X \cdot I) \tag{3}$$

It is clear, as can be seen by examining Equation 3, that a phase detected triangulation system in the small triangulation angle configuration is quite adaptable. Putting in some representative numbers for a long range system will yield some insight into the ways phase detected triangulation might be configured. Assuming a standoff S of 3 meters, a triangulation leg X of 50 mm, an image distance I of 200 mm, and a slit width (W) of 0.5 mm(1 degree triangulation angle); then the ambiguity measurement depth R is 900 mm. If we have a depth resolution of approximately 8 bits, this translates into a resolution of 3mm. If finer resolution is required, then the slit width could easily be reduced by a factor of ten, yielding 0.3mm resolution. Of course, the trade-off is a reduced ambiguity measurement depth. This ambiguity shortcoming could cost effectively be overcome by having a two detection system; one with course resolution and the other with fine resolution.

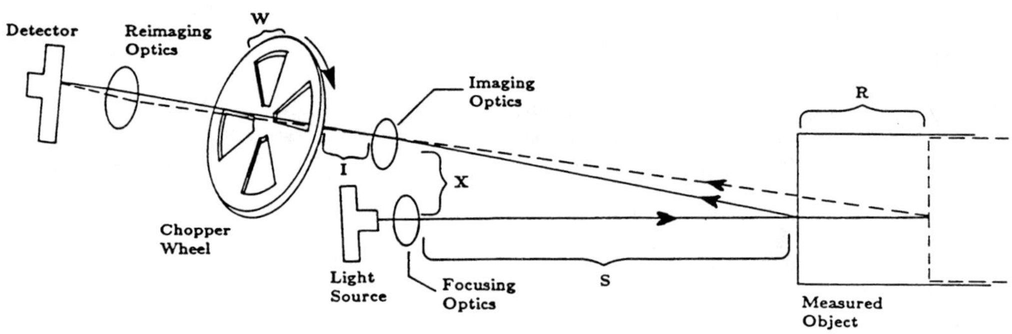

Figure 6. Small Angle Configuration for Phase Detected Triangulation

A desirable configuration for phase detected triangulation at larger angles is shown in Figure 7. This is the off-axis configuration which eliminates non-linearities associated with other commonly used triangulation configurations [8]. The major draw back with the off-axis configuration is reduced light gathering ability. This problem can be overcome by replacing the imaging lens with a mirror light collecting system. From the geometry of the off-axis configuration, the relationship between the ambiguity depth (R) and the radial grading slit width (W) can be established:

$$R = 2 \cdot W \cdot O / I \tag{4}$$

Phase triangulation, using the off-axis design, can be adapted for a wide variety of measurements. A system configured as shown in Figure 7 might be used to make high accuracy measurements on a polished surface. Such a surface at the triangulation angle shown would require a high signal to noise ratio for the detection system. As discussed in the experimental results section, this high signal to noise ratio can be obtained with phase detected triangulation system to a resolution of one part in 1000 or better. If we require a measurement range of 1mm(resolution of 1 micron), and the object distance (O) and image distance (I) are equal, then the slit width (W) is 0.5 mm. If a greater measurement range is required, then a two detection system (course and fine) could be used.

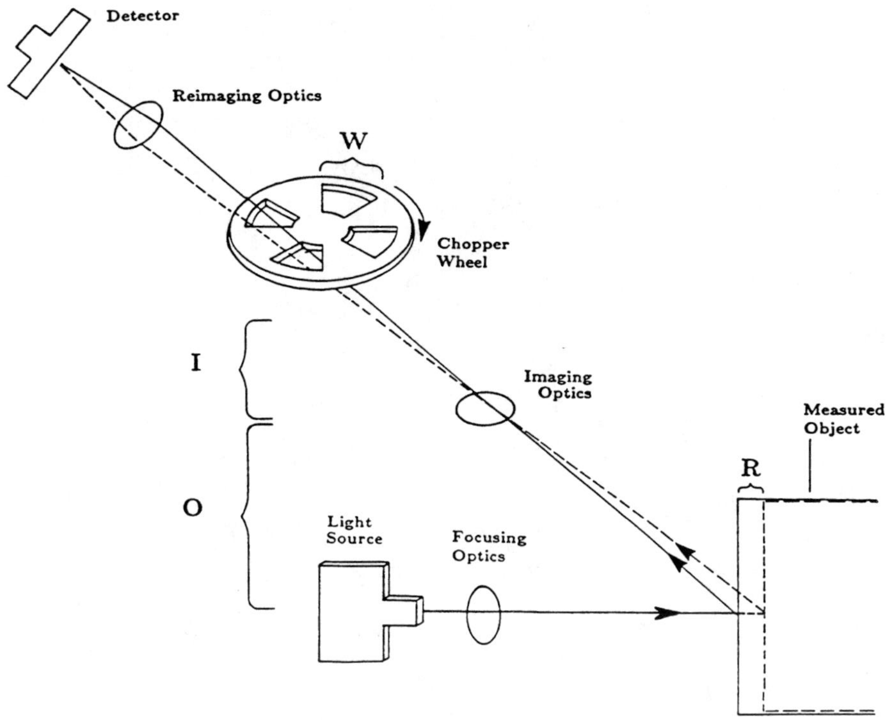

Figure 7. Large Angle Configuration for Phase Detected Triangulation

4. COMPARISON OF RANGING TECHNOLOGIES

Phase detected triangulation offers advantages over laser radar and traditional triangulation ranging sensors. Of course, there are trade-offs in using phase detected triangulation. This section will explore the advantages and disadvantages associated with the ranging techniques.

As a starting point for the comparison of ranging technologies, Table 1 was constructed. This table is a ball park estimate for each of the ranging techniques for comparative purposes. The data for laser radar and traditional triangulation range sensors represents a composite of information from vendor product descriptions. The information on phase detected triangulation was extracted from the previous section in this paper.

From Table 1 it can be seen that the performance of Phase Detected Triangulation falls between the performance of Laser Radar and traditional triangulation. The unit cost is lowest for phase detected triangulation and highest for laser radar. The major drawback with phase detected triangulation is the motion that is required for the detection scheme. However, with good design, this can be minimized.

The big advantage that phase detected triangulation has over laser radar is that the electronics do not have to operate at ultra high frequencies, as in the case of laser radar. Phase detected triangulation operates at frequencies under 1 megaHz whereas laser radar must operated at frequencies of 300 megaHz or better to achieve 1mm resolution. Operating at these ultra high frequencies increases cost and complexity for the system while reducing reliability and the signal to noise ratio of the detection system.

Phase detected triangulation has advantages over traditional triangulation including: better signal to noise ratio, potentially lower cost, faster measurement rate, and the ability to measure at smaller triangulation angles.

In summary, phase detected triangulation has advantages which allow this technology to tackle measurement problems that could not be addressed adequately by other techniques.

TABLE 1. RANGING TECHNOLOGIES

	PHASE DETECTED TRIANGULATION	LASER RADAR	TRADITIONAL TRIANGULATION
DEPTH RESOLUTION (# of Bits)	8-10*	8-10	8-11
MAXIMUM STANDOFF (Meters)	3-16	4	1
MEASUREMENT RATE (kHz)	50	300	1
UNIT COST (k$)	1-4	50-100	5-10

* Improved by 4-6 bits by using two detectors (coarse and fine).

5. REFERENCES

1. M. Hersman, F. Goodwin, S. Kenyon, and A. Slotivinski, "Coherent Laser Radar Applications to 3D Vision and Metrology", Vision '87 Conference Proceedings, SME, Dearborn, MI, 3-1 (1987).

2. D.J. Svetkoff, P.F. Leonard, R.E. Sampson, and R. Jain, "Techniques for Real-Time, 3D, Feature Extraction Using Range Information", Intelligent Robots and Computer Vision, SPIE 521, 302-311 (1984).

3. C.J. Bocchi, R.E. Keil, J.E. Bolbeom, S.K. Case, "Sensor Performance in Long Range 3D Triangulation Systems", Vision '87 Conference Proceedings, SME, Dearborn, MI, 3-13 (1987).

4. R.D. Lorence, "A Novel, High Range to Resolution Ratio, Optical Sensing Technique for High Speed Surface Geometry Measurement", Optics, Illumination, and Sensing for Machine Vision, SPIE 728, 152-159 (1986).

5. G. Marshall, Laser Beam Scanning, Marcel-Dekker Inc., New York, NY (1985).

6. M. Gottlieb, Electro-Optic and Acousto-Optic Scanning and Deflection, Marcel-Dekker Inc., New York, NY (1983).

7. D.J. Svetkoff, "Towards a High Resolution, Video Rate, 3D Sensor for Machine Vision", Optics, Illumination, and Image Sensing for Machine Vision, SPIE 728, 218 (1986).

8. K.H. Harding and D. Markham, "Improved Optical Design for Light Stripe Loges", Sensors '86 Conference Proceedings, SME, Dearborn, MI, 1-22 (1986).

Reprinted from *Optomechanical and Electro-Optical Design of Industrial Systems,* R.J. Bieringer, K.G. Harding, eds., Proc. SPIE 959, pp. 166-176 (1988). ©1988 SPIE.

ELECTRO–OPTICAL SENSORS IN INDUSTRIAL APPLICATIONS

Dr.–Ing. Rolf–Jürgen Ahlers

Fraunhofer–Institut für Produktionstechnik und Automatisierung
Eierstr. 46, 7000 Stuttgart 1

Quality Assurance plays a key role in modern flexible manufacturing systems. This is the reason why an increasing demand for highly adapted automated measuring and inspection systems is noticeable that have to assure the quality performance of the products leaving the production lines. Once proven to be reliable and economic, they will be introduced into production as systems for in–line quality control. In the following different electo–optical systems will be described that deal with topics as optical geometrical measurements or the automated visual inspection as they are applied for industrial use.

1. INTRODUCTION

The advent of microprocessor controlled electro–optical sensor systems has rapidly changed the face of production. They have been integrated into production lines, wherever possible, to improve the quality of the processes or products to be measured or inspected. Especially in the field of visual inspection tasks there is a great need for automated inspection systems.

In order to be technically acceptable and economically justifiable, such systems should in general meet the following requirements:
- high processing flexibility for the solution of different complex inspection and measuring tasks
- high processing speed for efficient on–line operation
- adapted "easy–to–handle" interface with human and production environment
- production adapted, ruggedized system concept.

Modern production systems consist of integrated manufacturing units using sophisticated software and hardware tools as well as different sensor equipment. This pilosophy of "Manufacturing and Inspection Cells" leads to distributed facilities with a very high level of integration. Thus "intelligent" sensor units become more and more important for industrial use in complex automated production areas /1/, most of them making use of electro–optical methods.

When using sensor aided equipment one will encounter a vast field of possible solutions. Among other things, economic considerations, flexibility, new technical developments, new means of technical information flow etc. of the automated sytem have to be dealt with. This is the reason why we will begin in the following with simple but effective principles of electro–optical sensors, ending up with complex and sophisticated sensor systems that make use of image processing equipment to cope with industrial requirements .

2. SIMPLE ELECTRO–OPTICAL SENSORS

"Simple" in the context of this chapter should not imply a contemtous judgement. "Simple" is just used as a terminus in contrast to "complex" as it is used in the chapters that follow.

2.1 ELECTRO–OPTICAL SENSORS FOR SURFACE MEASUREMENTS

Optical methods for the evaluation of surfaces are attracting increasing interest in the field of measurement techniques /7/. They avoid various problems characteristic for methods based on mechanical touching of the workpiece:

- errors resulting from the bending during the measurement cycle
- deformation of the workpiece in case of soft flexible materials
- danger of collision with the workpiece during scanning cycle
- wear of the mechanical probe.

Wherever surfaces contact each other in a defined manner, slide along each other or are to be interactive in some other way, it is necessary to test features as the roughness or microgeometry of each of the surfaces involved.

The key role of electro–optical methods was recognized very early, but it was not until the eighties that systems for broad industrial applications evolved, making use of different principles such as:

- moire interferometry
- light sectionning
- interferential evaluation
- reflection measurements
- dynamically focussing.

Some of these methods will be described in the following.

The first sensor represents a method especially developed for the optical determination of the roughness with the reflection method. The light of a photo diode (LED) is projected onto the rough surface and the backscattered component is covered in its spatial propagation with a linear array of light–sensitive semiconductors. In dependence on the surface profile, varying intensity characteristics will occur which permit an evaluation of the surface (Figure 1).

Closely related to speckle–contrast–methods different systems evolved, one of them is the white–light–method that makes use of the phase–contrast of a defocussed surface image with additional light superposition. If a surface is moved in front of this sensor, this will result in a relative scanning of the intensity pattern according to Figure 2. The contrast value C is determined with respect to

$$C = SI/\langle I \rangle$$

by appropriate signal evaluation, with SI representing the standard deviation and <I> the average value of the electro–optically converted signal. Figure 3 shows experimental results that were obtained. The slope can be changed through the constant light portion T. With T defined as the ratio of the light intensity coming from the surface, Is,to the light intensity coming from a mirror Io.

2.2 ELECTRO–OPTICAL SENSORS FOR DISTANCE MEASUREMENTS

The wellknown triangulation method is here adapted to the optical scanning of workpieces. A light beam is projected onto the surface of the workpiece. By means of an imaging optic

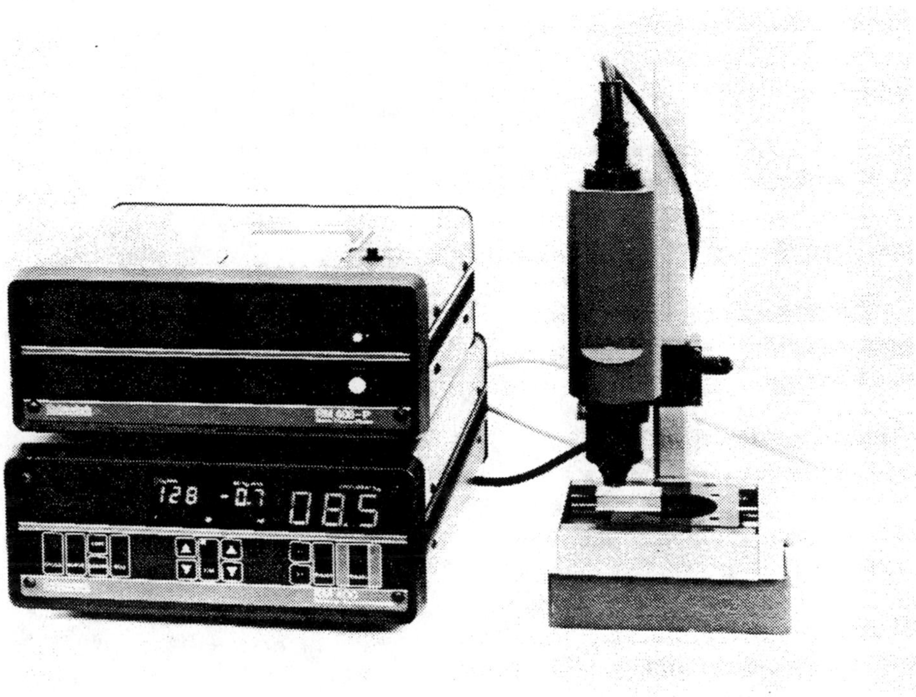

Figure 1: Reflection Sensor (Rodenstock)

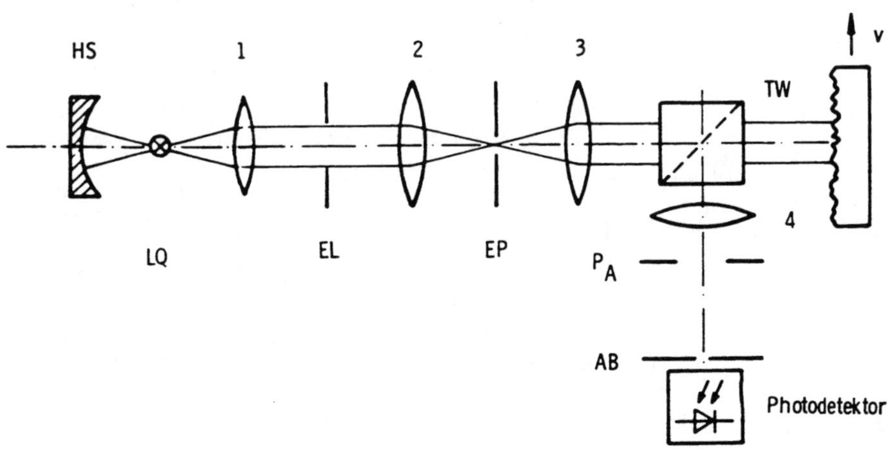

Figure 2: White Light Method

the lightspot is projected onto the sensor array. The measured distance is directly related to the displacement of the light spot on this array. The general principle is sketched in Figure 4.

For the application of this method there may be used either CCD–sensors or positition sensitive analog photosensors arranged in a linear or area array. Wheras the use of a position sensitive photosensors requires a focused spot the conditions are much less restricitivein case of a CCD–sensor because here the exact position of the light spot may be determined by digital signal processing of the sensorsignal, at the expense of measurement speed.

The triangulation method is not the one and only method. Other examples exist /3–5/. One of them will be described below. This system is used for high precision measurements that can not be reached by the triangulation method.

Until now there were no optical length measuring instruments commercially available offering an equal high resolution both in the direction of measurement, e.g. normal to the surface of the workpiece, and perpendicular to it. Now several methods have been discussed whichall have in common that the objective lens of the sensor follows the displacement of the surface. Without this feature the measurement range is very small depending on the surface structure. With objectives having a numerical aperture of 0.5 or larger the focus of the probing laser beam can reach around 1 um still maintaining a standoff of about 3 mm. A summary and discussion of various optical methods to measure displacements are given in /7/. Since the Foucault principle yields good results, and the necessary set–up being not too difficult to adjust, many instruments were built relying on this method. Figur 5 shows the principle /9/.

The light spot on the surface is imaged onto a differential photo diode where the beam has to pass a knife edge, which is adjusted such that the photodiodes are equally lit if the beam is precisely focussed onto the surface. A displacement of the surface in either direction results in a respective shadowing of one of the photodiodes from which the direction of the displacement can be deduced. It is obvious that only a small measurement range can be obtained which depends on the light reflecting properties of the surface. For a mirror surface the range will be around 5 um with a high sensitivity of the photodiode signal. This holds for a numerical aperture of the objective lens to at least 0.5. A weakly reflecting surface, e.g. glass, would yield a range of up to 100 um exhibiting a comparably low resolution. Due to the limited diameter of the lens a decrease in the measurement accuracy is observed if the test surface is not at the right angle to the lightbeam (Figur 6).

Therefore one has to be cautious interpreting the results of roughness or microprofile measurements with optical methods.

3. OPTICAL COORDINATE MEASUREMENT SYSTEMS

3.1 COORDINATE MEASUREMENT SYSTEMS WITH INTEGRATED IMAGE PROCESSING

The objective in developing instruments for opto–electronic coordinate measurement is to combine the advantages of high precision electro–optical sensor systems and those of mechanical coordinate measuring instruments. With the use of optical sensors, small objects can be measured directly while the large measurement range is due to the mechanical high precision tables of the coordinate measurement instruments.

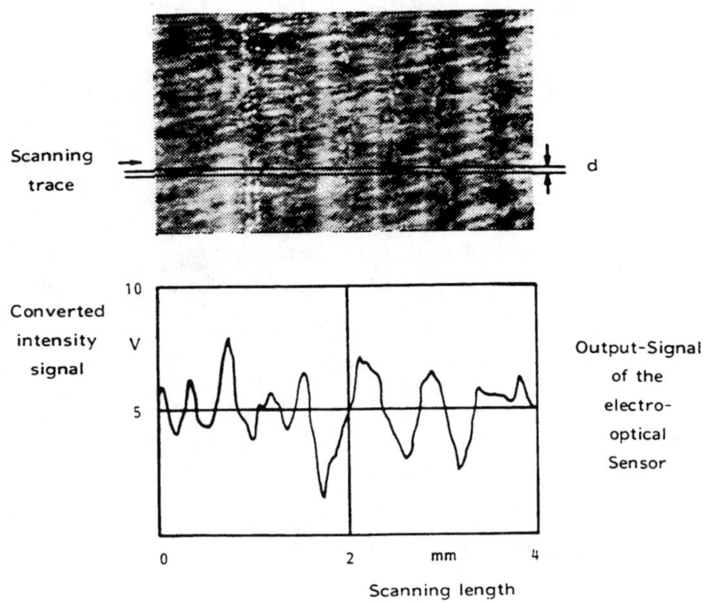

Figure 3: Measurement Results of the White–Light–Method

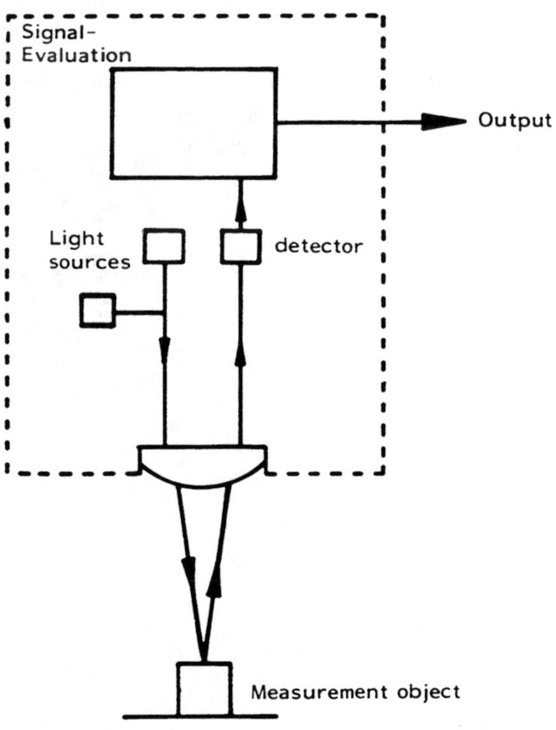

Figure 4: Triangulation Method (Honeywell)

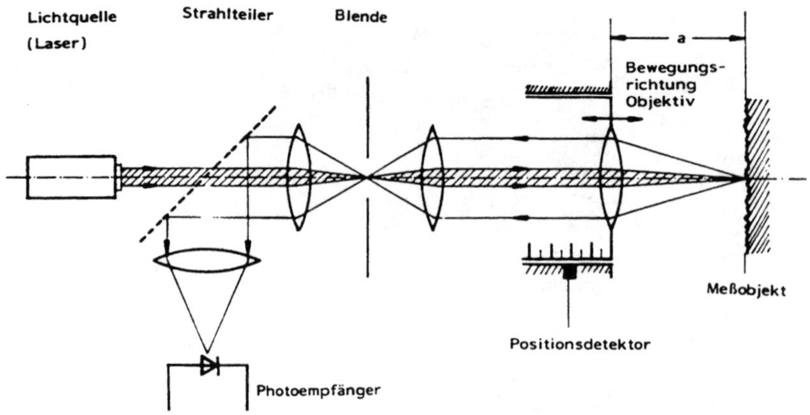

Figure 5: Dynamically Focussing System (DFS)

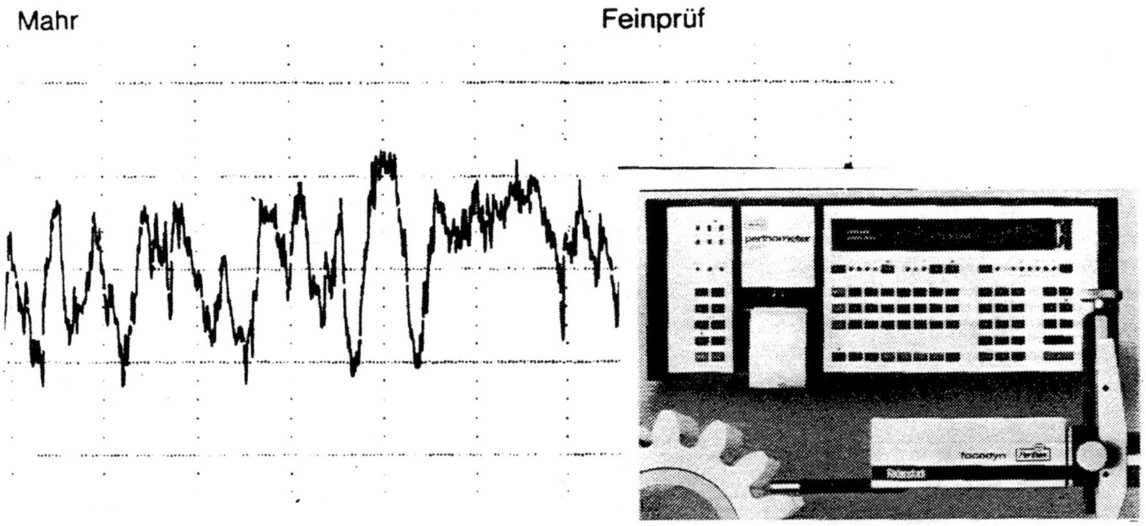

Figure 6: Results of the DFS

In general simple electro–optical sensors can be integrated into high precision mechanical displacement systems. In the following we will concentrate on integrated image processing systems.

Sensors with an area array are well suited for measurements. Both semiconductor cameras and tube cameras are in use and it would lead far beyond the scope of this contribution to judge on the use of the former or the latter.

By a measurement optic the object is imaged onto the target of the camera. The magnification of the objective determines the resolution as well as the field of view. The captured image is preprocessed (e.g.filtered) and then evaluated in the image processing system. The evaluation can be done either in binary or in grey–scale pictures. Coordinates of the objects are calculated by combining the coordinates of the axes of the coordinate measuring instrument with those obtained by the electro–optical system. Measurements in z–direction are realized by focussing on the object via contrast evaluation in the video signal, or an external autofocussing system.

A very important element for high accuracy measurements is the lighting. One should carefully devise and deliberately choose appropriate systems in order to reach for a well adapted system integration.Transmitted light illumination permits measurements of opaque or more or less transparent objects. For the integration a controllable incident light system is chosen.

To optimize the contrast for incident illumination dark–field (large angle of incidence) and bright–field (small angle) lighting arerecommended.

Figure 7 shows one examle of an electro–optical coordinate measurement system that has been developed for special applications related to the automobile industry.

3.2 LIGHT SECTIONNING METHOD

Often the three–dimensional measurement of workpieces is necessary where no direct mathematical description and thus geometry is available. For these objects, e.g. turbine–blades, drills, pump rotors, car bodies, different measurement systems exist /1/.

In order to assure accurate machining conditions of a drilling tool it is necessary to keep its cross–section within defined tolerances. The cross–section influences swarf removal, torsional characteristics of the tool etc..

Nowadays optical systems, e.g. profile projectors, are widely used to measure complex cross–sections, the disadvantage of which is that the tool has to be destroyed in order to measure the required geometry. Destructive testing of this kind is wholly unacceptable when dealing with small–batch production.

As a consequence a totally new measuring concept must include:

- nondestructive testing
- short measuring time
- automatic measuring process which eliminates subjective aspects of testing personnel.

When dealing with twist drills it is not only the cross–section which has to be measured. There are also several other characteristics like cutting edge centrality, chisel angle, lip height

Figure 7: Electro–optical Coordinate Measurement System

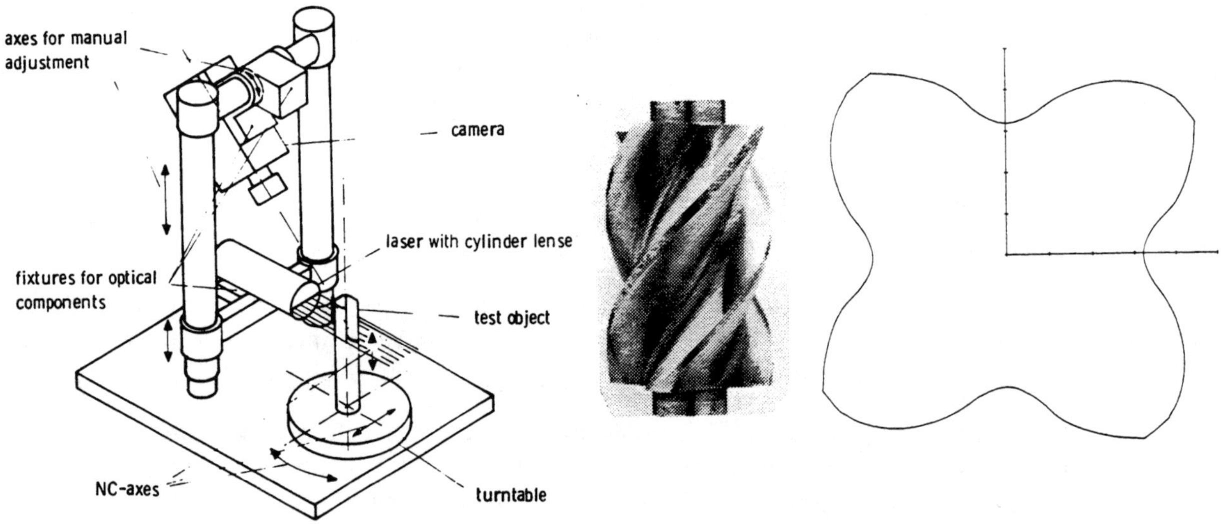

Figure 8: Light–Sectionning System

etc. which, even today, are being measured using optical comparators, tool microscopes and often even simpler testing gauges. The main characteristics of the cutting edges have to be considered with reference to the point angle, cutting edge centrality, and lip relief angle.

The applied optical set–up consists of a narrow collimated light pencil that is projected onto the workpiece. Distorted by the actual test object, the modulated light stripe represents the profile of the object that is recorded via the camera of the integrated image processing system (Fig.8).

3.3 ELECTRO OPTICAL COORDINATE MEASUREMENT

Another type of integration concerning electro–optical coordinate measurement systems for high precision measurements relates to systemsusing microscopic optics for high accuracy and precision displacement units with a big measurement range (Fig.9).

The measuring uncertainly can be pushed up to 1 um, whereas the optical accuracy depends on the optical set–up and the applied sensor–types. The accuracy is also strongly influenced by the lighting and the autofocus system /2/.

4. THE AUTOMATION OF VISUAL INSPECTION

The philosophy of "Inspection Cells" has become important when trying to cope with human skills in the inspection of products. The inspection personnel is handling the parts to improve the optical appearance for expected flaws, to get a better contrast.For an automated process, up to now, predominantly fixed cameras have been used which necessitate bringing the object to be tested into the defined visual field of the electro–optical sensor. A linked system involving an industrial robot and an image processing unit, which can be described as a "robot–guided camera" deserves attention.

For large objects with varying spatial dimensions, it is possible to take the camera to the test location. An alternative to an industrial robot can be achieved through the combination of the most varied transverse and rotational positioning units, however, it always leads to special and usually expensive one–off applications. Probably the most flexible system arrangement is based on the robot–guided image sensor, which allows the industrial robot to guide the camera to the test location. This type of camera positioning is no less futuristic in that, through the involvement of industrial robots, it can be extended to flexible standard products. Applications are manifold.Fig.10 just shows two aspects of the coupling of robots with image processing systems /8/.

5. CONCLUSION

Electro–optical sensors, together with efficient computer–systems, form a basis for flexible automation in the field of dimensional measurement techniques and the automated visual inspection. As shown integration of these systems into production lines has become an obvious fact. Furthermore, as production automation continues to increase, especially optical sensors will become dominant. This is founded on the vast amount of information which is contained in optical principles. Therefore some efforts have to be undertaken to reduce this information content where possible. Better and faster hardware will allow more and more manipulations of this information in order to compete with human intellectual capabilities.

Figure 9: Integrated electro–optical coordinate measurement system (Werth)

Figure 10: "Inspection Cell" for the automated visual inspection

The barrier "computer–system" or in general "signal evaluation system" will be crossed with the aid of appropriate user–friendly software structuring, as well as programming aids (menu technique, teach–in, etc.). Little by little the capabilities of such sensor systems will be enhanced and an increasing amount of applications using image processing devices for various tasks in automated production will be added.

6. REFERENCES

/1/ R.–J. Ahlers; K.W. Melchior; R.D. Schraft; L. Schreiber; H.J. Warnecke, Applied Pattern Recognition in Sensor Systems for Quality–Control and Automated Production Systems, Manufacturing Systems, Vol. 14, No. 3, 1986.

/2/ C. Keferstein; G. Pavel; H.J. Warnecke, Two– and Threedimensional Linear Measurement with Image Processing Systems, pp. 137–143, Measurement, Vol.3, April–Juni 1985.

/3/ B. Batchelor; D.A. Hill; D.C. Hodgson, Automated Visual Inspection,IFS Ltd, North Holland, Amsterdam, 1985.

/4/ L.H. Biemann, Three–Dimensional Machine Vision, pp. 81–86, Photonics Spectra, May, 1988.

/5/ L.H. Biemann; K.G. Harding; M. Michniewicz; R. Tait, Phase Detected Triangulation: a new twist on an old technolgy, SPIE Vol.850 (1987).

/6/ M. Rioux; F. Blais, Compact three–dimensional camera for robotic applications, Opt. Soc. Am. A./Vol.3, No.9/ Sept. 1986.

/7/ R.–J. Ahlers, Die optische Rauheitsmessung in der Qualitätstechnik, Springer–Publ. Co; New York (1986).

/8/ H.J.Warnecke,B.Oberdorfer,W.Rauh,Flexible Prüfzelle mit bildverarbeitungsgesteuertem Prüfroboter zur Einbindung in ein CIM–Konzeptwt–Z.ind.Fert. 75, Heft 4,5, p.225 228 (1986).

/9/ U.Breitmeier, R.–J.Ahlers, Dynamically focussing electrooptical sensor system for microprofilometry, SPIE Vol.802 (1987).

Reprinted from *Current Developments in Optical Design and Optical Engineering*, R.E. Fischer, W.J. Smith, eds., Proc. SPIE 1527, pp. 225-233 (1991). ©1991 SPIE.

Width gauge for hot steel plates by laser scanning range finder

Yuzi Adachi

Kashima System Planning Department, Sumitomo Metal Industries, Ltd.
3 Hikari, Kasima-cho, Kashima-gun, Ibaraki, 314, Japan

Kenichi Matsui

Instrumentation Technology Development Section, Sumitomo Metal Industries, Ltd.
1-8, fuso-cho, Amagasaki-city, Hyougo, 660, Japan

Abstract

We have developed a new advanced laser scanning range finder. The width gauge consists of a pair of these range finders. Conventional gauges cannot measure the effective width of complexly shaped plate edges. However, the gauge measures both side profiles of a plate and calculate the effective width with an accuracy of ±1.5 mm. This new type of width gauge has been developed and tested at the plate rolling mill at Kashima Steel Works of Sumitomo Metal Industries Ltd.

Introduction

In the plate rolling process, the dimensions of the product must be controlled during manufacturing in order to increase yield ratio. When the measurement accuracy in the rolling process is not high, bigger margins on the plates are needed to insure that the required dimensions of final product are achieved. Especially on hot plate rolling process we wanted to measure the width of plates more accurately, but conventional method of measuring plates width is inaccurate because it only measure the widest width of the plate.

With this in mind, We have developed a new advanced laser scanning range finder which can measures in the harsh environment of the steel industry. This laser scanning range finder measures the distance from sensor to side of the plate by scanning the laser beam, so that this finder can measure the edge profile of plates in spite of the rapid changes in reflection by an object (the plate) without a reflector. This range finder has following characteristics :

1)high speed measurement (40,000 time / sec)
2)high speed scanning (40 Hz)
3)a wide range of measurement (4000 mm)
4)high accuracy (±1.5 mm)
5)high stability (±1 mm)

The new width gauge reported in this paper measures width of plates more accurately because it can measure the edge profile of plates and calculate effective width.

Background

Plate rolling process

Fig.1 shows the hot steel plate rolling line at Sumitomo Metal Industries' Kashima Steel Works. Steel plates are rolled from steel slabs. The temperature of the slab is about 1000 degrees centigrade. The size of plate is from 0.8 m to 5.2 m in width, 4.5 mm to 200 mm in thickness and from 2 m to 55 m in length.

The continuous measurement of the effective width of a hot steel plates as it is rolled in the edging stand and the roughing stand is essential for reducing edge trim allowance, thus improving yield. Measurement must be made to determine that the incoming plate in the roughing stand will produce the correct finished width for the finishing stand.

Fig.2 shows the width change in rolling pass. If the plate edge is flat after the first pass (horizontal pass), the rise near the plate edge will be large after the second pass (vertical pass), and the width change (d_1) from second pass to third pass (horizontal pass) is also large. However, if the plate edge is caved after the first pass, the rise near the plate edge will be small after the second pass, and the width change (d_2) is also small. Therefore, the measurement of maximum width (W_0) does not produce plates with desired width (W).

It is important to measure the effective width, that is, the average width. If the effective width and the edge profile can be accurately measured, it is possible to estimate the rise after the second pass and thus predict the width change after the third pass. Then we can set the roll gap of edging stand and the roughing stand correctly to produce a product with the required width.

Conventional width gauge

The conventional width gauge for hot plates is installed over a rolling line and measures the maximum, that is, the widest width using the infrared radiation of hot plates or shadow from back-lighted plates. (Fig. 3) But when the profile of the plate edge is not flat, the maximum width is not equal to the effective width.
Sideguides are also used to measure the width , but they report only the widest width of the plate, and the curvature of a plate causes measurement error.

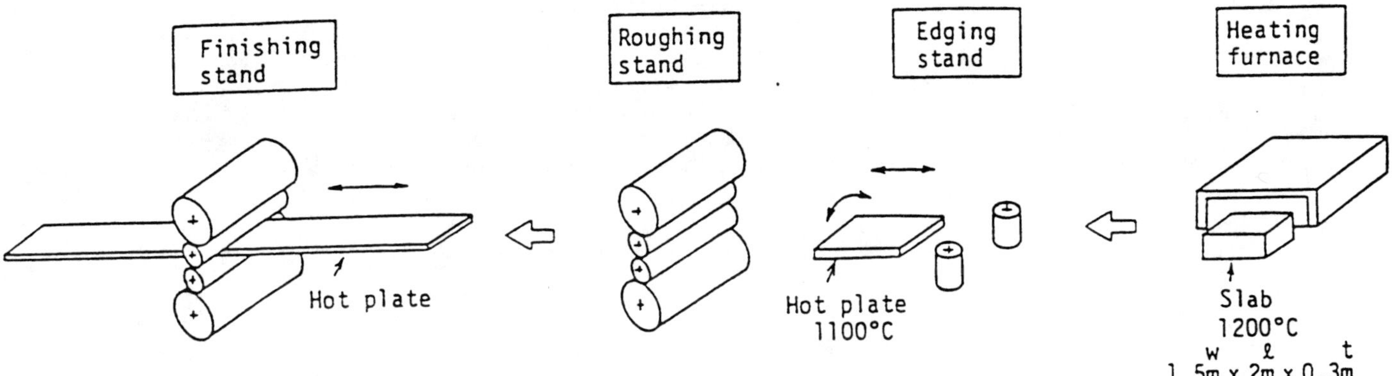

Fig . 1. Hot steel plate rolling process

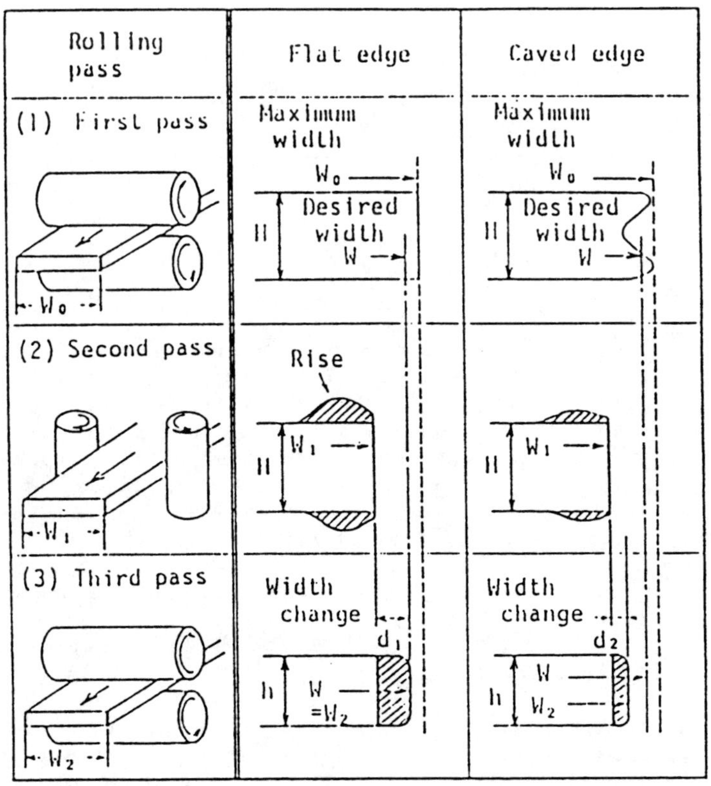

Fig. 2. Width change in rolling pass

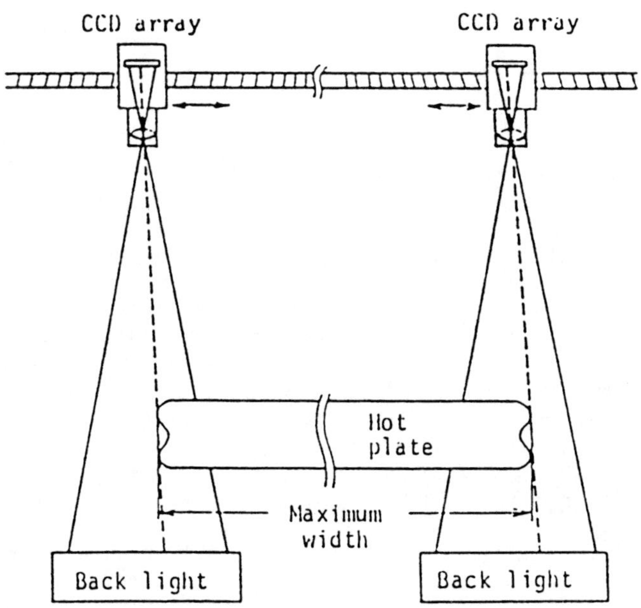

Fig. 3. Conventional Width gauge

Principle of range finder

The principle of the range finder is to measure the time taken for a ray of light to cover the distance from the instrument to the object (plate edge) and back to the instrument.
The range finder is built using a laser diode (LD) and avalanche photo diode (APD). The LD is modulated directly by an RF current whose frequency is F (150MHz). The modulated light beam emitted from LD is sent to the target. It is scattered at the surface of a target as shown in Fig 4, and as shown in Fig. 4, and scattered light beam is detected by an APD.
The heterodyne detection is done to measure the phase shift in the low frequency region.
The distance between the instrument and the target is calculated from the phase shift by using the formula

$$L = \frac{1}{2} \cdot \frac{C}{F} \left(N + \frac{\Delta \theta_1}{2\pi} \right) + \text{Constant} \qquad (1)$$

where C = light velocity, F = modulation frequency (150MHz) and N = an unknown integer.
In this case 1/2 modulation wavelength is 1000 mm, therefore use of single amplitude modulation frequency dose not permit determination of value of absolute range lager than 1000 mm. However, the difference between the plate width under rolling and the desired width is less than 100mm, so it is possible to predict the approximate distance and define the integer value N.

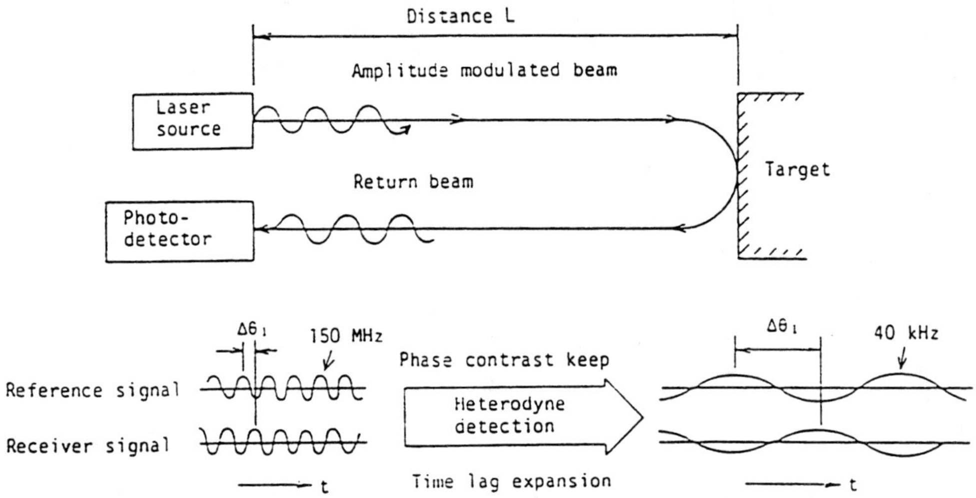

Fig. 4 . Principle of range finder

Optical schematic of sensor head

Fig . 5 shows the optical schematic of a sensor head. A laser diode (a) transmits an amplitude modulated continuous wave laser at a modulation frequency of 150 MHz. The laser output beam is directed through a intensity regulator (e) onto a optical scanner (b). A Pockels cell is used for the intensity regulator. The scanning frequency of the optical scanner is 40 Hz. Next a parabolic mirror (c) aligns a fan-shaped scanning beam to a parallel scanning beam. The laser beam is scanned up and down in a vertical line with a scanning range 200 mm. The laser beam then is divided by a beam splitter (f). The split beam is detected by APD (d_1) and detected signal is used as the reference signal. Finally the laser beam reflected from the plate edge is focus on the APD(d_2).

Fig . 6 shows the focus schematic of a sensor head. The lens (g) is moved in order to focus the laser beam on the target. Without this lens (g) the laser beam size on target will be bigger than 3 mm, so that it will not be able to measure small target (thin plate : 4.5 mm).
The laser beam reflected from the target is focused on the APD by lens (h), therefore in spite of a very low efficiency in collection of the scattered light returned from the surface of the target it has high sensitivity in the detection of the signal.

In addition all lenses and mirrors are AR coated. This was necessary in order to cut the laser beam leakage to the detector, because the leakage from emitted laser beam was the strongest noise source in the signal detection. Especially when the scattered laser from the beam splitter (f) returned to LD itself, the stability of LD become much worth due to back talk. So AR coating was necessary for the beam splitter too.

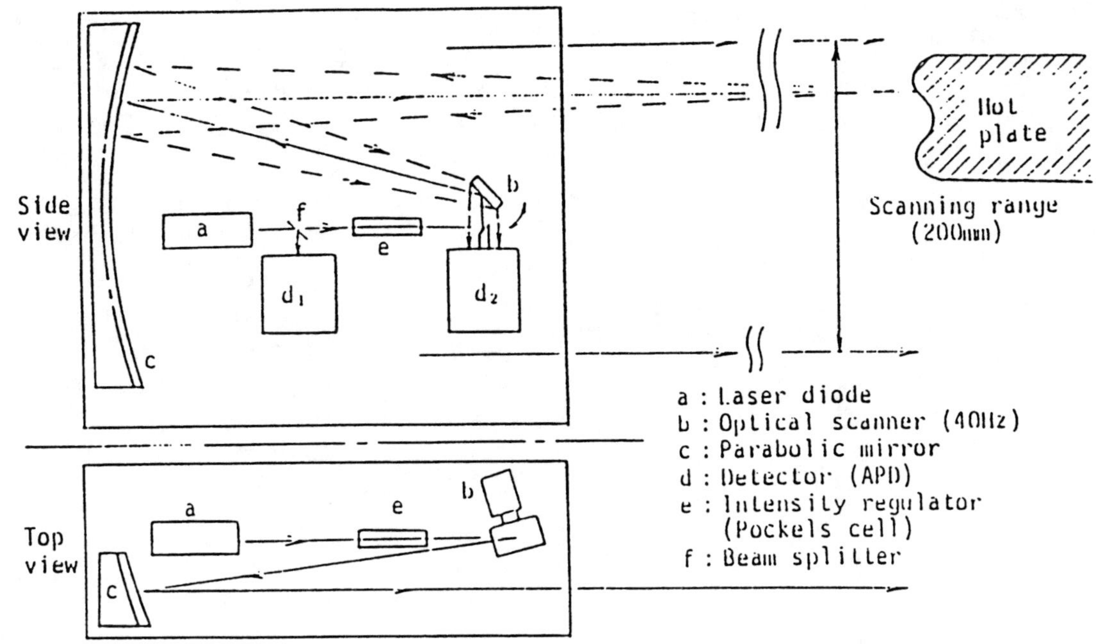

Fig. 5. The optical schematic of a sensor head

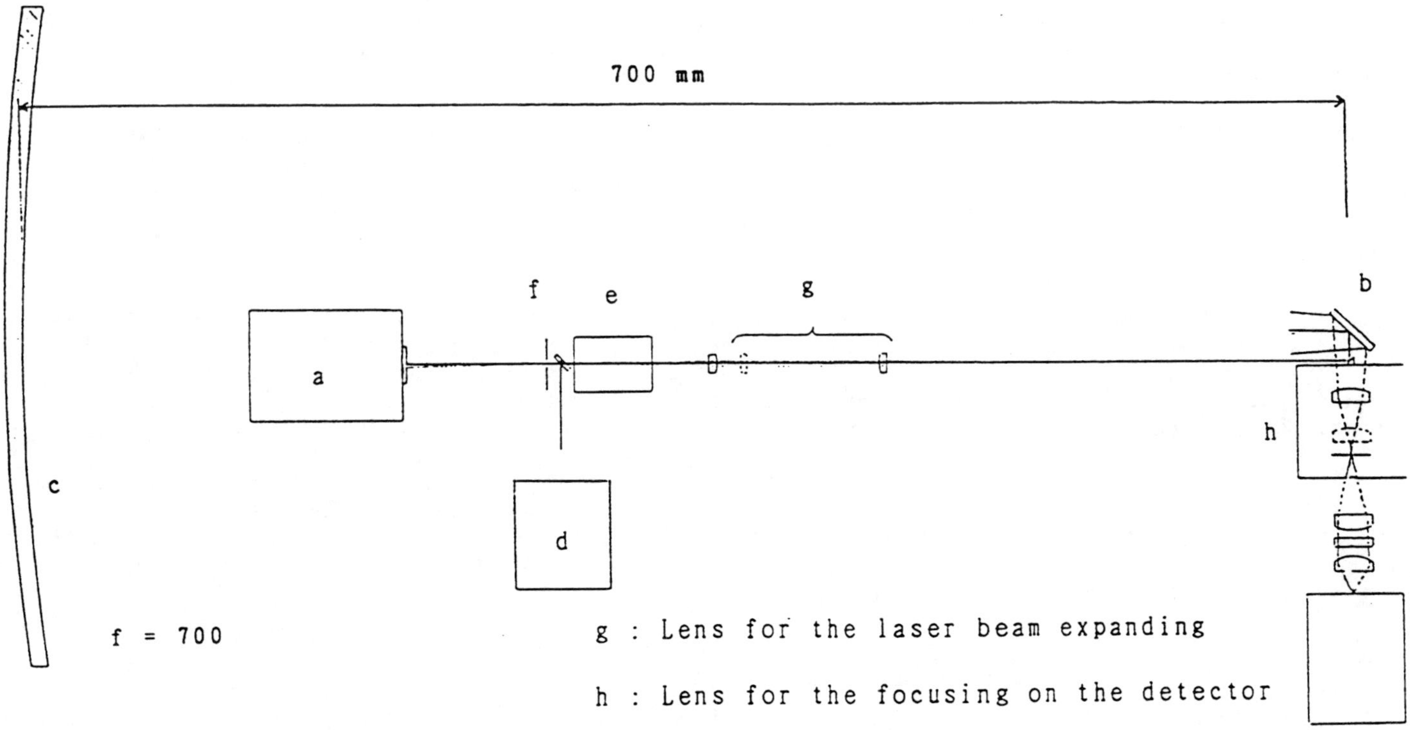

Fig. 6. The detailed schematic of a sensor head (focus lens)

Laser transmitter

The laser beam is amplitude modulated directly by current signal at frequency of 150 MHz . The temperature of the laser diode is controlled by a thermoelectric device (Perche device) within 24 ± 0.1 degrees centigrade to stabilize laser power and wavelength and phase shift. Wavelength is 780 nm and the maximum output power of the laser diode 20 mW.
Fig. 7 shows the phase shift in LD without a temperature control.

Detector

The temperature of the photo detector (avalanche photo diode) is also controlled by a thermoelectric device (Perche device) within 24 ± 0.1 degrees centigrade to stabilize their sensitivity and response time and phase sift.

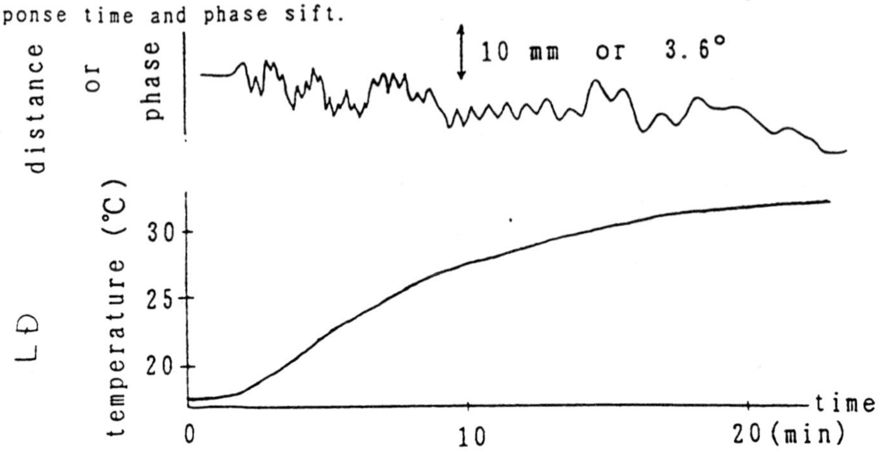

Fig. 7 The phase shift in LD without a temperature control.

Signal processing

Fig. 8 shows the detection and processing electronics. The phase difference* between the reflected laser signal and the reference signal is detected 40,000 times per second. The phase difference corresponds to the distance from the sensor unit to the plate edge.
To measure the absolute distance we adopted the following signal processing.

1)When the light intensity detected by APD varies drastically, the large change of the phase shift is caused by the intensity of the incident light beam and errors in the distance measurement occurs. In order to solve this problem the intensity of the incident light beam is kept constant. But the optical reflectivity of hot steel change drastically because of scale or rust. Therefore the Pockels cell is used to keep the power of the laser which is reflected from the plate edge constant.

2)When the frequency difference between RF frequency f1 (150 MHz) and f2 (149.96MHz) changes, large errors in the distance measurement also occur in the phase comparator.
In order to maintain the frequency difference between f1 and f2 another oscillator f3 (40 KHz) was used. The RF signal of frequency f2 was controlled so that it was between RF frequency f3 and the frequency difference f1-f2 . This electric circuit (phase locked loop)is shown in Fig. 9.

3)There is a large change of the phase shift between the RF current signal of LD and the amplitude modulated laser light beam. Its phase change cannot be stabilized by only temperature control of LD. Therefore the split beam is detected by APD (d_1) and detected signal is used as the reference signal in order to exclude the large change of the phase shift between the RF current signal of LD and the amplitude modulated laser light beam.

Installation of a new width gauge

The installation layout of the width gauge is installed in Fig.10 . A sensor head is installed on each side of a hot steel plate rolling line so that they face each other. Each sensor head continuously measures the distance (L1 , L2) from the sensor unit to the side edge of plate along the side in each down scanning. The microcomputer calculates the profile of the plate edge using each set of distance data and the position data of the optical scanner. The effective width is continuously measured as a plate runs through the sensor heads and the width profile for the entire plate length is also measured.

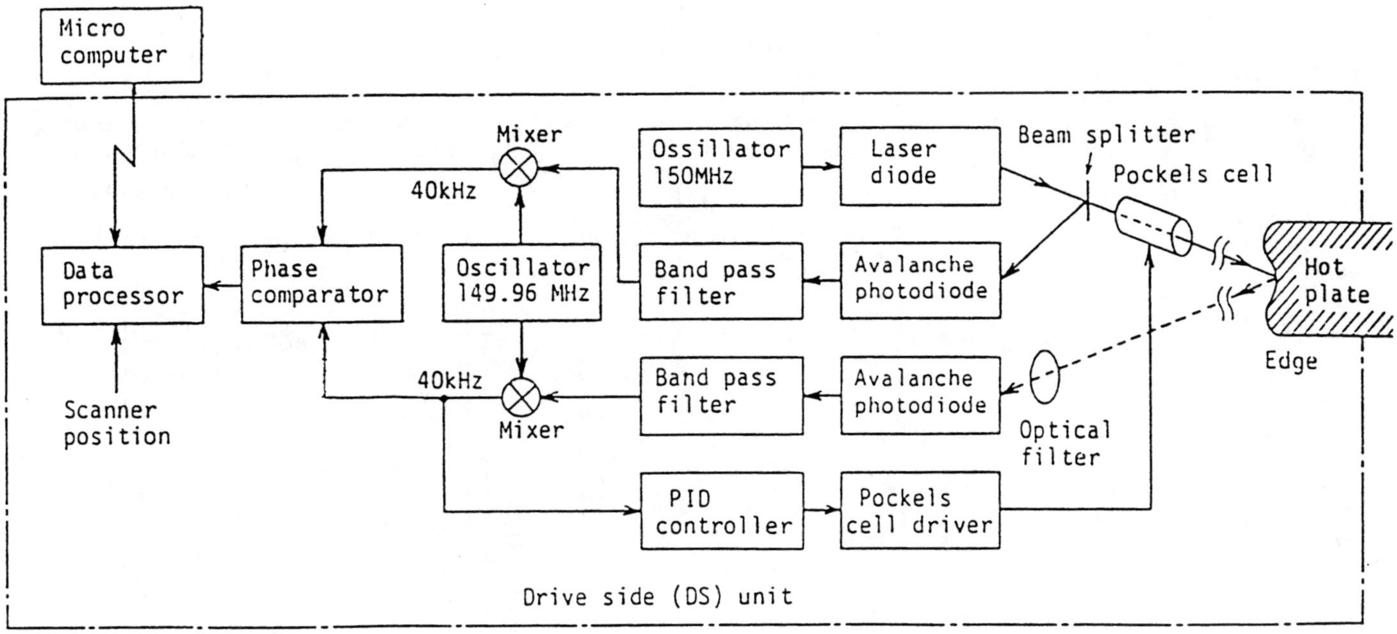

Fig. 8. Signal detection and processing electronics

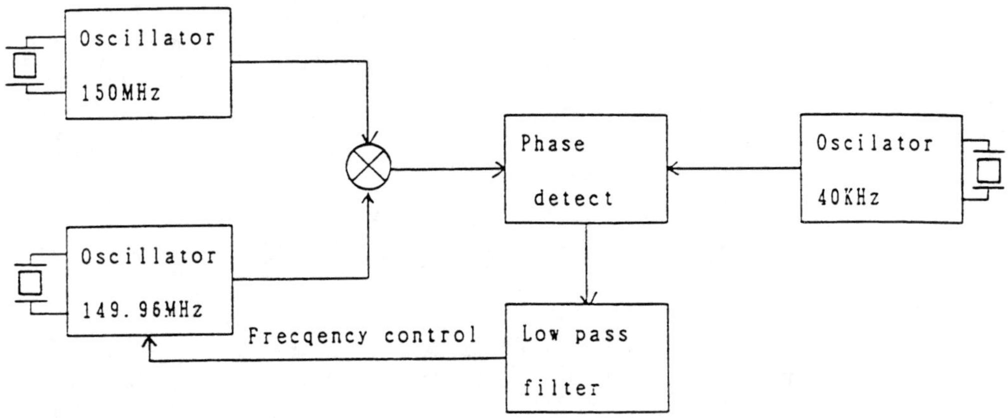

Fig . 9 . Phase locked loop circuit for the stabilizing of ossilator

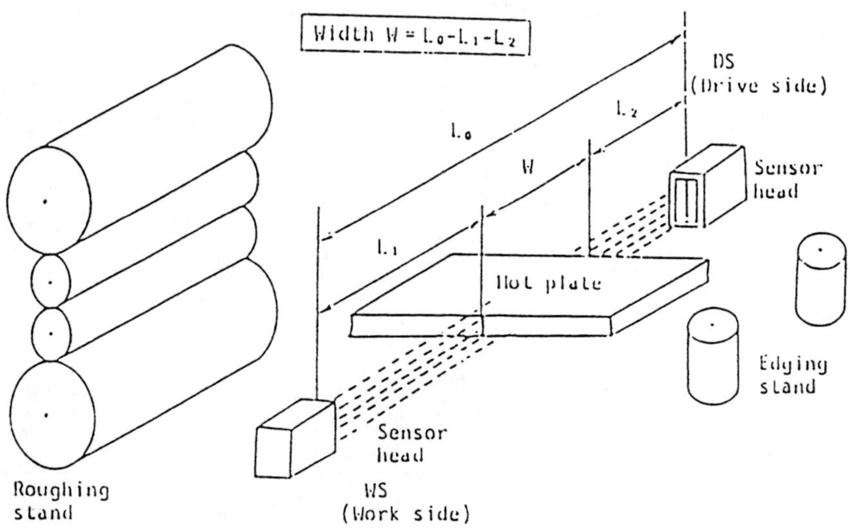

Fig. 10 Layout of the rolling stand and the sensor heads

Results

Profile Measurement

Fig . 11 (A) shows the profile of a test sample obtained with the new laser range finder. The profile looks very much like a staircase. This profile corresponds well with the sample's actual profile. Fig. 11 (B) shows the profile of the actual hot steel plate edge while rolling.

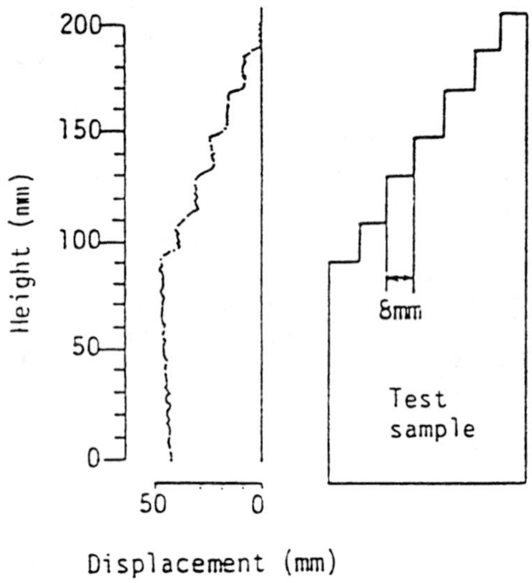

Fig. 11(A). Profile measurement on a cold steel test sample

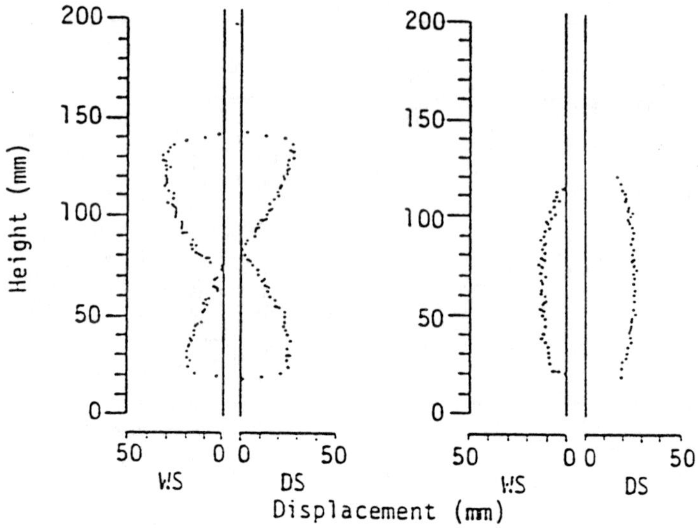

Fig. 11(B). Results of edge profile measurement on hot steel plates

Accuracy

The distance measuring accuracy of each laser range finder was confirmed to be ±1.0 mm in the entire range of gauge (3 m - 7 m). In addition, when two sensor units are used, the width measuring accuracy was confirmed to be ±1.5 mm on hot plates.

Stability

The distance measuring stability of each laser range finder was confirmed to be ±1.0 mm . This high stability in the distance measuring is obtained by following countermeasures.

1) Temperature control of the LD and the APD.
2) Reference signal from the additional APD.
3) Phase lock loop between RF frequency f3 and the difference frequency f1-f2 .
4) Reduction the emitted laser beam leakage and back talk.

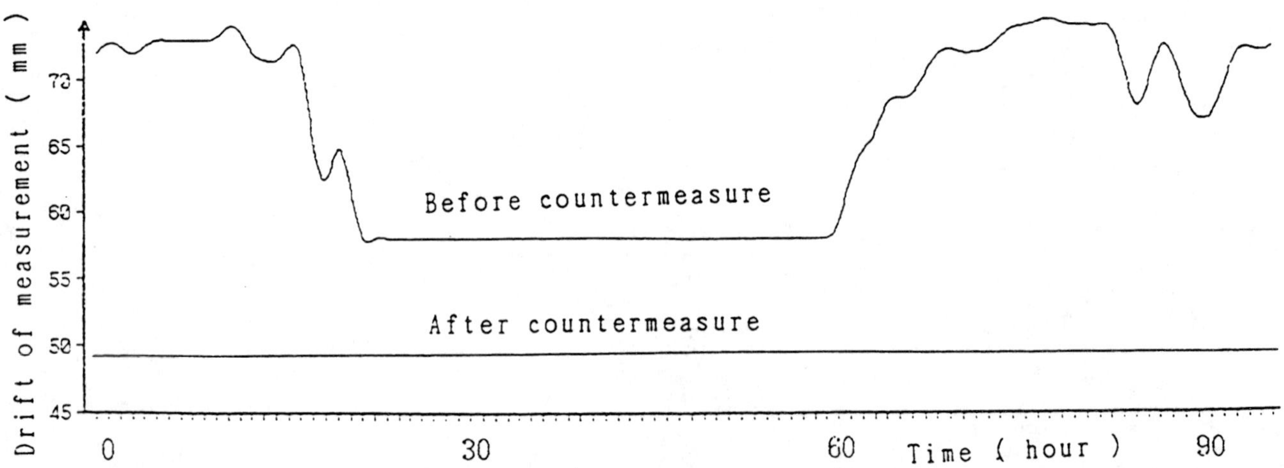

Fig . 12. A increase in the distance measuring stability

Response speed

In order to obtain high speed measurements, 40,000 time/sec or 0.025 msec/time, A high speed response in the intensity control of the incident light beam is required. Therefore a Pockels cell as the light modulator is used to control the power of the laser which is reflected from the plate edge. Fig. 13 shows the response speed of the intensity control. A high speed response such as the measurement speed noted above is obtained.

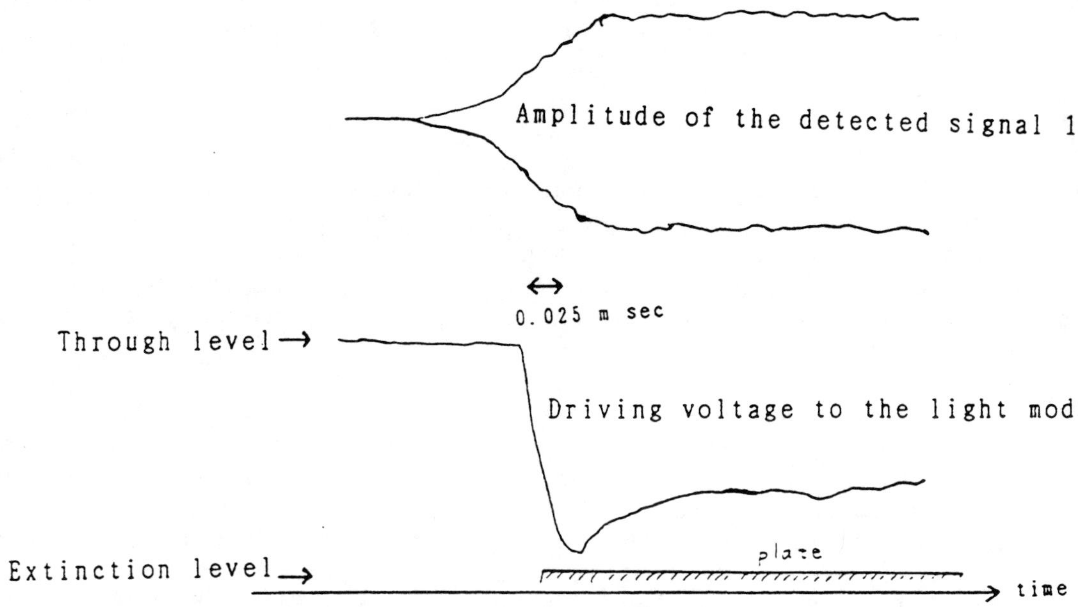

Fig . 13. The response speed in the intensity control of the incident light beam by the Pockels cell as the light modulator

Discussion for absolute distance measurement

In the absolute distance measurement, the drift constant in formula 1 is the biggest problem. It is easy to obtain a high resolution of about 0.5 mm, but it is difficult to stop the drift and to measure the absolute distance accurately.
However it is possible to cancel the phase drift by following countermeasures.

1) Temperature control of the LD and the APD.
2) Reference signal from the additional APD.
3) Phase lock loop between RF frequency f3 and the difference frequency f1-f2 .
4) Reduction the emitted laser beam leakage and back talk.

Conclusion

A high accuracy, a high stability, a high speed response and a high sensitivity are obtained in the distance measurement without reflectors. The new type width gauge consists of a pair of these range finders. The gauge measures both side edge profiles of hot steel plate and calculates their effective width with an accuracy of ±1.5 mm.

References

1. Tamura Y., Yamamoto T., Hiramoto K., Koyama A., Okada T., Morimoto E., "Development of Laser Profile Meter." SICE 86 ,p.479-480 (in Japanese)

2. Katuo Seta, Tdano Oh'ishi, Distance Measurements using a Laser Diode and by Heterodyne Detection with an Avalanche Photodiode" SPIE vol 740 - Laser Diode Optics (1987) p.60-65

3. Kenichi Matsui, Yuzi Adachi, Michio Mukoukubo, "Width Gauge for Hot Heavy Plates by Scanning Laser Range Finder" SICE 88 ,p.601-602 (in Japanese)

4. Kenichi Matsui, Yuzi Adachi, "Width Gauge for Hot Steel Plates using Scanning Laser Range Finder" IMEKO 91

Reprinted from *Laser-Diode Technology and Applications II,* D. Botez, L. Figueroa, editors, Proc. SPIE Vol. 1219, pp. 446-456 (1990). ©1990 SPIE.

DIODE LASER RADAR: APPLICATIONS AND TECHNOLOGY

J. Doyle McClure

Boeing Aerospace/Electronics High Technology Center
PO Box 3999, MS 7J-04, Seattle, Wa. 98124-2499

Abstract

Diode laser radar (ladar) offers unique advantages for applications that require ultra-high range resolution or measurement data rates. This paper describes several such applications, which include high precision metrology, high speed imaging, and fiber optic systems for aircraft avionics and/or flight control functions. Typical ladar requirements for these and other aerospace applications are discussed.

Several alternative modulation waveforms and associated device technologies available for diode laser radar are described. These include pulsed intensity modulation (pulsed IM), discrete frequency intensity modulation (discrete frequency IMCW), linear chirped optical frequency modulation (optical FMCW), and intensity modulation with linear chirp of the modulation frequency (linear chirped IMCW). Ladar performance relations that determine range-measurement resolution, absolute range capability, and measurement time or data rate in terms of the critical design parameters are discussed. Such design parameters include laser chirp bandwidth, modulation frequency, optical coherence length, laser and receiver noise sources, optical power, and receiver/processor data rate.

Strengths and weaknesses of alternative ladar technologies are summarized.

2. Laser Radar Applications and Requirements

Various applications for diode laser radar that are especially relevant to the aerospace industry are outlined in Fig. 1. As is discussed in Ref. 1 and in Sections 3. and 4. to follow, the capability for high intensity modulation frequency and/or extremely wideband chirp of IM or optical FM for semiconductor diode lasers compared to other types of lasers makes this type of ladar especially appropriate for applications that require extremely high precision range measurement and/or high data rate.

A major ladar sensor application is the large number of high-precision fiber optic position sensors for both military and commercial aircraft "fly-by-light" systems. With the usual triple-redundancy requirements on flight-critical control systems, 75 to 200 such sensors are typically required for each vehicle. Use of these sensors in high gain automatic flight control systems generally imposes requirements for 10-12 bit resolution on position measurement and update rates of several hundred Hz (associated distance resolution requirements vary from 0.5-50 μm depending on actuator stroke length). These performance requirements, combined with the need for small size and compatibility with operation in harsh aerospace environments, underlies the high challenge of developing aircraft fly-by-light capability. An advanced diode laser radar system for such applications is described below, and is treated in more depth in Ref. 2.

Configuration control and/or remote inspection of space assets also offer significant opportunities for diode laser radar applications. These applications include both high precision and high speed sensing requirements. One such high precision application involves use of a scanning diode laser radar to "map" the shape of large, light-weight space-based millimeter-wave radar antennas for precise configuration control. Scanning ladars of the type discussed below with 10 μm measurement resolution and absolute range capability of 50-100 m would provide the basis for dynamic control of the geometric shape of such antennas to the precision required for high performance radars.

Remote or robotic inspection of spacecraft typically involves high speed imaging requirements with a lesser precision requirement (e.g., millimeter to centimeter range resolution at absolute ranges of a few meters to a few hundred meters). For example, a two-dimensional range imager that provides individual "frames" having 256 x 256 range "pixels" and a rate of 10 frames/sec requires a data rate of 0.65 x 10^6 range measurements/sec (maximum measurement time of 1.54 μsec). Both coherent FMCW and fixed frequency IMCW diode laser radars have demonstrated centimeter and higher range resolution at data rates in excess of 10^5 pixels/sec, but with restriction to operational ranges of a few meters (Refs. 3 and 4), and pulsed diode laser radars have demonstrated data rates in excess of 10^4 pixels/sec with centimeter range resolution at ranges of over one hundred meters (Ref. 5).

Automated manufacturing and so-called "smart" structures and "health monitoring" applications provide further opportunities for development and use of diode laser radar. Measurement of structural dispacements as small as 0.1 μm and velocities of 0.5-5 μm/sec for typical vibrational frequencies of 1-10 Hz are typically required. A part of these requirements can be met by technologies discussed in this paper, others may require development of alternative technologies.

Figure 2 shows typical requirements on ladar operational range, range resolution, data rate, and optical powers for a number of aerospace applications, including those discussed above. As shown by the figure, diode laser radar is particularly appropriate for applications involving short range (and optical fiber propagation paths), ultra-high range resolution, and/or extremely high data rates. Long range imaging and wind measurements generally require higher optical powers than are currently available with diode lasers. However, Ref. 6 reports an IMCW diode laser that can provide terrain images at distances of 300-400 m or more. Diode laser radar can provide ranging capability at much longer operational range if "cooperative" retroreflectors can be used as targets. As will be shown below, the use of extremely wide chirp bandwidth for FMCW diode laser imaging ladars has the potential to greatly increase the range capability with low optical power, provided that adequate optical coherence can also be achieved.

Lidars and velocimeters to measure true air speed and wind velocities associated with gusts, clear air turbulence, etc., generally require higher laser optical power and better coherence properties than are currently available with commercial diode lasers. However, Ref. 7 reports use of short pulse (1-2 nsec) diode lasers operating at 75 W or higher peak power and very low duty cycle (average power) for aerosol measurements, and we are evaluating means to achieve single mode operation and high frequency stability of broadstripe diode lasers that can provide high average optical powers for coherent lidar/ladar applications.

3. Alternative Ladar Types and Waveforms

Figure 3 shows a number of laser radar waveforms, all of which have been demonstrated with diode lasers. This section includes a general discussion of these various types of ladar, and also provides more detailed descriptions of specific examples of coherent FMCW and chirped IMCW ladars. Section 4 to follow presents quantitative performance relations required to design ladars to meet specified performance requirements.

The pulsed IM ladar approach illustrated by the upper left inset of Fig. 3 is extremely simple conceptually. A short laser pulse is transmitted to the target and the reflected return pulse is detected on a photodetector (required transmit/receive optics not shown). A high speed reference clock or counter permits measurement of the two-way optical transit time to the target, which gives the target range for known propagation speed.

Typical high speed diode lasers and photo detectors permit use of 1 nsec or less pulse durations which can provide transit-time measurement accuracy of order 100 psec or less (limited, in part, by typical counter rates of 10-15 GHz), and corresponding range uncertainty of a few centimeters or less. Use of shorter pulses for improved range resolution is limited both by digital counter speeds, and by the reduced energy contained by short laser pulses which reduces the measurement signal-to-noise ratio (SNR).

As noted above, a pulsed IM lidar using a high peak power diode laser has been reported for aerosol measurements. Coherent and incoherent lidars for wind measurement have generally been developed for longer range applications which require longer pulses, higher power, and/or better coherence levels than are currently available with diode laser transmitters.

The discrete frequency IMCW technique shown at the upper right of Fig. 3 is also extremely simple in concept. For diode lasers with 10-20 GHz intensity modulation frequencies, the variation in phase delay of the sinusoidal photodetector signal with range permits measurement of transit time to less than 0.1 psec resolution , or 20-30 μm range resolution (Ref 8). This range resolution is many orders of magnitude higher than can be achieved with simple pulsed systems.

However, use of a single IM modulation frequency does not permit determination of values of absolute range larger than 1/2 the modulation wavelength due to phase "ambiguities", nor does the single frequency IM support ranging from multiple targets. Use of multiple discrete frequencies, transmitted either simultanously or sequentially, permits application to these more general cases, but with increased complexity of device hardware or reduced data rate. Design requirements for high precision, single-target/relative-ranging applications are discussed in Section 4 to follow. The fixed frequency, **coherent optical** counterparts of this fixed frequency IMCW technique are, depending on the application, variously known as Doppler velocimetry or optical interferometry which are not discussed here.

The linear chirped IMCW ladar approach illustrated at the lower left of Fig 3 corresponds to RF mixing of the frequency chirped output of the photodetector with a chirped local oscillator signal taken from the chirped RF laser driver. With appropriate filtering, the signal from the RF mixer is simply the difference-frequency term of the mixer inputs, i.e., a sinusoidal "heterodyne" signal whose frequency is $B/T \times \delta\tau$ where B is the chirp bandwidth, T the chirp duration, and $\delta\tau$

is the two-way delay time which is linearly proportional to the range. As is discussed below and in Ref. 2, high precision measurement of this heterodyne frequency (and thus the two-way time delay) provides the means for high resolution ladar range measurement. This technique provides both absolute ranging capability and the capability for simultanous multi-target ranging (each target return generates a different heterodyne tone whose frequency is proportional to the range to that target).

The coherent optical FMCW ladar illustrated at the lower right of Fig 3 is, in many respects, very similar to the chirped IMCW approach described above. In the FMCW case, however, the optical frequency is chirped, a fraction of the optically chirped laser beam is used as the local oscillator, and the heterodyne mixing of the local oscillator and return beams occurs at the photodetector. As with the IMCW case described above, the heterodyne frequency varies linearly with range, thus measurement of the frequency provides the desired range information. With diode lasers, the optical FMCW chirp bandwidth can in principle be many tens to hundreds of GHz which, as described in section 4, gives the potential for major performance benefits compared to the 5-10 GHz bandwidths that are practical for the chirped IMCW approach. However, the difficulty of developing coherent, wideband chirped FMCW diode lasers and the sensitivity of the technique to other application factors that result in coherence degradation limits the utility of the FMCW approach, at least for near-term applications.

A more complete schematic diagram of an FMCW diode ladar is illustrated by Fig. 4. This concept utilizes a "compressive" receiver/processor based on surface acoustic wave (SAW) devices and is appropriate for high speed imaging applications using laser chirp durations of a few microseconds, data rates of several hundred thousand pixels/sec, and moderately high range resolution (as noted below, a digital-processing receiver is required for ultra-high range resolution, albeit at much lower data rates than can be achieved with the compressive receiver).

The principal of operation of the compressive receiver is as follows. The heterodyne signal from the photodetector is mixed down to a convenient level with the output from a reference oscillator, then is combined in a second mixer with a down-chirped signal generated by an impulse generator and SAW expander. This chirped signal, which is effectively time-shifted by an amount that is linearly proportional to the heterodyne signal frequency (hence ladar range), is "loaded" into an up-chirp SAW compressor which outputs a transform-limited pulse at a time which is also linearly proportional to the heterodyne frequency and ladar range. A counter is used to measure the time of the output pulse, hence heterodyne frequency and ladar range (the complete compressive receiver is equivalent to a relatively high precision dispersive delay line processor). The FMCW ladar could alternatively be equipped with a high precision digital processing receiver, such as that discussed in connection with Fig. 5 to follow, for applications that require much higher range precision at lower data rate. Conversely, the SAW compressor just described could be used with the chirped IMCW ladar discussed in the following paragraph for applications that require high data rate with reduced measurement precision.

Fig 5 illustrates the concept for a chirped IMCW ladar system being developed by Boeing for high precision fiber optic position sensor applications (Ref. 2). The upper part of the figure illustrates a fiber network architecture in which the linear chirped IM signal is transmitted via a bus fiber and distribution network to a number of linear position sensors, each of which reflects a signal whose two-way time delay is governed by the sum of the fiber optical length and the free space distance to the retroreflector, which is placed on the actuator whose position is to be measured. With appropriate choice of fiber length to each sensor and filtering at the receiver, the individual heterodyne signal for each sensor can be separately processed to determine the frequency and corresponding actuator position. Reference reflections at the fiber ends can be used to null out effects of changes in fiber optical length due to temperature fluctuations etc.

The lower part of the figure illustrates an alternative fiber network in which the signals from individual sensors return via separate fibers to separate detectors (i.e., no signal multiplexing). In general, increased levels of multiplexing, as in the upper part of the figure, result in a lesser number of required components, but with increasingly demanding specifications on each. We are conducting the quantitative research required to optimize design of such systems as regards performance, reliability, development time, cost etc.

As is discussed in Ref. 2, high speed digital processor systems are required for the receiver of Fig 5 (or for a high precision version of the FMCW system shown by Fig 4). These processors are based on use of state-of-the-art processor chips to carry out high speed FFT operations, and are generally custom designed for each application to minimize the number of required processing steps which govern size, cost, and operating power of the receiver. The many processor design issues involve the characteristics and number of signals to be processed on individual receiver channels, the required precision and processing bandwidth, special purpose processing algorithms designed to take maximaum advantage of specific sensor characteristics, etc.

4. Ladar Performance and Design Considerations

Many factors govern the design and performance of ladar systems such as those illustrated by Figs. 4 and 5. The transmitter and receiver designs generally need to be "custom" tailored to meet specific application requirements. The performance relations discussed in the following paragraphs provide the basis for conducting ladar design/performance trades, and illustrate some of the principal issues and governing parameters.

Figure 6 presents a set of ladar performance relations taken from Ref. 1. The critical performance capabilities are calculated from the so-called Cramer Rao Lower Bound (CRLB) estimation relations shown in the upper part of Fig. 6 for both FM and IM linear chirped ladars (labeled FMCW), and for discrete frequency IMCW ladars. Detailed algebraic expressions for a critical signal-to-noise parameter, S/N_o, for the associated ladar return power (P_R), and for the background optical power (P_B) which are needed for application of the CRLB expressions are shown in the lower part of the figure for both coherent FMCW and direct detection IMCW ladars.

The CRLB equations for both chirped and discrete frequency ladars give the product of the expected variance (mean square statistical fluctuation) of the range uncertainty (δr^2) times the measurement or integration time-interval (δt) in terms of S/N_o and the FM/IM chirp bandwidth (B), or the discrete IM frequency (f_o). The CRLB provides a lower limit on the achievable range variance, which can generally be achieved by an unbiased estimator algorithm such as the FFT. These expessions clearly show the advantagous use of large values of B (or f_o) to achieve improved range resolution, reduced measurement time (required to support high data rate), or capability to operate with reduced S/N_o (e.g., resulting from operation at increased range with constant laser power or with reduced laser power at fixed range).

Fig. 7 illustrates results obtained from the equations of Fig. 6 for advanced diode laser radar applications. The curves labeled FM/CD correspond to optical frequency modulation with coherent heterodyne detection, and that labeled IM/DD to fixed frequency intensity modulation with direct (incoherent) optical detection. The IM/DD result was limited to a 2 GHz modulation frequency in order to permit advantagous use of a high power diode laser (P_o = 100 mW) and a high gain APD photodetector (G = 150). The FM/CD curves correspond also to a 100 mW laser power, a 10kHz laser linewidth, and a wideband PIN photodetector. The assumed single mode FM chirp bandwidth of 10-40 GHz is readily achieved by variation of the diode laser injection current. Other parameter values are typical of state-of-the-art components (see Ref. 1).

The ordinate of Fig. 7 is normalized such that the numerical scale is the measurement time (in seconds) for a high precision ladar with 0.001 cm resolution. Variation of required measurement times for other specified range resolution values, or variation of range resolution for specified measurement time, can also be read from the curves. The 10 GHz chirp FM ladar has about the same performance as the 2 GHz IM ladar at short range, but could support order of magnitude decrease in integration time (increase in data rate) at an operational range of 100m. The 40 GHz FM chirp reduces required measurement time by a factor of 16 for a fixed range resolution (or reduces range error by a factor of 4 for fixed integration time).

The curves of Fig. 7 clearly show the potential superiority of the coherent FMCW (FM/CD) approach, but the assumed linewidth of 10 kHz for a 10-40 GHz chirp bandwidth is far beyond the current state of the art. As is discussed in Ref. 1., the phase noise term shown in the FMCW S/N_o expression of Fig. 6 rapidly degrades the coherent ladar performance for larger laser linewidth.

The cause of the FMCW ladar performance degradation with increasing laser linewidth is graphically illustrated by the power spectra of diode laser heterodyne signals shown in Fig. 8 (see Ref. 9). The curve with the large spike corresponds to a (fixed frequency) laser whose linewidth was narrowed to a record low value of about 1 kHz (coherence length L_c is much larger than the effective round-trip range 2R, or coherence time much larger than the associated propagation delay time). The lower curve corresponds to a typical high quality commercial diode laser with a 10 MHz linewidth. For the narrow linewidth laser, the heterodyne signal power is about 42 dB above the receiver noise floor. For the commercial laser, the peak of the broad heterodyne signal (which corresponds to the "incoherent" beating of the laser signals) is about 15 dB above the receiver noise floor. Unfortunately, for ultra-high precision diode laser radar measurements, the heterodyne frequency must be measured to an accuracy of one to several Hertz, which is much smaller than the obtainable linewith. Consequently, the "incoherent" signal of Fig. 8 must be regarded as (phase) noise in the coherent FM ladar estimation and S/N_o equations of Fig. 6. Close examination of these phase noise effects shows that the coherent heterodyne signal decays exponentially with increasing laser line width, and that the phase noise rapidly increases at the same time. Thus, the linewidth of wideband FM diode laser must typically be reduced to 10 kHz or less to obtain the potential performance advantages of the coherent FMCW technique compared to the simpler direct detection IMCW approaches. It is relevant to note also that the results shown in Fig. 7 correspond to a smooth target surface that does not introduce speckle noise. For rough target surfaces, the speckle statistics would further degrade the coherent FM ladar performance compared to that for the direct detection IM approach.

5. Conclusions

As described above, CW diode laser radar generally provides capability for much higher range resolution (or data rate) for metrology, fiber optic, or object imaging applications than is possible with pulsed lasers. Both pulsed and CW lidars are required for atmospheric sensing, but available diode laser optical power and/or coherence levels are marginal for these applications.

Fig. 9 summarizes the relative advantages and disadvantages of coherent FMCW and (incoherent) chirped IMCW ladars. In general, the coherent FMCW approach has higher potential performance capability, but requires major advances in transmitter technology and, in certain applications, is subject to performance degradation due to coherence loss associated with motions of the target or propagation media. Thus, the simpler incoherent IMCW approach has significant advantages for some applications. Careful evaluation is usually required to determine the optimum ladar approach for specific applications, as determined by required performance levels, by required advances in component development, and by associated operational factors.

6. Acknowledgments

This paper has drawn on results from a number of research projects conducted at the Boeing High Technology Center. Individuals who have made major contributions to these efforts include Greg Abbas, Randy Babbitt, Ray Beausoleil, Mike deLaChapelle, Mark Fleshner, John McGarvey, and Ed Vertatschitsch. The work has also benefited from many discussions with Luis Figueroa.

7. References

1. M. deLaChapelle, J.D. McClure, E.J. Vertatschitsch, R.G. Beausoleil, J.G. Bull, and J.A. McGarvey, "Diode laser radar system analysis and design for high precision ranging", SPIE Proceedings on Laser Diode Technology and Applications, Luis Figueroa, Editor, vol. 1043, pp 228-237, 1989.

2. G.L. Abbas, W.R. Babbitt, M. deLaChapelle, M.L Fleshner, J.D. McClure, and E.J. Vertatschitsch, "High-precision fiber-optic position sensing using diode laser radar techniques", SPIE Proceedings on Laser Diode Technology and Applications, Don Botez and Luis Figueroa, Editors, this volume, 1990.

3. K.G. Wesolowicz and R.E. Sampson, "Laser radar range imaging sensor for commercial applications", SPIE Proceedings on Laser Radar, R.J. Becherer and R.C. Harney, Editors, vol. 783, pp152-161, 1987.

4. A. R. Slotwinski, "Utilizing GaAlAs laser diodes as a source for FMCW coherent laser radars", SPIE Proceedings on Laser Diode Technology and Applications,

Luis Figueroa, Editor, vol. 1043, pp 245-251, 1989.

5. R. Wangler, Private communication, Schwartz Electro-optics Inc., Orlando, Fla., 1990.

6. C.R. Hale, Private communication, Perkin-Elmer Electro-optics Group, Danbury, Conn. 1988.

7. W.J. Lentz, "High speed photon counting lidar", Proceedings of the OSA Topical Conference on Remote Sensing of the Atmosphere, Incline Village, Nevada, 1990

8. R.B. Hall, Private communication, Boeing Aerospace and Electronics Division 1989.

9. R.G. Beausoleil, J.A. McGarvey, R.L. Hagman, and C.S. Hong, "Narrow-linewidth semiconductor laser for FMCW ladar", Conference on Lasers and Electro-optics, Baltimore, Md., 1989.

- High precision measurement at high data rate
  - Fiber-optic sensor systems
  - Inspection/configuration control of space assets
  - Automated manufacuring/quality control
- High-speed imaging with high resolution
  - Automated systems/telerobotics for aircraft/space
  - Short/moderate range obstacle avoidance/target recognition
- Aircraft smart structures and health monitoring
- Air data sensors
  - Aircraft velocity, angle-of-attack, local winds

Figure 1. Diode Laser Radar Applications

Application	Range (m)	Resolution (cm)	Data rate (pix/sec)	Optical power (mW/W/kW)	Laser technology
• F/O Sensor – Smart Skins	3-100	0.001-0.1	100-1000	10-100 mW	Diode laser
• High Precision Metrology	3-100	0.001-0.01	1-1000	50-1000 mW	Diode Laser
• Remote/Automated Image System	10-1000	1-10	100-10^6	50-1000 mW	Diode laser (solid state or gas laser for longer range/lower data rate)
• Obstacle Avoidance/ Target Recognition	100-10^6	10-50	100-10^6	100mw - 100 kW	– Short range/high data rate - diode laser – Long range/low data rate - solid state or gas laser
• Atmospheric Winds	10-10^6	300cm - 150m	10-1000	1W-1 kW	– Short range air data – diode laser – Long range (500m-1000 km) – solid state or gas laser

Figure 2. Typical Laser Radar Requirements

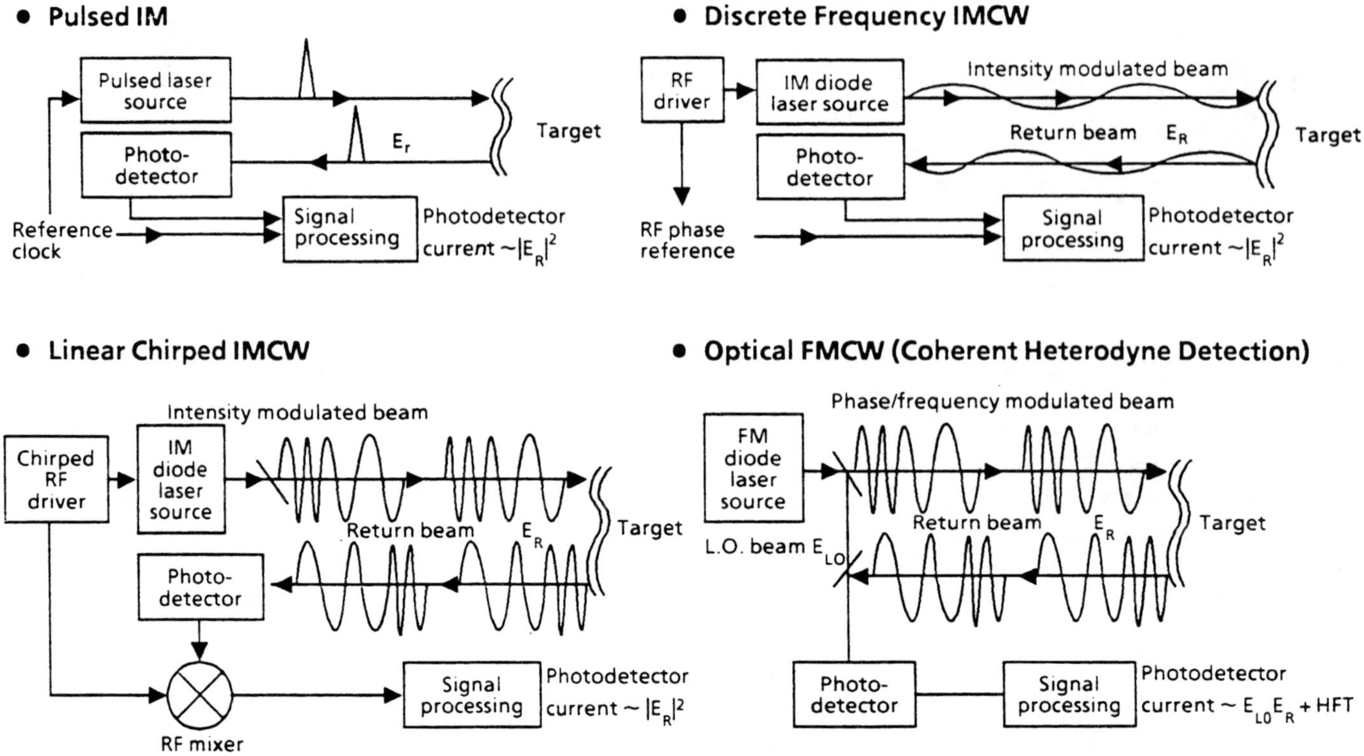

Figure 3. Alternative Laser Radar Waveforms

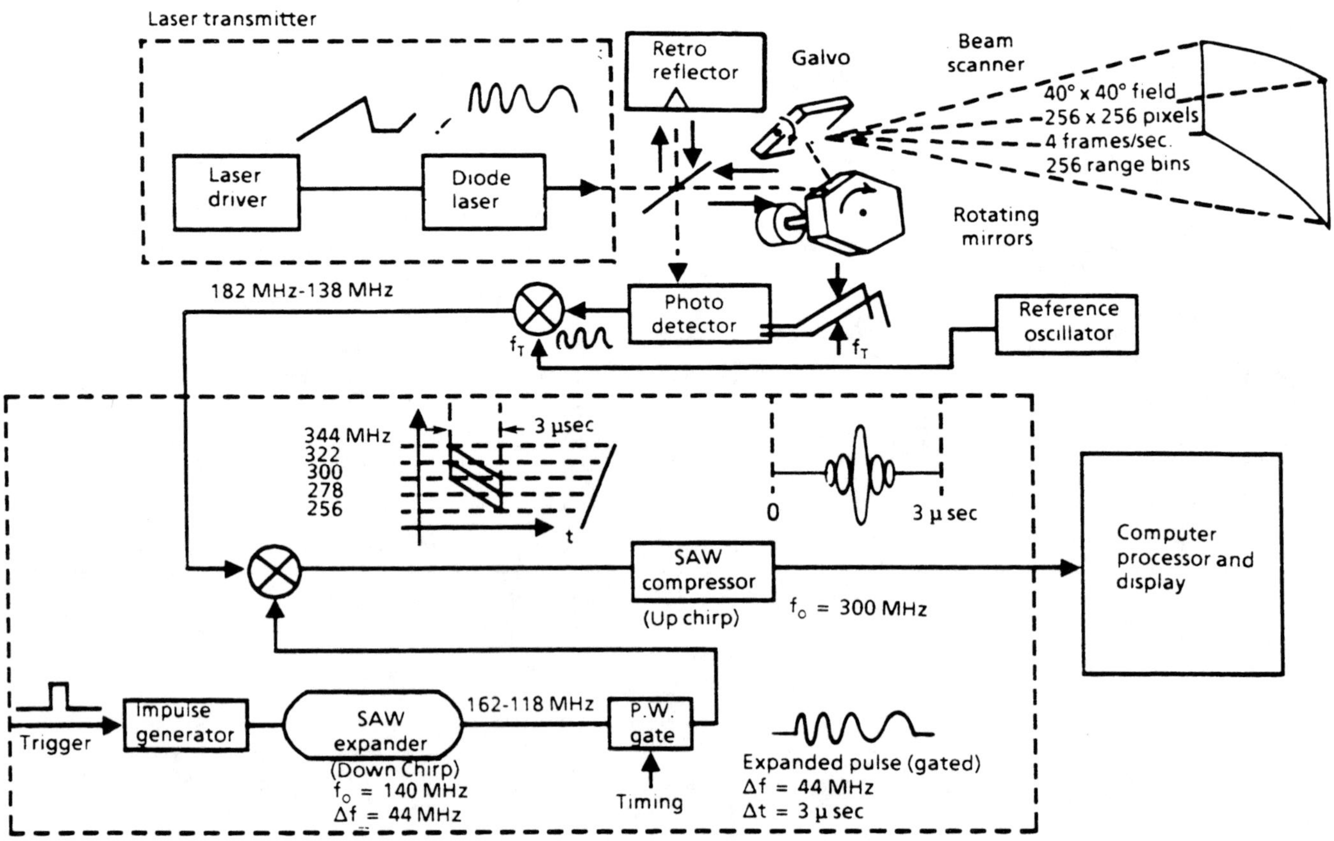

Figure 4. Coherent FMCW Ladar for High Speed Imaging

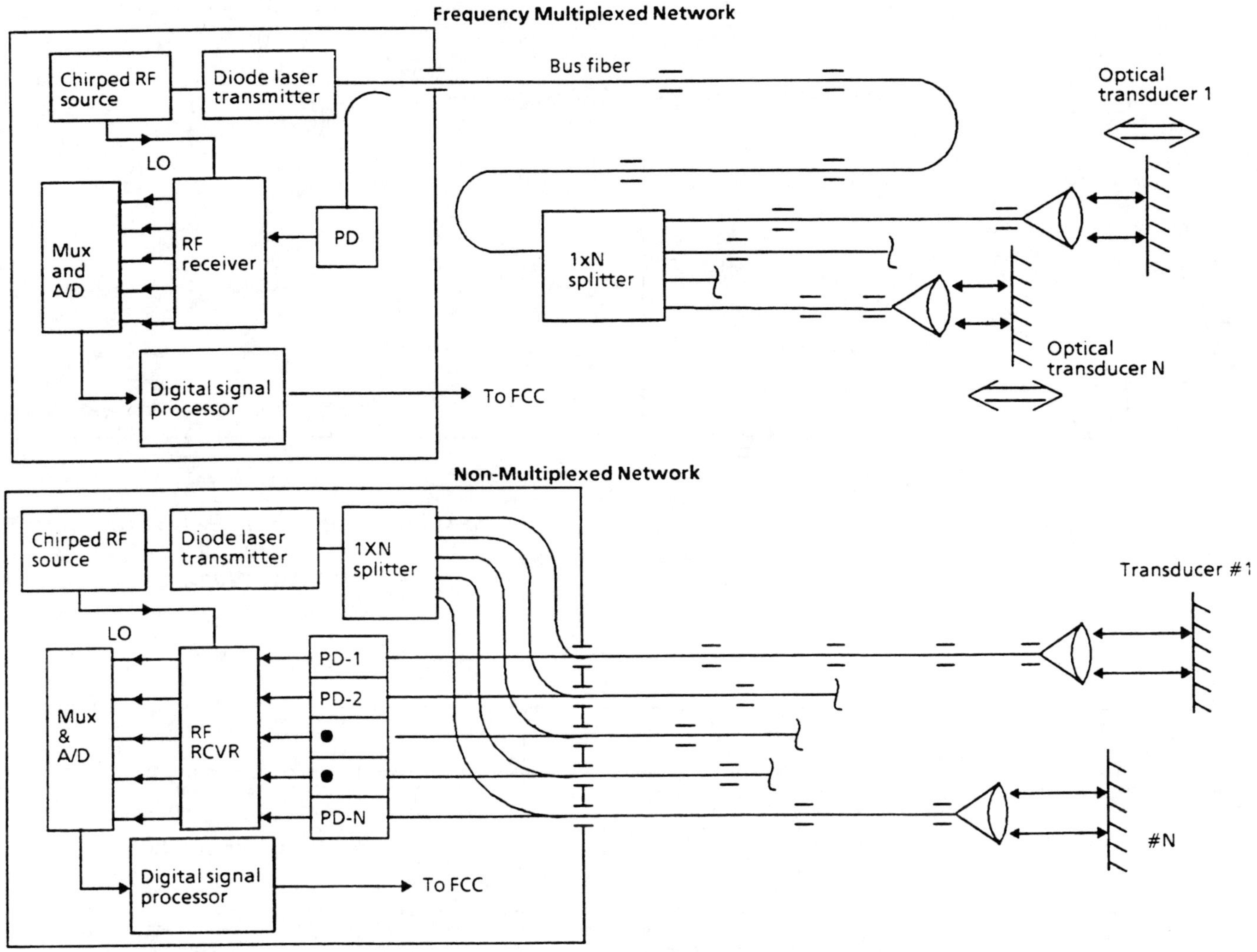

Figure 5. Chirped IMCW Ladar for Arrays of Fiber Optic Position Sensors

CRLB Estimation Relations:

FMCW	Discrete FrequencyIMCW
$\delta t\, \delta r^2 = \dfrac{3c^2}{4\pi^2 (S/N_0) B^2}$	$\delta t\, \delta r^2 = \dfrac{c^2}{16\pi^2 (S/N_0) f_0^2}$

Signal/Noise Per Unit BW-S/No:

FMCW	$\dfrac{m^2 r^2 P_R P_{LO} e^{-2\pi\Delta f t}}{e F_M m^2 [r(P_R + P_{LO} + P_B) + I_L + I_D] + \dfrac{2 K_B T}{R_o} + m^2 r^2 (P^2_{LO} + P^2_R) RIN + \dfrac{m^2 r^2 P_R P_{LO}}{\pi \Delta f} [1 - e^{-2\pi\Delta f t} (1 + 2\pi \Delta f t)]}$
IMCW	$\dfrac{\frac{1}{2} (m P_R r M)^2}{e F_M m^2 [r(P_R + P_B) + I_L + I_B] + \dfrac{2 K_B T}{R_o} + (m r P_R)^2 RIN}$

Optical Power:

$P_R = \dfrac{P_o \varepsilon_a^2 \varepsilon_o^2 \rho}{\pi} \left[\dfrac{d_r}{R}\right]^2$	$P_B = \dfrac{\pi}{32} \rho \varepsilon_a \varepsilon_o H I_f \dfrac{d_s^2 d_r^2}{R^2}$

Figure 6. Range Solution and Required Estimation Time for Alternative Ladars

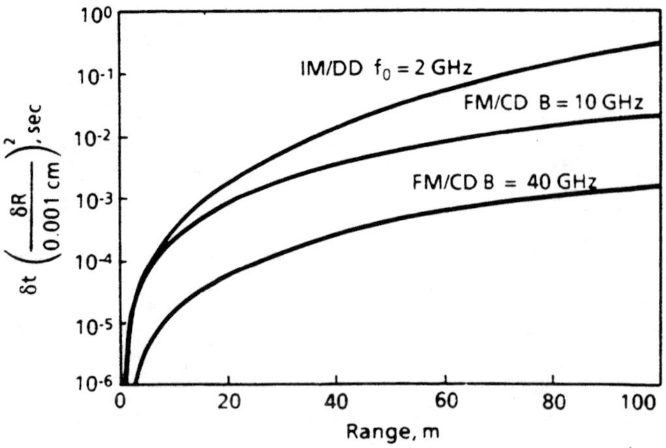

Figure 7. Range Resolution and Required Estimation Time for Alternative Ladars

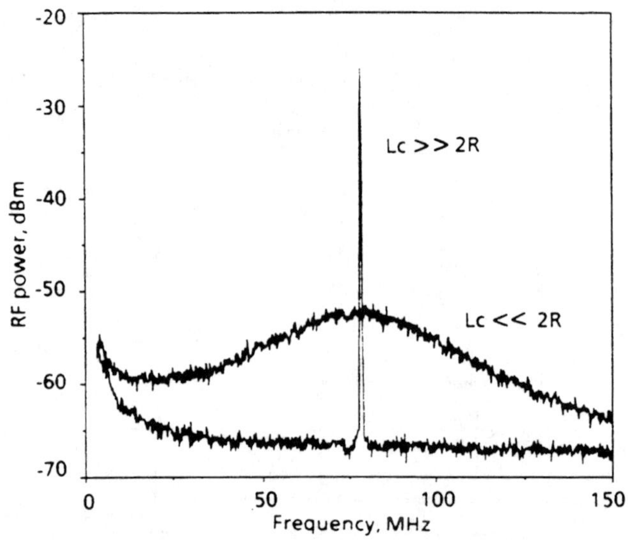

Figure 8. Effects of Diode Laser Coherence and Phase Noise on FMCW Ladar Signal Spectrum

	Coherent FMCW	Chirped IMCW
• Advantages	• Extremely wide chirp bandwidth (10-150 GHz) – Ultra-high resolution with low SNR – Ultra-high data rate at moderate resolution • High sensitivity with coherent optical detection – Resolution of weak targets at long range with low laser power	• High chirp bandwidth (2-20 GHz) – Good resolution/data rate trades • Reduced component complexity for near-term applications – No laser coherence requirement
• Disadvantages	• Requires complex laser transmitter development – Wide, single mode chirp bandwidth – High coherence (narrow linewidth) – High power (relative to conventional diode laser) – Stringent linearity of chirp • Subject to external coherence loss – Propagation path – Target motion	• Requires higher laser power(except F/O system) – Mode control and high frequency IM – Intensity noise/dynamic response • Requires high BW detection circuitry – Limits chirp BW/detectivity • Requires wideband RF laser driver development
• Summary	• Potential for extremely high performance with longer term development	• Potential for excellent performance with near-term development
	• Preferred approach is application-dependent • Parallel development reduces risk	

Figure 9. Summary of Factors Governing Choice of Diode Laser Radar

Section Three
Optical Techniques for Nondestructive Testing

Reprinted with permission from *Optics and Laser Technology,* Vol. 3(1), pp. 26-30 (February 1971).

Speckle pattern and holographic techniques in engineering metrology

J. N. BUTTERS & J. A. LEENDERTZ

Loughborough University of Technology, Loughborough, Leicestershire.

Some of the benefits of holographic interferometry in metrology are explored and related to difficulties experienced in practical methods. A new type of speckle pattern application is introduced which effectively samples the wavefront phase, enabling the interferometric measuring process to be carried out with low density data recording.

HOLOGRAPHIC INTERFEROMETRY has undoubtedly provided a new tool in engineering metrology which has been shown capable of use for precise surface measurements in circumstances where classical interferometry could never be carried out. The essential feature of the hologram is its ability to store wavefront phase information, and the process became practical with the development of highly coherent optical lasers producing a wave of known initial phase. When an object is illuminated with such light the phase distribution in the scattered wavefront is a measure of the surface geometry and metrological interpretation can be carried out in one or two ways, one of these being by the process of holographic interferometry. It is the purpose of this paper to point out in a preliminary form that another process for extracting required information is the use of coherent speckle pattern, and this is likely to show several advantages over holographic interferometry.

Speckle has not been well understood and in many cases treated as noise and something to suppress as far as possible. To date, a few workers have used speckle monitoring to detect surface stability and others have used speckle in spatial filtering to form a displacement sensitive system. A significant property of the speckle pattern, namely its ability to carry wavefront phase information, seems to have been missed until Leendertz published work at the I.C.O. Conference in 1969.

In contrast to this, applications of holographic interferometry have been progressively sought out, developed and used since the early discoveries in 1964. The hologram has promised great potential in engineering metrology but has been disappointingly slow in finding its way into system design and industrial application. In the authors' opinion this may be due to several basic application difficulties with holography, some of which can be overcome using the new speckle pattern methods. To establish this some of the metrological requirements must be considered together with the fundamentally different recording practices which can be used for holographic and speckle pattern interferometry. It would be desirable to carry out all data processing rapidly and preferably to avoid uncertainties of the photographic process.

AN APPRAISAL

For this purpose metrology is considered in its broadest sense of engineering measurement including processes where surface deformations are observed in the process of non destructive testing. The specific area of metrology to consider is that of optical interferometric measurement as is established for jobs such as the checking of slip gauges. Such measurements are basically attractive since they refer directly to the fundamental form of length standard which is expressed in terms of certain spectral wavelengths. There is good reason to re-define the standard in terms of a stabilised laser wavelength which would make possible multiple measurement with direct relationship to the standard. The limitation of direct application of classical interferometry is that the interfering optical wavefronts are formed from real geometric surfaces which have to be matched to a tolerance comparable to the wavelength of light. Such manufacture is only possible for simple geometric forms and practical application is almost entirely limited to the use of optical flats.

The basic process is not changed by the introduction of holography but the practical ability to utilise the method on any surface geometry is realised. The function of the hologram is here to record an exact replica of a surface of interest and then to use this as one element of a measuring interferometer, the real surface acting as the second element, interference effects being set up between them. Such an arrangement makes possible measurement or displacement between corresponding points in the surface as presented by the two elements of the interferometer. It can readily be shown[1] that this displacement d may be expressed in terms of the interferometer geometry and illuminating wavelength λ according to

$$d = n\lambda / 2 \cos\theta \ \cos\varphi$$

where 2θ is the angle between illuminating and viewing directions and φ is the angle between the displacement vector and the bisector of angle 2θ, n is an integer. This is simple in theory but in practice difficulties are faced in maintaining overall system stability throughout the measurement process. Since the basic measuring interval is $\lambda/2$, many experts would claim that relative positioning must be maintained to $\lambda/20$ for the instrument to be realistic in applications to metrology. In addition to maintaining this condition statically the process described with present holographic techniques requires the removal, processing and relocation of a photographic plate on which the hologram is recorded. The emulsion itself is far from stable as shown by Butters, Denby & Leendertz[2] and much of the precision is lost. The basic technique is not applicable to large displacements because the fringes become too numerous to resolve and accordingly the measuring system is severely limited.

Considerable improvements can be realised by using the so-called lapsed time techniques of holographic interferometry, but these require a full knowledge of circumstances in advance of recording. A full discussion of the applications of holography is given by Butters[3]. The permanent results are available after photographic processing in the form of an interferogram. No adjustment is possible and the optical system is equivalent to a common path interferometer. The results are precise and reliable and may be equally well applied for displacement and strain measurement as for vibration mode and amplitude plotting. The practical problems are reduced because many of the possible spurious system motions are cancelled out in the common path arrangement. The rig must however be stable to within $\lambda/8$ during the hologram recording process.

For metrological purposes the hologram is required to be of fairly high quality which demands a recording-medium capable of resolving about 3000 lines/mm. This is only attainable via photographic type processes. If one considers the basic information requirements of the hologram one finds that only the phase distribution in the recorded object wavefronts is significant. Most of the detail recorded on the photographic plate is therefore redundant. If the hologram could be sampled at low spatial frequency in such a way as to extract its phase information the metrological application could still be carried out but now a low definition record could be used. To date we have not discovered any satisfactory method for such a hologram sampling process, but we have observed and shown that the relevant information is carried by a speckle pattern. Such a pattern can be recorded at spatial frequencies compatible with electronic data recording and processing thus making possible an electro-optical system with the metrological facilities of holographic interferometry and with considerably greater practical convenience.

SPECKLE APPLICATIONS

Speckle pattern has almost certainly been qualitatively observed by all workers concerned with the use of lasers as optical sources. It can be observed without any special discipline of the eye and accounts for the 'frosty appearance' of stable objects illuminated exclusively with monochromatic laser light.

Speckle pattern are often associated with holography but there is no essential connection between the two processes. The association is almost certainly brought about because the aperture formed by any holographic viewing system gives rise to speckle effects in coherent light. This speckle represents noise on the reconstructed image and its removal for high fidelity systems has been the subject of considerable research effort particularly in micro-circuit applications.

A notable property of speckle patterns has been their use to enable an unaided eye to recognise whether or not a surface is stable, instability of a magnitude of 0.125λ being resolved. The effect is illustrated in that a human hand illuminated with laser light does not appear speckled but a stable metal object does. The significance of this observation is that a hologram can be formed from the stationary metal object but cannot be formed for a human hand. The rudimentary conclusion derived is that a hologram can be made from objects exhibiting speckle whilst this is not possible for objects which do not.

An application of this was realised at an early stage by Burch & Ennos at the National Physical Laboratory when they proposed a method of nodal detection in vibrating objects, based entirely on speckle pattern. Their work followed the discovery of the time-averaged hologram to display vibration amplitude contours. As their method did not involve phase processing the display was a rather imprecise nodal map giving no amplitude information. The principle was simply to show the stationary nodal regions as speckled areas and the remaining areas more or less uniformly illuminated. The work remained unpublished but was a starting point in the applications of speckle pattern in metrology. The actual size of the individual speckles in a speckle pattern is not uniform but the average size can be controlled by the viewing aperture dimensions. This can be easily observed by viewing a coherently lit surface through a small aperture formed between the thumb and finger. Some consideration of aperture size for most convenient viewing was given by Burch & Ennos and at the present time derivations of their interferometer are used as a kind of view finder for holographic vibration studies. This work is currently carried out by Stetson at N.P.L. who sets up his component for observation using real time viewing by a speckle pattern interferometry and when correctly set for a pertinent observation the time averaged hologram is formed. It is of course possible to set up a holographic recording rig and speckle pattern interferometer to view a single object at one and the same time.

Another area of application in which speckle patterns have been applied is that of spatial filtering and associated techniques. The principle is to record a reference speckle pattern and then to use the record as an optical mask. The processing concerned has some similarities to scatter-plate interferometry and was explored in some detail by Groh[4]. It was shown that by this method measurements of displacement and vibration could be made in terms of light intensity passing through the speckle mask. In addition to measurement of displacement some novel applications to non-destructive testing were proposed where a claim was made to detect fatigue at an early stage. The magnitude of motions was determined by calibration.

The next significant step in speckle pattern applications was made by Leendertz at Loughborough University of Technology. He showed at the International Conference on Optics (July 1960) that speckle pattern was not a completely random phenomenon as had previously been assumed, but was pseudo-random in a sense that although the speckle distribution could not be precisely predicted it was possible to extract information about the phase of the wavefront forming the pattern as a whole. This literally gave a new meaning to speckle pattern interferometry.

INTERPRETATION OF SPECKLE PATTERN

The general case of speckle pattern interferometry has been given by Leendertz[5]. When a non specular surface is illuminated with highly coherent laser light the phase of the contributions from the many scattering centres is not independent. Contributions from each scattering centre can be received over a wide viewing angle and within this angle contributions arrive with all phase angles which may differ significantly over a cycle between 0 and 2π. Complete multiples of $2n\pi$ are not significant at the present stage (n is an integer). The summation of the scattered contributions taken over all space simply gives the total scattered intensity but the phase condition gives distinct maxima and minima when the integration is confined to light emitted within a small solid angle.

A basic arrangement for seeing a speckle pattern is shown in Fig. 1. The viewing aperture is used to control the solid angle over which the surface scattered light is collected. A simple appreciation of the formation of speckle is to consider the summation of amplitudes at all points such as P = (x, y) in the viewing plane. Since the initial illumination is coherent the vectors representing contributions from each of the individual scattering centres will remain constant with time. They will be oriented at all angles between 0 and 2π (polarised light) in the plane of vibration. For a few scattering centres the vector diagram for point (x, y) may appear as an Fig. 2a, the resultant R giving the amplitude of light radiation at (x, y). In reality there are many scattering centres giving a diagram more like Fig. 2b. The intensity at a point in the speckle pattern is given by R^2 and this may vary between 0 and some maximum determined by the brightness of the laser. The intensity of this maximum is not important for speckle pattern interferometry as the information is carried as phase data.

If the surface in an arrangement similar to Fig. 1 is moved, the speckle distribution will change but one pattern will not be recognisable from the other by eye. An established method for making the system displacement-sensitive is to form a shadow mask photographically from one pattern, and measure changes in optical transmission arising from speckle re-orientation observed as a function of surface displacement. If the surface is arranged in such a way that the change in optical path of every scattering centre contributing to the mask is the same, the transmission will show a cyclic variation from zero to some maximum. When the phase of the light is

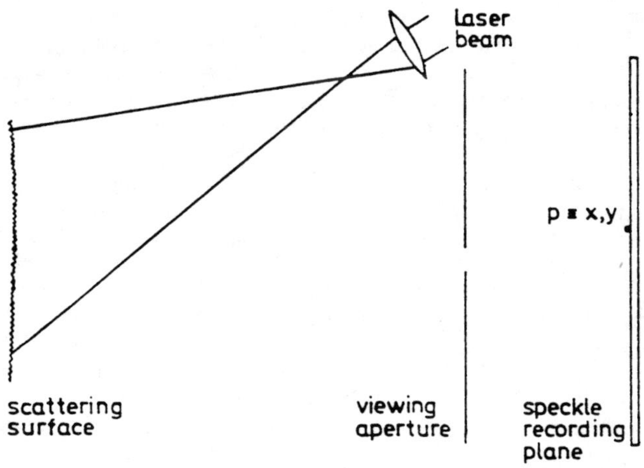

Fig. 1 Basic arrangement for viewing a speckle

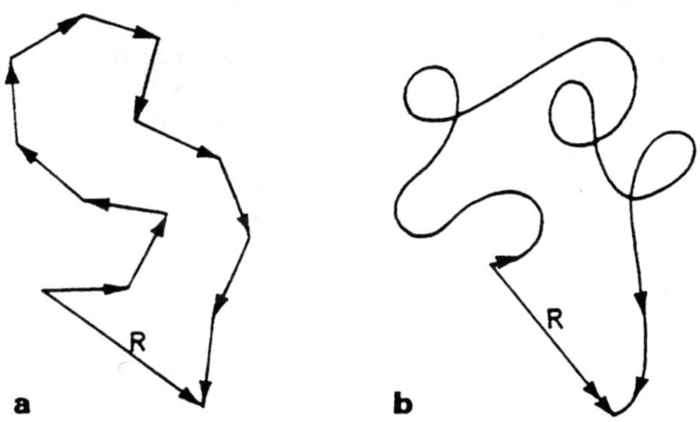

Fig. 2 (a) Possible vector diagram for the intensity at any point such as P(x, y), as a result of contributors from a few scattering centres.

(b) Possible development of vector diagram from a few scattering centres. In each case the intensity at P is R^2.

changed by 2π the speckle pattern for the displaced surface will fall exactly in coincidence with the speckle recorded on the mask giving rise to the zeros. The zeros can thus be used to measure the surface displacements in intervals of the wavelength of light as in ordinary interferometry.

The case described is rather special and it would not normally be true that all points on the surface had the same optical displacement. In the simple experiment described this would prevent the transmitted component from falling to zero. The situation can be conveniently corrected by focussing an image of the object surface onto the speckle pattern recording plane. This has the effect of forming speckles related to the localised surface conditions. If the previously described experiment is carried out with the focussing image, exact coincidence of speckles is experienced for all surface points giving optical phase changes of $2n\pi$ where ($n = 0, 1, 2, 3, ---$). These intervals of coincidence coincide exactly with interference fringes in holographic interferometry but at this stage are less easy to see.

The basic reason for observation difficulties is that the human eye does not easily recognise variations in spatial frequency, which is in this case variation in the number of dots per unit area across the field of view. To ease visualisation these spatial frequencies need to be converted to intensity variations. In practice this can be accomplished very simply by Fourier filtering which uses the laser source and an arrangement of lenses as shown in Fig. 3. The process is firstly to re-shape the laser beam to a parallel beam of sufficient size to uniformly illuminate the speckle pattern recordings. The recordings representative of the two states of the object are superimposed in the parallel beam where diffraction of the light takes place. The angle of diffraction at each speckle combination is proportional to the speckle size and hence the spatial frequency of the dots. When the light is refocussed by a subsequent lens the undiffracted component of light is concentrated exactly at its focus. The diffracted components come to a focus at different locations depending upon the spatial frequencies giving rise to them. If a circular stop is used at the focus of this lens to remove the undiffracted light the remaining light can be focussed by a further lens into a screen where an image in intensity distributions will be observed. The intensity distribution seen follow exactly the spatial frequency distributions in the original record but now a presentation easily appreciated by the eye is obtained (Fig. 4).

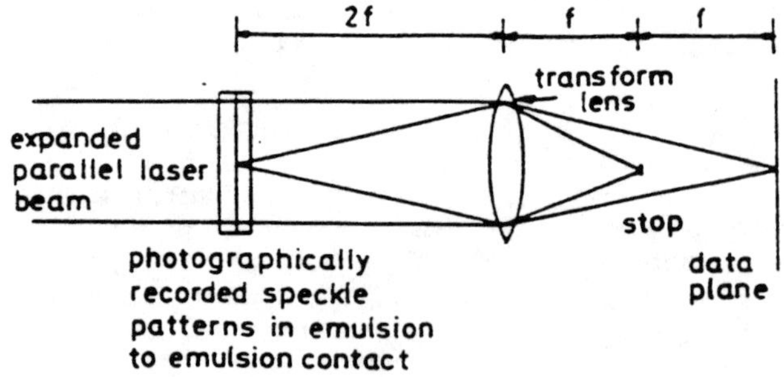

Fig. 3 Transform arrangement for comparing two speckle patterns and expressing the difference in terms of intensity variations.

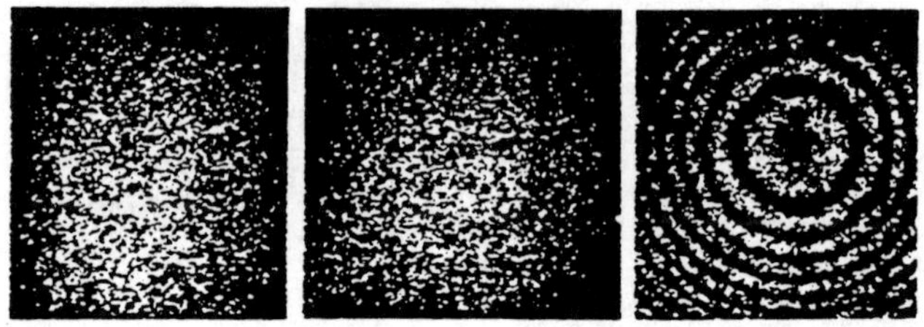

a b c

Fig. 4 (a) Two speckle patterns formed from a transducer diaphragm before and after pressure increment.

(b) Displacement fringe information obtained from the speckle patterns of (a).

COMPARATIVE ARRANGEMENT

In a general measuring system the process is based on the comparison of one object with another. For example if we desire to cut a one metre length of bar we first take a metre scale and cut off a length of bar corresponding to the distance between two marks on the scale. This principle is the basis of most engineering metrology. If we attempt to apply this process holographically, difficulties are experienced because the holographic process is sensitive to surface finish to a degree which is irrelevant to the engineer. The difficulties were highlighted by the work of Ennos[6] in an attempt to establish a routine inspection method for internal combustion engine cylinder bores.

The difficulties are less acute for speckle pattern applications because it is possible to form the mask with a speckle pattern obtained from combined portions of illumination received from the scattered wavefronts of two separated surfaces. One of the surfaces can be treated as a reference surface and held fixed. Changes in the other surface can then be observed relative to the first using the speckle pattern technique described. A diagrammatic arrangement of apparatus for this application is shown in Fig. 5. Component location difficulties are significantly eased by comparison with holography but so far the photographically recorded speckle mask has been assumed. Since a new mask is required for every pair of surfaces this is a disadvantage but in contrast to the hologram the speckle mask can be electronically stored. This is not possible with the high spatial frequencies of a hologram.

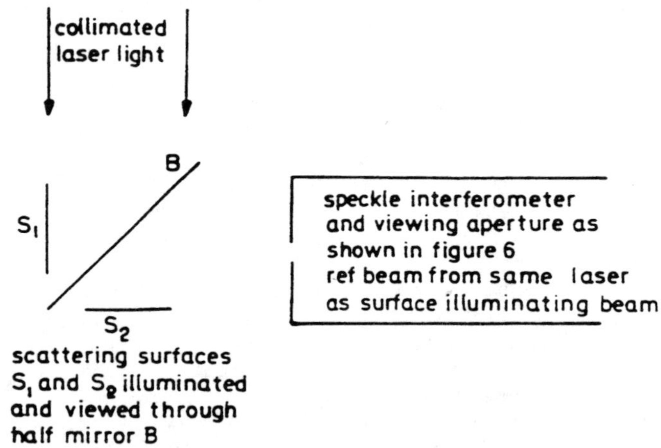

Fig. 5 Arrangement of speckle interferometer for comparative viewing.

VIDEO PROCESSING

With the processes so far described only one advantage over holographic interferometry is claimed and that is the use of low resolution photographic emulsions. The problem of photographic processing remains but fortunately the system can be extended to operate within the resolution capacity of an industrial vidicon giving some 350 × 175 resolved speckles.
In order to operate with static displacements either data store or delay line techniques must be used and investigations are currently in progress. The speckle pattern combination must still be carried out but now this is accomplished electronically together with the necessary filtering.

The initial investigations on the operation of the system have concentrated on applications with a time variable involving temporal frequencies instead of spatial frequencies and thus by-passing the need for data storage. Such an arrangement is obtained when speckle pattern interferometry is applied to vibration visualisation. A basic interferometer for this work is shown in Fig. 6 where a reference beam is introduced superimposed on the speckled image to satisfy measurement conditions. (Displacement measurements must be relative one point to another in a temporal frame of reference and accordingly a reference wavefront is required.) The image is scanned according to normal television practice and the resulting video frequency filtered to select regions where the optical phase change is some multiple $2n\pi$ ($n = 0, 1, 2, ---$). Fig. 7 shows typical vibration mode displays as photographed directly from a television monitor. The object was a circular disc excited with a crystal. The spurious fringes are parallel plate fringes formed by interference at the vidicon face arising from coherent illumination.

The display is exactly the same as would be obtained using time averaged holographic interferometry but in this case real time processing is involved. It is possible to watch the changing mode patterns with changing frequency just as with stroboscopic real time holographic interferometry, but there is no strobe frequency to synchronise and no switching frequency limit. Object location presents no problem and the vibrating object can be picked up during the test and adjusted if necessary. The modal pattern is immediately re-observed when the object is replaced in approximately the same position on the table. With present passive filters we are able to work up to the third amplitude fringe but we anticipate improvements with active filter arrangements.

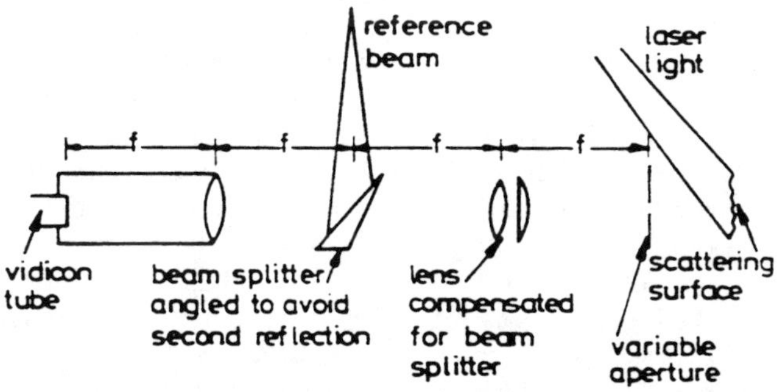

Fig. 6 Arrangement for speckle pattern interferometer.

SUMMARY

It has been established that many of the difficulties experienced in practical methods of holographic interferometry arise from the nature of the hologram recording and in particular from the high spatial frequencies involved. Much of the recorded information is redundant; the only metrological requirements being a record of wavefront phase. This information could be stored with sufficient accuracy using a coarsely sampled hologram but no satisfactory process for carrying out the sampling is known. A new facet type of speckle pattern application has been introduced which does effectively sample the wavefront phase enabling the interferometric measuring process to be carried out with low density data recording.

The early stages of application indicate that real time processing for displacement, strain and vibration measurement is possible using television type systems. Results are presented for vibration visualisation which give the essential modal information in a system suited to industrial application. Uses are likely to extend widely in metrology, inspection and non destructive testing.

REFERENCES

1 Ennos, A. E. (1968). Measurement of in-plane strain by hologram interferometry. J. Sci. Instrum. **11.** (2). 731-34

2 Butters, J. N., Denby, D. & Leenbertz, J. A. (1969). A method for reducing movement in holographic emulsions. J. Sci. Instrum. **2.** (2). 116-7.

3 Butters, J. N. (1970). Holography and its Technology. I.E.E. Monograph Series Chapt. 6. Cambridge University Press.

4 Groh, G. Engineering uses of laser produced speckle patterns. The Engineering Uses of Holography 1968 Symposium Report. 483-97 Cambridge University Press.

5 Lendertz, J. A. (1970). Interferometric displacement measurement on scattering surfaces utilizing speckle effect. J. Sci. Instrum. 3. (2). 214-18.

6 Ennos, A. E. (1967). Application of holography to the comparison of cylinder bores. J. Sci. Instrum. **44.**

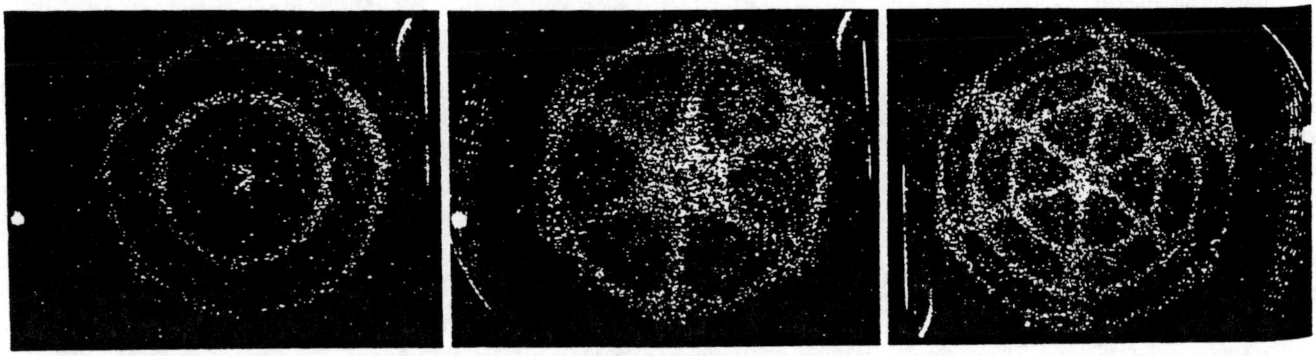

Fig. 7 Typical vibration mode displays.

Reprinted with permission from *Materials Evaluation,* Vol. 32, pp. 50-56 (March 1974).

Thermal Stress Studies Using Optical Holographic Interferometry*

By W. J. Harris and D. C. Woods†

Abstract

The application of holography to thermal stress studies is discussed. Interference fringes as produced by holograms and their interpretation are reviewed in relation to workpiece displacement. Three potential mechanisms are given to explain thermal displacement as detected by holographic methods. Results of some thermal stressing studies are reported, including tests on a live rocket motor.

Introduction

Optical holography is increasingly being used in nondestructive testing, and the use of thermal stressing for work-piece excitation has been a contributing factor.

An important advantage of thermal stressing is ease of application. Also, workpieces do not require fixturing prior to stressing. Thermal stressing periods often take only seconds involving temperature increments of 1 to 10°F (0.6 to 5.6K).

Since thermal stressing is dynamic, it is convenient to make observations in real-time. A continuous fringe motion is observed with real-time holography when thermal stressing is employed due to the continuous, and sometimes rapid, change in temperature. This fringe motion often makes interpretation difficult. This paper describes a technique for avoiding this problem.

The workpiece is heated to some temperature above ambient and allowed to start cooling. A hologram is recorded at some point after the cooling rate has decreased. Real-time observations are made at a later time when the temperature of the workpiece is below that at which the hologram was recorded. At this point, the cooling rate is negligible, and therefore the fringe motion is slow, permitting interpretation. This may be called reverse stressing to differentiate from the usual procedure of recording the hologram at ambient temperature and observing it when the workpiece is at a higher temperature.

There are basically three types of holographic interferometry: real-time, double-exposure and time-average. The results reported herein were obtained with real-time holograms. To provide a definitive understanding of holographic interferometry, a theoretical review has been included in the appendix. Special consideration is given in the review to a cylindrical geometry since much of the results involve a similar geometry.

Reverse Stressing

The usual procedure for applying real-time holography is to record a hologram of the undisturbed workpiece. The processed hologram is replaced in its original position, and observations are made while the workpiece is stressed. If the observer has control of the stresing and therefore the fringe motion, this technique works very well. However, in many cases where thermal stressing is employed, the continuous temperature changes produce very rapid fringe motion. Quite often the workpiece moves through the optimum stress point before the observer can photograph the data.

The fringe motion problem can be avoided if the reverse stressing procedure is used. With this procedure, thermal stressing is applied before the hologram is made. The hologram is subsequently viewed in real-time when the workpiece returns to ambient. (This technique was used to obtain results shown in Figs. 3, 5, 6, 7 and 8.

Differential Thermal Displacement of Separated Areas in a Bonded Structure

A theoretical model to explain deflection of unbonded areas produced by thermal excitation would aid in understanding and interpreting anomalies of holographic interference fields. A dependable model would also enhance the possibility of estimating success in related applications.

A promising model to explain deflection is thermal elongation. In this approach, unbonded zones are free to expand, linearly, but bonded surfaces are restrained

*Submitted for publication in March 1973.

†Mr. Harris is associated with the Lockheed Missiles and Space Co., Sunnyvale, Calif., and Mr. Woods with the National Aeronautics and Space Administration, Hampton, Va.

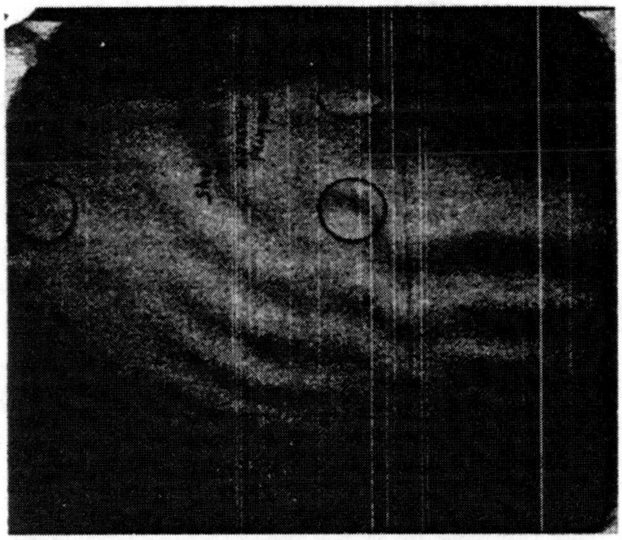

Fig. 1—View through real-time hologram of rocket nozzle having 1-in.-dia built-in disbonds, located by black rings, between casing and liner. Shrinkage of emulsion in developing hologram reveals casing convex distortion indication disbond before thermal excitation.

due to lower temperature or smaller coefficient of expansion. An analogy is the heating of a thin disc which is subsequently forced into an area of its original diameter. The deflection equation for this behavior is:

$$d = R\sqrt{2k\Delta T} - C\left(\sqrt{1 + 2k_1\Delta T_1} - 1\right)$$

where the second term is the expansion expected for the substrate of a cylinder of radius C, coefficient of expansion, k_1 experiencing a temperature change of ΔT_1.

In solving the above for a specific application, a solid rocket motor having a fiberglass casing and substrate of propellant, a displacement of 10^3 microinches (25 microns) results for $\Delta T = \Delta T_1 = 7°F$ (3.9K) and a diameter of 18 in. (0.456 m). Since the fringe resolution of real-time and reverse holograms ranges from about 20 to 300 microinches (0.5 to 7.7 microns), the estimated displacements are high by not quite two orders of magnitude.

A more realistic model of thermal elongation would be an expanded area having a fixed edge which should substantially decrease the predicted deflection.

A second model involves the deflection of a disc caused by a pressure differential produced from a temperature rise. In this model, the pressure, P_1, resulting from the temperature rise of the entrapped air in the disbonded volume exceeds the ambient pressure, P_0. However, this model was eliminated by a simple calculation. First, the elongation was computed from a known displacement, dp, the 1-in. (2.54 cm) diameter disbond shown in Fig. 1.

$$\Delta L = \Theta\,(r + dp) - L$$

Here, r is the radius and Θ the angle subtended by the disbond.

The pressure, P_1, required to produce the calculated ΔL can be determined from

$$\Delta L = \frac{L\,r\,P_1}{Y\,t}$$

where t = casing thickness and Y is Young's modulus.

The calculation showed that P_1, a differential pressure, was 15 to 18 psi (103.5 to 124.2 kN/m^2) over ambient pressure. This is much too great to expect for temperature rises of several degrees. For the temperature rises involved in thermal stressing, pressure increases of only 2 to 4 psi (13.8 to 27.5 kN/m^2) can be anticipated.

Another approach assumes a temperature differential through the casing thickness. In this case, a disbond acts not unlike a heated bi-metallic strip which deflects.

If the outer layer is a metal, maximum temperature differentials to expect are less than a few hundreths of a Kelvin for seconds of cooling time. The estimate of displacement for a typical metal, using $dp \approx \frac{L}{2}\sqrt{2k\Delta T}$, turns out to be about 50 microinches (1.3 microns) before considering edge restraint. However, for an epoxy matrix, such as fiberglass, 1000 microinches (26 microns) is computed.

But after observing displacement decay times, this model does not appear valid. For this model, calculations predict fringe anomalies, representing disbonding, will be visible for only several seconds, which is the time the temperature reaches equilibrium through the thickness of a thin-walled workpiece. Such indications actually persist for tens of seconds, which is approximately the time the entire workpiece cools to ambient temperature. This behavior, which has been observed for metals and nonmetals, supports the thermal elongation model.

Workpiece Preparation

In thermal stressing workpieces of less than 1 ft^2 (0.093 m^2) in area, the surface does not require a reflective coating. However, if spectral light presents intensity variance, a diffusive coating can be sprayed on the surface. Also, for large workpieces, a coating can be used to maximize illumination levels. A commercial aerospray, a penetrant developer, can be sprayed on in minutes and removed easily with a dry cloth. Figure 2 shows a rocket motor as viewed through a hologram plate holder, with a white coating of chlorinated hydrocarbons ~ 40 microns in diameter.

The coating also has the alvantage of providing a uniform emissivity, which minimizes hot spot formation that can generate confusing interference fringe anomalies.

Workpiece fixturing should be minimized. Any fixture has two purposes. One is to position the workpiece, and the other to eliminate cantilever vibrational modes. For thermal stressing, the workpiece must float to permit freedom of motion. If the workpiece has a compressive or tensile strain impressed on it by the fixture, this restricts thermal deflection.

Thermal Stressing Equipment

Hot Air

A manually operated hot air-gun creates sufficient temperature differentials in low-mass applications involving thin-gage materials such as honeycomb and composite-sandwich structures. Hot air guns providing air-stream temperatures from 100 to 125°F (311 to

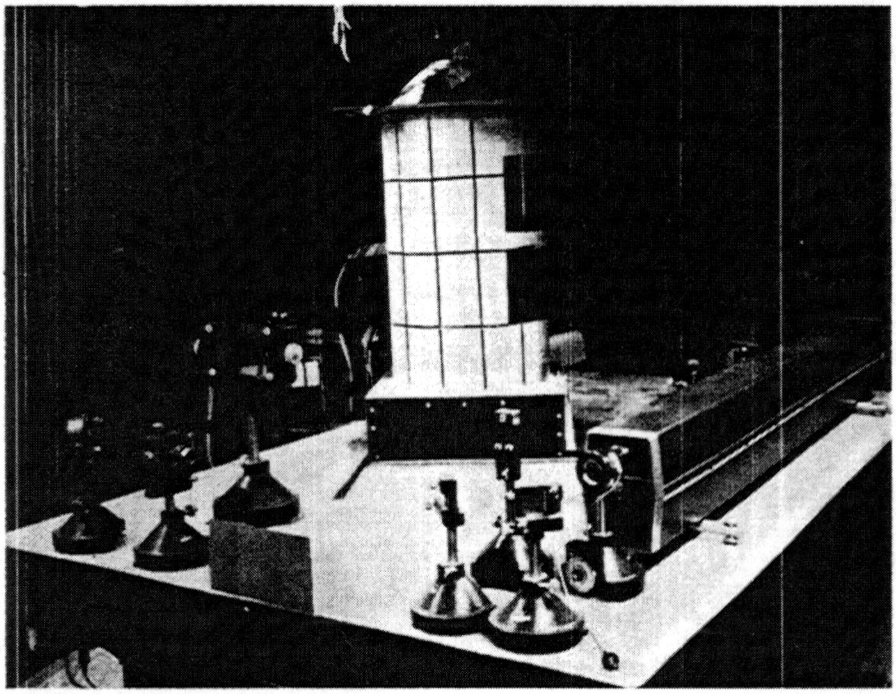

Fig. 2—Rocket motor shown mounted on air suspended holographic table for testing. Casing was sprayed with white coating before applying ¼-in.-wide black tape used for indexing.

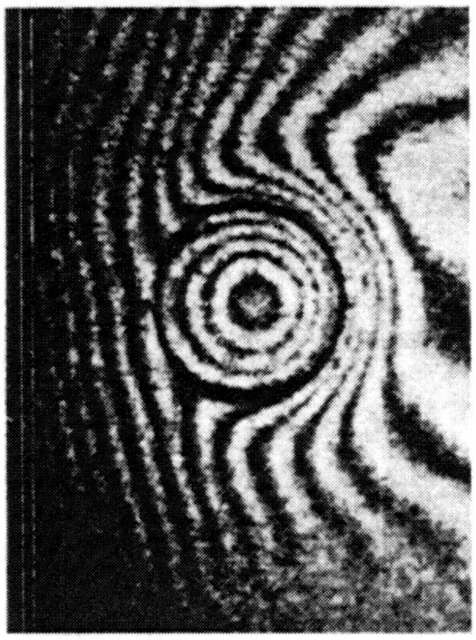

Fig. 3—Convex displacement of 1-in.-dia disband resulted from internal thermal stressing with a hot air gun.

325K) at a distance of 1 ft (0.30 m) are commercially available.

Care must be taken to prevent workpiece movement when directing hot air at the workpiece. In addition, a circular movement will reduce hot-spot formation, especially on the surface of poor thermal conductors.

Radiation

Quartz heaters offer an excellent radiative heat source. The energy can be uniformly distributed over a surface in seconds. Power levels of 1 to 3 kilowatts are adequate for most workpieces. The emissivity across a workpiece surface must be reasonably constant or a coating should be applied.

Evaporation

An alcohol spray can be used for cooling a workpiece surface by evaporation. This technique performs best in conjunction with reverse holograms. But, obtaining a uniform coating of alcohol can be difficult, making interpretation of fringe anomalies arbitrary.

Thermal Stressing of a Rocket Nozzle

Hot air was circulated inside the 6-in. (15.2 cm) dia rocket nozzle shown in Fig. 1. The outer casing, 0.150-in. (0.381 cm) thick D6AC, was bonded to a phenolic liner that varied in thickness from 0.25 to 1.25 in. (0.635 to 3.18 cm). A parting agent was applied for making the 1-in. (2.54 cm) dia disbonds.

To reveal the disbonds, it was necessary to heat the nozzle for 1 to 2 min with the hot-air gun. Both real-time and reverse holograms revealed convex displacement of the casing. The typical set of rings can be seen in Fig. 3. The convex displacement can be determined from the increased spacing between rings on the left, which was also to the left of the viewing center of the hologram.

An attempt was also made to flow hot air over the outer casing surface, but negligible straining and fringes resulted. The same was found for radiative stressing.

Radiative Stressing of a Solid Rocket Motor

A quartz lamp applied for periods up to 40 sec was successful on a rocket motor having a 0.055-in. (0.140 cm) fiberglass casing. The motor is shown mounted on a holographic table in Fig. 2.

This motor was 18 in. (0.457 m) in diameter and 30 in. (0.762 m) high in the cylindrical section. The motor was unrestrained, resting on the mount visible in the photograph.

The importance of the white coating can be appreciated by contrasting Figs. 4 and 5. Figure 4 was taken through a real-time hologram. Illumination of the entire cylindrical view was not possible with a 50-watt He-Ne laser. In addition, the cylindrical geometry precludes elimination of the spectral light reflection through the hologram. However, Figure 5 shows the view through a reverse hologram made after spraying the casing and indexing it with tape. Although the dark area at the top of the motor indicates a loss of coher-

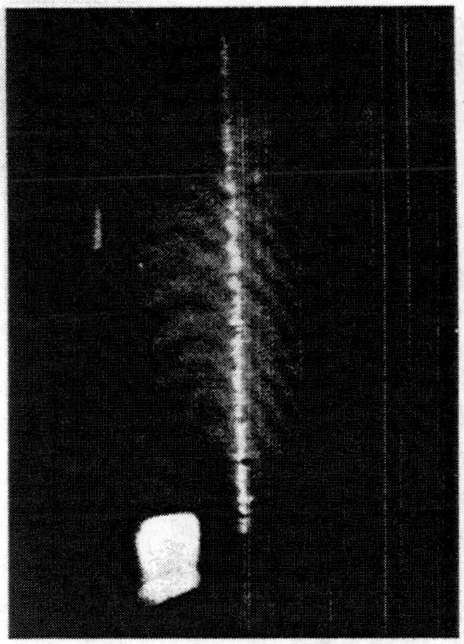

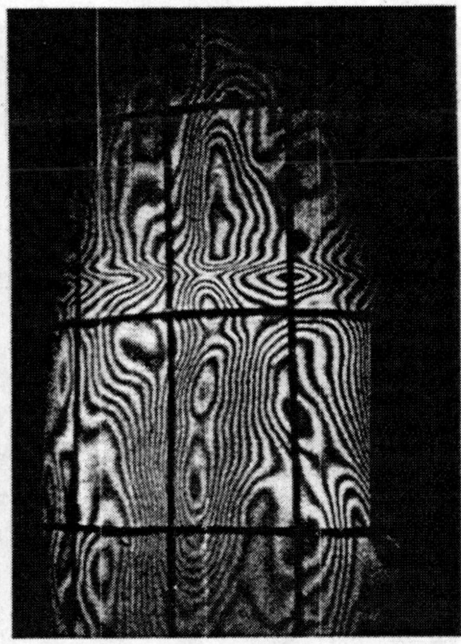

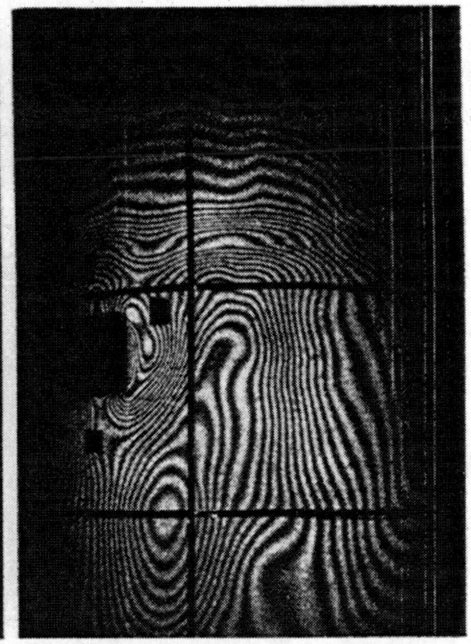

Fig. 4 (left)—View through real-time hologram of motor case prior to application of white coating. Square shape of light is a reflection from the plate holder. Fig. 5 (center)—Reverse hologram of rocket motor was exposed when the ambient temperature was at 80°F, and this photograph through it was made the following day when the temperature was 73°F. Fig. 6 (right)—Reverse hologram of solid rocket motor exposed 10 min after a 30-sec radiative stressing. The Polaroid above was taken 1 hour later.

ence, the entire casing is uniformly illuminated.

The myriad of fringe anomalies in Fig. 5 suggests subsurface defects revealed by the 24-hour reverse hologram. The bent and spread lines, as well as the rings, infer localized displacement. But looking at the interference patter in a general way, and forgetting the bent lines that show the termination of the liner which extends from the base between the fiberglass and propellant, two-fold symmetry is apparent. An imaginary line can be drawn down the center showing the pattern on the left to be similar to the right side. The same can be done for an imaginary horizontal line. In reverse holograms, the change in ambient temperature cannot be used as a thermal stressing technique. The workpiece displaces, but so does the holographic optics, and separating the two effects is not practical.

A more successful reverse hologram is depicted in Fig. 6. The summation of displacement in this figure occurred as a result of the case cooling approximately 1°F (0.6 K). Since the direction of surface displacement is perpendicular to a fringe line, the later displacement at the top of the motor is evident. The large rings represent a contraction after convex movement which was found in destructive testing to be a separation between the tapered rubber-type liner and the propellant. A composite of reverse holograms is depicted in Fig. 7. The concentric ellipses to the right of the separated zone again illustrate typical fringe behavior for a cylinder.

A real-time hologram (Fig. 8) revealed similar results as the reverse hologram. However, another set of rings to the right of the set corresponding to the reverse hologram rings indicated even a larger area of unbond. But these were discounted by plotting line density behavior in a real-time hologram upon cooling.

If the thermal elongation model is correct, then any relaxation of an unbonded area should occur exponentially over the period of cooling to room temperature. By graphing displacement with time, the set of rings on the right side of Fig. 8, which was the 210° positional of the motor, was discounted. The results are shown in Fig. 9.

The power output of the quartz lamp was 3000 watts. This converts to 715 calories/second. And for a 30-second period, the energy propagated amounts to 21,-450 calories. This was found to raise the casing temperature by 6°F (3.4K).

Thermal Stressing Other Materials

In general, radiative stressing seems to perform best on non-metals and convection heating on metals.

On honeycomb applications, radiative stressing works best if the skin is a thin gage, 0.005 to 0.050 in. (0.013 to 0.127 cm), metal. For heavy gages, convection or acoustical stressing should be considered. If the skins are non-metals, such as a fiber/resin systems, a quartz lamp can be applied for 5 to 10 sec. But fringe interpretation may be difficult with this material due to the large ratios of thermal coefficients that are possible between fiber and the resin matrix. The ratio can be as high as 10:1 of the resin to the fiber. Consequently, fringe anomalies may be indicating variations in resin content rather than subsurface defects.

The same thing that applies to honeycomb can be said for corrugation-stiffened structures and sandwich panels. One to two seconds of radiative stressing of multi-layer printed boards easily detects delaminations.

Conclusions

Calculations show that for thermally stressed workpieces, the principle mechanism of differential displacement is thermal elongation. This being the case, then

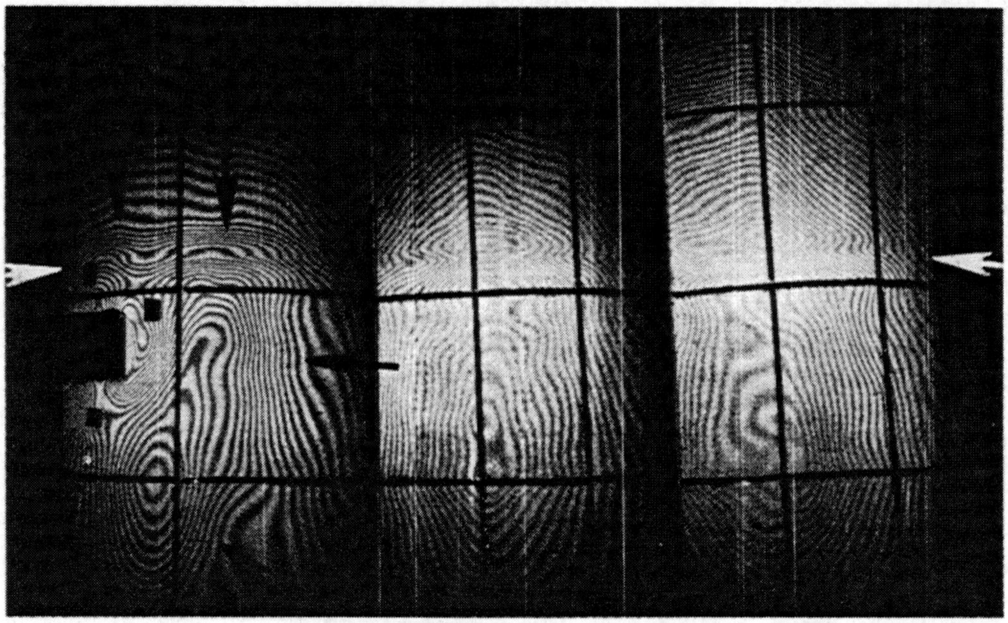

Fig. 7—Composite of photographs made through reverse holograms. Area defined by black arrows was subsurface separation. White arrows show plane of liner termination.

Fig. 8—Photograph through real-time hologram of solid rocket motor as it appeared 9 min after a 40-sec radiative stressing.

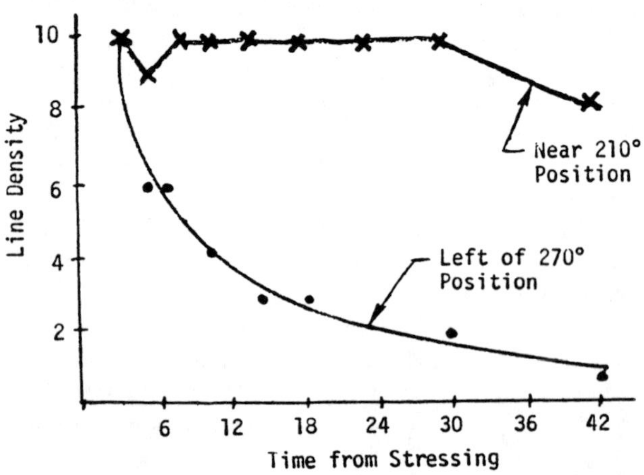

Fig. 9—Fringe line behavior with cooling time at two positions found to be separated at the line/propellant interface in destructive testing.

the greater the variation in coefficients of thermal expansion between a material and its substrate, the increased likelihood of success in revealing unbonding in holographic analysis.

To optimize illumination levels and eliminate spectral reflection from workpieces having surface areas of 2 to 3 ft^2 (0.186 to 0.279 m^2), a thin diffusive coating can be utilized. A commercial aerospray, a white penetrant developer, can be sprayed on in minutes and easily removed.

The technique of reverse stressing offers a static interference field for holographic analysis and photography of thermally stressed workpieces.

Radiative stressing of thin metallic or nonconductive workpieces, for periods under a minutes, provides sufficient thermal displacement to generate interference fringes for holographic analysis. For massive metal workpieces, a heat soak of a few minutes with a hot air gun is required.

APPENDIX

HOLOGRAPHIC INTERFEROMETRY

A hologram is a recording of the intensity and phase distribution that results from two interfering wavefronts, a reference wavefront and an object wavefront. The holographic recording contains both amplitude and phase information about the object which may be retrieved at a later time through the use of a coherent light source. If the hologram is illuminated with the original reference wave, normal image of the object may be seen by viewing through the hologram. The image is identical in every way to the original object in form, texture and position. Since the image is superimposed on the object, the two are indistinguishable. But, if the object is disturbed or moved by a very small amount (on the order of several wavelengths of light), then the object wavefront interferes with the image wavefront. Interference fringes will be observed just as in conventional interferometry.

The intensity distribution I (x, y) recorded on the photographic emulsion is given by: [7,8]

$$I(x,y) = |E_o|^2 + |E_R|^2 + E_o E_R^* + E_o^* E_R \quad (1)$$

where: $E_o = |E_o| e^{i\phi_o(x,y)} \equiv$ object wavefront,

$E_R = |E_R| e^{i\phi_R(x,y)} \equiv$ reference wavefront, and

$\phi_o(x,y)$ and $\phi_R(x,y)$ are phase functions and depend on position.

It has been shown that when the developed hologram is illuminated with the original reference wave, E_r, the transmitted amplitude wavefield is

$$E_r T(x,y) = E_r(t_o - KI), \text{ and} \quad (2)$$

$$E_r T(x,y) = E_r T_o - E_r K(|E_o|^2 + |E_R|^2 + E_o E_R^* + E_o^* E_R) \quad (3)$$

where: T (x,y) is the transmission of the developed hologram

t_o = amplitude transmittance of the unexposed plate

K = emulsion and development constant for a particular emulsion

I = intensity which is a function of exposure

Consider the 4th term on the R.H.S. of (3)

$$-KE_R E_R^* E_o = -K|E_R|^2 |E_o| e^{i\phi_o} = \alpha |E_o| e^{i\phi_o}$$

where: $\alpha = -K|E_R|^2$

The constant, α, times the original object wave represents the virtual image of the object.

The first three terms in (2) are zero order diffraction terms. The last term,

$$-K|E_R|^2 E_o^* = \alpha |E_o| e^{-i\phi_o}$$

is a constant times the complex conjugate of the object wave. This wavefront forms the real image of the object. It may be seen by turning the hologram around if a plane wave reference is employed. Since the term $\alpha E_o e^{i\phi_o}$ forms an image superimposed on the object, it is the only term in the equation that is useful for interferometry.

Let the wavefront representing the virtual image be written as $E_o e^{i\phi_{image}}$ where $|\alpha|$ is set equal to 1 for simplicity. The wavefront from the object is written as $E_o e^{i\phi_{obj.}}$. The total wavefield as seen by an observer, viewing the object and hologram virtual image simultaneously, is of the form,

$$E_o e^{i\phi_{image}} + E_o e^{i\phi_{obj.}} = E_o\left[e^{i\phi_{image}} + e^{i\phi_{obj.}}\right] \quad (4)$$

The intensity is given by

$$I = 2E_o^2(1 + \cos\Delta\phi \quad (5)$$

where $\Delta\phi$ equals the phase difference between object and image

When the image is precisely superimposed on the object $\phi_{obj.} = \phi_{image}$ and (5) reduces to

$$I = 4E_o^2, \quad (6)$$

there will be no interferences observed.

If the object is deformed or stressed so that $\phi_{obj.} = \phi_{image}$ the phase change $\Delta\phi$ results. The phase change $\Delta\phi$ is the factor effecting interference.

Considering the diagram below, it can be shown that

$$\Delta\phi = \frac{2\pi}{\lambda}\Delta l = \frac{2\pi}{\lambda} dp(\cos\beta + \cos\sigma) \quad (7)$$

where Δl is the optical path difference, dp is the object deformation and λ is the wavelength of light.

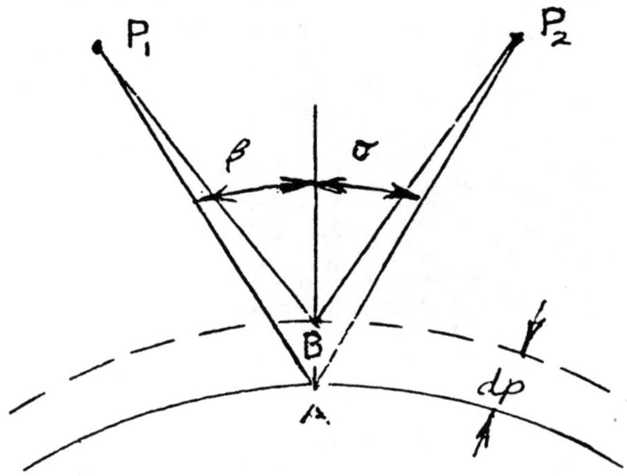

Substituting Eq. (7) into Eq. (5) gives

$$I = 2E_o^2\left[1 + \cos\left\{\frac{2\pi}{\lambda} dp(\cos\beta + \cos\sigma)\right\}\right] \quad (8)$$

The intensity is zero when $\cos\left\{\frac{2\pi}{\lambda} dp(\cos\beta + \cos\sigma)\right\}$ $= -1$, which implies,

$$\frac{2\pi}{\lambda} dp(\cos\beta + \cos\sigma) = (2n \pm 1)\pi, \; n = 1,2,3 \ldots \quad (9)$$

Solving (9) for dp gives,

$$dp = \frac{(2n \pm 1)\lambda}{2(\cos\beta + \cos\sigma)} \quad (10)$$

Equation (10) may be expressed in vector form as

$$\vec{dp} \cdot (\vec{N_1} + \vec{N_2}) = \frac{(2n+1)\lambda}{2} \quad (11)$$

N_1 and N_2 are unit vectors in the direction toward the source and the observer, respectively. The vector form represents the more general case in which the source point, P_2, the point on the object, A/B, and the observation point do not all lie in the identical place.

Zero intensity indicates dark fringes. Equation (8) can therefore be used to predict the fringe pattern from a know surface deformation, or by observing the fringe pattern one may determine something about the deformation.

Fringe patterns have been observed on a cylindrical surface by means of real-time holography while the cylinder was under thermal stress. The general fringe pattern, as shown in Fig. A1, approximates a set of concentric ellipses. This type of pattern is common for cylindrical surfaces under radial stress. There are a number of possible reasons for the deviations from perfect ellipses. Emulsion shrinkage contributes to a minor extent depending on geometry and number of fringes.

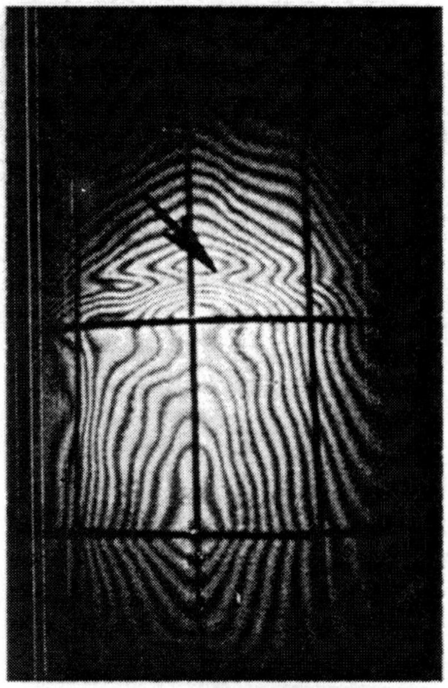

Fig. A1—Family of ellipses generated by a cylinder after an incremental surface displacement. Center of viewing through the hologram was above intersection of lower horizontal line and middle vertical line. Arrow points to bent lines indicating termination of liner.

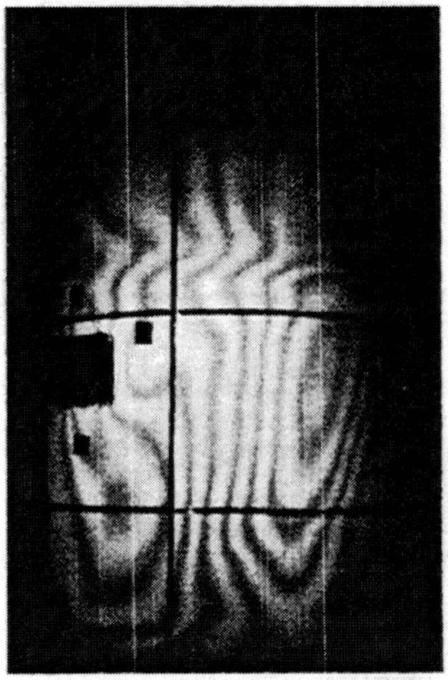

Fig. A2—Asymmetrical distortions in rocket motor cylindrical geometry are revealed in developed holograms. Bent lines, rings and ellipses suggest bulges in casing.

There may also be small thermal variations over the surface as well as the possibility that the surface may not be a perfect cylinder. Also, longitudinal as well as circumferential strains are present. These distortions are even more noticeable in Fig. A2. Distortions of this kind, while troublesome, are usually easy to distinguish from differences in fringe patterns due to anomalies in the surface having well defined boundaries. Figure 1 shows a circle which is clearly distinguished from the general pattern over the surface. The circle was caused by the presence of a 1-in. (2.54 cm) dia disbond between the steel casing and the phenolic liner of a rocket motor nozzle. The localized surface displacement resulted from longitudinal and circumferential strains in the casing due to contraction during cooling after the bondline adhesive cure.

References

1. K. A. Stetson and R. L. Powell, "Hologram Interferometry," *Journal of the Optical Society of America,* Vol. 56, No. 9, 1966, p. 1161-1166.
2. G. M. Brown, R. M. Grant, and G. W. Stroke, "Theory of Holographic Interferometry," *Journal of the Acoustical Society of America,* Vol 45, No. 5, 1969, p. 1166-1179.
3. L. O. Heflinger, R. F. Wuerker, and R. E. Brooks, "Holographic Interferometry," *Journal of Applied Physics,* Vol. 37, No. 2, 1965, p. 642-648.
4. K. A. Haines and B. P. Hildebrand, "Surface-Deformation Measurement Using the Wavefront Reconstruction Technique," *Applied Optics,* Vol. 5, No. 4, 1966, p. 595-602.
5. D. A. Evensen, R. Aprahamian, and J. L. Jacoby, "Holographic Measurement of Wave Propagation in Axi-Symmetric Shells," National Aeronautics and Space Administration, CR-2063, Washington, D.C., June, 1972.
6. J. P. Waters, "Holographic Inspection of Solid Propellant to Liner Bonds," *Applied Optics,* Vol. 10, No. 10, 1971, p. 2361.
7. R. J. Collier, C. B. Burkhardt, and L. H. Lin, *Optical Holography,* 1971, Academic Press, New York, N.Y.
8. J. W. Goodman, *Introduction to Fourier Optics,* 1968, McGraw-Hill, New York, N.Y.

HOLOGRAPHY IN A FACTORY ENVIRONMENT

ROBERT K. ERF, RONALD M. GAGOSZ AND JAMES P. WATERS

Abstract

A holographic technique, which has been developed for a particular inspection problem, should not have to be ruled out because holography is not adaptable to the factory environment. Specific investigations, to be reviewed herein, have established the tolerance of holography to levels of ambient illumination, vibration and media perturbations well above those existing in most manufacturing and maintenance facilities. Some special techniques have been developed to compensate for these deleterious effects in continuous-wave holography, but in general, pulsed holography offers a more versatile approach. The higher output energies available in closely spaced (10 to 500 microseconds), short duration pulses provide a capability for interferometrically recording the effects of a non-cyclic, or large amplitude, dynamic stressing of an object with a minimum of preparation. As examples, experimental investigations conducted in both a fatigue test laboratory and an experimental flight hangar at the operating divisions of United Aircraft Corporation will be described, and the results presented. Further, the use of dual reference beam techniques promises to provide an approach for employing double-pulse methods with the longer time interval (up to many seconds) necessary for interferometrically recording object deformation as a result of static loading.

1. Introduction

Adaptability of the holographic process to a physically hostile environment has been a question of continuing concern to the hardware oriented engineer who is constantly told of the virtues of the technique for facilitating many difficult measurement problems. In attempts to dispel his doubts, considerable effort has been devoted to the development of specialized methods to compensate for environmentally induced object motion during the relatively long continuous-wave (cw) laser holographic exposures. Foremost among these has been the use of local reference beam generation including: attachment of mirrors [1], or the holographic plate itself [2], to the object; optical processing of the reference beam [3]; and focused image [4] and speckle reference beam [5] methods. In addition, by sensing the object vibration, electronic feedback servo systems can be used with either phase[6] or frequency[7] modulation, to drive a reference leg mirror. Finally, scanning beam methods[8] have been developed which effectively provide for a reduced exposure time by virtue of an increased intensity. As an alternative to the use of these special methods, some have suggested the very real approach of providing special inspection areas isolated from the general environment, but nearby the facility where the component of interest is manufactured.

Pulsed holography however, with its extremely short duration high energy output pulses, probably offers the most straightforward approach for eliminating the deleterious effects of the factory environment. The short duration pulse (tens of nanoseconds) for single exposure holography, or brief pulse interval time (tens of microseconds) for double exposure interferometric holography, effectively eliminates problems of optical component or unintentional object movement due to vibration in almost all environments and, in addition, permits the use of shutters to prevent fogging of the film by the ambient light. Further, it freezes the motion of any surrounding media perturbations (density variations or particulate matter) thereby negating effects of random variations on the holographic phase fronts which can be troublesome in cw holography.

For vibration analysis work and/or HNDT with vibrational stressing, the time-averaging exposure requirement generally leads one to cw methods. However, even here, pulsed holography offers a potential solution. The use of non-Q-switched (open lase) ruby lasers for millisecond pulsed holography[9] provides a pulse short enough to negate the debilitating effects of environmental vibration, while time-averaging exposures of cyclically excited objects in the ten kilohertz range and higher, a frequency regime often utilized in HNDT. In addition, this mode of operation produces: (i) a significantly greater energy output for the same flash lamp power, providing larger surface area illumination; and (ii) a considerable decrease in the power density, permitting the use of conventional expander-spatial filter optics with a lower susceptibility to air breakdown effects.

For strain analysis work and/or HNDT requiring a static load stressing, the two exposures necessary for creating the interferometric hologram may be widely separated in time. This, together with application of the loading force itself, may result in displacement between exposures that is unrelated to the measurement, thereby precluding construction of a meaningful interferometric hologram. Dual reference beam holography, permitting fringe control in the reconstruction step [10], affords a method for obviating this problem. Thus, when taken in concert with pulsed lasers to eliminate ambient light and vibration effects in general, dual reference beam techniques have potential application in the factory environment.

Continuing investigations at UARL have been directed toward the development and demonstration of holographic techniques for application under non optical-laboratory conditions. Specific experiments, as reviewed in the following section, have been conducted to examine particular environmental parameters which can adversely affect the holographic process. The results of on-site tests, conducted at the Hamilton Standard and Sikorsky Aircraft Divisions of United Aircraft Corporation are then presented to graphically illustrate the capabilities of pulsed holography in the factory environment. Finally, the preliminary studies and the initial factory tests of a dual reference beam technique to compensate for rigid body displacement unrelated to the stressing force of interest are discussed.

2. Tolerance of pulsed laser holography to the factory environment

The pulsed laser studies to be reported herein have utilized: corporate developed and commercial lasers; low and high energy systems; Q-switched and non-Q-switched operation; active and passive Q-switching; and single and double pulsing modes. A typical low energy system had a coherence length of 1.5 meters, an output energy of 40 millijoules, and a pulse duration of 20 nanoseconds. It employed a 6 in by $\frac{1}{2}$ in diameter Czochralski-grown, ruby rod with Brewster angled end faces for the active medium. The laser cavity consisted of a dielectric-coated full reflector with a 10 meter radius of curvature and a Fabry–Perot etalon as a resonant reflector. In addition, a Pockels cell for synchronized, active Q-switching, and an aperture to control the transverse mode structure are mounted internally to the laser cavity. (The Pockels cell can be replaced with a bleachable dye cell for passive Q-switching.) The Pockels cell is driven by a high voltage power supply capable of delivering two negative pulses with pulse separations variable from 10 microseconds to a few hundred microseconds. The jitter in synchronizing to an external event is approximately 0.2 microseconds.

A typical high energy ruby laser system incorporates a ruby oscillator, producing 30–50 millijoules of energy, and one, two or three amplifiers to increase this value up to 4 or 5 joules. The commercial unit used in the UARL studies had a 3.5 joule capability, and a coherence length on the order of 5 meters. Pulse separations of 20 microseconds to 1 millisecond were available in the double-pulse mode, accompanied by a slight reduction in pulse energy (3.2 joules/pulse), with an individual pulse duration of 40 nanoseconds. Pockels cell Q-switching was utilized and pulse separations of less than 20 microseconds can be used with some decrease in both reliability and energy/pulse (approximately 2.7 joules with a 6 microsecond separation and perhaps 70% reliability).

2.1. *Ambient illumination compensation*

The inherent characteristics of pulsed lasers (high energy, short duration, single frequency output pulses) provide the necessary elements to eliminate the effects of film fogging, and subsequent loss of holographic information, due to high ambient light levels. Fast shutters, placed immediately in front of the holographic recording plane, synchronized to the laser output, and of sufficient duration to encompass both pulses in interferometric studies, are capable of holding film fogging to an acceptable level in the presence of several hundred foot candles incident on the photographic plate as reflected off the object.

The results of experimental studies, presented in fig. 1 illustrate a tolerance to ambient illumination of up to 200 foot-candles (a level described by the Health and Safety Code as hazardous, irritating and non-conducive to a good overall working environment). A 650 watt tungsten-halogen lamp was used to supply the ambient light. The holograms, whose reconstructions are shown, were recorded using a 75 millisecond duration shutter (conveniently achieved with the slight modification of a focal plane unit) and 40 millijoule laser with varying reference to object beam ratios; further, no preferential wavelength filtering was incorporated. This extremely excessive shutter open time (the interpulse interval would rarely exceed one millisecond), with more than adequate suppression of film fogging, conclusively demonstrates the negligible effect of ambient illumination, and emphasizes that the important parameter of interest is laser energy density at the film (a function of laser output, desired area coverage and object reflectivity characteristics). The object in fig. 1 was a one foot square aluminum plate and NBS resolution chart; both single and double-pulse recordings were made to permit examination of interferometric fringe contrast.

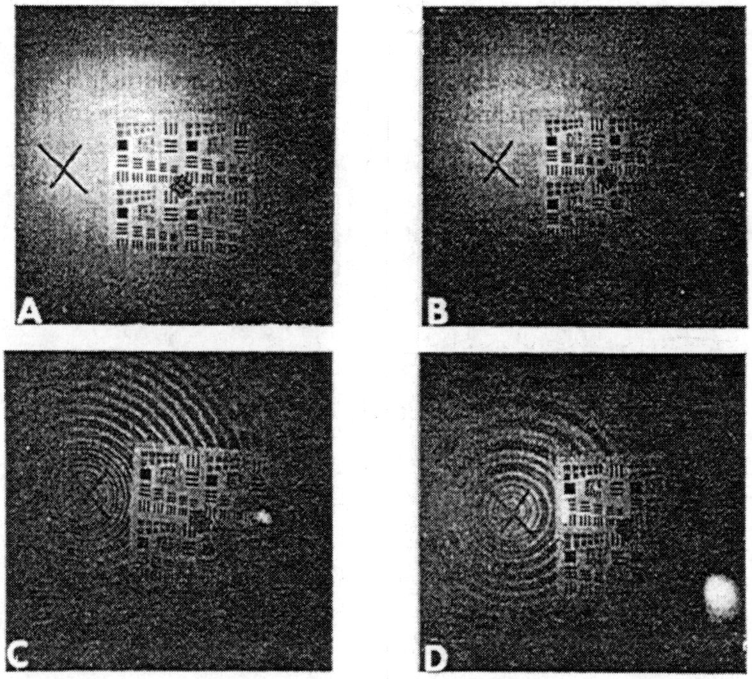

Fig. 1. Reconstructions illustrating the effect of (A) 0, (B) 30, (C) 150 and (D) 200 foot candles of ambient illumination, as measured at the photographic plate, upon the holographic process. Reference-to-object beam intensity ratios were (A and B) 5:1, (C) 30:1 and (D) 2:1. The double-pulse interferograms (C and D) were constructed with pulse intensity ratios of (C) 1:8 and (D) 1:1; excitation was provided by an acoustic pulse at the center of the 'X'.

For the double-pulse interferograms, acoustic excitation was used to provide a disturbance in the surface shape of the plate.

The monochromatic character of laser radiation often leads one to the use of narrow band interference filters as an apparently straightforward corrective measure. However, the angular dependency of the bandpass transmission characteristics of these filters can severely limit their application, particularly with regard to the angular separation between the reference and object beams in the holographic construction process. With especially intense ambient light, wider band red filters may be helpful, but are generally not required as suggested in the foregoing discussion of the experimental data.

Perhaps more germane to the problem is some brief consideration of the relative intensities of the holographic reference (R) and object (O)

beams. Hamasaki[11] showed that there is an optimum fraction of the laser power (greater than 50%) to be used for the reference wave in order to optimize the signal-to-noise ratio when reconstructing images from holograms constructed in the presence of a coherent source. The UARL studies, conducted with $R:O$ ratios varying between 40:1 and 1:20, showed that there is considerable latitude in this ratio, but that it should be maintained greater than unity. The hologram reconstructions of fig. 1 were recorded with $R:O$'s of 5:1, 5:1, 30:1 and 2:1 for figures A, B, C and D respectively. Of additional interest is the fact that when constructing double-pulse interferometric holograms, it is not necessary that the two pulse amplitudes be precisely matched. Although this was not a controlled variable in the experiments, the pulse energies were monitored for informational purposes; the energy ratios for figs. 1C and 1D were 1:8 and 1:1 respectively.

2.2. *Vibration compensation*

Again, the inherently short duration pulse of the ruby laser directly eliminates the effects of any mechanical vibration in the construction of a singly exposed hologram, immediately establishing its viability in the factory environment. Assuming a 20 nanosecond exposure, component velocities, along the optic axis, of up to approximately $\frac{1}{2}$ meter per second are tolerable without serious degradation of the reconstructed image. For interferometric work, however, the additional problems of unintentional object motion, and maintaning stability of position amongst the laser and holographic components themselves during the interpulse interval, must be considered.

Probably the most straightforward approach is the use of double Q-switching to generate both exposure pulses during a single optical pumping cycle of the laser. This permits selection of a time interval in the microsecond to millisecond range thereby avoiding any deleterious effects of vibration up to a few hundred hertz. Consequently, this mode of operation eliminates any restrictions placed on the holographic process by virtue of the physical environment existing in most industrial factory situations. However, for tests requiring an intentional stressing force, techniques must be provided for its rapid application; a sometimes impossible task. Consequently, test situations (e.g. application of a static load for strain analysis studies) not compatible with this double-pulse method do arise and, as with cw holography, special techniques must be contrived to cope with any inadvertent object or component motion.

One possible approach, to be discussed in a subsequent section, is the use of two reference beams in the construction step permitting application of fringe control procedures during reconstruction. The use of pulsed lasers for each of the individual holographic exposures, of course, still effectively isolates the object from surrounding vibrational influences.

One final item relating to vibrational effects, and alluded to earlier, is the necessity for good laser system design. It must be structurally stable and rigid in order to maintain the critical alignment required for reliable single frequency operation, while still being portable, since the large capital investment generally dictates multiple usage. Without such features, an inordinate amount of time would be consumed in re-establishing operational status after each relocation.

2.3. *Ambient media perturbations*

Another important consideration in the utilization of holographic techniques in an industrial environment is the state of the atmosphere through which the laser beams must propagate to construct the hologram. Of particular importance are the presence of scatterers in the air such as dust, smoke, etc. An obvious solution to this problem is the encasement of the holographic recording system to provide a uniform, non-scattering environment. Although this would not require as massive a structure as that required for vibration isolation, such a structure would reduce the portability of the system and also reduce the versatility. Such techniques (encasement) are quite effectively used for cw holographic work, especially to eliminate the effects of air currents even in relatively clean areas. However, pulsed lasers, again by virtue of their short pulse duration, eliminate deleterious effects of air motion, leaving particulate matter as the real problem to be considered.

The size of the aerosol will determine the particular effect it may have upon the holographic image quality. If the aerosol particles are less than 1 micron in diameter, scattering of the transmitted radiation will predominate. The effect should be to reduce the total transmission of laser energy to the photographic plate thereby possibly causing image degradation because of insufficient exposure. If the aerosol particles are of sufficient size (larger than 1 micron diameter) the particles can be resolved in the reconstruction, in addition to causing a loss of transmitted energy. Upon reconstruction, therefore, the particles will be seen causing an obstruction of the observed target.

In order to examine the effects of small (submicron) airborne aerosols upon the holographic image quality, a chamber was fabricated to house the holographic optical components and to contain the aerosols, which could be inserted and circulated to remove the effects of settling. The tests incorporated the use of cigar smoke, with a range of particle sizes approximating 0.01 to 1 micron. Even with smoke concentrations far in excess of the threshold limit value for oil vapors (5 milligrams/cubic meter, as established by the American Conference of Governmental Industrial Hygienists), quite satisfactory holograms could be constructed, although there was some loss in resolution and a noticeable decrease in contrast especially with concentration levels above 100 milligrams/cubic meter.

Experiments were also conducted to determine the effects of 1–100 micron diameter particles on the recording technique, and demonstrate the tolerance of holography to noise (i.e. light scattering centers which contain no target information). This work involved introduction of the particles into the holographic recording setup in discrete volumes (a cylinder approximately 25 cm high and 10 cm in diameter), rather than over the total volume, offering two important advantages: first, proven reliable seeding mechanisms could be used; and second, by utilizing smaller volumes, specific areas of interest in the optical setup could be examined separately, thus providing a means of evaluating the effects of the particles on different components. A simple, vibrating mesh screen was used as the seeder and cement dust was used for the particles. Dust particles were introduced in the target area, in the reference beam leg, and at the holographic plate. In addition to providing a means of isolating the effects of the dust particles, this technique of using a small seeded volume permitted greater uniformity of the particle density. In any one recording there were 0.25 ± 0.05 grams of particles in the enclosed volume giving a density of approximately 135 grams/cubic meter. This degree of particle concentration is generally higher than that which would be present in a typical industrial environment and is more than 100 times greater than the particle concentration reported to be typical in a dusty factory or mine-tunnel by the Mine Safety Appliance Company. Furthermore, it exceeds the lower explosive limit for airborne inflammable dust.

The effects of these dust particles on image quality are shown in fig. 2. The most ovbious effect of any particle on an image recording technique would be to either (i) scatter laser light back towards the

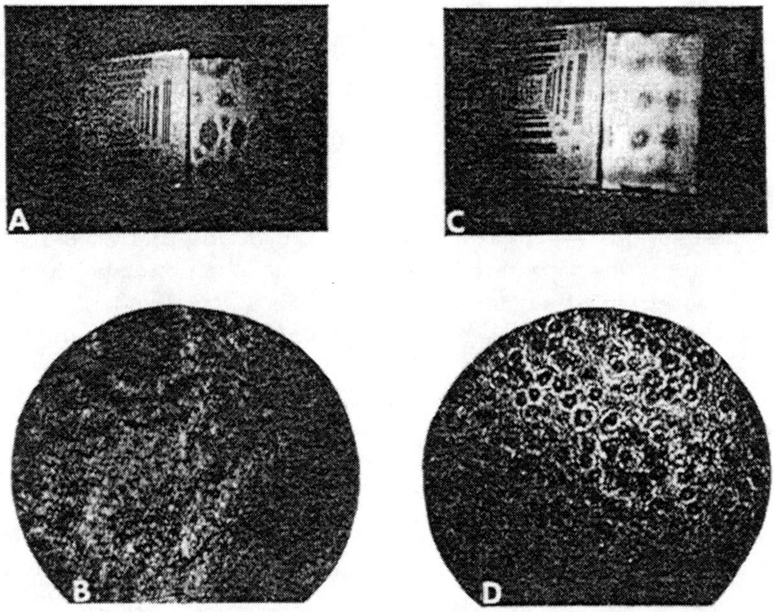

Fig. 2. Test results demonstrating the effect of 5–100 micron dust particles at the object (A and B) and holographic plate (C and D) upon holographic reconstructions (A and C) and plate exposures (B and D).

hologram thus reconstructing these particles; or (ii) shadow light from the object thus eliminating these portions of the target in the reconstruction. Both of these effects are in evidence here. With the dust at the target, a section of the reconstructed image is speckled with bright reflections from the particles. This section of the field was somewhat degraded (i.e. a loss of resolution) and for this reason the technique might be in question except that, for this concentration of particles, the field could not be seen with the human eye under normal lighting conditions either. With dust in the reference beam and at the holographic plate (the latter also shown in fig. 2), the plate was shadowed in part by the particles as seen in the plate exposure accompanying the reconstruction. The shadowing tended to make these holograms less efficient, but did not severely effect the resolution or clarity of the reconstructions.

3. Factory tests

The foregoing discussion has shown how the debilitating effects of the factory environment can be overcome primarily because of the short, 20–50 nanosecond duration of the pulsed ruby laser. Consequently, holography can be performed in a manufacturing or maintenance environment simply by installing a pulsed ruby laser system of sufficient output energy to illuminate the component of interest at its test site. Evidence of this, gathered at the Corporate divisions, is presented below to illustrate this capability. First, the study of a prototype propellor blade subjected to fatigue testing at the Hamilton Standard Division is reviewed; and second, application of double-pulse holography to a CH53 helicopter airframe at Sikorsky is discussed. The latter item also illustrates the large area coverage capability of the holographic process.

3.1. *Propellor blade holography in a Structures Fatique Laboratory*

Double-pulse holographic interferometry was successfully employed in a Structures Fatigue Laboratory at the Hamilton Standard Division giving evidence of its applicability to nondestructive testing in a physically hostile environment. The object under investigation was a prototype propellor blade, approximately 150 cm long and 45 cm wide, undergoing vibration fatigue testing at a frequency of 116.6 Hz with a peak-to-peak displacement amplitude of 0.57 cm at the tip. Successful recordings were made with the blade near its maximum stress points (*i.e.* position of maximum displacement) revealing structural nonuniformities.

A schematic diagram of the optical system as it was set up to record holograms of the propellor blade undergoing vibrational fatigue stressing is presented in fig. 3. The environment for the test was a conventional test cell and it was not necessary to institute any special procedures in order to perform the tests. The pulsed laser system employed was a Pockels cell, Q-switched ruby which included its own closed-cycle cooling and cw alignment systems. As shown in fig. 3, a thin glass plate was used to direct a sample of the beam to a pulse detector which provided a signal to synchronize the timing, and record the pulse separation. The main output beam was then divided by a partially reflecting mirror into the object and reference beams. The object beam was expanded by a negative lens and directed onto the blade at the area

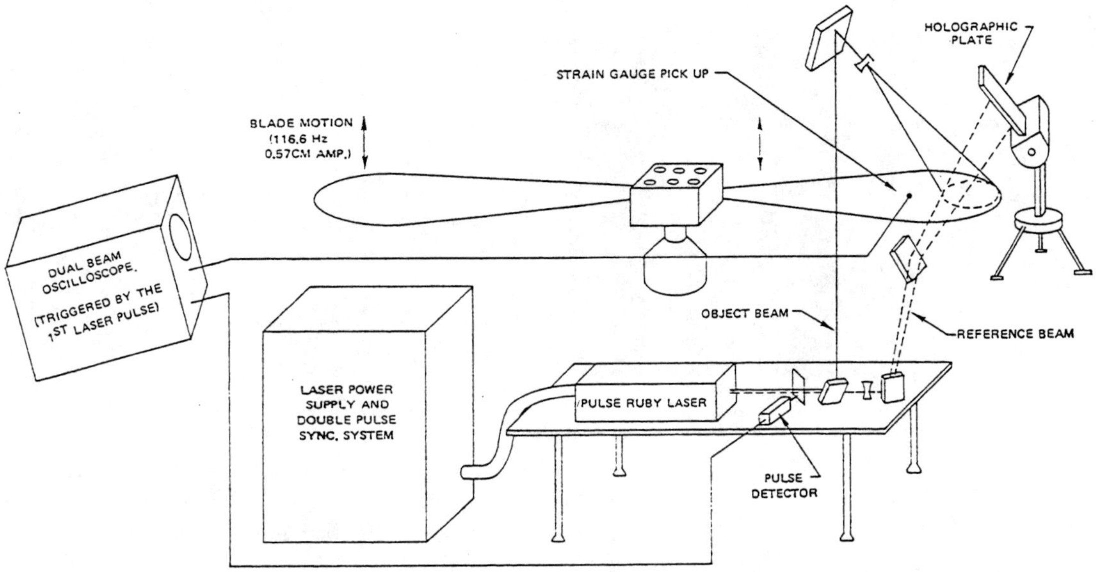

Fig. 3. Experimental apparatus for the recording of interferometric holograms of propellor blades as they are subjected to vibrational fatigue tests in a Structural Dynamics Facility.

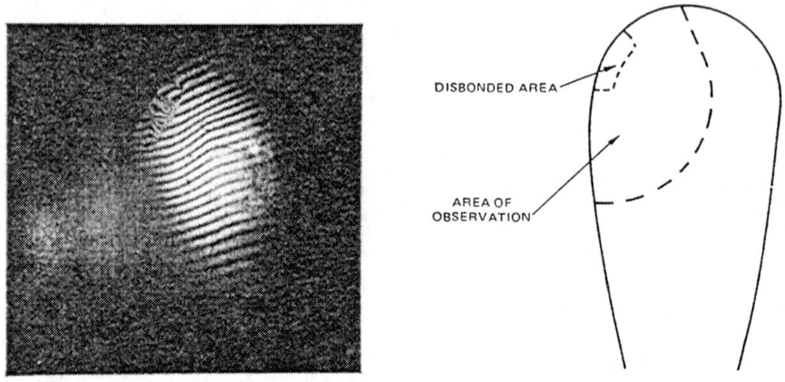

Fig. 4. Reconstruction and schematic diagram from a double-pulse interferometric hologram of a propellor blade undergoing cyclic fatigue at 116.6 Hz.

of interest, from which it was diffusely reflected to the holographic plate. The reference beam was also expanded with a negative lens and sent directly to the holographic plate.

Also shown in the schematic is a strain gauge pick-up which was used to determine the position of the blade when the exposures were recorded. The sinusoidal voltage output from the strain gauge (indicative of the blade displacement) was displayed on the upper channel of a dual beam oscilloscope, and the output from the laser pulse detector (beam sampler) was also fed into the scope for recording, and to trigger the sweep. Consequently, both the phase of the sine wave blade motion and the position of the two laser pulses were recorded for subsequent correlation with the holographic test results.

A typical test result is presented in fig. 4 showing the reconstructed holographic interferogram along with a schematic denoting the blade features. The oscilloscope recording for this test result showed that the recording was made near the blade's maximum displacement point. A laser pulse separation of 30 microseconds was found satisfactory for blade analysis at this point in the vibration cycle where the blade velocity is approaching zero. Unattainably shorter pulses would be required to acquire usable blade tip data as it neared its maximum velocity of just over 400 cm/sec at the zero displacement position. While the main thrust of these experiments was to demonstrate holographic viability in a most severe environment, two features in the reconstructed image are noteworthy. First, the area of fringe discontinuity at the blade tip represents a potential area of disbond, and the general curvature of the fringes across the central portion of the blade suggests the presence of some blade twist.

3.2. *Helicopter holography in a flight hangar*

Additional work at the operating divisions of United Aircraft has included the performance of holographic testing in the flight hangar at Sikorsky Aircraft to demonstrate applicability of the holographic process to large scale components (areas on the order of 10 square meters), as well as its adaptability to the factory environment. The object used for these studies was an unpainted CH53 helicopter airframe, and the laser system employed was a commercial unit manufactured by Quantrad, with a 3.5 joule output capability. Both the single and double-pulse operational modes were utilized.

A schematic illustration of the configuration used to construct both holograms and holographic interferograms of the airframe is presented in fig. 5. The laser output was directed to a dove prism which served as a beamsplitter for generating the reference and object waves. The portion of the beam reflected from the first surface of the dove prism served as the reference beam, and was reflected from a second glass surface (to further reduce its energy) before being expanded to fill the holographic plate. The beam transmitted by the dove prism was incident upon a second prism at its apex in a direction normal to the base, as shown in the insert to fig. 5, and thus was spatially split into two components; the angular displacement of the two components was approximately 60 degrees. This afforded an increase in the energy to the forward and aft portions of the aircraft while decreasing the energy in the central region. A diffuser, inserted after the second prism, spread the energy sufficiently to illuminate the center of the object area. This particular scheme was found to provide the greatest area coverage attainable (approximately 10 square meters) of the several tried. The plateholder and focal plane shutter system were mounted adjacent to the diffuser to permit visualization of the illuminated area normal to the surface of the object. All of the optical components, with the exception of the pulsed ruby laser were mounted on light, aluminum honeycomb-steel veneer plates which were supported by tripods standing on the hangar floor. The glass diffuser and holographic plate holder were approximately 3.5 meters from the helicopter airframe.

Fig. 6 graphically illustrates the holographic setup as installed in the

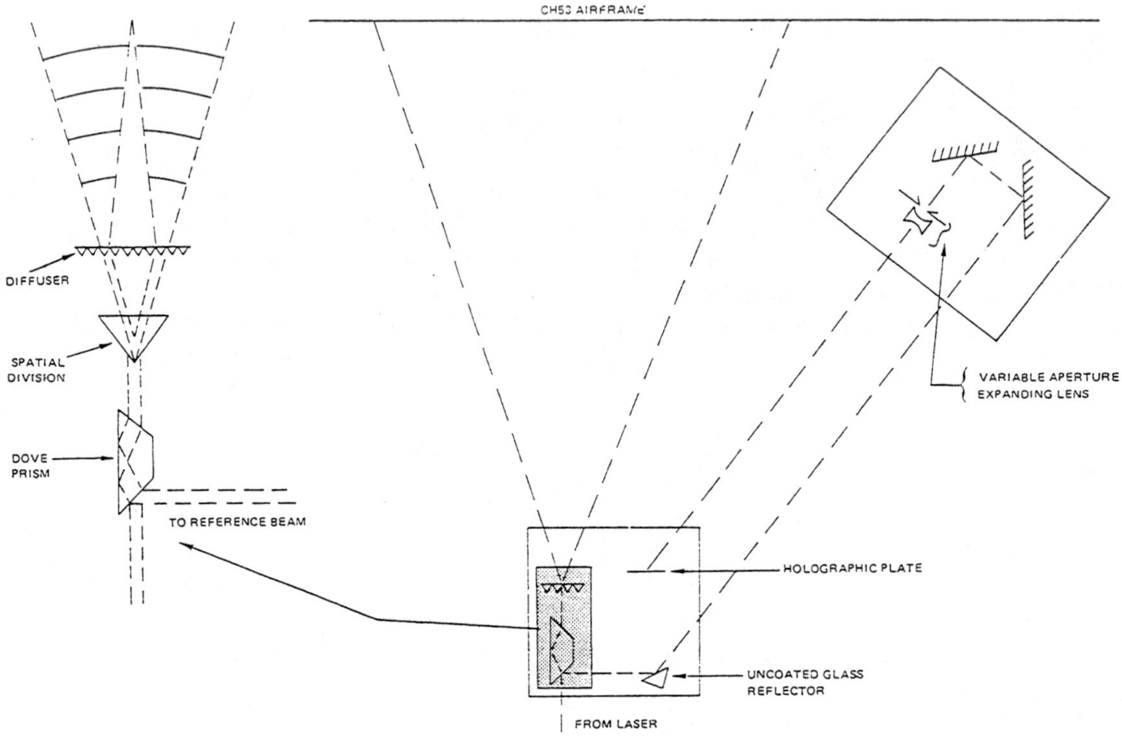

Fig. 5. Schematic diagram of the holographic configuration used for helicopter studies, showing an enlarged view (left) of the spatial division beam expander incorporated in the object beam to more evenly illuminate a large area of the airframe.

Fig. 6. General view of the test equipment for helicopter studies in a flight hangar, providing an indication of the existing ambient light level, and showing the pulsed laser holographic system as viewed from under the tail section.

flight hangar, providing an indication of the object size and the severity of the ambient illumination level; the holographic system can be seen under the tail assembly. All of the tests were performed during normal working hours under daylight conditions. The area of the airframe which was examined holographically is presented in fig. 7 along with a typical reconstruction (multiple views were necessary to attain a satisfactory photographic record of the entire reconstructed image from a single hologram). The photograph also shows the areas on the airframe which were painted to determine the effectiveness of such surface preparation on holographic visibility. All of the white areas visible in the figure were painted with a water-based diffuse paint (spotcheck developer) with the exception of a small square underneath the forward window. The material used at this location was 'CODIT', a retroreflective paint supplied by Minnesota Mining and Manufacturing. With the laser source and holographic plate located adjacent to each other, it can be

Fig. 7. Photograph (A) of helicopter airframe section to be examined, and typical reconstruction (B) from a single hologram covering approximately 10 square meters of surface area.

seen that the 'CODIT' provides a very noticeable increase in the reconstructed visibility whereas the painted areas actually appear darker, probably because they diffuse more light out of the holographic plate area than the unpainted surface.

The holographic interferometry studies were conducted in the double-pulse mode with a pulse separation of 20 microseconds. Typical test results are presented in figs. 8 and 9; the former providing more detail, and the latter illustrating large area coverage (again requiring more than one photograph to record the reconstruction). The stressing force used to provide object motion for the interferogram of fig. 8 was a rivet

Fig. 8. Reconstruction from a holographic interferogram of a helicopter airframe subjected to rivet gun excitation; a 20 microsecond pulse separation interval was used.

gun (the type commonly used to assemble these airframes) applied to the juncture of the main and aft sections of the fuselage, while impulse loading (hammer tap) was used for the result in fig. 9. Both reconstructions exhibit considerable buckling of the airframe skin within the unsupported regions.

It should be noted that the double-fringe character apparent at various locations in the reconstruction of fig. 8 (especially at the higher amplitude positions) is most probably the result of a third output pulse

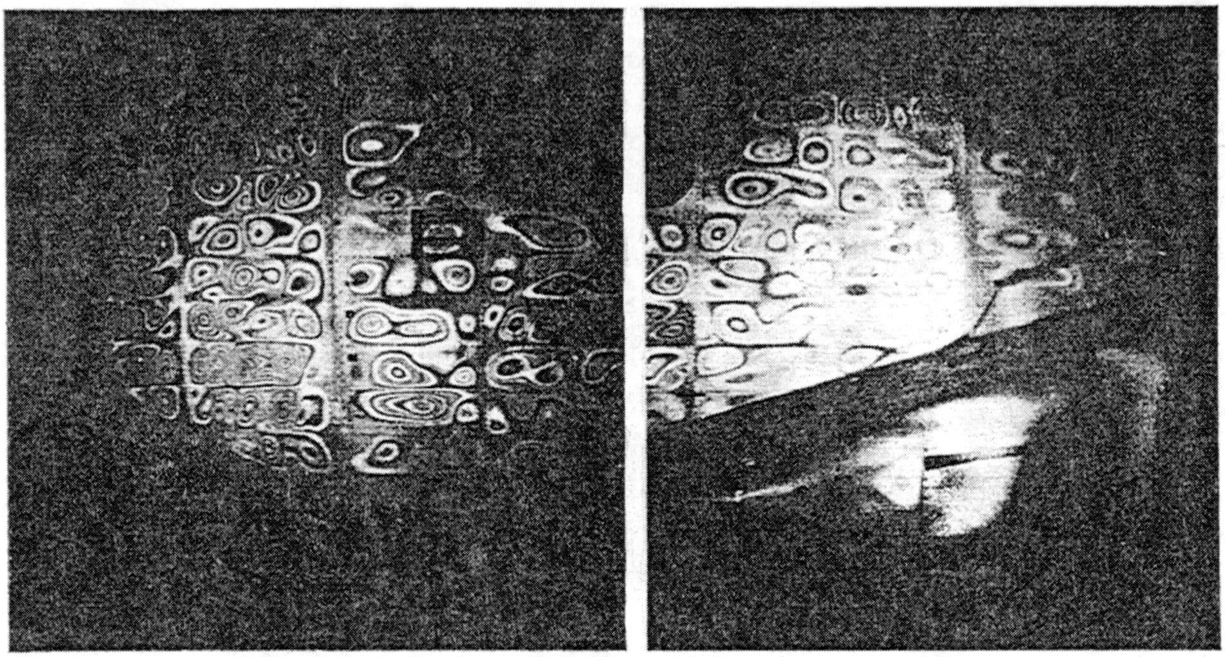

Fig. 9. Reconstruction from a holographic interferogram of a larger section of the helicopter airframe.

overcoming the laser cavity losses prior to Q-switching of the Pockels cell.

The foregoing studies graphically depict the important role holography can play in the factory environment. However, the work utilized only single and double-pulse laser operational modes. Even greater participation can be forecast with the development of practical dual reference beam systems, which would permit the use of two single-pulse exposures for recording interferometric holograms.

4. Dual reference beam holographic investigations

The dual reference beam technique is a method of holographic interferometry which employs a separate reference beam (*i.e.* separate in angle of incidence on the photographic plate) for each exposure[12]. It affords a distinct advantage for factory holography by providing a means for combining fringe control procedures, more generally associated with cw real-time holographic interferometry, with pulsed laser holography. Using both reference beams in the reconstruction step allows one to correct for lateral and axial object translation between the two images by manipulation of the optical components. Unfortunately, these adjustments are not simple to effect, and they are limited to correction of only the uniform (rigid body) motion of the object, since any bending cannot be compensated for in a uniform manner. Further, when using such a method, the area of examination must be limited to a small portion of the total reconstructed surface area. Thus, study of the entire surface requires multiple reconstructions.

4.1. *Laboratory investigations*

In order to investigate the dual reference beam technique, a holocamera which used the image plane holographic process, because of its necessity in fringe manipulation as reported by Champagne[10], was fabricated. A schematic diagram of the system, together with a photograph of its internal arrangement, is presented in fig. 10. As shown, the reference beam is directed into the holocamera chamber through two variable apertures. These apertures are provided to permit alignment of the pulsed ruby laser reference beam to the remainder of the holocamera components, as well as to permit introduction of the reconstructing He–Ne laser beam into the holocamera along nearly the same path so as to permit duplication of the constructing and reconstructing wavefronts. After passing through the alignment apertures the beam

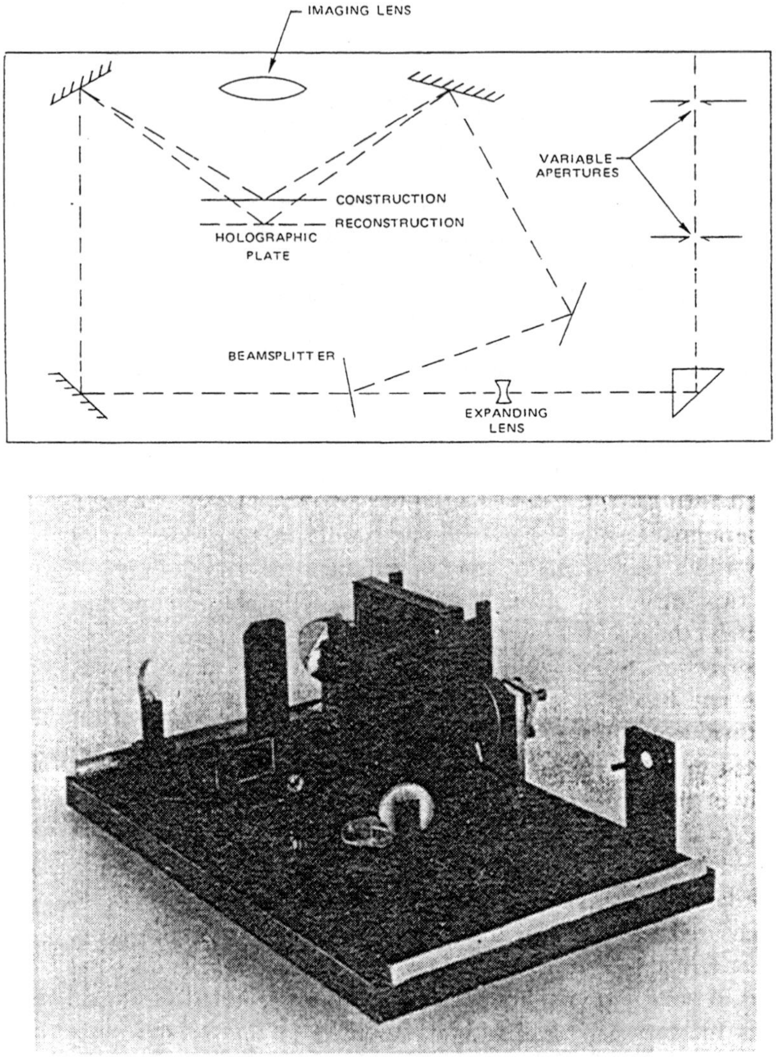

Fig. 10. Schematic diagram (top) of a dual reference beam holocamera, and photograph (bottom) showing the internal components.

is turned by means of a right angle prism and directed to a negative expansion lens ($f = -12$ mm) and then to the beamsplitter. The reflected beam is directed via a pair of flat mirrors to the holographic plate, while the transmitted beam is similarly directed but symmetrically

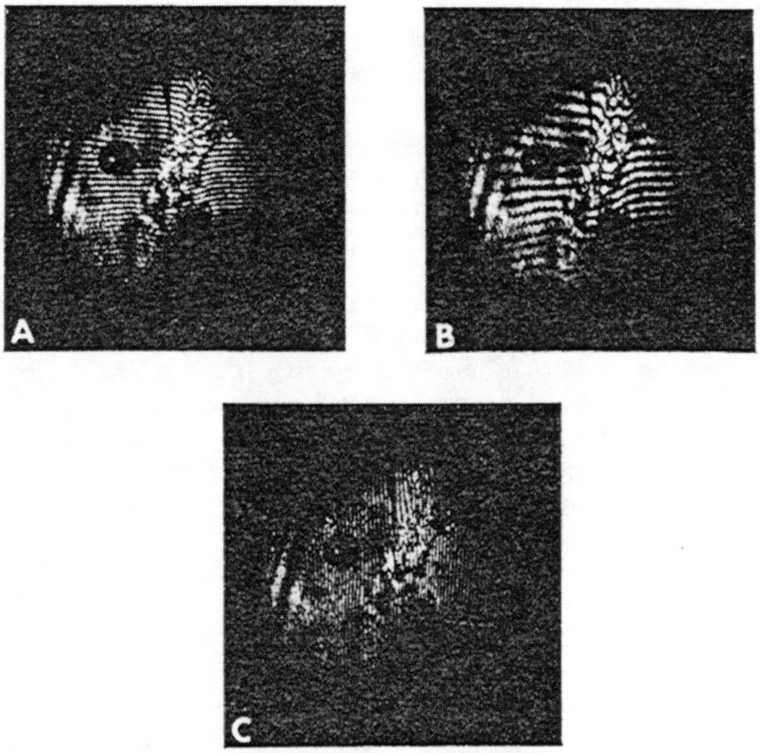

Fig. 11. Three reconstructions obtained from a single interferometric hologram recorded with a separate reference beam for each exposure, and illustrating the ability to control fringe spacing (A and B) and orientation (A and C). The object was a jet of freon blowing across a white plate, and the ball bearings and black arrow were used to aid in the image alignment.

disposed (relative to the first reference beam) about the normal to the holographic plate.

The results of laboratory, dual reference beam tests are presented in fig. 11, illustrating the fringe pattern control capability. A flat white plate, with a small jet of compressed freon to introduce an optical phase shift across the field, was used as the holographic subject. The ball bearings and black arrow were used to assist in aligning the two images, and the three photographs of different fringe spacing and orientation were taken from the same hologram, with a simple angular adjustment of one mirror in the reconstructing geometry.

4.2. *Flight hangar tests*

For field testing of the holocamera, an imaging lens of 75 mm focal length was mounted in a fixed position in order to image a 3 meter square object at a distance of approximately 3 meters. The system was then employed in the CH53 helicopter tests.

For this work the airframe was statically loaded by wedging the airframe (internally) or by an impulse at the juncture of the main and aft sections of the fuselage. A different reference beam was used for each exposure and the airframe was loaded between the two exposures (the first exposure being recorded with the airframe unloaded). The selection of the reference beam to be used was accomplished by the manual insertion of a dark slide into either reference beam path. A photograph of the reconstruction from perhaps the best dual reference hologram recorded in these first studies is presented in fig. 12. Immediately obvious in the figure is the reduced field coverage as a result of necessarily limiting study to a small portion of the reconstructed image. It is, however, most important to note the ability to successfully reconstruct an interferometric hologram recorded in this environment, using two separate single exposures separated by many seconds in time, of an object 3 meters removed from the photographic plate and subjected to a load between the exposures. The fine fringe structure is representative of the airframe deformation; the coarse parallel fringe system is caused by the beamsplitter in the reconstruction process. Of further note is the fact that the fringe character, in the reconstruction, could be altered by suitable adjustment of the reference beam parameters, although this capability was not demonstrated on all of the dual reference holograms. Further effort is most certainly required to more thoroughly define the capabilities and limits of applicability of this technique.

5. Conclusions

The above studies have attempted to show that the holographic process is directly adaptable to the factory environment through the use of pulsed laser systems. Successful tests, conducted at United Aircraft's corporate divisions, were reported that quite clearly demonstrated such a capability. In addition, the work amply illustrated the potential of holography for the rapid visualization of the full surface of a large object, a most important benefit for various holographic applications.

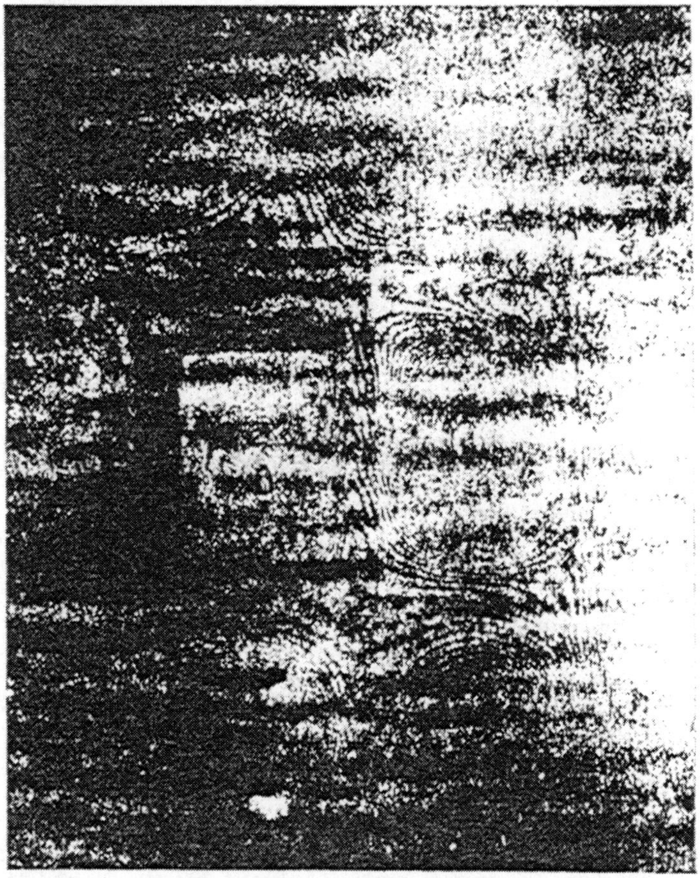

Fig. 12. Reconstruction from a dual reference beam hologram recorded of a helicopter airframe in a flight hangar.

In associated work, initial studies of a dual reference beam method showed promise of increasing still further the utility of holography in the factory, by providing a means to compensate for rigid body displacement in static testing situations requiring long pulse interval times.

In particular, investigations of the deleterious effects of ambient illumination gave evidence that good quality interferometric holograms can be constructed in well lighted areas, without the need for filtering, provided that the exposure time can be controlled. Further, the short duration pulses and brief pulse interval times (in double-pulse work) are effective in isolating the holographic process from most problems associated with environmental vibration. And finally, it was generally found that airborne particulates will only become a nuisance if the particle concentration is sufficient to affect human visual acuity or, in the case of submicron particulate matter, exceeds recommended working levels.

Acknowledgement

The work reported upon in this paper was partially supported by the United States Air Force Materials Laboratory, WPAFB under Contract F33615-71-C-1874.

References

[1] V. J. Corcoran, R. W. Herron, Jr. and J. G. Jamarillo. Generation of a hologram from a moving target. *Appl. Opt.*, 5, 668–669, 1966.

[2] D. B. Neumann and R. C. Penn. Object motion compensation using reflection holography (A), *J. Opt. Soc. Amer.* **62**, 1373, 1972.

[3] H. J. Caulfield *et al.* Local reference beam generation in holography. *Proc. IEEE* 55, 1758, 1967.

[4] G. B. Brandt, Image plane holography. *Appl. Opt.* **8**, 1421–1429, 1969.

[5] J. P. Waters. Object motion compensat'on by speckle reference beam holography. *Appl. Opt.* **11**, 630–636, 1972.

[6] D. B. Neumann and H. W. Rose. Improvement of recorded holographic fringes by feedback control. *Appl. Opt.* **6**, 1097–1104, 1967.

[7] H. W. Rose and H. D. Pruett. Stabilization of holographic fringes by FM feedback. *Appl. Opt.* 7, 87–89, 1968.

[8] J. C. Palais. Scanned beam holography. *Appl. Opt.* **9**, 709–711, 1970.

[9] J. P. Waters. Applications of millisecond pulsed holography (A). *J. Opt. Soc. Amer.* **62**, 1373, 1972.

[10] E. B. Champagne. Quantitative data reduction with the use of fringe control techniques in conjunction with holographic interferometry. *Proc. of the Symp. on Eng. Appl. of Hol.* SPIE, 133–145, 1972.

[11] J. Hamasaki. Signal-to-noise ratios for hologram images of subjects in strong incoherent light. *Appl. Opt.* 7, 1613–1620, 1968.

[12] T. Tsuruta, N. Shiotake and Y. Itoh. Hologram interferometry using two reference beams. *Jap. J. App. Phy.* 7, 1092–1100, 1968.

Reprinted from *Industrial Applications of Holographic Nondestructive Testing*,
J. Ebbeni, editor, Proc. SPIE Vol. 349, pp. 91-98 (1982). ©1982 SPIE.

Interference holography in aerospace technology

Rob. J. Schliekelmann
Fokker Technological Centre, Schiphol, The Netherlands

Abstract

Since 1969 interference holography has seen at the Fokker company a development from an exotic laboratory gimmic into a valuable method for quality control and evaluation of aerospace structures. It is used for inspection of adhesive bonded and advanced fibre composite structures as well as for evaluation of the deformation and resonance characteristics of structures of all kinds.

Evaluation principles.

Inspite of the fact that with special installations it is possible to measure deformations in three axis, the installation as used by the Fokker company records the deformation only in the direction of direct observation. This means that any evaluation of special characteristics of structural components can be based only on relative displacements in the direction of observation with magnitudes of a quarter wave length of the laser-light used(for Argon-lasers this is 0,00013 mm). The methods of deformation used in most cases are either thermal or by means of vibration, of which the former method is most popular for quality control of adhesive bonded and/or composite structures.

Means of operation

The Fokker Technological Centre developed in the early years a reliable installation for holographic interference evaluation. This installation is based upon a 3 Watt continuous wave Argon laser, of which a maximum of 1,2 Watt is usable for optimum coherence. The laser with the necessary optical components are placed on a heavy steel table, that is provided with an effective vibration damping suspension system. During the construction of the holography laboratory special noise and vibration damping provisions have been made in order to exclude external disturbance as much as possible. This was done in particular in view of the location of the laboratory only a few hundred meters away from the engine run-up and testarea of the Fokker flight test department. The mounting of the components takes place at the far end of the table. Maximum dimensions of the components to be inspected are 1.5 x 6 meters. Behind the mounted component is a battery with ceramic infra-red heater elements, that can be remotely control led by the observer behind the optical arrangements. The optics are provided with an effective means of fringe control and high power illumination of the full object surface. The laser is provided with a rotable beam splitter for continuous variation of the light intensities of object illumination and reference beam. Behind the plate holder an highly sensitive TV camera is placed for continuous observation of the interference patterns, that can take place on a monitor and which can be recorded on a professional video-tape recorder. Exposure times are automatically controlled by an exposure meter that operates the shutter in front of the laser. For deformation of the component by acoustical means a suitable amplifier provided with frequency generator and excitation equipment is available. The laboratory is provided with red light illumination which allows for all operations including the developments of the plates to be done by suitable lighting conditions.

Methods of inspection.

Adhesive bonded and composite structures are inspected by means of interference holography as follows.

The component concerned is placed to the support structure at the far end of the holographic table and is firmly clamped, taking care not to introduce mounting stresses. The optical components are adjusted in such a way that the full object surface to be inspected is equally illuminated. For components that have a high length/width ratio curved mirrors are available to ensure that all the laser energy is equally distributed over the full surface of the object and not adjacent to that. After the holographic plate has been put in the plate-holder an exposure is made and the plate is developed in the nearly chemical tanks. After drying it is placed back in the plate-holder. With help of the adjustments of the optical components the fringe pattern is corrected to the maximum possible fringe distance.

With help of the remote control a short infra-red heating is introduced, after which the fringe patterns are observed in real time either by direct observation through the plate or by means of the TV-camera and the monitor. The centre of the fringe pattern "the bull's eye", is moved by adjustment of the optical components over the whole surface of the component. During this operation continuous corrections are introduced eliminating any fringe deformations resulting from overall deformations of the component. During this scan carefully disturbancies in the fringe pattern are recorded. These anomalies are indications for irregularities in either the adhesive bonded or composite structures. The fact that during the scanning operation all the fringe patterns are continuously recordable on the video-tape-recorder has been found as to be extremely valuable. This allows for repeated reviewing the recorded fringe patterns at the various locations of the components. Typical configurations can be recorded also by a photographic camera, or by means of another hologram.

The availability of a powerful zoom lens on the TV-camera has provided the installation with a remakable sensitivity. Anomalies of about 3 mm diameter can be observated easily.

Methods of evaluation of structures.

In addition to the application of interference holography as a means of quality control for adhesive bonded or composite structures during which irregularities in deformation characteristics are searched and detected, the method has proven to be extremely valuable for the observation of structures or components. For the time being this has been limited so far to the vibration and resonance characteristics of structures or components. The evaluation takes place after a hologram has been made of the component in the stationary, undisturbed, situation. Then the component is excited at well defined frequencies. In pratice the excitation starts at a very low frequency and the component is observed in real time. During careful increase of the frequencies the resonance frequencies are noted down, in real time. That initial hologram is used to define precisely the resonance frequencies at which in each case a new hologram is made as a time average hologram. That allows for very precise evaluation of the resonance characteristics of the component. Successful applications of this method were numerous among which are resonance characteristics of transducers for ultrasonic testing equipment and acoustical characteristics of fuselage interior components. The resonance method is also suitable for determination of the quality of adhesive bonded structures. The stiffness of the edges of adhesive bonded joints has important effect on the clamping of not-bonded sheet metal areas. From the resonance frequencies of such areas the clamping effect of the bonded joint can be clearly seen, and an opinion of the cohesive quality of the adhesive bonded joint can be formed.

In the following a number of examples of typical aerospace technology applications of interference holography is given. There is no question that interference holography has grown into an indispensable means for inspection and evaluation of aerospace structures.

References.

1. Schliekelmann, R.J.,"Holographic Interference as a means for quality determination of adhesive bonded metal joints". ICAS, 8th Congress, August 28, 1972.
2. Erf, R.K. and all., Non destructive holographic techniques for structures inspection", April 1974, United Aircraft Corp., East Hartford, Conn., U.S.A.
3. Barbarisi, M.J. and all., "Initial feasibility study employing holographic vibrational analysis to locate non-bonds in thick ceramic to fiberglass composite", Picatinny Aersenal, Dover, N.J., U.S.A.
4. Schliekelmann, R.J., "Non-destructive testing of adhesive bonded joints", AGARD Lecture Series n° 102, Chapter 8, 1979.
5. Bunn, B.P. and all., "Materials Investigation for a failed spacecraft (MAROTS) antenne" ESA Journal. 1978. Vol. 2.
6. Schliekelmann, R.J., "Interference Holography, a means for non-destructive testing" Proceeding 6th International Symposium on Non-Destructive Testing, Antwerp, Belgium, 18-20 Sept. 1974, p. RS-1.

Fig. 1 - Fokker Holographic Tester in operation during inspection of a Space lab. hatch structure. Note the video observation and recording station at the lefthand side.

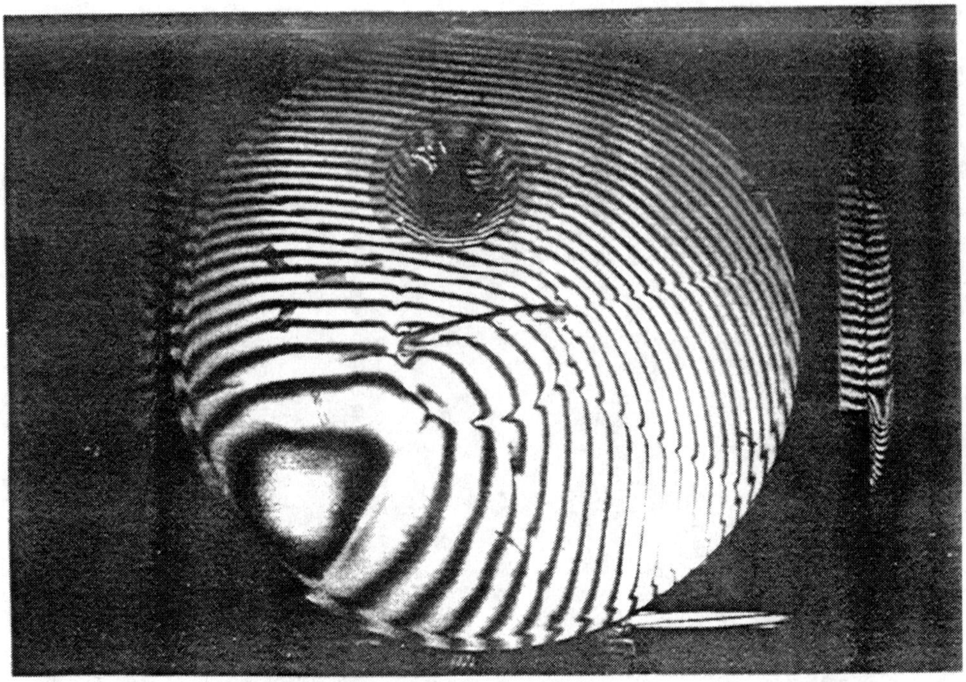

Fig. 2. The observation hatch of Space Lab. is an adhesive bonded honeycomb sandwich structure that has to fulfil the highest demands of quality and reliability. During the evaluation phase as well as actual production of the flight articles interference holography has played an important role for quality control of the honeycomb core, the core splices as well as the bonds with the face sheets and the edge members. Double hologram after inspection by Fokker is shown by the thermal deformation method.

Fig.3. Adhesive bonded sandwich wing leading edge structure shows large void area by means of closely spaced interference fringes.

Fig.4. Various degrees of damage as a result of impact inflicted on an honeycomb sandwich structure can be easily detected. Each individual cell in the honeycomb core is visible through the thin face sheet (photo: Fokker).

Fig.5. Advanced fibre reinforced composite component for the Fokker F28 "Fellowship" under periodic evaluation on the holographic installation of the Technological Centre.

Fig. 6. Ultra light solar energy panels built as carbon fibre reinforced sandwich panels for the MARECS satellite. The sandwich panels are inspected contactless by means of interference holography prior to bonding of the solar cells.

Fig. 7. Holographic interference test installation of the Fokker Technological Centre. On the foreground left the video-installations are visible. In the background a MARECS satellite panel is mounted for inspected in front of the battery of infrared heater elements.

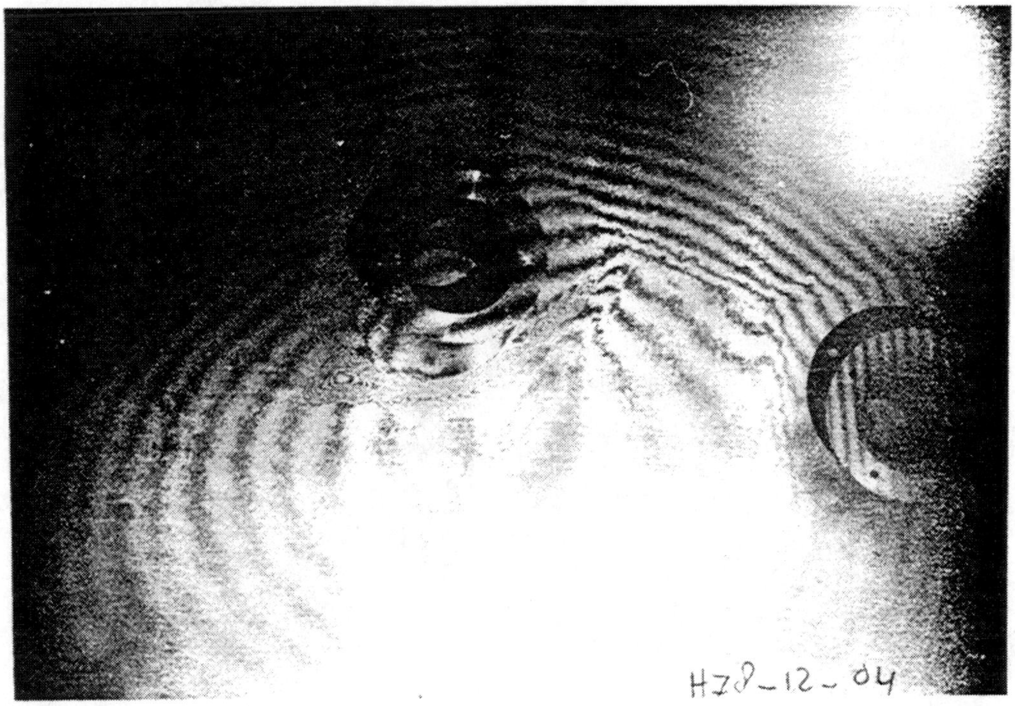

Fig.8. Interference holography is effectively used for evaluation of structural components for satellites before and after qualification testing.
This photograph is a detail of a damaged area of the MAROTS spacecraft antenne after subjection to severe structural loads. (Photograph: Fokker).

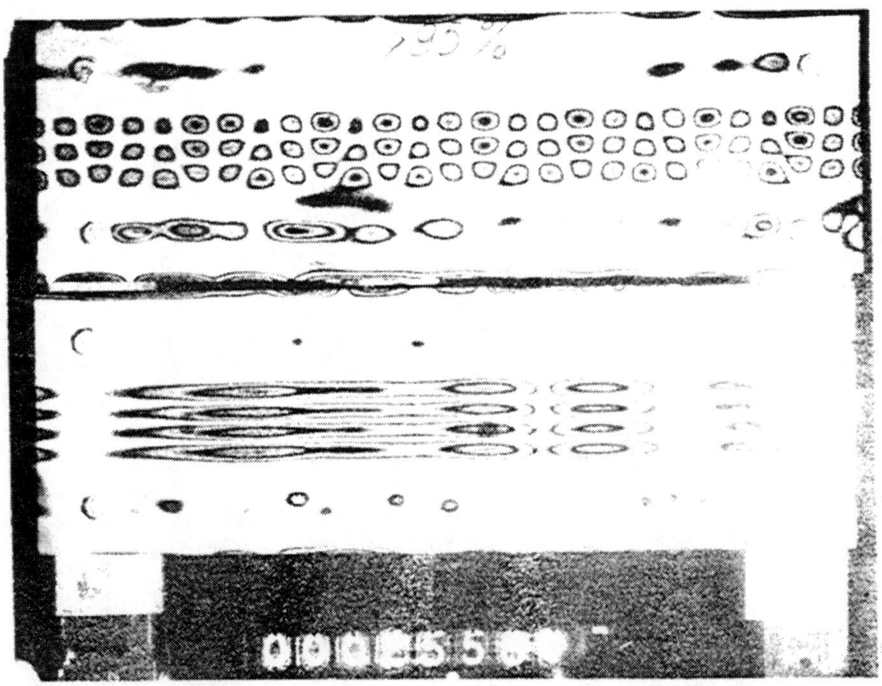

Fig.9. Evaluation of adhesive bonded panels by means of resonance characteristics as observed by interference holography. The shown fringe patterns concern the skin fields between two bonded stiffeners. At the given frequency of 25589 Hz the toppanel with high quality, stiff bonds, shows a distinctly different resonance pattern that the lower panel with low quality, low stiffness, bonded joints (Photograph: Fokker).

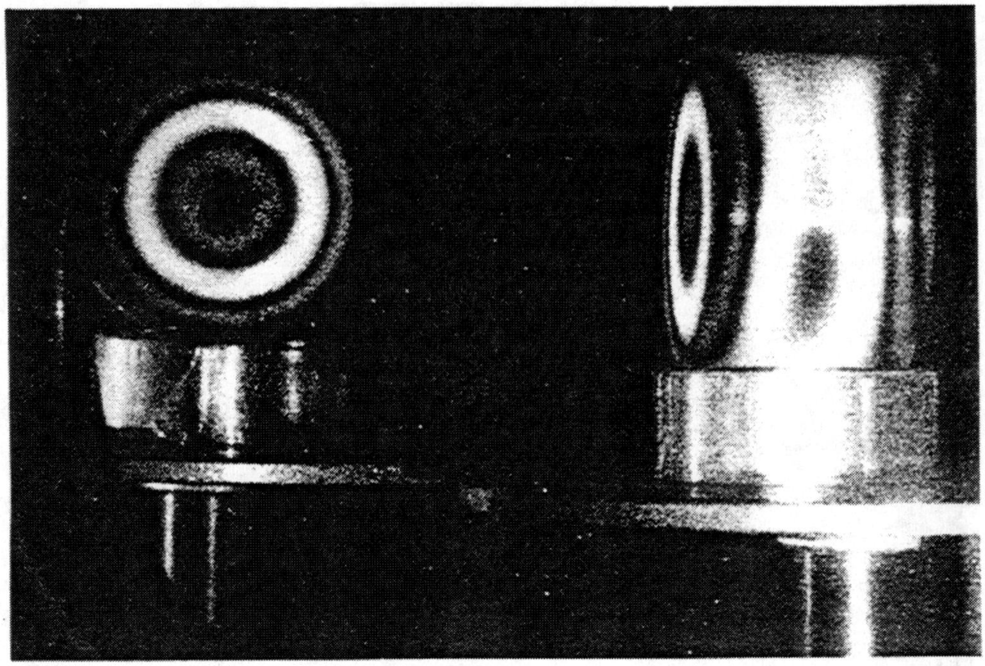

Fig.10. Evaluation of ultrasonic transducers can be done successfully by means of interference holography. This photograph shows the resonance characteristics of a bariumtitanate transducer of about 37 mm diameter and 25 mm thickness at resonance at a frequency of 90 kHz. The real time hologram of both the face- and the cylindrical surfaces of the transducer was made simultaneously with help of an effective mirror arrangement (Fokker).

Reprinted from *Optical Engineering,* Vol. 24(6), pp. 1086-1092 (November/December 1985). ©1985 SPIE.

Application of pulsed reflection holography to material testing

R. L. van Renesse
J. W. Burgmeijer*
Institute of Applied Physics TNO
P.O. Box 155
2600 AD Delft, The Netherlands

Abstract. When holography is applied outside the laboratory, some well-known problems are encountered: vibrations, rigid-body motion, and stray daylight. Pulsed holography can overcome the difficulties with vibrations, but the other problems are less easily solved. When the object area to be holographically tested is small, a very simple and convenient method may be employed, which was described independently by Boone (1974) and Neumann and Penn (1975). They fixed the hologram holder rigidly to the object under test, thus avoiding rigid-body motion of the object with respect to the hologram. In a similar configuration Denisyuk reflection holograms are made without any necessity of darkening the environment. The authors believe that the main reason this technique is not widely used is due to difficulties generally encountered in processing the Denisyuk hologram to good quality. This paper describes a simple processing technique that results in high quality reflection holograms that may be analyzed by microscope up to interference fringe densities of about 30 fringes/mm. As examples, the results of a projectile impact study and the study of early fatigue crack detection in a critical airplane structure are presented.

Subject terms: holographic interferometry; reflection holography; chemical processing; gelatin collapse; nondestructive testing; crack detection.

Optical Engineering 24(6), 1086-1092 (November/December 1985).

CONTENTS

1. INTRODUCTION

Surprisingly, the history of wavefront reconstruction by recording standing waves in a volume medium appears to date back as far as 1810, when Goethe's extensive work on chromatics was published.[1] In this work we read in the findings of Seebeck that silver chloride under certain circumstances tends to adopt the color of the light to which it is exposed. Along this principle the first color photographs were inadvertently made in about 1848 by Beckerel, who was as yet unaware of the underlying principle.

After Zenker's[2] explanation of the phenomena through the theory of standing light waves in 1868, it was Lippmann who in 1891 succeeded in making brilliant color photographs of excellent keeping quality that have stood the test of time and can still be found reconstructed in some museums.[3] It was soon recognized that color photography was to be developed along more practical lines, and thus the fascinating technique of Lippmann photography became of only academic interest. It must be emphasized, however, that at the turn of the century the fabrication of simple volume reflection holograms was practically possible and was in fact accomplished in 1908[4] by Ives, who actually made holographic optical elements using the green mercury line while applying a photographic bleach technique. In 1892 Lippmann had already demonstrated dichromated gelatin to be an excellent medium for volume color photography.[5] The real potential of the technique, apart from color photography, was not realized however, and practical applications had to wait until concrete form was given to Gabor's[6] ideas (1948) on wavefront reconstruction. It was Denisyuk[7] who in 1962 revived the old Lippmann process and demonstrated its suitability to wavefront reconstruction. While the simplicity of the optical principle has promoted its application to holographic interferometry, contrary to this, the need for extremely fine-grained emulsions and the difficulties connected with their processing have restricted the general application of the technique in this field.

The purpose of this paper is to discuss these difficulties and their convenient solution as well as the practical applicability of the technique to holographic interferometry.

2. NONDESTRUCTIVE INSPECTION

2.1. Crack detection

An investigation carried out by our Institute will serve as an introduction to the principles of the technique.[8] This investigation

*Present address: Philips Research Laboratories, P.O. Box 80.000, 5600 JA Eindhoven, The Netherlands.

Invited Paper 5163 received April 23, 1984; revised manuscript received March 7, 1985; accepted for publication March 25, 1985; received by Managing Editor April 15, 1985. This paper is a revision of Paper 398-18 which was presented at the SPIE conference on Industrial Applications of Laser Technology, April 19–22, 1983, Geneva, Switzerland. The paper presented there appears (unrefereed) in SPIE Proceedings Vol. 398.

concerns a nondestructive inspection (NDI) method that detects the initiation and growth of fatigue cracks at fastener holes in the wing of an aircraft. Using a periodic inspection procedure, fatigue crack growth has to be detected before the crack size has reached its critical value. The demand for reliable inspection methods comes from the need to fly the aircraft well beyond its initial design lifetime.

For most current inspection methods, the detection of cracks under installed fasteners forms an only partly solved problem. The inspection procedure must be appropriate for use in the aircraft hangar without the need to dismantle the wing or to dismantle fasteners. This last restriction is understood by taking into consideration that bad installation of fasteners is one of the principal causes of crack initiation in fastener holes.

One is most interested in the inspection of fastener holes in a highly stressed area, the so-called "critical area" where a higher probability of fatigue crack initiation exists. This critical area has a flat, diffuse metal surface and is approximately circular with a diameter of 60 mm. It is situated in the lower wing skin on a place not easily accessible for current NDI apparatus. Holographic techniques are known for the detection of fatigue crack growth,[9,10] but many of them have serious drawbacks for this application. For the detection of these flaws not only must a suitable loading mechanism be found, but also some extra conditions inherent to the application outside the optical laboratory must be satisfied. We will consider six principal items connected with the technique.

(1) A *loading method* has to be found that causes a detectable anomaly in the fringe pattern, which is a map of surface deformation, in the neighborhood of holes where cracks have reached a length much smaller than critical. An anomaly in the deformation pattern will be induced at the crack tip during mechanical loading. This can be caused by two phenomena: (a) The presence of a free-crack surface affects both the transverse displacement (Poisson effect) and the residual stress (assembling stress). (b) The sharp crack front will introduce singular terms to the equation of the stress field. Several mechanical loading methods can be considered, for instance, tensile loading, vibrational excitation, and impulse loading. For the latter two methods, a double-pulsed laser would be necessary. In a preliminary experiment in a test facility for aircraft model constructions, we have demonstrated that a tensile loading method is successful. For an inspection procedure in a hangar, however, this may be unpractical, and therefore other loading methods are still being sought.

(2) During hologram exposure *vibrations and motions* must not cause motions between the object and the holographic plate of more than a fraction of the wavelength. The method applied by most holographers is to shorten the exposure time such that motions during this time interval are restricted. To get sufficient exposure energy a high power pulsed laser is needed. We have used a ruby laser having a pulse width of 25 ns and a single-mode pulse energy of 30 mJ.

(3) *Rigid-body motions* of the object relative to the holographic plate between the two exposures of the hologram will give an unwanted extra number of fringes during reconstruction. There are many solutions to this problem, all of which have their own drawbacks. These fall into three categories:

(a) Using a double-pulsed laser, the pulse interval can be as short as 1 μs. The object motion in this interval will be small enough to prevent most of the formation of unwanted fringes. A serious drawback then is the restriction to dynamical loading techniques, for instance, vibration excitation and impulse loading.

(b) A compensation after the exposure of the interferogram is possible, e.g., using fringe control, sandwich holography, or holographic moire. At the present time these techniques are still too sophisticated for a routine inspection by untrained personnel.

(c) During the exposure of the hologram, rigid-body motions can be partly compensated for by a technique introduced by Boone[11] and Neumann and Penn.[12] They clamped the holographic plate holder to the object, with the holographic plate as near to the object as possible, and used a collimated laser beam to optimally compensate for rigid-body motion (Fig. 1). The holographic technique applied is Denisyuk reflection holography.[7] Although Denisyuk reflection holography has the advantage of simplicity, the impossibility of adjusting the intensity ratio between the object and reference beams limits its applications. Generally, objects holographed in this way must possess a high reflectance or otherwise be coated with a suitable white or metallic paint in order to render a sufficient intensity ratio.[11,12]

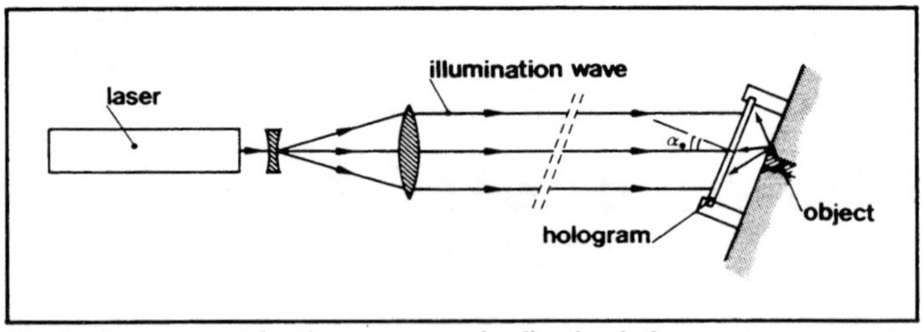

Fig. 1. Arrangement for the exposure of reflection holograms.

We have used the Denisyuk technique without any surface preparation of the object. Therefore, it is demanded that the holographic processing perform optimally with respect to diffraction efficiency and signal-to-noise ratio. This subject will be discussed in Sec. 3. A drawback of the technique may be that the size of the inspected area is restricted to the size of the holographic plate. For our application, however, the size of the area to be inspected was small enough to be covered by a 4 in. × 5 in. plate. The technique offers a very efficient compensation for machine vibrations and motions. Out-of-plane rotations of object and plate cannot be completely compensated for, and thus the use of a pulsed laser is still required for their elimination.

(4) An excess of *daylight exposure* on the holographic plate will seriously decrease the contrast of the interference fringes and along with it the brightness of the reconstructed image. The technique of Denisyuk pulsed holography, however, offers a good opportunity to expose holograms in a daylight environment. A plate holder was constructed that is shielded on one side by the object and on the other side by a mechanical shutter with a large aperture. During the opening of the shutter (0.1 s) the pulsed laser is fired, and the exposure to daylight will generally be negligible.

(5) *Simplicity of the setup* and its operation is demanded in case of operation by unskilled personnel. The need for a simple setup is all the more urgent since the location of the area to be inspected is not easily accessible. As is illustrated by Fig. 1, the Denisyuk holographic setup is very simple and easy to adjust. The plate holder can be manufactured to be compact so that it can be conveniently installed in all kinds of locations.

(6) *Interpretation of the fringe pattern* will be easiest if it can remain qualitative. In that case, the abrupt changes in the curvature of the fringes that appear near the crack tip allow the qualitative discovery of a crack before it can be detected by the usual means. If the anomaly in the fringe pattern is only slight, or if the crack size has to be evaluated quantitatively, automatic fringe readout and processing becomes necessary.

2.2. Effect of processing

Due to the chemical processing of the hologram, the emulsion will generally shrink, while its refractive index will probably change as well. These changes will affect the reconstruction angle as well as the reconstructed wavelength. This section evaluates these reconstruction parameters, while Sec. 3 discusses the processing parameters that are the cause of these changes.

The holographic plate is exposed by a coherent plane wave of wavelength λ_e under an angle α_e with the normal (within the emulsion). Part of this wave is transmitted and subsequently diffusely reflected by the object. Each object point will radiate a nearly spherical wave toward the plate, which will interfere with the original plane wave. The average period of the resulting interference pattern is d_e, in a medium of refractive index n_e. The optical path length between the fringes is then given by

$$n_e d_e = \frac{\lambda_e}{2\cos\alpha_e} . \quad (1)$$

After processing, the refractive index of the emulsion changes to n_r, while, due to gelatin collapse, the fringe period decreases to d_r. The factor denoting the change in optical path length between the fringes, causing a shift in reconstructed wavelength, is given by $s = (n_r d_r)/(n_e d_e)$. After processing, the optical path length between the fringes is obtained by substitution of the shift factor s in Eq. (1):

$$n_r d_r = \frac{s\lambda_e}{2\cos\alpha_e} . \quad (2)$$

For the reconstruction of the hologram a wavelength λ_r is used, which is not necessarily equal to λ_e. According to Bragg's law, constructive interference of the light reflected by the successive planes appears when

$$n_r d_r = \frac{\lambda_r}{2\cos\alpha_r} , \quad (3)$$

where the reconstruction angle α_r is again considered within the emulsion. Equations (2) and (3) may be considered in two ways, as follows.

(1) If the reconstruction angle α_r equals the angle of exposure α_e, an optimally bright reconstruction is obtained by selecting the proper wavelength λ_r of the reconstruction light source. From Eqs. (2) and (3) it follows that

$$\lambda_r = s\lambda_e . \quad (4)$$

Due to the processing of the emulsion, the reconstructed wavelength is shifted by a factor s with respect to the original wavelength. This shift factor s may be experimentally determined by measuring λ_r. The results of these experiments for different processing parameters are discussed in Sec. 3.

(2) For optimal resolution of reconstructed image detail, a monochromatic light source is needed. The reconstructed wavelength is then fixed, so that the angle α_r has to be optimized. From Eqs. (2) and (3) it follows that

$$\cos\alpha_r = \frac{\lambda_r}{s\lambda_e} \cos\alpha_e . \quad (5)$$

If the hologram is exposed applying a ruby laser (λ_e = 694 nm) and a He-Ne laser is used for its reconstruction (λ_r = 633 nm), we find $\lambda_r/\lambda_e = 0.91$. In practice α_e is generally small, so that α_r may be estimated without a large error if $\cos\alpha_e$ is omitted from Eq. (5). Therefore, if the path length change is more than 9% ($s < 0.91$), there is no solution for Eq. (5) and reconstruction using a He-Ne laser becomes impossible. As an example, without any path length changes between fringes ($s = 1$), the angle of reconstruction $\alpha_r = 24°$ within the emulsion, which becomes approximately 44° outside the emulsion, assuming a refractive index $n \sim 1.7$.

Considering the foregoing, it appears advantageous to find a holographic process that performs well enough to allow the making of Denisyuk holograms without the need for object surface preparation. Moreover, this process should control the shift factor s.

3. DIRECT BLEACH

3.1. Efficiency

The conversion of the developed image silver of an amplitude transmission hologram into a transparent silver compound image, after removal of the residual halide by the fixer, creates a phase hologram for which diffraction efficiency is considerably increased. Such a conversion, which is generally referred to as "direct bleach," results in a silver halide, a silver ferrocyanide, or a silver mercurochloride image. For a long time there has been a deeply rooted belief of many holographers that the conversion of the image silver into a compound with a refractive index substantially greater than that of the gelatin medium is demanded in order to prepare efficient holograms. The cause of the original misunderstanding that the refractive index modulation should increase along with increasing refractive index of the silver compound grains is probably due to the implicit supposition that we deal with grains that are large with respect to the wavelength of the light in the medium. In that case, the geometrical-optics formulation gives the resulting refractive index as the mean of the refractive indices of the respective components, weighted by volume. However, this supposition is obviously invalid in the case of extremely fine-grained holographic emulsions, in which the grain diameter amounts to a few tens of nanometers only. Such grains may be regarded as Rayleigh scatterers, and the refractive index of the composition is given by the Lorentz-Lorenz equation, which predicts an approximate linear relation between refractive index of the composition and the polarizability of the composite substances. Efficient phase transmission holograms therefore may be obtained independent of the bleaching agent applied, as long as differences in concentrations of polarizable silver compound molecules are optimized by adjusting the exposure energy. This explains why in past decades innumerable chemical formulations have been proposed—a research effort reminiscent of the alchemic quest of the philosophers' stone—not resulting, however, in anything like a "final" formulation.

In earlier work[13] it was set forth that instead of the refractive index, the molecular polarizability and the molecular volume of the silver compound are the crucial parameters governing the phase variations resulting from the variations in prebleach optical density. The—at first sight paradoxical—conclusion that follows from this proposition is that efficient phase transmission holograms may be obtained by conversion of the image silver into a silver compound of refractive index equal to that of the gelatin. This was experimentally confirmed by conversion of the image silver into silver ferrocyanide, for which the refractive index virtually equals that of the gelatin. This work has been extended to the reversal bleach process by Sirko and Slaby.[14]

In view of photographic sensitivity, the only consideration that follows from the foregoing is that we might wish to apply the bleaching agent that converts the least possible density variations into sufficient phase variations. Evidently, this is accomplished by the conversion of the image silver into the compound with the highest possible polarizability, silver mercurochloride.

It must be emphasized that the foregoing specifically applies to transmission holograms, in which, beyond spatial frequencies of some hundreds of lines per millimeter, the gelatin surface will generally no longer show thickness variations in accord with variations in concentration of the silver or its compounds (Ref. 13, Sec. 4). The contrary is true for reflection-type volume holograms, in which Bragg planes are oriented perpendicularly with respect to transmission-type holograms.

In this latter case, even at very high spatial frequencies, the gelatin is able to follow such variations in concentration to a large extent. Therefore, the foregoing considerations only apply to transmission holography.

3.2. Scatter

The one-sidedness of the previous considerations becomes evident

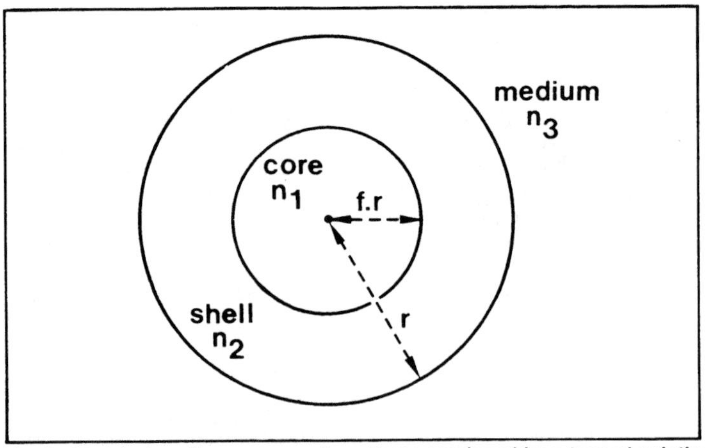

Fig. 2. Model of a silver-compound scatterer enveloped by a tanned gelatin shell in a nontanned gelatin medium.

once the scattering of the silver compound grains is taken into account. A bleaching agent that, for the smallest possible optical density modulation, renders optimal diffraction efficiency may well result in an objectionally high noise level. As it appears, the intensity of the light scattered by a particle that is small with respect to the wavelength is proportional to the square of its total polarizability.[15] This observation shows that diffraction efficiency is indissolubly linked to light scatter. The situation becomes more complex, however, because the grain may not in general be regarded as a scatterer in a homogeneous gelatin medium. On the one hand, the gelatin medium may be tanned through cross-linking of the gelatin molecules in the vicinity of the grain by the oxidation products of the developer; on the other hand, the gelatin may be locally compressed through the forces exerted by the grain, which increases in size when being bleached. These processes may change the mass density of the gelatin near the grain, thereby causing a change in the refractive index of the gelatin enveloping the grain. These considerations led to the proposition of the model of a scatterer, as depicted schematically in Fig. 2.

The spherical core of radius $f \times r$ consists of a silver compound of refractive index n_1 and is enveloped by a spherical gelatin shell of refractive index n_2, thus composing a scatterer of radius r immersed in the regular gelatin medium of refractive index n_3. The total polarizability of such a composite particle, small with respect to the wavelength, is given by van de Hulst.[16]

The validity of the shell model was confirmed experimentally by scattering measurements as well as by phase measurements, the results of both approaches being in close agreement.[17] It is a remarkable coincidence that the proposition of this shell model was accompanied by complementary experimental evidence collected by Joly and Vanhorebeek,[18] who applied developers of different tanning activity and showed, by means of electron photographs, that the average size of the bleached grains is independent of the type of developer. They concluded, therefore, that the observed differences in noise level must be attributed to gelatin distortions within as well as at the surface of the emulsion. As mentioned earlier, the growth of the grains may be expected to exert radial forces on the adjacent gelatin, forces that may compress the shell of tanned gelatin, thus increasing its molecular density and with it the total polarizability of the scatterer. Alternatively, however, these forces may indeed further enhance the amplitude of the surface irregularities coincident with the surface grains. A simple comparative experiment with a developer of low tanning activity reveals that it is not the collapse of the gelatin layer that accompanies the fixing step, but the bleaching that causes excess reticulation of the gelatin surface. Joly and Vanhorebeek have shown that surface irregularities decrease with increasing tanning activity of the developer. From the foregoing it follows that such a suppression of the surface scatter happens at the expense of an increased intraemulsion scatter, since the decrease in surface irregularities is compensated for by increased molecular density in the shells. Therefore, depending on the nature of the processing, the balance between intraemulsion scatter and surface scatter may deflect between extremes. These observations sustain our conclusion that the intensity of scattered radiation may vary widely with the type of processing applied and that it is therefore not adequate to relate results of various bleaching agents without investigating the influence of processing parameters on these separate contributions to the total emulsion scatter.

Such an investigation may be accomplished by utilizing an index-matching technique. Evidently, application of the direct bleach process will only yield optimal results if the total polarizability of the scatterers is minimized through a subtle processing that prohibits shell forming by allowing the emulsion to swell completely and if the surface is index-matched so as to eliminate surface scatter.

3.3. Gelatin collapse

The intricacy of the direct bleach process is further increased by the fact that the removal of the bulk of residual silver halide by the fixer causes the gelatin layer to collapse. For volume reflection holograms this involves a decrease in distance between Bragg planes, resulting in a generally unacceptable shift to shorter reconstructed wavelengths, an inconvenience that demands the application of a swelling agent such as triethanolamine.[19] This is illustrated by a series of experiments with holographic volume gratings exposed on Agfa Gevaert 8E75 HD emulsions at 633 nm wavelength. Two plane waves of about equal intensity impinged perpendicularly on either side of the plate, thus forming an interference pattern parallel to the surface. The holograms were developed in a pyrogallol developer of the following composition:

Solution A:	pyrogallol	20 g
	sodium sulfite sicc.	nil
	distilled water to make	1 liter
Solution B:	sodium carbonate sicc.	60 g
	distilled water to make	1 liter

Solutions A and B are mixed in equal proportions immediately prior to development. As the mixture is liable to fast oxidation, it should only be used once. Without sodium sulfite, solution A has a limited shelf life.

Holograms were developed for 4 min (at 20 °C with continuous agitation), washed for 2 min in running water, subsequently fixed in a 200 g/l sodium thiosulfate neutral fixer (10 min with intermittent agitation), washed for 10 min in running water, and finally bleached in formulations 2 or 4 (from Ref. 13) in order to convert the image silver into AgBr or $Ag_4Fe(CN)_6$, respectively. By means of a spectrometer, the reconstructed wavelength λ_r was measured at each stage of the processing as a function of optical density. The results of these experiments are given in Figs. 3 and 4.

Immediately after development, a decrease in reconstructed wavelength nearly proportional to optical density is established. This is due to a decrease in optical path length between the fringes during development. This decrease is caused by the removal of bromine from the emulsion through reduction of AgBr, which may result in gelatin collapse as well as a decrease in refractive index.

Subsequent removal of the residual AgBr by the neutral fixer drastically reverses this picture. Since most of the bulk halide is removed at low optical density, the wavelength shift is maximal in this region, as is indicated by the reverse in slope. A total shift of about 100 nm in the reconstructed wavelength is encountered for optical densities between 2 and 3. The final conversion of the image silver into a transparent silver compound results in a considerable growth of the grains, in accordance with the increase in molecular volume, which consequently causes a shift toward longer wavelengths. As the molecular volume of $Ag_4Fe(CN)_6$ exceeds that of AgBr by a factor of 2.4, the slope of the graph is expectedly higher.

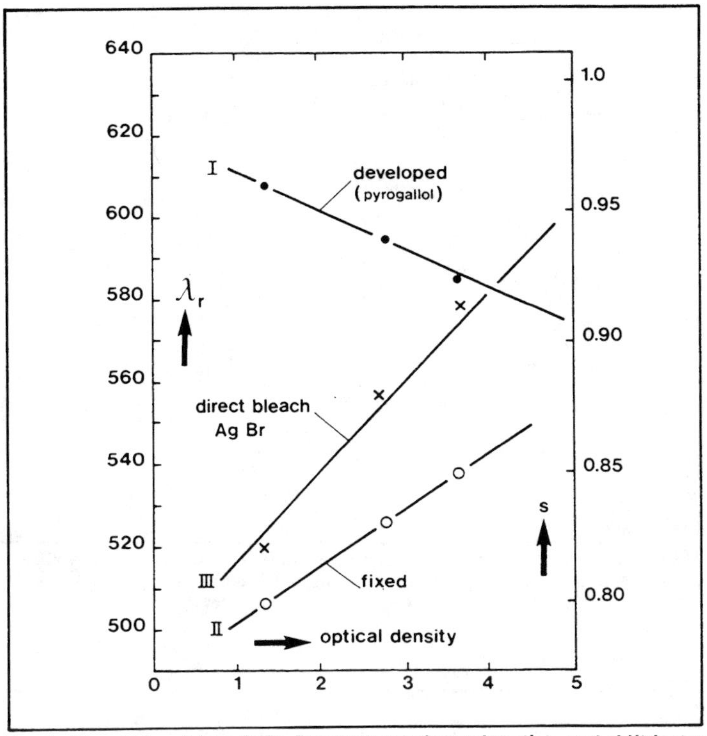

Fig. 3. Direct bleach into AgBr. Reconstructed wavelength λ_r and shift factor s as a function of optical density.

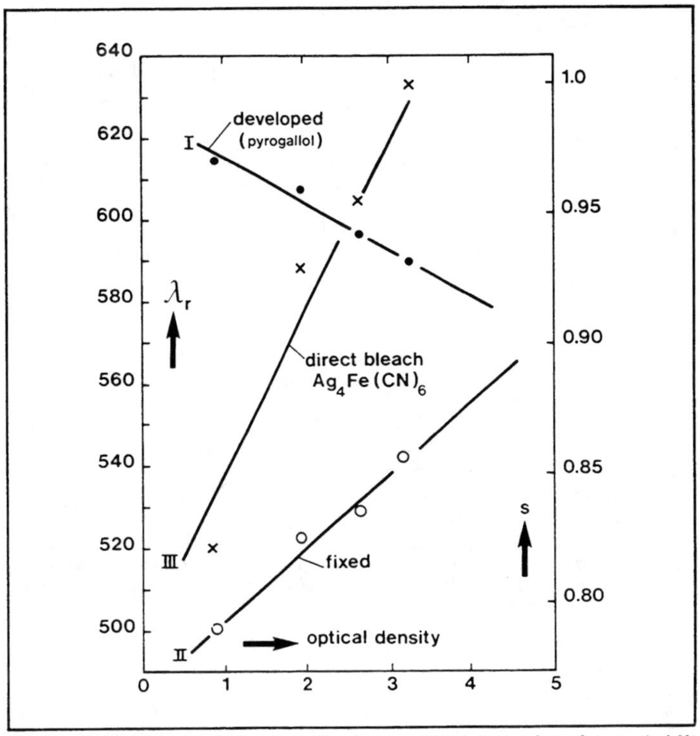

Fig. 4. Direct bleach into $Ag_4Fe(CN)_6$. Reconstructed wavelength λ_r and shift factor s as a function of optical density.

The large amount of alternate shrinking and swelling of the gelatin appears to be disadvantageous since it is only at the bleaching stage of the processing that serious surface scatter is induced. A considerable band broadening (up to 60 nm) is observed as well, indicating that with these large variations in gelatin thickness and grain size the smoothness and coherence of the Bragg planes diminishes at the expense of diffraction efficiency (maximum observed 10%).

Considering the difficulties apparently inherent to the rather circumstantial develop-halide solvent-rehalogenate procedure, an alternative processing that avoids these problems should be very useful. A more direct approach to be considered therefore is the develop-silver solvent process that modulates the concentration of the residual silver bromide.

4. REVERSAL BLEACH

4.1. General

If after development of an amplitude hologram, omitting the fixing step, a suitable silver solvent is applied that does not significantly remove any of the residual silver halide, a phase hologram is obtained as well. In this case, which is generally referred to as "reversal bleach," the concentration of the silver bromide that was originally deposited in the emulsion is modulated. This reversal process has a few advantages over the direct process: (a) The omission of the fixing step reduces the process to essentially two steps, which considerably curtails processing time. (b) The residual AgBr grains have not suffered any changes in size, nor has the gelatin adjacent to the grains undergone any differential tanning. Therefore, these grains are not enveloped by any shell structure, nor have they been the cause of any surface irregularities, and thus they may be expected to render the least scattering.

Conversely, it has been argued by Philips et al.[20] that removing irregular chunks of silver will leave an irregular vacancy in the gelatin host, and the scatter may thus be as severe as if the image were fixed and rehalogenated. For the same reason, however, one might argue that the fixing step, which removes the AgBr grains, must be expected to bring about a similar effect, which, as experiments show, does not appear. If indeed the silver grain removal would leave marked voids in the bulk gelatin, a considerable background scatter should be observed at higher prebleach densities. Our experiments, however, have invariably shown that the reverse is the case: scatter decreases with prebleach density and becomes negligible at densities beyond 2 or 3 in the case of reversal bleach. Evidently, the bulk of residual halide, which is inversely proportional to the optical density, is the main source of scatter. As will be shown in Sec. 4.2, the application of a strong hardening developer prohibits a further wavelength shift upon removal of the silver grains.

4.2. Experimental

We have conducted a series of experiments, similar to those described in Sec. 3.3, with the reversal bleach technique for the purpose of disclosing the influence of processing parameters on the wavelength shift.

The straight pyrogallol developer was selected again because of its strong hardening activity, which was expected to minimize gelatin collapse and thus to minimize wavelength shift. This is illustrated by a series of comparative experiments with a metol developer, the formulation of which is obtained by substituting metol for pyrogallol in solution A (Sec. 3.3) and adding 100 g/l of sodium sulfite.

Metol as a developing agent has a low tanning activity, while the high sulfite content further eliminates any remaining hardening activity by reducing the primary oxidation products of the developing agents so that these can no longer cross-link the gelatin molecules. The result of these comparative experiments is illustrated in Fig. 5. After development in the pyrogallol developer, a shift in reconstructed wavelength is observed that is virtually equal to that of the previous experiments, due to the liberation of bromine. Subsequent dissolution of the image silver by a 4 g/l potassium dichromate bleach with low pH (4 cm^3 H_2SO_4/l) surprisingly does not induce any further wavelength shift, as was already mentioned in Sec. 4.1. This is probably due to the fact that the hardening activity of the developer's oxidation products has a limited diffusion

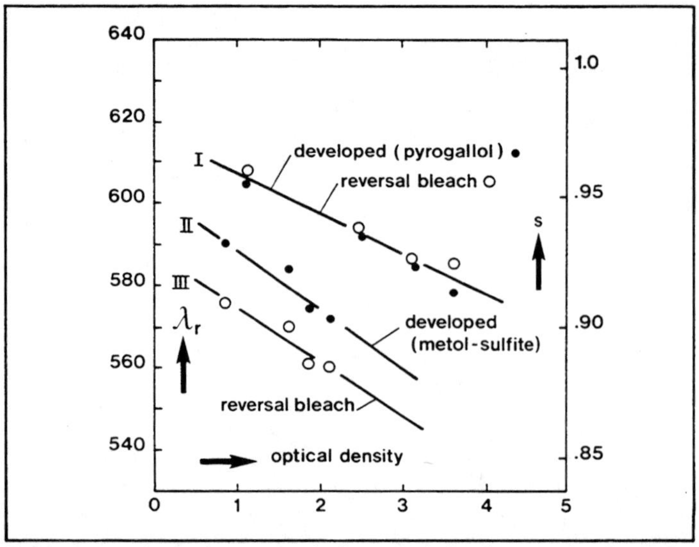

Fig. 5. Comparison of a nonhardening metol developer and a hardening pyrogallol developer. Reversal bleach: reconstructed wavelength λ_r and shift factor s as a function of optical density.

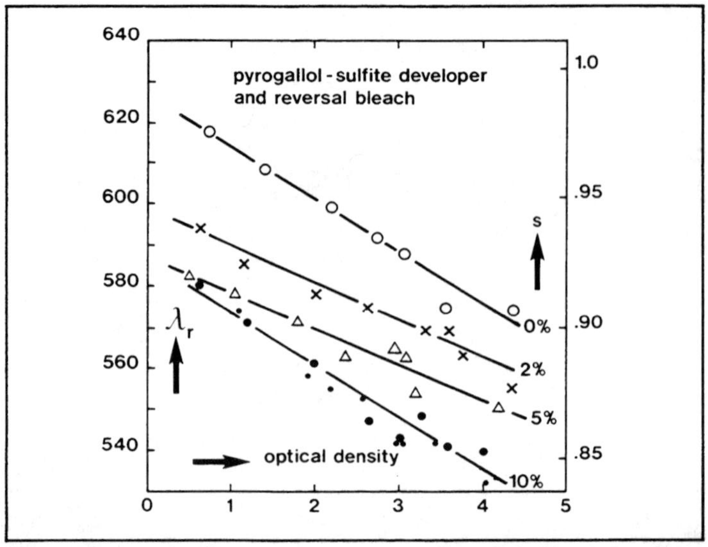

Fig. 6. Influence of sulfite content: o, x, Δ, and ● pyrogallol with 0, 2, 5, and 10% sulfite, respectively; • metol with 10% sulfite. Reversal bleach: reconstructed wavelength λ_r and shift factor s as a function of optical density.

lifetime so that the gelatin is cross-linked in the direct vicinity of the developed grains. Gelatin collapse by removal of these grains by the silver solvent is therefore prohibited by the rigidity of the locally hardened gelatin. Completely contrary to this, the removal of the undeveloped residual AgBr, consequently contained by unhardened gelatin, will cause a strong collapse of the gelatin layer, as shown in Figs. 3 and 4. The results with the nonhardening metol-sulfite developer shown in Fig. 5 sustain this explanation. Not only does the virtual absence of any hardening effect allow a considerably larger gelatin collapse during development, but subsequent dissolution of the silver by the dichromatic bleach now brings about an expected further gelatin collapse. It may be noted that the largest shrinkage is rendered by removal of the bromine, which has an atomic volume almost twice as large as that of silver.

These experiments show that phase holograms may be obtained by means of a simple reversal bleach procedure without the large disastrous shrinking and swelling movements of the gelatin that occur during the cumbrous direct-bleach procedure. Wavelength shifts are limited to a few tens of nanometers because of the hardening activity of the pyrogallol. These experiments further suggest that the amount of the wavelength shift may be manipulated to a large extent by adjusting the hardening activity of the developer formulations by changing the sulfite content.

Figure 6 shows the results of a series of experiments demonstrating this possibility of reconstructed wavelength manipulation by increasing the sulfite content from nil to 10% by weight. The wavelength shift increases linearly by about 10 nm/% sulfite up to a 3% concentration, beyond which the effect gradually saturates. At a concentration of 10% the shift appears to have reached maximum values, as no more difference between the pyrogallol-sulfite and the metol-sulfite developer can be observed. For a prebleach optical density between 2 and 3.5, the color of He-Ne exposed holograms may thus be varied between a deep orange-red and a neutral green.

Apart from the apparent advantages of the reversal bleach procedure combined with a tanning developer over the direct bleach procedure, the resulting diffraction efficiency of the holographic Bragg gratings thus obtained is of great interest. Diffraction efficiencies were measured from the spectrographs and are presented in Fig. 7. Optimal efficiencies of typically about 20% are obtained for prebleach optical densities between 2.5 and 3.5. Bandwidths are typically about 20 nm.

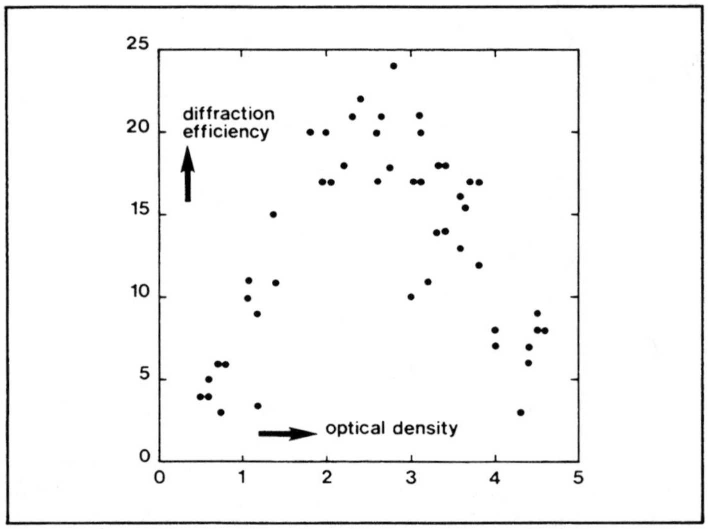

Fig. 7. Reversal bleach: diffraction efficiency (%) as a function of optical density.

Because the hardening pyrogallol development is accompanied by the formation of a considerable brownish stain of the gelatin, the transparency of the hologram suffers a decrease proportional to prebleach density. Evidently, the resulting absorption will cause a decrease in diffraction efficiency; in view of the simplicity of the related processing, however, this disadvantage seems of little interest for NDI holography. It should be mentioned, however, that this gelatin stain can be removed, for instance, by a 1 to 1½ min bath in a 0.05% potassium permanganate/0.4% sulfuric acid solution in water. As it appears that gelatin stain and gelatin hardening are independent to a high degree,[21] the resulting wavelength shift is only slight (10 to 15 nm). This stain removal, however, is accompanied by an increased background scatter, which is to be expected because this stain effectively absorbs multiple scattering by the halide grains. A further processing step, which consists of dying the gelatin with a suitable dye, is therefore demanded to suppress this scatter while preserving the gain in diffraction efficiency, which is increased to typically 30%.

5. RESULTS

The excellent performance of the reversal bleach processing

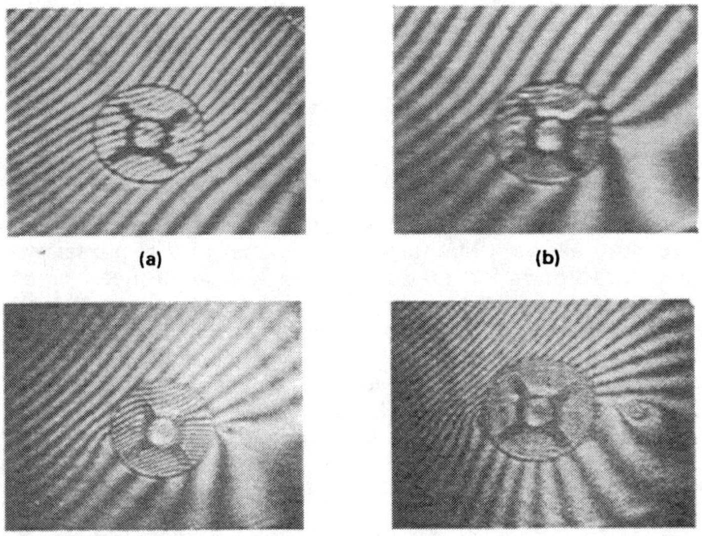

(a) (b) (c) (d)

Fig. 8. Double exposure interferograms of the critical fastener in a specimen of the wing. The mechanical load on the specimen is increased between the two exposures. The interferograms are made after an increasing number of stress cycles: (a) No anomaly in the pattern. (b) Indication of the development of a crack on the right side of the fastener. (c) The crack on the right side reaches the surface; crack development on the left side is also indicated in the pattern. (d) A few cycles before complete fracture of the specimen.

described rendered the possibility of making Denisyuk holograms of low reflectance diffuse objects. Although the diffraction efficiency considerably decreases with the reflectance of the object, reconstructions remained sufficiently bright to allow for their convenient evaluation. As an example, a few results are presented of our investigation of the fatigue crack growth discussed in Sec. 2.1. The area of interest had a diffuse reflectance of only 10% for the ruby wavelength of 694 nm. The holograms were made in an environment with machine vibrations and stray daylight. Figure 8 shows the double-exposure interferograms of the critical area for an increasing number of stress cycles. The interference pattern given in Fig. 8(b) significantly indicates the development of a crack well before it could be visually detected.

Generally, in a conventional setup for transmission holograms the object can only be observed under limited angles due to the rather large object-to-plane distance with respect to plate size. As a result, the sensitivity of such a setup for object deformations is practically normal to the object surface so that mainly out-of-plane deformations are detected. Contrary to this, the Denisyuk holographic setup allows minimal object-to-plate distances so that the object may be observed under rather large angles with the normal. Therefore, sensitivity to in-plane deformations is obtained as well.[8]

Moreover, an important advantage of the Denisyuk technique is that it is generally accessible for low magnification microscopes. Extremely fine details of the interference pattern may therefore be evaluated microscopically without difficulty. This advantage was utilized in a recent study of impact phenomena.[22] Figure 9(a) gives an example of an interferogram of a ceramic target 1 μs after impact. Figure 9(b) gives an enlarged portion of this interferogram showing the development of cracks after impact. Microscopic observation of such interferograms showed fringe densities up to 30 lines/mm. Fringe visibility in these areas was still sufficient to allow evaluation of the interferograms.

6. DISCUSSION

A simple processing technique has been developed that allows the making of low noise, high efficiency Denisyuk phase reflection holograms. Even of objects of relatively low diffuse reflectance, holograms are obtained of a quality sufficient for their convenient interferometric evaluation. Further advantages of the processing described are the low emulsion shrinkage as well as the possibility of shrinkage manipulation, and thus reconstructed-color manipulation, by varying the sulfite content of the developer.

Application of the processing allows the practical and easy utilization of Denisyuk holography outside the laboratory. This technique readily solves the problems connected with rigid-body motion if the plate holder is fixed to the object. Moreover, the reconstruction of the object in the immediate vicinity of the plate allows microscopic evaluation of the fringe pattern up to high spatial frequencies. Finally, the object may be observed under widely different angles so that sensitivity to in-plane as well as out-of-plane deformation is obtained.

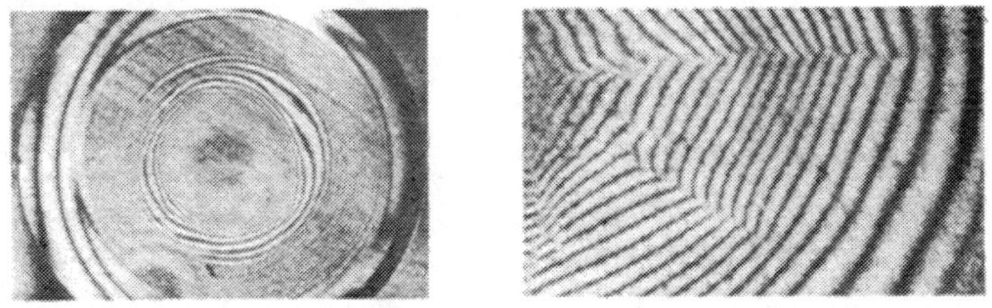

(a) (b)

Fig. 9. (a) Interferogram of ceramic target 1 μs after impact (enlarged 1.5 ×). (b) Detail of (a), showing crack development (enlarged 13 ×).

7. ACKNOWLEDGMENT

The authors wish to thank the National Aerospace Laboratory in Emmeloord for their collaboration in the experiments on aircraft wing specimens. The research work described in this paper has been supported by the National Defence Research Organization in The Netherlands.

8. REFERENCES

1. J. W. Goethe, *Naturwissenschaftliche Schriften,* Bnd. V, pp. 156–159, Ausg. R. Steiner Verlag, Dornach (1982).
2. W. Zenker, Lehrbuch der Photochromie, W. Zenker, ed., Berlin (1868).
3. M. G. Lippmann, C. R. Acad. Sci. 112, 274 (1891).
4. H. E. Ives, Astrophys. J. 27, 325 (1908).
5. E. Valenta, *Die Photographie in natürlichen Farben,* Encyclopädie der Photographie(2), W. Knapp, ed., Halle a.S. (1894).
6. D. Gabor, Nature 161, 777 (1948).
7. Yu. N. Denisyuk, Opt. Spectrosc. (USSR) 15, 279 (1963).
8. J. W. Burgmeijer and H. J. Raterink, Conf. Digest ICO-13, Sapporo, Japan, 404 (1984).
9. R. K. Erf, ed., *Holographic Nondestructive Testing,* Academic Press, New York (1974).
10. C. M. Vest, *Holographic Interferometry,* John Wiley & Sons, New York (1979).
11. P. M. Boone, Proc. Int. Opt. Computing Conf., Zurich, 75, IEEE (1974).
12. D. B. Neumann and R. C. Penn, Exp. Mech. 15, 241 (1975).
13. R. L. van Renesse and F. A. J. Bouts, Optik (Stuttgart) 38, 156 (1973).
14. L. Sirko and J. Slaby, Opt. Commun. 41(6), 407 (1982).
15. M. Kerker, *The Scattering of Light,* p. 39f, Academic Press, New York (1969).
16. H. C. van de Hulst, ed., *Light Scattering by Small Particles,* p. 73f, John Wiley & Sons, New York (1957).
17. R. L. van Renesse, Phot. Sci. Eng. 24(2), 114 (1980).
18. L. Joly and R. Vanhorebeek, Phot. Sci. Eng. 24(2), 108 (1980).
19. D. M. Rowley, J. Phys. E 12, 971 (1979).
20. N. J. Philips, A. A. Ward, R. Cullen, and D. Porter, Phot. Sci. Eng. 24(2), 120 (1980).
21. A. G. Tull, J. Photogr. Sci. 11, 1 (1963).
22. W. J. Kolkert, R. J. M. Amelsfort, and R. L. van Renesse, Proc. 6th Int. Symp. on Ballistics, Orlando, Fla., 22, 388, ADPA (American Defense Preparedness Assoc.) (1981).

Reprinted from *Optical Engineering*, Vol. 24(5), pp. 824-831
(September/October 1985). ©1985 SPIE.

Heterodyne and quasi-heterodyne holographic interferometry

R. Dändliker
R. Thalmann
Institut de Microtechnique
Université de Neuchâtel
Rue de la Maladière 71
CH-2000 Neuchâtel 7, Switzerland

Abstract. The use of heterodyne and quasi-heterodyne techniques in holographic interferometry allows a quantitative evaluation of the interference phase with high accuracy and offers the possibility of automated interferogram processing. In this paper, the fundamentals of these electronic methods are reviewed, the possibilities and limitations are discussed, and some applications in the field of displacement and strain measurement are presented. Heterodyne holographic interferometry is well suited for scientific applications in which very high accuracy (1/1000 of a fringe) is needed. Quasi-heterodyne holographic interferometry with video processing is adequate for industrial applications in which high speed and medium accuracy (1/100 of a fringe) are required.

Subject terms: holographic interferometry; automatic fringe analysis; optical phase measurement; heterodyne; video data acquisition; speckles; displacement and strain measurement.

Optical Engineering 24(5), 824-831 (September/October 1985).

CONTENTS

1. INTRODUCTION

The automated quantitative evaluation of holographic interferograms requires accurate interference phase measurement independent of fringe position and intensity variations in the reconstructed image. In many interferometric arrangements, heterodyne or quasi-heterodyne techniques have become a powerful tool for high resolution interference fringe interpolation.[1] In holographic interferometry, these techniques have been applied to real-time[2,3] as well as to double-exposure[2,4–6] holography. For double-exposure holography two reference beams are required.[7,8] During reconstruction a change of the relative phase between the two reference beams allows one to shift the fringes in the interferogram at will.

Invited Paper HI-105 received March 13, 1985; revised manuscript received April 29, 1985; accepted for publication May 31, 1985; received by Managing Editor June 25, 1985.

In heterodyne methods, the relative phase increases linearly in time, and the interference phase is measured electronically at the beat frequency of the reconstructed wavefields. Heterodyne holographic interferometry offers high spatial resolution and interpolation up to 1/1000 of a fringe. However, it requires sophisticated electronic equipment and mechanical scanning of the image by photodetectors.[4,9] In quasi-heterodyne methods, the relative phase is changed step-wise, using at least three different values. The interference phase then can be computed from the different measured intensity values. Quasi-heterodyne techniques are more adequate for digital processing and TV detection. Two-reference-beam holography with reference sources close together and video electronic processing allows one to measure the interference phase with an accuracy of 1/100 of a fringe at any point in the TV image.

2. INTERFEROMETRY AND FRINGE INTERPOLATION

Interferometry is used to transform phase differences of wavefields into detectable intensity variations, i.e., interference fringes. Assuming that the two wavefields to be compared are given by

$$V_1(\mathbf{x}) = a_1(\mathbf{x})\cos[\omega_1 t + \phi_1(\mathbf{x})] ,$$
$$V_2(\mathbf{x}) = a_2(\mathbf{x})\cos[\omega_2 t + \phi_2(\mathbf{x}) + \psi] , \tag{1}$$

where ψ is an additional constant phase, the local intensity $I(\mathbf{x})$ of their superposition becomes

$$I(\mathbf{x}) = |V_1 + V_2|^2$$
$$= a(\mathbf{x})\left\{1 + m(\mathbf{x})\cos[\Delta\omega t + \phi(\mathbf{x}) + \psi]\right\} , \tag{2}$$

where $a(\mathbf{x})$ is the local mean intensity, $m(\mathbf{x})$ is the fringe contrast, and $\phi(\mathbf{x}) = \phi_2(\mathbf{x}) - \phi_1(\mathbf{x})$ is the phase difference of the two wavefields.

In classical interferometry, the optical frequencies ω_1 and ω_2 are identical ($\Delta\omega = \omega_2 - \omega_1 = 0$) and the interference phase $\phi(\mathbf{x})$ is transformed into interference fringes independent of time. Interpolation of the phase $\phi(\mathbf{x})$ between dark fringes is not accurately possible because the mean intensity $a(\mathbf{x})$ and the fringe contrast $m(\mathbf{x})$ may change as well with position $\mathbf{x}$.

2.1. Heterodyne

In heterodyne interferometry the two optical frequencies are chosen to differ by a small amount $\Delta\omega = \omega_2 - \omega_1$.[10] The superposition intensity is then time dependent, and the interference phase $\phi(\mathbf{x})$ is transformed into the phase of the beat frequency signal. Since the beat frequency $\Delta\omega$ is chosen low enough (<100 MHz) to be resolved by the optoelectronic detector employed, the interference phase can be measured with high accuracy independently of the amplitudes a_1 and a_2 using an electronic phasemeter. This way, both the interpolation problem and the sign ambiguity of classical interferometry are solved. This method requires special equipment, such as acousto-optical modulators and phasemeters. The image is scanned mechanically by photodetectors to measure the interference phase $\phi(\mathbf{x})$ locally. Therefore, the speed is relatively low (~1 s/point), but the accuracy ($\delta\phi \simeq 0.3°$, or 1/1000 of a fringe) and the spatial resolution (>10^6 resolvable points) are extremely high.[4]

2.2. Quasi-heterodyne

For moderate accuracy and spatial resolution, quasi-heterodyne techniques[11] have been developed, which allow electronic scanning of the image by photodiode arrays (CCDs) or TV cameras and use microprocessor-controlled digital phase evaluation.[3] For this purpose, the linear increase of the relative phase ($\phi_t = \Delta\omega t$) due to the heterodyne frequency offset $\Delta\phi$ is replaced by a stepwise change of ψ. To determine the interference phase $\phi(\mathbf{x})$, the interferogram is analyzed at least three times, and the mutual phase ψ is changed each time between two analyses by a well controlled phase shift. The local intensities $I_k(\mathbf{x})$ are then given by

$$I_k(\mathbf{x}) = a(\mathbf{x})\left\{1 + m(\mathbf{x})\cos[\phi(\mathbf{x}) + \psi_k]\right\}, \quad (3)$$

which correspond for k = 1, 2, 3 to a system of three equations with three unknown values: the mean intensity $a(\mathbf{x})$, the fringe contrast $m(\mathbf{x})$, and the interference phase $\phi(\mathbf{x})$. For the interference phase $\phi(\mathbf{x})$ one gets from Eq. (3)

$$\tan\phi(\mathbf{x}) = \frac{(I_3 - I_2)\cos\psi_1 + (I_1 - I_3)\cos\psi_2 + (I_2 - I_1)\cos\psi_3}{(I_3 - I_2)\sin\psi_1 + (I_1 - I_3)\sin\psi_2 + (I_2 - I_1)\sin\psi_3}. \quad (4)$$

The other unknown values are given by

$$a(\mathbf{x})m(\mathbf{x}) = \frac{I_1 - I_2}{\cos[\phi(\mathbf{x}) + \psi_1] - \cos[\phi(\mathbf{x}) + \psi_2]}, \quad (5)$$

$$a(\mathbf{x}) = I_1 - a(\mathbf{x})m(\mathbf{x})\cos[\phi(\mathbf{x}) + \psi_1]\,. \quad (6)$$

The mean intensity $a(\mathbf{x})$ and the fringe contrast $m(\mathbf{x})$ determined in this manner allow one to control the quality of the interferogram evaluation. In particular, they can be used to distinguish between true interference fringes and other structures in the image, such as shadows, contours, holes, etc. Considerable fluctuations of the fringe contrast versus position may also indicate inaccurate reference phase shifts and intensity measurements or extraneous fringe patterns in the image.

2.3. Holographic interferometry

For both heterodyne and quasi-heterodyne fringe evaluation, independent access to the two wavefields is necessary in order to control their relative phase. For holographic interferometry, this implies independent recording of each wavefield by using different reference sources.[4,8] The frequency offset $\Delta\omega$ or the change of the mutual phase ψ can be introduced by acting on the respective reference beams.

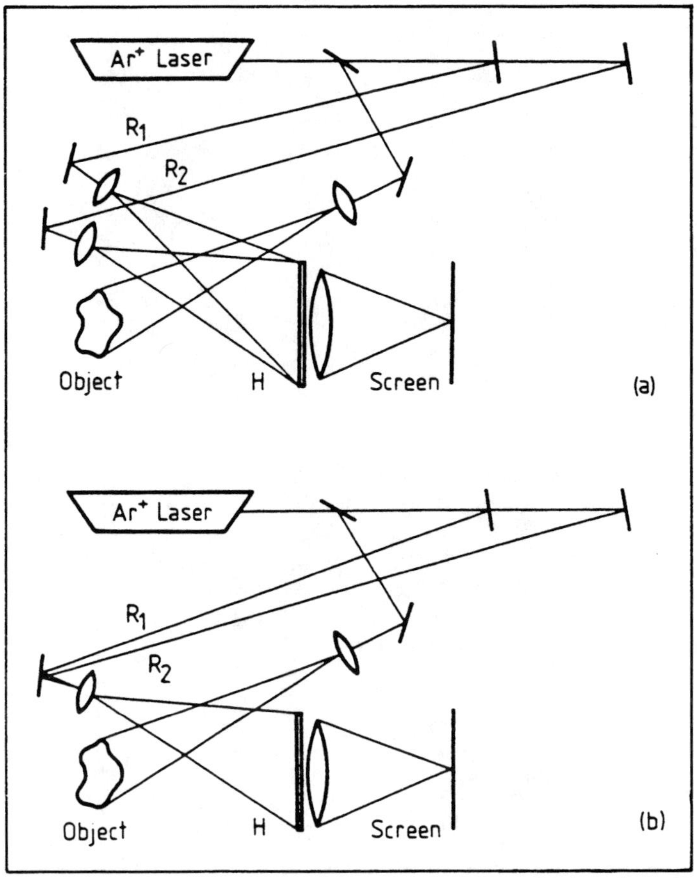

Fig. 1. Arrangements for two-reference-beam holographic interferometry. (a) Well separated reference sources; (b) reference sources close together.

3. TWO-REFERENCE-BEAM HOLOGRAPHIC INTERFEROMETRY

The optical arrangements for two-reference-beam holographic interferometry are sketched in Fig. 1(a) for well separated reference sources and in Fig. 1(b) for reference sources close together. Except for the two references, this setup is the same as for classical holographic interferometry. In both arrangements, the object is illuminated by a point source, and an imaging system permits one to observe the interferogram on a screen. Consecutively, the first object state O_1 is recorded by reference R_1 and the second object state O_2 by reference R_2 on the same hologram plate.

As described in Ref. 8, two-reference-beam holography requires special attention to the multiplicity of the reconstructed images and the influence of misalignment of the hologram with respect to the reference beams. Illuminating the hologram with both reference beams R_1 and R_2 yields eight reconstructions. Four of them are conjugate reconstructions. The other ones are the two desired primary self-reconstructions ($R_1R_1^*O_1$ and $R_2R_2^*O_2$), which give rise to the interference pattern, and the two undesired cross-reconstructions ($R_2R_1^*O_1$ and $R_1R_2^*O_2$). The direction of propagation of the various reconstructed waves depends on the geometry of the optical setup. The primary reconstructions are shown for both cases in Fig. 2(a) and Fig. 2(b).

To avoid disturbing overlapping of the different reconstructions [Fig. 2(a)], the two reference sources must be chosen on the same side of the object, with a mutual separation larger than the angular size of

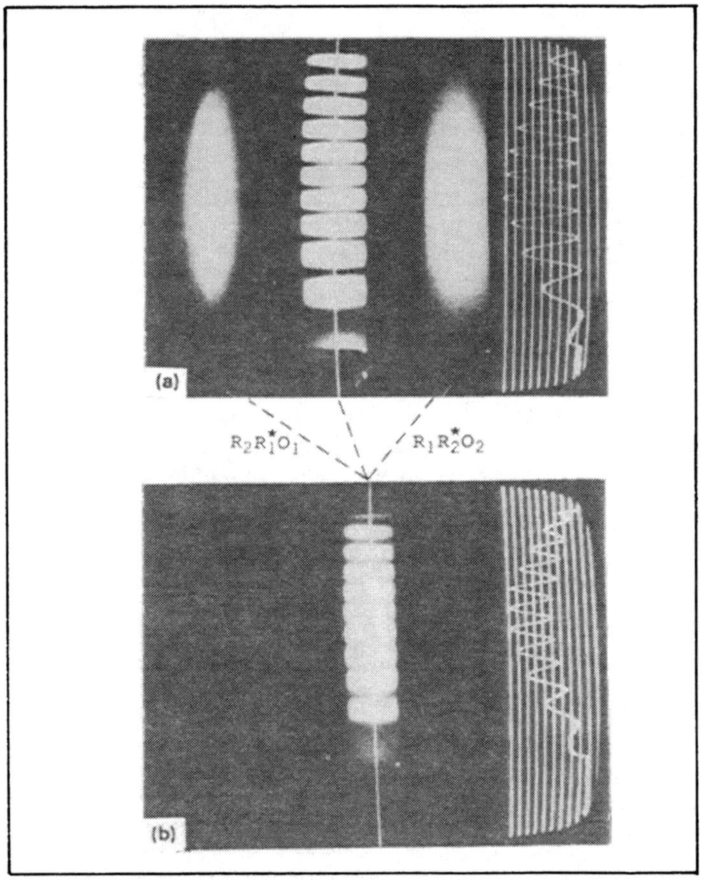

Fig. 2. Reconstructed images and intensity profiles on TV screen. (a) Separated cross-reconstructions $R_2R_1^*O_1$ and $R_1R_2^*O_2$; (b) overlapping reconstructions.

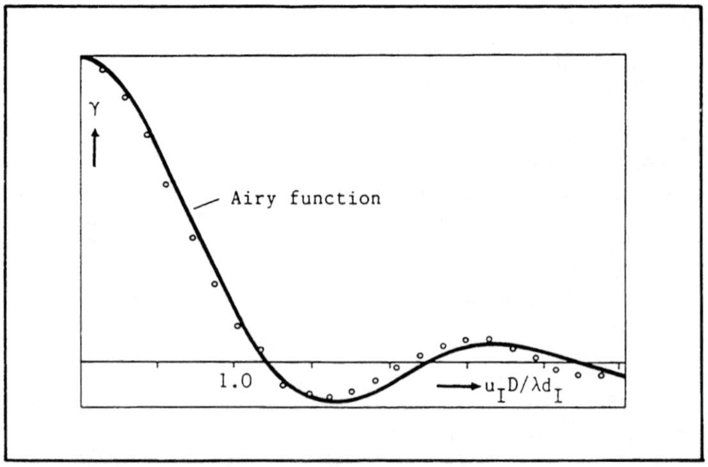

Fig. 3. Theoretical and experimental values of the fringe contrast γ versus transverse displacement u_I of the interfering speckle patterns, in terms of speckle size $\gamma d_I/D$.

the object in the corresponding direction [Fig. 1(a)]. However, the consequence of large separation of the reference sources is high sensitivity to repositioning errors, as described hereafter. Therefore, reference sources close together [Fig. 1(b)] would be preferred if, under certain conditions, overlapping of the cross-reconstructions [Fig. 2(b)] could be tolerated without undue loss of accuracy for the interference phase measurement.

3.1. Alignment and wavelength sensitivity

Two-reference-beam holographic interferometry is sensitive to repositioning errors and to wavelength changes, because the propagation directions of the reconstructed wavefields are differently affected by these alterations.[4,8] Wavelength changes or small rotations of the hologram plate result in an additional linear phase variation in the hologram plane, which is proportional to the difference $(\mathbf{k}_1 - \mathbf{k}_2)$ of the wave vectors of the two reference beams. If the imaging lens is placed close behind the hologram, this linear phase variation in the hologram plane produces a mutual transverse shift u_I between the two interfering reconstructions in the image plane. This results, as shown below, in a loss of fringe contrast in the interferogram.

The magnitude of the interference term depends essentially on the autocorrelation $C_h(u_I)$ of the impulse response function of the imaging system, which defines the speckle size in the image. The interference fringes are visible only as long as the mutual shift of the speckle patterns is smaller than the speckle size. For a circular aperture of diameter D, the fringe contrast $\gamma(u_I)$ is thus given by the well known Airy function:

$$\gamma(u_I) = C_h(u_I)\frac{2J_1(\pi D u_I/\lambda d_I)}{\pi D u_I/\lambda d_I} , \qquad (7)$$

where $J_1(\cdot)$ is the first-order Bessel function and d_I is the distance from the lens to the image plane. Figure 3 shows the decrease of the fringe contrast versus the normalized transverse speckle shift $u = u_I D/\lambda d_I = u_I/(\Delta x)_I$, i.e., u_I in terms of the diffraction-limited optical resolution $(\Delta x)_I = \lambda d_I/D$ (speckle size in the image plane).

Wavelength changes and misalignment of the hologram plate and the reference sources thus result in a loss of interference fringe contrast. For well separated reference sources, to prevent overlapping cross-reconstructions, the difference $(\mathbf{k}_1 - \mathbf{k}_2)$ is quite large, and therefore the sensitivity to wavelength changes and hologram alignment is very high. Typically, a rotation of the hologram of only 0.01° about its surface normal is enough to reduce the fringe contrast to 50%.

3.2. Overlapping cross-reconstructions

On the other hand, the overlapping of the cross-reconstructions $R_2R_1^*O_1$ and $R_1R_2^*O_2$ for reference sources close together does not produce macroscopic interference, as long as the lateral separation of the cross-reconstructions is much larger than the speckle size $(\Delta x)_I$ and, therefore, the speckle patterns of the desired and undesired images are uncorrelated. To avoid systematic interference phase errors, the angular separation between the two reference sources must be several times larger than the diffraction limit, since the higher order maxima of the Airy function (Fig. 3) still give rise to some correlation with the undesired reconstructions.[6] In other words, the two reference sources must produce many interference fringes across the lens aperture to avoid disturbing phase errors, although in a typical setup with reference sources close together [Fig. 1(b)] the sensitivity to misalignment is less than 10°.

However, the overlapping with the uncorrelated cross-reconstructions reduces the overall fringe contrast by a factor of two, as can be seen by comparing Figs. 2(a) and 2(b), and introduces an additional statistical error to the interference phase measurement.[6] The minimum value of this phase error is given by

$$\delta\phi^2 = \frac{5}{2(N+1)} , \qquad (8)$$

where N is the number of independent speckles or correlation cells within the detector area. Figure 4 shows theoretical and experimental values of the statistical phase error due to overlapping cross-reconstructions for different numbers N of speckles. For practical purposes N can be obtained from the F-number and the focal length f of the imaging lens through the relation

$$N = A_D\pi\left(\frac{f}{2F\lambda D_I}\right)^2 , \qquad (9)$$

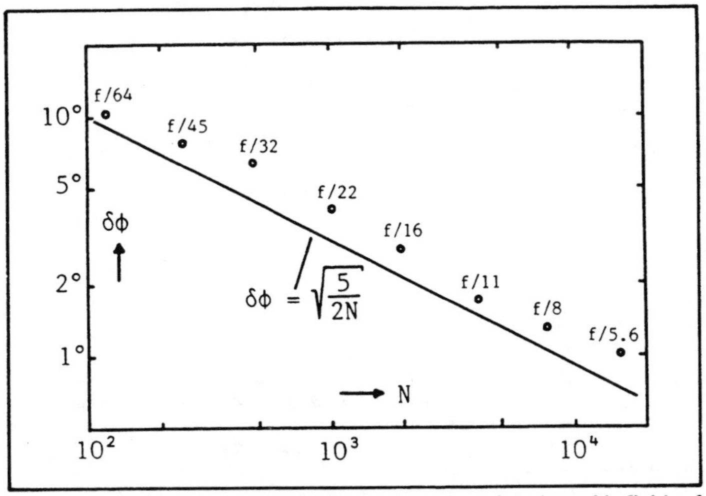

Fig. 4. Statistical phase error $\delta\phi$ due to the uncorrelated speckle fields of the overlapping cross-reconstructions versus the number N of speckles within the detector area. Experimental results for different lens apertures are shown.

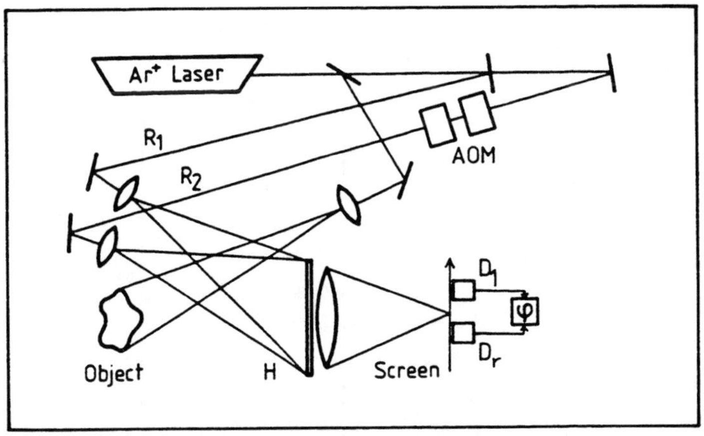

Fig. 5. Arrangement for heterodyne holographic interferometry using two acousto-optic modulators for the frequency offset.[4]

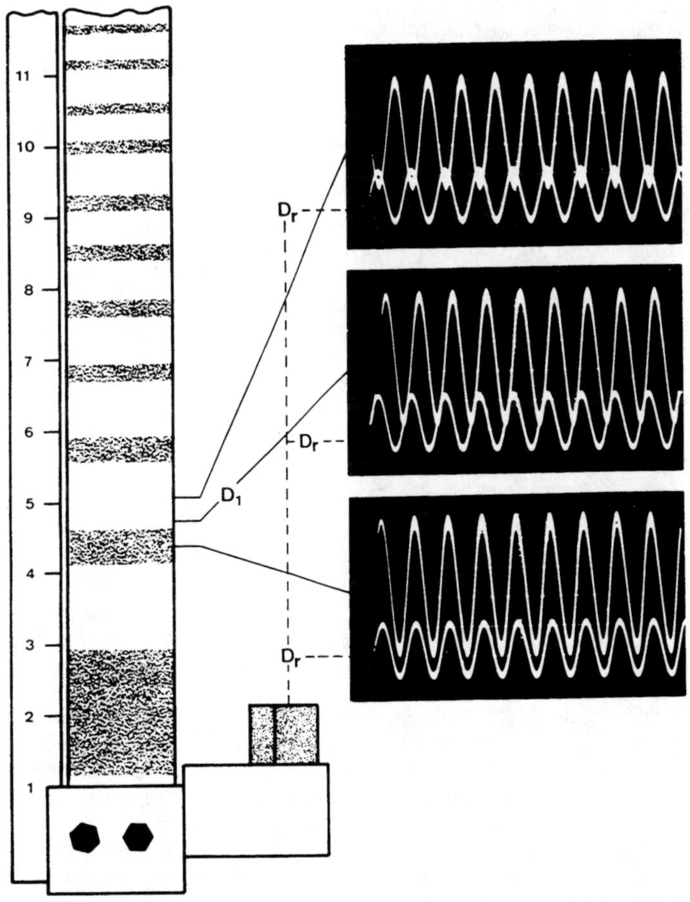

Fig. 6. Schematic representation of the interference fringes on a bent cantilever and the corresponding beat frequency signals from the two photodetectors D_r (fixed) and D_1 (scanning).[4]

where A_D is the detector area and d_I is the distance between the lens and the image plane.

3.3. Systematic errors

Additional errors of systematic nature may occur due to the fact that the hologram will usually not be exactly in the pupil plane of the imaging lens or due to nonlinear cross-talk in the hologram reconstruction.[4] The latter can be avoided by carefully cleaning the recorded scene of spurious reference light sources, such as parts of objects that remain unchanged between the two exposures.

4. HETERODYNE HOLOGRAPHIC INTERFEROMETRY

4.1. Experimental arrangement

A typical setup for double-exposure heterodyne holographic interferometry is shown in Fig. 5. The frequency difference $\Delta\omega/2\pi$ of about 100 kHz is realized by two commercially available acousto-optical modulators (AOMs) in cascade to give opposite frequency shifts. During recording, both modulators are driven with 40 MHz, so that the net shift is zero. During reconstruction, one modulator is driven with 40 MHz and the other with 40.1 MHz, so that the net shift is the desired beat frequency of 100 kHz. Figure 6 shows how the interference phase $\phi(\mathbf{x})$ is obtained by scanning the image of the object with a photodetector D_1 and measuring the phase of the beat frequency signal with respect to the reference phase obtained from a second detector D_r at a fixed point.

In practice an array of three detectors is used to determine the two phase differences $\Delta\phi_x$ and $\Delta\phi_y$ in the orthogonal directions x and y, rather than the interference phase $\phi(x,y)$ itself. The latter can be easily and accurately calculated by summation of the measured differences along a given path. All electronic amplifiers and filters in the signal path should be designed carefully to avoid phase distortion that could reduce the accuracy of the phase measurement.[12] The detector array can be realized by the ends of optical fibers of fiber bundles, which feed the light to three photomultiplier tubes. The beat frequency signals at 100 kHz are filtered with a bandwidth of 10 kHz, and the amplitudes are kept constant, independent of the intensity variations across the image, by a feedback control of the photomultiplier voltages. The phase differences $\Delta\phi_x$ and $\Delta\phi_y$ are measured with two zero-crossing phasemeters, which interpolate the phase angle to 0.1° and also count multiples of 360°, which correspond to the fringe numbers.[12] The detector array is mounted on a stepper-motor-driven stage to scan the image. Scanning and data acquisition is automated and computer controlled. The measuring time for one position, including displacement of the detector array, is about 1 s.

4.2. Accuracy

The overall accuracy and reproducibility of heterodyne holographic interferometry depend mainly on the specification of the phasemeter, the signal-to-noise ratio (SNR) of the detector signals, and the mechanical stability of the optical setup.[13] Using good experimental equipment, a phase measurement accuracy of 0.2° to 0.5° can be

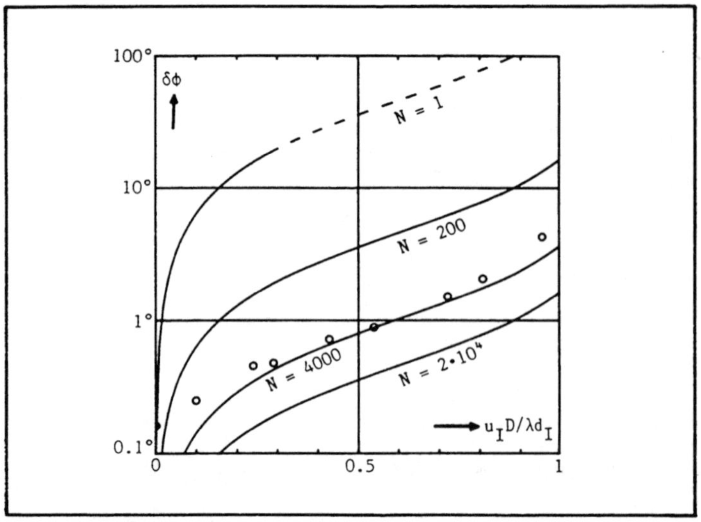

Fig. 7. Statistical phase error $\delta\phi$ due to partial speckle decorrelation versus transverse displacement u_I of the interfering speckle patterns, in terms of speckle size $\lambda d_I/D$. N is the number of speckles within the detector area. Experimental results for N = 4000 are shown.

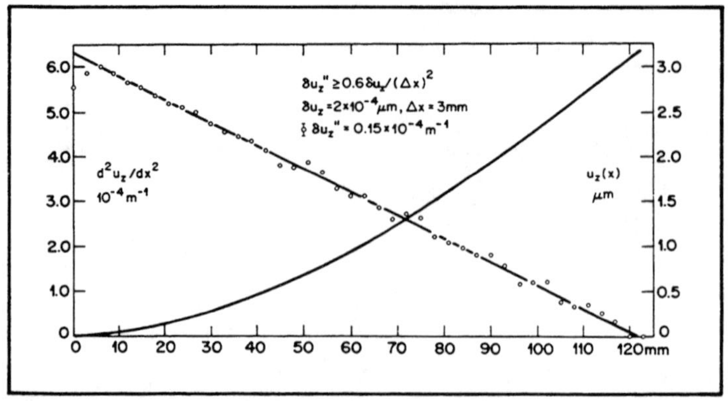

Fig. 8. Bending of a cantilever measured by heterodyne holographic interferometry. Comparison of the second derivatives d^2u_z/dx^2 of the normal displacement $u_z(x)$ with theory indicates an accuracy for the phase measurement of $\delta\phi = 0.3°$.[4]

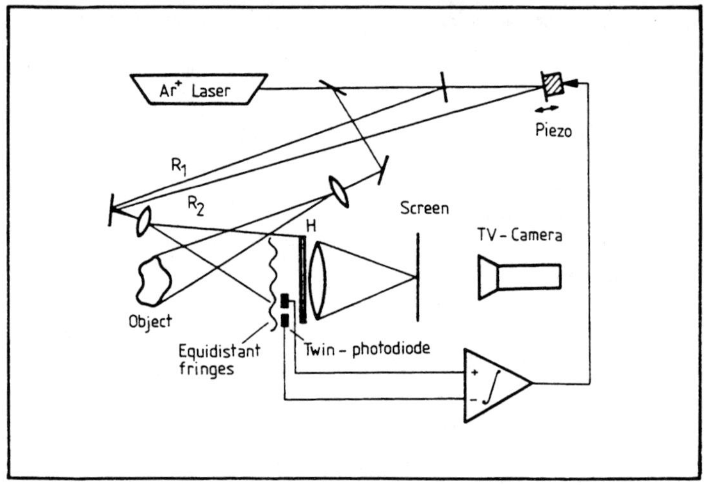

Fig. 9. Arrangement for quasi-heterodyne holographic interferometry with two reference sources close together, fringe stabilization, and TV observation.

obtained. The ultimate limits are due to loss of fringe contrast by speckle decorrelation and overlapping reconstructions, as already discussed in the preceding section. If high accuracy is required, the phase error due to overlapping cross-reconstructions from reference sources close together (Fig. 4) cannot be tolerated. For nonoverlapping cross-reconstructions, only a partial decorrelation of the interfering speckle fields (reduced fringe contrast, Fig. 3) due to a transverse component u_I of the object displacement gives rise to an additional random phase in the beat signal. The theoretical value of the statistical phase error $\delta\phi$ due to partial speckle decorrelation can be described by

$$\delta\phi^2 = \frac{1 - \gamma(2u)}{2(N+1)\gamma^2(u)} , \tag{10}$$

where $\gamma(u)$ is the fringe contrast [cf. Eq. (7) and Fig. 3] and $u = u_I D/\lambda d_I$ is the normalized transverse displacement in the image. Figure 7 shows the statistical phase error given by Eq. (10) for different speckle numbers N within the detector area. For completely correlated speckles ($u_I = 0$) this error becomes zero (no speckle noise).

The capability of heterodyne holographic interferometry is demonstrated by the experimental results given in Fig. 8.[4] The bending of a cantilever beam, clamped at the base and loaded at the end, was measured and compared with theory. The second derivative d^2u_z/dx^2 of the measured normal displacement $u_z(x)$ should follow a straight line, proportional to the bending moment M(x). The measurements were taken at intervals of $\Delta x = 3$ mm, and the comparison with theory indicates an accuracy for the interference phase of $\delta\theta = 0.3°$, corresponding to $\delta u_z = 0.2$ nm.

5. QUASI-HETERODYNE INTERFEROGRAM PROCESSING

5.1. Experimental arrangement

A typical setup for quasi-heterodyne holographic interferometry with two reference sources close together (c.f. Sec. 3.1) is shown in Fig. 9. The advantages of using reference sources close together are that, first, the requirements for correct repositioning of the hologram are low. This makes it possible to reconstruct the hologram anywhere in a similar optical setup. Even holograms recorded with a pulsed ruby laser can be realigned for reconstruction with a He-Ne laser. Second, the reference waves produce in the hologram plane a pattern of parallel macroscopic fringes of typically 1 to 2 mm separation, which can be used to stabilize the interference phase and to control the phase steps ψ very accurately. A twin photodiode placed just beside the hologram tracks the fringe maximum, and an integrating feedback loop to the piezoelectric mirror in one of the two reference beams keeps the fringes stable (Fig. 9). By translating the twin photodiode, the interference fringes will move with it. Accurate phase shifts of 120° are obtained by moving the photodiode, mounted on a stepper-motor-driven stage, in steps of 1/3 of the fringe separation.

Data acquisition and processing is carried out with a video system and a microcomputer. The interferogram is detected with a TV camera. It can be observed either directly through the hologram or by intermediate projection on a screen. A silicon target vidicon or CCD-array camera should be used to guarantee linear response for the intensity measurement. Three successive hologram reconstructions with different relative phases ψ of the reference beams are digitally stored and then analyzed pointwise by calculating the interference phase $\phi(\mathbf{x})$, the fringe contrast $m(\mathbf{x})$, and the average intensity $a(\mathbf{x})$ with the help of Eqs. (4) to (6). Video image processing systems with digital frame stores for fast data acquisition are readily available nowadays.

5.2. Accuracy

The limitations for the determination of the interference phase in quasi-heterodyne holographic interferometry as described above have been thoroughly investigated by the authors.[6] The possible sources of errors can be summarized as phase noise due to perturbations (mechanical, thermal), inaccuracy of the phase steps, speckle noise (mainly due to the overlapping cross-reconstructions), spurious interferences due to the overlapping cross-reconstructions, electronic noise of the detector (TV camera), and nonlinearity of the detector (TV camera).

To estimate the accuracy of the phase measurement, one has to distinguish between statistical and systematic errors. For the rms value of the statistical errors introduced by intensity and phase noise, $\delta I/\overline{I}$ and $\delta\psi$, respectively, the expression

$$(\delta\phi)^2 = \left(\frac{2}{3}m^2\right)\left(\frac{\delta I}{\overline{I}}\right)^2 + \frac{(\delta\psi)^2}{2} \tag{11}$$

can be derived from Eq. (4). The systematic errors will in general be periodic. Equation (11) is still valid for the rms value of the periodic phase errors.

5.2.1. *Phase errors*

The absolute value of the relative phases ψ_k between the two reference sources must be guaranteed with the same precision as desired for the interference phase $\phi(\mathbf{x})$. Any errors $\delta\psi$ in ψ_k cause systematic errors for $\phi(\mathbf{x})$ of the same magnitude as $\delta\psi$ [cf. Eq. (11)] and with the same periodicity as the interference fringes. With the stabilization scheme described above (Fig. 9), the phase noise was measured to be less than $\delta\psi = 0.4°$ rms for an integrating time of 20 ms (TV camera), and the deviations of the mutual phases ψ_k were found to be less than 1°.

5.2.2. *Overlapping cross-reconstructions (cf. Sec. 3.2)*

If the aperture of the imaging lens is too small compared with the angular separation of the two reference sources, the speckle patterns of the cross-reconstructions are not completely decorrelated [Eq. (7), Fig. 3]. This produces a spurious fringe pattern in the image, which will cause systematic phase errors with the same periodicity as the interference fringes. In addition, if the hologram is not exactly in the pupil plane of the imaging lens, the macroscopic fringe pattern of the reference beams on the hologram may be visible in the image. Its apparent periodicity depends on the distance between hologram and pupil plane; its visibility is determined by the aperture of the lens. To suppress these systematic phase errors, the lens aperture must be large enough to contain many fringes (at least seven fringes for a phase error of less than 2° rms).

The statistical phase error due to speckle noise is shown in Fig. 4 as a function of the number N of speckles integrated by the detector area. An estimation of the expected phase error for the described system with video electronic detection is difficult, because the resolution of the TV camera is not well defined. Assuming 300 resolved lines/TV-field, one obtains a spatial resolution of about 30 μm for the vidicon tube. For an aperture of the TV camera lens of $F = 2$, one gets from Eq. (9) $N \simeq 700$ speckles within a spatial resolution cell, and by applying Eq. (8), $\delta\phi \simeq 3.5°$. This value agrees well with experimental results. The error due to speckle noise can be further reduced by spatial averaging using a detector area that covers more speckles, either "hard" (electronically defocusing the TV camera) or "soft" (numerically averaging over neighboring sample points).

5.2.3. *Detector (TV camera)*

The detector introduces intensity measurement errors by electronic noise and nonlinear response. On the contrary, variations of the sensitivity across the image, or from one pixel to another in CCD cameras, do not cause any error, since the calculation of the interference phase is based on three consecutive measurements of intensity at the same position [cf. Eq. (4)].

For a silicon target TV camera, such as the RCA Ultricon, the electrical signal-to-noise ratio has been determined experimentally to be about 35 dB. Assuming a fringe contrast of $m = 0.4$, one

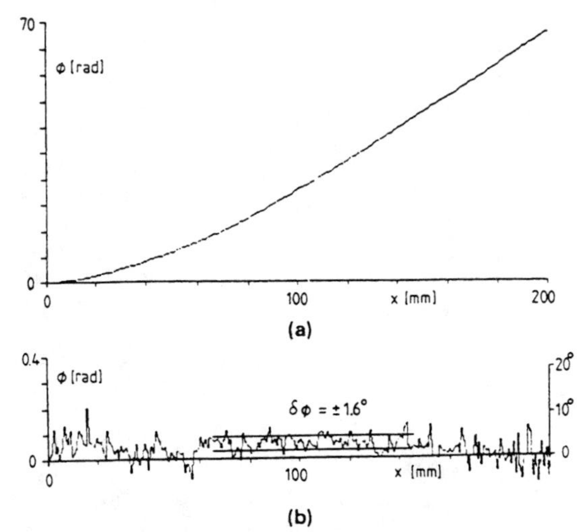

Fig. 10. Interference phase ϕ measured by quasi-heterodyne holographic interferometry. (a) Displacement of a bent cantilever versus length x. (b) Phase variations (statistical error) measured for the undeformed cantilever.[5]

obtains from Eq. (11) a statistical phase error of $\delta\phi = 2°$. Figure 10 shows the experimentally determined phase on an object without fringes. The intensity profiles have been averaged over three measurements, which reduces the expected error by a factor $\sqrt{3}$ to $\delta\phi = 1.2°$. The corresponding experimental value is found to be 1.6° (Fig. 10) in the central part of the image, where the average intensity is sufficiently high. (For this measurement, the speckle noise, which is normally of the order of some degrees, was further reduced by electronically defocusing the camera.)

A nonlinear response of the detector produces systematic errors of the same periodicity as the interference fringes. Again, the error can be estimated with Eq. (11) by taking for δI the deviation of the detector response from the best linear fit for the dynamic range of the measured intensities. The response of the TV camera mentioned above has been verified to be sufficiently linear to produce less than 1° error.

5.2.4. *Conclusions*

Quasi-heterodyne evaluation of holographic interferograms offers an accuracy of 1/100 of a fringe by using video electronic processing. The required two-reference-beam holography can be operated as easily as classical double-exposure holography by using two reference beams close together (about 0.5 mrad angular separation). The described system combines effectively the simplicity of standard double-exposure holography, the power of heterodyne holographic interferometry, and video electronic processing.

6. APPLICATIONS AND CONCLUSIONS

The described techniques for electronic processing of holographic interferograms can be applied to real-time, double-exposure, double-pulse, and stroboscopic holography. Both heterodyne and quasi-heterodyne methods offer high accuracy fringe interpolation, high spatial resolution, and the possibility for automated interferogram processing. They allow quantitative determination of displacement vector components on the surface of arbitrarily shaped objects. Using different illumination or observation directions, the high accuracy fringe interpolation also offers the possibility of automated determination of 3-D displacement and strain from a single hologram plate.[14,15] Moreover, using different illumination beams for each reference, holographic contouring can be performed in exactly the same setup.[6]

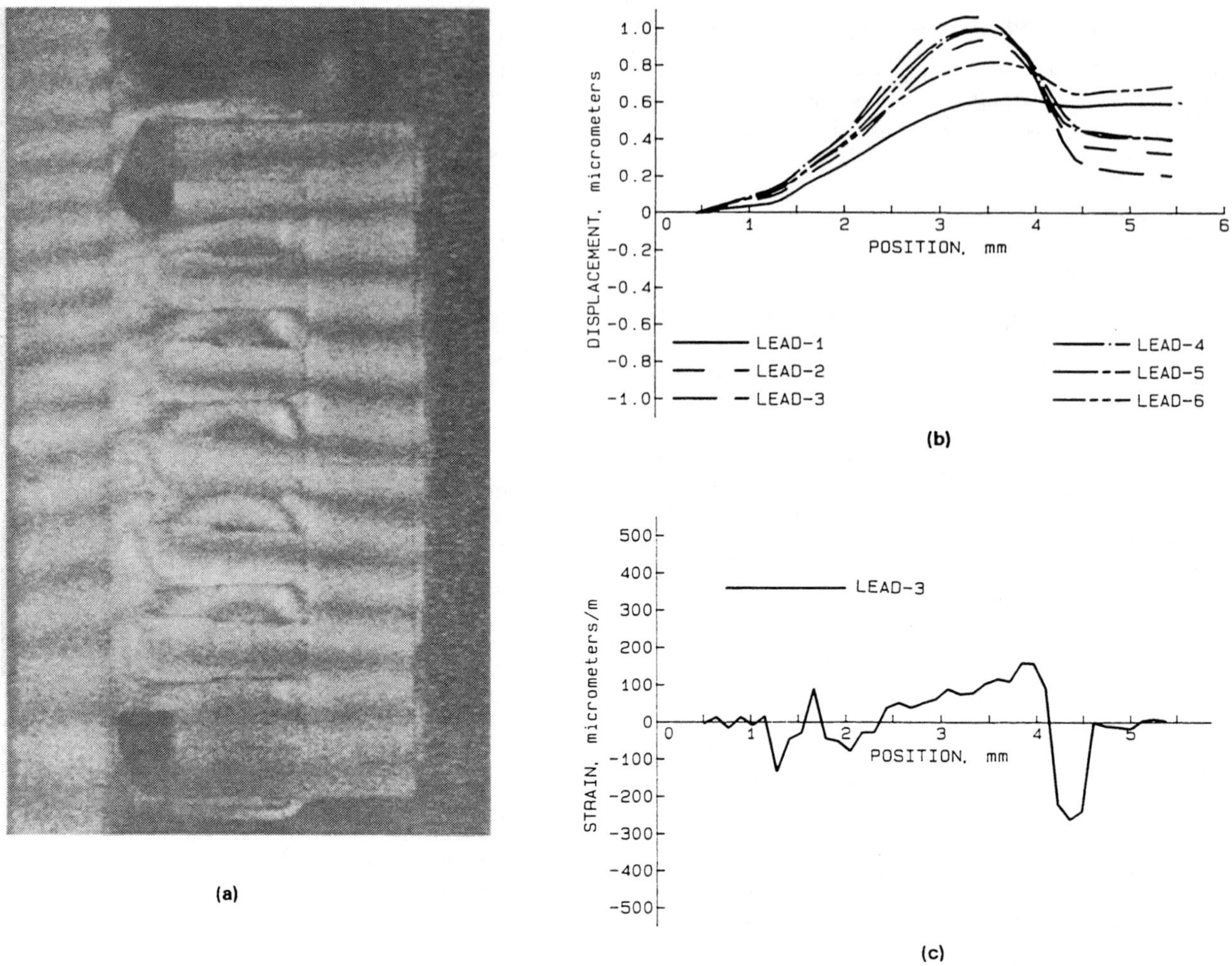

Fig. 11. Heterodyne holographic interferometry applied to displacement and strain measurements of computer chip carrier leads. (a) Typical fringe pattern. (b) Out-of-plane displacement component. (c) In-plane strain along one lead. (Figures courtesy of R. J. Pryputniewicz).

Heterodyne holographic interferometry requires sophisticated electronic equipment and mechanical scanning of the image by photodetectors. It is well suited for scientific applications in which very high resolution is needed. An application to quantitative displacement and strain measurements of computer chip carrier leads is presented in Fig. 11. The fringes were analyzed in the magnified image (about 8×) of the object with a heterodyne system similar to the one described in this paper (Sec. 4). Three different observation directions (about 20° difference) were used to determine the displacement vector at some 40 points along each lead, separated by about 0.13 mm. An accuracy of 0.5° for the measured interference phase would correspond to an error for the in-plane strain of about 10 μm/m.[15,16]

On the other hand, quasi-heterodyne holographic interferometry with TV detection is nearly as simple as standard double-exposure holography, and it does not require any special instrumentation apart from a video electronic data acquisition system. This method is very well suited for industrial applications in which high speed and medium accuracy are required. Applications to displacement and vibration measurements are shown in Fig. 12. The fringe patterns were analyzed with a quasi-heterodyne arrangement similar to the one described in this paper (Sec. 5.1) and a video image processing system with digital frame store. The deformation of the pressure-loaded CFC-panel of a radio telescope antenna [Fig. 12(a)], about 1.2 m by 1.4 m in size, was recorded by double-exposure, two-reference-beam holography using a cw argon laser. The vibration pattern of an oil pan [Fig. 12(b)] resonating at 1244 Hz was recorded with a synchronized double-pulse ruby laser in a holographic setup with reference sources close together. For the reconstruction and fringe analysis, however, a cw He-Ne laser was used. The effects of the wavelength change were compensated by readjusting the hologram and the angular separation of the reference sources.

7. ACKNOWLEDGMENTS

The authors would like to thank B. Breuckmann and R. Thieme of M.A.N. Neue Technologie, Munich, and R. Pryputniewicz, Worcester Polytechnic Institute, for making available some of their recent experimental results to illustrate the presented methods with actual applications. This work was supported in part by the Swiss Science Foundation.

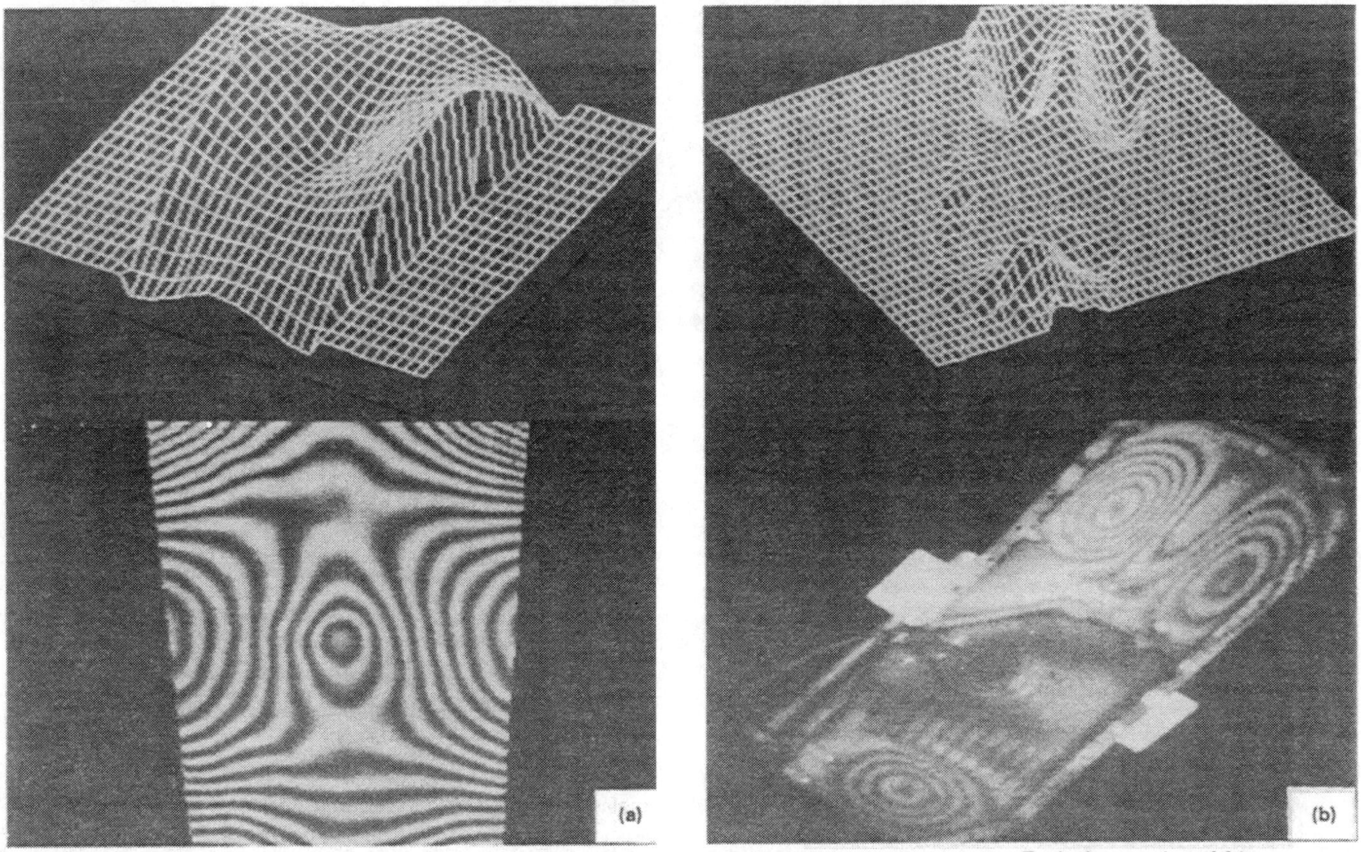

Fig. 12. Quasi-heterodyne holographic interferometry applied to displacement and vibration measurements. Typical examples of fringe patterns and deformation plots displayed on the monitor. (a) Out-of-plane displacement of a pressure-loaded CFC-panel of a radio telescope antenna. (b) Vibration amplitude of an oil pan resonating at 1244 Hz. (Figures courtesy of B. Breuckmann and W. Thieme.)

8. REFERENCES

1. Special issue on *Precision Surface Metrology,* J. C. Wyant, guest ed., Opt. Eng. 23(4), pp. 349-395 (1984).
2. W. Jüptner, T. M. Kreis, and M. Kreitlow, in *Industrial Applications of Laser Technology,* William F. Fagan, ed., Proc. SPIE 398, 22 (1983).
3. P. Hariharan, B. F. Oreb, and N. Brown, Appl. Opt. 22, 876 (1983).
4. R. Dändliker, in *Progress in Optics,* Vol. XVII, E. Wolf, ed., pp.1-84, North-Holland, Amsterdam (1980).
5. R. Dändliker, R. Thalmann, and J.-F. Willemin, Opt. Commun. 42, 301 (1982).
6. R. Thalmann and R. Dändliker, in *1984 European Conference on Optics, Optical Systems, and Applications,* B. Bölger and H. A. Ferwerda, eds., Proc. SPIE 492, 299 (1985).
7. G. S. Ballard, J. Appl. Phys. 39, 4846 (1968).
8. R. Dändliker, E. Marom, and F. M. Mottier, J. Opt. Soc. Am. 66, 23 (1976).
9. R. Dändliker, B. Ineichen, and F. M. Mottier, Opt. Commun. 9, 412 (1973).
10. R. Crane, Appl. Opt. 8, 538 (1969).
11. J. H. Bruning, D. R. Herriott, J. E. Gallagher, D. P. Rosenfeld, A. D. White, and D. J. Brangaccio, Appl. Opt. 13, 2693 (1974).
12. J. Mastner and V. Masek, Rev. Sci. Instrum. 51, 926 (1980).
13. R. Dändliker, in *Progress in Optics,* Vol. XVII, E. Wolf, ed., pp. 54-60, North-Holland, Amsterdam (1980).
14. R. Dändliker and R. Thalmann, in *Industrial Applications of Laser Technology,* William F. Fagan, ed., Proc. SPIE 398, 11 (1983).
15. R. J. Pryputniewicz, Int. Arch. Photogram. 24, 377 (1982).
16. R. Dändliker and B. Eliasson, Exp. Mech. 19(3), 93 (1979).

Reprinted from *Optical Engineering,* Vol. 14(5), pp. 482-489
(September-October 1975).

A Review of Speckle Photography and Interferometry

Karl A. Stetson

Instrumentation Laboratory, United Aircraft Research Laboratories, East Hartford, Connecticut 06108

Abstract

Since its discovery in 1969, speckle interferometry has become a very active area of research, so much so that there now exists quite a collection of techniques that are applicable to a wide variety of problems. Because of the surprising versatility offered by such a wide range of techniques, the potential user of such technology may become confused as to what can and cannot be done, and, further, what process is suitable for what task. This review is provided, therefore, to set the various techniques into context, point out their similarities, and delineate their differences.

Introduction

To me one of the most revealing comments ever made about laser speckle was by Dennis Gabor.[1] In a thorough exposition of the detrimental aspects of laser speckle in holography, and the various methods of dealing with it, he said: "The novelty in holography is speckle noise. This has really nothing to do with holography; it is a consequence of coherence. It has a special standing among noise phenomena because it is not really noise but unwanted information. When we put a sheet of white paper into laser light, the reflected light conveys to us information on the roughness of the paper in which we are not at all interested. I am glad that at last Dr. J. M. Burch has shown that we can make some use of this information." The reference was to a paper in the same conference.[2] What intrigued me, aside from the obvious thought that one man's bane is often another man's boon, was that even armed with a "holographic crystal ball" the scholar sometimes seems helpless to truly predict the future of science and technology. Who could have guessed that the universal nuisance of holography would have proved to be more useful, in some ways, than holography itself? And having missed the guess, how many scholars would be willing to say that the most practical thing holography has done is to have called our attention to laser speckle?

I certainly cannot fall into either category myself and like to regard myself as a partial convert and a partial skeptic. I could not have prognosticated the discoveries because I, like many others, was unaware of the history of interferometry with scattered light.[3] My own initial reaction to interferograms produced with laser speckle was that the contrast of the fringes was poor and the resolution of the objects was unsatisfactory. In short, the pictures looked unpleasing and, therefore, I felt that the whole field might become a passing fad. On that score, however, I had the good fortune to spend two years at one of the primary centers of activity in the field (The National Physical Laboratory, Teddington, England) and I was able to watch some of the developments first hand. As a consequence, I firmly believe that these techniques will have a clear place in the future of optical metrology. I believe, however, with equal firmness that there are many measurement problems that will always be handled better by the generalized interferometry that the hologram has made possible.

H-113 received April 10, 1975.

In this paper, I shall attempt to review the techniques of speckle photography and speckle interferometry so as to provide as comprehensive a guide as possible to these techniques. The techniques will not be treated in chronological order, because of the inevitably labyrinthine path of discovery, but rather in order of complexity. I shall, however, try to indicate the initial contributors for each major technique. A discussion of the properties of laser speckle patterns that are relevant to these techniques will begin the paper. These will include their distributions of brightness and spatial frequency and their movements relative to the motions of diffusely reflecting objects. Speckle photography with a single illumination beam will be considered next and the various measurements of object motion and strain that can be accomplished thereby. Some discussion will be given also to films to be used for this work. Next the subject of speckle interferometry will be considered wherein two or more illumination beams are used or where shearing of the image is employed. The utility of frequency plane processing of the information on speckle photographs will be discussed for both speckle interferometry and speckle photography. Finally, a number of concomitant systems will be discussed.

Before beginning the body of this paper, a brief discussion of the nomenclature to be used is in order. By this point, many may well wonder: what is speckle interferometry; what is speckle photography; and what is the difference? Strictly speaking, there really is no difference; the two terms are assigned somewhat arbitrarily. In photographing an object illuminated by laser light, interference occurs between the fields passing through various portions of the lens aperture, and it is this interference that causes speckles to appear in the image. Furthermore, practically all techniques in this paper involve making photographs of the object under study. Both interference and photography are always involved, therefore, but because there are generally two physical phenomena employed, I will make a distinction in terminology. Most of the techniques that will be considered here involve making two exposures, on a photographic film, of an image of the object under study. The resulting transparency is then examined in one of a number of ways that will yield fringes that indicate motions of the object. If, for the speckle pattern of one image, there is always a region on the second image where its speckle pattern is well correlated to that of the first, and the fringes result simply from translation of the two speckle patterns, then I will call the process speckle photography. If the fringes are derived from fluctuations in the correlation of the speckle patterns between the two images, whether or not there is a translation between the correlated portions of the patterns, then I will call the process speckle interferometry. These definitions will also hold if either multiple exposures or time-average exposures are used. I hope that the reasons for this distinction will become clear when the processes are described.

Pertinent Properties of Speckle Patterns

When laser light is incident on a finely irregular surface, the fine structure of the surface (which may even be small with respect to the wavelength of light) imparts a random phase modulation to the reflected light. In its propagation away from the surface, the effect of some type of aperture may be observed on the field. The aperture may be present due to any one of several possibilities; for example, the limited size of the object, a limited region of illumination, or the finite size of some lens system that images a particular plane in the object field. The effect of the aperture is to place a limit on the angle between the directions of propagation of the components of the field. The field at any point in space may be thought of as the sum of components arriving from a variety of directions that are determined by the size and shape of the aperture. The amplitudes of the field components may be nominally the same, but their relative phases will be random. If a large number of field points are considered, then the likelihood increases that all possible combinations of phases of field components will occur. The resulting irradiances of the field points will be randomly distributed, spatially, but will be distributed in brightness according to certain statistical rules. Although the statistics of laser speckle patterns have been discussed many times and in many contexts, the reader is referred here to two articles by J. M. Burch[2,3] because the treatments presented therein are specifically geared toward speckle photography and interferometry, and they are quite comprehensive. We shall extract, at present, one fact from those treatments; i.e., that the most probable brightness in a laser speckle pattern is zero and the probability of all other brightness decreases exponentially with increasing brightness. This presumes that all components in the field possess the same polarization, which is generally true only of reflections from metallic surfaces. A depolarized reflection could be described by the incoherent addition of two separate, orthogonally polarized, speckle fields. The statistical distribution of brightness in a simple speckle pattern means, however, that a photographic film exposed to such a field will possess a higher level of transmittance than might be expected after development owing to the considerable number of locations where the exposure will be zero, i.e., the black spots in the speckle pattern. This assumes, of course, that the film is capable of resolving these spots.

That last point brings us to another consideration in simple speckle patterns; the distribution of spatial frequencies. These are related directly to the size and shape of the aperture that governs the formation of the speckle. An excellent discussion of this subject was provided by H. J. Tiziani,[4] and from this we may extract the following relationship: the distribution of the spatial frequencies in the irradiance field of a laser speckle pattern may be obtained by convolving the associated aperture with itself. It should be emphasized that this yields only the envelope of the frequency distribution. The actual Fourier transform of a speckle irradiance pattern will itself have the same type of random fluctuations, or speckles, as the original speckle pattern. This confirms the notion that once you have speckle, it is quite difficult to get rid of it. The frequency distribution of greatest interest is that of a circular aperture, because most lens systems have circular exit apertures. That distribution is well known for analyses of diffraction-limited, optical transfer functions[5] and is

$$F = (2/\pi)[\cos^{-1} f - f(1-f^2)^{1/2}], \quad f < 1, \tag{1}$$

$$= 0, \quad f > 1,$$

where F is the frequency distribution and f is the radial spatial frequency normalized to the cut-off frequency. The cut-off frequency f_c is given in terms of the wavelength of light and the f/number of the cone subtended by the aperture and the field point by the approximation

$$f_c \approx 1/\lambda\,(f/n)\,. \tag{2}$$

The upper curve of Fig. 1 shows a radial plot of the frequency distribution in the speckle pattern generated by a circular aperture, and it can be seen that it deviates only modestly from a straight line between 1.0 and the cut-off frequency, f_c. To obtain a more accurate representation, however, it is common to take the cut-off frequency to be 1/1.2 of the value in Eq. (2) and represent the distribution as a straight line.

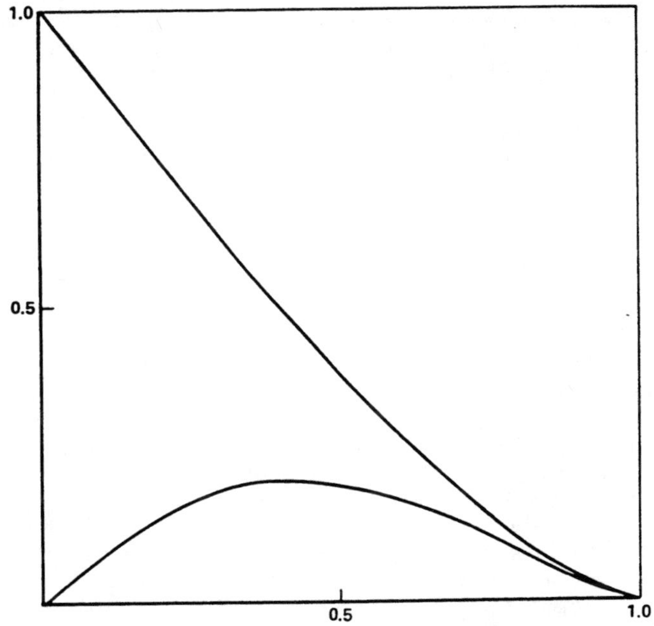

Fig. 1. Relative spatial-frequency distributions in simple speckle patterns for circular apertures. The upper curve is the envelope of two-dimensional frequency spectrum plotted versus normalized radial frequency. The lower curve is the relative concentration of spatial frequencies independent of angular orientation.

The motion of speckle patterns with motions of an object has been discussed extensively by Tiziani[4,6] and by Archbold and Ennos,[7] among others. From these discussions we may conclude that speckles in the image of an object will either shift their location as the object moves, or they will change their pattern, or both. Changes in the pattern of the speckles will cause a loss of correlation between the speckle patterns in subsequent photographs of the object. These changes in speckle pattern can occur in a number of ways, as categorized by Archbold and Ennos.[7] If the object moves toward the lens more than the focal depth of the imaging system, a different plane in the image field will be recorded in which the speckle pattern will be randomly different than the original one. If the object rotates about an axis in its own plane, the field in front of the camera lens will translate in a direction, and by a magnitude, that is related to the rotation of the object, and to the angles of illumination and observation relative to the surface of the object. (Near the specular reflection angle of the object surface, the field motion approximates that obtained by simple mirror reflection. At other angles, the vectorial grating equation must be used, wherein the field in front of the lens is assumed to be diffracted there by a grating-like structure on the surface of the object.) If the field translation is sufficient to cause a different portion of the scattered field to pass through the lens aperture, then the speckles in the image will be different. If the field penetrates into the object sufficiently, or if multiple reflections occur on the object surface, the field scattered by the object may be highly sensitive to changes in the angle of illumination. Thus, rotations of the object relative to the illumination will affect the speckles formed by the imaging lens. If none of these factors operate, then the speckles in the image of the object will retain their pattern with motions of the object, and, furthermore, if the object is in focus, they will retain their position on the object itself. If the object is not in focus, translations of the speckles relative to the object surface may be expected if the surface of the object rotates. Finally, any motion of the object that tends to increase or decrease the optical path length between the source of illumination and the point of observation will change the optical phase of the pattern. This, of course, will not be observable unless a second field is present with which the first interferes.

These, then, are the properties of speckle patterns that are primarily essential to speckle photography and to speckle interferometry.

Speckle Photography

Measurement of Object Translation

Let us consider a photograph in which a double exposure has been made of an object before and after a translation perpendicular to the optical axis of the camera. If correlation of the speckles is maintained between the two exposures, the speckles behave like a random moiré grid attached to the surface of the object. Unlike a real moiré grid, however, the fineness and contrast of the speckle pattern will not depend upon lens aperture aberrations. Further, and also unlike a real moiré grid, the speckle pattern will be aperiodic, and the conventional methods of observing moiré interference are not applicable. Consider a transparency of the photograph (e.g., the original photographic negative); it consists of two identical transmittance functions that are displaced from one another. If such a transparency is placed in a convergent beam of coherent, monochromatic light, the field at the focal plane of the convergent beam will be proportional to the two-dimensional Fourier transform of the transmittance function of the transparency. Because the transmittance function can be regarded as the sum of two identical, but displaced, transmittance functions, a simple application of the shift theorem in Fourier transform theory shows that the field in the transform plane will be multiplied by a cosinusoidal fringe pattern. The fringes will be aligned perpendicular to the direction of displacement between the two patterns and the fringe spacing will be inversely proportional to the magnitude of the displacement. These fringes are sometimes referred to as Young's fringes because they correspond exactly to those that would be obtained by taking any pair of corresponding points in the two speckle patterns as point sources of mutually coherent light. This construction makes it immediately obvious as to the relationship between the fringe spacing and the speckle displacements, which is

$$L_i/\lambda = f/d \qquad (3)$$

where L_i is the displacement of the speckle patterns, λ is the wavelength of the light in the observation beam, f is the distance between the transparency and the transform plane (assuming no intervening lens), and d is the fringe spacing.

If the beam that illuminates the transparency is made small enough so that essentially a region of uniform displacement on the image is being illuminated, then the speckle displacements of that region can be determined by the fringes observed in the transform plane. If the displacements vary across the surface of the object, a point by point application of this technique will allow a point by point determination of the object displacements. By noting both the spacing and orientation of the transform plane fringes, both the magnitude and direction of the object motion can be determined in a plane perpendicular to the optical axis of the camera that took the picture. Care must be taken to note the demagnification of the object in the recording of the image when the measured displacements are related back to the original object. If the total, three-dimensional motion of the object is required, a second photograph of the same surface, taken from a different angle, will allow the determination of the object motion in a different plane. This provides sufficient information for the complete determination of the vectorial motion to within a plus or minus sign. This, then, is the simplest technique of speckle photography, which has been discussed by Archbold, et. al.[7,8] Tiziani[4,6,9,10] and Kopf.[11] It would appear that the first publication of this method was Ref. 8.

Analysis of Object Vibration

The technique of speckle photography outlined above is also applicable to vibratory motions that are perpendicular to the axis of the recording camera. In this case, the two displaced speckle patterns become replaced by an ensemble of patterns. Any one point in the speckle pattern traces a line between the two points determined by the extremes of the vibration cycle, and the relative exposure at each point along the line will be proportional to the reciprocal of the velocity of the speckle at that point. The analysis of this case, first performed by Tiziani,[4] leads to the expected conclusion that the previously obtained cosine fringes become replaced by zero-order Bessel-function fringes when the double-exposure of two object positions is replaced by a time-average exposure of a sinusoidal vibration. The extension to generalized characteristic functions is straightforward, and recently, Archbold and Ennos[12] have published an excellent set of experiments for vibratory motions that are orthogonal and at rational frequencies. In these experiments, a single point on the speckle pattern traces a Lissajous figure.

Measurement of Object Strain and Rotation

Although the simple technique of speckle photography would, at this point, seem to provide everything we might wish to know about the motions of an object by yielding a complete vectorial determination of the object motion at any point on its surface, in fact a great deal is lacking. It is often desirable to determine strains and rotations of objects rather than total displacements. Determining rotations and strains by taking differential values of displacements can often lead to serious errors when the differentials begin to approach the error tolerances on the values of the displacements. Therefore, it is of great interest to determine strains and rotations as directly as possible. Object rotations about axes in the plane of the surface, sometimes called tilts, can be determined by photographing patterns formed in the far field of the object. Such patterns translate when the object is rotated by an amount that can be directly related to the object's rotation. The far field patterns can be obtained by means of a lens that will display the Fourier transform of the field scattered by the object, as suggested by Tiziani,[6] or by illuminating only a small region of the object, as suggested by Archbold and Ennos.[7]

Rotations in the plane of the object surface and strains of the surface itself can best be obtained by subjecting the speckle photograph to optical processing by means of a spatial frequency filter. Figure 2 shows what is perhaps the most practical arrangement for doing this. A laser beam from the left passes through a lens-pinhole smoothing filter which provides a diverging beam of nearly constant amplitude. This is followed by a lens (which focuses the beam at a convenient distance), a transparency of the speckle photograph, and a camera. The latter two (transparency and camera) are placed so that the speckle photograph is imaged on the film plane of the camera, and the entire fixture (containing the transparency and the camera) is gimbled to rotate about the center of the transparency. Rotation of the fixture will translate the aperture of the camera lens away from the point where the laser beam is brought to focus, and, by allowing two axes of rotation, the aperture can be translated in two orthogonal directions. During any such rotation, however, the transparency remains focused on the film plane of the camera.

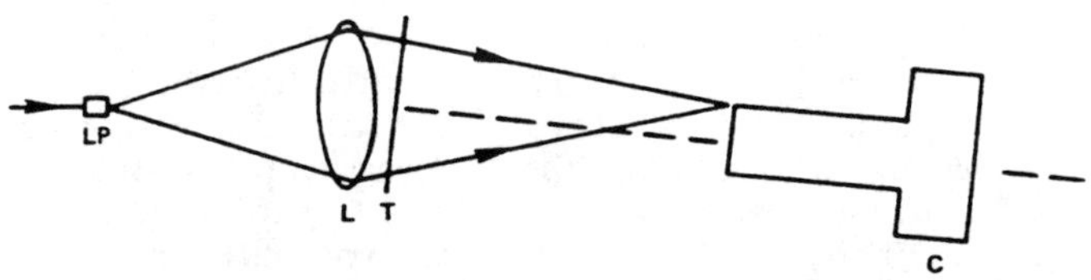

Fig. 2. An optical processing system for speckle photographs. LP is a lens-pinhole smoothing filter, L is a lens, T is the transparency of the speckle photograph, and C is a camera.

As discussed previously, the field at the plane where the laser beam is brought to focus will be proportional to the Fourier transform of the transmittance of the transparency. This remains true, to a reasonable approximation, when the transparency is tilted, or when the plane at the focus of the beam is replaced by a spherical surface centered on the transparency. (In fact, the spherical symmetry used in this processing scheme improves the accuracy of the Fourier transform representation rather than detracts from it.) The aperture of the camera lens transmits a small portion of the frequency spectrum of the transparency into the camera, and the camera performs an inverse Fourier transform on this segment of the spectrum. Thus, an image is formed of the speckle photograph that is derived only from a certain band of spatial frequencies in the original photograph.

If the camera aperture is set to admit only a certain portion of the transform of the speckle photograph, then those regions on the original photograph where the translation of the speckle pattern between the two exposures was such as to cause a dark fringe to fall on the aperture will appear dark in the processed image, and those regions where the translation was such as to cause a bright fringe to fall on the aperture will appear bright. Clearly, if all points on the object surface moved by the same amount, the image would appear uniformly dark, bright, or in between, depending upon the location of the aperture and the magnitude of the displacement. Variations in translations, whether achieved by strain, shear or in-plane rotation, will yield fringes in the resulting image; the strain, shear, or in-plane rotation will be inversely proportional to the fringe spacing, which, in turn, is inversely proportional to the angular displacement between the camera aperture and the axis of the converging laser beam.

The interpretation of the fringes obtained by spatial-frequency filtering of speckle photographs is reasonably simple; they are contours of constant displacement in the speckle photograph in the direction of displacement of the camera aperture from the point of zero-order diffraction of the laser beam. Thus, if the aperture is moved in the x direction, we measure contours of constant u, the displacement of the object in the x direction, and if the aperture is moved in the y direction, we measure contours of constant v, the displacement in the y direction. If a photograph is recorded that measures the x displacements, the fringe spacing in the x direction is inversely porportional to the rate of change of u with x, or symbolically u^x where we use a superscript to denote partial differentiation in the superscript variable. Similarly, the fringe spacing in the y direction for the same photograph gives u^y. From a photograph that measures the y displacements we may obtain v^y and v^x. From this data it is possible to separate strains, shears and in-plane rotations. This is done by putting the four values into a matrix, and expressing it as the sum of three matrices; the strain, shear, and rotation matrices. The strain matrix has only the two major diagonal terms; the shear matrix has only off-diagonal terms, but is symmetric; and the in-plane rotation matrix has only off-diagonal terms but is antisymmetric.

A very helpful variant of this technique was demonstrated by Duffy[13,14] wherein the aperture of the camera used to take the original speckle photograph was composed of two small apertures equidistant from the optical axis of the lens. Such a speckle has a high concentration of spatial frequencies in a band determined by the separation of the apertures in the taking lens. When the transform of this speckle photograph is displayed, a great deal of light is concentrated at the angles to the zero order that correspond to the angular separation of the taking-lens apertures. Spatial-frequency filtering with the filter aperture centered at either of these angles will yield a brighter processed image than if the entire taking-lens aperture had been used. Metrologically, however, there is no difference between this technique and that using the entire aperture. Furthermore, the technique works out to be more practical if the lens aperture is divided into four small apertures in the pattern of a square. This allows the same ability to determine x and y components of motion as with a single large aperture. The main disadvantages of the technique lie in a loss of light in the exposure of the original speckle photograph, and in a reduced ability to adjust the sensitivity of the measurements of x and y displacements. An additional trouble comes from the fact that virtually no lens is designed to yield good imagery with only the edges of its aperture open; in fact, they are usually designed to work better for the center of the aperture than for the edges. Use of the technique, therefore, becomes a matter of balancing pros and cons in any particular application.

Recording Films

The issue of the amount of light available for recording a speckle photograph leads logically to the choice of films for this work. Clearly, the film used should have high modulation transfer factor (MTF) for the range of spatial frequencies in the speckle pattern. To obtain a better understanding of the range of spatial frequencies in a speckle pattern, let us return to Fig. 1, and recognize that the two-dimensional spectrum of a speckle pattern is a figure of revolution about the vertical axis of the upper curve. Because the angular orientation of a spatial frequency is immaterial with regard to film resolution, this must be compensated for in reducing two-dimensional spectrum to a single magnitude for each spatial frequency involved. We can do this by simply multiplying the upper curve by the normalized radial spatial frequency to obtain the lower curve. It can be seen that the lower curve has appreciable value up to the reduced cut-off frequency, $.833f_c$. Thus we should require the film to have a high MTF up to the reduced cut-off frequency of the speckle pattern. If the film fails to have this resolving power, then the light energy in the higher range of spatial frequencies will merely act as background fogging the exposure.

For photographic films, speed and resolving power are inversely related, and choices run between high speed, low resolution films and low speed, high resolution films. The tradeoffs between these have been discussed quite adequately by Archbold, et. al.,[8] Who conclude that so far as shortening exposure times is concerned the advantage lies with the high resolution films. This is because resolution requirements only increase linearly with numerical aperture (approximately the inverse of f/number), whereas light admitted by the lens increases as the square of numerical aperture. There are situations, however, when the aperture must be reduced to avoid detecting bulk translations of the object. In these cases the advantage lies with the high speed film because it will compensate for the loss of light.

An additional aspect of recording speckle photographs, that is often overlooked, is that bleaching the negative transparencies to a silver halide will often render them more suitable to optical processing than leaving them in the black silver form. In the optical processor, the speckle photograph diffracts light into the pattern corresponding to the spectrum of the spatial frequencies of the speckle pattern. In the black silver form, the transparency does this by modulating the amplitude of the field that passes through it, whereas in the silver halide form it modulates the phase of the same field. The latter process is much more efficient with respect to diffraction than the former, and brighter images are formed at the output of the processor. Furthermore, variations in exposure of the speckle photograph tend to be less important when the silver is converted to a silver halide, and the processed image becomes more even in tone.

Speckle Interferometry

Having discussed the techniques of speckle photography we may now discuss speckle interferometry, whose history predates that of speckle photography. It is certain that the initial paper on this topic was presented by Leendertz in July 1969 at a conference of the ICO,[15] which was followed by a journal publication in 1970.[16] It was these papers that initiated the field of speckle interferometry and thereby led to speckle photography. Other papers quickly followed.[8,17,18]

The Basic Technique

To illustrate the basic idea of speckle interferometry, let us consider Fig. 3 which shows a Twyman-Green interferometer wherein the mirrors have been replaced by diffusely reflecting surfaces. If the light illuminating the interferometer comes from a linearly polarized laser, and the diffuse reflections do not depolarize the light, then the field where lens forms images of the two surfaces will contain a well modulated speckle pattern. The *character* of the speckle pattern at the image plane will be the same whether one, the other, or both diffuse surfaces are present. Furthermore, the *character* will not change if either surface is translated or rotated by a small amount. If, however, one surface is translated in a direction normal to its surface by

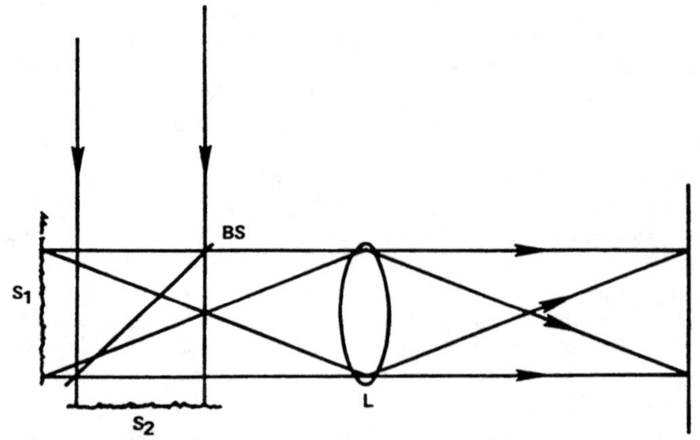

Fig. 3. A speckle interferometer based on the Twyman-Green interferometer. BS is a beam splitter, S_1 and S_2 are diffusely reflecting surfaces, L is a lens, and F is a photographic film.

one half of the wavelength of the laser light, not only the *character* of the speckle pattern at the image plane will be the same, but the *pattern* itself will be the same as before the translation. This results from the fact that the pattern at the image plane is the result of interference between two mutually coherent, but randomly speckled, fields. When the relative phase of the two fields changes by 2π they will interfere in exactly the same way as when the relative phase is 0 or 4π, etc. The phase referred to here is, of course, added in bulk to the random phase distributions of the speckled fields. When the relative phase is π, or an odd multiple thereof, the pattern resulting from the interference of the two fields will be randomly different than when the phase is an even multiple of π.

Now let us consider a double-exposure recording at the image plane of Fig. 3 before and after one of the surfaces has been rotated, by a small angle, about an axis in its surface. The rotation will introduce a linear variation in phase across the image plane of one of the two speckle fields, and, as a result, the speckle pattern of the second exposure will periodically match and differ from that of the first exposure. Where the two patterns differ, the modulation will be less than where they match. After the double-exposure recording has been developed, this periodic variation in spatial modulation may be extracted in at least two ways.

1. The nonlinearities of the photographic film can be employed. Because of the persistance of the black speckles where the patterns match, there will always be a residual transmittance in those regions even after considerably overdevelopment. Where the patterns do not match, there will always be some filling in of the black speckles of one pattern by the bright speckles of the other, and overdevelopment will render such regions quite low in transmittance. Thus, the regions where the speckles match will appear more transparent than where the speckles do not match, and the result will be light and dark bands in the photograph that correspond exactly to the light and dark fringes that would have been obtained where the diffuse surface replaced by plane mirrors, one of which was given the same rotation.

2. The photographic transparency could be bleached into a silver halide. The regions of matching speckle will be very effective in diffracting light that passes through the transparency relative to areas where the speckle patterns do not match. An optical processing system such as that of Fig. 2 can be used to detect this by placing the camera on axis, blocking off the zero-order beam, and widening the camera aperture to accept the light diffracted away from

the zero-order beam.

The preceding example should make it possible to see the difference between speckle photography and speckle interferometry. Speckle interferometry usually allows the construction of some equivalent optical interferometer, whereas for speckle photography such a construction would be more contrived. The key distinction, however, is the formation of fringes by fluctuation in correlation, or matching, of the two recorded speckle patterns.

Strain Analysis

If speckle interferometry were limited to the example given above, interest in it would not have become very widespread. Leendertz recognized, however, that the two speckle fields could be derived from the same object by illuminating its surface with two mutually coherent laser beams as in Fig. 4. The relative phases of the two scattered fields will change only for motions of the object parallel to the vector difference of the propagation vectors of the two fields, and, when the two beams make equal angles to the surface normal of the object, this direction lies in the surface of the object. Thus, fringes obtained in this configuration will measure variations in the in-plane motions of the surface, and they are suitable, therefore, for measuring in-surface strains.

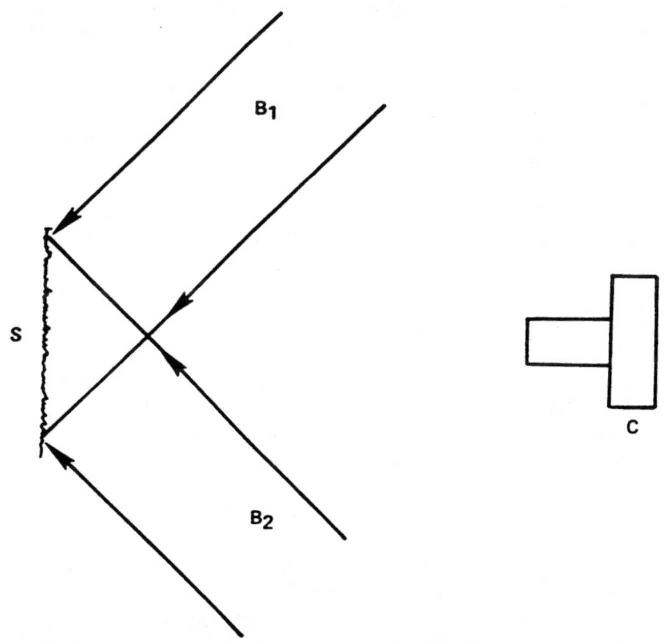

Fig. 4. A speckle interferometer sensitive to lateral motions of an object surface. S is the object surface, B_1 and B_2 are two illumination beams, and C is a camera.

In the foregoing, it was assumed that lateral motions of an object in Fig. 4 were so small that there was essentially no lateral shift of the speckle pattern in the image due to the motion of the object, or if so, it was small relative to the size of the speckles. It was inevitable that in some experiments this would not be the case, and when this occurred the effects associated with speckle photography were observed. It was noticed, however, that when the techniques of spatial-frequency filtering were used, as for processing speckle photographs, the effects of the speckle interferometry could still be seen. Butters and Leendertz discussed this from the point of view of introducing such shifts into the recording by translating the photographic film between exposures.[17] This enhanced the visibility of the speckle interference fringes. Archbold, et. al, discussed it from the point of view of separating large displacements from small strains.[8] That issue was also discussed analytically by this author, and certain of the relationships more clearly drawn.[19] To the best of my knowledge, only one application of this technique to more than two illumination beams has been tried,[20] and it was shown to allow measurement of complete strain tensors. Furthermore, the work reported by Kopf[21] indicates that it is not absolutely necessary to require that the photographs to be compared be a double exposure of the same photographic plate. This could suggest the possible application of speckle photography and speckle interferometry to high speed photography.

Analysis of Bending Strains

With respect to the issue of strain measurement, the two-beam method of speckle interferometry is much more sensitive than speckle photography because it is possible to obtain a far higher angle between two illumination beams than between two portions of a lens aperture. When strains occur due to bending of the object, however, they may be too small to measure by either technique if the object is too thin. Nonetheless, bending strains can be related to the second derivative of bending deflection, and can be determined at very low levels in this way. For this reason a technique was developed, apparently independently by Leendertz and Butters[21] and by Hung and Taylor,[23] to obtain fringes on an object surface that are related to contours of constant slope of out-of-plane deflection. The general idea is illustrated in Fig. 5. The object is illuminated with laser light, with as low a curvature as possible, and photographed through an optical device that will shear the field so that two images will be formed at the film plane of the camera that are displaced by a known amount. The image shear can be accomplished in one of several ways; for example, with a small Michelson interferometer, with a roof reflector where the roof angle is not exactly 90°, or even with a grating. The speckle in the image plane is formed by interference between the two sheared images. If the object translates as a whole in any direction, the relative phases between the two images will not change; however, if the object rotates about an axis perpendicular to the direction of shear, a phase change will occur. The rotation of the object imparts a linear phase change across its surface to the field it scatters. This imparts a linear phase change to both of the images in the camera. However, the two linear changes are offset by the shear of the images. The result is a bulk phase change between the two speckled fields that interfere to give the image speckle, and it is proportional to both the shear and to the rotation. If the object is bent to a cylindrical shape whose axis is perpendicular to the direction of shear, the rotation of the surface will vary linearly across the surface. This will impart a linear phase change between the two sheared fields and the result-

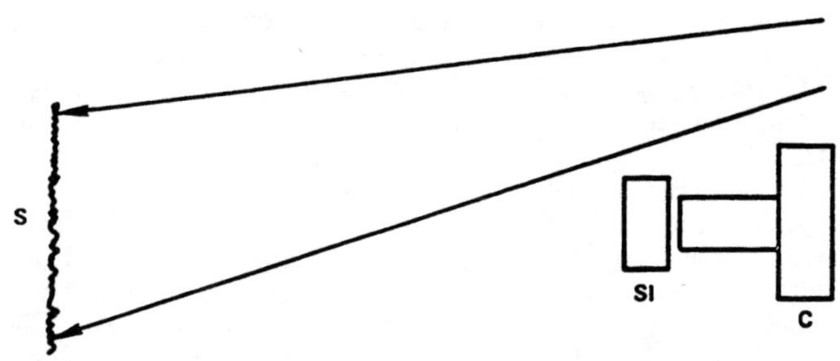

Fig. 5. A shear-speckle interferometer. S is an object surface, SI is a shearing interferometer, and C is a camera.

ing speckle pattern will go through a cyclic fluctuation of correlation with the speckle pattern that was formed before the object was bent. Therefore, a double exposure is made before and after the object is subjected to out-of-plane bending, and the resulting photographic transparency is optically processed to obtain the fringes of speckle correlation.

A practical innovation has been recently introduced by Hung, et. al.[24] It is obvious with the technique described above that it would be desirable to shear simultaneously in two orthogonal directions so that all bending strains and shears can be separated from in-plane rotations. This can be done, of course, with unidirectional image-shearing devices in tandem; however, a careful analysis of the resulting pattern indicates that there can occur a certain ambiguity in the interpretation of the pattern because the two orthogonal shears cannot be separated. The technique of Ref. 24 solves that problem by dividing the lens aperture into four small apertures, in the form of a square. If the apertures are small enough, the depth of focus of the images formed from the field passing through each aperture will be much greater than that corresponding to the original aperture. Defocusing the image relative to the total aperture will create four sheared images in the pattern of a square, and the amount of shear will be proportional to the amount of defocus. Not only does this accomplish the double shearing in a convenient way, but because the shears are associated with different apertures in the lens, they may be separated by optical frequency-plane processing. Imagine a double-exposure photograph made before and after the object is bent out of its plane. If the transparency is placed in an optical processor such as Fig. 2, diffracted light will be concentrated around points in the transform plane corresponding to the intersection of grid lines, one of which is centered on the zero order. This results from high frequency periodicities in the speckle photograph caused by interference between the fields coming from the four apertures in the original camera lens. If the aperture in the optical processing system is placed to transmit only one of these portions of the spectrum at a time, the resulting image will possess shear-speckle fringes due to the shear in that direction alone. In practicing this technique, it would be wise to employ camera lenses that are mounted in separable halves, between which is located the normal lens aperture. Replacing the normal aperture by one that has the required four holes will place the new aperture where it will be least affected by vignetting. It must also be mentioned that Ref. 24 presents the application of shear-speckle interferometry to vibration. Finally, we should note that if the object translates laterally between the exposures the effects of speckle photography will come into play here as in the previous cases of speckle photography. In fact, Leendertz and Butters mention the use of their film translation technique[17] to enhance the readout of the shear speckle fringes.[22]

Concomitant Techniques

In all of the preceding techniques, a photographic recording has to be prepared and processed in order to evaluate displacements or strains. In some cases it may be necessary to make these evaluations directly as the load is applied to the object. To do so requires systems for concomitant speckle interferometry, and, ironically, Leendertz initiated speckle interferometry with exactly such systems.[16] The idea is quite simple and analogous to concomitant hologram interferometry or concomitant moiré interferometry. A single-exposure speckle photograph is recorded of an object illuminated as in Fig. 4, preferably on a glass plate. The negative transparency is placed back in the image plane of the camera in exact register with the original image. Because of the contrast reversal, there will be a very noticeable null in transmittance when the correct registration is achieved. If a transverse stress is applied to the object, the resulting strain will introduce a phase shift between the two interfering fields that comprise the image, as in the previous discussions. This will cause the speckle pattern illuminating the negative transparency to alternately match and differ from the pattern of the transparency. Where the patterns differ, more light will be transmitted than where they match, and the result will be fringes that indicate contours of constant lateral displacement of the object surface that accompany the loading of the object. Exactly the same method can be used for the shear-speckle technique to allow concomitant inspection of bending strains. An additional possibility, which, although not concomitant, provides rapid access to the results of speckle interferometry is the use of television readout of speckle interferometers.[25] With video storage and electronic subtraction, it is possible to display speckle interferograms within seconds of applying a load. This can be important because it eliminates the need for processing a speckle photograph and replacing it in an image field.

The last system to be discussed in this paper will be that which chronologically was the first speckle interferometer ever to have been reported.[26,27,28] This can be described as a device for displaying to an observer an intensity field corresponding to a focused-image hologram, with an in-line reference beam, and a low spatial frequency bandwidth. What the observer sees is an image of the object as rendered by an optical system with a very high f/number, about f/64, so that the speckles are easily visible. A reference field of constant amplitude is introduced into the system so that its curvature and propagation direction exactly match that of the average of the image field. If the two fields are mutually coherent, their interference will yield a different speckle pattern covering the image, the speckles of which will be generally twice as large as those without the reference field. Motion of the object toward the observer so as to change the phase of the object field by π, will reverse the destructive and constructive interference between the image and reference fields. A slow, constant motion of the object will cause the speckles to twinkle.

This device is most practical for concomitant vibration analysis. When a vibrating object is observed through the instrument, the vibration nodes stand out distinctly as regions of high contrast speckle, whereas the vibrating portions appear smoothed out. The analyses of J. M. Burch[2,3] indicate that the reference beam should be four to five times stronger than the image beam for optimum observation of the fluctuations in speckle contrast, which is a considerable benefit owing to the great loss of light obtained by the very high f/number. This technique also has been adapted to television readout[25] to yield electronically processed images that correspond favorably to the fringes of hologram interferometry. These higher order fringes come from the fact that the speckle contrast is not entirely smoothed out in the vibrating regions, as pointed out by Ek and Molin,[28] but rather, the speckle contrast is proportional to the square of a zero-order Bessel function of the vibration amplitude of the object. With careful observation, these higher order fringes can be seen through a speckle interferometer as weak reappearances of speckle contrast. The utility of the speckle interferometer in vibration analysis is remarkable in that it allows immediate inspection of vibration modes of objects without the intervening preparation of a hologram, and it requires no long term stability of the object.

Epilogue

It has been my intention in the preceding discussion simply to review the many primary techniques of speckle photography and speckle interferometry. Virtually all of the mathematics that can be used to describe these phenomena has been omitted; however, it can be noted that perhaps the most endearing aspect of speckle techniques comes from the simplicity of the mathematics involved relative to hologram interferometry. The reader is encouraged to seek the references to gain what mathematical insights he or she may need. The reference list has been purposely limited to publications in the major optics and engineering journals and has excluded work that exists only in the form of unpublished reports. Thus what I discuss is merely representative of the work to date in a field where new techniques are being actively pursued.

At this point I could be expected to make comments about the future of speckle photography and speckle interferometry. I can imagine many aspects of laser speckle that could be pursued, but if the history of this field teaches us anything it is that we must expect the unexpected. Thus I hope I shall be forgiven if I limit myself merely to prognosticating that a great deal more benefit may come to mankind from the use of this unwanted information than from its suppression.

References

1. D. Gabor, *Proc. SPIE, 25*, 129-133 (1971).
2. J. M. Burch, *Proc. SPIE, 25*, 149-155 (1971).
3. J. M. Burch, Optical Instruments and Techniques, Ed. J. Home Dickenson (Oriel Press, Newcastle upon Tyne, 1970) p. 213-229.
4. H. J. Tiziani, *Opt. Acta, 18*, 891-902 (1971).
5. J. W. Goodman, Intro to Fourier Optics (McGraw-Hill, New York, 1968) p. 117-120.
6. H. J. Tiziani, *Opt. Comm., 5*, 271-276 (1972).
7. E. Archbold and A. E. Ennos, *Opt. Acta, 19*, 253-271 (1972).
8. E. Archbold, J. M. Burch, and A. E. Ennos, *Opt. Acta, 17*, 883-898 (1970).
9. H. J. Tiziani, *Optik, 34*, 442-455 (1972).
10. H. J. Tiziani, *Appl. Opt., 11*, 2911-2917 (1972).
11. U. Kopf, *Optik, 33*, 517- (1971).
12. E. Archbold, A. E. Ennos, *Opt. & Laser Tech, 7*, 17-21 (1975).
13. D. E. Duffy, *Appl. Opt., 11*, 1778-1781 (1972).
14. D. E. Duffy, *Exp. Mech., 14*, 378-384 (1974).
15. J. A. Leendertz, Optical Instruments and Techniques, Ed. J. Home Dickenson, (Oriel Press, Newcastle upon Tyne, 1970) p. 256.
16. J. A. Leendertz, *J. Phys. E: Sci. Inst., 3*, 214-218 (1970).
17. J. N. Butters and J. A. Leendertz, *J. Phys. E: Sci. Inst., 4*, 272-279 (1971).
18. Y. Y. Hung and J. D. Hovanesian, *Exp. Mech., 12*, 454-460 (1972).
19. K. A. Stetson, *J. Opt. Soc. Am., 64*, 857-861 (1974).
20. R. Jones and J. A. Leendertz, *J. Phys. E: Sci. Inst., 7*, 653-657 (1974).
21. U. Kopf, *Opt. Comm., 9*, 374-376 (1973).
22. J. A. Leendertz and J. N. Butters, *J. Phys. E: Sci. Inst., 6*, 1107-1110 (1973).
23. Y. Y. Hung and C. E. Taylor, *Exp. Mech., 14*, 281-285 (1974).
24. Y. Y. Hung, R. E. Rowlands, and I. M. Daniels, *Appl. Opt., 14*, 618-622 (1975).
25. J. N. Butters and J. A. Leendertz, *Appl. Opt., 11*, 1436-1437 (1972).
26. E. Archbold, J. M. Burch, A. E. Ennos, and P. A. Taylor, *Nature, 222*, 263-265 (1969).
27. K. A. Stetson, *Opt. Laser Tech., 2*, 179-181 (1970).

Reprinted from *Optical Engineering*, Vol. 21(3), pp. 391-395
(May/June 1982). ©1982 SPIE.

Shearography: a new optical method for strain measurement and nondestructive testing

Y. Y. Hung
School of Engineering
Oakland University
Rochester, Michigan 48063

Abstract. This paper presents a new optical method referred to as shearography. Shearography is an interferometric method which employs an image-shearing camera. It permits full-field measurement of derivatives of surface displacements. The object under study is illuminated by laser light, and it is imaged by an image-shearing camera. The image-shearing camera produces two laterally sheared images in the image plane. When a photographic film in the image plane is doubly exposed with the object being deformed between the exposures, the processed photograph yields a fringe pattern which depicts the surface displacement gradients. Theory of the method as well as its applications to strain meausrements and nondestructive testing are presented.

Keywords: coherent optical techniques in experimental mechanics; shearography; strain measurement; nondestructive testing; speckle interferometry.

Optical Engineering 21(3), 391-395 (May/June 1982).

CONTENTS

1. INTRODUCTION

In the past two decades, development of new methods of experimental mechanics seems to have been mainly in the area of optics. Optical methods enjoy the advantages of being extremely fast in response, noncontacting, and generally full-field. These optical methods may be classified into two categories, coherent light methods and noncoherent light methods. The noncoherent light methods are mainly moire techniques,[1-3] whereas the coherent light category includes holography[4] and speckle interferometry.[5] This paper describes a coherent optical method which will be referred to as shearography. Shearography is a speckle interferometric method allowing direct measurement of displacement gradients. It is superior to other speckle techniques in that the obtained fringes are of much better quality. Besides being a strain measuring technique, shearography has been developed into an industrial tool for nondestructive testing.

Invited Paper CO-102 received Nov. 16, 1981; accepted for publication Dec. 1, 1981; received by Managing Editor Jan. 7, 1982.

2. THE METHOD

The object to be studied is illuminated by a point source of coherent light as shown in Fig. 1. It is imaged by a camera equipped with a shearing device and is referred to as an image-shearing camera. The image-shearing camera produces a pair of sheared images in the film plane and thus, the method is named shearography. As the object is illuminated by coherent light, the two sheared images interfere with each other producing a random interference pattern commonly known as a speckle pattern. When the object is deformed, this speckle pattern is slightly modified. Superposition of the two speckle patterns (deformed and undeformed) by double exposure yields a fringe pattern depicting the surface displacement gradient of the deformed object. This fringe pattern is not readily visible on the processed photographic film; it is made visible by a high-pass Fourier filtering technique.

3. PRINCIPLE OF SHEAROGRAPHY

Shearography utilizes an image-shearing camera as illustrated in Fig. 1. A thin glass wedge located in the iris plane of the lens covers one half of the lens aperture. Without the wedge, rays scattered from an object point, say $P_o(x,y)$, and received by the two halves of the lens will converge to a point in the image plane; i.e., a point in the object is mapped into a point in the image plane. The glass wedge is a small angle prism which deviates rays passing through it. In the presence of the wedge, the rays from $P_o(x,y)$ in the object are mapped into two points P_1 and P_2 in the image plane. Thus, a pair of the laterally sheared images is observed in the image plane. In other words, the image-shearing camera brings the rays scattered from one point on the object surface to meet with those scattered from a neighboring point in the image plane. If the wedge is so oriented that the rays are deviated in the xz-plane, rays from a point P(x,y) in the object will be brought to meet the rays from a

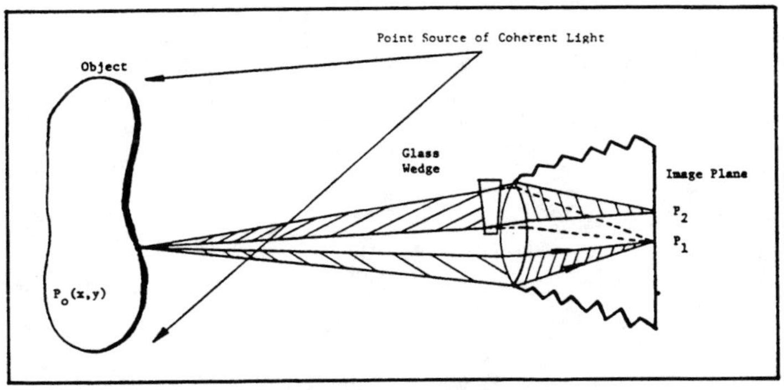

Fig. 1. Schematic diagram of shearography.

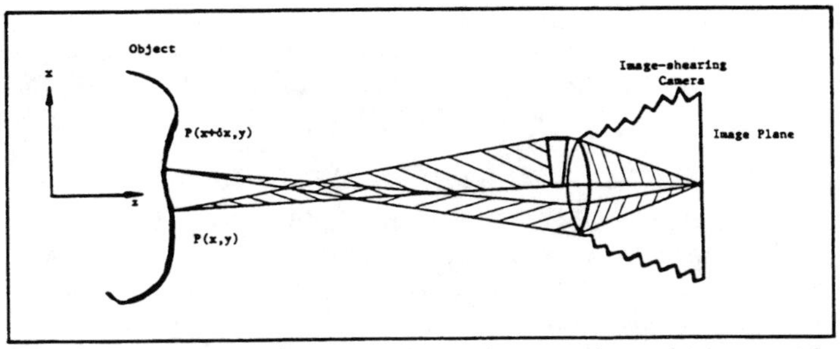

Fig. 2. Rays from two neighboring points are brought to meet in the image plane.

neighboring point in the x-direction $P(x+\delta x,y)$. This is illustrated in Fig. 2. δx is the amount of shearing in the object which is related to the shearing $\delta x'$ in the image plane and the magnification m by:

$$\delta x = \frac{\delta x'}{m} . \tag{1}$$

Assuming that the wedge is located at the iris plane of the lens, $\delta x'$ is given by

$$\delta x' = D_i(\mu - 1)\alpha , \tag{2}$$

where D_i is the image distance, μ is the refractive index of the wedge, and α is the wedge angle. Since m is the ratio of the image distance D_i to the object distance D_o, combining Eqs. (1) and (2) yields:

$$\delta_x = D_o(\mu - 1)\alpha . \tag{3}$$

For simplicity, hereafter, the magnification will be assumed to be unity so that no distinction is made between $\delta x'$ and δx.

4. ANALYSIS OF FRINGE FORMATION

Let the wave fronts of the two sheared images be represented by

$$U(x,y) = a \exp[\theta(x,y)]$$

and

$$U(x+\delta x,y) = a \exp[\theta(x+\delta x,y)] , \tag{4}$$

where $\theta(x,y)$ and $\theta(x+\delta x,y)$ represent the phase of the light from points $P(x,y)$ and $P(x+\delta x,y)$, respectively, and a is the light amplitude which is assumed equal for the two neighboring points.

The total light field U_T detected in the image plane is thus

$$U_T = U(x,y) + U(x+\delta x,y) , \tag{5}$$

and the intensity I detected in the image plane is:

$$I = U_T U_{T^*} = 2a^2[1 + \cos\phi] , \tag{6}$$

where $\phi = \theta(x+\delta x,y) - \theta(x,y)$ represents a random phase angle.

When the object is deformed, an optical path change occurs due to surface displacements in the object. This optical path change produces a relative phase change between the two sheared wave fronts. Thus, Eq. (6) becomes

$$I' = 2a^2[1 + \cos(\phi + \Delta)] , \tag{7}$$

where I' is the intensity distribution after deformation, and Δ is the relative phase change due to relative displacement between $P(x+\delta x,y)$ and $P(x,y)$. If a photographic film in the image plane is doubly exposed sequentially to I and I', the total energy recorded is proportional to the sum I_s of the two intensity distributions:

$$I_s = I + I' = 2a^2[2 + \cos\phi + \cos(\phi+\Delta)] . \tag{8}$$

ϕ in the above equation is fast varying, and the spatial frequencies of $\cos\phi$ and $\cos(\phi+\Delta)$ are generally beyond the eye's resolution capability. The eye sees an intensity averaged over an elementary area resolvable by the eye. Assuming equal probability of ϕ varying from 0 to 2π in the elementary area, the averaged intensity observed by the eye is given by:

$$I_{av} = \frac{1}{2\pi}\int_o^{2\pi} I_s d\phi = 4a^2 , \tag{9}$$

which shows that the eye sees a uniform intensity, i.e., no fringes. However, Eq. (8) can be rewritten in the following form:

$$I_s = 4a^2\left[1 + \cos\left(\phi + \frac{\Delta}{2}\right) \cdot \cos\left(\frac{\Delta}{2}\right)\right] . \tag{10}$$

The second term in Eq. (10) represents a high-frequency carrier [described by $\cos(\phi + \Delta/2)$] amplitude modulated by a low-frequency factor [$\cos(\Delta/2)$]. The high-frequency carrier is nulled when

$$\cos\left(\frac{\Delta}{2}\right) = 0 ; \tag{11}$$

that is,

$$\Delta = (2n+1)\pi ,$$

when $\pm$ n = 0,1,2,3,4, . . . Equation (10) represents a fringe patern of frequency variation type. Fringes of this type are identified as areas of nulling carrier. It should be emphasized that this type of fringe patten is different than the conventional fringe patterns. Conventional fringes are loci at which intensity is minimum. The conventional fringes are readily visible to the eye, whereas the present frequency variation type fringes are not visible. However, these invisible fringes can be converted to the visible intensity variation type by an optical high-pass Fourier filtering process described below.

Figure 3 shows the schematic diagram of the Fourier filtering setup. A point source of light is focused by the transforming lens (TL) onto a plane known as the Fourier filtering plane. The photographic transparency with an intensity transmittance represented by Eq. (10) is placed at the input plane. The light field appearing in the Fourier filtering plane is the Fourier transform of the transmittance. Fourier transform is a transformation from a spatial domain to a frequency domain. In other words, carriers of various frequencies recorded in the transparency are separated in the Fourier filtering plane. The areas of zero frequency (i.e., nulling of the carriers) are displaced at the focal point on the optical axis, whereas those of higher frequencies are located off-axis in the

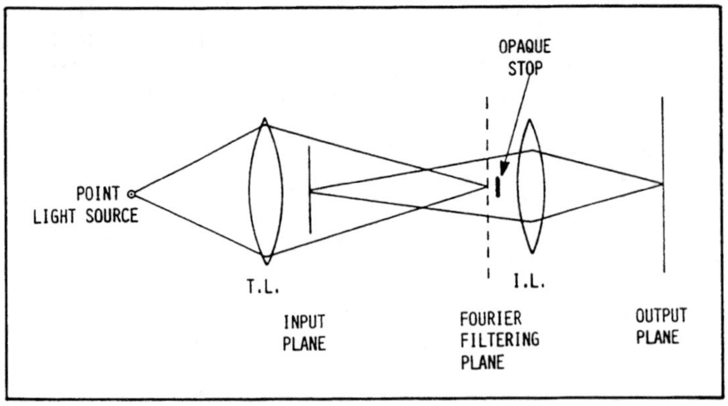

Fig. 3. The high-pass Fourier filtering setup.

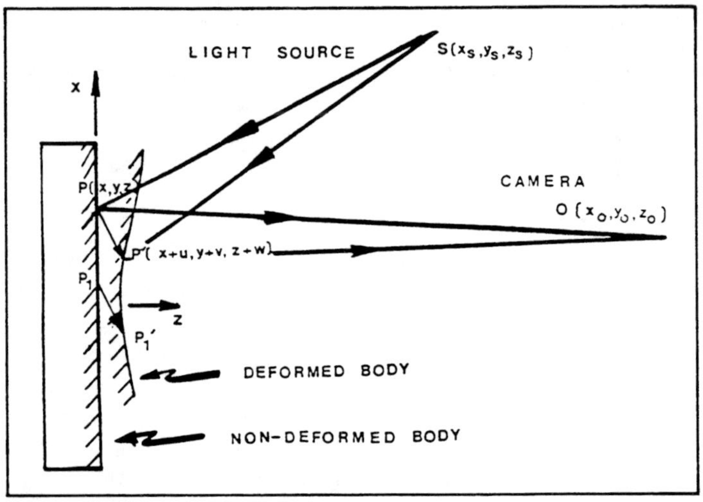

Fig. 4. Displacement optical path diagram.

Fourier filtering plane. By placing an opaque stop on the axis at the Fourier filtering plane to block the low frequency components, the contribution from the areas where the carriers are nulled (i.e., fringes) as described in Eq. (10) are stopped. This process is known as high-pass Fourier filtering. When reimaged by the imaging lens (IL) behind the Fourier filtering plane, those areas will appear dark in the image plane (output plane). Thus, the frequency variation type fringe pattern is converted to a visible fringe pattern of intensity variation. This fringe pattern depicts Δ. Δ is the relative phase change due to the relative displacements between the points P(x + δx,y) and P(x,y). Referring to Fig. 4, it can be shown that for the small shearing Δ is related to the displacement derivatives by:

$$\Delta = \frac{2\pi}{\lambda}\left[A\frac{\partial u}{\partial x} + B\frac{\partial v}{\partial x} + C\frac{\partial w}{\partial x}\right]\delta x\,, \tag{12}$$

where (u,v,w) represents the displacement vector of the point P(x,y); A, B, and C are related to the position of the illumination point $S(x_s,y_s,z_s)$ and the camera position $0(x_o,y_o,z_o)$ by[6]:

$$A = \left(\frac{x-x_o}{R_o} + \frac{x-x_o}{R_s}\right),$$

$$B = \left(\frac{y-y_o}{R_o} + \frac{y-y_s}{R_s}\right), \tag{13}$$

$$C = \left(\frac{z-z_o}{R_o} + \frac{z-z_s}{R_s}\right),$$

where

$$R_o^2 = x_o^2 + y_o^2 + z_o^2$$

and

$$R_x^2 = x_s^2 + y_x^2 + z_s^2 .$$

Note that z = z(x,y) describes the object surface.

Should the shearing direction be in the y-direction, Eq. (12) becomes

$$\Delta = \frac{2\pi}{\lambda}\left[A\frac{\partial u}{\partial y} + B\frac{\partial v}{\partial y} + C\frac{\partial w}{\partial y}\right]\delta y\,, \tag{14}$$

which describes the displacement derivative with respect to y; δy is the amount of shearing in the y-direction. It is possible to employ a multiple image-shearing camera[7] permitting displacement gradients in both x and y-directions to be recorded simultaneously.

5. APPLICATION TO STRAIN MEASUREMENT

Equation (12) indicates that the obtained fringes are generally functions of three displacement derivatives, namely, ∂u/∂x, ∂v/∂x, and ∂w/∂x. Therefore, three measurements with three different illuminating angles are required for the three displacement derivatives to be separated.

With normal illumination and normal viewing, i.e., $x_s = y_s = x_o = y_o = 0$, and for R_o and R_s to be large compared with the object size, Eqs. (12) and (13) are reduced to:

$$\Delta = \frac{4\pi}{\lambda}(\delta x)\cdot\frac{\partial w}{\partial x} \tag{15}$$

and

$$\Delta = \frac{4\pi}{\lambda}(\delta y)\cdot\frac{\partial w}{\partial y}\,. \tag{16}$$

The above equations show that fringes are whole-field representations of the loci of ∂w/∂x and ∂w/∂y. This technique is particularly suited to studies of plate deformation, as it is only necessary to differentiate the obtained data once to determine flexural strains. Figure 5 shows a fringe pattern depicting ∂w/∂x of a rectangular plate clamped along its four boundaries. The fringe pattern clearly illustrates the superiority of this technique over the speckle methods, namely, excellent fringe quality. The fringe density at the center of the plate is about 20 lines per cm before magnification.

Since the quality of the fringes is good, it is possible to obtain second derivatives of the plate deflection by moire effect of two identical fringe patterns of sufficient high fringe density, superposed and shifted with respect to each other.

Shearography can be employed to measure strains in an arbitrarily curved surface. Like ordinary photography, the tolerance to the depth of field is increased with the increased f-number of the lens. Figure 6 shows a fringe pattern depicting a deflection gradient around a crack in a cylindrical circular pressure vessel.

6. NONDESTRUCTIVE TESTING

Shearography permits full-field observation of surface strains in structures. Obviously, it is applicable to nondestructive testing of structural components. Flaws in structures usually create strain concentrations which will create anomalies in the shearographic fringe pattern. Figure 7 shows the presence of a crack in a pressure vessel. The crack which creates strain concentration is clearly revealed by the fringe anomaly in the crack area.

Shearography is particularly suited to detection of debonds or separations in structures of composite materials. It has already been developed into an industrial tool for evaluation of integrity of

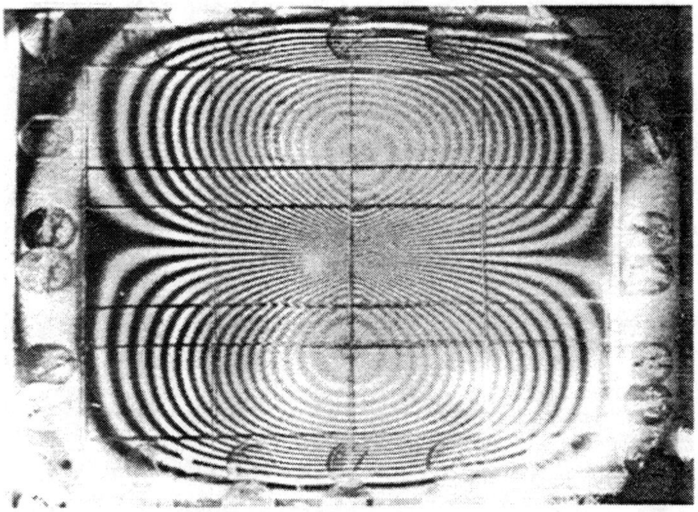

Fig. 5. Fringe pattern depicting $\partial w/\partial x$ of a rectangular plate clamped along the boundary and centrally loaded.

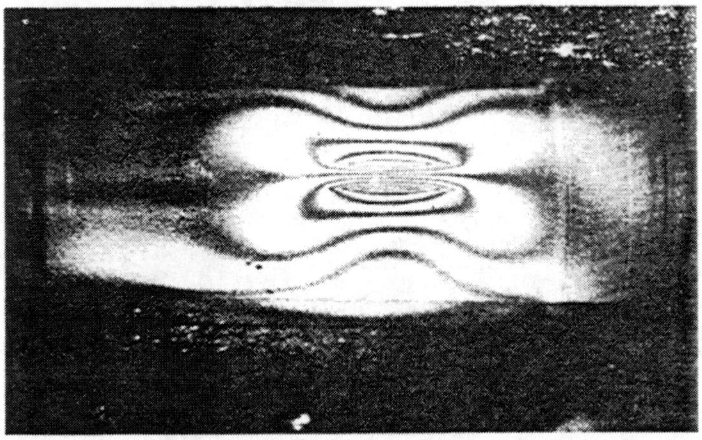

Fig. 6. Fringe pattern depicting the gradients of normal deflection in the crack area in a cylindrical shell.

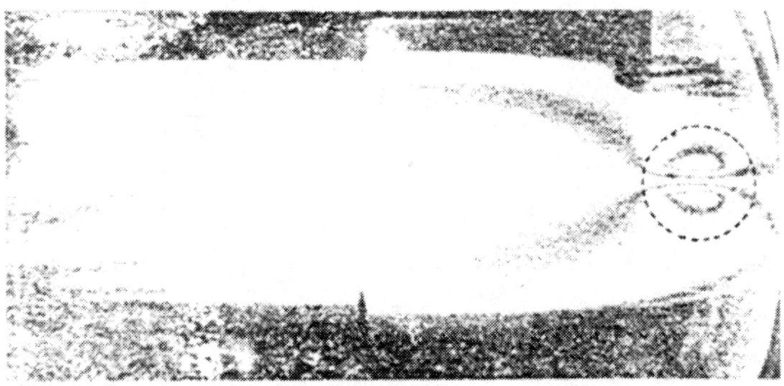

Fig. 7. A crack in a pressure vessel is revealed by the anomaly in the fringe pattern.

automotive and aircraft tires. Figure 8 shows that separation has occurred along the belt edge of a used truck tire. It also reveals an isolated ply separation. In the testing, the tire was placed in a chamber and a partial vacuum was used as a means of stressing.

7. SHEAROGRAPHY VERSUS OTHER NDT TECHNIQUES

One major difference between shearography and other nondestructive testing (NDT) techniques such as magnetic particle, dye

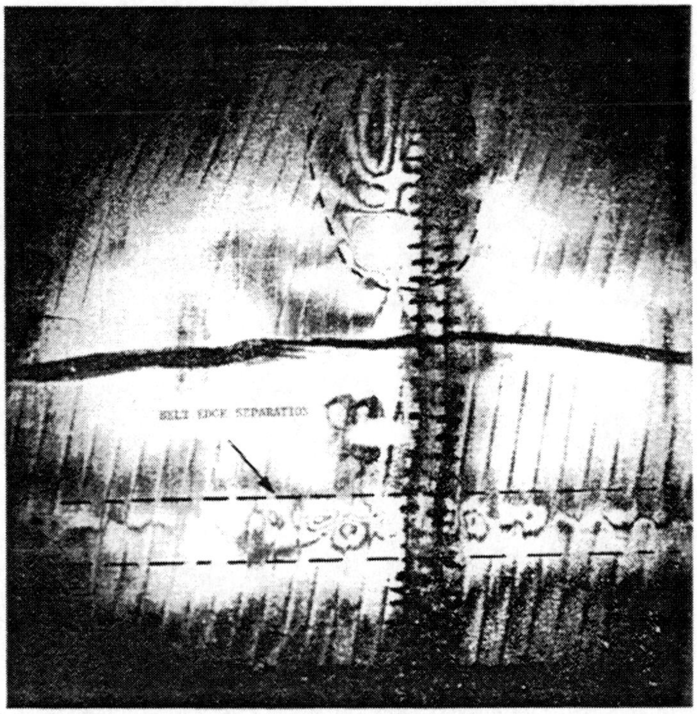

Fig. 8. Fringe pattern revealing an isolated ply separation and separations along the belt edge of a truck tire.

penetrant, x ray, ultrasonic, eddy current, etc., is the method of revealing flaws. Other NDT techniques reveal surface flaws by identifying abnormalities by enhanced visual means (such as magnetic particle, dye penetrant) and detect internal flaws by detecting nonhomogeneity (such as ultrasonics, x ray, etc.). Not all flaws are critical; the criticality depends on flaw size, flaw shape, and flaw nature as well as the location of the flaw. These techniques provide no information on whether or not these flaws will weaken the structures.

Premature failures of structures are usually caused by excessive stress levels developed by stress concentrations induced by flaws in structures. Shearography measures strains and thus, flaws are identified by identifying the strain concentrations induced by the flaws. If the structures under testing are loaded in a stress mode similar to the actual loading in service, shearography can be used to reveal critical defects only (i.e., those which create strain concentrations and thus will weaken the strength of the structures); cosmetic defects are ignored and false rejections can be avoided.

Furthermore, shearography is a full-field optical method which does not require scanning or contacting. Thus, inspection rate of shearography is inherently higher than other techniques.

8. SHEAROGRAPHY VERSUS HOLOGRAPHY

Being an optical method, it seems appropriate to compare shearography with holography. Both methods enjoy the advantages of being full-field and noncontacting. However, the image-shearing camera offers several advantages over holography. These can be summarized as follows:

(1) It requires a very simple optical setup, thus eliminating the optical alignment problem.

(2) It relaxes the vibration isolation requirement; thus, it is more suited for inspection in production environments.

(3) The coherent length requirement is greatly reduced, thus eliminating the problem of maintaining lasers running at single mode.

(4) It provides a wider and more controllable range of sensitivity, thus allowing larger deformation to be measured. Holography is too sensitive for many practical measurements.

(5) The requirement of recording media resolution is much

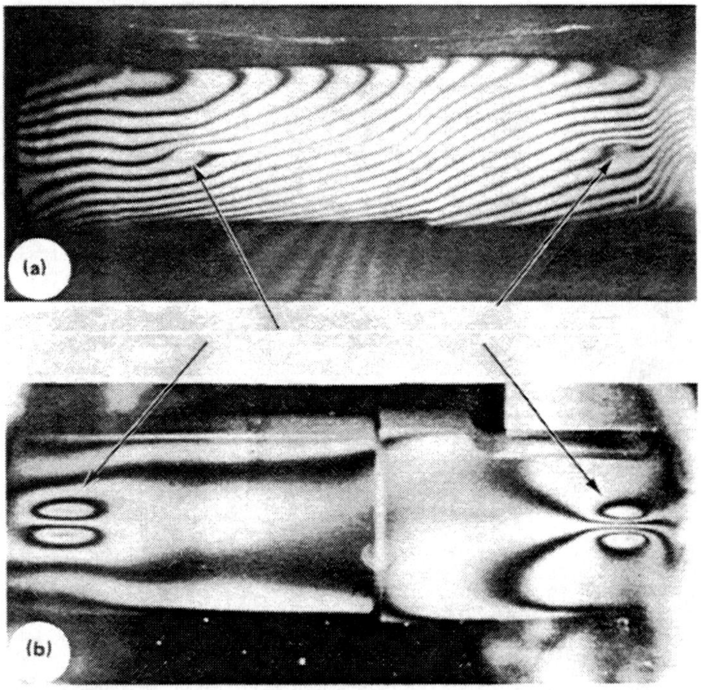

Fig. 9. Shearography versus holography in the detection of cracks: (a) Crack detection by holographic interferometry; (b) Crack detection by image-shearing camera.

lower, thus allowing faster and less expensive photographic films to be used. Even TV and instantly processible photoplastics may be employed.

(6) Shearography measures displacement gradients instead of displacements. This is perhaps the most significant advantage. Since defects normally create strain concentrations, it is easier to correlate defects with strain anomalies rather than displacement anomalies. Since rigid body motion does not produce strain, the image-shearing camera is insensitivie to rigid body motion. Note that rigid body motion produces confusing, if not misleading, fringes in holography. This advantage of shearography is clearly demonstrated in Fig. 9, which shows two cracks in a pressure vessel. Figure 9(b) shows the fringe pattern obtained by shearography, whereas Fig. 9(a) was obtained by holography, which does not reveal the cracks as prominently because of the rigid body motion.

9. CONCLUSION

Shearography, a new means for strain measurements and nondestructive testing, has been presented. Due to the numerous advantages of shearography over holography, it seems to possess the potential of being developed into a practical tool for measurements in industry. The full capability of shearography still awaits exploration.

10. ACKNOWLEDGMENTS

This investigation was supported by the National Science Foundation (Grant ME 79-20097). The author is grateful to C. C. Astill of NSF for his support.

11. REFERENCES

1. Theocaris, P.S., *Moire Fringes in Strain Analysis,* Pergamon Press, (1969).
2. Durelli, A. J. and Parks, V. J., *Moire Analysis of Strain,* Prentice-Hall, (1970).
3. Chiang, F. P., *Experimental Stress Analysis Techniques SESA Monograph,* edited by A. S. Kobayashi, Chap. 6, (1978).
4. Vest, C. M., *Holographic Interferometry,* John Wiley and Sons, (1979).
5. Efr, R. J., editor, *Speckle Metrology,* Academic Press, (1978).
6. Hung, Y. Y. and Liang, C. Y., Appl. Opt. 18 (7), 1046 (1979).
7. Hung, Y. Y. and Durelli, A. J., J. Strain Analysis 14 (3), 81 (1979).

Reprinted from *Optical Engineering*, Vol. 24(6), pp. 936-943
(November/December 1985). ©1985 SPIE.

Random (speckle) patterns for displacement and strain measurement: some recent advances

F. P. Chiang
D. W. Li
State University of New York
at Stony Brook
Laboratory for Experimental
Mechanics Research
Stony Brook, New York 11794

Abstract. Major advances in using coherent and incoherent speckle patterns for displacement and strain measurements are reviewed. Topics include the laws of spatial movement of laser speckle, the halo functions of coherent and incoherent speckle patterns, decorrelation factors and their effect on the fringe visibility and the upper limit of measurement, new techniques of holospeckle interferometry, real-time and electron microscopic speckle photographies, and the automatic processing of Young's fringes.

Subject terms: 3-D sensing for inspection; speckle; random patterns; displacement; strain; interferometry.

Optical Engineering 24(6), 936-943 (November/December 1985).

CONTENTS

1. INTRODUCTION

One of the major events in the history of measurement techniques for mechanics and metrology in the past two decades is the advent of random speckle methods. It is noteworthy in two distinctive aspects. First, a random collection of bright and dark dots (speckles) is used as a gauging device rather than as a well-defined geometric element. Second, quantitative information is not obtained by following the movement of individual speckles. Rather, it is retrieved from the correlation properties between the two displaced clusters of speckles. This branch of measurement technique has been variously referred to as speckle photography or speckle interferometry (one beam). Its basic principles and numerous applications can be found in a number of review papers[1-3] and books.[4,5]

Of particular interest is a family of techniques that can all be traced back to the 1968 classical paper of Burch and Tokarski,[6] wherein they reported the observation of interference fringes upon optically Fourier transforming superimposed speckle patterns. In the ensuing years, this phenomenon has been developed into a series of measuring techniques through the efforts of many researchers, including Archbold et al.,[7] Archbold and Ennos,[8] Chiang,[9,10] Stetson,[11] and Asundi and Chiang,[12] among others.

In the past few years, a flurry of papers have appeared in the literature dealing with various aspects and improvements of these techniques, for example, the development of objective speckle photography,[13,14] the combinational adaptation of speckle photography and holographic interferometry,[15-17] the development of remote sensing techniques using fiber optics,[18] the elimination of the effect of halo function[19,20] and speckle noise,[21] the development of automatic data processing techniques,[22-26] and statistical treatment.[27-29]

In the following sections, we review some of these important developments. The topics selected, of course, reflect the interest and bias of the authors. A review of this nature cannot possibly cover all activities, but we intend to offer a glance of this dynamic field from a particular angle.

Many of the recent important results derive from the adaptation of statistical optics in the treatment of speckle photography, in contrast to the deterministic approach previously used. Indeed, almost all the important features of speckle photography can be related to statistical concepts. For example, the autocorrelation function can be used for the determination of speckle size, and its Fourier transform, the Wiener spectrum, can be adapted for the calculation of the halo function.[30] Similarly, speckle length and speckle movement can be derived by applying the concept of cross-correlation.[31] While statistical optics in the treatment of speckle phenomena has long been used by many researchers,[30,32,33] its systematic adaptation in the analysis of the procedures in speckle photography is a fairly recent activity.[27-29,31] We shall first review

Invited Paper TD-103 received July 23, 1985; accepted for publication July 24, 1985; received by Managing Editor Aug. 9, 1985.

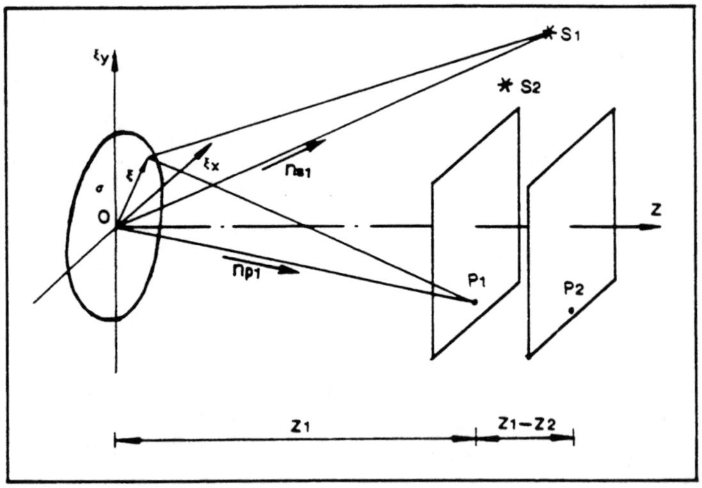

Fig. 1. Geometry for determining laws of speckle movement.

some of these results: speckle movement in space, halo functions for coherent and incoherent speckle patterns, decorrelation effect, fringe visibility, and upper limit of measurement. We shall then review the development of real-time speckle photography, speckle photography with an electron microscope, and automatic data processing.

2. SPATIAL MOVEMENT OF LASER SPECKLE

It is well known that laser speckles move according to the movement of their generating surface. The relationship that governs this movement is of obvious importance in speckle photography. A number of workers[34–40] have studied this problem both theoretically and experimentally. Two schools of thought exist, resulting in different field equations.

Figure 1 shows a speckle-generating surface σ, an illuminating point source S, and one of the speckles P. Denoting the incident and reflection distances OS and OP by L_S and L_P and letting their projections on the generating surface be $\boldsymbol{\ell}_S$ and $\boldsymbol{\ell}_P$, respectively, one school of thought[39] derives the following set of equations for describing the speckle movement in space:

$$\Delta(\boldsymbol{\ell}_S + \boldsymbol{\ell}_P) = 0 \ , \tag{1}$$

$$\Delta\left(\frac{1}{L_S} + \frac{1}{L_P}\right) = 0 \ . \tag{2}$$

On the other hand, a different approach[37,40] yields another set of equations:

$$\Delta(\boldsymbol{\ell}_S + \boldsymbol{\ell}_P) = 0 \ , \tag{3}$$

$$\Delta(L_S + L_P) = 0 \ . \tag{4}$$

While ample experimental evidence exists to support the validity of the vector equation, the question of which form of the scalar equation is more appropriate has not been put to sufficient test because of the difficulty in measuring longitudinal speckle movement. In a recent paper, Chiang and Li[41] set out to derive the equations of speckle movement by maximizing the cross-correlation between the initial and final states of the speckle field.

Let $A_1(P)$ and $A_2(P)$ be the amplitude distributions of the two speckle fields, and let $I_1(P)$ and $I_2(P)$ be their corresponding intensity distributions. The amplitude cross-correlation of the two fields is defined as

$$R_{CA}(P_1, P_2) = \langle A_1(P_1) A_2^*(P_2) \rangle \ , \tag{5}$$

where $\langle\ \rangle$ denotes the ensemble average and * denotes the complex conjugate. The corresponding intensity cross-correlation is

$$R_{CI}(P_1, P_2) = \langle I_1(P_1) I_2(P_2) \rangle \ . \tag{6}$$

The two correlation functions can be related by the following relation:

$$R_{CI}(P_1, P_2) = \langle I_1(P_1)\rangle \langle I_2(P_2)\rangle + |R_{CA}(P_1, P_2)|^2 \ . \tag{7}$$

It can be shown,[41] assuming that the microstructure of the generating surface is sufficiently fine, that

$$R_{CA}(P_1, P_2) = \frac{\exp[jk\Delta(L_S + L_P)]}{L_{S_1} L_{P_1} L_{S_2} L_{P_2}} \int_\sigma \exp\left\{ jk\left[-\Delta(\boldsymbol{\ell}_S + \boldsymbol{\ell}_P)\cdot\boldsymbol{\xi} + \Delta\left(\frac{1}{L_S} + \frac{1}{L_P}\right)\frac{|\boldsymbol{\xi}|^2}{2}\right]\right\} d\sigma_\xi \ , \tag{8}$$

where $\boldsymbol{\xi}$ is a position vector on the generating surface. From Eq. (8), it is seen that while $\Delta(L_S + L_P)$ influences the amplitude cross-correlation function R_{CA}, it has no effect on the intensity cross-correlation function R_{CI}, which is controlled by $|R_{CA}|$. Since it is the intensity rather than the amplitude of a speckle field that is recorded in speckle photography, it is reasonable to assume that Eq. (4) is not appropriate in describing the speckle movement. Furthermore, it can be shown by applying the Schwartz inequality that $|R_{CA}|$, and thus R_{CI}, is maximized when Eqs. (1) and (2) are satisfied, indicating that they are the equations describing the speckle movement.

An experiment was also designed to test the validity of the scalar equations with consistent results.[41]

3. HALO FUNCTIONS OF COHERENT AND INCOHERENT SPECKLE PATTERNS

When data analysis is performed on a specklegram, the halo function plays a vital role. In pointwise filtering the halo profile influences the position of Young's fringes, whereas in full-field filtering it affects the sensitivity and brightness of the isothetics. Earlier works[12,42] using a deterministic approach have assumed the halo function to be the same for both coherent (such as laser) and incoherent (white-light) speckle patterns. Recent works have shown that for a coherent speckle pattern, the halo function,[27,29,43] apart from a bright spot in the center, is

$$I_{ch}(\mathbf{U}) = \int |P(\mathbf{r})|^2 \left| P\left(\mathbf{r} + \frac{\lambda_0 q \mathbf{U}}{\lambda f}\right)\right|^2 d\sigma_r \ , \tag{9}$$

where $\mathbf{U}$ is a position vector in the halo plane, $P(\mathbf{r})$ is the pupil function with $\mathbf{r}$ being a position vector at the pupil plane, λ_0 and λ are the wavelengths of the beams used for recording and processing, respectively, q is the image distance, and f is the focal length of the transform lens.

On the other hand, the halo function for an incoherent speckle pattern is shown to be of the following form[44]:

$$I_{ih}(\mathbf{U}) = I_{ihI} I_{ihII} \ , \tag{10}$$

where

$$I_{ihI}(\mathbf{U}) = \left| \int P(\mathbf{r}) P^*\left(\mathbf{r} + \frac{\lambda_0 q \mathbf{U}}{\lambda f}\right) d\sigma_r \right|^2 \ . \tag{11}$$

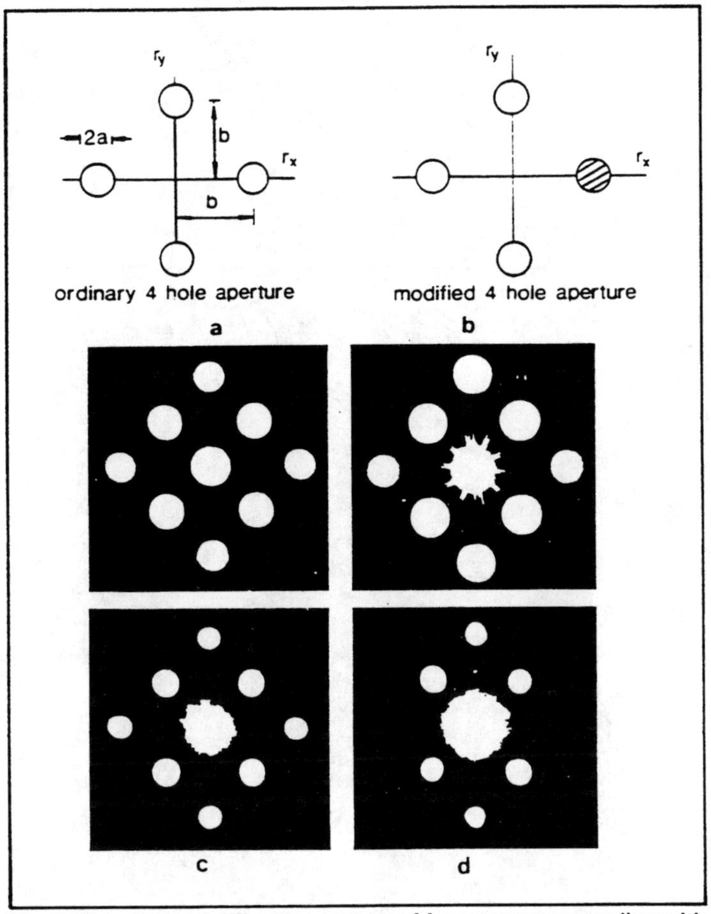

Fig. 2. Experimental diffraction pattern of four-aperture recording with and without random phase.

$$I_{ihII}(\mathbf{U}) = \iint \left\langle I_o\left(\frac{\mathbf{X}}{M}\right) I_o\left(\frac{\mathbf{X}'}{M}\right)\right\rangle \times \exp\left[-j\frac{2\pi}{\lambda f}(\mathbf{X} - \mathbf{X}')\cdot\mathbf{U}\right] d\sigma_X d\sigma_{X'} \ , \quad (12)$$

where $I_o(\mathbf{X})$ is the artificial speckle pattern on the object surface, M is the magnification factor, and $I_o(\mathbf{X}/M)$ is the geometric image of the artificial speckle pattern. It is seen that the diffraction halo of an incoherent speckle pattern is the product of two functions, I_{ihI} and I_{ihII}, where the former is the squared MTF of the recording system and the latter is the power spectrum density function of the demagnified artificial speckle pattern. It is only when the demagnified speckle pattern is so fine that $\langle I_o(\mathbf{X}/M)I_o(\mathbf{X}'/M)\rangle$ can be approximated by a δ function that the halo function becomes

$$I_{ih} = I_{ihI} = \left| \int P(\mathbf{r})P^*\left(\mathbf{r} + \frac{\lambda_o q\mathbf{U}}{\lambda f}\right) d\sigma_r \right|^2 \ , \quad (13)$$

which is the form used in earlier works.[12]

There are a number of differences between the two halo functions. First, while the boundary of the coherent halo function is determined solely by the pupil function of the recording system, the boundary of the incoherent halo function is controlled both by the pupil function and the cutoff frequency of the demagnified artificial speckle pattern. This has an effect on the sensitivity of the speckle methods. Second, the halo functions have different profiles, which in turn introduce different errors in the position of Young's fringes. The more interesting, albeit somewhat academic, difference between the two is the fact that a phase term introduced into the pupil function will influence the incoherent halo function while leaving the coherent halo function intact. An example is shown in Fig. 2. The top row shows two identical four-aperture arrangements except in one case one of the horizontal apertures is covered with a ground glass to introduce a random phase term. The second row shows the diffraction patterns from two laser specklegrams depicting the familiar nine-dot array. The third row shows those from an incoherent speckle pattern. It is seen that the random phase has affected the diffraction pattern to the extent that two diffraction orders are completely missing.

4. DECORRELATION, FRINGE VISIBILITY, AND UPPER LIMIT OF MEASUREMENT

In the earlier analysis of speckle photography,[42] it was tacitly assumed that the two speckle patterns, superposed by double exposure, were completely correlated in that the speckles were simply shifted without any microstructural change. In reality, displacement and deformation result in decorrelation, which manifests itself in reduced fringe contrast and thus affects the upper limit of measurement. Many factors, such as surface strain, tilt, rotation, change of illumination angle, etc., influence the microstructure of a speckle field in the recording stage, and the process of filtering contributes its share of decorrelation in the reconstruction stage.

Ennos et al.[45] studied the limiting case of total decorrelation. They proposed two criteria for it to occur under the conditions of surface tilt and out-of-plane displacement. Recently, Yamaguchi[27] performed an analysis of the recording process using statistical optics to show the general effect of object displacement on speckle decorrelation. He introduced a factor γ to describe the decorrelation associated with the recording process. Chiang and coworkers[28,29] investigated, also through statistical optics, the decorrelation effect due to spatial filtering and introduced two additional decorrelation factors, Ω and ψ. The effect of all these factors was studied in detail in Ref. 29, and this led to the establishment of a general criterion for fringe visibility and set the upper limit of measurable displacement. This latter work is briefly reviewed in the following.

Consider a linearly developed coherent specklegram with an amplitude transmittance $t(\mathbf{X})$ as follows:

$$t(\mathbf{X}) = I_1(\mathbf{X}) + I_2(\mathbf{X}) \ , \quad (14)$$

where I_1 and I_2 are the intensity distributions of the first and second exposures, respectively. In pointwise filtering, such a specklegram is placed at the front focal plane (restriction imposed for mathematical simplicity only) of a Fourier transform (FT) lens and illuminated by a narrow laser beam. A fringe pattern (the so-called Young's fringes) appears at the back focal plane of the FT lens. It is enclosed by a diffraction halo and modulated by the secondary speckle noise produced by the filter aperture, namely, the illuminated area in this case It can be shown that the Young's fringes are governed by the following equation:

$$I(\mathbf{U}) = 2\left[1 + \nu_d \cos\left(2\pi\frac{\mathbf{d}_{so}\cdot\mathbf{U}}{\lambda f}\right)\right] I_{ch}(\mathbf{U}) \ , \quad (15)$$

where $\mathbf{d}_{so}$ is the speckle displacement at the center of an illuminated area S and I_{ch} is the halo function given in Eq. (9). There are, apart from halo and speckle effects, a series of parallel cosine fringes perpendicular to $\mathbf{d}_{so}$, with fringe spacing being

$$S_f = \frac{\lambda f}{|\mathbf{d}_{so}|} \ . \quad (16)$$

The amplitude of the cosine term is the product of the three decorrelation factors; i.e.,

$$\nu_d = \gamma\psi\Omega \ . \quad (17)$$

The factor γ represents the decorrelation effect due to the microstructural change of the superposed speckle patterns during the recording process. It consists of the effect of out-of-plane displacement, surface tilt and rotation, strain, change of illumination angle, defocus, etc. The total effect may be divided into two categories: those that result in speckle longitudinal movement at the image plane, such as displacement, strain, change of illumination angle, and defocus, and those that result in the change of spectrum content of the wave that enters the pupil, such as surface tilt and rotation. Denoting these two effects by ϕ and $\mathbf{d}_r$, respectively, the expression for γ is

$$\gamma = \frac{\left| \int P(\mathbf{r})P(\mathbf{r}-\mathbf{d}_r)\exp(j\phi)d\sigma_r \right|^2}{\left| \int |P(\mathbf{r})|^2 d\sigma_r \right|^2} . \tag{18}$$

Lens aberration also contributes to decorrelation, as can be seen from the fact that a complex-valued pupil function $P(\mathbf{r})$ will reduce the value of γ in general. References 27 and 29 need to be consulted for more detailed information concerning γ.

In order to appreciate the effects of ψ and Ω, it is helpful to introduce a "twin speckle" physical model. A double-exposure specklegram may be considered as a random distribution of paired speckles. Each pair consists of one speckle due to the first exposure and its twin, displaced and slightly changed, from the second exposure. The separation between the twin speckles is a constant if only translation of the entire speckle pattern is involved. Otherwise, it varies from pair to pair. Under laser illumination, each speckle and its twin act as coherent point sources producing Young's fringes. On the other hand, light waves from different twins would not interfere to produce fringes. The observed fringe pattern is a result of incoherent superposition of various Young's fringes coming from various twin speckles.

Let us consider the case of a rigid-body displacement field within an illuminated area S, as shown in Fig. 3. After a displacement $\mathbf{d}_s$, the speckle set is displaced to region S" and recorded by the second exposure. (A symmetrical set is shown to be moved to region S' for a negative displacement.) Only those speckles in the overlapping region S_1 (or S_1') have their twins simultaneously illuminated within the region S. Thus, they alone contribute to the formation of overall Young's fringes. The decorrelation factor Ω, defined as

$$\Omega = \frac{S_1}{S} , \tag{19}$$

is therefore a measure of decorrelation due to this effect. It is easy to see that if the displacement is such that it exceeds the diameter of the illuminating beam, no Young's fringes will be observed. Figure 4 shows a plot of Ω versus $d_{so}/2r_x$, where r_x is the radius of the illuminated area.

If the separation between twin speckles is not uniform within the illuminated area, the Young's fringes produced by each twin will be different in orientation and spacing. The incoherent summation of all these fringes results in a blurred overall pattern. This effect is represented by the decorrelation factor ψ, for which the expression is quite lengthy.[29] For the special case of a circular illuminating beam of radius r_x and constant amplitude, assuming that the speckle displacement within this area is small compared to r_x, we have

$$\psi = \frac{2J_1(\rho)}{\rho} , \tag{20}$$

with

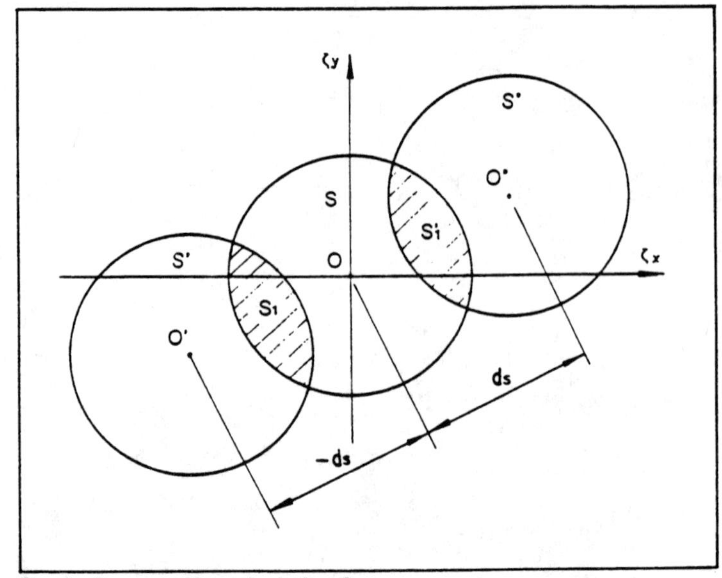

Fig. 3. Geometry for calculating Ω.

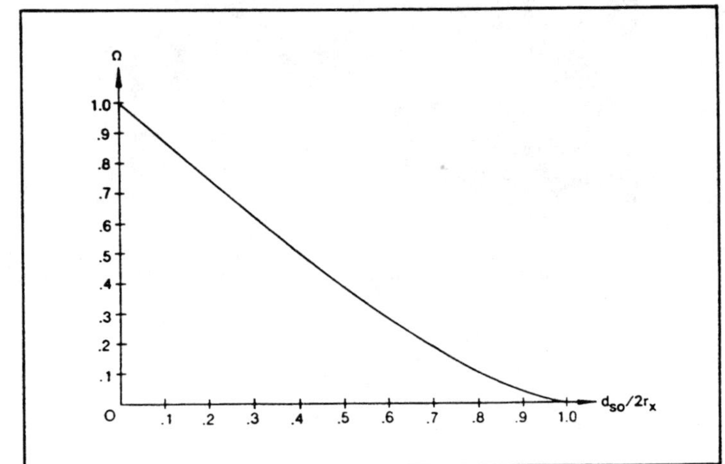

Fig. 4. Displacement factor Ω as a function of $|d_{so}|/2r_x$.

$$\rho = \frac{2\pi |\mathbf{U}| \, |\boldsymbol{\epsilon}| \, r_x}{\lambda f} , \tag{21}$$

where J_1 is the first-order Bessel function of the first kind,

$$\boldsymbol{\epsilon} = \begin{pmatrix} \epsilon_x \\ \epsilon_y \end{pmatrix} = \begin{bmatrix} \frac{\partial d_{sx}}{\partial x} & \frac{\partial d_{sy}}{\partial x} \\ \frac{\partial d_{sx}}{\partial y} & \frac{\partial d_{sy}}{\partial y} \end{bmatrix} \begin{pmatrix} \cos\alpha \\ \sin\alpha \end{pmatrix} , \tag{22}$$

$$\alpha = \tan^{-1} \frac{U_x}{U_y} . \tag{23}$$

Note that for a plane problem without defocusing, $\boldsymbol{\epsilon}$ is closely related to the two-dimensional strain tensor. Figure 5 shows a plot of ψ versus ρ. It is seen that Ω and ψ are competing factors. An increased r_x results in an increased Ω but a decreased ψ.

In the case of whole-field filtering, it can be shown that the ensemble average of the intensity distribution at the image plane of a spatial filtering optical system is

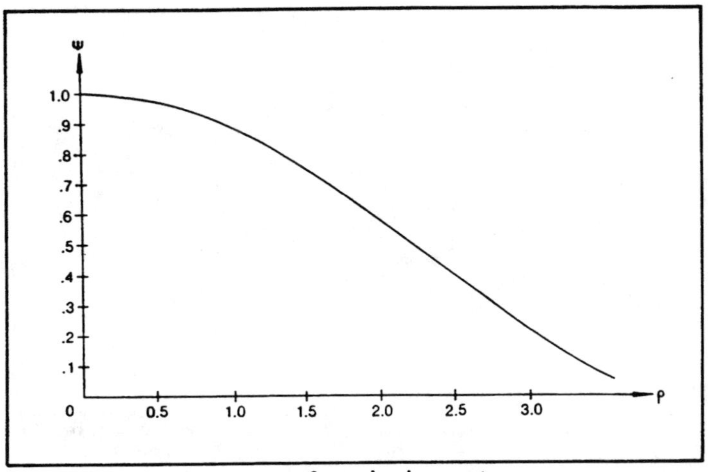

Fig. 5. Strain factor ψ versus ρ for a circular aperture.

$$I(X') = 1 + \nu_d \cos\left[2\pi \frac{d_s(X')\cdot U_o}{\lambda f}\right], \tag{24}$$

where U_o is the position vector of the filtering aperture at the FT plane. These are the so-called isothetic fringes representing contours of the speckle displacement component resolved along U_o. It is seen that the fringe's amplitude is also modulated by the product of decorrelation factors. The argument presented for the case of pointwise filtering can be extended to the whole-field filtering situation with some minor modification. The effect of the illuminated area in the specklegram in pointwise filtering is now replaced by the Airy disk generated by the filtering aperture at the output plane. The radius of the Airy disk approximated by its first zero point is $r_x = 0.61 f/r_u$, where r_u is the radius of the filtering aperture. With this replacement, the expressions for Ω and ψ for pointwise filtering can be directly adapted, while the factor γ remains unaffected. Both Figs. 4 and 5 for Ω and ψ can also be adapted. As pointed out earlier, Ω and ψ are competing factors. In the case of whole-field filtering, the competition is more direct because one always varies the filtering aperture in an effort to obtain better isothetics. A decreasing r_u results in increasing Ω but decreasing ψ. Thus, it is informative to plot the combined effect of $\Omega\psi$. Figure 6 shows such a plot for a special case.[29] The general trend is applicable to most situations. It is seen that while fringe contrast can be increased in general by reducing the filtering aperture, too small an aperture will result in drastically reduced contrast of the isothetic fringes.

While the global fringe contrast as determined by ν_d $(= \gamma\Omega\psi)$ obviously plays an important role in fringe visibility, it is not the only factor. As noted earlier, these fringes are modulated by the secondary speckles created by the filtering process. Indeed, it is easy to visualize that if the size of the secondary speckles is very small, a fringe pattern with a low ν_d would still be visible. On the other hand, if the secondary speckles are very large compared to fringe spacing, a high ν_d would not be sufficient for discerning the fringes.

For a given ratio of fringe spacing to the size of the secondary speckle (the FS ratio), there exists a value of ν_d below which the fringes will not be visible. This critical value of ν_d not only is a function of the FS ratio but also depends on the mechanism under which the fringes are observed (e.g., by the eye, by microdensitometer tracing, etc.). We shall define an ambiguity factor Q such that

$$Q = (\nu_d)_{critic} \ . \tag{25}$$

Q is a function of the FS ratio; i.e.,

$$Q = Q\left(\frac{S_f}{S_s}\right), \tag{26}$$

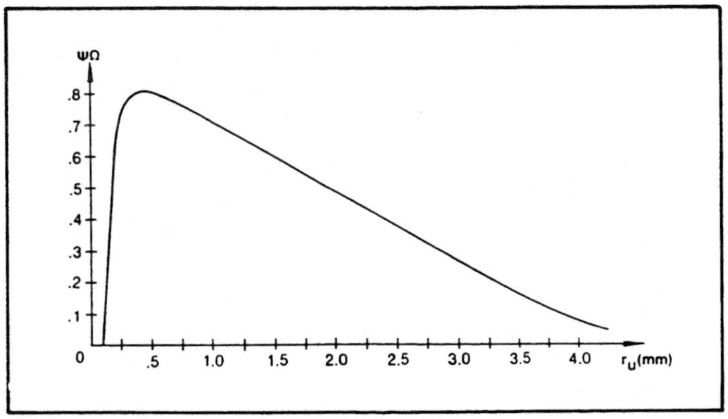

Fig. 6. Combined effect of factors Ω and ψ in full-field filtering.

where S_f and S_s are the fringe spacing and the secondary speckle size, respectively. With the help of the ambiguity factor, we may establish a criterion for fringe discernibility as

$$\gamma\Omega\psi - Q\left(\frac{S_f}{S_s}\right) > 0 \tag{27}$$

and the upper limit of displacement measurement as

$$\gamma\Omega\psi - Q\left(\frac{S_f}{S_s}\right) = 0 \ . \tag{28}$$

The FS ratio for pointwise filtering is

$$\frac{S_f}{S_s} = \frac{2r_x}{1.2\,|\mathbf{d}_s|}, \tag{29}$$

and for whole-field filtering is

$$\frac{S_f}{S_s} = \frac{2r_u}{1.2\,|\epsilon|\,|U_o|} \ . \tag{30}$$

The ambiguity factor Q is not easily obtainable through analytical derivation. Not only is it a function of the FS ratio, but it is also influenced by the instrumentation used to detect and process the fringe information. However, it is possible to obtain the functional form of the ambiguity factor by computer simulation through the plotting of secondary speckle fields modulated by fringes of various magnitudes and spacing; alternatively, experiments could be designed to generate the various patterns from which Q can be determined empirically.

5. HOLOSPECKLE INTERFEROMETRY AND REAL-TIME SPECKLE PHOTOGRAPHY

Holographic interferometry exploits the amplitude interference of two waves, whereas speckle photography employs the intensity interference of the two. These methods complement each other in that the former is best suited for the measurement of out-of-plane displacement and the latter is essentially used for in-plane displacement measurement. It has occurred to a number of workers[15,16] that the two methods should be combined to take advantage of their unique features. A recent entry is the scheme proposed by Chiang et al.,[17] which has some capabilities not available to the earlier schemes.

The experimental arrangement of the scheme is shown in Fig. 7. It is a conventional off-axis holographic system except that there are two reference beams whose polarization can be switched from parallel to orthogonal by a rotator. In the recording process, the first exposure is made with one reference beam and the second exposure

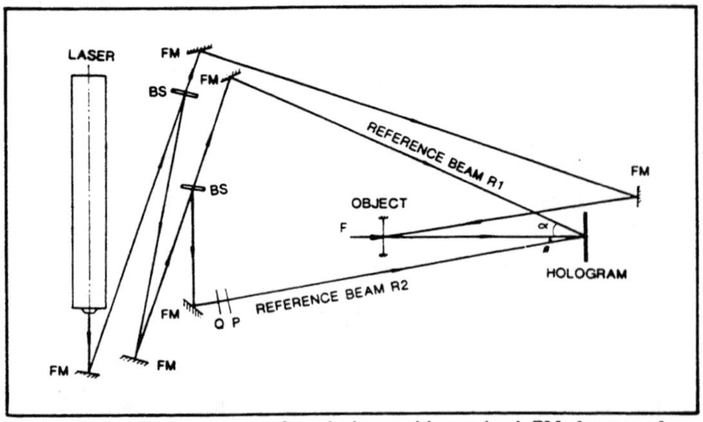

Fig. 7. Optical arrangement for a holospeckle method. FM: front surface mirror, BS: beam splitter, Q: quarter-wave plate, and P: plane polarizer.

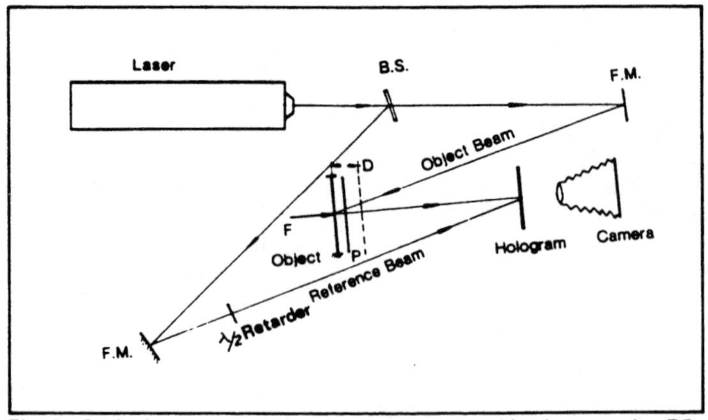

Fig. 8. Optical arrangement for real-time laser speckle photography. BS: beam splitter, FM: front surface mirror, P: polarizer, and D: 4.25 in. defocused distance.

(after object deformation) is made with the other. In reconstruction, both reference beams are utilized. They are polarized parallelly to view the holographic fringes and polarized orthogonally when the holographic fringes are to be destroyed and the remaining two fields are ready for speckle photography. It has been demonstrated[46] that speckle fields of orthogonal polarization remain highly correlated. Thus, when one uses a camera to photograph the two reconstructed speckle fields (one before and one after deformation), a specklegram is obtained; this specklegram can be processed in the conventional way to yield either Young's or isothetic fringes. This scheme has several advantages. First, only a single double-exposure holographic recording is needed to perform both holographic and speckle interferometries. Second, the reconstructed speckle fields interfere in real time. They can be sectioned by either focused or defocused photography to generate speckle fringes representing in-plane displacement and tilt, respectively. Third, one can vary the specklegram recording angle through the hologram window to generate more information, which is needed for a general three-dimensional deformation problem. Variations of holospeckle interferometry have found application in velocity measurement and crack-tip deformation determination.

Following this line of thought, Chiang and Li[47] proposed a scheme for real-time speckle photography. The experimental arrangement is shown in Fig. 8. It is an off-axis holographic system except that a provision is made to rotate the polarization direction of the reference beam and a linear polarizer is placed in front of the object surface. The object in its initial stage is holographically recorded and then reconstructed with a polarization orthogonal to that of the live object. As a result, when the object is deformed, no real-time holographic fringes will be seen. The "deforming" speckle field, on the other hand, can be photographed in real time as it interferes with the reconstructed "frozen" initial speckle field.

6. SPECKLE PHOTOGRAPHY WITH ELECTRON MICROSCOPES

The sensitivity of the speckle methods is determined by the smallest speckle that can be generated or resolved. When visible light is used, it is limited to $\lambda/2$, or about 0.3 μm, for a He-Ne laser. While coherent speckles substantially less than this size await the availability of new light sources such as x-ray lasers, incoherent speckles much smaller than this can be generated by a variety of means and viewed through electron microscopes. Such a method was recently demonstrated by Chiang.[48]

In the process of thin-film deposition, clusters of "islands" are formed before they coalesce into a thin film. These islands are an ideal pattern of random speckles, and these speckles are so small that they are not resolvable by visible light. The way to view them is through electron microscopes. Thus, if one deposits such a pattern onto a specimen's surface and has the speckle pattern photographed using a scanning electron microscope (SEM) before and after the deformation of the specimen, a double-exposure specklegram is formed. Alternatively, one can photograph the two states of the speckle pattern separately and then superimpose afterwards. Figure 9(a) shows a speckle pattern from the deposition of gold particles, and Fig. 9(b) shows the resulting Young's fringes from two such patterns superimposed.

An SEM not only has a higher resolution, of the order of 10^5 magnification, but it also has a 300 times better depth of field than an optical microscope. Thus, the difficulty often encountered in using an optical microscope, that under load the specimen surface is deformed out of focus, is virtually eliminated. This feature is especially useful when applying the speckle technique to crack-tip deformation measurement, wherein the out-of-plane deformation is quite pronounced. This "electronic" speckle technique may find wide application in the future. Apart from crack-tip deformation measurement, one can use it in determining strains in individual grains or grain boundary sliding in metallic materials. It is also feasible to apply this technique to the measurement of strains in microprocessor chips and other miniature elements.

The sensitivity of the method can be further increased if a transmission electron microscope (TEM) is used. The only drawback of using a TEM is that the specimen has to be extremely thin in order to be penetrable by electrons. Figure 10 shows a speckle pattern resolved by a TEM. Speckles smaller than those shown can also be prepared with special techniques.

7. AUTOMATIC YOUNG'S FRINGE ANALYSIS

A Young's fringe pattern (resulting from pointwise filtering) is the predominant mode of retrieving the displacement information carried by a double-exposure specklegram. The useful parameters are the fringe spacing and the fringe orientation. Since it is a pointwise approach, a great deal of work is involved if one has to analyze a number of specklegrams, which is often the case. Thus, in recent years a flurry of activities[23–26] have concentrated on how to automate the process. This is not a trivial pursuit because, as noted previously, the Young's fringes are modulated by secondary speckles and enveloped by the diffraction halo. Therefore, a good automatic fringe processor must be able to do the following: (1) fringe direction determination, (2) speckle noise suppression, (3) halo effect elimination, and (4) fringe spacing determination.

Ineichen et al.[23] suggested an automatic processor that can effectively perform tasks (2) and (4). The scheme proposed by Kaufmann et al.[24] satisfies requirements (2), (3), and (4). A computer algorithm recently designed by Robinson[25] fulfills the first need. Since it is impractical to review in detail all these schemes, we attempt to summarize the key points in the following.

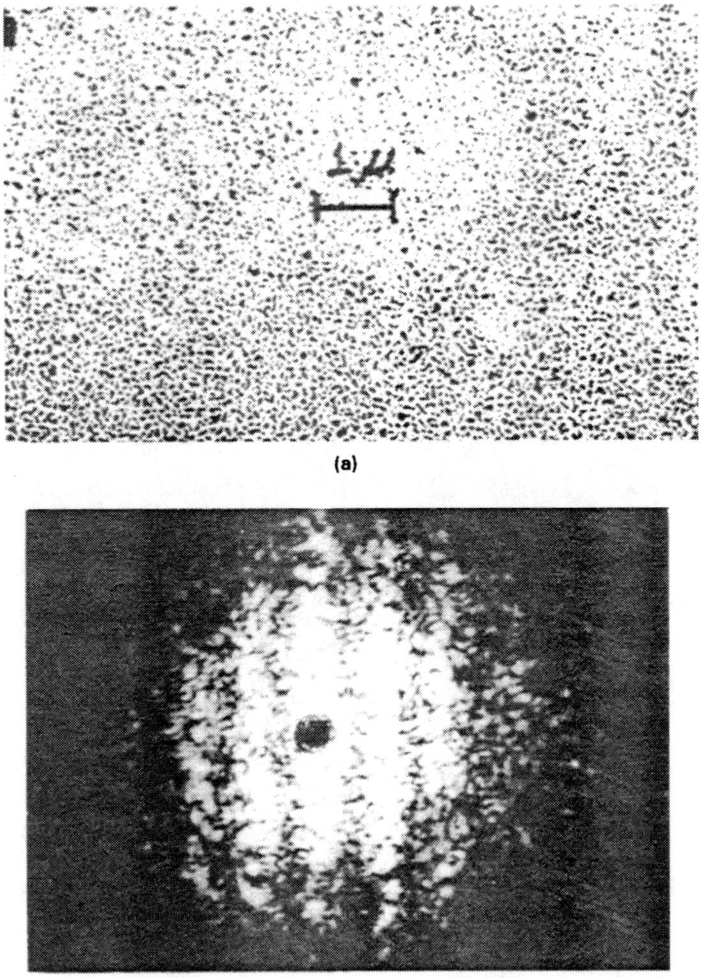

(b)

Fig. 9. Speckle photography with electron microscope: (a) speckle pattern of gold particles resolved by scanning electron microscope; (b) corresponding Young's fringes.

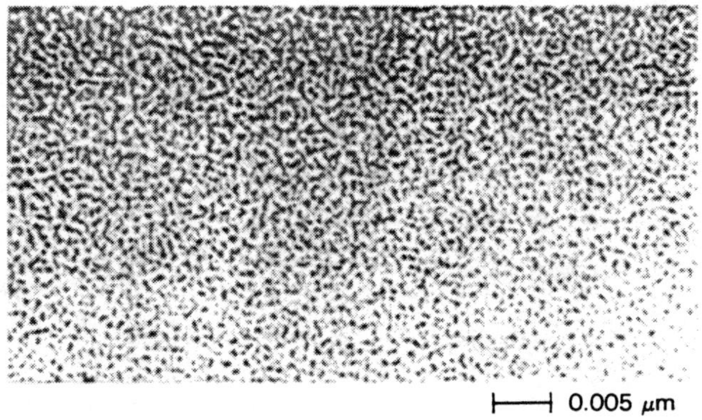

Fig. 10. Fine speckle pattern resolved by transmission electron microscope.

In general, a Young's fringe processor consists of three parts: an optical Fourier processor, a fringe pattern digitizer, and a fringe pattern analyzer. The first part is composed of a low-power laser, a stage to hold the specklegram, and a stepping motor to shift the specklegram so that it can be scanned. The second part may be a digitizing TV camera or other digitizing device. The last part is a computer with appropriate software to do all the tasks mentioned above. The sequence is usually as follows.

(1) Fringe direction determination: The Young's fringe that passes through the center of the diffraction halo is usually the brightest. A polar coordinate is established with its origin at the center of the halo. The intensity values along various radial sections are summed and compared. The direction along which the sum assumes its maximum is the direction parallel to the fringes.

(2) Speckle noise suppression: Once the fringe direction is determined, the intensity values along each fringe can be summed to produce a one-dimensional fringe pattern. The averaging process effectively suppresses the speckle noise present in the original fringes. (In the work of Maddux et al.,[26] line integration is done by a photodiode array.)

(3) Elimination of halo function: The intensity summation process is then carried out along sections perpendicular to the fringes, thus producing a speckle-noise-free one-dimensional halo function. Dividing the one-dimensional fringe pattern obtained in step (2) by the halo function thus obtained results in a one-dimensional fringe pattern with a nearly cosine profile.

(4) Fringe spacing determination: This task is carried out using either the process of autocorrelation or Fourier transformation. The former generates a further smoothed fringe pattern, and the average distance between fringe peaks is then determined. The latter produces the frequency spectrum of the fringe pattern, whose peak location is used to determine the fringe spacing.

8. EPILOGUE

Since Burch and Tokarski's[6] classical paper of 1968, speckle photography has come a long way. In addition to the development that has application to mechanics and metrology, there exists a parallel development called stellar speckle interferometry that has applications to astronomy.[49] Now that the basic theory that provides the foundation for the technique has been well developed, we have gained a deeper understanding as to what the technique can and cannot do. While most of the activity is in the fields of solid mechanics[2] and metrology in general,[5] applications of the technique have spread into such fields as fluid mechanics,[50] material science,[51] and ophthalmology.[52] But the potential of this technique is still enormous, and new applications await discovery.

9. ACKNOWLEDGMENTS

Author F. P. Chiang has been engaged in speckle method research since the early 1970s. In the ensuing years, his research would not have been possible without the generous support of the National Science Foundation's Solid Mechanics Program through the following grants: ENG7610750, ENG7808573 (capital equipment), MEA8009047, and MEA8402912. He and his coworkers gratefully acknowledge this support. They also acknowledge the support of the Office of Naval Research through Contract No N0001482K0566 in recent years for laser speckle work.

10. REFERENCES

1. F. P. Chiang, "A new family of 2D and 3D experimental stress analysis techniques using laser speckles," Solid Mechanics Archives 3(1), 1 (1978).
2. F. P. Chiang, J. Adachi, R. Anastasi, and J. Beatty, "Subjective laser speckle method and its applications to solid mechanics problems," Opt. Eng. 21(3), 379 (1982).
3. C. Froehly, "Speckle phenomenon and some of its applications," Proc. IUTAM Symposium on Optical Methods in Mechanics of Solids (1979).
4. M. Francon, *Laser Speckle and Applications in Optics,* Academic Press, New York (1979).
5. R. K. Erf, ed., *Speckle Metrology,* Academic Press, New York (1978).
6. J. M. Burch and J. M. J. Tokarski, "Production of multi beam fringes from photographic scatters," Opt. Acta 15(2), 101 (1968).
7. E. Archbold, J. M. Burch, and A. E. Ennos, "Recording of in-plane surface displacement by double exposure speckle photography," Opt. Acta 17(12), 883 (1970).
8. E. Archbold and A. E. Ennos, "Displacement measurement from double-exposure laser photographs," Opt. Acta 19(4), 253 (1972).
9. F. P. Chiang, "Multi-aperture speckle interferometry," Proc. Conf. on

Speckle Phenomena and Their Applications (1974).
10. F. P. Chiang, "A new three-dimensional strain analysis technique by scattered light speckle interferometry," Appl. Opt. 19(16), 2701 (1980).
11. K. A. Stetson, "Speckle photography and interferometry: a review," Opt. Eng. 14, 482 (1975).
12. A. Asundi and F. P. Chiang, "Theory and applications of the white light speckle method for strain analysis," Opt. Eng. 21(4), 570 (1982).
13. P. M. Boone, "Use of close range objective speckles for displacement measurement," Opt. Eng. 21(3), 407 (1982).
14. F. P. Chiang and C. C. Kin, "Strain determination on curved surfaces using far-field objective laser speckles," Opt. Eng. 21(3), 441 (1982).
15. F. D. Adams and G. E. Maddux, "Synthesis of holography and speckle photography to measure 3-D displacements," Appl. Opt. 13, 219 (1974).
16. P. H. Ruterbusch, J. A. Tome, and F. T. S. Yu, "Multiplexed speckle and holographic interferometry with color encoding by white-light processing," Opt. Eng. 22(4), 501 (1983).
17. F. P. Chiang, Q. B. Li, and J. B. Chen, "Holospeckle interferometry," presented at the 1984 SESA International Congress on Experimental Mechanics (1984).
18. T. D. Dudderar and J. A. Gilbert, "Fiber-optic measurement of the deformation field on a remote surface using numerically processed white-light speckles," Appl. Opt. 21(10), 3520 (1982).
19. G. H. Kaufmann, "On the numerical processing of speckle photograph fringes," Opt. & Laser Technol. 12, 207 (1980).
20. C. S. Vikram, "Simple approach to process speckle photography data," Opt. Lett. 7, 374 (1982).
21. M. E. Caluwe, "A simple device to reduce secondary speckle," Optics & Lasers in Eng. 3, 229 (1982).
22. W. H. Peters and W. F. Ransom, "Digital imaging techniques in experimental stress analysis," Opt. Eng. 21(3), 427 (1982).
23. B. Ineichen, P. Englon, and R. Dändliker, "Hybrid optical and electron image processing for strain measurements by speckle photography," Appl. Opt. 19(3), 2191 (1980).
24. H. Kaufmann, A. E. Ennos, B. Gale, and D. J. Pugh, "An electric-optical read-out system for analysis of speckle photographs," J. Phys. E 13, 579 (1980).
25. D. W. Robinson, "Automatic fringe analysis with a computer image-processing system," Appl. Opt. 22(4), 2169 (1983).
26. G. E. Maddux, S. L. Moorman, and R. R. Corwin, "A programmable data-retrieval system for in-plane displacement from speckle photography," AFFDL Report TM-78-109-FBE, Wright Patterson Air Force Base, Ohio (1978).
27. I. Yamaguchi, "Fringe formation in speckle photography," J. Opt. Soc. Am. 1(1), 81 (1984).
28. J. B. Chen and F. P. Chiang, "Statistical analysis of whole field filtering of specklegram and its upper limit of measurement," J. Opt. Soc. Am. A 1(8), 845 (1984).
29. D. W. Li, J. B. Chen, and F. P. Chiang, "Statistical analysis of one-beam subjective laser speckle interferometry," J. Opt. Soc. Am. A 2(5), 657 (1985).
30. J. W. Goodman, "Statistical properties of laser speckle patterns," in *Laser Speckle and Related Phenomena*, J. C. Dainty, ed., Chap. 2, Springer-Verlag, New York (1975).
31. I. Yamaguchi, "Speckle displacement and decorrelation in the diffraction and image fields for small object deformation," Opt. Acta 28, 1359 (1981).
32. L. I. Goldfisher, "Auto-correlation function and power spectrum density of laser produced speckle patterns," J. Opt. Soc. Am. 55(3), 247 (1965).
33. J. C. Dainty, "Some statistical properties of random speckle patterns in coherent and partially coherent illumination," Opt. Acta 17, 761 (1970).
34. H. J. Tiziani, "Analysis of mechanical oscillations by speckling," Appl. Opt. 11, 2911 (1972).
35. K. A. Stetson, "Problem of defocusing in speckle photography, its connections to hologram interferometry and its solution," J. Opt. Soc. Am. 66, 1267 (1976).
36. D. A. Gregory, "Basic physical principles of defocussed speckle photography: a tilt topology inspection technique," Opt. Laser Technol. 8, 201 (1976).
37. P. Jacquot and P. K. Rastogi, "Speckle motions induced by rigid body movements in free space geometry: an explicit investigation and extension to new cases," Appl. Opt. 18, 2022 (1979).
38. G. J. Jaisingh, "Studies in one-beam laser speckle interferometry," Ph.D. Thesis, State University of New York at Stony Brook (1979).
39. X. P. Wu, S. P. He, and Z. C. Li, "Movement of space speckle," presented at 1980 SESA Int. Congress on Experimental Mechanics (1980).
40. P. K. Rastogi and P. Jacquot, "Speckle metrology techniques: a parametric examination of the observed fringes," Opt. Eng. 21(3), 411 (1982).
41. F. P. Chiang and D. W. Li, "On the laws of laser speckle movement in space," in *International Conference on Speckle*, H. Arsenault, ed., Proc. SPIE 556 (1985).
42. R. P. Khetan and F. P. Chiang, "Strain analysis by one-beam laser speckle interferometry, 1: single aperture method," Appl. Opt. 15(9), 2205 (1976).
43. R. Megnard, "Diffraction halo in speckle photography," Appl. Opt. 23(14), 2235 (1984).
44. F. P. Chiang and D. W. Li, "Diffraction halo functions of coherent and incoherent random speckle patterns," Appl. Opt. 24(14), 2166 (1985).
45. E. Archbold, A. E. Ennos, and M. S. Virdee, "Speckle photography for strain measurement—a critical assessment," in *1st European Conf. on Optics Applied to Metrology*, Michel Grosmann and Patrick Meyrueis, eds., Proc. SPIE 136, 258 (1978).
46. F. P. Chiang and D. M. He, "Polarized speckles and their applications," Proc. 5th Int. Congress on Experimental Mechanics, Soc. for Experimental Mechanics (1984).
47. F. P. Chiang and Q. B. Li, "Real-time laser speckle photography," Appl. Opt. 23(24), 4469 (1984).
48. F. P. Chiang, "Speckle method with electron microscopes," Soc. of Experimental Stress Analysis spring meeting (1982).
49. J. C. Dainty, "Stellar speckle inteferometry" in *Laser Speckle and Related Phenomena*, J. C. Dainty, ed., Chap. 7, Springer-Verlag, New York (1975).
50. T. D. Dudderar and P. G. Simpkins, "The development of scattered light speckle metrology," Opt. Eng. 21(3), 396 (1982).
51. T. Asakura, "Surface roughness measurement," in *Speckle Metrology*, R. K. Erf, ed., Chap. 3, Academic Press, New York (1978).
52. E. Ingelstan and S. Ragnarson, "Eye refraction examined by aid of speckle pattern produced by coherent light," Vision Res. 12, 411 (1972).

Reprinted with permission from *Materials Evaluation,* Vol. 38(1), pp. 86-92 (1980).

Noncontact Material Testing Using Laser Energy Deposition and Interferometry

by C. A. Calder and W. W. Wilcox

Abstract

A technique is described for the noncontact testing of materials using laser energy deposition to generate a stress pulse and interferometry to record the transient surface displacement. The dilatational wave speed can be measured and, in the particular case of rod or plate specimens, sufficient information can be obtained to evaluate the two elastic constants of an isotropic material. Several applications illustrating the advantages of the approach are summarized.

Introduction

The accurate characterization of material behavior for use in computer models and calculations is continually required for the design of safe and efficient structures and machine components. Frequently, new materials or unique applications require measurements of material properties where conventional methods are unsuitable. In this work, an experimental technique is described which allows loading of a specimen and the recording of its dynamic response without physical contact with the specimen. The transient load is applied by rapid energy deposition from a single-pulse Q-switched, high energy laser. The rapid heating produces vaporization of the surface skin and the resulting blowoff drives a compressive stress pulse into the material. Stress wave arrivals are monitored using a dynamic displacement interferometer.

In a number of situations this technique has distinct advantages over the commonly used ultrasonic methods. These include the following:

1. The laser loading produces a large stress pulse amplitude of short duration so that tests of highly attenuating or very thin materials are possible.
2. The short measurement time of a few microseconds or less can be made at a precise instant in time.
3. The noncontact feature is especially useful for testing in severe environments and with toxic materials.
4. There are few restrictions on specimen size and configuration.
5. The specimen can be far removed from the instrumentation hardware.

In addition, flat and specular surfaces are often not necessary for adequate interferometry measurements when a relatively high power CW laser is available.

Specimen Loading by Laser Energy Deposition

The potential for generating stress waves in materials by laser energy deposition has been recognized since the early sixties following the invention of the pulsed ruby laser in 1960. Ready[1] gave one of the first treatments describing the effects of normal mode and Q-switched laser pulses absorbed at opaque surfaces. In the late sixties, Percival[2] conducted a study of laser generated stress waves in an absorbing rod and used Pochhammer-Chree theory to

Dr. Clarence A. Calder received his PhD in applied mechanics from the University of California, Berkeley. He is presently an associate professor of mechanical engineering at Oregon State University and serves as a consultant for the Materials Engineering Division of Lawrence Livermore Laboratory. Recent research interests include laser instrumentation and laser-induced stress wave applications in materials testing. Professional affiliations include ASME, SESA, and the American Society for Engineering Education. For inquiries concerning this work, contact the author at 503/754-2427.

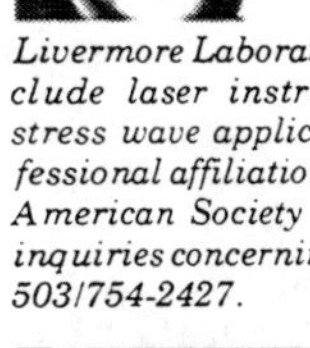

William W. Wilcox is a Technical Associate with the Lawrence Livermore Laboratory, Livermore, CA, and is presently working in the Nondestructive Evaluation Section of the Materials Engineering Division. He attended the University of Santa Clara, Santa Clara, CA, and is currently working toward a BS from the University of San Francisco. Since 1973 he has been involved in the research and development of laser-optical methods of materials evaluation and has been author or co-author of over 15 publications concerning this work. He is a current member of SESA and has been selected for inclusion in the 1979 edition of Who's Who in Technology Today.

describe the dynamic stress response. This approach was used to determine the elastic constants of aluminum rod specimens at elevated temperatures using a sandwiched piezoelectric tourmaline crystal.[3] The technique, simplified by Calder and Wilcox[4] to include a commercially available ultrasonic transducer, was used to evaluate the elastic constants of ten different materials including highly attenuating plastics, and to demonstrate the feasibility of flaw detection using laser induced stress pulses. Skeen and associates[5] used quartz stress gages to record stress pulse amplitudes created by the rapidly expanding plasma. A following study[6] explored the large plastic deformation of thin plates subjected to laser loading with the plasma confined by a surface overlay. Laser induced stress pulses with rise times of less than 1 ns and half-amplitude durations of 3 ns have been produced with a 3-ps pulsed laser.[7] Felix and Ellis[8,9] have produced laser generated ultrasonic single pulses of 200 ns and wave trains with frequencies from 1-25 MHz using a confined absorbing liquid layer. A similar arrangement was later used by von Gutfeld and Melcher for flaw detection applications.[10,11] Fairand and coworkers have conducted several studies[12–14] related to increased laser generated stress amplitudes by using surface overlays. They produced large plastic flow at the surface of an iron-silicon alloy specimen but found the stress amplitude decayed rapidly as it progressed into the interior. An extensive coverage of work done up to 1971 on material response to high power lasers is given by Ready.[15]

Much effort is currently being expended in laser-material interaction studies ranging from metal scribing at low energy to laser fusion at ultra high energy. At moderate laser-energy levels, stress pulse amplitudes on the order of the yield stress of metals can be reached. Such stress waves are induced by at least two mechanisms, and depending on the opacity of the material to the laser wavelength used, one or the other may predominate. In the first mechanism, the rapid deposition of laser energy into an absorbing material produces a state of unbalanced, compressive thermal stress as the heated material wants to expand but is restrained initially by its own inertia. This stress gradient propagates into the material and is immediately followed with a tensile tail arising from the front surface reflection. See, for example, Ref. 16.

In the second mechanism, an opaque target surface is rapidly vaporized by the high energy laser pulse, and the recoil from blowoff of the vapor causes a strong, transient pressure build-up. The pressure on the target surface induces a compressive stress pulse that propagates through the material. This mechanism is illustrated in Figure 1 which shows the laser generated plasma blowoff from a 6.35-mm-thick aluminum disk subjected to laser energy deposition from a neodynium-glass pulse of about 12 J and 30 ns acting on a 22-mm-diam area.

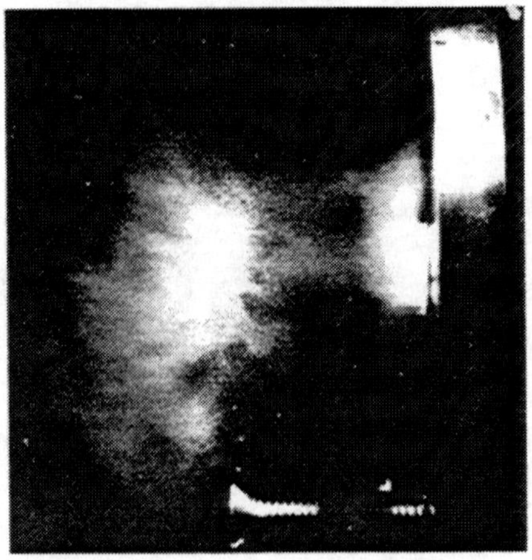

Figure 1—Plasma blowoff from 6.35-mm-thick aluminum disk. Pulse was 12 J, 30 ns from neodymium-glass laser on 22-mm-diam area.

The second mechanism can produce some surface damage but in most cases it is insignificant with the surface appearing slightly scorched. This is illustrated by the results of laser energy deposition on a 25-mm-diam aluminum disk at three energy densities shown in the upper left of Figure 2. The disk surface was first prepared with 600 grit sandpaper to provide a reasonably uniform surface roughness. Then the surface was exposed to 1.06 μm laser pulses of (A) 5 J and (B) 12 J over a 10-mm-diam area, and (C) 12 J focused to about 1 mm diameter. The latter produced a clearly evident crater in the surface. Scanning electron microscopy views of the original and exposed surface (B) are shown at 1000X. The vaporization and melting of the ridge peaks are evident in the upper right view of Figure 2.

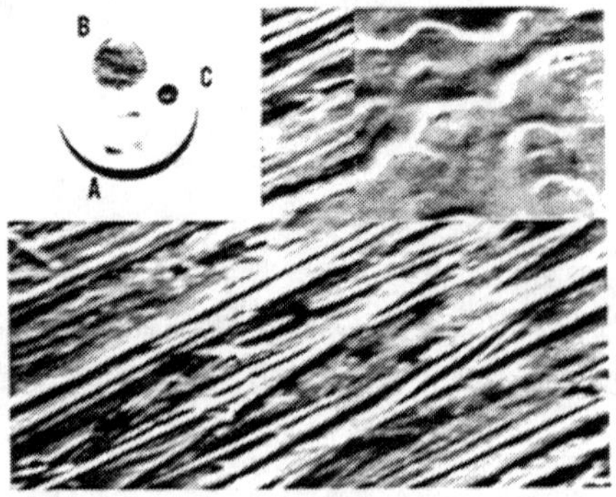

Figure 2—Effect of laser deposition on aluminum. Laser energy density was about (A) 6 J/cm². Background was unexposed surface at 1000X and upper right was area B at 1000X.

Measurements of actual material mass removed by laser deposition were taken for three samples each of 1018 steel, aluminum, tantalum, and nickel. The average surface depth removed was calculated assuming a uniform layer of the measured mass. The laser energy was 12 J at 1.06 μm and 30 ns over a spot about 10-mm diameter. The results, accurate to ± 2 μg, are tabulated in Table 1 and show the actual depth removed to be extremely small. The mass removed is a function of many parameters including melt and vaporization temperature, specific heat, heat of fusion, heat of vaporization and coefficient of absorption. For example, aluminum has a low melting temperature but little mass was removed due to its high specific heat, heat of fusion, heat of vaporization and reflectance.

TABLE 1 Mass of Surface Skin Vaporized in Laser Deposition Tests of Four Metals. Neodymium-glass laser, Q-pulsed with Irradiance of 15 J/cm².

Metal	Avg Mass Lost	Avg Depth Removed
1018 Steel	5 μg	0.008 μm
Aluminum	2 μg	0.010 μm
Tantalum	18 μg	0.014 μm
Nickel	25 μg	0.036 μm

Surface treatment or overlays, such as highly absorbing black paint or a transparent sheet of slide glass, are known to greatly effect the amplitude and duration of the stress pulse induced in the specimen. In cases where the specimen surface is made more highly absorbing, a larger percentage of the laser energy is coupled to the surface creating a greater blowoff pressure. When a transparent overlay (tamper) is used, the laser energy is absorbed at the specimen surface and the confined plasma produces a much higher pressure. These effects were explored by conducting laser energy deposition tests on a 6.35-mm-thick aluminum disk specimen with the following surface conditions:

1. bare and cleaned by previous shots,
2. blacked with a felt-tip pen,
3. tampered with cellophane tape,
4. blacked and tampered with tape.

Laser deposition from a Q-pulsed neodymium-glass laser operating with a 12 J and 30 ns pulse duration was spread approximately uniformly over an area of 22-mm-diam in each case. Stress arrival at the back face was monitored with an acoustically-matched quartz transducer with a 10-mm-diam guard ring using the technique described by Graham.[17] This one-dimensional strain experiment permitted recording of the complete laser induced compressive pulse before edge ef-

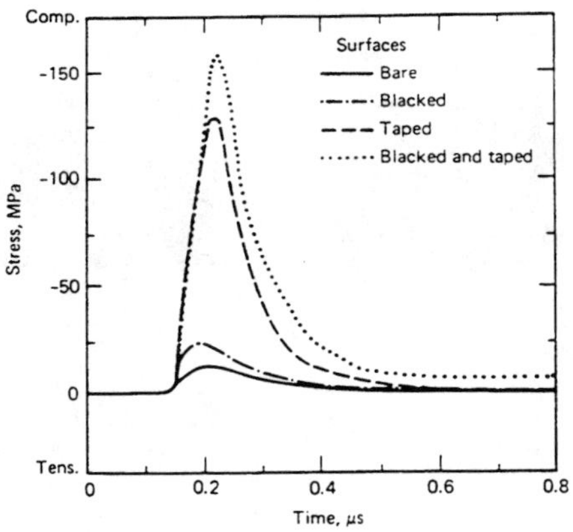

Figure 3—Stress histories for four surface conditions.

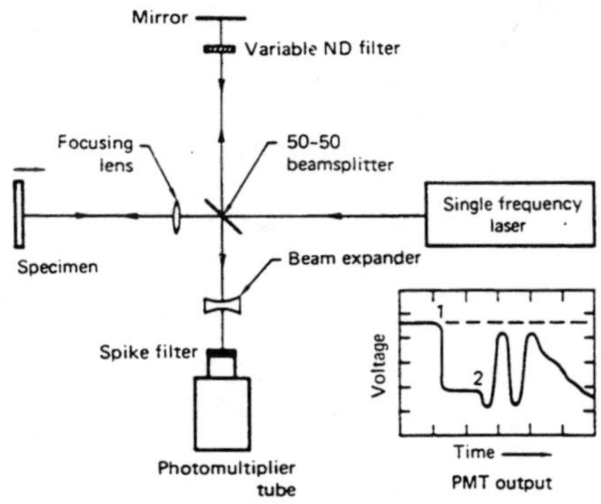

Figure 4—Displacement interferometer arrangement with simulated photomultiplier output trace.

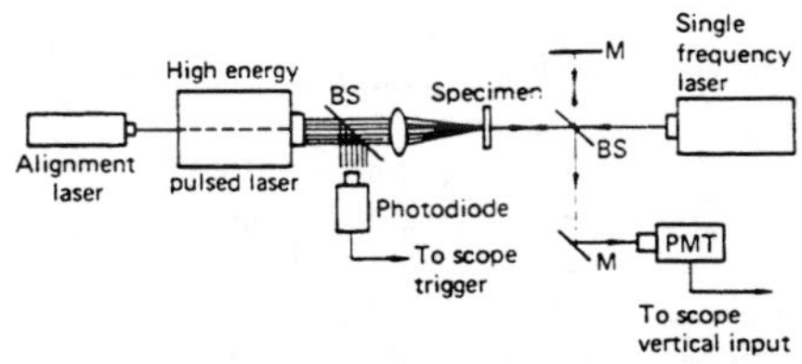

Figure 5—Schematic of noncontact testing arrangement using laser deposition loading and displacement interferometry.

fects invalidated the quartz transducer output.

Stress profile results for the four surface conditions are compared in Figure 3. While blacking the surface nearly doubled the peak stress of the bare surface, the taped overlay produced a ten-times increase in stress amplitude. The maximum stress of about 157 MPa was achieved with the combined blacked and taped surface. The half-amplitude stress pulse duration decreased from 240 ns for the bare surface to 95 ns for the blacked and taped condition. Rise time was about 45 ns in all cases. Earlier work[18] has shown the transparent cellophane tape to be as effective in enhancing the stress amplitude as more massive and less convenient transparent overlays of quartz, epoxy, and slide glass. Possibly the adhesive backing has a significant influence on the pressure developed.

Specimen Response By Laser Interferometry

Most work with laser deposition loading has been recorded using piezoelectric quartz transducers. These devices are very sensitive, have high frequency response, and in the case of one-dimensional strain or stress experiments, can provide accurate stress-time data. However, there are several obvious limitations to their use in general. First, the specimen must have flat surfaces to permit direct contact and good coupling with the transducer. Second, specimen size limitations are governed by the transducer size and possible overlapping of the original and echo stresses in the specimen. Third, in general the transducer is not acoustically matched to the specimen material requiring corrections to be made in the data to obtain accurate stress information. Finally, these transducers do not survive well in severe environments and their piezoelectric character disappears at high temperatures.

These limitations are overcome and completely noncontact testing made possible by using a Michelson displacement interferometer to sense the stress wave arrivals at the back face of the specimen. Although the normal stress of the free surface is always zero, the displacement and velocity are nonzero at the time the stress pulse is reflecting from the back face. The specimen surface becomes the moving mirror of the interferometer. When light reflected from the specimen surface is recombined with light from the reference mirrorn, a series of light and dark bands or fringes are produced. Surface motion of the specimen causes fringe movement which can be monitored by a photodetector.

A schematic of the interferometric arrangement used is shown in Figure 4 and is similar to that described by Barker and Hollenbach.[19] The output from a single frequency, Spectra Physics model 164 argon-ion laser, operated at a wavelength of 0.5145 μm, is directed to a beam splitter positioned at 45 degrees to the beam. The specimen and reference mirror, placed perpendicular to each other and equidistant from the beam splitter, reflect the split beams back to the beam splitter where they are recombined to form a pattern of concentric, circular fringes. The fringe pattern can be observed during setup by placing a screen at the photodetector.

The photodetector is an RCA model 7326 ten-stage photomultiplier tube and associated circuitry. The circuit is designed to provide up to 10 V of linear output from the photomultiplier tube. The tube is in an "off" mode when the high voltage is on until a turn-on trigger signal of 25 V is applied. This feature permits operation with the high ambient light level present and provides much higher sensitivity at the moment the desired signal occurs.

The fringe pattern is enlarged with an expanding lens so that the photodetector, equipped with a 0.7-mm-diam aperature, senses only one fringe at a time at the center of the bulls-eye pattern. Movement of the free surface a distance equal to one-half the wavelength of the laser ($\lambda/2$) causes the center of the concentric fringe pattern to change from light to dark to light. Thus, the detector output is a frequency modulated sine wave with each successive peak representing an additional surface displacement of $\lambda/2$. This is illustrated by the simulated trace in Figure 4 showing the photomultiplier circuit turn-on at (1) to a d-c voltage level representing the fringe position at that instant of time. Specimen movement causes a sine wave type output beginning at (2) with each complete cycle representing a specimen back surface displacement of 0.2572 μm. For perfect fringe contrast, the valleys of the output would return to the zero voltage level at (1) indicating complete light extinction. In cases where the specimen surface is not a good mirror this desired condition is difficult to achieve. In these cases fringe contrast can be improved by focusing the specimen beam to a point on the specimen surface and using a neutral density filter in the reference leg to equalize intensities of the recombined beams. A 0.5145 μm spike filter is placed in front of the photomultiplier tube aperture to filter out unwanted pulsed light from the high energy laser, flash tube, and laser generated plasma.

The interferometer system employed differed from the dual beam/stable path length system used by Palmer et. al[20,21] and Bondarenko and Drobot.[22] The photomultiplier tube was pulsed on just prior to the arrival of data and the recording time was short so that low frequency disturbances were not a problem. Also, it was not necessary to limit the displacement to the linearized portion of the interferometer output as only time-of-arrival information was of primary importance.

Noncontact Testing Arrangement

The combination of pulsed laser energy deposition for specimen loading and laser displacement interferometry for measuring specimen response provides a convenient arrangement for noncontact testing of materials in a number of situations which would be impossible or very difficult using more conventional approaches. A schematic of the noncontact test arrangement is shown in Figure 5. Specimen loading is accomplished by laser energy deposition on the specimen surface from a single-pulsed, Q-switched Holobeam series 6082 neodymium-glass laser at energies up to 15 J. Half-amplitude pulse duration is typically about 20-30 ns. A slide glass beam splitter is used to direct a small portion of the input beam to a 0.5-ns-rise time Monsanto silicon pin photodiode model MD2. The output of the photodiode circuit, about 10 V, is used to trigger the oscilloscope time base or a digital delay plug-in unit which triggers the trace after a preset time delay. The single-sweep signals were recorded on a Tektronix 7904 oscilloscope using a 7A19 plug-in amplifier and 7B92A dual time base resulting in a 500 MHz bandwidth capability. A 15 mW Spectra Physics model 124 helium-neon laser is used for precise alignment of the high energy laser and the specimen. The components of the interferometer are as described in the previous section with the exception that in some earlier applications the helium-neon alignment laser was used in place of the argon-ion laser. This particular model is not single frequency and more than one axial mode may be generated. This causes the laser output beam to be amplitude modulated at the beat frequencies of the active axial modes and results in "noisy" data traces.

In experiments requiring accurate measurement of stress wave transit time through the specimen, where the transit time may be only a few hundred nanoseconds, inherent delays in the scope triggering and vertical input amplifier, transit time through the photomultiplier tube, electrical and optical path travel time, and effective initiation time of the stress pulse must be known or excluded from affecting the data. The time of initiation of the stress pulse was measured by using a thin plate of high purity, oxygen-free copper of known dilatational velocity as the specimen in Figure 5. A second slide-glass beam splitter was placed at the high energy input to provide a fiducial marker superposed on the photomultiplier signal output from the interferometer. This procedure eliminated all other timing considerations except the difference in optical path lengths which could be easily determined. The optical path from the fiducial beam splitter to the photomultiplier tube is 0.46 m shorter than the air path via the interferometer, so the high energy laser pulse appears 1.5 ns early on the oscilloscope record shown in Figure 6. The transit time for the 0.91-mm-thick specimen is 192 ns based on a dilatational velocity of 4740 m/s for high purity copper.[23] Therefore, the time of initiation of the stress pulse is obtained by measuring back 190.5 ns from the break-out of the interferometer signal as shown in Figure 6. The

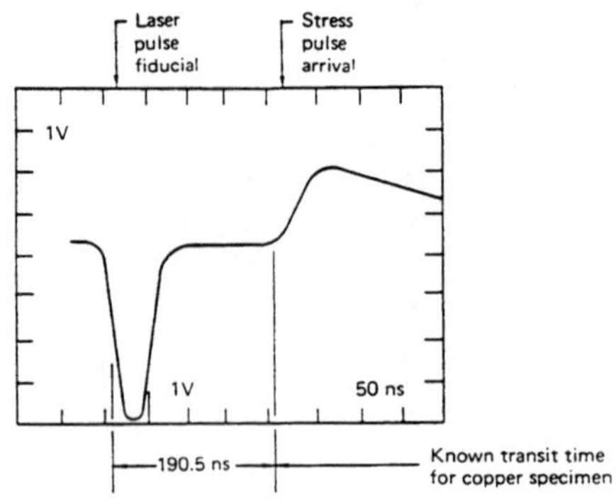

Figure 6—Data trace for determining time of stress initiation for 0.91 mm copper specimen.

time of stress initiation was found to be very close to the half-amplitude point on the rise of the high energy laser pulse fiducial. The same conclusion was found, within 5 ns, for similar tests on other metals.

Using the noncontact test arrangement, measurements of wave velocities accurate to 1-2 percent are possible. The scope sweep should be calibrated at the sweep rate used and corrections made for the various electrical and optical delays occurring in the particular setup. Transient displacements can be determined to within a fraction of wavelength of the laser light used provided tilt of the specimen surface is small. The determination of stress from the displacement record would be considerably less accurate due to errors and difficulties inherent in differentiating experimental data. Velocity interferometry might be considered as an alternative, but is somewhat less sensitive. Also, a recent technique called a "velocity interferometer system for any reflector" (VISAR)[24] would provide velocity information directly as well as reduce the necessity for a good surface mirror and small tilt.

Applications

Several examples of noncontact testing will be briefly discussed to illustrate the technique and give insight to potential applications which might be more conveniently done with this approach.

One-Dimensional Strain Experiment

Single-pulse laser loading of reasonably uniform intensity over a sufficiently large area of a disk specimen induces a one-dimensional strain pulse into the material. This strain pulse quickly deteriorates to a complex state due to unloading waves propagating inward from the edge of the loading. Displacement interferometry on the back face of the specimen will indicate the arrival of the main disturbance, the arrival of its echo off the front face, and the deterioration of the one-dimensional strain state. Data records for two tests at different laser intensities are shown in Figure 7 for a 53-mm-diam by 6.35-mm-thick

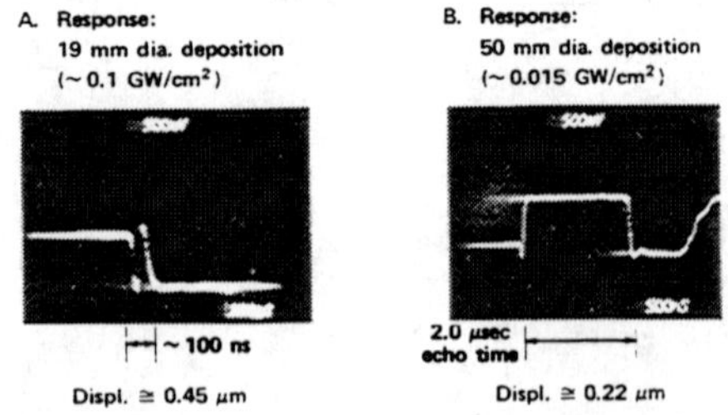

Figure 7—One-dimensional strain data for 53-mm-diam by 6.53-mm-thick aluminum disk at fluences of about (A) 0.1 GW/cm² and (B) 0.015 GW/cm².

aluminum disk. The laser energy of 9 J was spread over a 19-mm-diam in the first test and 50-mm-diam in the second. The expanded sweep in Figure 7A indicates a pulse duration of about 100 ns. The record of Figure 7B shows the first arrival, the echo or one-round-trip reflection 2.0 μs later, and the edge effect appearing an additional 1.0 μs later. The decrease in laser fluence from 0.1 to 0.015 GW/cm² reduced the back surface displacement by one-half.

Measurement of Elastic Constants

A longitudinal wave induced in one end of a rod will travel axially with grazing incidence to the rod lateral surface at the dilatational velocity with part of its energy continuously mode converted to shear waves. The shear wave travels at a slower velocity diagonally across the rod diameter to the opposite surface. There it mode converts back to a shear wave reflecting back into the material and to a dilatational wave again propagating at grazing incidence to the rod. This process is continually repeated resulting in a successive train of pulses arriving at the far end of the rod, the first appearing at the dilatational transit time with the following pulses delayed by a time related to the rod diameter and shear wave velocity. The measured time of arrival of the first two pulses along with the specimen dimensions and mass density provide sufficient information to calculate the elastic constants of the isotropic material using a single test.[25,26]

This ultrasonic test can be conducted using the noncontact approach where the compressional pulse is induced in one end of the rod by laser energy deposition and the stress wave arrivals at the opposite end sensed with the displacement interferometer. The noncontact feature is especially convenient for high temperature material property measurement with the addition of a clamshell oven surrounding the specimen.[27] An actual oscilloscope trace showing the displacement response of a 6.35-mm-diam by 101-mm-long tantalum rod at 200 C is illustrated in Figure 8. The sweep was delayed 23.5 μs

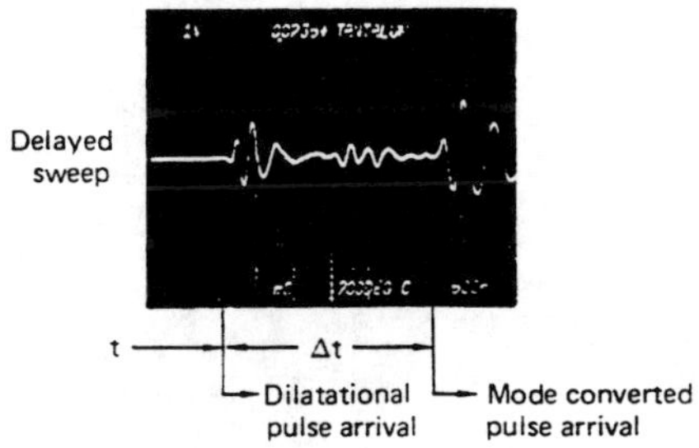

Figure 8—Displacement response of 6.35-mm-diam tantalum rod at 200C subjected to laser deposition on one end.

and swept at 500 ns per division. The arrival times of the dilatational and mode converted signals are clearly indicated. The values to t and Δt lead directly to the evaluation of the wave velocities and elastic constants of tantalum at 200 C. For the particular run shown, Young's modulus was found to be 176 GPa which agrees well with a published value of 180 GPa at 200 C.[23]

Acoustic Velocity Measurement in Liquid Metal Column

An isobaric expansion apparatus has been developed to measure various thermophysical properties of liquid metals at very high temperatures and pressures.[28] In this experiment, a specimen the diameter of a common paper clip (1 mm) and about 25-mm-long is subjected to rapid resistance heating by current dumped from a large capacitor band. Required measurements are taken during an instant a few microseconds after specimen melt but before instability and breakup of the molten column takes place. An additional measurement of sound velocity across the diameter of the specimen at this instant in time would permit a more complete evaluation of the thermodynamic state of the liquid metal. Such a measurement by traditional means would be impossible. However, the noncontact approach is ideally suited for obtaining this information provided the liquid metal remains sufficiently reflective for the interferometric measurements.[29]

The basic setup of Figure 5 was used with optical access to the specimen through tapered sapphire windows positioned in the pressure cell wall on opposite sides of the vertical, 1-mm-diam specimen. The high energy laser was focused to a spot about 0.2-mm-diam on the lateral surface of the specimen and, diametrically opposite, the interferometer laser beam was focused. The specimen and anvil holder are shown in Fig. 9 with the high energy beam and generated plasma on the left and the interferometer laser beam on the right. Lead specimens were selected to demonstrate the technique due to the relatively low energy required to obtain melt. Sonic velocity measurements were first made on solid lead rods and subsequently on lead rods liquified by the high current pulse from a 20 kV capacitor bank discharge. The stress transit time measurement was made a few microseconds following melt. Typical oscilloscope traces are shown in Figure 10 for the solid and liquid lead runs. As in the stress initiation time measurement. discussed in the previous section. the fiducial marker is simply a representation of the timing of the high energy laser pulse. The measured sonic velocity for the solid specimen was reasonably close to an average value 2.21 mm/μs taken from ref. 23. As would be expected, there was a considerable drop in velocity for the liquid lead test. The measured liquid lead sound velocity of

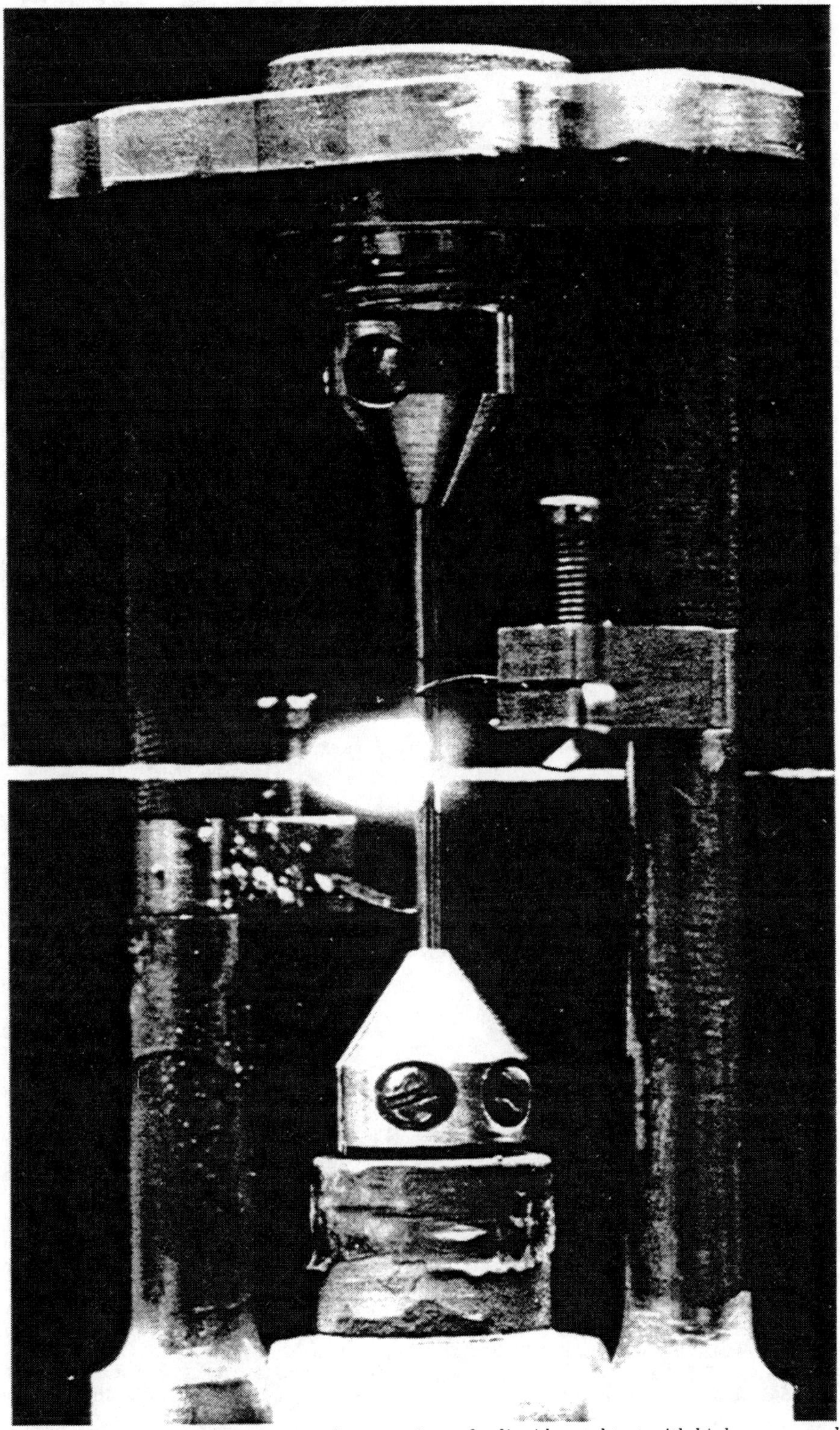

Figure 9—Anvil mount and 1-mm-diam specimen for liquid metal test with high energy and interferometer laser beams shown.

1.80 mm/µs agreed very well with a published value at melt of 1.78 mm/µs.[30]

Flaw Detection

As in conventional ultrasonic testing, the noncontact approach can provide data on the presence and location of flaws in the material. To show this, a 1.5-mm-diam hole was drilled at the midthickness of a 25-mm-thick aluminum plate to provide an artificial flaw as shown in Figure 11. The interferometer was set up using the unpolished, as received, back surface and a stress pulse was induced by single-pulse laser deposition. The oscilloscope trace clearly shows the arrival of the flaw induced echo as well as the first arrival and one-round-trip echo of the main pulse. Using a dilatational velocity for aluminum of 6.3 mm/µs, the disturbance due to echoes from the front and rear flaw surface should arrive at the rear face about 3.8 µs after the main pulse. This is the arrival time observed in Figure 11. The interferometry data has been used for time-of-arrival information only. The fringe order would have to be determined and surface tilt effects accounted for to determine accurate displacement amplitudes.

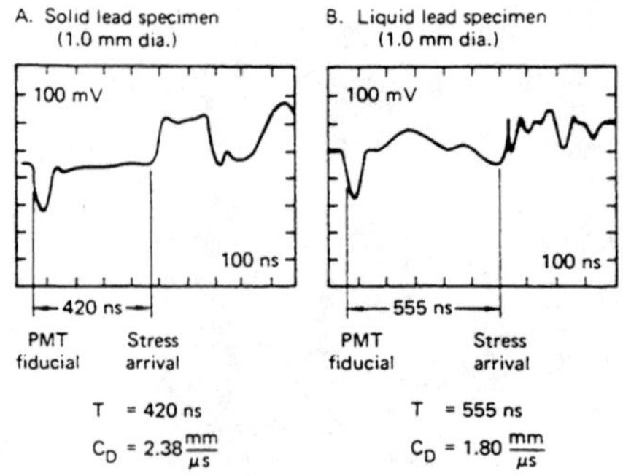

Figure 10—Data for sound velocity measurement across the diameter of the (A) solid lead and (B) liquid lead specimens.

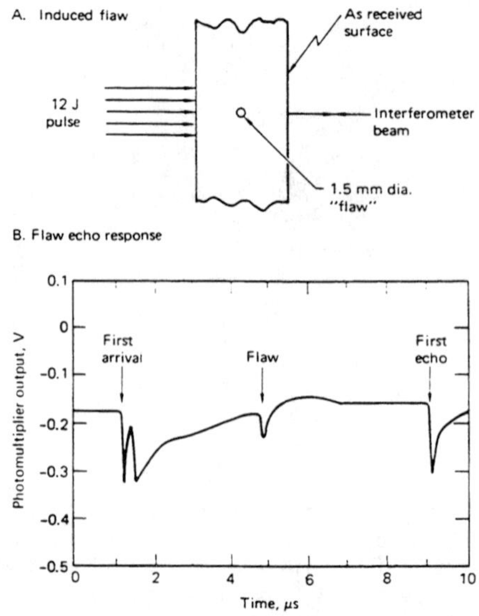

Figure 11—Flaw detection using noncontact testing.

Conclusions

Noncontact material testing using laser energy deposition loading and dynamic displacement interferometry can provide material response data and corresponding material properties accurate to within 1-2 percent. The large amplitude stress pulse induced along with the high sensitivity of the displacement interferometer used permits testing without requirement of a specular surface. The technique is especially advantageous in situations where more conventional ultrasonic methods are impossible or difficult to apply.

Acknowledgment

Work performed under the auspices of the U.S. Department of Energy by the Lawrence Livermore Laboratory under contract number W-7405-ENG-48.

References

1. Ready, J. F. "Effects Due to Absorption of Laser Radiation." *J. Appl. Phys.* 36, no. 2 (Feb. 1965): 462–68.
2. Percival, C. M., and Cheney, J. A. "Thermally Generated Stress Waves in a Dispersive Elastic Rod." *Exp. Mech.* 9, no. 2 (Feb. 1969): 49–57.
3. Brammer, J. A., and Percival, C. M. "Elevated-Temperature Elastic Moduli of 2024 Aluminum Obtained by a Laser-Pulse Technique." *Exp. Mech.* 10, no. 6 (June 1970): 245–50.
4. Calder, C. A., and Wilcox, W. W. "Applications of Pulsed Laser Induced Stress Waves." Lawrence Livermore Laboratory, UCID-16353, Sept. 1973.
5. Skeen, C. H., and York, C. M. "Laser-Induced 'Blow-off' Phenomena." *Appl. Phys. Letters* 12, no. 11 (June 1968): 369–71.
6. Skeen, C. H. et al. "The Deformation of Thin Plates by Stress Waves Generated in Confined Laser Heated Plasmas." Sandia Laboratories, SC-CR-69-3306, Nov. 1969.
7. Peercy, P. S. et al. "Ultrafast Rise Time Laser-induced Stress Waves." *Appl. Phys. Letters* 16, no. 3 (Feb. 1970): 120–22.
8. Felix, M. P., and Ellis, A. T. "Stress-Pulse Propagation in Solids: A Closer Look at Dispersion." *Appl. Phys. Letters* 21, no. 11 (Dec. 1972): 532–34.
9. Felix, M. P. "Laser-Generated Ultrasonic Beams." *Rev. Sci. Instrum.* 45, no. 9 (Sept. 1974): 1106–8.
10. von Gutfeld, R. J., and Melcher, R. L. "MHz Acoustic Waves from Pulsed Thermoelastic Expansions and Their Application to Flaw Detection." *Materials Evaluation* 35, (Oct. 1977): 97–99.
11. von Gutfeld, R. J., and Melcher, R. L. "20 MHz Acoustic Waves from Pulsed Thermoelastic Expansions of Constrained Surfaces." *Appl. Phys. Letters* 30, no. 6 (March 1977): 257–59.
12. Fairand, B. P. et al. "Quantitative Assessment of Laser-induced Stress Waves Generated at Confined Surfaces." *Appl. Phys. Letters* 25, no. 8 (Oct. 1974): 431–33.
13. Fairand, B. P., and Clauer, A. H. "Effect of Water and Paint Coatings on the Magnitude of Laser-generated Shocks." *Optics Comm.* 18, no. 4 (Sept. 1976) 588–91.
14. Clauer, A. H.; Fairand, B. P.; and Wilcox, B. A. "Pulsed Laser Induced Deformation in an Fe-3 Wt Pct Si Alloy." *Metall. Trans. A* 8A (Jan. 1977): 119–25.
15. Ready, J. F. *Effects of High-power Laser Radiation.* New York: Academic Press, 1971.
16. Percival, C. M. "A Quantitative Measurement of Thermally Induced Stress Waves." *AIAA J.* 9, no. 2 (Feb. 1971): 347–49.
17. Graham, R. A.; Neilson, B.; and Benedick, W. B. "Piezoelectric Current from Shock Loaded Quartz - A Submicrosecond Stress Gage." *J. Appl. Phys.* 36, no. 5 (May 1965) 1775–78.
18. Wilcox, W. W. "Enhances Stress Waves in Solids." Lawrence Livermore Laboratory, UCID-16447, Feb. 1974.
19. Barker, L. M., and Hollenbach, R. E. "Interferometer Technique for Measuring the Dynamic Mechanical Properties of Materials." *Rev. Sci. Instrum.* 36, no. 11 (Nov. 1965) 1617–20.
20. Palmer, C. H., and Green, R. E. "Materials Evaluation by Optical Detection of Acoustic Emission Signals." *Materials Evaluation* 35, no. 10 (Oct. 1977).
21. Palmer, C. H.; Claus, R. O.; and Fick, S. E. "Ultrasonic Wave Measurement by Differential Interferometry." *Applied Optics* 16, no. 7 (July 1977) 1849–56.
22. Bondarenko, A. N., and Drobot, Y. B. "Optical Excitation and Detection of Nanosecond Acoustic Pulses in Nondestructive Testing." *Sov. J. of NDT* 12, (1976): 655–58.
23. Simmons, G., and Wang, H. *Single Crystal Elastic Constants and Calculated Aggregate Properties: A Handbook.* 2nd Ed. Cambridge: MIT Press, 1971.
24. Barker, L. M., and Hollenbach, R. E. "Laser Interferometer for Measuring High Velocities of any Reflecting Surface." *J. Appl. Phys.* 43, no. 11, (No. 1972): 4669–75.
25. Calder, C. A., and Wilcox, W. W. "Technique for Measurement of Elastic Constants by Laser Energy Deposition." *Rev. Sci. Instrum.* 45, no. 12 (Dec. 1974): 1557–59.
26. Wilcox, W. W., and Calder, C. A. "Materials Testing Using Laser Energy Deposition." *Instrumentation Technology* 25, no. 8 (Aug. 1978): 63–68.
27. Calder, C. A., and Wilcox, W. W. "High Temperature Noncontact Material Testing." *Characterization of Materials for Service at Elevated Temperatures.* G. V. Smith, ed. ASME Series MPC-7, 1978, 169–81.
28. Gathers, G. R.; Shaner, J. W.; Calder, C. A.; and Wilcox, W. W. "Determination of Sound Velocity in Liquid Metals at Temperatures Greater than 4000 K." *Proceedings of the Seventh Symposium on Thermophysical Properties.* May 1977.
29. Calder, C. A., and Wilcox, W. W. "Acoustic Velocity Measurement Across the Diameter of a Liquid Metal Column." *Experimental Mechanics* 19, no. 5 (May 1979) 171–74.
30. **Webber, G. M. B., and Stephens, R. W. B. "Transmission of Sound in Molten Metals."** ***Physical Acoustics.*** **W. O. Mason, Ed. V. IV - Part B. New York: Academic Press, 1968, 53–97.**

Reprinted with permission from *Research Techniques in Nondestructive Testing,*
R.S. Sharpe, editor, Vol. 5, Chap. 8, pp. 281-327 (1982).

Laser Generation of Ultrasound in Metals

C. B. SCRUBY

Materials Physics Division, AERE, Harwell, Didcot, Oxon, UK

R. J. DEWHURST, D. A. HUTCHINS AND S. B. PALMER

Department of Applied Physics, University of Hull, Hull, UK

I. INTRODUCTION

Ultrasonic inspection techniques using piezoelectric transducers are established and accepted methods of nondestructive testing. It is therefore quite natural to ask why alternative methods of generating and detecting ultrasound, using relatively expensive and complicated laser systems, should be considered. In this chapter we shall try to answer this question and demonstrate the potential of this growing research area.

Laser generation of ultrasound in solids is caused by interaction of the optical wave train with the material. In the case of metals, which will be the main concern of this chapter, the interaction takes place at the metal's surface. The acoustic waves which are generated may propagate directly from the source at the solid surface, or may be the result of a shock front which at first changes the phase of the material and at a later stage of propagation weakens to become an acoustic wave. Two extreme situations can be identified, characterized by the laser intensity used. At very low power densities the light causes only a rise in temperature by conduction below the surface, with no change of material phase. When, on the other hand, the power density is very high, multiphoton ionization takes place on the surface within a few light cycles, causing a plasma to be formed. Between these two extremes, there is a wide variety of intermediate situations in which changes of phase, pressure due to vaporization, thermionic emission and shock wave generation may be important in providing the acoustic source.

It is clearly important to identify those features of laser-generated ultrasound that make it different from conventional piezoelectric generation, so that suitable application areas can be explored. For instance, the outstanding feature of laser-generation is that no physical contact with the sample is required. This is in marked contrast to conventional techniques where the piezoelectric transducer element must be bonded to the sample. Thus the problems of couplant response and variability are immediately eliminated with an optical technique, especially when scanning a sample. To take the most advantage of this feature, it is also desirable to have a noncontacting optical detector of ultrasound. Laser interferometers have been developed for this purpose, although at present their sensitivity tends to be low compared with conventional transducers.

In this chapter we shall first outline the historical development of the technique, before introducing some of the physical principles underlying the generation process. We shall then summarize recent experimental results that best demonstrate the characteristics of this ultrasonic source. We shall finally discuss areas where we believe laser-generated ultrasound could make a valuable contribution to nondestructive testing.

II. Historical Background

The first analysis of how ultrasound is generated by a laser pulse incident on a solid was carried out by White (1963). He considered in detail the way in which the temperature of the solid changes, both as a function of time and of depth into the material from the surface. Ready (1965) carried out a study along similar lines, with later work indicating the type of stress pulse likely to be developed from temperature gradients within the irradiated solid (Gournay, 1966; Tomulescu, 1975). However, no general theory has yet been published which describes ultrasonic waveforms expected in the far-field within solids. More is known for the case of liquids, where far-field predictions have been made (Kasoev and Lyamshev, 1977) and aspects of spherically heated regions described (Hu, 1969; Bunkin and Komissarov, 1973; Lyamshev and Naugol'nykh, 1976; Bozhkov *et al.*, 1976).

Experimental studies have been carried out in both solids and liquids. In liquids, such studies have attempted to verify theoretical work e.g. Hutcheson *et al.* (1971) have compared their results to the theories of Gournay (1966) and Hu (1969). Several variants on the method of generation have been employed. Techniques have been discussed (Westevelt and Carson, 1973) for the development of highly-directive sound beams in water, produced in a direction normal to the main absorption direction of the laser beam. Experimental work on these "broadside arrays" has been carried out by Muir *et al.* (1976). Another technique is the use of longitudinal mode beating from a hybrid CO_2 laser, which has been used to produce 60 MHz ultrasound in water (Sigrist *et al.*, 1978). A modulated, continuous-wave (CW) argon ion laser has also been used (Kohanzadeh *et al.*, 1975), but this resulted in a very low acoustic power level (Ivakin *et al.*, 1976). Continuous-wave lasers are also used in the related technique of photoacoustic microscopy (e.g. Rosencwaig, 1980). Other workers (Ivakin *et al.*, 1976) have used the interaction between two laser beams to produce beats, and have generated high frequency sound waves in glycerine.

In solids, investigations have tended to use high power (>10 MW) laser pulses (Penner and Sharma, 1966; Percival, 1967; Skeen and York, 1968; Anderholm, 1970; Peercy *et al.*, 1970; Jones, 1971; Kubota and Nakatani, 1973; Fairand and Clauer, 1979). These have included experiments to study the effect of constraining layers on the surface of the solid (Fairand, 1974), when enhancement over generation of ultrasound at free surfaces is observed. For example, enhancements of 46 dB have been demonstrated in certain cases (Von Gutfeld and Melcher, 1977). Enhancements have also been produced by the application of coatings to solid surfaces. O'Keefe and Skeen (1972) have used coatings of transparent, relatively volatile materials to produce order-of-magnitude increases in stress over bare targets, and it has also been shown (Fox, 1974) that a thin coat of black paint or drops of distilled water can produce results comparable to constraining techniques. Studies on free, unconstrained surfaces have been conducted mostly with Q-switched laser pulses giving light pulse durations in the range 10–50 ns. Solid media are capable of supporting transverse vibrations over large distances in addition to the compressional mode that propagates in liquids. Three types of acoustic pulse are therefore expected: shear, longitudinal and surface wave, and all have been generated by a pulsed laser (Ledbetter and Moulder, 1979). However, these authors used a focused 0·3–1·0 J Nd : glass laser, and damage to the specimen must have taken place. Surface (Rayleigh) waves have been reported elsewhere (Lee and White, 1968) but most work has concentrated on the production of longitudinal pulses with applications of the technique to nondestructive testing (Lee and White, 1968; O'Keefe and Skeen, 1972; Giglio, 1973; Kubota and Nakatani, 1973; Fairand, 1974; Fox, 1974; Calder and Wilcox, 1974, 1978; Bondarenko *et al.*, 1976; Von Gutfeld and Melcher, 1977; Kaule, 1978; Wellman, 1978; Bar-Cohen, 1979; Ledbetter and Moulder, 1979).

III. Principles of Ultrasound Generation by Laser

A. The Laser Pulse as a Stress Source

Ultrasonic generation by laser is a sufficiently new technique to warrant some consideration of the controlling physical principles. In this section we shall restrict our attention to generation at the surface of metals, to pulsed lasers, and to laser power regimes that are most likely to be suitable for NDT. Thus we exclude the very high power regimes ($>10^{15}$ W cm^{-2}) which are extensively studied in connection with fusion research.

1. *Absorption of Laser Energy*

A pulsed laser emits bursts of coherent electromagnetic radiation. When low intensity radiation ($\lesssim 10^6$ W cm^{-2}) is incident on a metal surface, the combination of electric and magnetic fields induces currents in the conduction band near the surface of the metal. Some of the incident energy is absorbed by resistive losses and converted into heat, while the remainder is re-radiated as a reflected pulse. The screening by conduction electrons is such that most of the absorption and reflection takes place very close to the surface, within what is called the "skin depth" and generally denoted by δ. At longer wavelengths in the infrared the following classical expression for δ can be used (Bleaney and Bleaney, 1965)

$$\delta^2 = \frac{1}{\pi \sigma f \mu \mu_0} \tag{1}$$

In this equation, σ and μ are respectively the conductivity and relative permeability of the metal, f is the frequency of the incident radiation, and $\mu_0 = 4\pi \times 10^{-7}$ H m^{-1} is the permeability of free space. At higher frequencies δ must be derived from quantum mechanics. For a Nd : YAG laser of wavelength 1·06 μm in the near infrared, the skin depth is ~ 5 nm in aluminium. In iron however, the skin depth is less, ~ 3 nm, because the permeability is higher.

Again, using classical electromagnetic theory (Bleaney and Bleaney, 1965) the reflectivity of the metal, R, can be calculated as the ratio of the reflected energy, E'', to the incident energy, E, i.e.

$$R = \frac{E''}{E} = \frac{2 - 2t + t^2}{2 + 2t + t^2} \tag{2}$$

where $t = \mu_0 \sigma c \delta$, c being the velocity of light. For a metal where $\sigma \sim 10^7$ $(\Omega - m)^{-1}$, $t \gg 1$ for all frequencies up to visible light, so that (2) can be approximated to

$$R = 1 - 4/t \tag{3}$$

Thus the absorbed energy E' is given by

$$E' = (1 - R)E = \frac{4}{\mu_0 \sigma c \delta} E \tag{4}$$

Substituting values for the material constants gives a theoretical absorption coefficient of 6% for aluminium and 36% for iron, for 1·06 μm radiation. Experimental values of absorption coefficients (Kaye and Laby, 1973) are generally somewhat higher, e.g. 7% for aluminium, because the surface reflectivity is reduced by surface roughness, oxidation or other contamination.

Thus in a low power regime in which other effects can be neglected the pulsed laser acts as a transient heat energy source at the metal surface (Fig. 1). If we now assume that the absorption of energy occurs so rapidly that thermal conductivity into the bulk of the metal can be neglected, then the deposition of energy, E', in a surface layer of area, A, and depth, δ, causes a rise in temperature, ΔT, given by

$$\Delta T = \frac{E'}{s\rho A \delta} \tag{5}$$

where s and ρ are respectively the specific thermal capacity and density of the metal.

Substituting E' from eqn (4) gives

$$\Delta T = \frac{4}{\mu_0 c \rho \sigma s \delta^2} \left(\frac{E}{A}\right) \tag{6}$$

i.e.

$$\Delta T \propto e \tag{7}$$

where the constant of proportionality $(4/(\mu_0 c \rho \sigma s \delta^2))$ is a material constant, and e is the mean incident energy density.

In a complete treatment, the spatial beam profile needs to be taken into

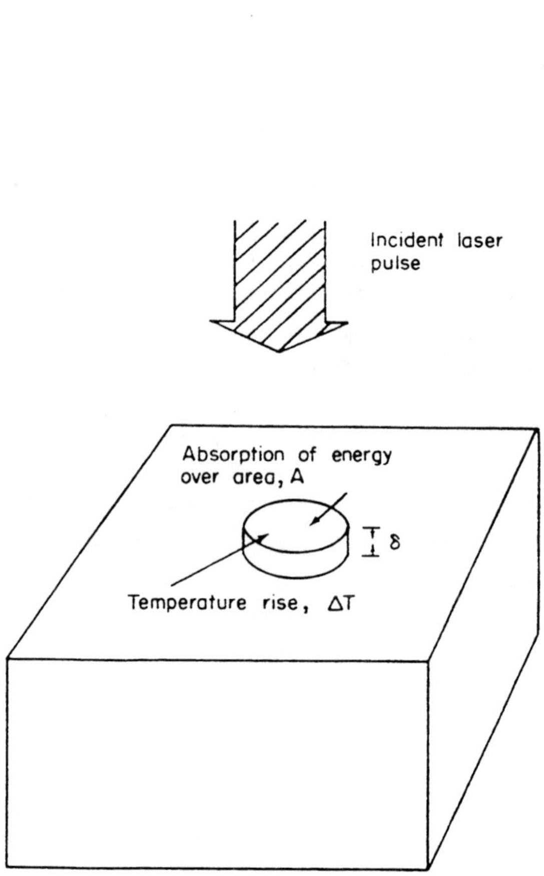

FIG. 1. Absorption of laser energy at metal surface takes place over area A within skin-depth, δ, causing rapid temperature rise, ΔT.

account; for a pure laser mode the profile is generally Gaussian. However, for most of the following it is adequate to assume that the beam has a fixed area, A, and constant energy density, e.

We have so far neglected thermal conductivity. Over the timescales of typical laser pulses (10 ns up to 100 ns), there is in fact appreciable thermal conductivity into the bulk of metals such as aluminium and copper. The problem has been tackled by a number of authors (White, 1963; Ready, 1965; Gournay, 1966; Tomulescu, 1975). We do not intend to give a detailed account of thermal conductivity in this treatment; however, the main effects as regards the generation of ultrasound are that the volume of heated material increases as a function of time, the largest being from the increase in thermal source thickness. Instead of being equal to the skin depth, δ, the source thickness increases with time as heat is conducted into the bulk of the metal. For instance, for a 20 ns pulse incident on aluminium, the mean effective thickness is about 3 µm, which is much greater than δ.

Ready (1971) has shown that the temperature at time, t, depth, z, and radial position, r, is given by

$$T(t,r,z) = \frac{F_{\max} a^2}{K}\left(\frac{k}{\pi}\right)^{1/2} \int_0^t \frac{p(t-t')}{t'^{1/2}(4kt'+a^2)} \exp\left[\frac{-z^2}{4kt'} - \frac{r^2}{4kt'+a^2}\right] \mathrm{d}t' \quad (8)$$

where F_{max} is the absorbed power density at the centre of the Gaussian spot, a the radius, $p(t)$ the normalized laser pulse profile, K the thermal conductivity and k the diffusivity. Because the heated volume is increased the result of including conductivity effects is also to lower the surface temperature increase, given by eqn (5). However, it should be noted that thermal conductivity redistributes the energy, so that it should not affect any measurement which is a function only of total deposited energy.

2. *Thermoelastic Effect*

The sudden rise in temperature of the surface layer caused by the absorption of thermal energy must be accompanied by thermal expansion. Thus if the volume $V = A\delta$ expands to a volume $V + \Delta V$, then

$$\Delta V = 3\alpha V \Delta T \tag{9}$$

where α is the coefficient of linear expansion. Substituting for ΔT from eqn (5) and for E' from (4) gives

$$\Delta V = \frac{3\alpha}{s\rho} E' \tag{10}$$

$$= \frac{3\alpha}{s\rho}(1 - R)E \tag{11}$$

Thus the absorption of laser energy causes thermoelastic strains equivalent to the sudden insertion of a volume ΔV of material immediately below the surface, that is proportional to the incident energy E. Note from eqn (11) that the magnitude of the extra volume, ΔV, is independent of both the beam area, A, and the thickness, δ, at constant R. The source is pure expansion and can generate only compressional waves if situated in bulk material. However, the close proximity of the surface introduces some conversion to shear modes, as discussed in a later section. For some purposes it is helpful to try and visualize the source in terms of the stresses it sets up, for these are modified by the surface. Compressive stresses parallel to the surface are unchanged, but the boundary conditions require there to be zero net stress normal to the surface. The source is just below the surface so that there is incomplete cancellation. However, in the far-field close to the normal to the surface, the source can be approximated to a set of dipolar stresses parallel to the surface (Fig. 2a). As the angle increases in the far-field this approximation becomes less valid. It can be seen from Fig. 2a that this thermoelastic source of ultrasound is different from a piezoelectric compressional transducer which principally generates a stress normal to the surface.

For many applications it is adequate to treat the thermoelastic source of

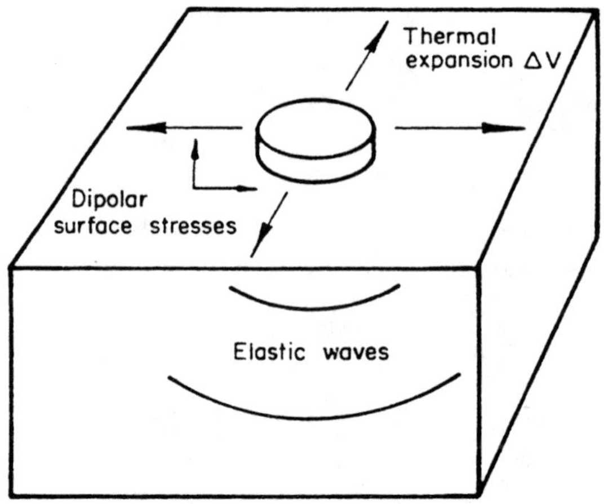

FIG. 2 (a). Temperature rise causes thermal expansion at the surface by an amount $\Delta V = 3\alpha A\delta\Delta T$, neglecting conductivity. Thermoelastic stresses can be approximated by dipolar stresses in surface.

ultrasound as a point source, provided small enough beam diameters are employed. Effects due to non-zero source dimensions will be pointed out where they arise. It is noted that, neglecting thermal conductivity effects, the thermoelastic strain has the same time dependence as the surface temperature.

3. *Ablation of the Surface*

From eqn (7) the surface temperature increases linearly with incident energy density and hence with power density. As the optical power density is increased, whether by increasing the energy per pulse, or by focusing a constant energy onto a smaller surface area using a converging lens, the temperature must rise until the melting point of the surface metal is reached. Some of the incident energy is now dissipated as latent heat. At yet higher incident energy densities the surface temperature reaches the boiling point of the metal, and ablation of metal from the surface takes place. Once ablation occurs, some of the incident energy is dissipated as kinetic energy of the metal vapour. It should be noted, however, that during transient heat pulses of short duration, thermal equilibrium is not reached, so that the surface can, for a short time, become superheated above the boiling point of the metal under conditions of high incident radiant power density.

As a consequence of the ablation of material and the high temperatures involved, a metal plasma forms at, and above, the surface of the metal. The interactions of the laser pulse, plasma and surface are too complex to be given detailed consideration here: (see e.g. Krehl *et al.*, 1975). For the purposes of ultrasonic generation, ablation will be considered solely as producing an impulsive recoil force, **F**, by the transfer of momentum from the surface (Fig. 2b).

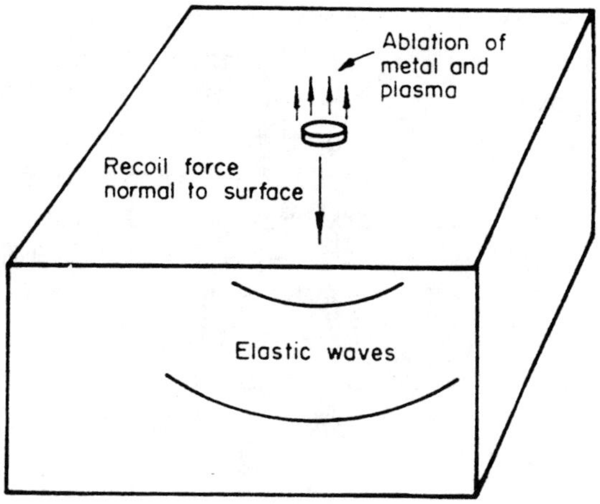

FIG. 2 (b). At high power density, ΔT is so large that metal vaporizes to form a plasma. There is now a recoil force F normal to surface.

4. *Radiation Pressure*

At this point it is appropriate to estimate the magnitude of radiation pressure caused by the incident pulse. The mean radiation pressure, p, is given by the energy density of the incident wave plus that of the reflected wave, i.e.

$$p = \frac{(1+R)E}{Ac\Delta t} \tag{12}$$

where Δt is the temporal width of the pulse. For a YAG laser pulse of energy 30 mJ and width 20 ns, focused to an area of 1 mm^2, the mean radiation pressure is 10^4 N m^{-2}. The mean force exerted on the surface during incidence is 10^{-2} N, which is much smaller than either thermoelastic or ablation forces.

B. *Ultrasonic Wave Generation and Propagation*

In both the thermoelastic and ablation (plasma) regimes, rapidly changing stress and strain fields are produced at and just beneath the surface of the metal, which act as sources of elastic waves. The strain energy is dissipated by both bulk and surface waves. For many purposes the irradiated area may be considered as a point source of ultrasound, with time dependence related to that of the laser pulse, i.e. only far-field terms generally need be considered. The displacement wave fields are best described in terms of Green's tensors, i.e. spatial and temporal transfer functions. If there is a disturbance at a point ($\mathbf{x}'$) at zero time that takes the form of a unit force with δ-function time dependence, then the displacement $\mathbf{u}(\mathbf{x}, t)$ at a point ($\mathbf{x}$) at time t is given by the Green's tensor. The finite dimensions and finite lifetime of a real source can be taken into account by a convolution of the Green's tensor with the time dependence and dimensions of the source. Detailed calculations of Green's tensors are beyond the scope of this chapter, and theoretical work will only be presented where it can be compared directly with experimental results. Green's tensors have only been calculated for simple bodies, an infinite isotropic elastic medium, and elastic half-space (Pekeris and Lifson, 1957), and an infinite parallel-sided plate (Pao *et al.*, 1979). Some of the laboratory experiments, to be described in Section IV, have been designed to approximate to situations where Green's function can be explicitly calculated, measurements of bulk waves being made on a plate at the epicentre, and those of surface waves on a block that approximates to a half-space.

1. *Bulk Waves*

Perhaps the most useful property to try and predict from a theoretical study is the angular dependence (directivity) of the bulk waves generated by the laser source in both the thermoelastic and plasma (ablation) regimes.

Thermoelastic source directivity

Since the thermoelastic source is a pure expansion, if it were sited in an infinite isotropic, elastic medium, it would radiate only compressional (longitudinal) waves equally in all directions. The directivity pattern for longitudinal (L) waves would be a sphere.

It is necessary to calculate what happens when this source is vanishingly close to a planar surface. Mode conversion of reflected waves at the surface causes shear (S) waves to be radiated in addition to L waves. The simplest method is to use reciprocity (Burridge and Knopoff, 1964) exchanging the positions of source (force) and generated displacement, and solving the boundary conditions for L and S plane waves from radial and transverse force sources incident on the surface that only generate dilatation on reflection. The wave equation has been solved for a free surface by a number of authors (e.g. Achenbach, 1973; Graff, 1975). Graff shows that if the L wave amplitude is A_1 incident at an angle of θ_1, there is a reflected L wave of amplitude A_2 at angle θ_1 and a reflected S_v wave at an angle θ_2, where

$$\frac{A_2}{A_1} = \frac{\sin 2\theta_1 \sin 2\theta_2 - k^2 \cos^2 2\theta_2}{\sin 2\theta_1 \sin 2\theta_2 + k^2 \cos^2 2\theta_2} \tag{13}$$

Similarly an S_v wave of amplitude B_1 incident at an angle θ_2 gives on reflection an Sv wave at angle θ_2, and an L wave of amplitude A_2 at angle θ_1, where

$$\frac{A_2}{B_1} = \frac{-2k^2 \sin 2\theta_2 \cos 2\theta_2}{\sin 2\theta_1 \sin 2\theta_2 + k^2 \cos^2 2\theta_2} \tag{14}$$

where in addition

$$k = \frac{c_1}{c_2} = \frac{\sin\theta_1}{\sin\theta_2} \tag{15}$$

c_1 and c_2 being respectively the L and S wave velocities. Thus, applying reciprocity, the thermoelastic source at the boundary generates in the far-field reflected L waves at θ_1 and S waves at θ_2, whose amplitudes are given by A_2 in eqns (13) and (14), together with a direct L wave at θ_1 of amplitude A_1, i.e.

$$U_r(\theta_1) \sim A_1 + A_2$$

$$= A_1 \frac{2\sin 2\theta_1 \sin\theta_1 (k^2 - \sin^2\theta_1)^{1/2}}{(k^2 - 2\sin^2\theta_1)^2 + 4\sin^2\theta_1(1-\sin^2\theta_1)^{1/2}(k^2 - \sin^2\theta_1)^{1/2}} \tag{16}$$

$$U_\theta(\theta_2) \sim A_1 \frac{-k\sin 4\theta_2}{k(1 - 2\sin^2\theta_2)^2 + 4\sin^2\theta_2(1-\sin^2\theta_2)^{1/2}(1 - k^2\sin^2\theta_2)^{1/2}} \tag{17}$$

These directivity patterns are shown for aluminium in Fig. 3. Note that there is no radiation normal to the surface; in this direction the source appears to cancel out with its reflection in the surface. Note also that both L and S directivities give lobes with cylindrical symmetry about the normal. The maximum longitudinal wave amplitude is radiated at an angle of $\sim 64°$ and the maximum shear at $\sim 30°$. The shear is zero at 45° and there is a second smaller lobe for $\theta_2 > 45°$. Above the critical angle $\theta_c = \sin^{-1}(1/k) \simeq 30°$, U_θ is complex; thus $|U_\theta|$ is plotted. This model predicts far-field ultrasonic directivity that is different from other well-known sources.

Ablation source directivity

The same approach (Lord, 1966) can also be applied to the calculation of the ultrasonic directivity of the ablation source. Alternatively, modelling the source as a time-varying force acting normal to the surface of the metal at a point, enables Miller and Pursey's theory (1954) to be applied. The far-field directivity can be derived from either author as

$$U_r(\theta_1) \sim \frac{\cos\theta_1(k^2 - 2\sin^2\theta_1)}{(k^2 - 2\sin^2\theta_1)^2 - 4\sin^2\theta_1(1-\sin^2\theta_1)^{1/2}(k^2 - \sin^2\theta_1)^{1/2}} \tag{18}$$

$$U_\theta(\theta_2) \sim \frac{\sin 2\theta_2(1 - k^2\sin^2\theta_2)^{1/2}}{k(1 - 2\sin^2\theta_2)^2 - 4\sin^2\theta_2(1-\sin^2\theta_2)^{1/2}(1 - k^2\sin^2\theta_2)^{1/2}} \tag{19}$$

These directivity patterns are shown for aluminium in Fig. 4; in contrast to Fig. 3, most of the L wave energy is directed about the normal, $\theta_1 = 0$, and shear energy is mostly directed into a lobe at an angle of $\theta_2 > 30°$, where there is no energy from the thermoelastic source. The ultrasonic field of the ablation source resembles that of a small, compressional piezoelectric transducer under pulse excitation.

Calculated waveforms

As far as is known theoretical bulk waveforms from a laser ultrasonic source have only been computed for one geometry, i.e. at the epicentre of a parallel-sided plate (Fig. 5a). The waveform calculated for the thermoelastic source is shown in Fig. 5b (Scruby *et al.*, 1980). The calculation neglected thermal conductivity and finite source dimensions and used a Green's tensor method (J. E. Sinclair, unpublished work). The waveform calculated for an ablation source is shown in Fig. 5c. This calculation modelled the ablation as a normal force with $\delta(t)$ time dependence. The two waveforms are quite different. Figure 5c is dominated by a fast rising positive pulse at the L arrival whose time dependence is the same to a first approximation as that of the model for the

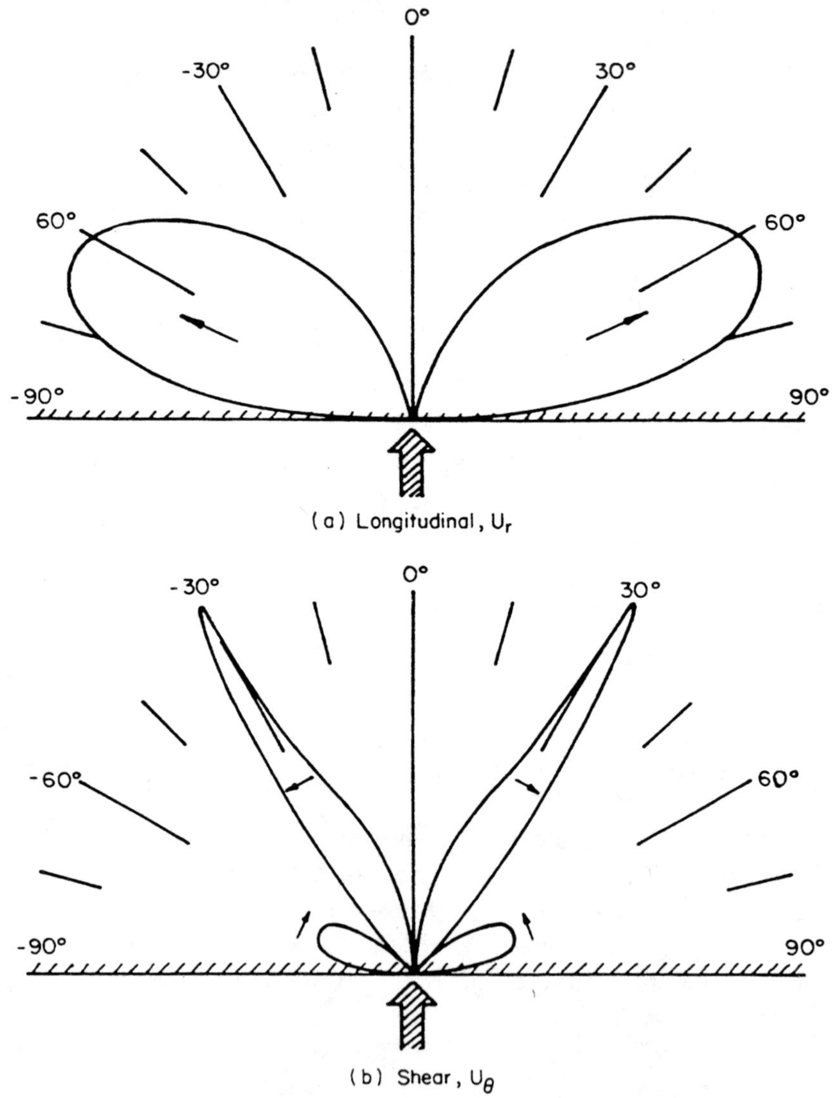

FIG. 3. Theoretical angular dependence (directivity) of (a) longitudinal and (b) shear waves for laser ultrasonic source in aluminium (ν = 0·34)—thermoelastic regime.

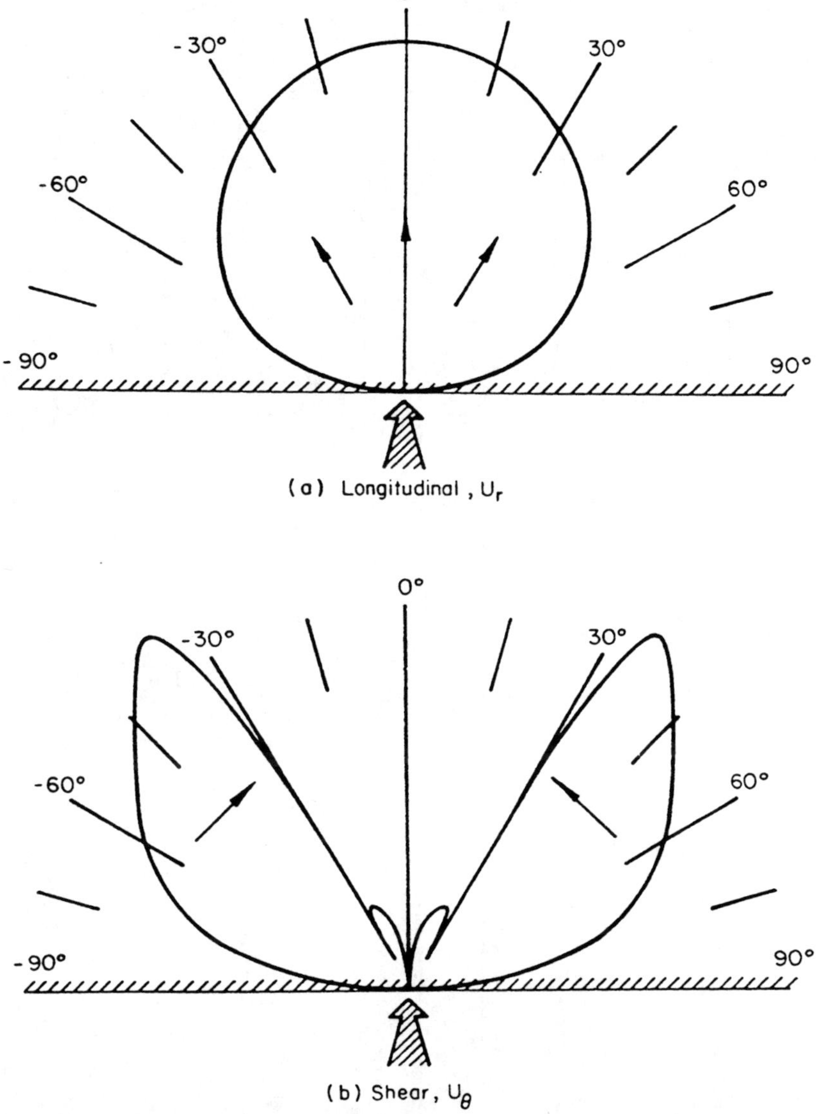

FIG. 4. Theoretical angular dependence (directivity) of (a) longitudinal and (b) shear waves for laser ultrasonic source in aluminium (ν – 0·34)—ablation regime (based on Miller and Pursey, 1954).

ablation force. Thus if alternatively the ablation force were a step (Heaviside) function, the L arrival would be a step also. The positive pulse implies that the surface of the metal rises away from the source. Figure 5b on the other hand, is negative, implying that the surface is depressed, and consists only of steps, with no fast pulses. This only holds for a thermoelastic source at the epicentre. Away from the epicentre there is again a positive pulse at the L arrival. The absence of the pulse at the epicentre is equivalent to the far-field directivity (Fig. 3), being zero for $\theta = 0$.

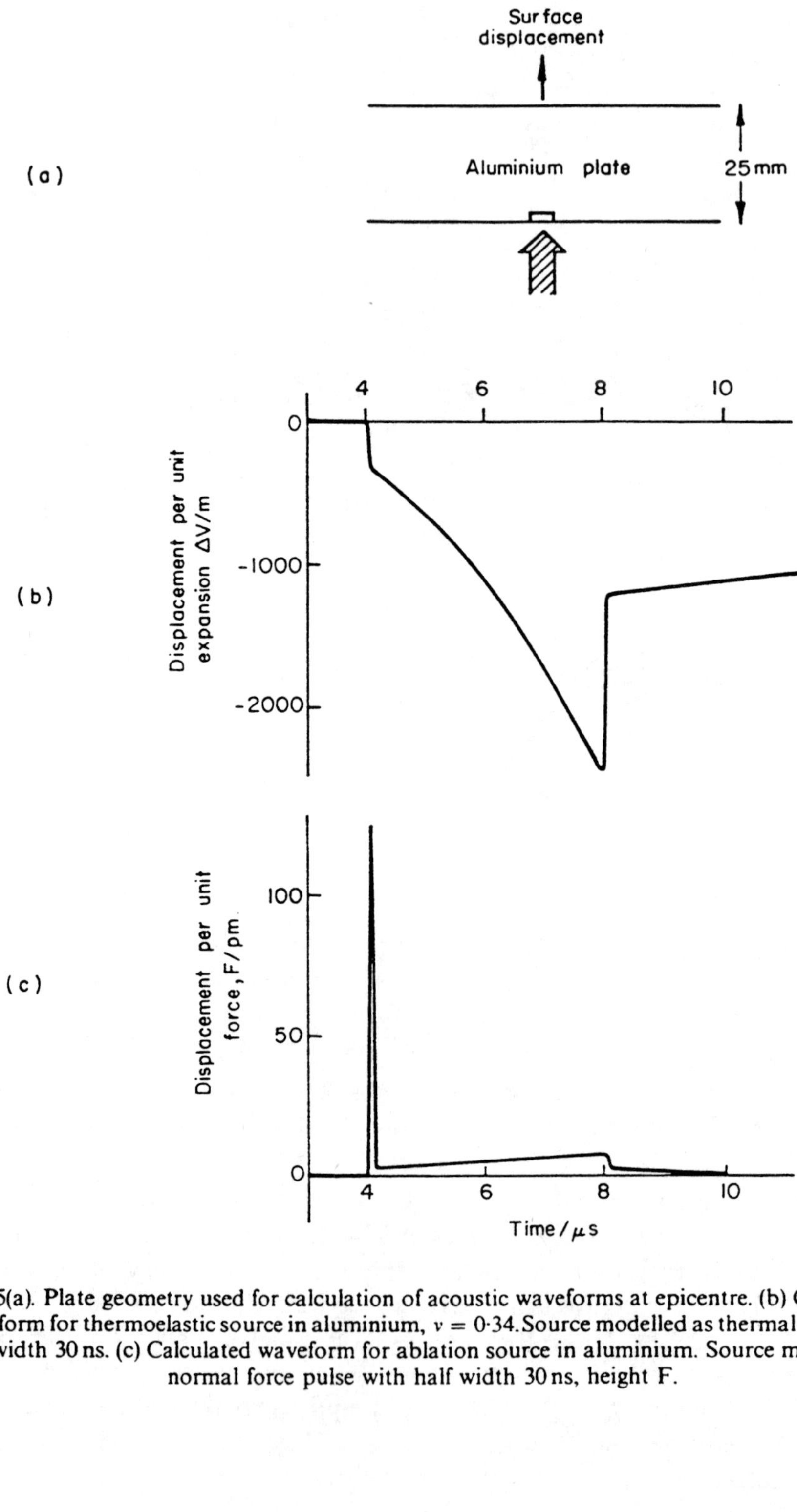

FIG. 5(a). Plate geometry used for calculation of acoustic waveforms at epicentre. (b) Calculated waveform for thermoelastic source in aluminium, $\nu = 0{\cdot}34$. Source modelled as thermal pulse with half width 30 ns. (c) Calculated waveform for ablation source in aluminium. Source modelled as normal force pulse with half width 30 ns, height F.

2. *Surface Waves*

The thermoelastic and ablation sources both have cylindrical symmetry about the normal to the surface, unless special lenses or apertures are used. Thus they would be expected to radiate surface waves equally through an angle of 360°.

It is instructive to compute the form of the surface wave that might be expected in each case. Assuming that the source and measurement points are both on the surface of a half-space, Pekeris's theory (1955) can be applied to calculate the displacement of the surface. Pekeris calculates the vertical and horizontal displacements for a normal point loading with step-function time dependence. It is relatively simple to obtain the solution for a source with a different time dependence, and the vertical displacement for a vertical force which has a Gaussian time dependence is shown in Fig. 6(a). It is not possible to use Pekeris's solution directly to compute the waveform generated by the thermoelastic source. However, an approximate solution can be obtained by assuming the thermoelastic source is a force dipole lying in the surface (cf. Fig. 2a). Then, assuming reciprocity of source and detector, the waveform can be calculated by differentiating Pekeris's solution for the horizontal displacement with respect to distance r. The result is given in Fig. 6(b). Note that for each waveform of Fig. 6 there is a small P arrival with the longitudinal wave speed. Note also that both are dominated by the Rayleigh wave, which is an asymmetrical bipolar pulse, whose polarity is reversed between thermoelastic and ablation waveforms. The polarity of the P arrival is the same for both sources.

3. *Effect of Surface Modification*

The effects of surface modification can be very complex, since the replacement of the metal–air boundary by a metal–liquid or metal–glass boundary completely changes the nature of the source. A number of features will be briefly noted in this section.

First let us consider the thermoelastic source. Suppose that the surface is covered with a transparent solid, such as glass, in intimate contact with the metal. As a result the boundary conditions for the reflection of ultrasonic waves at the interface are now different, since appreciable compressional and shear energy propagates into the glass. Let us just consider normal incidence ($\theta = 0$). Then the reflected L wave amplitude A_2 is given by:

$$A_2 = A_1 \frac{\rho' c_1' - \rho c_1}{\rho' c_1' + \rho c_1} \tag{20}$$

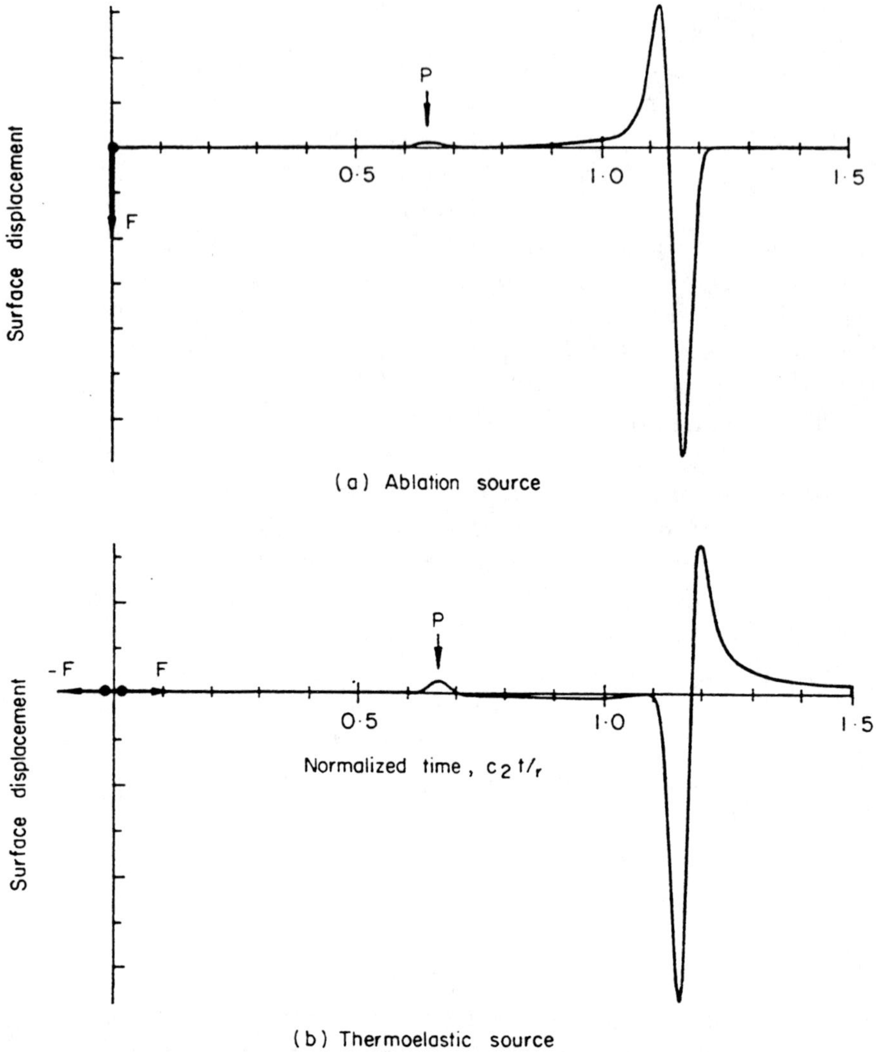

FIG. 6. Surface acoustic waveforms computed for metal with $\nu = 0{\cdot}25$, derived from Pekeris (1955); r is source-to-receiver distance; (a) ablation source (force pulse with half width 0·03); (b) thermoelastic source approximated by surface force dipole.

where ρ and ρ' are the densities and c_1 and c'_1, the L wave speeds of metal and glass respectively (Achenbach, 1973), therefore

$$U_r \sim A_1 + A_2 = A_1 \frac{2\rho' c'_1}{\rho' c'_1 + \rho c_1} \tag{21}$$

For air $\rho' c'_1 \simeq 0$, so that $U_r \simeq 0$ for $\theta = 0$ as before. For Pyrex® glass and aluminium, however, the impedance ratio $\rho' c'_1/\rho c_1 = 0{\cdot}73$, so that $U_r = 0{\cdot}84 A_1$. Thus there is now a large L wave amplitude at $\theta = 0$. The situation is approaching that of a buried dilatation, which radiates equally in all directions, and which would occur if $\rho' c'_1 = \rho c_1$.

A similar situation would be expected if the metal were covered with a substantial layer of liquid such as oil, although the impedance match is then rather worse. However there would be the added complication of the thermal properties of the liquid. It is likely that thermal expansion of the liquid would be the dominant source mechanism, exerting a force normal to the surface and changing the source characteristics. If the layer of liquid were thin, some would probably evaporate, in which case there would be a momentum transfer force exactly analogous to the case of ablation discussed above. In all these cases it is probable that the directivity characteristics would resemble those of the ablation (plasma) source rather than the thermoelastic source.

Surface constraints are also likely to modify the characteristics of the ablation source, although probably not so drastically. However since plasma formation is already a complex and only partially understood phenomenon, it is not intended to speculate on the detailed physical mechanisms under surface modification.

IV. Experimental Results

In this section we describe experimental studies of laser-generated ultrasound in metallic specimens which illustrate the different generation mechanisms outlined earlier. The production of both bulk and surface waves by a wide range of incident laser power densities is discussed, and the effects of modifying the metal surface examined. Finally, measurements of the directivity patterns of various acoustic modes are shown to be in keeping with the generation mechanisms outlined in Section III.

A. Bulk Waves

1. *Thermoelastic Generation*

A typical experimental arrangement for studying the laser generation of longitudinal and shear waves is shown in Fig. 7. A wide range of pulsed lasers can be used with wavelengths extending from the ultraviolet to the infrared. However, for ease of comparison and to provide some reference point, all results presented here will use the same laser, namely a Q-switched Nd : YAG laser operating in the near infrared at 1·06 μm, supplying single shot pulses of up to 60 mJ in energy with a half width pulse duration of typically 24 ns. Such lasers provide a reliable, reproducible output at a repetition rate of at least 10 Hz. Greater efficiency of generation could be achieved by using shorter wavelengths but convenience and reliability would suffer.

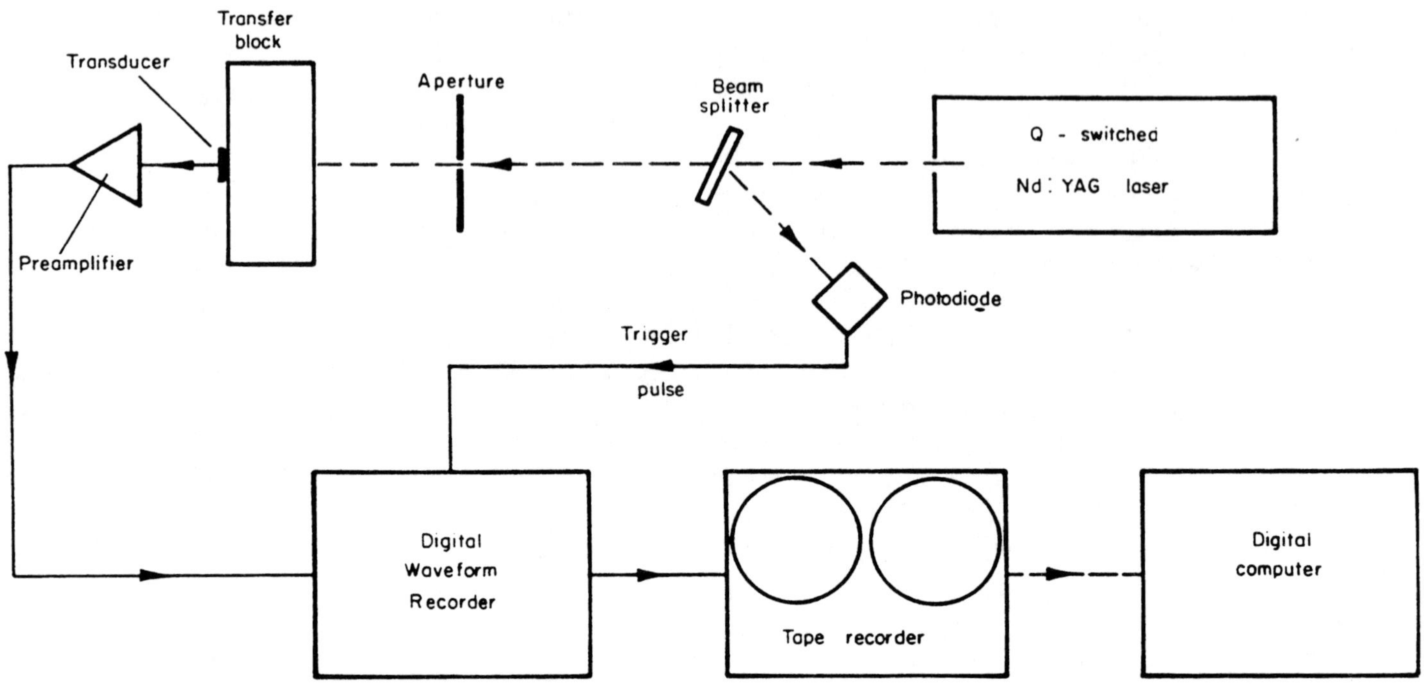

FIG. 7. Schematic diagram of apparatus to investigate epicentre waveforms. Piezoelectric and capacitance transducers may be used, and the signal recorded either digitally or on an oscilloscope.

In the arrangement shown (Fig. 7), the laser irradiates one face of a metallic disc, and the ultrasound is detected on the far face axial with the laser beam (i.e. at epicentre with respect to the acoustic source). The ultrasonic signals recorded by two different detectors will be described here. The first is the widely used piezoelectric transducer, using commercial PZT ceramic and designed to be resonant at a particular frequency. The second, less common, type is the capacitance transducer (Scruby and Wadley, 1976), which can provide quantitative measurements of surface displacements in addition to being a wide-band, non contact device. These latter two properties minimize any distortion of the ultrasonic signal. While PZT transducers are more widely used the capacitance detector is more useful for analysing generation mechanisms.

At low incident optical power density where the laser produces no discernible change in the metal surface, ultrasonic generation occurs through thermal expansion (Section III). The principal stresses in the acoustic source appear to act in directions parallel to the metal surface. Figure 8 shows the bulk wave generated in a 25 mm thick aluminium disc in the thermoelastic regime, as detected by the capacitance transducer (Scruby *et al.*, 1980). The waveform represents displacement perpendicular to the far surface, a positive deflection being a displacement away from the acoustic source. The longitudinal pulse is therefore a step-like depression of the far surface while the shear arrival is a larger elevation of the surface. Between the arrival of these two acoustic modes the surface moves gradually towards the source.

This waveform can be compared with the waveform calculated using wave propagation theory and shown in Fig. 5(b). Theory and experiment are in good agreement except for the small elevation of the surface that is observed experimentally at the start of the longitudinal arrival. Possible origins of this

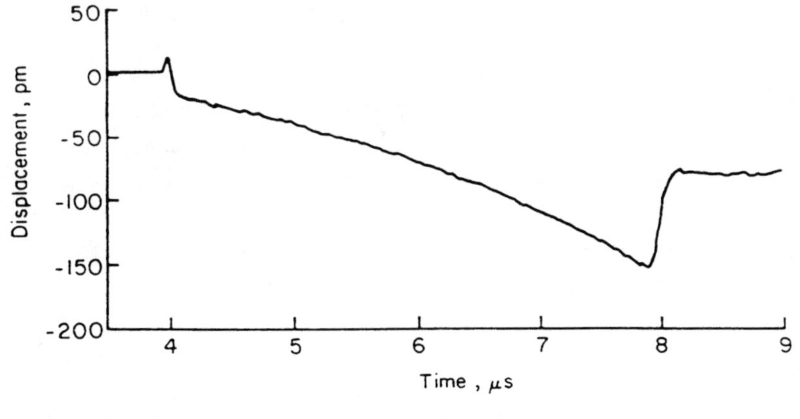

FIG. 8. Surface displacement waveform in thermoelastic regime recorded at epicentre by capacitance transducer (3 mm diameter, 18 mJ laser pulse incident on 25 mm aluminium disc), cf. theory Fig. 5(b).

small discrepancy are at present under investigation but, neglecting this one point, there is strong evidence that for power densities below 2×10^7 W cm^{-2}, thermoelastic mechanisms dominate the acoustic source. The almost complete absence of a large longitudinal pulse at the epicentre is confirmation that the model chosen for the source in Section III is substantially correct. If the same ultrasonic waveform is detected with a longitudinal 5 MHz, PZT transducer, the output shown in Fig. 9(a) appears on the oscilloscope (Aindow *et al.*, 1980a, b). Placing the same PZT transducer on a mild steel sample of similar thickness yields the output shown in Fig. 9(b). Although the velocities of the longitudinal and shear modes differ in the two samples, both modes are clearly detected with the displacement due to the shear arrival being larger in both cases.

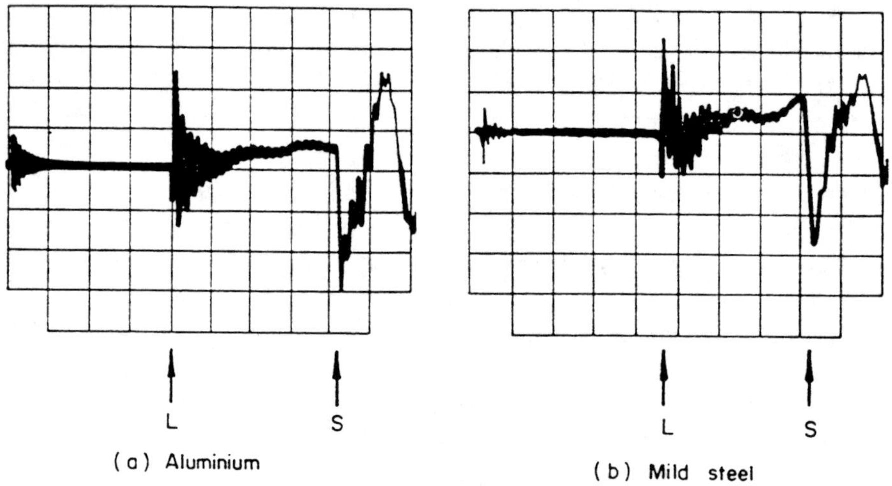

FIG. 9. Laser-generated acoustic waveforms, in thermoelastic regime detected at epicentre by longitudinal 5 MHz piezoelectric transducer; (a) 25 mm aluminium disc, (b) 25 mm steel disc (14 mJ laser energy, 1 μs/div horizontal, uncalibrated vertical scale).

In the thermoelastic regime all the experiments carried out so far indicate that the acoustic amplitude is proportional to the incident laser energy. The spatial profile of the laser beam is not critical provided that it is not larger than the diameter of the detector. Hence, for a fixed laser energy, variations in the optical power density play only a minor role in determining the acoustic amplitudes and waveforms in this regime. The experiments also indicate a high degree of reproducibility in the acoustic waveform.

2. *Plasma Regime*

The acoustic source that we have described above is altered drastically if the laser power density is increased sufficiently to produce a plasma at the metal surface. The ablating particles in the plasma produce, by momentum transfer, forces normal to the metal surface in addition to the thermoelastic stresses that have already been described.

An informative method to study the effects of plasma formation is to keep the laser energy constant while varying the power density using a suitable lens. Alternatively the pulse is focused down to a constant small area, and the power density varied by the insertion of neutral density filters. Initially, on increasing the power density into the plasma regime, the longitudinal arrival first becomes a pulse-like elevation of the surface and then at higher density a step-like elevation (Fig. 10). At the same time the shear arrival tends to diminish to a

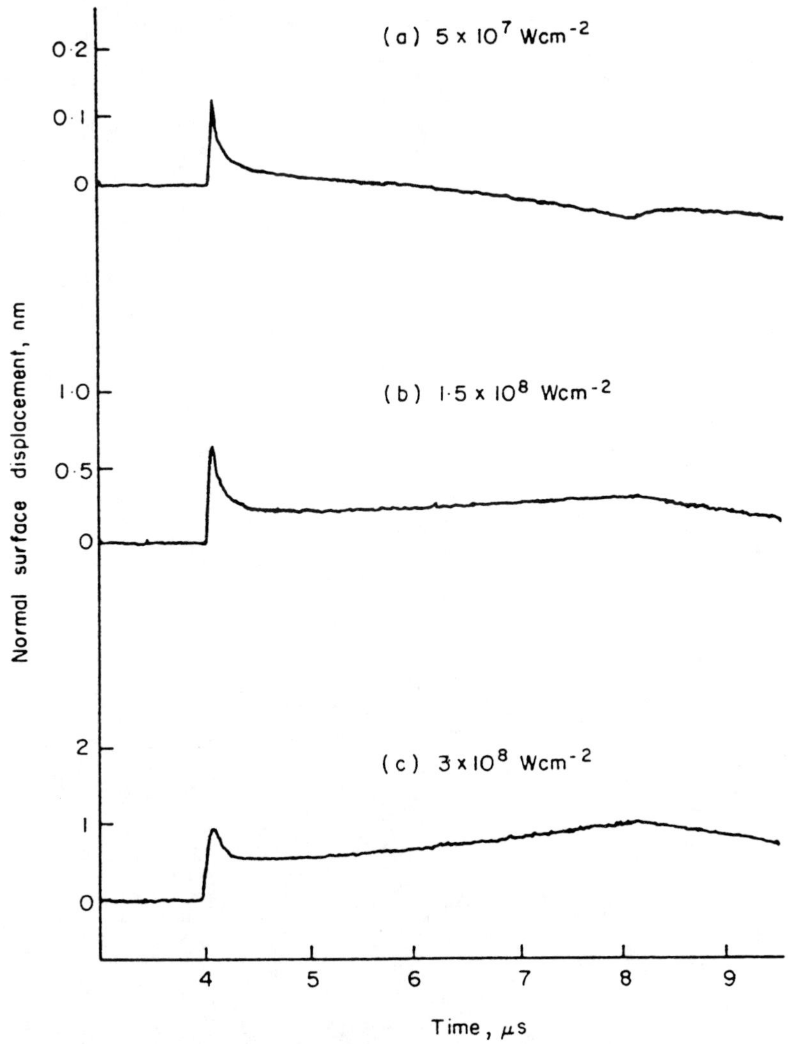

FIG. 10. Epicentre displacement waveforms in ablation regime recorded by capacitance transducer. At low-power density the L arrival is a pulse, (a). As the power density is increased the L arrival progressively becomes more step-like, (b) and (c).

change in slope. These measurements also indicate that the longitudinal displacements change sign and rapidly increase in amplitude as a plasma is formed on the surface.

Wave propagation theory can be used to predict the acoustic waveforms produced in the plasma regime as discussed in Section III. To obtain good modelling the normal forces due to material ablation are allowed to dominate the thermoelastic stresses at high optical power densities. Comparison of Fig. 10b with the theoretical waveform, Fig. 5(c), implies that the normal force generated by the plasma extends over longer time scales than the optical pulse

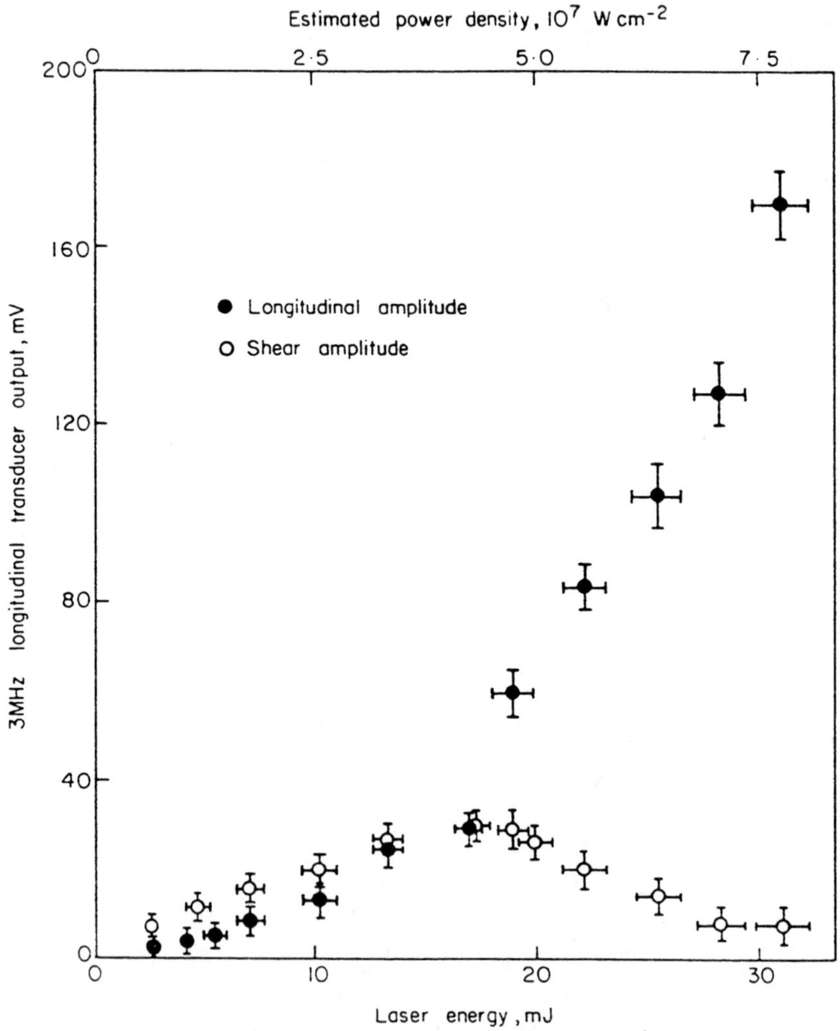

FIG. 11. Amplitude of L and S arrivals as detected by 3 MHz longitudinal transducer at epicentre as a function of laser power density.

length assumed in the theory. The particularly slow fall-time that is observed suggests that plasma effects persist for substantial fractions of a microsecond.

PZT transducers can also be used to observe the changes in acoustic waveform as the optical power density is increased beyond the plasma threshold (Aindow *et al.*, 1981). In Fig. 11 the amplitude of the longitudinal and shear arrivals are plotted as a function of laser power density using a 3 MHz longitudinal PZT transducer as detector. The generation efficiency of the longitudinal pulse is clearly enhanced while the amplitude of the shear arrival decreases markedly. The signals detected by the PZT transducer for an incident laser power density in excess of 10^8 W cm^{-2} are shown in Fig. 12. The amplitude of the longitudinal mode is greatly increased compared with Fig. 9, while the shear arrival is barely detected at this lower gain.

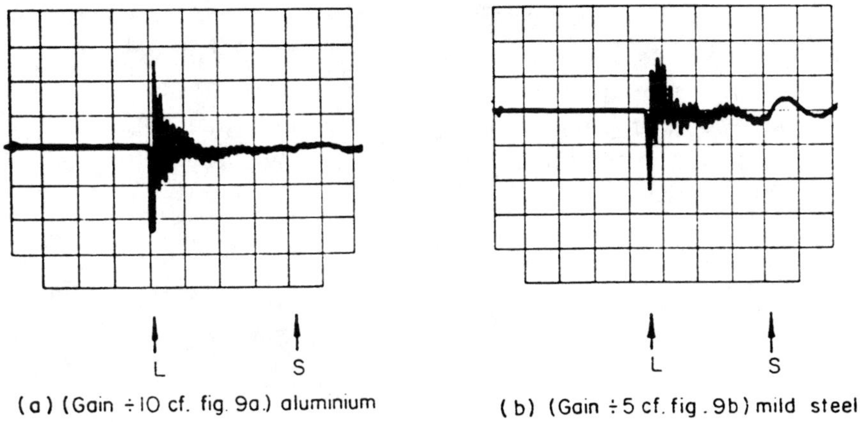

FIG. 12. Signals detected by 5 MHz longitudinal transducer at epicentre in ablation regime (power density > 10^8 W cm^{-2}). (a) 25 mm aluminium disc, (b) 25 mm steel disc (14 mJ laser energy, 1 μs/div. horizontal, uncalibrated vertical scale).

3. *Directivity Patterns at Free Surfaces*

The directivity patterns of longitudinal and shear waves generated in metals by pulsed lasers are most interesting. They depend, for example, on the ultrasonic frequency, the size of source and the stress distribution across it. A series of experiments has been carried out to determine the longitudinal and shear directivity patterns in aluminium at an ultrasonic frequency of 1 MHz. The laser source had a diameter of 1 mm, virtually a point source at this frequency, and the power density was varied to encompass both the thermoelastic and plasma regimes. The laser beam irradiated the centre of the face of a 10 cm diameter hemispherical sample while the piezoelectric transducer traversed the curved surface through the full 180°. Figure 13(a) shows the directivity patterns from the longitudinal mode in the thermoelastic regime while Fig. 13(b) shows the longitudinal mode in the presence of a plasma.

For the thermoelastic generation, longitudinal waves are mainly confined to a lobe at about 55° to the surface normal. The ablation source, on the other hand, produces a broad longitudinal lobe, centred on the surface normal. Measurements of shear directivity patterns are not yet complete, but preliminary results suggest that the shear energy experiences a strong directivity at an angle of about 30°. The experimental data (Fig. 13) can be compared with the predictions of the theoretical models (Figs 3 and 4). It can be seen that there is quite good overall agreement for the longitudinal directivity of both the thermoelastic and ablation sources. Discrepancies may be in part due to nearfield effects and the choice of a resonant transducer. Nevertheless, the agreement would suggest that the models of the ultrasonic generation process are substantially correct, i.e. thermoelastic generation can be modelled as a point expansion close to a free surface, and ablative generation as a normal force at a point on the surface.

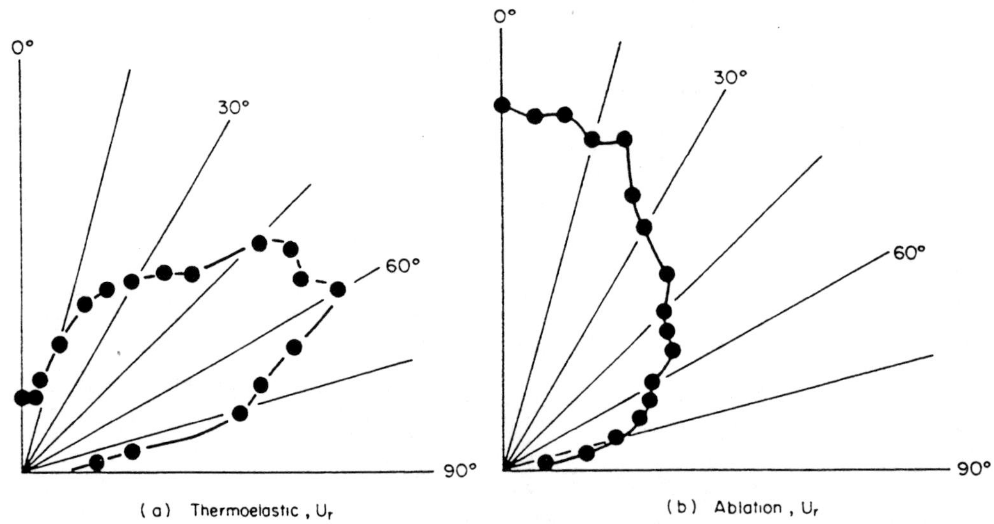

FIG. 13. Experimental directivity patterns for longitudinal wave field, detected by 1 MHz transducer; (a) thermoelastic source (1 mm diameter) cf. Fig. 3(a); (b) ablation source cf. Fig. 4(a).

4. *Surface Modifications*

All the results described so far have considered the laser irradiation of a clean, free metal surface. However, both the form of the generated acoustic waveform and its amplitude can be altered markedly with various surface modifications.

Early work carried out at high laser energies studied the effect of water on the irradiated metal surface (Fox, 1974). More recent measurements have been carried out over a wide range of incident power densities for a variety of liquids on a range of metallic substrates ((Hutchins *et al.*, 1981a). In all cases the laser pulse irradiated the surface of a 25 mm thick metallic sample after passing through the surface modification. Table I illustrates how the presence of these liquids strongly increased the longitudinal generation efficiency in the thermoelastic regime.

TABLE I
The Effect of Surface Modification on Longitudinal Wave Amplitude at Epicentre (Unfocused Laser Pulse)

Surface treatment	dB enhancement over unmodified surface		
	Aluminium	Brass	Mild steel
Constraining layers	30	28	33
Light oil	25	27	26
Silicon resin	21	24	26
Water	21	23	24
Acetone	25	24	32
Matt black paint	22	36	38
Matt white paint	25	29	22

In addition to pulse enhancement, acoustic waveforms in the thermoelastic regime are also noticeably changed by the modified surface. Quantitative measurements show that the longitudinal arrival is a positive-going pulse, similar to that produced by a plasma source, while the shear arrival is relatively unchanged. Figure 14 shows the waveform obtained when a 9 mJ laser pulse is incident through a 10 mm oil layer onto an aluminium surface. The acoustic waveforms for thin liquid coatings (Fig. 15a) resemble those for the plasma regime since the evaporation of the liquid produced by laser irradiation produces a normal force, very similar to that produced by material ablation. This is confirmed by almost identical directivity patterns in the two cases. A less convenient surface modification that has been discussed by several workers (Von Gutfeld and Melcher, 1977) is a solid transparent layer fixed to the metal surface. This "constraining layer technique" allows stresses to form normal to the metal surface by thermal expansion. The acoustic waveforms produced (Fig. 15b), are similar in amplitude to those shown in Fig. 14 with significant enhancement of the longitudinal mode (Table I). Multiple reflections in the oil or glass layer cause the additional pulses in Figs 15(a) and (b).

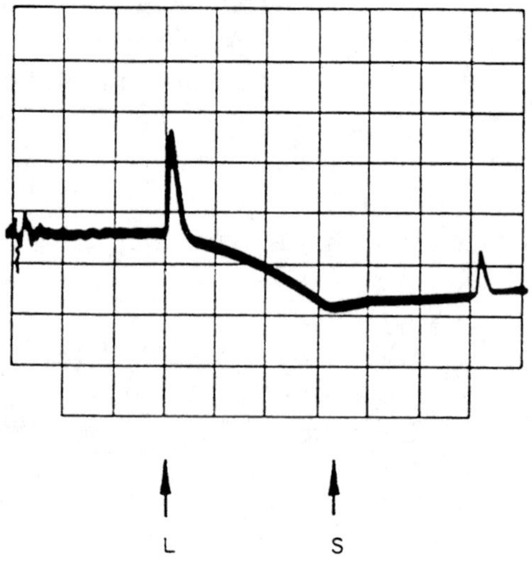

FIG. 14. Epicentre waveform detected by capacitance transducer. Unfocused laser pulse incident on aluminium surface through 10 mm of oil (0·5 μs/div., 0·68 nm/div.).

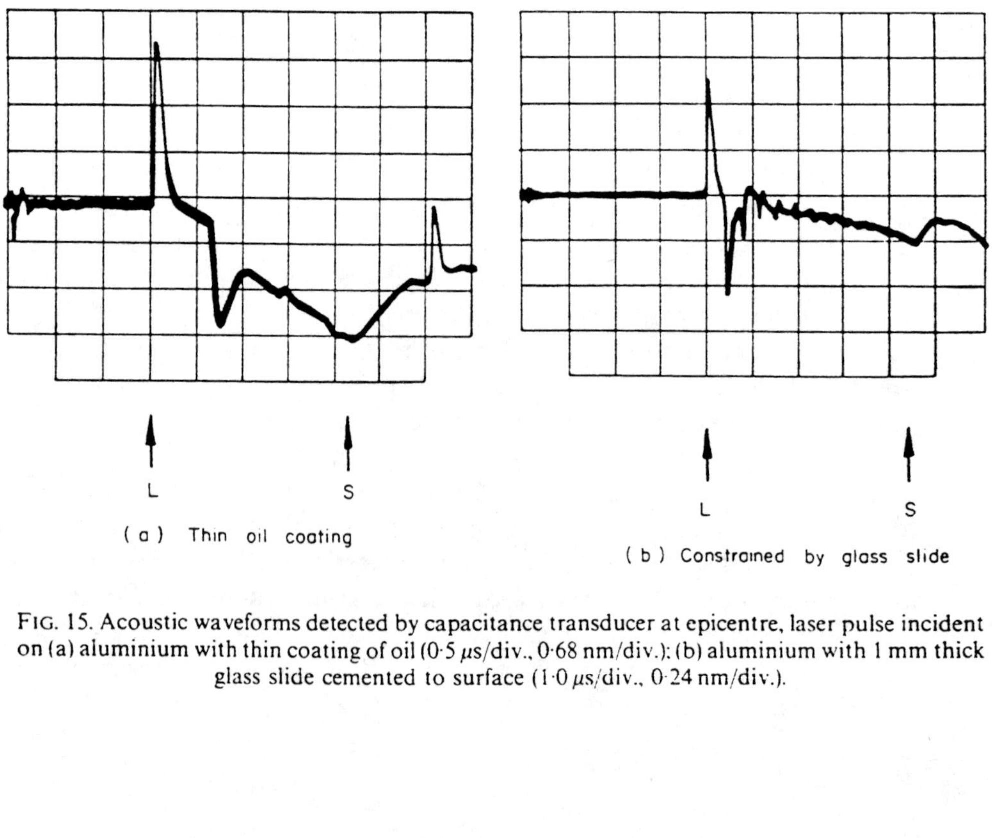

FIG. 15. Acoustic waveforms detected by capacitance transducer at epicentre, laser pulse incident on (a) aluminium with thin coating of oil (0·5 μs/div., 0·68 nm/div.); (b) aluminium with 1 mm thick glass slide cemented to surface (1·0 μs/div., 0·24 nm/div.).

B. *Surface Waves*

In addition to bulk waves, the pulsed laser will also generate surface acoustic waves, with the Rayleigh wave being most efficiently generated. A wide range of techniques is currently available to generate the Rayleigh wave: interdigital fingers, wedge probes and edge-mounted transducers. All have their particular areas of application but all have severe limitations in their use. The laser technique, being non contacting and producing either an omnidirectional or a unidirectional beam, provides an exciting alternative.

A 20 mJ pulse from a Nd:YAG laser generates both longitudinal and Rayleigh surface waves on an aluminium surface. Using a 3 MHz PZT slab at 45° to the edge of the sample as a detector, produces the oscilloscope trace shown in Fig. 16. The Rayleigh amplitude is much greater than the longitudinal (*P*) arrival, and the shear arrival is not separately observed since the shear wave travels at a velocity little different from that of the Rayleigh wave.

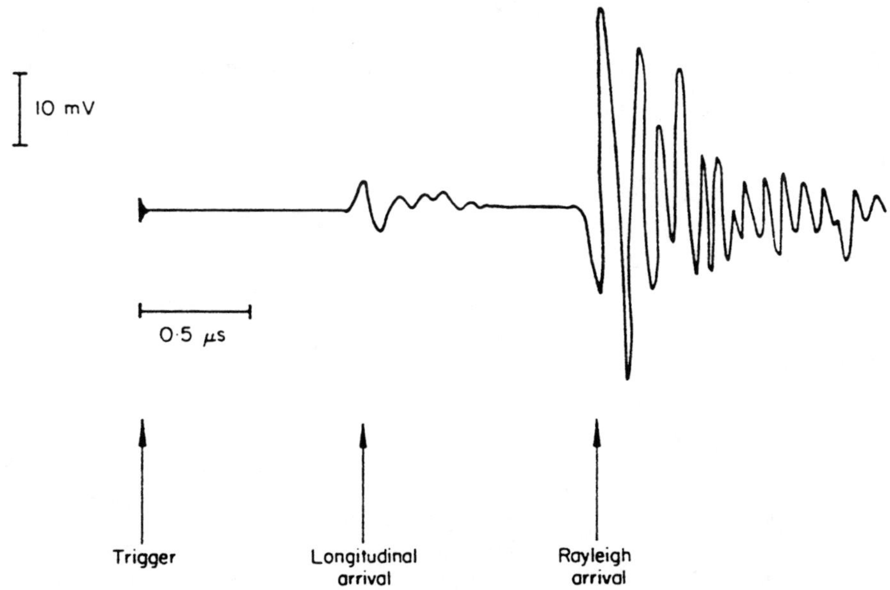

FIG. 16. Surface acoustic waveform generated by thermoelastic source and detected by 3 MHz edge-mounted transducer.

The edge transducer is highly resonant and therefore masks the true nature of the Rayleigh acoustic waveform. This can be remedied by using a much less resonant 3 MHz wedge transducer which reveals that the Rayleigh pulse is dipolar and has a time dependence proportional to the physical diameter of the laser beam. Figure 17 shows Rayleigh pulses produced by two diameters of laser beam, the longer pulse being produced by the wider beam.

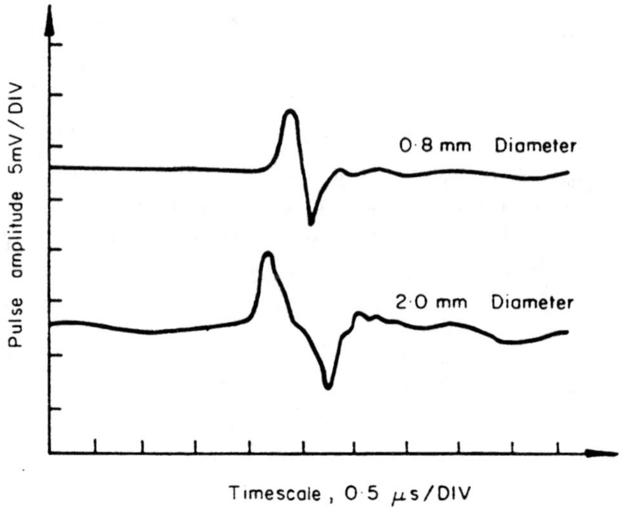

FIG. 17. Rayleigh waveforms generated by thermoelastic source of different diameters (3 MHz wedge transducer).

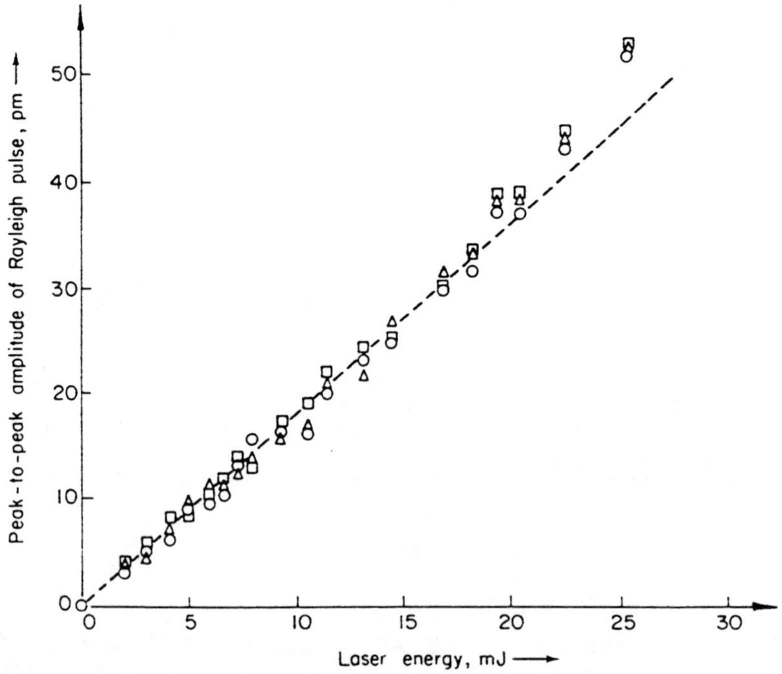

FIG. 18. Rayleigh wave amplitude as a function of incident laser energy; symbols ○, □, △ refer to three experiments (Aindow *et al.*, 1980a).

A wide-band capacitance transducer or laser interferometer can also be used to detect the laser-generated surface waves and to measure, for example, the scaling law. Figure 18 shows the linear relationship observed between Rayleigh pulse amplitude and incident laser energy in the thermoelastic regime (Aindow *et al.*, 1980a). If the laser energy is held constant and the power density increased by focusing down the beam, Rayleigh wave enhancement is observed in the plasma regime. Measured surface waveforms on steel ($\nu = 0{\cdot}29$) in both regimes (Fig. 19) can be compared with the theoretical waveforms, although calculated for a slightly different material ($\nu = 0{\cdot}25$). It can be seen that there is good general agreement, in particular that the Rayleigh wave is bipolar. Enhancement is also observed if thin liquid films are

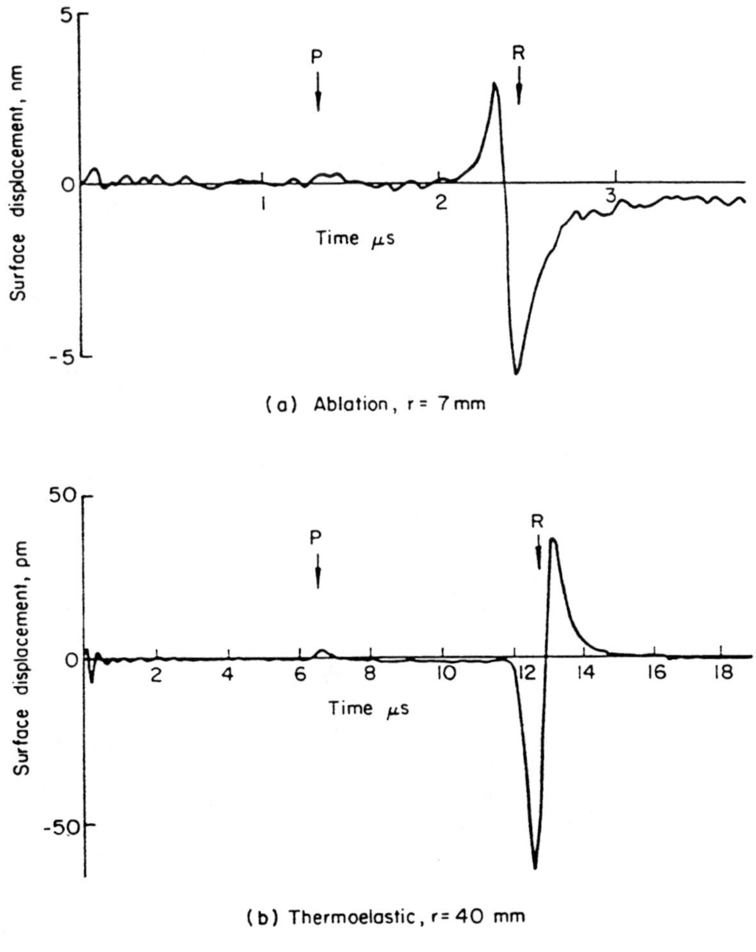

FIG. 19. Surface acoustic waveforms, measured using broadband detector; (a) ablation regime (by laser interferometer); (b) thermoelastic regime (by capacitance transducer). Compare theory, Fig. 6.

placed on the irradiated surface although care must be taken to ensure that the liquid film does not cover the whole acoustic wave path.

In order to achieve a directional surface acoustic wave the laser beam can be focused into a line source with a cylindrical lens. Figure 20 shows the experimental directivity pattern at 3 MHz for Rayleigh waves generated by a line source 4·5 mm × 0·1 mm, compared with an approximately circular source (Aindow *et al.*, 1980b). The directionality can be increased by increasing the length of line and the ultrasonic frequency of detection. The possible NDT applications of such a device are many and varied.

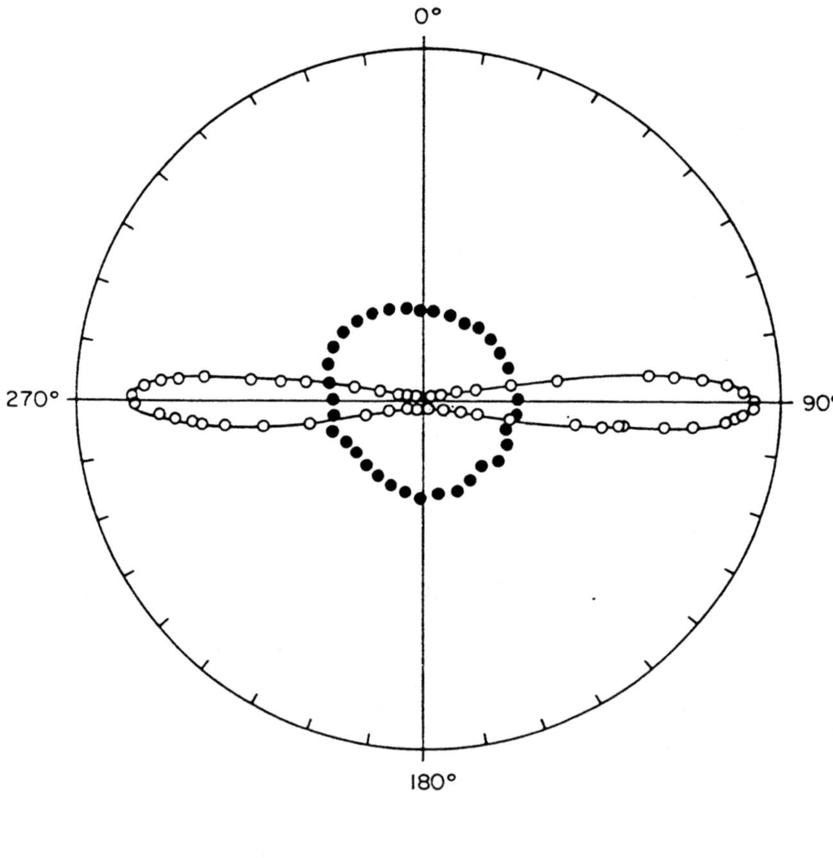

FIG. 20. Directivity of Rayleigh waves generated by thermoelastic line source, 4·5 mm × 0·1 mm, compared with nominal circular source, detected by 3 MHz surface wave transducer (Aindow *et al.*, 1980b); ○ 5 mJ Pulse focused by cylindrical lens, thermoelastic; ● 5 mJ Pulse focused by spherical lens, thermoelastic.

C. Summary of Source Characteristics

It is perhaps helpful at this stage to summarize some of the characteristics of laser-generated ultrasonic sources as determined experimentally.

(1) There are three major regimes, thermoelastic (incident power density $\lesssim 10^6$ Wcm^{-2}), plasma (power density $\gtrsim 10^7$ Wcm^{-2}) and constrained surface. These have different characteristics, but all generate longitudinal, shear and surface (Rayleigh) waves.

(2) Measured amplitudes can be greatly varied from ~ 10 nm (for a 60 mJ, 20 ns laser pulse) down to the detection limit, by simple optical means such as filters and lenses. The amplitudes and waveforms are highly reproducible in the thermoelastic regime, but rather less so in the other two regimes. The acoustic amplitudes are proportional to the absorbed optical energy in the thermoelastic regime. In the plasma regime the amplitudes are also a function of power density. Constraining the surface generally produces large enhancements in compressional wave amplitude.

(3) The far-field directivity pattern of the thermoelastic source is unusual in that the longitudinal energy is radiated into lobes directed at $\sim 55°$ to the surface normal. Unlike the conventional "piston-source", there is little energy along the normal. The directivity pattern in the plasma regime, on the other hand, is characterized by a single broad lobe centred on the normal, and is similar to that expected from a point piezoelectric transducer. The directivity when the surface is constrained is similar to that in the plasma regime. Surface waves can be generated with different directivities, according to the type of lens or aperture placed in the laser beam. Omnidirectional or highly directed beams can be readily produced.

(4) Different types of acoustic waveform can be generated. Potentially, all the waveforms are broadband and of rise-time comparable with the optical pulse. In the thermoelastic regime the L arrival is a step-function along the surface normal, and a pulse in other directions. In the plasma regime, a pulse or step can be generated according to the power density, while surface constraint generates a pulse, or series of pulses, from multiple reflection. The Rayleigh wave is bipolar in all regimes.

(5) The experimental data in the thermoelastic regime are consistent with a model in which the absorbed radiation causes a point expansion just below a free surface. Effects due to finite depth, finite lateral dimensions and laser beam profile are generally small and can be neglected in the far-field. In the ablation regime, the data are consistent with a model in which ablation of metal from the surface causes a time-varying normal point force. The constrained surface regime is more complex, but it can

often be approximated by a model similar to the plasma regime, with a force normal to the metal surface.

V. Applications of Laser-generated Ultrasound

A. *The Potential of the Technique*

When trying to assess the potential of a new NDT technique such as the generation of ultrasound by laser, it is valuable to study existing methods in order to pinpoint problems that might be solved by the technique. It is also important to examine the characteristics of the laser-generated source in the light of current NDT needs.

By far the most important method of generating and detecting ultrasound in NDT is by piezoelectric transducer. A range of piezoelectric materials is used, the most popular being lead zirconate titanate (PZT), but lead metaniobate, lithium niobate, quartz, cadmium sulphide and zinc oxide also have important applications. For much ultrasonic testing the transducers are designed to be resonant in a range 1–15 MHz and to have high efficiency. The piezoelectric transducer is so widely and successfully used that there is little chance of its replacement by an alternative except in those areas where it is less than ideal, and where its shortcomings become important.

Many of the problems with piezoelectric generation (and reception) lie, not in the transducer itself, but in the bond (coupling) which is necessary between transducer and specimen. The choice of couplant (often an oil or grease) and its method of application tend often to be an art rather than a science. Variability of couplant thickness or viscosity manifest themselves in loss of sensitivity, unwanted resonances and time-of-flight measurement errors. These problems are exacerbated at higher temperatures where fluid couplants lose viscosity and flow away. Rapidly demountable bonds have an upper operating temperature of $\sim 100\,^{\circ}C$, and at high temperatures more permanent bonding methods must be used. However at these temperatures ($\gtrsim 300\,^{\circ}C$) other problems arise: PZT is no longer suitable (its Curie temperature is 250–350 °C), and unless waveguides are used, less efficient lithium niobate (or quartz) transducers must be employed (Butler *et al.*, 1979). However even these materials cannot be used on red-hot metal. Piezoelectric transducers also have to be well protected against hostile environments, for highly corrosive or radioactive environments bring further problems.

Scanning with piezoelectric transducers generally involves the slow mechanical movement of single probes or pairs of probes, with care required to maintain good coupling with the sample. Area scanning rates are slow, even when a water bath is used for coupling. A further problem concerns access to regions of awkward geometry. Ideally the piezoelectric probe needs a local, fairly flat region of the specimen. It cannot so easily be used in regions of high curvature, for instance very close in to a right-angled nozzle on a pressure vessel. Finally, while the piezoelectric transducer with its finite area generates directed ultrasonic beams with high efficiency, it is less than ideal for situations that require a point ultrasonic source.

The laser generation of ultrasound provides at least a partial solution to all the problems outlined above. Some of the most important characteristics of the laser source have been summarized in Table II. It is first and foremost a noncontacting technique, eliminating the need for any coupling between source and sample. This clearly gives it an important advantage over piezoelectric generation techniques. The YAG laser can be pulsed at a rate of at least 10 pulses per second (higher rates are under development) and the laser beam can be readily scanned using optical methods, such as oscillating prisms and mirrors. Alternatively the sample can be scanned. Thus a relatively rapid continuous inspection of the test material is possible. The high repetition rate pulsed CO_2 laser would increase scanning rates considerably, although with some loss of ultrasonic generation efficiency. Noncontact generation, together with the low divergence of the laser beam, allows the laser to be placed well away from the test sample—even behind a transparent screen. It can therefore be used to generate ultrasound in materials and in situations that

TABLE II

Advantages and Disadvantages of Laser-generated Ultrasound Compared with Piezoelectric Techniques

Advantages	Disadvantages
(1) Noncontacting, hence no couplant problems.	(1) Signals in thermoelastic regime (no surface damage) have relatively low amplitudes.
(2) Remote, hence can be used in hostile environments.	(2) Ablation regime, with larger signal amplitudes, causes slight surface damage.
(3) Can be rapidly scanned by optical methods.	(3) Relatively expensive (at present).
(4) Can be made to operate on awkward-shaped testpieces.	(4) Requires Class IV laser safety precautions.
(5) Can be used as a point, ultrasonic source.	(5) Complete system less portable (at present) than piezoelectric transducer system.
(6) Broadband, precise time resolution possible.	
(7) Reproducible waveform and amplitude, especially in thermoelastic regime, hence a "standard" source.	
(9) Angular directivity (beam profile) of bulk and surface waves can be tailored by optical means.	

would be hostile to conventional probes, for instance when the material is at a very high temperature (e.g. a red-hot steel billet) or in a highly corrosive or radioactive environment.

These are the major gains to be made by using laser-generated ultrasound. There are other features of the source that may also prove useful. It can be made to operate on awkward surfaces of high curvature and it can be directed optically into very confined spaces. The ultrasonic characteristics of the source, notably its small dimensions, high bandwidth and reproducibility make it a candidate for applications where a standard source is required, e.g. in velocity or attenuation measurements and in calibration studies. It is also fascinating that characteristics of the source such as the beam profile and directivity can be changed optically, by the use of lenses, filters and apertures. It is probable that there are potential applications here that have not yet been explored.

It would be wrong not to mention some of the disadvantages of laser generation. Perhaps the most serious drawback is that there is an upper limit on the acoustic amplitudes that can be generated without causing visible damage to the metal surface. At distances of about 25 mm in aluminium, longitudinal amplitudes are limited to a few nanometres in the thermoelastic (no damage) regime. Unless the surface is in some way modified or constrained, these amplitudes are only reached along the maximum of the directivity pattern. Amplitudes in the plasma regime are somewhat higher for bulk and surface waves, reaching ≈ 10 nm for 100 mJ pulse. Here the penalty of some surface damage must be paid, although this is generally limited to a region $\sim 1\ \text{mm}^2$ in area and $\sim 10\,\mu$m deep. The other drawbacks are not acoustic, but concern cost, safety and portability. At the moment, laser generation is a research technique using relatively bulky and expensive all-purpose lasers. There is no reason, however, why costs and size could not be substantially reduced for specific applications. All pulsed lasers that are likely to be suitable for NDT are in Class IV for safety considerations, resulting in the need for protection against eye-damage. It is now proposed to discuss in more detail some areas of ultrasonics where laser generation is beginning to be applied.

B. Ultrasonic Flaw Detection

This is probably the most exciting potential application of laser-generated ultrasound, when combined with remote detection using a laser interferometer, constituting an entirely remote, noncontacting, ultrasonic flaw detection system. However, any detection system based on laser generation and detection looks at the moment to be a relatively expensive investment, so that initial applications are likely to be restricted to the inspection of high cost components or materials. There are also applications where structural integrity and safety are of paramount importance, where conventional techniques cannot provide a complete inspection, and where therefore the cost of a new alternative technique is relatively unimportant, provided it is reliable.

Several groups of workers have reported on the use of laser-generated ultrasound in a remote flaw detection system. One of the earliest published reports was that of Bondarenko *et al.* (1976). They generated ultrasound on one side of a steel block using a Q-switched ruby laser, and detected the transmitted waves on the other side with a dual-beam laser interferometer. The high-power pulsed laser source was focused so as to be in the ablation regime, and the minimum signal detected by the interferometer was 1 nm. As expected from focused pulses of several joules in energy, the detected direct wave amplitudes were greater than 10 nm. The acoustic pulse-widths were comparable with the optical pulse widths, i.e. ~50 ns. A planar defect was simulated by clamping two thin polished blocks together, with only part of the interface well-coupled with oil. They detected the unlubricated area by the disappearance of the transmitted ultrasonic pulse; they did not however report the scattering of ultrasound from defects.

In a paper that discusses a number of applications of laser-generated and laser-detected ultrasonics, Calder and Wilcox (1980) demonstrate the detection of an artificial flaw, a 1·5 mm diameter hole midway across a 25 mm aluminium plate. Their experimental arrangement (Fig. 21) was generally similar to that of Bondarenko *et al.* (1976), except that they used a focused 15 J Nd : glass laser system. This system had sufficient sensitivity to detect a strong ultrasonic "echo" from the artificial flaw (Fig. 22). This result looks very encouraging in terms of sensitivity, although it must be remembered that a 15 J laser pulse causes appreciable local surface damage, as the authors themselves show.

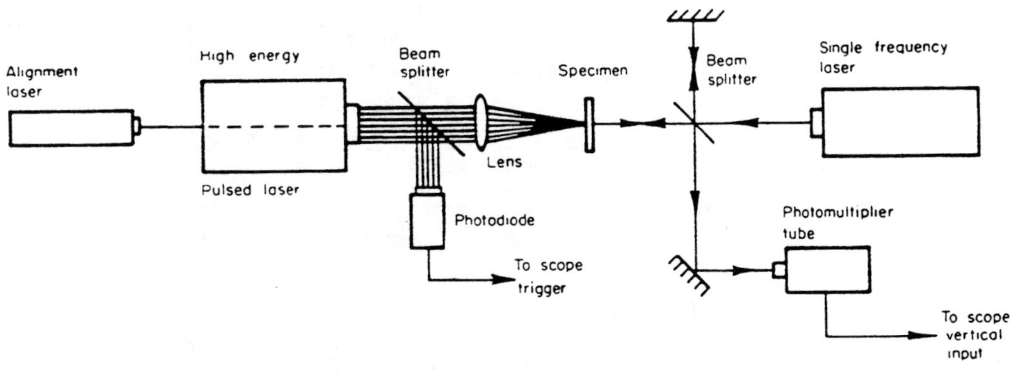

FIG. 21. Schematic diagram of noncontact transmission testing arrangement, using laser-generated ultrasonic source, with laser interferometer detector on opposite face of specimen (Calder and Wilcox, 1980).

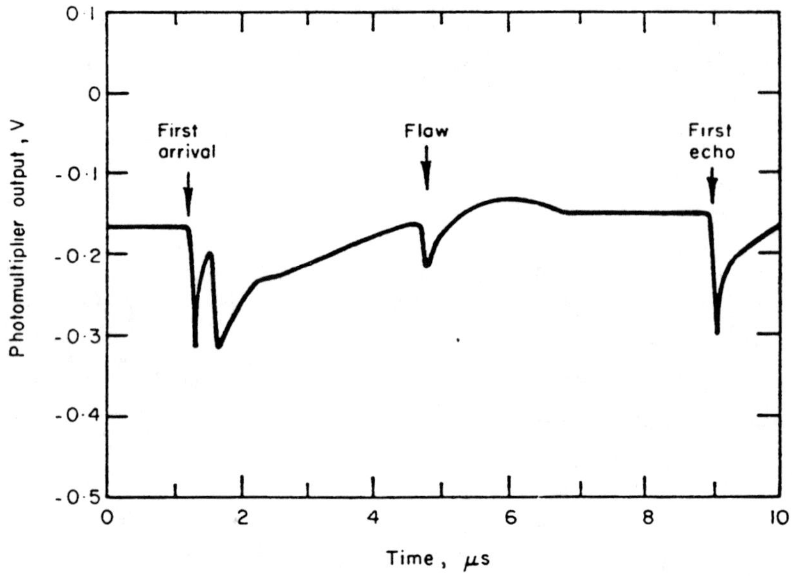

FIG. 22. Flaw detection using noncontact ultrasonic testing. Flaw was 1·5 mm diameter hole at mid-point of 25 mm aluminium plate (Calder and Wilcox, 1980).

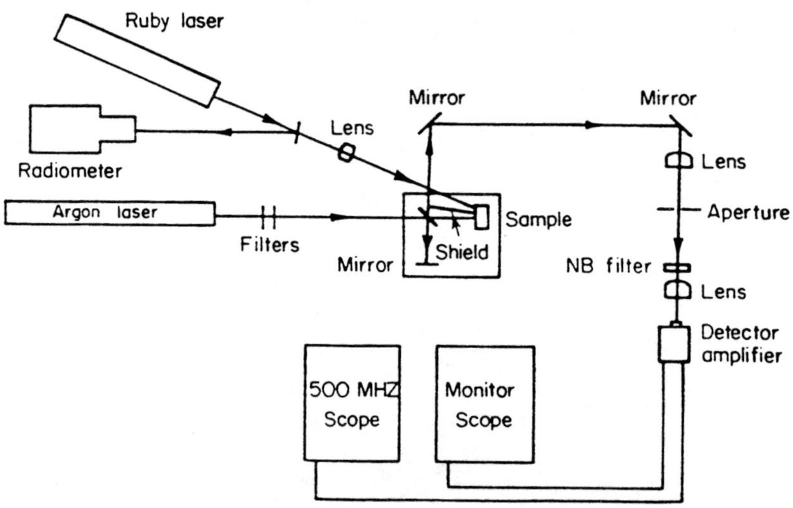

FIG. 23. Schematic diagram of experimental arrangement for noncontact ultrasonic testing with source and detector on same surface (Wellman, 1980).

The most comprehensive published study of laser systems for the detection of flaws in solids has been carried out by Wellman (1980) at the Harry Diamond Laboratory. In his report he first lays the physical foundations for his technique before continuing to demonstrate its applications to the detection of flaws. Wellman investigated two experimental arrangements, the

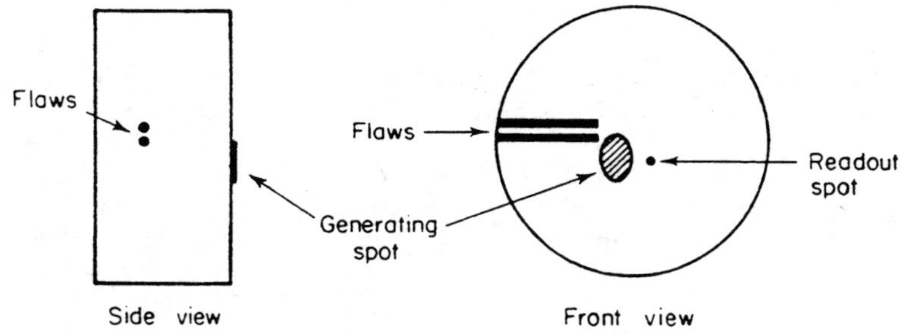

FIG. 24. Drawing of flawed sample, showing orientation of flaws (2 mm diameter holes, separated by 1 mm) and positions of source and detector (Wellman, 1980).

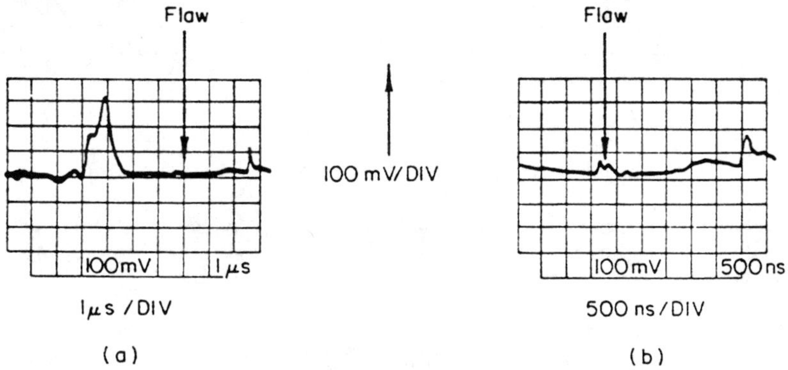

FIG. 25. Oscilloscope photographs of acoustic waves observed in flawed sample. Ultrasound scattered from flaw indicated by arrows (Wellman, 1980).

first with the laser source and detector on opposite sides of the sample (cf. Fig. 21) and the second with source and detector side by side on the same surface (Fig. 23). He used a ruby laser that was slightly less powerful (~ 1 J) and avoided power densities in excess of 500 MW/cm^2. Nevertheless ablation at this power density must have caused some pitting of the surface. When the source and detector are adjacent, filtering and shielding are necessary to reduce optical interference, and effects due to pressure waves in the air caused by the plasma formation. In one experiment (Fig. 24) Wellman simulated a flaw by two parallel 2 mm diameter holes, separated by 1 mm, and 16·5 mm below the surface of a polished steel block. His system was able not only to detect the presence of flaws by ultrasonic echo, but also to resolve the two holes (Fig. 25). He has also applied the technique to flaw detection in artillery shells. Here the interferometer detector had to be modified to collect as much light as possible from what is now an unpolished surface. Wellman (1980) was nevertheless able to demonstrate the difference between flawed and

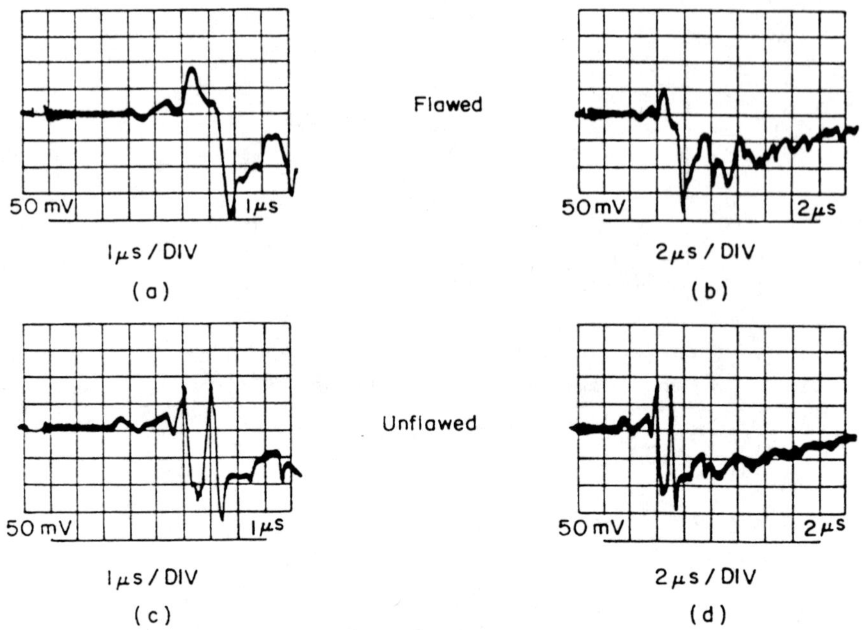

FIG. 26. Oscilloscope photographs of signal observed for 2 mm deep, 10 mm long by 0·5 mm wide flaw in 105 mm artillery shell casing compared with unflawed position (50 mV/div., Wellman, 1980).

unflawed shell cases as shown in Fig. 26. Budenkov and Kaunov (1979) have applied laser-generated ultrasound to hot (900 °C) steel samples, using electromagnetic acoustic (EMA) detection, but give no details of their results. These results, especially those of Wellman (1980), must surely give encouragement and impetus to the further exploitation of laser ultrasonics for flaw detection.

C. Measurement of Ultrasonic Velocity

The rise-time of the displacements produced by both the longitudinal and shear waves is comparable with the laser rise-time, and this sharp leading edge promises the possibility of high precision measurements of the velocity of these acoustic modes. The accuracy of absolute velocity measurements using standard ultrasonic techniques is limited to approximately $\pm 0{\cdot}2\%$ in solids with the major errors being incurred in the bond and the timing of the pulse transit. It is now possible to time ultrasonic transits to 1 ns with reference to a commercially available 500 MHz clock. With an ultrasonic rise-time better than 1 ns, as is possible using a laser pulse, the accuracy of the 500 MHz clock and associated electronic circuitry will be the major source of error if the ultrasonic pulse is detected with a noncontact technique. For a transit time of a few microseconds in materials a few centimetres thick this will result in an accuracy of a few parts in 10^4 rather than 10^3. The increase in absolute knowledge of velocity will have a wide range of application, not least in improving thickness and depth gauging, in addition to allowing a more accurate calculation of the elastic constants of materials.

Bondarenko *et al.* (1976) used the laser generation technique to measure the velocity of longitudinal waves in a steel rod. The elastic modulus, E, was then determined from the expression

$$E = \rho v_s^2 \tag{22}$$

where v_s is the velocity of sound and ρ the density. Unfortunately their paper did not present any quantitative results.

More recently Wilcox and Calder (1978) have used the technique to measure the velocity of longitudinal and shear waves in a range of materials. They then used these velocities to calculate the elastic constants of polycrystalline aluminium, stainless steel, copper, brass, tantalum, tungsten, steel and nylon. Agreement between elastic constants measured by standard ultrasonic techniques and by laser generation is generally better than 2%. It should be noted that the laser pulse used by Wilcox and Calder was obtained from a 15 J Nd : glass laser. This high energy pulse generated only longitudinal ultrasonic pulses, and the only detectable shear waves were produced by mode conversion of the longitudinal waves at the side wall reflections. Ledbetter and Moulder (1979) were the first to report that a Q-switched Nd : glass laser, incident on an aluminium surface, generated simultaneously longitudinal, shear and surface (Rayleigh) waves. They claimed an accuracy of ±1% for the velocity of their shear and surface waves and ±2% for the longitudinal modes. Their results compare favourably with accepted values when these errors are taken into account.

As far as the authors are aware, in this context only Krautkramer (1979) has taken advantage of the rapid rise-time of the ultrasonic pulse produced by a picosecond laser pulse from a mode-locked laser. The compressional wave pulse generated by the mode-locked laser had a pulse width of only 2 ns and a rise-time of less than 1 ns. Krautkramer demonstrated that a 2 ns acoustic pulse could be used to test a component as thin as a razor blade for faults and flaws. The acoustic transit time in the blade of ~30 ns ensured that there was no overlapping of echoes (Fig. 27). Calder and Wilcox (1978) have made use of the noncontact nature of the laser technique to generate compressional ultrasonic waves in liquid lead, and obtain a velocity at 200 C that is in good agreement (within 1%) of the accepted value.

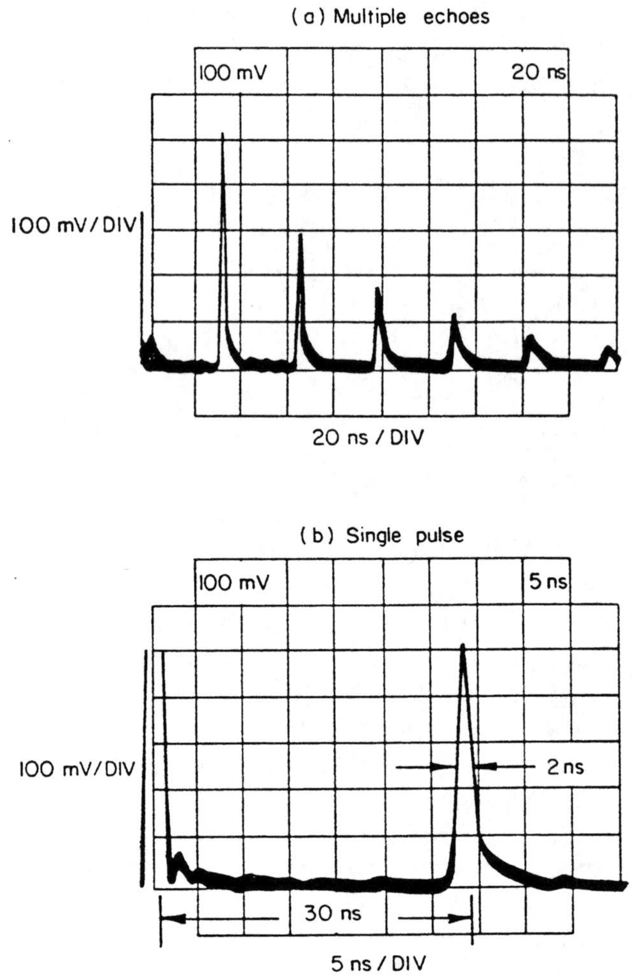

FIG. 27. Oscilloscope photographs of multiple echoes in razor blade, using mode-locked laser source and laser interferometer detector (Krautkramer, 1979).

D. Measurement of Ultrasonic Attenuation

The laser ultrasonic source, generating a sharp broadband pulse without couplant problems, must also be a candidate for ultrasonic attenuation measurement. In the megahertz frequency range, dispersion in acoustic velocity is negligible. However ultrasonic attenuation is frequency dependent. In the Rayleigh regime (wavelength greater than the scattering length, which is often the grain size) the attenuation is in fact proportional to $(\text{frequency})^4$, and varies from material to material. In other words the sample acts as a low-pass filter. The sharp pulses and steps associated with the arrival of the compressional and shear waves contain frequencies extending up to ~ 100 MHz. Thus frequency analysis of the wave arrivals after propagation through a known thickness of material should enable the ultrasonic attenuation to be measured. Ideally the measurement needs to be made with a broadband, noncontact detector. The laser interferometer is one possibility, although its sensitivity may be insufficient. Another possibility is to use an air gap capacitance transducer, which is also a noncontacting device.

As far as the authors are aware, there has been no reliable measurement of attenuation using laser-generated ultrasound. Bondarenko *et al.* (1976) report measurement of the decay of laser-generated ultrasonic pulses in stainless steel, deducing a value for the attenuation of $0{\cdot}143\ \mathrm{Np\,cm^{-1}}$. It must be noted however that this value is not referred to any particular frequency.

E. Ultrasonic Microscopy

Most scanning acoustic microscopes (SAMs) use sputtered ZnO transducers for both the generation and detection of the ultrasonic signal (see e.g. Atalar *et al.*, 1979). These thin film transducers can be used to generate ultrasonic waves with a frequency in excess of 2 GHz and consequently wavelengths of the order of 1 μm, depending on the medium. Using a sapphire lens enables the 2 GHz ultrasound to be focused to a spot size of the order of 1 μm, thus defining the resolution of the microscope. Two major problems associated with this technique could be alleviated by using a laser generation system. Firstly, the acoustic intensities generated at these high frequencies by the ZnO transducers are low, and this failing is not helped by ultrasonic attenuation of the order of 1500 dB/cm in the water mostly used as a coupling medium between sample and transducer. Secondly, the sample has to be mechanically scanned through the ultrasonic probe beam.

A scanned laser beam would immediately solve the scanning and couplant problems for the source, and an approach developed by Wickramasinghe *et al.* (1978) could well lead to an improvement in signal intensity. A mode-locked, Q-switched Nd : YAG laser was used with Q-switched pulses of 200 ns duration at a repetition frequency of 2·7 kHz. The mode-locked pulse train within each Q-switched pulse consisted of 0·2 ns pulses repeating at 210 MHz. When this probe train was incident on the sample (Fig. 28) ultrasonic signals were generated at the fundamental frequency of 210 MHz and at higher harmonics. Wickramasinghe *et al.* (1978) selected a frequency of 840 MHz by using a ZnO detector with a response centred near 800 MHz. This is by no means the upper limit for the Nd: YAG source and in addition, because of the acoustic amplitude scales with laser energy, improvements in signal intensity are also possible.

Another type of photoacoustic microscopy uses a scanned, intensity modulated, CW laser beam incident on a metal surface (Luukkala, 1980;

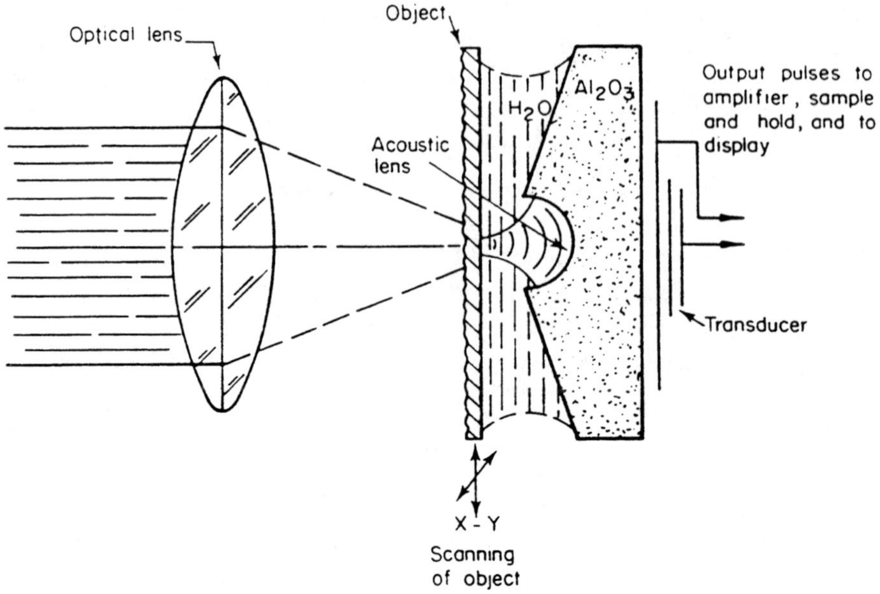

FIG. 28. Schematic diagram of photoacoustic apparatus. The optical source is a mode-locked and Q-switched Nd : YAG laser. The pulses repeat at 210 MHz and have a width of 0·2 ns. The acoustic lens has a radius of 200 μm and the transducer is tuned to 840 MHz (Wickramasinghe *et al.*, 1978).

Rosencwaig, 1980). The optical energy is absorbed and converted through the thermoelastic effect into elastic waves, which are detected either by a transducer attached to the back face of the specimen, or by an air microphone close to the surface. The absorption process and conversion to elastic waves is basically the same as for the nanosecond pulsed laser, except that the scales are different. In the case of the Q-switched Nd: YAG laser with a 30 ns pulse length there is a small thermal field in metals, i.e. penetration depths of a few micrometres, and a large elastic wave field. In the case of photoacoustic microscopy, with a modulation frequency of ~ 1 kHz, there is a far larger thermal field with dimensions in the range 0·1–1 mm. Thus in the latter case the image produced when the laser is scanned across the surface relates more to variations in the thermal properties than in the acoustic properties of the sample. The laser generation technique could be even more powerful in the field of microscopy if it were again coupled with laser detection. The low sensitivity and insufficient bandwidth of laser interferometry seem to make this development unlikely at present.

F. Standard Acoustic Source

Acoustic emission is now increasingly being used for detecting and locating flaws in a wide range of engineering structures (Spanner and McElroy, 1975;

Williams, 1980) and for the study of deformation and fracture processes in a variety of materials (Wadley *et al.*, 1980). Much work, however, is still carried out with equipment that is neither standardized nor calibrated. There is a clear need for a simulated emission source that generates elastic waves similar to those from a true emission event and can be used for calibration purposes.

A number of methods of calibrating acoustic emission systems have recently been reviewed by Sachse and Hsu (1980) including the use of a helium gas jet (McBride and Hutchison, 1976) and the fracture of pencil leads (Hsu *et al.*, 1977) or glass capillaries (Breckenridge *et al.*, 1975). The gas jet is essentially a continuous white-noise generator, with a different frequency spectrum from a true acoustic emission source, which is usually a short duration, transient event. The gas jet also gives relative rather than absolute calibrations. The fracture of a pencil lead or glass capillary, on the other hand, does constitute a transient event and can give an absolute calibration. However these sources are limited in high frequency content, are monopolar and of much greater strength than typical crack sources.

Several authors (Davidson *et al.*, 1976; Egle and Brown, 1976) have proposed the use of a pulsed laser as a simulated emission source. More recently Scruby *et al.* (1981) have demonstrated that laser generation is able to act as a standard acoustic emission source. For example, the stress field around the thermoelastic source, to a first approximation, appears to be of a similar form to that surrounding a vertical Mode I crack at the surface of a plate in that the latter is also composed principally of horizontal force dipoles, rather than the force monopole that is generated by the fracture of a pencil lead. When the laser pulse is incident on a transfer block with a 45° notch (Fig. 29), it generates a stress field that includes a vertical force dipole. This is

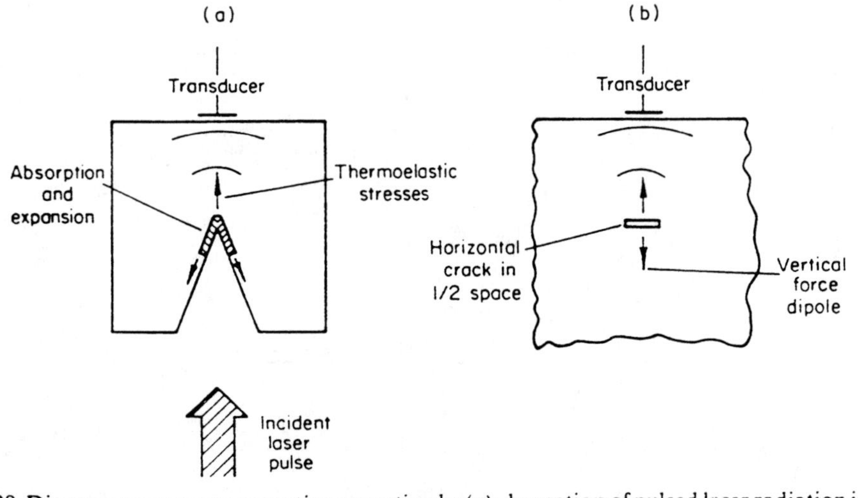

FIG. 29. Diagram to compare acoustic generation by (a) absorption of pulsed laser radiation in (a) a 45° notched specimen, and (b) a horizontal, Mode I crack in a half-space (Scruby *et al.*, 1981).

important if the laser source is to simulate a horizontal Mode I crack within the bulk of a material. In this orientation the simulated waveform (Fig. 30a) is observed to be very similar to that generated by cleavage microcracking (Fig. 30b; Scruby *et al.*, 1979)

Other types of standard source can be simulated by using the ablation regime, or by modifying the surface. In particular, monopolar sources with step function and δ-function time dependence can be approximated (Hutchins *et al.*, 1981b). The chief application of a standard acoustic source is in calibration, and Scruby *et al.* (1981) have demonstrated that the laser generation technique is ideally suited to the calibration of acoustic emission equipment. The laser-generated waveforms are so reproducible from shot to shot that quantitative comparisons can readily be made between the output from, for instance, a piezoelectric transducer and an absolute capacitance transducer. The development of quantitative flaw characterization by acoustic emission in engineering components is held back by the fact that the acoustic transfer functions of complex geometries are not known. Another application of a reproducible, standard acoustic source is therefore in the experimental measurement of acoustic transfer functions.

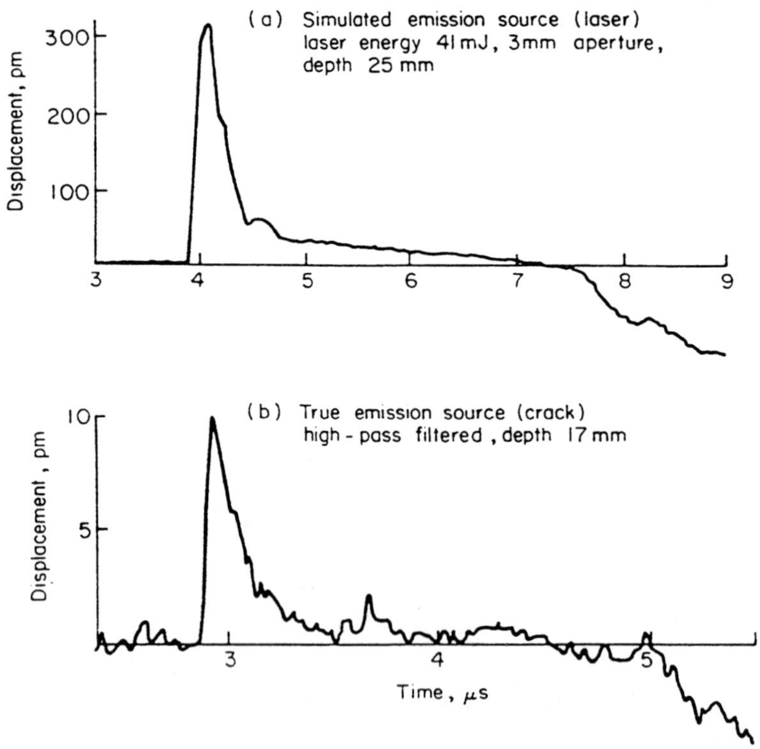

FIG. 30. Comparison of acoustic wave forms generated by (a) absorption of pulsed laser radiation (Fig. 29a) with (b) cleavage microcrack propagation in mild steel (Scruby *et al.*, 1981).

VI. Concluding Remarks

It has been our objective in this chapter to introduce the reader to ultrasonic wave generation by laser. The technique is still in its infancy, so we have devoted quite some time to explain the underlying physical principles and to introduce laboratory results that illustrate the characteristics of laser-generated ultrasound. We have however also discussed application areas where we believe the technique can make a valuable contribution in the future. A technique that offers the possibility of carrying out nondestructive inspection remotely and rapidly, under the conditions (e.g. high temperatures) that make conventional ultrasonics difficult, demands more than casual interest.

It is clear, nevertheless, that laser generation techniques still require development. Effort is needed to apply laser ultrasound to realistic problems and to develop compact, practical equipment. For many of the applications discussed laser generation only achieves its full potential when combined with laser detection. There is evidence that development is also necessary here, chiefly to improve sensitivity. While pulsed lasers are capable of generating ultrasonic amplitudes comparable with conventional methods, laser interferometer systems are still several orders of magnitude less sensitive than their piezoelectric counterparts.

In conclusion, we see piezoelectric techniques continuing to be used for the vast majority of NDT applications, with laser generation being developed for a number of special applications. NDT thrives on having an array of complementary techniques ready for an inspection problem when it arises. It is to be hoped that these exciting and promising laser methods will soon be generally available.

Acknowledgements

We are grateful to the Mechanical and Electrical Engineering Requirements Board of the Department of Industry for its funding of part of this work, through its support for the NDT Centre, AERE, Harwell.

References

Achenbach, J. D. (1973). "Wave Propagation in Elastic Solids". North-Holland Publishing Co., Oxford.

Aindow, A. M., Dewhurst, R. J., Hutchins, D. A. and Palmer, S. B. (1980a). *In* "Proc. Europ. Conf. on Opt. System and Applications, Utrecht".

Aindow, A. M., Dewhurst, R. J., Hutchins, D. A. and Palmer, S. B., (1980b). "Acoustics '80", Conf. Proc. 277.

Aindow, A. M., Dewhurst, R. J., Hutchins, D. A. and Palmer, S. B. (1981). *J. Acoust. Soc. Am.* **69**, 449.

Anderholm, N. C. (1970). *Appl. Phys. Lett.* **16**, 118.

Atalar, A., Jipson, V., Koch, R. and Quate, C. F. (1979). *Ann. Rev. Mat. Sci.* **255**.
Bar-Cohen, Y. (1979). *Brit. J. NDT* **21**, 76.
Bleaney, B. I. and Bleaney, B. (1965). "Electricity and Magnetism", 2nd edn. Oxford Press.
Bondarenko, A. N., Drobot, Y. B. and Kruglov, S. V. (1976). *Soc. J. NDT* **12**, 655.
Bozhkov, A. T., Bunkin, F. V. and Gyrdev, L. L. (1976). *Sov. J. Quant. Elect.* **6**, 809.
Breckenridge, F. R., Tschiegg, C. E. and Greenspan, M. (1975). *J. Acoust. Soc. Am.* **57**, 626.
Budenkov, G. A. and Kaunov, A. P. (1979). 9th World Conf. NDT, paper 4A-14.
Bunkin, F. V. and Komissarov, V. M. (1973). *Sov. Phys. Acoust.* **19**, 203.
Burridge, R. and Knopoff, L. (1964). *Bull. Seism. Soc. Am.* **54**, 1875.
Butler, B., Palmer, S. B. and Primavesi, G. J. (1979). *Ultrasonics* **17**, 249.
Calder, C. A. and Wilcox, W. W. (1974). *Rev. Sci. Instrum.* **45**, 1557.
Calder, C. A. and Wilcox, W. W. (1978). 6th Int. Conf. on Expl. Stress Analysis, Munich, p. 181.
Calder, C. A. and Wilcox, W. W. (1980). *Mat. Eval.* **38**, 86.
Davidson, G., Emmony, D. C. and Maloney, T. (1976). 2nd Acoustic Emission Conf AML, Dorset.
Egle, D. M. and Brown, A. E. (1976). *JTEVA* **4**, 196.
Fairand, B. P. (1974). *Appl. Phys. Lett.* **25**, 431.
Fairand, B. P. and Clauer, A. H. (1979). *J. Appl. Phys.* **50**, 1497.
Fox, J. A. (1974). *Appl. Phys. Lett.* **24**, 461.
Giglio, D. A. (1973). Harry Diamond Labs Report, HDL-TM-73-32, USA.
Gournay, L. S. (1966). *J. Acoust. Soc. Am.* **40**, 1322.
Graff, K. F. (1975). "Wave Motion in Elastic Solids". Clarendon Press, Oxford.
Hsu, N. N., Simmons, J. A. and Hardy, S. C. (1977). *Mat. Eval.* **35**, 100.
Hu, C. (1969). *J. Acoust. Soc. Am.* **46**, 728.
Hutcheson, L., Roth, O. and Barnes, F. S. (1971). 11th Symposium electron, ion and laser beam technology. p. 413. San Francisco Press Inc., San Francisco.
Hutchins, D. A., Dewhurst, R. J. and Palmer, S. B. (1981a). To be published in *Ultrasonics* **19**, 103.
Hutchins, D. A., Dewhurst, R. J., Palmer, S. B. and Scruby, C. B. (1981b). *Appl. Phys. Lett.* **38**, 677.
Ivakin, E. V., Lazaruk, A. M., Petrovitch, I. P., Rubanov, A. S. and Stepanov, B. I. (1976). *Sov. Tech. Phys. Lett.* **2**, 181.
Jones, E. D. (1971). *Appl. Phys. Lett.* **18**, 33.
Kasoev, S. A. and Lyamshev, L. M. (1977). *Sov. Phys. Acoust.* **23**, 510.
Kaule, W. (1978). 8th World Conf NDT, Cannes, pp. 1–7.
Kaye, G. W. C. and Laby, T. H. (1973). "Tables of Physical and Chemical Constants", Congress (London), 14th edition.
Kohanzadeh, Y., Whinnery, J. R. and Carroll, J. (1975). *J. Acoust. Soc. Am.* **57**, 67.
Krautkramer, J. (1979). 9th World Conf NDT, Plenary Lecture, Melbourne, Australia.
Krehl, P., Schwirke, F. and Cooper, A. W. (1975). *J. Appl. Phys.* **46**, 4400.
Kubota, K. and Nakatani, Y. (1973). *Jap. J. Appl. Phys.* **12**, 888.
Ledbetter, H. M. and Moulder, J. C. (1979). *J. Acoust. Soc. Am.* **65**, 840.
Lee, R. E. and White, R. M. (1968). *Appl. Phys. Lett.* **12**. 12.
Lord, A. E. Jr (1966). *J. Acoust. Soc. Am.* **39**, 650.
Luukkala, M. V. (1980). *In* "Scanned Image Microscopy" (E. A. Ash, ed.). Academic Press, London and New York.
Lyamshev, L. M. and Naugol'nykh, K. A. (1976). *Sov. Phys. Acoust.* **22**, 354.

McBride, S. L. and Hutchison, T. S. (1976). *Can. J. Phys.* **54**, 1824.
Miller, G. F. and Pursey, H. (1954). *Proc. Roy. Soc.* (*London*) **A223**, 521.
Muir, T. G., Culbertson, C. R. and Clynch, J. R. (1976). *J. Acoust. Soc. Am.* **59**, 735.
O'Keefe, J. D. and Skeen, C. H. (1972). *Appl. Phys. Lett.* **21**, 464.
Pao, Y. H., Gajewski, R. R. and Ceranoglu, A. N. (1979). *J. Acoust. Soc. Am.* **65**, 96.
Peercy, P. S., Jones, E. D., Bushnell, J. C. and Gobelli, G. W. (1970). *Appl. Phys. Lett.* **16**, 120.
Pekeris, C. L. (1955). Geophysics Proc. NAS **41**, 469.
Pekeris, C. L. and Lifson, H. (1957). *J. Acoust. Soc. Am.* **29**, 1233.
Penner, S. S. and Sharma, O. P. (1966). *J. Appl. Phys.* **37**, 2304.
Percival, C. M. (1967). *J. Appl. Phys.* **38**, 5313.
Ready, J. F. (1965). *J. Appl. Phys.* **36**, 462.
Ready, J. F. (1971). "Effects of High Power Laser Radiation". Academic Press, London and New York.
Rosencwaig, A. (1980). *In* "Scanned Image Microscopy" (E. A. Ash, ed.). Academic Press, London and New York.
Sachse, W. and Hsu, N. N. (1980). *Phys. Acoust.* **XIV**, 277.
Scruby, C. B. and Wadley, H. N. G. (1976). *J. Phys. D.* **11**, 1487.
Scruby, C. B., Wadley, H. N. G. and Sinclair, J. E. (1979). Proc. Conf. Periodic Insp. for Press Compts., I.Mech.E., London, p. 127.
Scruby, C. B., Dewhurst, R. J., Hutchins, D. A. and Palmer, S. B. (1980). *J. Appl. Phys.* **51**, 6210.
Scruby, C. B., Wadley, H. N. G., Dewhurst, R. J., Hutchins, D. A. and Palmer, S. B. (1981). To be published in *Mat. Eval.*
Sigrist, M. W., Fuchs, F. and Kneubuhl, F. (1978). Presented at ECOSA2, Brighton.
Skeen, C. H. and York, C. M. (1968). *Appl. Phys. Lett.* **12**, 369.
Spanner, J. C. and McElroy, J. W. (1975). "Monitoring Structural Integrity by Acoustic Emission", ASTM STP 571.
Tomulescu, R. (1975). *Rev. Roum. Phys.* **20**, 463.
Von Gutfeld, R. J. and Melcher, R. L. (1977). *Appl. Phys. Lett.* **30**, 357.
Wadley, H. N. C., Scruby, C. B. and Speake, J. H. (1980). *Int. Metals Reviews* **2**, 41.
Wellman, R. L. (1978). Conference on Laser and Electro-Optic Systems, San Diego, p. 106.
Wellman, R. L. (1980). Harry Diamond Labs Report, HDL-TR-1902, USA.
Westevelt, P. J. and Carson, R. S. (1973). *J. Acoust. Soc. Am.* **54**, 121.
White, R. M. (1963). *J. Appl. Phys.* **34**, 3559.
Wickramasinghe, H. K., Bray, R. C., Jipson, V., Quate, C. F. and Salcedo, J. R. (1978). *Appl. Phys. Lett.* **33**, 923.
Wilcox, W. W. and Calder, C. A. (1978). *Instr. Tech.* **25**, 63.
Williams, R. V. (1980). "Acoustic Emission". Hilger, London.

Reprinted with permission from *IEEE Transactions on Ultrasonics, Ferroelectrics, and Frequency Control,* UFFC-33(5), 485-499 (September 1986).

Optical Detection of Ultrasound

JEAN-PIERRE MONCHALIN

Abstract—**A review of the various optical methods to detect ultrasound (bulk and surface waves) at the surface of opaque solids is presented. The most useful techniques are thoroughly analyzed. Their performance when nonideal conditions are encountered, such as vibrations, air turbulence, and rough light scattering surfaces is evaluated. This review includes a description of knife-edge techniques, optical heterodyning, differential interferometry, and velocity (time-delay) interferometry methods, plus a mention of various less-important techniques.**

I. Introduction

EARLY EFFORTS to detect ultrasound by optical means were motivated by the need to visualize and image the ultrasonic field. Although this motivation is still present, as demonstrated by a survey of recent literature on the subject, the main thrust behind optical detection of ultrasound is now to fully exploit the potential of laser ultrasonic generation [1], [2] by coupling both techniques in order to make a powerful inspection tool for materials. Since in this case the sample could be located at a large distance (more than 1 m) from any optical element of the generating or receiving system, important applications are foreseen in the inspection of materials at high temperature (such as all metals and ceramics) for process and quality control. A technique based on laser generation and optical detection could be used, for example, to detect flaws as soon as they are created during processing, to measure production parameters (thickness, temperature ...), and to determine microstructural properties on-line (grain size, porosity ...). Ultrasonic inspection in hard to access areas (corners, edges) would also be greatly facilitated by this all-optical inspection system. These areas are often prone to the development of defects, during production or in service. Furthermore, the development of ultrasonic techniques for microstructural material characterization (grain size, magnetic domains, porosity ...) may benefit greatly from the broadband nature of ultrasound generated by the laser pulse. Since the poor sensitivity of optical detection is the main reason that limits the practical evolution of this technology to full commercial application, particular attention will be given to this question throughout this review.

In this review, we will try to cover all the techniques known to have been used to detect ultrasound produced in opaque solids. The detection of ultrasound in transparent solids or liquids is well established [3], [4] and will not be part of this review. In this paper, newer techniques will be analyzed thoroughly, whereas well-established techniques will not be detailed and the reader will be referred to previous reviews such as that by Stegeman [5] for further information. This review deals primarily with the detection of ultrasound and not the detection of thermal waves, although in many cases the distinction is unclear, since a stress and a strain is induced at the same time as the thermal wave and the same apparatus could be used. For these reasons, the references to works in photoacoustics which have used a method also applicable to ultrasonic detection will be given.

The optical detection techniques for ultrasound will be classified into noninterferometric techniques and interferometric techniques. The former are well developed or of limited application, while the latter are more general and are presently the object of active developments.

Manuscript received December 17, 1985; revised March 25, 1986.
The author is with the Physical Metallurgy Research Laboratories, CANMET, Energy Mines and Resources, Ottawa, ON Canada and the Industrial Materials Research Institute, National Research Council of Canada, 75 de Mortagne Boulevard, Boucherville, Québec, J4B 6Y4, Canada.
IEEE Log Number 8609260.

II. Noninterferometric Techniques

The various optical techniques to detect ultrasound which are not based on interferometry and which have been either considered or implemented are the following:

- knife-edge technique
- surface-grating technique
- technique based on reflectivity
- technique based on a light filter.

A. Knife-Edge Technique

The principle of this technique is shown in Fig. 1. When an ultrasonic wave impinges on a surface or propagates along it, a ripple is produced causing the deflection of a laser beam sent onto the surface. The beam deviation is measured with either a detector with a knife edge in front of it or a position-sensitive detector. The beam size on the surface has to be smaller than the ripple spacing. Surface waves have been visualized using this technique [5]–[9] as well as bulk waves. When bulk waves are used, the technique permits characterization of the sample microstructure and detection of internal defects [10]. It has matured into a commercial inspection device, the SLAM [11] (scanning laser acoustic microscope). Evaluation of the sensitivity of the technique can be found in a review by Whitman and Korpel [12] and a paper appearing in this issue [13]. The detection limit has been reported to be about a few 10^{-2} Å for a polished nonabsorbing surface, a detection bandwidth of 1 MHz, and a laser power of 10

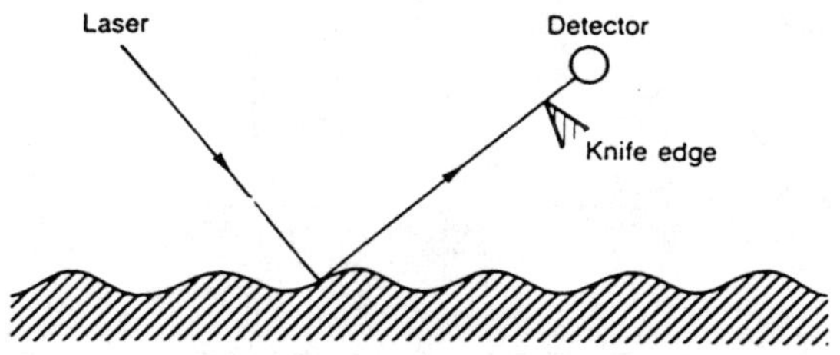

Fig. 1. Principle of the knife-edge technique.

mW. This corresponds to a detection limit of a few 10^{-6} Å $(W/Hz)^{1/2}$.

The technique is very insensitive to vibrations but requires a good surface finish and is hardly applicable to rough surfaces which scatter light. However, as it is done in the SLAM, this limitation can be avoided by mounting on top of the sample a transparent cover piece having a mirror at its bottom surface. A fluid ensures the transmission of ultrasound to the mirror surface, the deformations of which are probed by the laser beam.

The technique has been applied to the detection of ultrasound generated by laser [14] and also to the detection of the small "bumps" produced by photothermal heating [15], [16].

B. *Surface-Grating Technique*

This technique [17] has been applied to the detection of surface waves and, unlike the previous technique, uses a beam larger than an acoustic wavelength. The grating produced by the surface wave gives two first-order diffracted beams, which after interfering with the specularly reflected zeroth-order beam, produce a moving intensity corrugation at certain distances from the surface. The signal is detected through a Ronchi grid of suitable periodic spacing. This technique is mostly limited to continuous excitation on a clean polished surface and has been reviewed by Whitman and Korpel [12] and Stegeman [5].

C. *Reflectivity Technique*

The ultrasonic stress produces a small change of the sample refractive index which in turn causes a change of the specularly and diffuse reflectivity of the sample. The change of reflectivity is small and of the order of the ultrasonic strain. This technique has, however, been applied successfully to detect the ultrasonic echos in a thin film produced by laser generation with picosecond time resolution [18]. We should also note that a similar technique is used in photothermal detection [19], but the change of reflectivity being in this case caused by the temperature or carrier concentration changes.

D. *Use of a Light Filter Based on Absorption*

This technique is based on the frequency demodulation of frequency shifted by ultrasound scattered light (Doppler effect, see below) by the absorption band of an optical medium (solid, liquid, gas), the laser frequency being tuned to the slope of the band [20]. In practice, the absorption bands found in solids, liquids, and gases are too broad for normally used ultrasonic frequencies (<100 MHz) to give a sensitive device. However, in the case of liquids and gases the absorbing medium can be given easily a large cross-section and can be made long if necessary. Since the filtering action depends only on the pathlength through the medium, this technique is rather insensitive to the direction of incoming rays. This is combined with a large cross-section and enables a large gathering efficiency of scattered light (see below for gathering efficiency or étendue), which may make such a simple technique attractive in spite of the generally poor discrimination sensitivity mentioned above, but its use has not been reported so far. Light filtering action can be also easily obtained by interferometry and this is described below.

It is also possible to narrow the Doppler limited linewidth obtained in a gas cell by using saturated absorption [21]. A powerful laser beam burns a narrow hole through the atoms or molecules Maxwellian velocity distribution, which can be used to frequency demodulate the weaker beam received from the surface [22]. The sensitivity of such a technique depends on the chosen gas, the laser power, the interaction length, and the angular distribution of received scattered light that broadens the resonance feature by the residual Doppler effect. No experimentation to apply this technique to the detection of ultrasound is known at this time.

Finally, it should be noted that all these techniques give a filtering bandwidth that is fixed and determined by the medium, unlike interferometry, which enables easily to chose the most suitable bandwidth (see below).

III. Interferometric Detection: Background

In this section, we will present some background information which will be useful to understand the merits and limitations of the various interferometric detection techniques presented below. These techniques will be classified into the following three types, which can be identified by reference to Fig. 2.

1) *Optical Heterodyning:* This technique consists of beating the wave scattered by the surface, which can have a very complicated and distorted wavefront, with a reference wave.

2) *Differential Interferometry:* This technique consists of producing the interference of two waves issued from different points on the surface or from one single point illuminated by two distinct beams.

3) *Velocity or Time-Delay Interferometry:* This technique uses only one wave issued from the surface and generates a signal by beating it with itself delayed by a certain time.

If the pathlengths have been compensated the first two techniques can use a broadband source and do not require a single frequency laser. By opposition, the third technique, which involves necessarily a time delay, requires a narrow optical bandwidth source, in practice a single frequency laser, or to encode the spectrum of a broader band source (spectral modulation technique, see below).

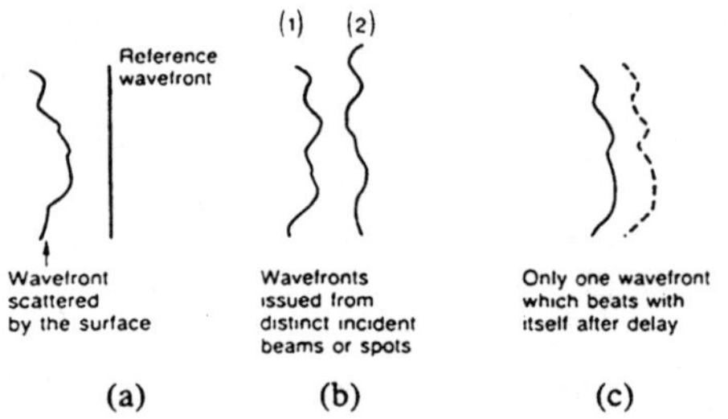

Fig. 2. Classification of the various interferometric techniques to detect ultrasound. (a) Optical heterodyning. (b) Differential interferometry. (c) Velocity or time-delay interferometry.

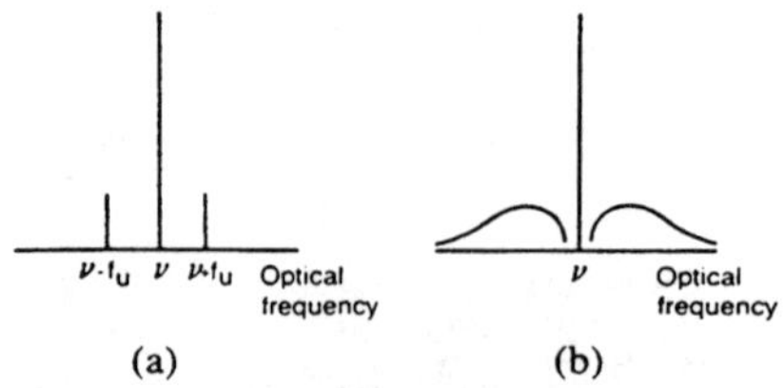

Fig. 3. Optical spectrum of the light scattered from the surface. (a) Continuous ultrasonic excitation. (b) Pulsed ultrasonic excitation.

A. *Effects of Ultrasound in the Time Domain and Frequency Domain*

Let us assume that the surface moves by $\delta(t)$ at a location where a light field $E_0 \cos \omega t$ is incident, $\omega = 2\pi\nu$, where ν is the optical frequency. In the backscattered direction, the light field is then proportional to $\cos[\omega t - 4\pi\delta(t)/\lambda] = \cos[\omega t - 2\omega\delta(t)/c]$, where λ is the optical wavelength and c is the speed of light. This phase factor can also be written $\int_0^t \omega(1 - 2v(t)/c)\, dt$, where $v(t) = d\delta(t)/dt$ is the surface velocity. This shows the well-known Doppler effect: the instantaneous frequency is $\nu(1 - 2v(t)/c)$.

If we assume, for simplicity, that $\delta(t) = U \cos(2\pi f_u t + \psi)$, where f_u is the ultrasonic frequency, U is the amplitude of the displacement (U is much less than λ), and ψ a phase constant (this is valid for continuous ultrasonic excitation, but pulse excitation can be readily handled by Fourier analysis), the back scattered light field can then be written as proportional to

$$\begin{aligned}
&\cos[\omega t - 4\pi U \cos(2\pi f_u t + \psi)/\lambda] \\
&\quad \approx \cos \omega t + 4\pi\delta(t) \sin \omega t/\lambda \\
&\quad = \cos(2\pi\nu t) + 2\pi U [\sin[2\pi(\nu + f_u)t + \psi] \\
&\qquad + \sin[2\pi(\nu - f_u)t - \psi]]/\lambda.
\end{aligned}$$

This last expression shows that ultrasound produces two weak sidebands on both sides of the optical frequency, the ratio of the amplitudes being $2\pi U/\lambda$ (see Fig. 3). In the case of pulsed ultrasonic excitation, these sidebands are broadened (see Fig. 3).

In conclusion, the effect of ultrasound can be described either as a phase modulation of the reflected light or a Doppler shift of the instantaneous frequency or the generation of sidebands. All these descriptions are equivalent and any of them can be used to analyze the response of an interferometric system used to detect ultrasound.

B. *Light Gathering Efficiency and Étendue*

Most of the surfaces found in practice are not mirror-like and scatter incident light in various directions. In order to have a sensitive receiver, a large portion of this scattered light should be collected. Generally, the light collection efficiency of an interferometric system is characterized by its *étendue* parameter (or throughput), defined as the product of its effective entrance aperture area by the solid angle limited by the rays of maximum inclination. The maximum inclination rays can be defined as those which produce a shift of the interference pattern by a quarter of a fringe. The importance of this *étendue* parameter stems from its invariance within the frame of geometrical optics. This invariance is equivalent to the conservation of radiance between object and image [23] or to the Lagrange–Smith–Helmholtz equation [24] (this is not valid for large angles from the optical axis, in this case an elementary *étendue* should be introduced and the angular distribution of scattering by the illuminated surface should be taken into account).

The general rule is to match the *étendue* of the interferometer to that in the object (surface) space. The *étendue* in the object space is defined as the product of the illuminated spot area by the solid angle under which the collecting lens aperture is seen from the surface. To fail to satisfy to this condition has the consequence that, either the interferometer is not used to its full light gathering capability or the front lens is too large without necessity and or part of the illuminated area is not viewed by the system.

Matching the *étendue* in the object space to that of the interferometer determines only the product of the spot size by the solid angle, but neither of these parameters. In some cases, either for the purpose of averaging the measured displacement or to be less sensitive to surface topography, which, for a sufficiently small spot could cause all the light to be scattered off the receiving aperture, a value of the spot size will be chosen. In other cases, there is no condition on the spot size and this size could be diminished, down to the limit given by diffraction. This increases in consequence the radiance of the spot and with the *étendue* kept constant to the value given by the interferometer, increases the received light intensity (in other words, this means that a larger solid angle is used). It should be noted that, in some cases, the spot size being diminished the solid angle will be limited by the maximum receiving lens size practically feasible and the *étendue* in the object space could not match that of the interferometer. The situation where the object space *étendue* is slightly below that permitted by the interferometer is the most desirable, since it ensures well-defined receiving parameters and a more tolerant alignment of the whole optical system.

C. *Speckle Phenomenon*

This well-known phenomenon is caused by the interference of the multiple wavelets scattered by various

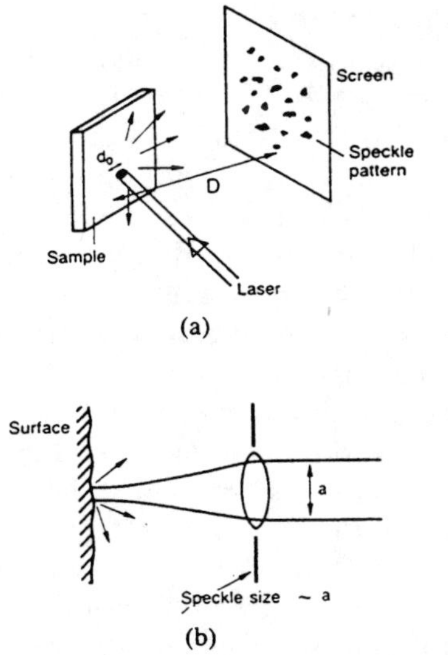

Fig. 4. Speckle phenomenon. (a) Observation of the effect. (b) Speckle formation when a beam is focused to a diffraction limited spot by a lens.

points of the illuminated spot [25]. It may be readily observed on a screen which receives the scattered light (see Fig. 4). The average speckle spot size on the screen is a function of the illuminated zone size diameter d_0 and of the distance D between the screen and the surface. It is approximately equal to $\lambda D/d_0$. When a beam of diameter a is focused to a diffraction limited spot on the surface with a lens, the speckle size on the lens aperture is then approximately equal to a. The intensity and the phase are roughly constant within one speckle spot and varies randomly further away.

If we let such a light field interfere with a reference wave or a second speckle field of the same average size, another speckle pattern is obtained. The intensity of each spot composing this speckle pattern can be written $a + b \cos(\Delta\phi(t) + \phi)$, where $\Delta\phi(t)$ is the phase change produced by the surface motion and where a, b, and ϕ are random parameters. This expression can be approximated by $a + b\cos\phi - b\Delta\phi(t)\sin\phi$ for small $\Delta\phi$. If N speckle spots are collected by the detector, a situation similar to a bidimensional random walk problem is encountered. The signal which is proportional to $\Delta\phi(t)$ has then a random magnitude. Its most probable value depends on speckle statistics. It can be crudely estimated to be $\sqrt{N}$ times the signal given by one speckle.

D. Signal-to-Noise Considerations

We will assume that the conditions for quantum limited detection are satisfied and that the performance is not limited by amplifier noise. In the case of low light levels, this could imply the use of a detector with internal gain such as a photomultiplier or an avalanche photodiode. A short discussion of noise problems can be found in another article of this journal [13].

The light intensity I_D received by the detector comprises two terms. The first one, noted I_S, to which the photocurrent signal i_s corresponds, is associated to the ultrasonic motion of the surface and vanishes when the ultrasonic displacement is zero. The second one, associated to the photocurrent i_0, is the residual part of the received intensity. In most cases encountered in practice, the ultrasonic displacements are much less than an optical wavelength, so I_s is small compared to I_0. The amplitude of the photocurrent signal i_s is given by

$$i_s = G\eta e\, I_s/h\nu \tag{1}$$

where η is the quantum efficiency of the detector, e is the electron charge, $h\nu$ is the energy of a quantum of light of frequency ν, and G is the internal gain of the detector. The i_0 is expressed by an equation similar to (1) with I_0 substituting I_s. The photocurrent noise (RMS) value is then given by (we assume that the amplification process does not introduce additional noise):

$$i_N = \sqrt{2\, GeBi_0} \tag{2}$$

where B is the detection bandwidth. The resulting signal-to-noise ratio is then

$$i_S/i_N = I_S\sqrt{\eta/(2Bh\nu I_0)}. \tag{3}$$

The limit of detection (δ or U minimum) is then determined by taking $i_s/i_N = 1$ and using the expression of I_S as a function of U or δ.

IV. Optical Heterodyning

As mentioned before, this technique consists in beating the wave reflected by the surface with a reference wave. The experimental configuration generally used is sketched in Fig. 5 and constitutes what is called a Michelson interferometer. Other configurations based on the other two-wave interferometers (Mach-Zehnder, Fiseau, ...) can be used as well. All of them have the common characteristic that the surface acts as one mirror of the interferometer. Without the frequency shifter, these probes are called homodyne and with it in either interferometer arm, these are called heterodyne (sometimes super-heterodyne).

A. Homodyne Interferometers

Without the frequency shifter in either arm (homodyne interferometer), the light intensity received by the detector can be written as follows:

$$I_D = I_L\{R + S + 2\sqrt{R}\sqrt{S}\cos[4\pi\,\delta(t)/\lambda - \Phi(t)]\} \tag{4}$$

where I_L is the laser power, R is the effective transmission coefficient in intensity for the beam in the reference arm, and S is that for the beam reflected off the surface (S is much less than unity for practical unpolished surfaces), $\Phi(t)$ is a phase factor that depends upon the interferometer path difference and is affected by vibrations. In practice, the reference mirror is mounted on a piezoelectric pusher, which enables to adjust the path difference in such a way that $\Phi = \pm\pi/2 + 2n\pi$ (n is an integer). It then follows that the signal is proportional to the *displacement*. In practice, electronic stabilization circuitry should be used to keep this adjustment in spite of ambient vibrations,

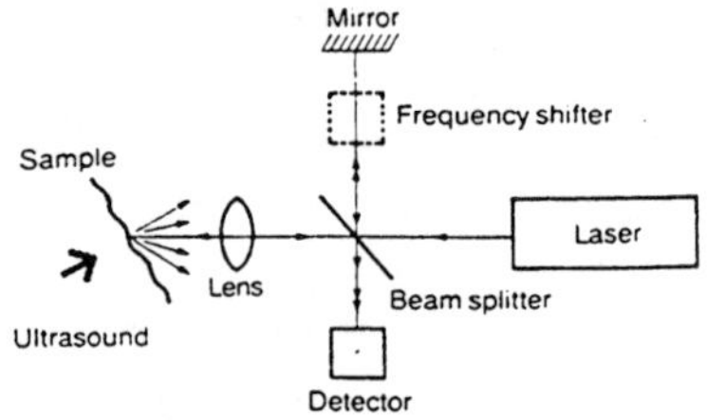

Fig. 5. Schematic of the Michelson interferometer probe. Without the frequency shifter, the probe is called *homodyne* and with it in either arm, it is called *heterodyne*.

which have amplitudes much larger than the ultrasonic displacements to be detected [26]–[34]. Faster compensation can be obtained with an electrooptic phase shifter [35]. The error signal for stabilization can be obtained by detecting the second harmonic of a dithering voltage applied to the piezoelectric pusher and using a lock-in amplifier tuned to the second harmonic (the zero of the second derivative of the signal corresponds to the $\phi = \pm\pi/2 + 2n\pi$ condition above). Another simpler stabilization scheme is based on the comparison of the dc output of the detector to that corresponding to the fringe midpoint ($\Phi = \pm\pi/2 + 2n\pi$, condition above).

These homodyne stabilized interferometers do not work well in presence of strong vibrations and often require for best operation to be located on a vibration-free table (except possibly the wider bandwidth compensation system of [35]).

Calibration is generally obtained in these systems by opening the stabilization loop and making the piezoelectric pusher to vibrate with an amplitude larger than $\lambda/2$. If the fringe pattern is not very stable, calibration should be performed frequently. This situation occurs with most surfaces (except mirrors) which scatter light and produce a speckle effect, except if a very stable setup is provided. A solution that has been experimented, but with additional complexity, is to dither constantly the piezoelectric pusher at a frequency above the response bandwidth of the stabilization loop [34]. It is also possible to obtain calibration without opening the stabilization loop by measuring the error voltage of the stabilization network [36]. This only requires to know the proportionality factor between displacement and observed voltage and vibrations amplitudes larger than $\lambda/2$.

Another method to eliminate the problem of vibrations is to vibrate continuously the reference mirror and to take a measurement at the proper time [37], but this is hardly applicable to short pulse ultrasound. It is also possible to have a system insensitive to vibrations by generating inside the Michelson interferometer two colinear optical beams in quadrature, which are detected by two separate detectors. As can be readily deduced by expanding (4) for small $\delta(t)$, two signals proportional to $\delta(t)\cos\Phi(t)$ and $\delta(t)\sin\Phi(t)$ are obtained by filtering out low frequencies. The $\delta(t)$ can be deduced by squaring the two signals [31], [38]. This system has a drawback to require critical detector adjustments (the detectors should see the same fringe pattern) and squaring at high frequency. Alternatively, it is possible to digitize the two signals and then, to compute the ultrasonic displacement [39].

In order to maximize the signal for a rough surface, the beam should be first focused to a diffraction limited spot on the surface. This gives a speckle spot size of the order of the incident beam diameter. Secondly, the reference mirror should be adjusted so that the reference wavefront is parallel to the average wavefront reflected by the surface. The first requirement can be understood as follows. When the surface is not at focus several speckles (let us say N) interfere with the reference beam. We have seen above that they add incoherently, so the response corresponds approximately at most to $\sqrt{N}$ times that of one speckle. Since in this case, the intensity of each speckle is on average $\approx 1/N$ that of the incident beam, the signal is smaller than at focus when $N \approx 1$.

We would like to mention that a solution to avoid the speckle problem would be to use a phase conjugate mirror [40] in association with a displacement interferometer such as described above. The phase conjugate mirror would receive part of the scattered light and send it back in such a way that light retraces its path and produces a uniform wave to interfere with the reference wave. No system using such a scheme is known to exist presently.

Finally, it should be noted that the detection bandwidth of such a detection scheme is given by the optical detector and extends from $\approx$ 50 KHz (displacements at lower frequencies are compensated by the stabilization loop) to 100 MHz and more. From the point of view of practical use, we note that the flexibility can be greatly enhanced by using a single-mode optical fiber in the probing arm [41].

B. Heterodyne Interferometers

If the detection is performed about a higher frequency by shifting the optical frequency in either arm (heterodyne interferometer), the signal from the detector ($-2\pi f_B t$ is added to the phase in (4)) includes a carrier at the shifting frequency f_B and two sidebands (as shown in Fig. 3, except that the spectrum is now centered on f_B). Calibration can readily be performed by comparing the sidebands amplitude with that of the carrier.

Many systems based on a Michelson heterodyne interferometer have been described in the literature and differ either by their optical configuration or electronic processing technique. They generally use a Bragg cell (acoustooptic cell excited at frequency f_B) to shift the optical frequency. The cell can be single pass, which gives a shift of f_B [42]–[49] or double pass (used as beam splitter and beam mixer), which gives a shift of $2f_B$ [50]–[54] (see Fig. 6).

By sending the two incident beams at proper angles, it is also possible to generate two angularly separated beams, one down-shifted and the other up-shifted [55]. Two Bragg cells can also be used: one is located in the reference arm and the other one is in the arm of the beam sent to the surface [56].

The simplest way to process the phase modulated signal at f_B, is to use a commercial spectrum analyzer [56] or a

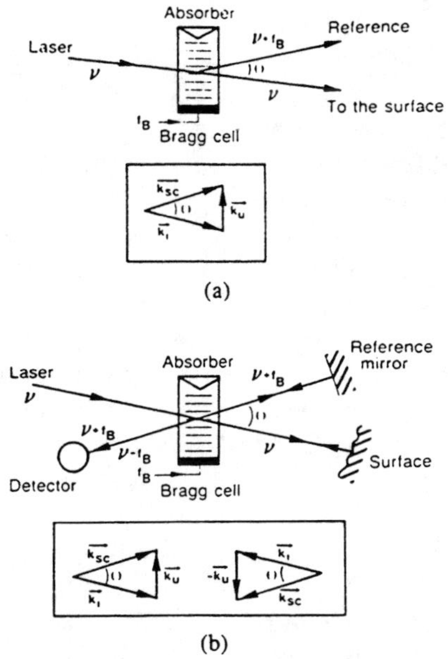

Fig. 6. Illustration of the use of a Bragg cell as beam splitter and beam splitter/combiner. (a) Single pass (beam combiner not represented, the insert shows the scattering diagram. (b) Double pass (the insert shows the scattering diagrams for the beam diffracted to the reference mirror and the beam from the surface diffracted onto the detector).

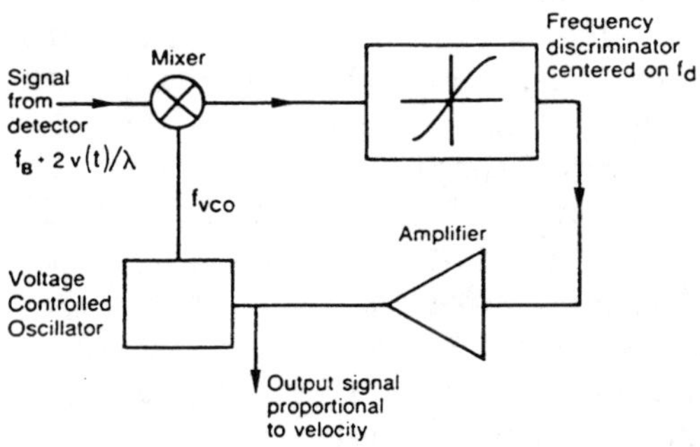

Fig. 7. Sketch of a frequency tracker giving a signal proportional to velocity.

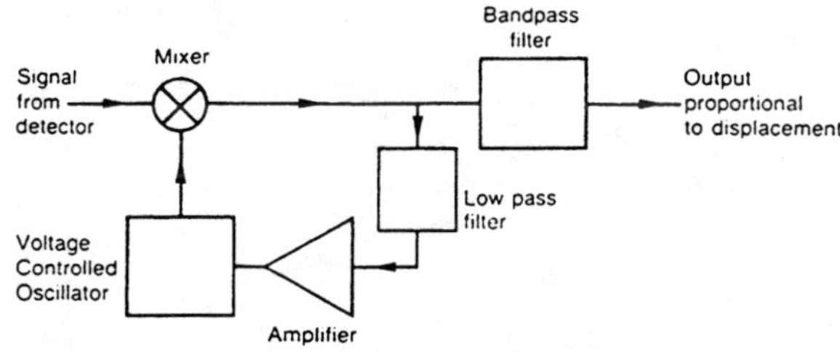

Fig. 8. Sketch of a frequency tracker giving a signal proportional to displacement (adapted from [48]).

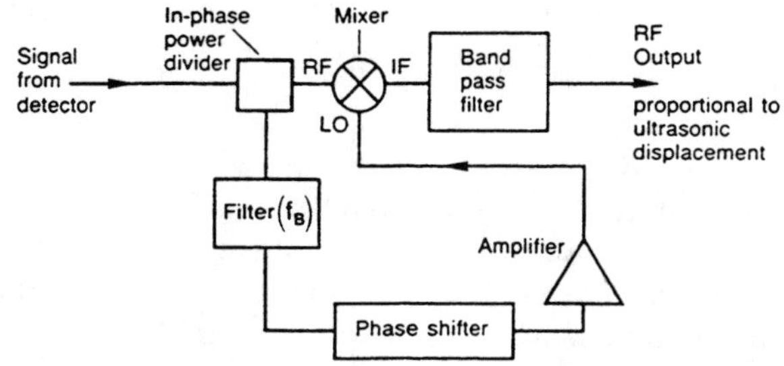

Fig. 9. Pulse broadband detection of both sidebands.

VHF receiver tuned to one of the sidebands followed by square law detection [44], [47]. In this case, the phase of the ultrasonic displacement is lost. Another way is to use standard FM demodulation [52], [53], (slope discrimination followed by diode detection and low-pass filtering) which gives a signal proportional to the surface velocity. The frequency response of such a detection scheme is limited approximately to the linear region of the discriminator. For higher frequencies, the response for constant displacement amplitude is not linear and levels off. A response proportional to surface velocity can be also obtained by a frequency-tracking technique [43], [57] described in Fig. 7. The frequency response of such a scheme is limited by the tuning capability of the voltage-controlled oscillator and the response time of the feedback loop. It appears for this reason mostly applicable to low ultrasonic frequencies and has been used for detecting vibrations [43].

It should be noted that the two modes of signal processing described above, which give an output proportional to the velocity, ensure in practice immunity to ambient vibrations since those occur only at low frequencies.

It is also possible with a frequency tracker to obtain a response proportional to the surface displacement [48], using the circuit described in Fig. 8. The operation of this mode of signal processing can be explained by expanding for small δ the equation for the heterodyne interferometer corresponding to (4):

$$I_D = I_L\{R + S + 2\sqrt{R}\sqrt{S}\,[\cos\,[(2\pi f_B t + \Phi(t)] + [4\pi\delta(t)/\lambda]\,\sin\,[2\pi f_B t + \Phi(t)]]\} \tag{5}$$

The voltage-controlled oscillator will lock to give zero dc signal at the output of the mixer, i.e., its output has to be proportional to $\sin\,[2\pi f_B t + \Phi(t)]$. Then the output bandpass filter (cuts low frequencies and frequencies around $2f_B$) gives a signal proportional to $\delta(t)$.

Another scheme to obtain a displacement signal for narrowband ultrasonic excitation [50], [55] consists of isolating the signal at f_B and one of the sidebands with filters and then mixing them. Since the vibrations have the same effect on the phase of the central frequency at f_B and the phase of the sidebands, a signal independent of the vibrations is obtained. It is also possible to use a commercial lock-in amplifier and to drive it with a reference signal derived from mixing the ultrasonic frequency with the central f_B frequency filtered out from the detector output [46]. If a signal derived from the Bragg cell generator at f_B is used instead of one derived from the detector, the output is observed to fluctuate constantly with vibrations.

It is also possible to detect both sidebands at the same time for broadband pulsed ultrasound with the scheme shown in Fig. 9 [58]. This scheme divides the detector signal into two in-phase components. The central peak at f_B is filtered out from one of them and then mixed to the other one after being given a phase shift of 90°. By reference to (5), it can be seen that, after eliminating with a filter high frequencies of the order of $2f_B$, a signal pro-

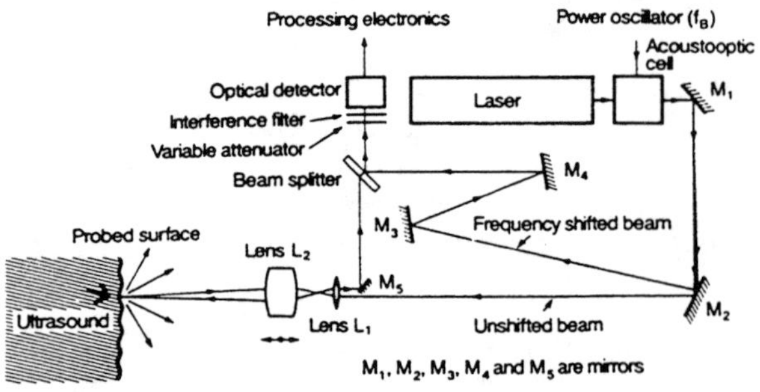

Fig. 10. Schematic of a heterodyne Michelson interferometer for the detection of ultrasonic displacements. The beam splitter is a bare glass slab which enables to maximize S (see (7)), for best sensitivity with a scattering surface (from [45], [46], and [58]).

portional to the displacement results without any effect of the vibrations. The f_B filter has a transient response at frequencies of the order of its bandwidth which is also eliminated by the output filter. The detection bandwidth of such a scheme is limited in practice by that of the output bandpass filter and by the value of f_B.

Finally, for all the schemes mentioned above, in order to have a response independent of the fringe signal amplitude and proper operation of the RF mixer, a constant amplitude signal is obtained from the detector output by amplification and hard clipping or automatic gain control.

An example of a heterodyne interferometer which maximizes the intensity sent to the surface for best signal-to-noise ratio (see below) and also enables to adjust the path difference (for multimode lasers the path difference should be $2n$ times, n being an integer, the laser cavity length) is shown in Fig. 10 [45], [46], [58].

C. Sensitivity and Detection Limit of Optical Heterodyning Techniques

We are considering below only the detection schemes which give an output proportional to the surface displacement. We discuss first the sensitivity of stabilized homodyne interferometers. Following (4), the detected signal intensity I_s, as defined above, is given by (stabilization at $\Phi = \pm\pi/2 + 2n\pi$):

$$I_S = 8\pi\delta(t) I_L \sqrt{R}\,\sqrt{S}/\lambda. \tag{6}$$

Since $I_0 = I_L(R + S)$, we obtain using (3) the following signal-to-noise ratio:

$$i_S/i_N = (4\pi\delta/\lambda)\,\sqrt{2I_L\eta/(Bh\nu)}\,\sqrt{RS/(R+S)}. \tag{7}$$

For a mirror-like surface with absorption coefficient A, this expression reduces to

$$i_s/i_N = (2\pi\delta/\lambda)\,\sqrt{I_L\eta(1-A)/(Bh\nu)} \tag{8}$$

where we have chosen the optimum reflection-transmission characteristics for the beam splitter, i.e., 50-percent reflection and transmission. Equation (8) gives for $I_L = 5$ mW, $\eta = 0.5$, $\lambda = 6328$ Å, $A = 0$, and $B = 10$ MHz a detection limit of $\sim$0.04 Å. This value corresponds approximately to 10^{-6} Å$(\mathrm{W/Hz})^{1/2}$.

For a rough surface, the reference beam is generally more intense than the beam reflected off the surface and the last factor in (7) reduces to $\sqrt{S}$. In this case the signal-to-noise ratio increases with S. Best sensitivity is obtained, therefore, when S has been maximized, i.e., when nearly all the laser intensity is sent onto the surface and only one highly transmitting beam splitter is used (see Fig. 10). We have seen that the signal is maximized when one speckle is detected. Using the notations of Section III-C, the solid angle of one speckle, is approximately equal to $\pi(a/2D)^2$. Assuming isotropic scattering and neglecting the losses produced by various optics, which should be small for best sensitivity, as mentioned above, we then find that the S parameter, which when multiplied by I_L gives the power in one speckle, is given approximately by

$$S \approx (1 - A)\,(a/2D)^2. \tag{9}$$

Equations (7) and (9) give for the preceding parameters with $a \simeq 4$ mm, $D \simeq 15$ cm a detection limit of $\simeq 1$ Å. In practice a surface rarely scatters isotropically and a speckle brighter than average can be found, so a lower limit can be expected. This limit can be lowered by using higher laser power (the detection limit scales as $1/\sqrt{I_L}$) or a closer working distance D (the detection limit scales as $1/D$), but closer focusing produces a smaller spot size and a shorter depth of focus.

These results are consistent with the antenna theorem for heterodyne detection [59], [60] which states that the effective *étendue* of the receiving system is λ^2 at most. For an illuminated spot radius w on an isotropically scattering surface the brilliance is $I_L(1 - A)/\pi^2 w^2$. This expression multiplied by the *étendue* λ^2 gives the effective received power for detection according to the antenna theorem. We see that the power is maximized for $w = w_0$, where w_0 is the diffraction limited spot radius. In this case, one speckle spot beats with the reference beam. Since $w_0 \sim 2\lambda D/\pi a$ we retrieve (9). It is useful for further comparison to rewrite (7) (in the limit $R >> S$) using the *étendue* parameter E:

$$i_S/i_N = (4\delta/\lambda)\,\sqrt{2\pi I_L(1 - A)E\eta/(Bh\nu S_I)} \tag{10}$$

where S_I is the illuminated spot area on the surface. Equation (10) can also be expressed with the solid angle subtended by the receiving aperture as seen from the surface. This solid angle Ω is equal to E/S_I. The detection limit calculated previously for a nonabsorbing isotropically scattering surface corresponds approximately to a limit of 10^{-6} Å$(\mathrm{W\ sr/Hz})^{1/2}$.

When a heterodyne interferometer is used a slightly reduced signal-to-noise is encountered. This originates mostly from the detection scheme (Fig. 8 or Fig. 9), which uses only half of the power produced by the mixing process by eliminating frequencies components around $2f_B$. It can be shown that the proper equation for the signal-to-noise ratio is (7) with $\sqrt{RS/(R+S)}$ replaced by $\sqrt{RS}/[\sqrt{2}(\sqrt{R} + \sqrt{S})]$, which reduces to $\sqrt{S/2}$ for $R >> S$. When a single sideband is detected [46], this expression should be further halved. A detection limit identical to the one found above for the homodyne interferometer is obtained for a mirror-like nonabsorbing surface and 50-percent reflecting-transmitting beam splitter. For the nonabsorbing isotropically scattering surface, the value found above should be multiplied by $\sqrt{2}$.

Finally, we want to mention that the limits of detection calculated in this section can be obtained in practice and they have been roughly verified [61], [62], [46]. When discrepancies have been reported [50], [55], it is suspected that the theoretical assumptions were not verified by the experiment system.

D. *Other Interferometric Methods*

The laser itself is made of an interferometer and can be used to detect ultrasonic motion by feedback upon it. A simple He–Ne laser can demonstrate easily the effect [63], which should exist with any laser, e.g., laser diodes [64]. It seems, however, difficult to make very quantitative measurements because of the complexity of the problem which involves three coupled cavities and a nonlinear medium (the laser).

Instead of using two-wave interference, a multiple-wave interference phenomenon can be used as well by making the surface one of the mirrors of a Fabry–Pérot interferometer [65]. Very large sensitivity can be obtained, but the surface has to be nicely polished and the alignment is critical.

V. Differential Interferometry

As mentioned previously, two wavefronts are made to interfere: they could be issued from one single point on the surface illuminated by two distinct beams (differential interferometry for in-plane measurement) or two distinct points (differential displacement interferometry).

A. *Differential Interferometry for In-Plane Measurement*

This technique is similar to differential Doppler anemometry [66] and is illustrated in Fig. 11. As seen in Fig. 11, the probe is made by having two beams, issued from the same laser and separated by an angle 2θ, focused at the same location on the surface. A grating is then produced on the surface, and part of the speckle field scattered from the surface is received by a detector. The grating is fixed when the frequency of the two beams is the same (homodyne probe) and moving when it is heterodyne. If N speckles of mean intensity I_{sp} are seen by the detector, the detected intensity in the case of the heterodyne probe can be written (the speckles add up incoherently):

$$I_D \sim \sqrt{N} I_{sp} \cos\left[2\pi f_B t + (\vec{k}_1 - \vec{k}_2) \cdot \vec{\delta}(t) + \phi(t)\right] \tag{11}$$

where f_B is the shifting frequency ($f_B = 0$, homodyne probe); $\vec{k}_1$ and $\vec{k}_2$ are the wavevectors of the two beams; and

$$(\vec{k}_1 - \vec{k}_2) \cdot \vec{\delta} = 4\pi \sin\theta(\sin\alpha\, \delta x + \cos\alpha\, \delta z)/\lambda$$

where α is the angle between $\vec{k}_1 - \vec{k}_2$ and normal-to-the-surface, and δx and δz are, respectively, the in-plane and out-of-plane displacements. As seen previously, when the frequency is shifted, this expression gives a carrier signal at f_B and two sidebands, which enables easy calibration. In-plane and out-of-plane displacements can be detected, but this probe is particularly useful to detect displacements parallel to the surface (in this case $\vec{k}_1 - \vec{k}_2$ is perpendicular to the surface and $\alpha = \pi/2$).

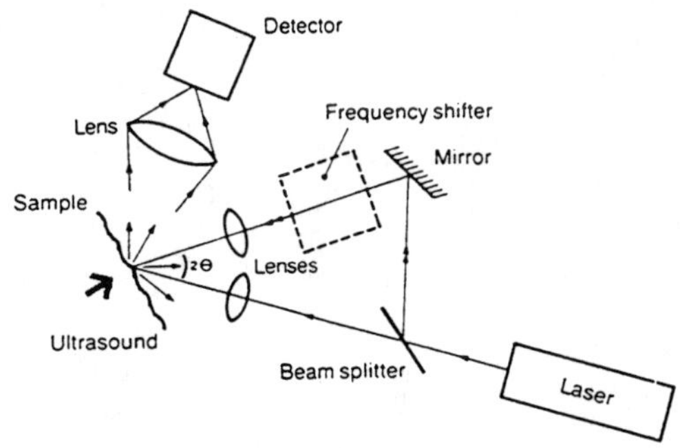

Fig. 11. Schematic of the differential interferometric probe (for in-plane detection). The probe is homodyne without the frequency shifter and heterodyne with it.

When both sidebands are detected, (or in the case of a homodyne probe) the detected signal intensity I_s, as previously defined, is given by

$$I_s \sim \sqrt{N} I_{sp} (\vec{k}_1 - \vec{k}_2) \cdot \vec{\delta}. \tag{12}$$

Since $I_0 \sim N I_{sp}$ and using (3), a signal-to-noise ratio approximately independent of number of speckles is obtained:

$$i_S/i_N = (\vec{k}_1 - \vec{k}_2) \cdot \vec{\delta} \sqrt{\eta I_{sp}/(2Bh\nu)}. \tag{13}$$

Equation (13), also shows that, in order to improve the signal-to-noise ratio, I_{sp} should be maximized, which means that the speckle size should be of the order of the detection aperture and consequently that this aperture should be large enough to collect a large part of the scattered light. Accordingly, this means very sharp focusing. When typical numerical values are used, a detection limit of the same order as the one found in Section IV is obtained. However, if we want the system not to be too bulky, θ should be small which could reduce the sensitivity.

It is possible to construct a probe which, by a simple change, could detect the normal displacement (as in Section IV) and the in-plane displacement (differential configuration) at the same point [56], [67]. A broadband probe with such a capability has been recently used to compare the normal and in-plane displacements of a Rayleigh surface wave, giving results in agreement with theory [68].

B. *Differential Displacement Interferometry*

This technique consists of focusing at two distinct locations on the surface, then to produce an interference pattern from the two point sources and to detect it through a Ronchi grid (see Fig. 12) [69], [70].

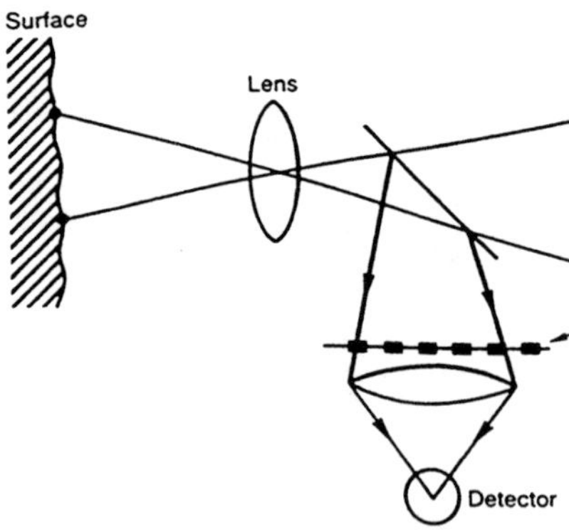

Fig. 12. Principle of differential displacement interferometry. The two point sources S_1 and S_2 are derived from the same laser.

This technique is very insensitive to ambiant disturbances, but detects only the differential normal displacement between the two illuminated points. Its detection bandwidth in the case of the detection of surface waves is a function of the spacing between the two points.

VI. Velocity or Time-Delay Interferometry

We have seen in Section IV-B that a signal proportional to surface velocity could be obtained by using a frequency discriminator, which could be a RF filter used on its slope, to demodulate the heterodyne signal. The same operation could be done directly at optical frequencies by using an optical filter. As mentioned in Section II-D, filters that are based on absorption of light do not generally have a sufficiently steep slope to be useful (except if saturated absorption is used). Steep slopes and high discrimination sensitivity can be obtained by optical interferometry. The basic setup is shown in Fig. 13.

As mentioned in the background section, the basis of the technique consists of making the wave scattered by the surface interfere with itself after some time delay. In practice, the delayed wavefront does not match exactly the initial wavefront and limitations on the size of the useful area of the wave and on its divergence are encountered. These limitations are related to the concept of *étendue* of the interferometer and a sensitive system will be such that they have been minimized to give a large light-gathering capability.

The interferometers which can be used can be classified into two types: 1) two-wave interferometers, e.g., Michelson and Mach-Zehnder and 2) multiple-wave interferometers, e.g., Fabry-Pérot. For all these devices, since the signal is (in the first approximation) proportional to velocity, there is excellent immunity to ambient vibrations.

A. Velocity Interferometry with a Two-Wave Interferometer

A typical setup is sketched in Fig. 14. We note the main difference between this setup and the one of Fig. 5: here the interferometer does not make use of the surface as a mirror, but views the light scattered by it. This type of

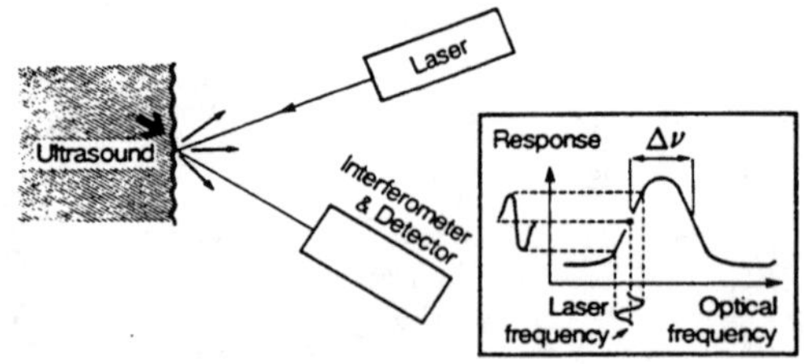

Fig. 13. Detection of ultrasonic motion using an interferometer working as an optical frequency discriminator. The insert indicates the principle of frequency demodulation.

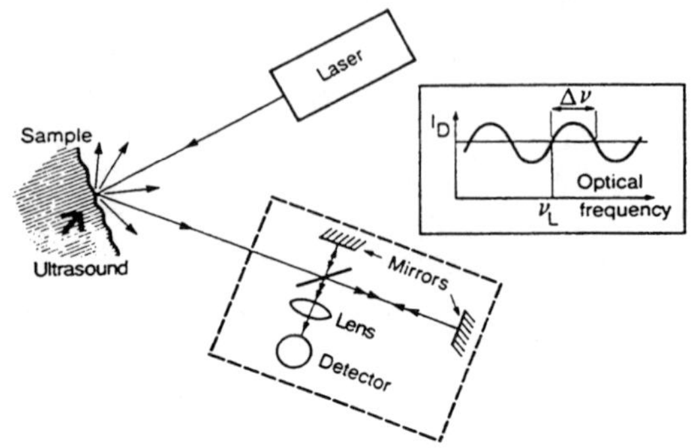

Fig. 14. Schematic of the Michelson velocity interferometer. The insert indicates the response of the interferometer to optical frequency.

system has been used in shock-wave research [71] and has been also considered for the detection of ultrasound [72].

The received intensity for a given direction θ of an incident ray is

$$I_D = A_1 + A_2 \cos (2\pi\nu/\Delta\nu_{FS} + \phi) \qquad (14)$$

where A_1, A_2, and ϕ are constant, the free spectral range $\Delta\nu_{FS} = 1/\tau = c/2\Delta d(\theta)$, τ is the delay time, and $\Delta d(\theta)$ is the difference of arm lengths for the incident ray inclined by θ. As sketched in the insert of Fig. 14, which plots the spectral response given by (14), frequency discrimination is obtained when the laser frequency (a single frequency laser source is necessary) is tuned to a zero crossing:

$$2\pi\nu/\Delta\nu_{FS} + \phi = \pm\pi/2 + 2m\pi$$

where m is an integer.

The surface being rough and irregular acts as an incoherent source, so the fringes are only observed at infinity [73] or at the focus of a lens. It can be readily shown that $\Delta d(\theta) = \Delta d \cos \theta$, where Δd is the path difference for the ray perpendicular to the mirrors. The fringes are then concentric rings, and in practice the central fringe is selected by using a circular hole at the lens focus (at the expense of a much higher complexity a mask pattern matching the fringes can be used). Taking a maximum path difference of $\lambda/4$, it follows a limit angle $\theta_M \simeq \sqrt{\lambda/4\Delta d}$. In order to be able to detect ultrasound in the range 1–10 MHz, the free spectral range $\Delta\nu_{FS}$ (the bandwidth $\Delta\nu$ is defined as $\Delta\nu_{FS}/2$) should not be too large: taking $\Delta\nu_{FS} = 25$ MHz yields $\Delta d = 6$ m (even if folding mirrors are used, the system will be large and bulky) and $\theta_M \simeq 2\ 10^{-4}$ rd, for $\lambda = 1.06\ \mu$m. Using an entrance aperture of 10 cm in diameter (and mirrors of this size), the system has an

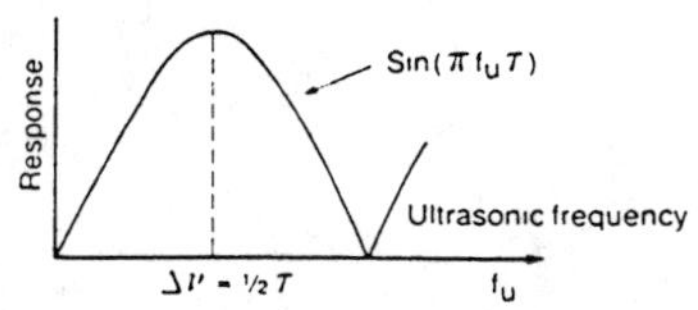

Fig. 15. Response of the Michelson velocity interferometer to ultrasonic frequency (the laser frequency is set at a zero crossing, as shown in the insert of Fig. 14). This response is periodic with period equal to $2\Delta\nu$.

étendue of $\simeq 10^{-3}$ mm^2 sr^2. This is much larger than the optical heterodyne systems described in Section IV, but not sufficient in many cases. A spot 4 mm in diameter viewed through a six-inch lens at 1.5 m requires an *étendue* of 0.1 mm^2 sr. Fortunately, there is a known way to increase the *étendue*, and it will be presented below after a discussion of the time and frequency response.

The time response of this interferometer can be understood by considering that the pathlength from the laser source is changed by $\delta(t)$ for one arm and by $\delta(t - \tau)$ for the other [74]. Then, when the laser frequency is properly adjusted to the zero crossing, the detected signal intensity I_s is given by ($\delta << \lambda$):

$$I_s \simeq A_2 4\pi[\delta(t) - \delta(t - \tau)]/\lambda. \quad (15)$$

After substitution of $\delta(t) = U \cos(2\pi f_u t + \psi)$, this expression shows that the frequency response of this probe to ultrasonic frequencies is proportional to $S(f_u) = 2 \sin \pi f_u / 2\Delta\nu = 2 \sin \pi f_u \tau$ (sketched in Fig. 15). The response is therefore only linear at low frequencies when $f_u << \Delta\nu$. In this case, it can readily be shown that the detected signal is proportional to the surface velocity and to the Doppler shift and that it can simply be obtained by differentiating (14). It should also be noted that the interferometer which produces the maximum response for a given ultrasonic frequency f_u has a bandwidth $\Delta\nu$ equal to f_u (i.e., a delay time equal to half the ultrasonic period). Identical results can also be obtained by an analysis in the frequency domain, where the interferometer is considered as a linear filter with $1 + \exp(-i2\pi\nu\tau)$ as transfer function and where the laser frequency plus the two sidebands are applied at the input. Other two-wave interferometers such as the Mach–Zehnder [75]–[77] can be analyzed in the same way and give similar results.

The *étendue* of the Michelson interferometer can be greatly enhanced by having in the longest arm a slab of index n (glass, liquid) in such a way that the two mirrors are imaged from each other through the beam splitter and the slab [78], [79]. In this case, the two emerging rays originating from the same incident ray are superimposed in first approximation (see Fig. 16). The path difference Δ is then equal to $2h(n - 1/n)$, where h is the slab thickness. The maximum ray inclination to the axis is limited by the third order spherical aberration of the slab (wavefront deformation $= \theta^4\Delta/8n^2$). This gives a maximum inclination $\theta_M = (2n^2\lambda/\Delta)^{1/4}$. Using the same numerical values as before ($\Delta\nu_{FS} = c/\Delta$) and $n = 1.5$, it follows $h = 7.2$ m and $\theta_M \simeq 1.5°$. If the mirrors have a size sufficient to accommodate this maximum inclination, it then

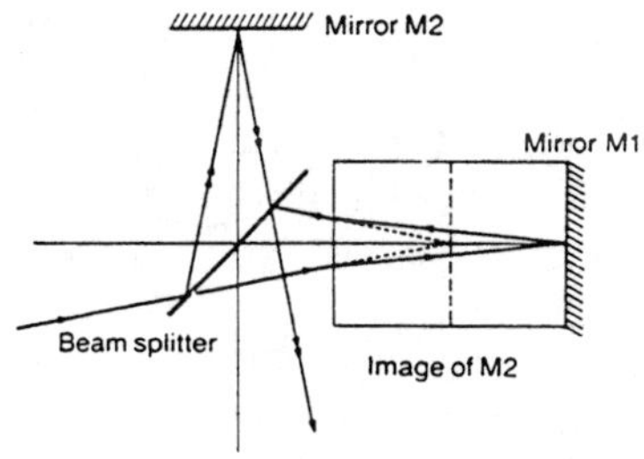

Fig. 16. Schematic of the Michelson interferometer with increased étendue.

results in a very large *étendue* (>>1 mm^2 sr). If the viewed area is a $\frac{1}{4}$ inch in diameter at 2 m, this *étendue* will then require in order to be totally used, a front lens too large to be feasible. Therefore, it can be calculated that the *étendue* will be limited by the viewed area and the front optics size ($\simeq 0.06$ mm^2 sr for the example chosen), and that there is no need to increase further that of the interferometer (as explained in [80]). Such wide-angle Michelson interferometers (WAMI) have been used in shock-wave research [81] and anemometry [82].

The sensitivity of a two-wave velocity interferometer is now evaluated. We assume a beam splitter with 50-percent reflection and transmission and neglect any other reflection losses in the interferometer. The average intensity received I_0, for stabilization at half maximum, is given by

$$I_0 = A_1 = A_2 = LE/2 \quad (16)$$

where L is the radiance (luminance) of the viewed area and E is the *étendue*. We find

$$L = I_L(1 - A)/\pi S_I \quad (17)$$

with the same notations as before. We have assumed isotropic scattering and that all the light from the illuminated spot passing through the input lens aperture is received by the detector. We assume further a surface displacement $\delta = U \cos(2\pi f_u t + \psi)$. The detected signal intensity at frequency f_u is

$$I_s = I_0 S(f_u)\,(4\pi U/\lambda) \quad (18)$$

where, in this case, from (15), $S(f_u) = 2 \sin(\pi f_u/2\Delta\nu)$. It then follows using (3) a signal-to-noise ratio:

$$i_s/i_N = 4 \sin(\pi f_u/2\Delta\nu)\,(U/\lambda)\,\sqrt{\pi I_L(1 - A)\,E\eta/(h\nu B S_I)}. \quad (19)$$

As noted previously, the sensitivity is best when $f_u = \Delta\nu$. For this value, we note that (19) is approximately the same as (9) of the heterodyne technique. By comparing those two equations, we can deduce that, for a given illuminated area S_I, the Michelson velocity technique is better since the *étendue* is not limited to λ^2. It is however possible to obtain about the same signal-to-noise ratio with the heterodyne technique if the incoming beam has the same size as the receiving aperture of the velocity interferometer and is focused onto the surface to a diffraction limited spot (i.e., the speckle size is of the order of the receiving aperture and $E/S_l = \Omega$, where Ω is the solid angle and is the same in both cases). It should be noted that, in this case, the heterodyne signal will be very sensitive to the position of the target surface and if it is moving, the signal will fluctuate constantly. Also because of the sharp focusing, the depth of field will be small. The velocity interferometric technique does not have these limitations since many speckles are collected and therefore should be preferred for industrial applications. Taking the same Ω as before (e.g., aperture 40 mm in diameter, distance = 1.5 m) and the same intensity (5 mW at 6328 Å), we find the same limit of $\simeq 1$ Å, which corresponds to 10^{-6} Å$(\text{W sr/Hz})^{1/2}$. For a mirror-like surface, the appropriate expression for the signal-to-noise is (19) multiplied by $\sqrt{\pi/\Omega}$. A detection limit approximately the same as the one found for heterodyne detection is also obtained.

This detection limit assumed a very stable single-frequency laser source, both in amplitude and frequency. In practice, the laser should be stabilized with respect to the interferometer. For a source which has good stability, simple piezoelectric control of the laser cavity length or the interferometer path difference is appropriate. For more unstable lasers, the fluctuations of amplitude and frequency could be compensated by dividing or subtracting electronically the signal coming from the sample with a signal representative of these fluctuations [83].

Other velocity two-wave interferometers, which could be used in principle for the detection of ultrasound have been reported in the literature: one is based on optical polarization interference in birefringent crystals [84] and the other on spectral coding or spectral modulation [85].

This technique consists of using a broad band source to illuminate an interferometer. The spectrum of the light emerging from the interferometer is modulated with a period $\Delta\nu_{FP}$ equal to $1/\tau$, where τ is the delay time of the interferometer. This light is reflected by the target surface into the same interferometer and then detected. When the target moves, the spectrum of the reflected light is shifted by the Doppler effect, and a signal indicative of the target velocity is detected. As might be expected, since the same signal should be obtained independently of the sign of the velocity, the response is quadratic for small velocities and practical delay times, which result in poor sensitivity.

Such a concept has been recently put into a useful device by using fiber optics technology [86], [87], which enables long delay times (see Fig. 17). By using either a phase modulator on a fiber arm or using nonreciprocal effects of fiber birefringence, a linear signal is obtained. The amplitude of the signal depends greatly upon the surface reflectivity into the fiber. This problem is solved by dividing the signal by the one at the modulator frequency [88] (or twice the modulator frequency). Being sensitive to velocity this device is immune from ambient disturbances. Single-mode fibers should be used and since their core diameter is less than 10 μm, the *étendue* is $\simeq$ or less than 10^{-4} mm^2 sr, which is more than given by optical heterodyning techniques but orders of magnitude less than provided by bulk *étendue*-widened Michelson or Mach–Zehnder interferometers (or confocal Fabry–Pérot, as follows).

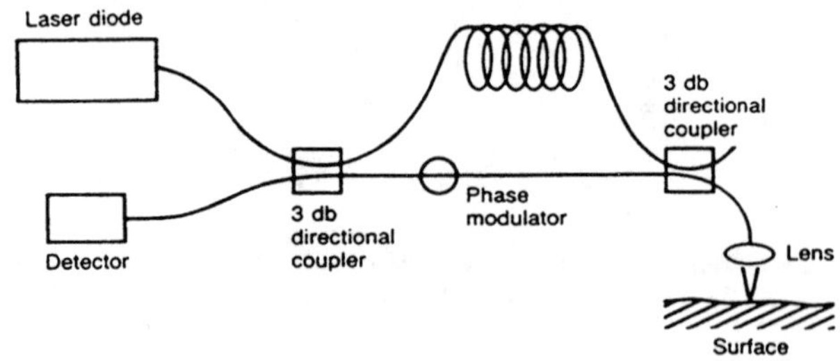

Fig. 17. Schematic of the All-Fiber Optic Sensor (adapted from [86] and [87]).

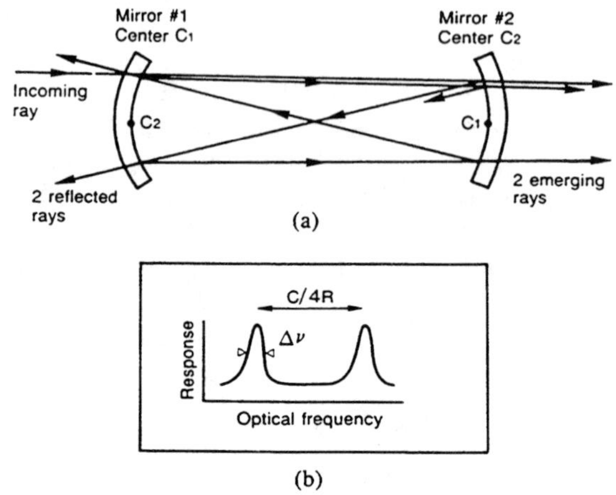

Fig. 18. (a) Schematic showing the principle of the confocal Fabry–Pérot. Only the two rays emerging first are represented for clarity. (b) Response to optical frequency.

B. Velocity Interferometry with Multiple-Wave Interferometers (Fabry–Pérot)

We have seen that in order to obtain adequate sensitivity at low ultrasonic frequencies with a two-wave velocity interferometer long path length differences are required (a bandwidth of 1 MHz requires a difference of 75 m). A solution is to use multiple-beam interference, i.e., a Fabry–Pérot interferometer. In this case, the interferometer bandwidth can be decreased nearly at will by choosing suitable mirror reflectivity (the higher the reflectivity, the narrower the bandwidth).

Fabry–Pérot velocity interferometers have been used previously in anemometry [89]–[91] and in shock-wave research [92]. For the narrow bandwidths of interest for the detection of ultrasound, a confocal Fabry–Pérot should be preferred to a planar one since it has a much larger *étendue* [93], [94]. Contrary to what has been stated (20), a planar Fabry–Pérot has approximately the same light gathering efficiency as a simple Michelson interferometer (not *étendue*-widened) for the same bandwidth. Large *étendue* is obtained by having the two mirrors confocal: as seen in Fig. 18, all the emerging rays overlap after two round trips (in the paraxial approximation). A large circular fringe is observed at the output, whose size is determined by the third order spherical aberration of the system.

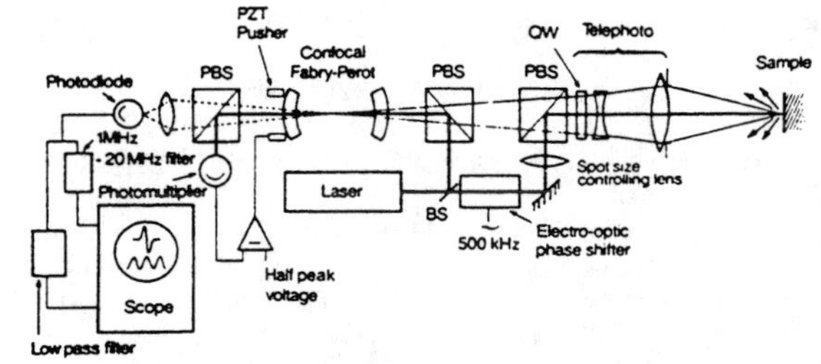

Fig. 19. Experimental setup to detect ultrasound using a confocal Fabry–Pérot interferometer: beamsplitter (BS); polarizing beamsplitters (PBS) (reflect nearly 100 percent of the light polarized perpendicularly to the plane of the drawing and transmit nearly 100 percent of the light polarized in the plane of the drawing); quarter-wave (QW) plate oriented at 45° to the laser polarization. A lens is used to produce the proper spot size on the sample through the telephoto (from [95]).

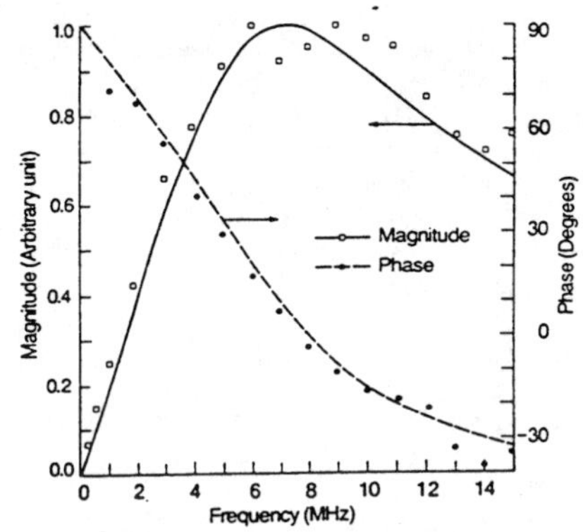

Fig. 20. Ultrasonic frequency response (magnitude and phase) of a Fabry–Pérot receiver (R = 50 cm, $\Lambda\nu$ = 10 MHz) with experimental verification data points (from [97]).

In the confocal Fabry–Pérot, the bandwidth and the *étendue* are interrelated [93], [94]: $E\nu/\Delta\nu \simeq 2\pi^2R^2$, where R is the radius of the mirrors. Therefore, adequate light gathering capability may require an interferometer sufficiently long. Setting $\Delta\nu$ = 10 MHz, R = 50 cm, λ = 0.6328 μm gives $E \simeq 0.1$ mm^2 sr^2, which enables to receive all the light through a lens 6 inches (15 cm) in diameter coming from a spot $\simeq$4 mm in diameter located on a target at 1.5 m from the lens. A confocal Fabry–Pérot system with these parameters has been realized [95] and is sketched in Fig. 19. In this system, real-time calibration is obtained by passing the incident beam through an electrooptic phase shifter, so the signal observed is independent of the target orientation and reflectivity.

The frequency response of the Fabry–Pérot velocity interferometer is not as simple to calculate as that of a two-wave interferometer. It has been calculated [96], [97] and is plotted for the parameters given above in Fig. 20, together with experiment verification data.

The sensitivity of such a receiver is now evaluated. When the laser frequency is stabilized at half height of a peak and when the two output beams are collected by the detector, the average received intensity is $I_0 \simeq LE/4$ (this assumes a mirror reflectivity close to 1 and that all the light from the illuminated spot passing through the input lens aperture is received by the detector). The detected signal intensity at f_u is given by (18), where $S(f_u)$ is equal to $2f_u/\Delta\nu$ for $f_u << \Delta\nu$. The signal-to-noise ratio is then given by

$$i_s/i_N = S(f_u)\,(U/\lambda)\,\sqrt{2\pi I_L(1-A)\,E\eta/(h\nu BS_l)}. \quad (20)$$

Since the maximum of S is $\simeq 1$, (20) gives a signal-to-noise ratio $\simeq 1/2.8$ of that given by the two-wave interferometer (19), i.e., for the same Ω as before (aperture 40 mm in diameter at a distance of 1.5 m) and the same intensity (5 mW at 6328 Å) the detection limit is $\simeq$3 Å for an isotropically scattering nonabsorbing target. This value corresponds to 3 10^{-6} Å(W sr/Hz)$^{1/2}$. We should note that, when the two emerging beams are made to interfere constructively the signal is doubled and the detection limit is reduced by $\sqrt{2}$. As mentioned above, in practice, many surfaces have a more directional scattering distribution, so the detection limit is smaller. For comparison, for a mirror-like nonabsorbing surface, this limit will be multiplied by $\sqrt{(1-A)\,\Omega/\pi}$, which is $\frac{1}{75}$ for the example given.

VII. Applications

Although the purpose of this review was not to describe all the applications of ultrasonic optical detection, but rather to focus on the detection techniques, several applications which have been described in the literature are mentioned below. It is expected that these techniques will be of wider use in the future owing to the development of more powerful, more stable, and higher repetition rate lasers.

Besides being at the basis of the operation of the SLAM [10], [11] mentioned earlier, optical detection of ultrasound has been used for the characterization of SAW devices [6]–[9], [12], [47], [50], [65], [87], the characterization of ultrasonic transducers [31], [35], [45], the study of the interaction of ultrasound with defects [52], [98], [99], the detection of acoustic emission [28], [30], and ultrasonic field mapping [12], [37], [58]. In association with laser generation, it has been used in the laboratory for nondestructive evaluation of materials, such as the measurement of elastic constants of composites [100], anisotropy and acoustoelectric coupling constant of piezoelectric materials [101] and for flaw detection [61], [102]–[105]. We note also the use of similar techniques in photothermal detection [15], [16], [106]. The application of ultrasonic optical detection in industrial environments where the surfaces are rough and absorbing, where vibration and turbulence levels are high, has not been so far reported, but appears possible if a detecting device with proper sensitivity and immunity to environmental disturbances is used with a laser source of adequate stability and sufficient power.

VII. Conclusion

Following this review of various optical techniques to detect ultrasound, the main conclusion is that all the techniques (knife-edge, optical heterodyning, differential interferometry for in-plane measurement, velocity interferometry, and other less important techniques having not been analyzed in depth) have approximately the same sensitivity. The detection limit amounts to about 10^{-6} Å (W/Hz)$^{1/2}$ for a mirror-like surface and about 10^{-6} Å (W sr/Hz)$^{1/2}$ for an isotropically scattering surface (except knife-edge). These figures apply for a nonabsorbing surface illuminated by visible light ($\lambda = 6328$ Å). When another wavelength λ is used, they are multiplied by $\lambda/6328$. When the surface absorbs the fraction A of incident light, they are multiplied by $1/\sqrt{1 - A}$. These figures also assume that all the techniques are used properly at their optimum capability for the intended use.

The choice of a particular technique for a given application is not therefore a matter of sensitivity. However, the various techniques we have described are not equivalent and some are better than others according to the intended use:

Laboratory Applications on Polished Surfaces: In this case, the knife-edge technique comes ahead, because of its simplicity, its insensitivity to vibrations, and its broad detection bandwidth.

Laboratory Applications on Scattering Surfaces: Optical heterodyning should be preferred because of its broad detection bandwidth. As we have seen, it is possible to build a system which permits the detection of normal and in-plane displacements at the same location on the surface, which may be necessary for some applications. Compensation for vibrations is better accomplished with a heterodyne probe (super-heterodyne technique) than with a stabilized homodyne one.

Industrial Applications: The requirements of good immunity from vibrations and air turbulence, and the requirement of insensitivity to precise alignment and system setting indicate that velocity interferometry which enables to receive many speckles, should be preferred. The system based on a confocal Fabry–Pérot has the advantage to enable to tailor easily the reception bandwidth to the ultrasonic frequencies to be detected. The systems based on two-wave interferometers which are very bulky for optimum detection in the MHz range are suitable for high frequency applications (>100 MHz).

References

[1] C. B. Scruby, R. J. Dewhurst, D. A. Hutchins, and S. B. Palmer, "Laser generation of ultrasound in metals," in *Res. Techniques in Nondestructive Testing* (vol. 5), R. S. Sharpe, Ed. New York: Academic, 1982, pp. 281–327.

[2] D. A. Hutchins, "Ultrasonic generation by pulsed lasers," in *Physical Acoustics*, vol. 18, W. P. Mason and R. N. Thurston, Ed. New York: Academic, to be published.

[3] D. M. Marsh, "Methods of visualizing ultrasound," in *Res. Techniques in Nondestructive Testing* (vol. 2), R. S. Sharp, Ed. New York: Academic, 1973, p. 317–367.

[4] P. A. Fleury, "Light scattering as a probe of phonons and other excitations," in *Physical Acoustics* (vol. 6), W. P. Mason, Ed. New York: Academic, 1970, p. 1–64.

[5] G. I. Stegeman, "Optical probing of surface waves and surface wave devices," *IEEE Trans. Sonics Ultrason.*, vol. SU-23, p. 33, 1976.

[6] R. Adler, A. Korpel, and P. Desmares, "An instrument for making surface waves visible," *IEEE Trans. Sonics. Ultrason.*, vol. SU-15, p. 157, 1967.

[7] G. Alers, M. A. Tennison, R. B. Thompson, and B. R. Tittmann, "Visualization of surface elastic waves on structural materials," *Ultrasonics*, vol. 11, p. 174, 1973.

[8] H. Engan, "Phase sensitive laser probe for high-frequency surface acoustic wave measurements," *IEEE Trans. Sonics. Ultrason.*, vol. SU-25, p. 372, 1978.

[9] —, "A phase-sensitive laser probe for pulsed SAW measurements," *IEEE Trans. Sonics. Ultrason.*, vol. SU-29, p. 281, 1982.

[10] L. W. Kessler and D. E. Yuhas, "Acoustic Microscopy-1979," *Proc. IEEE*, vol. 67, p. 526, 1979.

[11] Manufactured by Sonoscan, Bensenville, Illinois, USA.

[12] R. L. Whitman and A. Korpel, "Probing of acoustic surface perturbations by coherent light," *Appl. Opt.*, vol. 8, p. 1567, 1969.

[13] H. Sontag and A. C. Tam, "Optical detection of nanosecond acoustic pulses," *IEEE Trans. Ultrason., Ferroelectrics, Freq. Contr.*, vol. UFFC-33, no. 5, Sept. 1986.

[14] —, "Optical monitoring of photoacoustic pulse propagation on silicon wafers," *App. Phys. Lett.*, vol. 46, p. 725, 1985.

[15] M. A. Olmstead and N. M. Amer, "A new probe of the optical properties of surfaces," *J. Vac. Sci. Technol.*, vol. B1, p. 751, 1983.

[16] J. Opsal, A. Rosencwaig, and D. L. Willenborg, "Thermal-wave detection and thin-film thickness measurements with laser beam deflection," *App. Opt.*, vol. 22, p. 3169, 1983.

[17] A. Korpel, L. J. Laub, and H. C. Sievering, "Measurement of acoustic surface wave propagation characteristics by reflected light," *Appl. Phys. Lett.*, vol. 10, p. 295, 1967.

[18] C. Thomsen, H. T. Grahn, H. J. Maris, and J. Tauc, "Ultrasonic experiments with picosecond time resolution," *J. Physique*, colloque C10, suppl, vol. 46, p. 765, 1985.

[19] A. Rosencwaig, J. Opsal, W. L. Smith and D. L. Willenborg, "Detection of thermal waves through optical reflectance," *Appl. Phys. Lett.*, vol. 46, p. 1013, 1985.

[20] Wolf Bickel, "Method and apparatus for receiving ultrasonic waves by optical means," U.S. Patent #4,129,041, Dec. 1978, assigned to Krautkrämer-Branson Inc.

[21] A recent review is found in M. D. Levenson, *Introduction to Nonlinear Laser Spectroscopy*, Academic Press, New York, 1982.

[22] Wolf Bickel, "Method and apparatus for receiving ultrasonic waves by optical means, U.S. patent #4,345,475, Aug. 1982, assigned to Krautkrämer-Branson Inc.

[23] M. V. Klein, *Optics*, J. Wiley, New York, 1970, see section 4.1.E.

[24] Ref. 23, p. 92.

[25] A. E. Ennos, "Speckle interferometry," in *Progress in Optics* (vol. 16), E. Wolf, Ed. New York: North-Holland, 1978.

[26] J. A. Midgley and K. F. Sander, "High resolution laser homodyne interferometer," *Electron Lett.*, vol. 7, p. 117, 1971.

[27] A. N. Bondarenko, Y. B. Drobot, and S. V. Kruglov, "Optical excitation and detection of nanosecond acoustic pulses in nondestructive testing," *Sov. J. NDT*, vol. 12, p. 655, 1976.

[28] C. H. Palmer and R. Green, Jr., "Optical detection of acoustic emission waves," *Appl. Opt.*, vol. 16, p. 2333, 1977.

[29] M. Kroll and B. B. Djordjevic, "A laser stress-wave probe with subangstrom sensitivity and large bandwidth," in *Proc. 1982 IEEE Ultrason. Symp.*, p. 864.

[30] D. C. Emmony, M. W. Godfrey, and R. G. White, "A miniature optical acoustic emission transducer," *Rev. Prog. Quantitative Nondestructive Evaluation*, D. O. Thompson and D. E. Chimenti, Ed. New York: Plenum, p. 533, 1983.

[31] K. F. Etzold, "Transducer characterization by interferometry," in *Proc. 1982 IEEE Ultrason. Symp.*, p. 867.

[32] C. H. Palmer and R. E. Green, Jr., Report to the U.S. Army Research Office, Aug. 1979, unpublished.

[33] C. H. Palmer, "Circumventing laser relaxation oscillations in interferometers," *Appl. Opt.*, vol. 23, p. 3510, 1984.

[34] J.-P. Monchalin, Progress Report IGM 83-231-02, Appendix B (unpublished).

[35] L. E. Drain, J. H. Speake, and B. C. Moss, "Displacement and vibration measurement by laser interferometry," (1st European Congress on Optics Applied Metrology), *SPIE*, vol. 136, pp. 52–57, 1977.

[36] The author acknowledges the contribution of one referee of this paper for pointing out this possibility.

[37] R. Mezrich, D. Vilkomerson, and K. Etzold, "Ultrasonic waves: their interferometric measurement and display," *Appl. Optics*, vol. 15, p. 1499, 1976.

[38] D. Vilkomerson, "Measuring pulsed picometer-displacement vibrations by optical interferometry," *Appl. Phys. Lett.* vol. 29, p. 183, 1976.

[39] R. Riebold and W. Molkenstruck, "Laser interferometric measurement and computerized evaluation of ultrasonic displacements, *Acustica*, vol. 49, p. 205, 1981.

[40] D. M. Pepper, "Nonlinear optical phase conjugation," *Optical Eng.*, vol. 21, p. 156–183, 1982.
[41] A. Sarrafzadeh-Khoee and J. C. Duke, Jr., "Noncontacting detection in ultrasonic NDE of materials: simple optical sensor and fiber optics interferometric application," *Rev. Sci. Instrum.*, vol. 57, to be published.
[42] G. A. Massey and R. R. Carter, "Portable laser instrument for vibration analysis and transducer calibration," *Shock and Vibration Bull.*, vol. 37, part 2, pp. 1–6, 1968.
[43] P. Buchhave, "Laser Doppler vibration measurements using variable frequency shift," *DISA Informat.*, no. 18, p. 15–20, Sept. 1975.
[44] G. Bouchard and D. B. Bogy, "Experimental measurement of scattered surface waves using a laser—Doppler technique," *J. Acoust. Soc. Amer.*, vol. 77, p. 1003, 1985.
[45] J.-P. Monchalin, "Optical detection of ultrasound at a distance," *Can. J. Soc. Nondestructive Testing*, vol. 6, no. 1, pp. 48–53, 1984.
[46] ——, "Heterodyne interferometric laser probe to measure continuous ultrasonic displacements," *Rev. Sci. Instrum.*, vol. 56, p. 543, 1985.
[47] R. L. Whitman, L. J. Laub, and W. Bates, "Acoustic surface displacement measurements on a wedge-shaped transducer using an optical probe technique," *IEEE Trans. Sonics Ultrason.*, vol. SU-15, p. 186, 1968.
[48] M. J. Rudd, "Ultrasonic nondestructive evaluation using laser transducers," in *Rev. Progress in Quantitative Nondestructive Evaluation*, D. O. Thompson and D. E. Chimenti, Eds. New York: Plenum, pp. 1763–1782, 1983.
[49] D. Royer, E. Dieulesaint, and Y. Martin, "Improved version of a polarized beam heterodyne interferometer," in *Proc. 1985 IEEE Ultrason. Symp.*, p. 432.
[50] R. M. DeLaRue, R. F. Humphryes, I. M. Mason, and E. A. Ash, "Acoustic surface wave amplitude and phase measurements using laser probes," *Proc. IEE*, vol. 119, p. 117, 1972.
[51] S. Ameri, E. A. Ash, V. Newman, and C. R. Petts, "Photo-displacement imaging," *Electron. Lett.*, vol. 17, p. 337, 1981.
[52] S. Ayter, B. A. Auld, and M. Tan, "Optical probing of resonant ultrasonic scattering from machined flaws in plates," in *Rev. of Quantitative Nondestructive Evaluation*, D. O. Thompson and D. E. Chimenti, Eds. New York: Plenum, pp. 565–568, 1982.
[53] B. A. Auld, S. Ayter, and M. Tan, "Filter detection of phase-modulated laser probe signals," *Electron Lett.*, vol. 17, p. 661, 1981.
[54] H. K. Wickramasinghe, Y. Martin, D. A. M. Spear, and E. A. Ash, "Optical heterodyne techniques for photoacoustic and photothermal detection," *J. Physique*, colloque C6, suppl, vol. 44, p. 191, 1983.
[55] R. L. Jungerman, J. E. Bowers, J. B. Green, and G. S. Kino, "Fiber optic laser probe for acoustic wave measurements," *Appl. Phys. Lett.*, vol. 40, p. 313, 1982.
[56] R. Dändliker and J.-F. Willemin, "Measuring microvibrations by heterodyne speckle interferometry," *Optics Lett.*, vol. 6, p. 165, 1981.
[57] M. O. Deighton and E. A. Sayle, "An electronic tracker for the continuous measurement of Doppler frequency from a laser anemometer," *DISA Informa.*, no. 12, p. 5–10, Nov. 1971.
[58] J.-P. Monchalin, R. Héon, and N. Muzak, "Evaluation of ultrasonic inspection procedures by field mapping with an optical probe," *Canadian Metallurgical Quarterly*, 1986, to be published.
[59] A. E. Siegman, "The antenna properties of optical heterodyne receivers," *Appl. Opt.*, vol. 5, p. 1588, 1966.
[60] P. Buchhave, "Light collecting system and detector in Laser Doppler anemometer," *DISA Informa.*, vol. 15, p. 15, 1973.
[61] P. Cielo, F. Nadeau, and M. Lamontagne, "Laser generation of convergent acoustic waves for materials inspection," *Ultrasonics*, p. 55, Mar. 1985.
[62] D. A. Hutchins and F. Nadeau, "Non-contact ultrasonic waveforms in metals using laser generation and interferometric detection," in *Proc. 1983 IEEE Ultrason. Symp.*, p. 1175.
[63] J.-P. Monchalin, unpublished.
[64] A. Dandridge, R. O. Miles, and T. G. Giallonenzi, "Diode laser sensor," *Electron. Lett.*, vol. 16, p. 948, 1980.
[65] J. K. Thompson, H. K. Wickramasinghe, and E. A. Ash, "A Fabry–Pérot acoustic surface vibration detector-application to acoustic holography," *J. Phys. D.*, vol. 6, p. 677, 1973.
[66] L. E. Drain, *The Laser Doppler Technique*. New York: J. Wiley, chap. 5, 1980.
[67] A. Sarrafzadeh-Khoee and J. C. Duke, Jr., "Measurement of small acoustic displacement amplitudes on an object with a retro-reflective diffusing surface using laser speckle interferometry," *Experimental Techniques*, vol. 10, to be published.
[68] J.-P. Monchalin, R. Héon, C. K. Jen, to be published.
[69] C. H. Palmer, R. O. Claus, and S. E. Fick, "Ultrasonic wave measurement by differential interferometry," *Appl. Optics*, vol. 16, p. 1849, 1977.
[70] A. O. Garg and R. O. Claus, "Application of optical fibers to wideband differential interferometry," *Mat. Eval.*, vol. 41, p. 106, Jan. 1983.
[71] L. M. Barker, "Laser interferometry in shock-wave research," *Exp. Mech.*, p. 209, May 1972.
[72] W. Kaule, "Interferometric method and apparatus for sensing surface deformation of a workpiece subjected to acoustic energy," U.S. patent #4,046,477, Sept. 1977. (Assigned to Krautkrämer-Branson Inc.).
[73] M. Born and E. Wolf, *Principles of Optics*, 4th ed. New York: Pergamon, 1970, Chap. 7.5.4, p. 301.
[74] R. J. Clifton, "Analysis of the laser velocity interferometer," *J. Appl. Phys.*, vol. 41, p. 5335, 1970.
[75] W. Kaule, "Laser induced ultrasonic pulse testing," in *Proc. 8th World Conf. Nondestructive Testing*, 1976, paper 3J5.
[76] R. K. Mueller and R. L. Rylander, "New demodulation scheme for laser-scanned acoustical-imaging," *J. Opt. Soc. Amer.*, vol. 69, p. 407, 1979.
[77] Commercial note on the Laser Ultrasound System, Krautkrämer-Branson Inc., Lewistown, PA, unpublished.
[78] G. Hansen, "Die Sichtbarkeit der Interferenzen beim Michelson und Twyman Interferometer," *Zietschift für Instrumentenkurde*, vol. 61, p. 411, 1941.
[79] P. Connes, "Augmentation du produit luminosité x résolution des interféromètres par l'emploi d'une différence de marche indépendante de l'incidence," *Revue d'Optique*, vol. 35, p. 37, 1956.
[80] C. W. Gillard, "Optical differential interferometer discriminator for FM to AM conversion," U.S. Patent #3,503,012, Mar. 1970, assigned to Lockeed Aircraft Corp., CA.
[81] L. M. Barker and R. E. Hollenbach, "Laser interferometer for measuring high velocities of any reflecting surface," *J. Appl. Phys.*, vol. 43, p. 4669, 1972.
[82] G. Smeets and A. George, "Instantaneous laser Doppler velocimeter using a fast wavelength tracking Michelson interferometer," *Rev. Sci. Instrum.*, vol. 49, p. 1589, 1978.
[83] W. Kaule, "Method and apparatus for receiving ultrasonic waves by optical means," U.S. Patent #4,388,832, June 1983. (Assigned to Krautkrämer-Branson Inc.)
[84] S. E. Harris, "Conversion of FM light to AM light using birefringent crystals," *Appl. Phys. Lett.*, vol. 2, p. 47, 1963.
[85] P. Cielo and C. Delisle, "Vélocimétrie Doppler par modulation spectrale," *Can. J. Phys.*, vol. 54, p. 1199, 1976.
[86] J. E. Bowers, "Fiber-optical sensor for surface acoustic waves," *Appl. Phys. Lett.*, vol. 41, p. 231, 1982.
[87] J. E. Bowers, R. L. Jungerman, B. T. Khuri-Yakub, and G. S. Kino, "An all fiber-optic sensor for surface acoustic wave measurements," *J. Lightwave Tech.*, vol. LT-1, p. 429, 1983.
[88] R. L. Jungerman, B. T. Khuri-Yakub, and G. S. Kino, "Optical probing of acoustic waves on rough surfaces," *J. Acoust. Soc. Amer.*, vol. 73, p. 1838, 1983.
[89] D. M. Paul and D. A. Jackson, "Rapid velocity sensor using a static confocal Fabry-Pérot and a single frequency argon laser," *J. Phys. E: Scientific Inst.*, vol. 4, p. 170, 1971.
[90] D. A. Jackson and D. M. Paul, "Measurement of supersonic velocity and turbulence by laser anemometry," *J. Phys. E: Scientific Inst.*, vol. 4, p. 173, 1971.
[91] R. N. James, W. R. Babcock, and H. S. Siefert, "A laser-Doppler technique for the measurement of particle velocity," *J. Am. Inst. Aeronautics Astronautics*, vol. 6, p. 160, 1968.
[92] P. M. Johnson and T. J. Burgess, "Free surface velocity measurement of an impacted projectile by optical Doppler shift," *Rev. Sci. Inst.*, vol. 39, p. 1100, 1968.
[93] P. Connes, "L'étalon de Fabry-Pérot sphérique," *J. Phys. Radium,*, vol. 19, p. 262, 1958.
[94] M. Hercher, "The spherical mirror Fabry–Pérot interferometer," *Appl. Opt.*, vol. 7, p. 951, 1968.
[95] J.-P. Monchalin, "Optical detection of ultrasound at a distance using a confocal Fabry–Pérot interferometer," *Appl. Phys. Lett.*, vol. 47, p. 14, 1985.
[96] ——, "Detection at a distance of laser generated ultrasound using a confocal Fabry–Pérot interferometer," *Can. J. Phys.*, vol. 64, no.

9, Sept. 1986.
[97] J.-P. Monchalin and R. Héon, "Laser ultrasonic generation and optical detection with a confocal Fabry–Pérot interferometer," *Mat. Eval.*, vol. 44, to be published.
[98] R. L. Jungerman, B. T. Khuri-Yakub, and G. S. Kino, "Characterization of surface defects using a pulsed acoustic laser probe," *Appl. Phys. Lett.*, vol. 44, p. 392, 1984.
[99] D. A. Hutchins, F. Nadeau, and P. Cielo, "A pulsed photoacoustic investigation of mode conversion," *Can. J. Phys.*, vol. 64, no. 9, Sept. 1986.
[100] L. Piché and B. Champagne, "Characterization of the particle-matrix cohesion in composites by a sound propagation technique," in *Proc. 11th World NDT Conf.*, Nov. 1985.
[101] C. K. Jen, P. Cielo, X. Maldague, and K. El-Assal, "Noncontact ultrasonic characterization of piezoelectric ceramics," *J. Amer. Ceramic Soc.*, vol. 68, p. C-146, 1985.
[102] J. Krautkrämer, in Plenary sessions, *Proc. Ninth World Conf. Nondestructive Testing*, 1979, p. 2.
[103] A. M. Aindow, R. J. Dewhurst, S. B. Palmer, and C. B. Scruby, "Laser-based nondestructive testing techniques for the ultrasonic characterization of subsurface flaws," *NDT Int.*, vol. 17, p. 329, 1984.
[104] R. L. Wellman, "Laser system for the detection of flaws in solids," Harry Diamond Laboratories, rep. HD1-TR-1902, Adelphi, MD, 1980.
[105] P. Cielo, C. K. Jen and X. Maldague, "The converging-surface-acoustic wave technique: Analysis and applications to nondestructive evaluation," *Can. J. Physics*, vol. 64, no. 9, Sept. 1986.
[106] Y. Martin and E. A. Ash, "Imaging of micro-cracks by photo-displacement microscopy," in *Proc. 1984 IEEE Ultrason. Symp.*, p. 647.

Jean-Pierre Monchalin received a diploma in optical engineering in Paris in 1968 and the M.S. and Ph.D. degrees in lasers and optics from the Massachusetts Institute of Technology, Cambridge, in 1971 and 1976.

He is a Research Scientist in nondestructive evaluation at the Canada Centre for Mineral and Energy Technology (CANMET) and is presently detached at the Industrial Materials Research Institute of the National Research Council of Canada in Boucherville, Québec. His doctoral work consisted of very accurate interferometric measurement in the infrared and led to a new value for the speed of light. He was then employed by Ecole Polytechnique in Montréal and worked on chemical laser development, photoacoustic techniques and, in particular, on their application to the spectroscopy of powders. He has been involved with nondestructive evaluation since his association with CANMET in 1982. His present work includes the measurement of ultrasonic absorption in polycrystalline metals and essentially the development of laser and optical techniques to generate and detect ultrasound for industrial applications.

Dr. Monchalin is a member of the Optical Society of America, the Canadian Society for Non-Destructive Testing, the American Society for Non-Destructive Testing, the American Society for Metals, and Sigma Xi.

Reprinted with permission from *Research Techniques in Nondestructive Testing,*
R.S. Sharpe, editor, Vol. 5, Chap. 7, pp. 239-279 (1982).

Theoretical and Practical Aspects of the Thermal Nondestructive Testing of Bonded Structures

V. P. VAVILOV

Tomsk Polytechnic Institute, USSR

AND

R. TAYLOR

UMIST, Manchester, UK

I. Introduction

Practical realization of the idea of using a heat flux to investigate the inner structure of a solid has, in spite of its obvious simplicity, only become possible after the development of high speed infrared (IR) systems. Although there have been some successful studies using contact and semi-contact sensors in the area of thermal nondestructive testing (TNDT), most investigations in this field are carried out using noncontact IR apparatus. Naturally the first industrial applications using early IR techniques were concerned with the remote measurements of stationary temperature fields. The first applications of infrared to quality estimation can be traced to the 1940's in the USA— apparatus for detecting forest fires (Barker, 1934), temperature uniformity measured during the heating of strip in steel rolling mills (Nichols, 1935), evaluation of soldered seams in tinned can production (Gorrill, 1949), detecting hot joints on electrical transmission lines (Sandiford, 1952) and overheated journal bearings on railroad cars (Dzthubez *et al.*, 1958), etc. However, in 1960 IR techniques began to be used for the production of good quality thermograms in electrical engineering applications (Anon., 1961). Probably one of the earliest applications of TNDT to the detection of voids, disbonds and adhesive-free areas in laminates was the evaluation of rocket motor casings by Florant (1964) and Beller (1965). It is worth noting that the testing procedure was rather simple, including the placing of the motor casing in a heated room and then subsequently observing its cooling. Especially fruitful for TNDT was the period from 1960 to 1970 when fundamental investigations of the active testing of solid state devices for rockets, nuclear reactor fuel elements and aviation, space and electrical engineering components were carried out in the USA by Willburn (1960), Alzofon (1961), Day (1968), Ferguson (1968), Green (1968, 1971), Kubiak (1968), Vogel (1968), Kutzscher and Zimmerman (1968), Robichaud (1969), Stiefeld and Yoshimura (1971), Lavery *et al.* (1971), Vandamme and McGarvey (1972), Renius (1973), Green *et al.* (1973) and Bobo (1974). At the end of the 1960's, Bekeshko and Oshehepkov (1969) pioneered Russian research in this field. By 1973–1975 activity in the TNDT field reached a maximum and is now at the same level

The total number of published papers on this subject is now roughly 800. As regards the contribution from Soviet workers, it is worth emphasizing the comparatively large number of theoretical investigations which began to appear when attempts were made to put the hitherto empirical approach on a more formal quantitative basis. From 1973 onwards many workers studied the theoretical aspects of TNDT. Surface temperature signals in the presence of a defect were analysed and the connection between theoretical results and the optimum choice of the monitoring apparatus was evaluated (Bekeshko and Oshehepkov, 1969; Merhib and Taylor, 1972; Bekeshko, 1973; Vavilov *et al.*, 1973, 1978; Vavilov and Gorbunov, 1974, 1977; Gorbunov *et al.*, 1974; Popov *et al.*, 1975; Popov *et al.*, 1977; Potapov *et al.*, 1977; Rapoport, 1978; Singh *et al.*, 1978; Storozhenko *et al.*, 1978). At the present time a number of countries, firms and organizations are involved in TNDT investigations. A number of commercially available industrial systems have been marketed, e.g. AGA thermovisers (Sweden) and the TIRIS system for aviation construction inspection (Cohen and Sherman, 1973).

Infrared nondestructive testing in the USA has become an important component of some governmental and private programmes. Indeed IR testing groups and laboratories have been created at such firms as Boeing, General Electric, Lockheed Aircraft, Automation Ind., Raytheon and Goodyear. In 1965 40% of the papers presented at the annual conference of the American Society for NDT were concerned with TNDT.

Nevertheless at the present time there is a lack of information encompassing all the main aspects of TNDT, although the chronology of IR investigation and the design of IR systems are excellently described by Hudson (1969). Some problems of TNDT were discussed by Lawson and Sabey (1970). A bibliography of thermal and infrared nondestructive testing up to the end of 1972 is available (Merhib and Taylor, 1972).

In this paper we do not intend to review all aspects of TNDT. We shall, instead, attempt to describe the generation of a useful signal in bonded structures and the interrelation between the signal and the testing parameters.

II. Physical Aspects of Thermal Nondestructive Testing (TNDT) of Bonded Structures

A. General Features—Terminology and Quantitative Aspects

Western terminology for the method being described may be either "thermal" or "infrared" NDT, depending on the method of monitoring the temperature field. In the USSR the method is governed by the national standard 18353-73 for "thermal nondestructive testing", underlying the fact that the detection of subsurface defects is due to nonuniformity of a heat flux permeating through the specimen. In this chapter we shall adopt this terminology on the assumption that the temperature signal associated with the defect is not affected by the way it is monitored. Testing of objects with a temperature above ambient due either to technological or to functional reasons shall be called "passive testing", whereas "active testing" requires the provision of an additional heat source.

Over the past few decades there has been an increasing technological requirement to develop nondestructive testing methods for multilayer structures such as composites and honeycomb structures used in aerospace. Defects found in these composites, such as delamination, porosity, bad adhesion and abnormal glue thickness can, only with difficulty, be analysed by the more traditional nondestructive testing methods. X-rays as a rule are useless and ultrasonic inspection, whilst a possibility in certain cases, is characterized by low inspection rates and high noise level. That is why a number of firms concerned with the production and testing of various bonded structures consider TNDT, if not the main inspection method, a very useful additional technique.

B. Application to Aerospace Components

Although the technology to manufacture honeycomb and bonded structures for aerospace applications is well established there is still a lack of good, economical inspection procedures.

In the aerospace materials field the important requirement for accurate quality control and inspection becomes evident if one considers, for example, that the C-5 Lockheed aircraft contains up to 3400 m^2 of bonded structures. Ultrasonic testing was replaced by infrared testing at Lockheed, using the TIRIS system designed for inspection of honeycomb and bonded structures. This system is characterized by a high inspection rate, minimization of operator's errors, automation and continuous recording in real-time (Cohen and Sherman, 1973). It is possible to detect delaminations with diameters of about 25 mm and stored water in the titanium–aluminium honeycombs at a testing rate of 30 mm/s. The system incorporates a Bofors thermoviser and uses a powerful lamp as a heat source. Because of nonuniform heating with lamps, powerful scanning lasers have also been used (Renius, 1973). However, this change has produced negligible improvement in inspection sensitivity, mainly because of the low average heat density with laser scanning. The most successful TNDT methods use spot heating and a spot monitoring of temperature during the scanning process (Kutzscher and Zimmerman, 1968). Developing that idea, research on line heating and line scanning systems was carried out in the USSR for fibreglass tube inspection (Storozhenko *et al.*,

1978). The same apparatus including an operator console scanner and line heater, was used in the USA for the testing of honeycombs with an aluminium core and fibreglass skin (Robichaud, 1969). The possibility of detecting moisture in resin matrix composites and the influence of stored water on the inspection of delaminations were investigated by Singh *et al.* (1978). Similar research on honeycombs has been done by Popov *et al.* (1977) in the USSR. It is difficult to deduce any firm conclusions on the limit of TNDT in aviation and space technology because of the large variety of materials, defects, types and testing conditions of interest. Qualitatively the following conclusions may be drawn:

(1) Thermal testing is unique from the point of view of speed of inspection (up to $0{\cdot}1\ m^2/s$; Robichaud, 1969). It is not usually affected by the limitations of optico-mechanical scanning but by the power and geometrical parameters of the heating source.
(2) The noncontact character of the method and the ability to process the information in real-time allow the development of an automatic NDT system requiring a minimum of operator involvement (Day, 1968).
(3) The best NDT procedure for metal–nonmetal structures requires temperature monitoring from the side of the nonmetal and permits heating either from the metal surface (two-sided inspection) or from the nonmetal surface (one-sided inspection). As for honeycombs, most success has been achieved in the detection of delaminations between the fibreglass, graphite or titanium skin and phenolic or aluminium core. Thermograms of honeycombs sometimes contain a picture of the inner structure giving a better opportunity to exclude false signals than with say ultrasonic inspection.
(4) Moisture detection is still rather limited and TNDT merely gives an indication of whether or not water is present at a fairly high concentration.

C. Testing of Nuclear Reactor Components

A virtual monopoly in a field as difficult for TNDT as the fuel elements of nuclear reactors belongs to Green and his collaborators in the USA. They successfully used rather complicated systems of heating (plasma arc, inductive heating, electric current) and some interesting monitoring apparatus included a two-channel IR radiometer. Green has developed the ingenious idea of using two time delays during the temperature measurement which has allowed aluminium and zircaloy specimens to be tested without the need to paint the surfaces black (Green, 1968). In a later paper he describes an IR testing procedure for the electrical fuel pin simulators used in the nuclear reactor's thermal–hydraulic tests (Green *et al.*, 1973). This testing method was adopted by Hanford Laboratory and Oak Ridge National Laboratory after comparing cost estimates for IR techniques with X-ray and neutron radiography. Moreover TNDT can identify four main types of simulated defects: (a) low density of the boron nitride layer; (b) displaced (off-centre) core; (c) higher than normal heat production in core; (d) change in cladding thickness. Green's idea of double scanning was also applied by Day to bonds in niobium-carbide coated graphite support blocks of nuclear reactors (Day, 1968). In such tests, carried out at a speed of 25 mm/s, delaminations between the coating and the graphite block were detected. (Defects were simulated by using Teflon inserts.) The signal enhancement, relative to the standard level, was about 20% for defects of 3 mm diameter. The higher porosity of some blocks reduced the useful signal to 14% of normal.

D. Nondestructive Tyre Inspection

Modern tyres are complicated multilayer structures comprising cord, rubber, adhesive and other components. The combination of so many different materials such as natural and artificial rubbers, steel, nylon, fibreglass and polymers together with adhesives creates many difficulties for the development of reliable NDT methods. It is usually found that tyre destruction begins in an adhesive layer where small voids develop into big delaminations under the effects of high temperature and cyclic loading. The successful application of TNDT for bonded structures in aviation and space engineering has awakened interest in the application of IR techniques in the field of tyre manufacture. Now leading world firms such as Firestone, Goodyear, General Tyre and Rubber Company and Gatest Rubber Company are thermally testing tyres and viewing it as a potentially attractive method for use in conjunction with X-rays, ultrasonic and holographic interferometry (Willburn, 1960; Lavery *et al.*, 1971). The typical test procedure includes monitoring the thermal "signature" of a rotating tyre, initial interpretation of a thermogram by an operator and subsequent processing of the thermogram by computer. Current investigations in this field are directed to determining connections between tyre faults developed in service, type of defect and the parameters of the temperature pattern.

III. Determination of TNDT Sensitivity

At the present stage of development no TNDT researcher can predict the results of applying his apparatus to test a new class of specimens without preliminary experimental evaluation. The combination of objective factors (such as the variety of parameters influencing surface temperature) and subjective ones (absence of good theory, shortage of adequate standards) results in the fact that there is no simple way to define defect sensitivity for thermal defectoscopes. The available experimental data are also often of fragmentary character and do not readily allow extension of the results obtained to fit other testing conditions. The objective comparison of results obtained by different authors is difficult even for specimens of similar type because of variations in apparatus and analysis. The most usual "rule-of-thumb" estimate is that the minimum detectable defect size must be at least two to three times greater than the thickness of material above it. We believe, however, that such an estimate may be improved upon because these results are usually associated with rather simple testing conditions when judgement is being made by nonprofessional operators. In this context anyone with experience in the field of heat transfer is capable of detecting internal defects with minimum *a priori* information about the temperature distribution, e.g. using just the sign of the temperature signal. However, whilst there have been many investigations where smaller defects were detected it usually means that a single experiment with known defects was carried out. Sometimes increased sensitivity has occurred because of the special properties of the material (fibreglass, glass, rubber, metals with low conductivity, the presence of inner channels, etc.) or possibly uniform heating, e.g. with electric current heating in the whole volume of a specimen. More interesting are investigations where the enhancement of sensitivity was brought about because of a special test procedure (Day, 1968). In Table I are listed the experimental results for defect detection sensitivity on different materials from published information. We deliberately avoid giving detailed descriptions of test conditions for this comparison. As shown below the main limitations of TNDT are determined by the properties of the material if the other parameters are chosen in an optimum way. So we summarize as follows:-

(1) The combination of factors influencing TNDT results includes time, dimensions of specimen and defect, type of heat source, speed of inspection and the detectability of a useful signal against the background of noise.
(2) For small defects which do not emit energy during production or operation, a transient regime is always more sensitive than a steady-state one. The varying nature of the signal over a defect when working in the transient regime is such that it is possible to apply the method of double, consecutive or simultaneous, scanning in order to overcome the effects of fluctuations in surface emissivity (Day, 1968; Green, 1968). For nonmetals the time available for inspection is about 10–100 s, but for metals the smaller time for which a useful contrasty signal exists creates considerable practical difficulties. The time necessary for tuning a good

Table I

The Sensitivity of TNDT for Different Materials (the Data are obtained by Generalization of the Experimental Results from a Selection of Published Works).

Material and its thickness	Minimum cross-sectional area of detectable defects (mm^2)
Composites	
Plastic–titanium	100
Plastic (3 mm)–metal (5 mm)–plastic (3 mm)	100
Glass (0·2–1 mm)–Al (1·5 mm)	1
Al (2 mm)–asbestos (4 mm)	100
Steel (2 mm)–rubber (5 mm)	100
Al (2 mm)–honeycomb core (20 mm)–Al (2 mm)	140
Ferrite (1–4 mm)–metal	0·5–7
Metal (6 mm)–plastic (3 mm)–metal (6 mm)	400
Fibreglass (3·2 mm)–Al honeycomb core (9·1 mm)	100
Nonmetals	
Tufnol, Teflon, ceramics (2–10 mm)	1–25
Rubber	50–400
Wood, plastic (5 mm)	100
Fibreglass (0·3 mm)	0·75
Fibreglass (7 mm)	225
Fibreglass (12 mm)	100
Plastic (7 mm)–plastic (45 mm)	100
NbC-graphite	10
Metals	
Al–Al, brass–brass, steel–steel	0·1–7
Cu (0·3 mm)–Cu	0·5
Al, zircaloy (0·7 mm)–fuel element core	100
Titanium (0·5 mm)–Al (2–4 mm)	25
Babbit (0·8 mm)–steel (12 mm)	140

picture after heating stops is generally a few seconds but the temperature contrast over the defect may disappear completely during this time, especially for thin specimens with high conductivity.

(3) TNDT is applicable to practically all materials and their composites and may be either as two-sided or one-sided testing. The range of specimen thicknesses that can be examined is roughly from 0·1 to 10 mm for solid materials, if the defect is likely to occur anywhere within the material.

Minimum defect sizes should as a rule be two to three times greater than the layer of material above them (this condition may be obtained by dividing the square root of the defect surface by its depth).

(4) More and more it is being realized that progress in active TNDT will be linked to heater improvements; from snow, hot water, air and lamps to plasma arc and scanning laser beams—such has been the evolution of heating sources used. The main requirements are a high energy density, uniformity of heat emission, small size and minimization of background (noise) radiation. These parameters are discussed in more detail below. The practical development of IR systems has been described by Hudson (1969). A wide variety of optical materials, scanning systems and IR receivers is available, so these do not present any technical limitations for optimum design of apparatus. Also useful in this context would be a preliminary theoretical analysis of the TNDT problem for any particular case.

(5) As with other NDT methods the question of useful signal recognition is very important in TNDT. This is especially so for industrial application of IR apparatus. The presence of noise sources in factories, variation in surface condition of materials under test and the variety of natural defects leads to a situation where the operator cannot identify a defect with a high level of confidence. This aspect is, probably, the most backward in the field of TNDT.

IV. Apparatus for Thermal Testing

A. *Survey of Possible Test Procedures*

The most common active TNDT test procedures shown in Fig. 1 include (a) testing with a spot heat source (Fig. 1a, variant 1); (b) testing with a line heater (Fig. 1b, variant 2) and (c) the method of inspection using a pulsed area heater and thermoviser (Fig. 1c, variant 3).

Usually the energy profile inside the heat beam is a Gaussian distribution (Fig. 1, curves 7), although for a line heater placed near the specimen surface it could be quite uniform (Fig. 1, curve 8). Lasers are characterized by a specific circular distribution of energy with some peaks (Fig. 1, curve 9). Nonuniform and test results show amplitude temperature profiles (Fig. 2a) with a specific maximum occurring over the defect for the heated surface H and a minimum for the rear surface R. This is the so-called "rule of sign" which can be applied for the recognition of a defect. It is worth noting that for a thin specimen with a high conductivity the signal sign may be reversed. This effect has received virtually no investigation in TNDT. But usually the time behaviour of a spot temperature measurement over an experimental defect grows from zero (the

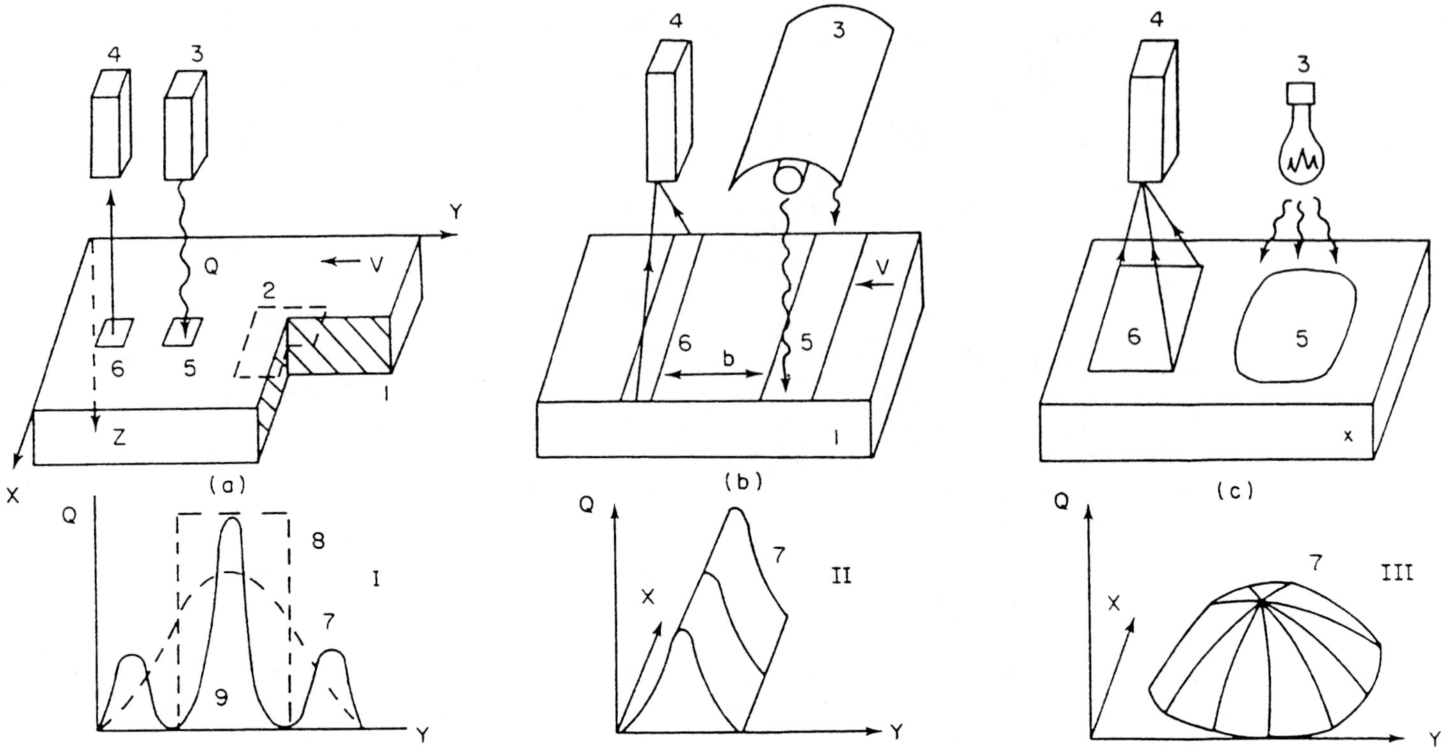

FIG. 1. Variants of active TNDT: (a) spot testing; (b) line testing; (c) area testing. I, II, III energy distribution inside the heat beam. 1 – specimen; 2—defect; 3—heat; 4—IR systems; 5—heated area; 6—scanned area; 7, 8, 9—Gaussian, uniform and circle distributions.

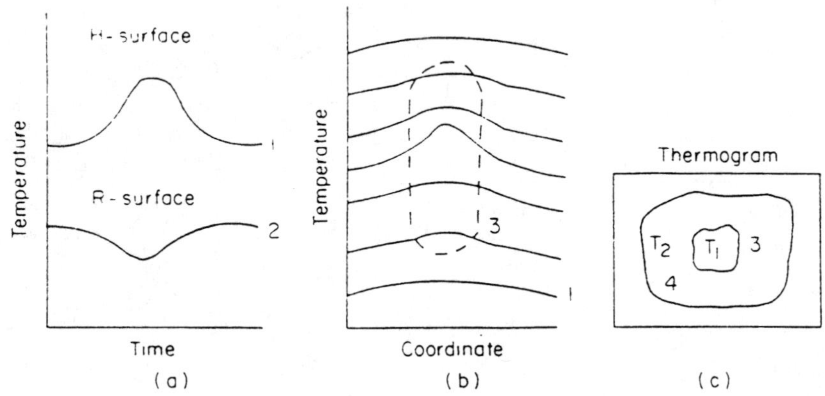

FIG. 2. Typical representations of TNDT results: (a) spot testing; (b) line testing; (c) area testing. 1, 2—temperature profiles; 3—defect area; 4—heated area.

moment of heating) to a maximum (corresponding to the optimum time delay) and then decreases to zero again some time later. Variant 2 (line testing) requires a uniform line energy input along coordinate X. If no correction is needed for energy distribution the temperature record has a bell-shaped form associated with the signal from a defect (Fig. 2b). But for both variants the optimum distance between the heated zone and the monitoring point b (or the optimum time delay $\tau_0 = b/V$) needs to be considered. For variant 3 representation of the results usually has the form of a thermogram. The example of a thermogram depicted in Fig. 2(c) shows two isotherms one of which outlines the heated area and the other the defect. The nonuniformity of heating makes it difficult to tune the thermoviser's picture for big and small defects simultaneously in this particular case. The optimum time delay for the third variant is the time gap between the end of heating and the instant of monitoring the temperature.

B. Spot Heating

Air gun, plasma arc systems, incandescent bulbs, open wire heaters and lasers may be used as spot heaters. Some of their properties are listed in Table II. Later we shall see that the smaller the heating area the greater the temperature contrast over the defect. From this point of view the plasma arc heater and the laser are particularly effective. However, the former has a high level of background radiation noise and time instability. We believe that lasers are the most effective because they possess a high energy density and background noise may be eliminated by a suitable choice of wavelength. At present experimental limitations are connected with their size, cost and cooling requirements.

TABLE II
Parameters of Some Heaters used in TNDT

Heater	Power density (W/cm^2)	Diameter of heating zone (cm)	Level of[a] radiation "noise"	Time stability[a]	Influence[a] of surface conditions	References
Air gun	0·05	0·5–50	0	0	0	Cohen and Sherman, 1973
Plasma arc	150	0·5–1	1	0	1	Green, 1968 Popov *et al.*, 1977
Incandescent filament bulb	0·5	0·5–50	1	1	1	Bekeshko, 1973 Popov *et al.* 1975
Heaters with open filament	10	0·5–5	1	1	1	Storozhenko *et al.*, 1978
Laser	[b]	0·5–1	1 or 0	1	1	Renius, 1973
Inductive heater	[b]	0·5–1	0	1	0	Green 1968
Contact heater	10	[c]	1	1	0	—

[a] 1 corresponds to the presence of a parameter, 0 to its absence; [b] Power density depends on the generator; [c] Contact heaters could have any size.

C. Line Heating

The most popular line heaters have an open filament. These are quite cheap and the energy distribution and choice of heating area may be quite simply regulated (Hsieh and Ellingson, 1977; Storozhenko *et al.*, 1978). On the other hand, Green has successfully used an inductive heater for cylindrical metal specimens using a 30 kW high frequency generator (Green, 1968).

D. Heaters with a Large Heating Area

The majority of investigations involving thermovisers use ordinary incandescent filament bulbs because of low cost, simple and direct energy supply and the flexibility associated with their combination with the scanning head of the thermoviser. However, everybody who has worked with such lamps as TNDT heaters is familiar with the situation when the operator watches the "thermal track" of the lamp instead of the defect signal. Whilst being aware that the direct application of such lamps hinders progress in TNDT it is difficult to propose an equally simple and reliable substitute that avoids this problem. Contact heaters are more uniform but their industrial application seems limited. For metals it is possible to use a pulsed electric current which provides uniform heating (Green *et al.* 1973).

E. *Apparatus for Monitoring Temperature Fields*

The technical parameters of industrial thermovisers and IR pyrometers have been widely discussed (Hudson, 1969; Vavilov *et al.*, 1975). The parameters of some specific systems have been discussed by Storozhenko and Gorbunov (1978). A full description is beyond the scope of this paper. Suffice it to note that IR systems engineering provides a spatial resolution in IR defectoscopes of about 5–100 μm (IR microscopes) or 0·1–5 mm (IR telescopes). Temperature resolution is usually in the range 0·05–0·5 °C and the productivity of scanning may be as high as 30 m^2/s (for the AGA thermoviser). However, in practice, the real rate of testing is usually limited by heating speed, not by optico-mechanical scanning. If the testing rate is defined as the area inspected per time unit P (cm^2/s), then for a square area $X_s \times Y_s$ which is scanned with speed V (see Fig. 1b) this parameter is given by

$$P = V.X_s \tag{1}$$

if there is no overlapping of the scanned lines. From (1) it is clear that the rate of inspection with TNDT, according to variant 2, is limited only by heating of the $X_s \times Y_s$ area by a heat pulse of duration τ_h, followed by a time delay of τ_d and then monitoring a temperature field during a period τ_m. So:

$$P = X_s . Y_s / (\tau_h + \tau_d + \tau_m) \tag{2}$$

if there is the consequent inspection of a whole specimen surface. Typical values of parameters mentioned above may, in practice, (Storozhenko and Gorbunov, 1978) be

Variant 2	Variant 3
$X_s \times Y_s = 250 \times 30\ mm^2$	$X_s \times Y_s = 200 \times 200\ mm^2$
$V = 50\ mm/s$	$\tau_h + \tau_d + \tau_m = 10\ s$
$P = 125\ cm^2/s$	$P = 40\ cm^2/s$

These results show that the thermovisers are not as rapid an inspection tool as their superior scanning system would lead one to suppose. Amongst other possible systems for monitoring the temperature field potentially the most interesting are liquid crystals described by Ferguson (1968) and the "transducer" introduced by Green (1971).

V. Solutions to Theoretical TNDT Problems

A. *Formulation of the Problem and Possible Approximations*

Strictly speaking TNDT problems are of no particular interest from the point of view of classic heat transfer theory. The problems of heat transfer in various situations have been summarized in many works, e.g. Carslaw and Jaeger (1959). More recent are the investigations by Soviet thermophysical workers, e.g. Lykov (1967), concerning heat transfer and Rykalin (Rykalin *et al.*, 1975), concerning moving heat beams. The first stage of TNDT theory development (Shultz, 1968) was characterized by the use of methods and ideas of the above two authors. Gradually this field, which has not received a lot of attention in heat transfer theory, has become more important. By this we mean the influence of small irregularities within a material on the time behaviour and topology of surface temperature fields. An intuitive understanding of the heat transfer process in such specimens needs to be replaced by a more quantitative treatment. At this stage the second part of the TNDT theory now becomes more important: the prediction of the TNDT limits (without consideration of the noise factors) and the selection, based on the theoretical calcuations, of the apparatus parameters. We are sure that, at present, this is now the most important role of TNDT theory. Disregarding the chronology of the various mathematical models of the TNDT problem (Shultz, 1968; Vavilov *et al.*, 1973; Vavilov and Gorbunov, 1974; Gorbunov *et al.*, 1974; Storozhenko *et al.*, 1977; Rapoport, 1978; Vavilov *et al.*, 1978), we believe an adequate formulation of this problem (Fig. 3) to be

$$\alpha\nabla^2 t_j = \partial t_j/\partial\tau; \quad j = 1 - n \tag{3}$$

$$\lambda\frac{\partial t_j}{\partial q_i} = -Q + \varepsilon\sigma\left[\left(\frac{t_j}{100}\right)^4 - \left(\frac{t_a}{100}\right)^4\right] + K(t_j - t_a) \tag{4}$$

$$\lambda_j\frac{\partial t_j}{\partial q_i} = \lambda_K\frac{\partial t_K}{\partial q_i} \tag{5}$$

$$t_j = t_K \tag{6}$$

$$Q = \begin{cases} Q(q_i,) & \text{inside the heat beam} \\ 0 & \text{outside the heat beam} \end{cases} \tag{7}$$

$$t_j\,(\tau = 0) = t_a \tag{8}$$

Here the various equations may be summarized as follows. Equation (3) is a transient three-dimensional equation of heat transfer; ∇^2 defines the Laplace operator with coordinates $i = 1 - 3$; j—layer number; n—the number of layers (the defect is one of them). Equation (4) deals with the conditions on the specimen surface including heating and cooling by convection and radiation. Equations (5) and (6) deal with the conditions of continuity of temperature and heat flux on the boundary between layers. Equation (7) is an equation describing time and space behaviour of the heat beam. Equation (8) is the initial boundary condition.

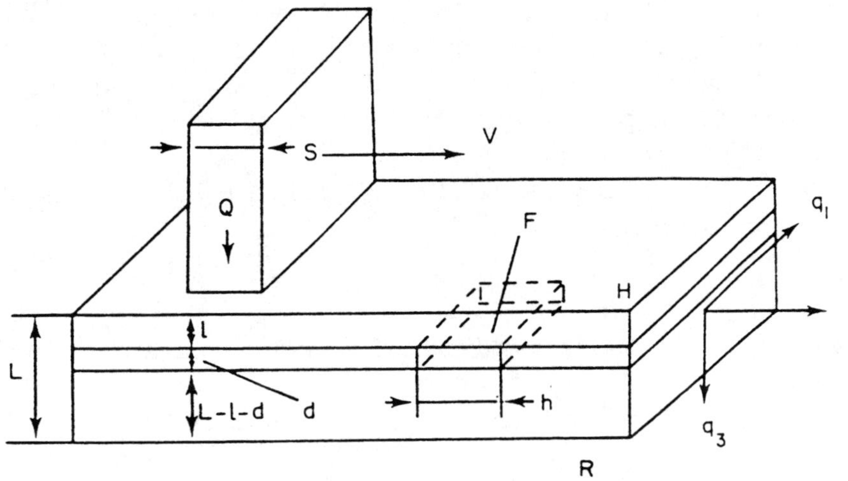

FIG. 3. General scheme of thermal nondestructive testing.

Investigations by western authors have been conducted with the objective of checking experimental results; specific mathematical models were not usually discussed (Green *et al.*, 1973). However a numerical analysis of the three-dimensional problem of stationary TNDT for cylindrical specimens was described by Hsieh *et al.* (1979) for a mesh of 14 × 13 × 6 points (IBM 370–195 computer).

B. Analytical Methods

Some particular problems which have been solved analytically will be discussed below. Here we must note that the analytical approach is nearly exhausted now because of the difficulties connected with a solution of finite-defect problems. Moreover if such solutions can even be obtained the results are usually so complex that a computer is needed to process the data anyway. Usually the Laplace transform is used in analytical methods of solution for one-dimensional problems. The simplest solution is obtained if there is no heat transfer through a defect and the boundary conditions, instead of (5) and (6), are as follows:

$$\frac{\partial t_i}{\partial q_i} = 0, \quad \text{on the surface of a defect} \tag{9}$$

otherwise even the one-dimensional solution becomes quite complex (Storozhenko *et al.*, 1977).

C. Numerical Methods and a Three-dimensional Problem

A detailed description of numerical methods will not be given in this paper, so we refer the reader to the in depth study by Tikhonov and Samarsky (1972) and the simpler treatment by Smith (1965). It is known that numerical methods deal in general with a four-dimensional network having three space and one time coordinates.

The three-dimensional second-order transient (eqn (3)) may be presented in the following general form:

$$\frac{\partial t}{\partial \tau} = \sum_{i=1}^{p} L_i t \tag{10}$$

where $L_i = \partial^2/\partial q_i^2$ is a differential operator and p is the number of coordinates.

The natural finite-difference approximation of (10) is

$$\frac{t^{j+1} - t^j}{\Delta \tau} = \theta(\Lambda t)^{j+1} + (1-\theta)(\Lambda t)^j \tag{11}$$

where $\Lambda = \sum_{i=1}^{p} \Lambda_i$; Λ_i is the finite-difference analog of L_i according to one form of representation of the derivatives; t is the temperature at the points of the network. θ is a constant specified for the numerical scheme; j and $j+1$ are moments of time which are divided by one time step $\Delta\tau$. $\theta = 0$ defines the explicit scheme, $\theta = \frac{1}{2}$ the Crank–Nicholson and $\theta = 1$ a fully implicit backward time difference method (Smith, 1965).

The explicit finite-difference representation seems to be the most attractive but it takes a great deal of computer time because the condition for its validity in Cartesian coordinates is

$$\alpha \,.\, \Delta\tau\left(\frac{1}{\Delta h_x^2} + \frac{1}{\Delta h_y^2} + \frac{1}{\Delta h_z^2}\right) < \frac{1}{6} \tag{12}$$

and it necessitates extremely small values for $\Delta\tau$.

The most efficient method for the solution of eqn (11) is the alternating-direction implicit method using the so-called "disintegration" of each time step $\Delta\tau$ into three substeps. The consequent calculation is conducted for each substep along one of the coordinates. So for a three-dimensional problem we use the following scheme:

$$\begin{aligned}
&\gamma_1 \,.\, T_{i-1,j,k}^{\tau+1/3} - (1-2\gamma_1)\, T_{i,j,k}^{\tau+1/3} + \gamma_1 \,.\, T_{i+1,j,k}^{\tau+1/3} = -T_{i,j,k}^{\tau} \\
&\gamma_2 \,.\, T_{i,j-1,k}^{\tau+2/3} - (1-2\gamma_2)\, T_{i,j,k}^{\tau+2/3} + \gamma_2\, T_{i,j+1,k}^{\tau+2/3} = -T_{i,j,k}^{\tau+2/3} \\
&\gamma_3 \,.\, T_{i,j,k-1}^{\tau+1} - (1+2\gamma_3)\, T_{i,j,k}^{\tau+1} + \gamma_3\, T_{i,j,k+1}^{\tau+1} = -T_{i,j,k}^{\tau+2/3}
\end{aligned} \tag{12a}$$

where $\gamma_n = \alpha\Delta\tau/\Delta h_n^2$; n is one of the coordinates; i, j, k are the numbers of points in the network. Specification of the boundary conditions is quite essential in such a scheme. There are different expressions for specimen and defect surfaces. They are quite complex and we shall show how one such expression may be determined. Let us suppose that a defect has a parallelopiped form and we need to obtain an expression to calculate the temperature at a certain point at the transition from one specimen layer into another, e.g. the boundary

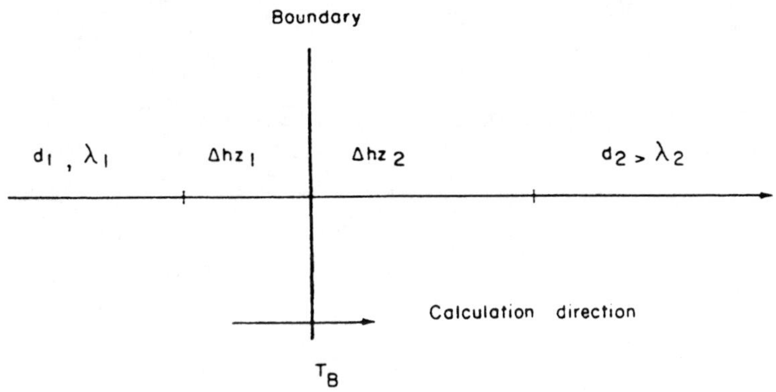

FIG. 4. Boundary condition between two layers expressed in numerical form.

conditions (5) and (6) (Fig. 4). An algorithm of the solution to the three-dimensional TNDT problem, boundary expressions and the programme "TEST" described below are obtained with the direct participation of Tanaseychuk of the Tomsk Polytechnical Institute. Let us assume that α_1, λ_1, α_2, λ_2 are thermophysical parameters of both layers; Δh_{z_1}, Δh_{z_2} are space steps along coordinate z; T_B the temperature at the boundary between the materials and T_{B-0} and T_{B+0} are the boundary temperatures for the approaches from left and right respectively. The general boundary conditions are

$$T_{B-0}(\tau) = T_{B+0}(\tau) \tag{13}$$

$$\lambda_1 \frac{T_{B-0}(\tau)}{\partial z} = \lambda_2 \frac{T_{B+0}(\tau)}{\partial z} \tag{14}$$

By Taylor's theorem the temperatures T_{B-0} and T_{B+0} may be obtained as follows:

$$T_{B-0} = T_B - \Delta h_{z_1} \cdot \frac{\partial T_{B-0}}{\partial z} + \frac{\Delta h_{z_1}^2}{2} \frac{\partial^2 T_{B-0}}{\partial z^2} \tag{15}$$

$$T_{B+0} = T_B + \Delta h_{z_2} \cdot \frac{\partial T_{B+0}}{\partial z} + \frac{\Delta h_{z_2}^2}{2} \frac{\partial^2 T_{B+0}}{\partial z^2} \tag{16}$$

Simultaneously the differential heat transfer equations valid for the two layers are

$$\frac{1}{\alpha_1}\frac{\partial T_{B-0}}{\partial \tau} = \frac{\partial^2 T_{B-0}}{\partial z^2} \tag{17}$$

$$\frac{1}{\alpha_2}\frac{\partial T_{B+0}}{\partial \tau} = \frac{\partial^2 T_{B+0}}{\partial z^2} \tag{18}$$

Substitution of (15) and (16) into (13) and (14) together with (17) and (18) gives, for the final expression:

$$T_B(\tau + \Delta\tau) = T_{B-\Delta hz_1}(\tau) + K T_{B+\Delta hz_2}(\tau) + (V - 1 - K)\, T_B(\tau) \tag{19}$$

where $K = \lambda_2 \Delta hz_1/\lambda_1 \Delta hz_2$; $V = (\Delta h^2_{z_1}/2\alpha_1 + \Delta h^2_{z_2}/2\alpha_2)$.

Similar ideas have been used to obtain the relevant expressions for boundary points sliding along the defect's side, moving perpendicularly to it and for the top points of the defect (in the last case we used the equation of heat balance for eight diagonal points which were not associated with the boundaries).

D. Programme "TEST"

This programme has been compiled, according to the above-mentioned algorithm, for the solution of three-dimensional nonlinear transient TNDT problems according to any of the testing variants shown in Fig. 1. It is assumed that the specimen being investigated consists of N regions, including an adhesive layer and a defect. Usually $N = 4$ (two plates are connected with an adhesive). The programme is concerned with either a moving heat beam or with a motionless large area heater. Each variant usually involves the calculation of the testing process for three to six different defects placed at three different depths in the specimen. As a result we obtain the temperature distribution over each defect, maximum temperature signal and contrast and respective time delays, both for the heated surface H and for the rear surface R. The programme "TEST" is written in FORTRAN and initially was composed for the CDC-7600 computer at Manchester University Computing Centre. The calculation time of one variant is usually about 120–1200 octal seconds for a standard mesh of $140 \times 10 \times 8$ points, depending on the testing conditions.

VI. Results

A. Analytical Expression

Dealing with analytical calculations we need to consider the so-called "infinite" defects. By this we mean defects of such large cross-sectional area in the X Y plane that further increase in size does not change the temperature signal over them. The reason for specifying this is that it is not usually possible to take into account the two- or three-dimensional character of the temperature field and results which may be obtained do not depend essentially on X-Y coordinates. That is why the analytical results merely represent an illustration to the real processes inside the specimen to be tested.

One of the first analytical attempts to relate the defect size and temperature signal was probably made by Vavilov *et al.* (1973) who considered TNDT of spot welds on an initially heated specimen. It was assumed that each welded point is just a single sphere which may or may not contain an internal air bubble. The problem was considered to be one-dimensional with no heat transfer through a defect. It did not take into account the influence of the ghost sheets so the boundary conditions were rather uncertain. The expression obtained showed that as the nondefect points cool exponentially the temperature difference between them and the defect is given by

$$\frac{\Delta T}{T_i} = \exp\left(-3\frac{K\alpha}{\lambda R}\tau\right) - \exp\left(-3\frac{K\alpha}{\Delta\lambda R}\tau\right) \tag{20}$$

where R is the diameter of the welded zone; T_i is the initial temperature difference between the point and ambient: Δ a function specified by the defect diameter (r_d) according to

$$\Delta = (1 - r_d/R)[3r_d/R + (1 - r_d^2/R^2)] \tag{21}$$

(Note that Δ tends to unity at small r_d).

It follows from (20) that the temperature signal is a function with one maximum, which may be obtained from:

$$\tau_{max} = \frac{\lambda R}{3K\alpha}\frac{\ln\Delta}{\Delta - 1} \tag{22}$$

The practical application of eqns (20) and (22) has an element of uncertainty dependent on the heat change coefficient K, which includes here not only convection and radiation but mostly the heat flux through the welded sheets. However, it is possible to find this constant experimentally and then it should prove possible to determine the influence of an air bubble on the temperature.

Another problem that has been treated concerns the uniform heating of plates (Vavilov *et al.*, 1978). Boundary conditions on the surface to be heated include uniform heating and cooling by convection; boundary conditions for the defect boundary and the specimen's rear surface are given by eqn (9), which means a lack of heat flux through both surfaces. In this case the temperature expression is the same for either defect or nondefect areas:

$$t/t_0 = 1 - \sum_{n=l}^{\infty} A_n \cos\mu_n \exp(-\mu_n^2\tau) \tag{23}$$

where A_n, μ_n are functions of either specimen thickness L or defect depth, l. Usually only a few members of the series in eqn (23) are needed. For thin nonmetal plates

$$t/t_0 = 1 - \cos\sqrt{Kl/\lambda} \, . \exp(-K\alpha\tau/\lambda l) \tag{24}$$

The optimum time for temperature monitoring if heating continues may be obtained by differentiation of the respective expressions given by eqn (24) for defect and nondefect areas:

$$\tau_{max} = \frac{\lambda(\ln L/l)L \, l}{K\alpha(L-l)} \tag{25}$$

This expression is similar to eqn (22) revealing a similar mechanism of useful signal creation. The most important conclusion from eqns (22) and (25) is that the maximum temperature contrast $A = t/t_0$ occurs earlier the smaller are λ, L, l and the larger are K and α. Unfortunately separate consideration of defect and nondefect areas is possible only for large defects in low conductivity materials. Metals are characterized by a high level of heat transfer in the $X-Y$ plane so the problem becomes a multidimensional one. There have been attempts to consider the parameters of defects by Vavilov *et al.* (1973) and Storozhenko *et al.* (1977). The results obtained seem to be quite complex and merely illustrate that the neglect of heat transfer through the defect causes the largest errors when the defect thickness is small and the conductivity of the material being tested is high.

Among other analytical expressions we should indicate the formula for the temperature t of a semi-infinite specimen heated by a moving heat beam (Rykalin *et al.*, 1975),

$$t = \frac{Q\sqrt{\alpha D/V}}{0{\cdot}885\lambda} \tag{26}$$

where D is the diameter of the heat beam and V is the speed of heating.

B. *Numerical Solution*

1. *Investigated Materials*

For calculations we have chosen four materials (representing different classes from the point of view of thermophysical parameters) and two adhesives (Table III). The range of conductivity is from 0·2 W/m.K to 100 W/m.K which covers most known materials. We tried to identify some common features to find roughly the TNDT parameters for any particular problem. The values of heat beam size cover the range from 16 mm^2 (4 × 4) to large area heaters. Testing speed ranged from 2·6 to 10·6 mm/s and specimen thicknesses from 2·4 to 8·3 mm. Most of the results, unless otherwise stated, are concerned with an epoxy-glue adhesive layer of thickness 0·05–0·5 mm.

Some preliminary remarks are perhaps necessary. Unfortunately we have failed to find any general relationships between all the parameters. This is why we give most results for a fixed NDT condition: testing speed $V = 5{\cdot}2$ mm/s;

TABLE III
The Thermophysical Properties of the Materials Investigated (Robichaud, 1969)

Material	Thermophysical properties: Conductivity (W/m . K)	Thermophysical properties: Diffusivity (m²/s)
Air (as a defect)	0·027	$2{\cdot}36 . 10^{-5}$
Plastic	0·2	$2{\cdot}54 . 10^{-7}$
Teflon	3·5	$1{\cdot}59 . 10^{-6}$
Zircaloy	12·5	$0{\cdot}624 . 10^{-5}$
Brass (65% Cu, 35% Zn)	103	$3{\cdot}17 . 10^{-5}$
Silver	430	$8{\cdot}17 . 10^{-4}$

heat beam size $S = 40$ mm (line heater); the defect being an air bubble inside an epoxy adhesive layer of thickness 0·1 mm. Some results merely compare the particular parameters for materials with very different thermophysical properties, e.g. plastic and brass. Although we have investigated defect sizes in the range 2–40 mm most results are concerned with 4×4 mm defects at differing depths. In spite of using such precisely defined conditions we discuss variations of all the parameters and deduce whether it is necessary to take these variations into account. Subscript H specifies the surface which has been heated, subscript R the rear surface. With A_H and τ_H we shall define the maximum temperature contrast and optimum time delay on the H-surface. The R-surface is characterized by a maximum temperature signal Δt_R at time τ_R but for generalization we shall also give the temperature contrast. A_R corresponds to τ_R (as will be shown below the R-contrast has no maximum and increases constantly when moving nearer to the heated zone).

2. *Influence of Boundary Conditions*

The question whether nonlinear conditions on both surfaces of the specimen have a pronounced effect on TNDT parameters is of interest because if it is not so, all the results obtained do not depend on heat flux density Q and the theoretical calculations can be simplified. The results presented in Table IV show that the TNDT problem can be viewed as a linear one until the temperature is at least 150 °C (for this particular case this is the threshold of thermal destruction of plastic at $Q = 20\,\mathrm{kW/m^2}$; $S = 40$ mm; $V = 5{\cdot}3$ mm/s). The errors for all variants are less than 2% for the contrast and less than 5% for the time delay, having a tendency to decrease with diminution of defect size. Moreover what is most important for the analytical calculations, the heat change constant h, could be assumed equal to zero with only a small increase of contrast and extension of the time delay.

TABLE IV

Influence of Boundary Conditions on the Main Parameters of TNDT for Plastic ($L = 8{\cdot}1$ mm; $S = 40$ mm; $V = 5{\cdot}3$ mm/s; $d = 0{\cdot}1$ mm; $l = 2$ mm)

TNDT Parameters	Defect size (mm²)	Boundary conditions: Linear $K = 0$ $Q = 0$–$20\,\mathrm{kW/m^2}$	Boundary conditions: Linear $K = 10\,\mathrm{W/m^2\,K}$ $Q = 0$–$20\,\mathrm{kW/m^2}$
A_H	12 × 12	0·149	0·144
	4 × 4	0·109	0·107
τ_H (s)	12 × 12	15·8	14·2
	4 × 4	11·7	10·9
A_R	12 × 12	0·081	0·050
	4 × 4	0·028	0·028
τ_R (s)	12 × 12	23·4	23·0
	4 × 4	21·1	20·8
		Boundary conditions	
		Non-linear $K = 0$ $Q = 0$–$20\,\mathrm{kW/m^2}$	Non-linear $K = 10\,\mathrm{W/m^2\,K}$ $Q = 0$–$20\,\mathrm{kW/m^2}$
A_H	12 × 12	0·143	0·142
	4 × 4	0·107	0·106
τ_H (s)	12 × 12	13·6	13·6
	4 × 4	10·5	10·5
A_R	12 × 12	0·050	0·050
	4 × 4	0·028	0·028
τ_R (s)	12 × 12	23·0	23·0
	4 × 4	20·4	20·4

3. *Topology of Temperature Field. Determination of Defect Sizes*

The computed temperature profiles, curves 1, 3 presented in Fig. 5, describe the TNDT of plastic with a line heater ($Q = 10\,\mathrm{kW/m^2}$, an easily obtained value) at the optimum heater–radiometer distances (when Δt is a maximum). Some features of TNDT are clearly shown from this result: (a) for a defect depth

$l = 2$ mm, defect sizes of 30×30 and $12 \times 12\,\text{mm}^2$ produce the same signal, so these defects could be classified as "infinite" for these conditions; (b) there is the influence of large neighbouring defects on each other (note the change of nondefect level t_{nd} between defects).

At the present time the real problem is the determination of the defect size and form using thermograms or temperature profiles. Looking at curves 1, 3 we could visualize an appropriate level of isotherm from which the real defect form and hence size could be defined on the thermoviser screen or recorder tape. But this operation requires a knowledge of the material and at least the depth of defect; then a suitable choice of time and monitoring level, etc. could characterize the defect. We have found the criterion which seems to be independent of all the TNDT parameters. This is the surface derivative $\partial t/\partial Y$ of the temperature profile along any line (curves 2, 4 in Fig. 5). For any variations of L, l, V, S, d, h, λ and a, the extremes of the derivative lie on the projection of the defect edge on the specimen surface. The amplitude derivative $\partial t/\partial Y$ depends on the useful signal amplitude (about 0·5 °C/mm and

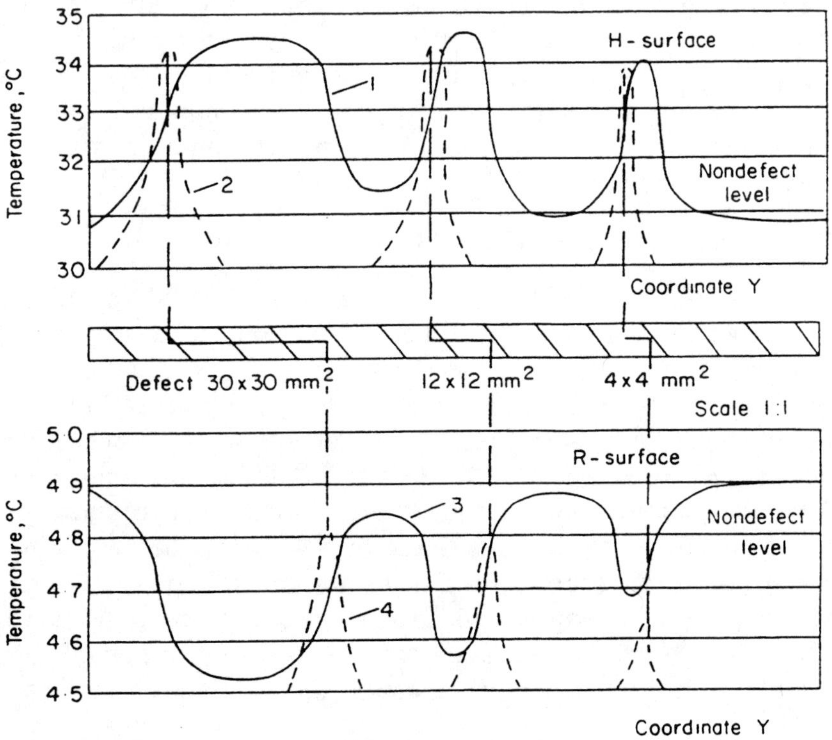

FIG. 5. Temperature profiles (curves 1, 3) and derivatives $\partial t/\partial Y$ (curves 2, 4) on the H, R surfaces for plastic. $L = 6{\cdot}1$ mm; $l = 2$ mm; $V = 5{\cdot}3$ mm/s; $Q = 10\,\text{kW/m}^2$; $S = 40$ mm (line heater): 1 — $\tau_0 = 7{\cdot}25$; 3 — $\tau_0 = 12{\cdot}8$ s. (Note that the derivative maximum corresponds to a rising temperature signal, conversely the minimum occurs for a decreasing temperature signal.)

0·05 °C/mm for the H and R surface respectively) and its sign depends on the sign of Δt and the direction of scanning. Physically this effect could be explained by saying that the heat flux $\lambda \partial t/\partial Y$ at any depth has its maximum near the defect edge and its projection on the surface. The effect of some variables will now be briefly considered.

4. *Second-order Parameters*

A researcher who attempts to find common relationships between some TNDT parameters soon encounters a difficult task because the contrast A and the optimum time delay τ_{opt} depend on the following set of parameters: α, λ (for specimen and defect), L, d, h, S, V, τ and even the configuration of the defect. In an earlier study (Storozhenko *et al.*, 1978) it was thought that certain parameters do not significantly affect A and τ_{opt} (e.g. the independence of τ_{opt} on V and independence of A on defect thickness for thermally insulated defects, etc.). But usually such conclusions have only been derived for particular materials or for a restricted variety of TNDT conditions. Our calculations show that, in general, there are no absolute and invariant parameters or their sets. However calculation of errors for various cases allows some assumptions to be accepted. The aim of such an operation is to facilitate approximate determinations of TNDT sensitivity for new materials and new testing requirements. (We believe that at the present time an accuracy of about 20–50% is acceptable).

Cross-sectional area of defects

Excluding from consideration the configuration of the defect, we have calculated some variants for defects of the same cross-sectional (and whole) area, F, but with different sizes along the X, Y coordinates. (Note: since the heat beam moves along the Y coordinate, the Z coordinate is inside the specimen.) The results are presented in Table V. As can be seen for even a three-fold change of defect size, h, the errors are usually only about 10–25%, depending on the defect depth (there is just one exception for the defect 4 × 36 mm at $l = 6$ mm, when the error is 45%). So, below, all our results are concerned with square form defects assuming that, firstly, very stretched defects are rather rare for bonded structures and, secondly, as we mentioned above, TNDT needs some standard defects. In experimental works they are usually drilled holes; here we assume a square shape.

Heat Beam Size

Comparing the NDT of plastic and brass with a large range of heat beam size from line heater to spot heater, we have obtained the results shown in Table VI, for a defect of depth 2 mm and defect size 4 × 4 mm^2.

TABLE V

Influence of Configuration of Defects on TNDT Parameters for Teflon ($L = 8{\cdot}1$ mm, $d = 0{\cdot}1$ mm, $V = 5{\cdot}3$ mm/s, $S = 4 \times 4$ mm^2)

TNDT parameters		Cross-sectional area of defect $F = 144$ mm^2: 12 × 12	24 × 6	4 × 36	Cross-sectional area of defect $F = 16$ mm^2: 4 × 4	8 × 2
$l = 2$ mm	A_H	0·878	0·827	0·635	0·508	0·434
	A_R	0·526	0·440	0·470	0·364	0·297
	τ_H(s)	10·9	11·7	8·3	6·4	6·8
	τ_R(s)	7·6	6·8	6·8	5·6	5·6
$l = 6$ mm	A_H	0·092	0·067	0·051	0·029	0·030
	A_R	0·546	0·519	0·470	0·440	0·427
	τ_H(s)	13·2	13·2	12·1	10·6	10·6
	τ_R(s)	7·9	7·5	6·8	6·4	6·4

TABLE VI

Influence of Heat Beam Size on Temperature Contrast A and Optimum Time Delay τ ($L = 8{\cdot}1$ mm; $d = 0{\cdot}1$ mm; $V = 5{\cdot}3$ mm; $l = 2$ mm; $h \times h = 4 \times 4$ mm^2)

	Heat beam size							
	$S = 40$ mm (line heater)				$S = 4 \times 4$ mm^2 (spot heater)			
Material	A_H	A_R	τ_H(s)	τ_R(s)	A_H	A_R	τ_H(s)	τ_R(s)
Plastic	0·12	0·05	14·0	24·9	0·11	0·023	10·9	20·8
Brass	0·06	0·04	− 3·8[a]	−0·4[a]	0·11	0·1	1·5	0·7

[a] Negative value of τ means that the optimum is inside the heat beam.

The reduction of S causes an increase in contrast A and a longer time delay τ, especially for metals. For nonmetals it occurs mainly on the rear surface. But in general it could be pointed out that a wide variation of beam size does not lead to more than a doubling of the signal.

Heating Speed

The results presented in Fig. 6 for $S = 40$ mm (line heater) show that the growth of V causes higher values of contrast and longer time delays. If the time of heat input, $\tau_h = S/V$, is introduced it can be shown that, beginning from the certain value of τ_h. the H-contrast does not depend on τ_h. In our example for

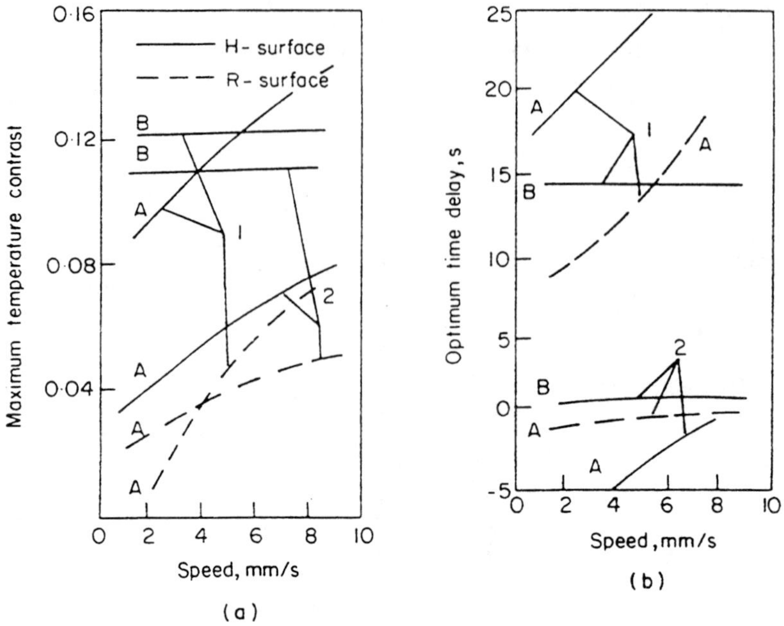

FIG. 6. Influence of heating speed on the main TNDT parameters; $L = 8{\cdot}1$ mm; $l = 2$ mm; defect size $= 4 \times 4\ \mathrm{mm}^2$; 1—plastic; 2—brass. A-s = 40 mm (line heater); B-s = $4 \times 4\ \mathrm{mm}^2$ (spot heater).

spot heating, with $S = 4 \times 4\ \mathrm{mm}^2$, the TNDT parameters do not vary with V so the choice of suitable speed is dictated, firstly by τ_h and a necessary value of A and, secondly, by the adequate level of heating according to eqn (26). It is important to note that for brass (Fig. 6b) the increase of speed allows the optimum monitoring point to be moved from inside the heat beam to outside it. This is a very important practical conclusion which means that faster speeds are favoured especially for metals.

5. *First-order Parameters*

Time

The dramatic influence of time on the signal in TNDT has been demonstrated by a number of authors who applied IR techniques to active TNDT (Alzofon, 1961; Ferguson, 1968; Kubiak, 1968; Vogel, 1968; Green, 1968, 1971; Kutzscher and Zimmerman, 1968; Stiefeld and Yoshimura, 1971). A certain distance b between the heat zone and the monitoring point is required in any thermal test especially with metals. The influence of b on the temperature distribution surrounding defects is shown in Fig. 7 for Teflon. These results are associated with a spot heater $S = 4 \times 4$ mm, $Q = 5{\cdot}10^5\ \mathrm{W/m}^2$ (e.g. a laser beam with absorbed energy 9 W) moving with speed $V = 5{\cdot}3$ mm/s. The specimen temperature above ambient is of the order of 25–60 °C for the H-surface

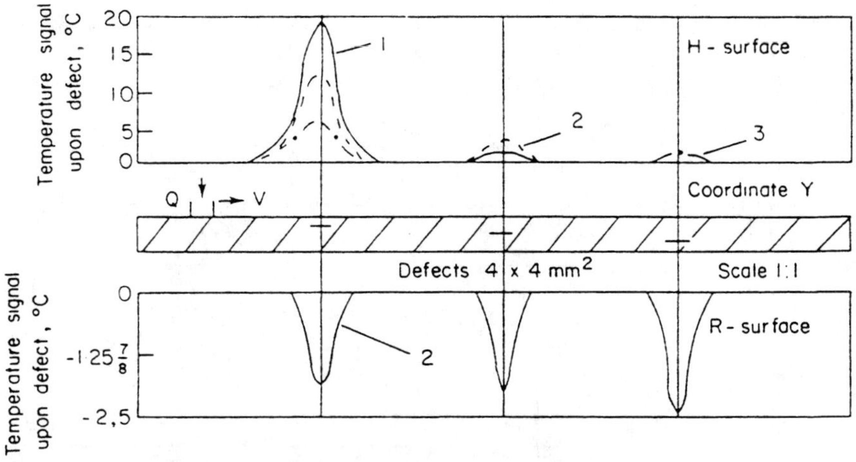

FIG. 7. Temperature profiles for different radiometer-heater distances. Teflon $L = 8{\cdot}1$ mm; $S = 4 \times 4$ mm^2; $Q = 5{\cdot}10^5$ W/m^2; $V = 5{\cdot}3$ mm/s; 1—b = 14 mm; 2—b = 30 mm; 3—b = 46 mm. Defect depths are 2, 4, 6 mm, respectively.

(b = 46–14 mm) and about 5 °C for the R-surface (b = 30 mm). For the R-surface the optimum value of b does not change significantly for different defect depths (this is typical for any TNDT conditions) but the signal on the H-surface depends strongly on b. Each of the three b values is optimum for one value of the defect depth. If we need to detect defects at any depth the best value of b is the optimum one for the deepest defect (b = 46 mm for l = 6 mm). Although we could create better conditions for the detection of subsurface defects (e.g. by decreasing the b value) deeper defects would be missed.

Another point in this consideration is a comparison of the actual behaviour of the absolute temperature signal and the temperature contrast vs b for the H and R surfaces, which is presented in Fig. 8. On the H-surface, as contrast A and temperature signal ΔT reveal, the specific maxima occur at different b values. Outside the heat beam, contrast occurs later than the temperature signal whereas inside the heat beam the converse is true. This difference reduces for smaller specimen thicknesses and larger defect depth but in any case it could reach 20–30 mm both for metals or nonmetals. Hence, it has to be taken into account. On the R-surface the temperature signal behaves similarly but the contrast increases constantly in the direction of the point situated opposite the beam, so for any particular case the apparatus arrangement needs to be based on a compromise between higher contrast but lower heating level. These results confirm the earlier conclusion that sensitivity enhancement for this regime is connected with higher temperature resolution of the IR system (Karpelson *et al.*, 1975).

Next we shall operate with maximum contrast on the H-surface A_H and the optimum time $\tau_H = b/V$ corresponding to it. For the R-surface, an optimum

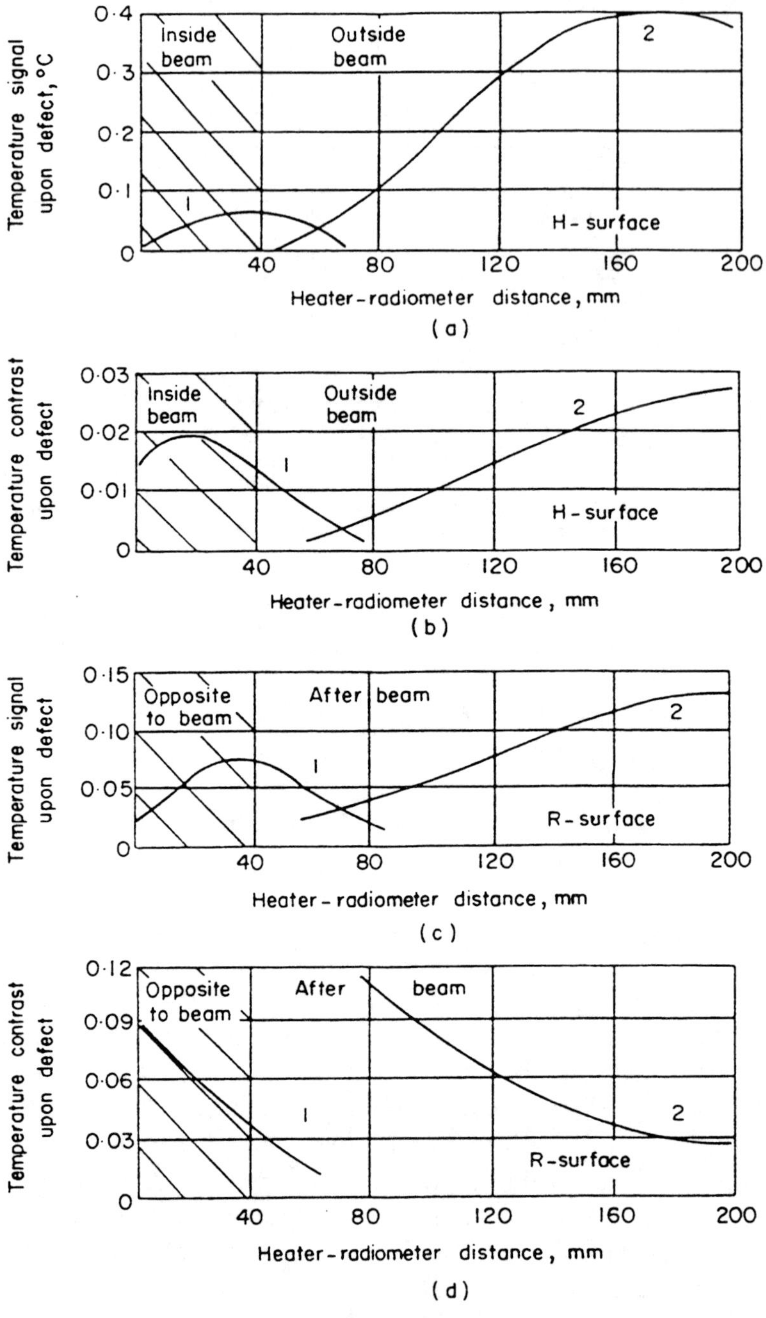

FIG. 8. Relationship between temperature signal and temperature contrast vs heater–radiometer distance for H and R surfaces; $L = 8{\cdot}1$ mm; $l = 4$ mm; defect size $= 4 \times 4\ mm^2$; $S = 40$ mm (line heater); $Q = 10\ kW/m^2$; $V = 5{\cdot}3$ mm/s; 1—brass; 2—plastic.

time τ_R is valid for the maximum temperature signal and contrast A_R relates to the same point. In addition it is interesting to note that for large values of τ the so-called inversion of signal sign can occur but the level of signal in this case tends to be low and this probably explains why this effect has not been described experimentally.

Finishing this section we present in Table VII the optimum time delays defined as $\tau = h/V$ for all investigated materials and different defect depths for fixed TNDT conditions; specimen thickness 8·1 mm, defect size 4 × 4 mm, beam size 40 mm, testing speed 5·3 mm/s. The influence of different L, h, d will be shown below. Reduction of heat beam size and increase in speed causes a slight increase of optimum time within a range of a few seconds. This effect is most important for metals because it allows the monitoring point to be moved from within the heat beam area where testing is impossible because of reflected radiation (negative values of τ_H corresponding to inside beam points occur in TNDT of metals with relatively low heating speeds). Table VII illustrates clearly the influence of the material (specifically its diffusivity) and defect depth on τ. Thus far we are unable to explain the maximum values of τ_R for $l = 4$ mm (R-surface) but this effect seems to be quite stable.

TABLE VII

Time Conditions for TNDT of Defects 4 × 4 mm^2 inside Various Materials of Thickness 8·1 mm, $S = 40$ mm, $V = 5·3$ mm, $d = 0·1$ mm

Material	Defect depth (l) (mm)	Optimum time delay (s)	
		H-surface	R-surface
Plastic	2	10·9	20·8
	4	32·1	29·4
	6	>40[a]	25·5
Teflon	2	3·3	4·1
	4	6·0	4·5
	6	8·2	4·9
Zircaloy	2	1·1	0·7
	4	1·9	1·1
	6	1·8	0·7
Brass	2	−3·8	−0·7
	4	−3·8	−0·4
	6	−3·8[a]	−0·4[a]

[a] These values are approximate.

Defect size and depth

Most of the results were obtained for a defect thickness $d = 0{\cdot}1$ mm. The influence of different d values is shown in Fig. 9. It is evident that for researched materials heat transfer through the defect is not negligible for $0{\cdot}1 < d < 0{\cdot}3$ mm. Its change causes an approximately linear change of contrast and time delay. But more dramatic is the influence of defect size in a plane perpendicular to the main heat flux direction. We have found the parameter $L.l/(L-l)$ to be useful since this allows the H-contrast for various testing conditions to be fitted to one of the curves shown in Fig. 10 with an error of less than 15%. These curves (Fig. 10) represent the joint influence of cross-sectional area of defect, its depth

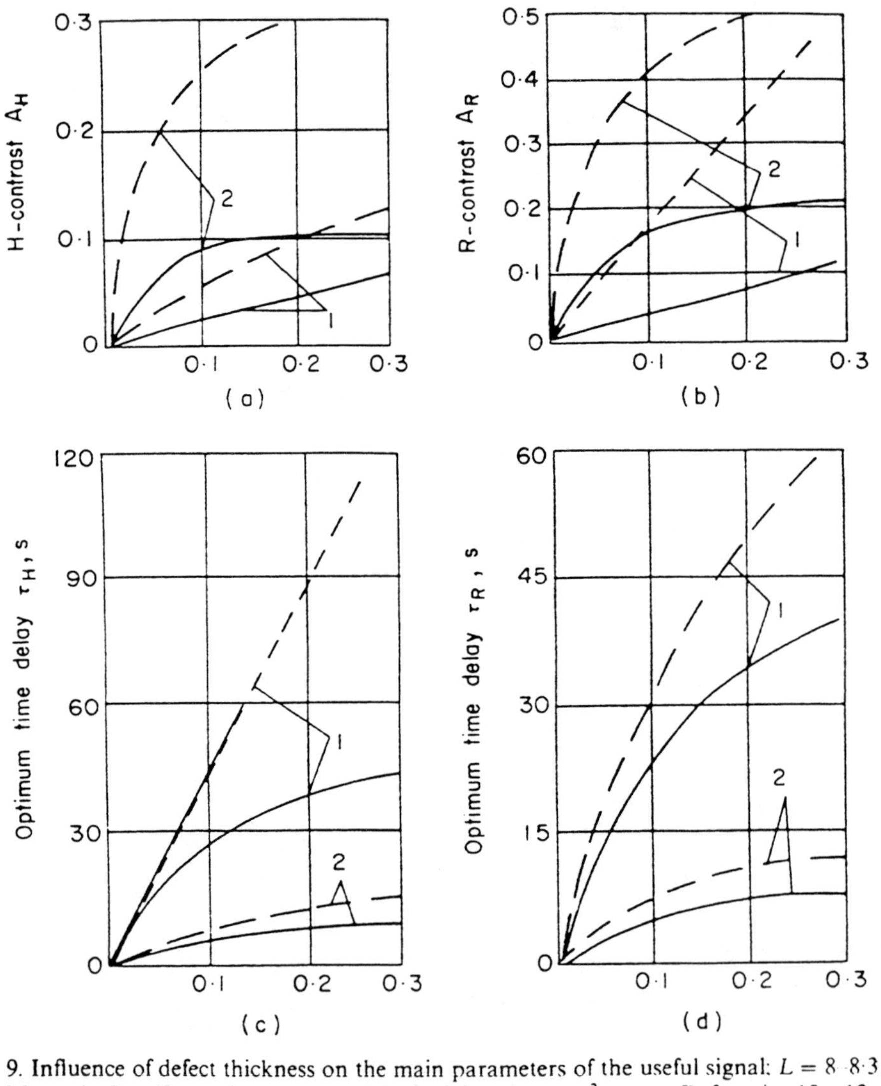

FIG. 9. Influence of defect thickness on the main parameters of the useful signal: $L = 8$–$8{\cdot}3$ mm; $V = 5{\cdot}3$ mm/s; $S = 40$ mm; $l = 4$ mm. ——Defect size 4×4 mm^2. - - - - - -Defect size 12×12 mm^2.

and specimen thickness and illustrate the difficulties of detecting small defects in materials having a high thermal conductivity (curve 4 for brass). The *R*-contrast, according to Fig. 10(c), does not vary much with defect depth for an *L* value. The variation of optimum time delay with *l* and *h* can also be illustrated by the data in Table VII.

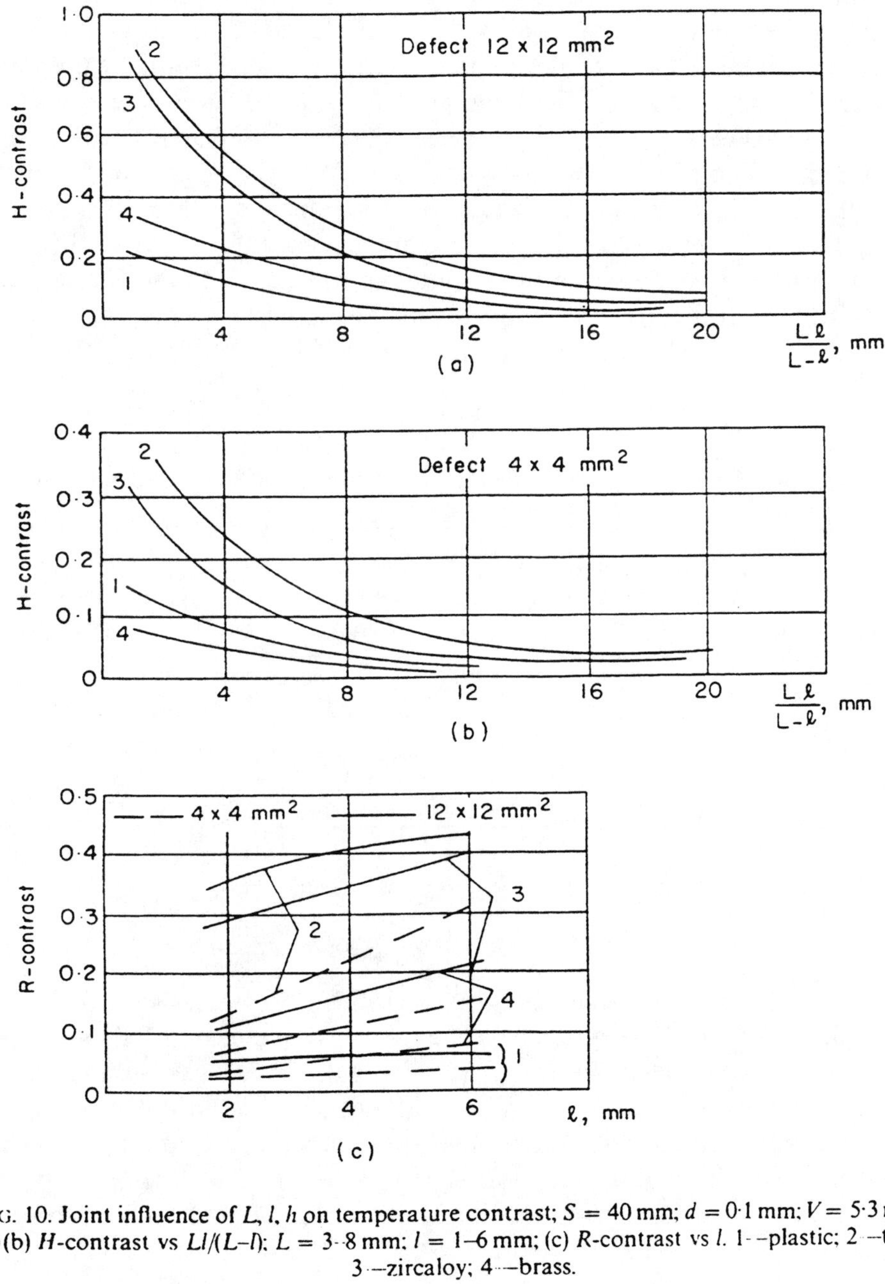

FIG. 10. Joint influence of *L*, *l*, *h* on temperature contrast; $S = 40$ mm; $d = 0{\cdot}1$ mm; $V = 5{\cdot}3$ mm/s; (a)(b) *H*-contrast vs $Ll/(L-l)$; $L = 3{\cdot}8$ mm; $l = 1$–6 mm; (c) *R*-contrast vs *l*. 1—plastic; 2—teflon; 3—zircaloy; 4—brass.

6. *Comparison of Spot, Line and Uniform Heating*

One of the variants for Teflon was calculated for TNDT by uniform heating of the specimen for 7·5 s with further monitoring of the temperature field. The comparative results given in Table VIII show that from the point of view of signal level the spot heating regime is more suitable than line and uniform testing, which provide similar values for A. But testing rates, according to (1) and (2), are equal to 42 mm^2/s for spot heating and 2650 mm^2/s for line heating if the size of the heat beam along the Y coordinate is 250 mm; the possibility of such a line heater has been reported by Storozhenko and Gorbunov (1978). If it is assumed that the processing time for the thermovision method assumed is about 12 s (Table 8) the rate of inspection in this case will depend substantially on the heating area (see expression (2)). For an area 200 × 200 mm it has a potential productivity of 3300 mm/s but taking into account the time for apparatus rearrangements we shall obtain a value close to the line heating one. It is important that the rate of line testing be increased by higher testing speed and this is much easier to do than to provide uniform and powerful heating of a large area.

TABLE VIII

Parameters of Various TNDT Regimes for Teflon (L= 8·1 mm; d = 0·1 mm; defect 4 × 4 mm^2)

Defect depth (l) (mm)	Spot heating $S = 4 \times 4$ mm^2 V= 10·6 mm/s)		Line heating (S = 40 mm; V= 10·6 mm/s)		Uniform heating for 7·5 s			
	A_H	A_R	A_H	A_R	A_H	A_R	τ_H (s)	τ_R (s)
2	0·5	0·3	0·35	0·15	0·34	0·13	3	4·5
4	0·12	0·38	0·1	0·22	0·1	0·18	6	4·5
6	0·03	0·44	0·02	0·33	0·02	0·25	7·5	4.5

VII. EFFECT OF NOISE ON RESOLUTION

A. *Interpretation of Thermograms*

For a long time, when signal amplitudes in TNDT systems were of the order of 10–30 °C, very little attention was paid to decision making. This was especially true for stationary temperature fields where the defect was itself a heat source. But the transition to low level signals associated with passive TNDT of multilayered structures, and competition with more traditional methods of NDT, have altered this situation. A high level of false alarm has become the price to be paid for sensitivity enhancement. In connection with this some of the problems that have arisen are:

(a) Instructing operator staff in the art of thermogram interpretation. Whilst some "dark" areas may be disbonds, others result from scratching the black painted surface. In order to exclude such spurious signals, detailed visual examination is necessary. Even more complicated is the temperature profile shown in Fig. 11. Its interpretation will be given in Table X.

(b) Direct or indirect computerizing of TNDT systems allows temperature information to be presented in a form suitable for an operator to improve information access and to analyse isotherms, etc. (Stiefeld, 1973; Hsieh and Ellingson, 1977; Whitcomb, 1978). A unique distance where a computer was used for the analysis of a growing crack front during cyclic unloading of specimen was described by Whitcomb (1978). Standard programmes for three-dimensional graphics, plane evolution of cylindrical specimens and contouring of defects have been developed by Sandia Laboratories (Stiefeld, 1973).

(c) Developing physical methods to subdue noise (e.g. optimum choice of the spatial resolution of the IR device cancels the noise from surface nonuniformities of tyres, Lavery *et al.*, 1971).

The choice of optimum criteria for defect identification is an important consideration for all NDT methods, and there is an urgent need to collect and collate statistical information about each class of material investigated by TNDT. As an example of this we could cite the work of Day (1968) who has shown that a suitable choice of monitoring level enables defects to be classified as "delamination" or "porosity". Application of pattern recognition techniques in TNDT described by Hsieh and Ellingson (1977) for ceramic materials illustrates the complicated character of three-dimensional temperature fields even for relatively big defects.

B. Noise Sources. Problems of Emissivity

For a scanning IR radiometer with a chopper the signal from the receiver will be

$$U \sim \phi_s + \phi_r + \phi_i - \phi_c \tag{27}$$

where ϕ_s is radiation from the specimen; ϕ_r is radiation from outside sources which is reflected by the specimen's surface; ϕ_i is internal radiation of the IR radiometer; ϕ_c is reference radiation reflected by the chopper blade (Hudson, 1969).

Technically it is possible to reduce the last three terms in (27) to a minimum. Then signal fluctuations will be as follows:

$$\Delta U/U = \Delta\phi_s/\phi_s \tag{28}$$

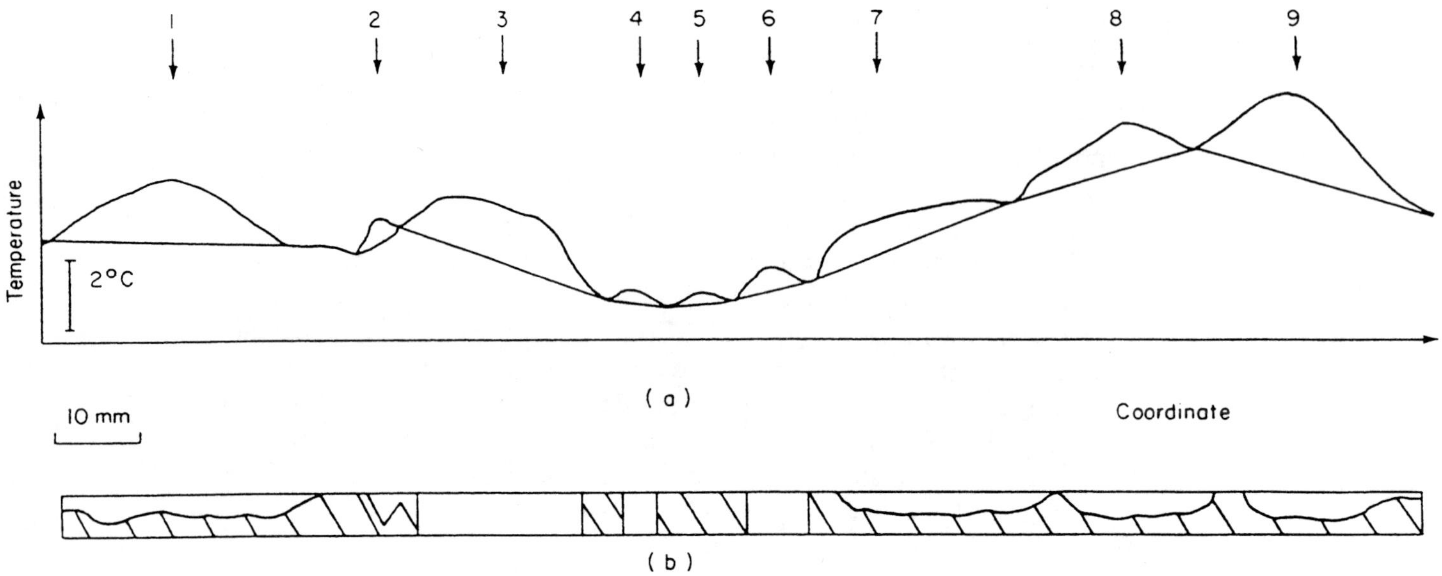

FIG. 11. Temperature profile for a Tufnol (5 mm)–aluminium (2 mm) specimen and the situation inside the adhesive layer, an epoxy glue. $Q = 5\,\mathrm{kW/m^2}$; $V = 1$ mm/s. Specimen temperature 35 °C.

We assume below that (a) an infrared sensor monitors the total specimen radiation (this is true for the most widely used detectors InSb, Ge(Au), CdTeGe); (b) fluctuations of the average temperature could be due to the presence of a defect and variations of thermophysical constants α, λ; speed V; specimen absorptivity for the incident heat beam ε_h, and heat beam density Q.

Total radiation from the specimen is given by the Stefan Boltzmann law

$$\phi_s = \varepsilon_r \sigma T^4 \tag{29}$$

where ε_r is the total average emissivity; T is the absolute temperature of the specimen; $T = t_a + t_h$; t_a is the ambient temperature; t_h is the heating level. For t_h expression (26) may be used:

$$t_h = \frac{Q\varepsilon_h\sqrt{\alpha S/V}}{0{\cdot}885\lambda} \tag{30}$$

where ε_h differs from ε_r because of the different radiation spectra of heater and specimen.

From (28) to (30):

$$\frac{\Delta U}{U} = \frac{\Delta\varepsilon_r}{\varepsilon_{rr}} + \frac{4t_h}{t_a + t_h}\left(\frac{\Delta Q}{Q} + \frac{\Delta\varepsilon_h}{\varepsilon_h} + \frac{\Delta\alpha}{2\alpha} - \frac{\Delta\lambda}{\lambda} - \frac{\Delta V}{2V} + \frac{\Delta T}{t_h}\right) \tag{31}$$

where ΔT = signal from the defect.

If the criterion for defect detection is the contrast, $\Delta T/t_h$, the requirements for possible fluctuations of all the parameters can be found from (31). Let us assume that the specimen is quite uniform ($\Delta\alpha = \Delta\lambda = 0$), heating time fluctuations are rather slow and can be excluded by frequency filtering and ΔV is reduced to zero. In this case

$$\frac{\Delta U}{U} = \frac{\Delta\varepsilon_r}{\varepsilon_r} + \frac{4t_h}{t_a + t_h}\left(\frac{\Delta\varepsilon_h}{\varepsilon_h} + \frac{\Delta T}{t_h}\right) \tag{32}$$

It is obvious that emissivity fluctuations are quite dramatic from the point of view of sensitivity enhancement. If $t_a = 300$ K, $t_h = 100$ °C, $\Delta\varepsilon_h = 0$, the temperature resolution of the IR system is equal to 0·2 °C; if $\Delta T = 1$ °C the signal-to-noise ratio assumed equal to 5 requires that fluctuations of ε_r must be less than 0·2%. Such a result is hardly possible in practice. This example illustrates then that surface conditions, rather than temperature resolution, play a key role in TNDT, so reduction of this kind of noise is always crucial in practical testing. There are two main directions for solving this problem. The first one requires preliminary information about surface conditions or controlled preparation of the surface before inspection. Another group of methods includes rational processing of the radiation flux, usually by spectral filtering. So far methods of the first group are more effective. Among them the most popular ones are the comparison of results with standard thermograms and the more commonly used one of black painting. But even in the latter case, as we shall see below, the noise level is rarely less than 3–4%. An interesting solution to the emissivity problem was devised by Green (1968) using two-channel scanning. A modification of this method consisting of the double consequent testing could, we believe, be effective for cases where black painting is neither desirable nor practicable.

C. *Concept of Local Signals and Determination of TNDT Possibilities*

The empirical approach to TNDT could include a count of the so-called "local signals" for any temperature data set (e.g. measuring of signal amplitude between neighbouring extremes). The simplest decision making requires a comparison of these local signals with those for a nondefect specimen. *A priori* some statistical data would be necessary in this case. We have made some experiments with the objective of finding the maximum noise signal for various surfaces. Testing procedure requires uniform heating of the specimen up to 40–80 °C and then monitoring temperature profiles with an IR radiometer. A typical distribution is presented in Fig. 12 and the maximum values of local contrast are given in Table IX. It is necessary to note the subjective character of local signal choice which depends on operator experience and influence on the character of distributions. We have chosen the largest signal as the level for decision making. Returning to Fig. 11 it may be stated that if nine local signals are found and all of these are viewed as originating over defects then quite a small adhesive-free area (up to 10 mm^2) could be detected and also a change of adhesive thickness. But in this particular case it will lead to false defects (see signal 5 and the data in Table X). If we could take into account that the maximum noise signal for the Tufnol could reach 3·5% (Table IX) only three among eight real defects will be revealed in this case (signals 1, 3, 9) although a false alarm signal will not occur. This example illustrates the difficulties of thermogram interpretation when the detection of small defects is required. In practice more complicated criteria for decision making should be used, e.g. the

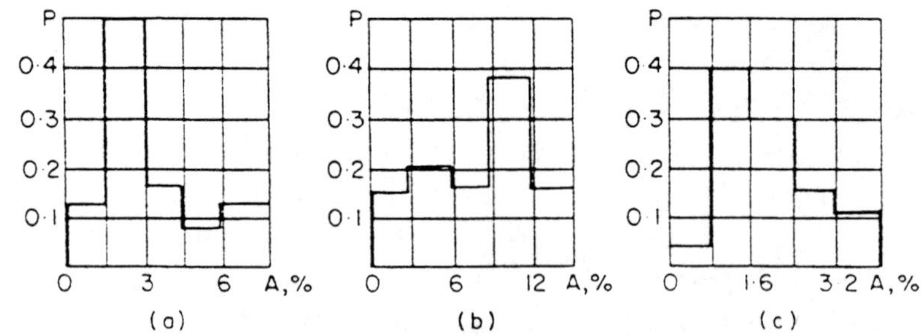

FIG. 12. Probability distribution for the local temperature contrasts; (a) steel oxidized; (b) steel coated with boron nitride; (c) steel covered with Aquadag.

TABLE IX
Maximum Noise Local Signals for Different Materials

Material	Maximum noise signal (%)
Cardboard	2
Plastic	3·5
Tufnol	3·5
Carbolite	1·9
Rubber	2
Teflon	2
Steel after welding	50
Steel oxidized	7·5
Steel with boron nitride	15
Steel covered by Aquadag	4
Brass covered by Aquadag	4
Brass covered by scratched Aquadag	11

TABLE X
Analysis of Local Signals for the Temperature Profile of Fig. 11; Tufnol (5 mm)–aluminium (2 mm) specimen; heating level—35 °C

Local signal	Amplitude (°C)	Contrast (%)
1	1·8	5·1
2	0·6	1·7
3	1·6	4·6
4	0·4	1·1
5	0·4	1·1
6	0·6	1·7
7	1·0	2·9
8	1·2	3·4
9	2·4	6·9

Neuman–Pearson criterion minimizing false alarm probability (Cox, 1970). The choice of such criteria, knowledge of noise statistics for the particular material being tested and theoretical results described above will allow the possibilities of TNDT to be calculated for any practical problem. We shall illustrate it with the following example. Assume a composite of plastic ($L = 9$ mm) and steel ($L = 3$ mm) is to be inspected by one-sided testing. The minimum size of defect to be detected is $4 \times 4\,mm^2$. Realistic values of the testing speed and heating parameters may vary substantially. We shall assume that line testing is preferable at a speed of about 5 mm/s. If the noise level is about 4% (black painted metal and natural nonmetal*) and the signal-to-noise ratio is three, a defect $4 \times 4\,mm^2$ could be detected up to a depth of 1·6 mm in plastic and 1·9 mm in steel according to Fig. 10. (Zircaloy data could be substituted for steel without affecting this conclusion). Without black painting, TNDT possibilities for steel are reduced to at least a depth of 1·3 mm and problems due to reflected radiation could also arise in this case.

VIII. Conclusions

We believe that active thermal nondestructive testing, TNDT, is not merely an unusual testing method useful only as an adjunct to traditional NDT methods but that it is, in its own right, a useful tool for obtaining additional and sometimes unique information about the situation inside layered composites and bonded structures. The level of IR systems development is now high enough to merit paying more attention to statistical procedures necessary to decision making and theoretical investigations of heat transfer through the specimens to be tested. An optimum choice of the parameters of TNDT systems and computerizing the testing process enhance the possibility of making TNDT an important industrial method in the not too distant future.

IX. Acknowledgements

We would like to thank Mr R. Sharpe (Harwell NDT Centre) for allowing us to use the Harwell NDT information store and Dr R. Procter (UMIST, Manchester) for assistance with computer programming.

References

Alzofon, P. E. (1961). "Application of Infrared Technology in NDT)). Proceedings of Conference, Missiles and Rockets Symposium, Concord, California.

Anon. (1961). Infrared camera spots malfunctions. *Electronic Design* 9, 12 (6 December).

Anon. "Bibliography of Infrared and Thermal Techniques for Nondestructive Testing". The Boeing Co., Philadelphia, Pa (distributed by the Infrared and Thermal Committee of ASNT).

Barker, G. A. (1934). "Apparatus for Detecting Forest Fires", US Patent No. 1959703.

Bekeshko, N. (1973). Nondestructive testing of glass–metal joints using the thermal method (in Russian). *Sov. J. of NDT (Defectoscopy)* 5, 118.

* Industrial plastic specimens are often multilayer constructions and substantially anisotropic, so this case requires further consideration.

Bekeshko, N. and Oshehepkov, P. K. (1969). Apparatus for recording thermal fields at temperatures above 150 °C (in Russian). *Sov. J. of NDT (Defectoscopy)* **1**, 94.

Beller, W. S. (1965). Navy sees promise in infrared thermography for solid case checking. *Missiles and Rockets* **16**, 22.

Bobo, S. (1974). Recent developments in the use of infrared for nondestructive tire testing. *Mat. Eval.* (July), p. 147.

Carlslow, H. S. and Jaeger, J. C. (1959). "Conduction of Heat in Solids", 2nd edition, p. 122. Clarendon Press, Oxford.

Cohen, J. and Sherman, E. (1973). The design and application of the traversing infrared inspection system---TIRIS. *Nondest. Test.* **6**(2), 74.

Cox, C. W. (1970). "Data handling techniques", *In* "Research Techniques in Nondestructive Testing" (R. S. Sharpe, ed.), Vol. 1. Academic Press, London and New York.

Day, C. K. (1968). "Infrared Testing of Bonds or Niobium Carbide Coated Graphite Support Blocks". Rep. BNWL–787, Battelle Pacific Northwest Labs., Richland, Washington, USA.

Dzthubez, R. K. *et al.* (1958). *In* "Apparatus and Method for Detecting Overheated Journal Boxes", US Patent No. 2856539.

Ferguson, J. L. (1968). Liquid crystals in nondestructive testing. *Appl. Optics* **7** (9), 1729.

Florant, L. E. (1964). Testing using infrared. *J. Am. Instrum. Soc.* **11**, 61.

Gorbunov, V., Vavilov, V. and Parvatov, G. (1974). Solution of two-dimensional problem of thermal NDT for a metal plate with a rectangular internal defect (in Russian). *Sov. J. of NDT (Defectoscopy)* **5**, 74.

Gorrill, W. S. (1949). Industrial high speed infrared pyrometer. *Electronics*, **22**, 112.

Green, D. R. (1968). Principles and application of emittance-independent infrared nondestructive testing. *Appl. Optics* **7**(9), 1779.

Green, D. R. (1971). Thermal and infrared nondestructive testing of composites and ceramics. *Mat. Eval.* **29** (11), 241.

Green, D. R. *et al.* (1973). Infrared NDT of electrical fuel pin simulators used in LMFBR thermal hydraulic tests. *Mat. Eval.* **41** (12), 247.

Hsieh, C. K. and Ellingson, W. A. (1977). "Feasibility of using Infrared Scanning to Test Flaws in Ceramic Materials", Paper to the 15th Int. Thermal Conductivity Conf., Ottawa, Ontario, Canada. 24–26 August, 1977.

Hsieh, C. K., Ellingson, W. A. and Su, K. S. (1979). A model study for quantitative flaw detection using infra-red scanning. *Brit. J. NDT* **21** (4), 185–192.

Hudson, R. D. (1969). "Infrared System Engineering". Wiley (Interscience).

Karpelson, A. E. *et al.*, (1975). Determination of optimum conditions of active thermal inspection of objects possessing a discontinuity. *Sov. J. of NDT (Defectoscopy)* **2**, 47.

Kubiak, E. J. (1968). Infrared detection of fatigue cracks and other near-surface defects. *Appl. Optics* **7**, 1743.

Kutzscher, E. W. and Zimmerman, K. H. (1968). A scanning infrared inspection system applied to nondestructive testing of bonded structures. *Appl. Optics* **7**(9), 1715.

Lavery, A. L. *et al.* (1971). "Survey of Nondestructive Tire Inspection Techniques", Rep. DOT–TSC–NHTSA–71–5 (July).

Lawson, W. D. and Sabey, J. H. (1970). "Infrared techniques". *In* "Research Techniques in Nondestructive Testing" (R. S. Sharpe, ed.), Vol. 1. Academic Press, London and New York.

Lykov, Ju. V. (1967). "Heat Conductivity Theory", Science and Technology (in Russian). Minsk Press, Moscow.

Merhib, C. P. and Taylor, L. (1972). "A Report Guide to Thermal Testing Literature". Techn. Rep. AD 740654 (April). Army Materials and Mechanics Research Center, Watertown, Mass.

Nichols, J. T. (1935). "Temperature Measuring". US Patent No. 2008793.

Popov, Ju. and Ketkovich, A. *et al.* (1975). Detection of separations in three layer products using a high speed thermal defectoscope (in Russian). *Sov. J. of NDT (Defectoscopy)* **6**, 62.

Popov, Ju. and Karpelson, A. *et al.* (1977). Possibility of active thermal inspection of honeycomb structures (in Russian). *Soc. J. of NDT (Defectoscopy)* **4**, 59.

Potapov, A. *et al.*, (1977). Automated infrared inspection of plastic products (in Russian). *Sov. J. of NDT (Defectoscopy)* **2**, 116.

Rapoport, D. (1978). "Solution of the two-dimensional problem applied to the inspection of multilayered products" (in Russian). *Sov. J. of NDT (Defectoscopy)* **8**, 86.

Renius, O. (1973). Laser illumination for infrared nondestructive testing. *Mat. Eval.* (May), p. 80.

Robichaud, R. E. (1969). "Development of a Thermal Nondestructive Inspection System to Detect Corrosion in Aircraft Structures". Techn. Rep. AD–863490, Naval Air Systems Command, Washington, USA.

Rykalin, N. N., Uglov, A. A. and Kokora, A. N. (1975). Laser processing of materials (in Russian). Moscow.

Sandiford, P. J. (1952). Simple method for detecting hot joints on electrical transmission lines. *Rev. Sci. Instr.* 23, G44.

Shultz, A. W. (1968). An infrared transient method for determining thermal inertia, conductivity and diffusivity of solids. *Appl. Optics* 7(9), 1845.

Singh, J. J., Andronicos, C. K. and McErlean, E. A. *et al.* (1978). "Effects of Moisture in the Infrared Thermography of Resin Matrix Composites". NASA, Techn. Memorandum 78649, NASA-TM, 78649 (January).

Smith, G. D. (1965). "Numerical Solution of Partial Differential Equations". Oxford University Press.

Stiefeld, B. (1973). "NDT and Interactive Data Display". USA Techn. Rep. SLA–74–5054. Sandia Lab., Albuquerque.

Stiefeld, B. and Yoshimura, R. H. (1971). A capability and limitations study of thermography of carbon–carbon cones. *Mat. Eval.* **12**.

Storozhenko, V. A. and Gorbunov, V. I. (1978). Principal problems of an active thermal non-destructive inspection. *Sov. J. of NDT (Defectoscopy)* **8**, 76.

Storozhenko, V. *et al.* (1977). Theoretical analysis of the optimum conditions of inspection of materials having low thermal conductivity by a thermal method. *Sov. J. of NDT (Defectoscopy)* **4**, 70.

Storozhenko, V. and Vavilov, V. *et al.* (1978). Estimate of the principal parameters of single-side thermal testing of layers-parting in epoxyglass (in Russian). *Sov. J. of NDT (Defectoscopy)* **9**, 41.

Tikhonov, A. and Samarsky, A. (1963). "Equations of Mathematical Physics", translated by A. R. M. Robson and P. Besu. Pergamon Press, Oxford.

Vandamme, G. E. and McGarvey, J. W. (1972). "Infrared NDT of Laminated Structures and Electrical Circuits". US Army Command Techn. Rep. SWERR–TR–72–35 (June).

Vavilov, V. and Gorbunov, V. (1974). Analytical calculation of temperature fields for thermal inspection of laminated products (in Russian). *Sov. J. of NDT (Defectoscopy)* **2**, 100.

Vavilov, V. and Gorbunov, V. (1977). Influence of radiation on the results of measurements of temperature fields (in Russian). *Sov. J. of NDT (Defectoscopy)* **1**, 65.

Vavilov, V. *et al.* (1973). Theoretical investigation one of the problems of thermal control of spot welds (in Russian). *Sov. J. of NDT (Defectoscopy)* **1**, 21.

Vavilov, V. *et al.* (1975). Some theoretical and experimental problems of thermal NDT methods. *Sov. J. of NDT (Defectoscopy)* **6**, 67.
Vavilov, V. *et al.* (1978). Comprehensive approach to the design of a thermal inspection system for soldered joints (in Russian). *Sov. J. of NDT (Defectoscopy)* **10**, 63.
Vogel, P. E. J. (1968). Evaluation of bonds in armor plate and other materials using infrared nondestructive testing techniques. *Appl. Optics* **7** (Sept.), 1739.
Whitaker, S. (1977). "Fundamental Principles of Heat Transfer". Pergamon Press, Oxford.
Whitcomb, J. B. (1978). "Thermographic Measurements of Fatigue Damage". NASA Techn. Memorandum (78693) (May).
Willburn, D. K. (1960). "An Infrared Diagnostic Technique for Evaluation of Automotive Tires". Tacom Propulsion Systems Lab. TACOM TR 11154 (AD 719692).

Reprinted from *Infrared Technology and Applications,* L.R. Baker, A. Masson, editors, Proc. SPIE Vol. 590, pp. 277-284 (1985). ©1985 SPIE.

TRANSIENT THERMAL TECHNIQUE FOR INFRARED NON DESTRUCTIVE TESTING OF COMPOSITE MATERIALS

Hervé TRETOUT - Jean-Yves MARIN

AVIONS MARCEL DASSAULT - BREGUET AVIATION
78 Quai Carnot - 92214 Saint-Cloud - FRANCE

Abstract

With the fast growing utilisation of composite materials in aircraft structures, manufacturers have to acquire appropriate non destructive testing systems. A large effort have been devoted for the last ten years to the improvement of two major techniques : ultrasonics and radiography. Up to now, these methods have been fully automated and their efficiency proven. However a demand exists for new methods which can offer a higher inspection rate than ultrasonics and can detect a wider range of defects than radiography. These considerations led us to look to thermography. Infrared non destructive testing is attractive since it requires no physical contact. It is potentially very rapide and easy to set up. It can be applied when only one side of a surface is accessible. Real time display and recording associated with image treatment make the interpretation of thermograms easy. A large investigation programme is carried out in order, to define the field of application of thermography to composite materials evaluation, to set up a method which can be industrialised, to write down specifications for an industrial installation. Thermophysical properties of the materials, different heat sources, infrared detectors and image processing systems are investigated. Calibration of the apparatus is also considered. An original method using the combination of a line heater, a line scanner and a real time digital signal processing and recording system is presented.

Introduction

The rapidly increasing use of fibre reinforced composite materials in structural aircraft components has led Aircraft manufacturers to acquire large inspection installations in order to assess the quality of their production. In effect, defects can occur during manufactoring, such as material porosity, disbonds and delaminations, and after service from fatigue, over loading, impact damage and environmental degradation. Inspection is required before assembly, on completed structures and in service. Several methods already exist for the detection of significant defects. They include Ultrasonic inspection, Television radioscopy, and various proprietary devices such as the Fokker Bond Tester. Ultrasonic inspection and Television radioscopy are already fully computerized and are complementary for defects detection. However a requirement exists for a technique which will detect the same defects but :

- is much more rapid,
- is easier to set up,
- does not require physical contact or coupling media such as water,
- can be applied where access is difficult or possible from one side only,
- provides a permanent record suitable for storage and image analysis,
- is relatively easy to interpret.

Different techniques have been investigated for this purpose, including optical holography, acoustic emission and thermography. All have not been generally adopted in the industries concerned, for various reasons. From preliminary studies and recent hardware developments, it appears infrared thermography must satisfy all the above requirements. These consideration have led AMD-BA to carry out an industrial assessment programme. The aim of the programme is to :

- assess the applicability and reliability of infrared thermography as a method for inspection of composite and bonded materials,
- develop the technique for an industrial utilisation,
- specify the installation required for its application.

Infrared Thermography principle

By use of IR thermography one can visualize the spatial distribution of infrared radiations emitted by an actively heated or self radiating object. The infrared pattern is monitored by an infrared camera which provides a real time video display of the surface temperature. The radiant emittance monitored by the camera (See Figure 1) is expressed by the "fondamental" equation of thermography :

$$L' = \tau_o \cdot \varepsilon_o \cdot L_o + \tau_o (1 - \varepsilon_o) \cdot L_a + (1 - \tau_o) \cdot L_{atm} \quad (1)$$

where τ_o is the atmospheric transmission coefficient, ε_o is the object emissivity, Latm is the atmospheric radiant emittance, La is the ambient radiant emittance and Lo the object radiant emittance.

For short distances, less than 2 meters, $\tau o = 1$ and equation (1) yields.

$$L' = \varepsilon o\ Lo + (1 - \varepsilon o)\ La \qquad (2)$$

From this equation, it appears that emissivity and ambient temperature must be measured if one wants to know the absolute temperature of the object. The output signal of the IR camera is proportionnal to the monitored radiant emittance. Calibration of the camera with a black body radiation source yields a relation ship between the output signal and the surface temperature. We can distinguish between passive and active thermography. Passive thermography is the surface temperature monitoring of a self radiating object with an IR camera. Defects appear with a different temperature gradient. For active thermography, the object is heated with an external heating source. A defect in the material disturbs the heat wave propagation resulting in a different temperature gradient on the object surface monitored by the IR camera.

Active thermography

In this case, a very short pulse of radiant energy such as a pulsed laser beam or a flash lamp is incident on the surface of the inspected part and absorbed by the material. The temperature of the local heated area rises above ambient and then decays (1). The presence of subsurface defects can only be observed as transient temperature gradient corresponding to hot peaks when the surface is monitored on the same side the applied heat, and cold peaks when monitoring is on the opposite side of the applied heat. The heat distribution in the material is governed mainly by two quantities (2). The amount of heat to be deposited per unit area in order to obtain a required degree of contrast, and the time taken for that contrast to develop on the scanned face. Several authors (3, 4) have presented analytical solutions for the second quantity of interest. Their computed values confirm our observations and the requirement for a specific well controlled heating source. With composite materials the time necessary to reach maximum contrast is in the order of a few seconds depending on the depth of the defect. This relatively long time has led us to develop an industrial pulsed photothermal technique. The pulse duration is produced by the linear motion of the inspected part, at a constant speed in front of a line heater. Temperature Monitoring is performed with a line scanner positionned at a precise distance of the line heater. This distance is set in relation to the thickness of the inspected part and defect location (See Figure 2). The main advantage of a line scanner is that the thermogramme of the inspected part is uniform, therefore enhancing the contrast between good and bad areas (See Figures 3 and 4). This method should enable us to inspect 30 square meters per hour compared to 3 square meters with the most advanced ultrasonic installations. The Flash method is also investigated for on site non destructive testing. (4)

Thermal characteristics evaluation

Measurements of the following thermal characteristics have been carried out for the different materials inspected .

k : thermal conductivity	α : diffusivity	ε : emissivity
c : specific heat	e : effusivity	μ : thermal propagation depth

Emissivity

Surface emissivity of a material is an essential parameter if accurate temperature measurement has to be performed. Emissivity is dependent on the material, on the surface finish and the temperature of the test part, on the spectral sensitivity of the infrared detector and on the view angle. The more the emissivity is important, the more accurate is the temperature measurement and the less the reflectance is. Therefore if the reflectance is important, the IR camera detects spurious radiations from surrounding objects and sources which are difficult to eliminate. Great care must be taken to get rid of these radiations. High emissivity paint is used when it is possible. Otherwise, the effects of variations in emissivity over the surface of a test part are minimized by using the temperature difference between one time interval and the next (time domain reference technique) (1, 5) or between the part prior to heating and after. Directionnal spectral emissivity measurements are performed in the Service Universitaire d'Energetique, Université de Reims (France) with the "indirect method" (See Figure 5). The indirect method considers the radiation exchange which is formulated as :

$$\varepsilon + \rho + \tau = 1$$

Absorbed radiations + Reflected radiations + Transmitted radiations = 1

For an opaque, smooth and polished surface sample. $\varepsilon + \rho = 1 \qquad \varepsilon = 1 - \rho$

With this method, material emissivity can be obtained from measurement of the reflected radiations. For real, rough surfaces the incident radiations are scattered ; therefore bidirectionnal spectral reflectance measurement will be performed.

Results obtained (See Figure 6) :

Spectral directionnal reflectance measurements have been carried out on flat and polished CFRP and epoxy resin samples of different thicknesses over the spectral domain ranging from 3,25 up to 13 . Emissivity of polished CFRP samples decays slowly from 0,8 at 3,25 down to 0,72 at 13 . Variations other the sample area are less than 3 % at 5 (0,8 - 0,82), therefore the measured values are attributed to the fibers emissivity.

For polished epoxy resin samples, 0,6 mm thick and above, emissivity is constant over the spectral domain ranging from 5,5 up to 13 , the measured value is 0,94. In this spectral region, epoxy resin samples are opaque and are equivalent to dielectrics. Over the spectral domain ranging from 3,25 up to 5,5 , the 0,6 mm thick epoxy resin sample is semi-transparent, transmission is in the order of 0,06 à 5 . As a conclusion, emissivity of real CFRP samples will vary from 0,8 (emissivity of the fibers) up to 0,94 (emissivity of the epoxy resin) in the spectral domain ranging from 3,25 up to 13 . However thinner epoxy resin samples (0,2 mm) have to be made to confirm opacity in the spectral domain ranging from 5,5 to 13 .

The other thermal characteristics will be measured in an attempt to establish an analytical model which would predict the detectability of defects by thermal non destructive testing of parts of various materials and geometries.

Experimental rig (See Figure 7)

The test system has been developed for three main functions :

- IR cameras and heating sources evaluation and comparison
- development of the method on test pieces
- testing mechanical devices for specification of a final installation.

The mechanical parts of the rig consist of the following major sub-systems :

- A mechanical structure on to which is mounted a carriage supporting the test piece which performs the two motorized scan (X) and/or index (Z) motions. The necessary motions to inspect the parts are based on a combination of a continuous motion (scan) and a discontinuous motion (index) where off the index motion takes place at the end of each scan motion. The index motion displacement is related to the length of the line scanner. The scanned area is presently of 3 m^2 (3 x 1 m). Scanning speed is continuously adjustable from 0,5 to 10 m/mn.
- An IR camera carriage. It permits manual vertical positioning of the detector and motorized alignment in relation to the line heater.
- A line heater carriage. It permits motorized positioning of heater focal line in relation to the test part.
  Safety devices are included in order to prevent the test part from stopping in front of the lighted heater. The rig must be free of vibrations.

Heat sources investigation

Heating is the most important factor in the process. The best compromise must be found between, radiated intensity, heated area, spectral emission characteristics and heat pulse duration. Specially developed sources will be required for particular problems (6). A product survey of available heat sources has been carried out. Three different types of source have been tested.

Hand held flash bulb

A commercially available photographing lighting device has been tested. This hand-held single flash bulb delivers up to 5 KJ per flash in 600 ms, this energy is sufficient for most applications. This type of source is easy to set up, it is of moderate cost and can be used with direct energy supply. However, the heated area is small and non uniform. Positioning of the source in combination with the detector is cumbersome. Therefore its industrial application is limited, but on site inspection can be considered.

Line heater

The 14 KW infrared line heater choosen (up to 400 Watts per linear inch) generates highly concentrated radiant heat flux density along a thin line at the external focal axis of an elliptical reflector. The radiant energy emitted from a high temperature (5400°F) tungsten filament is dependent on voltage applied to the heater and can be regulated. The spectral energy peak is around 1 micron. Rapid heating rates can be obtained. The lighted length is 400 mm. This type of heating is relatively cheap, easy to regulate, however characterisation of the radiant energy emitted is difficult but necessary in order to, check filament and reflector ageing, monitor power supply fluctuations. Important drawbacks are : high level of background reflective noise in the spectral range detected by a short wave infrared camera, lack of uniformity of the heated line (See Figure 8) and important dependence on the positoning of the focal distance with regards to the inspected parts.

Laser

We believe that lasers would be most effective and therefore, we are developping that type of heating. Laser heating provides high energy densities which can be perfectly characterized and regulated. Background noise may be eliminated by a suitable choice of wavelength (6). Positioning of the laser beam as regards to the inspected part is not crucial. Line testing with a laser can be produced by different optical devices. The fact that a 200 W CO_2 laser without power regulation was only available in our laboratory constrained us to select a scanning system in order to regulate the energy density and the length of the heated line (See Figure 9). Some experiments have been carried out with a mirror scanning system but has been abandonned since it is not possible to obtain a uniform enough heated line. Present limitations for an industrial application of laser heating would be connected with size and cost requirements.

Selection of the optimal system

Eight commercially available systems were tested. These tests were carried out on specially manufactured CFRP test samples with artificial defects (See Figures 8, 10). Systems spatial and temperature resolution have been evaluated. For this purpose, 35 different material disks were introduced in one sample, and PTFE bar targets in the other. It has not been possible to select the optimal working wavelengths band of the infrared camera. In effect, at the observed temperatures (40°C to 70°C) detectivity is not really dependent on the wavelength (7). Material emissivity and spectral emission characteristics of the heating sources will guide the choice of a SW or a LW infrared camera for maximum sensitivity.

From the experiments, the main characteristics of the optimal system were selected :

Infrared camera	- Remote controlled functions (focusing, sensitivity, level, line scanner and frame modes of operation, scanner frequency and frames rate). - Line scanner and frame modes of operation. - Adjustable line scanner frequency - 9 hours minimum operation without refilling
Acquisition and visualisation unit	- 8 to 12 bits A/D conversion - Real time colour display and digital storage - 8 or 16 bits // interface - Image processing functions such as zoom, isotherm

The two "CIRCE" IR cameras LW and SW selected have been recently introduced by the french manufacturer IRIS. They are interfaced either with a 12 bits unit or a 8 bits acquisition and visualisation unit both manufactured by the Laboratoire National d'Essais (LNE). These units can be interfaced to a host computer. The two selected configurations (See Figures 11, 12), will allow for real time digital image display, storage and processing in order to avoid cumbersome and inadequate utilisation of video tape recording and play back. For an industrial application the 12 bits acquisition unit enables us to acquire 4 external signals such as the coordinates of the part, ambient temperature, instant radiated power in the case of laser heating. Image processing methods will be required and developped such as filtering, convolution, image averaging, space-domain reference and time-domain reference subtractions (See Figures 13, 14, 15 and 16). A dual line scanner mode can be selected on the remote controlled unit. Distance between the two lines is adjustable. This system offers the possibility to apply the time domain reference technique (1).

Inspection means calibration

Equipment calibration is of primary importance in NDT whatever the technique used (X rays, ultrasonics, thermography). Calibration is required at different stages of an industrial application. At first the equipment must be characterized on receipt, then the selected characteristics have to be checked periodically. Calibration of thermographic equipment is a necessity if one wants to carry out accurate temperature measurements, which are needed for a reproductible evaluation. The main characteristics investigated are :

- S.R.F (Slit Response Function)
- N.E.T.D (Noise Equivalent Temperature Difference)
- Uniformity in the fied of view
- Thermal drift
- Signal/temperature correlation
- M.R.T.D (Minimum Resolvable Temperature Difference).

These characteristics are choosen since they are an indication of the temperature resolution of the system correlated to the spatial resolution. For our application, temperature resolution, lack of thermal drift and spatial response uniformity are the main criterias.

The instrumentation adopted, consists of an accurate temperature controlled Black Body radiation source with a resolution of 0.1°C, and an extended area hot black body. Signal acquisition is carried out on a digital oscilloscope interfaced to a micro-computer. The acquired signal for NETD measurement is presented (See Figure 17). Heat sources calibration is also required in order to check the spatial uniformity and the radiated power of the thermal impulse. In the case of laser heating, influence of fluctuations in the beam power profile upon the quality result of the heated line is considered. Measurement of laser beam characteristics, can be easily carried out with commercially available systems. Calibration of the other heat sources such as : heat gun, flash tubes and IR line heater is more difficult. For the two latter types of heat sources, the radiated power is a function of emitter and reflector lifetimes, and power supply fluctuations. A Pyranometer has been used to characterize the radiant energy of IR line heater. The plot (See Figure 18) simulates the on and off working conditions.

Field of application assessment

Four main phases are defined covering a wide range of samples :

- feasability study on flat samples,
- extended study on flat monolithic samples of various materials,
- extended study on bonded, cocured and semicured samples,
- extended study on complex geometries.

Well characterized samples have been choosen with due regard to the various materials, structural configurations, defects and pass/fail criterias. Results obtained so far show that both transmission and reflection techniques are complementary. Whith the transmission observation mode thicker samples can be inspected than the reflection observation mode, however only large defects 20 mm diameter, compared to 5 mm can be detected. Comparative test are carried out using ultrasonics and radiography on the same samples.

Conclusion

In order to industrialize thermography as a reliable non destructive testing technique, several parameters must be assessed. These factors are related to the material thermal and infrared cameras characteristics, types of heating source, positioning and setting of the different systems. The undertaken programme will lead to the definition of a specific installation, which implies development in optics, robotics, signal and image digital processing. Quality control requirements in aircraft industry obliged us to carry out and develop calibration of the heating and monitoring systems. It is very important to begin calibration from the starting point of the programme in order to have a really good grip on the process. We are seeking a theoretical model which would likely be a most valuable practical aid to reduce the development time required to optimize inspection conditions for specific applications.

Acknowledgements

The authors would like to express their thanks to P. Bru, C. Lambert and D. Legros for valuable contributions to this programme, M. Egee and C. Bissieux from Université de Reims for supplying the emissivity measurement results.

References

(1) P. CIELO, R. LEWAK and D.L. BALAGEAS, Thermal sensing for industrial quality control.
SPIE thermosense VIII Symposium, Cambridge, MA - Sept. 15-20, 1985.

(2) W.N REYNOLDS and G.M. WELLS, Video - Compatible Thermography.
British Journal of NDT, January 1984.

(3) PARKER W.J, JENKINS R.J, BUTLER C.P. and ABBOTT G.L.
"Flash method of determining thermal diffusivity, heat capacity and thermal conductivity".
J. Appl. Phys., 32, 1679-84, 1961.

(4) D. BALAGEAS, Méthode photothermique impulsionnelle.
Application aux milieux stratifiés solutions analytiques exactes.
ONERA, Rapport technique 1/7254 SN, 19 février 1985.

(5) C.M. SAYERS, Detectability of defects by thermal Non-Destructive Testing.
British journal of NDT, January 1984.

(6) V.P VAVILOV and R. TAYLOR, Theoretical Practical aspects of the thermal non destructive testing of bonded structures.
Research Techniques in non destructive testing, Vol. V, R.S. SHARPE, 239 - 279.

(7) G. GAUSSORGES, La Thermographie infra-rouge.
Technique et documentation Lavoisier.

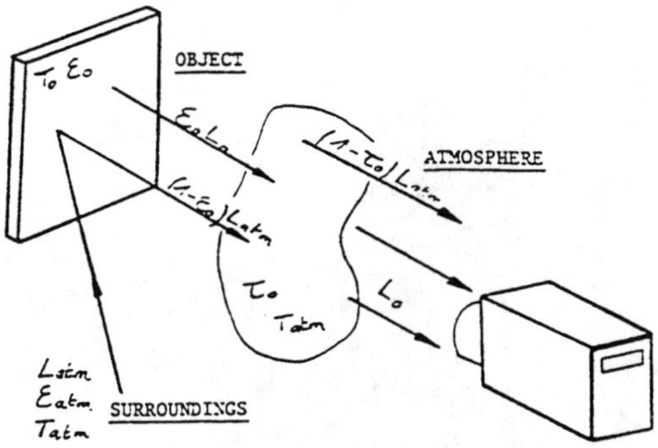

Fig. 1 : Infrared thermography principle

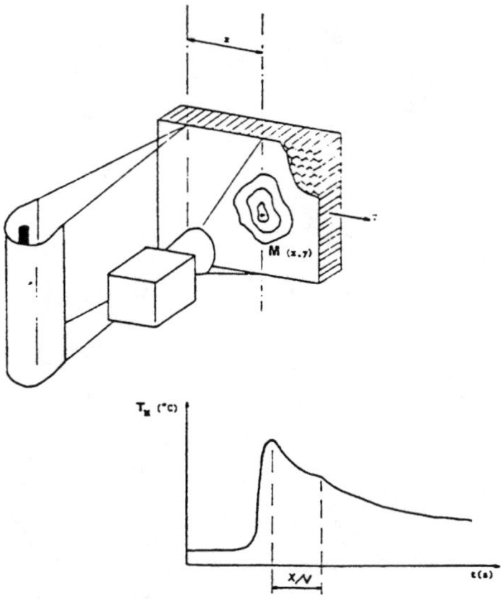

Fig. 2 : Active thermography principle. Line scanner and line heater setting. Temperature plot as a function of time.

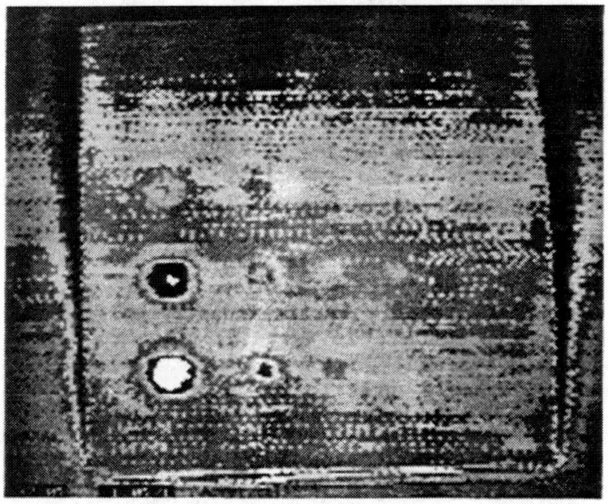

Fig. 3 : Video Frame scanner. Colour thermograph of a CFRP sample with flat bottom holes at different depth

Fig. 4 : Line scanner

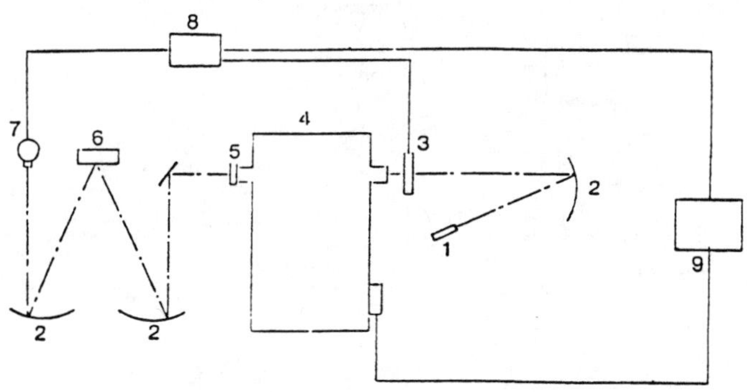

Fig. 5 : Reflectance measuring equipement.
1. Source 2. Mirror 3. Modulator 4. Monochromator
5. High pass filter 6. Sample 7. Detector
8. Synchronous amplifier 9. Micro-computer.

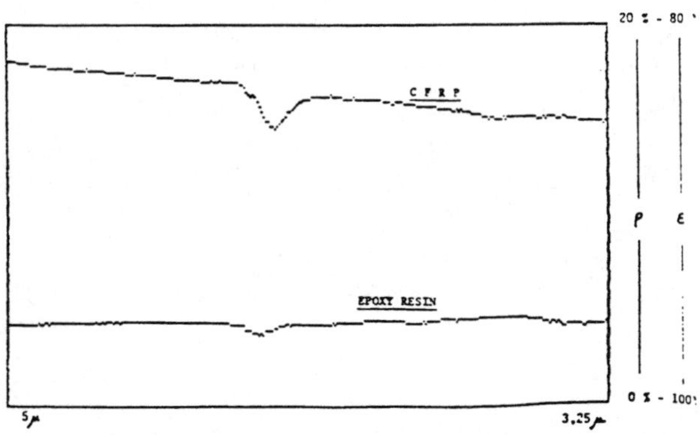

Fig. 6 : Spectral directionnal reflectance curves of polished CFRP and Epoxy resin samples.

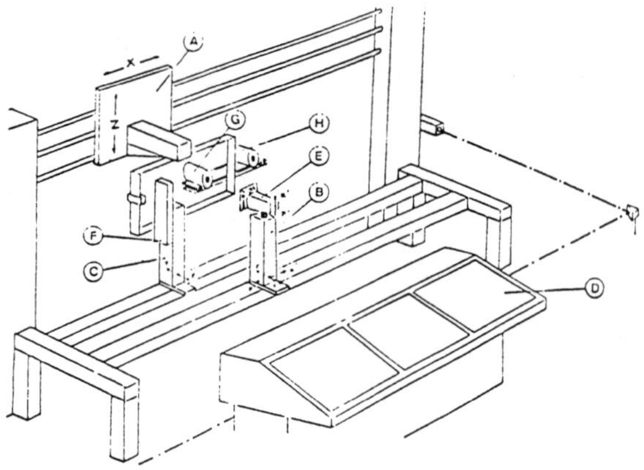

Fig. 7 : Experimental rig.
A. Test part carriage B. IR camera carriage
C. Line heater carriage D. Control panel
F. IR camera F. Line heater G. Black body
H. Extended hot black body

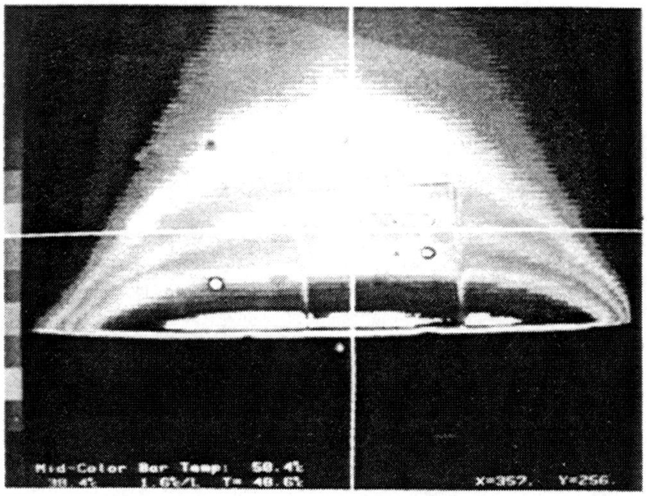

Fig. 8 : Colour thermograph of test sample
with artificial defects of different materials.

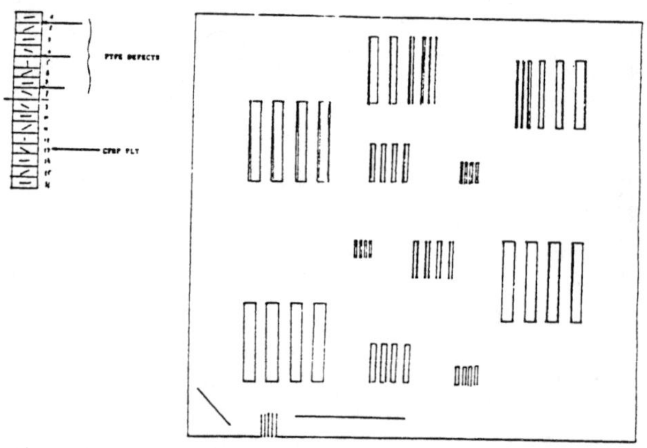

Fig. 10 : Test sample with bar targets
of different spatial frequencies.

Fig. 9 : Laser heating system whith line scanning
optical device

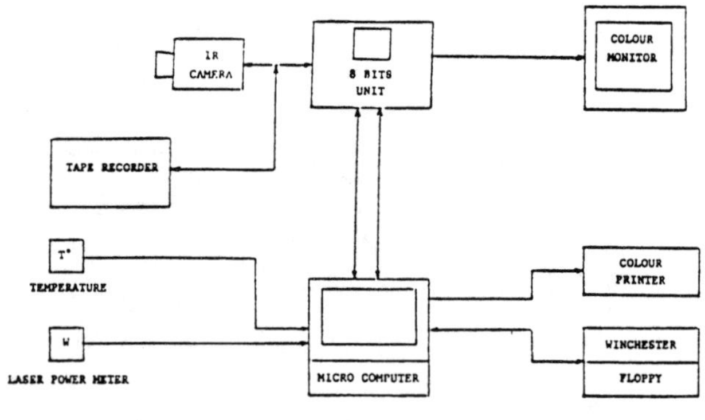

Fig. 11 : 8 bits IR system configuration

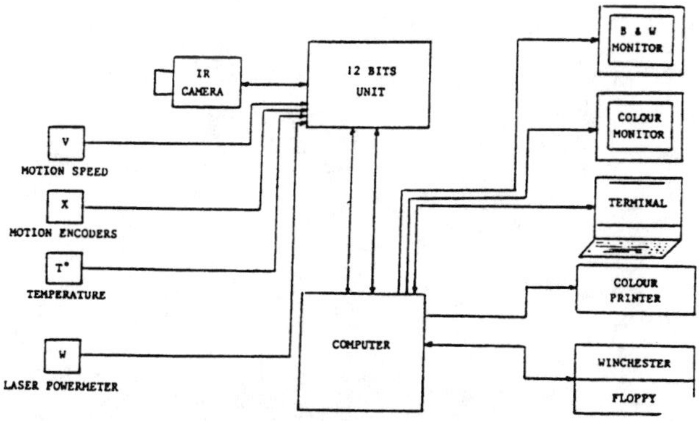

Fig. 12 : 12 bits IR system configuration

Fig. 13 : Thermal image enhancing of PTFE bar targets introduced in a CFRP sample. Depth 1 mm

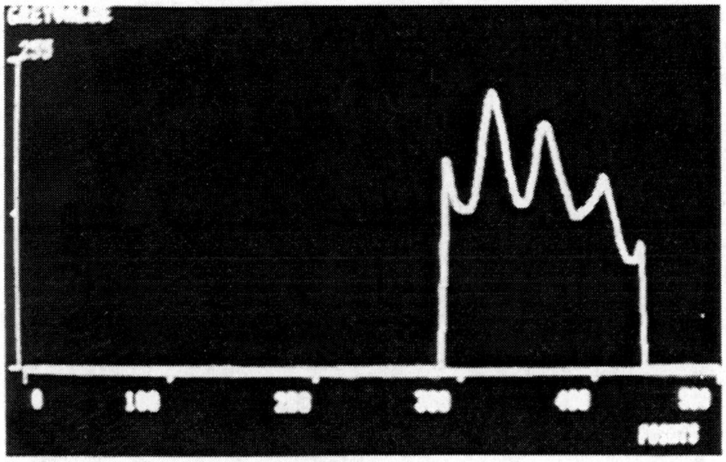

Fig. 14 : Thermal profile of the brightened line displayed in fig. 13.

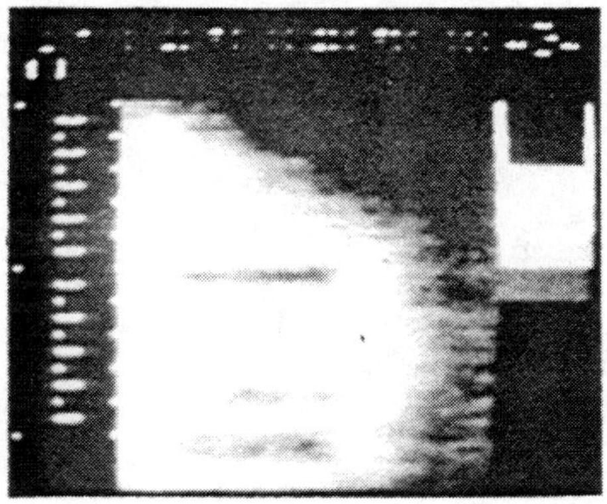

Fig. 15 : Contrast enhancement

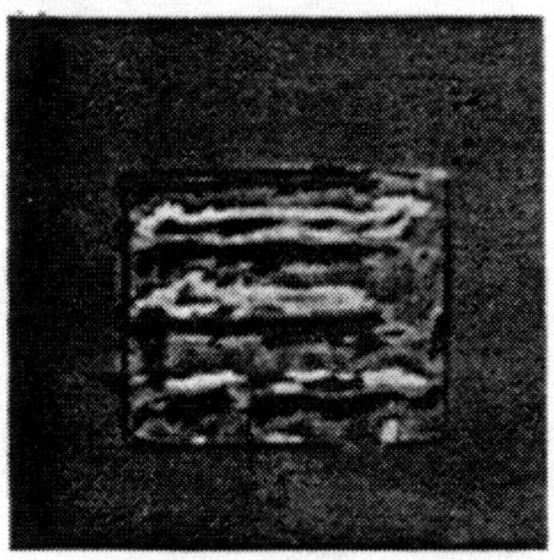

Fig. 16 : Thermal image enhancing. Coloration

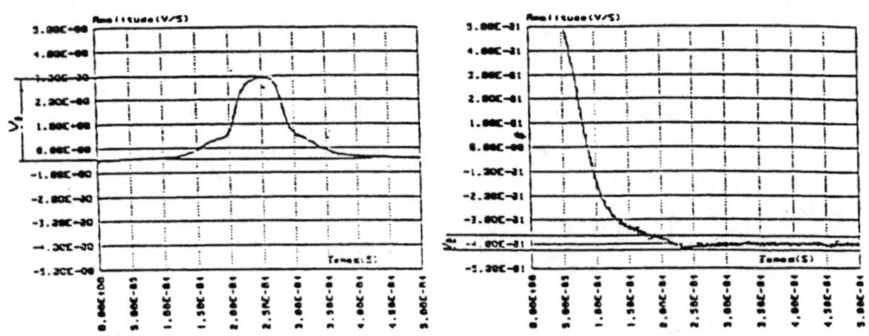

Fig. 17 : NETD measurement. Signal digitalisation

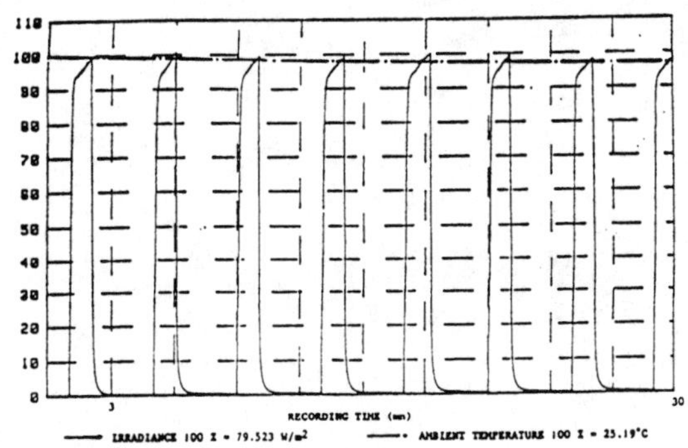

Fig. 18 : IR line heater calibration on and off working conditions are simulated.

Section Four
Optical Techniques for Process Control

Reprinted with permission from *Temperature: Its Measurement and Control in Science and Industry,* Vol. 3, Part 2, pp. 391-403 (1962). ©1962 Van Nostrand Reinhold.

36. The Place of Photovoltaic Detectors in Industrial Pyrometry

R. Barber and T. Land

Land Pyrometers Ltd., Sheffield, England

INTRODUCTION

The thermopile has been the basis of nearly all industrial radiation pyrometry. The silicon "solar" cell can now be used in place of the thermopile, and for some measurements it is much better. Other photovoltaic detectors such as the selenium "barrier-layer" cell and the indium antimonide photoelectromagnetic cell can also be used for certain special measurements. These photoelectric devices have their limitations and it is clear that the thermopile will not be supplanted entirely at the present time. The advantages of the photoelectric detectors are so striking that it is necessary to review the whole field of radiation pyrometry to see where they can be used with advantage.

This paper deals only with photovoltaic cells. These solid-state devices generate an electric current when illuminated with visible light or infrared radiation, without the aid of an external voltage source. We shall not deal with photoemissive cells, photomultiplier tubes, or photoconductive devices; nor shall we consider the class of pyrometer which depends on "chopped" radiation, which has its own, quite different, field of application.

The advantages of photovoltaic cells arise from their robustness, their great sensitivity, and their high speed of response, and from the fact that they respond only to radiation of certain wavelengths.

The construction of a silicon solar cell has been described by Wolf and Prince[1] and is shown in Fig. 1. A thin layer of *p*-type silicon is laid on a thicker layer of *n*-type silicon. The negative contact is deposited on the back surface of the silicon, while a metallized strip on the top surface forms the positive contact. The silicon is sufficiently strong to need no physical support. The impurity layers and hence the polarity can be reversed.

The selenium "barrier-layer" cell described by Veszi[2] has a similar layer type of construction, but the negative contact is normally a small iron plate, on which the selenium is deposited. The surface layer is usually of cadmium oxide and the positive contact strip is deposited on the cadmium oxide surface.

The indium antimonide photoelectromagnetic cell described by Hilsum and Ross[3] is a different kind of device. It consists of a thin strip of indium antimonide in a strong magnetic field, as shown in Fig. 2. The contacts are taken from the ends of the strip. When infrared radiation falls on the strip, an electric current is generated in the circuit to which the contacts are connected.

The most important difference between the different types of photovoltaic detectors is the range of wavelengths to which each is sensitive. These are shown in Fig. 3. The sensitivities are not, of course, plotted on the same scale.

A thermopile is sensitive almost equally to radiation of all wavelengths, but the incident radiation is restricted by the material chosen for the optical system. The transmission of glass is shown for comparison in Fig. 3 as a broken curve.

Sensitivity

The most obvious advantage of the silicon cell is its great sensitivity. To compare the silicon cell, which produces a current, with the thermopile, which produces a voltage, we must specify the load resistance. A resistance of 400 ohms has been found convenient for silicon cells, and using this as a basis for comparison we find the following ratios of sensitivity for pyrometers with similar optical systems at different temperatures. The comparable figures for a selenium cell with a load resistance of 1000 ohms have also been included in Table 1.

It will be seen that the silicon cell is more sensitive than the thermopile over the whole range of temperatures in which radiation pyrometers are generally used. The selenium cell is also more sensitive than the thermopile but over a more limited range of temperatures. The extra sensitivity can be used to take the pyrometer further from the hot object, or to sight on a smaller target, or to use a simpler or less sensitive indicating instrument. A millivoltmeter, which could be used with a total radiation pyrometer only for full scale deflection of at least 1500°C, could be used with a comparable silicon cell pyrometer over the range 800 to 1100°C.

TABLE 1. SENSITIVITY OF PHOTOVOLTAIC CELLS RELATIVE TO A MOLL-TYPE THERMOPILE

Temperature °C	Ratio of sensitivities	
	Silicon cell	Selenium cell
700	1	
1000	6	0.1
1500	24	3
2000	60	22

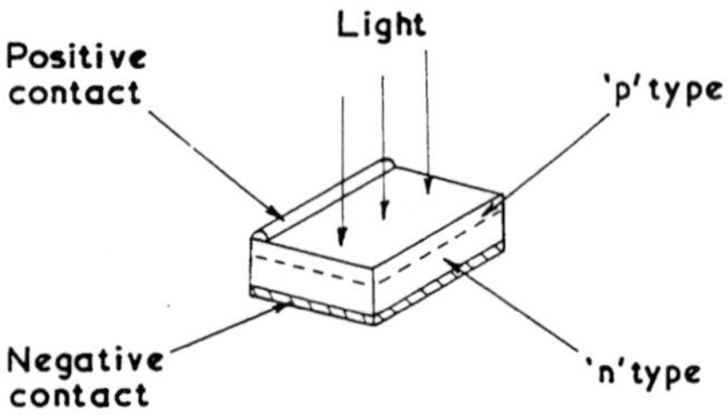

FIG. 1. Construction of silicon solar cell.

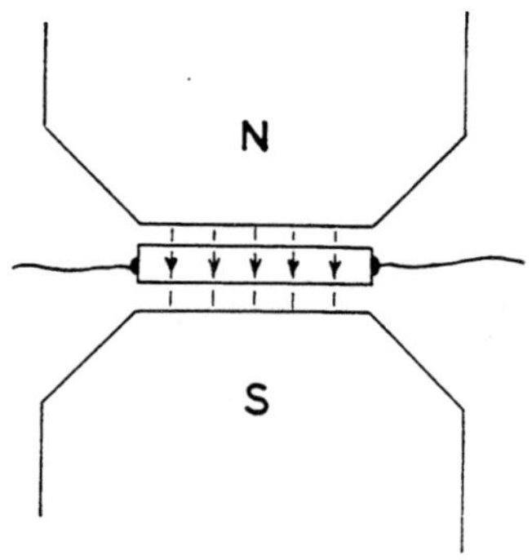

FIG. 2. The indium antimonide photoelectromagnetic cell.

Speed of Response

The speed of a normal industrial pyrometer is usually about 2 seconds to 98% of the final reading. A response time of less than one second is considered to be something rather special. But the photovoltaic detectors have a response time of less than a millisecond, much faster than any industrial recorder. A pyrometer using a silicon cell in conjunction with a ¼-second re-

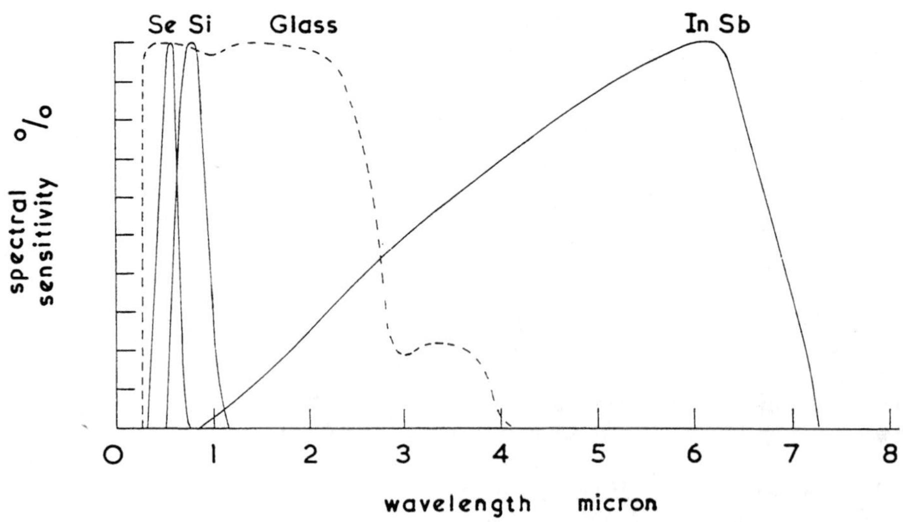

FIG. 3. Spectral sensitivities of some photovoltaic cells. The broken line shows the transmission of glass.

corder will handle most industrial problems adequately. There are many industrial applications in rolling mills, in induction heating, in glass manufacture, and elsewhere, where this high speed of response is most welcome.

Target Size

The ability to sight on small targets has already been mentioned, but there are two additional features that help in this respect. The limitation for small target applications with a thermopile is usually the chromatic aberration of the lens. The silicon and selenium cells are really monochromatic detectors, so this imperfection is of little importance. In addition, the thermopile is a nonuniform receiver, being more sensitive to radiation falling near the junctions. There are also an integral number of junctions, and a reduction in aperture size may result in the outer junctions becoming redundant. This limitation does not apply to photocells.

Stability

When a pyrometer using a photovoltaic cell is tested in the laboratory, two features are particularly pleasing and very evident. Firstly, the zero, when the incident radiation is interrupted, is absolutely stable. Secondly, there is no atmospheric absorption, so that it is possible to design a pyrometer giving exactly the same reading at a wide variety of distances. This is important in industrial work where a path length of 6 feet in air may lead to a variation of $\pm 10°C$ from day to day due to variations in the humidity of the atmosphere. This feature is, of course, only true for silicon and selenium cells which miss the water vapor absorption bands, and not for the indium antimonide cell.

Reduction of Emissivity Errors

We come now to the principal advantage of the silicon and selenium detectors. This is very important in industrial pyrometry, but as it is not immediately evident in the same sense as the other advantages mentioned above, it will be explained in some detail.

It is well known that an optical pyrometer is a much more accurate instrument for the measurement of surface temperature than a total radiation pyrometer. This is because the optical pyrometer measures only visible radiation of short wavelengths. Silicon and selenium cells also respond only to radiation of short wavelength and, therefore, share the advantages of the optical pyrometer.

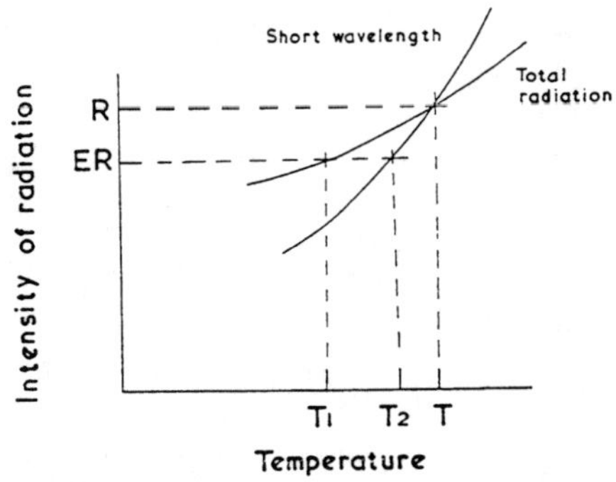

FIG. 4. The effect of a change in emissivity on the apparent temperature indicated by pyrometers sensitive to different parts of the spectrum.

The brightness of a hot object increases with temperature much more rapidly than does the total radiation which it emits. Between 1000 and 1500°C the brightness increases about a hundred times, while the total radiation increases by a factor of less than 4. The short wavelength radiation is clearly a much more sensitive measure of the temperature than the total radiation. The accuracy of the optical pyrometer is likewise less affected by any loss of radiation during its passage from the hot body to the pyrometer, whether this be by variations in emissivity of the surface or by absorption by smoke or spray.

This feature is illustrated in Fig. 4 in which the changes in intensity of total radiation and of short wavelength radiation over a small range of temperature are compared. If the radiation R is reduced to ER, the temperature indicated by a total radiation pyrometer will drop to T_1. But if the pyrometer measures only radiation of short wavelength, the indicated temperature will fall much less, to T_2.

To express this numerically we can say that while the total radiation from a hot blackbody is known (Stefan's law) to vary in proportion to the fourth power of the absolute temperature, the intensity of radiation of short wavelength varies in proportion to a higher power of this temperature. For example, a pyrometer using a silicon cell gives an output at temperatures around 1000°C which varies in proportion to the 13th power of the absolute temperature. In consequence a 13% change in output due to some cause, for example due to a change in the surface emissivity, results in only a 1% (in °K) change in the temperature indicated. A "total" radiation pyrometer with a glass lens follows a 6th power law at the same temperature. Consequently the effect on the indicated temperature of variations in emissivity of the measured surface are less, when a silicon detector is used, by a factor of 6/13.

The power laws for various types of radiation pyrometers at 1000 and at 1500°C are given in Table 2. The numerical value of the power can be shown to be equal to $C_2/\lambda T$ where C_2 is a constant having the value of 1.438 cm°K and λ is the wavelength of the radiation. The "effective" wavelength of each type of pyrometer is also given.

TABLE 2. POWER LAW AND EFFECTIVE WAVELENGTHS OF DIFFERENT PYROMETERS

Type of pyrometer	1000°C		1500°C	
	Effective wave-length (micron)	Power law	Effective wave-length	Power law
Optical	0.65	17th	0.65	13th
Selenium cell	0.62	18th	0.60	14th
Silicon cell	0.91	13th	0.86	10th
"Total" radiation	—	6th	—	5th

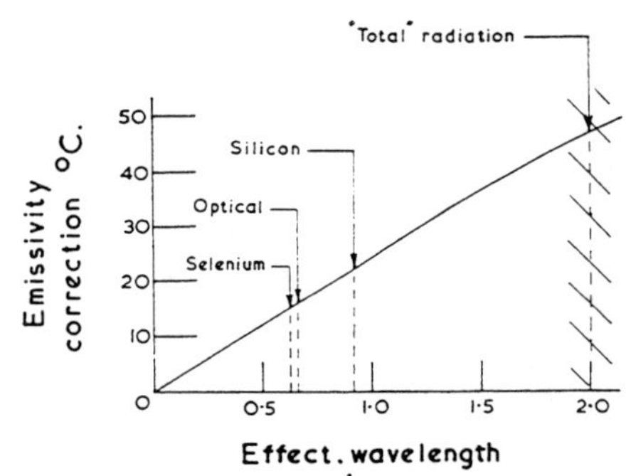

FIG. 5. Emissivity correction, for a surface of emissivity 0.8 at a temperature of 1000°C, as a function of the effective wavelength of the pyrometer used.

The effect of emissivity variations on indicated temperature is in fact almost proportional to the effective wavelength of the pyrometer used for the measurement. This is illustrated in Fig. 5 where the emissivity correction for a surface at 1000°C (emissivity 0.80) is plotted against the effective wavelength of the pyrometer.

The success of silicon and selenium detectors in reducing errors in the measurement of surface temperatures is often enhanced by the fact that many surfaces have higher emissivities at shorter wavelengths. This is universally true of metals and sometimes true of oxidized surfaces.

In addition, a very thin oxide layer may be opaque to short wavelength radiation but almost transparent to longer wavelength infrared radiation. On one installation, where the temperature was being measured of brass section leaving a rolling mill, large differences in emissivity were noted due to this effect. A thermopile detector gave readings that implied that the emissivity of the surface was about 0.4 to 0.5. A selenium photocell detector was substituted, and the readings were reproducible and suggested an emissivity of 0.8. The thermopile detector was quite unsatisfactory but the pyrometer with a photocell detector has been in successful use for several years.

Disadvantages of Photovoltaic Detectors

The most obvious disadvantage of a silicon or selenium detector is that the output changes with temperature so rapidly that only about 300 degrees can be accomodated on the scale. Scales for silicon and selenium detectors are compared with the scale for a "total" radiation pyrometer in Fig. 6. The result is a little ridiculous on a long scale, where 10°C is often repre-

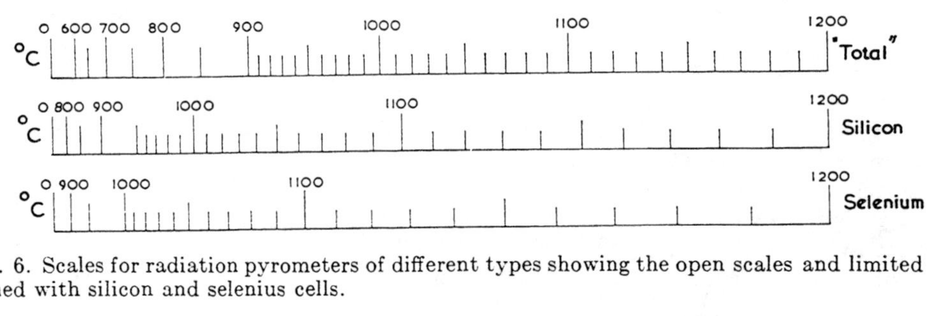

FIG. 6. Scales for radiation pyrometers of different types showing the open scales and limited ranges obtained with silicon and selenius cells.

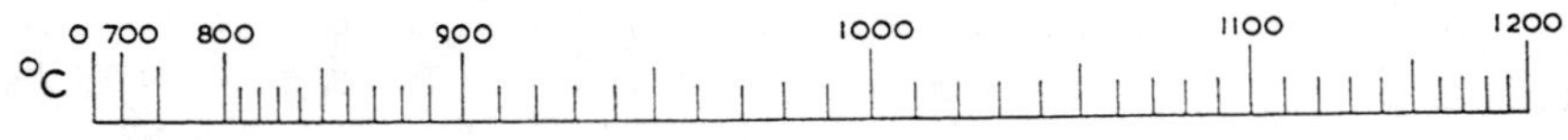

FIG. 7. Scale for silicon detector using the modified potentiometer circuit shown in Fig. 18.

sented by one inch at the top of the scale. This is the price that is paid for a reduction in the errors due to changes and uncertainties in the emissivity of the surface.

Fortunately, it is possible to make a minor modification to the circuit of a potentiometer which makes the scale shape much more nearly linear. This circuit modification is explained later in the paper. The resulting scale for a silicon detector would appear as in Fig. 7.

Another disadvantage which applies to selenium and indium antimonide detectors is that they must be kept cool or they suffer permanent changes in calibration. Water cooling is usually essential. Although a reliable supply of cooling water is not often particularly difficult to find in an industrial installation, it is a complication which would be better avoided. Silicon cells on the other hand are no more affected by heat than thermopiles and will stand temperatures well over 100°C.

Selenium detectors do not have large temperature coefficients. As in addition they must be water-cooled, temperature compensation is not a problem. The indium antimonide photoelectromagnetic cell has a large negative temperature coefficient and special circuits are necessary to provide compensation even when the instrument is water-cooled.

Silicon cells need to be compensated for changes in output with ambient temperature as they have (positive) temperature coefficients of current output. It is not possible to provide perfect compensation for all temperatures of the measured hot body as the cause of the variation in output is a shift in the cutoff wavelength, not a change in sensitivity of the cell. A residual error of the order of ±0.2°C per °C may be encountered in extreme cases, but the error is negligible in most applications.

Measurements in a furnace are usually better made with a conventional radiation pyrometer, simply because of its more convenient scale-shape. In addition, it can be shown that a pyrometer measuring short wavelength radiation will give greater errors than a total radiation pyrometer if other surfaces or flames are present which are at very much higher temperatures. Consequently, total radiation pyrometers are preferable for measurements such as that of the roof of an open hearth steel melting furnace. If the measurement is made through the back wall, flame reflection may lead to serious errors if a selenium detector is used.

APPLICATIONS

The use of selenium cells for the measurement of the temperature of the roof of a steel melting furnace was described 20 years ago by Larsen and Shenk,[4] and later by Land.[5] Over this period pyrometers using selenium cells have been used continuously under arduous industrial conditions and have proved fully reliable. At the time when these pyrometers were developed the accuracy of total radiation pyrometers was not always satisfactory, and large blackbody furnaces of high precision were not available. The improvement of the total radiation pyrometer and the development of better calibrating furnaces, such as the spherical furnace described by Burton and Mayorcas,[6] have reduced the advantages of the photoelectric pyrometer for this particular application. Nevertheless, the experience obtained makes it clear that such pyrometers are entirely reliable for industrial measurement.

Liquid Iron Temperature

An application of the selenium detector which is proving very successful is the measurement of the temperature of the liquid iron stream from a cupola. A successful trial has recently been reported by Rew[7] at the British Cast Iron Research Association. Attempts have often been made to measure the temperature of the liquid metal with a "total" radiation pyrometer, but it seems to be generally recognized that the accuracy of such measurements is not adequate for proper control of the process. Immersion thermocouples are often used, but they are rather expensive to use and need regular servicing and some skill in use. The following calculations suggest why the selenium cell makes a more suitable pyrometer for this measurement.

We assume that the emissivity of liquid iron at 0.6 micron is 0.42 and its total emissivity is 0.30. We suppose that two causes of error may exist. First, there may be ripples and turbulence and we will assume that on 10% of the surface the pyrometer can "see" reflections of other parts of the liquid metal surface. Second, we will assume that at times as much as 5% of the surface is covered with iron oxide having an emissivity, at all wavelength, of 0.7. The temperature of the iron is 1500°C, and we consider the temperatures as indicated by a "total" radiation and a pyrometer using a selenium cell. Both pyrometers are calibrated to read correctly when sighted on a clear metal surface. The indicated temperatures under the different conditions have been calculated and are set out in Table 3.

Rolling Mills

A typical application of a pyrometer with a silicon detector is the measurement of the temperature of sheet steel during rolling. The installation is outwardly identical with a conventional radiation pyrometer installation, but no adjustment of the recorder was necessary during installation. The recorder had been calibrated for an emissivity of 0.8, and it agreed at once with the optical pyrometer normally used to set up such installations in the works.

TABLE 3. APPARENT TEMPERATURES OF LIQUID IRON

Condition of surface (true temperature 1500°C)	Apparent temperature	
	Total radiation pyrometer, °C	Selenium cell pyrometer, °C
Clear flat metal	1500	1500
Ripples affecting 10% of surface	1525	1508
Oxide on 5% of surface	1524	1505

TABLE 4. EFFECT OF STEAM AND WATER IN THE OPTICAL PATH

Water flow onto hot steel sheet	Indicated temperature	
	Total-radiation pyrometer, °C	Silicon cell pyrometer, °C
None	1100	1100
A little	1080	1098
Some	1055	1075
A lot	850	1030

In this type of application there is often a certain amount of steam and spray in the optical path as water is used to loosen the scale on the hot metal. Water may also run on the hot surface in globules across the line of sight. We were interested to compare the pyrometer with a total radiation pyrometer under such conditions, so a jet of water (estimated at a maximum of 5 gallons/minute) was allowed to fall on the hot sheet steel just outside the lines of sight of the two pyrometers. This was a very crude experiment, but it was sufficient to indicate that the pyrometer using the silicon cell was definitely less affected than the conventional pyrometer. The results are shown in Table 4.

Laboratory experiments have confirmed that a cloud of steam has less than a third of the effect on a silicon cell pyrometer when compared with a "total" radiation pyrometer. The percentage loss of radiation is only a little less with the silicon cell pyrometer, but this represents a much smaller change in temperature. Other experiments showed that liquid water in the optical path has very little effect on the silicon cell. Even an inch of water causes a drop of only about 30°C.

Induction Heating

A light pipe, consisting of a solid rod of transparent material, may be used to convey the radiation from a small area of a hot surface to the detector. Sapphire has been used very effectively with total radiation pyrometers. Silica is equally satisfactory if a silicon detector is used, and it is of course much less expensive. A small pyrometer with a silicon detector and a silica light pipe is illustrated in Fig. 8. It has been used successfully to measure temperatures of steel rolls during induction heating. The light pipe slips between the coils of the induction heater. An interesting feature of this installation is that the steel rolls are water-quenched, and a globule of water often forms on the end of the light-pipe, but this has no noticeable effect on the reading of the pyrometer.

Other pyrometers of more orthodox construction, but using silicon detectors, have been used to measure the temperatures of non-ferrous metals during induction heating. Measurements are sometimes made through the gap between the turns of the induction coil, sometimes on the ends of the billets. The materials measured include brasses, bronzes, copper-nickel, and alumi-

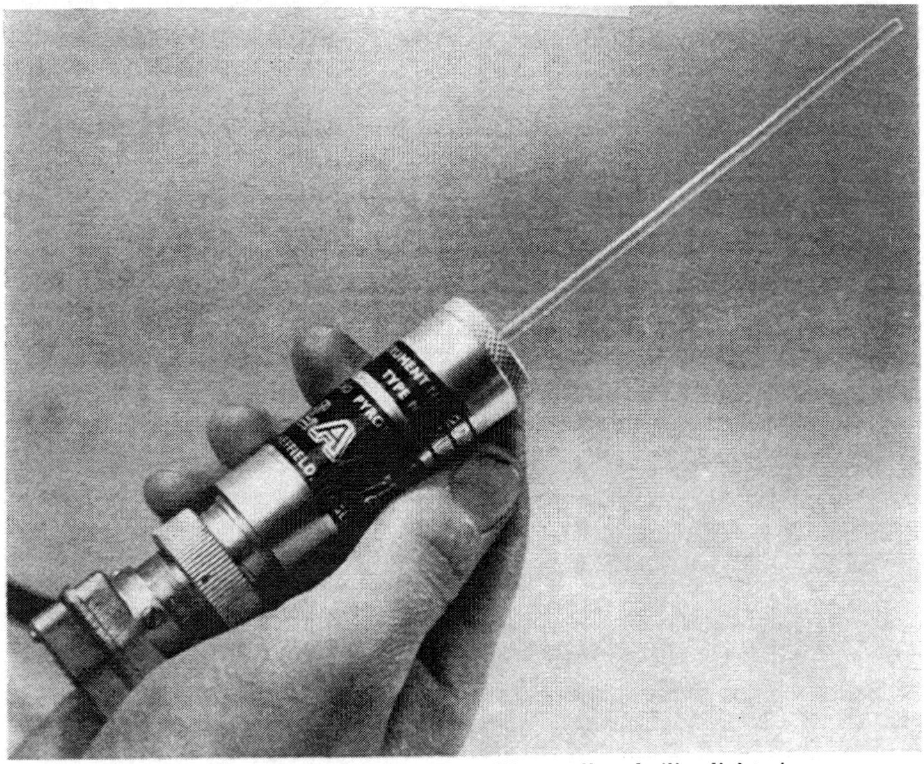

FIG. 8. Miniature pyrometer using silicon cell and silica light pipe.

num-nickel alloys. On one freshly cut surface of aluminum bronze a total radiation pyrometer registered a temperature about 100°C lower than on an adjoining surface which was more heavily oxidized. A pyrometer with a silicon detector registered the same temperature on both surfaces, corresponding to an emissivity of 0.7. The total emissivities of the two surfaces were found to be about 0.2 and 0.4 respectively.

The small size and high sensitivity of the silicon cell make it possible to design very small pyrometers, and units much smaller than the one shown in Fig. 8 have been made and used successfully. One has been placed transversely in a water-cooled tube to sight through a hole in the side of the tube. This unit has been put into a furnace and used to measure the temperature of glass, which would otherwise have been inaccessible. A similar probe has been inserted through the roof of a steel melting furnace to measure the roof temperature.

Glass Surface Temperature

An interesting application of the indium antimonide photoelectromagnetic cell is in the measurement of the surface temperature of glass. During the toughening process the surface temperature of the glass changes extremely rapidly. Pyrometers have been made for this type of measurement using the indium antimonide cell as the detector with a filter of lead selenide to select a suitable band of wavelengths for the measurement. This is not an easy application. The pyrometer has to be fitted in a water-cooled jacket, and a dc amplifier is used to match the impedances of the cell (about 7 ohms) and the recorder. The amplifier output current is equal to the input current from the cell, but the amplifier allows a large load to be used without detriment to the cell characteristics.

Two-Color Pyrometers

Two photovoltaic cells can be used to make a simple and effective "two-color" or ratio pyrometer. The principle of such pyrometers is well known. Instead of measuring the radiation emitted in a particular range of wavelengths, the pyrometer measures the radiation in two different bands of wavelengths. The ratio of the two intensities depends on the temperature of the source. The blackbody temperature which would yield the same ratio is called the "color temperature" of the hot body.

To measure the color temperature, a silicon cell and a selenium cell are illuminated from the same optical system in the pyrometer. The output current from one cell is passed through a fixed load resistance, and the potential across the load is applied to an electronic potentiometer. The current for the slide-wire of the potentiometer is obtained not from a battery but from the output of the second photovoltaic cell. In this way the recorder indicates directly the ratio of the two cell currents and hence the color temperature of the hot body.

A pyrometer of this type has been made and tested in the laboratory, and at the time of writing is waiting to be tested on a steel melting furnace. If a ratio pyrometer is correctly designed, the indicated temperature is independent of the size of the hot object on which it is sighted.

If the surface which is being measured were perfectly "gray" a color pyrometer would indicate the true surface temperature without any "emissivity correction." A gray surface is defined as one which has the same emissivity at all wavelengths. In practice, very few surfaces are truly gray and so we have a residual error, which may, however, be considerably smaller than for other types of radiation pyrometer.

A convenient way of comparing the errors of a color pyrometer and a conventional "brightness" pyrometer is shown in Fig. 9.

The difference between the true temperature T and the "brightness" temperature T_B, indicated by a pyrometer having an effective wavelength λ, is given by the equation:

$$\frac{1}{T} - \frac{1}{T_B} = \frac{\ln E}{C_2/\lambda}$$

where E is the emissivity of the surface at wavelength λ. For a ratio pyrometer, the error of the color temperature T_c is given by the corresponding equation:

$$\frac{1}{T} - \frac{1}{T_C} = \frac{\ln E_1 - \ln E_2}{C_2/\lambda_1 - C_2/\lambda_2}$$

where E_1 and E_2 are the emissivities at wavelengths λ_1 and λ_2.

Consequently, if the logarithm of the emissivity of a surface is plotted against C_2/λ, the slope of the line joining the origin to any point on the curve gives the error for a brightness pyrometer at that particular wavelength. The error of the color pyrometer is given by the slope of the chord joining the two points on the curve corresponding to the two wavelengths chosen in designing the pyrometer.

An analysis of such emissivity data as are available suggests that the color pyrometer may be the most accurate type of pyrometer for many surface temperature measurements. A

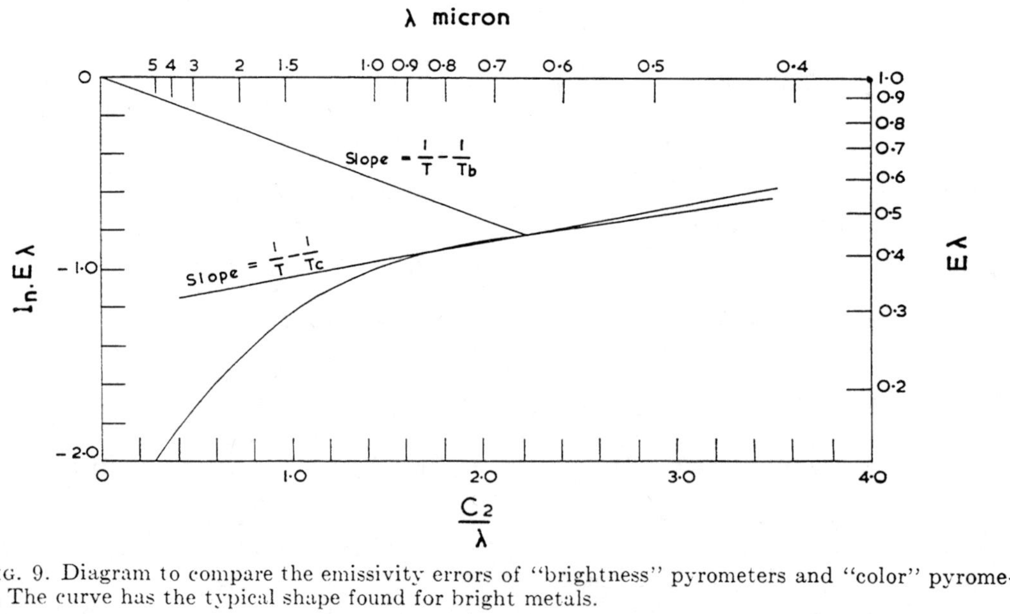

FIG. 9. Diagram to compare the emissivity errors of "brightness" pyrometers and "color" pyrometers. The curve has the typical shape found for bright metals.

ratio pyrometer using two photovoltaic cells is much simpler and less expensive to make than the usual type of two-color pyrometer and may well have an important future.

PROPERTIES OF PHOTOVOLTAIC CELLS

The operation of a selenium or silicon photovoltaic cell can be explained by reference to the equivalent circuit shown in Fig. 10 which must be taken as a simplified model, not a perfect representation. The current generator represents the physical fact that photons release electrons. The light generated current is proportional to the intensity of illumination of the cell. It must be emphasised that whereas a thermopile produces an emf proportional to the incident radiation, a photovoltaic cell produces a proportional *current*.

Diode Characteristics

The second feature of the cell is that the current generator is shunted by a diode. Since the properties of this diode determine the electrical properties of the cell, it is necessary to discuss the diode characteristics before we look at the cell as a photoelectric device.

A diode has a resistance which varies with the magnitude and direction of the current passing through it. The relationship between current and voltage is, therefore, not linear, and may be determined by the circuit shown in Fig. 11. The voltage V across the diode is measured with a voltmeter and the current I passing through it is varied by means of the resistor R and measured by the low-resistance ammeter A. The type of curve obtained for a silicon diode is shown in Fig. 12. If the diode is a photovoltaic cell, then the characteristic curve is displaced when the cell is illuminated. The light-generated current cannot change the shape of the diode characteristic; it merely provides a source of extra current. The characteristic of the illuminated cell is shown in Fig. 12 as a broken curve.

The series resistance is present due to the electrical resistance in the top layer between the contact strip and the point of release of the electron. The effect of this resistance is small at the low currents obtained in radiation pyrometers and its presence can usually be neglected. Typical values would be 50 ohms in a silicon cell and 1000 ohms in a selenium cell.

The effect of increasing the illumination of the cell is shown in Fig. 13. In this diagram the light generated current is shown positive instead of negative as in Fig. 14. It is evident that the open-circuit voltage is not proportional to the intensity of illumination. In fact, above a certain level of illumination the voltage is approximately proportional to the logarithm of the intensity of the illumination. This logarithmic relationship might be turned to advantage

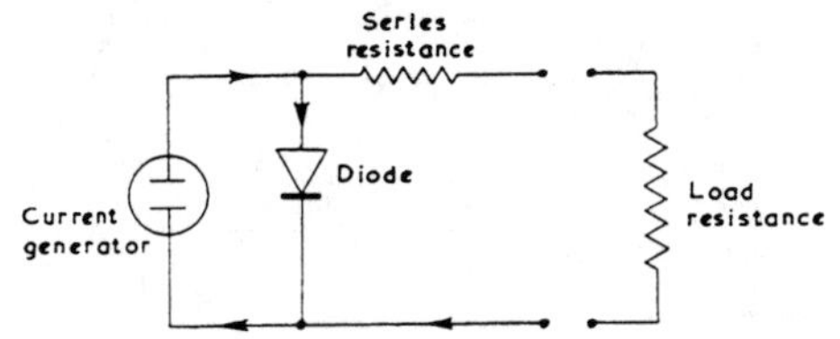

FIG. 10. Equivalent circuit of a silicon or selenium photovoltaic cell.

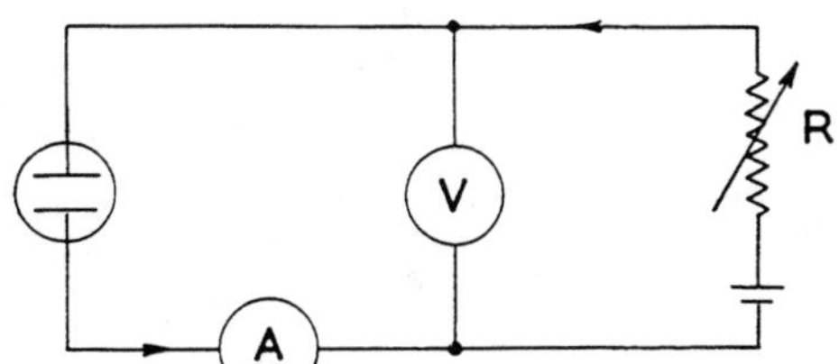

FIG. 11. Circuit for measuring the characteristics of a diode.

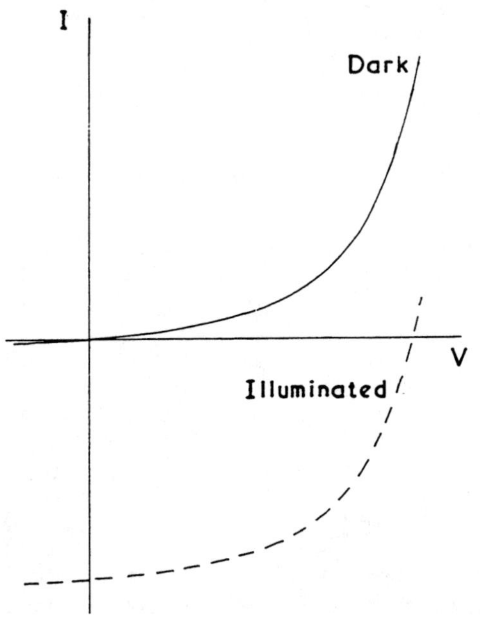

FIG. 12. Current-voltage characteristics of a photovoltaic cell, dark and illuminated.

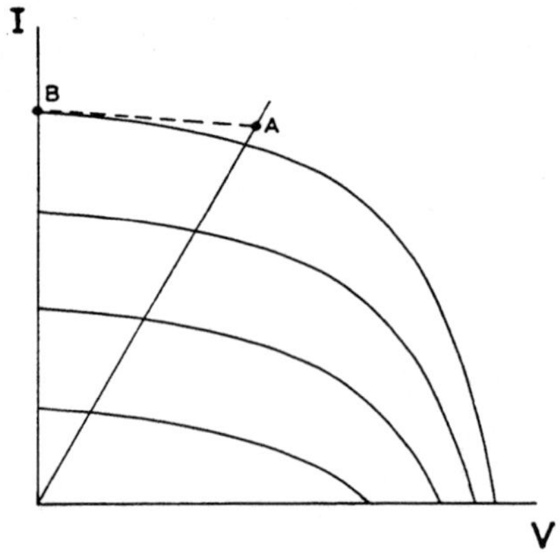

FIG. 13. Characteristics of a cell at different levels of illumination.

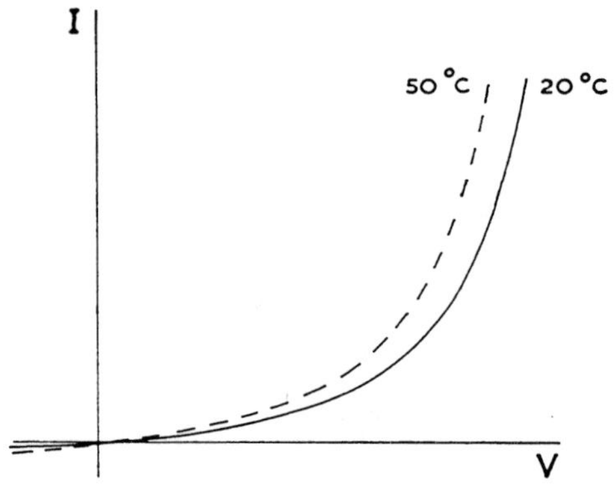

FIG. 14. Change in the diode characteristic of a dark cell on heating.

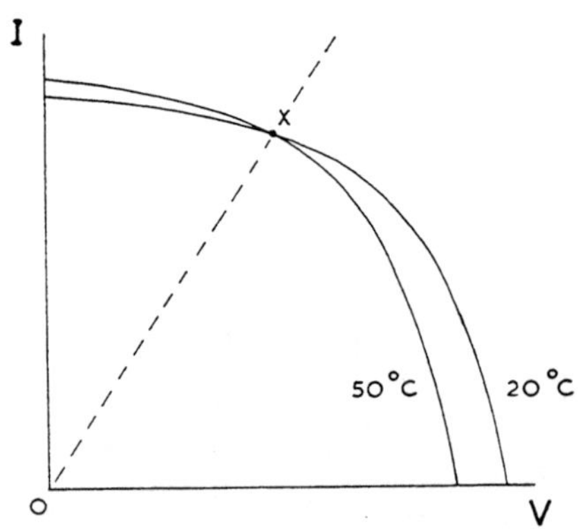

FIG. 15. Characteristics of an illuminated cell at vo temperatures. At load OX the temperature coefficient is zero.

in pyrometry. We have preferred to use a voltage proportional to the intensity of illumination which we obtain by passing the cell current through a small resistance.

The line OA represents the resistance of the measuring circuit, and the deviation of the curve BA from the tangent at B represents the departure of the cell from linear response to illumination at that particular load. The curves are nearly parabolic close to the origin. A high quality silicon cell with a load of 400 ohms is substantially linear up to an output of about 50 millivolts. An input resistance of 400 ohms is about as high as a normal electronic potentiometer will comfortably tolerate, and seems to be about the right load for a silicon cell from all points of view. A load of 1000 ohms may be preferable for a selenium cell; there is room for a diversity of opinions on the best compromise for any situation.

Temperature Coefficient

When a photovoltaic cell is heated, the diode characteristic of the dark cell is displaced to the left without greatly changing its shape, as shown in Fig. 14. In addition to this, the cutoff wavelength of the spectral sensitivity curve is displaced a little towards longer wavelengths so that the light generated current is increased.

The effect on the output of this combination of events is shown in Fig. 15. There is a load, shown by the line OX, for which the temperature coefficient is zero. But this load varies with the temperature range, with the illumination, and between cells, so the effect is of no great practical value.

The relation between the short circuit current and the temperature is substantially linear, but as the load is increased the line falls away in a series of curves as shown in Fig. 16. The

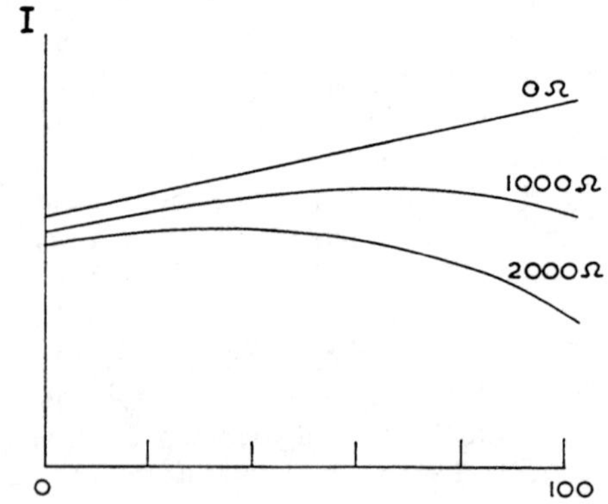

FIG. 16. Variation of the cell current with temperature at different load resistances; silicon solar cell.

slope of the zero-load line depends on the color temperature of the source. The temperature coefficients of silicon cells at low loads have been found to lie between about 0.2% per °C at 1200°C and 0.6% per °C at 700°C, varying considerably from cell to cell. A value of about 0.4% per °C seems to be a good working mean.

Temperature compensation for a silicon cell can be obtained with a shunt circuit combining thermistors and menganin resistors. The circuit shown in Fig. 17 is a useful one for general purposes. It has a substantially linear increase of conductance between 20 and 60°C. Above 60°C the temperature coefficient of conductance decreases, and so matches the curvature of the 400-ohm characteristic of an average silicon cell.

A little care in the mounting of the pyrometer will usually avoid excessive temperature in most applications for which a silicon cell would be chosen in preference to a thermopile. The small imperfections in compensation which are unavoidable are not often of an important magnitude.

The selenium barrier-layer cell must always be water-cooled, so the need for temperature compensation does not normally arise. The general picture given for silicon cells seems to apply also to selenium cells, but temperature coefficients are usually about 0.1 to 0.3% per °C at zero load. The indium antimonide photoelectromagnetic cell has a large negative temperature coefficient of about 2% per °C, which must be compensated even when water cooling is used.

The temperature coefficient of a thermopile depends on the material chosen to make it, and on the length and thickness of the wire or strip which is used. An average value for the coefficient would be minus 0.15% per °C. Compensation is often obtained by a simple shunt resistor of nickel, which has a high positive temperature coefficient of resistance. It might appear that the uncompensated temperature coefficient of a silicon cell is much larger than that of a thermopile, but the comparison must be made in terms of degrees change in indicated temperature for each degree change in the temperature of the detector. The comparative figures at 1000°C are approximately −0.3° per degree for the thermopile and +0.4° per degree for the silicon cell.

Linearity and Effective Wavelength

All industrial pyrometers of a given type must give the same "scale shape"; that is to say that

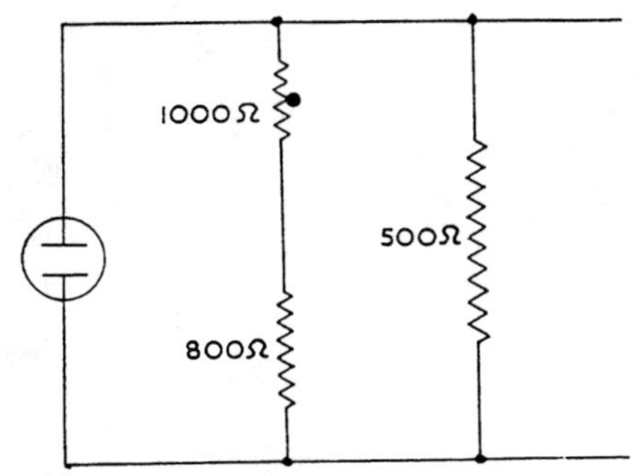

Fig. 17. Temperature compensation circuit for a silicon cell, using a 1000-ohm thermistor.

the ratio of outputs at any two selected temperatures must be the same for every individual pyrometer of that type. This will be true if the detectors have a linear response to the intensity of incident radiation, and if they all have the same effective wavelength.

A large batch of silicon cells was recently tested for linearity. Each cell in turn was illuminated from a blackbody furnace, and the intensity of illumination was then reduced, first by a factor of 5, and then by a factor of 25. The actual output was compared with the output expected from a proportional relationship. The actual output for a reduction of 5 times in illumination was always less than the output predicted, the ratio lying between 0.81 and 0.99. For a reduction in illumination of 25 times the ratio lay between 0.68 and 0.98. Cells having a ratio greater than 0.95 for a reduction in illumination of 5 times were considered acceptable. Selenium cells tested some years ago by Land[8] were more nearly linear and the smaller illumination gave a rather greater output than predicted. Recent experience has confirmed that the deviation is in this direction: the deviation is usually very small, but a bad cell is occasionally found.

An optical pyrometer responds to a narrow band of wavelengths and it is usually treated as if it were a monochromatic pyrometer measuring radiation of a certain "effective wavelength." The calculation of the effective wavelength has been described by Forsythe.[9] A pyrometer using a silicon or selenium cell can similarly be considered to have an effective wavelength which is calculated in a similar manner to the optical pyrometer, as described by Land.[8] If the logarithm of the photocurrent is plotted against the reciprocal of the absolute temperature of the source, the result is a line

TABLE 5. EFFECTIVE WAVELENGTHS FOR PYROMETERS USING SILICON AND SELENIUM CELLS

Temperature, °C	Effective wavelength (micron)	
	Silicon cell	Selenium cell
500	0.98	—
1000	0.91	0.62
1500	0.86	0.60
2000	0.83	0.59

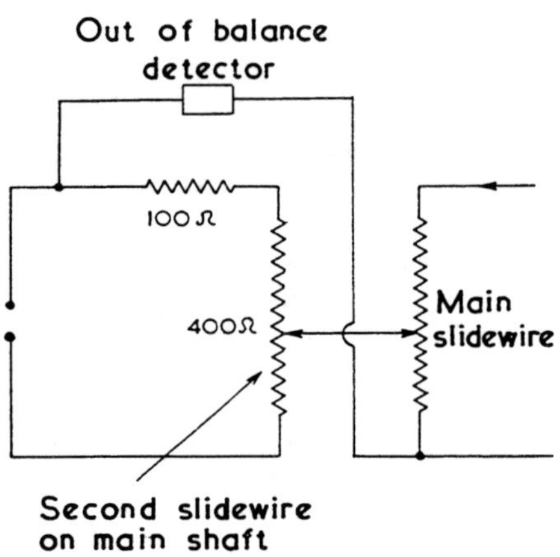

FIG. 18. Modified potentiometer circuit for radiation pyrometers.

which is slightly curved. The slope of the curve is equal to C_2/λ_e, where C_2 is the usual constant from Wien's formula and λ_e is the effective wavelength of the pyrometer. The effective wavelength of an optical pyrometer varies very little with temperature, but a photoelectric cell responds to a wider band of wavelengths and consequently, the effective wavelength changes rather more. Values for silicon and selenium cells are given in Table 5.

Photoelectric pyrometers of these types will have different scale shapes if the effective wavelength differs from cell to cell. The effective wavelength depends primarily on the optical properties of the silicon or selium, such as the optical transmission. Certain variations have been noticed in both silicon and selium cells, but the variations are rarely of a serious magnitude. The best test is a simple comparison of calibrations at two widely different temperatures.

Measuring Circuits

The circuit shown in Fig. 17 has been designed to provide temperature compensation for a silicon cell. The thermistor and the 800-ohm manganin resistor are incorporated in the pyrometer, while the 500-ohm resistor is incorporated in the indicator or recorder. It can be either a 500-ohm millivoltmeter or a resistor in parallel with the input terminals of a potentiometer.

Many makes of electronic potentiometer can be simply modified to give a more linear scale for a radiation pyrometer, and the circuit shown in Fig. 18 is designed with this possibility in mind. A second slide-wire is fitted to the same shaft as the main slide-wire. The pyrometer current is passed through the auxiliary slide-wire and a fixed series resistor. When the contact to the main slide-wire is taken off through a parallel slide-wire, no extra slide-wire is necessary. The second slide-wire becomes a variable potential divider which changes the proportion of the total potential difference which is applied to the potentiometer. The modified circuit using a 400-ohm slide-wire is shown in Fig. 18. It will be seen that near the top of the scale only one-fifth of the full potential is applied to the potentiometer, but near the bottom of the scale the full potential is applied. The improvement in scale shape has already been mentioned and illustrated in Fig. 7. Higher ratios than 5:1 can be used, but an automatic adjustment of the amplifier gain is then usually necessary, and the scale becomes rather cramped near the top.

Low Temperatures and Reflected Light

At temperatures below about 700°C, the silicon cell is less sensitive than the thermopile, but it becomes increasingly effective in eliminating the effect of emissivity variations. A change from a bare metal surface of emissivity 0.4 to an oxidized surface of emissivity 0.8 could change the indicated temperature by about 120°C at 500°C if a thermopile were used as the detector. With a silicon cell the change would be 30°C. The lowest range that can be measured is something like 400 to 500°C. The output at 400°C is less than one-tenth of the output at 500°C, half the scale would be covered by the range 470 to 500°C.

At temperatures below 700°C it is necessary to keep an eye on reflected light. It is well known that reflected light increases the apparent temperature of a surface measured with an optical pyrometer. When sighted on a piece of white paper, an optical pyrometer will read something like 800°C in strong artificial light (3 feet from a 100-watt tungsten lamp) or 900°C in front of a window (not direct sunlight). With a silicon cell, the corresponding temperatures would be about 600°C and 700°C respectively. It is only rarely that this causes any error, and the effect can readily be overcome by shielding the surface from direct illumination.

Other Detectors

The three detectors which we have used so far do not by any means exhaust the possibilities of photovoltaic cells in radiation pyrometry. It does not appear that any device so far developed is likely to approach the silicon solar cell in general usefulness. We need a cell which is sensitive to longer wavelengths, up to about 3 microns. Germanium takes us a little way, and the germanium phototransistor has photovoltaic properties which can be used to investigate the general possibilities of such a cell. Gold-doped germanium is sensitive to longer wavelengths, but has not yet been tested. Lead sulfide photovoltaic cells have been described by Smith, Jones, and Chasmar[10] and appear to have a useful range of spectral sensitivity. Other possible cells are numerous, and continue to be developed. The present review must, therefore, be considered as an early and incomplete review of a large and growing field of work.

Conclusions

There seems no doubt that the silicon solar cell will be important in radiation pyrometry. It seems logical to prefer it to the thermopile for nearly all surface temperature measurements at temperatures above about 700°C. The effect of variations in emissivity of the hot surface is reduced by a factor of 2 or 3, atmospheric absorption by water vapor is eliminated, and steam and water in the optical path have much less effect on the temperature measurement. The cell gives a larger output than a thermopile so that smaller targets can be measured and smaller pyrometers can be designed. The high speed of response of the cell is often a benefit in industrial pyrometry. It is simple and robust and will withstand high ambient temperatures without damage. Its disadvantages are the limited range of temperature which can be accommodated on a single scale, and an unavoidable imperfection in temperature compensation. These seem a reasonable price to pay for the advantages offered.

The selenium cell and the indium antimonide photoelectromagnetic cell have been found useful for special measurements where it is necessary to use either short wavelength or long wavelength radiation, and they clearly have a place in industrial pyrometry. It seems quite possible that other similar devices will take a place alongside these three cells as new types become commercially available.

References

1. M. Wolf and M. B. Prince, *Solid State Physics in Electronics and Telecommunications* (Academic Press, Inc., London and New York, 1960), Vol. 2, Part 2, p. 1180.
2. G. A. Veszi, J. Brit. I.R.E. **13,** 183 (1953).
3. C. Hilsum and I. M. Ross, Nature **179,** 146 (1957).
4. B. M. Larsen and W. E. Shenk, *Temperature, Its Measurement and Control in Science and Industry* (Reinhold Publishing Corporation, New York, 1941), p. 1150.
5. T. Land, J. Iron Steel Inst. (London) **155,** 568 (1947).
6. E. J. Burton and R. Mayorcas, J. Iron Steel Inst. (London) **168,** 151 (1951).
7. R. Rew, Brit. Cast Iron Research Assoc. J. **8,** 821 (1960).
8. T. Land, J. Iron Steel Inst. (London) **149,** 481 (1944).
9. W. E. Forsythe, *Temperature, Its Measurement and Control in Science and Industry.* (Reinhold Publishing Corporation, New York, 1941), p. 1115.
10. R. A. Smith, F. E. Jones, and R. P. Chasmar, *The Detection and Measurement of Infrared Radiation* (Oxford University Press, London, 1957), p. 160.

Reprinted with permission from *Temperature: Its Measurement and Control in Science and Industry,* J.F. Schooley, editor, Vol. 5, pp. 485-489 (1982).

Steel surface temperature measurement in industrial furnaces by compensation for reflected radiation errors

J. E. Roney

Jones and Laughlin Steel Corporation, Pittsburgh, Pennsylvania 15227

Optimum heating control is achieved when steel temperatures rather than zone temperatures are measured. Traditionally, zone temperatures were used to control furnaces because reflected radiation from flames, heating tubes, and hot furnace walls caused large errors (> 100 °C) in the measurements made with radiation pyrometers aimed directly at the steel surface. Dual pyrometer methods are described which compensate for the error caused by reflected radiation. These pyrometers are positioned at a safe distance from the product passline to insure long term reliability. Accurate measurement can be made of slab and billet temperatures in reheat furnaces and strip temperature in galvanizing lines. Dual pyrometer installations on these process lines are described. Equations with sample calculations are shown to estimate the effect of various parameters on the measurement accuracy.

Until recently furnace zone temperatures were used to control the heating in our furnaces. Operating rates were based on experience for the product size and zone temperature. Zone temperature measurements rather than steel temperatures were used because of the difficulty in obtaining accurate surface temperature measurements in a furnace. For a pyrometer aimed directly at the steel surface measurement errors of over 100°C could be caused by radiation from flames, hot furnace walls and heating tubes.

With zone control, productivity and fuel rate are not at the optimum because the steel temperature is usually higher than required. Quality may also be inconsistent because of the variability in the actual steel temperature.

Direct measurement of steel surface temperature can be made by aiming the pyrometer through a cooled sight tube with a cooled flange near the steel surface. The intensity of the furnace radiation is reduced with each reflection between the flange and steel surface. Analysis of this method shows that two of the many parameters that affect the pyrometer's error are the ratio of flange height above the surface to the flange radius and an experimentally determined number that is a measure of the specularity of the surface [1]. Because the flange tube is near the steel passline it may be subject to damage and it interferes with maintenance.

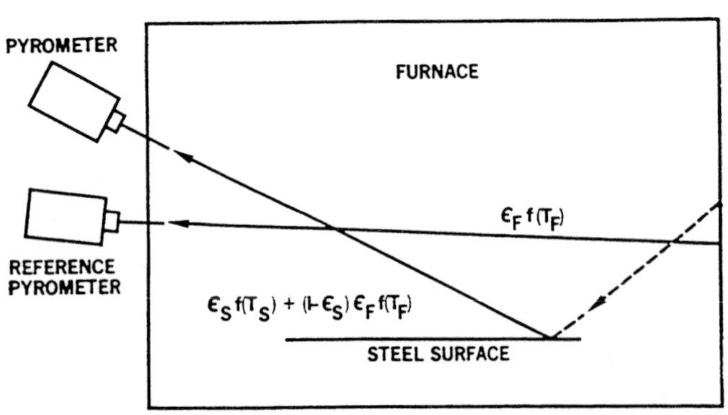

FIG. 1. DUAL PYROMETER - BACK-OFF METHOD

In order to reduce the error caused by the reflected radiation we use the signal from a second pyrometer to compensate for the error of the pyrometer aimed at the steel. By this method the pyrometers and their accessories can be placed far away from the passline to achieve long term reliability. Several different versions of the dual pyrometer systems will be described. The version selected depends on the particular measurement conditions.

BACK-OFF METHOD

As shown in Fig. 1, one pyrometer is aimed at the furnace wall [2]. Then a portion of the signal of the pyrometer aimed at the wall is subtracted from the signal of the pyrometer aimed at the steel surface. The relative magnitude of each of these signals is shown on the rays representing the radiation received by the dual pyrometer. The three terms correspond to the emitted and reflected radiation from the steel surface and the radiation from the furnace wall.

Conceptually:

$$Wa = \frac{\varepsilon_S f(T_S)}{\varepsilon_p} + \frac{(1 - \varepsilon_S)\, \varepsilon_F f(T_F)}{\varepsilon_p} - \frac{\varepsilon_F f(T_F)}{\varepsilon_R} \qquad (1)$$

where:

Wa	=	Net radiation received by the dual pyrometer
ε_S and ε_F	=	Effective emissivity of the steel surface and furnace wall, respectively
$(1 - \varepsilon_S)$	=	Effective reflectivity of the steel surface
ε_p and ε_R	=	Emittance settings of the pyrometers aimed at the steel surface and at the wall, respectively
$f(T_S)$ and $f(T_F)$	=	Radiation emitted by a black body at the steel temperature T_S and wall temperature T_F, respectively (Wien's approximation).

To simplify the equations used to describe the dual pyrometer concepts, the two pyrometers are assumed to have identical response characteristics. In addition, the solid angle subtended by the target area of the steel surface is assumed to be exactly equal to that of the reference area on the furnace wall. The furnace itself is assumed to be a black body cavity. Then to cancel the reflected error, the second and third terms of equation (1) must be equal. This requires that the reference pyrometer's emittance setting be at:

$$\varepsilon_R = \frac{\varepsilon_p}{1 - \varepsilon_s} \tag{2}$$

With the reference pyrometer's emittance set at the value given by equation (2), the radiation detected by the reference pyrometer is the same as the radiation reflected from the steel surface. They may not always be the same if the products of the view factor and area for the reflected radiation from the steel surface and from the wall are not equal. These products which can be considered to be contained in ε_F may not vary equally in slab reheat furnaces where over 95% of the heat transfer is by radiation from the flames. For best accuracy, the steel temperature measurement should be near its end close to the wall area. But at the slab end the heating and cooling may be quite different than near the center of the furnace.

COOLED REFERENCE TARGET

To get improved compensation for the reflected radiation, we have placed a cooled reference plate in our reheat furnace near the measured area of the slab [3]. This target plate reflects radiation from both the flames and the furnace wall. As shown in Fig. 2, one pyrometer is aimed at the slab surface and the reference pyrometer is aimed at the cooled reference target. The oxidized target surface has nearly the same reflectance as the oxidized slab surface in the furnace.

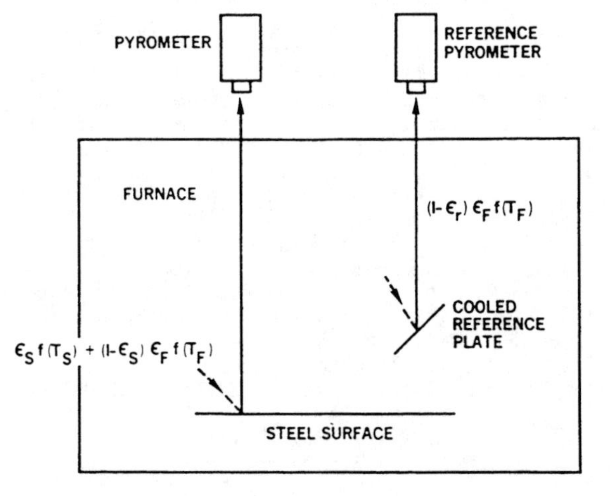

FIG. 2. DUAL PYROMETER - COOLED REFERENCE TARGET METHOD

The equation for this method is:

$$Wa = \frac{\varepsilon_s f(T_s)}{\varepsilon_p} + \frac{(1 - \varepsilon_s)\,\varepsilon_F f(T_F)}{\varepsilon_p} - \frac{(1 - \varepsilon_r)\,\varepsilon_F f(T_F)}{\varepsilon_R} \tag{3}$$

where:

ε_r = Effective emissivity of the reference target

To cancel the reflected radiation error:

$$\varepsilon_R = \varepsilon_p \; \frac{1 - \varepsilon_r}{1 - \varepsilon_s} \tag{4}$$

When the reference target emittance equals the steel surface emittance, $\varepsilon_R = \varepsilon_p$, both pyrometers are set at the same emittance. If the target surface temperature is too high it can cause an error in the measurement. This error will be less than 5°C if the target temperature is lower than 650°C at a steel temperature of 815°C. For convenience we use water cooling which keeps the surface temperature below 100°C. An experimental furnace with overhead firing similar to reheat furnaces was used to determine the dual pyrometer error caused by tilting the reference target from the plane of the slab surface. A slab with a thermocouple embedded in its surface was placed in the furnace with the pyrometer aimed at 45° to the slab surface. The cooled reference target could be tilted up to 45° from the plane of the slab surface with only a 5°C error in the dual pyrometer temperature indication with the error increasing to 55°C with the reference target at 90° to the slab. Thus, to keep the error from target position less than 5°C, we install the reference targets at an angle of 45° or less. At this angle dust and dirt slide off the target surface.

These dual pyrometer systems have been installed on three zones of each of the three reheat furnaces at both our Cleveland and Indiana Harbor Works. They can be mounted on the side walls to aim underneath the flames or they can be mounted on top the furnace to aim between the flames. Even if a pyrometer wavelength is selected that can look through clear flames excess fuel mixtures or incomplete mixing or burning of the fuel will produce luminous carbon particles which will interfere with the measurement.

Fig. 3 shows a cut-a-way view of a furnace zone with the pyrometer heads mounted on the roof. One pyrometer is aimed at the cooled reference target mounted on the burner wall, the other pyrometer aims at the slabs as they pass beneath it. Each pyrometer's signal is equally amplified, then the signal of the pyrometer aimed at the reference target is subtracted from the signal of the pyrometer aimed at the slab. Finally, the resulting signal goes to the meter, recorder, and control system. All our dual pyrometer systems use a similar subtraction circuit. Pyrometer emittance setting, ε_p and ε_R are fixed by the gain adjustment of the amplifier for each pyrometer.

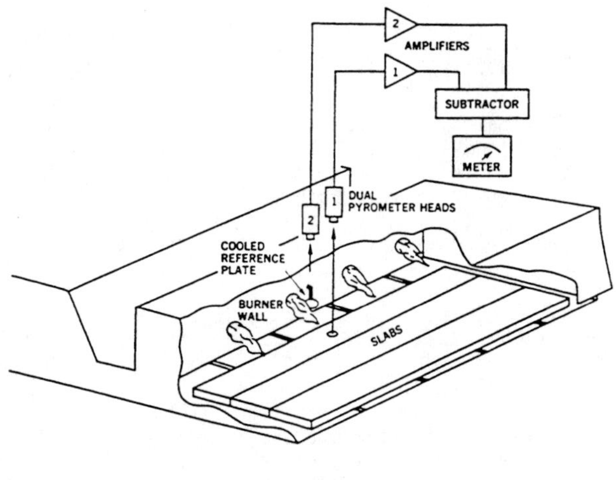

FIG. 3. DUAL PYROMETER ON TOP ZONE OF REHEAT FURNACE

To withstand the severe environment of the reheat furnace, the cooled reference target is ruggedly built and then mounted against the burner wall as shown in Fig. 4. Mounting the target against the burner wall provides additional support and an out of the way location. Yet it is still close to the measurement area of the slab.

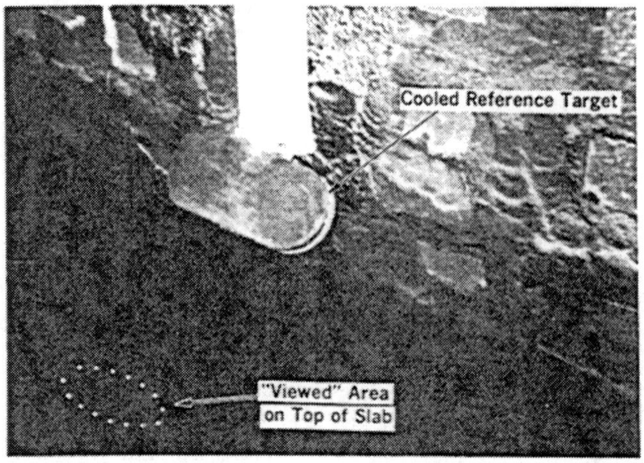

FIG. 4. COOLED REFERENCE TARGET MOUNTED ON FURNACE BURNER WALL

Bottom as well as top slab surface temperatures are measured with the dual pyrometers. Firing of the zones can thus be controlled independently to get the desired slab temperature. No specially designed cooled reference plate is needed in the bottom zone because a section of water cooled support pipe with the insulation removed is used as the reference target.

On-line performance is checked with an embedded thermocouple in the slab surface. As the slab passes through the furnace, the thermocouple is fed through the rear charging door. When the thermocouple junction passes near the pyrometer target, the two measurements are compared. Further checks were also made with a thermocouple embedded in a small test slab. This test slab was attached to a water cooled pipe and then pushed through a side door during the heat-up when the slabs were not moving. Gas flow, air flow, and air/fuel ratios were varied to confirm the dual pyrometer's performance for various firing conditions. The pyrometer's indication of the surface temperature varied less than ±11°C for gas flow variations of 90% of range, air flow changes of 65% of range, and air/fuel ratios from 0.6 to 1.6. Emittance differences of the oxidized surface for various grades of steel were measured. These differences will cause the pyrometer's indication to vary only ±6°C.

Estimated measurement inaccuracy of ±18°C is the root-sum-square (RSS) of the individual errors in the system. These errors are the individual pyrometer inaccuracy in the temperature range we use of ±9°C for each of the two pyrometers, the errors caused by furnace firing variations of ±11°C and the errors caused by emittance variations of ±6°C. Although the inaccuracy of each pyrometer is ±9°C, its reproducibility is ±4°C. Furnace firing variations, fuel and air flow and air/fuel ratios are normally within one-half of those described above. As a consequence, the reproducibility of the measurement of the slab temperature in the furnace is estimated to be ±10°C.

Surface scale temperature is measured by the dual pyrometer instead of the desired steel surface temperature underneath the scale. Slabs enter the furnace with a thin tight scale which grows to a thickness of about 3 mm at the exit end of the intermediate zone. In the first (primary) zone, the scale is thin and the rate of heat flow is high. Computer modeling analysis indicates that the temperature of the scale surface is an average of 55°C higher than the steel surface. In the intermediate zone, the scale is thicker, and the rate of heat flow into the slab is low, the temperature difference averages 28°C. When using the dual pyrometers to control the furnaces, these differences should be used to calculate the heat flow into the slab and to determine control strategy.

Dual pyrometer measurements have improved control and operation of the furnaces. Slabs of different thicknessess in the furnace at the same time cannot be heated to the same temperature, but the dual pyrometer will show when the thick ones are too cold and the thin ones are too hot. Dripping, melted scale from overheating the surfaces damages the insulation on the support pipes, causes build-up on the furnace bottom, and causes defects because melted scale is difficult to remove from the steel surface during rolling. The melting of the scale can be prevented by not allowing the surface temperature to go over 1370°C. During mill delays the furnace firing is reduced. Before the use of dual pyrometers for steel temperature monitoring and control, the firing cutback was not enough because the "Heater" was concerned that the slabs would be too cold to roll. Now with the dual pyrometer measurement, he knows and maintains the slabs at a predetermined temperature. This change alone has resulted in a 4% fuel savings.

Dual pyrometer signals cannot be used in a standard feedback control system because of the variability of slab thickness in the furnace. A thick cool slab would cause high firing while a thin hot slab would reduce the firing. To control this process a list of the dimensions of the slabs in the furnace and the desired dual pyrometer temperatures as a function of operating parameters are stored in a microprocessor. If the slabs are not at the desired temperature when measured by the dual pyrometer, the furnace firing is changed. This microprocessor control is now being installed in the furnace control system. When it is in operation, further improvements in furnace control and fuel rate should result.

CLOSED-END TUBE

Cooled reference targets cannot be placed in some furnaces because of their design, nor are cooled targets needed in all furnaces. If the zone temperature is uniform; no flames, no heating tubes, etc., only a measurement of the zone temperature is needed to correct the steel temperature measurement for the error caused by furnace radiation that is reflected from the surface. This zone temperature measurement can be made with a thermocouple and then the correction calculated. Instead of using a thermocouple to obtain a correction measurement, a correction signal can be obtained from a pyrometer aimed into a closed-end tube as shown in Fig. 5. This correction signal given by Wien's approximation can then be used to subtract the reflection error from the signal of the pyrometer aimed at the steel surface.

The equation for the closed-end tube method is:

$$Wa = \frac{\varepsilon_S f(T_S)}{\varepsilon_p} + \frac{(1 - \varepsilon_S)\ \varepsilon_F f(T_F)}{\varepsilon_p} - \frac{f(T_F)}{\varepsilon_R} \quad (5)$$

To cancel the reflection error:

$$\varepsilon_R = \frac{\varepsilon_p}{(1 - \varepsilon_S)\ \varepsilon_F} \quad (6)$$

We use the closed-end tube method at our Indiana Harbor Works to measure strip temperature on two galvanizing lines. Strip temperature is measured just beyond the heating zone exit. There are no heating tubes beyond the exit at the pyrometer position, but the zone temperature at this measurement location is high enough to cause a reflection error.

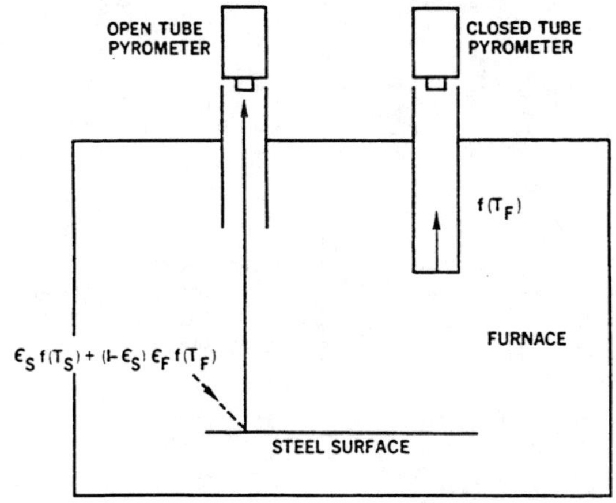

FIG. 5. DUAL PYROMETER - CLOSED TUBE METHOD

By accurately measuring and controlling the strip annealing temperature to the minimum needed, both the heating and cooling requirements are minimized. This lower strip temperature allows us to maximize productivity by operating at the fastest possible line speed.

SPECULAR REFLECTION

The last dual pyrometer method to be described can be used for specularly reflecting (mirror-like) surfaces. For this method one pyrometer is aimed at an angle to the smooth strip. Then because the reflected radiation's angle of reflection is equal to its angle of incidence, the area (reference area) of the furnace wall from which the reflected radiation is coming is known. Another pyrometer as shown in Fig. 6 measures the radiation from this reference area. It is assumed that the furnace wall radiation is equal both in the direction of incidence to the steel surface and in the direction of the reference pyrometer.

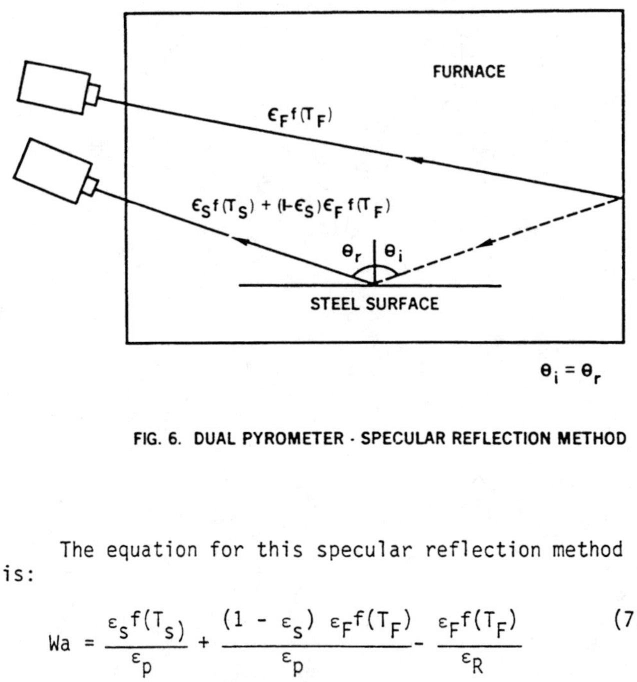

FIG. 6. DUAL PYROMETER - SPECULAR REFLECTION METHOD

The equation for this specular reflection method is:

$$Wa = \frac{\varepsilon_s f(T_s)}{\varepsilon_p} + \frac{(1 - \varepsilon_s)\ \varepsilon_F f(T_F)}{\varepsilon_p} - \frac{\varepsilon_F f(T_F)}{\varepsilon_R} \qquad (7)$$

To cancel the reflection error:

$$\varepsilon_R = \frac{\varepsilon_p}{1 - \varepsilon_s} \qquad (8)$$

By placing a black body furnace at the reference area both strip temperature and its emittance can be measured [4]. However, if the strip emittance is known, a reference furnace is not needed.

Another modification which we have used on a pilot line is to make the reference area cool and non-reflecting by installing a water cooled conical cavity in place of the black body furnace. This change makes the second term of equation (7) insignificant and eliminates the need for the reference pyrometer.

ERROR CALCULATION

When planning a dual pyrometer system it is essential to be able to estimate measurement accuracy before the installation. For the Back-Off method, let S be the differential signal obtained from the dual pyrometer. Using Wien's approximation for the spectral radiance of the steel surface and the furnace wall, S can be represented to a close approximation by:

$$S = G_p/\varepsilon_p \{\varepsilon_s \exp[-C_2/\lambda_p T_s] + (1-\varepsilon_s)\ \varepsilon_F \exp[-C_2/\lambda_p T_F]\} - G_R/\varepsilon_R \{\varepsilon_F' \exp[-C_2/\lambda_R T_F']\} \qquad (9)$$

where:

G_p and G_R	=	The response coefficients for the pyrometer aimed at the steel surface and for the pyrometer aimed at the furnace wall, respectively. Calibration information as well as other measurement constants of the pyrometers are contained in these factors.
λ_p and λ_R	=	The effective wavelengths of the pyrometer aimed at the steel surface and the pyrometer aimed at the furnace wall, respectively
ε_p and ε_R	=	The emittance setting of the pyrometer aimed at the steel surface and the pyrometer aimed at the furnace wall, respectively
ε_s and ε_F	=	The effective emissivities of the steel surface and the furnace wall, respectively
T_s and T_F	=	The temperature of the steel surface and the furnace wall, respectively
ε_F' and T_F'	=	The effective emissivity and temperature of the furnace wall which may not be the same as ε_F and T_F, respectively

Equation (9) can be rearranged in a more useful form for error analysis.

$$S = G_p/\varepsilon_p \{\varepsilon_s \exp(-C_2/\lambda_p T_s)\} \cdot \{1 + \frac{(1-\varepsilon_s)\ \varepsilon_F}{\varepsilon_s} \exp[-\frac{C_2}{\lambda_p}(\frac{1}{T_F} - \frac{1}{T_s})] - \frac{G_R}{G_p}\frac{\varepsilon_p}{\varepsilon_R}\frac{\varepsilon_F'}{\varepsilon_s} \exp[-C_2(\frac{1}{\lambda_R T_F'} - \frac{1}{\lambda_p T_s})]\} \qquad (10)$$

Similar equations can be derived for the other three methods. Reference pyrometer emittance setting for the method used as given by equations (2), (4), (6), and (8) should be substituted for ε_R. As can be seen in equation (10), to get accurate reflection error compensation, the magnitude of the last and next-to-last terms should be small compared to one. These two terms are smaller for a high effective emissivity, ε_s, than for a low effective emissivity. When $T_s = T_F$ the exponential term is one. When the steel temperature, T_s, is much higher than the furnace temperature, T_F, the magnitude of these terms is small compared to one. The following examples show the error in dual pyrometer compensation that can occur for some installations.

1. Slab temperature in a reheat furnace using the cooled reference method.
   where: $T_s = 1260°C,\ T_F = T_F' = 1620°C$

   $$\varepsilon_s = 0.85,\ \varepsilon_p = \varepsilon_r = \varepsilon_R = 0.80$$

   $$\varepsilon_F = \varepsilon_F' = 0.2,\ G_p = G_R$$

   $$\lambda_p = \lambda_R = 0.9\ \mu m$$

The dual pyrometer temperature measurement is 1256°C, for an error of -4°C. T_F and ε_F represents the flame temperature and emittance, because most of the radiation is from the flame rather than the furnace walls. If the wavelength is changed to 3.9 µm the error is 27°C for the same conditions. For other flame and slab temperatures, the error may be less at the longer wavelength. We chose the short, 0.9 µm wavelength because the silicon detector used at this wavelength is more rugged and stable than the longer wavelength detectors.

2. Strip temperature in the heating zone of a galvanizing line where the heating tubes and furnace walls are much hotter than the strip. Specular Reflection method.
   where: $T_s = 540°C,\ T_F = 980°C,\ T_F' = 974°C$

   $$\varepsilon_s = \varepsilon_p = 0.55,\ \varepsilon_R = \frac{\varepsilon_p}{1 - \varepsilon_s}$$

   $$\varepsilon_F = \varepsilon_F' = 1,\ G_p = G_R$$

   $$\lambda_p = \lambda_R = 0.9\ \mu m$$

The dual pyrometer measurement is 744°C, an error of 204°C. This large error occurs because over 99% of the radiation received by the pyrometer is reflected radiation. A slight error in the measurement of the reflected radiation (less than the 1% pyrometer temperature measurement accuracy) causes a large error in the dual pyrometer measurement. Obviously a dual pyrometer measurement is not feasible at this location.

3. Strip temperature beyond the exit of the heating zone of the galvanizing line using the Closed-End Tube method.
   where: $T_s = 700°C,\ T_F = 700°C,\ T_F' = 692°C$

   $$\varepsilon_s = \varepsilon_p = 0.55,\ \varepsilon_R = \frac{\varepsilon_p}{(1 - \varepsilon_s)\ \varepsilon_F}$$

   $$\varepsilon_F = \varepsilon_F' = 1,\ G_p = G_R,\ \lambda_p = \lambda_R = 0.9\ \mu m$$

The dual pyrometer temperature measurement is 707°C an error of 7°C. The small compensation error of 7°C occurs in this example because only 45% of the radiation received by the pyrometer is reflected radiation.

CONCLUSION

Dual pyrometer systems have provided temperature measurement in many locations where accurate measurements could not be obtained by other means. These steel temperature measurements improve quality, productivity, and reduce the fuel rate. Nevertheless, as shown in the example error calculations, dual pyrometers should not be used where the reflected radiation is many times higher than the emitted radiation. To estimate the installed accuracy, error calculations should be made which determine the effects of pyrometer wavelength, emittance variations, and pyrometer accuracy.

REFERENCES

1. T. Iuchi, et al, Transactions ISIJ, Vol. 16, 1976, "Temperature Measurement System of Steel Strips in a Continuous Annealing Furnace".
2. R. Barber, Industrial and Process Heating, February, 1967, "Furnace Load Temperature Measurements During the Heating Process".
3. J. E. Roney, U.S. Patent No. 4,144,758, "Radiation Measurement of a Product Temperature in a Furnace", Assignee: Jones and Laughlin Steel Corporation.
4. T. Iuchi, U.S. Patent No. 4,172,383, "Method and Apparatus for Simultaneous Measurement of Both Temperature and Emissivity of a Heated Material, Assignee: Nippon Steel Corporation.

Reprinted with permission from *Journal of Physics E: Scientific Instruments,* Vol. 13, pp. 306-310 (1980).

Multi-wavelength radiation pyrometry where reflectance is measured to estimate emissivity

J L Gardner and T P Jones
CSIRO Division of Applied Physics, National Measurement Laboratory, PO Box 218, Lindfield 2070, Australia

Received 6 July 1979, in final form 19 September 1979

Abstract A new technique of multicolour radiation pyrometry is described. The directional emissivity of the surface whose temperature is being sought is estimated from measurement of the surface reflectance factor. Temperature measurements of a tungsten strip lamp in the range 1180–1680 K using this technique agree within 0·3% with those obtained from its spectral radiance temperature and published emissivity data of tungsten.

1 Introduction

Radiation pyrometry, in which radiation emitted by a body is correlated with its temperature using Planck's radiation law, holds the promise of a convenient non-contact method of measuring surface temperatures. However, radiation pyrometers measure only the radiance temperature and the surface emissivity must be known to determine the true surface temperature. The emissivity in general is a function of radiation wavelength, temperature, viewing geometry and surface conditions; consequently these pyrometers require calibration for each application. A summary of pyrometer types and methods has been provided by Heimann and Mester (1975).

Various techniques have been applied to obviate the need to know the surface emissivity. One type of pyrometer (Land and Barber 1954) covers the surface with a domed reflector to increase the surface emissivity, but thereby loses the advantage of remote sensing. Two-colour pyrometers (e.g. Ruffino 1975) calculate the temperature from the ratio of radiance measurements at different wavelengths, assuming the emissivity to be constant. Kelsall (1963) and Toyota *et al* (1972) have developed instruments in which an auxiliary heater irradiates the surface being measured. The heater temperature is adjusted to equal that of the surface by comparing the direct heater radiation with that emitted and reflected by the surface. Similarly, the polaradiometer (Murray 1967) examines the polarisation of black-body radiation reflected from a surface, in order to maintain the black body at the same temperature as the surface. This technique has been proven as a practical remote sensing method for non-diffuse reflectors, but it has a slow response time (Murray 1972).

A laser absorption pyrometer has been developed (De Witt and Kunz 1972; Kunz 1975) in which a high-power dual-wavelength laser pulse heats the surface of interest and the emissivity ratio required by a two-colour ratio pyrometer is measured. Svet (1972a, b, 1976) has described a pyrometer in which a multicolour measurement of the emitted radiation, together with prior determination of the variation of emissivity with wavelength, is used to correct radiance temperature to true temperature. In principle (Svet 1967, 1975; Svet and Naidich 1971), the emissivity variation with wavelength can be deduced from the emitted radiation itself. This is discussed further in §6.

This paper describes the basis of a multicolour radiation pyrometer in which the emissivity of the surface is calculated from a measurement of the spectral variation of the surface reflectance, sampled in the specular direction. The basic assumption is made that the variation of this sampled reflectance factor with wavelength is the same as the variation of the total hemispherical reflectance with wavelength. Measurement of absolute reflectance is not required and only the wavelength dependence of the sampled reflectance is necessary to compute temperature. Results of good accuracy are reported for the determination of the temperature of a tungsten strip lamp, in a laboratory atmosphere. The effect of scattering between the lamp and the pyrometer is discussed.

2 Technique

When the temperature of the surroundings is low so that reflected or transmitted background radiation is insignificant, the response $R(\lambda)$ at wavelength λ of a radiation detector viewing a surface of spectral emissivity $\epsilon(\theta, \phi, \lambda)$ in a given direction (θ, ϕ) in the absence of scattering or absorption, is given by

$$R(\lambda)=s(\lambda)\ \epsilon(\theta, \phi, \lambda)\ L_b(\lambda, T) \qquad (1)$$

where $L_b(\lambda, T)$ is the spectral radiance of a blackbody at temperature T and $s(\lambda)$ is the responsivity of the system at the wavelength λ, including the system geometry.

By invoking the Helmholtz reciprocity relationship and conservation of energy, it can be shown (Nicodemus 1965) that

$$\epsilon(\theta, \phi, \lambda)=1-\rho_H(\theta, \phi, \lambda)$$

for opaque materials, where $\rho_H(\theta, \phi)$ is the reflectance of the surface for radiation incident from the direction (θ, ϕ) reflected into the total hemisphere, or equivalently, the reflectance into the direction (θ, ϕ) of radiation incident uniformly over the total hemisphere. This hemispherical reflectance is difficult to measure in a system where the main aim is a non-contact measurement of surface temperature. In general, for radiation incident from the direction (θ, ϕ) we will only sample the radiation reflected into a limited solid angle in the direction (θ', ϕ'), i.e. the reflectance factor $\rho(\theta, \phi, \theta', \phi')$ will be measured. Then equation (1) may be written as

$$R(\lambda)=s(\lambda)[1-G\rho(\theta, \phi, \theta', \phi', \lambda)]\ L_b(\lambda, T) \qquad (2)$$

where G is the ratio of the hemispherical reflectance to the measured reflectance factor.

Suppose now that measurements of the emitted radiation and the reflectance factor are made at a number of wavelengths and that G has the same value at each of those wavelengths. This is clearly true if the surface is a specular reflector, viewed in the specular direction ($G=1$) and is also true if the surface is a perfect Lambertian diffuse reflector. For practical surfaces intermediate between these two limiting cases, the polar distribution of the reflected radiation tends to become more diffuse as the wavelength decreases, depending on the surface roughness (Beckmann and Spizzichino, 1963). Little information on the wavelength dependence of the polar distribution of reflected radiation exists in the literature. However, in many cases of practical interest to near infrared pyrometry the variations are expected to be small over the chosen wavelength range, and the assumption that G is constant may be sufficiently valid to introduce only a small systematic error in the technique.

Provided the system responsivity $s(\lambda)$ is known, equation (2) represents a family of curves with two unknowns, G and T. Hence it is sufficient to record measurements of emitted and reflected radiation at two wavelengths to uniquely determine G and T (provided the reflectance factor is non-zero). Practical measurements are subject to random noise, however, and a more reliable estimate of G and T is made by recording data at a greater number of wavelengths and then performing a non-linear least squares fit of the data $R(\lambda)$, $\rho(\lambda)$ to the functional form of equation (2) (Bevington 1969). A further advantage of this fitting procedure is that the effect of scattering between the apparatus and the surface being measured can be included by parameterisation of its wavelength dependence (see §6).

Measurement of absolute reflectance is not required for the determination of temperature: the wavelength dependence of the reflectance is sufficient, with any scaling factors transferred into the evaluation of G.

The Appendix derives the error relationship for the temperature determined by the above technique. The reciprocal of the instrument effective wavelength Λ is the square root of the sum of squares of reciprocals of the wavelengths used in the measurement, leading to a short effective wavelength and much reduced uncertainty compared with the long effective wavelength of the ratio pyrometer, where $1/\Lambda = 1/\lambda_1 - 1/\lambda_2$ (Ruffino 1975).

3 Apparatus

A gas filled tungsten strip lamp was chosen as the test object. The spectral radiance temperature at 0·66 μm of a 1 mm^2 target area on the filament was calibrated against a gold-point black body, as a function of the lamp current. Using the published emissivity of tungsten (Latyev *et al* 1970) and allowing for the spectral transmittance of the glass envelope, the temperature of the tungsten could be determined to better than 2 K.

The apparatus is shown schematically in figure 1. The tungsten strip lamp filament was imaged with magnification of 1:2 on to a 1 mm diameter aperture and the beam chopped at a frequency of 346 Hz by a series of apertures near the outer edge of the chopping wheel. The radiation passing through the aperture was collimated by a further lens (not shown in figure 1) before passing through an interference filter to the silicon photodetector (United Detector Technology PIN-10D), used in photovoltaic mode. Six interference filters, with transmittances centred at 0·75, 0·8, 0·85, 0·95 and 1·0 μm were used: the filter bandwidths (FWHM) were 0·10, 0·10, 0·09, 0·09, 0·05 and 0·05 μm, respectively.

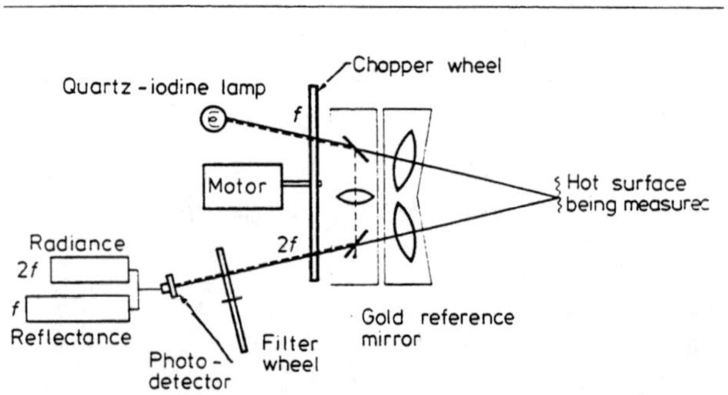

Figure 1 Schematic diagram of the apparatus. The modulation frequencies f and $2f$ refer to the reflected and emitted signals respectively.

The filament of a quartz-iodine lamp was imaged with 2:1 magnification at the strip lamp, with the beam chopped on its outward path at a frequency of 173 Hz, half that of the returning beam, by a series of apertures near the inner edge of the chopping wheel. The strip lamp was positioned so that specularly reflected radiation returned to the detector; the angle between the beams was 10°, near normal in order to minimise polarisation effects. The reflected signal was measured directly with a phase-sensitive detector operating at 173 Hz which was insensitive to the emitted radiation chopped at 346 Hz. Simultaneously, a second phase-sensitive detector operating at 346 Hz yielded the emitted signal from the strip lamp, after the silicon detector output was gated at 176 Hz to remove the reflected component. This gating reduced the signal by a factor of two, but it enabled the weaker emitted signal to be measured independently rather than as a difference between the composite signal and the reflected signal.

Figure 1 also shows a reference module whose purpose was to determine the spectral distribution of the radiation falling on the strip lamp. This system, containing two gold-coated mirrors and a lens to image the quartz-iodine lamp on to the detector aperture, was calibrated against an external front-surface gold mirror, placed in the position of the test object. The reflectance of the external gold mirror was found to be spectrally non-selective, in the wavelength range of interest, by comparing its reflectance with that of evaporated aluminium, whose reflectance has been tabulated (Bennett *et al* 1962, Shaw and Blevin 1963).

4 Calibration

The system must be calibrated to determine its responsivity $s(\lambda)$. In principle, only the relative spectral responsivity need be measured, with the normalisation constant fitted as a third parameter in equation (2). This is discussed further in §6. In practice, the fitting is more reliable if the number of degrees of freedom is minimised, i.e. the absolute responsivity of the system is determined.

Calibration was effected by measuring the radiance from a black-body aperture (Infrared Industries Model 464) placed at the position of the strip lamp. The response measured was compared with that calculated for the black body whose temperature was determined with a thermocouple.

The black-body radiance $L_b(\lambda, T)$ varied strongly across the bandwidth of the filters at the wavelengths and temperatures considered; consequently, a more exact form of equation (1) was employed, namely

$$R(\lambda_0) = \epsilon(\lambda_0, T)\, s(\lambda_0)\, \frac{\int s'(\lambda)\, L_b(\lambda, T)\, d\lambda}{\int s'(\lambda)\, d\lambda} \qquad (3)$$

where $R(\lambda_0)$ is the measured response at the nominal wavelength, $s(\lambda_0)$ is the required responsivity at the nominal wavelength, $s'(\lambda)$ is the product of the spectral response of the silicon detector and the transmittance of the interference filter at wavelength λ, and it is assumed that the variation in emissivity is small across the bandwidth of the filter.

The ratio of the two integrals in equation (3) is the mean black-body radiance, weighted by the spectral response of the filter/detector pair. We then defined an effective wavelength λ_e by equating this weighted black-body radiance to the black-body radiance at the effective wavelength and the same temperature at which the mean radiance is evaluated. That is,

$$L_b(\lambda_e, T) = \frac{\int s'(\lambda) L_b(\lambda, T)\, d\lambda}{\int s'(\lambda)\, d\lambda} \quad (4)$$

The effective wavelength of the filter/detector system for each filter was calculated over a wide range of temperatures. The filter spectral transmittances were measured with a double-beam spectrophotometer. These data were multiplied by the silicon detector spectral response (measured relative to a gold-black thin-film bolometer with a silica window to obtain $s'(\lambda)$) and numerical integration was used to solve equation (4) for the effective wavelengths at the various temperatures. As an example, for the filter with a nominal wavelength of 900 nm and a bandwidth of 90 nm, the effective wavelength varied by 3 nm for the temperature range 900–1600 K, equivalent to approximately 3% variation in radiance in this temperature range.

It was noted empirically that over an extensive range of temperatures, the reciprocal of effective wavelength was linear in the reciprocal of temperature, i.e.

$$1/\lambda_e = A/T + B \quad (5)$$

This same relationship was observed in single-colour pyrometry (Righini *et al* 1972), but with a different definition of effective wavelength. The linear coefficients were evaluated for each filter used; the temperature dependent wavelength λ_e was then applied to calculate the black-body radiance in the determination of the system responsivity $s(\lambda_0)$ and also the black-body radiance $L_b(\lambda_e, T)$ during the fitting procedure.

The system responsivity was determined for a number of black-body temperatures in the range 1000–1450 K, and for a number of positional changes in the black body. No systematic changes with temperature could be seen. The standard deviation in the mean responsivity, which closely followed that of the silicon photodetector because of our choice of filter bandwidths, was of the order of 2%. Most of this variation occurred because of positional changes affecting the absolute value of the radiance; the variation in the relative spectral responsivity was considerably less.

5 Results

Radiance and reflectance data were recorded for several strip lamp currents, corresponding to temperatures in the range 1184 to 1680 K. The radiance signals ranged from 60 μV to 10 mV after the preamplifier and were repeatable to better than 1%. The reflected signals, of the order of 100 mV, were repeatable to better than 0·3%. The reflectance factor reference was measured at each current setting, although this was unnecessary as the relative spectral distribution of the quartz-iodine lamp output was stable to better than 0·1%.

Figure 2 shows a plot of the measured reflectance factor at the two temperature extremes. Also included in the figure are reflectance values calculated from the tungsten emissivity data of Latyev *et al* (1970), assuming the surface to be a specular reflector and allowing for the glass envelope. It can be seen that the strip approaches within 10% of a perfect specular reflector, although the measured reflectance factor systematically departs from the slope of the calculated curve, dropping by 2% at the short wavelength end.

The radiance of the tungsten filament at each wavelength was computed from the measured radiant flux by correcting for the system response and the transmittance of the glass envelope. These radiance values and the simultaneously measured reflectance factors were submitted to an iterative non-linear least-squares fitting procedure as described by Bevington (1969), using the function in equation 2. An initial estimate of the temperature was made by assuming the surface grey and taking the ratio of the spectral radiance data at two wavelengths, using Wien's law. An initial estimate for the parameter G was made by assuming an emissivity of 0·5. Planck's radiation law was used to calculate the black-body radiance, recalculating the effective wavelength after each iteration. In general, 3–4 iterations were required for each set of data.

The fitted temperatures are plotted in figure 3 as differences from the known temperatures. The data shown include variations caused by repositioning of the strip lamp and also by adjustments to the optical system, followed by re-calibration with the black body. Most of the differences are less than

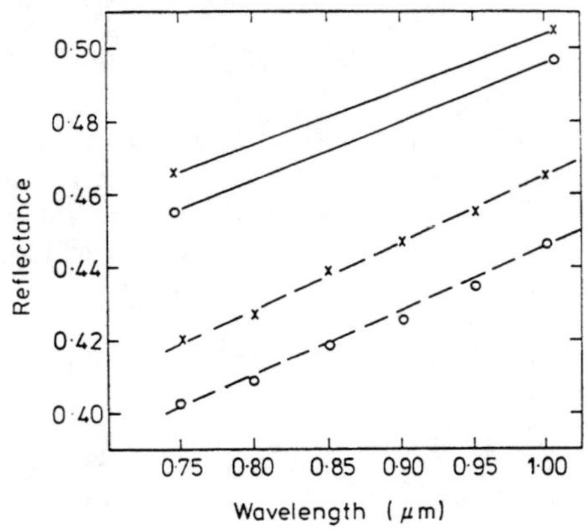

Figure 2 Measured reflectance factor (broken lines) of the tungsten strip. ×, 1680 K; ○, 1184 K. The solid lines are the specular reflectance calculated from the emissivity of tungsten.

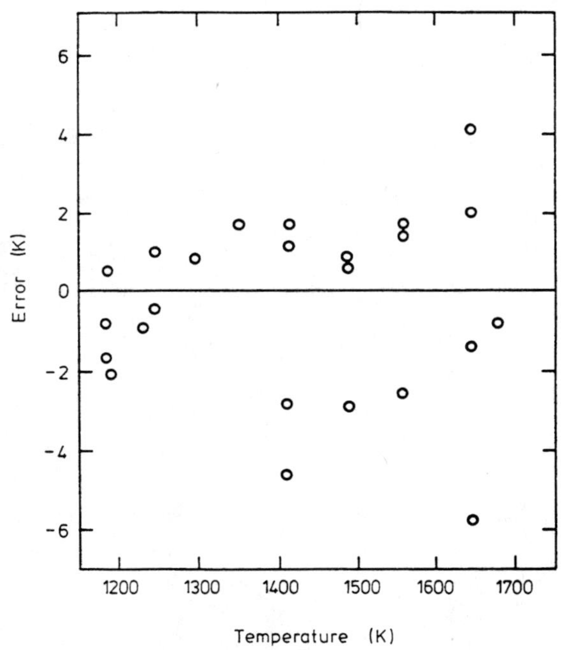

Figure 3 Measured temperature of the tungsten strip, plotted as a difference from the corrected spectral radiance temperature.

5 K, representing an error of about 0·3% at these temperatures. The black-body calibration data were also fitted as a consistency check, assuming a uniform reflectance. The fitted temperatures all agreed within 1 K of the measured temperature.

The reflectance data were modified artificially to check the effect of systematic errors in the measurement of reflectivity, equivalent to a variation in the value of G. In the temperature range of interest, a systematic linear variation of 8% in the reflectivity values over the wavelength range covered gave a 1% change in the temperature.

6 Effect of scattering

In many instances, radiation scattering (or absorption) may occur between the object of interest and the pyrometer. The reflected beam is attenuated in both the outward and returning paths, whereas the emitted beam is attenuated once. It is easily shown that equation 2 is then modified to

$$R(\lambda)=s(\lambda)\,e^{-\alpha}[1-Ge^{2\alpha}\rho(\theta,\phi,\theta',\phi',\lambda)]\,L_b(\lambda,T) \qquad (6)$$

where α is the scattering factor, which may be wavelength dependent.

In the simplest case where the scattering is constant, solution of equation 6 involves three unknowns, T, G' ($\equiv Ge^{2\alpha}$) and s_0 ($\equiv e^{-\alpha}$), i.e. it is equivalent to solving equation 2 with renormalisation of the responsivity curve. By adjusting three parameters compared with two previously, a more exact fit to the data is obtained, but the resulting parameter values are more susceptible to random errors in the data. The same data as used above, obtained under conditions where scattering was negligible, were subjected to the three-parameter fit. The resulting fitted temperatures are plotted in figure 4, again as differences from the known temperature. The errors are some five times greater than for the two-parameter fit (figure 3).

The same data were also treated in the manner suggested by Svet (1976) where the emissivity, its logarithm assumed linear in wavelength (an excellent approximation for tungsten in our wavelength range), is deduced from the radiance data in temperature independent form. Again, the procedure is susceptible to random errors in the data and although the residual sum of squares was low, the resulting temperatures for the strip lamp were up to 70 K in error.

Increased error arises when the data is fitted in shape, with freedom to also adjust the magnitude, as in both the above cases. The effect of random errors can be reduced by using more wavelengths. Choice of the wavelength range covered may also influence the accuracy in a given temperature range.

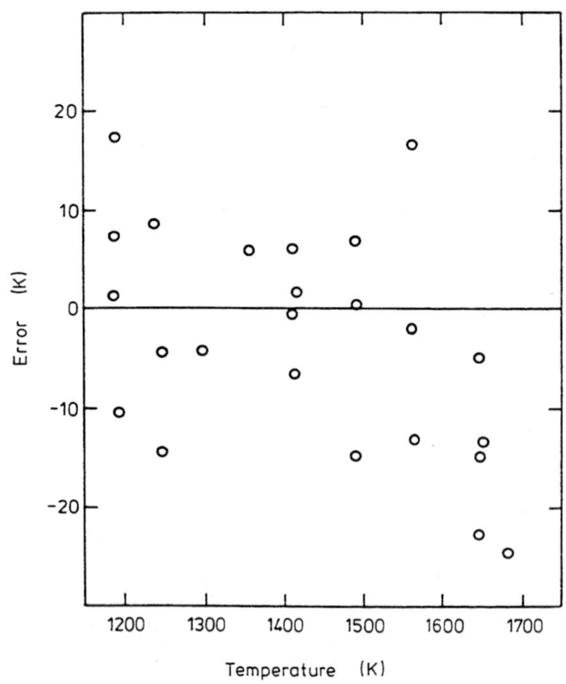

Figure 4 Temperature errors where the fitting procedure allows for spectrally non-selective scattering.

7 Conclusion

Radiation pyrometry using reflectance measurement to estimate emissivity, as described above, has been shown to be accurate for the tungsten strip lamp measured. Although the test object was a reasonably specular reflector, the technique is valid for perfectly diffuse reflectors.

For cases between these two extremes, the measured temperature will depend on the relative spectral variation of the reflected radiation; if the polar radiation distribution remains constant with wavelength, or approximately so, then the technique is still valid, without the need to measure the absolute value of the total hemispherical reflectance. Preliminary measurements of the reflectance factor and total hemispherical reflectance of diffusely reflecting roughened aluminium surfaces are encouraging toward more general application of the technique.

In conditions where radiation scattering and absorption are not significant, a multicolour pyrometer as described in this paper can measure surface temperatures with an uncertainty of less than 0·5% in temperature in the range 1200–1700 K. The technique is not suitable for temperature measurement when scattering or absorption is significant and cannot be included in the calibration. In the simplest case of non-selective absorption, an additional fitted parameter is required and the uncertainty in temperature measurement increases to 1·5% for a six-filter instrument covering the wavelength range 0·75 to 1·0 μm.

8 Acknowledgment

It is a pleasure to acknowledge the assistance of S Meszaros and J Devries in the mechanical design and construction of the equipment.

Appendix

Derivation of the variance in the fitted temperature. Consider N values of radiance R_i ($R(\lambda)/s(\lambda)$, equation 2), measured at wavelengths λ_i.

R_i is a sample from the distribution described by (cf equation 2)

$$f_i=(1-G\rho_i)\,L_b(\lambda_i,T) \qquad (A1)$$

$$=(1-G\rho_i)\frac{c_1}{\lambda_i^5}\exp(-c_2/\lambda_iT) \qquad (A2)$$

using Wien's law for simplicity.

The parameters T and G are fitted by minimising χ^2, defined by

$$\chi^2=\sum_{i=1}^{N}\frac{1}{\sigma_i^2}(R_i-f_i)^2 \qquad (A3)$$

where σ_i^2 is the variance in R_i.

The variance in the temperature T is given by (Bevington)

$$\sigma_T^2=\frac{2}{\partial^2\chi^2/\partial T^2}. \qquad (A4)$$

Using (A3) and (A2), we find

$$\partial^2\chi^2/\partial T^2 = -2\sum_{i=1}^{N}\frac{1}{\sigma_i^2}\frac{c_2 f_i}{\lambda_i T^3}\left[(R_i - f_i)\left(\frac{c_2}{\lambda_i T}-2\right)-\frac{c_2}{\lambda_i T}f_i\right]. \tag{A5}$$

Near the minimum,

$$R_i \approx f_i \qquad \text{and} \qquad \frac{c_2}{\lambda_i T}\approx 10 \approx \left(\frac{c_2}{\lambda_i T}-2\right)$$

Hence

$$\frac{\partial^2\chi^2}{\partial T^2}\approx 2\sum_{i=1}^{N}\frac{1}{\sigma_i^2}\frac{c_2^2 R_i^2}{\lambda_i^2 T^4} \tag{A6}$$

and

$$\sigma_T^2 \approx \frac{T^4}{c_2^2}\frac{1}{\sum_{i=1}^{N}(R_i/\sigma_i)^2(1/\lambda_i^2)}. \tag{A7}$$

Note that in the case of constant relative error (σ_i/R_i) in the radiance, it follows that

$$\left(\frac{\sigma}{R}\right)=\frac{c_2}{\Lambda T}\left(\frac{\sigma T}{T}\right) \tag{A8}$$

where

$$\frac{1}{\Lambda}=\left(\sum_{i=1}^{N}\frac{1}{\lambda_i^2}\right)^{1/2}. \tag{A9}$$

Λ is the 'instrument effective wavelength', by analogy with the differentiation of the black-body radiation law, which yields

$$(\sigma/R)=\frac{c_2}{\lambda}\frac{\sigma_T}{T}, \qquad \text{where} \qquad \sigma \equiv \mathrm{d}R \quad \text{and} \quad \sigma_T \equiv \mathrm{d}T.$$

Equation (A9) shows the statistical advantage that is gained by taking measurements at a number of wavelengths.

References

Beckmann P and Spizzichino A 1963 *The Scattering of Electromagnetic Waves from Rough Surfaces* (Oxford: Pergamon)

Bennett H E, Bennett J M and Ashley E J 1962 Infrared reflectance of evaporated aluminium films
J. Opt. Soc. Am. **52** 1245–50

Bevington P R 1969 *Data Reduction and Error Analysis for the Physical Sciences* (New York: McGraw Hill)

DeWitt D P and Kunz H 1972 Theory and technique for surface temperature determinations by measuring the radiance temperatures and the absorption ratio for two wavelengths
Temp. Meas. Control in Science and Industry, ed H H Plumb (Pittsburg: Instrument Society of America) **4** part 1 599–610

Heimann W and Mester U 1975 Non-contact determination of temperatures by measuring the infrared radiation emitted from the surface of a target
Temperature Measurement 1975: *Inst. Phys. Conf. Ser. No* 26 219–37

Kelsall D 1963 An automatic emissivity-compensated radiation pyrometer
J. Sci. Instrum. **40** 1–4

Kunz H 1975 On the state of ratio pyrometry with laser absorption measurements
Temperature Measurement 1975: *Inst. Phys. Conf. Ser. No* 26 273–7

Land T and Barber R 1954 New pyrometers for glass and other surfaces
J. Soc. Glass Technol. **38** 45–53

Latyev L W, Chekhovskoi V Ya and Shestakov E N 1970 Monochromatic emissivity of tungsten in the temperature range 1200–2600 K and in the wavelength range 0·4–4 μm
High Temp.–High Press **2** 175–81

Murray T P 1967 Polaradiometer – A new instrument for temperature measurement
Rev. Sci. Instrum. **38** 791–8

Murray T P 1972 The polarised radiation method of radiation thermometry
Temp. Meas. Control in Science and Industry, ed H H Plumb (Pittsburg: Instrument Society of America) **4** part 1 619–23

Nicodemus F 1965 Directional reflectance and emissivity of an opaque surface
Appl. Opt. **4** 767–73

Righini F, Rosso A and Ruffino G 1972 Temperature dependence of effective wavelength in optical pyrometry
Temp. Meas. Control in Science and Industry, ed H H Plumb (Pittsburg: Instrument Society of America) **4** part 1 413–21

Ruffino G 1975 Increasing precision in two-colour pyrometry
Temperature Measurement 1975: *Inst. Phys. Conf. Ser. No* 26 264–72

Shaw J E and Blevin W R 1963 Instrument for the absolute measurement of direct spectral reflectances at normal incidence
J. Opt. Soc. Am. **54** 334–6

Svet D Ya 1967 On one form of the spectral distribution of thermal radiation energy density
Sov. Phys. Dokl **11** 894–5

Svet D Ya 1972a Some new methods and systems of pyrometry and their application
Temp. Meas. Control in Science and Industry, ed H H Plumb (Pittsburg: Instrument Society of America) **4** part 1 587–97

Svet D Ya 1972b Optimal utilisation of redundant information in thermal radiation in thermophysical measurements
High-Temp.–High Press. **4** 715–22

Svet D Ya 1975 Independent determination of emissivity from natural thermal radiation spectra
Sov. Phys. Dokl **20** 214–5

Svet D Ya 1976 Determination of the emissivity of a substance from the spectrum of its thermal radiation and optimal methods of optical pyrometry
High Temp.–High Press. **8** 493–8

Svet D Ya and Naidich M E 1971 Application of computers to measure actual temperature of bodies from self radiation
Meas. Tech. **14** 1858–61

Toyota H, Yamada T and Mariai Y 1972 Improved radiation pyrometry for automatic emissivity compensation
Temp. Meas. Control in Science and Industry, ed H H Plumb (Pittsburg: Instrument Society of America) **4** part 1 611–18

Reprinted with permission from *Temperature: Its Measurement and Control in Science and Industry,* J.F. Schooley, editor, Vol. 5, pp. 491-503 (1982).

Two methods for simultaneous measurement of temperature and emittance using multiple reflection and specular reflection, and their applications to industrial processes

T. Iuchi and R. Kusaka[a]

Fundamental Research Laboratories, Central R & D Bureau, Nippon Steel Corporation, 1618 Ida, Nakahara-ku, Kawasaki 211, Japan

For radiation thermometry, the two problems of emittance and stray radiation from the surroundings always exist. The authors have developed two methods which can simultaneously measure both the temperature and emittance. One method, named "MA," utilizes the multiple reflection of radiation within a cavity. The second method, named "TERM," utilizes the specular reflection characteristics of the object being measured. The latter method has a special feature of eliminating stray radiation at the same time. These methods are closely related to each other from the viewpoint of interreflection. Both methods have resulted in new instruments which give not only the true temperature, but also surface information associated with the emittance of the object examined. In this paper, the principle of each method and experimental results are described.

INTRODUCTION

Radiation thermometry can be widely used where the measurement by a contact method such as with a thermocouple is impossible or undesirable, because the surface temperature of an object can be measured quickly and without contact. However, the reliability and accuracy of radiation thermometry is not yet high for low emittance materials or in the case of varying emittance. This is due to the fact that problems of emittance and stray radiation from the surroundings always exist. In order to reduce the emittance problem, several methods or techniques have been suggested. For example, Drury et al. [1] developed a pyrometer with a hemispherical reflector which makes the surface of the object to be measured into a pseudo-blackbody. Kelsall [2] suggested an emittance-compensated pyrometer utilizing specular reflection. Svet [3] investigated a new ratio pyrometer which eliminated the need for actually measuring the ratio, which in turn greatly improved the precision and dependability of the temperature measurements. These methods have been aimed at making radiation thermometry free from emittance variation, or reducing emittance effects.

On the other hand, the authors have focused their efforts on the development of simultaneous measurement of both temperature and emittance. Two methods of measurement have been developed. One method, named "MA" utilizes the multiple reflection of radiation within a cylincrical cavity. The second method, named "TERM", utilizes the specular reflection characteristics of the object being measured. The latter method has the special feature of eliminating stray radiation at the same time. The two methods are closely related to each other from the viewpoint of interreflection. Both methods have resulted in new instruments which give not only the true temperature, but also surface information associated with the emittance of the object examined.

In this paper, the principle of each method and experimental results are described and discussed. Online systems which apply these methods to several steel processes are introduced and several examples of the online measurement associated with the operation and control of processes are described.

PRINCIPLES OF MEASUREMENT

Multiple Reflection Method

The unknown quantities of temperature and emittance can be uniquely determined if two simultaneous equations, each including the two quantities, are available. The authors obtained these equations by utilizing the multiple reflection of radiation in a cylindrical cavity [4] which has a specularly-reflecting gold-plated interior surface. The fundamental phenomena are illustrated in Fig. 1. In Fig. 1(a), the open cylindrical cavity is placed above the object to be measured. There is a gap, H, between the object and the bottom opening of the cavity. In Fig. 1(b), the cylindrical cavity with a base surface

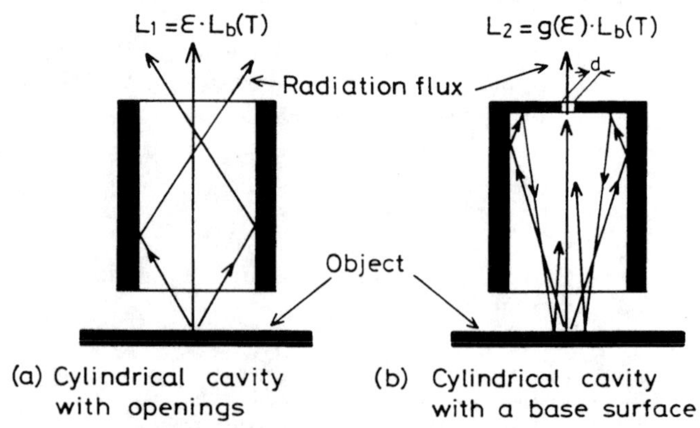

Fig. 1 Fundamental phenomena of interreflection of radiation between a cylindrical cavity and an object.

having a small aperture of diameter, d, is placed above the object. The aperture is much smaller than the diameter of the cavity opening. Identical radiometers located at the upper end of each cavity detect the radiance which comes through the cavity opening in Fig. 1(a) and the aperture in Fig. 1(b), respectively.

In Fig. 1(a), the radiance, L_1, is proportional to the emittance, ε, of the object as given in Eq.(1).

$$L_1 = \varepsilon \cdot L_b(T) \tag{1}$$

where T is the surface temperature of the object and $L_b(T)$ represents the blackbody radiance at temperature T.

On the other hand, in Fig. 1(b), the radiation from the object in multiply reflected between the object and the cavity so that the radiance increases. Accordingly, the detected radiance, L_2 is

$$L_2 = g(\varepsilon) \cdot L_b(T) \tag{2}$$

where $g(\varepsilon)$ represents the apparent emittance of the object, which is a function of ε.

The following assumptions are made in order to formulate $g(\varepsilon)$:

(i) The radiation by the cavity surface can be neglected if its temperature is kept much lower than that of the object. (This is possible by water-cooling or air-cooling.)

(ii) The effective reflectance, ρa, of the cavity with a base surface is constant if the reflectance of the interior surface of the cavity and the cavity dimensions (diameter 2R and length L) are constant.

(iii) The emittance of the object is independent of direction.

(iv) Reflected radiation from the cavity is reflected by the object. The fraction reflected back into the cavity is $(1-\varepsilon)\cdot q$ where the coefficient, $q(0<q<1)$, depends upon the diameter of the cavity, the distance H, and the reflection characteristics of the object surface.

(v) Only a negligible amount of radiation from the region surrounding the object and cavity reaches the detector.

From the above assumptions, L_2 is derived as

$$L_2 = \varepsilon \cdot L_b(T) \cdot \sum_{n=0}^{\infty} (\rho a \cdot (1-\varepsilon) \cdot q)^n \tag{3}$$

where the zeroth term in the series corresponds to the radiance of the object itself and successive terms correspond to the increased radiance due to nth interreflection between the cavity and the object. Because the product $\rho a \cdot (1-\varepsilon) \cdot q$ is less than one, Eq.(3) can be written as

$$L_2 = \frac{\varepsilon}{1-\rho a \cdot (1-\varepsilon) \cdot q} \cdot L_b(T) \tag{4}$$

$$= \frac{(\alpha+1) \cdot \varepsilon}{\varepsilon+\alpha} \cdot L_b(T) \tag{5}$$

where

$$\alpha = \frac{1-\rho a \cdot q}{\rho a \cdot q} \ (\geq 0) \tag{6}$$

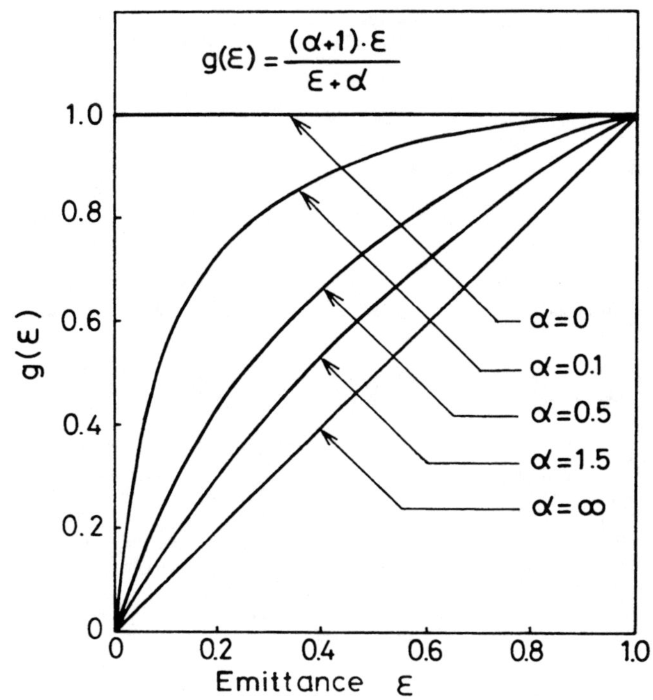

Fig. 2 Relation between $g(\varepsilon)$ and ε. (Cf. Eq. (7)).

Accordingly, $g(\varepsilon)$ becomes

$$g(\varepsilon) = \frac{(\alpha+1) \cdot \varepsilon}{\varepsilon+\alpha} \tag{7}$$

The parameter, α, is independent of ε, but depends upon the gap, H, the reflectance, ρa, the cavity dimensions, and the reflection characteristics of the object.

In Fig. 2, the characteristic curve of $g(\varepsilon)$ is shown for various values of α. It is obvious that, for a given ε, $g(\varepsilon)$ increases as α becomes smaller. That is, $g(\varepsilon)$ increases with increasing ρa and q.

The most important point of this method is that α remains constant even though ε changes. By solving Eqs.(1) and (5) simultaneously, both the temperature and the emittance of the object can be obtained.

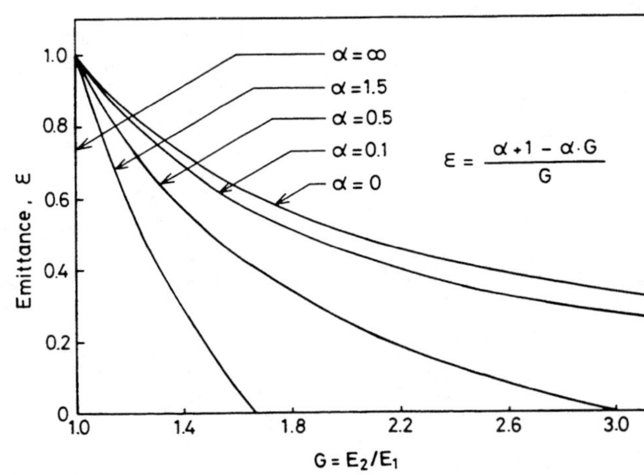

Fig. 3 Relation between ε and G. (Cf. Eq. (9)).

This can be done as follows: Let G equal the ratio L_2/L_1. Then the following equation results.

$$G = \frac{\alpha + 1}{\varepsilon + \alpha} \tag{8}$$

Eq.(8) can be rewritten as

$$\varepsilon = \frac{\alpha+1 - \alpha \cdot G}{G} \tag{9}$$

The emittance is obtained from Eq.(9). The emittance is generally a nonlinear function of G as shown in Fig. 3.

The temperature can then be obtained from Eq.(1) upon substituting the value of ε obtained above.

The principle of the measurement is realized by the apparatus illustrated in Fig. 4. A cylindrical cavity with an open top is placed above the object to be measured. A fan-shaped rotating chopper is placed close to the open top of the cavity. There is a slit of width, d, in the chopper blades. A radiometer detects the radiance from the object through this slit. The specular surfaces of the cavity interior and the chopper blade are gold plated.

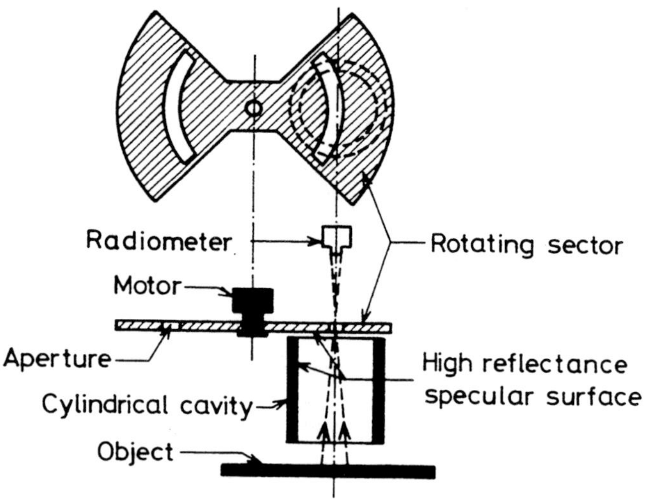

Fig. 4 Schematic diagram of the measurement apparatus. (MA).

When the upper opening of the cavity is not covered by the chopper, the situation is equivalent to Fig. 1(a). At that time the signal from the radiometer, E_1, is given by

$$E_1 = K \cdot \varepsilon \cdot F(T) \tag{10}$$

where K is the response coefficient for the radiometer. This factor contains the constant of calibration as well as other measurement constants including the solid angle of the target and the effective wavelength and bandwidth of the instrument, ε is the emittance of the object surface, and F(T) is the quantity $\exp[-C_2/\lambda_e T]$ from the Wien approximation of the spectral radiance of a blackbody. (C_2 is the second radiation constant and λ_e is the effective wavelength of the radiometer.)

When the opening is covered by the chopper, the signal from the radiometer, E_2, is given by

$$E_2 = K \cdot g(\varepsilon) \cdot F(T) \tag{11}$$

The two signals, E_1 and E_2, are alternately detected as the chopper rotates. Temperature and emittance can be simultaneously derived from these signals. A block diagram of the measuring system is shown in Fig. 5.

The method is named "MA" (Modulation Aided).

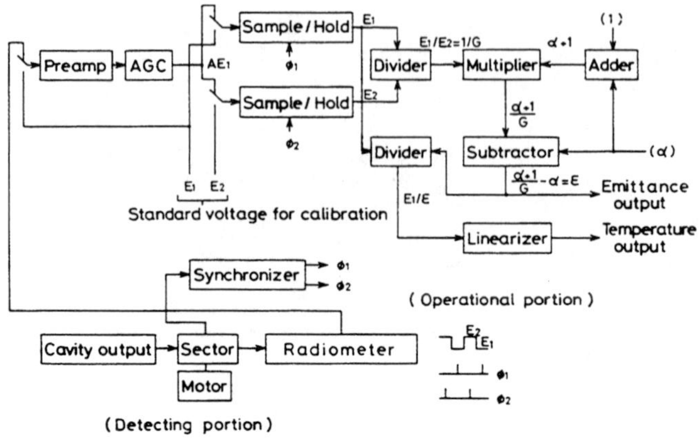

Fig. 5 Block diagram of a measuring system.

Specular Reflection Method

When the surface of an object is optically smooth and flat, namely a perfect specular reflecting surface, the following equation is satisfied

$$\varepsilon_\theta + \rho_\theta = 1 \tag{12}$$

where ε_θ is the emittance of the object in the direction of angle θ with respect to the normal to the object surface and ρ_θ is the reflectance of the object for the flux incident from the direction $-\theta$. The principle of the method will be described with reference to Fig. 6.

A radiometer and a blackbody are disposed symmetrically at an angle θ with respect to the normal to the surface of the object. If the temperatures of

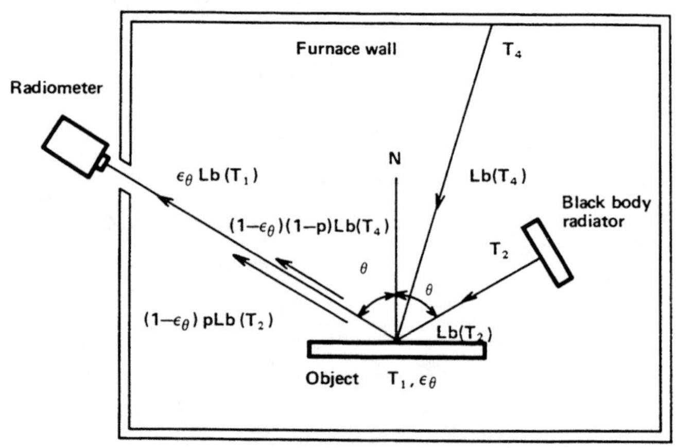

Fig. 6 Schematic diagram of the measuring method. (TERM).

the object and the blackbody radiator are T_1 and T_2, respectively, the output signal from the radiometer can be expressed as

$$E_2 = K\cdot\{\varepsilon_\theta\cdot F(T_1)+(1-\varepsilon_\theta)\cdot F(T_2)\} \tag{13}$$

where the quantities K, $F(T_1)$, and $F(T_2)$ are the same as for Eq.(10) above. The first term on the right side of Eq.(13) is associated with the radiance from the object itself. The second term is associated with the radiance of the blackbody, at temperature T_2, which is specularly reflected to the radiometer from the surface of the object.

When the temperature of the blackbody radiator is changed to T_3, the output signal from the radiometer, E_3, is given by

$$E_3 = K\cdot\{\varepsilon_\theta\cdot F(T_1) + (1-\varepsilon_\theta)\cdot F(T_3)\} \tag{14}$$

Equations (13) and (14) can be solved simultaneously for ε_θ and $F(T_1)$.

$$\varepsilon_\theta = 1 - \frac{E_2 - E_3}{K\cdot\{F(T_2) - F(T_3)\}} \tag{15}$$

$$F(T_1) = \frac{F(T_3)\cdot E_2 - F(T_2)\cdot E_3}{(E_2-E_3) - K\cdot\{F(T_2) - F(T_3)\}} \tag{16}$$

From equations (15) and (16) it will be seen that, for the case of the perfect specularly reflecting plane surface, the emittance and the temperature of the object surface can be determined independently by this method.

In general, however, because the surface of a real object has a certain amount of roughness, the condition of the measurement is not ideal. This situation causes the specular componants of the reflectance to be reduced and a diffuse component to appear. The fraction of reflectance that is specular can be expressed through the use of a "specular reflection factor", p ($0\leq p\leq 1$), such that the real specular reflectance is

$$\rho_\theta\cdot p = (1 - \varepsilon_\theta)\cdot p \tag{17}$$

(The factor (1-p), correspondingly, represents the fraction of the reflectance that is diffuse.)

For the real object, which is placed in a furnace having a uniform wall temperature of T_4, equations corresponding to Eqs.(13) and (14) above are

$$E_2 = K\cdot\{\varepsilon_\theta\cdot F(T_1) + (1-\varepsilon_\theta)\cdot p\cdot F(T_2) + (1-\varepsilon_\theta)\cdot(1-p)\cdot F(T_4)\} \tag{18}$$

$$E_3 = K\cdot\{\varepsilon_\theta\cdot F(T_1) + (1-\varepsilon_\theta)\cdot p\cdot F(T_3) + (1-\varepsilon_\theta)\cdot(1-p)\cdot F(T_4)\} \tag{19}$$

As before, the first term on the right side of Eq.(18) and (19) is associated with the radiance from the object itself. The second term is associated with that fraction of the radiance of the blackbody at temperature T_2 or T_3 which is specularly reflected to the radiometer from the surface of the object. The third term is associated with the stray, background radiance that is diffusely reflected to the radiometer from the surface of the object. The background radiance is assumed to be that of a blackbody at the furnace temperature T_4.

Solving equations (18) and (19) simultaneously for ε_θ and $F(T_1)$ will give the following expressions

$$\varepsilon_\theta = 1 - \frac{E_2-E_3}{K\cdot p\cdot\{F(T_2)-F(T_3)\}} \tag{20}$$

$$F(T_1) = \frac{p\cdot[F(T_3)\cdot E_2-F(T_2)\cdot E_3] + (1-p)\cdot F(T_4)\cdot(E_2-E_3)}{(E_2-E_3) - p\cdot K\cdot\{F(T_2) - F(T_3)\}} \tag{21}$$

It can be seen from Eq.(20) that ε_θ can be determined independently of T_1 and T_4 provided that these temperatures and p remain constant over the time interval required to obtain E_2 and E_3.

Although the value of T_1 calculated by means of Eq.(21) is independent of the emittance of the object (provided that p remains constant over the time interval required to obtain E_2 and E_3), it is apparent that the temperature will be in error by an amount which depends upon p, T_4, and temperatures selected for the two blackbodies, T_2 and T_3.

Substituting the expression for ε_θ in Eq.(20) into Eq.(19) will result in

$$F(T_1) = \frac{E_3}{K\cdot\varepsilon_\theta} - \frac{1-\varepsilon_\theta}{\varepsilon_\theta}\cdot p\cdot F(T_3) - \frac{1-\varepsilon_\theta}{\varepsilon_\theta}\cdot(1-p)\cdot F(T_4) \tag{22}$$

which is a useful form for computation. Fig. 7 is a block diagram of the computation for effectively implementing the principle of the measurement.

This method is named "TERM" (Temperature and Emittance measurement be Reflection Method).

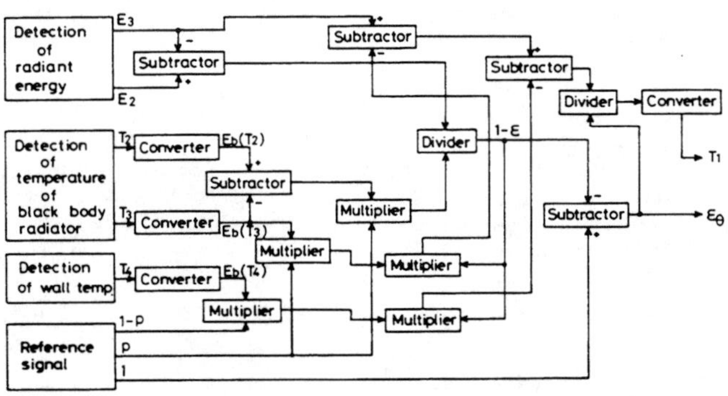

Fig. 7 Block diagram of a measuring system.

EXPERIMENTS FOR MA

Method of Measurement

An experimental apparatus shown in Fig. 4 was manufactured. The brass cavity (inside diameter of 50 mm and length of 150 mm) is a co-axial double cylinder through which cooling water flows. The diameter of the fan-shaped chopper is 220 mm and the width of the slit in the blade is 5 mm. The detector is PbS which is sensitive over the wavelength region of 1.8 to 3.0 μm.

A sample to be measured (diameter of 100 mm) was heated on a furnace. The temperature of each sample was measured using a Type K thermocouple spot-welded to its surface. The two signals, E_1 and E_2, were detected and their ratio, G, was computed. The emittance of the sample at temperature, T, was calculated from the following equation.

$$\varepsilon = \frac{E_1}{K \cdot F(T_1)} \tag{23}$$

The parameter α was selected so that the ε's of Eqs. (9) and (23) were equal.

Confirmation of Principle of the Method

Samples of cold rolled steel sheet, stainless steel sheet, and aluminum plate were prepared with surfaces having various conditions of roughness. The relation between ε and G was obtained over the temperature range 200 to 900 °C.

It is clear from Table I that the parameter, α is nearly constant for samples of a given material in spite of large variation in ε. On the other hand, different materials exhibit different α's because the surface conditions are different. The principle of the method was experimentally confirmed.

Table I Experimental confirmation of the measurement principle.

Sample	Temperature range (°C)	Emittance range ε	$\alpha + \Delta\alpha$	RMS α (μm)	$\lvert\frac{\Delta\varepsilon}{\varepsilon}\rvert$	$\lvert\frac{\Delta T}{T}\rvert$
Cold rolled steel (dull)	200 - 600	0.19 - 0.81	0.53 ± 0.04	1.329	0.11	0.01
Stainless steel (BA1)	300 - 900	0.20 - 0.74	0.25 ± 0.02	0.078	0.06	0.01
Stainless Steel (BA2)	300 - 900	0.21 - 0.73	0.34 ± 0.02	0.190	0.06	0.01
Aluminum (specular)	300 - 500	0.06 - 0.08	0.38 ± 0.02	0.351	0.28	0.026
Aluminum (shot blast)	300 - 500	0.30 - 0.38	0.75 ± 0.04	8.541	0.05	0.005

Analysis of Error

The error due to fluctuation of the measured values is analyzed. When α and G fluctuate by amounts $\Delta\alpha$ and ΔG, respectively, the error, $\Delta\varepsilon$, of ε is computed from Eq.(9) as follows:

$$\Delta\varepsilon = -\frac{G-1}{G} \cdot \Delta\alpha - \frac{\alpha+1}{G^2} \cdot \Delta G \tag{24}$$

The temperature error, ΔT, can be expressed as

$$\Delta T = \frac{-\lambda_e \cdot T^2}{C_2} \cdot \frac{\Delta\varepsilon}{\varepsilon} \tag{25}$$

By substituting the expression for $\Delta\varepsilon$ in Eq.(24) into Eq.(25), the error, ΔT, can be expressed as

$$\Delta T = \frac{\lambda e \cdot T^2}{C_2} \cdot \left[\frac{G-1}{\varepsilon \cdot G} \cdot \Delta\alpha + \frac{\alpha+1}{\varepsilon \cdot G^2} \cdot \Delta G\right] \tag{26}$$

In Eqs.(24) and (26), the first term on the right side shows the error due to variation of α, and the second term shows the variation due to G, respectively. While the former increases with G, the latter decreases with G^2. Accordingly, larger G causes less error. This implies that the selection of the cavity dimensions is important.

It is obvious from Eq.(26) that the temperature error is reduced as λ_e becomes shorter.

In Table I, the relative errors of ε and T are computed as a function of $\Delta\alpha$. Errors $|\Delta\varepsilon/\varepsilon|$ and $|\Delta T/T|$ are within 10% and 1%, respectively, except for the specular aluminum plates which have very low emittances.

Characteristics of the Cavity

The reflection characteristics of the cavity are closely related to its dimensions, surface condition of its interior and of the object to be measured, and the gap, H. The cavity should have the following characteristics.

1) Large G
2) Small variation of G with H
3) A shape that is easily constructed and an inside surface that is easily maintained.

As described below, a long cylindrical cavity is highly suitable.

In order to investigate the effect of cavity shape, both cylindrical and hemispherical cavities were constructed. The disk type upper base of the cylindrical cavity could be moved vertically inside the cavity. A coated steel sample was selected. The emittance of the sample remained constant during the measurements.

In Fig. 8, measured values of G are plotted as a function of the normalized gap distance, H/R. It is clear that the cylindrical cavity is superior to the hemispherical cavity. It produces a larger G and a smaller variation of G with H.

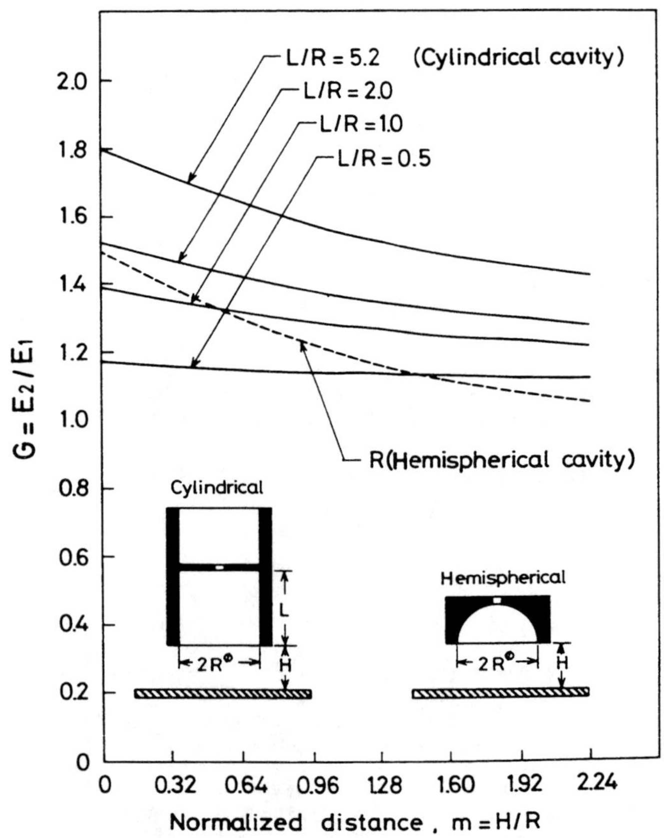

Fig. 8 Experimental relation between $G = E_2/E_1$ and $m = H/R$ for cylindrical and hemispherical cavities.

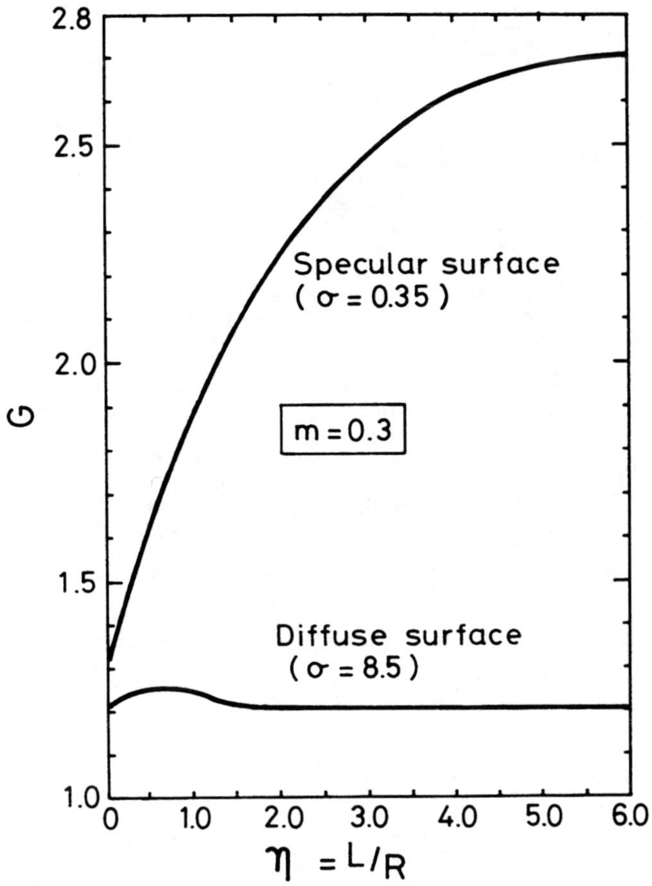

Fig. 9 Experimental relation between G and η =L/R for specularly reflecting and diffusely reflecting samples.

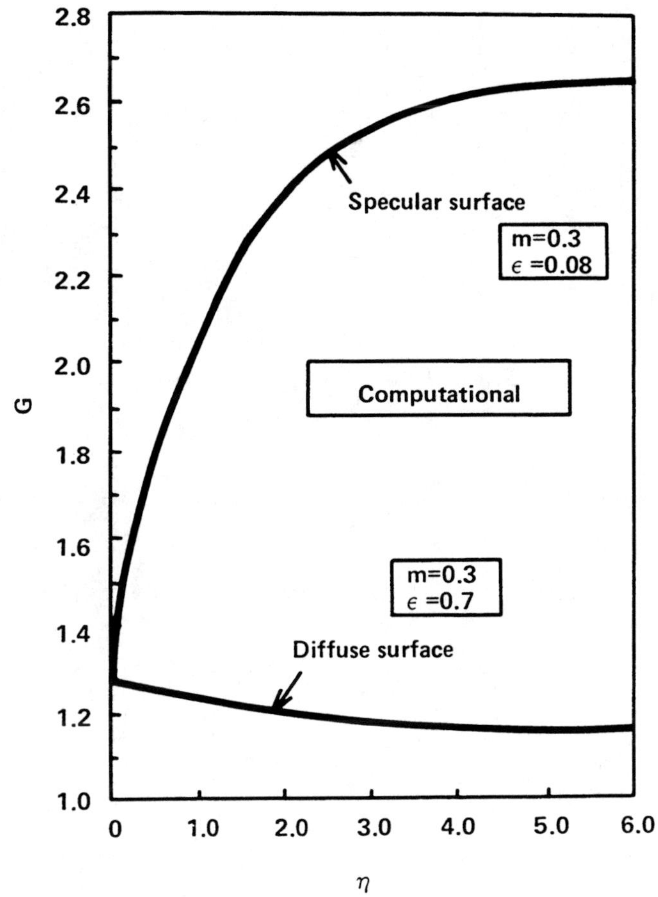

Fig. 10 Computational relation between G and η for specularly reflecting and diffusely reflecting surfaces.

In Fig. 9, measured values of G for two aluminum samples are plotted as a function of the normalized cavity length, η=L/R. One sample is an aluminum plate having specular reflection characteristics (root-mean-square roughness, σ = 0.35 μm), and the other is a shot-blasted aluminum plate having diffuse reflection characteristics (σ = 8.5 μm).

For the specular sample, the value of G increases rapidly as η increases from 0. The rate at which G increases gradually diminishes in the region of η greater than 4. On the other hand, for the diffuse sample, the value of G gradually decreases with increasing η.

Fig. 10 shows the values of G as a function of η calculated for a specular object surface and a diffuse object surface. (See the appendix for details of the calculations.) The computed curves correspond well with the experimental results shown in Fig. 9.

It is clear that the reflection characteristics are strongly affected by the surface roughness. RMS roughness [5], σ, was measured for every specimen. The relation between α and σ is plotted in Fig. 11 together with the empirical relation

$$\sigma = 7.8 \times 10^{-3} \exp(9.3\alpha) \tag{27}$$

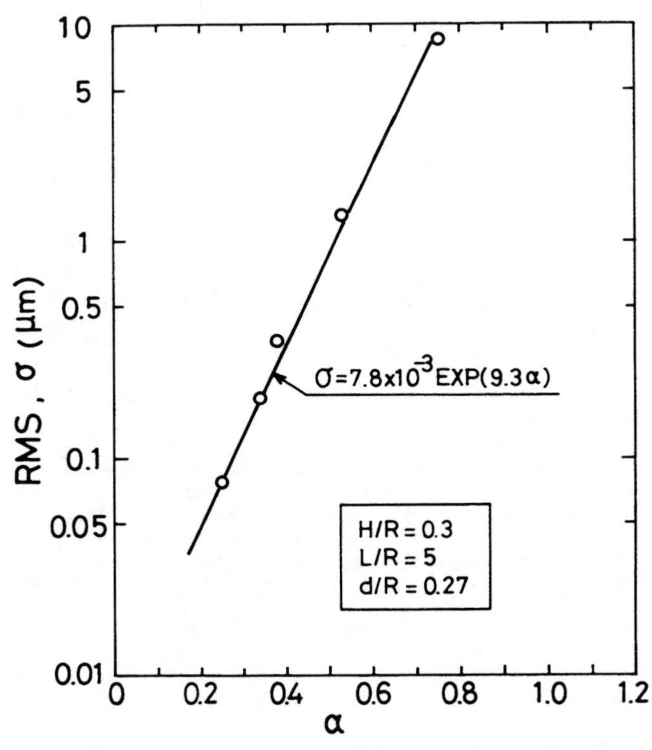

Fig. 11 Experimental relation between α and σ.

EXPERIMENTS FOR TERM

Determination of p

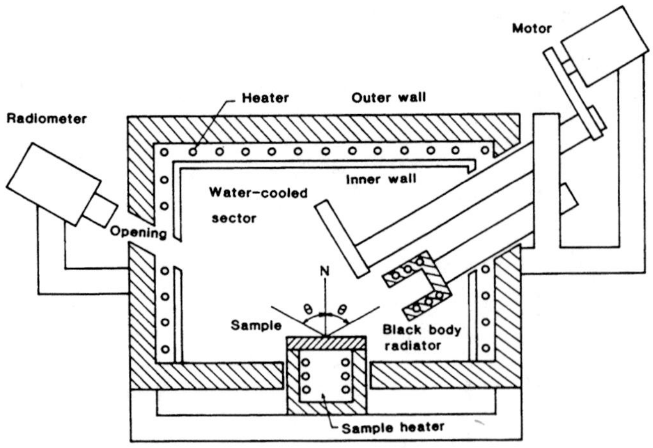

Fig. 12 Simulation furnace to measure p and f (TERM).

A simulation furnace was set up. Fig. 12 schematically shows the experimental arrangement. The inner wall of the furnace is made of thin steel plate and is of boxlike shape. The entire surface of the inner wall was coated with black paint to have an emittance of 0.95. The ceiling and side wall portion of the inner wall are heated by three separate heaters. The inner wall temperature, T_4, can be controlled independently so that it is substantially uniform over the entire surface. The temperature of the inner wall was monitored with Type K thermocouples placed at three locations on the ceiling and side wall portions.

An opposite pair of side walls of the furnace were each provided with an opening 50 mm wide and 100 mm long. A blackbody radiator was inserted through one opening and a radiometer through the other. The two were symmetrically disposed at an angle of 56° with respect to the normal to the sample surface.

The blackbody radiator was a hollow graphite cylinder having an inner diameter of 50 mm and a depth of 125 mm. It was maintained at 368 ± 1 °C using a three-mode controller. Its temperature was detected with a Type K thermocouple mounted in the bottom surface of the radiator. A blackened, water-cooled sector was placed between the sample and the radiator in order to cover the opening of the radiator as required. The solid angle subtended by the sector from the sample surface was 0.05π sr.

Three radiometers were used. The detecting elements were PbS ($\lambda_e = 2.2$ μm), InSb ($\lambda_e = 5.0$ μm), and a thermistor bolometer ($\lambda_e = 8.0$ μm).

To determine the factor p in the absence of significant background radiation, a series of experiments were performed with T_4 kept near 40 °C. The experimental procedure was as follows. Each sample was mounted on the sample heater and inserted into the unheated furnace. The surface temperature of the sample (100 mm in diameter) was measured with a Type K thermocouple which was spot-welded to the sample surface. Each sample was set at an arbitrary temperature, T_1, in the range from 250 to 450 °C. To determine the emittance, ε_θ at T_1, the water cooled sector was rotated in front of the opening of the blackbody radiator. The measured value of ε_θ under these conditions is

$$\varepsilon_\theta = \frac{E_1}{K \cdot F(T_1)} \tag{28}$$

When the sector is rotated away from the blackbody opening, the output signal of the radiometer is given by Eq. (18).

$$E_2 = K \cdot \{\varepsilon_\theta \cdot F(T_1) + (1 - \varepsilon_\theta) \cdot p \cdot F(T_2)\} \tag{18}$$

Combining Eqs. (28) and (18) by eliminating $E(T_1)$ and solving for ε_θ gives

$$\varepsilon_\theta = 1 - \frac{E_2 - E_1}{p \cdot K \cdot \{F(T_2)\}} \tag{29}$$

The specular reflection factor, p, is determined such that the value of ε_θ computed from Eqs. (28) and (29) are the same for each sample at every temperature.

Table II lists the results of measurements of various samples of cold rolled sheet, stainless steel sheet, and aluminum plate with a roughened surface. It will be seen that, for each sample, the calculated values of p are nearly constant for a given wavelength and tend to increase with increasing wavelength.

The variation in calculated values of emittance that is caused by small variations in p can be estimated from these data.

By rearranging Eq. (29), p can be expressed as

$$p = \frac{E_2 - E_1}{(1 - \varepsilon_\theta) \cdot K \cdot F(T_2)} \tag{30}$$

By differentiating Eq. (30) with respect to ε_θ one can express the variation $\Delta\varepsilon$ with Δp as

$$\frac{\Delta\varepsilon}{\varepsilon_\theta} = \left(\frac{1}{\varepsilon_\theta} - 1\right) \cdot \frac{\Delta p}{p} \tag{31}$$

Table II Experimental results of p for cold rolled steel sheet, stainless steel sheet and aluminum plate.

Sample	λ(μm)	T_1(°C)	ε_θ	p	$\bar{p}$	Δp	$\Delta\varepsilon/\varepsilon_\theta$	$\Delta T/T_1$
Cold rolled steel sheet	2.2	326 327 369 390 410 410	0.316 0.443 0.477 0.495 0.690 0.820	0.90 0.91 0.94 0.94 0.94 0.89	0.92	±0.03	0.07	0.007
Stainless steel sheet (bright annealed)	2.2	306 327 390 410 390 369	0.261 0.300 0.305 0.356 0.338 0.339	0.94 0.99 1.00 1.00 0.98 0.98	0.98	±0.02	0.06	0.006
Shot-blast treated aluminum plate (#250)	2.2	306 368 410 368	0.383 0.424 0.440 0.434	0.091 0.098 0.109 0.097	0.098	±0.009	0.15	0.015
Cold rolled steel sheet	5.0	221 275 326 368 430	0.144 0.146 0.151 0.170 0.287	0.97 0.96 0.97 0.97 0.98	0.97	±0.01	0.07	0.007
Stainless steel sheet (bright annealed)	5.0	221 326 368 430	0.155 0.181 0.178 0.181	0.98 0.99 0.98 0.99	0.99	±0.01	0.06	0.013
Shot-blast treated aluminum plate (#250)	5.0	221 326 368 430 326	0.412 0.423 0.438 0.462 0.453	0.14 0.14 0.15 0.15 0.15	0.15	±0.006	0.25	0.057
Cold-rolled steel sheet	8.0	296 326 368 431 296	0.138 0.139 0.147 0.195 0.376	0.99 0.99 0.98 0.99 1.00	0.99	±0.01	0.06	0.024
Stainless steel sheet (bright annealed)	8.0	296 326 368 431	0.184 0.176 0.164 0.174	1.01 1.00 1.00 1.01	1.00	±0.005	0.03	0.009
Shot-blast treated aluminum plate (#250)	8.0	326 368 431 368 326	0.451 0.464 0.483 0.464 0.463	0.27 0.26 0.26 0.26 0.26	0.26	±0.005	0.02	0.009

The relative variation of the temperature with respect to the emittance can be expressed by the well known formula

$$\frac{\Delta T}{T_1} = \frac{-\lambda_e \cdot T_1}{C_2} \cdot \frac{\Delta \varepsilon}{\varepsilon_\theta} \qquad (32)$$

Substitution of Eq. (31) into Eq. (32) will give

$$\frac{\Delta T}{T_1} = \frac{-\lambda_e \cdot T_1}{C_2} \cdot \left(\frac{1}{\varepsilon_\theta} - 1\right) \cdot \frac{\Delta P}{P} \qquad (33)$$

Taking $\bar{p}$ to be the mean of the calculated values of p for each sample at a given λ_e, and Δp to be the value given in Table II, the values of $\Delta\varepsilon/\varepsilon_\theta$ and $\Delta T/T_1$ appearing in columns 8 and 9 were calculated using Eqs. (31) and (33) and the values of ε_θ obtained from Eq. (29).

From Table II it will be seen that p approaches 1 with increasing wavelength, λ_e, and that the relative variation, $\Delta p/p$, decreases as λ_e increases.

Because the emittance and temperature determined by this method depend upon ε_θ and λ_e in addition to $\Delta p/p$ as shown in Eq. (31) and (33), the measurement accuracy at long wavelengths is not necessarily excellent. For example, for the data in Table II the relative error of the emittance is least at λ_e = 8 μm, while the relative error of the temperature is least at λ_e = 2.2 μm.

Fig. 13 shows the relation between the true emittance (Eq. (28)) and emittance obtained by this method (Eq. (29)) for measurements made at 2.2 μm on many samples of cold rolled steel sheet. In this experiment, $\bar{p}$ of 0.92 was used. It can be seen that the two values of emittance are in good agreement in spite of the large variation in emittance and the formation of an oxide film on the sample surface during heating.

Fig. 14 similarly shows the relation between the true temperature and that obtained by this method. The uncertainty is about ± 5 °C.

From Table II, and Figs. 13 and 14, we have shown that, for the same kind of steel plate, once the value of the p is established, the temperature and emittance can be measured simultaneously and accurately, regardless of wide variations in emittance.

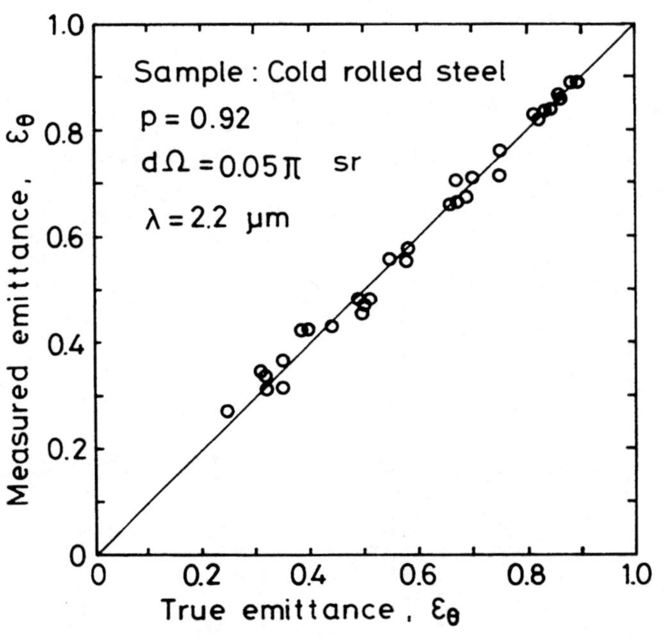

Fig. 13 Comparison of the true emittance and the emittance measured by TERM.

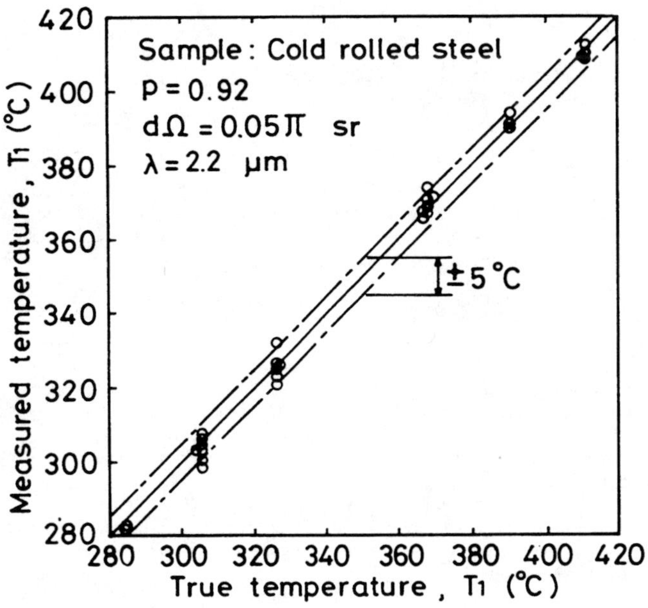

Fig. 14 Comparison of the true temperature and the temperature measured by TERM.

Next the temperature, T_4, was controlled at 390°C. The opening of the blackbody radiator was covered with the water-cooled sector. For this series of measurements, each sample was water-cooled so that its surface temperature remained near room temperature. Under these conditions the output signal from the radiometer, E_s, provides information on the fraction of the radiation from the furnace wall that is received by the radiometer after being diffusely reflected from the surface of the sample.

$$E_s = (1-\varepsilon_\theta) \cdot (1-p) \cdot \phi \cdot F(T_4) \qquad (34)$$

where ϕ is a function of the solid angle, dΩ subtended by the sector from the measured area on the sample surface.

Because the emittance of each sample had been previously measured, the "background factor", f, defined as

$$f = (1-p) \cdot \phi \qquad (35)$$

could be computed from the measurements.

The results of a series of measurements in which the solid angle subtended by the cooled sector was varied from 0.05π to 0.20π sr are given in Table III and plotted in Fig. 15. It will be seen that for each sample, f approaches zero as λ_e increases, or as the sample approaches a more specularly reflecting surface. In addition, an increase in dΩ makes f small. It is therefore possible to control the error in the temperature determined by this method by choosing the optimum position for the cooled sector.

Now let us describe the method for determining the placement of the sector. When the sector covers the blackbody radiator and the true temperature of the sample is T_1, the apparent temperature, T_a, can be determined from

$$F(T_a) = F(T_1) + \frac{1-\varepsilon_\theta}{\varepsilon_\theta} \cdot f \cdot F(T_4) \qquad (36)$$

If the maximum acceptable error in the temperature is ΔT, where $T_a = T_1 + \Delta T$, then

$$F(T_1 + \Delta T) = F(T_1) + \frac{1-\varepsilon_\theta}{\varepsilon_\theta} \cdot f \cdot F(T_4) \qquad (37)$$

Solving Eq.(36) for f, and substituting the minimum value expected for the emittance, ε_{min}, will yield the maximum value, f_{max} of f, as follows

$$f_{max} = \frac{\varepsilon_{min}}{1-\varepsilon_{min}} \cdot \frac{F(T_1 + \Delta T) - F(T_1)}{F(T_4)} \qquad (38)$$

Giving specific values of T_1 and T_4, and selecting λ_e will determine the f_{max} for the temperature measurement error within $\pm\Delta T/2$. Then the solid angle, $d\Omega_{min}$, corresponding to f_{max} can be determined from data of f vs $d\Omega$.

Practically, the maximum value of T_4 should be selected because the temperature of a furnace wall is not uniform.

Specific computations for cold rolled steel sheet and stainless steel sheet will be described. Table IV lists the f_{max} and $d\Omega_{min}$ necessary for restricting ΔT to within 10°C for T_1 = 700°C and T_4 = 600, 700, and 800°C. From Fig. 15, it will be seen that f decreases as λ_e increases, thus reducing the effect of the background radiation. However, the $d\Omega_{min}$ necessary for restricting ΔT to a specific value becomes smaller as λ_e decreases. The capability of reducing $d\Omega_{min}$ is very important from the technical view point. For stainless steel sheets, it is possible to make $d\Omega_{min}$ very small.

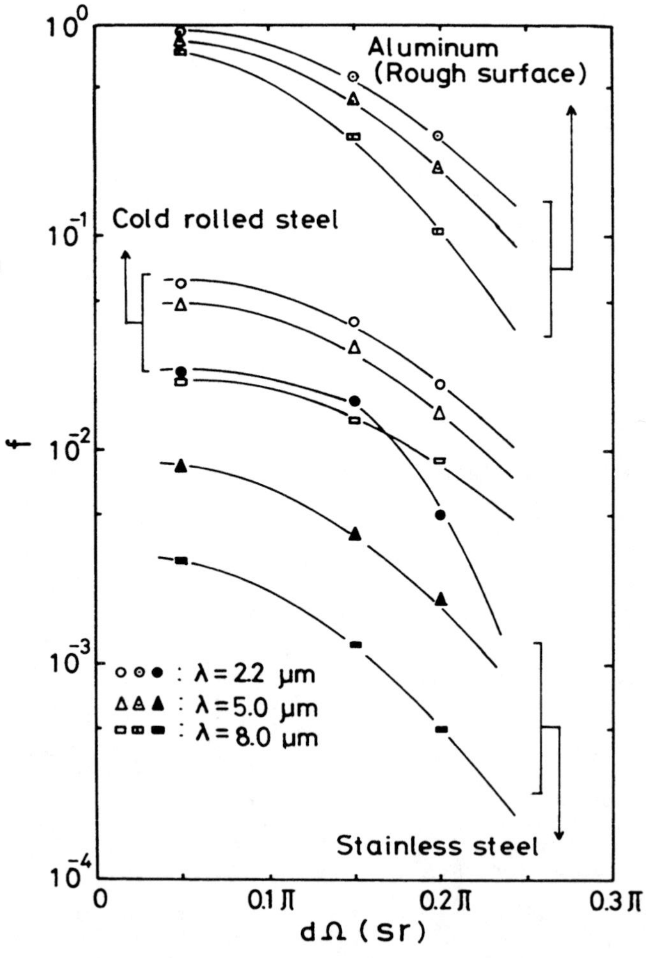

Fig. 15 Experimental relation between f and dΩ.

Table III Experimental results of the relation between f and dΩ for cold rolled steel sheet, stainless steel sheet and aluminum plate.

Sample	Effective wavelength λ (μm)	Emittance ε_θ	Solid angle dΩ (sr)	Diffuse reflection factor $f = (1 - p)\phi$	Specular reflection factor, p
Cold rolled steel	2.2	0.25 ~ 0.82	0.05 π 0.15 π 0.20 π	0.061 ± 0.01 0.04 ± 0.003 0.03 ± 0.005	dΩ = 0.05 π sr 0.92
	5.0	0.15 ~ 0.40	0.05 π 0.15 π 0.20 π	0.048 ± 0.005 0.03 ± 0.004 0.015 ± 0.003	0.97
	8.0	0.08 ~ 0.38	0.05 π 0.15 π 0.20 π	0.021 ± 0.001 0.013 ± 0.001 0.009 ± 0.001	0.99
Stainless steel	2.2	0.30 ~ 0.36	0.05 π 0.15 π 0.20 π	0.023 ± 0.007 0.017 ± 0.004 0.005 ± 0.002	0.98
	5.0	0.15 ~ 0.19	0.05 π 0.15 π 0.20 π	0.008 ± 0.002 0.004 ± 0.002 0.002 ± 0.001	0.99
	8.0	0.15 ~ 0.19	0.05 π 0.15 π 0.20 π	0.003 ± 0.001 0.0012± 0.0005±	1.00
Aluminum (rough surface)	2.2	0.23 ~ 0.26	0.05 π 0.15 π 0.20 π	0.902 ± 0.009 0.550 ± 0.005 0.299 ± 0.003	
	5.0	0.25 ~ 0.28	0.05 π 0.15 π 0.20 π	0.850 ± 0.006 0.440 ± 0.004 0.212 ± 0.003	
	8.0	0.27 ~ 0.29	0.05 π 0.15 π 0.20 π	0.740 ± 0.00 0.290 ± 0.00 0.105 ± 0.00	

The experimental relation between f and θ_a is plotted in Fig. 16, where θ_a is an average slope angle which is one of the definitions of surface roughness as shown in Fig. 17. This shows that if θ_a of a sample is known, the value of f can be determined.

Table IV Computational results of f_{max} and $d\Omega_{min}$ necessary for restricting ΔT within 10°C.

Specimen	Effective wavelength λ (μm)	ε_{min}	Temperature of specimen, T_1(°C)	Temperature of wall, T_4(°C)	f_{max}	$d\Omega_{min}$ (sr)
Cold rolled steel	2.2	0.25	700	600	0.051	0.10 π
				700	0.024	0.19 π
				800	0.0127	0.22 π
	5.0	0.15	700	600	0.012	0.21 π
				700	0.005	0.25 π
				800	0.004	0.26 π
	8.0	0.08	700	600	0.004	0.25 π
				700	0.0017	0.27 π
				800	0.0014	0.28 π
Stainless steel	2.2	0.30	700	600	0.066	0
				700	0.031	0
				800	0.016	0.15 π
	5.0	0.15	700	600	0.0076	0.07 π
				700	0.0054	0.12 π
				800	0.0046	0.14 π
	8.0	0.15	700	600	0.0043	0
				700	0.0037	0
				800	0.0031	0.05 π

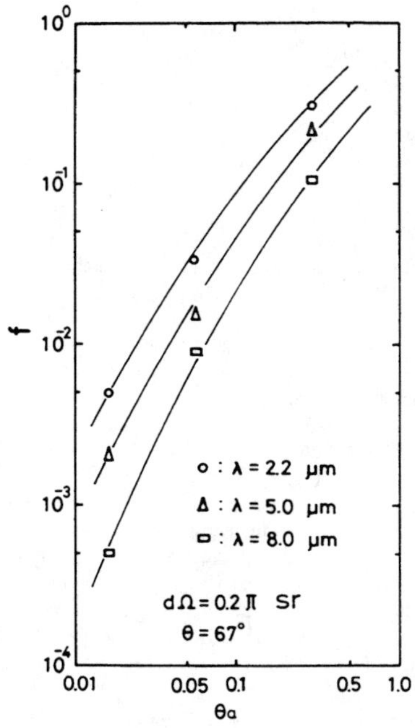

Fig.16 Experimental relation between f and θa in case of $d\Omega = 0.2\pi$ sr, $\theta = 67°$.

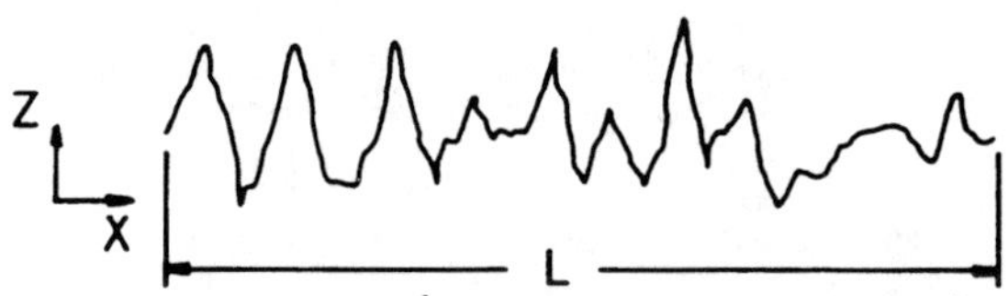

$$\theta a = \frac{1}{L}\int_0^L |\theta_c - \theta m|\,dx, \quad \theta m = \frac{1}{L}\int_0^L \left(\frac{dz}{dx}\right)dx$$

Fig.17 Definition of average slope angle, θa.

ONLINE APPLICATIONS TO STEEL PROCESSES

Online use of MA

The application of this method to the temperature measurement of a coated steel strip is an appropriate example, though there are many processes to which this method is applicable. For this process, the parameter α is very constant because the coated strip has a uniform surface roughness.

The outline of the process for making the coated steel strip is shown in Fig. 18 and is briefly described as follows. After passing through a continuous annealing furnace, the steel strip enters the zone where the coating is applied and dried. The control of the coating process was based upon the temperature of the steel strip at the outlet of the dryer. Accurate measurement by a conventional radiation thermometer was impossible because of the striking changes in the emittance of the coating. Therefore, a contact-type thermometer was used intermittently and the dryer temperature was manually controlled. As a result, the condition of the coating was unstable and the operation of the process was difficult.

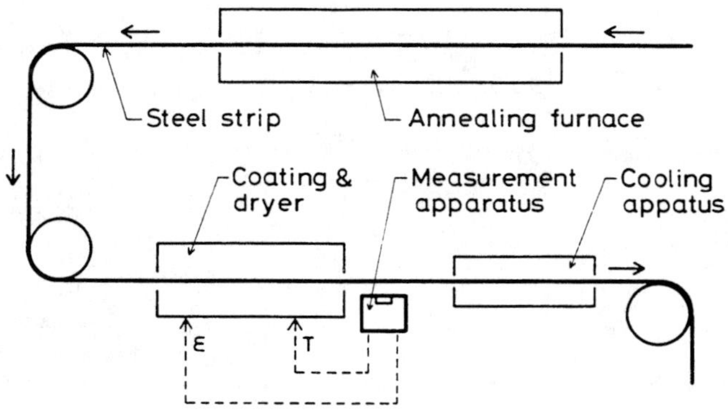

Fig. 18 Outline of the film coated steel making process.

To such a process, an online MA system has been applied, making possible continuous measurement of both temperature and emittance. The online system was installed at the outlet of the dryer.

A fused silica window was placed across the opening of the cavity to keep the interior surface clean. The outside of the window was purged with dry N_2 gas to prevent an accumulation of dust. The transmittance of the window was checked every two months, causing no trouble for practical use.

To prevent infiltration of dust, all of the detection apparatus, including the cavity, radiometer, and optical system were placed in a water-cooled box in which the pressure was kept several tens of torrs above atmospheric.

In Fig. 19, the results of typical online experiments are collected, where (a) indicate the readings of the strip temperature, Tr, by the contact-type thermometer, (b) are the emittance readings determined with the online MA system, and (c) are differences, ΔT, between T and Tr, where T is the temperature indicated by the MA system. At the lower portion of (c), the coil number of the strips is given, where S1-S5 and Th1-Th6 identify the coated steel strips and the threading strips, respectively.

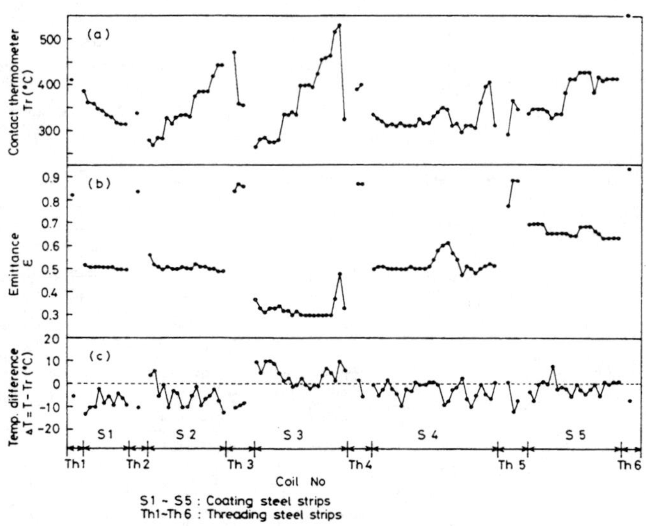

Fig. 19 Results of online experiments in the film coated steel making process.

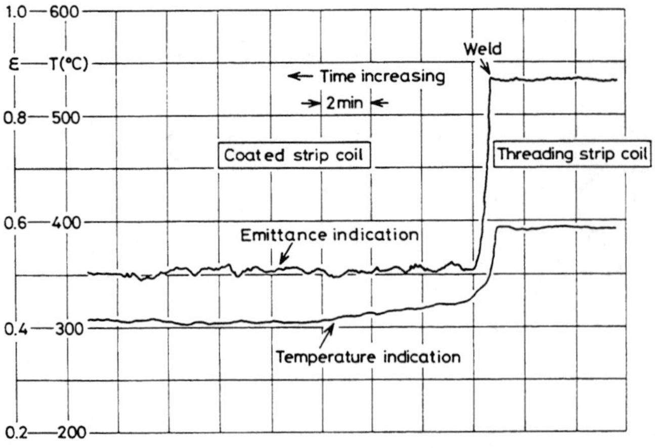

Fig. 20 An example of online measurement near a welded joint portion between threading and coated steel sheets.

The authors have previously shown [6], that the total estimated measurement error for the contact-type thermometer, at a line speed of about 100 m/min, was ±5°C when used by a skillful engineer.

Fig. 19 clearly shows that while the strip temperature ranged between 300°C and more than 500°C (see (a)), the emittances indicated large changes from 0.3 to 0.9 for all sorts of coils (see (b)). On the other hand, the temperature difference, ΔT, was almost within ±10°C. Each ΔT of the coil S3 in the temperature region below 300°C was slightly higher than the others. This was caused by the low sensitivity of the radiometer in the range of low emittance and low temperature. The emittances of the threading coils were about 0.8 because their surfaces were blackened from repeated use. From these results it was concluded that the online system succeeded in practical use.

Two additional examples of the online measurements, which represent typical features of the MA method, will be introduced. Fig. 20 shows the measured data near a welded joint between a threading coil and a coated steel coil. The values of ε and T changed abruptly in the vicinity of the joint. The ε of the threading steel was 0.87 in this case. Because of its high emittance, the threading steel was a good absorber of heat and its temperature is higher than the coated strip. On the other hand, the ε of the coated steel was about 0.51 and its temperature is lower by about 80°C than the threading steel. These measurements provided reliable information for the stable operation of the dryer.

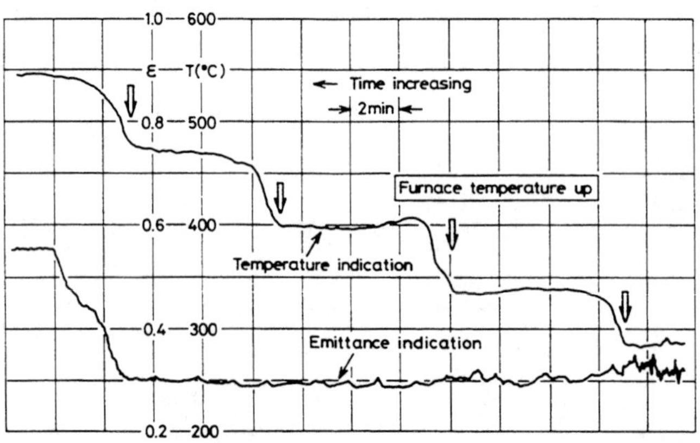

Fig. 21 An example of online measurement in case of the dryer temperature increasing.

Fig. 21 is an example of the measurement where the dryer temperature increases by several steps, each ranging from 50 to 80°C. The temperature of the coated steel increased with increasing dryer temperature. The emittance of the coated steel, however, showed no change until the strip temperature reached 470°C. Above that temperature, the ε rapidly incresed. The phenomenon corresponded with an abrupt change in the surface condition of the coating material. This information was a control variable for the quality control of the coating.

Online use of TERM

Recently, new furnaces for annealing of continuous steel sheet have been constructed. These furnaces were designed to lower energy costs through improved thermal efficiency. Unlike conventional furnaces in which the sheet is heated in an atmosphere of reducing gases, the new furnaces use rapid heating by direct heating burners in the first zone. This zone is called the Non-Oxidizing Furnace (NOF). However, the weak oxidizing atmosphere in the NOF causes wide variations of the emittance of the steel sheet. The control of the burners of the NOF is based upon the temperature of the sheet as it passes into the second zone, the Reducing Furnace (RF).

Fig. 22 shows schematically the four zones of the new furnace for annealing continuous steel sheet. In the last three zones the sheet is annealed in an atmosphere containing H_2, N_2 and CO so that the oxide film on the steel sheet is reduced and the emittance is stabilized when it leaves the annealing furnace.

A TERM system was installed at the throat between the NOF and the RF as shown in Fig. 23. The two blackbody radiators were controlled to temperatures T_2 and T_3, respectively. The solid angle subtended from the sheet by the opening of each radiator was 0.08π sr. The radiometer alternately produced signals E_2 and E_3 by optical scanning. The detecting element of the radiometer was PbSe ($\lambda_e = 3.45 \pm 0.5$ μm).

Online experiments were done for four kinds of cold rolled steel sheet. The desired temperature for the sheet as it entered the RF was from 250 to 650°C. The two blackbody radiators were set to temperatures

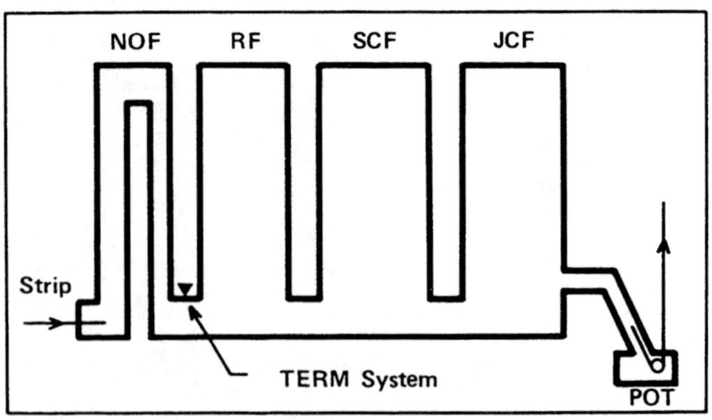

Fig. 22 A continuous annealing furnace composed of four zones, where TERM was introduced.
NOF : Non-Oxidizing Furnace,
RF : Reducing Furnace,
SCF : Slow Cooling Furnace, and
JCF : Jet Cooling Furnace.

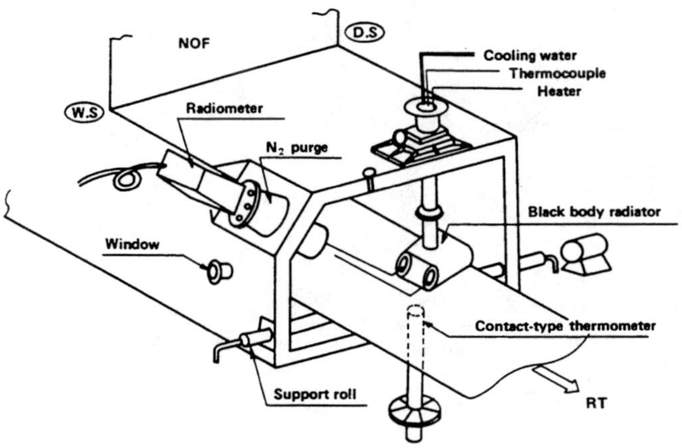

Fig. 23 TERM system installed at a throat between NOF and RF.

450°C and 40°C. The temperature of the surroundings, T_4, was assumed to be 400°C. Values for the emittance and the specular reflection factor, p, were obtained in such a manner that the temperatures determined by the TERM system coincided with the readings of the contact-type thermometer used to intermittently monitor the temperature of the bottom surface of the steel sheet.

Table V shows the relative variation in p, range of emittance, and the temperature error resulting from the use of a constant value of p for each kind of steel sheet. Except for the steel sheet specified as A, the temperature error remained within ±10°C. If a conventional radiometer having the same PbSe detecting element is used, the variation of emittance from 0.2 to 0.4 will cause a temperature error of about ±50°C.

The use of the TERM system for the control of the NOF temperature has significantly reduced the effects of such disturbances as changes in line speed or sheet thickness. Moreover, the measurement of the emittance provided useful information about the oxide film formation on the sheet as it passed through the NOF. By contributing to more stable control of the new furnace for annealing continuous steel sheet, it has helped to improve the quality of the steel product and to lower the energy costs required to produce it.

CONCLUSIONS

By utilizing interreflection of radiation, two methods of simultaneous measurement of temperature and emittance have been developed.

The one method, named "MA", is a passive method using multiple reflection between a cylindrical cavity and the surface of the object to be measured. The other method, named "TERM" is an active method utilizing the specular reflection characteristics of the surface to be measured and blackbodies as standard radiators. For the MA method it is necessary that a parameter, α, be kept constant within some tolerance regardless of emittance change. The same situation exists for the TERM method where a parameter, p, must remain constant.

It was confirmed that these two parameters were closely correlated with the surface roughness of an object. TERM has a special feature of eliminating stray radiation at the same time, so that this method especially meets the requirements for radiation thermometry in an environment such as the inside of an industrial furnace filled with stray radiation. Online systems of both methods were applied to the measurement of temperature and emittance in steel making processes where the temperature errors were within 1% and the emittance errors were within 10%.

Table V Online measurement results of Δp/p and ΔT for sheets of a different kind.

Sample	Δp/p	Emittance	Temperature error ΔT (°C)
A	± 0.051	0.2 - 0.4	±13.7
B	± 0.027	0.2 - 0.4	± 3.6
C	± 0.025	0.2 - 0.4	± 2.1
D	± 0.039	0.2 - 0.4	± 5.8

Acknowledgements

The authors would like to thank Dr. J. Ohno, Senior Researcher, and Mr. F. Tanaka, Research Engineer of the Fundamental Research Laboratories of Nippon Steel Corporation for their useful suggestions and discussions during the basic study of this paper. They would like to express their hearty thanks to Messrs. H. Hirata, Manager, and K. Watanabe, Assistant Manager, of Plant Engineering & Technology of Nippon Steel Corporation, and T. Kawamura, Senior Assistant to General Manager, of Research & Development Office of Plant & Machinery Division of Nippon Steel Corporation, and T. Shibata, Assistant Manager, of Engineering & Design Office of Nagoya Works of Nippon Steel Corporation for their contribution to the development of online systems of these methods.

APPENDIX

The principle of the MA method is based upon condition that parameter, α, is constant regardless of the value of emittance of the surface of the object. As shown in Eq. (a), α is a function of both ρa and q.

$$\alpha = \frac{1-\rho a\cdot q}{\rho a\cdot q} \quad \text{(a)}$$

where ρa is the effective reflectance of the cylindrical cavity and q is the fraction of radiation from the cavity that is reflected from the object surface back into the cavity.

The effective reflectance of a cylindrical cavity having specularly reflecting interior surface was derived from Lin's analysis [7].

$$\rho a = \rho-2\cdot\eta\cdot(1-\rho)^2\cdot\sum_{\ell=1}^{\infty}\rho^{\ell}\cdot\left[\sqrt{1+\left(\frac{\eta}{\ell}\right)^2}-\left(\frac{\eta}{\ell}\right)\right] \quad \text{(b)}$$

where

ρ = the reflectance of the interior surface of the cavity,

η = L/R, where L is the length of the cavity and R is the radius of the cylinder, and

ℓ = the number of reflections of radiation within the cavity.

When the object has a perfect diffusely reflecting surface, the coefficient, q, can be expressed as follows [8]

$$q(\text{diff}) = F_{1H} = \tfrac{1}{2}\{2 + m^2 - \sqrt{(2+m^2)^2 - 4}\} \qquad (c)$$

where

F_{1H} = geometric factor which represents the fraction of radiation that leaves the surface of the object and arrives at the cavity opening, and

m = H/R, where H is the distance between the surface of the object and the opening of the cavity.

The curve, shown in Fig. 10, of G for a diffusely reflecting object surface was calculated from

$$G = \frac{\alpha + 1}{\varepsilon + \alpha} \qquad (d)$$

together with Eqs. (a), (b) and (c) using $\rho = 0.95$, $m = 0.3$ and $\varepsilon = 0.7$.

When the object has a perfect specularly reflecting surface, it is assumed that the main portion of the detected radiance comes from only a few interreflections which are in the direction near the normal to the object surface. As a consequence, q can be expressed as

$$q(\text{spec}) = q(\text{diff}) \cdot \cos(\theta)$$

$$= F_{1H} \cdot \cos[\tan^{-1}(\frac{1}{m+n})] \qquad (e)$$

where tan (θ) is defined as R/(L+H).

The curve, shown in Fig. 10, of G for a specularly reflecting surface was calculated from Eq. (d) together with Eqs. (a), (b), and (e) using $\rho = 0.95$, $m = 0.3$, and $\varepsilon = 0.08$. In Eq. (b), the upper limit for interreflections, ℓ, was 4.

REFERENCES

a) Present Address: School of Energy Engineering, Toyohashi University of Technology, Tenpaku-cho, Toyohashi, 440 Japan.

1. M.D. Drury, K.P. Perry and T. Land, J.I.S.I. 169, 245 (1951).

2. D. Kelsall, J. Sci. Instrum. 40, 1 (1963).

3. D. Ya Svet, Temperature, Its Measurement and Control in Science and Industry (Instrument Society of America, Pittsburgh, 1972), Vol. 4, Part 1, p. 587

4. T. Iuchi, 8th ACTA IMEKO, 249 (1979).

5. P. Beckmann and A. Spizzichino, The Scattering of Electromagnetic Waves from Rough Surface (Pergamon Press, 1963), p. 187.

6. T. Iuchi, J. Ohno and R. Kusaka, Trans. I.S.I.J. 16, 195 (1976).

7. S.H. Lin, J. Heat Transfer C87, 299 (1965)

8. E.M. Sparrow and R.D. Cess, Radiation Heat Transfer (Bocks/cole Publishing Company, California, 1966) Chapter 4.

Polaradiometer—a New Instrument for Temperature Measurement

Thomas P. Murray

United States Steel Corporation, Applied Research Laboratory, Monroeville, Pennsylvania 15146

A new automatic instrument for the measurement of surface temperatures is described. The instrument examines for polarization the sum of (1) radiation emitted by the test surface and (2) radiation emitted by a blackbody and reflected from the test surface. In measurements up to 450°C on metallic surfaces with emittances of 0.05 to 0.47, the determination of equality of sample and reference temperatures was accurate within about ±2%, with no emittance corrections. Further tests are required to apportion this error among sample surface characteristics, instrument characteristics, and experimental error. Away from the null, the error signal depends on the difference of the emittances ϵ_p and ϵ_s.

INTRODUCTION

Various attempts have been made in the past to develop a radiation pyrometer that would indicate the temperature of an object by measuring its thermal radiation uninfluenced by the surface characteristics of the target. One approach to this problem has been the ratio pyrometer.[1,2] Other attempts[3] at a solution have used variations on the basic approach of measuring the sum of (1) radiation emitted by the surface of interest and (2) radiation emitted by a blackbody and reflected from the surface. This latter approach depends on the assumption that, from Kirchoff's law, whatever energy the sample absorbs from the blackbody beam is just compensated for by radiation emitted from the sample, when sample and blackbody are at the same temperature.

The defect in the latter approach is that in the complementary relation derived from Kirchoff's law (that reflectance plus emittance equals unity), the reflectance referred to comprises all the reflected radiation—that reflected diffusely as well as that reflected specularly. Unless a hemispherical collector to collect all the reflected radiation is used (which is not generally a practical solution) the measurements will be in error. Some preliminary tests of this method showed the existence of the scattering error, as was expected. They also showed that for targets of low emittance, the method is relatively insensitive. Figure 1 shows a test of this method on a low emittance sample. Note that the output indication varies only slightly as the sample temperature varies from 175 to 207°C. Although the scattering effect might possibly be calibrated out of the system, at least for certain ranges of roughness, there is no obvious way to overcome this defect of lack of sensitivity.

The method of measuring temperature presented in this paper consists of examining the reflected plus emitted beam not for intensity, but for polarization. The belief that this method might be generally usable on other than highly polished surfaces was based on the hypothesis that the diffusely reflected radiation is unpolarized, at least for many surfaces. If this were true, then any polarization of the reflected beam would be caused by actual absorption by the sample, but this would be complementary to the polarization of the emitted beam, by Kirchoff's law. That Kirchoff's law holds for the separate components of polarization was shown in 1902 by Pflueger.[4]

The polarization of the radiation emitted by metals was discussed by Arago[5] in 1824 and was later investigated more thoroughly by Worthing[6] and others.

A search of the literature, which may be incomplete, shows that the basic method of examining the combined radiation beam for polarization was described by Tingwaldt[7] in 1952 and again[8] in 1960. Tingwaldt used the ribbon filament of a tungsten lamp as a target, in order to maintain a clean polished metal surface for the measurements. Apparently the method has never been seriously considered as the basis for an automatic pyrometer capable

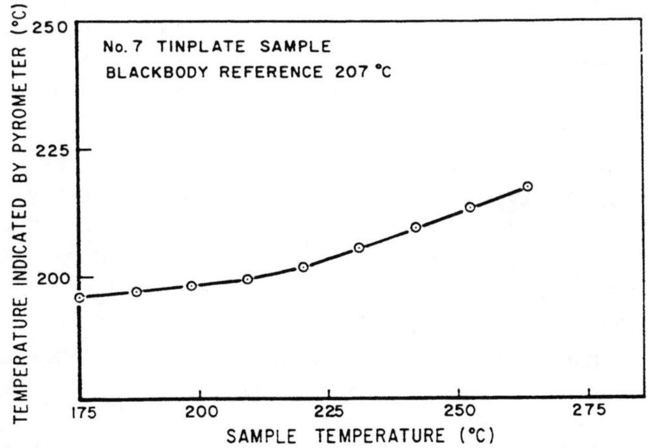

Fig. 1. Temperature indicated by a lead sulfide pyrometer measuring the sum of emitted radiation and reflected blackbody radiation.

[1] H. W. Russell, C. F. Lucks, and L. G. Turnbull, *Temperature, Its Measurement and Control in Science and Industry* (Reinhold Publishing Corporation, New York, 1941), Vol. I, p. 1159.

[2] T. P. Murray and V. G. Shaw, ISA J. 5, (No. 12) 36 (1958).

[3] W. G. Fastie, J. Opt. Soc. Am. 41, 872 (1951).

[4] A. Pflueger, Ann. Physik 7, 806 (1902).

[5] F. Arago, Ann. Chem. et Phys. 27 (2), 89 (1824).

[6] A. G. Worthing, J. Opt. Soc. Am. and Rev. Sci. Instr. 13, 635 (1926).

[7] C. Tingwaldt, Optik 9, 232 (1952).

[8] C. Tingwaldt, Z. Metallk. 51, 116 (1960).

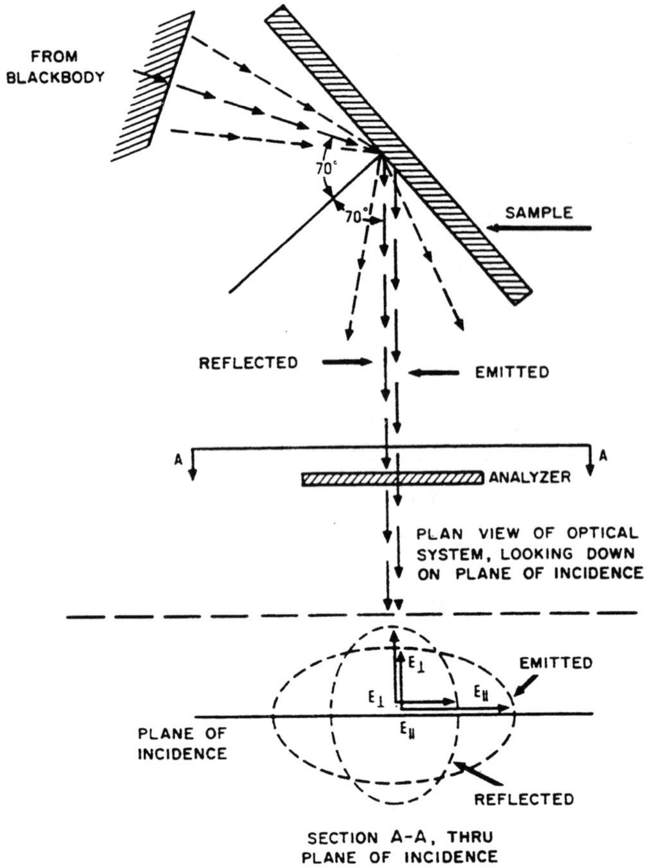

FIG. 2. The polarization of emitted and reflected radiation.

of widespread application. This is probably because of the common misconception that polarization techniques are applicable only to smooth, clean metallic surfaces. This author showed[9] in 1960 that an elliptically polarized light technique could be used to measure the thickness of films in the monomolecular range deposited on rough, oxidized metal surfaces. In view of this, it is not surprising that the polarization technique, as is shown in this paper, has ample sensitivity for the much less demanding application of measuring the temperatures of such surfaces. The belief that the use of polarization might result in better sensitivity, and higher accuracy, than the use of unpolarized radiation was, as already stated, a hypothesis that required confirmation by experiment.

The essential nature of this approach is presented in Fig. 2. The radiation emitted by a metal is such that the dominant component of polarization is $E_\parallel$ (in the plane of incidence) with a lesser component $E_\perp$ (perpendicular to the plane of incidence). Since Kirchoff's law holds for each component, the originally unpolarized blackbody radiation, when reflected, has $E_\perp$ as the dominant component with the lesser component $E_\parallel$. A rotating analyzer viewing the sum of the emitted and reflected beams sees the resultant of two elliptically polarized radiations, as indicated in Fig. 2. This resultant is comprised of a steady component and an alternating component. The alternating signal reaches its maxima as the plane of transmission of the analyzer passes through the plane of incidence, if the sample is hotter than the blackbody. If the sample is colder than the blackbody, the maxima occur perpendicular to the plane of incidence. If sample and blackbody are at the same temperature, the alternating signal vanishes.

In what follows, the theory of the method is presented, a fully automatic instrument embodying the principle is described, and experimental results on metal surfaces with varying degrees of roughness and varying oxide coatings are presented.

THEORY

Consider first a collimated beam of roughly monochromatic radiation of frequency f from the blackbody incident on the sample at an angle θ_0, and reflected at an angle θ_0 to the detector. This blackbody radiation is unpolarized and thus has polarization components making all angles χ_i with the plane of incidence. Each element of polarization of the reflected radiation can be resolved into components parallel (X) and perpendicular (Y) to the plane of incidence,

$$X_{Ri}=A_{XRi}\cos\chi_i\cos(\omega t+\delta_{XRi})$$
$$Y_{Ri}=A_{YRi}\sin\chi_i\cos(\omega t+\delta_{YRi}),$$

where $\omega=2\pi f$, the phases δ_i are arbitrary, and the amplitudes A_i will be defined later.

The thermal radiation emitted by the sample at an angle θ_0 likewise contains all components of polarization and can be resolved in a similar fashion,

$$X_{\epsilon i}=A_{X\epsilon i}\cos\chi_i\cos(\omega t+\delta_{X\epsilon i})$$
$$Y_{\epsilon i}=A_{Y\epsilon i}\sin\chi_i\cos(\omega t+\delta_{Y\epsilon i}).$$

The amplitude transmitted by an analyzer placed perpendicular to the beam and oriented so that its plane of transmission makes an angle ϕ with the plane of incidence would then be

$$A=\cos\phi\sum_{\chi_i=0}^{\pi}(X_{Ri}+X_{\epsilon i})+\sin\phi\sum_{\chi_i=0}^{\pi}(Y_{Ri}+Y_{\epsilon i}),\quad (1)$$

and the intensity transmitted by the analyzer is the time-averaged value of A^2.

In squaring the expression for A, the cross-products vanish, and the expression for A^2 reduces to

$$A^2=\cos^2\phi\sum_{\chi_i=0}^{\pi}(X_{Ri}^2+X_{\epsilon i}^2)+\sin^2\phi\sum_{\chi_i=0}^{\pi}(Y_{Ri}^2+Y_{\epsilon i}^2).\quad (2)$$

Since the optical frequencies involved are of the order of 10^{14} cycles per second, a rotating analyzer at any instant sees a time-averaged value (equal to $\frac{1}{2}$) of the factors $\cos^2(wt+\delta_i)$.

[9] T. P. Murray, Rev. Sci. Instr. 33, 172 (1962).

Therefore, the time-averaged value of

$$\sum_{\chi_i=0}^{\pi} X_{R_i}^2$$

is

$$\overline{\sum_{\chi_i=0}^{\pi} X_{R_i}^2}^{\,t} = \tfrac{1}{2}\sum_{\chi_i=0}^{\pi} A_{XR_i}^2 \cos^2\chi_i,$$

and similar results hold for the other summations in (2) so that

$$I=\overline{A^2}^{\,t}=\tfrac{1}{2}\{(\sum_{\chi_i=0}^{\pi} A_{XR_i}^2\cos^2\chi_i+\sum_{\chi_i=0}^{\pi} A_{X\epsilon_i}^2\cos\chi_i)\cos^2\phi$$
$$+(\sum_{\chi_i=0}^{\pi} A_{YR_i}^2\sin^2\chi_i+\sum_{\chi_i=0}^{\pi} A_{Y\epsilon_i}^2\sin^2\chi_i)\sin^2\phi\}. \quad (3)$$

The amplitudes A can be related to the temperatures of the sample and blackbody as follows:

$$\sum_{\chi_i=0}^{\pi} A_{XR_i}^2\cos^2\chi_i=R_pE_B^2(T_R),$$

$$\sum_{\chi_i=0}^{\pi} A_{X\epsilon_i}^2\cos^2\chi_i=\epsilon_pE_B^2(T_S),$$

$$\sum_{\chi_i=0}^{\pi} A_{YR_i}^2\sin^2\chi_i=R_sE_B^2(T_R), \quad (4)$$

$$\sum_{\chi_i=0}^{\pi} A_{Y\epsilon_i}^2\sin^2\chi_i=\epsilon_sE_B^2(T_S),$$

where R_p and R_s are the sample reflectances, and ϵ_p and ϵ_s are the sample emittances, for radiation polarized parallel and perpendicular to the plane of incidence, respectively. E_B^2 is the square of the (total) amplitude of the radiation of frequency f from a blackbody at temperature T, and the subscripts R and S refer to the reference and sample sources. With this definition of E_B^2,

$$R_p+\epsilon_p=0.50 \quad R_s+\epsilon_s=0.50. \quad (5)$$

Substituting (4) in (3) [using $E_{BR}^2\equiv E_B^2(T_R)$, etc.] results in

$$I=\tfrac{1}{2}\{(R_pE_{BR}^2+\epsilon_pE_{BS}^2)\cos^2\phi+(R_sE_{BR}^2+\epsilon_sE_{BS}^2)\sin^2\phi\}$$

and, making use of the appropriate trigonometric identity,

$$I=\tfrac{1}{4}[R_pE_{BR}^2+\epsilon_pE_{BS}^2+R_sE_{BR}^2+\epsilon_sE_{BS}^2$$
$$+(R_pE_{BR}^2+\epsilon_pE_{BS}^2-R_sE_{BR}^2-\epsilon_sE_{BS}^2)\cos2\phi]. \quad (6)$$

The Eqs. (5) are strictly true only when blackbody and sample are at the same temperature (apart from the assumption that scattering can be neglected) but are true to a good approximation near the null, which is the region of interest in any case. Substituting for the reflectances from Eqs. (5), the following expression is obtained for the intensity as seen by the rotating analyzer:

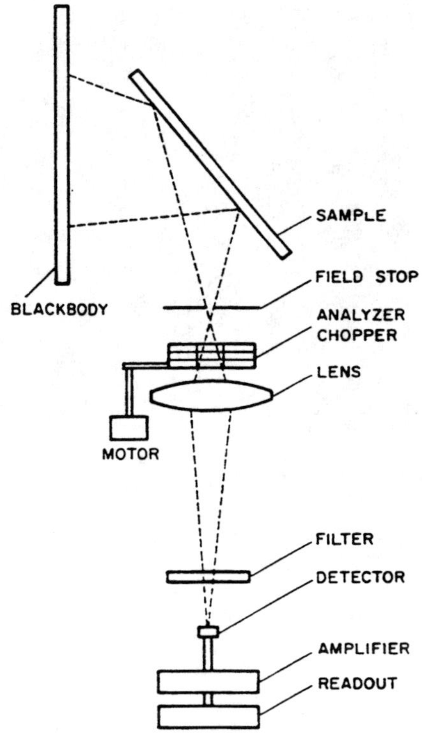

FIG. 3. Schematic of polaradiometer.

$$I=\tfrac{1}{4}[E_{BR}^2+(\epsilon_p+\epsilon_s)(E_{BS}^2-E_{BR}^2)$$
$$+(\epsilon_p-\epsilon_s)(E_{BS}^2-E_{BR}^2)\cos2\phi]. \quad (7)$$

This derivation was carried out for a collimated, roughly monochromatic beam. If a cone of incident angles is used, then the component of polarization of orientation χ_i is due to the superposition of all the elements of radiation reflected at somewhat different angles with this polarization. If a wide band of frequencies is used, rather than a roughly monochromatic beam, then the intensities must be summed over the band of frequencies. Equation (7) will again result, with the emittances now being those for the cone of angles and band of frequencies used. The apparatus to be described actually uses a noncollimated incident beam and a relatively wide band of frequencies.

The dc component of Eq. (7) (the first two terms) represents essentially the same signal (apart from a constant attenuation factor) as would be seen without the use of polarization. These two terms represent the signal plotted in Fig. 1. For small emittances, the output depends much more strongly on the reference than on the sample.

The method of measurement used in these investigations ignores this dc component and uses only the third term (the alternating component) for the measurements. If we represent the intensity curves for blackbody and sample over a short temperature interval near the reference temperature by curves of the form of T^n, then the third term of Eq. (7) can be expressed explicitly in terms of tempera-

FIG. 4. Optical system of polaradiometer. A close-up of optical arrangement; A—blackbody; B—sample heater assembly; C—field stop; D—rotating analyzer; E—reference voltage potentiometer; F—lens; G—optical filter; and H—detector.

ture,

$$I(\phi)=\frac{K_1}{4}(\epsilon_p-\epsilon_s)(T_S{}^n-T_R{}^n)\cos 2\phi, \tag{8}$$

where K_1 is a constant. The exponent n depends on the temperature range and the wavelength interval involved. (If the latter are such that Wien's law holds, then n is given approximately by $n=c_2/\lambda T$, where c_2 is the second radiation constant, and T is the mean absolute temperature of the target.) For the wavelengths and temperatures used in these experiments, n has values of about 10 to 15.

The alternating signal represented by Eq. (8) vanishes when sample and reference are at the same temperature, independent of sample emittance. Away from the null, the scale factor of the instrument depends on the difference in emittances $\epsilon_p-\epsilon_s$. The magnitude of Eq. (8) depends about equally on sample and blackbody temperatures, whatever the value of the emittances, so that the lack of sensitivity using the unpolarized approach has been overcome. The experimental results to be described demonstrate this, and confirm that the effects of scattering can be largely eliminated, as was hypothesized initially.

EXPERIMENTAL APPARATUS AND PROCEDURES

Figure 3 is a schematic view of operable apparatus using the principle of polarization for temperature measurements. Figure 4 is a photograph of the apparatus, and Fig. 5 shows an over-all view of the experimental arrangement.

Referring to Fig. 4, radiation from a heated blackbody, A is incident on a sample B and the reflected-plus-emitted radiation from the sample passes through an opening in a diaphragm C. The field stop C ensures that the field of view is completely filled by both the sample and the blackbody (viewed by reflection from the sample).

The analyzer D is made from "HR" sheet, a polarizer for the near-infrared region made by the Polaroid Corporation. This has its peak transmission in the region of 1.8 to 2.7 μ, and is used here with an interference filter G, which has a bandpass from 1.8 to 2.7 μ. The analyzer is rotated at 5 cycles/sec by a drive-motor and gear-and-belt assembly.

If the radiation beam is polarized, a radiation signal of twice the frequency of rotation is generated by the analyzer. The resultant alternating electrical signal generated by detector H is passed through an amplifier and, in the first version of the apparatus, was read out on an oscilloscope.

The oscilloscope trace was observed as the sample was heated up to and through the reference temperature. The temperatures of the sample and blackbody at the instant the signal passed through the null (or, more strictly, the minimum) were obtained from thermocouples attached to the sample and blackbody.

In instrumental arrangements preceding that shown here, a lens was placed in front of the blackbody and another in front of the analyzer, and an attempt was made to use a collimated beam from the blackbody. In those tests, it was found that the arrangement was very sensitive to the angle of viewing. In commercial applications, strip wobble, for instance, would cause sizeable errors. With the arrangement shown here, the detector assembly can be moved through an angle of at least ±5° without affecting the reading. In laboratory applications, of course, motion of the target is usually not a factor.

This embodiment of the instrument also has the merit that there are no optical components intervening between the analyzer and the two sources of radiation to introduce extraneous polarization.

EXPERIMENTAL RESULTS

Figure 6 shows experimental data taken with the apparatus just described on samples of black plate (low-carbon steel), melted and unmelted tinplate (tin-coated steel),

FIG. 5. General view of laboratory apparatus. A—temperature controller for blackbody source; B—recorder for readout of polaradiometer; C—optical bench; D—potentiometer for thermocouple readout; E—electronics filters; F—phase discriminator circuitry; and G—oscilloscope.

and crystalline and remelted galvanized (zinc-coated) steel. The plotted values are the random differences between the set-point temperature of the blackbody reference and the temperature of the sample, as indicated by thermocouples, at the instant the alternating output of the oscilloscope went through the null. All plotted values were corrected for a systematic error of 7 C°, which was taken as the mean of the total errors.

The additional data shown in Fig. 7, taken over a wider temperature range at a later time, have been corrected for a systematic error of 6 C°.

These data were taken without any correction for emittance, and without any adjustment of the instrument, on samples with emittances ranging from 0.05 to 0.47. (These values are known from work previously carried out on the measurement of emittances in the same wavelength band.)

The results amply demonstrated the feasibility of the method, and showed it to be independent of emittance within the accuracy of the measurements.

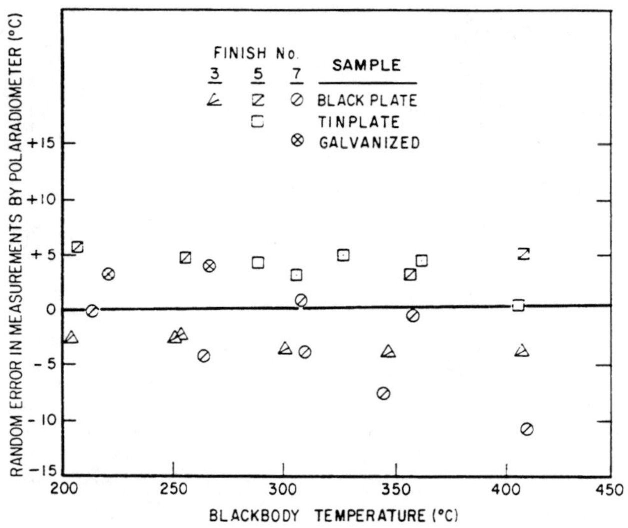

FIG. 7. Further experimental data over wider temperature range.

FULLY AUTOMATIC APPARATUS

To obtain a fully automatic instrument, the following components were added to the apparatus already described: (1) a reference voltage source, the output of which is a function of the angular position of the analyzer, and (2) a readout system that can detect the phase as well as the magnitude of the output of the radiation detector (see Fig. 8).

The wiper arm of the potentiometer M is attached to a rotating shaft driven by the same motor that drives the rotating analyzer. A source of voltage K is provided across the potentiometer; as the wiper arm rotates, a sinusoidal voltage is generated between the wiper tap and another tap L on the potentiometer. The angular position of the potentiometer is adjusted on its drive shaft so that the positive maximum of its alternating output occurs when the transmission axis of the analyzer is parallel to the plane of incidence. If the sample is hotter than the reference blackbody, the maximum in the signal generated by the rotating analyzer then occurs at the same time as the maximum in the signal generated by the potentiometer, so that the two signals are in phase. If the sample is colder than the blackbody, then the two signals are out of phase.

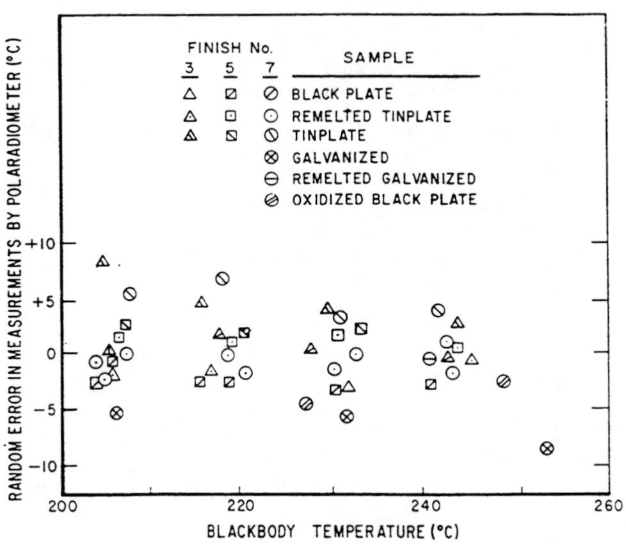

FIG. 6. Experimental data taken with polaradiometer.

The signal from the rotating potentiometer is fed to the voltage divider ABC, and the signal from the rotating analyzer is fed to BE in the same figure from the amplifier. When the instrument is adjusted as just described, the voltage to which the capacitor across DE charges (neglecting leakage through the resistors) is given by

$$V_{DE}=\frac{M}{2}+KI_M,$$

where M is the peak value of the reference voltage, I_M is the peak value of Eq. (8), and K is an instrumental constant. Similarly, the voltage across FE is then

$$V_{FE}=\frac{M}{2}-KI_M,$$

so that the output voltage, taken as V_{DF} is

$$V_{DF}=2KI_M=\frac{KK_1}{2}(\epsilon_p-\epsilon_s)(T_S^n-T_R^n). \tag{9}$$

The output is thus positive, with the conventions adopted, if the sample is hotter than the blackbody, and is negative if the sample is colder than the blackbody.

EXPERIMENTAL RESULTS WITH AUTOMATIC APPARATUS

Recorder charts of the output obtained with the fully automatic apparatus are shown in Figs. 9 to 12. This out-

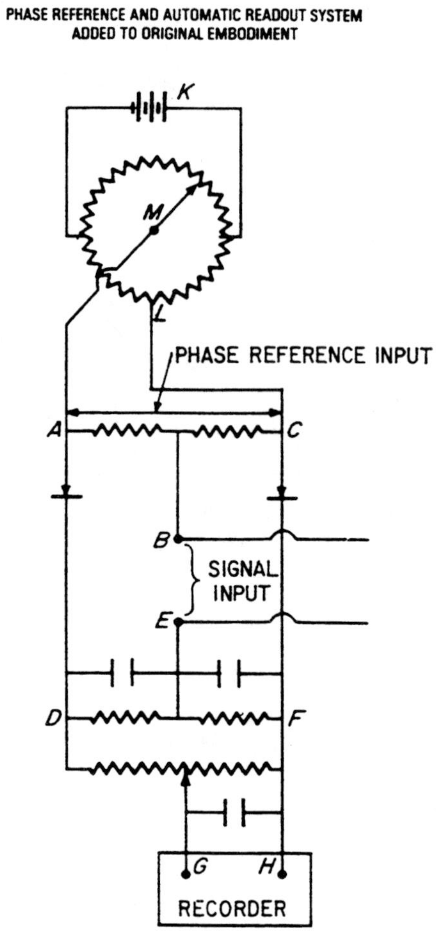

FIG. 8. Phase discriminator circuit for automatic instrument.

put is a continuous indication of the difference between the set-point of the reference blackbody and the actual sample temperature. The recorder has a mid-chart zero, so that if there are no errors in the measurement, the recorder graph should pass through zero when reference and sample are at the same temperature. For example, Fig. 9 shows results obtained on a sample of No. 5 finish melted tinplate, the reference blackbody temperature being set at 210°C. As the sample was heated up through the reference temperature, the sample temperature was continuously monitored by means of a thermocouple attached to its surface. As the sample heated, the error signal continually decreased and the indication passed through zero at a sample (thermocouple) temperature of about 212°C.

Figures 10 and 11 show similar data taken on a crystalline galvanized sample and on an oxidized steel sample at about 260°C.

Figure 12 shows the results obtained while monitoring the temperature of a crystalline galvanized sample as it was taken through the melting point. The emittance of the zinc changes from about 0.09 to 0.11 on going through the melting point and then changes to about 0.45 as the steel substrate reacts with the molten zinc and the zinc iron alloy "grows through" the surface. The spike on the curve is the transient change in the instrument output as the emittance changes from about 0.11 to about 0.40. The relatively small departure from a smooth curve as the emittance changes by 300% demonstrates directly the insensitivity of the instrument to emittance changes.

The responsivity shown in these results (of the order of 1 mV for a 5 C° temperature difference) would suffice for most measurements. No attempt has been made as yet to determine the optimum performance in this respect.

DISCUSSION

It has been demonstrated that this principle of temperature measurement is feasible and that it can be embodied in an automatic instrument for use on metallic surfaces with varying finishes and oxide coatings.

The method does not work if the sample is viewed at or near normal incidence or if the sample exhibits no polarization (which would be true for a blackbody or for a surface that acts as a completely diffuse reflector).

The systematic error in the experimental results is of the order of 6 to 7 C°. One source of such error is the fact that a perfect blackbody is not available. To a lead sulfide detector with the same bandpass filter as was used in these measurements, the blackbody sources used in these experiments had apparent emittances of about 0.90 to 0.92. Another source of error in the opposite sense is that, since the distances used in these experiments are not large compared to the focal length of the lens, the illumination of the detector by the blackbody and target is a function of distance, which differs for the two sources. Contributions to both the systematic and random error result from (1) temperature gradients in the blackbody and the sample and variations in these gradients and (2) differences between the conduction errors in the blackbody and sample thermocouples and variations in these differences. There are

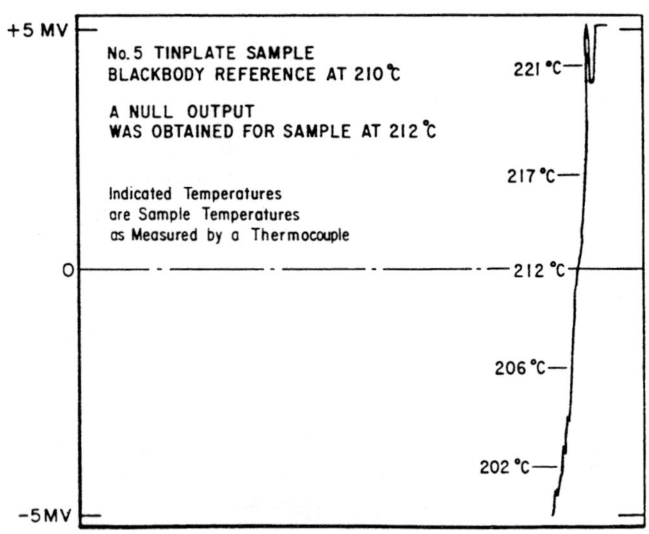

FIG. 9. Recorder chart of automatic instrument–tinplate sample.

other sources of possible error, but these are the major sources.

The systematic error can be calibrated out of the system, with the restriction that the emittance of the "blackbody" source must then remain constant as a function of time.

The experimental results demonstrate that the determination of the null is independent of emittance, as would result from Eq. (7), to an extent that is certainly adequate for most measurements. If an absolute error exists, it is sufficiently small to be masked by the other uncertainties referred to. The magnitude of such error could be determined by more careful experiments in which the blackbody source used should be of the melting-point type used by the National Bureau of Standards, the samples being in contact with another melting-point bath. The dependence of this error (particularly the scattering contribution) on wavelength should be established.

Whether the output away from the null follows Eq. (7) has not yet been established experimentally. Determination of ϵ_p and ϵ_s for the various types of samples used will be made. The difference $\epsilon_p - \epsilon_s$ is a function of the angle of incidence used. Whether there is any advantage so far as uniformity of the scale factor is concerned in working at certain angles of incidence requires further study.

The real advantage of this type of instrument over existing pyrometers lies in the fact that the null is independent of emittance, at least to a very high degree. In its present form, the instrument would best be used to drive the heating means for the sample or process in order to arrive at and maintain the desired temperature, at which the reference blackbody would be set. For the surface conditions discussed here, the error-voltage signal can be made sufficiently large that the variations in its magnitude due to different values of $\epsilon_p - \epsilon_s$ become relatively unimportant.

The experimental work presented is in essence a feasibility study, and no attempt has been made as yet to optimize the various parameters of the system. Work is now

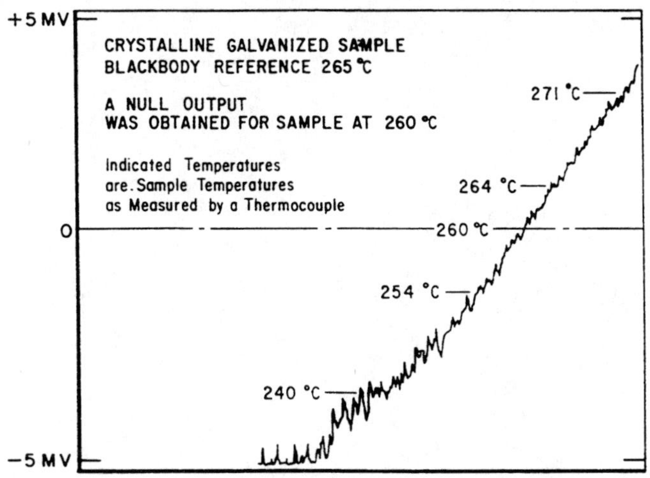

FIG. 10. Recorder chart of automatic instrument–galvanized sample.

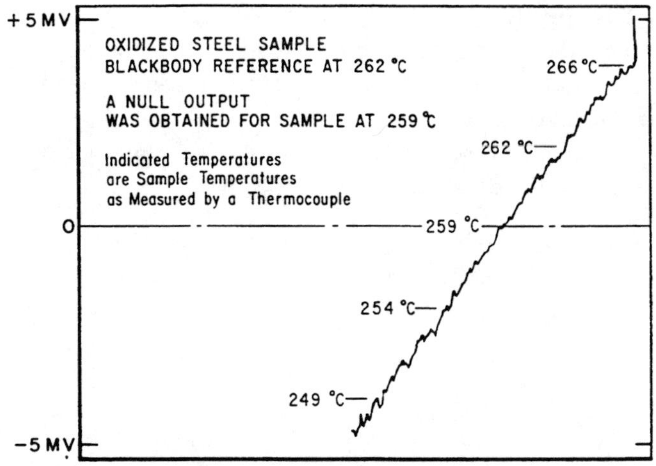

FIG. 11. Recorder chart of automatic instrument–oxidized steel sample.

in progress to construct a finished instrument to replace the "breadboard" version used in these experiments. In this process some of the components will be replaced by more sophisticated or more suitable elements. A single compact unit to rotate the analyzer and at the same time generate the reference voltage will be used in this new version, to replace the motor-belt-gear assembly used here.

Determination of the detectivity can be made following these improvements in the system.

The instrument has been used in the range of 177 to 538°C on metals. Extension to lower temperatures will certainly be possible when the detectivity is improved, as it can be. For use on many nonmetals, this extension to lower temperatures is a prerequisite. Measurements have been made on a Micarta sample to show that the method will, in general, work on nonmetals also. Extension to higher temperatures involves no particular problem, although the field of view would be made smaller in order to use a smaller blackbody, so that the power requirements for the blackbody do not become prohibitive.

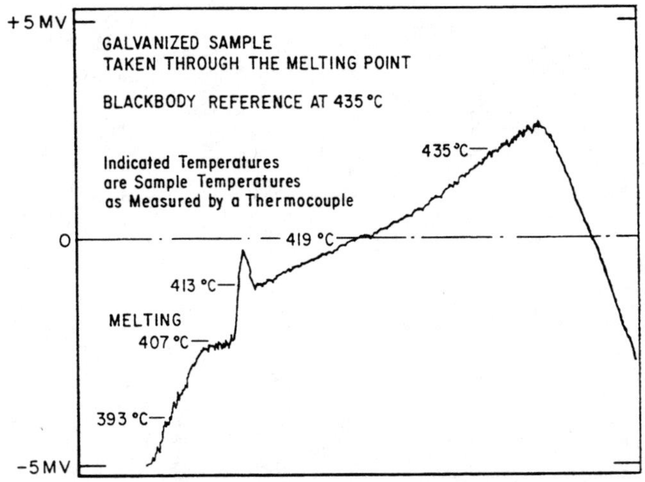

FIG. 12. Recorder chart of automatic instrument–galvanized sample taken through melting point.

The advantages of using electro-optical components rather than mechanical components to modulate the signal, and at the same time generate the reference voltage, will be investigated.

The United States Steel Corporation has applied for a patent on the developments described here.

ACKNOWLEDGMENTS

The author wishes to thank the United States Steel Corporation and in particular Samuel B. Prellwitz for support of this work and permission to publish these results, and Carl Willis for his aid in making the measurements.

Reprinted with permission from *International Journal of Thermophysics,* Vol. 11(1), pp. 269-281 (1990).

Dual-Wavelength Radiation Thermometry: Emissivity Compensation Algorithms[1]

B. K. Tsai,[2] R. L. Shoemaker,[2] D. P. DeWitt,[2] B. A. Cowans,[3] Z. Dardas,[3] W. N. Delgass,[3] and G. J. Dail[4]

The traditional contact methods of temperature measurement for metal processing applications provide accuracies of ±10 K. Noncontact methods based upon emissivity compensation techniques have the potential for improved accuracy with greater ease of use but require prior knowledge of the target emissivity behavior. The features of the basic spectral and ratio methods and five dual-wavelength methods are reviewed. Experiments were conducted on a series of aluminum alloys with different surface treatments characterized by x-ray photoelectron spectroscopy in the temperature range 600 to 750 K. Compensation algorithms that account for surface characteristics are required to achieve improved accuracy.

KEY WORDS: dual-wavelength radiation thermometry; emissivity; pyrometry; radiation thermometry; radiometry; surface characterization; temperature measurement; x-ray photoelectron spectroscopy.

1. INTRODUCTION

Accurately measuring the temperature of aluminum alloys during processing is a challenging problem. The commonly used contact methods employ two-pronged thermocouples or other contact devices which can achieve accuracies of ±10 K but are difficult to use. Direct noncontact or radiometric methods have the potential for increased accuracy and ease of use but necessitate some knowledge of the spectral emissivity of the target

[1] Paper presented at the Tenth Symposium on Thermophysical Properties, June 20–23, 1988, Gaithersburg, Maryland, U.S.A.

[2] School of Mechanical Engineering, Purdue University, West Lafayette, Indiana 47907, U.S.A.

[3] School of Chemical Engineering, Purdue University, West Lafayette, Indiana 47907, U.S.A.

[4] Lafayette Works, Aluminum Company of America, Lafayette, Indiana, U.S.A.

material. Spectral emissivity data are difficult to acquire since the emissivity of aluminum alloy surfaces depends on many variables, including temperature, wavelength, surface conditions, process conditions, and thermal history. Further, the inherently low emissivity and high reflectivity of aluminum alloy surfaces make property measurements difficult under laboratory or simulated mill conditions. In addition, environmental factors in the mill, such as vibrations and electrical noise, contribute greatly to the complexity of radiometric design necessary to make reliable observations. These problems emphasize the difficulty of achieving reliable temperature measurements for automatic control and metallurgical quality requirements, increasing product reproducibility and reliability, and reducing cost [1].

In applying any radiation thermometry method, some assumptions must be made on the emissivity of the target material. The spectral method, the simplest and most widely used approach, requires that the target have a constant and known emissivity. The ratio method, which requires approximately gray surfaces, finds special utility with highly oxidized metallics and nonmetallics but does not work well with low-emissivity metals. These methods can be satisfactorily used for numerous applications where the errors due to emissivity uncertainties are known and acceptable. Foley [2] has demonstrated a modified ratio method which allows for nongray behavior. The cross-correlation method of Svet [3] had good accuracy with near-black surfaces (cavities) experiencing small emissivity changes. The method of Watari et al. [4, 5] employed an algorithm involving emissivity as an exponential function of wavelength with oxidized stainless steels. Anderson [6, 7] used a linear combination of the ratio and spectral radiance temperatures for low-emissivity metallic surfaces.

The adjustable parameters of these dual-wavelength methods have not been related to surface conditions, and hence, the methods have found limited utility. An understanding of the effect of changing surface conditions on emissivity is essential in identifying the proper emissivity function or compensation algorithm. Because of the attractions of simplified, mill-hardened instrumentation, there are strong incentives to investigate whether dual-wavelength methodology, with appropriate algorithms, can satisfy an accuracy of ± 3 K for aluminum alloy processes. The objectives of the study are (1) to review dual-wavelength emissivity compensation algorithms, (2) to evaluate the effectiveness of these algorithms in predicting the true temperature of aluminum surfaces using laboratory data, and (3) to examine surface conditions for a relationship between emissivity and surface microstructure.

2. PRINCIPLES OF THE METHODS

Dual-wavelength radiation thermometry is based upon two relations which are written below for a spectral condition. Because our interest is generally restricted to wavelengths shorter than the peak of the blackbody curve, Wien's law

$$L_{\lambda,\mathrm{b}}=\frac{c_1\lambda^{-5}}{\exp(c_2/\lambda T)} \tag{1}$$

is a good approximation to Planck's law for the spectral radiance distribution. The spectral emissivity can be expressed as the ratio of the blackbody spectral radiance at the spectral radiance temperature (T_λ) to the blackbody spectral radiance at the true temperature (T),

$$\varepsilon_\lambda=\frac{L_{\lambda,\mathrm{b}}(\lambda, T_\lambda)}{L_{\lambda,\mathrm{b}}(\lambda, T)}. \tag{2}$$

From these relations, the *spectral temperature equation* is

$$\frac{1}{T}=\frac{1}{T_\lambda}+\frac{\lambda}{c_2}\ln\varepsilon_\lambda \tag{(3)}$$

In the *spectral* method, the true temperature is inferred by measurement of the spectral temperature, T_λ, and knowledge of the spectral emissivity, ε_λ, at a specified wavelength.

For the *ratio* method, using the same basic relationships, but written for two spectral conditions, the *ratio temperature equation* is

$$\frac{1}{T}=\frac{1}{T_\mathrm{r}}+\frac{\Lambda_\mathrm{r}}{c_2}\ln\left(\frac{\varepsilon_1}{\varepsilon_2}\right) \tag{4}$$

where the ratio temperature, T_r, and the equivalent wavelength, Λ_r, are defined such that

$$\frac{1}{T_\mathrm{r}}=\Lambda_\mathrm{r}\left[\frac{1}{\lambda_1 T_{\lambda 1}}-\frac{1}{\lambda_2 T_{\lambda 2}}\right] \tag{5}$$

$$\Lambda_\mathrm{r}=\frac{\lambda_1\lambda_2}{\lambda_2-\lambda_1} \tag{6}$$

To achieve emissivity compensation, the second term on the right-hand side of Eq. (4) needs to be zero or a known constant independent of temperature and surface conditions; that is, the ratio method requires that

$$\frac{\varepsilon_1}{\varepsilon_2}=\text{constant} \tag{7}$$

The *Foley* method [2] (FOL) is a variation of the ratio method and has a temperature equation of the form

$$\frac{1}{T}=\frac{1}{T_{mr}}+\frac{A_r}{c_2}[(1-K_f\lambda_1)\ln\varepsilon_1-(1-K_f\lambda_2)\ln\varepsilon_2] \tag{8}$$

where T_{mr} is the modified-ratio temperature defined by

$$\frac{1}{T_{mr}}=A_r\left[\frac{1-K_f\lambda_1}{\lambda_1 T_{\lambda 1}}-\frac{1-K_f\lambda_2}{\lambda_2 T_{\lambda 2}}\right] \tag{9}$$

Emissivity compensation is achieved when the term in paren theses in Eq. (8) is zero, namely,

$$\frac{\ln\varepsilon_1}{\ln\varepsilon_2}=\frac{1-K_f\lambda_2}{1-K_f\lambda_1} \tag{10}$$

where K_f is the parameter which is adjusted to account for nongray behavior. Note that when $K_f=0$, T_{mr} becomes the ratio temperature (T_r) and the FOL method reduces to the *ratio temperature equation*, Eq. (4).

The *Watari* method [4, 5] (WAT) is yet another variation of the ratio method and has a temperature equation of the form

$$\frac{1}{T}=\frac{A}{T_{\lambda 1}}+\frac{B}{T_{\lambda 2}} \tag{11}$$

where

$$A=\frac{1}{1-(\lambda_1/\lambda_2)^3}, \qquad B=\frac{1}{1-(\lambda_2/\lambda_1)^3}. \tag{12, 13}$$

That is, the reciprocal of the temperature is a specified linear combination of the two spectral radiance temperatures. Alternatively the temperature equation can be written as

$$\frac{1}{T}=\frac{1}{T_r}+\frac{A_r}{c_2}\alpha(\lambda_1^2-\lambda_2^2) \tag{14}$$

from which it is evident that the emissivity compensation algorithm has the form

$$\varepsilon_\lambda=\exp(\alpha\lambda^2) \tag{15}$$

The preceding temperature equations have their origins in simply prescribed emissivity functions, Eqs. (7), (10), and (15). An alternative approach is to postulate empirical forms of the temperature equations. The *ratio-with-offset* method (RWO) is a variation of the ratio method and relates the true temperature to the ratio temperature plus a constant offset.

$$T = T_r + B \tag{16}$$

This relationship is commonly used in commercial ratio instruments, and emissivity compensation is achieved if B is a known constant.

Anderson [6, 7] recognized that for metallic surfaces the spectral radiance and ratio temperatures are systematically lower and higher, respectively, than the true temperature. He postulated an empirical temperature equation of the form

$$T = (1 - x_0) T_r + x_0 T_{\lambda 2} \tag{17}$$

referred to as the *linear spectral-ratio* (LSR) method, in which the true temperature (T) is expressed as a linear combination of the ratio temperature (T_r) and a spectral temperature ($T_{\lambda 2}$). It can be shown that x_0 is related to emissivity by

$$\frac{\ln \varepsilon_1}{\ln \varepsilon_2} = 1 - \frac{x_0}{1 - x_0} \frac{\lambda_2}{\Lambda_r} \left(1 - \frac{\Lambda_r T}{c_2} \ln \frac{\varepsilon_1}{\varepsilon_2}\right) \Big/ \left(1 - \frac{\lambda_2 T}{c_2} \ln \varepsilon_2\right) \tag{18}$$

A variation of this approach, referred to as the *inverse spectral-ratio* (ISR) method, uses reciprocal temperatures, giving

$$\frac{1}{T} = \frac{(1 - y_0)}{T_r} + \frac{y_0}{T_{\lambda 2}} \tag{19}$$

for which the parameter y_0 can be expressed as

$$\frac{\ln \varepsilon_1}{\ln \varepsilon_2} = 1 - \frac{y_0}{1 - y_0} \frac{\lambda_2}{\Lambda_r} \tag{20}$$

By comparing Eqs. (10) and (20), it is evident that the emissivity compensation requirements of the FOL and ISR methods are identical. The LSR method contains a parameter x_0, which is dependent upon T, but for short-wavelength and low-temperature conditions, the LSR method compensates for emissivity in approximately the same manner as the ISR method.

3. EXPERIMENTAL APPARATUS AND PROCEDURES

An apparatus was developed to observe both the spectral radiance and the true temperatures of aluminum alloy samples. Its main functions included calilbration of the radiometer against a reference blackbody, measurement of spectral radiances from the sample, control and measurement of sample temperature, and computation of spectral radiance temperatures. The experimental arrangement included a two-band radiometer (Williamson,[5] modified Model 12200, $2.1 \pm 0.05\ \mu m$, $2.4 \pm 0.05\ \mu m$) calibrated against a blackbody (Williamson, Model 45), a gold mirror which allowed sighting either the blackbody or the sample, a 10-channel digital thermometer (Fluke, Model 2176A), an interface for amplification and integration of radiometer signals, and a microcomputer (Zenith, Model ZW-248-82). The samples were heated by two embedded electrical heating elements, which were powered by a digital temperature controller (Watlow, Model 808C-0100-0000). The *true* surface temperature of the sample was determined from a thermocouple positioned parallel to and at a depth of 1.6 mm from the surface. A heat transfer analysis of the heater-sample arrangement showed the sample temperature to be uniform within $\pm 0.25°C$ [9]. At the start of each experiment, the radiometer was calibrated against the blackbody in the range of 550–750 K, and a linear regression of spectral radiance vs signal output for both spectral bandpasses was obtained. Following completion of the planned heating schedule, the regression result was checked against the true temperature to confirm satisfactory radiometer performance.

Radiometric measurements were made on four types of aluminum alloys (1100, 2224, 5052, 7075) with different surface conditions (C, as cast; E, as extruded; P, as polished; S, as cut). Cast samples were cut from the surface of an ingot that had left a direct chill casting mold. Extruded samples was passed through an extrusion die and had surfaces with a degree of roughness in the range of 0.1 to 0.5 μm. Polished surfaces were defined as the condition obtained by standard metallurgical polishing procedures ending with 800 grit Al_2O_3. Saw-cut samples were made by a circular cutoff saw resulting in a directional surface with rms roughnesses ranging from 4 to 12 and 0.15 to 2.4 μm across and along the sawing direction.

Three sets of experimental data were examined. The first set was comprised of the results from seven samples (1100E, 1100S, 5052E, 5052S, 7075C, 7075E, 7075S) at temperatures in the range of 620–760 K and heating times of 7–62 h. No special effort was made to expose these alloy samples systematically to similar heating schedules. The second data set was taken on the 1100E, 1100S, 5052E, and 5052S alloys and was measured at a fixed temperature of 740 K. Two different samples of the 5052E alloy were used with short (1-h) and long (4-h) heating exposures but for the other three

[5] The mention of a commercial product is provided for completeness of description of the experimental procedure but does not constitute an endorsement.

alloys only one sample of each alloy was used. The third data set, also at 740 K, included observations on the 2224P and 2224S alloys, each with long-term (4.5-h) and short-term (1-h) heating schedules. Attention was given to these samples to assure similarity in heating schedules, temperatures, etc.

The spectral emissivities were calculated from Eq. (2), while the *inferred* temperatures, as well as the parameters, for the LSR, ISR, and the RWO methods were evaluated using a least-squares technique, in which the sum of the squares of the temperature errors, the true minus the inferred temperature, $(T - T_{inf})$, was minimized. The first data set was also analyzed with the FOL and WAT algorithms. The *true* temperature was determined from thermocouple observations.

Angle-resolved x-ray photoelectron spectroscopy, ARXPS, provided information on the composition of the metal surface and the oxidation states of the elements present. Spectra were collected on the 1100, 2224, and 5052 alloys and recorded on a Perkin–Elmer PHI 5300 spectrometer. The Mg (1253.6 eV)/Al (1486.6 eV) dual-anode x-ray source was operated at 15 kV and 300 or 400 W, respectively. The spectrometer was calibrated by setting the binding energies of the Au $4f_{7/2}$ and Cu $2p_{3/2}$ levels to 84.0 and 932.7 eV, respectively. The full width at half-maximum for the Ag $3d_{5/2}$ level (368.4 eV) was 0.86 eV at 489 kHz for the pass energy at which data were collected. Background pressure in the analysis chamber was always less than 1×10^{-9} Torr. Data accumulation times to measure major and minor species at a given angle were roughly 16 h, and data reduction was accomplished using Perkin–Elmer software.

4. DISCUSSION

The results of the first data set in Fig. 1 indicated that the ISR method was identical to the FOL method as expected from the earlier discussion. In addition, the ISR method was moderately more accurate than the LSR method. For the types of alloys used and the temperature range selected the WAT method proved to be the last accurate. For the full range of temperature from 620 to 740 K, the best precision with any method was ± 10 K. The results for the extruded samples displayed a temperature error $(T - T_{inf})$ that increased with temperature, whereas the temperature error associated with either the cast or saw-cut samples decreased with temperature. This phenomenon may be linked to the emissivity ratio behavior of the types of samples because, as the emissivity ratio increases, the inferred temperature (T_{inf}) increases and the temperature error $(T - T_{inf})$ decreases in the RWO, LSR, and ISR methods. More data and analysis would be necessary to substantiate this statement clearly. The least-squares

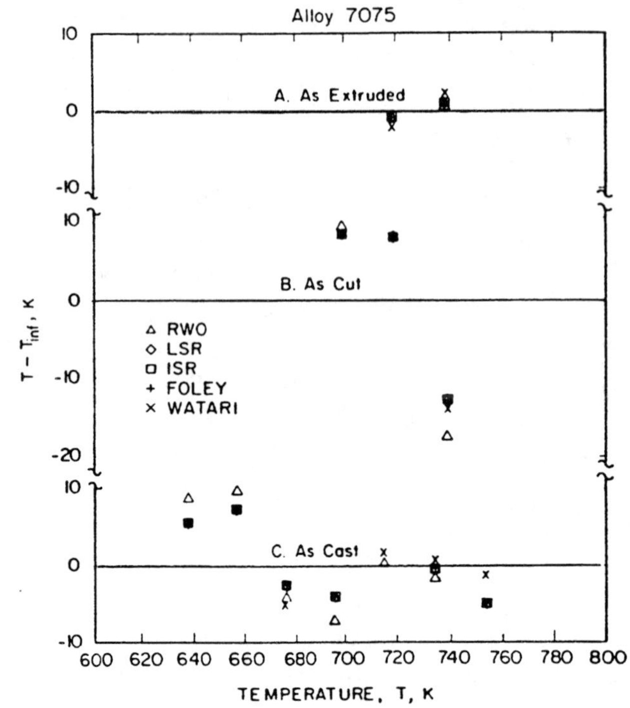

Fig. 1. Difference between true temperature and inferred temperature, $T - T_{inf}$, using the five emissivity compensation methods for the aluminum alloy 7075 under the (A) as-extruded, (B) as-cut, and (C) as-cast conditions.

analysis revealed that the RWO method was the best method for extruded samples, while the ISR and the FOL methods were most appropriate for the cast and saw-cut samples.

The second data set, shown in Fig. 2, revealed that the emissivity of the 1100 series alloys remained constant throughout the 5-h heating schedule, while the emissivity of the 5052S and 505E alloys doubled in the same time interval. Note that the sample-to-sample variation was approximately 0.02. As expected from surface roughness effects, the emissivity of the 1100S alloy was consistently higher than that of the 1100E alloy. The emissivity of the 5052S alloy was higher than that of the 1100 series alloys and increased with time as did the 5052E alloy. Again, the accuracy of the emissivity compensation methods was not better than ± 10 K.

The third set of results showed that the emissivity of the 2224S and 2224P alloys exposed to long-term (Fig. 3) and short-term heating schedules increased only slightly with time. In addition, the emissivity of

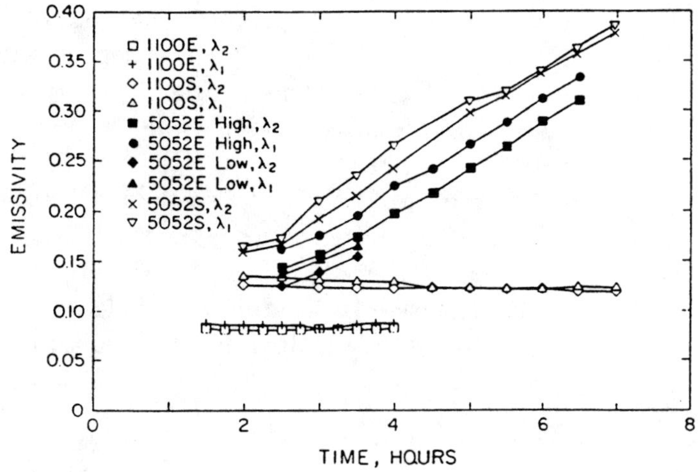

Fig. 2. Emissivity at 740 K for the aluminum alloys 1100 and 5052 under the saw-cut and extruded conditions.

the polished surfaces was less than the emissivity of the saw-cut surfaces. The accuracy of the compensation methods was not better than ± 10 K.

In the temperature difference or compensation results (Fig. 1) and the emissivity determinations (Fig. 2), the major sources of measurement error were due to uncertainties in observed spectral radiances, in radiometer calibration, and in the target temperature. The radiometer calibration uncertainty at the lower spectral radiance temperatures (about 600 K) was unacceptably large. Above 600 K, the error in temperature difference was

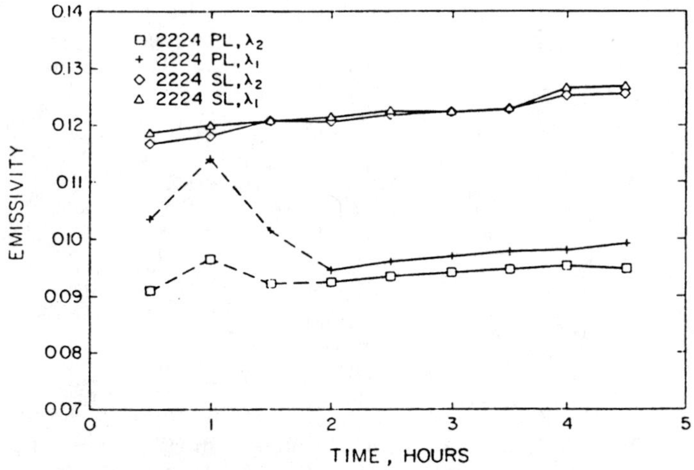

Fig. 3. Emissivity at 740 K for the aluminum alloy 2224 under the saw-cut and polished conditions with long-term heating.

about ± 5 K for the full range of temperatures. The differences among the methods were not altered by this random error, because the results were derived from the same set of experimental data. The error band for the temperature difference calculations is ± 5 K, while the error band for the emissivity determinations is about ± 0.005.

The ARXPS results were analyzed to provide a measure of the oxide layer thickness and composition. The oxide layer thickness was calculated from the relative intensities of the Al 2p lines for the Al and Al_2O_3 states as shown in Fig. 4. If we assume a uniform oxide layer of thickness covering an infinitely thick metal substrate, the equation relating the relative intensities has the form

$$\frac{I_{Al_2O_3}}{I_{Al}} = \frac{S_{Al_2O_3}\sigma_{Al_2O_3}\eta_{Al_2O_3}\lambda_{Al_2O_3}[1-\exp(-d/\lambda_{Al_2O_3}\sin\Theta)]}{S_{Al}\sigma_{Al}\eta_{Al}\lambda_{Al}\exp(-d/\lambda_{Al}\sin\Theta)} \tag{21}$$

where I is the intensity, S is the spectrometer term, which includes the analyzer transmission function, the detector efficiency, etc., σ is the photoelectron cross section [10], η is the atomic density, λ is the photoelectron mean free path [11], and Θ is the angle between the sample surface and the electron path to the analyzer. Since the Al and Al_2O_3 lines are very close in energy, we assume that the spectrometer terms are equal and solve Eq. (21) for the oxide layer thickness, d. The results of these calculations are shown in Fig. 5 and are discussed below. Note that the maximum thickness measurable by this method for this system is approximately 15 nm.

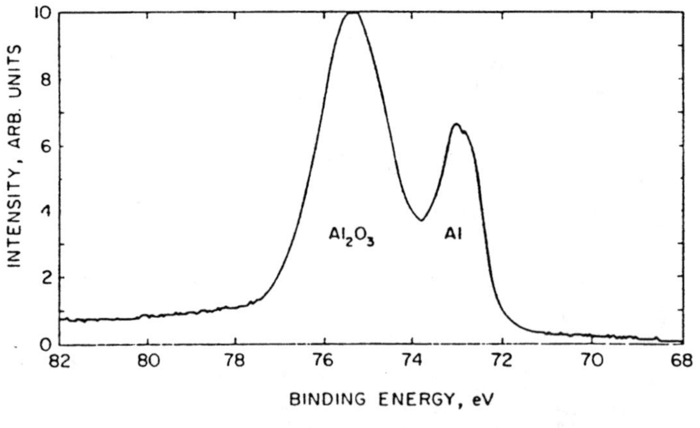

Fig. 4. X-ray photoelectron spectrum of the Al(2p) core level from 2224E at an electron takeoff angle of 60°. The two peaks show the chemical shift of Al_2O_3 from Al. An increase in the intensity ratio of $I_{Al_2O_3}/I_{Al}$ at a lower electron takeoff angle (spectrum not shown) confirms that the Al_2O_3 layer lies above the unoxidized aluminum.

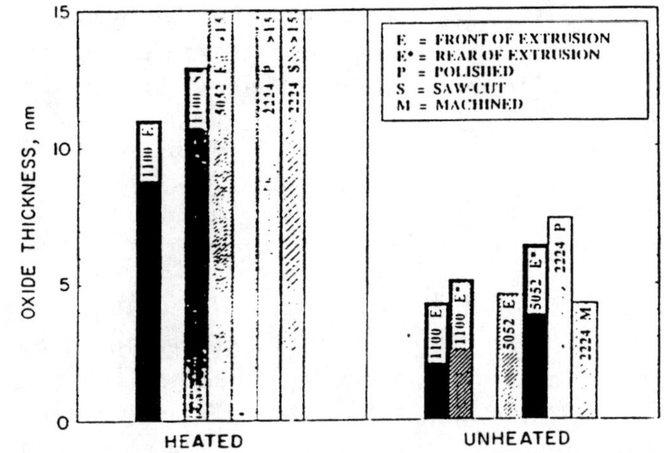

Fig. 5. Oxide layer thickness as determined by ARXPS for various alloys and conditions.

The fresh 1100E alloy possessed a relatively thin Al_2O_3 layer of roughly 4 to 5 nm. Prior to heating the sample, the oxide layer was slightly enriched in MgO relative to the nominal bulk Mg concentration. This result was also observed on the 2224 and 5052 alloys. Heating the 1100E alloy for 7 h at 740 K produced an Al_2O_3 layer approximately 11 nm thick with a MgO layer less than one monolayer thick. Note that the oxide thickness for the 1100S sample is 12.9 nm. The emissivity of this alloy remained constant at 740 K, as shown in Fig. 2.

Both machined (M) and polished (P) surfaces of the 2224 alloy were examined. The machined surface had an Al_2O_3 layer of 4.3-nm thickness and was similar in composition to the 1100 alloy, while the polished surface had an oxide layer 7.4 nm thick that was righ in Mg. The effect of polishing on oxide thickness has been noted previously [12]. The 2224 alloy, which contains slightly less Mg but much more Cu than the 5052 alloy, produced a MgO layer approximately 15 nm thick when heated to 740 K for only 1 h. Although some Al_2O_3 could be observed, the total oxide thickness of this alloy became too thick to measure with ARXPS. Extended heating for 4.5 h did not appear to increase the thickness of the MgO layer significantly. Thus, heating the 2224 alloy produces a surface oxide layer which is MgO for the first 15 nm and Al_2O_3 for an unspecified depth below the MgO. The emissivity of this alloy increased slightly over a 4.5-h period.

The oxide thickness of the unheated 5052E alloy changed from 4.6 to 6.4 nm from the front to the rear of the extrusion. Further, the surface at the rear of the extrusion contained nearly three times the amount of MgO as the front of the extrusion. When heated for 1 h at 740 K, the 5052 alloy produced a MgO layer greater than the 15-nm limit of ARXPS, with no

Al_2O_3 observable. We interpret this dramatic change in surface composition as indicative of the unlimited oxide growth reported in the literature [13] for this alloy. The emissivity of this alloy increased linearly from 0.16 to 0.32 over a 7-h heating period.

5. CONCLUSIONS

The objectives of this study were to evaluate dual-wavelength emissivity compensation methods for aluminum alloys subjected to different surface preparations and heating schedules. It was shown that the ISR method is slightly better than the LSR method for all the alloys studied. The ISR method produced the same results as the FOL method since, they in fact have the same emissivity compensation requirement. However, different algorithms seemed to work better for different surface conditions. For example, the RWO method worked best for the extruded samples, while the ISR and FOL. methods provided the best results with the cast and saw-cut samples. The temperature errors for any of the methods were still above ± 10 K.

The 1100 series alloys had a constant emissivity for 5 h while being held at a constant temperature of 740 K. This correlated with the fact that Al_2O_3 was the main constituent on the surface of the 1100 series alloys after heating and more importantly that the oxide layer was relatively thin, of the order of 10 nm. The emissivities of the 2224 alloys increased only slightly while heating at a constant 740 K. Although the surface of this alloy, when heated, became greatly enriched in MgO to a depth of roughly 15 nm, Al_2O_3 could still be observed. Thus the surface oxidation showed signs of being limited. The 5052 alloys, on the other hand, were characterized by a rapidly increasing emissivity at a constant temperature. This large change in emissivity appeared to be related to the dramatic restructuring of the alloy surface. Upon heating, the full observation depth, 15 nm, of this alloy became MgO with no Al_2O_3 observed. Thus, the large change in emissivity correlated with the growth of a very thick oxide layer.

From this study, it was determined that the dual-wavelength methods of emissivity compensation do not provide satisfactory performance to achieve ± 3 K for the prescribed temperature and heating schedule conditions. We conclude that the emissivity functions are inappropriate for the behavior of the alloys and that to develop useful dual-wavelength methods, an improved understanding of the spectral emissivity of the materials will be required.

ACKNOWLEDGMENTS

The authors acknowledge the efforts of Mr. A. S. Anderson, Williamson Corporation, for engineering the modifications of the Model 12200 to suit our requirements and for providing valuable advice on experimental procedures and data analysis. Professor D. R. Gaskell, School of Materials Engineering, contributed to developing procedures for appropriate preparation of test samples and our understanding of surface microstructure effects on emissivity. This work was supported in part by the National Science Foundation under Grant CDR 8803017 to the Engineering Research Center for Intelligent Manufacturing Systems, Schools of Engineering, Purdue University. Funding by the Aluminum Company of America, Kaiser Aluminum and Chemicals Corporation, and Inland Steel Company is gratefully acknowledged as well.

REFERENCES

1. M. J. Haugh, in *Theory and Practice of Radiation Thermometry*, D. P. DeWitt and G. D. Nutter, eds. (Wiley, New York, 1988), pp. 905–971.
2. G. M. Foley, *High Temp. High Press.* **10**:391 (1978).
3. D. Y. Svet, *High Temp. High Press.* **4**:715 (1972).
4. M. Watari, Y. Watanabe, S. Chigira, and Y. Tamura, *Yokogawa Tech. Rep.* **29**:25 (1985).
5. T. Andoh, F. D. Banta, T. Kawano, I. Fujimoto, and M. Watari, *Yokogawa Tech. Rep.* **31**:8 (1987).
6. A. S. Anderson, *Proceedings of Aluminum Association Workshop on Sensors*, p. 91 (1986).
7. A. S. Anderson, *Adv. Instrument.* **40**:1337 (1985).
8. B. K. Doloresco, MSME thesis (School of Mechanical Engineering, Purdue University, Lafayette, Ind., August 1986).
9. G. J. Dail, M. G. Furhman, and D. P. DeWitt, *Proc. Fourth Int. Alum. Extrus. Technol. Semin.* **2**:281 (1988).
10. J. H. Scofield, *J. Electron Spectrosc.* **8**:129 (1976).
11. C. J. Powell, *Surf. Interface Anal.* **7**:263 (1985).
12. E. A. Gulbransen and W. S. Wysong, *J. Phys. Chem.* **51**:1087 (1947).
13. C. N. Cochran and W. C. Sleppy, *J. Electrochem. Soc.* **108**:322 (1961).

Reprinted from *Optical Techniques for Industrial Inspection,* P.G. Cielo, editor, Proc. SPIE Vol. 665, pp. 75-88 (1986). ©1986 SPIE.

Thermographic Sensing For On-Line Industrial Control

Dag Holmsten

AGEMA Infrared Systems, 550 County Avenue, Secaucus, New Jersey 07094

Abstract

It is today's emergence of thermoelectrically cooled, highly accurate infrared linescanners and imaging systems that has definitely made on-line Infraread Thermography (IRT) possible. Specifically designed for continuous use, these scanners are equipped with dedicated software capable of monitoring and controlling highly complex thermodynamic situations.

This paper will outline some possible implications of using IRT on-line by describing some uses of this technology in the steel-making (hot rolling) and automotive industries (machine-vision). A warning is also expressed that IRT technology not originally designed for automated applications e.g. high resolution, imaging systems, should not be directly applied to an on-line measurement situation without having its measurement resolution, accuracy and especially its repeatability, reliably proven. Some suitable testing procedures are briefly outlined at the end of the paper.

Introduction

Infrared thermographic techniques (IRT) have been applied for more than 20 years to such established industrial applications as electric power-line and equipment inspection, industrial preventive maintenance, building insulation control, flat-roof moisture checks and heating equipment refractory verification. Practical field inspection techniques, as well as sophisticated quantitative heat loss measurement methods have been developed using portable scanners in the 2-5 or 8-14 micron range.

In parallel, the increased use of IRT in thermodynamic research applications, often looking at very rapid events (thermally, spatially or both), as well as its use to develop practical NDE-methods, has created a large amount of data indicating many potential uses of IRT for automated on-line product quality control and process monitoring.

Until this year, however, the general application of IRT to "around-the-clock" continuous operations in often hostile industrial environments had been hampered by the lack of suitable technology. The recent availability of low cost, powerful microcomputers and software capable of rapidly taking high frequency, dynamic data through complex temperature measurement and image processing algorithms was an important factor in improving this situation. The stage was now set for an important thermographic innovation.

On Line Thermography is Measurement

On-line thermography, or continuous thermal mapping, whether 1-, 2-, or 3-dimensional, refers to the ability of an infrared sensing system to detect (relative location and relative temperature of e.g. hot-spot), identify (relative location + absolute radiance level) and measure (absolute location and absolute temperature) the thermal characteristics of an object. Furthermore, this has to be done in such a manner as to make all relevant thermal data comparable in space and time.

Note that only true hot-spot detection is really a relative measurement method since both the identification and the measurement of an object has to be repeatable, to be of any value in on-line operation. If not, the information cannot be compared to a reference (identification) nor used as a consistent feedback (as absolute temperature) for readjustment of production parameters. All of this may seem rather obvious to anyone used to working with input from discrete sensors such as pressure gauges, thermocouples and spot radiometers. Remember, however, that off-line, portable thermography has, to the industrial user, mainly been identified with thermal imaging and relative temperature measurements in such applications as electric hot-spot detection and thermal insulation verification in buildings.

The advantage of using thermography in process monitoring and product quality control is largely related to its capacity of analyzing a large amount of time and/or space resolved thermal data. In other words, the more thermally and spatially dynamic the object/process, the greater the advantage of using Total Surface Temperature Monitoring or TSTM as we shall call it in the following. This is readily demonstrated by looking at a few typical linescanning applications (see Table 1).

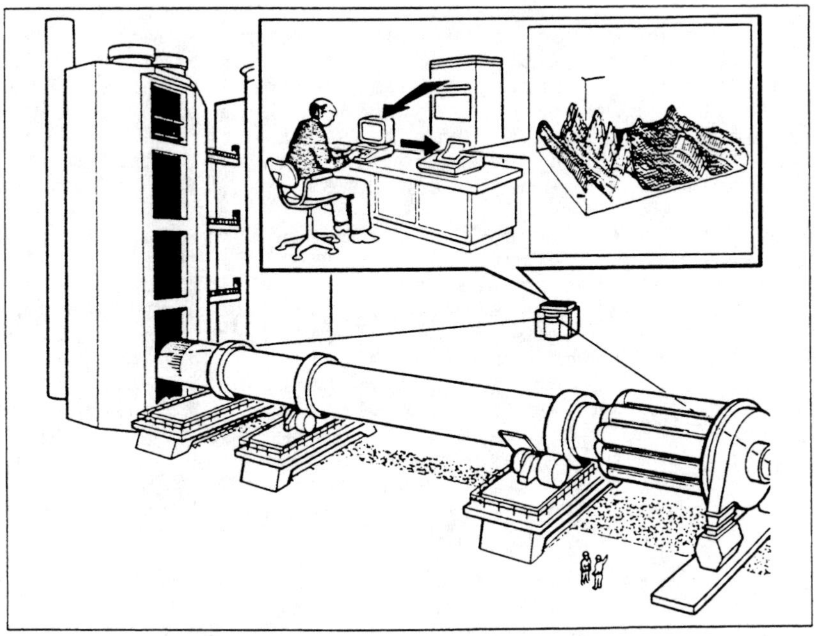

Fig. 1: Rotary-kiln inspection in the cement industry

	HOT SPOT DETECTION	PATTERN ID	TEMPERATURE MEASUREMENT
CEMENT AND GYPSYM			
- Monitoring of cement kilns	X		
- Detection of voids in plaster boards			X
- Curing of plaster boards	X	X	
IRON & STEEL			
- Hot strip mill		X	X
- Cold strip mill			X
- Continuous annealing			X
- Continuous casting		X	X
- Continuous galvanizing	X		
- Sinter strand	X		X
- Lime kilns			
GLASS			
- Monitoring of sheet glass		X	X
MANUFACTURING INDUSTRY			
- Re-heat furnace	X	X	
- Automotive on-line quality control		X	X
- Arc welding, resistance welding			X
- Induction hardening		X	X
- Friction welding, weld integrity control		X	
- Uniformity control of surface coating, painting, etc.		X	
PLASTIC			
- Extrusion of plastic film		X	X
- Lamination			X
- Thermo forming			
PULP AND PAPER			
- Moisture control of paper webs		X	X
- Mesa kilns	X		X
NON-FERROUS METALS			
- Hot Mills	X		
- Aluminum melt ovens	X		
- Cu and Zn refineries	X		
- Continuous Casting	X		

Table 1: Typical Linescanning applications

The cement industry was first to extensively adopt TSTM, in this case mainly to monitor the refractory lining of rotary kilns and prevent e.g. unplanned shutdowns due to brick failure or build-up of internal deposits ("cold rings") hampering the efficiency of the process (fig. 1). A truly eventful process, this application necessitates the dynamic scanning of a large, slowly moving object (1-3 RPM) internal to which a highly dynamic thermal process takes place. The amount of data-points required is high (400-800 points per scan-line 480 scan-lines per kiln revolution). Kiln applications are also important in pulp (mesa kilns) and iron making (lime kilns) processes.

The manufacturing industry is well represented by automotive applications related to welding (mobile, high-temperature applications), induction hardening (very thermodynamic application on rotating or otherwise mobile objects, painting (uniformity verification compared to a reference) and on-line quality control. This quality control often involves the detection, identification and rejection of inferior products, e.g. containing cracks, voids, thermal stresses or simply foreign materials (more about that later). All of these latter products are inspected on a moving conveyor belt and decision about their (thermal) status has to be made rapidly. They are true thermographic "machine-vision" applications.

Flat-glass, plastic-film and paper manufacturers have one thing in common: they have to monitor wide sheets or rolls of thermally non-uniform, rapidly moving objects. Both longitudinal and transverse temperature variations are important, so here again the total surface has to be monitored.

There are, of course, a large amount of other applications as well, but we shall today continue our discussion of TSTM by mainly concentrating on a few steel-making applications and finish by outlining a few challenges that await us in the automotive industry. The reason for this is clear. No other industries display such a diversity of applications as steel-making and automobile-component construction and assembly. We are confident that those in the audience that are more interested in e.g. aluminum, textile or food-processing usages of TSTM will find that the reasoning that we shall now apply to steel-making can readily be used in these earlier mentioned application as well.

In order to clear the scene for more detailed on-line measurement discussions, however, it is first necessary to outline the recent technology that has made TSTM possible.

On-Line IR-Technology Designed for Measurement

Today's state of the art, TSTM-technology, is represented by a new generation of high resolution thermoelectrically cooled scanners that are greatly superior for measurement tasks. Fig. 2 is showing the internal design of the Thermoprofile®-5 as a good example. The linescanner, already introduced at last year's Thermosense, is not a derivative from any military or television technology but is specifically designed for measurement. The heart of the system, can be said to be two internal black-body references that are scanned by the thermoelectrically cooled detector during each revolution. Years of development has made it possible to produce proprietary references with a nominal accuracy of 0.1%. Something quite unusual for an infrared system today! Great data repeatability is here assured.

(Fig. 3) The THP-5 typically scans a line in 30 milliseconds, producing some 800 discrete data-points. The large amount of data, notwithstanding, a linescanner is often sampled at a maximum rate in order to enable the software to zoom in ("changing lenses") on any particular part of the scan line and still maintain enough resolution per viewing sector. By adding consecutive lines, two dimensional thermal maps are produced of moving objects. (Fig. 5)

Looking at fig. 5 and comparing to the spot-pyrometer in fig. 3, it is easy to understand why a linescanner can be better understood as a thermographic total surface measurement system than simply as "many" spot-pyrometers. With similar temperature measurement accuracy, the linescanner would "replace" 200-300 pyrometers, but as we shall see, the purpose of using the former is rarely to "just" measure discrete temperatures.

In the domain of two-dimensional scanners, much less experience is available in on-line applications. The Infrared Burden Inspection System (IRBIS) that we designed for temperature monitoring inside the blast furnace, is one of the few examples available up until today of applying a thermal imager to process control. Recently, however, automated systems have been designed for the continuous inspection and quality control of such diverse products as PCB's, turbine blade cooling systems and composite aeronautical or automotive components. Based upon liquid nitrogen cooled scanners, however, these systems will need some evolution before they can be called truly autonomous on-line systems.

Only, a few days ago, we finally saw the world-wide introduction of the first really continuously operating imager, AGEMA's Thermovision® 870. Thermoelectrically

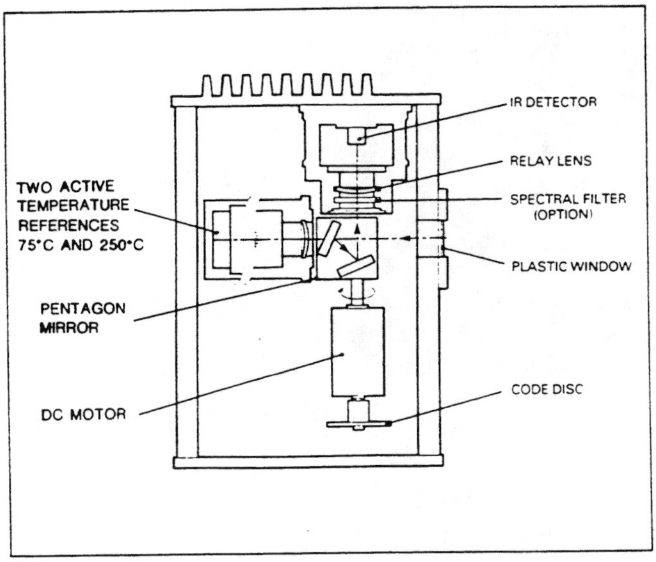

Fig. 2: Internal optical layout of a measurement scanner

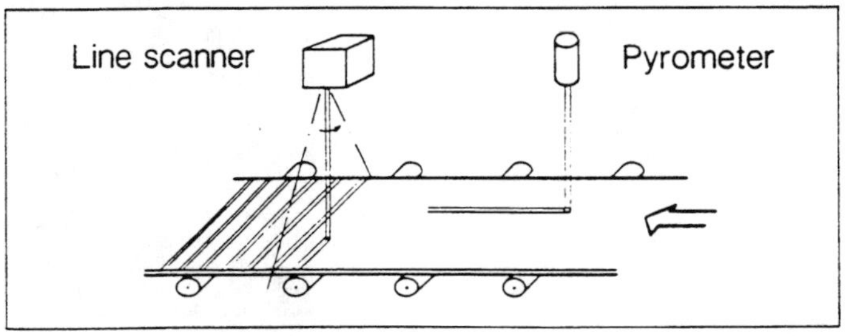

Fig. 3: Spot-Radiometer versus linescanner. A linescanner produces approcimately 400-800 "spot measurements" per scan line.

Fig. 4: The recent introduction of x-y scanning, thermographic measurement system designed for on-line applications, leads the way to the wider use of infrared for machine-vision.

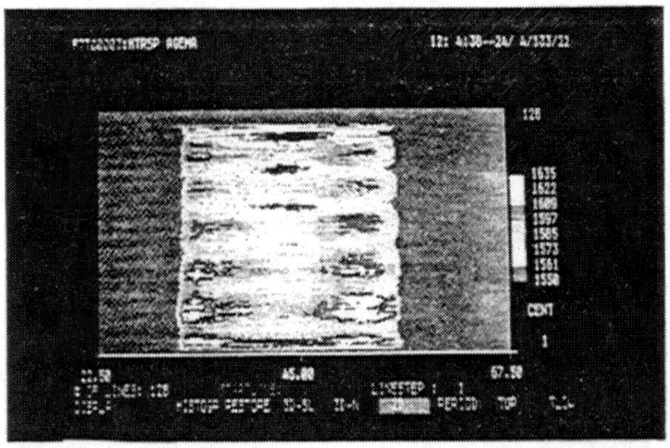

Fig. 5: Two-dimensional thermal image of strip in a hot rolling mill produced by a line-scanner (x - axis = linescan; z - axis = movement of steel sheet).

Fig. 6: The environment of a steel mill is the best possible proving ground to validate the capabilities of a well designed infrared measurement system. Note the aiming spot radiometer.

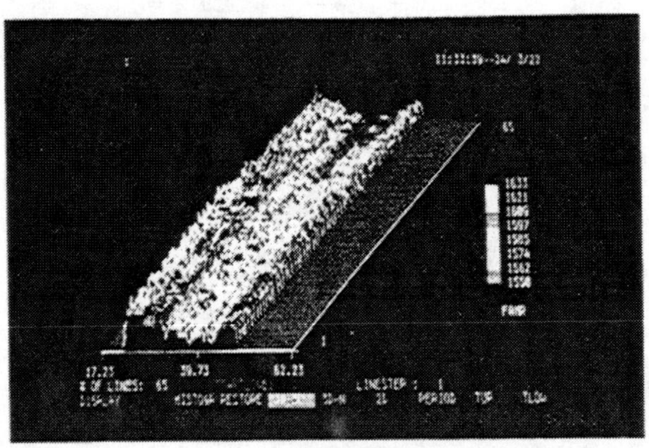

Fig. 7: Typical thermograms from end of finishing stands. The decrease in temperature from beginning to end of the strip as well as the remarkably lower temperature (/T of close to 90°F) in the middle of the strip compared to the edges is unexpectedly common in many hot strip mills.

cooled, as the THP-5, this instrument will be instrumental in quickly developing on-line TSTM applications in all industries. The entire technology of this 2.5 kHz scanner is geared towards "intelligent" process-monitoring, from the high resolution SPRITE detector, its innovative scanning concept, the two internal micro blackbodies, to the flexible real time digital image processing system (see. fig. 4).

The autonomous cooling is only one important feature. Here, again, we stress the importance of looking at total design rather than at a single feature. Many other systems will soon incorporate thermoelectric cooling, it's part of the future. This doesn't automatically transform an imaging system into a reliable on-line TSTM system. Durability, environmental ruggedness, and especially measurement accuracy/repeatability are much more important features. Let's now illustrate this by looking at some typical on-line steel applications.

Steel Applications

The steel industry (see Fig. 6) is certainly a good proving ground for evaluating the measurement capabilities of an infrared system. Not only does the measurement environment often contain vapors and gases that absorb in the infrared (C02, (NO)x, H20 etc.), is richly endowed with sources of electromagnetic interference for sometimes sensitive electronic components but this industry can also display a large variety of emissivity/reflectivity values for its hot and cold products.

It should be noted that there are many parallels to be made with other industries. In flat-glass production, transverse analysis (and lateral temperature variations) is as important as in the hot rolling mill. In a similar fashion, edge to edge variations in drying efficiency in paper-production is at least as important as uniform cooling is in hot rolling before coiling. Variations in temperature in extruded plastic film means uneven thickness and quality. A plastic sheet that has been unevenly heated before shooting out of the infrared oven will result in uneven stress in the e.g. thermoformed automotive dashboard in the same way as an unevenly reheated slab will be rolled unevenly and may even damage the rollers (by not being warm enough in some areas).

Fig. 4 is a 2-D image of part of a 48" sheet in the last stages of hot rolling. The eight colors correspond to temperature variations on the sheet surface with black areas being off-range. Note the uneven borders due to the side ways movements of the strip. Now, a spot-pyrometer would usually look at the slab or strip somewhat like the radiometer in Fig. 6. The idea is usually to aim at the centerline of the slab/strip, sometimes adding a few pyrometers on each side of the center (this is common in Japan) for "good measure".

Unfortunately, once a pyrometer is far off center, it is difficult to determine exactly where to put it. If it is placed too close to the edges, it is possible it will be looking partially outside the strip due to the often large spot-size of such an instrument, especially if products of very different widths are being rolled. Closer to the middle, the readings of a pyrometer will produce data, the relevance of which is more difficult to determine - it's not edge data and it's not centerline temperature - and therefore, seems a less useful alternative. Back then to the centrally placed sensor, at least here we can be fairly sure of reading (monitoring) the warmest part of the product and therefore be able to directly relate readings to desired production parameters?

In this respect, Fig. 7 may be of some interest. It is a typical three dimensional presentation of the end of a strip. Not all lines are displayed simultaneously since it might be problematic to present an entire 2000 feet long strip on a screen with only 512 lines. The color as well as the height of the profile (Y-axis) indicates temperature level. The X-axis is width and the Z-axis length of strip. What is especially noteworthy here is that the middle of the sheet is sensibly colder than the rest of the strip. The image is recorded right at the end of the last finishing stand before final water spraying and coiling. And it is typical of thermograms taken in many hot-strip mills both in the U.S. and in Europe.

Now where do we aim our spot pyrometer again? In the middle or at the edges? Obviously, temperature variations actually taking place during hot rolling and subsequent cold rolling operations are often much too complex to be handled by arbitrarily placed spot-measurement systems. Looking again at Fig. 7, there are many reasons for the temperature variations displayed.

The cold middle, to start with, in some cases seems to be due to the fact that sheets rolled are not completely flat and more water from earlier spray coolers and roll sprays simply stay a longer time in the middle before evaporating thereby cooling more then the rest of the strip. Also note the decrease in temperature along the length of th sheet. As a matter of fact, the decrease in temperature from the middle of this sheet to the end is more than 120°F and one of the edges constantly maintains a much higher temperature than the centerline. The red area in Fig. 8a is another example of a hot "spot" which may or may not be sufficiently cooled down by the subsequent water sprays.

Some automotive manufacturers experiencing serious problems in spray painting and in coating sheet steel, have been able to trace some of these difficulties back to a temperature non-uniformity in the sheet just before coiling: Hot areas make the sheets stick like glue. When uncoiling the steel, serious surface defects have then been caused which have remained all the way to the automotive paint shop. Fig. 9 is a typical thermogram from another mill after cooling and just before coiling. By using the linescanner just at the end of the finishing stand to indicate the transverse and longitudinal location of any hot areas, this information can be fed to the water sprays and the result checked by a linescanner positioned at the recoiler (see also fig. 10).

Of course, an early warning system e.g. at the end of the roughing stands could also provide useful information as to the existence or development of thermal stresses and surface defects. Fig. 4 discussed earlier is a good example.

To summarize, there are several inspection stations along the line of a hot strip where a fixed thermographic system could be installed to monitor both edge to edge and longitudinal sheet (slab-, strip-, plate-, bar-, beam-, etc.) temperature. Catching a lack of product thermal uniformity at an early stage means that potential uneven rolling and poorly distributed cooling can be adjusted for from pass to pass or, depending on software used, during the rolling itself. Result: less equipment wear, less frequent roll changes, less waste and above all much more even product quality (surface structure, shape and metallurgical composition).

The same principle applies for cold rolling, annealing, galvanizing and rolling of non-ferrous metals. Here, shape and surface structure are often even more important production parameters. These latter applications, furthermore have a well established reputation as "difficult". Low emissivity, uneven surface structure, emissivity changing with temperature (<u>and</u> with spectral response of sensor used) have sometimes created the impression that IR cannot measure temperature very effectively. Today, fortunately, many excellent solutions are available due to the increased competence in infrared measurement and the flexibility of lower cost, micro computer based process-control software. Intelligent software can be made to have a linescanner actually "see" surface structure dependent emissivity variations and automatically correct for these. The "blindly" aiming spot-pyrometer can now be used to produce spectrally well defined single wavelength and dual ratio measurements of a spot as a reference input to a total surface monitoring systems scanning a larger area incorporating this spot. An interesting result of this development is also the use of an IR scanner to actually check surface structure variations e.g. on a galvanizing line by measuring emissivity variations and not temperature!

That a thermal scanner can be used for other things than measuring surface temperature is of course not new. The dramatic change is often instead the efficiency and accuracy of such methods when they are combined with rapid image processing software. Look, for example, at the detailed horizontal scale of the profile in fig. 10. Knowing the distance scanner to object and the speed of the strip, a known fov. can be translated into inches of <u>strip width</u> or <u>slab position</u>. Today, using an appropriate software, it may take a TSTM system 2-4 seconds to convert rolling speed change from an earlier stand into horizontal product width and longitudinal position of the scan line along the strip, as it passes the last finishing stand.

Future Trends In On-Line TSTM

Two key phrases in future developments are <u>input comparison</u> and <u>time resolved techniques</u>. Using an input from a spot pyrometer, another linescanner or a stored reference thermogram are all examples of comparison to the object actually scanned that can be made rapidly to determine its status. Even digital inputs from completely different techniques such as ultrasound, visual radiation or x-ray can be compared with TSTM. Similarly, rapidly comparing the thermal signatures of an object at short time intervals during heating or cooling will reveal many characteristics of an object not easily discernible from a "steady state" thermogram. Two examples of combining these techniques will be related below. In both cases, the purposes of using thermography is <u>not</u> to measure surface temperature....

Prices on raw material being what they are, avoiding wastes in the primary mill as well as in the continuous caster are commendable opportunities for TSTM. Fig. 11a has been recorded by an X + Y scanning infrared system. Note especially the fish tail at the end of the ingot slab. The question here is how to determine exactly where to cut off the ends of

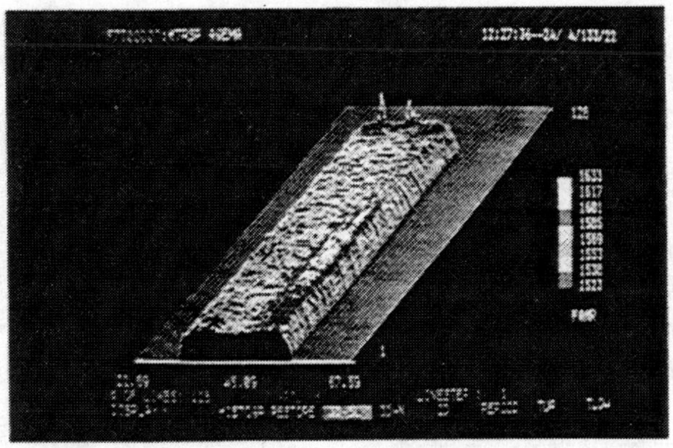

(8A)

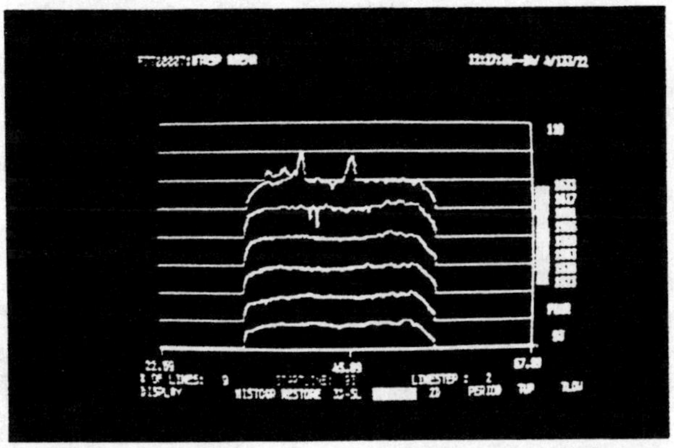

(8B)

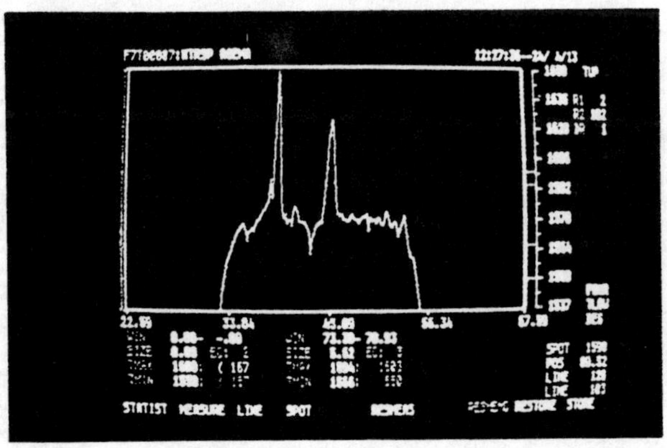

(8C)

Fig. 8: Analysis of a hot spot: Thermograms from the end of a steel strip, taken right after the last finishing stand with a linescanner.

A) 3-D image consisting of 120 consecutive linescans. Note the temperature peaks at end of strip (top of image).

B) The suspicious hotter areas are pinpointed by looking more closely at the last lines of the strip.

C) The exact extent, temperature and location of the hot spots are identified in this single profile. Alarm windows will be set to watch for similar spots in the next passes of the same product.

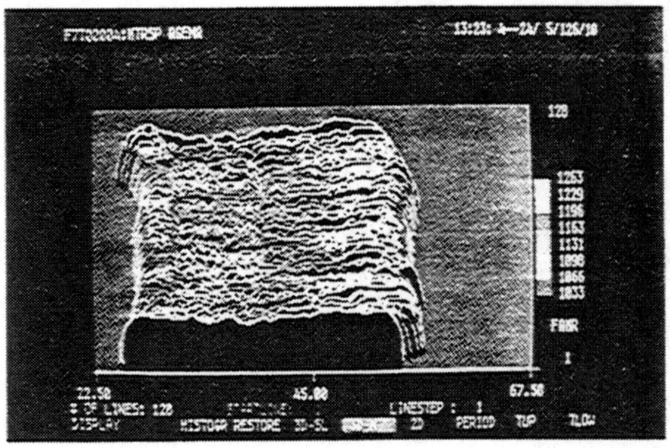

Fig. 9: Thermogram taken after water spraying and before the recoiler in a hot strip mill.

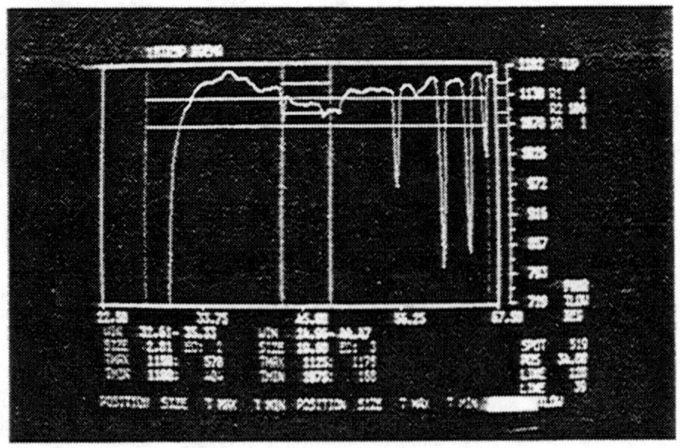

Fig. 10: In this thermal profile, only half of the total 90°fov is being utilized. Fov and temperature range 719°F to 1182°F. Note the two alarm windows.

Fig. 11A

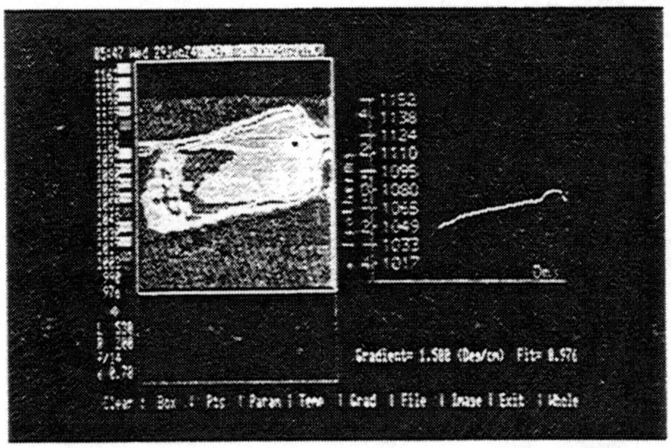

Fig. 11B

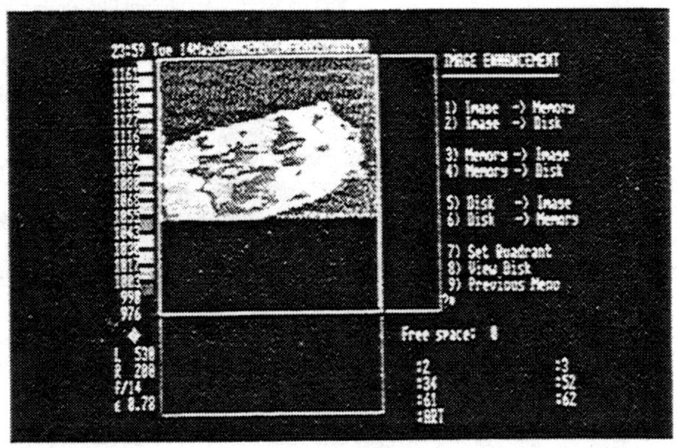

Fig. 11C

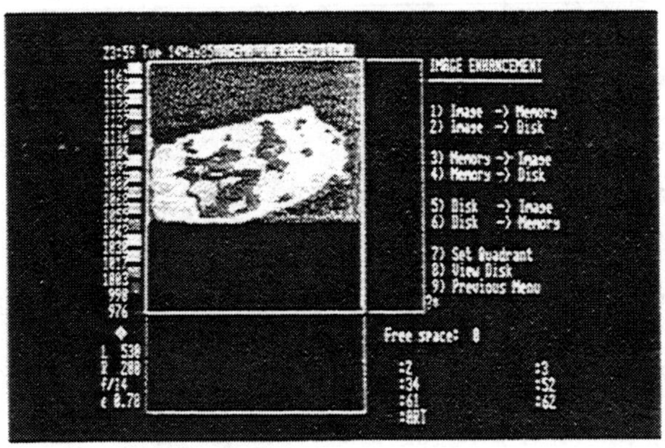

Fig. 11D

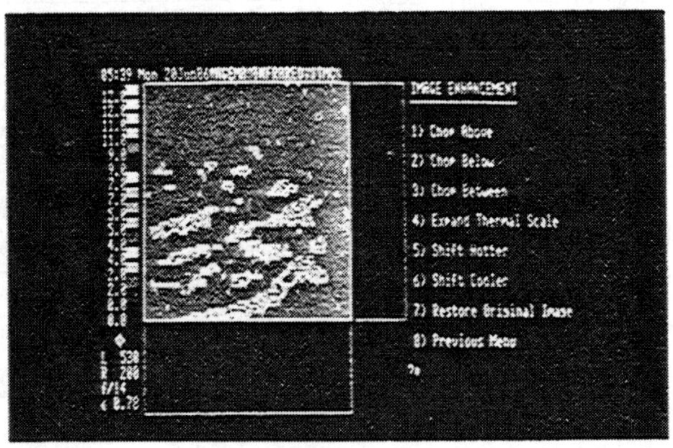

Fig. 11: Thermal imaging of ingot slabs. (degrees of centigrade)

A) Two-dimensional scan slab/ingot in the primary mill. Note fish-tail shape and uneven thermal signature close to tail.

B) Thermogram of end of slab

C) Thermogram of same slab 3 seconds later

D) Subtraction of image c from image b produces a completely different thermal pattern, earlier marked by preventing gross temperature variations in both image b and c. This time resolved substraction method may reveal otherwise hidden thermal patterns related to internal temperature distributions. Applied at larger time intervals, it can also show the growth of scale (ferrous oxides).

the slab in order to make sure that the remainder has enough uniform metallurgical quality but still not waste good steel. Previous image processing of this thermogram was made to filter out apparent temperature variations due to scale. It reveals an earlier hidden thermal pattern depicting an inhomogenous area close to the end of the slab. The uneven area could be due to higher porosity or the existence of piping, that is cavities in the upper end of the slab caused by shrinkage during solidification. Due to the large variations in the size and importance of such pipes, however, much additional testing is necessary before it can be definitely proven that small enough piping and/or porosity problems can be consistently detected in this manner.

Due to the large thermal mass of the ingot relative to the size of the potential cavities and the extensive amount of thermal diffusion in other directions than towards the scanner (scanning perpendicular to the upper surface of the ingot) it is virtually impossible to discover any significant temperature variations once you are getting close to thermal equilibrium. A time-resolved approach therefore seems the most feasible.

As indicated in earlier papers e.g. by Celio and his group at the Research Council here in Boucherville[19,20], sub-surface defects in highly conductive materials are more easily depicted by closely observing the change in slope of the temperature versus time curve for each individual pixel on the surface of the scanned object. Two thermograms taken at a time interval of say 3-5 seconds to allow a change in heat flow to have taken place could e.g. be substracted from each other and compared to a reference substraction showing a "perfect" slab with acceptable levels of porosity/piping recorded in the same manner from the same mill. We think further investigations along these lines could show great promise.

Ultrasound Versus(?) Thermography

In this application as well as in the measurement of internal slab temperature, ultrasonic techniques are presently being developed at the NBS in Washington[21]. The principle involved is simple: In the case of pipe/porosity measurement, an ultrasonic pulse is measured. Knowing the distance between the transmit and receive transducers, one can calculate the average velocity of the ultrasonic wave. By generating tables of velocity temperature dependence for different types of steel, the earlier mentioned average velocity can then be related to temperature. Unfortunately, it is not as easy as it sounds (!). Highly accurate measurements, in the order of 0.1%, have to be made of both the distance transmit-to-receive and the time-of-flight of the ultrasonic pulse in order to meet the standards formulated by a recent AISI workshop. Furthermore, the closeness to the moving surface of a few millimeters necessitated by the otherwise large loss of sensitivity of the EMATS (electromagnetic acoustic transducer) that are employed to receive the sound wave, can cause severe mechanical and thermal survivability problems. To create and transmit ultrasonic waves, piezo electric transducers have been found to be impractical in this environment. Instead NBS has suggested using a 850 mJNd:YAG Laser. Again, a very costly solution not to talk about the safety problems created by beams from a high-power laser that could accidentally bounce off the surface of the steel sheet etc. So why not combine ultrasound and thermography!

Due to the high energy level of an ingot or slab, the effective sensitivity of a thermal scanner becomes extremely good (better than 0.01°C). Repeatable time-resolved techniques based on thermography should therefore be feasible both for internal temperature measurements and for internal liquid/solid interface detection in ingots. It is also suggested that such measurements would be emissivity- and thus scale-independent since the same surface is substracted from itself. During the time-lapse in question, it is assumed that temperature has changed, but not enough to affect emissivity (but scale will increase). Ultrasound could then be used to "calibrate" the internal temperature measurements of the infrared scanners when installing the latter, thus combining the accuracy of the ultra sound approach with the use of the more practical TSTM system for on site continuous measurements.

Thermographic "Machine-Vision" Applications

I mentioned earlier (see also Table 1) that many applications do not require absolute temperature measurements, but still require a knowledge of absolute radiance levels. As an example, most machine-vision-type applications only require of infrared systems to detect cracks, foreign materials, malfunctioning components etc. These applications are based on the technique of comparing a scanned object with a reference thermogram, in order to identify the object and its status (good, bad, to be rejected etc.) Fig. 12 will illustrate this. Hot spots in a transmission cover seem to indicate the presence of unwanted chips remaining from the machinery operation used to produce the cover. In order to correctly establish if the hot spots are due to artifacts in the procedure used to heat the cover, or are the actual emission from heated chips, the thermogram is compared to a reference taken under the same heating conditions but on a transmission without chips. In the difference thermogram, the actual chips stand out nicely from the substracted background. For such an identification procedure to work, temperature measurements per se are not necessary, but the same thermal signature has to be reproducible, scan after scan, day after day, year after year for analogous measurement conditions. How do we make sure of that? Using appropriate testing procedures could help.

A Few Thermographic Measurement Procedures For On-Line Systems

A thermographic system on-line has to accurately at least identify both temperature and space and preferably also measure these parameters. And the key word is consistency. Repeatedly, and continuously...

Thermographic systems do not lack standards and testing procedures, far from it. In especially the military field and the building industry, several excellent, mostly rather complex testing methods have been laboriously put together. But not a single one of these tests is concerned with absolute measurement. As well in insulation void check as in enemy detection, the main requirements of the thermographic system is to well image the thermal scene and detect low-contrast temperature differences as rapidly (early) as possible. The objective is to resolve small temperature differences at very high spatial frequencies.

A continuous thermographic measurement task, however, requires a fairly large temperature difference in order to assure accuracy and repeatability of the information, and a drift free operation for the same reason. Given that these two tasks are satisfactorily performed, but only then, a high enough spatial measurement resolution should still be obtained to be able to monitor the required spot-size for a given object width. The imaging versus measurement controversy is summarized in Fig. 13.

In an attempt to increase the public interest in providing measurement capability tests, Appendix I describes the Slit Response Function (SRF), its capability to better define the spatial measurement resolution of any thermographic system and a simple testing procedure that can be performed by the user of IR equipment himself without the need of any sophisticated apparatus. This test is today used by us to define the imaging and measurement resolution of our system and consequently the necessary spotsize for repeatable detection as well as accurate non contact temperature measurements (see also fig. 14).

Conclusion

Thermography can only be seriously introduced to the on-line market if infrared systems are made available that are specifically designed for continuous, reliable, absolute measurements. The existence today of stable, thermoelectrically cooled, IR systems equipped with internal black body references is a good sign in the right direction. The increased knowledge of conditions necessary for good in-situ measurements, coupled with the real-time image/measurement processing capabilities of state of the art digital hardware and software, should further improve the possibilities of correcting for such traditionally problematic parameters as unknown emissivities, reflection problems and variations in object size/shape/position.

It is certainly this author's experience that the large amount of reliable thermal data that can now be provided by linescanners and imaging systems will substantially contribute to improve industrial process and production efficiency. Today's often very crude and nearly always discrete (spot) temperature measurement techniques will increasingly be replaced by thermal signatures of each individual process based on continuous (total) surface monitoring of the products in question at strategic locations. These signatures will generally be used as reference maps to which the day to day thermal events will be compared. Hot spots, unexpected thermal variations etc., will all then be compared to this uniformity index describing thermal events that are "within tolerance" and will not seriously effect the (uniform) quality of the finished product. Out of tolerence events will automatically trigger a correction in the process. These reference maps or uniformity indices will then, depending on the process, be expressed in terms of temperature, emissivity, shape, thickness, humidity, stress or other IR radiation intensity related variations and compared to the desired absolute levels of these parameters for the process in question.

Appendix I

The Slit Response Function, SRF

Traditionally, the geometrical resolution of an imaging system is evaluated by defining its ability, in one way or another, to reproduce the spatial frequency content of the scanned object (line). Several assumptions have to be made in order to adapt such tests to thermal linescanners or imaging systems. The aim of these tests are mainly to obtain a measure of the best possible spatial resolution of a system at high radiance levels (as in

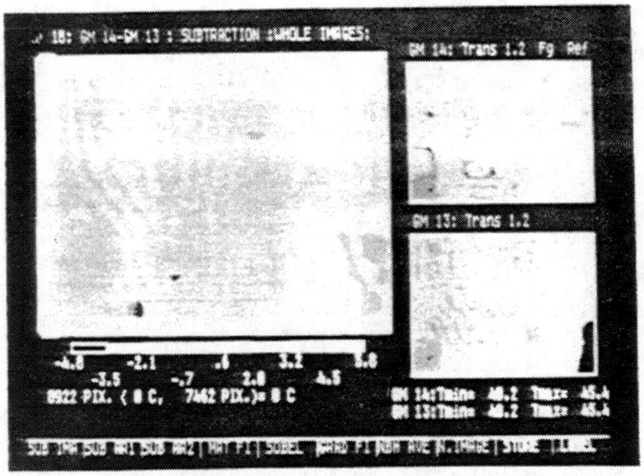

Fig. 12: Detection and identification of transmission chips using a reference substraction method.

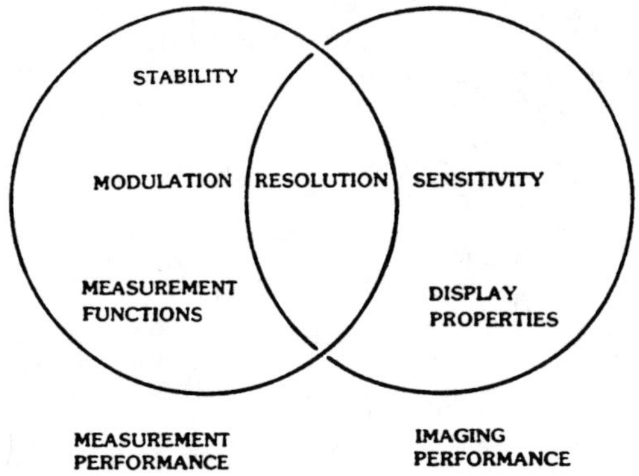

Fig. 13: Imaging versus measurement

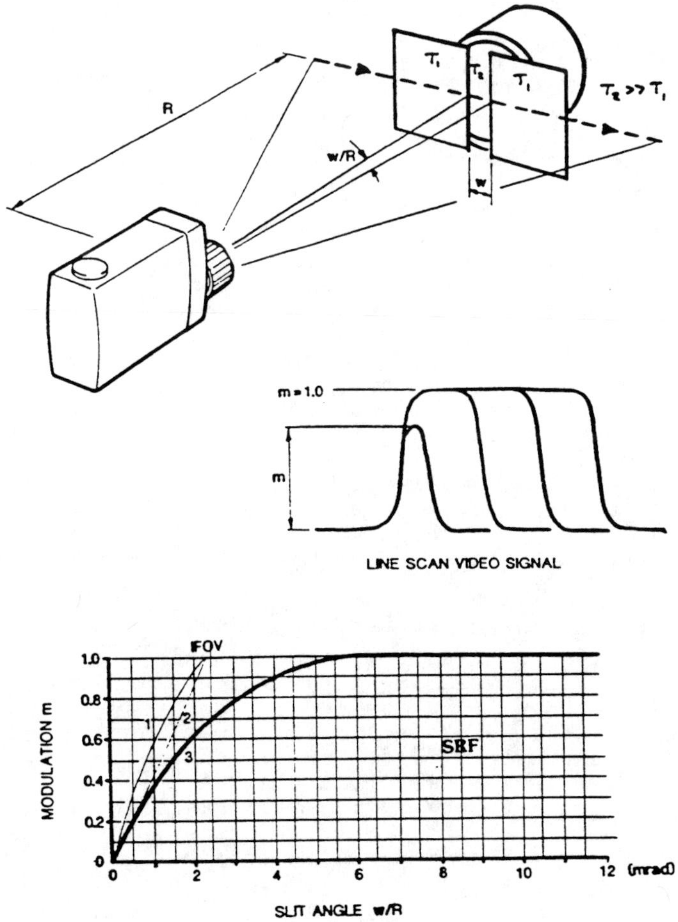

Fig. 14: Slit response function: testing procedure and typical curves. Note that an imaging systems measurement resolution (90 or 95% modulation) can be anything from 2.5 to 10 times more than its imaging/detection resolution (50% modulation). Typical spatial resolution spec's (1 FOV, cycles per milliradian, "elements per line" etc.) therefore reveal discouragingly little about the system's true measurement spot-size if modulation is not specified.

the Line Spread Function test*) looking at objects much smaller than the IFOV of the system or to obtain the best possible thermal sensitivity at high spatial frequencies (MRTD* at high spatial frequencies). The one-dimensional Fourier transform of the Line Spread Function would be a good approximation of the imaging systems Optical Transfer Function.* The modulus of the Fourier transform then determines the amplitude response of the imaging system in question.

A few excellent facilities are now available for testing industrial thermographic systems in this respect both in the U.S. (National Bureau of Standards in Gaithersburg, Maryland) and in Canada (Public Works Canada, Ottawa). Unfortunately, none of these tests are applicable to an IR scanners measurement precision. In terms of the modular transfer function, we are looking for the best possible spatial resolution that will still assure a given minimal and repeatable precision using a specific IR scanner. We want to have a strong enough signal (large enough temperature difference) to make the information repeatable.

This calls for looking at the sensors performance at high modulation rates ("high contrast") and not at the frequency limited modulation.

For measurement applications, therefore, we suggest that a simple diagnostic test that can be performed by any potential user, is to check the slit response function (SRF) of the IR-sensor to be used.

Fig. (14) describes the measurement procedure for the SRF and shows a typical curve. To represent the desired resolution, a slit of variable width is positioned between the IR-sensor and a high temperature source. The amplitude of the signal pulse created by the IR-system when looking at the "warm" slit is measured with an oscilloscope. A very broad slit will give a pulse with a flat top. This amplitude corresponds to full modulation, w = 1. With decreasing slit width, w, the pulse height expressed as the modulation factor w versus slit angle subtended, forms the SRF curve. Not only is this test very simple to perform, but it is also highly repeatable and applicable to individual system by system comparison. By not requiring the user (tester) to "observe" contrasts on a screen, it also lacks the element of subjectivity inherent in the more common MRTD tests.

It is recommended that the (sampling rate determined) spatial resolution be related to the slit response function of the scanner used. As an example, the Thermoprofile®-5 specifies both an SRF at 50%, which has been our standard for detection ("imaging") resolution for more than 10 years and an SRF at 90% (10% precision which in most applications translates into a measurement accuracy of better than 2%). Some manufacturers, especially those working with IR systems that have a peak response beyond (at longer wavelengths than...) the spectral peak of the object surveyed, may want to specify a higher SRF. Typically, 95% or even 99% would be desirable for some LW systems used in higher than ambient temperature application. This is to assure sampling large enough spots to be able to obtain accuracies comparable to other measurement (SW) systems.

It is now possible for the user to devise his own test requirements based on the precision that his application may require. As an example, the resolution obtained at 95% modulation, specifies the minimum spot-size that the system can resolve and still limit the precision error to 5% or less. This 5% is in terms of signal amplitude not in degrees F or C.

It should also be noted that the availability in addition to SRF spec's of individual calibration (system-response) and attenuation curves for all IR systems intended for measurement would definitely make the job of comparing measurement systems easier. As an example, it can generally be assumed that a SW system will have a more dynamic range (requiring less precision for a given accuracy) than a LW system in industrial applications. On the other hand, differences in detector response/quality, video amplification etc. may make a difference. Only individual system response curves will tell the whole story. In the same way a 3.9 micron peaked system may seem ideal for using in environments with high CO2 and/or H20 content. On the other hand, if the system optics are not coated correctly, internal attenuation (reflection) problems may occur. System specific attenuation curves are therefore necessary for on-line systems.

With an IR equipment designed for measurement, a carefully selected spot-size (SRF!) and a good understanding (= control) of actual measurement conditions (choice of optimal IR wave-length-band, measurements of actual emissivities and transmission factors, controlling reflections, environment enclosure etc.) absolute temperature measurement accuracies in the order of 2% to 5% should be achievable with a TSTM system designed for on-line use.

*See e.g. Ref. 1 and 2 for a detailed description of these functions and tests.

References

1. Wood, J., "Description of a Facility for Evaluating Infrared Imaging Systems for Building Applications", Proceedings of Thermosense V, SPIE, Vol. 371, 1982.
2. Chang, Y., Grot, R., Wood, J., Laboratory Design and Test Procedures for Quantitative Evaluation of Infrared Sensors to Assess Thermal Anomalies, June 1985
3. Gaussorgues, G., La thermographie infrarouge, Technique et Documentation, 1981
4. Lloyd, J., Thermal Imaging Systems, Plenum Press, New York, 1975.
5. Ohman, C., "Imaging Versus Measurement: A Review of Performance Characterization Parameters for Thermographic Systems', Infrared Information Exchange, October 1985.
6. Kaplan, H., "Thermal Signatures of Electronics Modules for Automatic Testing" - Progress Report, Infrared Information Exchange, October 1985.
7. Cox, C., "Automated Infrared Inspection of Linear Welded Thermoplastic Bumper Assemblies", Infrared Information Exchange, October 1985.
8. Cohen, J., Elements of Thermography for Nondestructive Testing, NBS TN 1177, U.S. Government Printing Office, Washington, D.C. 20402, 1985, 35 pp.
9. Levitt, R., Rao, N., Fisher., "Pyroelectric vidicon camera systems: status and capabilities", Proceedings of Thermosense IV, SPIE, Vol. 313 1981.
10. Frederickson, D., "New Developments in Thermal Imaging Systems", Proceedings of Thermosense V, SPIE, Vol. 371, 1982.
11. Elabd, H., Kosonocky, W., "Silicon based Schottky barrier infrared sensors for power system and industrial applications", Proceedings of Thermosense VI, Vol. 446, 1983.
12. Hunter, G., Allemand, C., Eagar, T., "An iomproved method of multi-wavelength pyrometry", Proceedings of Thermosense VII, Vol. 520, 1984.
13. Holmsten, D., "Industrial thermographic process monitoring and control: a new modular, line-scanning concept for continuous temperature measurement", Proceedings of Thermosense VII, Vol. 520, 1984.
14. Ding, K., "Test of jet engine turbine blades by thermogrpahy", Proceedings of Thermosense VII, Vol. 520, 1984.
15. Bantel, T., Bowman, D., Halase, J., "Automated infrared inspection of jet engine turbine blades", Thermosense VIII, Vol. 581, 1986.
16. Orlove, G., Madding, R., "Void detection in metals using heat flow parallel to the target surface", Thermosense VIII, Vol. 581, 1986.
17. Kelly, M., Abney, L., "A thermographic look at plasma spraying", Thermosense VIII, Vol. 581, 1986.
18. Holmsten, D., "Precision infrared on-line scanning for process control", Thermosense VIII, Vol. 581, 1986. 19. Cielo, P., Lewak, R., Balageas, D., "Thermal sensing for industrial quality control", Thermosense VIII, Vol. 581, 1986.
20. Cielo, P., Malgadue, X., Lemak, R., Lamantagne, M., "Recent Progress in NDE at IMRI"
21. Linzer, M., Droney, B.E., Martinez, F., Mauer, S.J., Norton and H.N.G. Wadley, "Progress in Development of Ultrasonic sensors for Monitoring Hot Steel Product", 5th Process technology Converence Detroit, Mighigan 1985

Reprinted from *Applied Optics,* Vol. 11(11), pp. 2603-2612 (November 1972).

Measurement of Particle Size, Number Density, and Velocity Using a Laser Interferometer

W. M. Farmer

A method for determining particle size, number density, and velocity utilizing a laser interferometer is analyzed. The results show that when the fringe spacing is comparable to a particle diameter, size can be estimated; and when the fringe spacing is much greater than the average particle diameter, number density can be measured. Since the optical arrangment for the interferometer is identical to that for a number of laser velocimeters, the effects of particle size on the velocimeter signal are discussed.

Introduction

The purpose of this paper is to present an analysis of a method for determining sizes and number densities of dynamic small particle fields. The method is an extension of the real fringe laser interferometer arrangements that are commonly used to measure the velocities of small particles.[1–5] At least two methods of determining the sizes of small particles from scattered light observed with a laser Doppler velocimeter have been reported.[6,7] In the first method it was shown that if a number of assumptions were made regarding the spatial and kinetic properties of very small particles, spectral analysis of the homodyne or heterodyne signal used in conjunction with the Rayleigh-Gans scattering theory would indicate particle size.[6] In the second method, an attempt was made to determine the particle size by measuring the absolute value of the scattered intensity and correlating this value with that predicted from a Mie cross-section calculation.[7]

The method to be presented here takes advantage of the nonuniform illumination of a particle by a set of spatially well-defined interference fringes generated by two or more crossed laser beams. For purposes of clarity and discussion, an experimental arrangement for generating such fringes is shown in Fig. 1. The laser beam (assumed to be in a TEM_{00q} mode) is split into two beams of equal intensity that are brought to a simultaneous cross-focus point by a lens. Henceforth the cross-focus region will be called the *probe volume.* In the region of focus the wavefronts are planar and tilted at an angle α. A Huygens diagram shows that the interference fringes exist as planes that are perpendicular to the plane of the beam center lines and parallel to the bisector of the beams. As Rudd has suggested, the Doppler frequency of light scattered from the probe volume may be interpreted in terms of the rate at which a scatter center (SC) crosses alternate bright and dark interference fringes.[8] Precise knowledge of the fringe spacing and time period of the scattered light allows a straightforward calculation of the velocity of the particle. Previous work has also suggested that the observed signal should be a function of the particle size and shape relative to the spacing of the interference fringes.[8] Thus, the problem to be addressed is the determination of the functional dependence of the scattered light on the particle size and shape as it crosses the fringe set. The results of the analysis show that particle size and number density can be determined through analysis of the signal. Furthermore, the effects of particle size on the velocity determination can be estimated, and it is shown how the signal-to-noise ratio for velocity measurements is corrected to account for size and shape of the particle or particles crossing the interference fringe set.

Signal Dependence on the Mie Scatter Cross Section

Consider the geometry shown in Fig. 2. In the general case, a spherical particle of radius a sits at the center of an orthogonal coordinate system (X,Y,Z). It is illuminated by two plane waves with their respective K vectors lying in the YZ plane and separated by an angle α. The bisector of α is coincident with Z. The spherical coordinates of some observation point in (X,Y,Z) are r, θ, ϕ. Each illuminating beam also has a coordinate set (X_i,Y_i,Z_i) $(i = 1, 2)$ centered on the particle. Z_i is coincident with K_i, while the X_i is coincident with the orientation of the polarization vector (assumed linear) for K_i. Each (X_i,Y_i,Z_i) coordinate system has corresponding spherical coordinates r, θ_i, ϕ_i for the observation point.

The author is with ARO, Inc., Arnold Air Force Station, Tennessee 37389.

Received 17 January 1972.

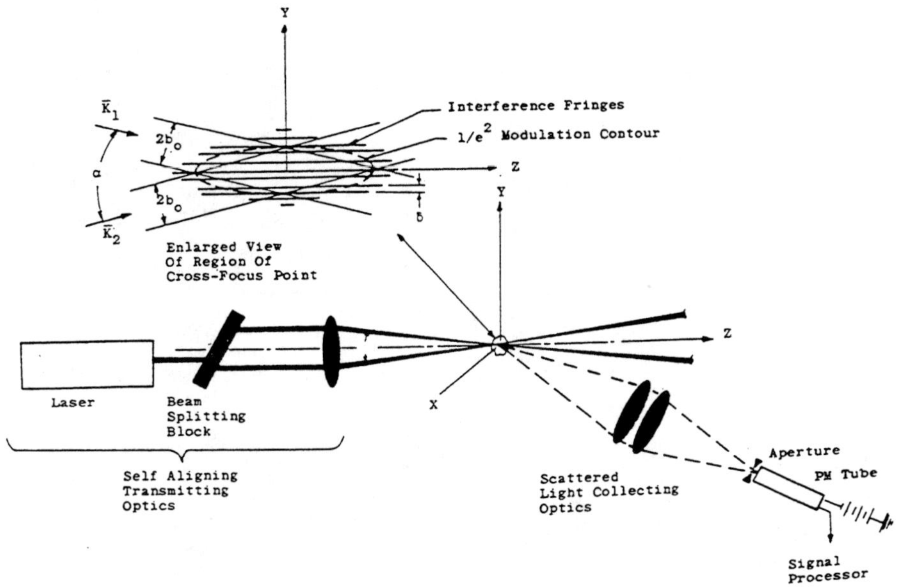

Fig. 1. Typical optical arrangement for generating a set of well-defined interference fringes.

The observed scattered intensity is the magnitude of the time-averaged Poynting vector $\mathbf{S}$:

$$\langle \mathbf{S} \rangle = (c/8\pi)\mathrm{Re}(\mathbf{E}\wedge\mathbf{H}^*), \tag{1}$$

where $\mathbf{E}$ is the electric field vector, $\mathbf{H}$ the magnetic field vector, and $\wedge$ the vector product. In this analysis it will be assumed that the dielectric constant and magnetic permeability of the medium surrounding the particle are approximately 1, that the scattered intensity is observed in the far field of the particle, and that the angle between the incident beams is such that

$$\sin(\alpha) \simeq \alpha, \tag{2a}$$

$$\cos(\alpha) \simeq 1. \tag{2b}$$

The $\mathbf{E}$ and $\mathbf{H}$ at the point of observation are the sums of the respective incident and scattered fields:

$$\mathbf{E} = \mathbf{E}_i(1) + \mathbf{E}_i(2) + \mathbf{E}_s(1) + \mathbf{E}_s(2), \tag{3a}$$

$$\mathbf{H} = \mathbf{K}_i(1) \wedge \mathbf{E}_i(1)/K + \mathbf{K}_i(2) \wedge \mathbf{E}_i(2)/K + \mathbf{K}_s(1) \wedge \mathbf{E}_s(1)/K_s + \mathbf{K}_s(2) \wedge \mathbf{E}_s(2)/K_s, \tag{3b}$$

where the i and s subscripts denote the incident and scattered fields, respectively; the numbers in parentheses denote the fields due to beams 1 and 2; K is the magnitude of the K_i vectors; and K_s is the magnitude of the scattered K vectors (see Appendix A). It is of interest to examine only the scattered power in $\langle \mathbf{S} \rangle$. Appendix A indicates the appropriate transformations and approximations required to calculate $\mathbf{E}_s \wedge \mathbf{H}_s^*$ in r, θ, ϕ coordinates. For paraxial front or back scatter from the incident beams with their respective polarization vectors parallel and oriented perpendicular to the plane of the incident beams, the value of the scattered intensity is

$$I_s = \left|\frac{c}{8\pi}\mathrm{Re}(\mathbf{E}_s \wedge \mathbf{H}_s^*)\right|,$$
$$I_s = \frac{c}{8\pi}\left\{|\mathbf{E}_s(1)|^2 + |\mathbf{E}_s(2)|^2 + 2\mathrm{Re}[\mathbf{E}_s^*(1)\cdot\mathbf{E}_s(2)]\right\}. \tag{4}$$

Under the assumption of small α and paraxial front or back scatter, the only differences in the $\mathbf{E}_s$ fields for the two beams are those due to differences in the incident field amplitudes and phases. The scattered field amplitudes can therefore be written as[9] (also see Appendix A):

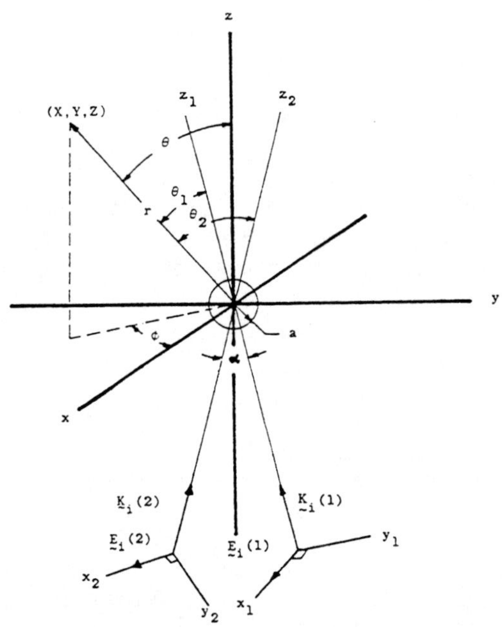

Fig. 2. Scattering geometry for determination of Mie cross-section dependence of $\mathbf{E}_s$ in a two-beam interference fringe system.

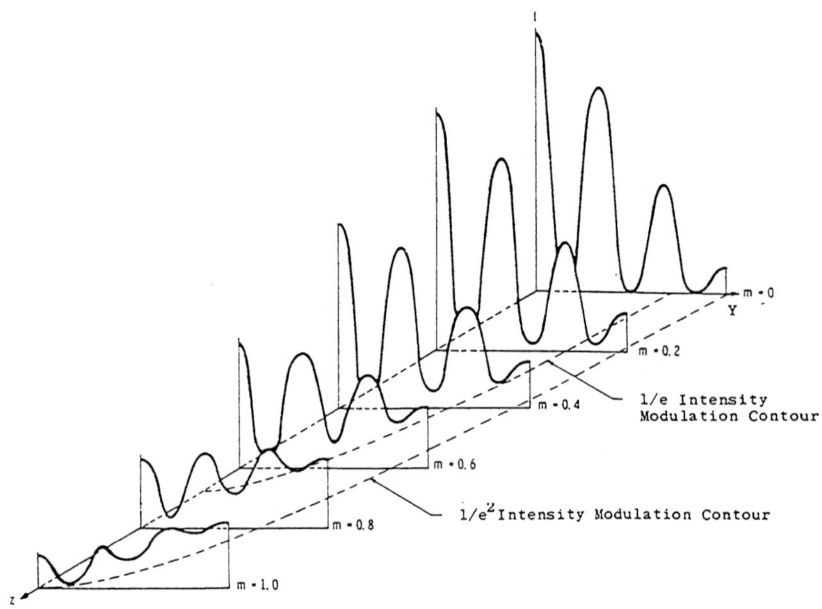

Fig. 3. Intensity distribution in plane of crossed beams (assumes beams of equal intensity and polarization vectors linear and parallel).

$$\mathbf{E}_s(1) = [i \exp(iK_s r)/K_s r]\{\cos(\phi)S_2[\cos(\theta)]\mathbf{e}_\theta + \sin(\phi)S_1[\cos(\theta)]\mathbf{e}_\phi\}E_i(1), \quad (5a)$$

$$\mathbf{E}_s(2) = [i \exp(iK_s r)/K_s r]\{\cos(\phi)S_2[\cos(\theta)]\mathbf{e}_\theta + \sin(\phi)S_1[\cos(\theta)]\mathbf{e}_\phi\}E_i(2), \quad (5b)$$

where θ, ϕ subscripts denote θ, ϕ vector components and where S_1, S_2 are the so-called amplitude functions,[9] which consist of a series of Legendre polynomials. On using Eqs. (5a), (5b) in Eq. (4), the scattered intensity can be written as

$$I_s \simeq c\{\cos^2(\phi)|S_2[\cos(\theta)]|^2 + \sin^2(\phi)|S_1[\cos(\theta)]|^2\}\{|\mathbf{E}_i(1)|^2 + |\mathbf{E}_i(2)|^2 + 2\mathrm{Re}[E_i^*(1)E_i(2)]\}/8\pi K_s^2 r^2, \quad (6)$$

where it is understood that $\theta \simeq 0$ or π. The second term in braces is approximately the intensity distribution in the probe volume. When the center line intensities in each incident beam are identical and the respective polarization vectors parallel, the expression is exact.[10] This requirement is highly desirable experimentally in order to optimize fringe contrast in the probe volume and will be assumed for the rest of this paper. Calling this term I_0, it can be shown[10] that the intensity distribution in the probe volume is

$$I_0 = 2I \exp\{(-2/b_0^2)[X^2 + Y^2 + (Z^2\alpha^2/4)]\} \times [\cosh(2YZ\alpha/b_0^2) + \cos(K\alpha Y - g)], \quad (7)$$

where I is the center-line intensity of one beam at the geometric center of the probe volume and g is a function that accounts for relative phase differences in the beams due to differences in pathlength in arriving at the probe volume. For the rest of the paper, g will be assumed zero (again for optimum fringe contrast in the probe volume). b_0 is given by[11]

$$b_0 = 2\lambda F\#/\pi, \quad (8)$$

where $F\#$ is the f/number of the beam.

Figure 3 illustrates the intensity distribution for the positive quarter of the YZ plane for the previous assumptions. Z has been defined in terms of a dimensionless variable m such that the relative intensity in Z has fallen to $1/e^2$ of the peak value when $m = 1$ [see Eq. (16b)]. The $1/e^2$ and $1/e$ modulation contours are indicated and represent the probe volume limits (the volume from which signals are detected) as defined previously in the literature.[2,4,10,12]

Inspection of Eq. (7) leads to the obvious conclusion that the illumination of the SC is nonuniform unless the SC is much smaller than a fringe spacing. To account for nonuniform particle illumination, it will be assumed that a simple average of I_0 is sufficient to determine the mean illuminating intensity. Defining $(-)$ as the ensemble average, the mean illumination intensity is

$$\bar{I}_0 = \frac{1}{A_p}\int_{A_p} I_0 dA_p, \quad (9)$$

where A_p is the cross-sectional area of the SC. Using Eqs. (8) and (9), the scattered intensity may be written as

$$I_s = \frac{c}{8\pi K_s^2 r^2}\{\cos^2(\phi)|S_2[\cos(\theta)]|^2 + \sin^2(\phi)|S_1[\cos(\theta)]|^2\}\bar{I}_0. \quad (10)$$

Equation (10) is the desired result. It shows that in the forward and back scatter directions with the angle between the incident beams small, the scattering cross section will be identical with that for a single beam. Any calculations of the scattered intensity made off-axis must account for the two-beam scattering cross section as indicated in Appendix A. The equation also shows that account must be taken of the distributed nature of the intensity in the illumination of the scatter center.

Determination of Particle Parameters from Signal Visibility

Since the observed scattered signal is from interference fringes, it is natural to define the ratio of ac to dc in the signal as the signal visibility V. V is fully equivalent to the Michelson visibility function and is defined as

$$V = (I_{s_{max}} - I_{s_{min}})/(I_{s_{max}} + I_{s_{min}}), \tag{11}$$

where $I_{s_{max}}$ is the maximum scattered intensity from a bright fringe and $I_{s_{min}}$ is the corresponding minimum scattered intensity from the next consecutive dark fringe. Using Eq. (10) in Eq. (11), V can be written approximately as (see Appendix B)

$$V = \int_{A_p} \exp\left[\frac{-2}{b_0^2}(X^2+Y^2)\right]\cos\left(\frac{2\pi Y\alpha}{\lambda}\right) dA_p \bigg/ \left\{\cosh\left(\frac{Z\lambda}{2b_0^2}\right) \times \int_{A_p} \exp\left[\frac{-2}{b_0^2}(X^2+Y^2)\right]\cosh\left[\frac{2Z\alpha}{b_0^2}\left(Y \pm \frac{\lambda}{4\alpha}\right)\right] dA_p\right\}. \tag{12}$$

The $\pm\lambda/4$ term appearing in the hyperbolic cosine term in Eq. (12) is due to the spatial phase shift in going from a bright to a dark fringe. The plus sign is chosen if the fringe contrast is decreasing (i.e., Y increasing), while the minus sign is chosen if the contrast is increasing (i.e., Y decreasing). Straightforward manipulation of Eqs. (10) and (11) also shows that

$$V = \langle P_s^1\rangle / \langle P_G^1\rangle, \tag{13}$$

where p_s^1 represents the ac signal power for a single SC as detected by the collection optics, P_G^1 represents the dc signal power for a single SC, and the time average is taken over one cycle of information. The interpretation of V given in Eq. (13) will prove most valuable in indicating how signal-to-noise power ratios should be corrected to account for particle size and shape. It is significant to observe that V as defined in Eq. (12) is independent of the Mie scattering cross section. Thus, as will be shown, particle sizing by this method requires only a single measurement at a fixed angle of the relative scatter intensity between a bright and dark fringe and not a pair of different angular measurements and the absolute scatter intensity at one angle as required by the usual process of determining sizes using light intensity measurements and Mie theory. Furthermore, no *a priori* knowledge of the index of refraction of the particle is required.

To calculate V for a particle, it is convenient to transform the coordinates in Eq. (12) into two sets—one that locates the position of the particle in the probe volume (X_0, Y_0) and one that describes the coordinates of the particle cross section (X', Y'), i.e.,

$$X = X_0 - X', \tag{14a}$$

$$Y = Y_0 - Y', \tag{14b}$$

$$r'^2 = X'^2 + Y'^2. \tag{14c}$$

Substitution of Eqs. (14a), (14b), and (14c) into Eq. (12) gives

$$V = \left(\int_{A_p} \exp\left\{\frac{-2}{b_0^2}[r'^2 - 2(X'X_0 + Y'Y_0)]\right\} \times \cos\left[\frac{2\pi}{\delta}(Y_0 - Y')\right] dA_p\right) \bigg/ \left(\cosh\left(\frac{m\delta}{b_0}\right) \int_{A_p} \exp\left\{\frac{-2}{b_0^2} \times [r'2 - 2(X'X_0 + Y'Y_0)]\right\} \cosh\left[\frac{4m}{b_0}\left(Y_0 - Y' \pm \frac{\delta}{4}\right)\right] dA_p\right), \tag{15}$$

where δ and m are defined as

$$\delta = \lambda/\alpha, \tag{16a}$$

$$Z = 2mb_0/\alpha. \tag{16b}$$

Note that δ is the interference fringe spacing and m was defined previously and is taken to be constant for the calculation. V as written in Eq. (15) will be called the generalized one-dimensional visibility function. The term *one-dimensional* follows from the fact that the two-beam intensity distribution can measure only one-dimensional velocities.

V will be evaluated in the general case for a spherical particle and in special cases for a spherically and a cylindrically shaped particle to illustrate some of the limitations in sizing particles with this technique.

To obtain a closed-form solution in V for spherical particles, it will be assumed that $2a \ll b_0$ so that the r'^2 exponential dependence can be neglected in the integration (numerical integration shows this is an accurate approximation for sizes up to $a \sim \delta$). A transformation is first made to cylindrical coordinates:

$$Y_0 = r_0\cos(\beta), \tag{17a}$$

$$X_0 = r_0\sin(\beta), \tag{17b}$$

$$Y' = r'\cos(\psi), \tag{17c}$$

$$X' = r'\sin(\psi). \tag{17d}$$

Then V is approximately given by

$$V \simeq \left(\int_0^{2\pi}\int_0^a \exp\left[\frac{4r_0r'}{b_0^2}\cos(\psi-\beta)\right]\cos\left\{\frac{2\pi}{\delta}[r_0\cos(\beta) - r'\cos(\psi)]\right\} r'dr'd\psi\right) \bigg/ \left(\cosh\left(\frac{\delta m}{b_0}\right)\int_0^{2\pi}\int_0^a \exp\left[\frac{4r_0r'}{b_0^2} \times \cos(\psi-\beta)\right]\cosh\left\{\frac{4m}{b_0}[r_0\cos(\beta) - r'\cos(\psi)]\right\} r'dr'd\psi\right). \tag{18}$$

Using integral formulas found in Refs. 13 and 14, we can write V for a spherical particle as

$$V \simeq [\exp(i2\pi Y_0/\delta)I_1(\zeta a)/\zeta + \exp(-i2\pi Y_0/\delta)I_1(\zeta^* a)/\zeta^*] \bigg/ \cosh\left(\frac{\delta m}{b_0}\right)\left\{\exp\left[\frac{4m}{b_0}\left(Y_0 \pm \frac{\delta}{4}\right)\right] I_1(\varphi_+ a)/\varphi_+ + \exp\left[\frac{-4m}{b_0}\left(Y_0 \pm \frac{\delta}{4}\right)\right] I_1(\varphi_- a)/\varphi_-\right\}, \tag{19}$$

where

$$\zeta = [(16r_0^2/b_0^4) + (i16\pi Y_0/b_0^2\delta) - (4\pi/\delta^2)]^{\frac{1}{2}} \tag{20}$$

and

$$\varphi_\pm = [(16r_0^2/b_0^4) \mp (32mY_0/b_0^3) + (16m^2/b_0^2)]^{\frac{1}{2}}. \tag{21}$$

I_1 is a first-order hyperbolic Bessel function of the first kind. V in Eq. (19) predicts the visibility variations for a SC whose size is much less than a beam waist radius located at any position in the fringe pattern. When used with Eq. (13) it can be used to predict all signal shape characteristics as a function of particle size and passage trajectory through the probe volume. This can be accomplished by writing the total signal power $\langle P^1 \rangle$ as

$$\langle P^1 \rangle = \langle P_s^1 \rangle + \langle P_G^1 \rangle, \tag{22}$$

or, using Eq. (13),

$$\langle P^1 \rangle = \langle P_G^1 \rangle (1 + V). \tag{23}$$

As written in Eq. (19), it is difficult to observe the size dependence of V, since the equation has taken into account the variation in contrast in the illuminating intensity distribution. There is no variation in fringe contrast at $r_0 = 0$, $m = 0$ (i.e., at the geometric center of the probe volume). Under these constraints, V reduces to

$$V \simeq 2J_1(2\pi a/\delta)/(2\pi a/\delta), \tag{24}$$

where J_1 is a first-order Bessel function. It follows that V has zeros at 1.22π, 2.233π, 3.238π. . . . This means, for example, that when the diameter of the particle is 1.22δ, no Doppler or ac signal can be observed with this type of velocity measuring system nor can the particle size be determined.

The process of obtaining the visibility for a particle as expressed in Eqs. (19) and (24) is fully equivalent to centering a particle first on a bright fringe, then on the next consecutive dark fringe, calculating the respective amounts of scattered intensity and using these forms in Eq. (11). Thus, it can be seen that V can change sign when $\bar{I}_{s_{min}} > \bar{I}_{s_{max}}$, i.e., when the SC scatters more light when centered over a dark fringe than when centered over a bright fringe. This effect would be observed experimentally when a *normally* observed peak in the signal is shifted from a maximum position to a minimum position as sketched in Fig. 4 for two differently sized SC. While such an effect appears to be of no significance for determining the velocity of a single SC, it does play a significant role in determining, unambiguously, the size of the particle. This is most easily demonstrated through Fig. 5, which plots V [Eq. (24)] as a function of $2\pi a/\delta$ for a spherical particle and a long, narrow ($X' \ll 1$) cylindrical particle with major axis moving perpendicular to the fringe planes. V is the type of data that could be extracted from the signal without knowledge of the phase dependence of V. For the spherical SC, it is clear that the size of the SC cannot be determined unambiguously unless $V \gtrsim 0.15$, while for the cylinder, $V \gtrsim 0.23$. At least two possibilities can be suggested for determining the sign of V:

(1) Determine the phase shift in the signal through a velocity-size calibration.

(2) Determine the approximate size range of the SC through measurement of the magnitude of the scattered intensity.

It remains to determine experimentally the most feasible method to use in practice for ascertaining the sign of V.

The visibility function for the cylindrically shaped particle shown in Fig. 5 can be written as [same assumptions as for Eq. (24)]

$$V \simeq \sin(2\pi a/\delta)/(2\pi a/\delta), \tag{25}$$

where a is the half-width of the major axis of the SC, and it is assumed that the major axis of the particle is normal to the fringe planes. From Eq. (25), it is immediately seen that V changes significantly with the shape of the SC.

Determination of Number Density from the Visibility

Number density of particles in the probe volume can be determined in the following manner. It will be assumed that all SC have the same velocity although it has not been established that this is a necessary condition for the analysis. In analogy to Eq. (13), the n particle visibility function $\bar{V}_n$ will be defined as

$$\bar{V}_n = \langle \bar{P}_s^n \rangle / \langle \bar{P}_G^n \rangle, \tag{26}$$

where $\bar{P}_s^n$, $\bar{P}_G^n$ are the respective ac signal and dc signal powers for n particles. It has been shown[15] that

$$\langle \bar{P}_s^n \rangle = \sqrt{n} \langle \bar{P}_s^1 \rangle, \tag{27a}$$

$$\langle \bar{P}_G^n \rangle = n \langle \bar{P}_G^1 \rangle. \tag{27b}$$

Then using Eqs. (27a), (27b), and (13), we can write Eq. (26) as

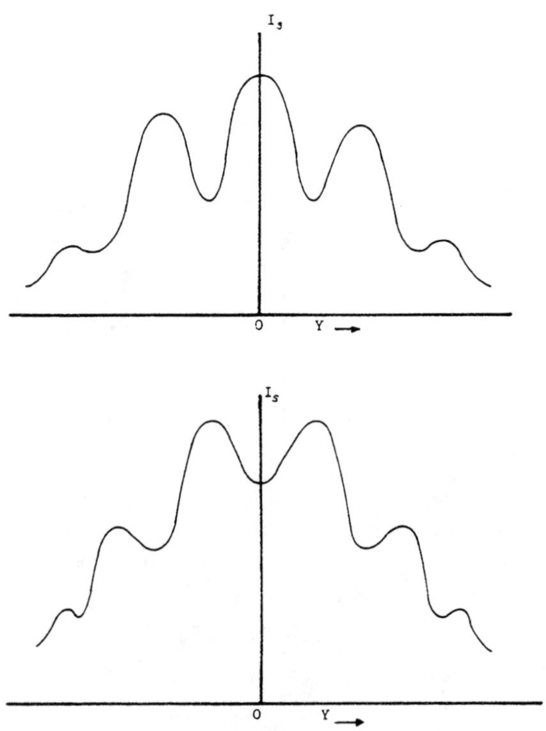

Fig. 4. Phase dependence of the signal on size of a spherical particle.

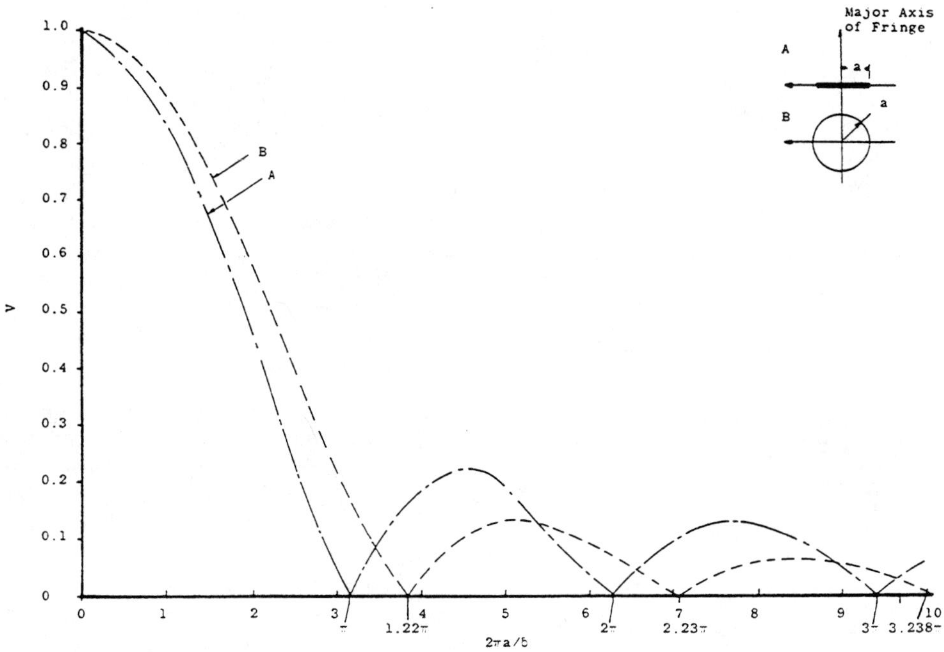

Fig. 5. Fringe visibility vs particle size for a sphere and cylinder.

$$\bar{V}_n = \bar{V}/\sqrt{n}, \qquad (28)$$

where $\bar{V}$ is the ensemble average one-particle visibility function for the size distribution of the n particles. The signal visibility decreases as n increases. From Eq. (28), it is clear that n can be determined experimentally by adjusting δ until $\bar{V}$ is constant (i.e., approximately 1) for the size distribution so that any visibility measurement can reflect only the number of SC contributing to a signal. It is suggested that the size distribution can be determined by observing the variation of $\bar{V}_n$ with δ once n is determined.

Correction to S/N Power Ratios for Finite Size SC

The effects of finite size SC on the signal-to-noise, (S/N) power ratio have been neglected in previous calculations involving the velocity determination of the SC.[4,12,15] This section indicates how the signal visibility enters into the S/N determination.

For systems using photomultiplier (PM) tube detection devices, the shot noise limited S/N at the output of the PM tube can be written as

$$\mathrm{S/N} = (q\lambda/hc\Delta f)\{1/[1 + (\langle P_{BG}\rangle/\langle \bar{P}_{G^n}\rangle)]\}[(\langle \bar{P}_{s^1}\rangle)^2/\langle \bar{P}_{G^1}\rangle], \qquad (29)$$

where q = quantum efficiency of the PM tube,
h = Planck's constant,
Δf = shot noise limiting electronic bandwidth, and
P_{BG} = stray background power.

On using Eqs. (27) and (28) in Eq. (29), the S/N becomes

$$\mathrm{S/N} = (q\lambda/hc\Delta f)\{1/[1 + (\langle P_{BG}\rangle/\langle \bar{P}_{G^n}\rangle)]\}\langle \bar{P}_{G^1}\rangle \bar{V}^2, \qquad (30)$$

which is identical to the result in previous calculations[15] when $\bar{V}$ is 1 (i.e., $a \ll \delta$). Equation (30) points out the explicit need to design laser velocimeter systems carefully such that particle sizes are not detrimental to the S/N.

Limitations and Sources of Error

Several limitations on this particle analysis technique are apparent. Examination of Eq. (19) shows that precise knowledge of the particle's position in the probe volume is required if the probe volume fringe contrast variations (see Fig. 3) are observable in the signal. Since identical signal shapes can come from different portions of the probe volume,[12] the scattered light collection optics must be focused only in that portion of the probe volume near geometric center or in those regions of the probe volume where the fringe contrast is less than or equal to a predetermined constant. Figure 6 sketches the possible shapes of particle sizing probe volumes as determined by the $1/e^2$ modulation contour[10] and a constant contrast constraint. Another possibility for elimination of this problem is construction of electronic filters that could filter the probe volume fringe contrast variations out of the signal.

Comparison of Eq. (24) with Eq. (25) and examination of Eq. (15) shows that for one-dimensional fringe systems, *a priori* knowledge is required about the particle's shape if its size is to be determined from the signal visibility. This suggests that simultaneous measurements from two or three orthogonal fringe sets could give considerable information about asymmetrically shaped particles and could possibly specify shapes to within a resolution limit determined by component fringe spacing. This, of course, assumes that the SC are not rotating between any two consecutive bright and dark fringes. That this should be a good assumption is clear if one calculates the required rotational rate for a particle in air to rotate one revolution be-

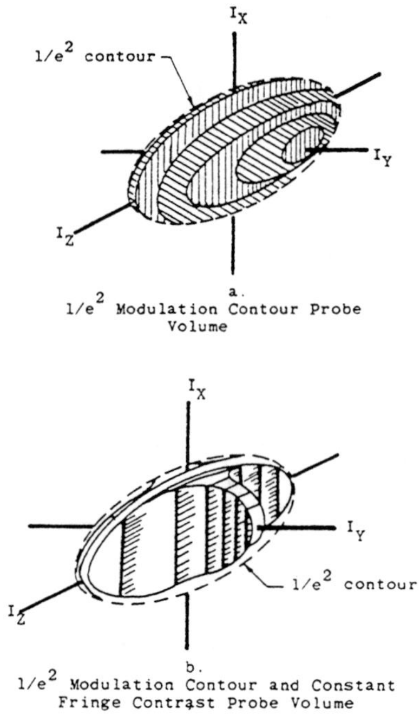

Fig. 6. Shapes of the probe volume for different signal constraints.

tween a bright and dark fringe. For example, a particle with a velocity of 1 m/sec passing through a probe volume with a fringe spacing of 20 μm would have to be rotating at a rate of 10^5 rev/sec in order to revolve once in passing from a bright to a dark fringe. Energy considerations show that this would happen only under highly turbulent or unusual circumstances.

In evaluating particle size from the visibility, it was tacitly assumed that a simple average of the illuminating intensity could be performed over the cross section of the particle. This should be a good assumption for all scatter centers in which the gross scattering process is not different for bright vs dark fringe scattering. The details for the signal visibility in this case have not yet been determined.

The determination of number density from the visibility assumed that all scatter centers on a particular fringe have the same velocity and phase. In a number of practical cases this will not be the case. However, it is assumed that the number density using the visibility function can be determined if the distribution function for the velocities is known. The arguments for this assumption are as follows. The dc level of an observed signal from n SC results from two sources: (1) the inherent dc due to the size of the SC and (2) cancellation of the ac signals due to the random phase relationships between the various particles. Properly adjusting δ removes the first source of unknown dc. When a signal is observed due to n SC with a velocity distribution, the ac signal for each set of n_i particles with velocity increment dv_i is present on the gross dc level due to n. It is believed that $\sqrt{n_i}$ contributes to the respective ac signal, although this is yet to be proved analytically or experimentally. The velocity distribution for the n SC observed is required in order to determine the relative numbers of SC that contribute to each dv_i after each $\bar{V}_{n_i}$ is measured. A summation or integration for the relative n_i and their additive relationships would then be involved in computing the gross number contributing to the signal. A further complication exists when a particle with one velocity can generate a local oscillator type signal for a scatter center with a different velocity. However, such beat signals are usually much weaker than the single SC signals and can virtually bc neglected when $\Omega/\lambda^2 n^{2/3}$ (Ω is the scattered light solid collection angle) is large.[16]

Summary and Conclusions

A technique for sizing dynamic particle fields utilizing a laser interferometer has been examined analytically. The basic parameter chosen to evaluate particle size is the ratio of ac to dc signal or the signal visibility that, for forward and back scatter measurements and small angular illumination, is independent of the Mie scattering cross section and absolute values of the scattered intensity. In this context the method appears to be easier to use than Mie analysis, both from calculational and measurement viewpoints. The requirements for the measurement are an ability to observe real-time signals and a knowledge of the general shape of the SC.

From the analysis, the following conclusions may be drawn:

(1) Scatter cross sections for single-beam illumination should be used in laser velocimeter calculations only for small angle forward and back scatter measurements with small angles between the illuminating beams.

(2) One-dimensional fringe systems may be used to size symmetrical slowly rotating (small rotation angle in passage between consecutive fringes) dynamic particles passing through the probe volume.

(3) For asymmetric slowly rotating particles of unknown shapes, multidimensional fringe systems must be used.

(4) Signal visibility depends on particle size, shape, and number density. Specifically, the following predictions can be made:

(a) When scatter centers pass near the geometric center of the probe volume, no *Doppler* signal will be seen for spherical particle diameters of 1.22δ, 2.23δ, 3.23δ . . . .

(b) Peaks in the ac portion of the signal will undergo relative phase shifts of π when the diameter lies between 1.22δ and 2.23δ or 3.23δ and 4.24δ, etc.

(c) Unless the sign of the phase shifts can be measured, the diameter of the particle cannot be determined unambiguously unless the visibility is $\gtrsim 0.15$ (spherical SC).

(d) For single spherical particles, diameters as small as $\delta/2\pi$ can be determined assuming that the visibility can be measured accurately within $\pm 2\%$. The size uncertainty for a particular value of V depends strongly on both the measurement accuracy for V and the specific value of V, the best resolution occurring in the interval between the first zero of V and $V = 1$.

(e) If the velocity distribution is known, number density can be determined from the visibility.

(5) When n particles (with identical velocities) exist simultaneously in the probe volume, the signal visibility varies as the average one-particle visibility and inversely as the $\sqrt{n}$.

(6) Signal-to-noise power ratios for dual scatter laser velocimeter systems should be corrected to account for signal visibility.

(7) The accuracy of the method depends primarily on how accurately the interferometer system parameters are known. This requires that observations be restricted to particular regions of the probe volume.

Statements (2) and (3) suggest that the requirement of knowledge of particle shape can be considerably relaxed through the use of a three-dimensional fringe system. Multiple-particle coherent scattering effects on the signal were neglected on the basis that the system could be adjusted to eliminate them.[16] The calculation of signal visibility offers a means whereby the shape of the signal for two particles an arbitrary distance apart can be determined analytically.

The author wishes to acknowledge the many fruitful discussions with D. B. Brayton and W. H. Goethert on this subject and the support given by the Experimental Research Group in the Office of the Managing Director, ARO, Inc. The author is especially indebted to A. E. Lennert for his continued support and encouragement while this work was being performed.

The research reported in this paper was sponsored by Arnold Engineering Development Center, Air Force Systems Command, Arnold Air Force Station, Tennessee, under Contract F40600-72-C-0003 with ARO, Inc. Further reproduction is authorized to satisfy the needs of the U.S. Government. This research supported by National Aeronautics and Space Administration, Langley, Virginia, under L-52080.

Note added to proof: It has recently come to the author's attention that related work has been performed by W. P. Altman in a master's thesis, "Incoherent Disdrometer Employing CO_2 Laser Radiation," Univ. of Pa., School of Elec. Eng., Philadelphia, Pa., and by F. Durst in a Ph.D. dissertation, "Development and Application of Optical Anemometers," Heat Transfer Section, Dept. of Mech. Eng., Imperial College, London, S.W.1.

Appendix A

The problem to be resolved here is to calculate the scattered electric field at a point with coordinates r, θ, ϕ from a spherical scatter center simultaneously illuminated by two plane electromagnetic waves that are monochromatic and coherent. The general approach to this problem is beyond the scope of this paper. Thus a number of simplifying assumptions will be made that reduce the mathematical complexity considerably. Chief among these are that (1) the polarization vectors of the illuminating waves are linear and parallel; (2) the scattered light is observed in the far field of the scatter center; and (3) the angle between the illuminating beams is small. With these assumptions, the problem reduces to one primarily of calculating transformations and making accurate approximations.

The coordinate systems for the problem are shown in Fig. 2. The nonsubscripted coordinate system is that with its Z axis coincident with the bisector of the angle between the illuminating waves and with its X axis rotated to coincide with the polarization vectors of the beams. Subscripts for beams 1 and 2 are placed on the coordinates accordingly. In general, the scattered far-field electric fields for each beam can be written as[9]

$$\mathbf{E}_s(1) = [i \exp(iK_s r)/K_s r]\{\cos(\phi_1)S_2[\cos(\theta_1)]\mathbf{e}_{\theta_1} + \sin(\phi_1)S_1[\cos(\theta_1)]\mathbf{e}_{\phi_1}\}E_i(1), \quad (31a)$$

$$\mathbf{E}_s(2) = [i \exp(iK_s r)/K_s r]\{\cos(\phi_2)S_2[\cos(\theta_2)]\mathbf{e}_{\theta_2} + \sin(\phi_2)S_1[\cos(\theta_2)]\mathbf{e}_{\phi_2}\}E_i(2), \quad (31b)$$

where it is assumed that

$$|\mathbf{K}_s(1)| = |\mathbf{K}_s(2)| = K_s. \quad (32)$$

Inspection of Eqs. (31a), (31b) reveals the problem's transformation requirements—expression of the scattered amplitudes in terms of θ and transformation of $\mathbf{e}_{\theta_i}, \mathbf{e}_{\phi_i}$ $(i = 1,2)$ to $\mathbf{e}_\theta$, $\mathbf{e}_\phi$.

As to the expression of $\cos(\theta_1)$ and $\cos(\theta_2)$ in terms of $\cos(\theta)$, it is straightforward to show that

$$\cos(\theta_1) = \cos(\theta)\cos(\alpha/2) - \sin(\theta)\sin(\phi)\sin(\alpha/2), \quad (33a)$$

$$\cos(\theta_2) = \cos(\theta)\cos(\alpha/2) + \sin(\theta)\sin(\phi)\sin(\alpha/2). \quad (33b)$$

It is clear that to first order in $\alpha (\alpha \ll 1)$, $\cos(\theta_1)$ and $\cos(\theta_2)$ are approximately $\cos(\theta)$ and strictly so when θ or ϕ is either zero or π. This will be assumed here. Similar relationships hold for $\cos(\phi_i)$, $\sin(\phi_i)$, and the same types of assumptions will be made in this case also. Thus the remaining difficulty is the transformation of $\mathbf{e}_{\theta_i}, \mathbf{e}_{\phi_i}$ to $\mathbf{e}_\theta$, $\mathbf{e}_\phi$. This can be accomplished in the following manner.

The transformation from Cartesian to spherical coordinates is given by a transformation matrix $\hat{M}$:

$$\hat{e}(r,\theta,\phi) = \hat{M}\,\hat{e}(X,Y,Z), \quad (34)$$

where the $\hat{e}$ are column matrices of the vector components indicated in parentheses. $\hat{M}$ is given by[17]

$$\hat{M} = \begin{bmatrix} \sin(\theta)\cos(\phi) & \sin(\theta)\sin(\phi) & \cos(\theta) \\ \cos(\theta)\cos(\phi) & \cos(\theta)\sin(\phi) & -\sin(\theta) \\ -\sin(\phi) & \cos(\phi) & 0 \end{bmatrix}. \quad (35)$$

$\hat{M}$ is orthonormal. The transformation of X_2, Y_2, Z_2 coordinates to those of the XYZ system $(\gamma = 0)$ is given by a matrix $\hat{T}$:

$$\hat{e}(X_2,Y_2,Z_2) = \hat{T}\hat{e}(X,Y,Z). \quad (36)$$

$\hat{T}$ is given by

$$\hat{T} = \begin{bmatrix} 1 & 0 & 0 \\ 0 & \cos(\alpha/2) & -\sin(\alpha/2) \\ 0 & \sin(\alpha/2) & \cos(\alpha/2) \end{bmatrix}. \quad (37)$$

It may also be shown that

$$\hat{e}(X_1,Y_1,Z_1) = \tilde{\hat{T}}\hat{e}(X,Y,Z), \quad (38)$$

where ($\tilde{\ }$) indicates the transpose of the matrix. Using Eqs. (34), (36), and (38), it is easily shown that

$$\hat{e}(X_1,Y_1,Z_1) = \hat{T}\tilde{\hat{M}}\,\hat{e}(r,\theta,\phi), \tag{39a}$$

$$\hat{e}(X_2,Y_2,Z_2) = \hat{T}\tilde{\hat{M}}\,\hat{e}(r,\theta,\phi). \tag{39b}$$

Let the transformation from the specific Cartesian set to the specific spherical coordinate set be written as

$$\hat{e}(r_1,\theta_1,\phi_1) = \hat{M}_1\hat{e}(X_1,Y_1,Z_1), \tag{40a}$$

$$\hat{e}(r_2,\theta_2,\phi_2) = \hat{M}\hat{e}_2(X_2,Y_2,Z_2), \tag{40b}$$

where $\hat{M}_i$ is identical in form to $\hat{M}$ and the subscript indicates that the angular variables are denoted with appropriate integer identification. From the same arguments that made the scattering amplitude functions identical, the $\hat{M}_i$ can be written as $\hat{M}$. Then using Eqs. (39a), (39b) in Eqs. (40a), (40b), the desired transformations are

$$\hat{e}(r_1,\theta_1,\phi_1) = \hat{B}\,\hat{e}(r,\theta,\phi), \tag{41a}$$

$$\hat{e}(r_2,\theta_2,\phi_2) = \tilde{\hat{B}}e(r,\theta,\phi), \tag{41b}$$

where

$$\hat{B} = \hat{M}\hat{T}\tilde{\hat{M}}. \tag{42}$$

Define the $\hat{E}_s$ matrix as

$$\hat{E}_s = \begin{Bmatrix} \cos(\phi)S_2[\cos(\theta)] \\ \sin(\phi)S_1[\cos(\theta)] \end{Bmatrix}. \tag{43}$$

Then the scattered electric fields may be written as

$$\hat{E}_s(1) = \hat{B}\hat{E}_{s1}[i\exp(i\,K_s r)/K_s r]E_i(1), \tag{44a}$$

$$\hat{E}_s(2) = \tilde{\hat{B}}\hat{E}_{s1}[i\exp(i\,K_s r)/K_s r]E_i(2). \tag{44b}$$

Poynting's theorem can now be used to calculate the scattered intensity.

From Eqs. (3a) and (3b), $\langle\mathbf{S}\rangle$ can be written as

$$\begin{aligned}\langle\mathbf{S}\rangle = \frac{c}{8\pi K_s}\,&\mathrm{Re}\big([|\mathbf{E}_s(1)|^2\mathbf{K}_s(1) + |E_s(2)|^2\mathbf{K}_s(2)] \\ &+ \mathbf{K}_s(1)[\mathbf{E}_s{}^*(1)\cdot\mathbf{E}_s(2)] + \mathbf{K}_s(2)[\mathbf{E}_s{}^*(2)\cdot\mathbf{E}_s(1)] \\ &\times\{\mathbf{E}_s{}^*(2)[\mathbf{K}_s(2)\cdot\mathbf{E}_s(1)] + \mathbf{E}_s{}^*(1)[\mathbf{K}_s(1)\cdot\mathbf{E}_s(2)]\}\big).\end{aligned} \tag{45}$$

The last two terms are approximately zero due to small α and to the fact that respective wave and K vectors are perpendicular. The third and fourth terms in Eq. (45) can be rewritten using Eqs. (41) and (44) as

$$\begin{aligned}&[\mathbf{K}_s(1)/\mathbf{K}_s][\mathbf{E}_s{}^*(2)\cdot\mathbf{E}_s(1)] + [\mathbf{K}_s(2)/K_s[\mathbf{E}_s{}^*(1)\cdot\mathbf{E}_s(2)] \\ &\quad = 2\mathrm{Re}[\hat{E}_s{}^+\hat{B}\hat{B}\hat{E}_sE_i{}^*(1)E_i(2)][\hat{B}\,\hat{e}(r,\theta,\phi)/K_s{}^2r^2],\end{aligned} \tag{46}$$

where the $(\ \ )^+$ notation indicates the transpose of the complex conjugate. The first two terms can be rewritten using Eqs. (41) and (44) as

$$\begin{aligned}&|\mathbf{E}_s(1)|^2[\mathbf{K}_s(1)/K_s] + |\mathbf{E}_s(2)|^2[\mathbf{K}_s(2)/K_s] \\ &\quad = [|\hat{E}_s|^2|E_i(1)|^2\hat{B} + |\hat{E}_s|^2|E_i(2)|^2\tilde{\hat{B}}][\hat{e}(r,\theta,\phi)/K_s{}^2r^2].\end{aligned} \tag{47}$$

Using Eqs. (45)–(47), the Poynting flux may be written as

$$\begin{aligned}\langle\mathbf{S}\rangle = (c/8\pi K_s{}^2r^2)\{&|\hat{E}_s|^2|E_i(1)|^2\hat{B} + |\hat{E}_s|^2|E_i(2)|^2\tilde{\hat{B}} \\ &+ 2\mathrm{Re}[\hat{E}_s{}^+\hat{B}\hat{B}\hat{E}_sE_i{}^*(1)E_i(2)]\hat{B}\}\,\hat{e}(r,\theta,\phi).\end{aligned} \tag{48}$$

It can be shown that $\hat{B}\hat{B}$ is simply $\hat{B}$ with α substituted for $\alpha/2$. Calculation of $\hat{B}$ shows that the off-diagonal elements may be neglected when α is small. By neglecting the off-diagonal elements in $\hat{B}$, $\hat{B} \simeq \tilde{\hat{B}}$, so that using Eq. (43) and (48), $\langle\mathbf{S}\rangle$ is expressed as

$$\begin{aligned}\langle\mathbf{S}\rangle = (c/8\pi K_s{}^2r^2)\big(&|\hat{E}_s|^2[|E_i(1)|^2 + |E_i(2)|^2] \\ &+ 2\{\cos^2(\phi)|S_2[\cos(\theta)]|^2b_{22} \\ &+ \sin^2(\phi)|S_1[\cos(\theta)]|^2b_{33}\}\mathrm{Re}[E_i{}^*(1)E_i(2)]\big)\hat{B}\,\hat{e}(r,\theta,\phi),\end{aligned} \tag{49}$$

where the b_{ij} are the components of $\hat{B}\hat{B}$. The three b_{ij} components of interest can be written as

$$b_{11} = 2\sin^2(\theta)\cos^2(\phi)\cos^2(\alpha/2) + \cos^2(\theta)\cos(\alpha), \tag{50a}$$

$$b_{22} = \cos^2(\theta)[\cos^2(\phi) + \cos(\alpha)\sin^2(\phi)] + \cos(\alpha)\sin^2(\theta), \tag{50b}$$

$$b_{33} = \sin^2(\phi) + \cos(\alpha)\cos^2(\phi). \tag{50c}$$

The b_{22} and b_{33} components are each approximately 1 with a maximum error of order α^2. Thus the magnitude of the Poynting vector is given by

$$I_s = |\langle\mathbf{S}\rangle|,$$

$$I_s = (c|\hat{E}_s|^2/8\pi K_s{}^2r^2)\{|E_i(1)|^2 + |E_i(2)|^2 + 2\mathrm{Re}[E_i{}^*(1)E_i(2)]\}, \tag{51}$$

which is the result presented in Eq. (6).

Appendix B

The method for obtaining Eq. (12) is indicated here. Using Eqs. (10) and (51), Eq. (11) is written as

$$V = (\bar{I}_{0_{\max}} - \bar{I}_{0_{\min}})/(\bar{I}_{0_{\max}} + \bar{I}_{0_{\min}}). \tag{52}$$

Let Y be the coordinate of the center of a bright fringe in I_0; then using Eq. (7), $I_{0_{\max}}$ is written as

$$\begin{aligned}I_{0_{\max}} = 2I\exp(-2m^2)\exp[-2(X^2 + Y^2)/b_0{}^2] \\ \times[\cosh(4Ym/b_0) + \cos(2\pi Y/\delta)].\end{aligned} \tag{53}$$

$I_{0_{\min}}$ is found by shifting $I_{0_{\max}}$ through half an interference fringe period $\delta/2$. Hence

$$\begin{aligned}I_{0_{\min}} = 2I\exp(-2m^2)\exp\{-2[X^2 + (Y \pm \delta/2)]^2/b_0{}^2\} \\ \times\{\cosh[(4m/b_0)(Y \pm \delta/2)] - \cos(2\pi Y/\delta)\}.\end{aligned} \tag{54}$$

It will be assumed that $\delta \ll b_0$, which allows neglect of the exponential variation in Y in Eq. (54). The plus sign is chosen for decreasing probe volume fringe contrast and the minus sign for increasing contrast. Using Eqs. (53) and (54),

$$\begin{aligned}I_{0_{\max}} - I_{0_{\min}} \simeq 2I\exp(-2m^2)\exp[-2(X^2 + Y^2)/b_0{}^2] \\ \times\{\cosh(4mY/b_0) - \cosh[(4m/b_0)(Y \pm \delta/2)] \\ + 2\cos(2\pi Y/\delta)\}.\end{aligned} \tag{55}$$

Similarly,

$$\begin{aligned}I_{0_{\max}} + I_{0_{\min}} \simeq 2I\exp(-2m^2)\exp[-2(X^2 + Y^2)/b_0{}^2] \\ \times\{\cosh(4mY/b_0) + \cosh[(4m/b_0)(Y \pm \delta/2)]\}.\end{aligned} \tag{56}$$

Using standard identities,

$$\begin{aligned}&\cosh(4mY/b_0) + \cosh[(4m/b_0)(Y \pm \delta/2)] \\ &\quad = 2\cosh[(4m/b_0)(Y \pm \delta/4)]\cosh(m\delta/b_0),\end{aligned} \tag{57a}$$

$$\cosh(4mY/b_0) - \cosh[(4m/b_0)(Y \pm \delta/2)]$$
$$= 2\sinh[(4m/b_0)(Y \pm \delta/4)]\sinh(m\delta/b_0)$$
$$\simeq 0 \quad (m\delta \ll b_0). \qquad (57b)$$

Using Eqs. (57a), (57b), and (9) in Eq. (52), the visibility is

$$V \simeq \int_{A_p} \exp\left[\frac{-2(X^2+Y^2)}{b_0^2}\right]\cos\left(\frac{2\pi Y}{\delta}\right) dA_p \Bigg/ \cosh\left(\frac{m\delta}{b_0}\right)\int_{A_p} \exp\left[\frac{-2(X^2+Y^2)}{b_0^2}\right]\cosh\left[\frac{4m}{b_0}\left(Y \pm \frac{\delta}{4}\right)\right] dA_p, \qquad (58)$$

which is identical to Eq. (12) with appropriate changes in the variables [see Eq. (16)].

References

1. C. M. Penny, IEEE J. Quantum Electron. **QE-5,** 318 (1968).
2. D. B. Brayton and W. H. Goethert, Trans. Instrum. Soc. Am. **10,** 41 (1971).
3. L. Lading, Appl. Opt. **10,** 1943 (1972).
4. A. E. Lennert, D. B. Brayton, F. L. Crosswy, *et al.*, AEDC-TR-70-101.
5. F. H. Smith and J. A. Parsons, AEDC-TR-70-119.
6. J. W. Dunning Jr., "Application of Laser Homodyne Spectrometer to Particle Size Measurements," Ph.D. Thesis (School of Engineering, Case Western Reserve University, Cleveland, Ohio, 1967).
7. D. G. Andrews and H. S. Seifert, "Investigation of Particle-Size Determination from the Optical Response of a Laser Doppler Velocimeter," SUDAAR No. 435, November 1971.
8. M. J. Rudd, J. Phys. E **2,** 55 (1969).
9. M. Kerker, *The Scattering of Light* (Academic Press, New York, 1969), p. 46.
10. D. B. Brayton and W. M. Farmer, "An Analysis of the Probe Volume in a Dual Scatter Laser Doppler Velocimeter," to be submitted to Applied Optics.
11. H. Kogelnik, Bell Syst. Tech. J. **44,** 468 (1965).
12. R. S. Adrian and R. S. Goldstein, J. Phys. E **4,** 505 (1971).
13. I. S. Gradshteyn and I. M. Ryzhik, *Tables of Integrals, Series, and Products* (Academic Press, New York, 1965), p. 488.
14. Ref. 13, p. 683.
15. W. M. Farmer and D. B. Brayton, Appl. Opt. **10,** 2319 (1971).
16. C. P. Wang, Appl. Phys. Lett. **18,** 522 (1971).
17. M. Born and E. Wolf, *Principles of Optics* (Pergamon Press, New York, 1965), p. 633.

Reprinted with permission from *Combustion and Flame,* Vol. 23(1), pp. 57-71 (August 1974).

Velocity Measurements in the Recirculation Region of an Industrial Burner Flame by Laser Anemometry with Light Frequency Shifting

R. J. BAKER,* P. HUTCHINSON,† and J. H. WHITELAW

Mechanical Engineering Department, Imperial College of Science and Technology, London, England

Detailed velocity measurements within the highly turbulent recirculating region of an industrial-burner flame, by laser anemometry, are reported. The results extend those previously obtained in the same burner by using a light frequency-shifting technique to resolve the directional ambiguity of the Doppler signal and seeding by solid nonreacting particles to enhance the light scattering.

1. Introduction

Previous values of mean velocity and turbulence intensity, measured in an industrial burner flame by laser anemometry, have been demonstrated by Baker, Bourke, and Whitelaw [1]. These measurements were restricted by the instrumentation employed and inadequate seeding to the regions of the flame where recirculation was absent. The present paper supplements these results by providing detailed measurements of 3 components of the instantaneous velocity in the highly turbulent and recirculating zone of the flame. These results were made possible by use of an electrooptic modulator to shift the frequency of one of the light beams and by seeding with titanium dioxide particles dispersed in a fluidized bed.

Many of the turbulent diffusion flames currently used in industrial combustion processes are designed to possess highly turbulent reverse flow regions which provide the mixing, heat, and low velocities required to sustain the flame at high total-mass flow rates. Swirl, which provides one method by which a recirculating flow may be formed, has been studied extensively by many investigators. Leuckel [2], for example, emphasizes the importance of the turbulence properties as a controlling influence in the combustion mechanism of swirling flames. At the present time, however, little or no turbulence data have been obtained in industrial burner flames. The majority of the measurements have been in laboratory flames with predominant flow directions. The measuring techniques have usually been indirect, such as the static pressure method of Becker and Brown [3] and the fluctuating emission method of Günther and Simon [4]. Only recently has laser anemometry been successfully applied to instantaneous velocity measurements in laminar and turbulent diffusion flames, by Baker, Bourke, and Whitelaw [5], and in premixed flames by Durst, Melling, and Whitelaw [6]. Günther [7], in a review of turbulence measurements in flames, and Schwar and Weinberg [8] in a review of laser techniques in combustion research, emphasize the importance of optical anemometry techniques for detailed localized measurements in flames. It is clear, however, that at the moment there is a scarcity of reliable data for the more complex combustion systems.

*Now with Electro-Watt Engineering Services, 19 Cavendish Square, London, W1, England.

†Theoretical Physics Division, Atomic Energy Research Establishment, Harwell, England.

A certain amount of mean velocity data have been obtained by water-cooled pitot probes in the Ijmuiden furnaces. Single-burner tests were carried out by Fricker and Leuckel [9] and multiple burner tests by Fricker, Van Heyden, and Michelfelder [10]. The geometries of all the burners were similar to the present burner; several examples of the mean velocity profiles, from these data, are presented in Ref. [13].

The purpose of the present research program is to develop and test instrumentation which can be used in combusting situations for diagnostic purposes and to aid the development of design procedures based upon solution of conservation equations [e.g., 11]. The electrooptic modulator and signal-processing instrumentation described here represent significant improvements on previous reported results in laboratory scale experiments of [1]; they will also facilitate the extension of measurements in furnace environments such as those described in [12 and 13].

The specific purposes of the present paper is to report measured values of the mean velocity and turbulence intensity in regions of a burner flame which were previously inaccessible to such measurements, to indicate the important implications of the results, and to evaluate the instrumentation employed.

2. Equipment and Experimental Procedure

The equipment used for the present measurements is shown in Fig. 1. The burner had an exit diameter of 87 mm and was identical to the multihole arrangement used for the measurements in [1] and described in more detail in [13]; it operated at a swirl number of 0.5 with 0.13 l/sec of methane and 8.5 l/sec of air, at approximately 6% of the design flow rate, and was mounted in a 3-dimensional traversing arrangement. The total air supply was seeded with TiO_2 particles dispersed in a fluidized bed. Additional information on seeding is given by Melling [14].

The optical arrangement, Fig. 2, consisted of an argon-ion laser, operating multimode at 488 nm and 200 mW output power, an electrooptic frequency shifter, integrated optical unit, receiving optics, and photomultiplier (EMI 9635Q). Only the frequency shifter and optical unit differ from

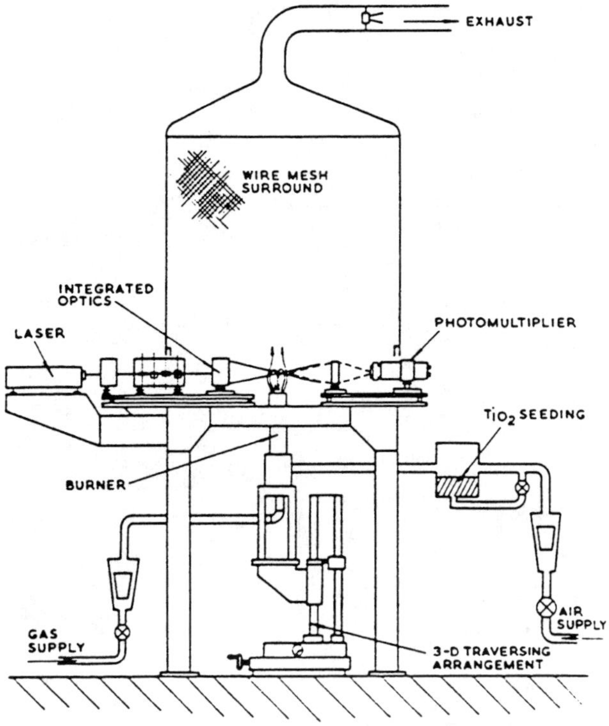

Fig. 1. Experimental arrangement.

the system described in [1] and are discussed in more detail below.

Several techniques for frequency-shifting laser beams are available; the method considered most applicable for combustion situations utilized the electrooptic effect. Figure 2 shows the arrangement of polarizing elements that must be used in conjunction with the electrooptic modulator.

Circularly polarized light was obtained by passing the beam through a polarizer and quarter wave plate ($\lambda/4$). The electrooptic modulator consisted of a 4-plate Kerr Cell operating at 3.25 MHz; the principle of operation has been described by Drain and Moss [15]. At fairly low modulation, the Kerr Cell shifts the frequency of the input circularly polarized light (f) by the applied frequency (f_s), to give $f + f_s$, with almost complete suppression of unwanted sidebands, i.e., $f - f_s, f + f_s$, etc; the shifted frequency is reversed in polarization. By use of a second $\lambda/4$ plate and the Wollaston Prism mounted in the integrated optical unit, it is thus possible to split the shifted and unshifted beams. With the Kerr Cell operating at a conversion efficiency of about 50%, equal intensity beams are produced by the above arrangement. The final $\lambda/2$ plate ensures that the polari-

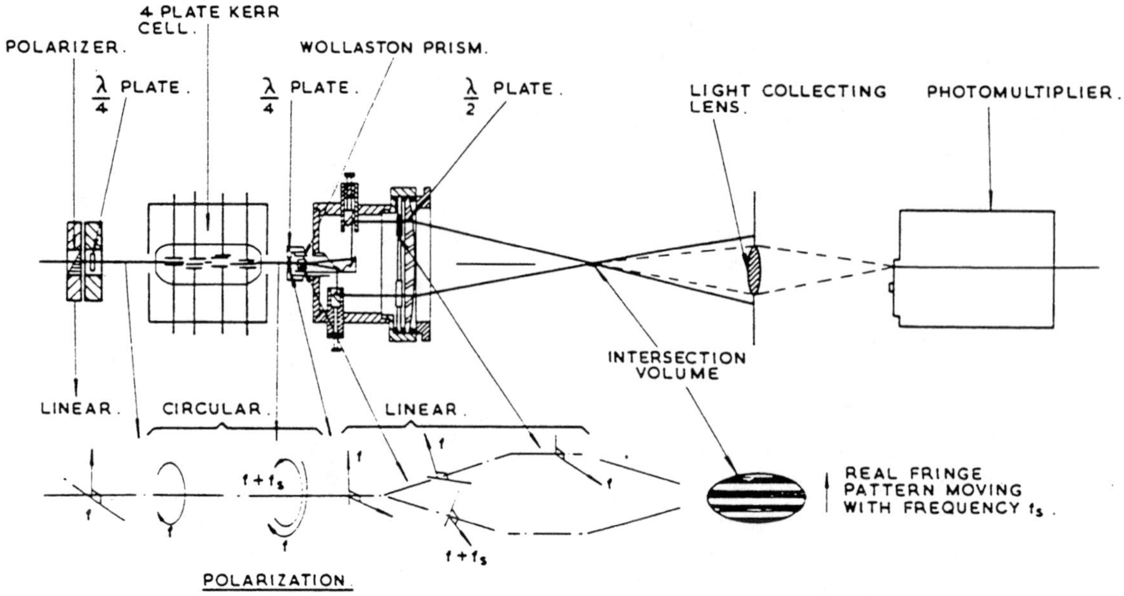

Fig. 2. Optical arrangement.

zation direction of the two intersecting beams is the same and a real fringe pattern is formed; in this case, however, the fringe pattern is moving in space with a velocity of $\lambda f_s/2 \sin \psi$. The complete system has the advantage that the integrated optical unit may be rotated to measure any component of velocity, with or without frequency shift, while maintaining alignment. The real fringe system is also advantageous in low particle concentration situations [see 16].

The two beams were focused by a 300-mm lens in the integrated optical unit to approximately 0.17-mm diameter. The magnification of the receiving optics was such that the measuring control volume diameter (d_c) was reduced to about 0.12-mm by a 0.25-mm aperture at the photomultiplier.

The signal from the photomultiplier was processed with the instrumentation shown in Fig. 3. The wide-band amplifier removed the d.c. component from the signal before it was analyzed by the Computing Counter. The counter measures the frequency of the burst within a preset measurement time: initiation of the measurement was caused by the delay trigger mechanism of the oscilloscope; the actual time for an integer number of positive zero crossings was then used to determine the frequency. The operating cycle for the counter is shown in Fig. 3; the minimum repetition rate was about 3 msec, e.g., for a continuous 1 kHz signal, 1000 samples, and no delay trigger, a processing time of 4 sec may be achieved. The counter operated for a preset number of bursts and computed mean and rms frequencies. With frequency shifting these are related to the derived turbulence quantities as follows:

$$f = f_s + \frac{2u \sin \psi}{\lambda}, \tag{1}$$

$$\bar{u} = \frac{\lambda}{2 \sin \psi}(\bar{f} - f_s), \tag{2}$$

$$\frac{\sqrt{\overline{u'^2}}}{\bar{u}} = \frac{\sqrt{\overline{f'^2}}}{(\bar{f} - f_s)} \tag{3}$$

where f is the measured frequency, $\bar{u}$ and $\bar{f}$ the mean velocity and frequency, $\sqrt{\overline{u'^2}}$ and $\sqrt{\overline{f'^2}}$ the rms components of the instantaneous velocity and frequency, respectively, λ the laser wavelength, and ψ the half angle of the beam.

Each measurement with the counter was based upon 1000 samples; the preset measuring time varied from 2–5 μsec and the delay time was

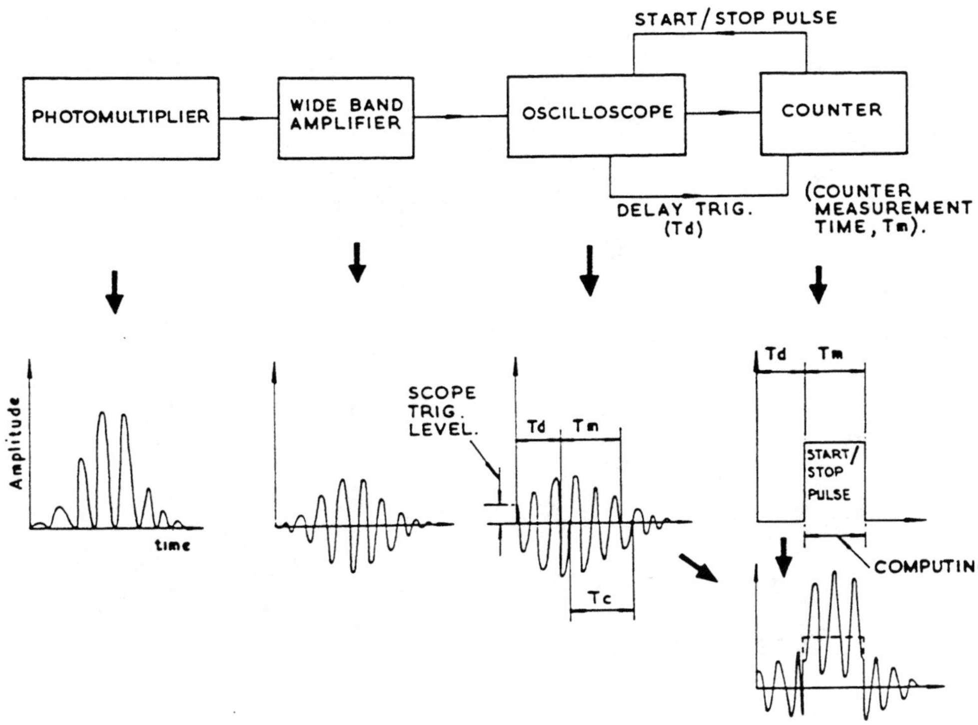

Fig. 3. Signal-processing instrumentation.

10–20 μsec (i.e., measuring about 10 cycles in the center of a burst).

3. Analysis

The application of frequency shifting to highly turbulent flows introduces several effects, not encountered without frequency shifting. The number of cycles in a Doppler burst becomes dependent upon the magnitude of the frequency shift in addition to particle velocity; improved high-pass filtering of the signal may also be achieved. The signal-processing time which is unaffected by frequency shifting is, however, dependent upon the particle intermittency and intensity of the turbulent flow.

3.1 Signal-processing Time

With a knowledge of the particle concentration, and hence intermittency, η [see 13, Sec. 2.6, for more details] it is possible to calculate the signal-processing time, SPT, for a given number of Doppler bursts. The SPT depends upon the rate of passage of particles through the control volume and hence the particle transit times. For simplicity, therefore the following assumptions are made:

(a) The velocity probability distribution is normal.
(b) The control volume has a Gaussian light-intensity distribution (of diameter d_c at the $1/e^2$ pts).
(c) Particles are uniformly distributed and monodispersed.

The SPT may thus be calculated as follows:

$$\mathrm{SPT} = \sum_{i=1}^{N} \left| \frac{d_c}{u_i} \right| (1 - \eta), \tag{4}$$

where the velocity probability distribution $P(u)$ is given by

$$P(u) = \frac{1}{\sigma\sqrt{2\pi}} e^{\frac{-(u - \bar{u})^2}{2\sigma^2}}, \tag{5}$$

where

$$\sigma = \sqrt{\overline{u'^2}}, \tag{5}$$

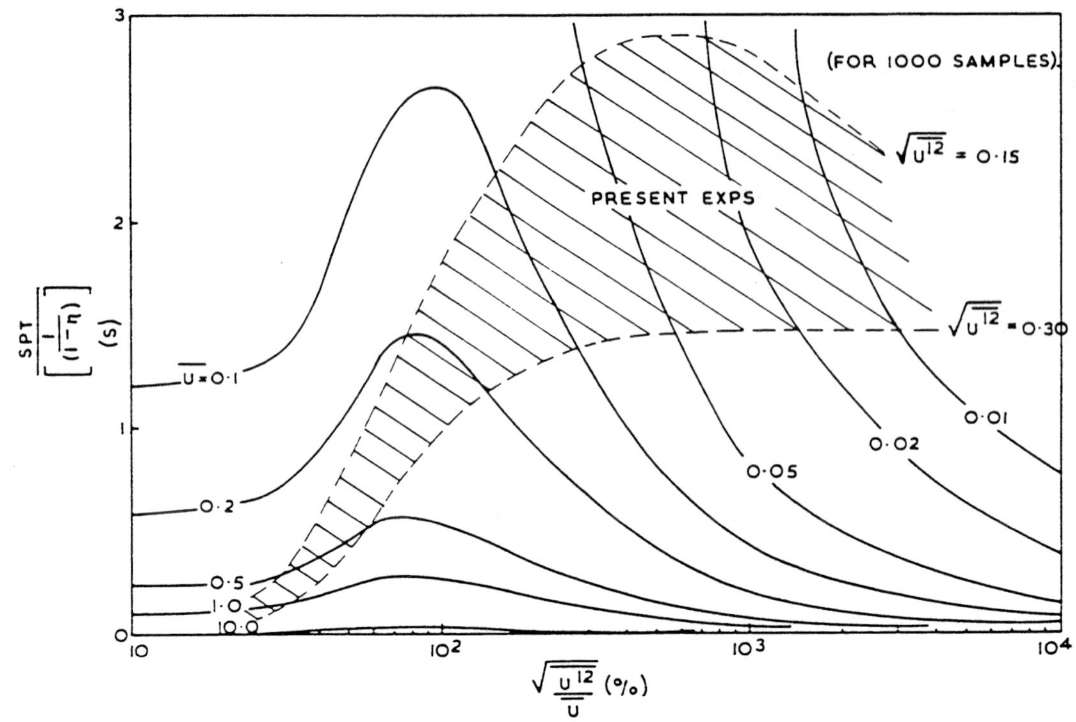

Fig. 4. Signal-processing time as a function of mean velocity and turbulence intensity.

hence

$$\mathrm{SPT} = \frac{N}{(1-\eta)} \int_{-\infty}^{\infty} P(u) \left| \frac{d_c}{u} \right| du, \tag{6}$$

provided that

$$\left| \frac{d_c}{u} \right| \leqslant T_R, \tag{7}$$

where T_R is the relaxation time of the turbulent flow, i.e., the maximum time any particle will spend in the measuring control volume. Numerical integration of Eq. (6) gave the results shown in Fig. 4. As to be expected, the signal-processing time increases with decreasing mean velocity but indicates a maximum at about 100% turbulence intensity. The results for the present case are indicated by the dotted lines which represent rms values from 0.15 to 0.30 m/sec.

3.2 Influence of Shifting on the Number of Cycles

With frequency shifting, the number of cycles in any burst is related to the velocity of the particle by

$$n = fT_b = (f_s + f_d)\, T_b, \tag{8}$$

where T_b is the duration time of the burst $= d_c/|u|$ and f_d the doppler frequency $= 2\, u \sin \psi/\lambda$ so that

$$n = \frac{(f_s + f_d)}{|f_d|} \frac{d_c\, 2 \sin \psi}{\lambda}, \tag{9}$$

where the number of cycles without frequency shifting, $n_{f_s=0}$ is given by the constant $d_c\, 2 \sin \psi/\lambda$.

Typical Doppler bursts, with and without shifting are shown in Fig. 5. It may be seen that as $f \to f_s$ (i.e., $f_d \to 0$), the number of cycles in the burst increases; the maximum, however, is governed by the relaxation time for change in the local flow velocity, so that

$$T_b|_{\max} \leqslant T_R. \tag{10}$$

The criteria for choice of the magnitude of the frequency shift must, therefore, be based upon the maximum expected negative Doppler frequency, $f_{d_{\max,\,\mathrm{neg}}}$ such that

$$n_{\min} > n_{f_s=0}; \tag{11}$$

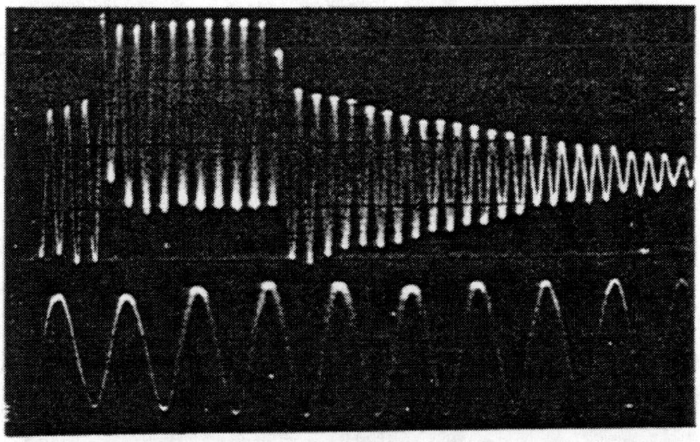

(a) WITHOUT SHIFTING

X = 20 mm
Y = 45 mm
T_s = 20 μs (scan time)
(top trace)

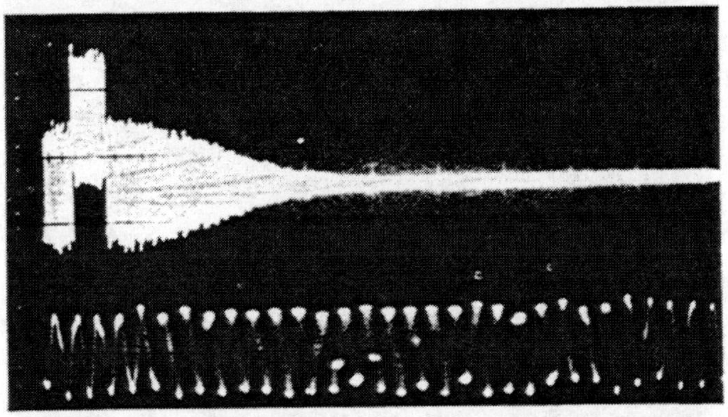

WITH SHIFTING fs = 3.25 MH_z

(b)

X = 20 mm
Y = 0
T_s = 200 μs
fd = −.202 MHz

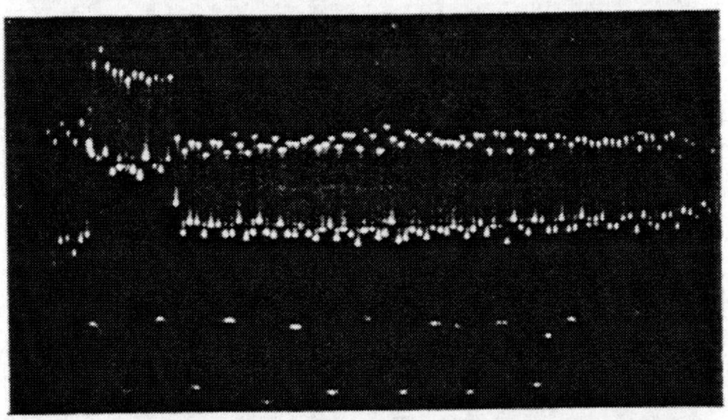

(c)

X = 20 mm
Y = 45 mm
T_s = 20 μs
fd = 1.51 MHz

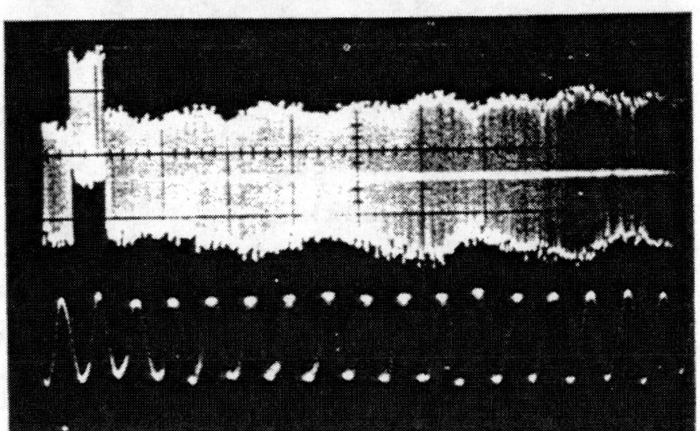

(d)

X = 85 mm
Y = 0
T_s = 100 μs
fd = 0.032 MHz

Fig. 5. Typical Doppler bursts with frequency shifting.

application of Eq. (9) then gives

$$f_s > 2 f_{d\,\mathrm{max,\,neg}}. \tag{12}$$

3.3 Influence of Shifting on High-pass Filtering

In turbulence measurements using Counting techniques, it is necessary to remove the low-frequency component of the Doppler signal by high-pass filtering. A flow with only 30% turbulence level has velocity fluctuations which typically vary by a factor of 20. At higher turbulence levels, the frequency variations are much greater, and it is impossible to separate the low frequency component of a high-velocity burst from the high-frequency component of a low-velocity burst. In addition, the filter characteristics combined with the constant preset triggering level of the Counter cause suppression of the tail of the probability distribution. With suitable frequency shifting, however, these errors are almost totally removed, as biasing the frequency measurements by a constant amount (f_s) reduces the relative range of the frequency variations.

This is demonstrated in Fig. 6(a) where a typical high-pass filter characteristic (F.C.) is plotted together with a velocity probability distribution of 30% turbulence intensity. The position of the high-pass filter (i.e., the frequency, f_{hp}, at the 3*db* point of the characteristic) is usually set so that the D.C. (low frequency) component of the doppler signal is completely removed. The maximum D.C. component, f_{dc}, may be estimated from the maximum expected doppler frequency and the number of cycles in the burst, as follows:

$$f_{dc}|_{\max} \approx f_d|_{\max} / n_{f_s=0}. \tag{13}$$

A typical wide-band filter characteristic is given by

$$(\mathrm{F.C.}) = \frac{1}{[1 + (f_{hp}/f)^2]^{1/2}}, \tag{14}$$

where the filter time constant $T_f \equiv 1/2\pi f_{hp}$, so that for (F.C.) $\ll 1$, f may be written as

$$f \approx f_{hp} \cdot (\mathrm{F.C.}) \tag{15}$$

and if the D.C. component of the *maximum* doppler frequency, $f_d|_{\max}$, is attenuated by at least 95%, then the value of the high pass filter, f_{hp}, may be estimated, as follows:

$$f_{hp} \approx \frac{20}{n_{f_s=0}} f_d|_{\max}. \tag{16}$$

The effect of a constant triggering level upon the velocity distribution shown in Fig. 6(a) may now be calculated if it is assumed that the amplitude probability, *Pa*, of the doppler signals, at any frequency, is Gaussian, as shown in Fig. 6(b). Assumptions (a) and (b) from Sec. 3.1 are still retained, and in addition the average value of the light-intensity distribution (equivalent to a uniform intensity distribution of diameter d_c) corresponds to a particle size and triggering level of unity, i.e., a particle of size unity passing through part of the beam of intensity unity will just be detected by a triggering level of unity. The equivalent triggering level with high-pass filtering will be given by the reciprocal of the filter characteristic, Eq. (14), and is shown by the dotted line in Fig. 6(a). The probability distribution is thus given by

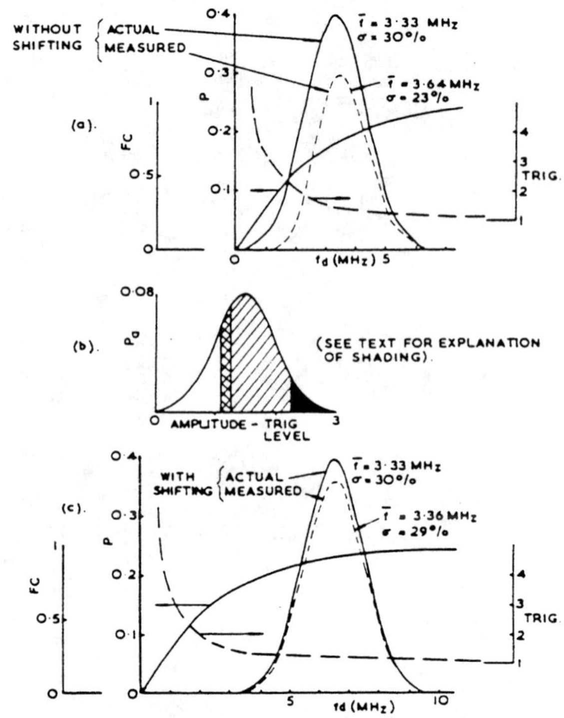

Fig. 6. The effect of high-pass filtering and trigger level.

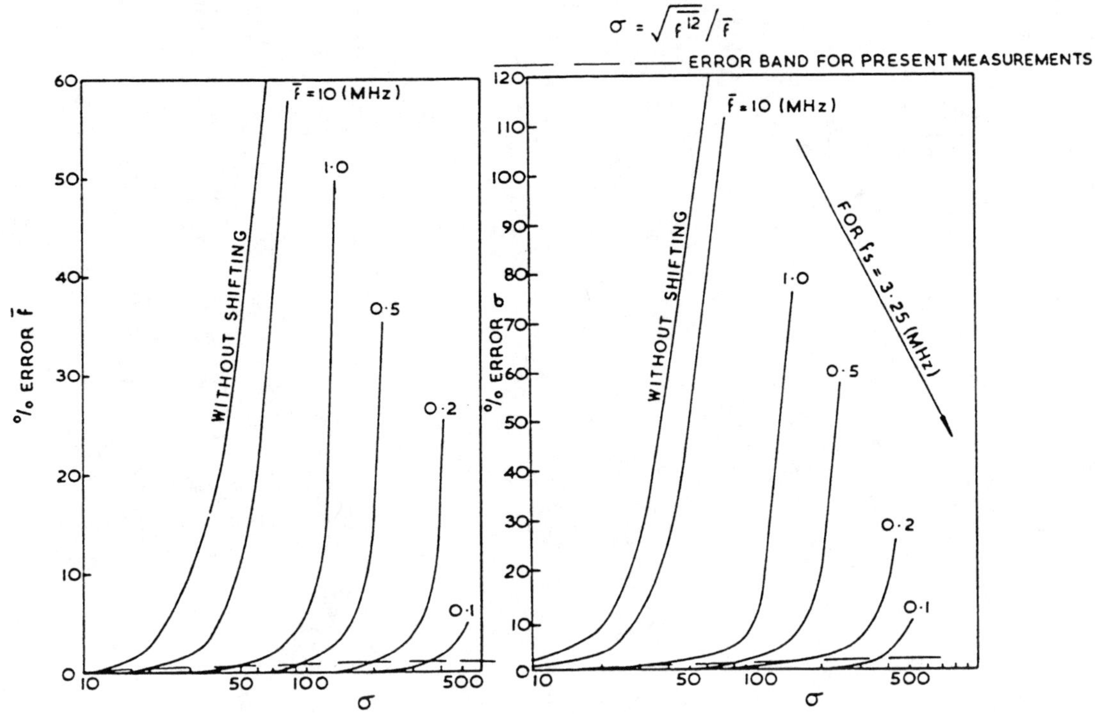

Fig. 7. Error in $\bar{f}$ and σ due to high-pass filtering.

$$P|_{\text{measured}} = P|_{\text{actual}} \frac{1 - \int_0^{\text{trig}} P_a da}{1 - \int_0^1 P_a da} . \quad (17)$$

It is assumed that below an amplitude of unity, corresponding to the product of light intensity and particle size, no triggering occurs, thus giving an approximate variation of 5 to 1 in the amplitudes of triggered bursts. It is possible to see that to fulfill the filtering requirement of Eq. (16) the triggering amplitude at f_d = 1.5 MHz will suppress many signal bursts which at f_d = 5.5 MHz would be processed [indicated by the shaded and hatched areas in Fig. 6(b), respectively] ; hence erroneously skewed velocity probability distributions result. The same calculation may be performed with frequency shifting, so that

$$f_{hp} \approx \frac{20}{n}(f_s + f_d|_{\max}), \quad (18)$$

which by substitution in Eq. 5 may be shown to be identical to Eq. 16, i.e., the same filter time constant may be used with and without shifting. This is indicated in Fig. 6(c). If f_s is put equal to $\bar{f}$ (to represent the present data), it may be seen from Fig. 6(b) that the effect of triggering level at f_s = 4.5 MHz and f_d = 8.5 MHz is now greatly reduced [indicated by the cross hatch area in Fig. 6(b)]. More generally, the errors in σ and $\bar{f}$ due to triggering and high-pass filtering are shown in Fig. 7 as functions of the turbulence intensity and shifted frequency.

The preset triggering level has two further effects; it biases the results towards the larger particles and those passing through the center of the control volume, and thus increases the spatial resolution. Based upon the previous assumptions, calculations have been performed to quantify these effects; the results are shown in Fig. 8.

3.4 Triggering Considerations for the Computing Counter

A problem which may occur with Computing Counting techniques, when used to measure high-turbulence levels, is associated with omitting the final zero crossing at the end of the measurement time, T_m, and recording an erroneously low fre-

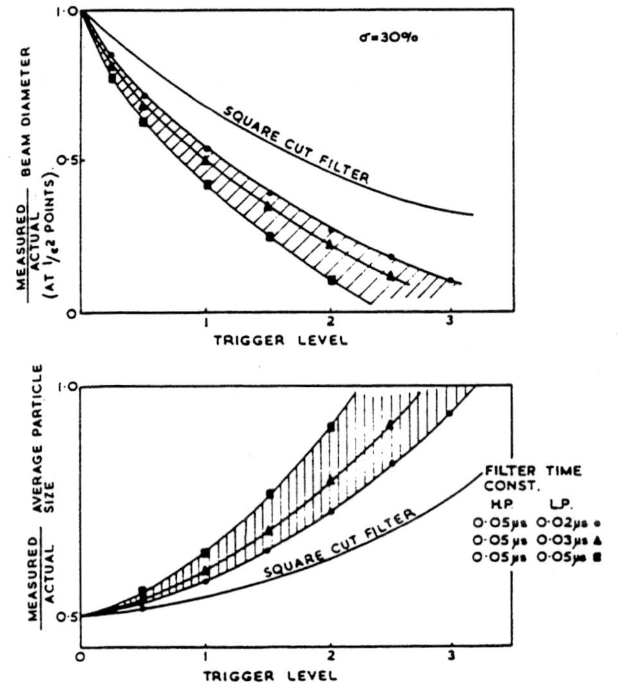

Fig. 8. Effect of trigger level on detected particle size and effective beam diameter.

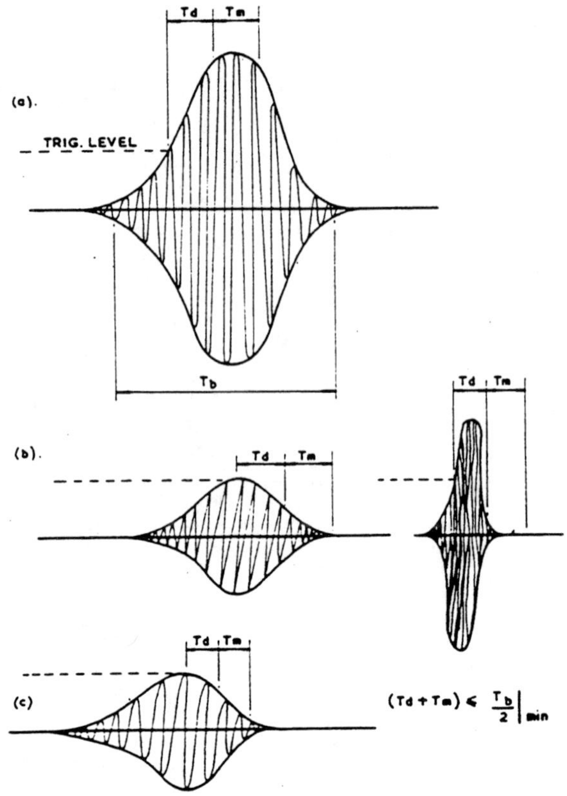

Fig. 9. Counter triggering condition.

quency. This is shown diagrammatically in Fig. 9. Figure 9(a) represents the ideal case with the measurement time in the middle of the burst. Figure 9(b) shows two cases where erroneous measurements occur due to too long delay, T_d, and measurement times, T_m. Figure 9(c) shows the worst case that may be expected, i.e., triggered in the middle of the burst; provided $(T_d - T_m)$ is always less than the minimum expected burst time, $T_b/2$ min, the above effect will be suppressed.

There may also be an error in the frequency measurements, δ_f, obtained on either side of the center of the Doppler burst, due to the shape of the envelope and the finite discrimination level, a_d, inherent in the computing counter. This is given by

$$\frac{\delta_f}{f_d} = \frac{a_d}{a \tan \dfrac{n_1 \pi}{n}}, \tag{19}$$

where a is the maximum amplitude of the doppler burst, and n_1 the measurement position in the burst. Equation (19) is shown in Fig. 10 together with experimental results obtained with the computing counter on a simulated Doppler burst with 100 cycles. It may be observed that the error passes through zero at the center of the burst.

4. Results

The measured profiles of mean axial velocity and the corresponding turbulence intensity, for three axial positions, are shown in Fig. 11; profiles of mean radial and circumferential velocity and the corresponding turbulence intensities are shown in Fig. 12.

Where comparisons are possible, i.e., outside the recirculation zone, the present measurements are in close agreement with those of [1]; such discrepancies as exist stem from the difficulties in interpreting the nonshifted velocity probability distributions recorded in [1]. The advantage of the frequency shifter is demonstrated in Fig. 13 where a probability distribution obtained in the stabilization region of the flame ($Y/R = 0.78$, $X = 2$ cms) is shown. The local turbulence intensity at this point is 112% (the hatched area indicates the negative velocity portion of the probability distribution). Also shown is the same

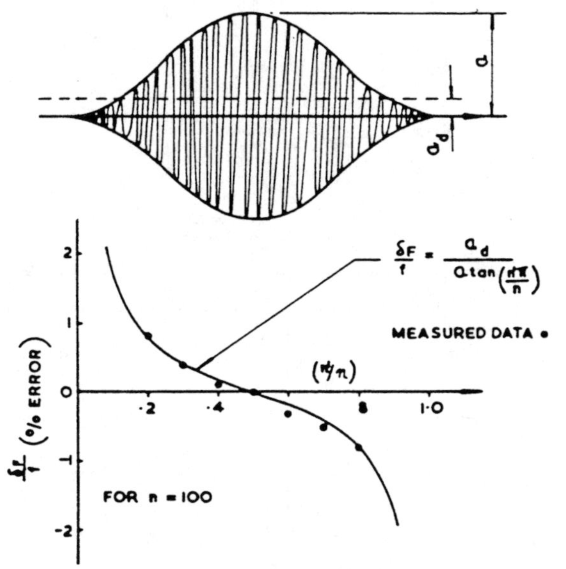

Fig. 10. Measurement error due to Doppler burst envelope for single cycles.

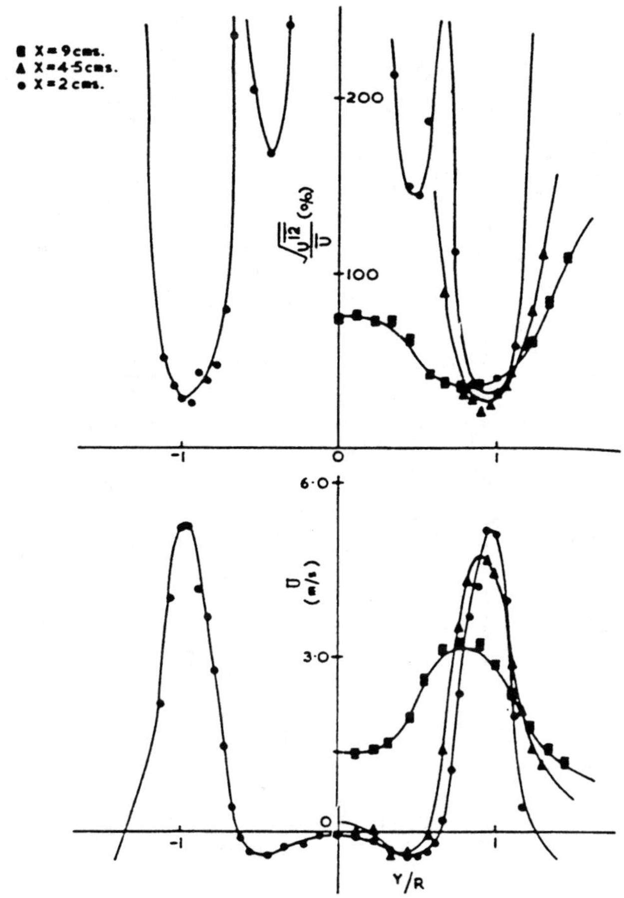

Fig. 11. Mean and fluctuating axial velocity.

probability distribution as though measured without frequency shifting (and with the wide-band filter at 1 MHz center frequency. The resultant spectrum is now positively skewed, with a 101% higher mean velocity and 331% lower rms turbulence level.

The results shown in Fig. 11 indicate both the symmetry of the burner flame and that the maximum axial velocity occurs around the edge of the quarl; this maximum moves towards the center line with increasing distance from the burner exit. The zero velocity locations mark the edge of the recirculation zone and indicate an annular reverse-flow region within the burner quarl which has disappeared at 9 cm downstream.

The reverse-flow velocities are small in magnitude but give rise to singularities in the turbulence intensities when normalized with the local mean velocities. The absolute values of the rms fluctuations, however, are remarkably constant across the flow, indicating only a variation from 15% to 35% when normalized with the bulk velocity. These are shown in Fig. 14 for the three axial radial and tangential fluctuating velocity components. It is worth noting, however, that even at a distance of 90 mm from the burner exit the local turbulence intensities are very high, e.g., a value of 75% on the center line which is well outside the capabilities of alternate instrumentation such as the hot-wire anemometer. The radial and circumferential measurements shown on Fig. 12 also reveal high turbulence intensities but, as expected the circumferential component does not reverse.

5. Discussion

The present measurements were prompted by two problems which were encountered during an initial study of the same burner arrangement [1]. The problems were a result of inadequate intensity of scattered light in the combusting region of the flame and high turbulence levels in regions of recirculation, which produced overlapping velocity spectra.

The present results demonstrate that these problems may be adequately overcome by seeding the burner airflow with solid, nonreacting TiO_2 particles and optically frequency shifting one of the beams from the integrated optical unit to resolve the directional ambiguity of the Doppler signal.

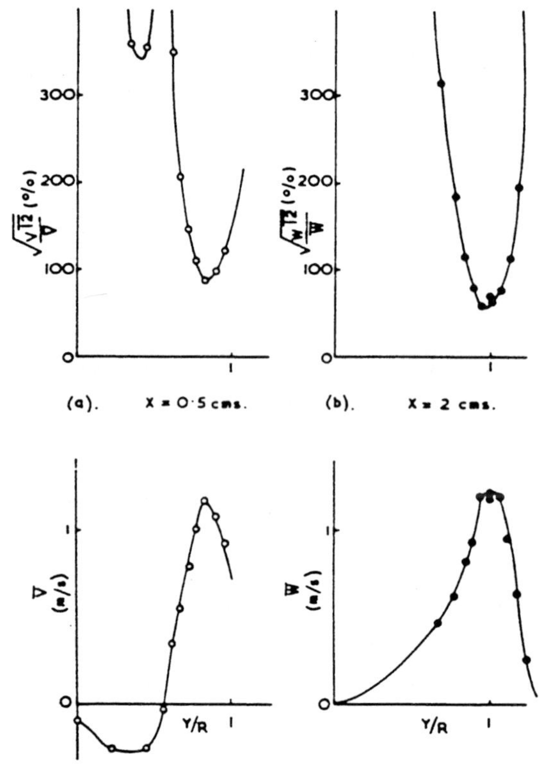

Fig. 12. (a) Radial and (b) tangential (swirl) mean and fluctuating velocity components.

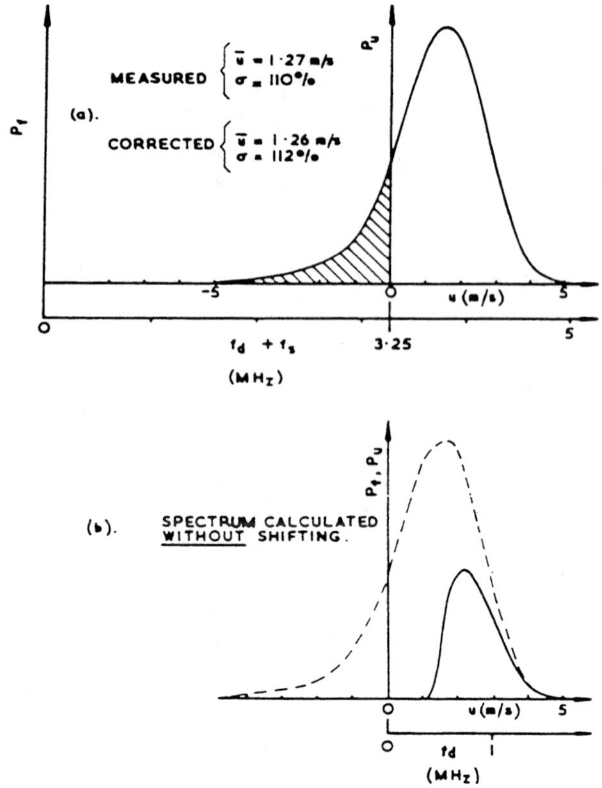

Fig. 13. Probability distributions (a) with and (b) without frequency shifting.

In addition, a computing counter was employed to process the signals, thus making data accumulation less time consuming and more accurate than by Spectrum analysis.

Little other data exists with which to compare the present measurements: mean velocities are in agreement with those previously obtained by Spectrum Analysis [1], and indicate the same trends as data from the Ijmuiden Furnace [9, 10]. Fluctuating velocity components, outside the recirculation zone, also compare well with those indicated in [1], but no other data exists with which to compare the high turbulence intensities measured in the center of the burner.[1] The present experimental arrangement employed frequency shifting and computing counting techniques to obtain measurements at points in the flow where the instantaneous velocity was both positive and negative. The large velocity variations together with the instrumentation employed, introduced effects which it has been necessary to quantify in order to estimate the accuracy of the measurements and to optimize the experimental arrangement. Parameters such as signal processing time, filter characteristic, triggering level, and errors due to a finite number of samples, have been analyzed and are discussed in more detail below.

With frequency-measuring techniques by zero crossing detection, such as the computing counter, the removal of the low-frequency component of the Doppler signal is important, and hence any effect the filter characteristic has upon the frequency measurement should be determined. The number of cycles in a Doppler burst has been shown to increase to a large but finite number (due to the turbulence relaxation time) as the absolute velocity decreases, and then to rapidly fall to zero at $-f_s$. An interesting outcome of this effect is the value of the high-pass filter time constant, T_f, which becomes independent of the

---

[1] At $Y/R = 1$ and $X = 2$ cms, a mean velocity of 5.23 m/sec and turbulence level of 22% was obtained without shifting; within 5% of the same measurements with shifting.

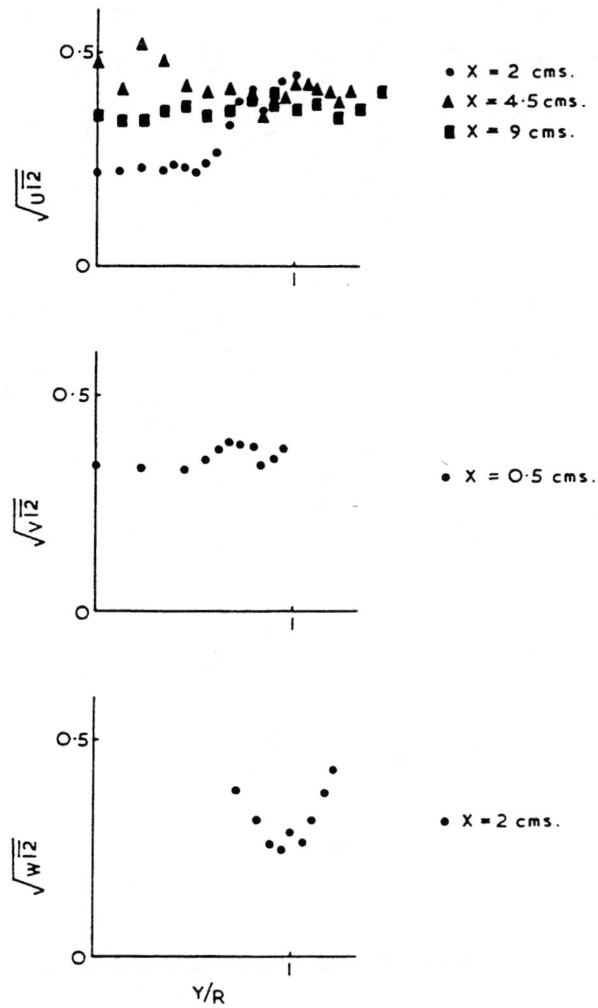

Fig. 14. rms fluctuating axial, tangential, and radial velocities.

shifted frequency, and may be calculated from an unshifted optical arrangement. Figure 5 shows several photographs of typical Doppler bursts, with and without frequency shifting. Comparison of Fig. 5(a) and (b) shows the considerably greater number of cycles in the frequency-shifted burst. The top trace shows the part of the burst measured by the computing counter, and in each case is shown on a stretched time base in the lower trace. Figure 5(d) shows a burst at almost zero velocity (0.1 m/sec); almost 400 cycles are present in the scan time shown (which is only one tenth of the total transit time of the particle).

It is apparent from Fig. 5(a) that the envelope of the Doppler burst without shifting is considerably smoother than those with shifting [Figs. 5(b), (c), and (d)]. The amplitude modulation of the Doppler envelope with shifting is attributed to incomplete suppression of the carrier frequency of the shifted beam, causing beating between the carrier and shifted frequency (f_s), as shown in Fig. 5(c), and beating between the carrier and Doppler (f_d), as shown in Fig. 5(d). The $\lambda/4$ plates were inferior in quality and produced eliptic rather than circularly polarized light, thus preventing complete separation of the shifted and unshifted light beams by the beam splitter (Wollaston Prism). The lower traces in each photograph, indicate that the amplitude modulation has no significant effect upon the accuracy of the Computing Counter; spectrum analysis, however, would result in additional spectra at these frequencies and hence greater difficulty in determining the mean and rms velocities accurately.

Signal-processing time was computed to give some indication of the variations to be expected from the widely varying mean and rms velocities. It had been observed during the experimental measurements that a considerably longer processing time was required in the highly turbulent recirculating regions of the flame than at the outer edge. Figure 4 shows that the processing time at different positions through the flame varied by a factor of 50. If a typical particle intermittency of 99% is assumed then processing times of the order of 5–6 min result; this is compatible with the observed measurements. A knowledge of the signal-processing time can become particularly important (to ensure good statistics) when estimating the integrating/averaging time required for signal processing by spectrum analyzer or filter bank.

It should be noted that the analyses given in Sec. 3 for n and SPT considered only one velocity component, i.e., 1-dimensional; in general a 3-dimensional analysis results in a smaller value of the particle transit time, T_b, due to the additional velocity components, and consequently lower values of n and SPT. However, the assumption of a monodispersed particle distribution gives a lower signal processing time than would occur in practice; due to the triggering level only the large particles would tend to be processed.

The effect of high-pass filtering and trigger level on velocity probability distributions has been demonstrated in Fig. 6 and applied to a measured distribution in Fig. 13. This error analysis was extended in Fig. 7 over a mean frequency range of 10 MHz to 0.1 MHz, turbulence levels from 10% to 500%, with and without shifting. It may be seen that the error in the mean and rms frequencies, without shifting, are independent of mean frequency; the error increases rapidly from a turbulence level of about 20%. At a turbulence level of 30% the mean velocity is 9% high and the rms 28% low. With 3.25 MHz shifting the errors decrease with decreasing mean velocity; the error band for the present measurements is indicated by the dotted line, and in no case exceeds 1% for the mean, and 2% for the rms frequencies.

Due to the finite number of velocity samples at any location, statistical errors result in the mean and rms quantities; these errors have been calculated in the Appendix. If the measurements are all assumed uncorrelated and of normal distribution it may be shown that for N samples the error in the mean is given by $(100/\bar{u})\sqrt{\overline{u'^2}/N}\,\%$ and the rms by $100\sqrt{2/N-1}\,\%$. For 1000 samples the rms fluctuating component is obtained to 7% precision and the mean 1% at the maximum velocity.

6. Conclusions

(a) Measured values of three components of mean and fluctuating velocity have been obtained, by laser anemometry, in an industrial burner flame.

(b) An electrooptic frequency shifter and TiO_2 seeding have allowed these measurements to be extended to the highly turbulent recirculation zone of the flame.

(c) Quantitative analyses of the various errors involved in the present measurements are provided.

(d) The measurements provide physical information suited to the testing of prediction procedures which incorporate turbulence models and assist burner designs.

Appendix 1. Error Analysis

Due to the finite number of samples, N, (i.e., instantaneous velocity measurements) processed by the Computing Counting technique, statistical errors will arise in the determination of the mean and rms fluctuating velocity components. If all the measurements are assumed uncorrelated and of normal distribution the variance in the determination of $\bar{u}$ and $\sqrt{\overline{u'^2}}$ may be calculated as follows:

The measured mean

$$\bar{u} = \frac{1}{N}\sum_{i=1}^{N} u_i, \quad \langle\bar{u}\rangle = \langle u_i\rangle.$$

The variance of $\bar{u}$

$$\sigma_{\bar{u}}^2 = \langle\bar{u}^2 - \langle\bar{u}\rangle^2\rangle^2,$$

$$= \frac{\sigma^2}{N}$$

$\therefore$ Error in $\bar{u}$

$$= \frac{\sigma_{\bar{u}}}{\bar{u}} = \frac{1}{\sqrt{N}}\frac{\sigma}{\bar{u}},$$

$$= \frac{1}{\sqrt{N}}\frac{\sqrt{\overline{u'^2}}}{\bar{u}}.$$

The measured variance

$$\overline{u'^2} = \frac{1}{N}\sum_{i=1}^{N}(u_i - \bar{u})^2, \quad \langle\overline{u'^2}\rangle = \frac{N-1}{N}\sigma^2.$$

The variance of $\overline{u'^2}$

$$\sigma_{\overline{u'^2}}^2 = \langle(\overline{u'^2} - \langle\overline{u'^2}\rangle)^2\rangle,$$

$$= \langle\overline{u'^2}^2\rangle - \langle\overline{u'^2}\rangle^2,$$

and

$$\langle\overline{u'^2}^2\rangle = \frac{1}{N^2}\left\langle\left(\sum(u_i - \bar{u})^2\right)^2\right\rangle,$$

$$= \frac{1}{N^2}\left\langle\sum_i\sum_j(u_i - \bar{u})^2\right\rangle,$$

$$= \frac{1}{N^2}\left(\sum_i(u_i - \bar{u})^4\right.$$

$$\left. + \sum_{i\neq j}(u_i - \bar{u})^2(u_j - \bar{u})^2\right),$$

$$= \frac{1}{N} \langle (u_i - \bar{u})^4 \rangle$$

$$+ \frac{(N-1)}{N} \langle (u_i - \bar{u})^2 \rangle^2.$$

Hence

$$\sigma^2_{\overline{u'^2}} = \frac{2(N-1)}{N^2} \sigma^4.$$

∴ Error in $\overline{u'^2}$

$$\frac{\sigma_{\overline{u'^2}}}{\sqrt{\overline{u'^2}}} = \sqrt{\frac{2}{N-1}}.$$

Nomenclature

a	amplitude of Doppler burst
d_c	diameter of measuring control volume at $1/e^2$ pts
f	measured frequency
f_d	Doppler frequency (due to velocity only $2u \sin \psi/\lambda$)
f_{hp}	high pass filter frequency at $3db$ point
f_s	optical frequency shift
F.C.	filter characteristic
n	number of cycles in a Doppler burst
$n_{f_s=0}$	number of cycles without shifting (i.e. number of fringes in a conventional unshifted optical arrangement $2d_c \sin \psi/\lambda$)
N	number of measured Doppler bursts
P_u	velocity probability of u
P_a	amplitude probability of a
SPT	signal-processing time
T_b	time duration of Doppler burst
T_f	filter time constant ($\equiv 1/2\pi f_{hp}$)
u	axial velocity component
$\bar{u}$	mean axial velocity
u'	fluctuating axial velocity component
v	radial velocity component
w	tangential (swirl) velocity component
λ	wavelength of laser beam
ψ	half angle of beam intersection
η	signal intermittency
σ	standard deviation of probability distribution (i.e., $\equiv \sqrt{\overline{u'^2}}$)

The authors are extremely grateful to the following persons without whose help the present measurements would not have been possible: Dr. T. M. Lowes of the International Flame Research Foundation for the burner; Dr. L. E. Drain of Materials Physics Division, Harwell, for the electro-optic modulator; Mr. A. Hare of British Titan Products for supplying the TiO_2. *Thanks are also due to colleagues at Harwell and Imperial College who gave helpful advice on many occasions and particularly to Mssrs. B. Moss, A. Glastonbury, and R. Pass who gave valuable assistance during the experimental work.*

References

1. **Baker, J. R., Bourke, P. J., and Whitelaw, J. H.,** 1973 Symposium on Combustion.
2. **Leuckel, W.,** The effects of swirl on ignition and combustion behaviour of turbulent diffusion flames, Int. Flame Res. Foundation, Doc. nr K20/a/44, June 1969.
3. **Becker, M., and Brown, A.,** Velocity fluctuations in turbulent jets and flames, 1969, p. 1059.
4. **Günther, R., and Simon, M.,** Turbulence intensity, spectral density functions, and eulerian scales of emission in turbulent diffusion flames, Twelfth Symposium on Combustion, The Combustion Institute, Pittsburgh, Pa., 1969, p. 1069.
5. **Baker, R., Bourke, P., and Whitelaw, J.,** Measurements of instantaneous velocity in laminar and turbulent diffusion flames using an optical anemometer. *J. Inst. Fuel*, December 1973.
6. **Durst, F., Melling, A., and Whitelaw, J.,** The Application of optical anemometry to measurements in combustion systems. *Combust. Flame*, in press.
7. **Günther, R.,** Measurements of flame turbulence: aims, methods and results, *J. Inst. Fuel* **43**, 187 (June 1970).
8. **Schwar, M., and Weinberg, F.,** Laser techniques in combustion research, *Combust. Flame* **13**, 335 (August 1969).
9. **Fricker, N., and Leuckel, W.,** Draft report on trials NG-1, Int. Flame Res. Foundation, Doc. nr. F35/a/4, June 1970.
10. **Fricker, N., Van Heyden, L., and Michelfelder, S.,** Investigations into the combustion of natural gas in multiple burner systems–A report on the MJ-1 trials series, Vols. I and II, Int. Flame Res. Foundation, Doc. Nr. F35/a/5, June 1971.
11. **Patankar, S. V., and Spalding, D. B.,** *Int. J. Heat Mass Trans.*, 1972.
12. **Baker, R. J., Hutchinson, P., and Whitelaw, J. H.,** 1973, A.S.M.E. Paper No. 73-HT-34. *J. Heat Trans.*, in press.

13. **Baker, R. J.**, AERE Harwell Report R. 7137, 1972.
14. **Melling, A.**, Scattering particles for laser anemometry in air: selection of criteria and their realisation, Imperial College Mech. Eng. Rept. ET/TN/B/7.
15. **Drain, L. E., and Moss, B.**, *Opto-Electronics 4* (1972).
16. **Drain, L.**, Coherent and non-coherent methods in Doppler optical beat velocity measurements, *J. Phys. D.* 5, 481 (1972).

Received February 1, 1974; revised April 6, 1974

Reprinted from *Applied Optics*, Vol. 18(10), pp. 1539-1545 (May 15, 1977).

Laser Doppler anemometry measurements in an engine

J. B. Cole and M. D. Swords

A simple and effective technique is reported which allows measurement of the flow field in the cylinder of a motored internal combustion engine with good temporal and spatial resolution. The experiment, a real-fringe laser Doppler anemometer, uses photon counting digital autocorrelation for data capture. The experimental autocorrelation functions have been processed using curve-fitting to a Gaussian velocity profile and by direct Fourier transformation without prior assumption of the form. The ability of such a technique to tolerate poor SNRs makes it a likely choice for measurement in firing engines.

Introduction

The flow patterns of air and fuel in an internal combustion engine play an important role in determining many of the primary operating characteristics such as economy, emission, and lean burning operation. A lack of detailed knowledge of the flow has meant that design changes are based largely on empiricism and intuition rather than on well-established scientific principles. The data we report are for a motored single-cylinder research engine, i.e., one where all the functions of a four-stroke spark-ignition engine operate, but the fuel is absent. However, the techniques are directly applicable to fired engines and similar combustion devices. (Since this paper was submitted we have obtained similar measurements with the engine firing.)

There have been several studies on motored engines using hot-wire anemometry[1–5]; however, this technique is extremely sensitive to vibration and contamination from fuel and lubricant. What is more, it becomes decreasingly reliable at high relative turbulence intensities. A further restriction of hot-wire measurements is that they cannot be extended to the fired engine.

Laser Doppler anemometry does not suffer from any of these restrictions although it does need reasonable optical access to the engine cylinder. We report here the application of the technique to engine measurements and discuss some of the problems associated with processing and validating our data.

The authors are with Shell Research, Ltd., Thornton Research Centre, P.O. Box 1, Chester CH1 3SH, England.

Received 26 October 1978.
0003-6935/79/101539-07$00.50/0.

Engine Modifications

The engine used in the work is a Ricardo E6 variable-compression single-cylinder research engine connected to a dynamometer, which allows both motored and fired experiments to be performed. A full description of the engine and the associated instrumentation has been given elsewhere by Allwood *et al.*,[6] and so only the modifications made to allow optical access will be described here.

The principal modification was the fitting of a spacer plate which was inserted between the cylinder head and block of the engine and into which optically flat quartz windows mounted in holders were fitted (Fig. 1). The object of our experiment was to study the flow in the region of the spark plug, as it is the early stages of combustion that seem to have the most marked effect on over-all engine performance. Consequently, the spacer geometry was such as to give maximum access in this region. The holes and slots through the spacer are connections between head and block for studs and coolant and have the effect of severely restricting the window area available. The broken lines on Fig. 1 indicate the access volume available for windows. The thickness of the spacer is dictated by the necessary compromise between incorporating the widest possible windows and keeping the attendant reduction in compression ratio to a minimum. The variable compression facility of the engine is helpful here, and the spacer eventually fitted resulted in a compression ratio of 7:1.

The windows were given an antireflection coating of magnesium fluoride and were tapered so that compression pressure in the engine pushed them into the tapered recess. Scattered light inside the engine cylinder was minimized by painting all internal nonsliding surfaces matt black.

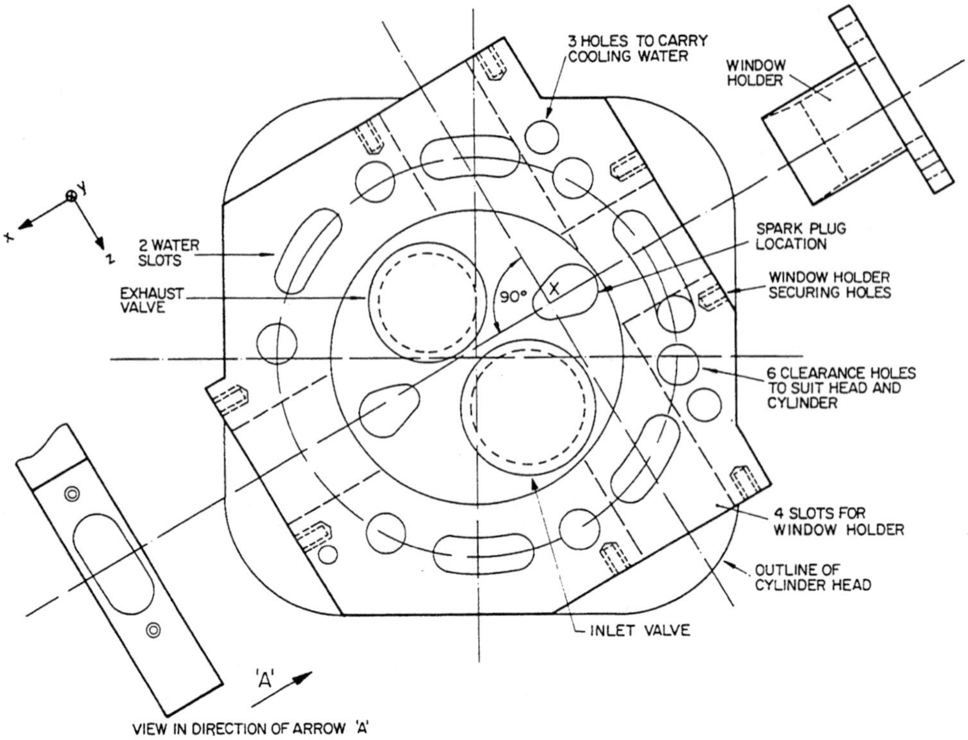

Fig. 1. Optical slice: plan view.

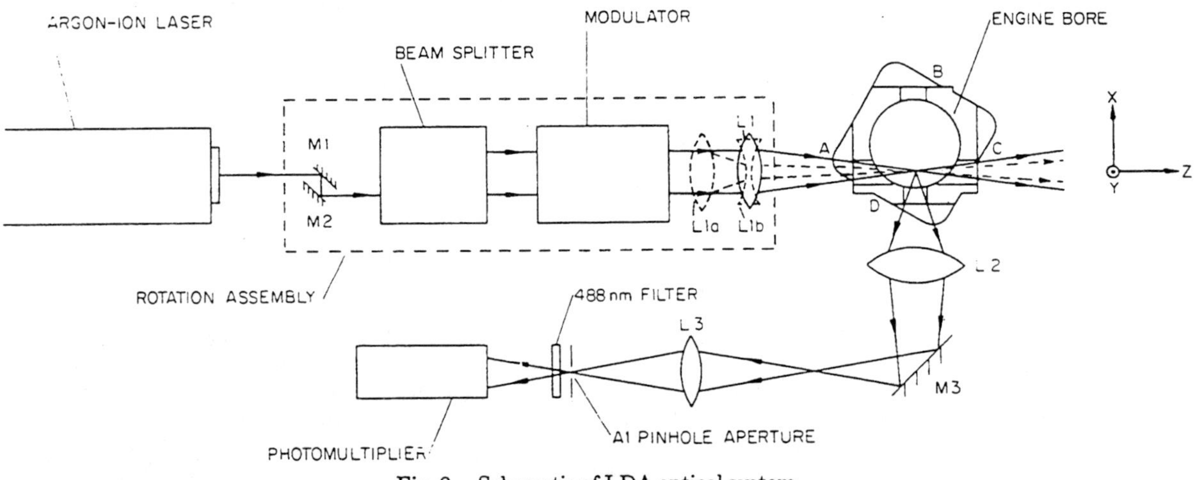

Fig. 2. Schematic of LDA optical system.

Laser Doppler Anemometry (LDA) System

The LDA system used photon counting and is based on the Malvern Instruments K7023 correlator. The outstanding advantage of photon correlation,[7,8] over frequency tracking and burst counting systems, is that it provides useful data over a very wide range of seeding conditions, in respect of both particle concentration and size distribution, so that we have been able to obtain data over the whole four-stroke cycle of the motored engine without artificial seeding. The photon count rate observed is typically 10^6 sec^{-1}. It increases to typically 3×10^6 sec^{-1} during the expansion stroke, apparently owing to lubricating oil and water condensation associated with adiabatic cooling.

The optical system is shown diagrammatically in Fig. 2. The two input beams pass through one of the engine windows to form a real fringe system in their elongated volume of overlap. The light scattered from the fringe volume by particles in the air flow is collected at right angles to the incident beams through a second window. There are two advantages of this arrangement over the more usual forward scatter and backscatter geometries. First, the laser light scattered from the input windows does not directly enter the receiving optics, and with no special precautions the intensity of the background light is very low. Second, the aperture $A1$ can be used to limit the part of the fringe volume viewed by the receiving optics, so that the effective measurement volume

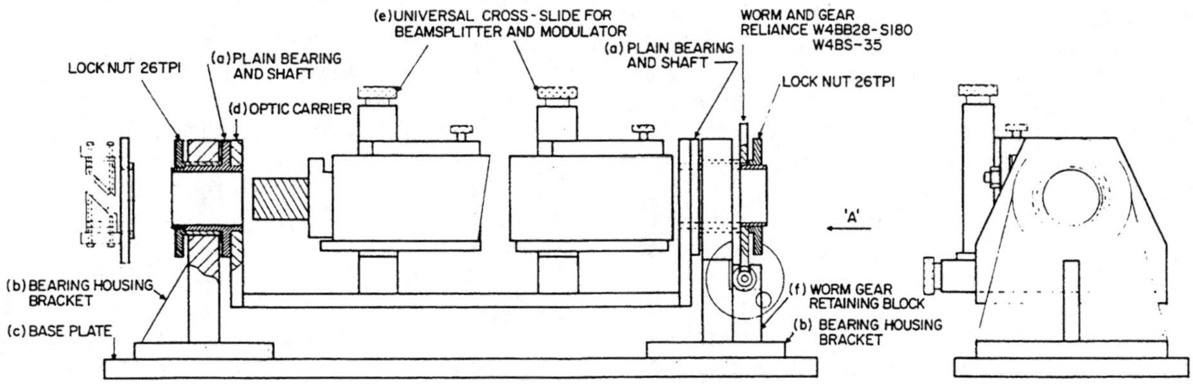

Fig. 3. Rotation assembly.

is not unduly elongated. An interesting further possibility, which has not been pursued, is to replace the aperture $A1$ by a slitlike mask with a Gaussian transverse transmittance function. For a suitable magnification in the receiving optics and TEM_{00} input beams, the measurement volume can be given spherical symmetry with a Gaussian radial distribution of sensitivity. The orientation shown measures the x component of velocity in the laboratory coordinate system, and the optics can be moved to measure in turn all three components throughout a volume of a few cm^3 around the spark plug. The mirrors $M1$ and $M2$, the beam splitter, and the modulator are mounted on the assembly shown in Fig. 3, which can be rotated about the axis of the original laser beam to measure the x and y components of velocity. The cruciform arrangement of the engine windows allows the y and z measurement directions to be obtained by moving the whole system through 90° with respect to the engine.

Since the alignment of the LDA system is critical and time consuming all the optical components are mounted on a single table equipped for translation in three perpendicular directions, allowing the measurement point to be scanned within the engine without the optical components moving relative to each other. The necessity for frequent realignment is further reduced by ensuring that the input beam crossover does not move in space as the rotation assembly revolves. This latter condition depends on a number of adjustments so that a systematic analysis of their effects is worthwhile. Such an analysis shows that translational motions of the laser have no (first-order) effect and that at least one tilt adjustment of the laser with respect to the rotation axis is necessary because misalignment here introduces an axial movement of the crossover during rotation. Once axial movement has been eliminated, the motion of the crossover follows a circle centered on the rotation axis, which may be slightly displaced from the nominal axis. Therefore, the position of the crossover is measured for two orientations separated by 90° of rotation, and the center of the circle is calculated by coordinate geometry. Once the crossover is made to coincide with the center by transverse adjustments to $L1$ (or $L1b$), the measurement point remains stationary during rotation to better than 0.1 mm.

The argon-ion laser is the Spectra-Physics 164/03, giving up to 1.3 W at 488 nm. The beam splitter and modulator are commercial components (Malvern RF307 and K9023) similar to those described by Foord *et al.*[9] The modulator is indispensable for measurements within the engine because of the very high relative turbulent intensities that are observed. The fringes are given a constant high velocity in order to reduce the apparent relative turbulent intensity. A linearly increasing phase difference between the input beams is introduced, followed by a rapid fly-back each time the phase difference has changed by 2π rad. The finite time for fly-back limits the usable fringe velocity to about 5×10^5 fringes/sec. With a 400-mm focal length lens at $L1$ the fringe spacing is 9.8 μm, giving a maximum fringe velocity of about 5 $msec^{-1}$. Since the observed turbulence velocities increase with engine speed, a higher value is required for measurements at 800 rpm and above. The single lens $L1$ is then replaced by the double lens combination $L1a$ and $L1b$, shown in broken lines in Fig. 2. The fringe spacing is then 24 μm, and the maximum fringe velocity is about 12 $msec^{-1}$, which is sufficient for the maximum speed of the engine. The diameters of the measurement volume calculated from conventional TEM_{00} mode propagation are about 0.3 mm and 0.7 mm for the single and combination lenses, respectively. However, at maximum laser power the diameters increase twofold or threefold. Ideally the waists of the measurement beams should occur at their crossover, and this will happen (to a good approximation) if the wavefronts of the beams incident on $L1$ form part of a single sphere. This condition can be achieved by giving the beams a slight mutual divergence, or by placing a weak converging lens between the laser and the rotation assembly to produce a beamwaist at $L1$ (or $L1a$). However, the variation of the fringe spacing through the measurement volume when these precautions are not taken ($\lesssim$5%) is of little consequence for the present measurements involving very large relative turbulent intensities. Note that it is the measured flow velocity, not the flow relative to the moving fringes, that varies in proportion to the fringe spacing.

A block diagram of the electronic system is shown in Fig. 4. The velocity information is contained in the statistical properties of the pulses from the D307LC photomultiplier, which are presented to the photon correlator via amplifier and discriminator circuits. The correlator also totalizes one count per cycle of the electrooptic modulator for subsequent subtraction from the observed LDA frequencies. The system for gating the discriminator circuit during a chosen portion of the four-stroke cycle is based on the crank-angle selector described by Allwood *et al.*[6] The crank-angle selector receives signals at each degree of rotation and at tdc from optical transducers on a toothed wheel attached to the crankshaft. An additional signal from the ignition circuit is used to distinguish the same crank angle in exhaust-inlet and compression-expansion crankshaft revolutions. The measurement is then repeated at different crank angles through the four-stroke cycle. Consequently, the measurements presented here for consecutive intervals of crank angle are statistically independent with respect to cyclic dispersion.

This method of taking data is to be distinguished from an experiment in which every crank-angle interval is measured during the same period of engine operation. There are three methods of outputting data from the correlator. A live display is provided on an oscilloscope, a slow analog output is provided for an x-y chart recorder, and digital information is output to a cassette tape recorder for subsequent analysis on the Shell time-sharing computer network.

Data Validation and Analysis

The LDA technique is undoubtedly less prone to systematic errors than any of the other methods available for in-cylinder flow measurement. Nevertheless, it is important to bear in mind that the measurement depends inherently on interference fringes being formed within the measurement volume, and that degradation of the regularity and visibility of the fringes from whatever cause will degrade the measurement also. Figures 5(a) and 5(b) show fringes formed by the input beams before and after an oil-contaminated window was interposed in their path. Despite considerable scattering of light from the beams a reasonably good fringe system remained. Such observations are important to validate the results of LDA experiments. Note that when the modulator is used, the presence of periodic signals at frequencies around the modulator frequency does not confirm that the system is operating correctly. As an extreme example, consider a fringe system in the measurement volume that has become so scrambled as to appear like a laser speckle pattern. Nevertheless, the illumination of each point will vary as the relative phase of the input beams changes, so that the light scattered by a stationary particle will be periodic at the modulator frequency. The same result would be observed for perfect fringes, and the signal will be correctly interpreted as zero velocity flow. If now the particle is given a constant velocity, the periodic photon autocorrelation function $F(\tau)$ will decay with increasing τ. The decay arises because of the random phase-modulation of the

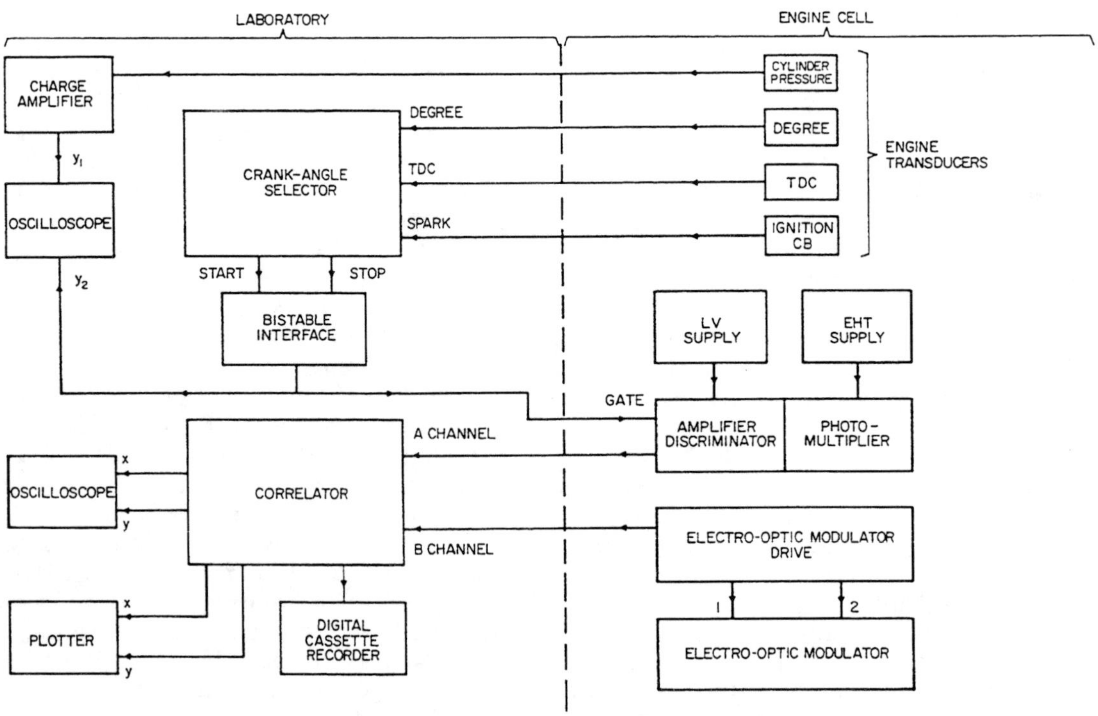

Fig. 4. Block diagram of the electronic system.

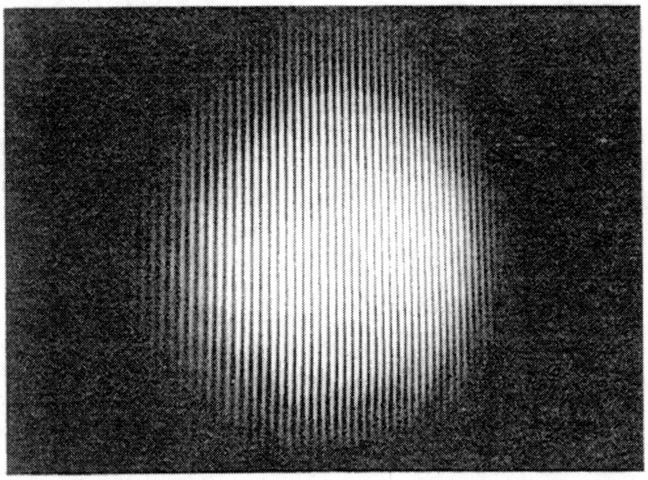

(a)

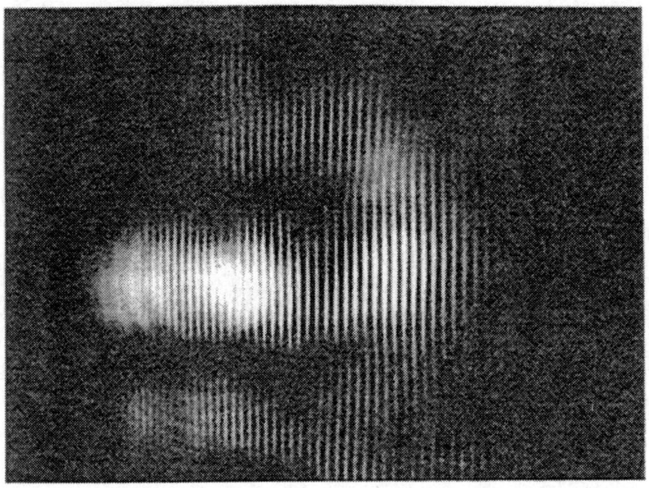

(b)

Fig. 5. (a) Projected fringe system; (b) fringes of (a) with an oil-contaminated window.

radio frequency amplitude modulation of the scattered light as the particle moves through the light and dark regions of the pattern; but it will be interpreted as being due to turbulence, which is not present, superimposed on an apparently zero mean flow.

Two methods have been used to analyze the experimental correlation data to obtain the incidence of each velocity during the measurement period, which (when normalized) approximates to a point on the true ensemble-averaged probability distribution of velocities, $P(u,v,w)$. The first involves a Gram-Charlier curve-fitting technique[9,10] using a Fortran program by A. D. Birch. Since in general the Gram-Charlier distribution has not resulted in a notable improvement in the quality of fit, the measurements presented in Figs. 6 and 7 have been obtained using the same program for a five-parameter fit, corresponding to a Gaussian distribution of velocity. The fitted function then has the form

$$F(\tau) = A\exp(-B\tau^2)\left\{1 + C\exp\left[\frac{-8\pi^2\tilde{u}^2\tau^2}{\lambda^2}\sin^2\left(\frac{\theta}{2}\right)\right] \times \cos\left[2\pi\tau\left(\frac{2\bar{u}}{\lambda}\sin\left(\frac{\theta}{2}\right) + f_0\right)\right]\right\}, \quad (1)$$

in which λ is the laser wavelength, θ is the angle of crossover, and f_0 is the modulator frequency. The fitted parameters are $A, B, C, \bar{u}$ (the mean velocity normal to the fringes), and $\tilde{u}$ (the corresponding rms turbulent velocity). The first exponential in Eq. (1) describes the decay of the cosine and its unmodulated pedestal due to the finite size of the measurement volume. Strictly, to use the exponential form here is to make assumptions that are not necessarily valid, i.e., that the laser mode is accurately Gaussian, that the measurement volume is long compared with its width, and that the fluid velocity is uniform throughout the measurement volume. However, the number of fringes within the beam diameter is sufficiently large ($\geqslant 40$) that the effect of the second exponential in Eq. (1), arising from the spread of velocities, is dominant, the first providing only a small correction over the portion of the function that is fitted.

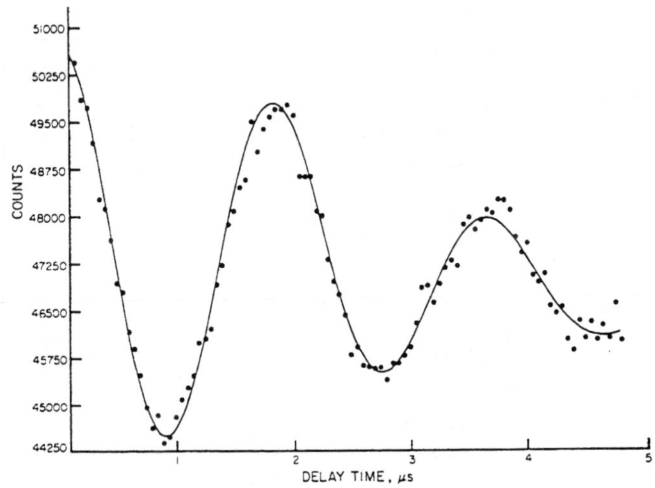

Fig. 6. Experimental and computed curves.

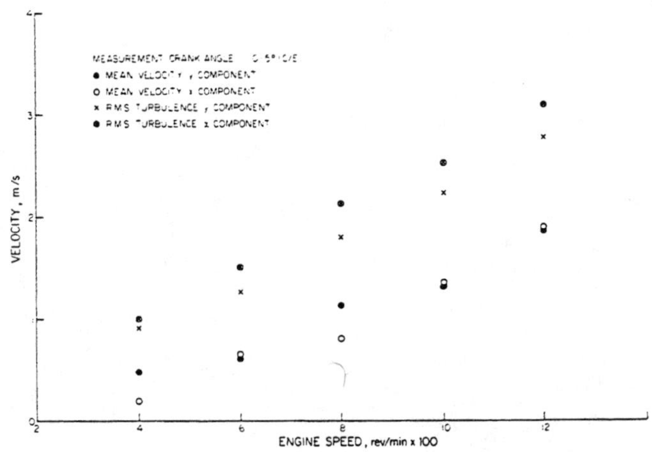

Fig. 7. Mean and rms turbulent velocity vs engine speed.

A typical set of data points and the fitted curve are shown in Fig. 6. The data were obtained at 400 rpm for crank angles between 40° and 50° after tdc on the expansion stroke, indicating values for $\bar{u}$ of +0.21 $msec^{-1}$ and for $\tilde{u}$ of 0.64 $msec^{-1}$. Besides the flow parameters the computer fit also gave a value for C corresponding to an effective fringe visibility of 2.2 from the depth of modulation of the scattered light. Such a high effective visibility could occur only if a number of conditions were satisfied simultaneously. Evidently, the actual fringe visibility was ≥2.2, the naturally occurring seed particles were small compared with the 9.8-μm fringe spacing, the intensity of stray light reaching the photomultiplier was small, and the measurement volume was usually occupied by no more than a few significant particles. Mean and rms turbulent velocities vs engine speed, measured and processed in the manner outlined above, are shown in Fig. 7. The measurement crank angle is 0–5° on the expansion stroke. The previously predicted[5] linear relationship, at least over the engine speeds studied, seems to be verified.

The second method of analysis is to take the discrete Fourier transform of the autocorrelation values, which according to the Wiener-Kintchine theorem yields the velocity distribution without prior assumption of its form. It has been found useful before transformation to subtract the mean of each data set from each value and to weight the resulting values by a Gaussian function centered on zero delay. Thus from the correlator values for a delay time τ_0 per channel $F_n = F(n\tau_0)$; $n = 1,2\ldots 96$, we obtain for a suitable value of T

$$\phi_n = \left[F_n - \left(\sum_{n=1}^{96} F_n\right)\bigg/96\right]\exp\left(\frac{-n^2\tau_0^2}{T^2}\right), \tag{2}$$

and hence the discrete cosine transform Φ_r; $r = 1,2\ldots 96$. The weighting of the correlation function is equivalent to a convolution in the velocity domain with a Gaussian of width (FWHM)

$$\lambda(\ln 2)^{1/2}/[\pi T\sin(\theta/2)],$$

which provides smoothing and avoids the undesirable effects of the truncation at a finite number of channels that were noted by Moore and Smart.[11] Provided that the artificial envelope is narrow compared with the envelope produced by the finite size of the measurement volume, a well-defined probability of each velocity can be calculated. Since the value of τ_0 is chosen to be approximately one quarter of the modulator period, zero velocity corresponds to $r \simeq 48$, and the fringe spacing is chosen so that the velocity information is contained within $25 \leqslant r \leqslant 72$. The Φ_r for $1 \leqslant r \leqslant 24$ may contain baseband contributions from particle motion in the input beams, but the values for $73 \leqslant r \leqslant 96$ can be used to eliminate the arbitrary constant that is contained in all the Φ_r owing to the absence of a zero delay channel. The normalized probability distribution is then

$$P(u_k) = \frac{\Phi_k - \sum_{r=73}^{96}\Phi_r/24}{\sum_{r=25}^{72}\Phi_r - 2\sum_{r=73}^{96}\Phi_r}; \quad k = 25,26\ldots 72 \tag{3}$$

in which

$$u_k = \frac{\lambda}{2}\left(\frac{k}{192\tau_0} - f_0\right)\bigg/\sin\left(\frac{\theta}{2}\right). \tag{4}$$

The reduction in the measured value of F_1 due to discriminator dead-time in principle produces some degree of baseline distortion of the probability distribution. A series of experiments without modulator drive showed that the effect is negligible and that the scatter of the measured probability is well accounted for by the counting statistics. Figure 8 shows a 3-D surface of the

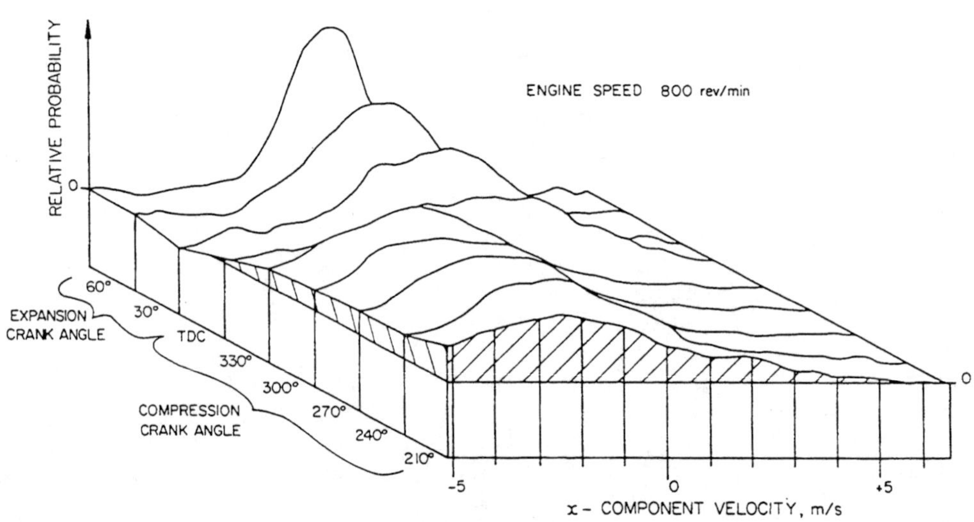

Fig. 8. Velocity distributions by Fourier analysis.

velocity distributions for a variety of crank angles of the compression and expansion strokes in a cylinder, generated in the manner outlined above. Each trace was obtained for a crank-angle range of 10° centered on the indicated crank angle. The relatively slow development between successive 30° increments of crank angle justifies the use of a 10° range for each measurement, except possibly in the vicinity of tdc. It should be noted, however, that in the parts of the cycle where the flow varies rapidly with crank angle (for instance immediately after valve opening), the use of too wide a range would give a spurious increase in the apparent width of the distributions.

The Gaussian smoothing function convolved with the data has a width (FWHM) of 0.5 $msec^{-1}$. The general features are initially of a fairly broad distribution resulting from the turbulence of the inlet process, narrowing as the turbulence decays with the mean of the distribution tending toward zero. This relaxation of the flow is seen to extend beyond the end of the compression stroke (tdc) into the expansion. However, in a fired situation the developing flame front would cause a rapid increase of turbulence rather than the reverse.

The authors gratefully acknowledge the technical assistance of R. J. Wade, the computational assistance of R. Caradine, and helpful discussions with P. T. Davies.

References

1. E. S. Semenóv, Studies of Turbulent Gas Flow in Piston Engines, NASA Tech. Trans. F97 (1963).
2. P. O. Witze, Hot Wire Turbulence Measurements in a Motored i.c. Engine, AIAA 14th meeting, Washington, D.C. (1976).
3. P. O. Witze, Measurement of Spatial Distribution and Engine Speed Dependence of Turbulent Air Motion in an i.c. Engine, SAE Paper 770220 (1977).
4. J. C. Dent and N. S. Salama, The Measurement of the Turbulence Characteristics in an i.c. Engine Cylinder, SAE Paper 750886 (1975).
5. R. J. Tabaczynski, Prog. Energy Combust. Sci. **2**, 143 (1976).
6. H. I. S. Allwood, G. A. Harrow, and L. J. Rose, SAE Automotive Paper 700063, Detroit (1970).
7. J. B. Abbiss, T. W. Chubb, and E. R. Pike, J. Phys. D. **5**, L100 (1972).
8. W. T. Mayo, Appl. Opt. **16**, 1157 (1977).
9. R. Foord, A. F. Harvey, R. Jones, and J. M. Vaughan, J. Phys. D. **7**, L36 (1974).
10. A. D. Birch, D. R. Brown, and J. R. Thomas, J. Phys. D. **8**, 438 (1975).
11. C. J. Moore and A. E. Smart, J. Phys. E. **9**, 977 (1976).

Reprinted with permission from *Journal of Physics E: Scientific Instruments,* Vol. 14, pp. 7-19 (1981). ©1981 IOP Publishing Ltd.

Correlation in instruments: cross correlation flowmeters

M S Beck
Department of Instrumentation and Analytical Science, University of Manchester Institute of Science and Technology, Manchester M60 1QD, England

Abstract The basic principles are briefly explained, showing how cross correlation can be used to identify dynamic characteristics of a wide range of telecommunication, structural and process systems. The main emphasis of the paper is on cross correlation flowmeters, which are developing to a stage where they can successfully solve industrial and environmental measurement problems, ranging from the flow of highly polluted liquids in pipes to the flow of gas from volcanic jets.

Cross correlation flowmeters are based on measuring the transit time of a tagging signal (turbulence, clumps of particles, etc.) in the flow between two axially separated sensors. The transit time is measured by a cross correlator. The design of the various subsystems is discussed in some detail and the various sensing techniques required for specific applications are described. Cross correlation flowmeters have become a realistic proposition because of the reducing cost of large scale integrated circuits and microprocessors from which the correlator may be realised, a number of different designs of cross correlator are described and their relative mertis discussed.

1 Introduction

Statisticians have used cross correlation to estimate the cause and effect relations in sets of population data (e.g. the dependence of crop yield on previous rainfall). Engineering applications of cross correlation have increased during the past few decades, no doubt because of the availability of inexpensive computers to produce cross correlation results in 'real time' (e.g. for recovery of information in weak radar echoes, identifying the dynamics of chemical processes, etc.). Lange (1967) outlines a wide range of cross correlation applications.

The basic concept is simple. The dynamic relationship is the way in which values of the input signal affect future values of the output signal. The response to a unit (dirac) impulse is the simplest embodiment of this, unfortunately many systems cannot have unit impulses applied which give a clearly discernible output response (Eykhoff (1974) presents an excellent account of system identification problems). By regarding the normal (randomly time varying) input as a series of impulses, merging to give the continuous input signal $x(t)$ and convoluting (i.e. multiplying and integrating with suitable time shifts) this with the system output $y(t)$, we obtain the cross correlation function R_{xy} which, assuming the input signal has a wide bandwidth, contains all the information about the dynamics of the system. The equation for this cross correlation is:

$$\text{cross correlation } R_{xy}(\tau)=\frac{1}{T}\int_0^T x(t-\tau)y(t)\,dt.$$

A specific advantage of cross correlation is that any spurious signals present as interference on the measured output $y(t)$ are strongly rejected because they have no correlation with the input signal $x(t)$. The process by which the spurious signals are rejected will now be considered in more detail. Let the measured output $y(t)$ be equal to the sum of a (desired) noise free signal $y'(t)$ and an (undesired but unavoidable) spurious noise interference $z(t)$. i.e.

$$y(t)=y'(t)+z(t).$$

Substituting in the cross correlation equation gives:

$$R_{xy}(\tau)=\frac{1}{T}\int_0^T x(t-\tau)[y'(t)+z(t)]\,dt.$$

This can be split into two integrations hence:

$$R_{xy}(\tau)=\frac{1}{T}\int_0^T x(t-\tau)y'(t)\,dt+\frac{1}{T}\int_0^T x(t-\tau)z(t)\,dt.$$

Now the second integral represents the cross correlation between the system input $x(t)$ and a spurious signal $z(t)$. Since these are not related their cross correlation tends to zero provided that the integration time T is sufficiently long. This rejection of spurious interference is the most important advantage of cross correlation.

The cross correlation method of determining system dynamics has many similarities to the frequency response method. The former determines the relationship for specific time delays, whereas the latter determines the relationship for specific frequencies. The cross correlation and the frequency response functions are related by 'Fourier type' transforms, so provided one function is known, the other can be readily computed. The reader should refer to the book by Bendat and Piersol (1971) for a complete treatment of these procedures for random data analysis.

In practice, direct calculation of the frequency response has been used mainly for structural vibration analysis, control system design by Bode, Nyquist diagrams, etc. (Shinners 1973) and telecommunication channel response measurement. Cross correlation is more directly used where measurements requiring some knowledge of time delays are required, such as velocity measurement of steel (Butterfield *et al* 1961), paper strip (Kashiwagi 1968), road and rail vehicle speed (Zimmer *et al* 1976, Idowaga and Ono 1976), range measurement by sonar and radar (Arthur *et al* 1979). This paper will describe one of the major uses of cross correlation, its application to flow measurement, and in so doing will illustrate the advantages and problems associated with the use of correlation for measurement purposes. General design principles, methods of implementation, means of on-line computation, and accuracy of measurement will be discussed.

1.1 *Cross correlation for flow measurement of difficult fluids*

The process industries handle a wide range of fluids from clean liquids and gases to aggressive and difficult fluids such as hydraulically and pneumatically conveyed solid materials, sewage etc. Measurement of flow and other process variables in industries handling clean fluids has been comparatively easy and this has led to a great deal of automatic control in such processes. On the other hand industries handling solid materials and difficult fluids have found that process measurement has been very difficult and this has hampered the use of automatic control in these industries. It seems certain that the main effort in process automation during the next decade will be in industries handling solid materials. This trend will be accentuated by an increasing use of solid fuels as our liquid and gas fuels become depleted.

Essential features of flowmeters for measuring flows of difficult fluids in pipes are that they should not obstruct the flow, which would give a risk of blockage and they should not have any parts which wear out by being in abrasive contact with the fluid (Beck *et al* 1978). In other cases the measuring instrument must even be truly remote from the fluid, consider for example the measurement of the flow of hot gases from steel making furnaces (Webb *et al* 1976).

Referring to figure 1 cross correlation flowmeters measure flow by sensing the passage of some disturbance or tagging signal in the flow at one position, and measuring how long it takes that disturbance to travel to a downstream point. This is basically a very simple principle and fortunately the tagging signals can be detected by sensors which do not contact the fluid.

The cross correlation function as defined above is:

$$R_{xy}(\tau)=\frac{1}{T}\int_0^T x(t-\tau)y(t)\,\mathrm{d}t.$$

It can be shown that this function has a maximum value when the cross correlation lag τ is equal to the transit time τ^* of the tagging signals, hence the flow velocity is given by

$$u=l/\tau^*$$

where l is the spacing of the sensors A and B.

A proof that the time delay of the cross correlation peak is equal to the fluid transit time τ^* is given in an early publication by Beck *et al* (1968).

The cross correlation calculation has in the past required quite elaborate and expensive computing systems; now, fortunately, modern large scale integrated circuits enable cheap cross correlators to be made. The design of cross correlators will be reviewed in §4. Before this we will consider alternative flowmeters (§1.2), total system design (§2), and the selection of suitable sensors (§3).

1.2 *Alternatives to cross correlation flowmeters for difficult fluids*

The flow of a fluid is often measured by putting some obstruction into the flow (Open University T291). Orifice plates, venturis and vortex flowmeters all fall into this category. They

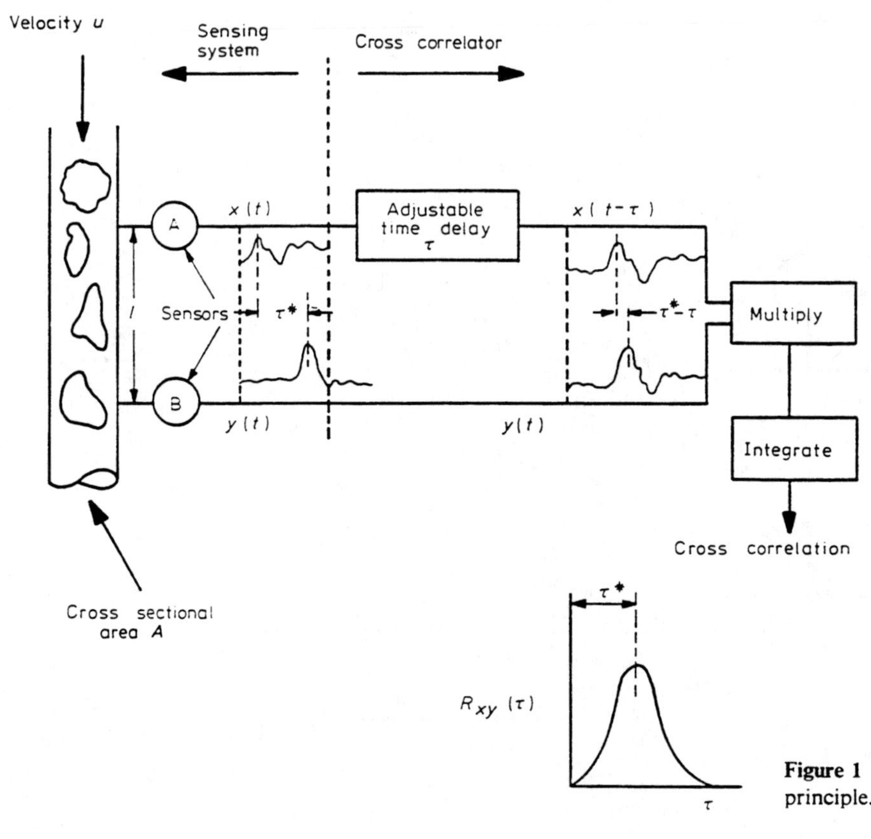

Figure 1 Cross correlation flowmeter – basic principle. Velocity $u=l/\tau^*$; flowrate $Q=kAu$.

are often unreliable due to wear by abrasion when the fluid contains suspended solid particles; they can also cause solids to drop out, leading to a complete stoppage of flow. The pressure loss caused by the obstruction leads to an increase in the energy required to pump the fluid, this can be significant with large flows of water, sewage, natural gas etc. In contrast there is no pressure loss with cross correlation flowmeters.

Ultrasonic pulse transit flowmeters (Cousins 1978) seem to be promising for difficult fluids. The generally preferred arrangement uses piezoelectric transducers located diagonally across the pipe which transmit ultrasound pulses upstream against the flow and downstream with the flow. The difference between the upstream and downstream pulse transit times is a measure of the flow. However, because the velocity of sound is much greater than the velocity of the fluid these devices involve measuring a small difference between two large quantities which has in the past been a cause of considerable error, although it is claimed that newer systems have reduced this error (Watts 1980). In order to obtain an accurate transmission path the piezoelectric transducers have to be in contact with the fluid, in contrast with ultrasonic cross correlation flowmeters in which the transducers can be clamped onto the outside of the pipe.

The electromagnetic flowmeter does not obstruct the flow and is possibly the only commercially available instrument which has been proved successful in measuring the flow of difficult fluids (De Jong 1978). However, it can only be used with electrically conducting fluids, and it is very expensive for the larger pipe sizes. In contrast, with suitable sensor design, cross correlation flowmeters can measure virtually any fluid and their cost does not increase significantly as the size of the pipe increases. In addition, cross correlation flowmeters rely on the fundamental measurements of time and distance and do not require calibrating for individual applications, this contrasts strongly with electromagnetic flowmeters which have to be individually calibrated for the specific pipe size in use and which may require recalibration when a fault is suspected.

2 Design of cross correlation flow meters

Our initial development of cross correlation flow meters was done on a largely empirical basis and a fuller theoretical framework was built at a later stage. In my opinion, the quickest way to develop a new technological area – such as cross correlation flowmeters – is to develop experimentally instruments for a range of applications. After the early development phase, the sight of a few experimental prototype instruments actually working helps to stimulate the more theoretical approach which is required to design the optimum cost-effective systems.

In the following section we will model the whole measurement system and break it down into subsystems in order to illustrate the more fundamental aspects of designing cross correlation flowmeters. Further details of modelling correlation flow measurement systems are given by Fischer and Davies (1964) and Kippham and Mesch (1978).

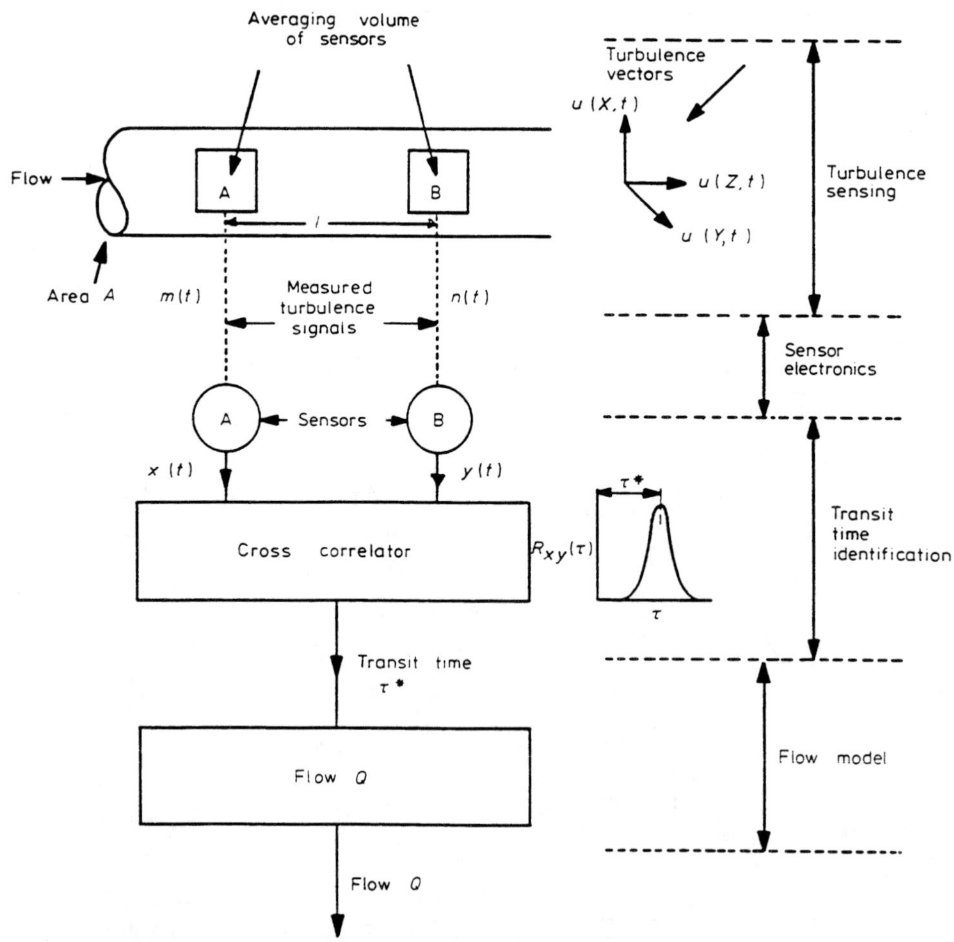

Figure 2 Cross correlation flowmeter – schematic of subsystems.

2.1 *Model of the measurement system*

Referring to figure 2, a cross correlation flowmeter measures the flow velocity by obtaining the transit time of a tagging signal between sensors at locations A and B. The following flow model is used to derive the measured flow Q from the measured transit time τ^*—

$$Q = k(l/\tau^*)A \qquad (1)$$

where l = spacing of sensors, A = cross sectional area of pipe, k = a calibration factor, mainly to allow for non-uniformity of the velocity profile (see §5.1).

The tagging signal is generated by the fluid turbulence, the turbulent eddies can be detected, for example, by the modulation they superimpose (by doppler effect) on a beam of ultrasound perpendicular to the direction of flow. Alternatively particles can become entrained in the turbulent eddies to form clouds (just as clouds are formed in the atmosphere by the effects of atmospheric turbulence on the water droplets); these clouds can form the tagging signal and they may be detected by ultrasonic or optical beams, electrical conductivity, electrical capacitance effects etc. The design of sensors for these signals is discussed in §4. The book by Hinze (1959) gives a comprehensive account of the structure of turbulence.

The flow turbulence eddies can be considered as vector velocities in the X and Y directions perpendicular to the flow axis and the Z direction along the flow axis (figure 2). The three component velocities X, Y and Z can be measured in the laboratory by laser velocity meters or by hot wire probes. They are all stationary ergodic variables as shown in figure 3(*a*). The mean values of the X and Y components are zero, because there can be no flow of the fluid through the pipe wall. The mean value of the Z component is of course equal to the transport velocity of the fluid. An 'observer' moving at the same velocity as the flowing fluid would note that the eddies at position A gradually changed their shape as they moved down the pipe, however; provided that the downstream sensing point B is reasonably close to A (say less than five pipe diameters), the eddies are sufficiently similar to be recognised by the correlator in order to enable the transit time to be measured. Clouds due to atmospheric turbulence behave in a similar manner, they keep a reasonably recognisable shape over a large arc of the sky so we can estimate wind speed from them.

The turbulent eddies can be represented by their frequency spectrum, figure 3(*b*). The spectrum is a band-limited white noise process, with its upper frequency dependent on viscous damping of the turbulent eddies. Typically the upper frequency limit may be as high as 10 kHz. Ideally the flow sensors in a cross correlation flowmeter should respond to the whole range of these turbulent frequencies, this would give the maximum statistical data into the correlator and enable accurate flow measurement with a very short response time. Unfortunately this wide bandwidth is only possible when using point sensing devices such as laser probes, hot wires etc., which are much too delicate for use in industrial environments. The sensors for industrial use, such as ultrasonic beams, capacitance and conductivity sensing plates etc., interrogate (spatially average) a substantially sized volume of the flow, and hence they will not respond to the higher frequency turbulent eddies. A prime need in system design is therefore to consider the averaging effects of the sensors used, because these fundamentally limit the information bandwidth to the cross correlator and hence limit the accuracy and response time of the system.

2.2 *Spatial averaging effect of the cross correlation sensors*

Referring to figure 2 the spatial averaging effect can be defined as a transfer function relating the turbulence $u(X, Y, Z, t)$ at a point in the pipe to the signals $m(t)$, $n(t)$ into the sensors. The calculation of this relation is complex for sensors having a non-uniform field, such as capacitance and electrical conductivity sensors. Even with ultrasonic sensors which can focus beams across the pipe, the calculation is complicated by the effects of beam reflection and standing waves.

The following analysis will be based on the use of optical sensors which detect the absorption of a beam of parallel light by solid particles in the flowing fluid. Optical beams are well defined so that the analysis is simplified. Even though optical transducers are not often used in industry (because of problems with soiling of optical surfaces) the following model will serve to illustrate the major factors involved in designing any type of sensing system.

Referring to figure 4, turbulently developed clouds of particles pass at velocity u through an optical beam of

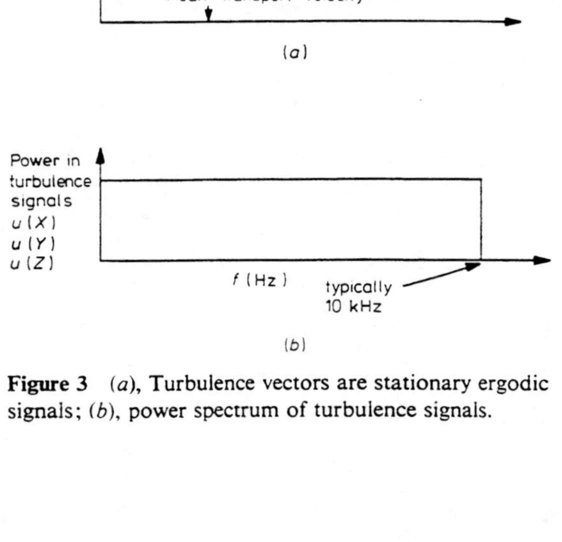

Figure 3 (*a*), Turbulence vectors are stationary ergodic signals; (*b*), power spectrum of turbulence signals.

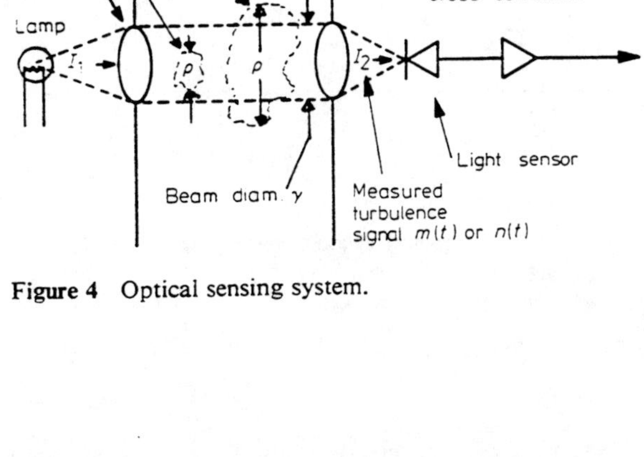

Figure 4 Optical sensing system.

diameter γ. The incident light I_1 lumens is attenuated by the particles to allow the emergent light I_2 to pass to the receiving optics. The emergent light I_2 corresponds to the input signals $m(t)$ and $n(t)$ to the sensors shown in figure 2.

Now consider a turbulent cloud of size ρm which is smaller than the beam diameter γ. In this case the emergent light is –

$$I_2 = I_1[1-(\rho/\gamma)^2 r] \qquad (2)$$

where r is the attenuation coefficient of the cloud.

Any cloud which is larger than the beam diameter ($\rho > \gamma$) will for a period of time obscure the beam so that–

$$I_2 = I_1[1-r]. \qquad (3)$$

Examination of equations (2) and (3) suggests that a cloud of size $\rho > \gamma$ is fully detected, but small clouds of size $\rho < \gamma$ have relatively little effect on the emergent light which is received by the photodetector. Now a cloud of size γ m travelling at a velocity of u m s^{-1} will take γ/u s to pass across the beam. Hence the highest unattenuated turbulence signal which can be detected corresponds to a signal having a period γ/u s as shown in figure 5(*a*). This signal has a fundamental frequency of $1/(2\gamma/u)$ Hz.

Summarising the above argument, the cut-off frequency (f_1) at the averaging volume is

$$f_1 = \frac{1}{2\gamma/u} = \frac{1}{2}\frac{u}{\gamma} = \frac{1}{2}\,\frac{\text{velocity } (u)}{\text{length } (\gamma)}$$

e.g. for a liquid flowing at 4 m s^{-1}, sensor beam diameter 20 mm, cut-off frequencies $f_1 = 4/(2\times 2\times 10^{-2}) = 100$ Hz, the corresponding frequency response of the averaging volume is shown in figure 5(*b*).

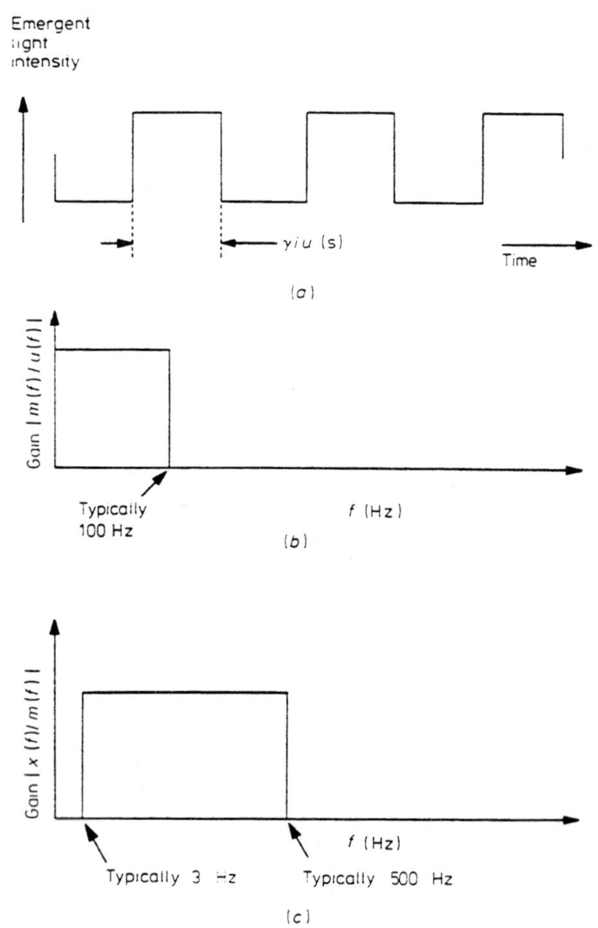

Figure 5 (*a*), Highest frequency signal unattenuated by spatial averaging; (*b*), frequency response of averaging volume; (*c*), frequency response of sensor electronics.

2.3 *Frequency response of sensor electronics*

The sensor electronics can be easily designed to have a wider bandwidth than the spatial averaging bandwidth derived above. Taking the particular system discussed above, which gave a spatial averaging bandwidth of 100 Hz, it would be reasonable to make the bandwidth of the sensor electronics say 500 Hz. This would allow the sensors to detect all the turbulent information and allow for higher frequency signals, which would be generated if the flow velocity of the fluid increased. A frequency response above 500 Hz may allow unwanted electronic noise to enter the cross correlator, which would reduce the statistical accuracy.

A typical response for the sensor electronics is shown in figure 5(*c*). The low frequency cut-off point is usually arranged at about 3 Hz so that low frequency signals which cannot be considered statistically stationary within the correlation averaging time (typically 2 s) are not passed to the cross correlator.

3 Sensing the tagging signal

The sensors detect the tagging signal at points A and B, which are located along the flow axis as shown in figure 1.

The candidate sensing principles fall into three broad categories: modulation of radiation by the flowing fluid, emission of radiation by the flowing fluid, electrical and thermal measurement of properties of the flowing fluid. Often several alternative techniques could be used for a practical measuring problem, the following descriptions present various sensing devices to help the reader make an appropriate choice for a particular situation. The choice should be principally based upon reliability and cost of the sensing devices. The gain stability of the sensors is not important, because the cross correlator simply measures the time delay of signals between the sensors, and this time delay is not dependent on the gain.

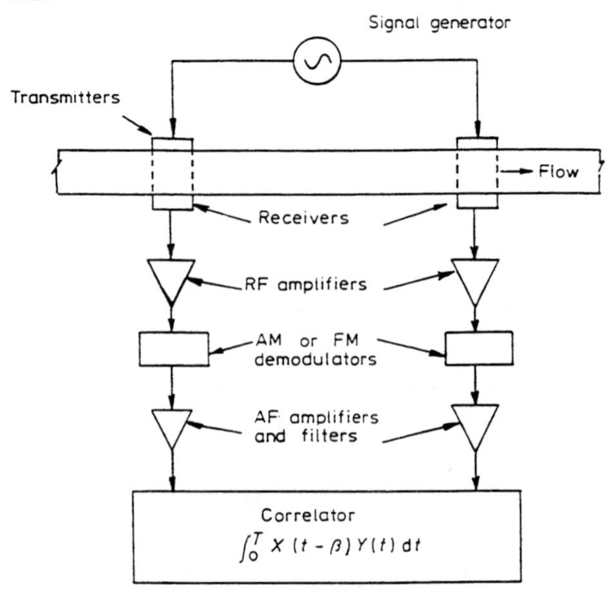

Figure 6 Ultrasonic correlation flowmeter.

3.1 *Modulation of radiation by the flowing fluid*

Radiation techniques are attractive because they can interrogate the flow across a large part of the pipe section, radiation sensors can frequently be mounted externally to the pipe.

(i) *Modulation of acoustic radiation*

When a beam of ultrasound (figure 6) is passed across a flowing fluid, the received ultrasound is modulated by the turbulent eddies in the fluid (doppler effect), or by absorption and scattering effects if particles are entrained in the fluid (Coulthard 1973, Beck and Ong 1975b). For liquids the transmitting and receiving sensors can be mounted on the outside wall of the pipe, frequencies in the region of 1 MHz can pass readily through the pipe wall and this enables a compact and easily maintainable instrument to be constructed, that is likely to find widespread industrial application. With gases the problem is not so simple, the acoustic mismatch between the pipe wall and the low density gaseous medium leads to the requirement of a transducer being in contact with the gas; although ultrasonic cross correlation flowmeters for gases have been successfully demonstrated under laboratory conditions it may be some time before they find their way to industrial development.

(ii) *Modulation of light*

Optical devices find only a very limited application for fluid flow in pipes because of soiling of the optical surfaces. However, optical cross correlation devices have proved successful for measuring flow in open channels using the system illustrated in figure 7, in this case the optical device is mounted well above the flowing fluid and the optical windows can be cleaned by wind-screen wipers if required (Beck and Kaghazchi 1977). Another successful application of optical cross correlation is in measuring the flow velocity of gases from chimney stacks. When steam or particles are present in these gases, turbulent clouds are generated and the velocity of the gas can be measured from a remote point by cross correlating the output signals from two photodetectors which view through telescopes two positions along the axis of the plume (Llewellyn 1976).

(iii) *Modulation of gamma rays etc.*

When solid particles are present in a fluid flow, they can be used to modulate a beam of radiation which is directed perpendicular to the flow. If two such sensors are spaced axially along the flow the received radiation signal can be cross correlated to give the flow velocity. Unfortunately rather strong sources of radiation are required, because it is essential for the count rate registered by the ionising radiation detector to be at least an order above the period of the highest frequency signal to be correlated (Stamford 1980). Hence these techniques may only be worth considering in situations where

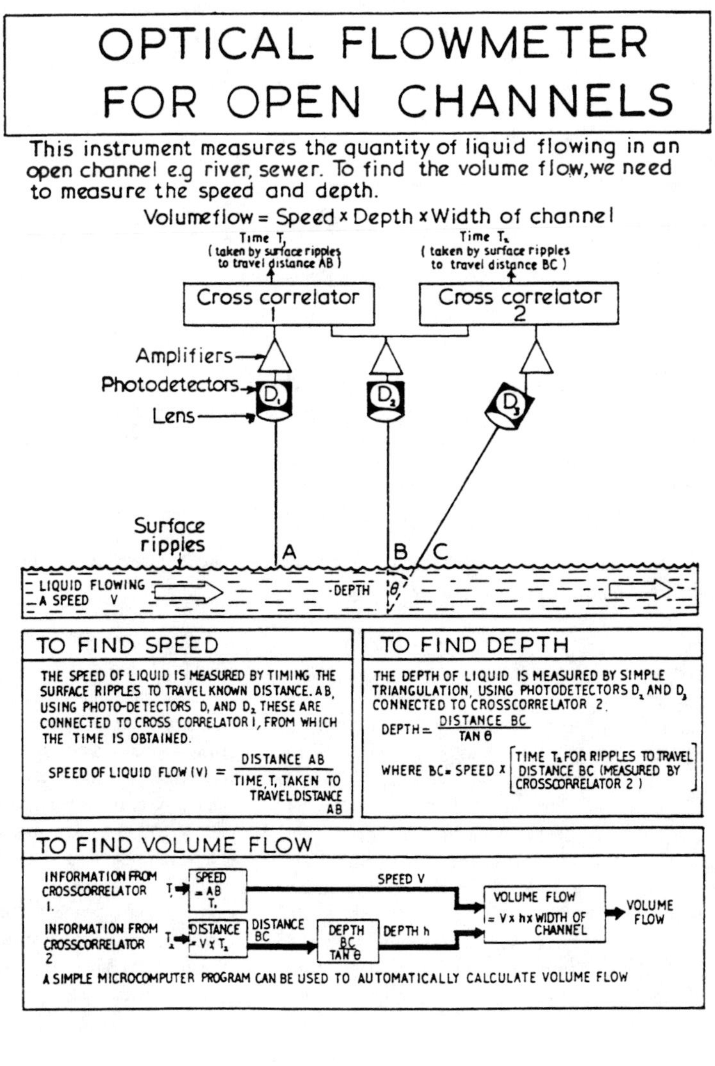

Figure 7 Optical flowmeter for open channels.

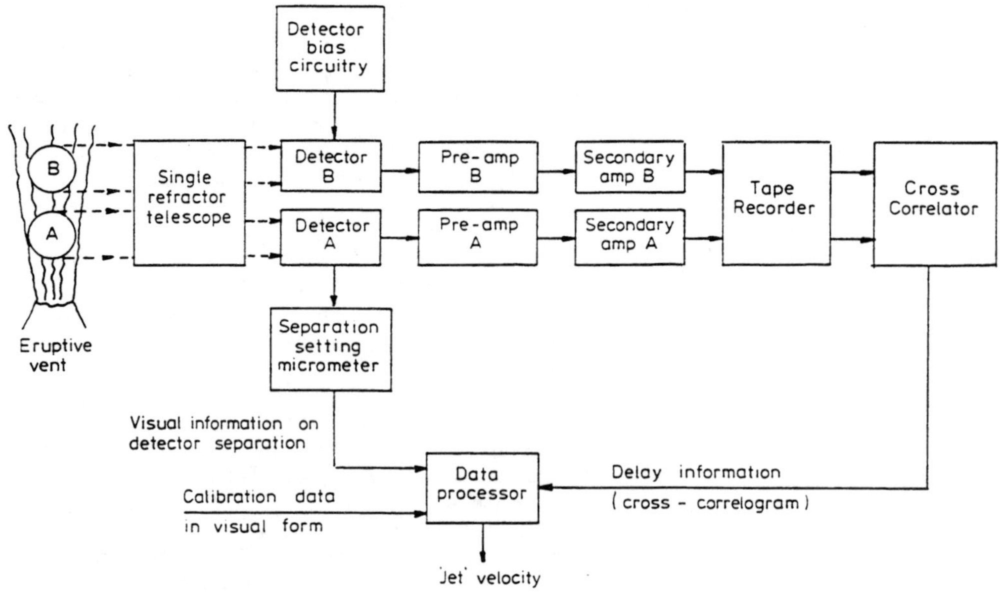

Figure 8 Volcanic jet velocity – antarctic expedition instrument system (from Llewellyn).

other devices could not be used, as might arise in measuring the flow in pipes located inside power station boilers etc.

3.2 *Emission of radiation by the flowing fluid*

(i) *Thermal radiation*

The most significant applications of radiation techniques have used the thermal radiation emitted by hot gases. The turbulence in hot gas flow inevitably leads to spatial variations of temperature and these provide good tagging signals. They can be detected by infrared sensors using suitably focused telescopes and located at a safe distance from the hot flowing gas.

Referring to figure 8, our first application was concerned with measuring the velocity of hot gas from volcanic jets (Llewellyn 1976). The telescope for these measurements was located over 200 metres from the gas jet and well away from the danger zone caused by the fall out of hot rock and magma which was also emitted from the jet. A similar application (figure 9) was by British Steel who measured the flow of hot gases from oxygen steel-making vessels (Webb *et al* 1976). This latter technique seems to work well and will almost certainly lead to less pollution, better quality steel and better operating efficiency in steel production plants.

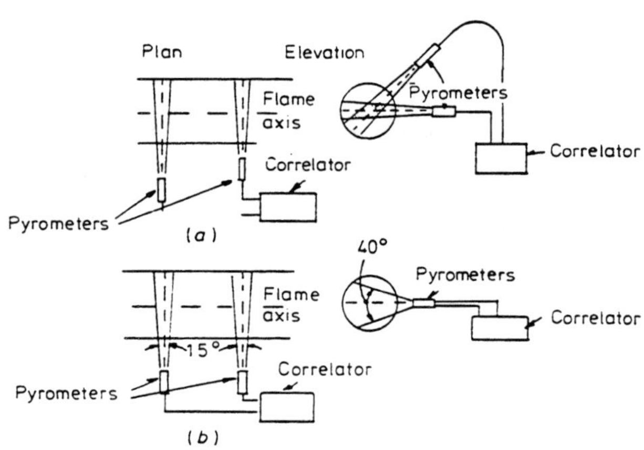

Figure 9 Flow of hot gases from oxygen steel-making vessels (from Webb *et al*). (*a*), Velocity profile sighting; (*b*), mean velocity sighting. The shaded area represents the correlation area.

(ii) *Ionising radiation*

A technique is being investigated (Stamford 1980) for measuring the flow of highly radio-active liquor in nuclear fuel separation plants. Variations in the emitted radiation are caused by either solid particles or gas bubbles in the flowing fluid and it is thought that these variations could be detected by scintillation detectors mounted outside the primary radiation shield, but viewing the pipe through very small holes in the shield. The net effect should be a flowmeter without parts inside the radiation zone where maintenance would be impossible. The scintillation detectors would be located in bulges to which standard maintenance procedures apply.

3.3 *Instantaneous measurement of electrical and thermal properties of the fluid*

Sensors for these instantaneous measurements have proved simpler to develop than the radiation type devices described above, hence there is already considerable information known about their performance.

(i) *Capacitance sensors*

Capacitance sensors are particularly useful for measuring the flow of gases with entrained solids. The solids form clouds due to turbulence and these clouds modulate the capacitance of a plate which is exposed to the flowing fluid. By cross correlating the capacitance modulation of two plates, suitably spaced along the axis of the pipe, the flow velocity can be measured (Beck *et al* 1968, Kippham and Mesch 1978). The capacitance sensing electrodes (figure 10) are very hard wearing, they do not protrude into the flow and the necessary insulating materials are available to withstand most conditions of temperature, corrosion or abrasion which may be encountered. Capacitance sensors have been used for several years in shot blasting machines, where they have been exposed to extremely abrasive

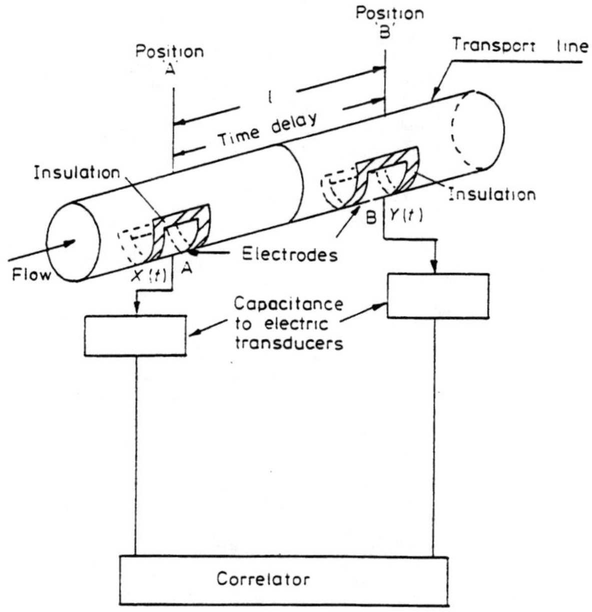

Figure 10 Schematic diagram of pneumatic conveyor, transducers and cross correlator.

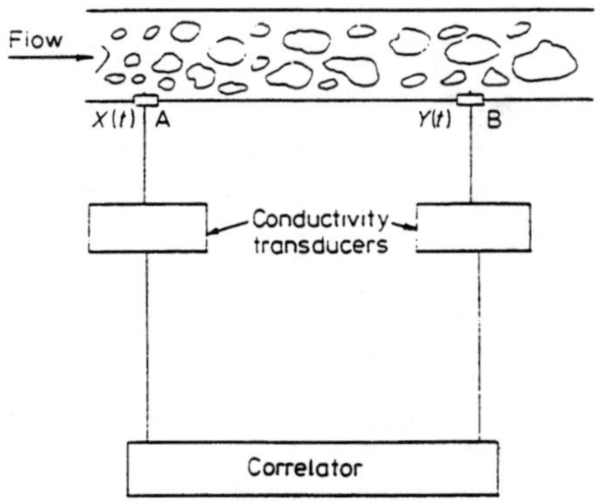

Figure 11 Conductivity cross correlation flowmeter.

conditions with no deterioration in performance (Beck *et al* 1978).

(ii) *Electrodynamic sensors*
Whenever solid particles are entrained in a flowing stream of gas a considerable electrostatic charge accumulates on the particles. This electrostatic charge can be detected by sensors using plates somewhat similar to the capacitance sensors described above. The voltage on the plate is amplified by a relatively simple AC amplifier and the output can be cross correlated to give the flow velocity (King 1973). In many ways they are similar to the capacitance sensors, but it should be possible to make them even more sensitive for measuring the flow of gases with very small amounts of entrained solids; such situations are likely to occur in suction lines feeding industrial gas cleaning plants.

(iii) *Electrical conductivity sensors*
The electrical conductivity sensors (Beck *et al* 1974) for liquid/solid mixtures and for liquid/liquid mixtures are analogous to the capacitance sensors for gas/solid mixtures. The discontinuous phase in the liquid causes instantaneous variations in the electrical conductivity of an electrode exposed to the flowing fluid, and these changes can be measured and cross correlated to give the flow velocity. The electrodes can be very robust (figure 11); the electronic circuits are very simple and hence these devices could be a strong contender for use as a low cost flowmeter for slurries. They do not have the advantage of clipping on the outside of the pipe, which ultrasonic sensors do, and they will not work reliably if the slurry contains a large amount of grease (as in some sewage slurries) because the grease will deposit on the conductivity electrodes and stop them working. On the other hand they are very simple and will certainly be much cheaper than ultrasonic systems.

(iv) *Electrolytic and injected bubble cross correlation sensors*
Attempts have been made to modify the electrical conductivity sensors so that these will work with single phase liquids. The first attempts used a DC measurement of electrical conductivity. Referring to figure 12, a polarising layer of gas bubbles was formed on the conductivity electrode and these bubbles were swept away at a variable rate according to the velocity of the turbulent eddies as they passed the electrode. Hence a variable current passed into the electrode to replenish the polarising bubble layer and this current was able to be cross correlated to give the velocity. Unfortunately all the practical realisations of this system suffered from chemical deposition on the electrode which eventually stopped it working. More promising, however, is a recent development (Kirkby 1979) in which the electrode is replaced by a stainless steel needle with a very small flow of fine bubbles emerging from its end (figure 13). These bubbles behave in the same way as the polarisation layer mentioned earlier, but in this case the electrode itself can be energised by an AC conductivity meter and it appears that the problem of chemical deposition on the electrode is overcome.

(v) *Thermal sensors to detect turbulent eddies*
Turbulent eddies in fluids cause corresponding random variations in the rate of heat exchange with a heated body. A heated thermistor provides a convenient way of detecting these eddies. The thermistor can be run at constant voltage, or constant current, and the corresponding variations in current

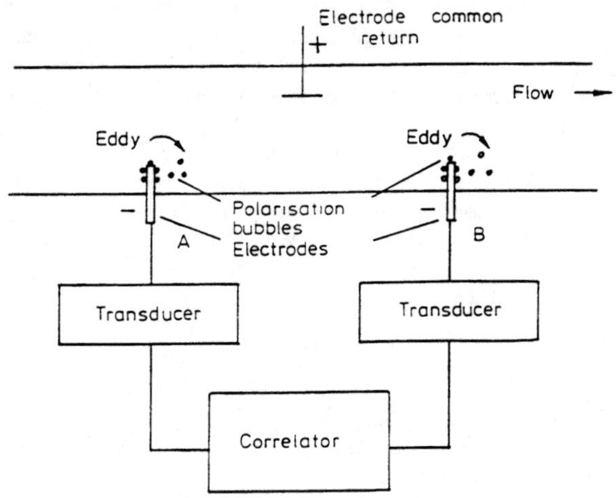

Figure 12 Electrolytic correlation flowmeter (from Kirkby).

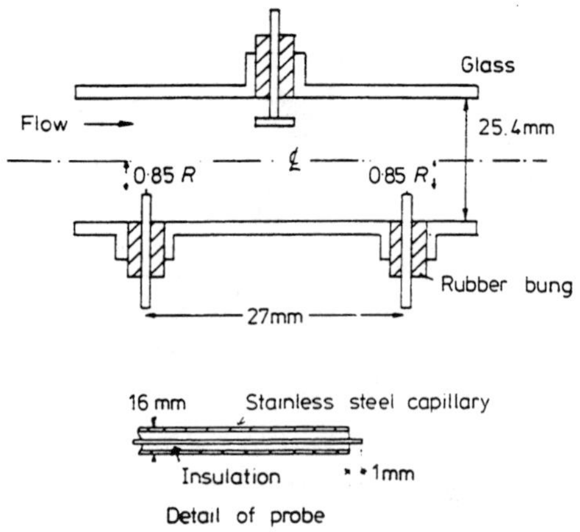

Figure 13 Probe arrangement for electrolytic flowmeter (from Kirkby).

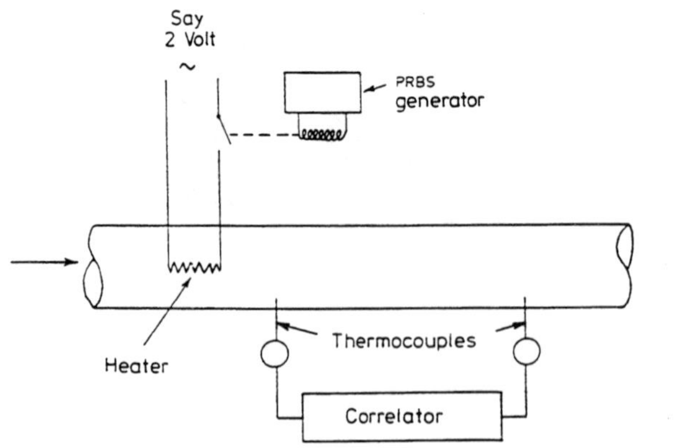

Figure 14 Cross correlation of injected heat pulses.

or voltage due to the temperature changes can be cross correlated. Alternatively the thermistor can be operated as a constant temperature anemometer by including it in a suitable feedback circuit, so that its resistance is maintained constant by varying the current through it; the voltage signal across the thermistor can then be used as an input to the cross correlator. It is claimed (Goh and Bentley 1977) that the constant temperature anemometer mode reduces to some extent the thermal time constant (and hence limitation in frequency response) which is inherent in thermistors used in the constant current or voltage modes. The thermistor is probably one of the very few alternatives to the ultrasonic system for a single phase fluid. It is a cheap and simple device to use, but it does intrude into the flow and it does suffer from a limited frequency response unless a very small thermal mass device is used, unfortunately the small thermal mass devices would be too fragile for industrial flows.

(vi) *Thermal sensors to cross correlate injected heat pulses*
Referring to figure 14, when a fluid is flowing under laminar conditions it may be necessary to inject a tagging signal into the flow (Beck and Abeysekera 1972). This can be done by switching on and off in a pseudo-random manner a small heater in the pipe. The resultant temperature pattern in the flow is detected by thermocouples or thermistors etc. at two points along the flow axis; the detectors outputs are cross correlated to give the flow velocity. When the fluid happens to pass over a heat exchanger, sufficient temperature variations for cross correlation purposes may be generated by the natural randomness of heat exchange. This has been exploited to measure the water circulation of internal combustion engines and to measure the coolant flow in the tubes of a nuclear reactor (Bentley and Dawson 1966). A paper by Boonstoppel *et al* (1968) evaluated the potential accuracy of the technique.

4 On-line cross correlation computation
The calculation of the cross correlation function involves the delaying of one signal with respect to the other, the multiplying of the two signals together, and the averaging of the product over a suitable period of time. The continuous (direct correlation) equation for the above operation is:

$$R_{xy}(\tau)=\frac{1}{T}\int_0^T x(t-\tau)y(t)\,dt. \tag{4}$$

A digital cross correlation computes the sampled data form of the above equation:

$$R_{xy}(j)=\frac{1}{N}\sum_0^N x(k-j)y(k). \tag{5}$$

Initially an expensive general purpose digital computer was used to perform this correlation. An analogue correlator using a magnetic tape delay was next used (Butterfield *et al* 1961) then a general purpose digital correlator (Anderson and Perry 1969). All these systems have been too expensive for general on-line industrial use.

The data sampling rate for flow cross correlation can be as high as 1 kHz. At this speed many digital computers and currently available microprocessors are unable to solve equation (5) in real time. A reduction in the amount of computation required can be obtained by cross correlating only the polarity of the input data and solving the equation:

$$R_{xy}(j)=\frac{1}{N}\sum_0^N \operatorname{sgn} x(k-j)\operatorname{sgn} y(k) \tag{6}$$

where sgn denotes the sign of the data with respect to its mean value. In a polarity cross correlator the data is read as a series of binary values as shown in figure 15. By reading binary values the amplitude information in the data is lost and the statistical accuracy of the cross correlation will suffer.

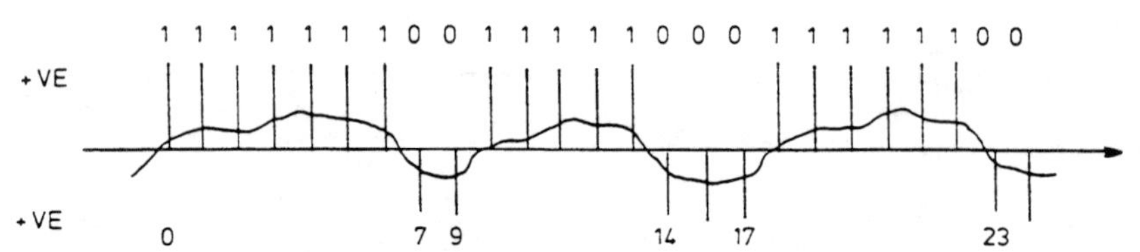

Figure 15 Zero crossing encoding.

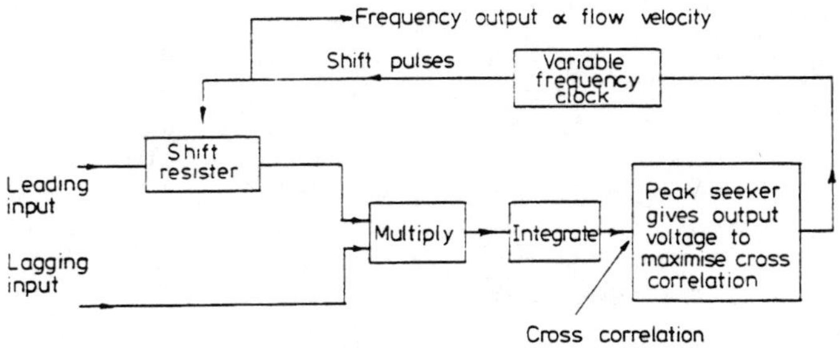

Figure 16 Peak seeking cross correlator.

This can be compensated for by increasing the data observation time by a factor of about 2·2 over the time required for the direct solution of equation (5) (Jordan 1979).

In order to avoid this approximate doubling of response time necessary with a polarity cross correlator, 'relay' cross correlators have been designed in which one data stream is polarity quantised and this is correlated with the actual value of the other data stream using the equation

$$R_{xy}(j)=\frac{1}{N}\sum_{0}^{N}\operatorname{sgn} x(k-j)y(k). \qquad (7)$$

Relay correlation requires the data observation time to be increased by a factor of 1·4 over the time required for the direct solution of equation (5).

4.1 *Correlators designed specifically for flow measurement*

(i) *Peak seeking optimisers*

Early designs of flow cross correlators used a hill climbing arrangement (Mesch *et al* 1971) so that the correlator time delay was automatically adjusted to give the maximum output of the cross correlation integrator as shown in figure 16. This arrangement appears attractive for minimising the hardware required, but unfortunately when the flow changes suddenly, or when there are certain disturbances in the flow, there are small spurious peaks on the cross correlation function and the peak-seeking circuit can lock on to a subsidiary peak, giving a gross error in flow measurement. This could be overcome by using logic circuits which direct the correlator to periodically rescan the whole function and reselect the major peak. These additional circuits increase the cost and complexity to a level where it seems preferable to use special large scale integrated circuits or microprocessors to inspect the whole cross correlation function and obtain the highest peak.

(ii) *Special purpose large scale integrated circuits*

This system uses the fact that the correlation function grows with increasing integration time (figure 17). If an arbitrary level is chosen, and the correlation function is allowed to grow through this level, then the first part of the function to reach this level is the peak. This principle can be implemented in a circuit having a regular repetitive form that can be readily realised in an integrated circuit (Jordan and Kelly 1976).

(iii) *Microprocessor cross correlators*

Microprocessors are attractive because they are cheap and becoming even cheaper. They are too slow to solve even the polarity cross correlation in real-time, but a number of systems using external circuits for delay and high speed multiplication have been described (Jordan 1979, Coulthard and Keech 1979). Unfortunately these add complications and expense over the attractive concept of a single chip microprocessor correlator. This problem was considered by Henry who devised a new

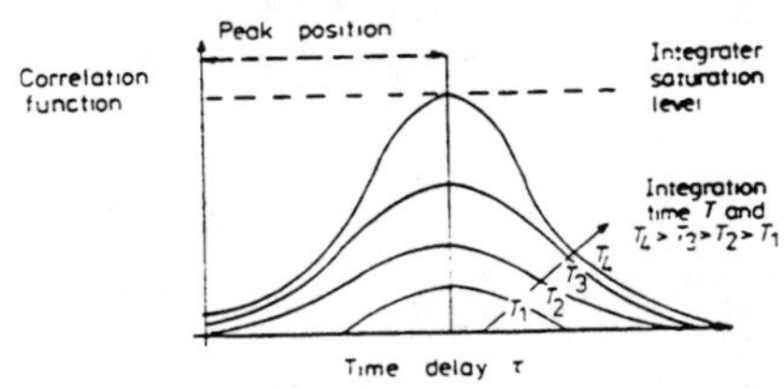

Figure 17 Peak detection in LSI correlator chip.

cross correlation algorithm for a microprocessor (Henry 1979) which increased its speed by well over ten times and eliminated the need for the additional circuits.

Henry's arrangement is based on the principle that the only significant information used by a polarity cross correlator is the actual zero crossing times of the signals as shown in figure 18. This information can be stored in the computer as

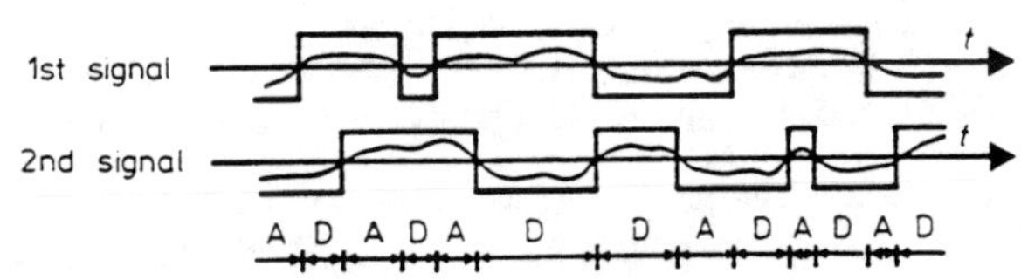

Figure 18 Cross correlating from zero-crossings. A, agreement; D, disagreement.

binary data denoting agreement or disagreement between the signals as shown. This avoids the need for sampling the data at close intervals that were previously thought necessary to give adequate resolution (say 0·2%) in flow measurement as shown in figure 15. Henry noted that the zero crossings are randomly distributed in time, hence the observation by the correlator of their time of occurrence gives data with a very fine time resolution. This seems to be an optimal way of handling polarity data and accounts for the large savings in computer time.

4.2 *Future developments in flow cross correlators*

In the immediate future it seems almost certain that most requirements will be met by microprocessor cross correlators using polarity correlation and based on Henry's method for maximising computer efficiency.

Polarity correlation (equation (6)) does have the penalty of slower response time compared with direct correlation (equation (5)). Intermediate systems using two or three level amplitude quantisation could speed up the response without excessive computational effort. However, looking further into the future one can see the possibility of microprocessors being constructed using charge coupled devices, storage and computation logic in a single package (Arthur *et al* 1979). The charge coupled device would offer the advantage of analogue data handling and would give the possibility of making correlators with the best possible response time at a realistic cost for industrial flow measurement.

5 Accuracy of cross correlation flowmeters

The measured flow rate is obtained using the flow model (equation (1)) derived in §2.1. The measured length (l) and measured cross sectional area (A) are subject to normal geometrical constraints. However, uncertainties in the measurement of the time delay τ^* and the calibration factor k affect the flow measurement accuracy in a more complex way.

5.1 *Calibration factor k*

It is convenient to consider the calibration factor k as a product of three factors, k_1, k_2 and k_3

$$k = k_1 k_2 k_3$$

k_1 = flow profile factor

k_2 = velocity/time trade-off factor

k_3 = sensor phase error factor.

Now examining these factors individually

(i) *Flow profile factor k_1*

The flow profile across a pipe is not uniform. An ideal sensor would take an average of the turbulence information across the whole cross section of the pipe, hence one would expect the value of k_1 to be equal to unity. Unfortunately the sensors currently available do not average the whole cross section, for example an ultrasonic beam averages over a diameter of the pipe and for turbulent flow geometrical analysis shows that a flow profile correction factor of $k_1 = 0{\cdot}94$ would be required to allow for the effects (Beck and Ong 1975a).

(ii) *Velocity/time trade-off factor k_2*

This factor allows for the basic fact that eddies in the flow which travel most quickly (usually those near the centre of the pipe) have less time to decay between the sensing zones A and B (figure 2) than have the slower moving eddies nearer to the wall of the pipe. This causes the measurement to be biased towards the faster moving eddies. We have not been able to determine the factor k_2 from theoretical considerations, but we think it may be sufficiently close to unity to be reasonably confident that it can be determined when calibrating the flowmeter, and that its value will not change significantly in service.

(iii) *Phase error in sensors*

Sensors which instantaneously interrogate the flow (e.g. optical, capacitance, conductivity etc.) can be designed to cause virtually no problems with phase errors. All that is necessary is for both the sensors to be reasonably well matched.

A potentially serious problem can be caused by mismatch of the acoustic beams when using ultrasonic sensors. The problem here is that reflections at the fluid to pipe wall interface cause a high standing wave ratio (say a ratio of 5/1) across the acoustic cavity in the pipe; the turbulent eddies are detected by their modulating effect on the standing wave pattern. If the standing wave patterns are mismatched, by say a half wavelength, the cross correlation peak will be inverted. If mismatched by a quarter wavelength the peak will be shifted by a 90° phase shift. The resultant errors would be intolerable, so it is essential to match the standing wave patterns to within a few degrees. Now for measuring liquid flows ultrasound frequencies as high as 1 MHz have to be used for the ultrasound to penetrate the pipe wall. At this frequency the acoustic wave length is about 1·5 mm. Even with the best physical and electronic matching of the sensors, it would not be possible to maintain the standing wave pattern sufficiently constant for long periods of time in a practical process situation. Fortunately, recent work indicates that this problem can be overcome by using phase control schemes and adaptive frequency trackers to control precisely the phase and acoustic wave pattern across the cavity (Leach 1978).

5.2 *Errors in determining cross correlation time delay τ^**

The cross correlator itself is a digital device, with its timing controlled by a crystal clock. Therefore there is no significant fundamental error in the cross correlation measurement of time delay. There could be errors due to quantisation of the time delay measurements by the digital device, however, these errors can be overcome by suitable design of the digital system (Jordan and Kelly 1976).

A fundamental source of statistical error is the limited bandwidth of the signals input to the correlator and the limited time available to make a measurement. In considering these limitations, the need for flow measurement can be sub-divided into two separate areas:

(i) Total flow measurement for accounting and custody transfer purposes etc. In such cases the averaging time of the

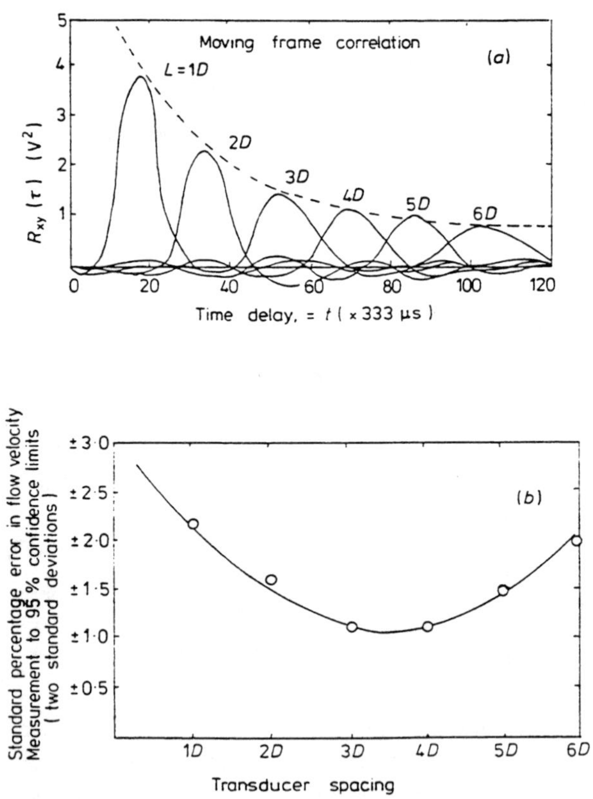

Figure 19 (a) Measured cross correlation function against transducer spacing; (b) normalised standard error in flow velocity measurement against transducer spacing. Experimental data: water flow in one inch (25·4 mm) diameter copper pipe. Ultrasonic probe diameter = 10 mm; carrier frequency = 1 MHz; AM signal processor filter bandwidth = 1–5 kHz; integration time = 20 s; mean velocity $V_m = 4{\cdot}2$ m s^{-1}; L = transducer spacing; D = pipe diameter (25·4 mm).

flow measurement is very long and there would be negligible statistical error provided that the flow is steady during the averaging time of the cross correlator. If the flow varies, the averaging time may have to be quite short and therefore the accuracy considerations discussed in (ii) below will apply.
(ii) Process flow control where the integration time of the cross correlator should be between say 1 and 5 s in order to enable the control loop to control against load changes. In these cases the statistical error could be significant and this will be discussed below to show which of the system parameters can be adjusted to minimise the error.

5.2.1 *Statistical errors in cross correlation flowmeters.* The standard error $\epsilon(\tau)$ of the transit time measurement is a function of signal bandwidth (B), correlator integration time (T) and the normalised cross correlation (ρ^*). Several investigators have developed relationships of the form (Beck and Ong 1975a):

$$\epsilon(\tau)=\frac{k_4}{B^{1\cdot 5}T^{0\cdot 5}}[1+(1/\rho^*)^2]^{0\cdot 5} \qquad (8)$$

where k_4 is a constant.

The flow rate Q is related to the transit time measurement as shown in equation (1), hence the standard error in flow measurement $\epsilon(Q)$ is given by:

$$\epsilon(Q) \propto \epsilon(\tau)/l. \qquad (9)$$

Now turbulence decays as the length l increases, hence the standard error $\epsilon(\tau)$ (equation 8) increases in a nonlinear way as the length l increases. Equation (9) shows that increasing the length l can reduce the error in flow measurement. The overall result is that excessively short length (very short transit time) and excessively long length (poor correlatability of signals) can result in poor flowmeter performance, but there is a range of values of between say one pipe diameter and six pipe diameters which gives an acceptable performance. The way in which the cross correlation function decays as l is increased is shown in figure 19(a) and the corresponding way the standard error varies is shown in figure 19(b).

6 Conclusions

We have seen how cross correlation flowmeters illustrate the advantages of the use of correlation for measurement purposes. The principle features being firstly that calibration of the sensors is not required because the time delay is measured with reference to a crystal controlled time standard in the cross correlator, secondly cross correlation rejects the effect of spurious interference on the signals. This is clear when observing the records from the two sensors of a cross correlation flowmeter, the turbulence pattern breaks up between the sensors so that the eye cannot determine the time delay relationship, fortunately the cross correlator can recognise and precisely determine this time delay.

We can expect cross correlation flowmeters to be developed to solve a wide range of industrial and environmental flow measurement problems. Until recently the commercial development of these flowmeters was inhibited by the high cost and complexity of cross correlators. We have seen how recent developments in large scale integrated circuits and microprocessors should enable simple, fast and reliable cross correlators to be designed at a cost that should be acceptable for industrial use. A few of the possible applications of cross correlation flowmeters are being developed by industry, but undoubtedly there is still a considerable amount of basic research work to be done concerned with sensor design and obtaining a better understanding of the way in which the sensors interrogate the flow.

In future, new cross correlation measurement systems will be developed and some of these will make even greater demands on data processing speed, fortunately we can expect recent developments in data processing such as the use of charged coupled devices (Arthur *et al* 1979) and optical signal processing (Murison *et al* 1979) to provide the required increase in data processing speed.

Acknowledgments
I am grateful to all the Bradford University students, staff and technicians and also my industrial colleagues whose enthusiasm and painstaking effort has enabled the work to be done. I also thank the Science Research Council for providing so much of the money needed for the work and the National Research Development Corporation for guiding the applications of the techniques.

References

Anderson G and Perry M A 1969 A calibrated real time correlator/averager/probability analyser
Hewlett Packard Journal, vol 21, no 3, pp 9–15

Arthur J W, Dix J F, Harland E, Widdowson J W, Denyer P B, Mavor J and Milne A D 1979 Large time-bandwidth product CCD correlators for sonar
IEE Conference Publication 1979/180 pp 62–68

Atomic International 1965 Development of noise-analysis techniques for measuring reactor coolant velocities
NAA-SR-11193

Beck M S, Drane J, Plaskowski A and Wainwright N 1968 Particle velocity and mass flow measurement in pneumatic conveyors
Powder Technology **2** 269–77

Beck M S and Abeysekera S A 1972 Liquid flow measurement by cross correlation of temperature fluctuations
Measurement and Control **5** T143–7

Beck M S, Lee K T and McKeown K J 1974 An on-line instrument for measuring small quantities of dispersed non-conducting liquid in a conducting liquid
Measurement and Control **7** T65–9

Beck M S and Ong K H 1975a Volume flow measurement of liquids and slurries by cross correlation of natural turbulence signals
Proc. of Conference on Fluid Flow Measurement in the Mid 1970's National Engineering Laboratory, April 1975

Beck M S and Ong K H 1975b Slurry flow velocity, concentration and particle size measurement using flow noise and correlation techniques
Measurement and Control **8** 453–63

Beck M S and Kaghazchi B 1977 Remote level and velocity measurement of rivers and open channels by analysis of random signals from surface waves
IEE Conference Publication No 159, April 1977

Beck M S, Green R G, Kwan H K and John R 1978 A low cost instrument for solids flow measurement in industry
J. Phys. E: Sci. Instrum. **11** 1005–10

Bendat J S and Piersol A G 1971 *Random Data: Analysis and Measurement Procedures* (New York: Wiley)

Bentley P G and Dawson D G 1966 Fluid flow measurement by transit time analysis of temperature fluctuations
Trans. Soc. Instrum. Technol. **18** 183–93

Boonstoppel F, Veltman V and Kergrouwen F 1968 The measurement of flow by cross correlation techniques
IEE Conference, Publication no 43, June 1968

Butterfield M H, Bryant G F and Dowsing J 1961 A new method of strip speed measurement using random waveform correlation
Trans. Soc. Instrum. Tech. **13** 111–23

Coulthard J and Keech R P 1979 A six channel microprocessor controlled correlator
IEE Conference Digest no 1979/32 pp 4/1–6

Cousins T 1978 The doppler ultrasonic flowmeter
FLOMEKO 78 (Amsterdam: North-Holland) pp 513–8

Davies D E N 1979 Signal processing for radar applications
IEE Conference, Publication no 1979/180 pp 145–50

De Jong J 1978 Comparisons of some 500 mm diameter electromagnetic flowmeters
FLOMEKO 78 (Amsterdam: North-Holland) pp 209–13

Durrani T S and Greated C A 1977 *Laser Systems in Flow Measurement* (New York: Plenum Press)

Dyer D and Stewart R M 1977 Detection and rolling element bearing damage by statistical vibration analysis
Symposium on On-line Surveillance and Monitoring of Process Plant (London: The Society of Chemical Industry) pp 3.1–3.13

Eykhoff P 1974 *System Identification* (London: Wiley)

Fischer M J and Davies P O A L 1964 Correlation measurements in a non-frozen pattern of turbulence
Journal of Fluid Mechanics **18** 97–116

Goh C K and Bentley O P 1977 Investigation into cross correlation techniques to measure low velocity flue gases
FLOW-CON 77 (London: The Institute of Measurement and Control) pp 409–32

Hayes A M 1976 *PhD Thesis* School of Electrical Engineering, Bradford University

Henry R M 1979 An improved algorithm allowing fast on-line polarity correlation by microprocessor or minicomputer
IEE Conference Digest no 1979/32 pp 3/1–4

Hinze J D 1959 *Turbulence* (New York: McGraw-Hill)

Idowaga T and Ono T 1976 A new method of vehicle-speed measurement using cross correlation
Proc. IMEKO VII, London, May 1976 (London: Institute of Measurement and Control) pp BLV 218/1–9

Jordan J R and Kelly R G 1976 Integrated circuit correlator for flow measurement
Measurement and Control **9** 267–70

Jordan J R 1979 Correlation circuits for measurement systems
IEE Conference Digest no 1979/32 pp 1/1–4

Kashiwagi H 1968 Paper strip speed measurement using correlation techniques
Trans. Soc. Instrument and Contr. Engrs., Japan **4** 304–12

King P W 1973 Mass flow measurement of conveyed solids by monitoring of intrinsic electrostatic noise levels
Pneumotransport 2, 5–7 Sept. 1973 (Cranfield: British Hydrodynamics Research Association) pp D2–9

Kipphan H and Mesch F 1978 Flow measurement systems using transit time correlation
FLOMEKO 78 (Amsterdam: North-Holland) pp 409–16

Kirkby H J A 1979 *MPhil Thesis* School of Control Engineering, Bradford University

Krause F R, Betz H T and Lysobey P H 1973 Pollution detection by digital correlation of multispectral, stereo-image pairs
Atmospheric Environment **7** 455–76

Lange F H 1967 *Correlation Techniques* (London: Iliffe Books)

Leach K 1978 *PhD Thesis* School of Control Engineering, Bradford University

Llewellyn G J 1976 Development of an instrument for measuring volcanic jet-velocity
Proc. IMEKO, V. 11, London, May 1976 (London: Institute of Measurement and Control) pp 247/1–6

Mesch F, Daucher H H and Fritsche R 1971 *Correlation Method of Velocity Measurement. Part I, Principles, Application to Two Phase Flows* (Germany: Meestechnik) Vol 79, no 7, pp 152–7 (CEGB Information Services, CE Trans. no 5890)

Murison A, Grant P M and De La Rue R M 1979 Application of optical techniques to signal processing
IEE Conference, Publication no 1979/180 pp 34–45

Open University, *Transducers 2: Flow, Instrumentation Unit 10* (Milton Keynes: The Open University Press) T291/10

Shinners 1973 *Modern Control System Theory and Application* (New York: Addison–Wesley)

Stamford B 1980 *PhD Thesis* School of Control Engineering, Bradford University

Watts F W 1980 Ultrasonic flowmeters: their reliability and accuracy in industrial plants
I. Chem. E. Symposium Series no 60 pp 147–60

Webb P J, Ashton R M and Kenahan T 1976 Remote monitoring of the Properties of Exhaust Fume from Oxygen Steelmaking Processes
Proc. IMEKO, vol 11, London, May 1976 (London: Institute of Measurement and Control) pp AML/148/1–10

Zimmer C, Ryser H and Meyr H 1976 An instrument for non contact speed measurement by image correlation
Proc. IMEKO, vol 4, London, May 1976 (London: Institute of Measurement and Control) pp BLV/217–1 to 10

Maurice Beck is the Professor of Instrumentation and Analytical Sciences at The University of Manchester Institute of Science and Technology (UMIST). He is at present engaged in setting up this new department which aims to integrate research and teaching in the previously separate disciplines of Instrumentation and Analytical Science. This will become a 'complete systems approach' and will make maximum use of the new possibilities provided by micro-electronic systems. Professor Beck is a graduate of Leeds University and was, from 1963 until his appointment at UMIST, at the University of Bradford in the School of Control Engineering. Professor Beck is author and co-author of over sixty publications.

Reprinted from *Fibre Optics '84,* J.M. Tait, editor, Proc. SPIE Vol. 468, pp. 190-200 (1984). ©1984 SPIE.

The Applications of Fibre Optics in Gas Turbine Engine Instrumentation

Ian Davinson

Electronics and Measurement Techniques Department,
Rolls-Royce Limited, P.O.Box 31,Derby,England

Abstract

Instrumentation in Gas Turbines must operate in extremely harsh environments. Electro-optical methods are being increasingly used to measure such variables as displacement, temperature and gas flow and fibre optics are often required to enable sensitive electronic components to be placed remote from the hostile region. This paper reviews applications of fibre optics in Rolls-Royce up to the present. In addition the case for using fibre optic sensors for the measurement of other parameters in future will be presented, along with a discussion of the prospects for fibre optic data transmission on the next generation of digitally controlled engines.

Introduction

Increasing fuel prices have generated greater demand for fuel efficiency in aero-engines. Figure 1 shows the performance parameters of a typical large civil engine. It can be seen from the fuel flow figures that improvements in fuel consumption can result in large financial savings, for example a 5% improvement could save an airline operating Boeing 747's £500,000 per year on each plane.

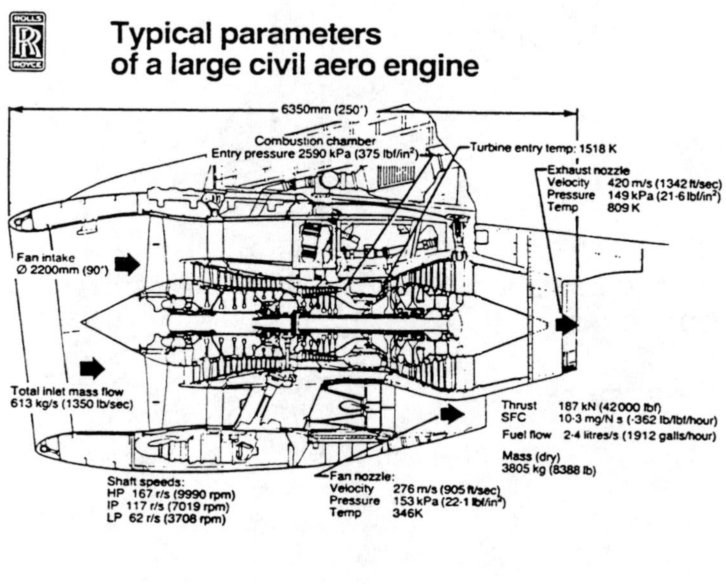

FIGURE 1

The development of gas turbine engines is generally a process of gaining small improvements in a large number of individual components and more accurate instrumentation is required to monitor this. Unfortunately improved performance often means increased operating temperatures and we therefore have the situation where more accurate instrumentation must be developed to work in even harsher conditions than those encountered at present. Figure 2 shows a typical test bed environment.

Aero engine running on reheat in sea level test bed

FIGURE 2

The standard instrumentation points on the 04 (hottest) module of an RB 211 are shown in Figure 3. These are mostly temperatures and pressures with a few flow and vibration sensors. Not all of these would be measured on every test,typically about 500 pressures and 300 temperatures covering the whole engine would be monitored on a development test. To these must be added any special instrumentation specific to the particular engine build, e.g. strain gauges. It can be seen that the amount of information involved is already considerable but there is a great incentive to increase it further. If the amount of data obtained from each engine build can be increased to the point where one less engine is needed in a development program this can save several million pounds. Packing more instrumentation onto each engine means that the sensors must be more compact and immune from crosstalk and that the data transmission and acquisition systems must be capable of handling larger quantities of data at higher rates. Reliability is also an important criterion even on development engines as the cost of stripping and rebuilding an engine to replace a defective transducer could be high.

Standard instrumentation points on RB211 HP system module

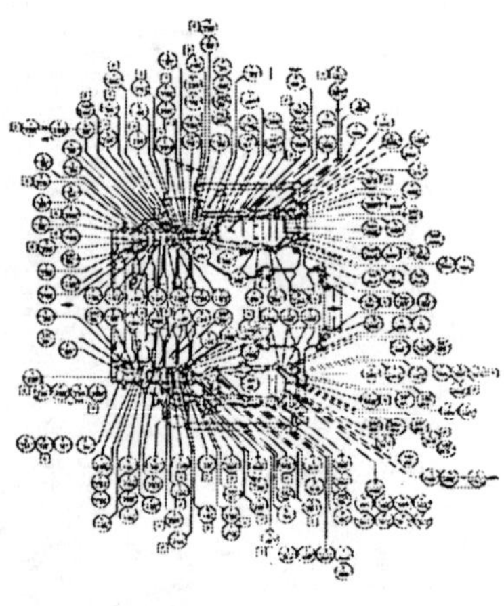

FIGURE 3

The amount of instrumentation on flight engines, where of course, reliability is vital, is also increasing as more parameters need to be sensed and controlled. In this case the factors such as weight, price maintainability and immunity to lightning strike (and for military engines, electromagnetic interference and nuclear electromagnetic pulse) must also be considered.

It is against this background that those involved in the manufacture of aero-engines have become interested in the possible advantages offered by optical and in particular fibre optical sensors. The activity of airframe manufacturers in this field is also a driving force as obviously at some stage it is necessary to interface the engines with aircraft systems.

A lot of preliminary work has been done in the USA on the concept of an integrated system of optical sensors for propulsion control (Refs: 1,2) and Rolls-Royce have some interest in this approach but this paper will concentrate on reviewing the use of fibre optics within Rolls-Royce up to the present and indicate the lines along which we see, or would like to see, fibre optical and related transducers developing. As in other fields these uses can be divided into data transmission and sensing although as will be seen later the boundary between these is likely to become blurred.

DATA TRANSMISSION APPLICATIONS

The majority of Rolls-Royce engine test beds and rigs are supported by an automatic data recording system based on distributed mini computers linked to a large mainframe complex and thence to distributed display terminals. Large amounts of data are handled and the use of fibre optic transmission is attractive,however this will not be discussed in detail here as the requirements are largely similar to any other distributed computing network.

One example will be discussed because of its possible implications for the future.This is a data acquisition facility installed on an open air test bed used for engine noise measurements. The engine is surrounded by microphones, meteorological sensors etc. and signals from these are transmitted via conventional wiring to an underground room beneath the test stand where they are multiplexed into a fibre for transmission to the engine control room 300m away. Figure 4 is an aerial view of this facility with the route of the fibre marked. The link was custom designed for Rolls-Royce by a sub-contractors and operates at 27 MB/s using 200μcore PCS fibre with an LED transmitter and PIN diode receiver.

Open air noise test site showing fibre optic link

FIGURE 4

This approach could be extended to all development engines with a multiplexer mounted on the engine gathering all the instrumentation information. If this eliminated the large amount of conventional wiring in test beds it could produce significant cost savings and the replacement of the current complex multiway connector used to link an engine into the test bed (Figure 5) with a single fibre optic connector has obvious benefits,speeding up the installation of the engine on the bed and enabling it to be easily connected to any data acquisition facility that had spare capacity.

Multiway connector plate carrying instrumentation and control data between engine and test bed

FIGURE 5

The collection of data into the multiplexer from conventional transducers via fibre optic links could be considered, but the short runs involved and the modular construction of some engines (necessitating many connectors) make this approach unattractive unless it can be combined with the advantages of passive optical sensors.

For certain instrumentation the fibredyne data highway (Ref 3) offers a possible compromise. A number of conventional sensors distributed around the engine would each be equipped with an ultrasonic modulator running at a different carrier frequency. During the later stages of engine build a single multimode fibre would be clipped into each transducer in turn and then led to a signal recovery unit and demultiplexer mounted either on or off the engine. Most of the advantages offered by the fibre are lost unless power for the sensors and modulators can be supplied optically or from batteries, however, the very low power requirements of the modulators make this a possibility.

Another data transmission problem is the acquisition of signals from transducers mounted on rotating components. Currently this is done via conventional slip-rings or radio telemetry At first sight an optical slip ring appears attractive for higher rates but there are a few problems. The temperatures encountered on rotating components are usually too high for semiconductor light sources and on multishaft engines transmission access to the axis of rotation is not possible, implying some sort of fibre optic hoop collector wrapped round the engine at a diameter of about 30 cm.

SENSOR APPLICATIONS

As was mentioned in the introduction the environment on a jet engine can be very harsh and in particular high temperatures are encountered. This has led to the development of high temperature optical sensor constructions and some unusual applications of fibres. The following subsections detail some of these.

(a) Endoscopy

In the past aero engines were serviced on a regular overhaul basis and were not designed for inspection on the wing. Any unscheduled in-service inspection that was required had to be done with flexible fibre endoscopes through any available access points. This is still necessary in special cases but routine examination of flight engines is now done through purpose designed ports with rigid endoscopes as these are cheaper and give a better image. These ports have been introduced to permit regular internal inspection as the current generation of civil engines are only removed from the aircraft for servicing when their condition indicates the need.

Components such as turbine blades with complex cast in cooling channels are inspected before they get to the engine with rigid needle endoscopes using Selfoc fibres.

(b) Flame Detection

Requirements for the detection of flames in engines or combustion test rigs arise occasionally and naturally the areas where flames can occur are often too hot for electronic components to survive, although some use has been made of high temperature ultraviolet sensitive Geiger tube detectors. Fibre optics can therefore be useful to lead the short wavelength radiation from flames out to where an electronic detector can be situated.

A requirement arose in the mid 1970's for a sensor to be used to investigate "flame-out" in an annular combustor. The design used a short length of nichrome plated quartz rod to relay the flame radiation to a conventional glass bundle when then led it to a blue filtered photodiode. The overall transmission of this system was obviously quite poor due to the quartz rod but at the time this appeared to be the only type of light guide that would survive the very high temperatures encountered near the combustor. This particular sensor was never run on an engine but the same idea has recently been used (Ref 4) to introduce light into an engine to enable component movement to be measured by an image correlation technique. In this case the light guide used was a nichrome plated sapphire rod.

Another flame detector is shown in Figure 6. This has been successfully used to monitor ignition in a high temperature, high pressure combustion rig. The light guide is an air cooled glass clad rod.

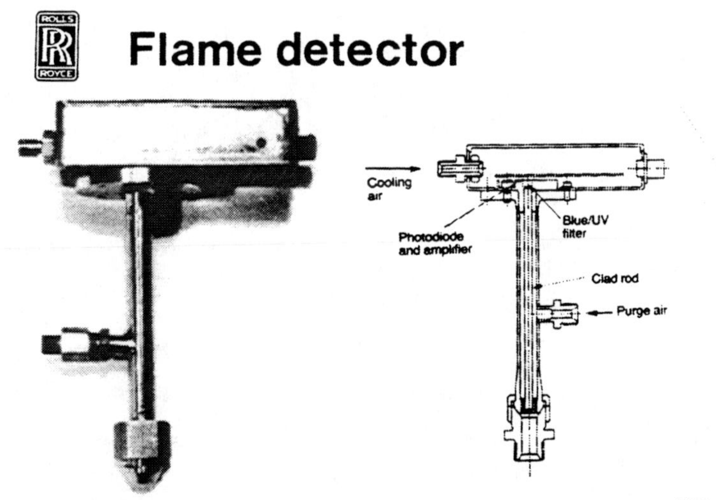

FIGURE 6

(c) Metal Temperature Measurement by Radiation Pyrometry

The use of radiation pyrometry for the measurement of turbine blade temperatures dates from the 1960s. The detector used is almost universally a silicon photodiode and the early pyrometers used a simple lens to concentrate radiation from the blades onto it. As a pyrometer is fundamentally an intensity sensitive device requiring response down to DC, stabilisation of the photodiode and its amplifier is critical. The most important turbine stage from an efficiency point of view is the high pressure stage and the environment above this can be at 500 - 600 °C. Optical systems using sapphire lenses brazed into metal holders have been developed to withstand these temperatures and as soon as fibre optics became widely available attempts were made to use them to enable the detector to be situated remotely. Some early work was done on a combustion can using epoxied glass fibre bundles in hypodermic tubing and turbine blade pyrometers also used epoxy terminated fibres initially. This limited them to areas where the termination could be kept below 250 °C and even here problems were encountered with the epoxy outgassing and/or creeping over the end of the fibres thus affecting calibration. Modern versions use hot formed termination which avoids these problems and also improves transmission because of the higher packing density. Figure 7 shows a typical fibre optic coupled turbine blade pyrometer. The pyrometer used by the control system of the RB 199 engine,which powers the Pan Avia Tornado,is probably the only example of a fibre optic sensor currently in use on a production gas turbine engine.

Typical fibre optic coupled pyrometer

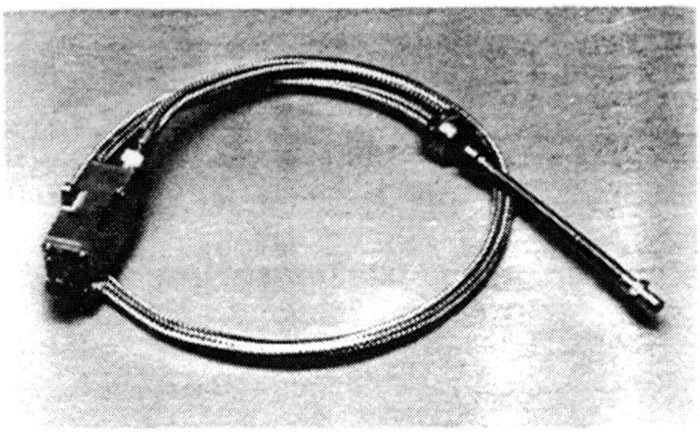

FIGURE 7

(d) <u>Laser Anemometry</u>
Laser two spot anemometry is an accepted method of studying airflows which has been widely used in turbomachinery applications. The basic principle (shown in Figure 8) involves timing the passage of small particles in the flow between two intense spots of light by detecting the light scattered by the particles as they pass through the spots. The flow patterns of interest are such that particles less than one micron in diameter must be used in order that they will adequately follow the flow direction. To give sufficient scattered signal the spots must be very small and very intense requiring the use of a high power laser source with diffraction limited focussing optics. A commercial system has been in use on fan test rigs for a number of years where it has proved very effective, but it is too large and delicate for applications where the anemometer must be rigidly mounted on an engine. It was therefore decided to have a miniature head extension to this system built containing only those components (mainly the focussing optics) which need to be attached to the engine and couple this to the main head containing the laser, photomultipliers etc.using fibre optics. The need to obtain diffraction limited spots dictated the use of monomode fibres for coupling the laser to the miniature head but the specification for the collection fibres transmitting the scattered light back to the photomultipliers is not so critical and PCS fibres were used here. Figure 9 shows the original anemometer head with the miniature extension (inside a water cooling jacket) in the foreground.

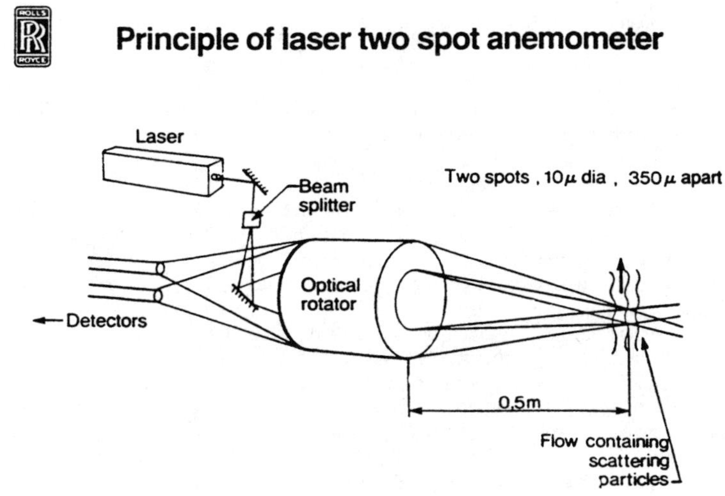

FIGURE 8

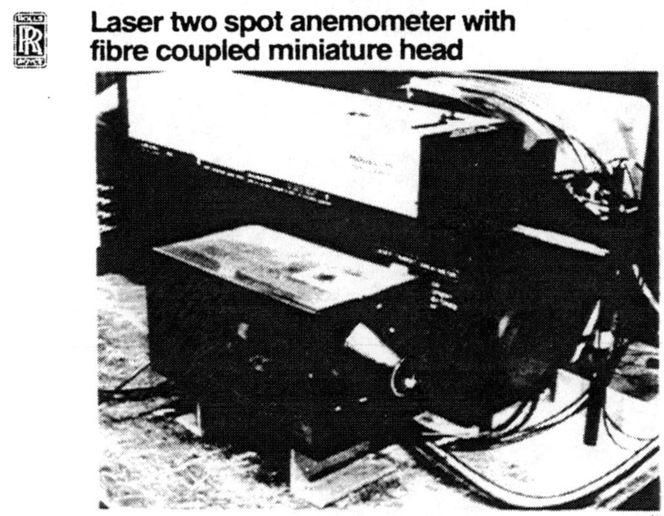

FIGURE 9

(e) Blade Tip Clearance Sensors

The clearance between rotating blades and the shroud surrounding them affects aerodynamic efficiency quite dramatically particularly in the high pressure turbine stage. Due to the hot supersonic gases and the presence of soot, unburnt fuel and intense thermal radiation from the blades, this is a particularly difficult parameter to measure. Some success has been obtained by American manufacturers using an optical triangulation technique (Refs 5,6,7). The principle involved (Figure 10) is to focus a spot of light onto the end of the blade at an angle and monitor the movement of the spot across the blade as the blade moves in and out. Figure 11 shows the folded version of the system which must be used to penetrate the multilayer casings around a turbine. Fibre optic coupling of the light source and detector is usually used.

Although effective this approach has some disadvantages, particularly if the sensor is to be used for control purposes on flight engines. The prism in figure 11 must be quite close to the gas stream making it difficult to keep clean, the inherent asymmetry of the system makes it expensive to manufacture and the linear position sensor required can lead to complex electronics or drift problems. In view of possible requirements for a flight standard sensor it was decided to make a preliminary study of an alternative fibre optical sensor which could use the constructional methods already proven by flight standard pyrometers. This is the astigmatic sensor (Ref.8), the princple of which is shown in Figure 12. The light is focussed into a spot on the end of the blade by an astigmatic lens system which then re-images the scattered light back into the plane of the fibre end. Due to the astigmatism this image and the spot on the blade will change shape as the blade moves along the probe axis. (Figure 12). If four collection fibres are placed around the input fibre as shown the ratio of the light outputs from diagonally opposite pairs of fibres will give an output dependant on blade position but independant of source intensity, lens contamination or blade reflectivity. The fibre optic termination of a prototype engine sensor is shown in Figure 13, where the four 400 micron core PCS collector fibres can be seen. More difficult to see is the central 250 micron core PCS emitting fibre. This sensor works in the laboratory and is awaiting evaluation of an engine. A calibration curve is shown in Figure 14.

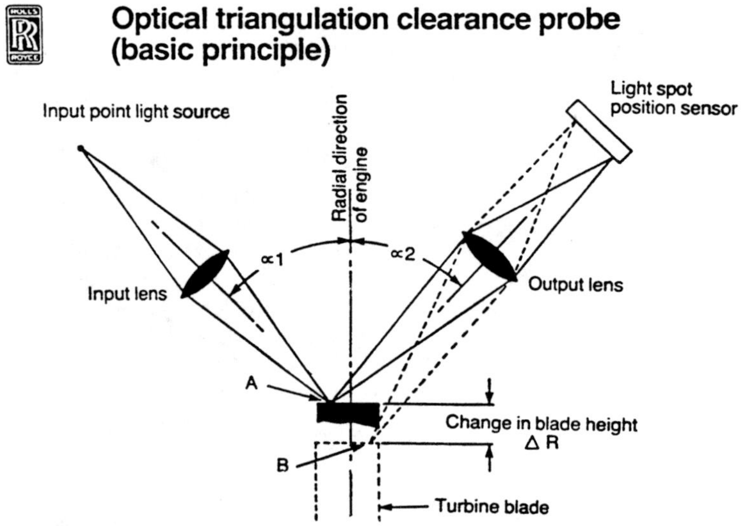

FIGURE 10

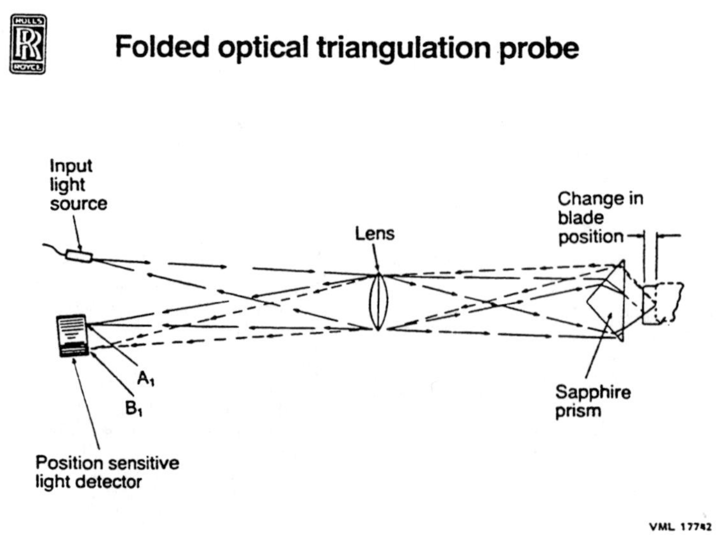

FIGURE 11

Astigmatic clearance probe schematic

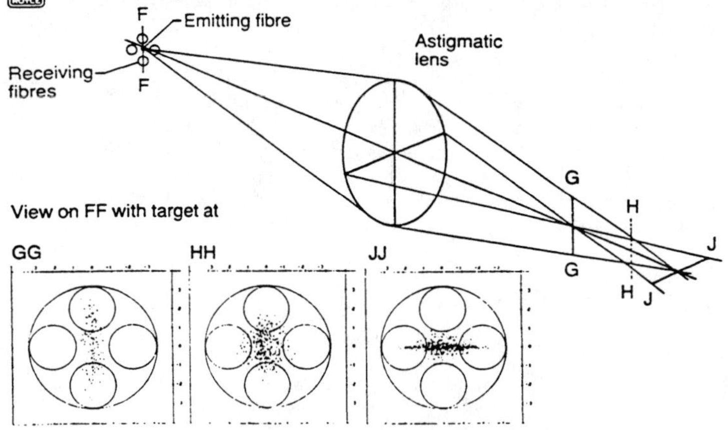

Output $= \dfrac{A - B}{A + B}$ where $A = 1 + 3$, $B = 2 + 4$

FIGURE 12

Fibre termination of astigmatic probe

FIGURE 13

Astigmatic probe calibration curve

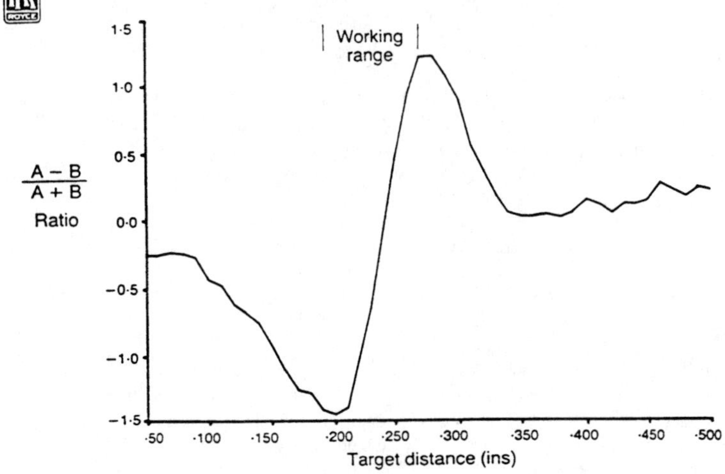

FIGURE 14

(f) <u>Fibre Optic Pinhole Camera</u>

Early in 1979 a requirement arose for monitoring the formation of carbon on the fuel burners of an RB 211. As access to the burners was very restricted, temperatures very high and contamination by fuel, soot and other combustion products very likely it was decided to manufacture a fibre optic coupled pinhole camera which could be installed in place of one of the burners to view the adjacent burner in the combustion can. The use of a pinhole as an imaging element has the advantages that it can be made to withstand high temperatures and has no optical surfaces to get contaminated. A rigid glass fibre image conduit was used behind the pinhole to transmit the image away from the hot area near the combustor. Cooling air flowing down the outside of the image conduit was vented through the pinhole to prevent the ingress of contamination. The image was transferred from this device to a video camera by a flexible coherent fibre bundle. The light losses inherent in this system were not important in view of the high brightness inside of the combustor. The sort of result obtained after image processing is shown in Figure 15. A similar idea has recently been used in the USA to study rocket exhausts (Ref 9).

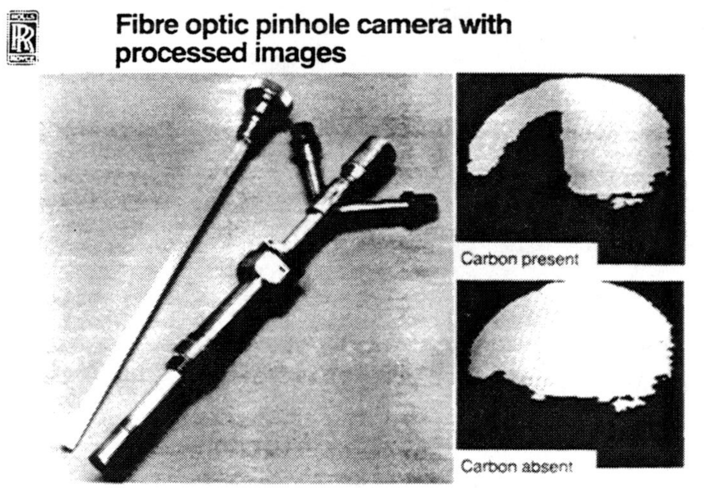

FIGURE 15

FUTURE DEVELOPMENTS

The future for passive fibre optical sensors on gas turbine engines looks promising in the long term although there are still many problems to resolve.

On production engines the modular construction mentioned in section 2 implies that sensors which are tolerant of multiple connectors will have to be developed and a suitable connector design proven. Exceptional cases such as pyrometers, fundamentally requiring an intensity measurement are unlikely ever to involve connectors and this may restrict their application especially on civil engines.

More information is required on the behaviour of fibres in conditions outside those encountered in telecommunications uses and recent developments, for example the temperature sensor usable up to 2000 $^{\circ}$C developed at the United States National Bureau of Standards (Ref 10) are of considerable interest.

A lot of effort will have to be put into the reliability and maintenance requirements of fibre optic equipment under the vibration levels and temperature excursions encountered on production engines. Whilst reliability and maintainability may be important in ground based applications they are crucial to the acceptance of fibre optic sensors on aircraft, particularly in the civil market. Airline operators will not appreciate having to employ a staff of fibre optic specialists at every airport they fly into and it will be necessary to prove to them that such systems can be supported easily on a world-wide basis.

On development engines the situation is different. Optical measurement techniques have been used for many years and fibre optics are increasingly being used to extend them into previously inaccessible areas. Often the main design problem is the lack of suitable information on fibre performance, for instance a possible application for the remote monitoring of optical structure in 30 ns Q switched ruby laser pulses would require information the dispersion properties of fibres at 694.3 nm which is not readily available.

Despite the above problems the use of light guides in gas turbine instrumentation can be expected to grow and as has been shown in this paper this can be expected to involve the full range of products from simple rigid clad rods to sophisticated monomode fibres.

REFERENCES

1. Baumbick R.J. Alexander J. Katz R and Terry J.
   Optical Sensors for Aeronautics and Space NASA Technical Memorandum 81407.

2. Baumbick R J.
   Fibreoptics Technology and its Application to Propulsion Control Systems AIAA-83-0534.

3. Culshaw B, Ball P.R. Pond J C and Sadler A.A.
   Optical Fibre Data Collection. Electronics and Power - February 1981 P 148-150.

4. Tuxworth R W
   A High Temperature Optical Displacement Measuring System. Transducer Tempcon 1983

5. White S D
   Turbine Tip Clearance Measurement. USARTL-TR-77-47 (Avco Lycoming Division).

6. Baker L C, Grady G.E, Mench H R.
   Turbine Tip Clearance Measurement. USARTL-TR-78-4 (General Electroc Co.)

7. Barranger J P and Ford M J
   Laser-optical Blade Tip Clearance Measurement System. ASME Gas Turbine Conference 1980. (P & W Aircraft Co.)

8. Davinson I. (Rolls-Royce Ltd.)
   UK Patent - GB 2066449, US Patent 4357104.

9. Electro Optical Systems Design - December 1982, P.22

10. Dils R R
    High Temperature Optical Fibre Thermometer
    J Appl. Phys. 54(3) March 1983.

Reprinted with permission from *IEEE Journal of Quantum Electronics*, Vol. QE-18(10), pp. 1624-1633 (October 1982). ©1982 IEEE.

Fiber-Optic Fabry–Perot Interferometer and Its Sensor Applications

TOSHIHIKO YOSHINO, KIYOSHI KUROSAWA, KATSUJI ITOH, AND TERUZI OSE

***Abstract*—Fiber-optic Fabry–Perot interferometers using monomode fibers are fabricated and their basic properties of finesse, polarization, and thermal response are studied. Fiber-optic Fabry–Perot interferometers are applied to the sensors of temperature, mechanical vibration, acoustic wave including human voice, ac voltage, and ac and dc magnetic fields. It has been demonstrated that a fiber-optic Fabry–Perot interferometer can simplify the interferometric fiber sensor system and that high measurement sensitivity can be obtained by using a high-finesse and/or long-distance fiber Fabry–Perot interferometer.**

I. Introduction

THE advent of monomode fibers has opened doors for the coherent use of optical fibers. As monomode fibers can carry light coherently, they can be used for arms in interferometers. Until now, two different forms of the fiber-optic interferometer using monomode fibers have been well known. One is the Mach–Zehnder form, which is used for the interferometric fiber sensor of acoustic wave [1], temperature [2], and magnetic field [3]. The other is the Sagnac form, which is used for fiber gyro [4]. Both of the interferometers utilize the two-beam interference. Recently, the Fabry–Perot form of fiber-optic interferometer, which utilizes the multiple-beam interference, has been demonstrated [5] and followed by some authors [6], [7]. The concept of using the fiber-optic Fabry–Perot interferometer for a sensor is noted in [8]. The fiber-optic Fabry–Perot interferometer (FFPI) developed by the authors consists of a single monomode fiber with dielectric-coated high-reflectance end faces, not with butt-coupled mirrors used by other authors [6], [7]. FFPI has many attractive features not only as the spectrum analyzer, but also the interferometric fiber sensor. In this paper, experimental and theoretical studies of FFPI made by the present authors are presented. In Section II, basic properties of FFPI are investigated. In Section III, thermal characteristics of FFPI are studied. In Section IV, FFPI's are applied to the sensor of temperature, vibration, acoustic wave, voltage, and magnetic field.

II. Basic Properties of FFPI

A FFPI is schematically shown in Fig. 1. Two end faces of a monomode fiber are optically polished and coated with the multilayer of dielectric films by the vacuum evaporation method. The monomode fiber used consists of a 4 μm diameter silica core and a 125 μm diameter silica cladding, covered with a 0.9 mm outer diameter nylon jacket. The optical loss of the fiber is 7.5 dB/km at a wavelength λ of 633 nm. FFPI has a 30 mm long metal sleeve on each end part, but is equipped with no mode stripper. Fig. 2 shows the experimental setup. The light source is an internal mirror He-Ne ($\lambda = 633$ nm) laser, normally operated at two axial modes with a frequency separation of 640 MHz. The output power is 2.4 mW. The two lasing modes are linearly polarized but have orthogonal azimuths to each other. Using the polarization properties, the laser frequency is stabilized by a method described in [9]. The single mode laser light or the two-mode laser light with an arbitrary mode-intensity ratio was obtained by passing the output laser light through a suitably oriented polarizer. The laser light is passed through a 20X microscope objective and injected into FFPI. In order to avoid the undesirable optical coupling between FFPI and the laser cavity, the incident light was somewhat defocused from the end face of FFPI; thereby the laser instability caused by the optical coupling could be smaller than 1 percent. Part of FFPI, typically 5 cm lengths, was wound about a 100 mm diameter and 15 mm thick piezoelectric transducer (PZT) driven by 50 Hz sinusoidal voltages. The fiber length is then modulated, so FFPI operates as a scanning Fabry–Perot interferometer. The use of FFPI as the spectrum analyzer has also been very recently reported by [7]. A number of FFPI's are studied as parameters of fiber length L of 1 cm to 100 m and end reflectance R of 60–90 percent. Fig. 3(a)–(c) shows the light intensity transmitted through three different FFPI's, where the incident light is the single mode. The horizontal axis δ represents the phase delay which the fiber mode undergoes per pass through FFPI and is given by

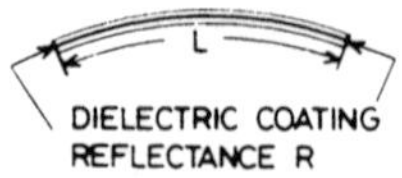

Fig. 1. Geometry of fiber-optic Fabry–Perot interferometer (FFPI).

$$\delta = kL \approx 2\pi nL/\lambda \tag{1}$$

where k is the phase velocity of the fiber mode and n is the refractive index of silica, equal to 1.46. Successive peaks of the transmission curves occur at $\delta = m\pi$, m being integers. Fig. 4 shows the output mode patterns of some FFPI's. The output mode pattern involves more or less the speckle pattern, which stems from the cladding mode. As is seen from Fig. 4,

Manuscript received March 9, 1982; revised June 4, 1982.

T. Yoshino and T. Ose are with the Institute of Industrial Science, University of Tokyo, Roppongi, Minato-ku, Tokyo 106, Japan.

K. Kurosawa is with the Tokyo Electric Power Corporation, Ltd., Tokyo, Japan.

K. Itoh is with the Furukawa Electric Corporation, Ltd., Chiba, Japan.

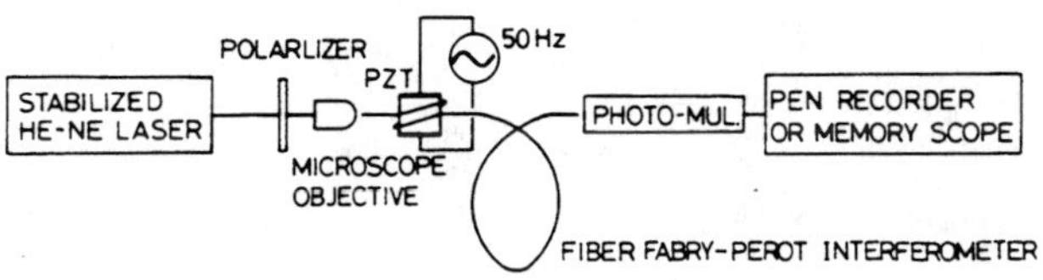

Fig. 2. Experimental setup for FFPI.

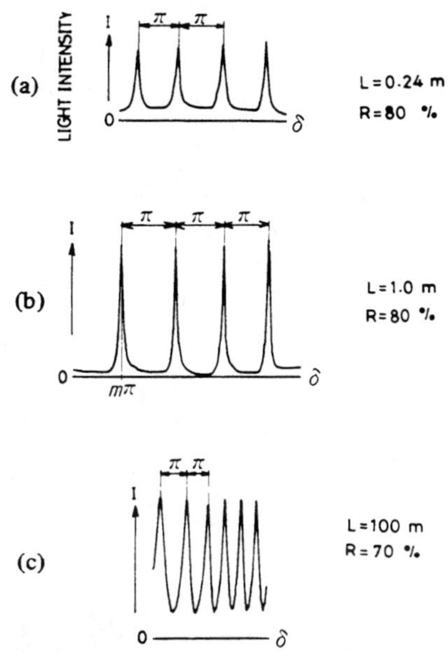

Fig. 3. Output light intensity I of scanning FFPI. δ represents the phase delay due to single pass through FFPI.

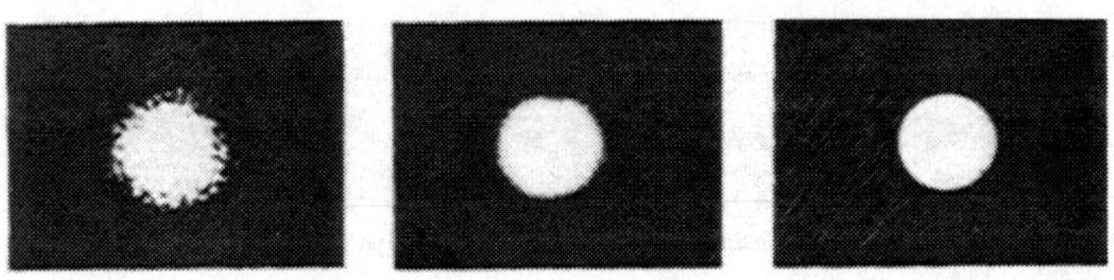

(a) L= 24cm, R=60%. (b) L= 1 m, R= 80%. (c) L=100m, R=70%.

Fig. 4. Output pattern of FFPI.

the speckle pattern component is generally larger for shorter FFPI. The speckle component involved in the output light intensity gives a bias in the transmission curve. In order to represent the transmission characteristics of FFPI without being affected by such spurious bias levels, we introduce an effective finesse $\mathfrak{F}_e$, defined as

$$\mathfrak{F}_e = \pi/\Delta\delta \tag{2}$$

where $\Delta\delta$ represents the full-width of δ yielding half the peak-to-peak transmittance (not half the maximum transmittance in the conventional definition [10]), as shown in Fig. 5. Theoretically, the transmittance of FFPI is given by

$$\tau = 1/(1 + F \sin^2 \delta) \tag{3}$$

in accordance with the conventional etalon [10]. Here

$$F = 4R/(1 - R)^2. \tag{4}$$

Considering that $\tau_{max} = 1$ and $\tau_{min} = 1/(1 + F)$, it can be shown that

$$\mathfrak{F}_e = (\pi/2)/\sin^{-1} (1/\sqrt{F+2}). \tag{5}$$

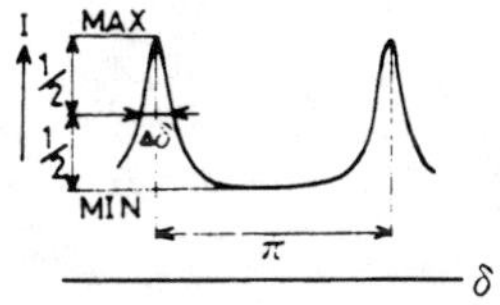

Fig. 5. Relationship for representing effective finesse $\mathfrak{F}_e$ of FFPI.

The free-spectral range of FFPI is, from (1), given by

$$\Omega_f \approx c/(2nL), \tag{6}$$

c being the light velocity in vacuum. Since the linewidth of the single mode of the He-Ne laser is as small as 2 kHz [11], the transmission curves as shown in Fig. 3 may be considered as those for a perfectly monochromatic light. From the transmission curves as shown in Fig. 3, the effective finesse $\mathfrak{F}_e$ was determined for various FFPI's. Some of the measured results are listed in Table I, together with the theoretical values. Comparing the experimental and theoretical values of $\mathfrak{F}_e$, both are generally in fairly good agreement. In some of the fabricated FFPI's, however, the experimental values are a few times smaller than the theoretical ones. The most probable cause for this is that the end faces of such FFPI's are tilted to the fiber axis. In that case, the fiber mode which is internally reflected on the end face of the fiber, cannot perfectly couple with the fiber mode. Letting the tilt angle be θ, the internal coupling efficiency, from the result for tilt losses in connecting fibers [12], is calculated to be

$$\eta = 1 - K_\theta (2n\theta/\sqrt{2\Delta})^2 \tag{7}$$

where $\Delta = 0.002$, which is the relative refractive index difference between the core and the cladding, and $K_\theta = 1.8$, which is the angular loss coefficient. If the coupling efficiency is η, the end reflectance of the fiber effectively becomes ηR. In order to achieve $\eta = 0.9$, it is required from (7) that the tilt angle θ should be smaller than $0.29°$, the fulfillment of which is not technically difficult. It should be noted that, in the present experiment, no particularly exact polishing was conducted so as to make the tilt angle as small as possible. The effective values of end reflectance R_e were also calculated by corresponding the measured $\mathfrak{F}_e$ values to the theoretical ones given by (5) and the results are also shown in Table I for comparison. In general, in the conventional Fabry-Perot interferometer or etalon, the tilt of the reflection surface causes the light beam to walk off the interferometer during the multiple passes through the interferometer, thereby causing the walk-off loss of light power. This walk-off loss is a very serious problem in the conventional form of Fabry-Perot interferometer. The walk-off loss increases with an increase in the interferometer length, so that long-distance Fabry-Perot interferometers, say longer than several meters, are hardly possible to construct. On the contrary, in FFPI, owing to the light trapping capability of optical fibers, there exists no walk-off loss; the presence of the tilt of the reflection surfaces simply reduces the effective end reflectance. As a consequence, it is much easier to fabricate FFPI than to construct the conventional form of Fabry-Perot interferometer. The spectral resolution of FFPI is given by $\Omega_f/\mathfrak{F}_e$ and increases with increasing Ω_f if $\mathfrak{F}_e$ is constant. For the 100 m

TABLE I
CHARACTERISTICS OF FFPI

FFPI length L	Surface reflectance R	Effective finesse $\mathfrak{F}_e$ Exp.	Theo.	Effective reflectance Re	Free-spectral range Ω_f
0.01m	90%	20	29	85%	1GHz
0.24m	80%	10	14	71%	0.43GHz
1.0m	80%	13	14	77%	103MHz
10m	80%	6	14	54%	10MHz
100m	70% (59%)	3	9 (6.3)	28%	1MHz

The number in the parentheses represents the theoretical value taking into consideration the optical transmission loss in fiber.

FFPI shown in Table I, $\Omega_f = 1$ MHz and $\mathfrak{F}_e = 3.3$, so that the spectral resolution is as high as about 300 kHz. Such a high spectral resolution is mainly owing to the very small free-spectral range of the FFPI. Fig. 6(a) and (b) shows the transmission curves in the cases that the input light involves two axial modes f_1 and f_2 ($f_1 - f_2 = 640$ MHz) with different intensity ratios. The transmission curve is given by the two independent transmission curves corresponding to each mode f_1, f_2.

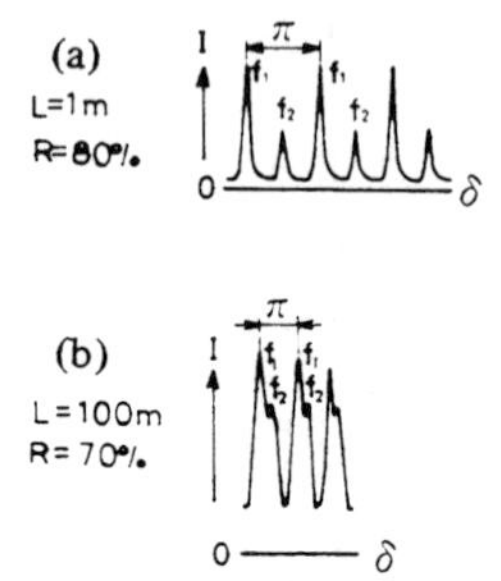

Fig. 6. Output light intensity I of scanning FFPI in the case that input light involves two modes f_1 and f_2 ($f_1 - f_2 = 640$ MHz).

The polarization properties are also studied. Generally, a monomode fiber, when placed in a coil form without twisting, has linear birefringence. Therefore, a coiled FFPI has the eigenstate of polarization consisting of orthogonal linear polarizations; their azimuths of polarization are either parallel or orthogonal to the plane where the fiber is placed. Each polarization mode has different phase velocity. Fig. 7 shows the light intensity transmitted through the 1 m FFPI (R = 80 percent) when two polarization modes s and p are simultaneously excited. It was achieved by passing the input light through a 45° polarizer. The input light is the single mode in Fig. 7(a) but is the two modes in Fig. 7(b). In the latter case, the transmission curve is given by the superposition of four different curves corresponding to the four combinations of two polarization modes s and p and two frequency modes f_1 and f_2. Naturally, such polarization-related complexity of the transmission characteristics of FFPI can be eliminated by using the polarization-maintaining fiber for the etalon fiber.

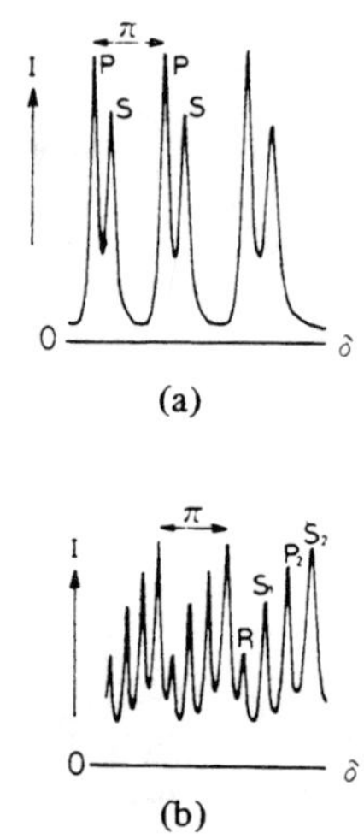

Fig. 7. Output light intensity I of scanning 1 m FFPI (R = 80 percent) in cases where input light involves single frequency but two polarization modes s and p, (a), and two frequencies f_1 and f_2 and two polarization modes s and p, (b).

From the above study, it has been verified that 1) FFPI can have high finesse, 2) FFPI makes it possible to achieve very long distance Fabry-Perot interferometer, thereby achieving very high spectral resolution, 3) FFPI requires much less precision for the tilt angle of the reflection surface than the conventional form of Fabry-Perot interferometer or etalon, and 4) FFPI is particularly well suited for a scanning Fabry-Perot interferometer because the optical path length in fiber can be readily modulated by appropriate external perturbations such as mechanical forces or temperature.

III. THERMAL CHARACTERISTICS OF FFPI

The temperature characteristics of FFPI are important for the practical use of it. The temperature induced phase shift in FFPI was measured with an FFPI of L = 24 cm (R = 60 percent). In order to prevent the undesirable effect of mechanical vibrations, the FFPI was inserted into a hollow glass tube of 19 cm length and 3 mm inner diameter. The glass tube was slowly heated or cooled by blowing hot or cool air on it. The output light intensity from the FFPI then changed with time as shown in Fig. 8(a); the input laser light is the single mode. One interference fringe or pulse corresponds to the change of $\lambda/2$ in the optical path length of FFPI. When the FFPI was cooled from 65 to 30°C, the number of pulses was counted as a function of the fiber temperature change ΔT, which was monitored by a thermistor

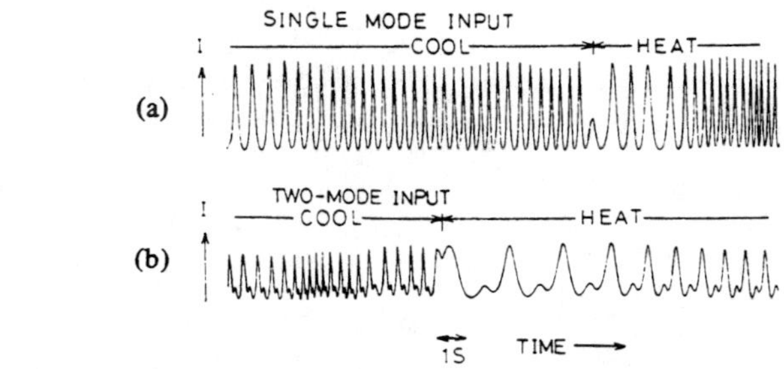

Fig. 8. Output light intensity I of 24 cm FFPI (R = 60 percent) when it was heated or cooled; input light is (a) single mode and (b) two modes.

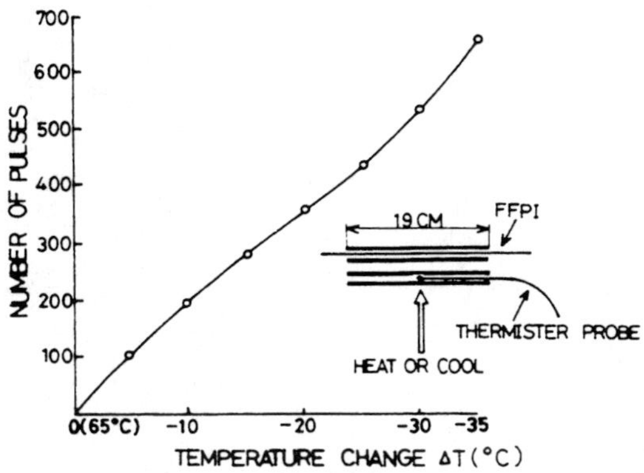

Fig. 9. Number of light pulses counted as a function of temperature change of 19 cm lengths of FFPI.

probe placed inside the glass tube. Fig. 9 shows the measured results, together with the experimental scheme. From Fig. 9, the average temperature sensitivity of the FFPI is obtained as $S_{\text{jacket}}(\text{pulse}) = 100/°\text{C} \cdot \text{m}$, which yields the temperature sensitivity in the relative phase shift as $S_{\text{jacket}}(\text{phase}) = \Delta\phi/(\phi\Delta T) = 2.2 \times 10^{-5}/°\text{C}$, where $\phi = nL$. Similar experiments were done in the case that the FFPI was unjacketed. The measured temperature sensitivity is $S_{\text{unjacket}}(\text{pulse}) = 36/°\text{C} \cdot \text{m}$, yielding $S_{\text{unjacket}}(\text{phase}) = 7.9 \times 10^{-6}/°\text{C}$. Thus, the temperature sensitivity for jacketed fibers is by a factor of 2.8 larger than that for unjacketed ones. Similar results have been recently reported [13], [14].

Theoretical considerations of the temperature sensitivity are done on a simple model. The fiber is composed of the core and the cladding with typically similar properties of silica, the cushion layer made of silicone, and the nylon jacket. The parameters of each component are listed in Table II. When the fiber system undergoes a temperature change of ΔT, the following relative phase shift is induced in the core

$$S(\text{phase}) = \frac{\Delta\phi}{\phi\Delta T} = \frac{1}{n}\frac{\partial n}{\partial T} + \frac{(\Delta n)_{\text{strain}}}{n} + \frac{\Delta L}{L} \tag{8}$$

where ΔL is the net expansion of the fiber length. The quantity $(\Delta n)_{\text{strain}}$ represents the strain induced change in the refractive index of the core, which is, by an ordinary photoelastic calculation, shown to be

$$(\Delta n)_{\text{strain}} = \frac{n^3}{2}\,[(p_{11} + p_{12})\nu - p_{12}]\,\sigma_1/E_1 \tag{9}$$

where p_{11} and p_{12} are the photoelastic constants of silica and ν is Poisson's ratio; σ_i and $E_i (i = 1, 2, 3)$ are, respectively, the axial stress and Young's modulus of each fiber component numbered in Table II. Assuming that all the fiber components are uniformly deformed without slip at the boundaries, it follows that

$$\Delta L/L = \alpha_i + \sigma_i/E_i \qquad (i = 1, 2, 3) \tag{10}$$

under

$$\sum_{i=1}^{3} \sigma_i S_i = 0 \tag{11}$$

where α_i and S_i are the linear expansion coefficient and cross-sectional area of each fiber component, respectively. Using (10), (11),

$$\sigma_1 = (\beta - \alpha_1)\,E_1$$

where

$$\beta = \frac{\Delta L}{L} = \sum_{i=1}^{3} (\alpha_i E_i S_i) \Big/ \sum_{i=1}^{3} (E_i S_i). \tag{12}$$

Hence, (8) becomes

$$S(\text{phase}) = \frac{1}{n}\frac{\partial n}{\partial T} + \frac{n^2}{2}\,[(p_{11} + p_{12})\nu - p_{12}]\,(\beta - \alpha_1) + \beta \tag{13}$$

which is in substantial accordance with the result in [14], neglecting tensor relations among the stresses and strains.

The numerical values of $E_i S_i$ and $\alpha_i S_i E_i$ are also shown in Table II. The numerical calculation of (12) gives

$$S_{\text{jacket}}(\text{phase}) = [(0.076) + (-0.014) + (0.24)] \times 10^{-4}/°\text{C} \tag{14a}$$

$$= 3.0 \times 10^{-5}/°\text{C}. \tag{14b}$$

Each parenthesis in (14a) corresponds to each term in (8). It is seen from (12) and (13) that S depends on the parameters of the silicone layer and nylon jacket through β, which depends on $E_i S_i$ and $\alpha_i E_i S_i$. As is seen in Table II, $E_1 S_1$ and $E_3 S_3$ have comparable magnitudes to each other, but are much larger than $E_2 S_2$; in addition, $\alpha_2 E_2 S_2 \ll \alpha_1 E_1 S_1$ or $\alpha_3 E_3 S_3$. Consequently, the effect of the silicon cushion layer on S is negligibly small. The calculated value of $S_{\text{jacket}}(\text{phase})$ of $3.0 \times 10^{-5}/°\text{C}$ agrees fairly well with the measured result of $2.2 \times 10^{-5}/°\text{C}$. The calculated value of $S_{\text{unjacket}}(\text{phase})$ is $8.0 \times 10^{-6}/°\text{C}$, which agrees well with the measured value of $7.9 \times 10^{-6}/°\text{C}$.

IV. Application of FFPI to Fiber Sensors

A. Temperature

The temperature change can be measured by counting the number of pulses as shown in Fig. 8(a), where one pulse corresponds to the temperature change of 0.04°C. For the practical temperature sensor, it must be distinguished whether temperature is rising or falling. In order to supply this ability for FFPI, the two-mode laser light was used as the input light. Fig. 8(b) shows the output light signal in that case. The output signal involves two peaks, in one period, corresponding to the two mode input whose intensity ratio is 5:1 in this case.

TABLE II
FIBER PARAMETERS

	Diameter d (mm)	Young's modulus $E (10^8 N/m^2)$	Cross-section $S (mm^2)$	Linear expansion coeff. $\alpha (10^{-4}/^\circ C)$	ES $(10^2 N)$	αES $(10^{-2} N/^\circ C)$
1. Core and cladding (Silica)	0.125	730	0.012	0.004	8.8	0.035
2. Cushion layer (Silicone)	0.4	0.01	0.11	2.5	0.0011	0.0028
3. Jacket (Nylon)	0.9	5.5	0.52	1	2.8	2.8

$\partial n/\partial T = 1.1 \times 10^{-5}/^\circ C$, $P_{11} = 0.121$, $P_{12} = 0.270$, $\nu = 0.17$

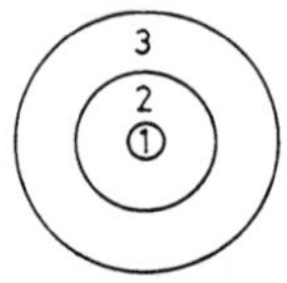

It is manifested in Fig. 8(b) that the order in which the two peaks appear is reversed between whether the temperature is rising or falling, i.e., the optical path length of FFPI is increasing or decreasing. Using this phenomenon, the direction of temperature change can be discriminated.

Since the nylon-jacketed FFPI with a length $L(m)$ has the temperature sensitivity-in-pulse-number of 100 $L/^\circ C$ as shown in Section III, very high temperature sensitivities can be obtained with long FFPI. This was demonstrated by some experiments. Fig. 10(a) shows the output light signal when a human came near a 10 m long FFPI (R = 70 percent), which was coiled in 10 turns; the input light is the two-mode. The fiber temperature was then changed by the termal radiation from the human body and convection. Fig. 10(b) shows the output light signal in the case where a human hand was made close to the FFPI; in this case, the fiber was most dominantly heated by the thermal radiation. It is clearly seen in both figures that small temperature changes caused by the human body were very sensitively and quickly detected by the FFPI. The pulse height reduction seen in Fig. 10 is only apparent, stemming from the relatively slow response of the pen recorder. Fig. 11 shows the output light signal in the case where a human hand was made near the 100 m FFPI coiled in 100 turns. The output signal was recorded with a memory scope. In the FFPI, if the temperature change is uniform over the entire length of it, one pulse of the light signal corresponds to a temperature change as small as 0.1 m · °C. Such a high temperature sensitivity is an attractive feature of a long FFPI and would find interesting applications.

The thermal response time of FFPI is one important factor for the temperature sensor. The thermal response time of a cylinder structure, defined by the time required for attaining to $1/e$ of the initial temperature difference, is given by $\tau_h = 0.12a^2 \rho C/K$ [15], where a, ρ, C, and K are the radius, density, specific heat, and thermal conductivity of the cylinder, respectively. For a silica cylinder with $2a$ = 125 μm, as ρ = 2.22 g/cm^3, C = 0.20 cal/g · °C, and $K = 3.5 \times 10^{-3}$ cal/cm · s · °C, τ_h is calculated to be 0.6 ms, which is very small. The response time of the jacketed fiber can be reasonably approximated by that of a nylon cylinder of $2a$ = 0.9 mm. As for nylon, ρ = 1.12 g/cm^3, C = 0.46 cal/g · °C, and $K = 5.2 \times 10^{-4}$ cal/cm · s · °C, τ_h is calculated to be 0.2 s, which is also suf-

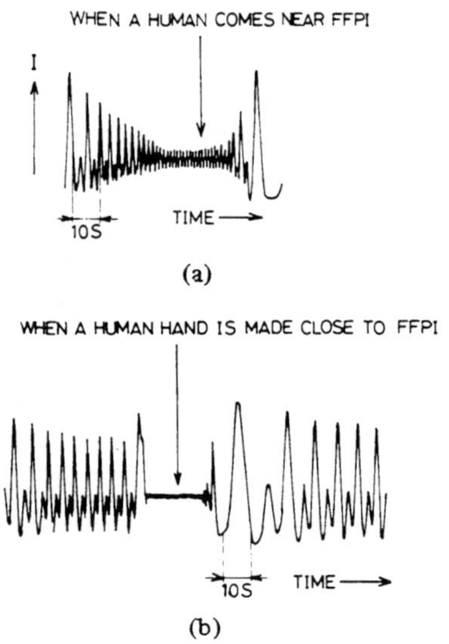

Fig. 10. Output light intensity I of 10 m FFPI (R = 70 percent) subject to temperature changes (pen recorder output); input light is two modes.

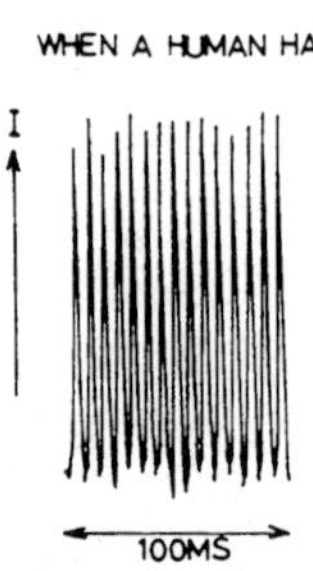

Fig. 11. Output light intensity I of 100 m FFPI (R = 70 percent) subject to temperature change (oscilloscope trace); input light is single mode.

ficiently small for most practical use for a temperature sensor. Since the optical path length in fiber is sensitive to mechanical vibration, FFPI as a temperature sensor has to be sufficiently well isolated from the ambient mechanical vibration.

B. Mechanical Vibration

Because FFPI has no reference arm in itself, FFPI can be not only a simple, but also a reliable, fiber sensor for vibrations. The 10 m FFPI ($R = 70$ percent) was placed in a coil on an optical bench. An impact was given to the bench. Fig. 12 shows the output light signal. In this case, the impact was so strong that phase modulation of several π was induced in the fiber. Fig. 12(a) and (b) corresponds to the cases that the input light is the single mode and two modes with an intensity ratio of 5:1, respectively. In the latter case, the output signal has an asymmetrical structure consisting of two peaks with different heights as shown in Fig. 12(b). From the order in which the two peaks appear, it can be discriminated whether the fiber is expanding or contracting. The change in the ambient temperature drifts the point of operation of FFPI. This is, however, not a significant problem if the vibration amplitude is large enough to cause the phase modulation larger than π, which is the case in most practical measurements of vibration.

C. Acoustic Wave

The acoustic sensitivity of FFPI is proportional to the fiber length. The 100 m FFPI was placed on a metal plate in a coil of 30 cm diameter. Acoustic waves of different frequencies of $f_A = 1$ kHz, 5 kHz, and 10 kHz, were radiated on the metal plate. Fig. 13(a)–(c) shows the corresponding output signals; the input light is the single mode. The operation point was located near the middle between the maximum and minimum transmittance of the FFPI. The incident acoustic pressure was 0.1 Pa in amplitude. The phase modulation induced in the fiber was about $\pi/8$ in amplitude. The corresponding sensitivity was about 4 rad/Pa. This sensitivity is one to two orders of magnitude greater than that by the free jacketed fiber [16], [17]. This is because the fiber was placed on the metal plate, which acted as a resonator and enhanced the acoustic vibration of the fiber. The operation point of the FFPI, however, drifted naturally. The stabilization of the operation point required a suitable feedback system as mentioned below. Fig. 14(a) and (b) shows the output signals in the cases that two human hands were clapped at distances of 1 m and 10 m from the FFPI, respectively. It is demonstrated that the FFPI can detect acoustic signals with high sensitivity.

An attractive use of FFPI is to exploit its high finesse, i.e., its sharp transmission characteristics. In general, the sensitivity of the Fabry-Perot interferometer is proportional to $d\tau/d\delta$. It can be derived from (3) that if $F \gg 1$, the sensitivity takes its maximum value of $0.65\sqrt{F}$ at $\delta = \pm 1/\sqrt{3F}$ (mod. π), i.e., at the point of $\tau = 75$ percent. This maximum sensitivity is by a factor of about $1.3\sqrt{F}$ larger than the sensitivity with a two-beam interferometer such as the Mach-Zehnder interferometer. A fiber-optic microphone system using FFPI was constructed as shown in Fig. 15(a). The 1

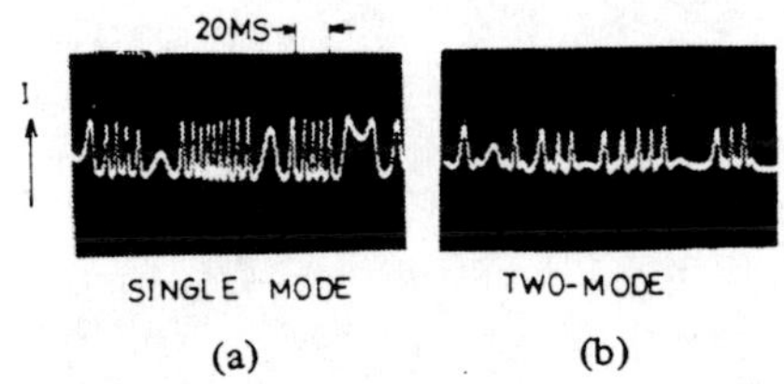

Fig. 12. Output light intensity I of 10 m FFPI ($R = 70$ percent) subject to mechanical vibrations in cases that input light is (a) single mode and (b) two mode.

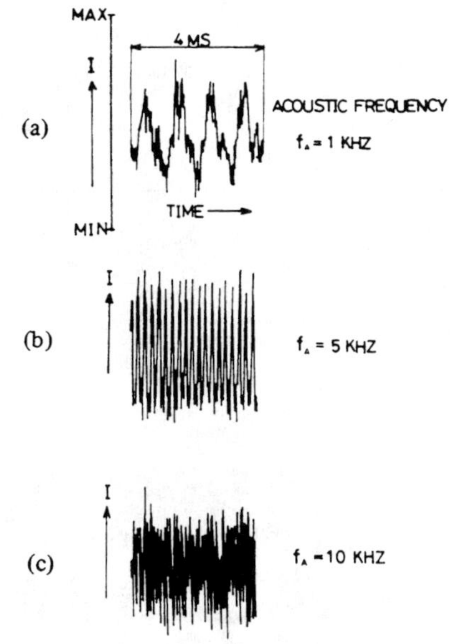

Fig. 13. Output light intensity I of 100 m FFPI ($R = 70$ percent) subject to acoustic waves of various frequencies.

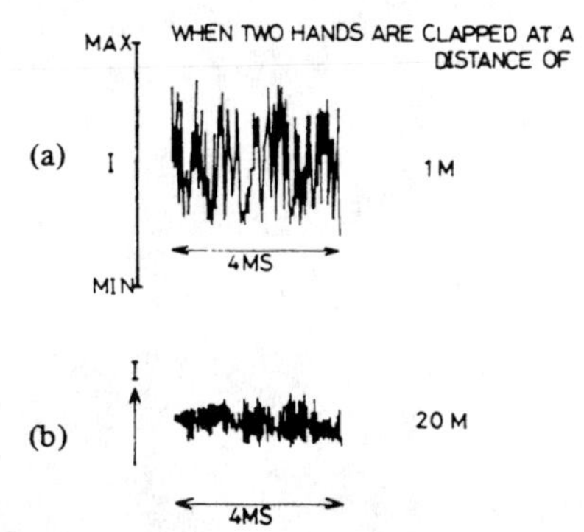

Fig. 14. Output light intensity I of 100 m FFPI ($R = 70$ percent) subject to sounds generated by clapping.

m FFPI ($R = 80$ percent) of Fig. 3(a) was employed for the sensor. In order to set the operation point at the steepest slope of the transmission curve, the following servo system was employed. The output light intensity of the FFPI was passed through a low-pass filter LF and compared, in a differential amplifier DA, with a constant voltage V_o corresponding to the point to which operation point should be stabilized [see Fig. 15(b)]. The differential signal was then fed, through a power amplifier PA, to the voice coil for driving a speaker cone. A part of the FFPI was attached on the cone so that the fiber length could be modulated by the displacement of the cone. By this servo system, the temperature-induced phase shift in

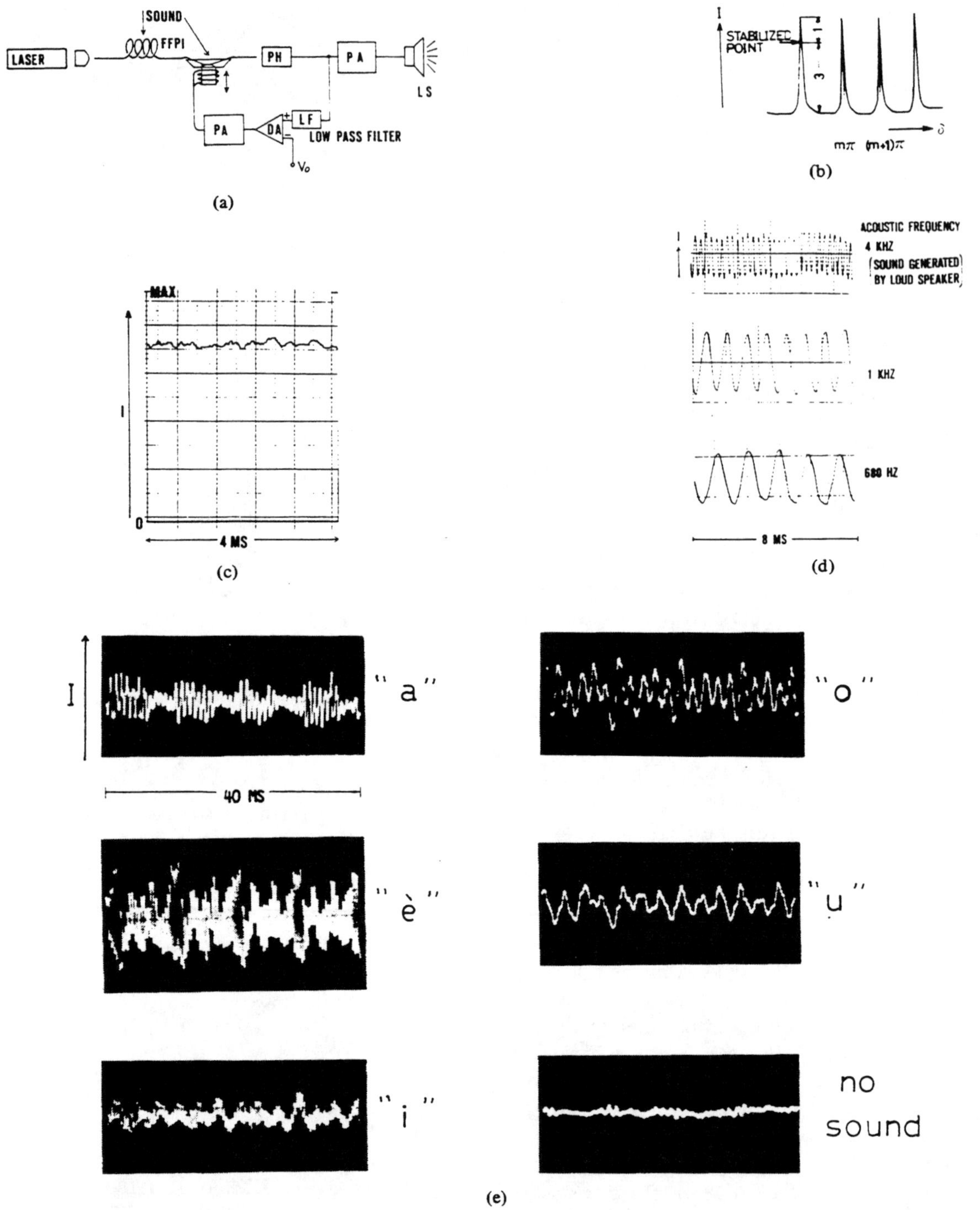

Fig. 15. Optical microphone system using FFPI: (a) entire system, (b) operation scheme (I output light signal), (c) stabilized result, (d) output light signal for the incidence of sinusoidal acoustic waves, (e) output signal for the incidence of vowels.

the fiber was compensated so that the operation point was stabilized to the steepest slope of the transmission curve of Fig. 15(b). Fig. 15(c) shows the typical result of stabilization. Somewhat intensity fluctuation involved in the stabilized result is attributed to the ambient acoustic noise. In order to increase the acoustic sensitivity, the acoustic wave was incident on the speaker cone, on which the FFPI was attached. Fig. 15(d) shows the output light signal of the FFPI for the incidence of sinusoidal acoustic waves with different frequencies. The modulation depth of the output light signal was about 20 percent of the peak-to-peak transmittance of the FFPI. Good linearity is obtained between the input acoustic wave and the output light signal. Fig. 15(e) shows the output light intensity of the FFPI for the incidence of the vowels generated by a man. The modulation depth of the output light was typically 20 percent. The frequency spectrum of each vowel pattern shown in Fig. 15(e) accords well with the well-known formant. By means of this microphone system, human voice would be well regenerated through a loud speaker LS. In this FFPI, since $R_e = 77$ percent (see Table I), the sensitivity enhancement factor of $1.3\sqrt{F}$ is 9.9. The minimum detectable phase shift by using the FFPI scheme is proportional to the above sensitivity enhancement factor but also depends on the light transmittance [6]. Concerning this, more detailed experimental study is demanded. One great advantage of using FFPI as an acoustic sensor is that it requires no reference arm and so it is free from any spurious signals associated with the reference arm.

D. Voltage

By the pulse counting method, ac electric voltages were measured. The measurement scheme is substantially the same as shown in Fig. 2. A part of FFPI was attached on PZT's as illustrated in Fig. 17(a) and (b). Fig. 16 shows the typical output signal when 50 Hz voltages were applied to the PZT of Fig. 17(a). Many light pulses were generated even for relatively low voltages. The pulse number was counted with an electronic counter. Fig. 17(a) and (b) shows the counted pulse number N as a function of the applied 50 Hz voltage V_{ac}. By using these V_{ac} - N curves, the ac electric voltages can be optically measured. In the case of Fig. 17(b) in particular, N is proportional to V_{ac}, which is favorable for voltmeter. The frequency characteristic of this voltmeter is directly related to that of PZT. The PZT's used in Fig. 17(a) and (b) have the resonance frequencies of 1 kHz and 92 kHz, respectively. The measurement sensitivity of the voltmeter increases around the resonance frequency.

E. Magnetic Field

Both ac and dc magnetic fields were measured by the pulse counting method. Fig. 18 shows the measurement system. The 10 m FFPI was wound about a magnetostrictive cylinder of ferrite, which has a hollow form of 4 cm inner diameter, 5.5 cm outer diameter, and 8 cm length. The magnetic fields were generated by a coil placed in front of the magnetostrictive cylinder, and was applied to the cylinder in its axial direction. Fig. 19 shows the typical output light signal when 50 Hz magnetic fields were applied to the cylinder; the input light was the single mode. It is observed in Fig. 19 that the modulation of the output light intensity becomes small around the zero levels of applied fields. This corresponds to the fact that the magnetostriction is an even function of the applied magnetic field. The number of the light pulses generated was measured as a function of the amplitude of the applied 50 Hz magnetic field H_{ac} and is shown in Fig. 20 by the circles. The pulse number was changed by the additional application of dc magnetic fields. The dependence of the pulse number N on dc magnetic field H_{dc} was measured as a parameter of the amplitude of the 50 Hz magnetic field H_{ac}. The measured results are shown in Fig. 21 by the circles.

In Fig. 20 it is found that N increases with an increase in H_{ac}, depending on its square. In Fig. 20, the solid line represents the calculated curve of $N = 1.6H_{ac}^2$. Good agreement is shown between the calculated and measured results. This indicates that, in the present case, the magnetostriction of the used magnetostrictive cylinder is proportional to the square of the applied magnetic field. Consequently, in the presence of both ac and dc magnetic fields, the magnetostriction is proportional to $(H_{ac} + H_{dc})^2$, and hence, for the modulation term in particular, to $H_{ac} \cdot H_{dc}$, if $H_{dc} \gg H_{ac}$. In Fig. 21, the solid lines represent the calculated curve of $N = 0.8\ H_{ac} \cdot H_{dc}$. Good agreement is obtained between the cal-

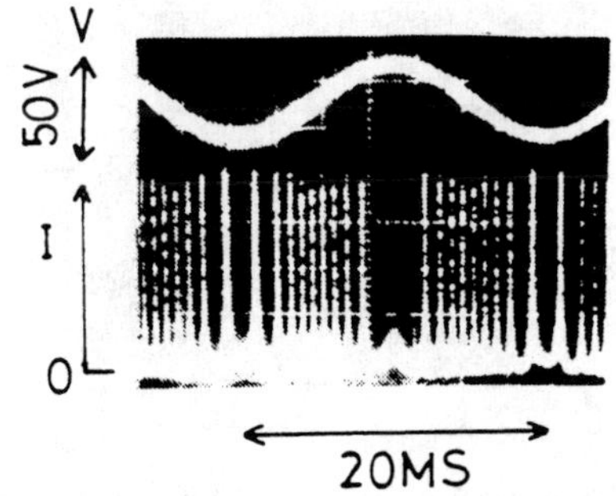

Fig. 16. Output light intensity I of 1 m FFPI (R = 80 percent) attached on PZT subject to 50 Hz voltage V.

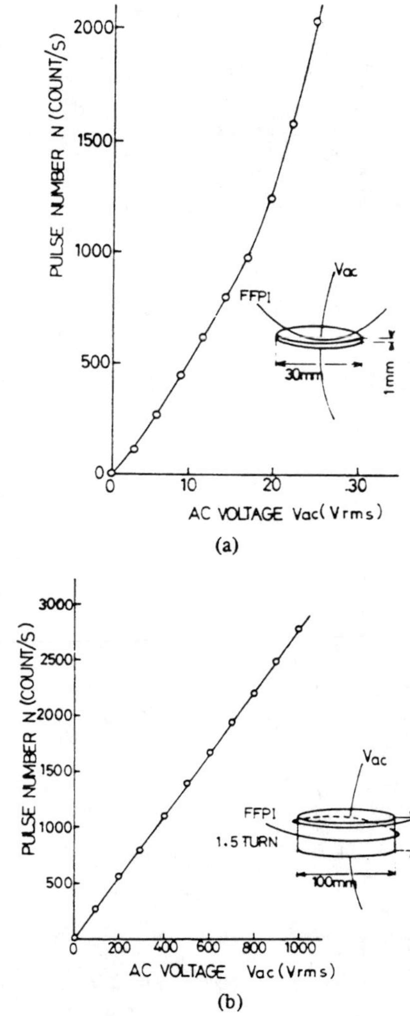

Fig. 17. Measured dependence of pulse number N on PZT 50 Hz voltage V_{ac}.

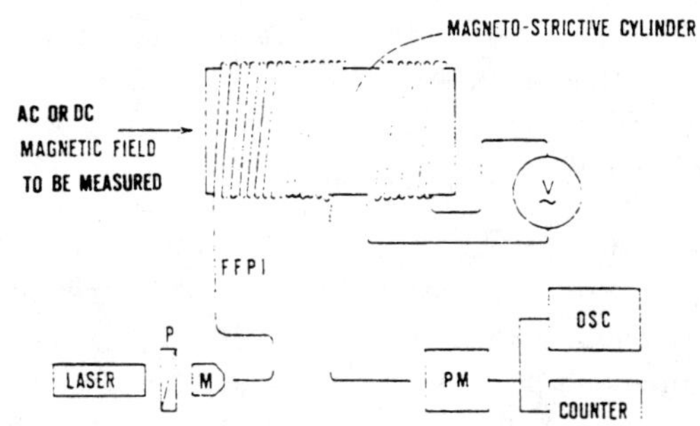

Fig. 18. Magnetic field measurement system using FFPI.

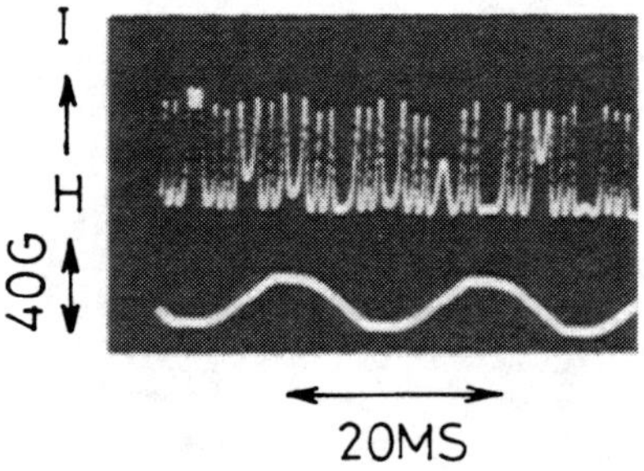

Fig. 19. Output light intensity I of 10 m FFPI wound about magnetostrictive cylinder subject to 50 Hz magnetic field H.

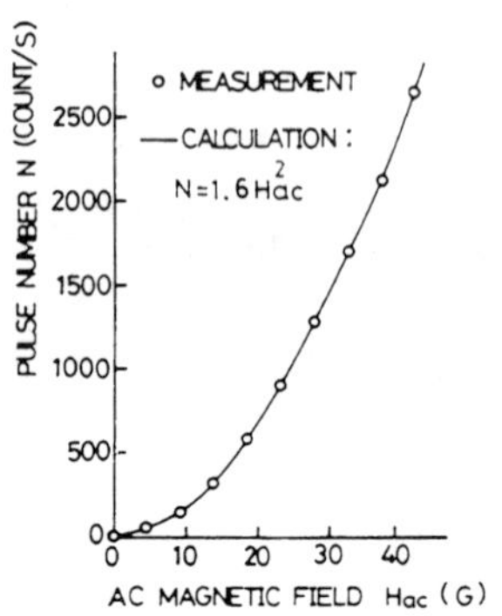

Fig. 20. Measured and calculated dependence of pulse number N on amplitude of 50 Hz magnetic field applied to the magnetostrictive cylinder H_{ac}.

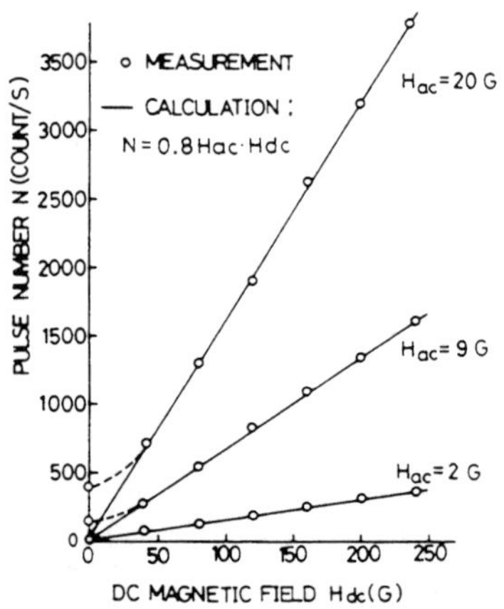

Fig. 21. Measured and calculated dependence of pulse number N on dc magnetic field applied to the magnetostrictive cylinder H_{dc} as a parameter of amplitude of 50 Hz magnetic field H_{ac}.

culated and measured results except the region where H_{dc} is near H_{ac}; in this region, the present calculation is invalid. The experimental results as shown in Fig. 21 make it possible to measure dc magnetic fields, which is generally difficult by the optical method.

In general, a digital sensing scheme undergoes the measurement error due to the quantization of a signal. The quantization error is significant for small signals because it increases in inverse proportionality to the number of generated pulses. Obviously, such quantization error can be decreased by increasing that part of FFPI which is attached on a PZT or a magnetostrictive material. Another useful approach to decreasing the quantization error is to scan the operation point across one or multiple periods of the pulse in such a way that one or multiple scans are involved in the measurement time but that the scanning speed is sufficiently slower than the modulation frequency of the signal. In the present measuring scheme using FFPI, such scanning action is actually, although not perfectly, performed by the naturally occuring temperature drift of the optical phase in fiber. The measurement error was typically ±3 percent or less in both the measurements of electric and magnetic fields.

V. Concluding Remarks

The fiber-optic Fabry-Perot interferometer (FFPI) has been developed for the first time. It has been shown that FFPI can have high finesse. It has been shown that, unlike the conventional Fabry-Perot interferometer, FFPI is flexible, can be very long, requires relatively low precision of fabrication, and moreover, it can be readily scanned. The polarization and thermal characteristics of FFPI are investigated. It is shown both experimentally and theoretically that the temperature sensitivity of jacketed FFPI is about three times larger than the unjacketed one. FFPI has been applied to the sensors of temperatures, mechanical vibrations, acoustic waves, ac voltages, and ac and dc magnetic fields. The fiber-optic microphone system using FFPI is also constructed. The analog or digital sensing scheme has been applied for these purposes. It has been demonstrated that FFPI can achieve high sensing sensitivity by using its high finesse and/or by making its fiber length very long. Unlike the conventional fiber-sensor scheme using the Mach-Zehnder interferometer, FFPI has a very simple form, can easily produce the interferometric signal, and is free from the false signal due to the reference arm. Consequently, FFPI can be a very attractive interferometer not only as the spectrum analyzer but also as the interferometric fiber sensor.

References

[1] J. A. Bucaro, H. D. Darty, and E. F. Caron, "Optical fiber acoustic sensor," *Appl. Opt.*, vol. 16, no. 7, pp. 1761-1762, 1977.

[2] G. B. Hocker, "Fiber-optic sensing of pressure and temperature," *Appl. Opt.*, vol. 18, no. 9, pp. 1445-1448, 1979.

[3] A. Dandridge, A. B. Treven, G. H. Sigel, Jr., E. J. West, and T. G. Giallorenzi, "Optical fiber magnetic field sensors," *Electron. Lett.*, vol. 16, no. 11, pp. 408-409, 1980.

[4] A. Vali and R. W. Shorthill, "Fiber ring interferometer," *Appl. Opt.*, vol. 15, no. 5, pp. 1445-1448, 1979.

[5] T. Yoshino, "Fiber Fabry-Perot interferometers," in *Proc. 3rd Int. Conf. on Integrated Optics and Optical Fiber Commun.*, San Francisco, CA, Apr. 1981, paper WL2.

[6] S. J. Petuchowski, T. G. Giallorenzi, and S. K. Sheem, "A sensitive fiber-optic Fabry-Perot interferometer," *IEEE J. Quantum Electron.*, vol. QE-17, pp. 2168-2170, Nov. 1981.

[7] D. L. Franzen and E. M. Kim, "Long optical fiber Fabry-Perot interferometer," *Appl. Opt.*, vol. 20, no. 23, pp. 3991-3992, 1981.

[8] P. G. Cielo, "Fiber-optic hydrophone: improved strain and environmental noise protection," *Appl. Opt.*, vol. 18, no. 18, pp. 2933-2937, 1979.

[9] T. Yoshino, "Frequency stabilization of internal-mirror He-Ne ($\lambda = 633$ nm) lasers using the polarization properties," *Japan. J. Appl. Phys.*, vol. 19, no. 11, pp. 2181-2185, 1980.

[10] M. Born and E. Wolf, *Principles of Optics*, 2nd ed. Oxford: Pergamon, 1964, p. 327.

[11] P. J. Nagill and T. Young, "Measurement of He-Ne laser linewidth utilizing the Doppler effect," *Appl. Phys. Lett.*, vol. 5, no. 1, pp. 13-15, 1964.

[12] Y. Murakami, I. Hatakeyama, and H. Tsuchiya, "Normalized frequency dependence of splice losses in single-mode optical fibers," *Electron. Lett.*, vol. 18, no. 9, pp. 277–278, 1978.

[13] M. Tateda, S. Tanaka, and Y. Sugiyama, "Thermal characteristics of phase shift in jacketed optical fibers," *Appl. Opt.*, vol. 19, no. 5, pp. 770–773, 1980.

[14] N. Lagakos, J. A. Bucaro, and J. Jarzynski, "Temperature-induced optical phase shifts in fibers," *Appl. Opt.*, vol. 20, no. 13, pp. 2305–2308, 1981.

[15] H. S. Carslaw and J. C. Jager, *Conduction of Heat in Solids*, 2nd ed. Oxford: Clarendon, 1959, p. 230.

[16] G. B. Hocker, "Fiber-optic acoustic sensors with composite structures: An analysis," *Appl. Opt.*, vol. 18, no. 21, pp. 3674–3683, 1979.

[17] B. Budiansky, D. C. Drucker, G. S. Kino, and J. R. Rice, "Pressure sensitivity of a clad optical fiber," *Appl. Opt.*, vol. 18, no. 24, pp. 4085–4088, 1979.

Toshihiko Yoshino was born in Tokyo, Japan, on April 27, 1939. He received the B.S., M.S., and Ph.D. degrees in applied physics from Tokyo University, Tokyo, Japan, in 1963, 1965, and 1968, respectively.

Since 1968 he has been with the Institute of Industrial Science, Tokyo University, where he has been engaged in research and education on optics. From 1971 to 1973 he was a Guest Researcher with the Technical University, Hannover, West Germany, where he worked on laser spectroscopy.

Dr. Yoshino is a member of the Japan Society of Applied Physics.

Kiyoshi Kurosawa was born in Nagano Prefecture, Japan, on September 6, 1952. He received the B.E. degree in electrical engineering from Nihon University, Tokyo, Japan, in 1978 and graduated from the Department of Technology, University Course, Tokyo Electric Power Institute, Tokyo, Japan, in 1980.

Since 1971 he has been with the Tokyo Electric Power Corporation, Ltd., Tokyo, Japan, where he has been engaged in research and development in application of optics in electric power systems.

Katsuji Itoh was born in Tokyo, Japan, on February 15, 1942. He received the B.S. and the M.S. degrees in electrical engineering from the Yokohama National University, Yokohama, Japan, in 1966 and 1968, respectively.

Since 1968 he has been with the Furukawa Electric Corporation, Ltd., Chiba, Japan, where he has been engaged in research and development in distribution and application of optics to high voltage.

Mr. Itoh is a member of the Institute of Electrical Engineers of Japan and the Japan Society of Applied Physics.

Teruzi Ose, photograph and biography not available at the time of publication.

Reprinted from *Journal of Applied Physics,* Vol. 54(3), pp. 1198-1201 (March 1983).

High-temperature optical fiber thermometer

R. R. Dils
National Bureau of Standards, Washington, D.C. 20234

A high-temperature optical fiber thermometer made from single crystal sapphire has been developed for use from 600 to approximately 2000 °C. The device consists of a small blackbody cavity which is sputtered on the end of a thin (0.25–1.25-mm diameter, 0.05–0.30-m length) sapphire fiber, a connecting low-temperature glass fiber, and a conventional optical detector. The radiance from the cavity is used to measure its temperature. The present instrument is calibrated at a single temperature and uses the fundamental radiation laws to extrapolate to other temperatures. It is accurate and has a high sensitivity and rapid temporal response. There appear to be a number of applications of the device both in science and industry.

During the past several years, a rapidly developing optical fiber sensor technology has emerged to replace existing measuring techniques or to extend measurement capabilities.[1–5] The devices use the optical fiber either as a signal generator[1–3] or transmitter.[4,5] In most instances, glass optical fibers with low-temperature coatings are used and the measurements are limited to relatively low temperatures. This letter describes an optical fiber sensor which is made from single crystal aluminum oxide (sapphire) and can be operated above 600 °C. The device concept is straightforward. A blackbody cavity is formed on the tip of a thin (0.05–0.3 m long) single crystal aluminum oxide (sapphire) fiber, and the radiance emitted from the cavity in a narrow wavelength band is used to measure its temperature. The high-temperature sapphire fiber and a low-temperature connecting fiber transmit the signal to an interference filter and optical detector.

A typical thermometer consists of three basic elements: a signal generator, one or more transmitting optical fibers, and an optical detector, as shown in Fig. 1. The signal generator at the tip of the high-temperature fiber can be any effective blackbody configuration. A simple blackbody cavity can be created by sputtering a thin metallic or opaque oxide coating on the surface of the fiber. For example, the initial experiments have used platinum or iridium films which are coated with a thin film of aluminum oxide to reduce oxidation of the metallic films. The equations describing the radiant emission from such a cavity are long established. The Planck equation[6] describing the spectral distribution of radiance for an ideal blackbody is used in conjunction with the Gouffé equation[7] describing the apparent emittance of a cavity of finite dimensions and isotropically diffuse reflecting walls in order to determine the radiance at the exit of the cavity. The flux per unit wavelength entering the fiber is obtained by integrating the internally reflected flux from the critical angle to the axis of the fiber.[8,9] The total spectral flux per unit wavelength entering the fiber is simply

$$\Phi(\lambda_0) = \frac{a\epsilon_0 c_1}{\lambda_0^5[\exp(c_2/\lambda_0 T) - 1]} \quad [\mathrm{W/m}], \qquad (1)$$

where a is the area of the cavity exit, m^2, ϵ_0 the apparent emittance of the cavity, c_1 the first radiation constant 3.7418×10^{-16} W m^2, c_2 the second radiation constant 1.43879×10^{-2} m K, λ_0 the wavelength in vacuum m, and T the temperature K. λ_0 is selected to provide a maximum sensitivity to temperature changes with a measurable flux and low absorption in sapphire at elevated temperatures. For example, from 600 to 1300 °C a 0.1-μm-wide band centered at 0.6–0.7 μm is sufficient; at 1000 °C and 0.6 μm, a 0.25-mm-diam cavity will emit $\sim10^{-7}$ W of optical power in a 0.1-μm bandwidth and a 1% change in temperature will cause a 20% change in flux. Above 1300 °C, a 0.1-μm-wide band centered at 0.4–0.5 μm will provide similar sensitivity and flux.

The apparent emittance of an iridium cavity is not a strong function of temperature, even for relatively small length-to-diameter (L/D) ratios. As shown in Table I,[7,8] for λ_0 from 0.5 to 0.7 μm, a cavity with an L/D ratio of 2 has an apparent emittance that varies less than $\pm1\%$ over the prescribed temperature ranges; this corresponds to a temperature change of less than $\pm0.05\%$. Accordingly, even a short cavity has a sufficiently constant $\epsilon_0(T)$ for most engineering applications, and there is an attendant improvement in the spatial resolution of the device.

The second element of the thermometer consists of the high- and low-temperature optical fibers which transmit the signal to the detector. In an engineering device, it is only required that the transmission losses are reproducible within approximately 1% and are not a function of temperature. Reflection at the ends of the high-temperature fiber and at both ends of the low-temperature fiber is understood[9,10] and can be evaluated experimentally. Typically, a 4–5% loss is observed at a simple crystal/air surface. Using a fixed air gap between the fibers, these losses can be reproduced within

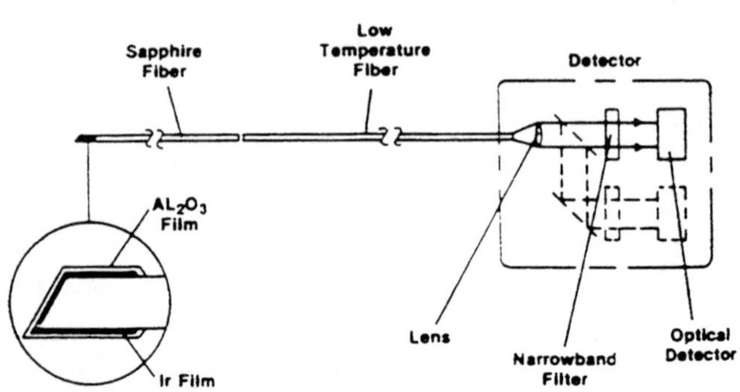

FIG. 1. High-temperature optical fiber thermometer.

TABLE I. Apparent emittance of an iridium cavity.[7]

λ_0 (μm)	Temperature (°C)	$L/D=2$	$L/D=5$	$L/D=20$
0.5	827	0.8205	0.9419	0.9848
	1227	0.8330	0.9473	0.9863
	1727	0.8350	0.9492	0.9869
	2127	0.8378	0.9494	0.9869
0.6	827	0.8106	0.9375	0.9835
	1227	0.8148	0.9410	0.9845
	1727	0.8163	0.9413	0.9846
	2127	0.8224	0.9428	0.9850
0.7	827	0.7886	0.9273	0.9804
	1227	0.7931	0.9296	0.9811
	1727	0.7937	0.9300	0.9812
	2127	0.7970	0.9314	0.9817

1%.[11] The coupling losses due to fiber misalignment can also be reproduced within 1% by using high- and low-temperature fibers with dissimilar diameters. For an interpolative engineering device, the low-temperature fiber can be either larger or smaller than the high-temperature fiber.

Transmission losses due to absorption in the low-temperature fiber are negligible since the fiber is cooled during operation and is typically less than 100 m long. Losses in the low-temperature portion of the sapphire crystal are also negligible. For example, it has been found that even imperfect crystals 1.25 mm in diameter, which are grown by the edge-defined growth method[12] and contain varying amounts of optical scattering defects within the crystal and surface imperfections, attenuate a signal less than 1% in 30 cm.[11] At temperatures above 1100 °C, scattering, absorption, and reemission at internal defects and impurities and surface imperfections are potential problems, and high quality crystals must be used. At present, selected flame polished fibers are of sufficient quality that, in tests from 1100 to 1700 °C in gas turbine combustor exhaust gases at 0.27–0.47 Mach (240–345 m/s), the total scattered, absorbed, and re-emitted radiant intensity measured in a 1.25-mm-diam, 5-cm-long uncoated fiber was less than 2% of the radiance from an identical fiber with the blackbody sputtered on its end. No evidence of fundamental lattice absorption has been observed and therefore it can be anticipated that with superior crystals, the amount of scattering and absorption will be further reduced. The majority of the scattering appeared to be due to surface imperfections. Exposure of the fibers to a hot gas stream for several hours at these temperatures has not caused significant surface degradation.[11] However, it can be anticipated that as the melting temperature of sapphire (2050 °C) is approached the surface topography will change due to surface diffusion or contamination and the practical maximum temperature and lifetime of the device will depend on the response of the surface to the environment. The lifetime can be improved through designs that limit the length of fiber exposed to the high-temperature environment and protect the fiber from direct impact damage or contamination.

The third element of the thermometer is the detector system which can be constructed from a wide range of conventional components whose operation is well known. The principle components are a light gathering lens, a narrow band filter, and a photomultiplier or silicon detector. Neutral density filters can be included to control the light intensity at the detector. Two wavelengths can be used for calibration and periodic system checks. Using standard photometric practices, a 1% level of uncertainty can be maintained[13,14]; advanced techniques permit the measurement of light intensity to better than 0.1%[15,16]; the respective uncertainties in temperature measurement are 0.05% and 0.005%.

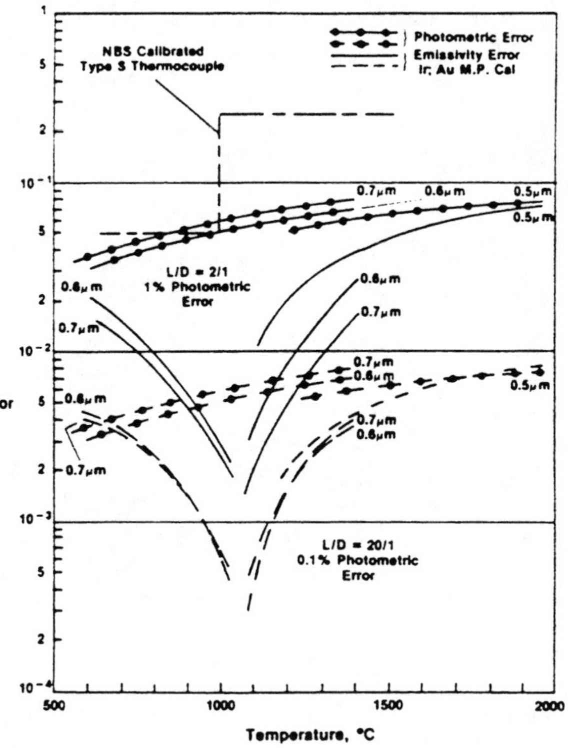

FIG. 2. Fundamental errors in high-temperature optical fiber thermometer measurements.

The relative errors encountered in optical fiber thermometers using cavity L/D ratios of 2 and 20 and several wavelengths are presented in Fig. 2 where it is observed that the photometric error introduces a greater inaccuracy than the emissivity error. In Fig. 2, the 1% photometric accuracy is a relative accuracy for an interpolative device which has an L/D ratio of 2 and can be calibrated at a single known temperature in a melting point furnace or a calibrated large blackbody furnace. Such a device has been operational for several months and the indicated accuracies have been realized. Note that even such a simple engineering device exhibits approximately one-fourth the inaccuracy level of an NBS calibrated ANSI type S thermocouple standard above 1000 °C. The 0.1% photometric errors and radiance errors due to the apparent emissivity of a cavity with a 20/1 L/D ratio are those anticipated for an advanced laboratory thermometer. Once the transmission and detector efficiency of a device is known, the device also can be self-calibrated by measuring the radiance at two wavelengths at any two unknown but different temperatures. For example, at 1000 °C, using 0.6- and 0.7-μm wavelengths and maintaining a 0.1%

TABLE II. Initial test data: Pt film $L/D = 2/1$; 1.25 mm × 30 cm crystal.

Furnace temperature (K)	λ_{eff} (μm)	Flux ratio Φ/Φ_{cal} Theor	Exp	Apparent (or measured) optical fiber temperature (K)	Flux Φ	(%) Error temperature
Uncoated fiber						
973.70	0.6075	0.0400	0.0399	973.63	0.13	0.007
	0.7050	0.0622	0.0632	974.42	1.61	0.074
1023.1	0.6072	0.1286	0.1274	1022.77	0.92	0.039
	0.7047	0.1700	0.1717	1023.60	1.00	0.049
1073.6	0.6069	0.3782	0.3697	1072.48	2.20	0.100
	0.7044	0.4321	0.4294	1073.24	0.62	0.034
1123.6	0.6055	...	...	CAL	...	...
	0.7040	...	...	CAL	...	...
1174.4	0.6062	2.474	2.510	1174.24	1.52	0.072
	0.7037	2.185	2.211	1175.20	1.19	0.068
1225.7	0.6058	5.709	5.739	1226.03	0.53	0.027
	0.7033	4.500	4.500	1225.70	0.00	0.032
1276	0.6055	12.173	12.252	1275.81	0.64	0.015
	0.7030	8.644	8.620	1275.80	0.27	0.016
				σ0.6	1.20	0.054
				σ0.7	0.95	0.048
Coated fiber						
969.79	0.6075	0.0380	0.0386	970.46	1.58	0.069
	0.7050	0.0595	0.0604	970.54	1.51	0.077
1018.9	0.6072	0.1223	0.1224	1019.27	0.08	0.012
	0.7047	0.1600	0.1643	1019.36	0.80	0.045
1070.4	0.6069	0.3708	0.3732	1070.75	0.65	0.033
	0.7044	0.4247	0.4235	1070.28	0.28	0.011
1121.2	0.6065	...	...	CAL	...	...
	0.7040	...	...	CAL	...	...
1171.1	0.6062	2.448	2.493	1172.23	1.84	0.096
	0.7037	2.166	2.174	1171.45	0.38	0.030
1223.6	0.6058	5.785	5.788	1223.71	0.04	0.009
	0.7033	4.549	4.548	1223.66	0.03	0.005
1275.2	0.6055	12.590	12.612	1275.33	0.02	0.010
	0.7030	8.898	8.686	1273.30	2.35	0.149
				σ0.6	1.02	0.050
				σ0.7	1.20	0.072

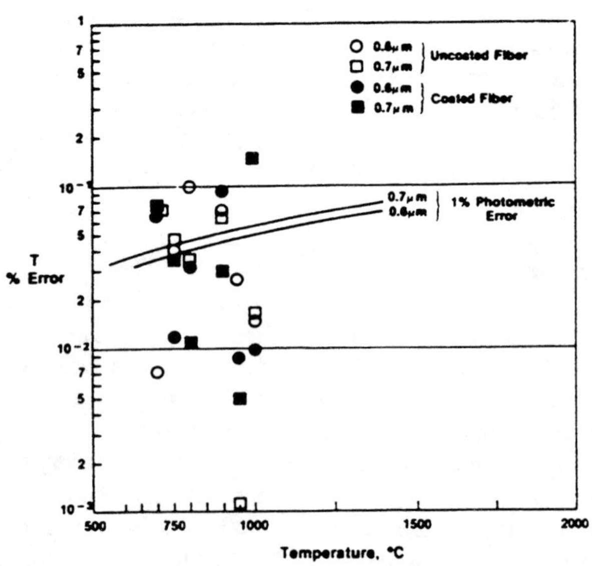

FIG. 3. Initial blackbody furnace tests: 700–1000 °C.

photometric error, the inaccuracy level of the self-calibration will be less than 0.05%.

There are two other features of the device which are unusual. First, the sensitivity is quite high. For example, using a silicon detector and amplifier with a $10^5/1$ dynamic range and a 0.2 μV equivalent input noise within a 0–1-kHz bandwidth, a 10^{-3} °C change can be measured at 1000 °C with a signal-to-noise ratio of 10/1. Because of the small thermal mass of the optical fiber tip, only a 10^{-8} cal heat input is required to change the temperature of the cavity by 10^{-3} °C. Second, the useful frequency bandwidth is much larger than comparable metallic thermocouples. The increased bandwidth is due to an improved thermal response of the fiber as well as the high signal-to-noise ratio. In the optical fiber thermometer, the thermal waves are attenuated at the surface, the high-frequency gain decreases only 3 dB per octave, and a two- to eightfold increase in the transfer function is obtained. The high signal-to-noise ratio permits small cavity temperature oscillations to be detected and further increases the bandwidth relative to that obtained with a fine wire thermocouple. The actual practical bandwidth is determined by the temperature fluctuation power spectral density functions and the spatial resolution considerations of a specific application. As an example, in a gas turbine combustor, the optical fiber thermometer will provide useful information in a 0–100-kHz bandwidth, which is a 100-fold improvement over present fine wire thermocouple techniques.[17,18]

The initial experiments reported here were conducted from 700 to 1000 °C in a commerical blackbody furnace. The furnace temperature was measured with a calibrated thermocouple. Experiments were conducted with uncoated and coated 1.25 mm, 30-cm-long sapphire fibers. The uncoated fiber was inserted directly into the blackbody furnace to evaluate furnace calibration, transmission, and photometric errors. In a separate test, a 10-μm-thick platinum tip with an L/D ratio of 2/1 was sputtered on the other fiber and this fiber was inserted to the end of the conical blackbody cavity. The sapphire fibers were coupled across a fixed air gap to a 0.6-mm-diam 10-m long low-temperature glass optical fiber. The detector system, which was designed and constructed at the Sandia (Livermore) Laboratory, consisted of a lens, neutral filters with optical densities of 0.3–2.0, beam splitters, 0.1-μm-wide filters centered at 0.6 and 0.7 μm, and two photomultipliers. The photomultiplier output current was measured across a 50-kΩ resistor with a voltmeter.

The data are presented in Table II and Fig. 3. In Table II λ_{eff} is defined by the convolution integral of the narrow band filter transfer function and the spectral radiance computed from Eq. (1), and the relative theoretical value of the flux at this wavelength is computed from Eq. (1). For this demonstration, the 850 °C measurement was used as the calibration point. Inspection of the data shows that over a 300/1 range of radiance, the uncoated fiber data display standard deviations of 0.054% and 0.048% from the values expected from Eq. (1), while the coated fiber data display standard deviations of 0.050% and 0.072%. The similarity of the standard deviations of the uncoated and coated fiber data shows that the errors are primarily due to the furnace calibration or photometric errors rather than errors in the radiance from the tip of the sapphire fiber and transmission through the fibers. However, even this simple test demonstrates that unusual accuracy can easily be obtained with an elementary device and suggests that a more accurate device can be constructed.

Research is continuing to evaluate the potential accuracy of a more sophisticated device. As indicated in the preceding analyses, a blackbody cavity with a L/D ratio of 20/1 sputtered on a long sapphire crystal which is directly coupled to a detector may enable interpolative or absolute temperature measurements to be made with an uncertainty of less than 0.01%. Sapphire fiber growth and characterization studies are under way and investigations of the component inaccuracies associated with flux from the blackbody and transmission through the sapphire fiber are to be conducted in parallel with measurements of the overall device accuracy in several melting point furnaces.[19]

The author wishes to thank R. Dibble and D. Tichenor of the Sandia (Livermore) Laboratory for many useful discussions and development of the initial optical detector, J. C. Richmond and J. Geist for their help on a wide range of radiation problems, P. Huang for his experimental assistance, and G. Manning and J. Hord for their support during the critical initial stages of the project.

[1]A. R. Teko, Electro. Opt. Syst. Des. **15**, 39 (1982).
[2]M. Gottlieb and G. B. Brant, Appl. Opt. **20**, 3408 (1981).
[3]M. Gottlieb and G. B. Brant, Appl. Opt. **20**, 3867 (1981).
[4]K. A. Wickersheim and R. B. Alves, Ind. Res. Dev. **21**, 82 (1979).
[5]D. A. Christensen, J. Biomed. Eng. **1**, 541 (1977).
[6]M. Planck, Ann. Phys. **4**, 553 (1901).
[7]A. Gouffé, Rev. Opt. **24**, 1 (1945).
[8]Y. S. Touloukian and D. P. Dewitt, eds., *Thermophysical Properties of Matter* (IFI/Plenum, New York, 1970), Vol. 7, p. 288.
[9]J. W. T. Walsh, Dept. Sci. and Ind. Res. Illumination Research Tech. Paper No. 2 (1926), p. 10.
[10]B. Rossi, *Optics* (Addison Wesley, Massachusetts, 1957).
[11]R. R. Dils (unpublished).
[12]Saphikon, 51 Powers Street, Milford, New Hampshire 03055.
[13]J. C. Richmond (private communication).
[14]B. Woerner, *Temperature, Its Measurement and Control in Science and Industry* (AIP, New York, 1982), Vol. 5, p. 429.
[15]J. Geist, W. K. Gladden, and E. F. Zalewski, J. Opt. Soc. Am. **72**, 1068 (1982).
[16]J. Geist (private communication).
[17]R. R. Dils and P. S. Follansbee, ISA Aerospace Inst. Proc. **22**, 307 (1976).
[18]R. R. Dils and P. S. Follansbee, NBS Spec. Publ. No. 561 (1979), p. 1027.
[19]J. Geist, M. Reilly, and R. R. Dils (unpublished).

Reprinted from *Fiber Optics Reliability: Benign and Adverse Environments IV,* R.A. Greenwell, D.K. Paul, eds., Proc. SPIE Vol. 1366, pp. 284-293 (1990). ©1990 SPIE.

Reliability Improvement Methods for Sapphire Fiber Temperature Sensors

by: C. Schietinger, B. Adams

Accufiber Division of Luxtron Corporation
9550 S.W. Nimbus Avenue, Beaverton, Oregon 97005

ABSTRACT

Novel and practical methods of improving reliability of high temperature measurements in harsh environments are discussed. These improvement methods include: mechanical, optical, electrical and software designs. Discussed are mechanical designs for combustion chambers, gas turbines, and especially corrosive environments. Systems are described which use multi-wavelength algorithms for determining the trustworthiness of measurements. These systems use average and difference calculations of the two wavelengths. Also, hybrid optical and thermocouple sensors are presented.

BACKGROUND

The concept of high temperature Optical Fiber Thermometry (OFT) was pioneered at the National Bureau of Standard around 1984[1]. This fiber optics based temperature measurement technology has three fundamental elements:

1. Controlled light generation and transmission,
2. Linear radiometry, and
3. Fundamental signal analysis using independently measured optical components and the appropriate application of the Planck Equation.[2]

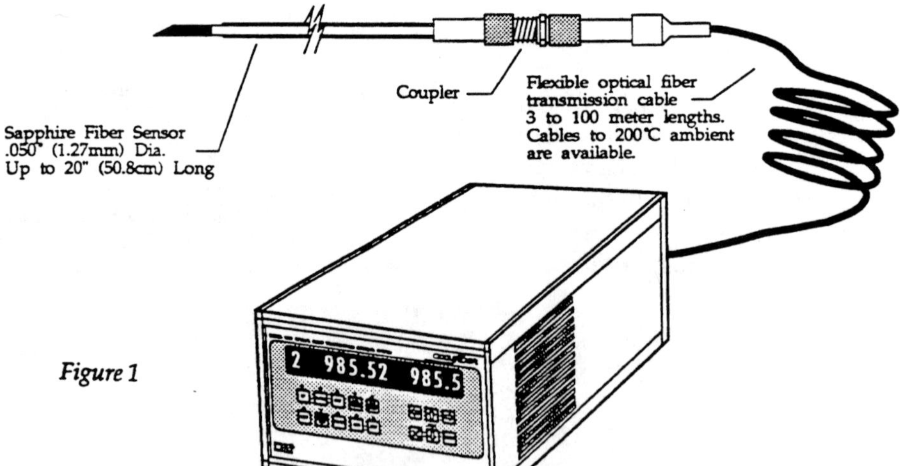

Figure 1

The complete OFT instrument has three components. *See Figure 1.*

1. The sapphire optical fiber sensor,
2. The optical transmission cable, and
3. The microprocessor based signal analyzer.

In measurement applications where physical or good thermal contact is desirable, a "blackbody sensor" can be used. *See Figure 2*

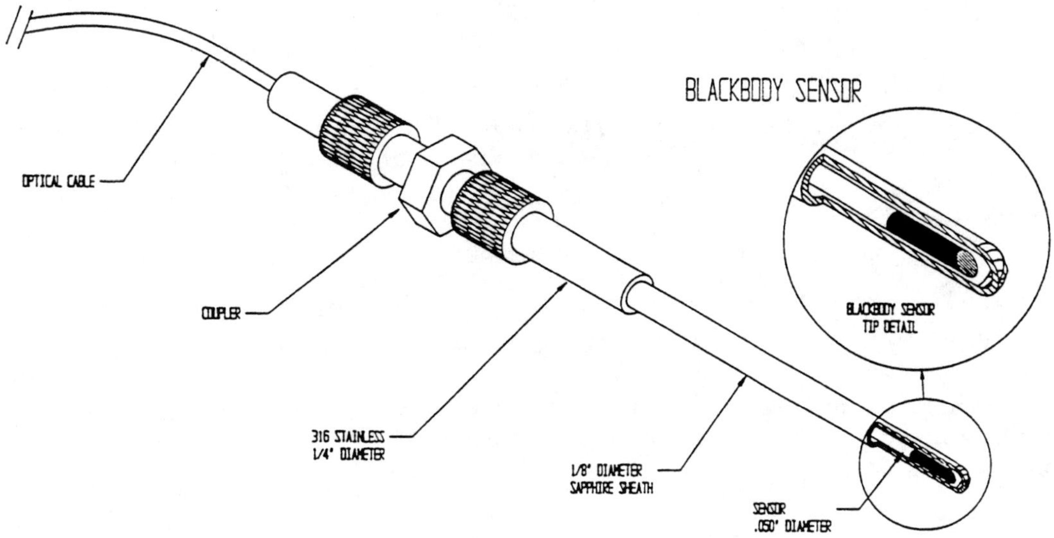

Figure 2

The blackbody sensor consists of an optically polished .03" to .08" diameter rod of single crystal aluminum oxide (sapphire), which is coated at its tip with a thin film of precious metal to form a blackbody emitter. Forming the blackbody at the end of the crystal controls the source of radiant energy and its emissivity, and the crystal controls transmission of that energy to the low temperature optical fiber.

OFT technology can also be applied in situations where physical contact between the blackbody cavity and the heat source is not practical. Where the sensor can be placed into a process, a sapphire "lightpipe sensor" is used to read light energy directly from a sample's surface, or ideally from a dark cavity within the sample. *See Figure 3.*

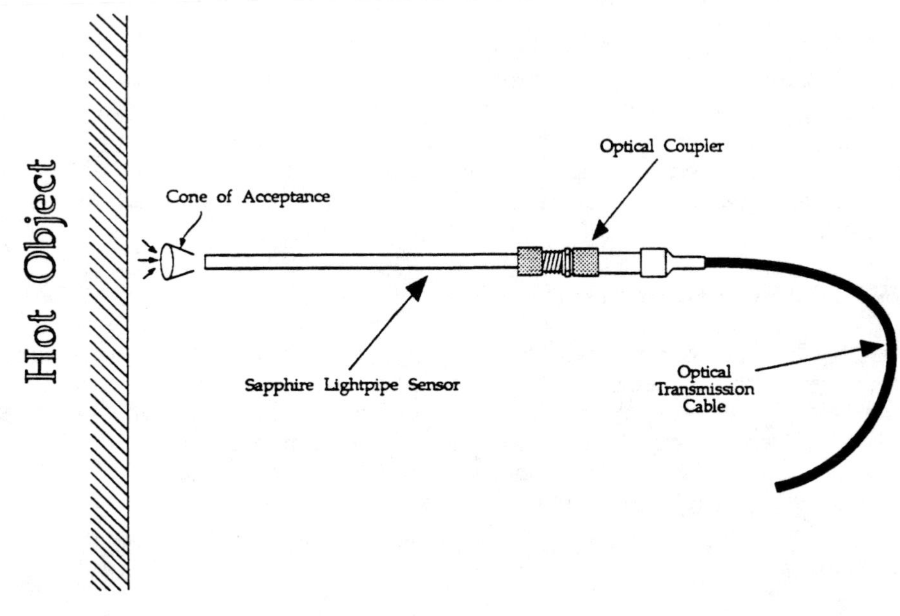

Figure 3

Both types of sapphire sensors transmit light energy to an optical fiber which in turn transmits it to a remote optical detector. At the detector, the light energy is converted into an electrical signal which is then amplified and digitized. Finally, a microprocessor converts the digitized signal into a temperature reading by use of the Planck Equation.

$$E_b(\lambda,T) = \frac{C_1 * \lambda^{-5}}{e^{\frac{C_2}{\lambda * T}} - 1}$$

Where: $E_b(\lambda, T)$ = Energy emitted by a blackbody cavity per unit area per unit wavelength, at temperature T and wavelength λ. $\left(\frac{W}{m^2 * \mu m}\right)$
λ = Wavelength (μm).
T = The Absolute Temperature (K).
C_1= Constant: $3.743 * 10^8 \left((W * \mu m^4) / m^2 \right)$
C_2= Constant: $1.4387 * 10^4 \left(\mu m * K \right)$

The temperature readings are then frequently used as inputs for control algorithms to provide closed loop process control, to levels unattainable by any other technology, as precise as ±0.03°C.[3]

SAPPHIRE AS AN OPTICAL FIBER

The choice of sapphire as a sensor material is based on its long term stability and survivability in harsh environments. Single crystal sapphire (Al_2O_3) has many features that offer advantages over other optical fiber materials:

1. High melting point over 2000°C,
2. A very hard material that allows for fine optical polish and low abrasion in high flow applications, i.e., auto engines and gas turbines,
3. Extremely low optical transmission losses over a wide wavelength band,
4. Good chemical corrosion resistance,
5. Index of refraction which give a large numeric aperture to the lightpipe sensor, and
6. Immunity to interference from electro magnetic radiation.

MECHANICAL DESIGNS FOR SENSORS

Optical Fiber Thermometry (OFT) is in use today in applications where the fiber sensors are routinely subjected to extreme mechanical stresses.[4] These applications include, auto engine cylinders, turbines, rocket exhaust testing, and wind tunnels to name only a few. The technology is also being used on the factory floor in steel mills and fiberglass factories. The reliability problems related to sensor breakage is overcome by good mechanical design of the sensors and sheaths for the specific application. A few of these designs are included in *Figures 4, 5, 6*.

Sensors for Corrosive Environments

There are basically three techniques for installing optical fiber temperature sensors into highly corrosive environments. The simplest is to make the fiber out of a corrosive resistant material such as sapphire. The second method is to sheath the fiber in a material such as silicon carbide, sapphire, boron nitride, or platinum to protect it. The other technique is to sheath the fiber in an open end sheath and purge that sheath with a low flow of gas, thereby keeping the tip free of corrosion or deposition.

Hybrid Sensors

To provide additional reliability to temperature measurements, hybrid sensors combine optical and thermocouple technologies into one sensor. These OFT/TC's provide backup in case of failure of either technology. They also provide a wider temperature measurement range. (0°C to 1900°C) The final advantage of these hybrids is that they practically separate the heat transfer modes. For example, the lightpipe sensor is solely measuring radiation and the thermocouple is mostly measuring temperature by convection and conduction.

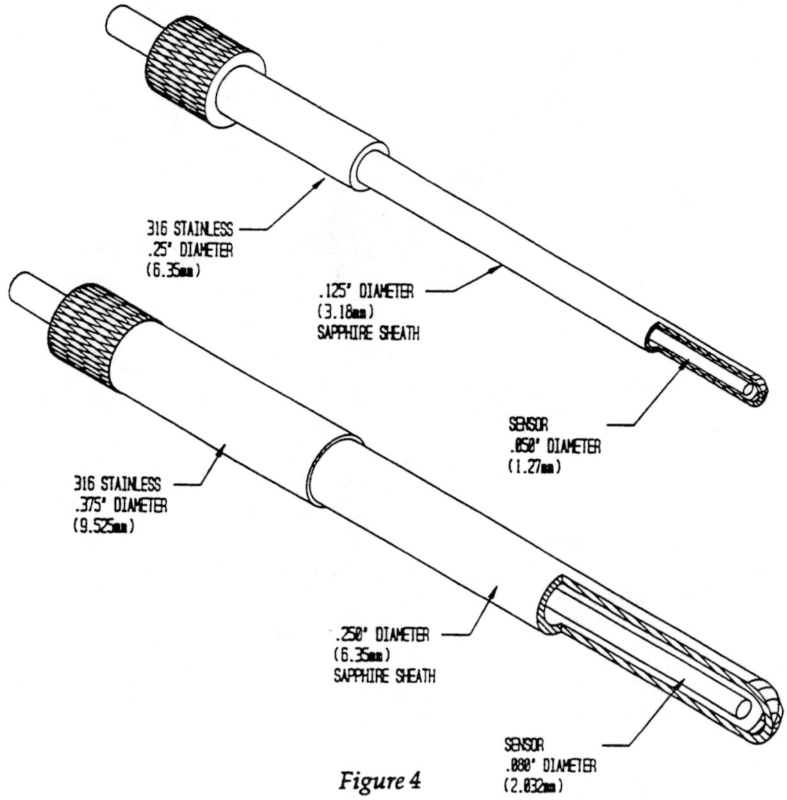

Figure 4

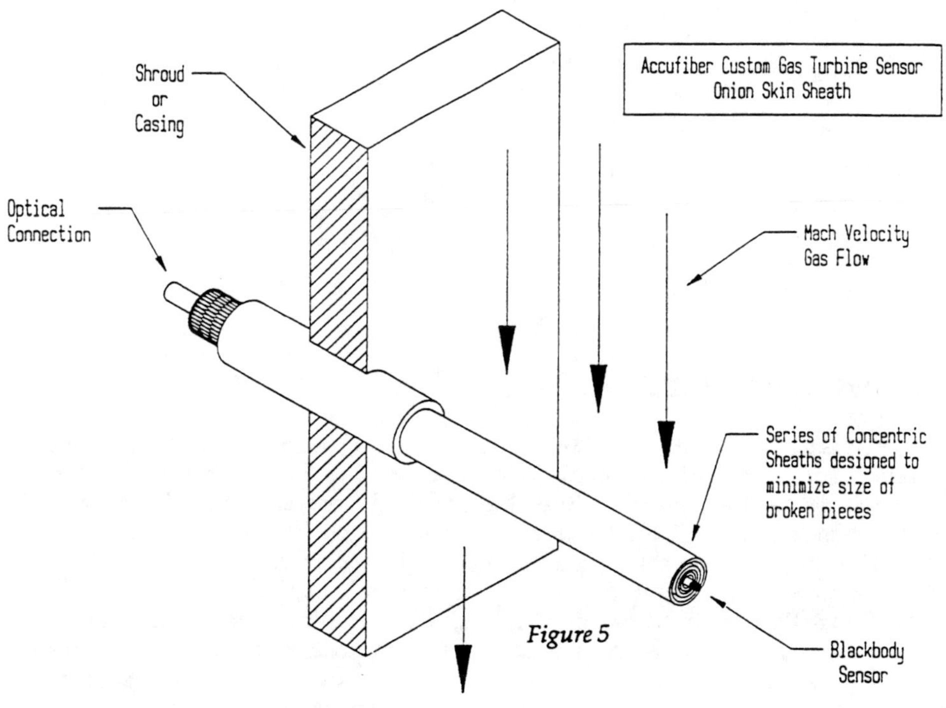

Figure 5

Figure 6

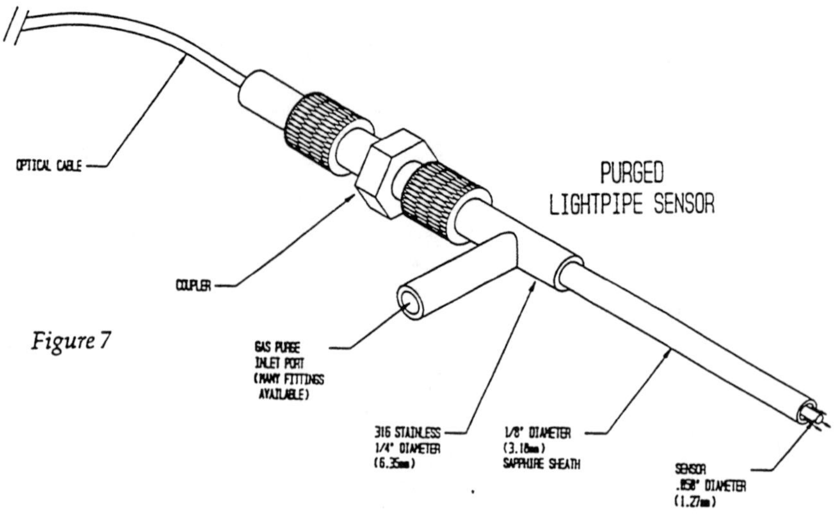

Figure 7

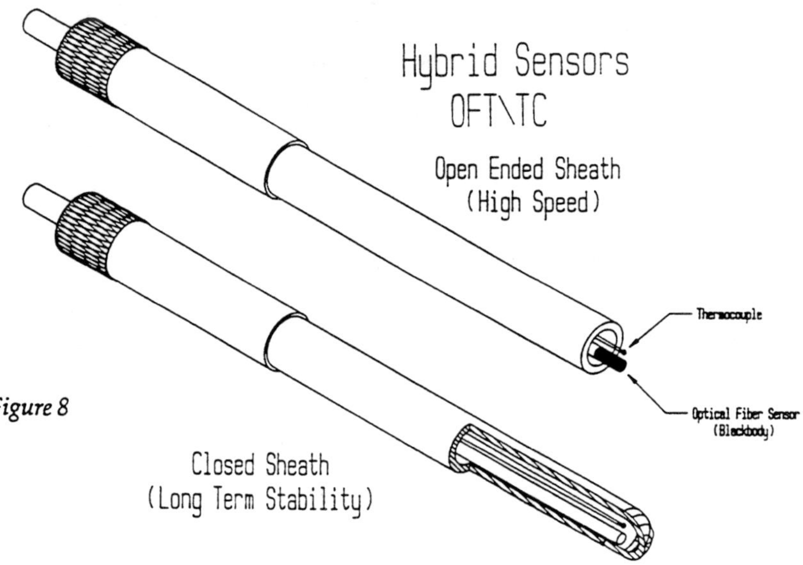

Figure 8

TWO WAVELENGTH MEASUREMENTS

Two different wavelengths are often used when making non-contact pyrometric temperature measurements. This technique usually calculates the ratio of the two intensities to determine the object's temperature from the slope of the Planck Equation. This ratio technique is independent of total light intensity. This method is therefore applied to non-contact measurements where the object does not fill the field of view. An example is a fine wire filament viewed by a lightpipe sensor. This technique is also useful for looking through dirty or partially blocked view ports. In most cases it is incorrect to assume two wavelength measurements eliminate emissivity as a measurement variable. If an object's emissivity is not known at one wavelength it is usually not known at two wavelengths. Why incorrectly assume it is the same at the two wavelengths that are measured?[5,6] Instead, the OFT technique provides a curve fitted second order polynomial (as a function of temperature) for each of the two wavelengths to be used in the calculation of ratios of the intensities. The polynomial for each wavelength's emissivity takes the form, $\epsilon = C{+}CT{=}CT^2$.

In addition to accounting for "real" emissivities at each wavelength, OFT provided for temperature measurement to be carried out as dual wavelength independent, ratio, or the average of the two wavelengths and their difference. In other words, the system can calculate temperature separately at each wavelength and combine that as an average and difference or the system can calculate the ratio of the two wavelengths.

$$T_{ratio} = f\left(\frac{\epsilon_{\lambda 1} E_{b\lambda 1}}{\epsilon_{\lambda 2} E_{b\lambda 2}}\right)$$

Where: E_b is the flux emitted by a blackbody and ϵ_λ is the emissivity function at a wavelength.

$$T_{average} = \frac{T_{\lambda 1} + T_{\lambda 2}}{2}$$

$$\Delta T = T_{\lambda 1} - T_{\lambda 2}$$

The two wavelength techniques described above can give the user better insight into not only the object's temperature, but also the quality of the optical transmission system, real time during measurement. The relationship of the two wavelengths can be used as a tool for checking the optical integrity of the optical signal transmission.

The difference between the two wavelengths, 950nm and 800nm in this case, can seem to change as the optical transmission system is degraded by any of a number of degradation mechanisms. *(See Graphs 1, 2, 3, 4.)* In Graphs number 1 and 2, the flexible optical

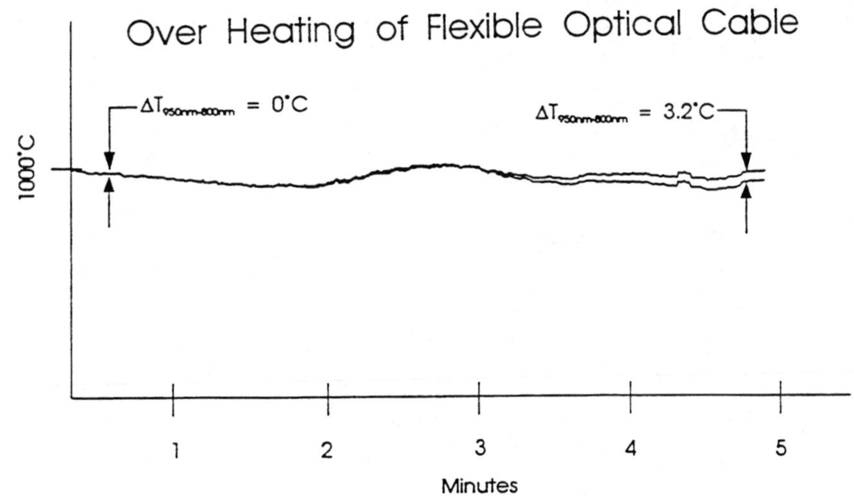

Graph 1

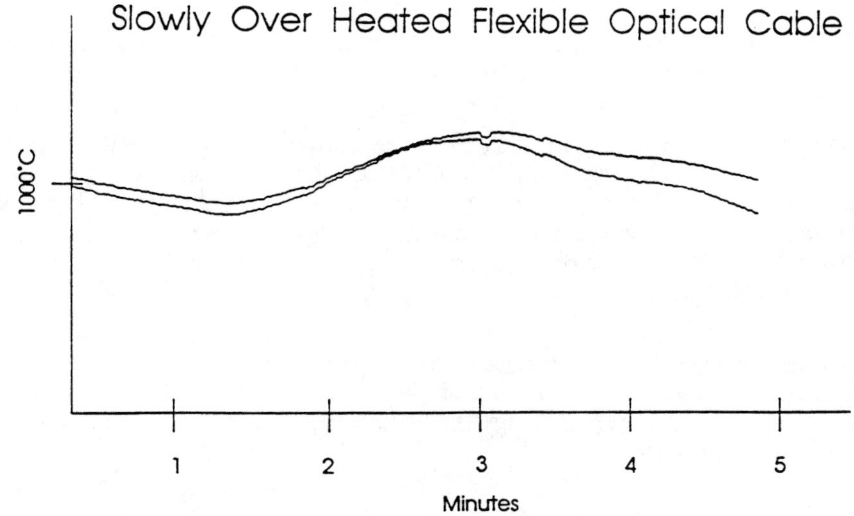

Graph 2

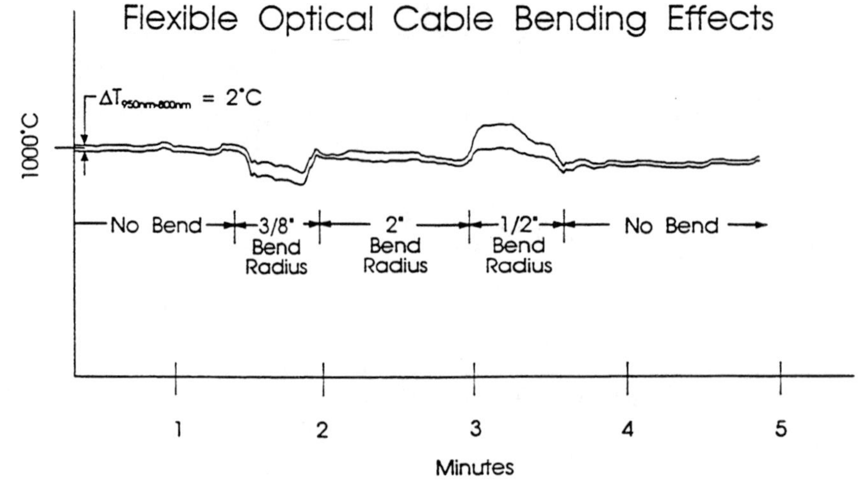

Graph 3

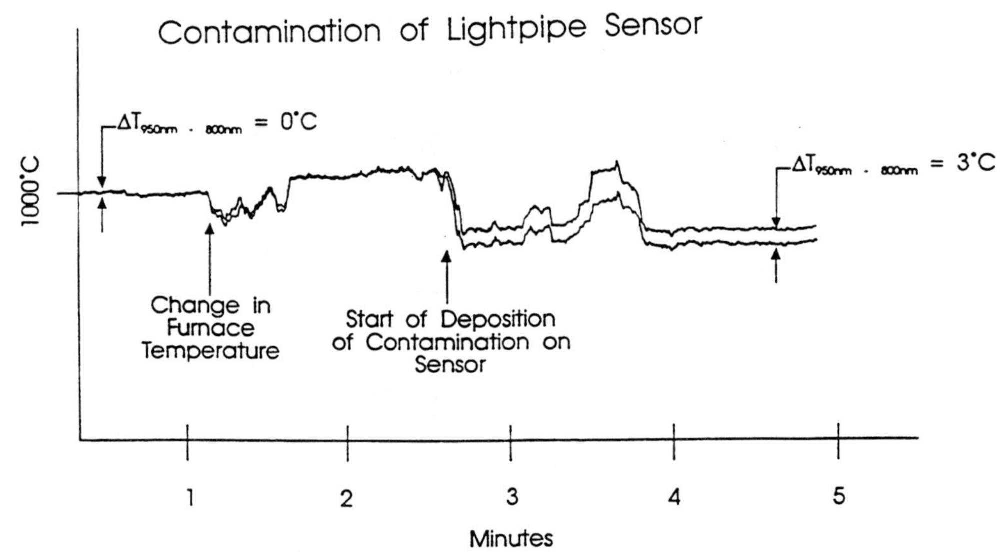

Graph 4

transmission cable is overheated and this affects the transmission of the 950nm and the 800nm light differently. In Graph 3, the flexible cable is bent to well beyond its normal operating conditions. Here again, the ΔT at the two measured wavelengths varies greatly as a function of severe bending causing micro bending losses. The final Graph, number 4, shows how gross contamination of the lightpipe sensor can cause a change in transmission at the two wavelengths.

These techniques are especially useful where total inspection of all the components of the system is routinely not possible, i.e., in an in-flight aircraft or power plant. *Note, that the system has a resolution of 0.01 °C, so that changes of approximately 3 °C are two orders of magnitude larger than the system resolution.*[7]

CONCLUSION

The long term reliability of fiber optics based temperature sensors are one of the major hurdles to wider acceptance and implementation. The need for improved temperature measurement over the thermocouples or conventional pyrometers is clear.[8] Improved temperature control leads to yield and quality improvements plus potential energy savings. This quality improvement can be realized in processes as diverse as fiberglass production to the growth of semiconductor films.

Today, there are about two thousand fiber optics based temperature measurement systems in the field. The number of customers and new applications are also growing steadily. Whether a temperature measurement system is on the factory floor or is an in-flight measurement system, reliability is critical. OFT technology can provide substantial advantages, including improved reliability when compared to conventional techniques.

ACKNOWLEDGMENT

The authors would like to thank the many creative customers of Accufiber and Luxtron. These innovative engineers and scientists have truly advanced the technology of Optical Fiber Thermometry (OFT).

REFERENCES

1. R.R. Dils, *Journal of Applied Physics*, Volume 59, Number 4, (1986).

2. M. Planck, *Annals of Physics*, 4(3), 553, (1901).

3. Accufiber, Furnace Control Application Note AN-10, (1988).

4. R.R. Dils, *EPRI Field Test Program Report*, (RP2102), (1987).

5. M.C. Lee, *NASA Conference Publication Number 2503*, (1988).

6. Grum Becherer, *Optical Radiation Measurements*, (1979).

7. Accufiber, Resolution Application Note AN-04, (1988).

8. *Business Week*, December 5, 1988, page 168A.

Fibre optic sensors for the chemical industry

R.T. MURRAY

ICI Electronics Group, Runcorn, Cheshire, U.K.

Abstract: A description is given of fibre optical sensors employed in the industrial measurement of pressure, temperature, level and safety. The questions of standardization and economics are cited as important considerations for widespread commercial acceptance.

The chemical and process industries have been major users of sensors for many decades. They fulfil two main functions:

(1) Indication that the process is straying from previously established optimum conditions.

(2) Detection of the onset of hazardous situations within the area of the plant. In that many of the products handled emit flammable vapours subject to accidental leakage it is important that the power used by the sensors is not able to cause ignition. Sensing loops rigorously satisfying this requirement are said to be intrinsically safe.

In both of the above roles it is important that the sensors (a) are reliable under process conditions, (b) require infrequent checking either for accuracy or safety, and (c) are cheap to instal.

The realization that greater hydrocarbon efficiency could significantly reduce the cost of a product became an important element of management thinking subsequent to the years of the oil crisis. The most significant advances towards optimum efficiency have been brought about by improved control systems in

(1) Energy management and the detection of heat leaks.

(2) The control of the process variables to be close to optimum for a high proportion of the plant running time.

More complex control systems under computer control have been installed, and since the goals are purely economic this has led purveyors of sensing systems to find technologies offering higher reliability and lower life costs.

Analysis of the costs involved in installing a sensor linked to the control computer reveal that for conventional sensors transmitting analogue signals over twisted pair wiring the most significant items are[1]

(1) The cost of installing the wiring over an EMI-free route.

(2) The cost of the anti-ignition housing for the sensor.

(3) The power control barrier devices used to achieve 'Intrinsic Safety' standards.

During the lifetime of the system, annual checks of insulation, earth continuity, etc, are required by the Safety Authorities and this certification is a significant ongoing cost. For those involved in fibre optics it is immediately apparent that each of these elements can be tackled by optical sensors with fibre links, i.e. extrinsic

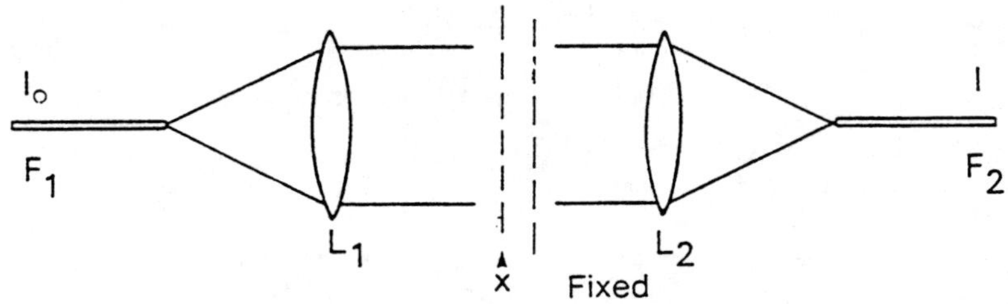

FIG. 1. (a) Optical principles of a fibre optic Ronchi grating reader. I_o: input light; I: output light; LI: collimator; L2: launch lens; X: sensing grating.

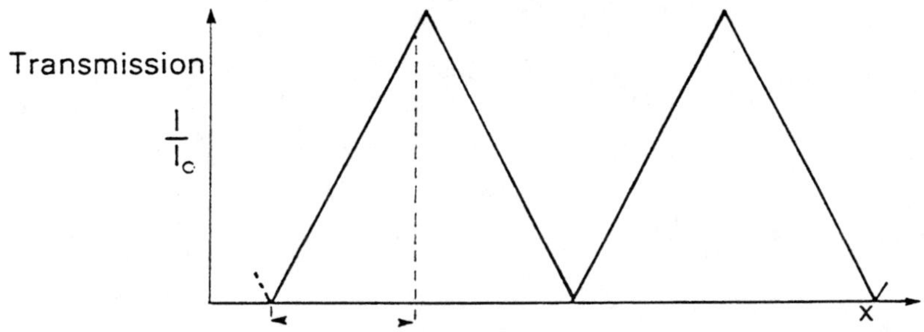

FIG. 1. (b) Transmission function expected from the reader. The linear region between the arrows gives good performance.

fibre sensors, but at a penalty. Almost all conventional electrical sensors for process use, work to a standard analogue signal of 4–20 m amps driven by 24 volts d.c. and are freely interchangeable between manufacturers. The achievement of standardization has been painful politically but was facilitated by the suppply of significant amounts of 'safe power' at the sensor head. Passive optical sensors will find standardization more difficult but it will be crucial within the economic framework I have constructed.

To illustrate progress towards plant-worthy systems I will describe work within ICI and then on a broader basis review other European offerings. This will be followed by a discussion of the standardization problem.

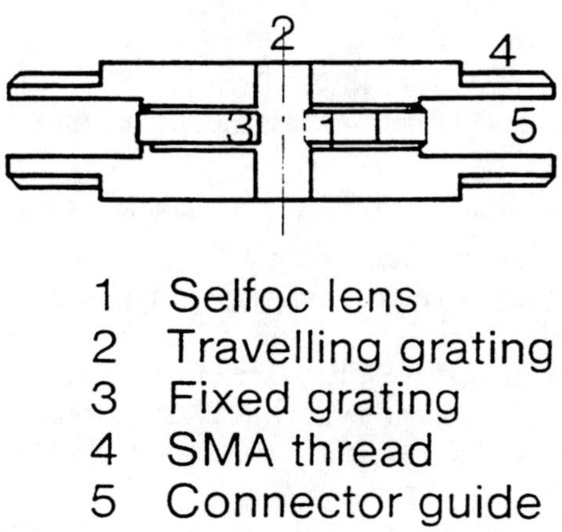

FIG. 2. Design of the optical train using selfoc lenses 1 and 3; the fixed grating is cemented to lens 1. 4-SMA, and 2-space for sensor grating.

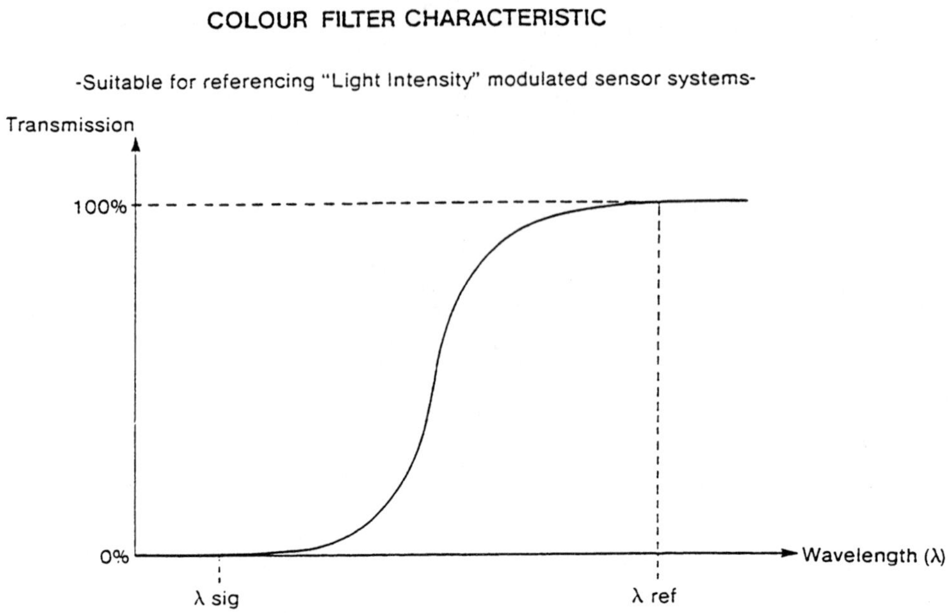

FIG. 3. Desired filter characteristics. For best referencing λ sig and λ ref should be as close as possible.

Grating pressure sensor

When a pair of square wave (Ronchi) gratings are translated relative to one another, the transmission function is a sawtooth (Fig. 1). Thus a simple optical displacement unit may be designed (Fig. 2) to form the basis for a family of sensors — pressure, temperature, flow rate or level.

Advantages of this approach are

(1) Linearity of transmission versus displacement

(2) Tailorability of grating period to the displacement range.

The design adopted for the displacement element seeks to marry these features in a simple design. However the approach

(3) Is purely analogue and subject to extraneous fluctuations in the transmission path.

(4) Requires precision alignment of the gratings and optical train.[2]

The problems arising from the sensitivity of the total optical path to stray variables has been tackled by colour referencing; in this approach one conventional Ronchi grating is combined with a dichroic grating in which the bars absorb up to λ_T and transmit efficiently above this wavelength. The inter-bar spaces transmit over a broad range on either side of λ_T. Thus as illustrated in Fig. 3 at λ_S the grating performs conventionally whereas at λ_R it acts as a window. Thus with a thermal source the displacement signal

$$x = k\,(I_s/I_R)$$

is insensitive to transmission changes due to connector variability etc. However, unless λ_S and λ_R are closely adjacent, fluctuations in bulb temperature will only be partially referenced. Several simple embodiments of the gratings have been tried:

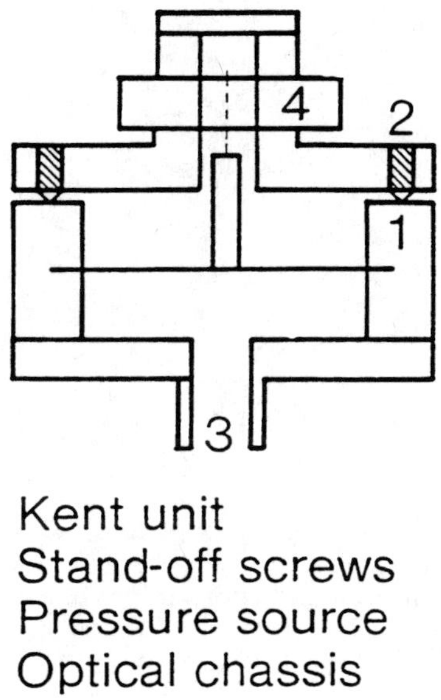

FIG. 4. Diagram of a pressure sensor constructed around a Kent Industrial Measurements low hysteresis diaphragm.

(1) Photographic colour film[3]

(2) Ge semiconductor band edge

However, neither of these gives a sufficiently fast transition to reference source fluctuations. Interference filters may well be the cheapest effective construction.[4]

The displacement element has been coupled to a taut diaphragm to form a fibre optic pressure gauge as illustrated in Fig. 4. The displacement of the diaphragm over its working range is 50 μm and therefore a grating period of 120 μm has been selected to maximise sensitivity and linearity. The characteristic obtained is illustrated in Fig. 5 giving reproducibility at ambient temperatures of 0.1%. However, the system has not been properly designed to minimise the effects of temperature. An improved design will minimise the separation of the diaphragm from the optical train and will compensate for the change in diaphragm compliance by mounting it upon a ring of higher thermal expansion. Judicious choice of material can reduce the temperature drift to negligible proportions for the range 0–100°C.

Thus we believe a robust and accurate plantworthy sensor which is insensitive to interference may be constructed for reasonable cost, and several U.K. instrument companies are building upon our experience.

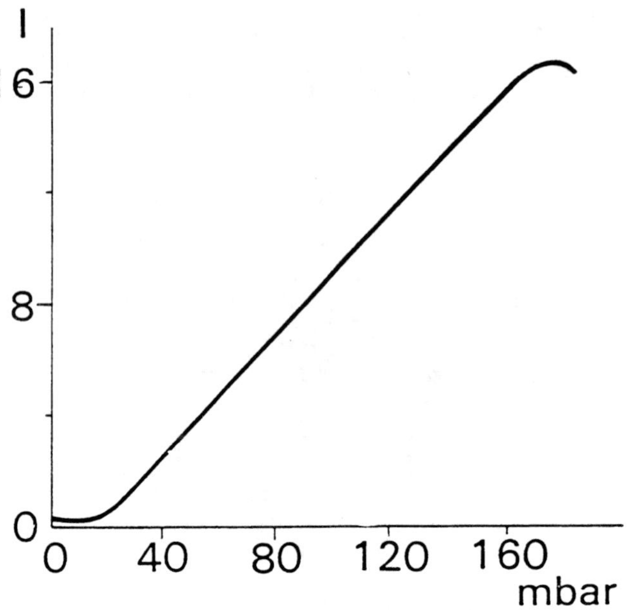

FIG. 5. Characteristics of the pressure sensor. Optical flux is in arbitrary units. The non-linear toe and head are caused, respectively by deviation from 1:1 mark to space ratio and slight imperfections to the bar edges.

Optically powered sensors

The passive optical sensor is a very attractive proposition but, as we remarked in our introduction, the lack of any signal processing capacity may be a real limitation. In addition, process plant designers with a high cost project in their care are prone to be conservative and are unwilling to commit their future to a totally alien technology. We have, therefore, pursued a hybrid approach in which a conventional transducer is powered by optical energy received along a fibre and transmits its signal in the form of an interpulse interval on a second fibre, thus incorporating the security of time as the signal medium.[5]

The simplest transducer to convert to the optical domain is the ubiquitous resistance thermometer (RT). The approach adopted involves the retention of as many features of present commercial RT transmitters as possible and to this end the primary sensor, its mounting system and protective housing have all been used without change. For reliability the optical power source is a Honeywell SE 3352/4 LED operating at 820 nm, able to launch 3.5 mW into a 600 μm fibre of numerical aperture 0.4, rather than a bulb. An array of Si solar diodes coverts the optical power to a 5 volt d.c. signal able to drive the necessary switching comparators and optical pulse electronics in a stable regime. Figure 6 illustrates the scheme in which a reference resistor R_R within the transmitter head is used to correct the effect of ambient temperature on the electronics. Because of the very low duty cycle of the signal LED the electronics requires only 1 mW of optical power to measure the resistance of the R_T once per second.

The internals of the head and the complete unit are illustrated in Figs 7 and 8. The pulses are returned to a clock circuit and a microprocessor performs the necessary algebra to calculate the temperature to ±0.5°C.

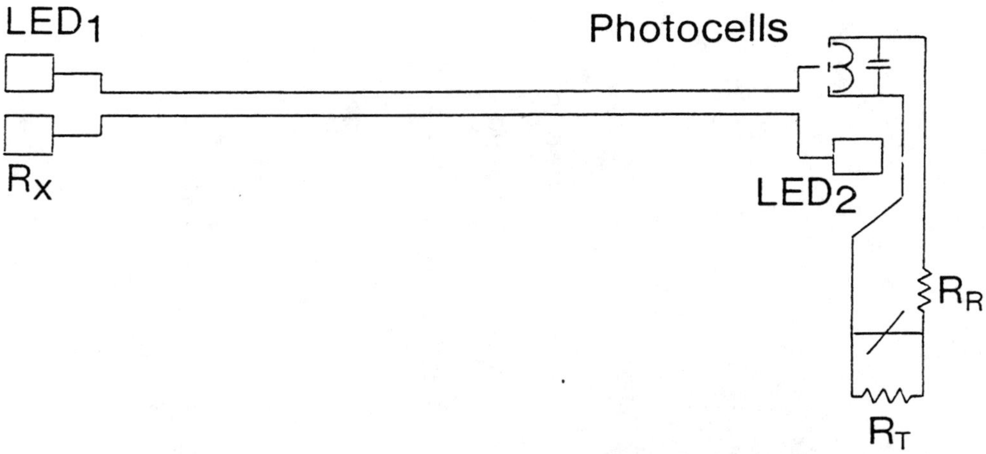

FIG. 6. Diagram of the optically powered resistance thermometer. R_R reference resistor which is in the head mounted transmitter, R_T sensor resistor.

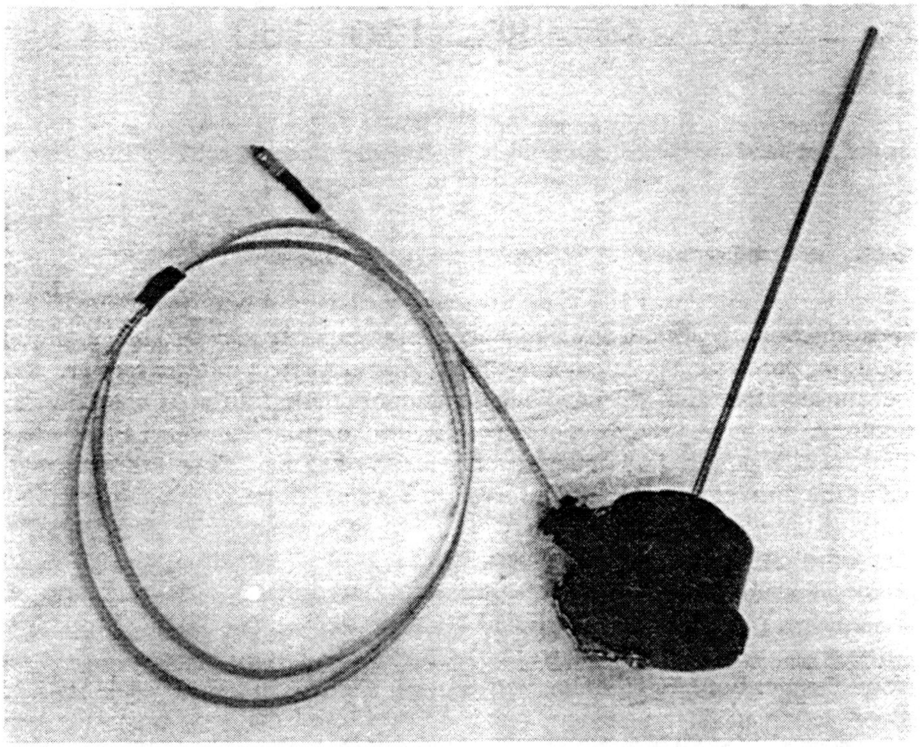

FIG. 7. View of the assembled optically powered resistance thermometer.

FIG. 8. Internals of the head showing the fibre optic leads and the optics block which sits on top of the electronics timing circuits.

The nature of the time generation circuitry should enable a range of other transducers to be similarly addressed so that a broad family of variables may be transmitted using a common language. Sensors based on this technology will appear on the market in the near future.

European developments

Other European corporations have also been active in the development of optical sensors and a range of devices are now available in the market place. The problems of incompatibility are well illustrated by the examples quoted.

Optronics pressure gauge*

Following the philosophy of winning customer acceptance by incorporating as much conventional sensor technology as possible into a fibre optic system, Optronics have incorporated an optical Grey code shaft encoder into a Budenburg pressure gauge.[6] Advantages include

(1) Digital signal of high integrity

(2) Visual presentation of the data at the point of measurement is combined with passive fibre optic telemetry.

However the linear to rotary motion converter inside the gauge is a source of hysteresis, limiting accuracy to about 1%.

---

*Optronics Ltd, Cambridge, U.K.

FIG. 9. Internals of the Optronic modified pressure gauge with absolute shaft encoder. The optical fibre cable enters from the bottom and contains 2×8 fibres for an 8-track encoder.

The shaft encoder principle involves eight binary tacks arranged circumferentially around a disc attached to the rotary shaft (Fig. 9). The angle of the shaft is deduced from the set of transmission values. A comparator level determines whether each channel is in a zero or one state and a lower threshold transmission is available to warn of breaks in the link. As installed in several gas and water pumping stations around the U.K. the device is linked passively to the control centre by a 16-core 200 μm plastic coated silica (pcs) cable to distances up to 300 metres.

Photoluminescent thermometer (ASEA)†

GaAs may be excited to luminesce efficiently by radiation in the 700–800 nm range as may be provided by conventional LED sources. Ovren and his colleagues at ASEA Innovation have demonstrated that the shift of the luminescence peak with temperature can be used to read the temperature of a tiny chip of the semiconductor mounted on the end of an optical fibre (Fig. 10).[7] The wavelength shift signal is obtained by splitting the luminescence signal equally onto two photodiodes with markedly different colour responses. The ratio of the two diode currents is then a measure of the peak position (Fig. 11) and should give temperature signals immune to connector or source fluctuations. However, drift in receiver temperature will affect the photodiodes differentially and this possibility should be minimized.

†ASEA, Vasteras, Sweden.

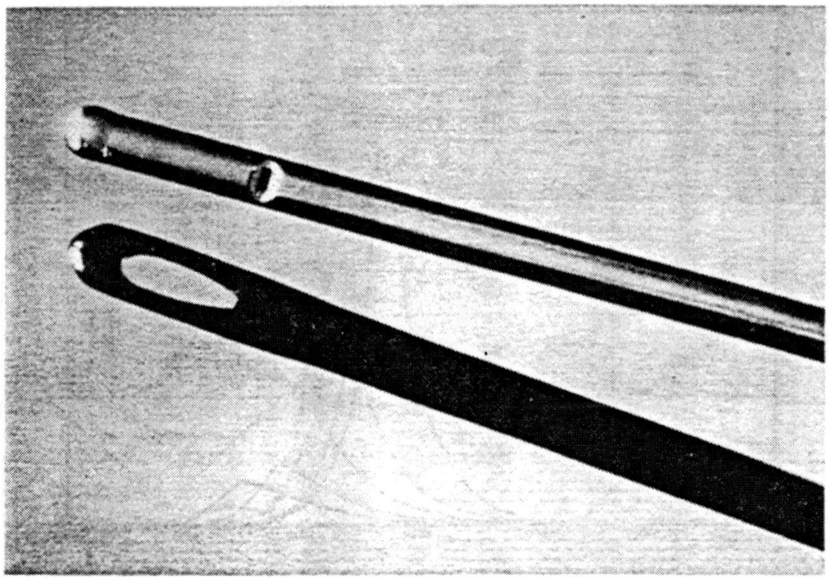

FIG. 10. Sensor head of ASEA photoluminescent thermometer compared with a domestic needle.

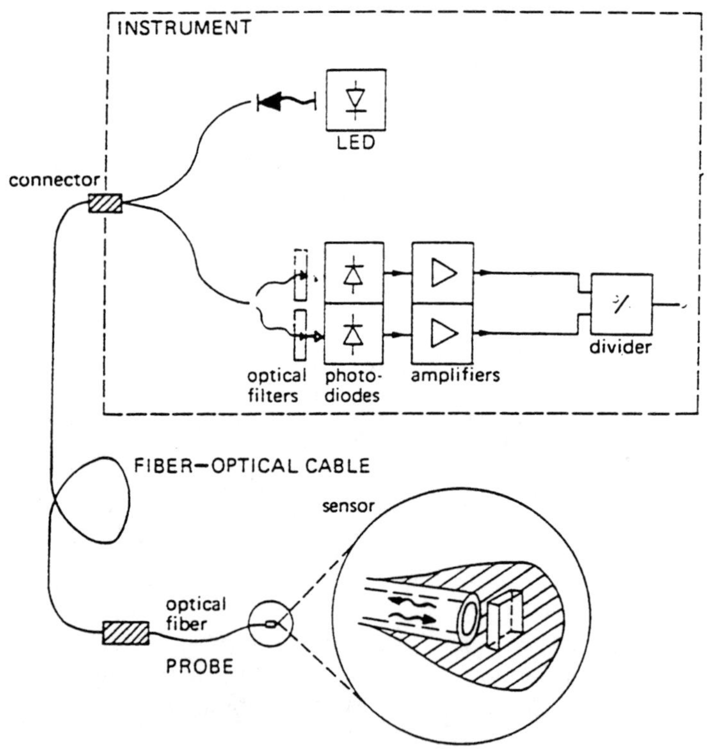

FIG. 11. (a) Schematic of the sensor components. Light emitted by the phosphor is split at the connector and passes through two optical filters before being detected by identical detectors. The ratio of the signal from the two detectors is a measure of the spectral shift and hence of the phosphor temperature.

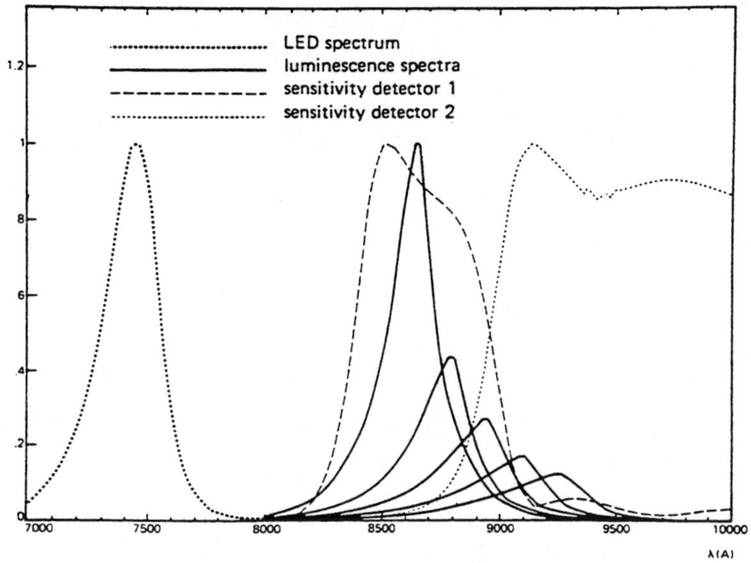

FIG. 11. (b) Spectral distribution of exciting and phosphorescent light and the filter characteristics.

The thermometer has found a range of applications, but from our point of view the ruggedized installation in the Cl_2 cells at Norske Hydro is the most exciting evidence of the fitness of this technology for our harsh world. However, in our introductory framework economics was a major plank and it must be admitted that at present a single thermometer system is expensive. ASEA are tackling this barrier by multiplexing a number of thermometer heads into an integrated electronics system.[9]

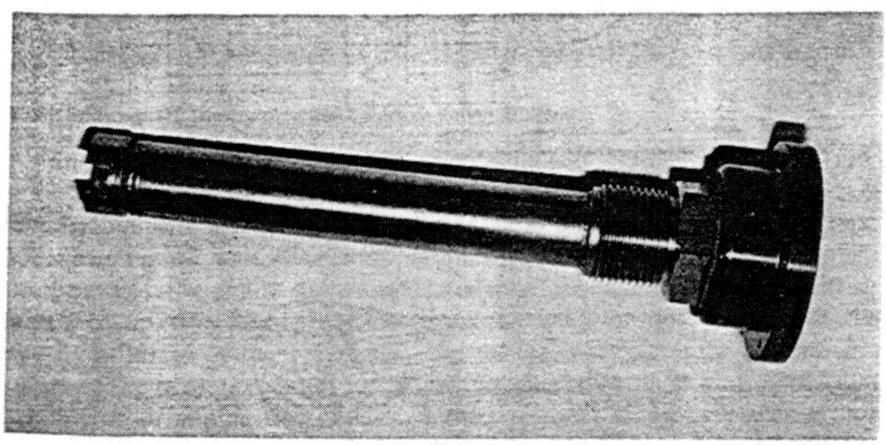

FIG. 12. Ruggedized prismatic level gauge with fibre optic leads [Delta Controls].

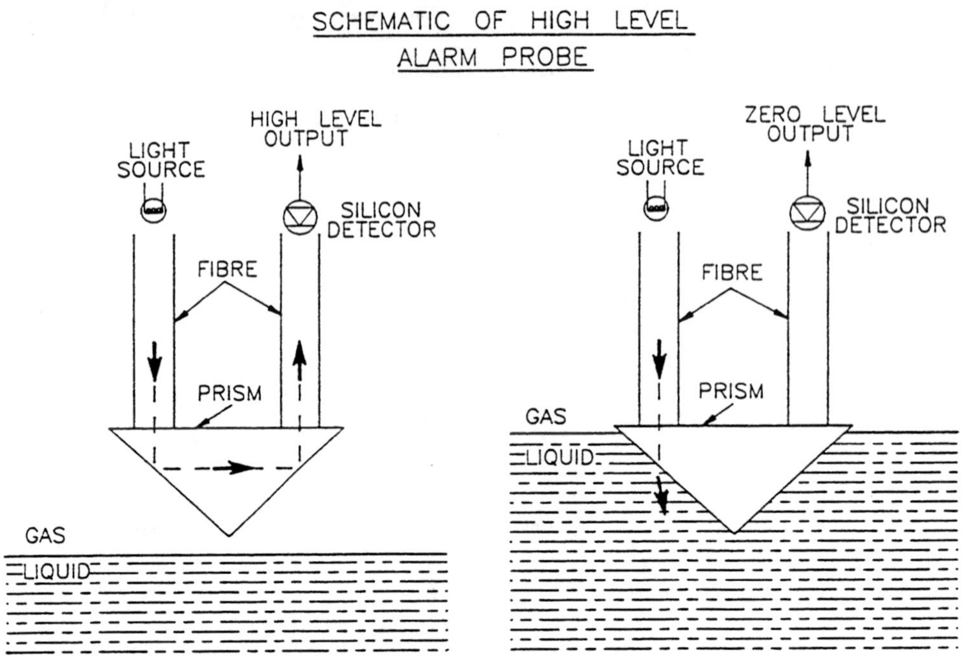

FIG. 13. Schematic illustrating the principles of the prism gauge. When the prism is immersed in a fluid of similar refractive index very little light is internally reflected thus signalling a fail safe warning.

Optical safety systems

In addition to their use in efficiency improvement, optical sensors/switches have been developed for hazard monitoring. This survey, which is restricted to existing commercialized European items, reveals a quite broad range from which we draw your attention to the following.

Optical switch for level or pressure (Delta Controls)*

The Delta optical level switch (Fig. 12) is designed to prevent overfilling of stock tanks, particularly where they contain hazardous fluids such as petroleum. The active element is a prism set to give total internal reflection when in air, but to pass light into the fluid when wetted by it (Fig. 13). The prism is addressed by two fibres with collimating optics for efficient launch and good contrast. Several industrial installations are already in satisfactory operation with link lengths up to 2 km being available.

The same company make an optical pressure trip in which an optical microswitch is incorporated into their standard units. The microswitch is based upon a conventional electrical device but misaligns abutting fibres rather than effecting contact. To distinguish the off state from cable faults, the device is arranged to pass a low level signal in this state. The optical signal is returned to an electro optic relay in the safe area.

*East Molesey, Surrey, U.K.

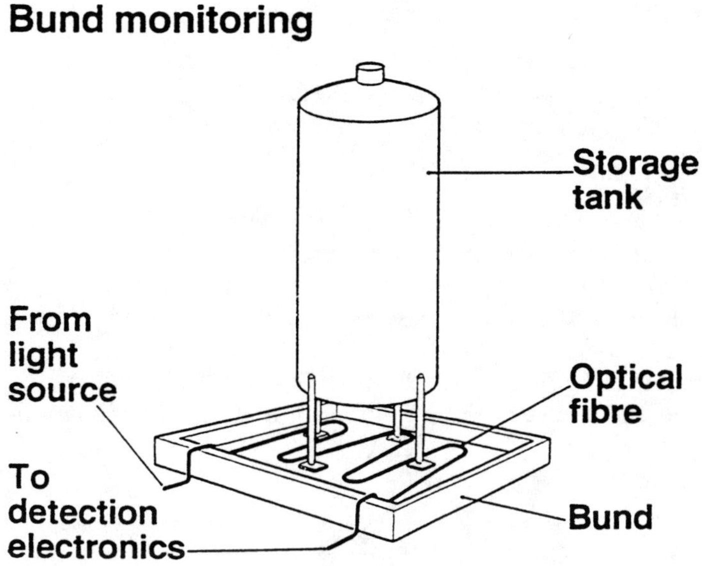

FIG. 14. Outline of the cryogenic sensors as installed on a British Gas installation.

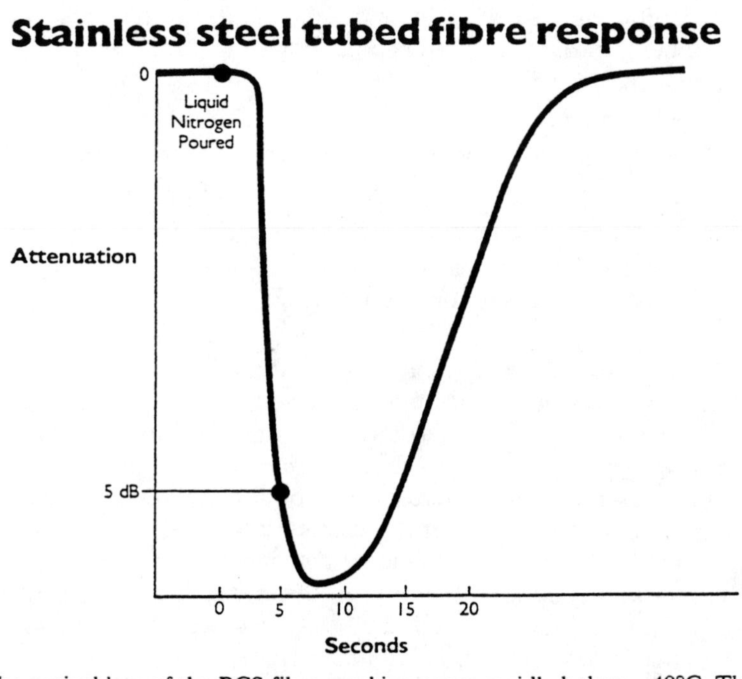

FIG. 15. The optical loss of the PCS fibre used increases rapidly below −40°C. The temporal response to immersion in liquid nitrogen is illustrated below.

Cryogenic leak switch (Pilkington Security Systems)†

A commonly cited disadvantage of pcs cable is the rapid change in numerical aperture of the fibre as ambient temperature falls through the glass transition for the polymer. Pilkington Security Systems have taken advantage of this phenomenon to design a cryogenic leak detector with applications to liquified petroleum gas storage.[10]

The detector fibre, which is encased in a stainless steel jacket, may be snaked back and forth under a storage tank or valve array to give a single distributed sensor (Fig. 14). Onset of a spillage rapidly lowers the temperature of the encased fibre and when it falls to −60°C the increased attenuation causes the system to trip an alarm (Fig. 15).

The prototype has been successfully installed on a liquified methane tank of British Gas over the last year and a number of additional installations are in the planning stage.

System standards

For optical sensor technology to fulfil its potential in process control, network standards will be necessary. We are now at the stage where consideration of the options and establishment of standards is necessary. Why? The customer engineer in designing a plant must ensure ease of installation and maintenance, compatibility between components of the system and continuing supply from several sources. He will therefore demand

(1) A single fibre standard

(2) A small range of cable designs fitted to the harsh environment

(3) Standard connectors which are intermatable between manufacturers

(4) A common transmission language whether analogue or digital

(5) For reasons of cost he will probably push for systems which offer in addition cheap multiplexing of sensors onto a fibre loop.

The range of sensors described above illustrates just how big a problem will arise if standardization is not achieved. Thus several are analogued in signal level, one uses wavelength shift, one uses multi-fibre cable with binary code and one uses time interval as the encryption process. The latter sensor also requires a larger diameter fibre than is necessary for the other techniques in order to supply adequate optical power. What then is the most profitable direction for the first standards to take? I believe the essential conservatism of the customer will dictate that he be offered devices with which he is familiar and comfortable incorporated into a system which offers him economic advantages. Hybrid sensors in which a conventional element such as a resistor or capacitor are varied by the measurand to give a time constant or time interval becomes very attractive under this regime.

A satisfactory base standard might then be as Table 1 (a). To accommodate the probability of multiplexing an overlay standard as in Table 1 (b) would permit addressing individual sensors by the central station.

†St Asaph, Clywyd, U.K.

TABLE 1. (a) Hybrid sensor standards.

Fibre: 200 μm core with 250 μm cladding and numerical aperture of 0.4
Cable: All dielectric
Connectors: SMA intermatable and with protective boot
Power: Optical 820 nm
Signal: Optical 1300 nm. Time interval 100–500 μs. Internal repetition rate 1 s

TABLE 1. (b) Multiplex overlay.

Power: Optical 820 nm
Address code: Carried as negative pulses + error code

Because of the rapid advances still being made in sensor types, e.g. distributed sensors, gas sniffer systems and chemical sensors, these standards will inevitably have a limited lifetime. Without them however there will be no future for fibre sensors in industry since funding sources need commercial success to maintain interest.

Safety

It is often assumed that fibre optics carries no detonation risk and is therefore inherently safe. Recent studies by Tortishell at SIRA suggest that future high powered laser diodes such as might power a large multiplexed system may be capable of causing ignition under special (but conceivable) circumstances. To retain the economic advantages of fibre optics it may therefore be necessary to seek alternative sources of power in multiplexed arrays.

Conclusion

The dawn of fibre optic sensors has arrived but before they replace conventional sensors in the economic and physical climate of the process industries it will be necessary to establish transmission standards and to address the problem of system cost. Multiplexing undoubtedly offers cost savings but will itself call for standardization. Since *de facto* standards are not emerging, the industry itself must act. The U.K. Optical Sensors Collaborative Association has started to examine the options.

References

1. BOND, D. *Industrial Growth of Optical Sensors in Optical Sensors and Optical Techniques in Instrumentation.* I.M.C., London (1981).
2. MURRAY, R.T. Optical fibre sensors. *IEE Conference Optical Sensors*, p. 221 (1983).
3. JONES, B.E. and SPOONER, R. *J. Phys. E.* **16,** 1124 (1983).
4. Brit. Patent No. 2138588 (1983) to Barr and Stroud W O Patent 84/01824 (1982) to FOXBORO.
5. CPW Boeder Optical Techniques in Process Control BHRA/SIRA. London (1983).
6. BRADLEY, N.J. *Control and Instrumentation,* p. 39. March 1982.
7. OVREN, C., ADOLFSSON, M. and HOK, B. *Optical Techniques in Process Control,* p. 67. BHRA/SIRA, London (1983).
8. BERGSTROM, J., FAHLANDER, M. and LARSON, L. *OFS 85 Digest,* p. 140. San Diego (1985).
9. LINGE, A., HOK, B. and THOREN, E. *OFS 85 Digest,* p. 148. San Diego (1985).
10. PINCHBECK, D. and KITCHEN, C.A. *Proceeding of Conference on Electronics in Oil and Gas Industries,* London, Jan 1985.
11. TORTISHELL, G. *Fibre Optics 85.* London (1985).

Section Five

Optical Techniques for Materials Analysis

Reprinted from *Electro-Optical Instrumentation for Industrial Applications,* F.M. Zweibaum, R.A. Carella, eds., Proc. SPIE Vol. 411, pp. 72-79 (1983). ©1983 SPIE.

Process Gas Analyzers Using Non-Dispersive Infrared

Rodney M. Durham

Instrumentation Division, Infrared Industries, Inc.
P.O. Box 989, Santa Barbara, California 93102

Abstract

The use of NDIR techniques for the monitoring of gases in industrial processes is discussed. Gas analyzer configurations utilizing microphone detectors and solid state detectors are reviewed. Typical applications are presented.

Infrared Spectroscopy

The manner in which electromagnetic radiation interacts with matter is a function of the wavelength of the radiation. Infrared energy interacts with matter at the molecular level. It is this molecular interaction which makes the infrared radiation a valuable probe in studying the properties of molecules. The infrared wavelength region from 3 µm to 10 µm is especially valuable for detecting the presence of molecular species. Infrared radiation at these wavelengths has sufficient energy to excite molecular vibration. The frequency of the vibration is a function of the weights of the atomic elements bound in the molecule and the strength of the molecular bond. This characteristic vibration frequency results in an infrared absorption band which is indicative of the presence of a particular molecular bond. For example, the carbon-hydrogen bond is excited by infrared radiation at the wavelength of 3.4 µm while the carbon-oxygen bond in the carbon monoxide molecule is excited by 4.7 µm radiation. Scientists have long used infrared absorption as a means of probing the structure of molecules. Studying the manner in which specific wavelengths of infrared energy excite vibration and rotation in molecules reveals information about the molecule that can be used to determine what and how many molecules are present. The Infrared Spectrophotometer is the principal instrument used by the scientist for these measurements. Most laboratory spectrophotometers are of a dispersive design, that is, a prism or grating is used to separate the spectral components in the source radiation. Modern infrared spectrophotometers have a wide wavelength range from 2 µm to 50 µm. They find use in research, quality control, and analytical service laboratories.

Infrared Process Analyzer

The infrared process analyzer has evolved from the laboratory spectrophotometer to satisfy the specific needs of industrial process control. While dispersive instruments continue to be used in some applications, the work horse infrared analyzers in process control are predominantly non-dispersive infrared (NDIR) analyzers. The NDIR analyzer can be used for either gas or liquid analysis. For simplicity, the following discussion addresses the NDIR gas analyzer but it should be recognized that the same measurement principle applies to liquids. The use of infrared as a gas analysis technique is certainly aided by the fact that molecules, such as nitrogen (N_2) and oxygen (O_2) which consist of two like elements, do not absorb in the infrared spectrum. Since nitrogen and oxygen are the primary constituents of air, it is frequently possible to use air as a zero gas.

Many different analyzer configurations have been developed to address the diverse needs of the industrial process control industry. The basic constituents of an NDIR analyzer are: 1) a source of infrared radiation; 2) a means of restricting the wavelength range of the source radiation; 3) a means of detecting the infrared radiation; 4) a sample chamber to hold the gas or liquid to be measured; 5) a means of modulating the source radiation; and 6) electronics to process the signal generated by the source energy falling on the detector.

NDIR Analyzers Using Microphone Detectors

The majority of infrared process analyzers installed over the last twenty years have utilized microphone detectors. These detectors are generally the Veingerov single-sided microphone system or the Luft balanced condenser microphone system. [1)]These are shown schematically in Figures 1 and 2. The microphone detector uses an absorbing gas as its detecting medium. When radiation reaches the detector that the sensitizing gas will absorb, the gas heats up and expands. This causes a diaphragm to distend. The diaphragm movement varies the condenser microphone capacity which is part of an electrical circuit which generates an electrical output signal. Both analyzers use dual sources which are chopped to alternately allow energy to pass through a sample cell and a reference cell. If the sample cell contains a non-absorbing zero gas, such as nitrogen, the modulated beams reaching the detector through the two paths are of equal amplitude. In the case of the Veingerov single-sided detector, the chopper is configured so that at any given time the sum of the cross-sectional areas of the two beams as seen by the detector equals the total cross-sectional area of a single beam so that when no absorbing sample is present, the sample and reference path signals become imbalanced and a signal at the chopper frequency is developed. The amplitude of this signal is a function of the concentration of the gas present in the sample cell.

The Luft detector operates similarly but the detector has two chambers separated by a diaphragm. The signal generated by the presence of an absorbing gas in the sample cell is at twice the chopping frequency. This is an advantage over the single-sided microphone detection system since it is less susceptible to vibration caused by imbalance of the chopper motor. Having separate chambers does, however, allow for the possibility of a change occurring in one-half of the detector and not the other and thus resulting in zero drift. Recently, infrared process analyzers have been introduced which use a Luft type detection system but replace the diaphragm with flow sensors. The flow of gas from one chamber to another is sensed by the flow sensor rather than by using a capactiance detection technique. This is claimed to eliminate one of the major modes of detector failure-failure of the thin diaphragm.

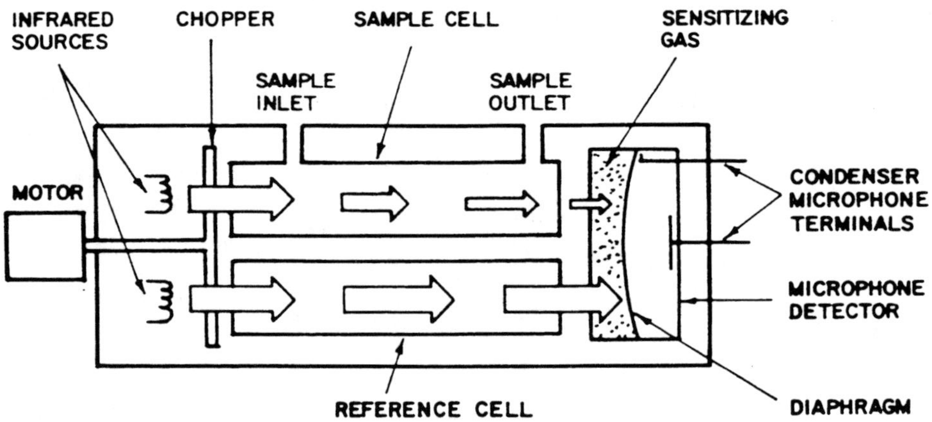

NDIR ANALYZER WITH VEINGEROV TYPE DETECTOR

FIGURE 1

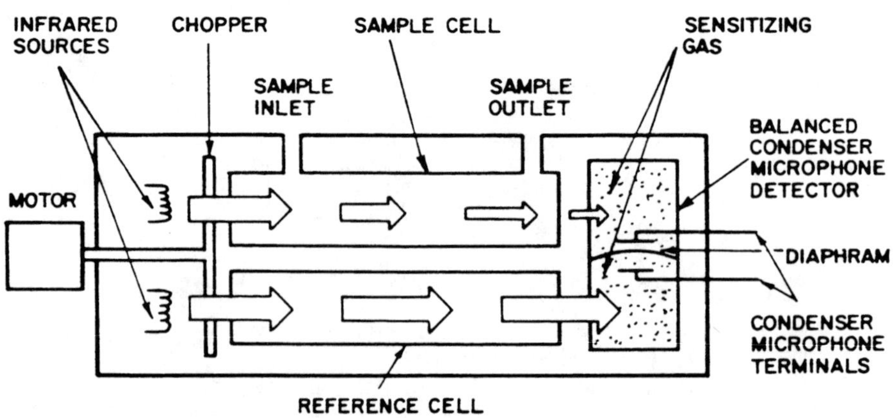

NDIR ANALYZER WITH LUFT TYPE DETECTOR

FIGURE 2

For a given path length, process analyzers which use the microphone detector are more effective than those analyzers which use solid state detectors and optical filters at measuring low concentrations of gases which have a lot of structure in their absorption band. This structure results from the molecular rotation spectrum being superimposed on the vibration spectrum and is easily resolved in the simpler molecules such as carbon monoxide, methane and ammonia (See Figure 4).

Disadvantages of the use of microphone detection for NDIR analysis include sensitivity to vibration and the high cost of detector replacement. A Luft type detector typically costs over $600. The diagrams in Figures 1 and 2 show that two sources are required in these systems. To avoid zero drift, these sources must have a stable balanced output. Also note that no lenses or mirrors are used to direct the energy from the sources to the detector. These systems rely on the reflections of the infrared beams off the interior walls of the sample and reference cells. To achieve high reflectivity, these walls are highly polished and are frequently lined with gold foil. This makes the replacement of a sample cell expensive. Contamination of the cell walls by the sample will also result in zero drift. Since the microphone detector uses a gas as its detecting agent, NDIR analyzers using this detection mechanism are generally offered for only a limited number of gas species. Unstable or corrosive gases cannot be used to charge the cell, although it is possible to use an alternative gas that has similar infrared absorption properties.

NDIR Analyzers Using Solid State Detectors

A new generation of NDIR analyzers has evolved to satisfy the frequently harsh industrial environment. These analyzers utilize solid state sensors for the detection of infrared radiation. The sensors most frequently used are lead selenide (PbSe), thermopiles or pyroelectric detectors. The gas analyzers generally are configured as single path instruments, dual beam with a reference path, or dual channel with a reference filter.

The single beam instrument finds application where low cost is important but stability requirements are not stringent. Generally, this is true when the measurement period is short and frequent rezeroing is practical. Changes in source intensity due to power variations or changes in detector sensitivity due to temperature fluctuations are reflected directly in the output as zero drift. To avoid this, a reference path is commonly used. The dual beam configuration is shown in Figure 3. The source energy is modulated by a chopper blade. This allows the source to alternately pass through the reference and sample paths. The reference path is always free of absorbing gas so the detector is exposed to the source through a path unaffected by the presence of the sample. This signal level is monitored by an automatic gain control (AGC) circuit which holds the reference signal level constant. If the source intensity or detector sensitivity should change, the AGC would correct for it in both the reference and sample channels. Sync pick ups monitor the chopper position and inform the electronics when the sample or reference path is irradiated. A narrow bandpass optical filter is located in the front of the detector to limit the infrared energy and sensitize the analyzer to a particular gas absorption band. The signals generated by the two optical paths are synchronously demodulated. The amplitude of the reference signal is used by the AGC circuit. When an absorbing gas is introduced, the signal reaching the detector through the sample path is attenuated and the magnitude of the detected signal corresponds directly to the concentration of the sample gas present in the sample cell. The use of a reference path results in stable operation with effectively reduced drift due to power line or temperature fluctuations. The synchronous detection is an effective means of providing a very narrow bandpass filtering of the chopped signal, resulting in good noise rejection. The use of collimating optics eliminates the reliance on internal reflections in the sample and reference cells to transmit the source energy to the detector. The sample cell construction is thus simplified and the cell is easier to clean, less susceptible to damage, and less costly to replace.

A reference optical filter can be used as an alternative to the reference path. This requires that a spectral window exist where there is no interference from the sample. The 3.8 μm to 3.9 μm spectral region is frequently used in NDIR gas analyzers for this purpose. The spectral curves shown in Figure 4 show that this spectral window would work well as a reference for measuring carbon monoxide, carbon dioxide, methane, or ammonia. It would be unsuited for nitrous oxide.

The dual channel (reference filter) configuration is outlined in Figure 5. The reference filter and the sample filter, which defines the spectral region of interest, are mounted on a spinning chopper wheel. As the chopper spins, it alternately positions the filters in the optical path. The signal is demodulated in a similar manner to the dual beam approach. In order for the reference filter to be effective, it is important that the performance of the source and detector behave the same in the sample and reference spectral regions. Unfortunately, this is not normally the case. The spectral properties of the source and detector are both strong functions of temperature. They must be

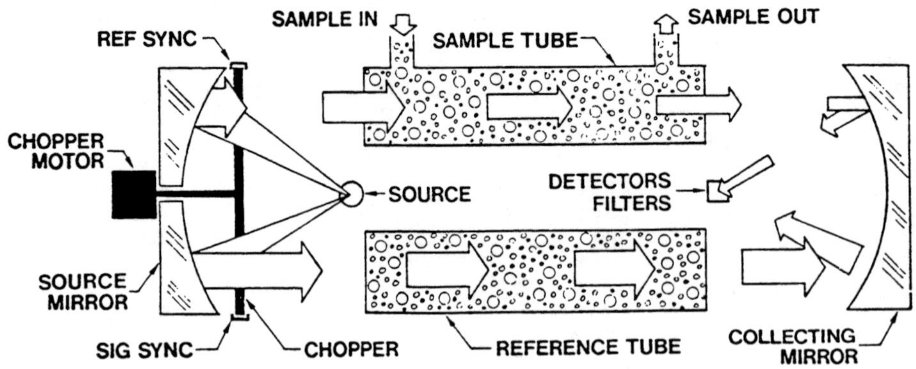

DUAL BEAM NDIR ANALYZER WITH SOLID STATE DETECTOR

FIGURE 3

controlled precisely or zero and span drift will result. Similarly, the optical interference filters used to isolate the spectral region of interest will shift with temperature. These too, should be temperature controlled. An advantage of this approach is that drift due to accumulation of dirt or film on the windows is avoided since this is generally non-spectral in nature and corrected by the AGC electronics. The analyzer is effectively zeroed each revolution of the chopper wheel.

Another attribute of both the dual-beam and dual channel (reference filter) process analyzers is that they can be configured for measuring more than one gas.

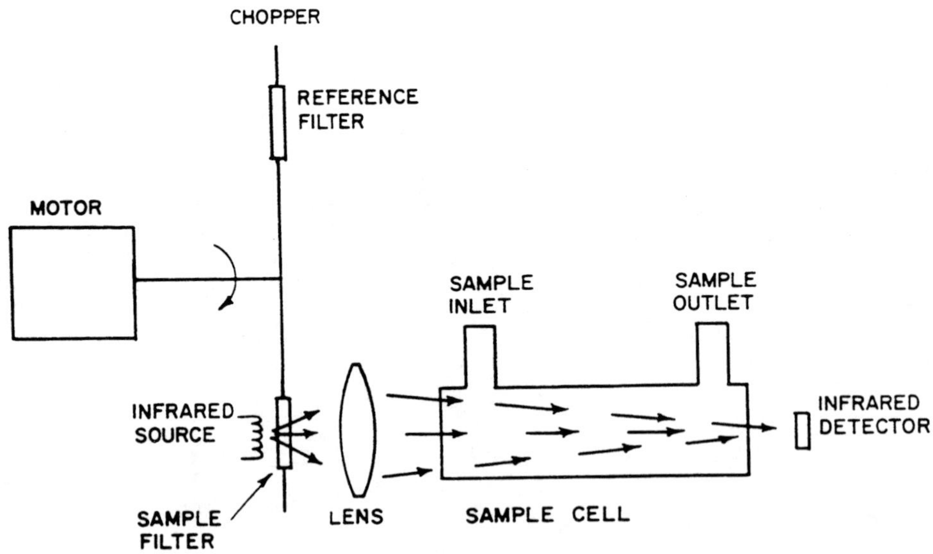

DUAL CHANNEL NDIR ANALYZER WITH SOLID STATE DETECTOR

FIGURE 5

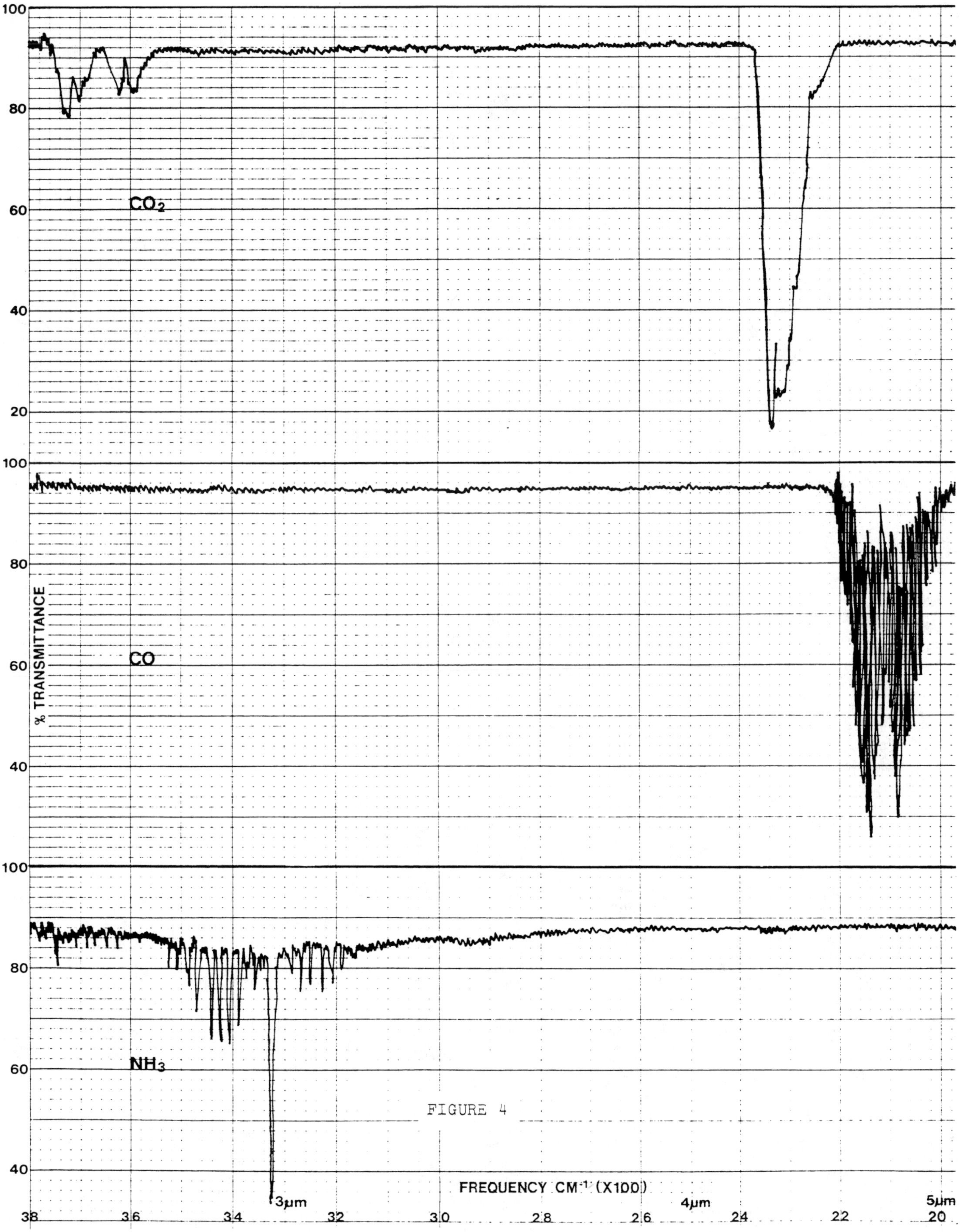
100
80
60
40
20
CO$_2$
100
80
60
40
CO
% TRANSMITTANCE
100
80
60
40
NH$_3$
FREQUENCY CM^{-1} (X100)
3μm
4μm
5μm
38
36
34
32
30
28
26
24
22
20

FIGURE 4

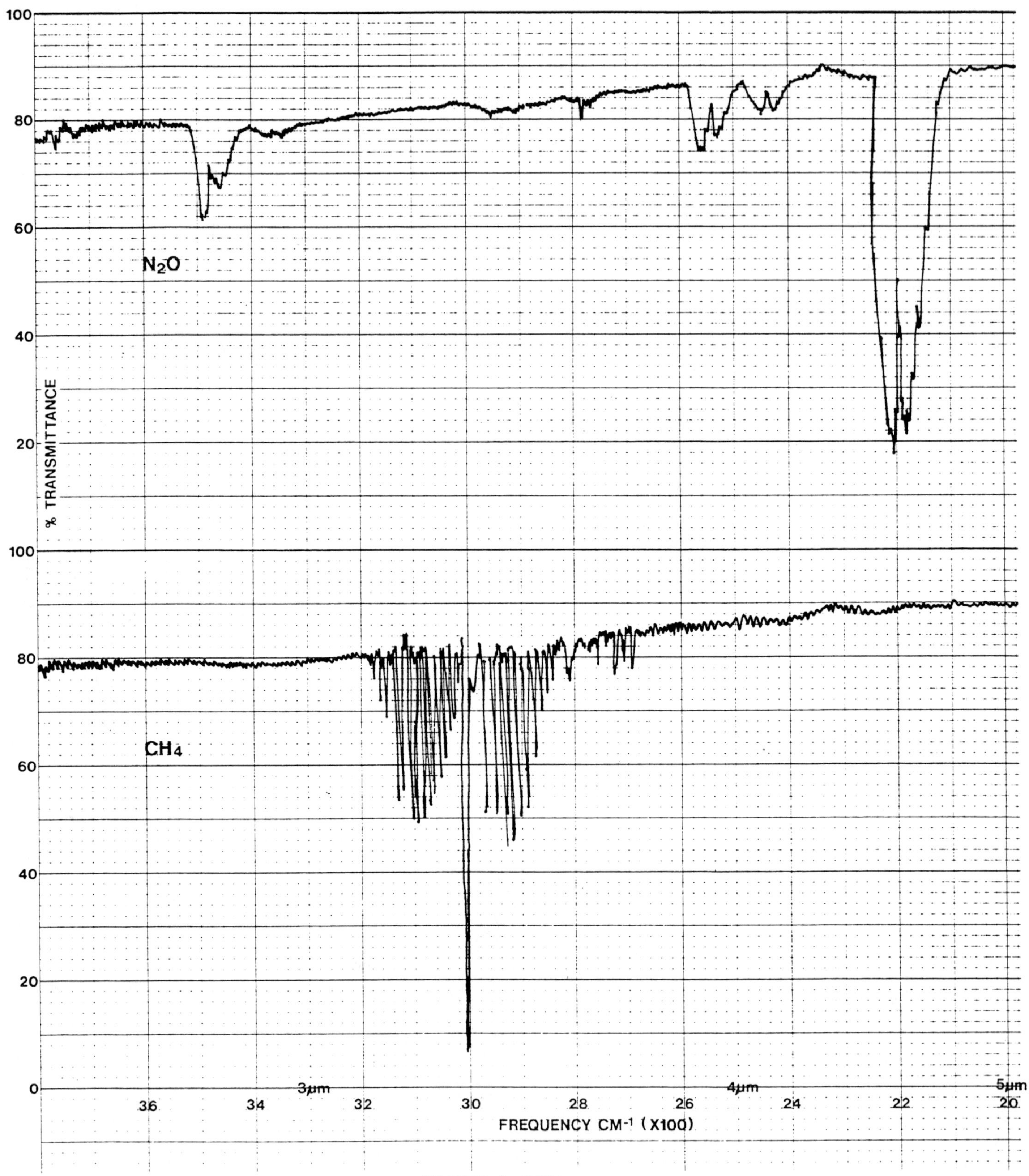

FIGURE 4 CONT.

Applications

The availability of rugged NDIR process analyzers at moderate cost has resulted in wide acceptance as a process control tool. Table 1 lists typical gases and concentration ranges generally available from NDIR analyzer suppliers. Literally hundreds of different gases can be monitored with infrared process analyzers. The Petrochemical Industry utilizes infrared process analyzers to monitor CO_2 in the manufacture of ethlyene oxide and ammonia. Acetylene is monitored during the manufacturing of acetylene and vinyl chloride. The Metals Industry uses infrared analyzers to monitor CO_2 in steel converting, soaking pits and in the heat treat process. Carbon monoxide is monitored during heat treating, aluminium powder processing and tin plant annealing. The Food Industry monitors CO_2 in greenhouses, storage facilities and during fermentation. The analyzers find use in monitoring for explosive and toxic hazards. Infrared analyzers are used to monitor stack gases for pollution control and are becoming increasingly important in flue gas monitoring for combustion efficiency.

TABLE 1

	Range (Full Scale Concentration)	
Gas	Minimum	Maximum
Ammonia	1000ppm	100%
Butane	300ppm	100%
Carbon Monoxide	500ppm	100%
Carbon Dioxide	200ppm	100%
Ethylene	1000ppm	100%
Ethane	500ppm	100%
Ethylene Oxide	1500ppm	100%
Hexane	400ppm	5%
Methane	400ppm	100%
Nitrous Oxide	150ppm	100%
Propane	400ppm	100%
Sulphur Dioxide	400ppm	100%
Water Vapor	1500ppm	10%

References

1. D.W. Hill, T. Powell, "Non-Dispersive Infra-Red Gas Analysis in Science, Medicine and Industry", Plenum Press, Plenum Publishing Corporation, New York, 1968.

2. The Sadtler Standard Spectra, Gases and Vapors, Sadtler Research Laboratories, Inc., Philadelphia, PA, 1972.

3. W.D. Perkins, "Infrared Spectroscopy", Process Instruments and Controls Handbook, McGraw-Hill, 1974.

4. The material presented in this paper has been derived from an article "Infrared Process Analyzers" written by R.M. Durham for the Process Instruments and Controls Handbook, third edition, to be published this year by McGraw-Hill Book Co.

Reprinted from *Electro-Optical Instrumentation for Industrial Applications,* F.M. Zweibaum, R.A. Carella, eds., Proc. SPIE Vol. 411, pp. 83-91 (1983).

Study of boiler optimization by means of spectral emission analysis

Robert W. Astheimer

Barnes Engineering Company
44 Commerce Road, Stamford, Connecticut 06904

Abstract

Spectral radiance measurements in the 1.7μm - 15μm region were made on the burner flames of several industrial boiler facilities using different fuels over a wide range of operating conditions. From these data, it was found that the ratio R_O of the flame radiance in two spectral bands is closely correlated to the excess air or fuel/air ratio at the burner. This ratio is dependent upon the load or fuel rate and it must be compensated for load changes. Under some conditions, another spectral ratio R_L was found to be proportional to the load and can be used to correct R_O. By means of a simple algorithm involving R_O and R_L (or the fuel rate), the excess air at an individual burner can be calculated.

A simple prototype burner monitor was built which measured these two ratios simultaneously. This was tested on boiler facilities burning oil, gas and coal. It was found that this instrument would indicate the excess air at a burner to an accuracy of 1.5% (0.3% O_2). In some cases, such as when firing with gas, the R_L ratio does not adequately correct for load changes and an external input of fuel or steam rate must be used instead of R_L.

Introduction

The large increase in the cost of fuels, which has occurred in recent years, has greatly increased the economic benefits to be gained by optimizing the combustion process in boiler furnaces. Stack instrumentation measuring O_2, CO and CO_2 in the exhaust gas is being increasingly used, and permits much better control of the combustion process and thus better combustion efficiency. Most large boiler facilities, however, employ multiple burners and stack exhaust gas analysis only gives the average for all burners and is a poor indicator of individual burner performance. In, say a 10 burner boiler, one burner could be fouled and operating at poor combustion efficiency but its affect on stack measurements would only represent 10% of the readings. Furthermore, there is a significant time delay between a change in burner conditions and its influence on stack gas content. If there is any leakage in the furnace or stack casings, "tramp" air will leak in and cause stack oxygen content to read erroneously high, which may be interpreted as too much air to the burners.

For these reasons, it would be desirable to monitor combustion efficiency at each individual burner rather than depend only on stack measurements. The more conventional means of gas analysis are not feasible because of the high temperatures and the non-equilibrium conditions which prevail in the flame region. Optical probing techniques such as Coherent Antistokes Raman Scattering (CARS) are useful for research purposes, but are far too complex and expensive for on-line monitoring and control.

It was thought that the infrared spectral radiance of the flame may contain characteristics indicative of combustion efficiency and that it may be possible to devise an algorithm using measured values of flame spectral radiance at several discrete wavelengths which would provide a burner optimization control signal. To this end, a joint research program was established by Barnes Engineering Company and the Hays/Republic Division of the Milton Roy Corp. Infrared spectral radiance measurements were made in the range from 1.6μm to 15μm on burner flames for a variety of fuels, furnaces and burner operating conditions. These measurements were made on fully instrumented boiler facilities, with only a single burner operating, whenever possible, so that the actual combustion conditions were accurately known. Barnes Engineering Company was responsible for making the spectral radiance measurements and the data analysis, while Hays Republic made the arrangements with appropriate steam generating plants, provided in-stack instrumentation, and obtained the reference data, or to borrow a term from the remote sensing community "Boiler Truth."

Experimental procedure

The first step was to schedule measurement programs with boiler facilities of the right type, who could and would be willing to operate their plant over a wide range of loads and burner combustion conditions. This is not easy to arrange, since these boilers are on-line and the load is established by the plant requirements not the desires of the researchers. Secondly, operating outside of normal conditions can cause problems such as flame-outs, smoke, internal explosions, etc. which are unwelcome in a production plant. Ideally, the boiler should have only one large burner which can be operated over a wide load range, and the facility should be well instrumented to provide reference data. In some cases, Hays/Republic installed additional stack instrumentation to support the research program. The boiler parameters measured and recorded were:

Steam Rate (lbs./steam/hr.)
Fuel Rate (gals./hr., CFH)
Stack Temperature
Stack Concentrations of O_2, CO and CO_2

The spectral radiance measurements were made with a Barnes Model 12-550 Spectral Radiometer directly coupled to a Hewlett Packard 85 computer. This instrument, shown in Figure 1 employs a rotating, circular variable filter wheel which scans the spectrum from 1.6µm to 5.2µm with a resolution of 0.02µm. A second filter wheel was available which extended the spectral range to 15µ but this was not used beyond the first few tests, since it was found that most of the spectral features of significance occurred below 5µm. When coupled to the HP 85, the computer controls the spectral scanning time and it required several minutes for a complete scan.

Use of the computer in the field had several advantages. The complete flame spectrum was stored on tape cassettes which made it very convenient to try various control algorithms at a later time. Also, the calibration table for each wavelength increment of the filter wheel was stored and, after each run, the computer could print out a fully calibrated spectral radiance curve.

The spectral radiometer had a 1" diameter objective lens and a 1° field of view, which could be focussed over a range of about 2' to ∞. The radiometer was mounted on a tripod several feet away from the burner view port and aimed through the port so that the flame filled its field of view. Burner view ports are generally covered with a blue colored glass window about 3" in diameter. This window was replaced with one of rock salt (NaCl) of the same size in order to transmit the infrared radiation.

Figure 2 shows a typical flame viewing arrangement on an oil burner assembly. Most oil burners have a diffusing plate on the nozzle and it requires a little maneuvering to aim the field of view between this diffusion plate and the insulated housing. In some cases, it was not possible to fully fill the field with the flame, however, since all results are expressed as ratios of spectral radiance, this is not significant.

Figure 1. Barnes Model 12-550 Spectral Radiometer

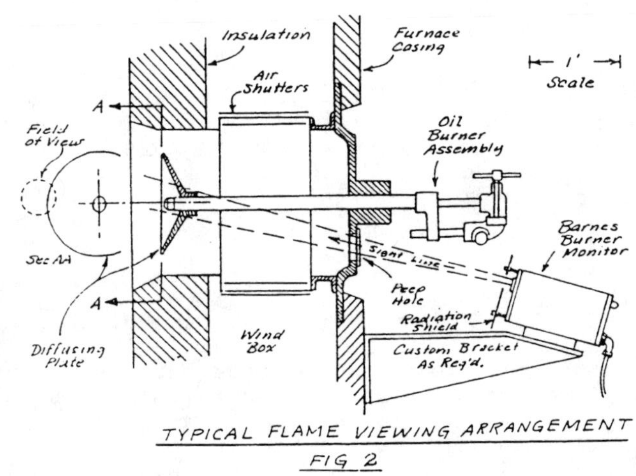

Figure 2. Typical Flame Viewing Arrangement

Flame spectral radiance data was taken at two different boiler facilities, one burning oil and the other burning either oil or natural gas. In general, the oxygen (or excess air) in the stack gas was taken as the independent variable over the range 0 to 10% oxygen (84% excess air). Figure 3 shows a computer printout of a typical oil flame spectrum and Figure 4 that of a typical gas flame spectrum. These were made on the same boiler at approximately half load. Full load for this boiler was 100,000 lbs. steam/hr. The peaks of the gas flame spectrum correspond to CO_2 and H_2O emission bands. The most obvious difference between the two flames is that below 3µm, the oil flame has a much higher continuum between the emission bands than the gas flame.

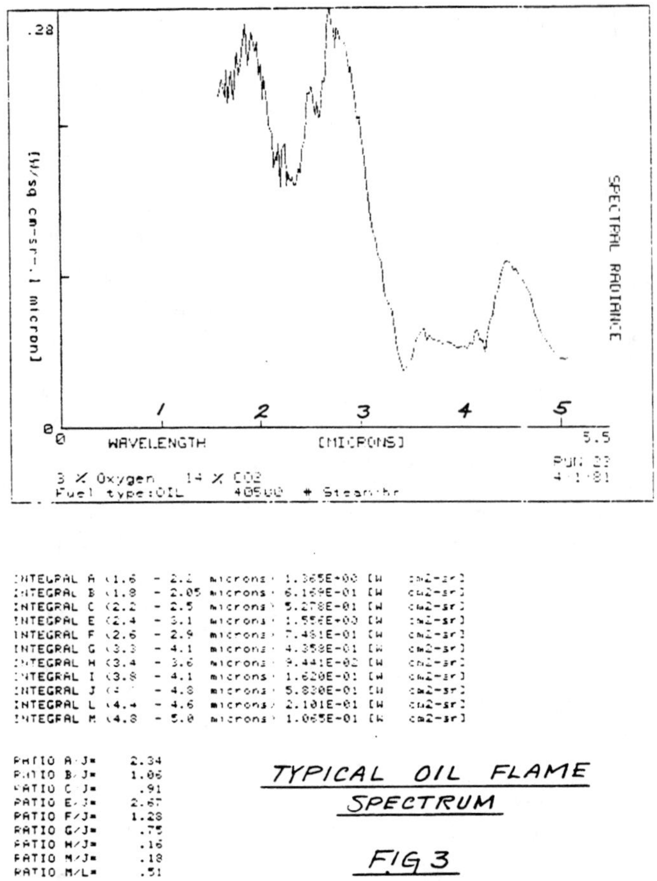

Figure 3. Typical oil flame spectrum

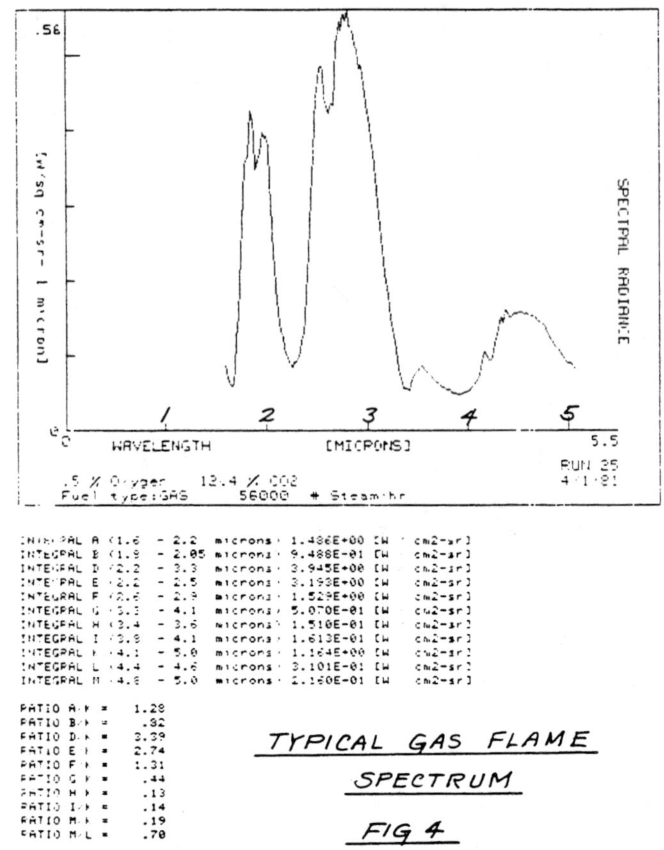

Figure 4. Typical gas flame spectrum

Below each spectral curve is a tabulation of the integrated radiance in a set of spectral bands and their ratios with respect to portions of the 4.1μ - 5.0μ CO_2 band (J, L, K). These ratios were plotted vs stack oxygen content and load (or fuel rate) in order to detect correlations between these parameters and the spectrum. The 4.1μ - 5.0μ band was taken as a reference for the ratio, since it appeared to show the least variation with burning conditions. It embraces the temperature broadened emission of the 4.3μ CO_2 band which is very strong and it probably is a good measure of flame temperature.

Data analysis

Ideally for maximum combustion efficiency, just enough air should be mixed with the fuel to oxidize all of it to CO_2 and H_2O. When this occurs, there will be zero oxygen content in the stack gas. Any excess air wastes energy heating the unneeded air going up the stack, while insufficient oxygen results in unburned fuel and causes sooting and smoke. Since mixing of the fuel and air is never perfect, best efficiency generally occurs at about 2% to 3% oxygen (10%-15% excess air) in the flue gas rather than zero. The oxygen content of the flue gas was, therefore, considered to be the criterion for combustion efficiency and spectral characteristics of the flame were sought which would correlate with stack oxygen. Our objective was to develop, if possible, some algorithm of flame characteristics which would be applicable for on-line monitoring and control of individual burner performance, therefore, it was decided to work with ratios of spectral radiances rather than absolute values. This avoids errors from such practical problems as dirt on windows, the flame not filling the field of view, variations in detector response with temperature, aging, etc.

The basic approach was to compute ratios of various features of the spectra and compare them with stack gas composition - particularly excess air. At first, no consistent correlations could be found, however, if the boiler load or fuel burning rate was held constant, a good correlation appeared between excess air and the ratio of the integrated radiance in the 3.8μ - 4.1μ band to that in the 4.4μ - 4.6μ band. These are the I and L bands in Figures 3 and 4.

Figure 5 shows plots of this ratio, which has been called the oxygen ratio R_o, vs stack oxygen for two different oil-fired boilers. The circled dots labelled P & W points (for Pratt & Whitney plant) were for one of the boilers, while all of the other points are for the second boiler. These plants were operational and the load at any one time was determined by the plant requirement, therefore, sets of points at exactly the same load could, in general, not be obtained. The points are grouped in load ranges expressed as lbs. steam per hour which is proportional to fuel burning rate. In spite of the fact that the loads in each group were not identical, the points fall fairly close to a smooth curve.

If it is assumed that the load is known, it is possible to make a correction to the R_o ratio which results in a load independent ratio R_c as shown in Figure 6. The correction amounts to dividing the ratio R_o by a load factor as follows:

$$R_c = \frac{R_o}{C_1 L + C_2}$$

where C_1 and C_2 are constants and L is the fraction of full load. With this correction, all of the points fall close to a smooth curve with the exception of the pair at 7% O_2.

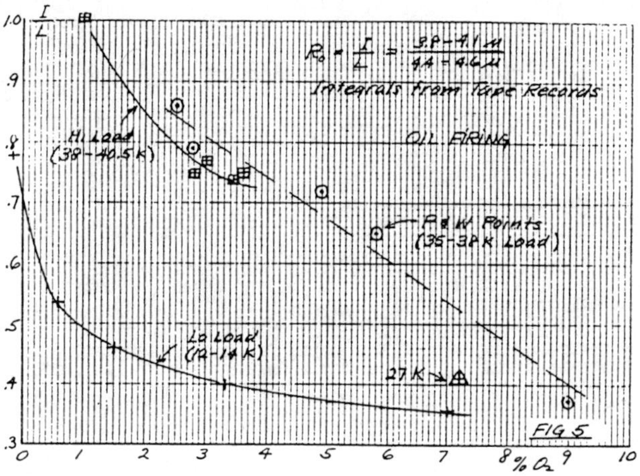

Figure 5. R_o vs % O_2 for two different oil-fired boilers

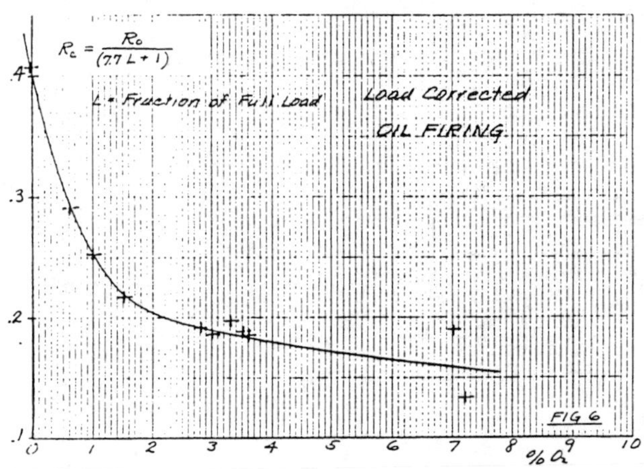

Figure 6. R_c vs % O_2 (oil firing)

It would be desirable to derive the load correction from the spectral data rather than to require a separate load or fuel rate input and, therefore, a search was made for another ratio, which could be used as a measure of load. Such a ratio was found for oil-firing in F/L (Figure 3), which is 2.6μ - 2.9μ/4.4μ - 4.6μ and is shown plotted in Figure 7. This load ratio R_L provides a good measure of the load which can be used in Equation 1 to correct the oxygen ratio R_o. It should be noted that the points in Figure 7 are at all different fuel/air ratios but R_L is primarily dependent only upon load.

The oxygen ratio R_o is also a good measure of excess air when gas-firing except the shape of the curve is different. This is seen in Figure 8 which also shows the same strong dependence on load. However with gas-firing, the load ratio R_L was not sufficiently well correlated with the load to permit using it to correct R_o and the actual load must be used.

The R_o and R_L ratios were originally arrived at empirically by trying to correlate various ratios of spectral bands to fuel air ratio and fuel rate. However, there appear to be physical reasons for their behavior. It is believed that the "oxygen" ratio R_o is a measure of unburned particles in the flame. The numerator is the radiance in the 3.8μm - 4.1μm band which is a region of minimum emission from the major combustion products, CO_2 and H_2O. The emission in this band, which we have called "I" will then be primarily black body radiation from particulates in the flame, and will be proportional to the product of the particulate concentration C, the view path through the flame or flame depth D, and a power of the absolute flame temperature T^n.

$$I \sim C\,D\,T^n$$

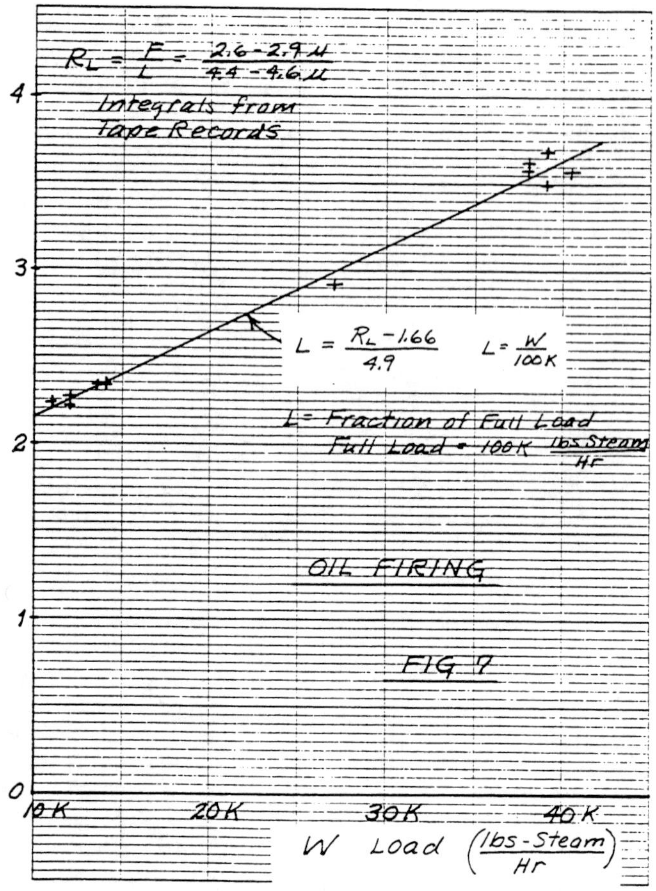

Figure 7. R_L vs load (oil firing)

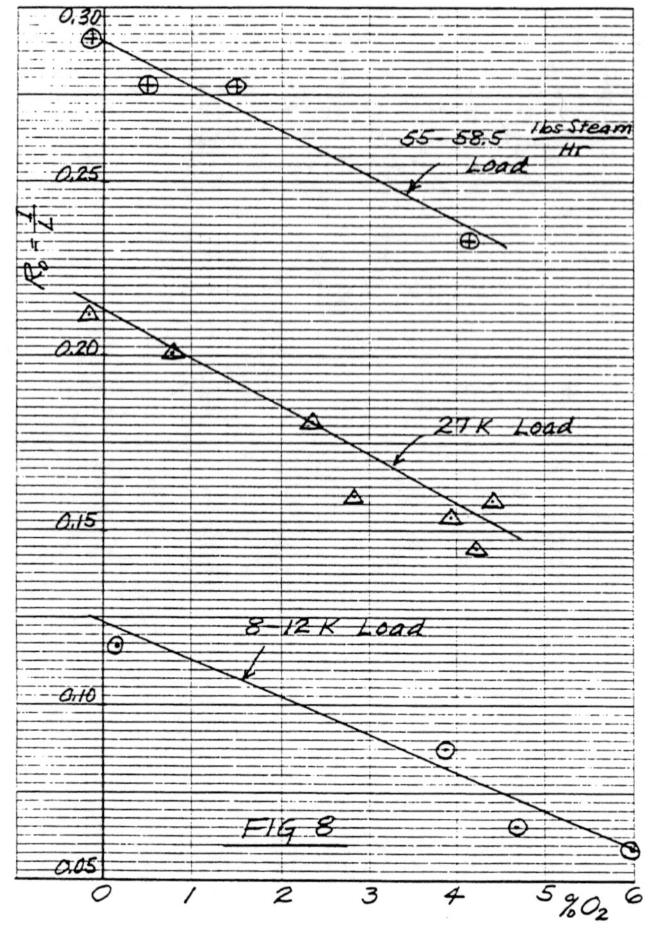

Figure 8. R_o vs % O_2 (gas firing)

The denominator L is the radiance in a very strong CO_2 absorption band (4.4µm - 4.6µm). In this region the flame will be opaque and radiate as a black body at flame temperature.

$$L \sim T^n$$

Since the L band is close to the I band, the power n will be about the same in the two bands. Thus when we take the ratio R_o, the temperature terms cancel and we have:

$$R_o = \frac{I}{L} \sim C\ D$$

The particulate concentration C may be assumed to be the factor which depends on the excess oxygen and is what we wish to measure, thus $C = f(O_2)$ and we have:

$$R_o = D \times f(O_2)$$

The flame depth D is the load-dependent factor which must be corrected for. As the fuel burning rate or load increases, the flame will grow in size and D will increase. The load ratio R_L has for its numerator F the radiance in a weaker band (2.6µm - 2.9µm) where the flame is not opaque. The emission in this band will depend upon flame depth D and temperature. When divided by band L to form the ratio R_L, the temperature will cancel as previously to give:

$$R_L \sim \frac{D\ T^n}{T^n} \sim D$$

When the ratio R_O/R_L is taken, the flame depth D cancels to give a control ratio R_C dependent upon oxygen only:

$$R_C = \frac{R_O}{R_L} \sim \frac{D\ f\ (O_2)}{D} \sim f\ (O_2)$$

In actuality, the proportionality between F band emission and flame depth D will be exponential rather than linear and in addition the F band will contain some particulate emission, although reduced because of the gas absorption. Thus the control ratio R_C will be a more complex function of R_O and R_L as indicated in the actual algorithms rather than their simple ratio.

$$R_C = f\ (R_O,\ R_L) = f\ (\%\ O_2)$$

The foregoing discussion has been greatly oversimplified, but it indicates qualitatively the physical significance of these spectral ratios. Since the numerators of R_O and R_L are in spectral regions where the flame is semitransparent, the field of view must be directed so that the flame is viewed against the water-cooled walls of the boiler and not brickwork. The temperature of a brickwork background will be much higher than that of water-cooled walls and will also change with load. If this is viewed, large errors will result.

Prototype burner monitor

The aforementioned results were quite encouraging, however, they represented only a limited amount of data on two boilers. It was also time-consuming to obtain complete spectral scans. Several minutes would elapse between measurements at the different spectral bands of interest and burner flicker or load changes could cause ratio errors. It was decided that a more efficient data collection instrument was needed which would measure radiance in the three spectral bands of interest simultaneously. To this end, a small filter wheel radiometer was built using a PbSe detector. The filter wheel contained the three bandpass filters admitting the I, L and F bands mentioned previously and rotated at 50 RPS. The signal from the detector was decommutated into separate channels for each spectral band and averaged with a time constant of 1 second. This gave essentially simultaneous readings in the three spectral bands. By means of a 3-position switch, the radiance in each of the bands (I, L and F) could be read on a meter mounted on the back of the radiometer. In addition, the three outputs were brought out separately to jacks so that they could be simultaneously recorded on a chart recorder. Figure 9 shows this prototype burner monitor mounted on an oil burner.

Data was obtained with the prototype burner monitor on three boiler facilities burning oil, gas and coal. The results were consistent to that described previously except that the data scatter was considerably less than with the spectral radiometer. This is because the averaging time was longer and the measurements were taken simultaneously, both of which minimize flame flicker which is the largest source of noise.

Example plots of R_O vs % O_2 for oil and gas firing are shown in Figures 10 and 11 respectively. From observed values of R_O and R_L (or Load), an algorithm can be formed which will compute stack oxygen concentration, i.e., % O_2 = f (R_O, R_L). Figures 12 and 13 show the results of such a calculation for both oil and gas firing. In these plots, the oxygen (or excess air) computed by algorithm from spectral radiance measurements is plotted against the actual oxygen content measured in the stack. The plotted points are for a wide variety of loads and fuel/air ratios. The 45° lines represent perfect results. It is seen that the points all fall close to this line. The standard deviation in Figure 14 is 0.3% O_2. In the algorithm for calculating excess air when gas firing, the fuel rate was used rather than the radiometric load ratio R_L, since as mentioned previously, this does not give an adequate load correction with gas firing.

Figure 14 shows a plot of R_O vs % O_2 coal firing. Again, the correlation is good. This was a grate type furnace with an automatic stoker and no changes in load occurred during the test.

Figure 9. Burner monitor installation

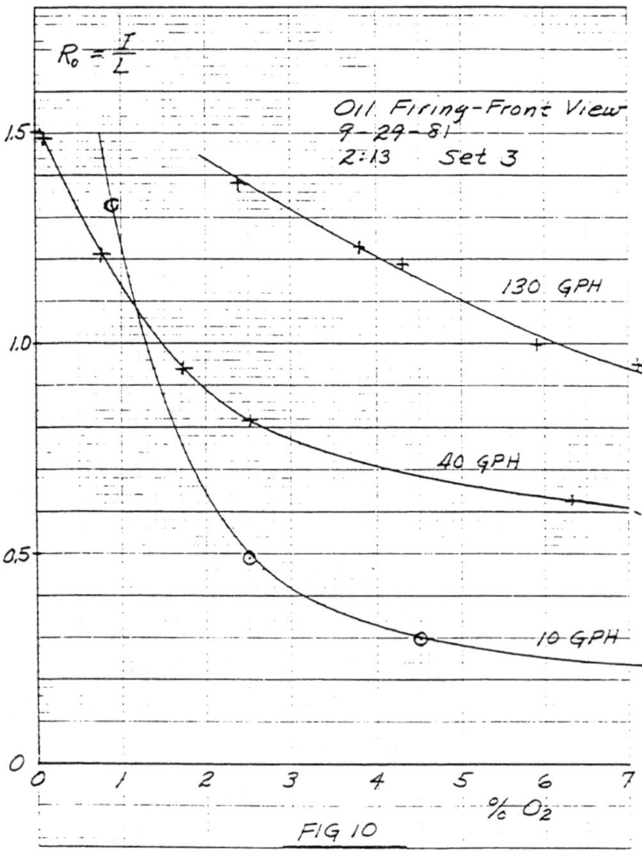

Figure 10. R_o vs % O_2 from burner monitor (oil firing)

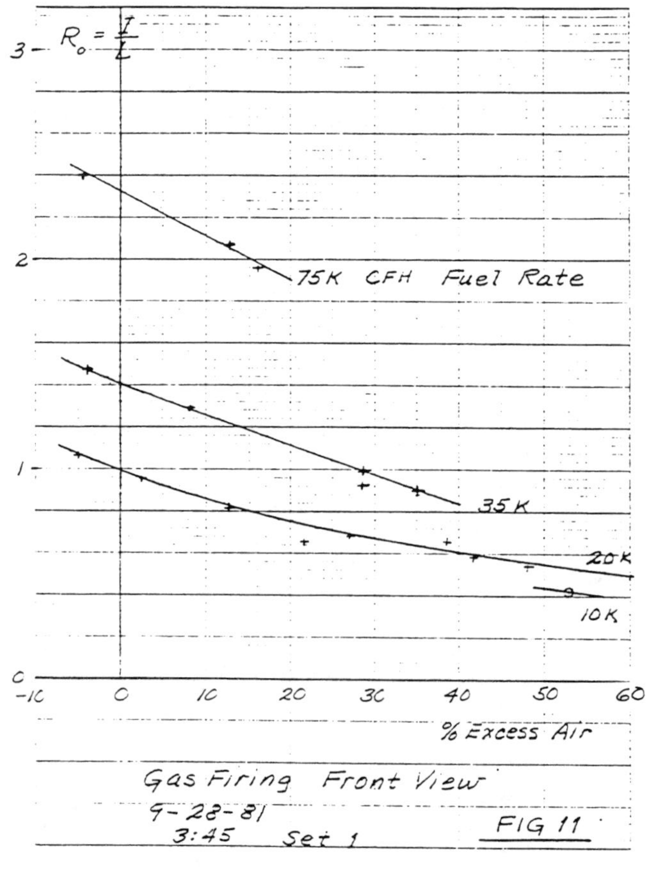

Figure 11. R_o vs % O_2 from burner monitor (gas firing)

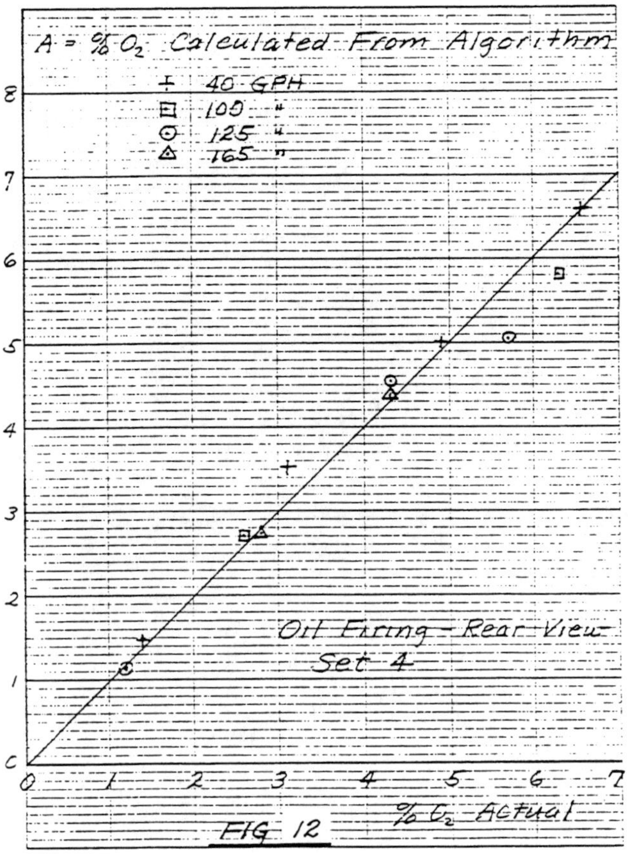

Figure 12. Indicated % O_2 vs measured % O_2 (oil firing)

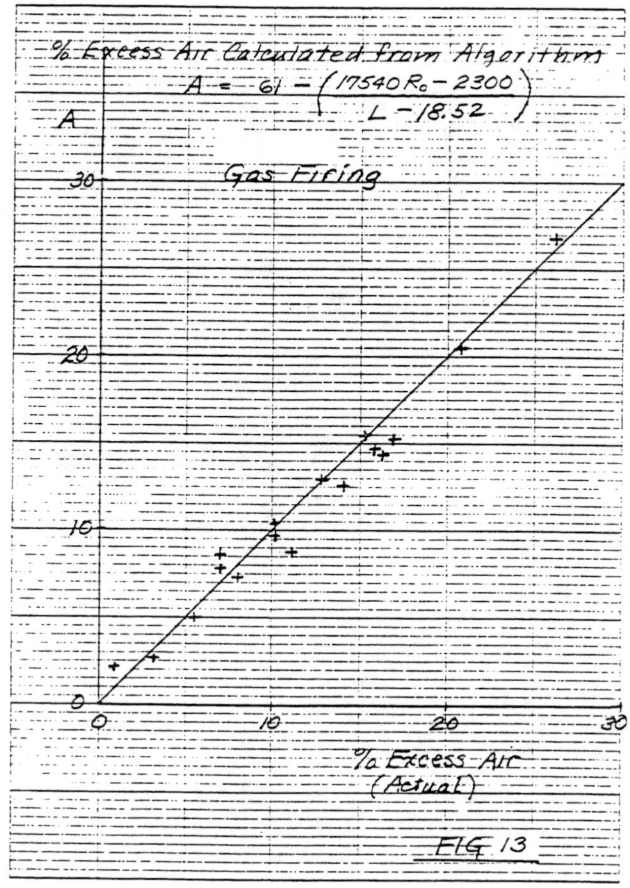

Figure 13. Indicated excess air vs measured (gas firing)

Conclusions

It has been found that the ratio R_o of the radiance in the spectral bands 3.8µ - 4.1µ and 4.4µ - 4.6µ is an accurate indicator of the fuel/air ratio of the flame. The fuel/air ratio can be calculated by a simple algorithm involving R_o and a few constants. This ratio is dependent upon the load or fuel burning rate. The affect of variations in load on R_o can be eliminated by adding another term to the algorithm which may be either the fuel rate, the steam rate or a second radiometric ratio R_L, which is the ratio of spectral bands 2.6µ - 2.9µ and 4.4µ - 4.6µ. The use of R_L does not give adequate load compenation in some situations such as when gas firing. In this case, the fuel or steam rate must be used. This algorithm enables the excess oxygen content to be measured at the burner to an accuracy of 0.3% O_2. A prototype burner monitor has been constructed to measure the three required spectral radiances simultaneously. The technique has been shown to be applicable to oil, gas and coal firing.

The need for an external load compensation input from measurement of fuel or steam rate, when R_L is inadequate, is the greatest complication in applying the technique. However, many large multi-burner installations compensate for load changes by cutting burners in or out and, when a burner is on, its load remains constant. A load compensation input would not be necessary in such installations.

Data on a limited number of boiler facilities indicates that the technique is applicable to oil, gas and coal firing. The form and the constants in the algorithm for computing excess oxygen are not dependent upon the installation but do depend upon the fuel type.

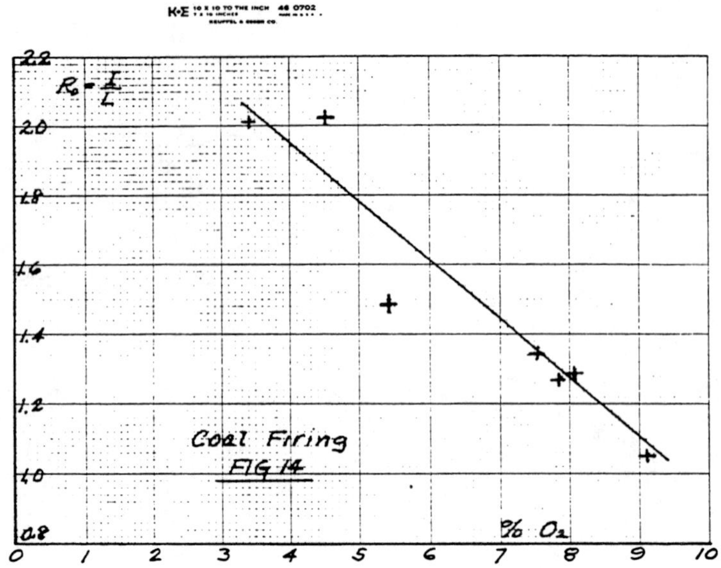

Figure 14. R_o vs % O_2 from burner monitor (coal firing)

Reprinted with permission from *Applied Spectroscopy*, Vol. 42(6), pp. 972-979 (1988).

In Situ FT-IR Analysis of a Composite Curing Reaction Using a Mid-Infrared Transmitting Optical Fiber*

DAVID A. C. COMPTON,† STEPHEN L. HILL, NORMAN A. WRIGHT, MARK A. DRUY, JOSEPH PICHE, WILLIAM A. STEVENSON, and D. WARREN VIDRINE

Bio-Rad, Digilab Division, 237 Putnam Ave., Cambridge, Massachusetts 02139 (D.A.C.C., S.L.H., N.A.W.); Foster-Miller Inc., 350 Second Ave., Waltham, Massachusetts 02254 (M.A.D., J.P.); IRIS Fiber Optics Inc., 40 Nagog Park, Acton, Massachusetts 07120 (W.A.S.); and Measurex Corporation, 1 Results Way, Cupertino, California 95014 (D.W.V.)

The use of a mid-infrared transmitting fiber to carry the infrared beam of an FT-IR spectrometer outside of the optical bench is reported. In addition it is demonstrated that it is possible to analyze samples using the fiber as an internal reflection element. The fiber is covered with a protective coating which can be removed for a short region, to allow contact with the sample over a controlled length. Two examples of the use of an optical fiber for remote sampling are discussed. The first shows that a spectrum of a liquid sample (2-butanone) can be easily measured, and the second shows how a fiber can be used to monitor the progress of curing reactions in thermoset composite materials. The spectrum of a resin was recorded before the cure by embedding the optical fiber in the graphite fiber/polyimide matrix resin prepreg, and then the progress of the cure was monitored during the curing process. This type of remote sampling shows tremendous potential for opening totally new areas of usage for FT-IR spectrometry, including the studies of hazardous materials, enclosed reactions, and processes that do not allow samples to be taken inside the spectrometer.

Index Headings: **FT-IR; Infrared spectroscopy; Fiber optics; Internal reflection spectroscopy; Polymer curing; Composite prepregs.**

INTRODUCTION

The use of infrared spectrometry is well established for the general-purpose analysis of a wide range of samples. It may be stated that this technique is the single most powerful analytical tool that can be brought to bear for the identification of an unknown sample, as long as that sample is not metallic in nature. This is because of the ability of infrared analysis to examine samples that are solids, liquids, or gases, whether these samples are pure or are present as a mixture.

Early in the development of dispersive infrared spectrometers, it was realized that the simple manner of holding a thin sample in the infrared beam for analysis by transmission was a limitation of the technique. In order that this problem might be overcome, a number of accessories have been designed which assist the analyst in examination of samples that are difficult to study by transmission. Such samples can be, for example, opaque, difficult to prepare as a thin sample, available as coatings on a surface, or existing at temperatures other than ambient. A major advance in sample handling for infrared analysis was made in the early 1960s, when Attenuated Total Reflectance (ATR) was used in the design of sampling accessories. This technique, now known as Internal Reflection Spectroscopy (IRS), uses the phenomenon whereby a light beam reflecting at a surface with high refractive index actually penetrates a short distance into the adjoining medium with a lower refractive index. The use of suitable crystals (Internal Reflection Elements—IREs) having a high refractive index, such as germanium, allows for an infrared beam inside the IRE to be used to analyze a sample pressed against the surface. For a detailed study of the theory and application of IRS, the reader is referred to the treatise on the subject, and a subsequent volume, by Harrick.[1,2]

Many potential applications for infrared analysis were impractical or even impossible, due to the requirement that a suitable sample be brought inside the spectrometer. This limitation was particularly true for dispersive spectrometers, which base their measurements on the comparison of sample and reference beams of energy. The use of FT-IR spectrometers has eased this restriction somewhat, since such instruments are single-beam in operation. A certain amount of remote sensing has been accomplished by directing the infrared beam out of the spectrometer through a sample, and then onto a detector. This can be done as long as a suitable reference single-beam spectrum can be recorded without the sample being present.

There are still many potential applications that remain outside the scope for infrared analysis, because of the

Received 18 February 1988.
* This work was sponsored by the Phase 1 SBIR Contract NAS1-18420 from NASA, at the Langley Research Center, Hampton, VA 23665.
† Author to whom correspondence should be sent.

difficulty in getting the infrared beam to the sample. These include the examination of samples that are too hazardous or too large to bring into a laboratory, industrial processes inside reactors or furnaces, and reactions that cannot conveniently be monitored by conventional IRS techniques, which may require an artificial environment.

An example of the last application is the direct monitoring of the state of cure of thermoset composite materials. Current methodology for the monitoring of such processes involves primarily dielectric or ultrasonic principles. These techniques measure secondary resin effects, such as the dielectric loss or ultrasound attenuation. The chemistry of the matrix resin is thus only inferred from these effects. Recent studies on resins in the absence of graphite fiber which utilize the diffuse reflectance sampling technique[3-5] have established the chemical changes that the matrix resin undergoes during cure. Diffuse reflectance infrared spectrometry is a powerful method for observing the spectra of powdered samples in particular, but suffers from a problem for this type of work. It is not possible to apply the technique to the real-time study of fiber/resin matrix prepregs in a composite manufacturing environment.

Fiber optic probes are in use for a number of different types of analytical techniques, because they allow for the examination of samples outside of the instrument. In order to characterize the optical properties of near-infrared transmitting fibers near their long wavelength cutoff point (typically about 4 μm), research groups have examined such fibers using FT-IR spectrometry.[6-7] In addition, it has been demonstrated that a near-infrared transmitting fiber can be used as an internal reflection element for IRS work.[8] Researchers at British Telecom have published a spectrum of liquid isopropanol, recorded by the use of a fluorozirconate fiber.[8] This fiber was both the transmission medium, to bring the beam outside a dispersive near-infrared spectrometer, and also the IRE used to record the spectrum of the O-H and C-H stretching regions of the isopropanol spectrum.

The ability to record the infrared absorbance spectrum of a compound out to a wavelength of about 4.6 μm (approximately 2200 cm^{-1}) is very exciting. However, most of the chemical information of the spectrum that can be used to monitor reactions lies below 1800 cm^{-1}, and this region is inaccessible to the type of fiber previously examined. One attempt to use a very short piece (less than 10 cm in length) of silver halide in a "fibercell" has been reported.[9] This type of material can be used as an IRE, but does not lend itself well to being fashioned into a long fiber.

We report here our studies using a mid-infrared transmitting fiber to carry the infrared beam outside the FT-IR spectrometer and demonstrate the analysis of samples using the fiber as its own IRE. The fiber used was a chalcogenide glass fiber of about 120-μm diameter, having a silicone coating of about 90-μm thickness. The transmission properties of similar fibers have recently been published.[10] These chalcogenide fibers are manufactured with the use of varying glassy compositions of Ge, As, Sb, Se, and Te. The very high optical throughput and sensitivity of a modern FT-IR spectrometer have been coupled with the information-rich region of the

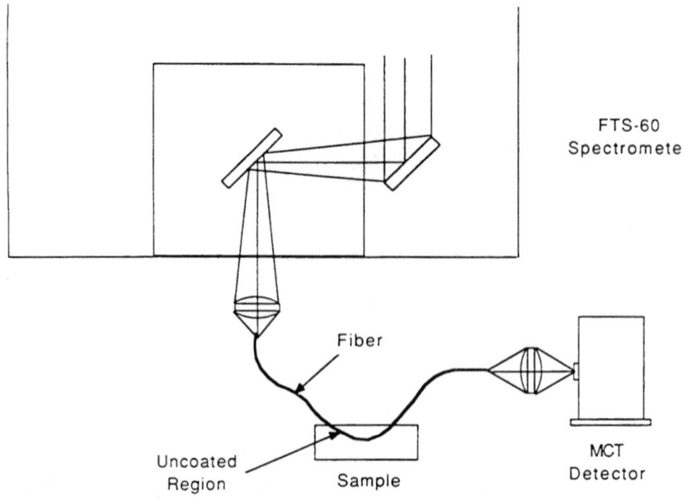

FIG. 1. Schematic showing one of the experimental arrangements used for the work. Each end of the fiber is mounted in an XYZ microtranslational stage.

mid-infrared spectrum extending down to about 1200 cm^{-1}, with the use of such a fiber. This combination of a powerful instrument and remote sensing device has great potential to alter the way infrared analysis is carried out. The use of such a combination allows for many of those potential applications listed above to be finally realized.

EXPERIMENTAL

Infrared spectra were recorded by the use of either a Digilab FTS 40 or an FTS 60 spectrometer equipped with a KBr/Ge beamsplitter and standard high-intensity ceramic source. Interferograms were coadded at 8-cm^{-1} resolution, with a data collection speed of 20 kHz, apodized with the use of a triangular function, and then transformed to spectra. At this data rate, about 250 scans are coadded per minute of observation time. The actual number of scans coadded for each example is described in the figure captions.

The mid-infrared transmitting fiber was supplied by IRIS Fiber Optics, Inc., and manufactured by British Telecom Laboratories, Martlesham Heath, Ipswich, England. The fiber had a 120-μm core of an arsenic germanium selenide glass, coated with a 90-μm thickness of silicone polymer. This fiber was used because of its availability, but cannot be considered fully optimized at this point. The attenuation loss of 10–15 dB per meter restricted use to about 3 m. The manufacture of lower-loss fibers would allow for longer lengths of fiber to be used. In addition, the silicone coating is not ideal, since it has significant infrared absorptions, as will be shown later.

Several experimental arrangements were tried for the work described herein. A schematic of one such arrangement is shown schematically in Fig. 1. The infrared beam was directed out of the sample compartment and focused onto the inlet end of the fiber. Each end of the fiber was mounted in an XYZ microtranslational stage, to allow sensitive adjustment of the fiber position for maximum energy throughput. The outlet end of the fiber was directed towards a high-sensitivity MCT detector, again with the use of suitable focusing optics.

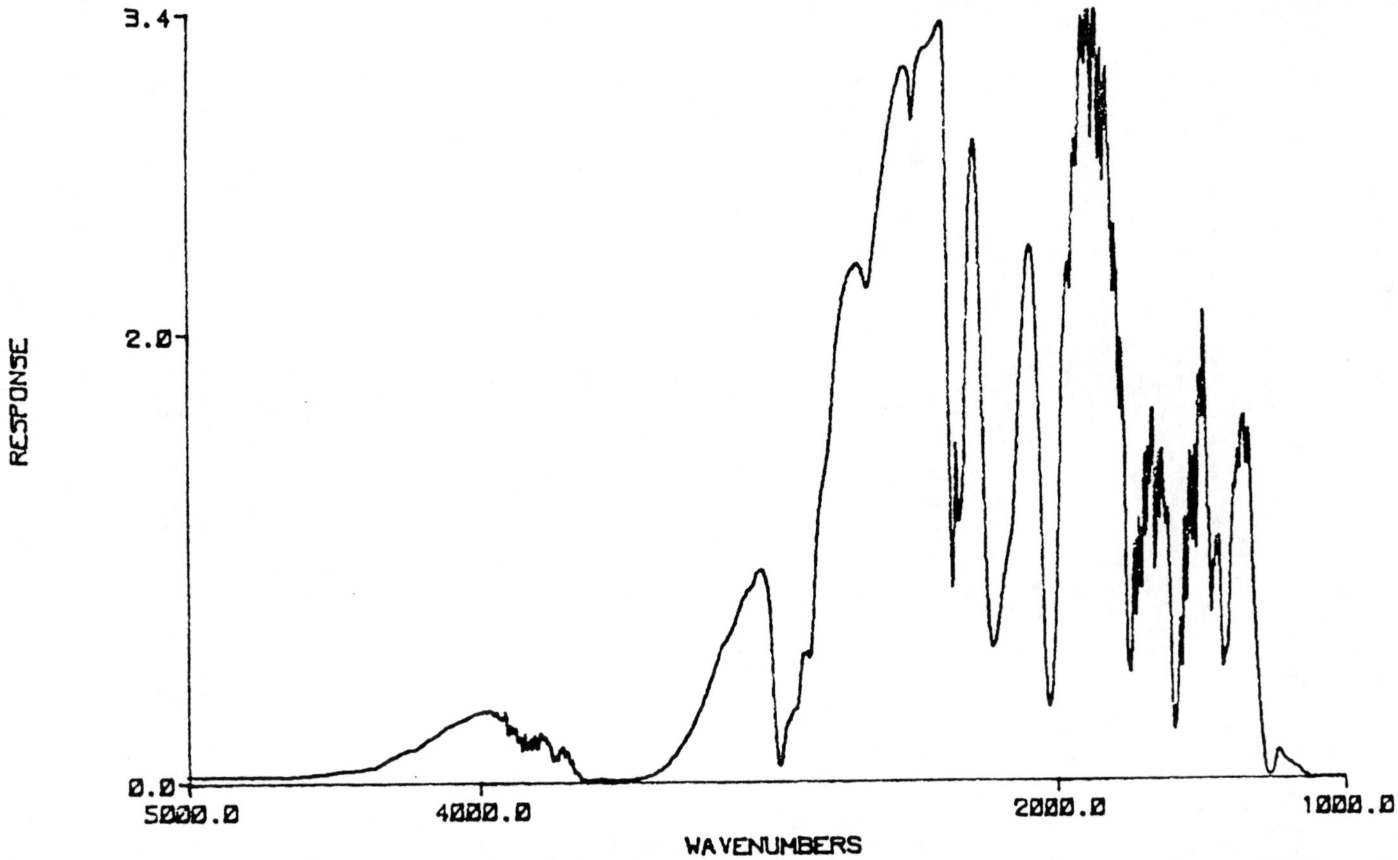

FIG. 2. Single-beam spectrum (5000–1000 cm^{-1}) recorded through a single two-meter length of fiber, with the use of the standard FTS 40 mid-infrared optics and a high-sensitivity MCT detector. The absorption bands in the spectrum arise mainly from the silicone coating on the fiber.

The coating of the fiber was carefully removed from a selected region of the fiber with a decladding tool. It was found that an uncoated length of 3–6 cm gave a reasonable level of infrared absorbance when brought into contact with a sample. In all cases, the background reference spectrum was recorded through the same length of fiber to be used for the experiment, in order to cancel out the significant absorptions due to the fiber and the cladding material.

To perform the first curing reaction, we made a simple experimental arrangement. A small sample of the resin was coated onto the declad fiber, and then the fiber was laid down on a heating plate. The reference spectrum for the experiment was recorded with the fiber coated with the room-temperature resin. The temperature was raised to each monitoring temperature, and then the single-beam spectrum was recorded for that temperature. By ratioing that spectrum to the one recorded at room temperature, we obtained an effective difference spectrum, which shows only bands due to chemical species that have changed during the cure at each point.

RESULTS

Optical Characteristics. The first experiment to be accomplished was to measure the spectral response of the total equipment including the fiber. To do this, we used a fiber of about two meters in length. The position of the fiber was set grossly, with the aid of the red spot of the instrument reference laser, since this is centered in, and colinear with, the infrared beam. The input end of the fiber was manipulated in three dimensions to maximize the observed interferogram intensity. The outlet end was also similarly manipulated near the detector face until the signal was further optimized. For these initial measurements we were able to record a signal of more than one volt on the detector.

The resulting single-beam spectrum is shown as Fig. 2. The spectrum shows the numerous absorption bands due to the silicone coating superimposed on the general response curve from 4500 to 1200 cm^{-1}. However, even with a fiber having a 2-m length, the only silicone absorptions that are totally absorbing are the broad O-H stretching region from 3600 to 3300 cm^{-1} and sharp bands at 2950 and 1260 cm^{-1} that arise from the Si-OMe groups. In addition, atmospheric absorptions of water (sharp bands in the 1800–1400 cm^{-1} region) and carbon dioxide (the broad doublet at about 2350 cm^{-1}) can be observed, since no attempt was made to completely exclude the atmosphere during the experiment.

By ratioing two successive single-beam spectra together, we recorded the 100% line shown in Fig. 3. This spectrum shows a noise level of less than 1%, except in the region of the silicone hydroxyl absorption.

The observed response curve through the fiber is significant from three respects:

1. The low-frequency cutoff of 1200 cm^{-1} is low enough to allow the carbonyl region and much of the fingerprint region to be examined. Many chemical processes have absorptions that can be monitored in this range.

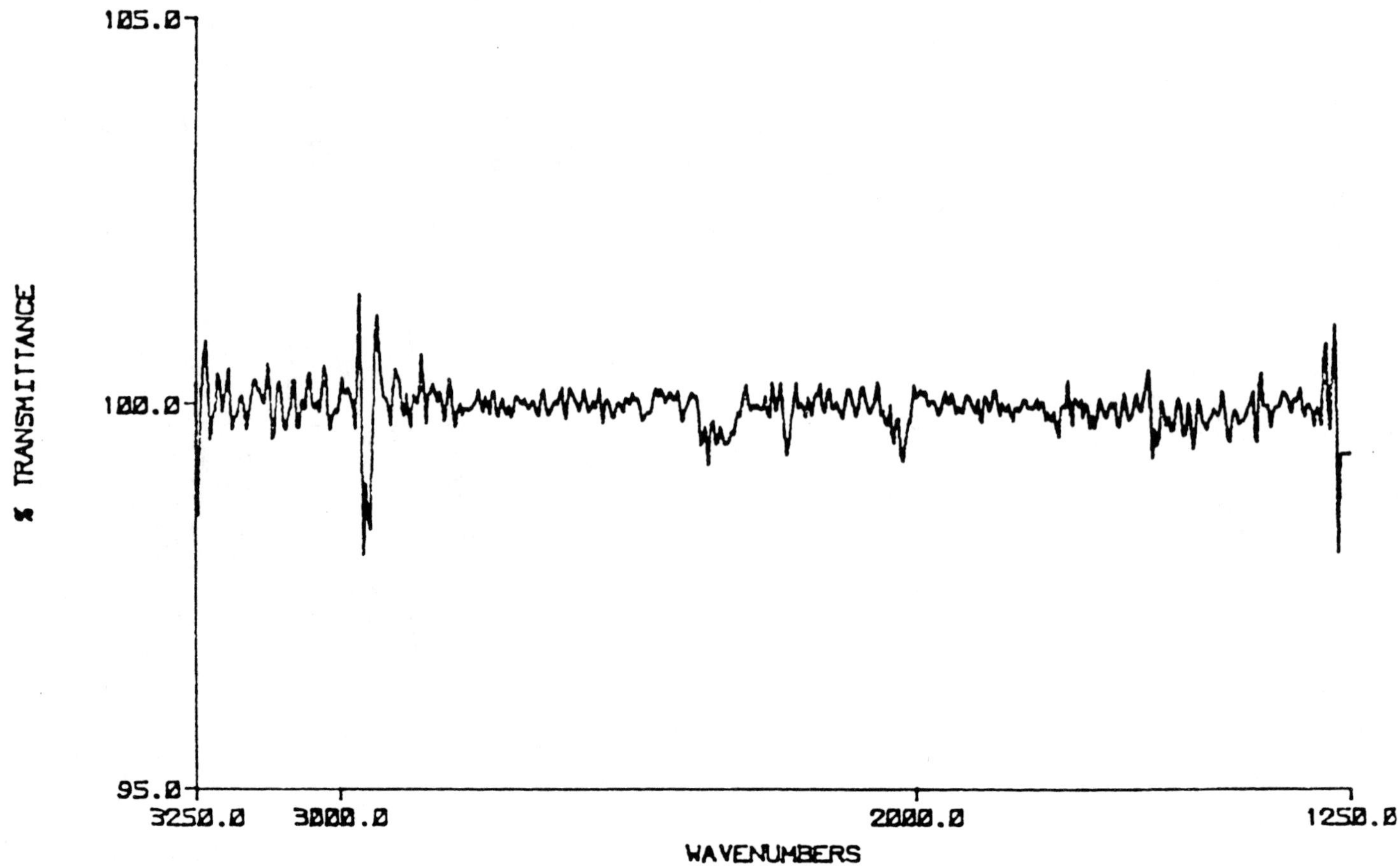

FIG. 3. One hundred percent line obtained by ratioing two successive single-beam spectra (as in Fig. 1); 500 scans were coadded for each spectrum at a resolution of 8 cm^{-1}.

2. Even with this long length of fiber, the majority of the silicone absorption bands were not totally absorbing. This result indicates a low penetration of the signal into the clad material, and allowed us to use this fiber to obtain experimental data.
3. The magnitude of the interferogram signal obtained was very encouraging, since it allowed for the spectrum to be recorded in about 2 min with excellent signal-to-noise ratio. The observed 100% line shows a very low noise level.

Spectrum of Test Solvent. Using the fiber sample mentioned above, we made a test to find the length of declad fiber region that would give a reasonable spectrum when brought into contact with a sample. A section of the fiber about 5 cm in length was declad, and the reference background spectrum recorded. The declad portion was dipped into a sample of 2-butanone (methylethylketone), and the transmittance spectrum of the sample recorded by coaddition of 100 scans. The observed spectrum between 3100 and 1200 cm^{-1} is shown as Fig. 4. The strongest absorption, at about 1740 cm^{-1}, has an intensity of about 20% *T*, almost an ideal value. The spectrum agrees well with reference spectra, and is easily identifiable. Even though no correction has been made to the relative intensities for varying penetration depth of the IRS beam, this spectrum was verified by being searched against a spectral library.[11]

Preliminary Study of Resin Cure Process. The above experiments have demonstrated that an optical fiber can be coupled to an FT-IR spectrometer and used as an IRS in order to measure the mid-infrared spectrum of an organic compound. We now wished to extend this result into the realms of an experiment that not only is in the "real world" but also solves an important engineering problem. To do this, we demonstrate the use of a fiber to monitor how the spectrum of a composite prepreg changes with the degree of cure.

A preliminary study of a resin curing reaction was carried out to demonstrate the feasibility of the proposed experiment. For this preliminary work, the spectrum of LARC-TPI (a thermoplastic polyimide developed by NASA at the Langley Research Center) was recorded, first as an uncured resin and then again after a simple curing cycle had been performed.

Prior to our running the spectra of the uncured LARC-TPI, an identical fiber with the same total length (~1 m) and same uncoated sensor area (~5 cm) was aligned as before and its spectrum was measured. This spectrum was stored as the single-beam background spectrum for the experiment, thus compensating for the absorptions due to the fiber and silicone coating. The sample fiber was then coated with the LARC-TPI and heated to 40°C to ensure that all diglyme solvent had been removed.

The transmittance spectra obtained for the uncured sample at 40°C and the cured sample at 260°C are shown in Fig. 5 as the lower and upper spectra, respectively. Comparison of the two spectra in Fig. 5 reveals that imidization of the polyamic acid to the polyimide has occurred. Specifically, the peak at 1535 cm^{-1} (associated

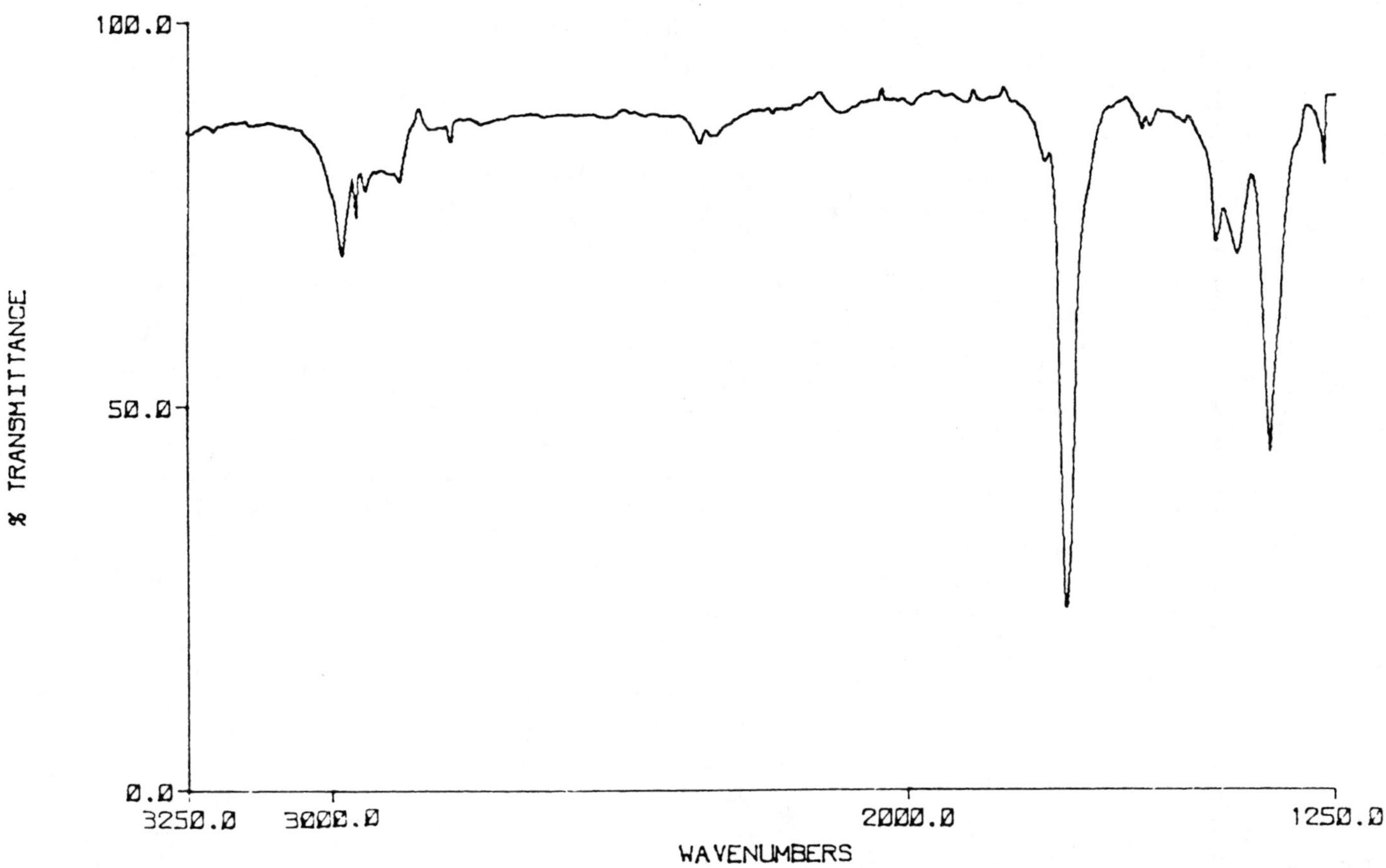

FIG. 4. Mid-infrared spectrum (3100–1100 cm^{-1}) of 2-butanone in units of % transmittance recorded by dipping an uncoated region of the fiber, approximately 3-cm long, into the solvent; 100 scans were coadded at a resolution of 8 cm^{-1} for this spectrum.

with the presence of the amide) is observed in the lower spectrum but disappears after heating to 260°C, as noted by its near absence in the top spectrum. Conversely, peaks associated with the imide functionality, namely at 1780, 1710, and 1370 cm^{-1}, are present in the spectrum recorded at the higher temperature. Thus, for the first time the ability to use a mid-IR transmitting optical fiber for measuring the curing reactions of a thermoplastic adhesive has been demonstrated. Because the fibers were coated with a silicone material, the spectrum of the cured sample shows evidence for the degradation of this cladding, which was also heated close to the reaction zone.

Detailed Study of Prepreg Curing Reaction. In order to further show the potential practical utility of this concept, we attempted the monitoring of the cure reactions and the kinetics of the curing process *in situ* on a graphite fiber/composite prepreg, PMR-15.[5] In this case the amount of resin is less than that in a bulk sample.

The spectrum of a chalcogenide fiber was recorded after a section had been declad and wrapped with the prepreg resin, as previously described. An aluminum block with a thermocouple was placed on top of the fiber and prepreg, which were resting on the surface of a hot plate equipped with a thermocouple. We determined the difference spectrum of the prepreg, starting at 22°C and continuing to 270°C once every 5 min, while the temperature was ramped 10–15°C during a 4-min period. The observed single-beam spectra were all ratioed directly against that recorded at 98°C, chosen since it is at the onset of the stage when curing reactions in PMR-15 are known to occur.[5] While a spectrum was being recorded, the temperature of the prepreg was within 1°C of the desired temperature.

Figure 6 shows the difference spectra of the prepreg at 22°C (lower trace) and after curing at 263°C (top spectrum). Because these spectra are plotted in Differential Absorbance units, chemical species that react during the cure process give rise to negative spectral bands, whereas the reaction products give rise to positive bands. Selected peaks associated with the curing reaction are listed in Table I. It should be noted that additional peaks associated with partial degradation of the fiber coating were observed, since the coating nearest the reactive region was also heated somewhat. The chemistry involved in the curing of the PMR-15 resin is complex.[12] The monomers involved have been described[12] as:

NE: monomethyl ester of 5-norbornene-2,3-dicarboxylic acid,
BTDE: dimethyl ester of 3,3′,4,4′-benzophenonetetracarboxylic acid,
MDA: 4,4′-methylenedianiline.

The first step of the process is the reaction of the amine group of the MDA and the ester group of the BTDE to form a secondary amide with loss of methanol, according to:

$$\text{-NH}_2 + \text{H}_3\text{COOC-} \rightarrow \text{-NH-CO-} + \text{CH}_3\text{OH}.$$

The NE acts as a terminating group, also by reacting with the MDA to form a secondary amide. Both of these

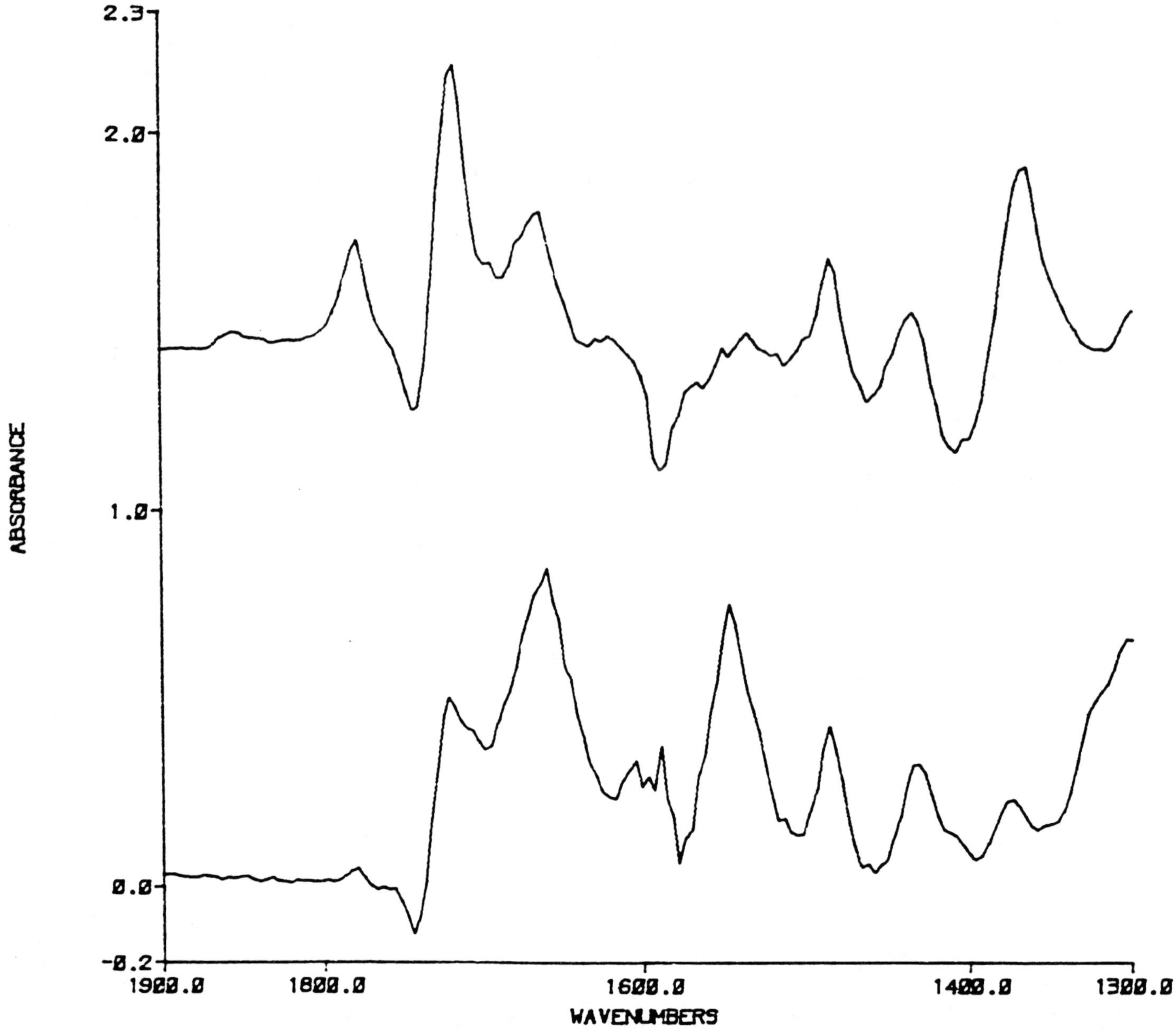

FIG. 5. Absorbance spectra of LARC-TPI resin (2000–1100 cm^{-1}) recorded by covering an uncoated portion of the fiber with the resin. The background spectrum was that of the uncoated fiber at room temperature. The spectra were recorded at 40°C (bottom) and after cure at 260°C (top). The top spectrum has been offset upwards, for clarity.

amide functionalities are next to a carboxylic acid group, and further heating causes the amide and acid groups to dehydrate and cyclize to give the imide group.[12]

Thus, during the curing process we could examine the spectral data for evidence of the following changes:

1. Loss of amine and ester groups at low temperature.
2. Build up of intermediate amide groups as step 1 progresses.
3. Conversion of the intermediate amide and acid groups to imide groups at higher temperatures.

Clearly, the complexity of the reaction may lead to some ambiguity in the observed spectral data, as all of these functional groups give rise to a number of bands in the optimum region of study, 1800–1300 cm^{-1}. To make the assignments even more complicated, both reactants and products have conjugation, which is known to shift the frequencies of these bands markedly. The bottom spectrum in Fig. 6 shows negative bands at about 1730, 1700, 1625, and 1510 cm^{-1}, which may be assigned to loss of ester, acid, aniline, and aniline, respectively. The change at the frequency of 1510 cm^{-1} is surprising, since the aromatic amide product also has a similar frequency, but the product band may be relatively weaker. The sharp rise in the 1590 cm^{-1} region is also hard to explain.

A convenient way to monitor the progress of such a reaction is to plot the intensity of the observed spectral bands as a function of temperature. We obtained these values automatically, using the Quant-32 software package, by defining a peak region and two baseline values for each band. Because of the crowded nature of the spectrum, the choice of baseline points was not always optimal for each band, but the general trends for the data are still evident. Figures 7–9 show how some of the

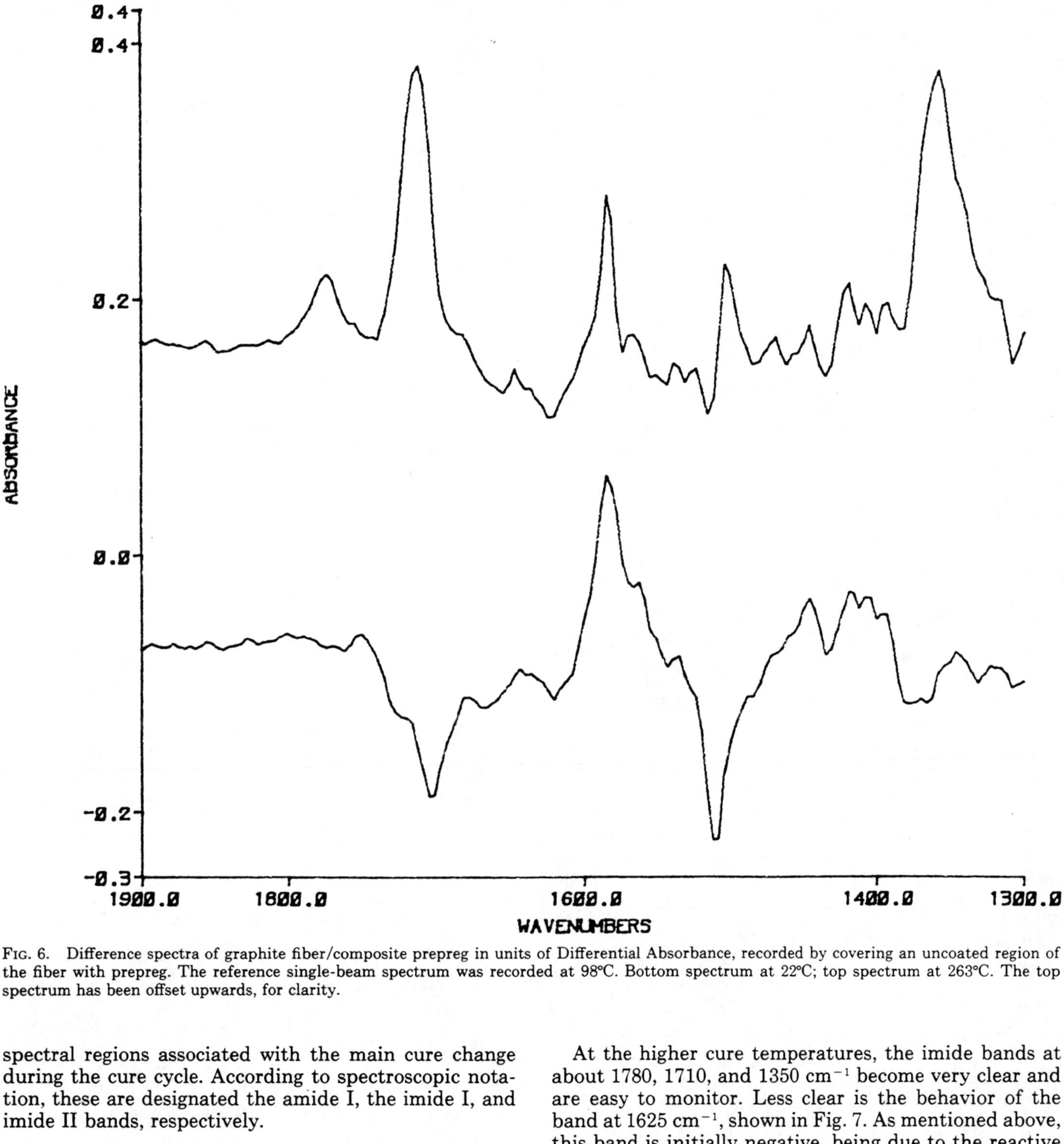

FIG. 6. Difference spectra of graphite fiber/composite prepreg in units of Differential Absorbance, recorded by covering an uncoated region of the fiber with prepreg. The reference single-beam spectrum was recorded at 98°C. Bottom spectrum at 22°C; top spectrum at 263°C. The top spectrum has been offset upwards, for clarity.

spectral regions associated with the main cure change during the cure cycle. According to spectroscopic notation, these are designated the amide I, the imide I, and imide II bands, respectively.

TABLE I. Peaks associated with curing reaction in PMR-15.

Absorbance, cm^{-1}	Assignment
1625	Amide I[a]
1510	Amide II[a]
1710	Imide I[a]
1785	Imide II[a]
1355	Imide[a]
1730	Ester
1700	Carboxylic acid
1625	Anilide
1510	Anilide

[a] Frequencies taken from Ref. 5.

At the higher cure temperatures, the imide bands at about 1780, 1710, and 1350 cm^{-1} become very clear and are easy to monitor. Less clear is the behavior of the band at 1625 cm^{-1}, shown in Fig. 7. As mentioned above, this band is initially negative, being due to the reactive aniline group. However, it increases in intensity as the cure starts, since this is also the absorption frequency of the amide intermediate. The intensity at 1625 cm^{-1} rises sharply up to about 100°C, and then drops down again as the intermediate amide is converted to imide. The behavior of Fig. 7 should be compared with that observed in Figs. 8 and 9. The imide production clearly coincides with the loss of the amide.

One observation may be made about the measured spectral baselines. These are not exactly at zero absorbance units, which seems to indicate that the contact

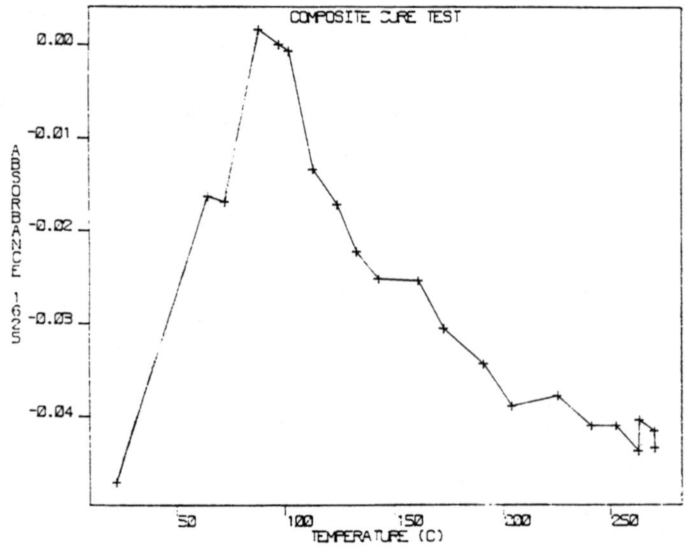

FIG. 7. Intensity of composite prepreg amide I absorbance band in difference spectra as a function of temperature.

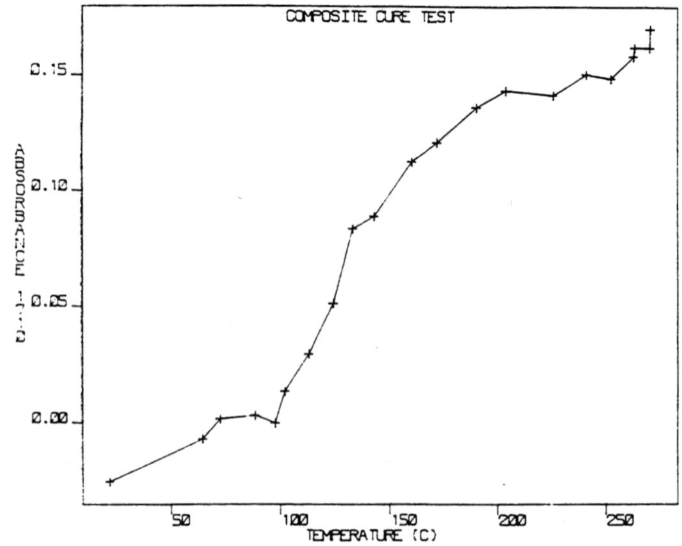

FIG. 9. Intensity of composite prepreg imide II absorbance band in difference spectra as a function of temperature.

between the fiber and the resin increases as the temperature, and thus viscosity, changes.

CONCLUSIONS

The feasibility of using a mid-infrared transmitting fiber as a spectroscopic probe has been demonstrated. Even with the relatively low-grade fibers used in this experiment, we have been able to transfer the infrared energy up to three meters to a remote detector. By selectively removing the coating of the fiber over a short portion, and bringing this portion into contact with a desired chemical species, one can conveniently obtain the spectrum of that chemical.

The study of the prepreg curing reaction led to the production of a large amount of information. Although the spectroscopic assignments needed to be made with care and are somewhat ambiguous, because of the complexity of the system, bands due to the various starting reactants, the intermediate, and the final product have been identified. It would be possible from this information to determine the kinetics of the curing reaction during processing and therefore to make immediate modifications in the curing cycle. Using real-time, *in situ* FT-IR monitoring of composite cure cycles will advance the state of the art in composite processing by correlating the spectral changes associated with cure to the actual mechanical properties of the part being fabricated. In this manner it will be possible to measure the actual cure process and unambiguously determine the extent of cure.

There are a whole host of other potential applications for this type of infrared sampling technology. We expect a number of such applications to be realized in a short period of time.

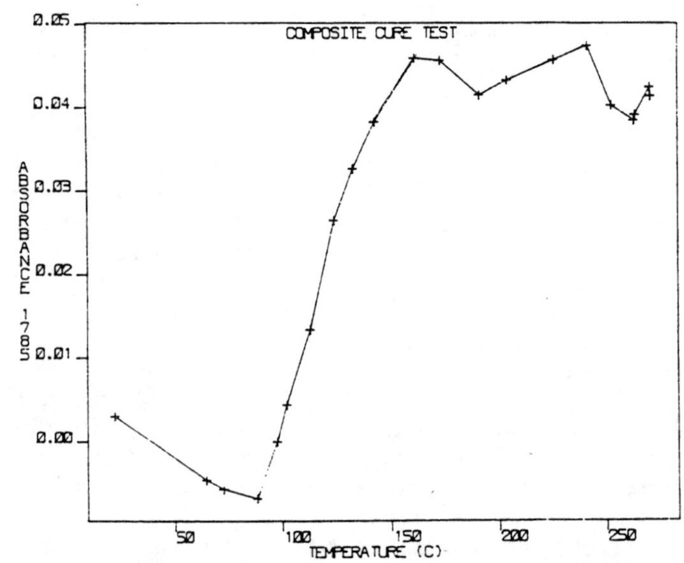

FIG. 8. Intensity of composite prepreg imide I absorbance band in difference spectra as a function of temperature.

1. N. J. Harrick, *Internal Reflection Spectroscopy* (John Wiley, New York, 1967).
2. F. M. Mirabella, Jr., and N. J. Harrick, *Internal Reflection Spectroscopy: Review and Supplement* (John Wiley, New York, 1985).
3. P. R. Young and A. C. Chang, Nat. SAMPE Tech. Conf. **16,** 136 (1984).
4. K. C. Cole, D. Noel, and J. J. Hechler, Proc. of the Nat. SAMPE Symp./Exhib. **30,** 624 (1985).
5. P. R. Young and A. C. Chang, Nat. SAMPE Tech. Conf. (1986).
6. W. F. X. Frank, W. Goertz, and H. H. Belz, Appl. Spectrosc. **41,** 323 (1987).
7. P. M. Fredericks, P. J. Samson, and A. D. Stuart, Appl. Spectrosc. **41,** 327 (1987).
8. P. W. France, S. F. Carter, M. W. Moore, J. R. Williams, and C. R. Day, In press.
9. S. Simhony, E. M. Kosower, and A. Katzir, Biochem. and Biophys. Res. Comm. **142,** 1059 (1987).
10. D. J. McEnroe, M. J. Finney, P. H. Prideaux, and P. C. Schultz, *Proceedings of the Conference on Infrared Optical Materials and Fibers, V,* SPIE Symposium on Fiber Optics and Integrated Electronics, San Diego, California (1987).
11. Sadtler Solvents Digital Library, Bio-Rad, Sadtler Division, Philadelphia, Pennsylvania.
12. G. D. Roberts and R. D. Vanucci, SAMPE Journal, **March/April** 24 (1986).

Comparison of Infrared Spectroscopic Methods for the Quantitative Analysis of Epoxy Resins Used in Carbon-Epoxy Composite Materials

K. C. COLE,* A. PILON, D. NOËL, J.-J. HECHLER, A. CHOULIOTIS, and K. C. OVERBURY

National Research Council Canada, Industrial Materials Research Institute, 75 bd. de Mortagne, Boucherville, Québec, Canada J4B 6Y4 (K.C.C., A.P., D.N., J.J.H.); and Canadair Ltd., P.O. Box 6087, Montréal, Québec, Canada H3C 3G9 (A.C., K.C.O.)

Three different methods of obtaining the infrared spectrum of carbon-epoxy prepreg for quality control purposes are critically evaluated with respect to their suitability for quantitative analysis. They are: transmission using a cast film of resin, attenuated total reflection (ATR) using prepreg, and diffuse reflection (DR) using prepreg. The cast film method is subject to significant error resulting from nonuniform film thickness; careful sample preparation is necessary in order to obtain acceptable results. The ATR method is faster and easier to use than the transmission method and gives more reproducible results; the use of a 45° germanium prism is recommended. The DR method is not suitable for quantitative work because of complications arising from front-surface reflection effects; it can be very useful, however, as a qualitative method.

Index Headings: **Infrared spectroscopy; Fourier transform infrared spectroscopy; FT-IR; Quantitative analysis; Epoxy resins; Composites; Attenuated total reflection; Diffuse reflection.**

INTRODUCTION

Composite materials based on continuous-fiber reinforcement embedded in a polymer matrix are becoming increasingly common. In "high-performance" applications, such as in the aerospace industry, the most widely used components are epoxy resins together with carbon or polyamide fibers. As the requirements for parts made from these materials become ever more demanding, there is a growing need for rapid and reliable analytical methods for quality control of the chemical composition of the raw materials. Infrared spectroscopy, along with liquid chromatography and thermal analysis, is one of the most useful methods.[1]

The raw material usually consists of "prepreg," or fiber reinforcement impregnated with partially cured resin. The latter is typically a complex mixture containing more than one epoxy resin, one or more hardeners, and sometimes catalysts and diluents. These are mixed and partially reacted to give the desired viscosity properties. The reactive system must then be stored at freezer temperatures in order to prevent further reaction. Variations in either the initial composition or the degree of advancement can have detrimental effects on the processing behavior of the material and on the final properties of parts made from it; hence the need for strict quality control.

Infrared spectroscopy has long been known as a versatile method for both qualitative and quantitative analysis. The increased sensitivity provided by Fourier transform instruments has opened up new possibilities, particularly through the use of reflection techniques. This paper compares three different measurement techniques for the quantitative analysis of carbon-epoxy prepreg, namely transmission, attenuated total reflection, and diffuse reflection. These three methods have been applied to a typical commercial prepreg in order to determine the easiest and most reliable way of characterizing such products.

EXPERIMENTAL

Materials. The system studied was Narmco Rigidite 5208/WC3000, manufactured by Narmco Materials, Inc. The reinforcement consists of Celion 3000 carbon fibers, woven in an 8-harness satin pattern. The woven fabric is impregnated with Narmco 5208 resin to a nominal resin content of 42% by weight, and the prepreg is supplied on a plastic backing sheet in rolls. Structural parts are typically made by laying up several layers of prepreg in a mold and curing the epoxy resin through application of heat and pressure. The recommended cure temperature is 177°C. It has been reported[2] that Narmco 5208 contains three main components: an epoxy resin based on tetraglycidyldiaminodiphenylmethane (TGDDM), about 69% by weight; a second epoxy resin, of the bisphenol A Novolac type, about 9% by weight; and the aromatic amine curing agent diaminodiphenylsulfone (DDS), about 22% by weight. From the respective companies, samples of the following commercial products corresponding to these components were obtained: Ciba-Geigy Araldite MY 720 (TGDDM epoxy resin), Celanese Epi-Rez SU-8 (bisphenol A Novolac epoxy resin), and Ciba-Geigy Hardener HT 976 (DDS).

Instrumentation and Methods. Infrared spectral measurements were done on a Nicolet 170SX Fourier transform instrument equipped with a mercury-cadmium-telluride detector. All spectra were the result of an accumulation of 128 scans recorded with a resolution of 4 cm^{-1}. For transmission measurements, the resin was extracted from prepreg with acetone, and thin films were cast on potassium bromide windows. The detector response recorded with the sample in place was ratioed against the "background" recorded with an empty beam to give the transmittance spectrum. Attenuated total reflection spectra were measured with a twin parallel mirror reflection attachment (Variable Angle Manual) from Harrick Scientific Corporation, with the use of single-pass parallelogram prisms (50 × 10 × 3 mm). Three types of prism were tested: KRS-5 (thallium bromoio-

Received 17 September 1987; revision received 21 December 1987.
* Author to whom correspondence should be sent.

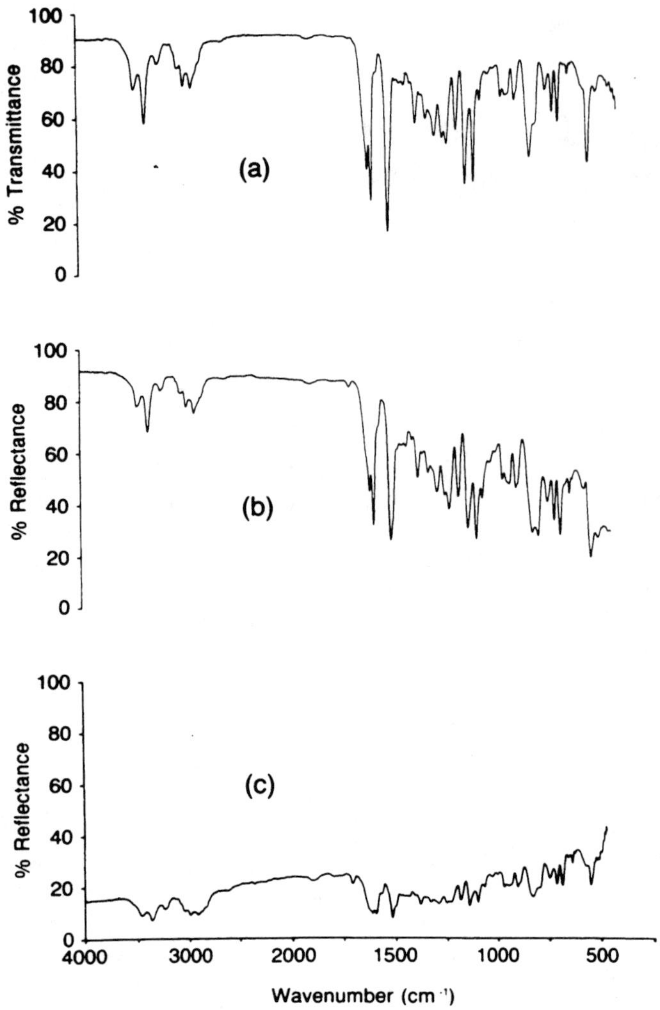

FIG. 1. FT-IR spectra of Narmco 5208/WC3000 prepreg as recorded by: (a) transmission (cast film of resin); (b) attenuated total reflection (prepreg on 45° KRS-5 prism); and (c) diffuse reflection (prepreg).

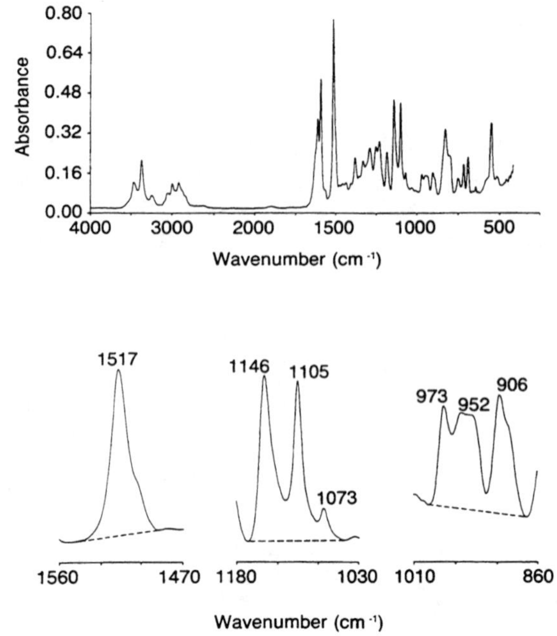

FIG. 2. (Top) Transmission spectrum of Fig. 1 expressed in absorbance units. (Bottom) Bands used for quantitative analysis.

dide), angle of incidence 45°; KRS-5, angle 60°; germanium, angle 45°. Strips of prepreg about 5 cm × 1 cm in size were cut with scissors and clamped against the prism, with the use of finger pressure to tighten the screw. In addition to the spectra recorded with the prepreg in contact with the prism, a second set was recorded for the residual film of resin left on the prism when the prepreg was removed. The spectrum obtained with the prism alone was used as a "background." In the case of the 45° KRS-5 prism, the spectra recorded with prepreg on both sides were overly intense, so prepreg was applied to only one side of the prism; in the other cases it was applied to both sides. After each specimen was measured, the prism was cleaned with a 50:50 mixture of toluene and acetone. Diffuse reflection measurements were done with a "praying mantis" type accessory, also from Harrick Scientific. Samples consisted of small squares of prepreg (about 1 cm × 1 cm). For the 8-harness satin fabric weave used in the system studied, on a given surface seven-eighths of the total area is covered by fibers aligned in one particular direction. With the reflection accessory used, it was found that better-quality spectra resulted if samples were oriented so that this direction corresponded to the beam direction of the instrument. Powdered potassium bromide, compacted with a spatula, was used as a reference material for the diffuse reflection spectra.

RESULTS AND DISCUSSION

The Narmco 5208 Spectrum. Figure 1 compares the spectra recorded for Narmco 5208/WC3000 prepreg (Batch 1010) with the use of the three techniques. The differences will be discussed later. For analysis, the transmission spectrum is converted to absorbance, as shown in Fig. 2. Figure 3 shows the spectra of the three components reported to be present in Narmco 5208. The major peaks in the TGDDM and DDS spectra have been assigned previously.[3] The most useful peaks for quantitative analysis purposes in the Narmco 5208 spectrum are shown in detail in the lower part of Fig. 2. The first is the strong band at 1517 cm^{-1}, which has been assigned to a stretching mode of the benzene rings. It arises mainly from the TGDDM, but small contributions from the other two components are responsible for the shoulder around 1500 cm^{-1}. This peak is useful as a reference, as explained below. The second peak of interest is really a group of three peaks located at 1146, 1105, and 1073 cm^{-1}. These all arise from the DDS hardener, and the two strongest ones have been attributed to vibrations of the sulfone group.[3] By dividing the area of these peaks by that of the aromatic ring peak at 1517 cm^{-1} it is possible to calculate an index which is independent of the sample thickness and which is a good indicator of the hardener-to-epoxy ratio in the resin. Thus:

$$\text{Sulfone index} = \frac{\text{Area of peaks at 1146, 1105, 1073 cm}^{-1}}{\text{Area of peak at 1517 cm}^{-1}}. \quad (1)$$

Since both the aromatic rings and the sulfone groups remain intact throughout the polymerization reaction, this index would not be expected to be particularly sensitive to the degree of resin advancement. However, it has been shown that the sulfone band intensities are

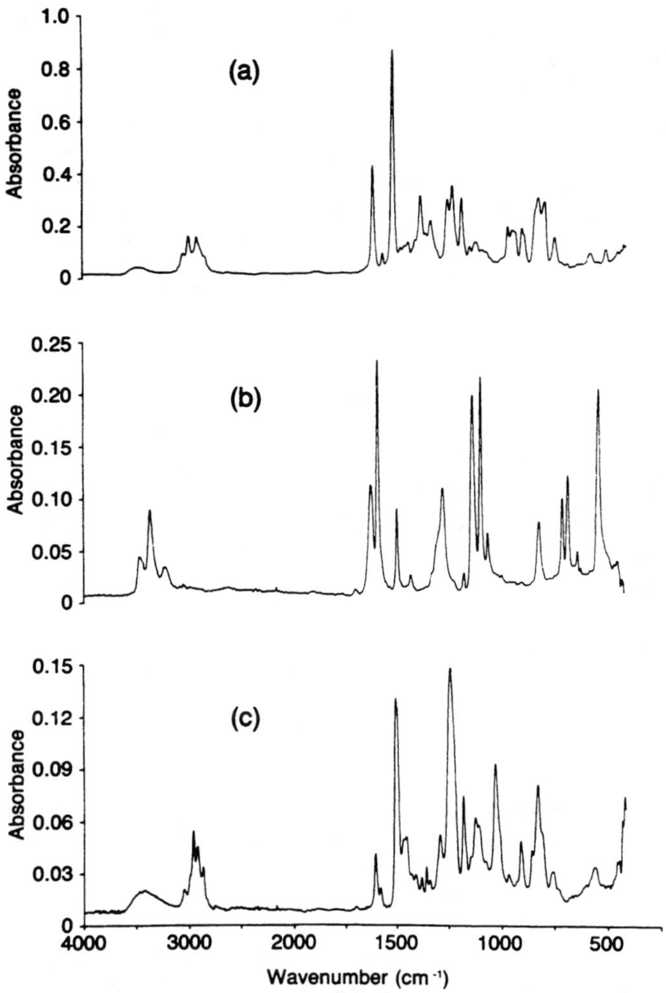

FIG. 3. FT-IR spectra of components believed to be present in Narmco 5208 resin: (a) Araldite MY 720 (TGDDM); (b) Hardener HT 976 (DDS); (c) Epi-Rez SU-8. Thin cast films in transmission.

affected somewhat by the transformation of the primary amine groups of the DDS to secondary amine groups,[4] as well as by the presence of moisture.[3] The third set of peaks of interest occurs at 973, ≈952, and 906 cm^{-1}. These are due to vibrations of the epoxide rings, mainly in the TGDDM, although the minor epoxy contributes to the peak at 906 cm^{-1}. When the resin cures, all these peaks disappear at much the same rate, confirming that they all arise from epoxide groups. The peak near 910 cm^{-1} has long been used to follow epoxy cure.[5] As above, an index can be calculated as follows:

$$\text{Epoxide index} = \frac{\text{Area of peaks at 973, 952, 906 cm}^{-1}}{\text{Area of peak at 1517 cm}^{-1}}. \quad (2)$$

In this case, since the epoxide groups disappear upon polymerization, while the benzene rings do not, this index is an indicator of the number of *unreacted* epoxide groups in the system. Thus the two indices together provide a means of checking both the hardener-to-epoxy ratio and the degree of polymerization (or "advancement") of the prepreg. The remainder of the paper is concerned with finding the most effective way to determine these two parameters.

Transmission Results. Generally speaking, quantitative IR analysis is best performed on solutions of the analyte, with the use of a cell of known pathlength and calibration with solutions of known concentration. This method has been used, for example, to determine the DDS content in epoxy formulations.[6–8] However, its applications to epoxy analysis are limited, because the few solvents which have good IR transmission properties are often poor solvents for the resins, particularly if the latter are partly polymerized, as in the present case. On the other hand, solvents which have good dissolving properties, such as tetrahydrofuran or acetone, tend to absorb strongly in the IR and interfere with the peaks of the compound being analyzed. In this work the solution method was rejected in favor of the cast film technique. The resin is extracted from the prepreg with solvent; a small quantity is placed on a KBr window; and the solvent is evaporated in order to leave a thin film, which is measured in transmission. Since the film thickness is difficult to determine, the amount of sample absorbing is unknown. This problem is overcome by using one peak in the spectrum as an "internal standard" or reference to which other peaks are compared, as described in the previous section.

TABLE I. **Quantitative analysis results obtained in transmission experiments.**

Expt. no.[a]	Prepreg batch	No. of specimens	Sulfone index Avg.	Sulfone index RSD (%)	Epoxide index Avg.	Epoxide index RSD (%)
1	1010	10	1.338	10.1	0.670	28.5
2	1010	5	1.097	5.5	0.417	8.2
2a	1010	4	1.072	2.5	0.402	2.4
3	1014	5	1.073	1.9	0.413	1.8
4	1014	2	1.310	—	0.623	—

[a] See text for details.

The following solvents were tested as candidates for extraction of the Narmco 5208 resin: tetrahydrofuran, acetone, acetonitrile, dichloromethane, and chloroform. Extractions were performed with the use of 100 mL of solvent with a piece of prepreg measuring 5 cm × 5 cm. Although tetrahydrofuran is the best in terms of resin solubility, it is not suitable for IR purposes because it associates strongly with the resin and is very difficult to remove completely without heating. Traces left in the resin give peaks at 1070 and 925 cm^{-1} which interfere with the peaks used for analysis. The next best solvent in terms of solubility, and the best for IR work, is acetone. It readily dissolves the resin, it is easily removed by evaporation under vacuum, and even if traces remain, the major peaks do not coincide with the peaks used for analysis. The remaining three solvents were found to be unsuitable because they gave slightly turbid solutions, indicating limited resin solubility.

An experiment was done to check the reproducibility of the cast film technique. A square of prepreg measuring 5 cm × 5 cm, from Batch 1010, was extracted with 20 mL of acetone in two stages (15 mL + 5 mL). We prepared ten different specimens by depositing a few drops of the solution on a KBr window, blowing gently with a stream of air from a rubber bulb to remove most of the solvent, and drying under vacuum (without heat, which would cause polymerization of the resin) to remove the remainder of the solvent. The IR spectra were recorded

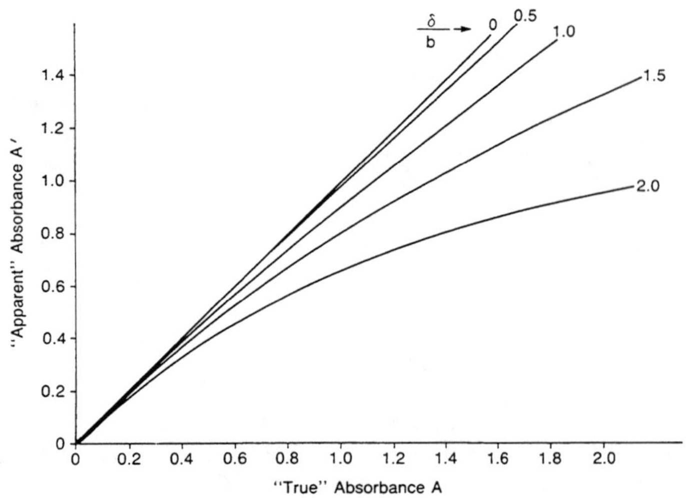

FIG. 4. Effect of nonuniform sample thickness on measured absorbance for "wedge-shaped" sample.

and converted to absorbance. The amount of resin deposited on the window was controlled to give spectra with a maximum absorbance of about 1. Peak areas were calculated by computer integration of the spectra with Nicolet software using the baselines shown in Fig. 2. The approximate limits of integration were 1550–1485 cm^{-1} for the aromatic ring band, 1166–1048 cm^{-1} for the sulfone bands, and 993–875 cm^{-1} for the epoxide bands. Sulfone and epoxide indices were calculated according to Eqs. 1 and 2. The average values, together with relative standard deviations, are given in the first line of Table I. A surprisingly high variation is observed, with relative standard deviations of 10% for the sulfone index and 29% for the epoxide index. This is much higher than expected and obviously unacceptable for quality control purposes.

A number of possible causes of the variation were investigated. To check the effect of using peak heights for the calculation instead of peak areas, we reanalyzed the spectra using the heights of individual peaks at 1517, 1146, 1105, and 907 cm^{-1}. No improvement was observed, and in fact, the relative standard deviations were greater. When the experiment was repeated, with the use of chloroform and dichloromethane as solvents, similar variations were observed, so the choice of solvent does not appear to be a contributing factor. Likewise, the presence of weak residual solvent peaks in a few of the spectra did not correlate with the variation of the indices. When specimens are prepared with the use of dichloromethane, the rapid evaporation of the solvent sometimes cools the sample to the point of causing condensation of atmospheric moisture, but this factor also had no detectable effect on the results.

The cause of the variation appears to be nonuniform film thickness. This problem has been discussed by Koenig[9] and Hirschfeld.[10,11] It is sometimes referred to as the "wedging effect," since it occurs in liquid cells in which the windows are not parallel and the sample takes the shape of a wedge. For a sample of uniform thickness b, the "internal transmittance," or the transmittance of the sample itself (neglecting reflection effects at the boundary surfaces), is given by the well-known Lambert absorption law:

$$T = e^{-\alpha b} \tag{3}$$

where α is the absorption coefficient of the material. Suppose, however, that within the area covered by the IR beam the sample thickness varies from the average value b over a range δ (i.e., from $b - \frac{1}{2}\delta$ to $b + \frac{1}{2}\delta$), and that all thicknesses within this range are equally probable. This will be the case, for example, for a wedge-shaped sample in a rectangular beam. It can be shown that the transmittance will be given by:

$$T' = e^{-\alpha b}\left\{\frac{\sinh\left(\frac{1}{2}\alpha\delta\right)}{\frac{1}{2}\alpha\delta}\right\}. \tag{4}$$

In terms of absorbance ($A = -\log_{10}T$), the measured absorbance A' is related to the "true" absorbance A (which would apply if there were no thickness variation) by the equation:

$$A' = A - \log_{10}\left\{\frac{\sinh\left(\frac{1}{2}\frac{\delta}{b}A \ln 10\right)}{\frac{1}{2}\frac{\delta}{b}A \ln 10}\right\}. \tag{5}$$

The relationship is shown graphically in Fig. 4 for several values of δ/b. The apparent absorbance A' is always lower than the true absorbance A, and the deviation is greatest when the absorbance is high (i.e., for strongly absorbing bands and for thick samples) or when the thickness variation is high. The effect on the appearance of the spectrum is to produce an apparent suppression, or "flattening," of the stronger bands. This is illustrated in Fig. 5, which shows two spectra from the series used to obtain the results previously described. Spectrum "a" is for a specimen whose thickness variation is believed to be low. Many of the recorded spectra closely resembled this one in shape, and it is believed to be a good representation of the "true" absorbance spectrum. Spectrum "b," on the other hand, shows strong evidence of nonuniform film thickness. The weaker bands, such as the epoxide bands near 1000–900 cm^{-1}, have the same shape as in spectrum "a," but the stronger bands, such as the aromatic ring bands around 1600 and 1500 cm^{-1}, show substantial distortion. In Fig. 6, the peak heights from spectrum "b," measured with respect to a baseline extrapolated from the 2500–1700 cm^{-1} region of the spectrum, are plotted against the corresponding peak heights from spectrum "a." If both specimens were of uniform thickness, the two spectra would have the same appearance, and the plot would be a straight line with slope equal to the ratio of the two thicknesses. Instead, a smooth curve is obtained whose form is similar to that of the curves shown in Fig. 4. An exact correspondence is not observed, since the two films differ in average thickness, the nonuniform sample is probably not exactly wedge-shaped, and the infrared beam is circular rather than rectangular. Nevertheless, the general shape of the curve in Fig. 6 provides strong evidence that the deviations in the spectra are the result of nonuniform film thickness. At low absorbance values, the slope of the curve is 2.3,

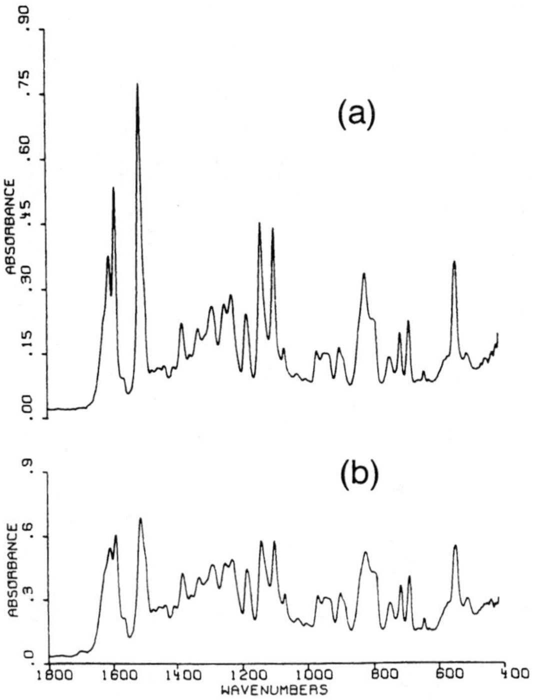

FIG. 5. Transmission spectra of films of Narmco 5208 resin: (a) film of uniform thickness; (b) film of varying thickness.

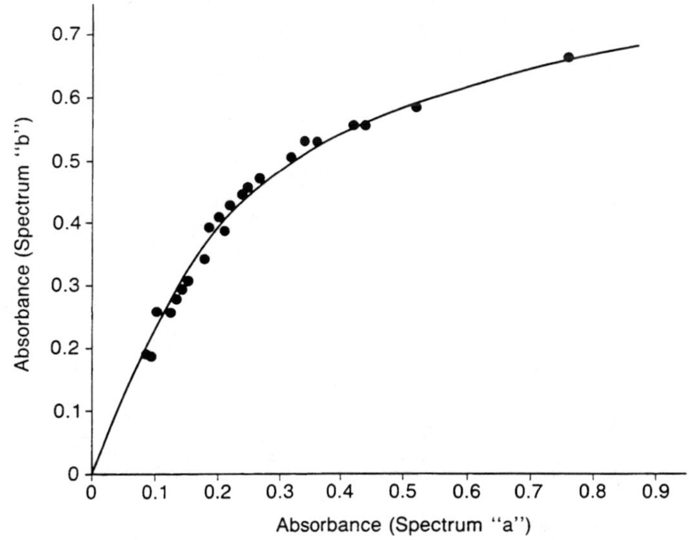

FIG. 6. Relationship between peak absorbance values for the spectra of Fig. 5.

indicating that the average thickness of specimen "b" is 2.3 times that of specimen "a." However, the absorbance of the strongest peak in spectrum "b" is actually lower than that of spectrum "a," because of the band suppression caused by nonuniform thickness.

Since stronger peaks such as the aromatic ring band at 1517 cm^{-1} are attenuated more than weaker peaks such as the sulfone and epoxide bands, the indices defined in Eqs. 1 and 2 will always be artificially high if there is a nonuniform thickness effect. Consequently, the average of several measurements will also be higher than the "true" value, and increasing the number of specimens measured will not improve the accuracy very much. A truer value will be obtained if unusually high values in the set are eliminated from the averaging, on the assumption that these correspond to specimens of nonuniform thickness. One should minimize the number of such values by careful sample preparation in order to obtain films that are as uniform as possible.

To check this assumption, we repeated the initial experiment, using greater care in sample preparation. A square of prepreg about 5 cm × 5 cm was extracted with 100 mL of acetone in three stages (50 + 30 + 20 mL), and the resulting solution was concentrated to about 25 mL on a rotary evaporator. This gives a resin concentration of about 25 mg/mL. Using a microsyringe, we prepared specimens by depositing 150 μL of this solution on KBr windows 25 mm in diameter. This was done slowly and carefully in an area protected from air currents so that the solvent could evaporate slowly enough to leave as uniform a film as possible. When most of the solvent had evaporated, the samples were placed under vacuum for at least 20 min to remove remaining traces. The spectra were then recorded and analyzed as previously. The results, based on the average of five specimens, are given as "Experiment 2" in Table I. The standard deviations are considerably lower than in Experiment 1, but are nevertheless higher than expected. One particular specimen, with indices of 1.197 and 0.476, respectively, was responsible for most of the error. When it was eliminated from the averaging process (Experiment 2a), the standard deviations dropped to the more acceptable level of 2.5%. A third experiment was done in a similar manner, but with the use of a different lot of prepreg (Batch 1014). This time the standard deviations were under 2%. In this case, two additional specimens were prepared from the same solution but with the use of an air stream to accelerate solvent evaporation (Experiment 4). As seen in Table I, these samples gave average indices as much as 50% higher, because of the effects of nonuniform thickness.

Taken together, the results of Table I show that the transmission method using cast films is acceptable for quality control of prepreg formulations only if great care is taken to prepare films of uniform thickness. With good technique and windows of good quality, a precision of 2% is obtainable.

Attenuated Total Reflection Results. The attenuated total reflection (ATR) technique, also known as internal reflection, has been discussed in a monograph by Harrick[12] and in a recent review.[13] The spectrum of the surface of an opaque solid sample is obtained by pressing the sample against an IR-transmitting prism of high refractive index, through which the IR beam passes while undergoing multiple internal reflections. At each reflection, the beam is "attenuated" by a partial absorption of the evanescent wave existing at the prism/sample interface. This technique is particularly interesting for prepreg analysis, because the prepreg can be measured "as is." This benefit not only eliminates the time-consuming solvent extraction and film preparation procedures required for transmission work, but also avoids the errors associated with nonuniform film thickness. However, ATR measurements are not without complications.

Figure 1b shows the ATR spectrum of Narmco 5208/WC3000 prepreg obtained with the use of a KRS-5 prism with an angle of incidence of 45°. This spectrum corre-

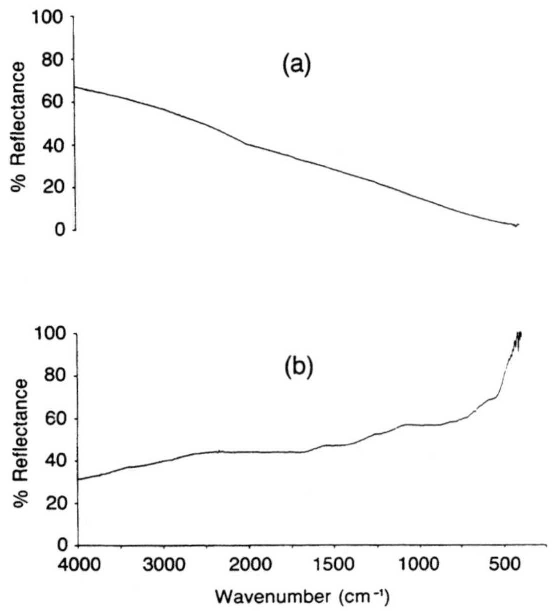

FIG. 7. FT-IR spectra of woven cloth made from carbon fibers, as recorded by (a) attenuated total reflection with 45° KRS-5 prism; (b) diffuse reflection.

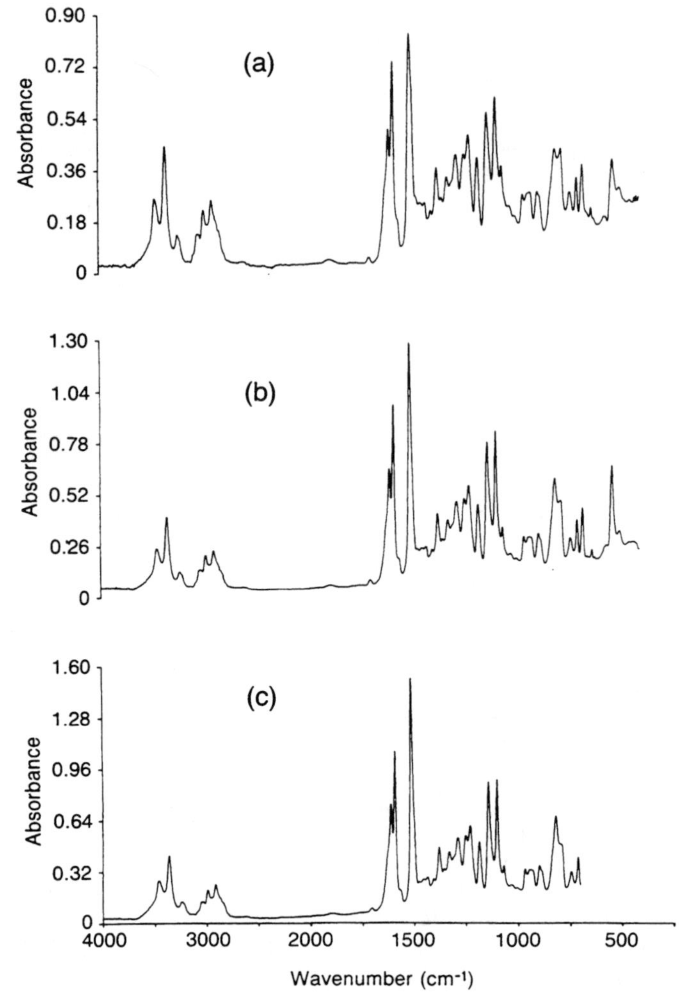

FIG. 8. Attenuated total reflection FT-IR spectra (corrected for wavelength dependence of sampling depth) of Narmco 5208/WC3000 prepreg, recorded with the use of three different prisms: (a) KRS-5, 45°; (b) KRS-5, 60°; (c) germanium, 45°.

sponds mainly to the thin surface layer of resin, with some contribution from the underlying carbon fiber fabric. Comparison with the transmission spectrum of Fig. 1a shows that although the two are quite similar, there are certain differences. In the ATR method, the "effective sampling depth" of the evanescent wave is proportional to the wavelength. As a result, there is an apparent increase in peak intensity toward longer wavelength (lower wavenumber), in comparison to the results for the transmission spectrum. A second effect observed in the ATR spectrum is a continuous "background" absorption due to the carbon fibers. These absorb at all frequencies, without distinctive peaks, as shown in Fig. 7a. The absorption is greater at lower wavenumber because of the sampling depth effect mentioned above.

These effects can be corrected mathematically to a large extent. Figure 8a shows the ATR spectrum of Fig. 1b after conversion to absorbance units, baseline correction to remove some of the graphite absorption, and multiplication by $\bar{\nu}/1000$, where $\bar{\nu}$ is the frequency in wavenumbers. The latter step corrects for the wavelength dependence of the sampling depth. However, when the resulting spectrum is compared with the absorbance spectrum measured in transmission (Fig. 2), certain differences are still apparent in the form of variations in band shape. These are most obvious in the case of stronger peaks such as the aromatic ring band at 1517 cm^{-1}, the sulfone bands at 1146 and 1105 cm^{-1}, and the peaks near 830 and 554 cm^{-1}. The reason for this lies in optical effects. In ATR, the reflectance at the prism/sample interface is determined by the optical constants of the prism and the sample, as well as by the angle of incidence and the polarization of the radiation. Both the refractive index n and the absorption coefficient α play a role. The exact relationship, which is rather complicated, is given by Fresnel's equations.[12] In order for total internal reflection to occur, the prism geometry must be in such a configuration that the angle of incidence is greater than a certain critical angle given by $\sin \theta_c = n_s/n_p$, where n_s and n_p are the refractive indices of the sample and prism respectively. Generally speaking, when the angle of incidence is much larger than the critical angle, the refractive index contribution is small and the ATR spectrum resembles the transmission spectrum. If the angle of incidence is near the critical angle, the refractive index contribution becomes more important and "distorts" the spectrum by introducing some dispersion-type character into the usual Lorentzian-type peaks.

In the Narmco 5208 spectrum, in regions of strong absorption peaks, the refractive index of the resin also increases, and becomes closer to that of the prism. The critical angle increases and approaches the angle of incidence, with the result that the refractive index has a greater effect on the spectrum, causing peak distortion. These effects can be reduced by either increasing the angle of incidence or lowering the critical angle (with a prism of higher refractive index). The results are illustrated in Fig. 8b and 8c, which show "corrected" ATR spectra obtained with two other prisms, one of KRS-5 (n = 2.37) with an angle of 60°, and one of germanium (n = 4.0) with an angle of 45°. In both cases they bear a closer resemblance to the transmission spectrum, but the latter is the better of the two.

A set of experiments was performed to investigate the feasibility of using the ATR method for quantitative analysis of the prepreg. Spectra were measured for the three types of prism, both with prepreg (Batch 1010) and with the residual films left behind by the prepreg. For each prism, ten prepreg specimens were measured, five with the principal fiber direction parallel to the long axis of the prism and five with it perpendicular. Spectra were converted to absorbance and treated in the same way as the transmission spectra. They were *not* corrected by multiplying by $\bar{\nu}/1000$, since this is an extra step which is needed only if they are to be visually compared with transmission spectra. When the results were examined, the fiber orientation was found to have no significant effect, so the figures for different orientations were combined, giving a total of ten results for averaging. These results, along with their relative standard deviations, are given in Table II. Spectra measured for the case of the 45° KRS-5 prism with prepreg were not analyzed, because some bands were overly intense, due to the high effective sampling depth occurring under these circumstances.

TABLE II. Quantitative analysis results obtained in ATR experiments.

		Sulfone index		Epoxide index	
Prism type	Sample form	Avg.	RSD (%)	Avg.	RSD (%)
KRS-5 (45°)	Prepeg	†[a]	†	†	†
KRS-5 (45°)	Residual film	1.836	1.9	0.938	4.0
KRS-5 (60°)	Prepeg	1.561	0.4	0.729	0.8
KRS-5 (60°)	Residual film	1.523	0.6	0.698	0.8
Ge	Prepeg	1.436	0.5	0.669	1.2
Ge	Residual film	1.403	0.5	0.636	1.2

[a] (†) Spectra were too intense for quantitative analysis.

The 45° KRS-5 prism is obviously the least acceptable for quantitative analysis. The reproducibility is poorer and the calculated indices are significantly different from those obtained with the other prisms. This result is no doubt due to the greater distortion of peaks by refractive index effects, as discussed above. Of the other two prisms, although there is not much difference in terms of reproducibility, the 45° Ge prism is preferred, as refractive index effects are less important and the band shapes are closer to those of the transmission spectrum. It is interesting to note that the indices obtained from residual film spectra are slightly lower than those obtained from prepreg spectra on the same prism. Since the precision is similar in both cases, and the residual film spectra are not affected by carbon fiber absorption, it might be expected that the film spectra would be preferred. This is not the case, however, since there is another factor which must be considered, namely the film thickness. If the film is sufficiently thin, the effective sampling depth may exceed the film thickness, causing an apparent decrease in absorption intensity. The effect will be more pronounced at lower wavenumber, since the sampling depth is proportional to wavelength. This is the most probable explanation for the deviations observed in the present case. Because the sulfone and epoxy peaks occur at lower wavenumber than does the aromatic ring reference band, they are more affected, and the ratio with respect to the reference peak is lower than it is in the case of the prepreg spectra, where the resin layer is thicker. The fact that the discrepancy for the epoxide index (about 5%) is greater than that for the sulfone index (about 2%) offers further evidence, since the epoxide peaks occur at lower wavenumber. In the present work, the good reproducibility observed for the residual film spectra indicates that the film thickness is reasonably reproducible. However, a different lot of prepreg, with more viscous resin, might leave a much thinner layer. In order that potential problems related to such variations might be avoided, it is recommended that the spectra be recorded with the prepreg present.

The preferred results in Table II are therefore those for the 45° Ge prism with prepreg. One slight disadvantage of the Ge prism is that the spectrum ends at 700 cm^{-1}, because Ge absorbs strongly beyond this point. However, a possible advantage is that with Ge the sampling depth is about four times lower than with KRS-5, so that the absorption of the carbon fibers is less apparent.

The indices given in Table II are considerably higher than those obtained in transmission, because the ATR spectra used for the calculations were not "corrected" for the wavelength dependence of the sampling depth. In order to compare the results from the two methods, we multiplied the 45° Ge ATR indices by $\bar{\nu}$(peak)/$\bar{\nu}$(reference), where $\bar{\nu}$(peak) is the wavenumber corresponding to the center of the group of peaks in question and $\bar{\nu}$(reference) is the wavenumber of the aromatic reference peak. This procedure gave 1.065 and 0.410 for the sulfone and epoxide indices, respectively. These figures are in excellent agreement with the values of 1.072 and 0.402 obtained in transmission.

In view of the results just described, the ATR method is recommended over the transmission method for the quantitative analysis of uncured carbon-epoxy prepreg. Not only is sample preparation faster and simpler, but the results are just as reproducible, if not more, since the problem of nonuniform sample thickness associated with the transmission method is avoided. A further advantage of the ATR method is that it gives the spectrum of the prepreg as received, whereas the manipulations required by the transmission method can change the resin composition. Thus in Fig. 8, a small peak can be seen at 1708 cm^{-1} which does not appear in the transmission spectrum. This result was found to be due to a volatile product which was collected and identified as acetone—apparently residual solvent from the prepreg manufacturing process. It is not seen in the transmission spectrum because acetone is used to extract the resin and is then evaporated, together with any other volatiles present, before the spectrum is recorded.

Diffuse Reflection Results. Although diffuse reflection (DR) spectroscopy has been known for a long time, it did not come into widespread use in infrared spectroscopy until the highly sensitive FT-IR instruments were developed. Since then it has proved extremely useful in a number of applications.[14–16] An advantage over ATR is that no physical contact with the sample surface is required. Since ATR requires intimate contact between the sample surface and the prism, it is generally limited to samples which are either deformable or else have a very smooth surface. Thus, although it is useful for pre-

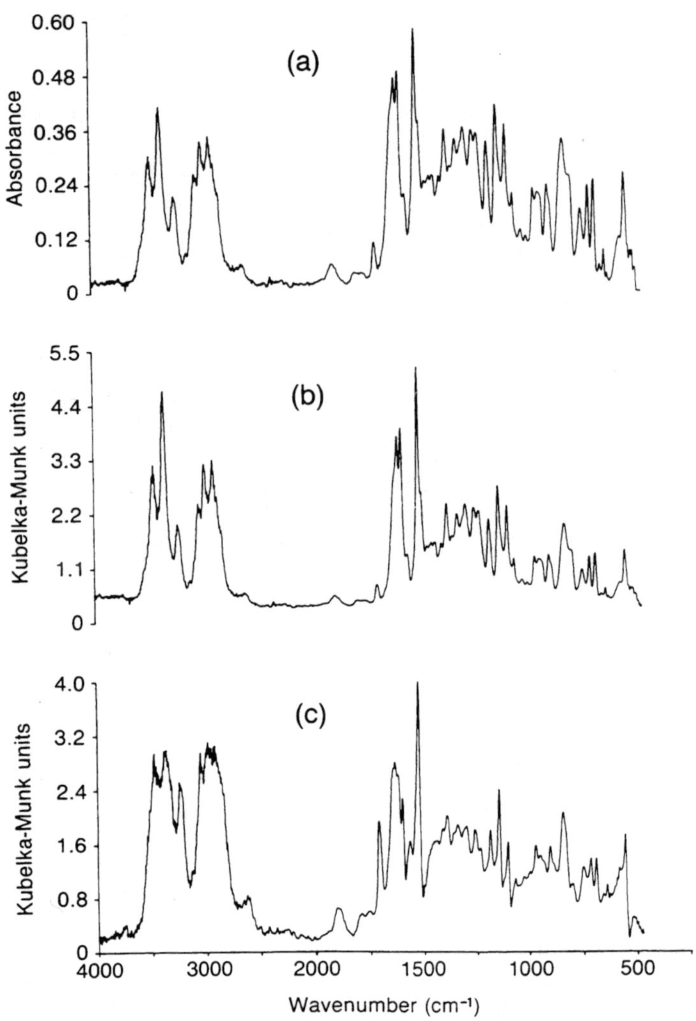

FIG. 9. Diffuse reflection FT-IR spectra of Narmco 5208/WC3000 prepreg: (a) spectrum of typical sample, converted to absorbance units; (b) same spectrum, converted to Kubelka-Munk units; (c) spectrum of resin-rich specimen, converted to Kubelka-Munk units.

preg analysis, it can be used for cured composites only if care is taken to prepare samples with a smooth surface.[3] Diffuse reflection, on the other hand, can be applied to composites at all stages of cure.[17–21]

Figure 1c shows a typical DR spectrum of Narmco 5208/WC3000 prepreg. The general features are the same as those in the transmission spectrum. Unlike the case for ATR, where the effective sampling depth is only a few micrometers, in diffuse reflection the IR beam penetrates the surface layer of resin completely and is then reflected back by the resin/fiber interfaces. At the same time there is scattering and partial absorption by the fibers. This is responsible for the strong continuous background absorption of Fig. 1c, which prevents the overall reflectance from exceeding about 40%. The diffuse reflection spectrum of woven carbon fiber cloth is shown in Fig. 7b. Since the sampling depth is independent of frequency in this case, the absorption does not increase at lower wavenumber. Instead the reverse is observed, probably because of increased scattering losses at shorter wavelength.

For further analysis, the diffuse reflection spectrum must be suitably transformed. This is done with the use

TABLE III. Quantitative analysis results obtained in diffuse reflection experiments.

Expt. No.	Pre-preg batch	No. of speci-mens	Units used	Sulfone index Avg.	Sulfone index RSD (%)	Epoxide index Avg.	Epoxide index RSD (%)
1	1010	18	K-M	1.118	11.4	0.567	15.0
1	1010	18	Abs.	1.486	6.5	0.986	8.9
2	1014	7	K-M	0.855	7.4	0.541	28.1
2	1014	7	Abs.	1.272	4.1	1.005	19.3
3[a]	1014	7	K-M	1.758	4.8	1.552	25.6
3[a]	1014	7	Abs.	1.893	7.1	1.721	20.1
4[b]	1014	7	K-M	1.863	7.4	1.484	23.2
4[b]	1014	7	Abs.	1.934	4.6	1.622	16.2

[a] KBr sprinkled on surface.
[b] KBr embedded in resin.

of the Kubelka-Munk relationship,[14] commonly represented as:

$$F(R_\infty) = k/s = (1 - R_\infty)^2/2R_\infty \qquad (6)$$

where R_∞ is the diffuse reflectance of the sample, which is assumed to be "infinitely thick," k is the absorption coefficient (denoted above by α), and s is a scattering coefficient. Since s varies only gradually with frequency, the Kubelka-Munk spectrum is usually similar, although not identical, to the transmission spectrum expressed in absorbance units. The Kubelka-Munk equation was derived for the case of an isotropically scattering particulate sample. However, it is sometimes applied to other types of sample, such as the present one, for which an appropriate model is lacking. Figure 9a shows the diffuse reflection spectrum after conversion to absorbance units and baseline correction, while Fig. 9b shows the result obtained with the Kubelka-Munk transformation. Comparison with Fig. 2 shows that the Kubelka-Munk transformation gives a better match to the "true" absorbance spectrum and is therefore preferred. All the details of the transmission spectrum can be seen in the diffuse reflection spectrum. However, the match is not exact for at least two reasons. The first is the contribution of the scattering coefficient s, which causes a certain variation in relative peak intensities. The second is the effect of front-surface reflection.[22] When the IR beam impinges on the sample surface, part of it is reflected at the air-resin interface before it ever enters the sample. This component may be labeled "surface" reflection R_s to distinguish it from the "volume" reflection R_v, which corresponds to radiation which has entered the sample and left again after undergoing further reflection and partial absorption. The front-surface reflection is governed by the laws of Fresnel, and its frequency dependence is quite different from that of R_v; in particular, the refractive index makes a greater contribution. Both R_s and R_v contribute to the diffuse reflection spectrum, and the relative importance of each will vary, depending on the sample. Figure 9c shows the DR spectrum of a prepreg sample with a surface layer of resin that was noticeably thicker than usual. As a result, the beam entering the sample is highly absorbed and the R_s component is much stronger than usual. This causes severe distortion of the band shapes, due to the refractive index contribution.

In spite of these complications, it was felt that the

diffuse reflection method might give sufficiently reproducible results to allow it to be used as a quantitative method, even if only rather empirically. To test this, we measured spectra for eighteen specimens cut from prepreg Batch 1010. For analysis, they were converted with the use of both Kubelka-Munk and absorbance units. Indices were calculated by peak integration in the same way as for transmission and ATR, and the results are given as Experiment 1 in Table III. The reproducibility is very poor, with relative standard deviations as high as 15%. Although the results obtained with the use of absorbance units are somewhat more reproducible, the Kubelka-Munk transformation gives indices closer to those obtained in transmission. Examination of the spectra indicated that the main cause of the poor reproducibility was band shape variations due to front-surface reflection effects.

It has been demonstrated[23–25] that the effects of fiber orientation and front-surface reflection can be reduced if the sample is covered with a layer of powdered KBr. This solution was tried with the Narmco prepreg. It was found that an opaque layer of KBr reduced the spectral contrast considerably, and since the contrast was low to begin with, because of the strong absorption of the carbon fibers, the resulting spectra were unusable for quantitative work. Instead, two other approaches were tried. In the first, the prepreg was covered with powdered KBr, which was then shaken off in order to leave a light sprinkling on the surface. In the second, the prepreg was covered with KBr, a piece of aluminum foil was placed on top, and the KBr was pressed into the resin by finger pressure. The sample was then inverted and tapped, to shake off excess KBr. Although the resulting spectra are of lower contrast than those obtained without KBr, they are more reproducible with respect to band shape, and the fiber orientation is found to have little or no effect on the quality. However, the band shapes are still somewhat different from those obtained in transmission, suggesting that front-surface reflection effects have not been eliminated.

To check the effects of these treatments on the quantitative analysis, we measured seven specimens from Batch 1014 for each of three conditions: no KBr, sprinkled with KBr, and KBr embedded. The results are given in Table III as Experiments 2, 3, and 4, respectively. The treatments with KBr do not improve the poor precision to any significant extent. Hence it would seem that diffuse reflection is of very limited use as a quantitative method for prepreg analysis. The main reason for this appears to be a variation in the thickness of the surface layer of resin, which results in a varying contribution of front-surface reflection effects in the spectrum.

CONCLUSIONS

Of the three methods tested for quantitative IR analysis of carbon-epoxy prepreg, the best is attenuated total reflection using a 45° germanium prism. This method involves practically no sample preparation and gives the most reproducible results. The more common transmission method, involving a cast film of resin, is subject to substantial error caused by nonuniform film thickness, which can only be avoided by very careful and time-consuming sample preparation. Diffuse reflection, although easy to use, appears unsuitable for quantitative work, because of complications caused by front-surface reflection effects.

1. G. R. Thomas, B. M. Halpin, J. F. Sprouse, G. L. Hagnauer, and R. E. Sacher, Nat. SAMPE Symp. Exhib. [Proc.] **24,** 458 (1979).
2. H. S. Chu and J. C. Seferis, Polym. Compos. **5**, 124 (1984).
3. R. L. Levy, D. L. Fanter, and C. J. Summers, J. Appl. Polym. Sci. **24,** 1643 (1979).
4. A. Gupta, M. Cizmecioglu, D. Coulter, R. H. Liang, A. Yavrouian, F. D. Tsay, and J. Moacanin, J. Appl. Polym. Sci. **28,** 1011 (1983).
5. H. Dannenberg and W. R. Harp, Anal. Chem. **28,** 86 (1956).
6. C. A. May, D. K. Hadad, and C. E. Browning, Polym. Eng. & Sci. **19,** 545 (1979).
7. M. Cizmecioglu, S. D. Hong, J. Moacanin, and A. Gupta, Polym. Preprints **22,** 224 (1981).
8. D. Crozier, G. Morse, and Y. Tajima, SAMPE J. **18(5),** 17 (Sept./Oct. 1982).
9. J. L. Koenig, Anal. Chem. **36,** 1045 (1964).
10. T. Hirschfeld, Anal. Chem. **51,** 495 (1979).
11. T. Hirschfeld, "Quantitative FT-IR: A Detailed Look at the Problems Involved," in *Fourier Transform Infrared Spectroscopy: Application to Chemical Systems,* J. R. Ferraro and L. J. Basile, Eds. (Academic Press, New York, 1979), Vol. 2, Chap. 6, pp. 193–242.
12. N. J. Harrick, *Internal Reflection Spectroscopy* (Harrick Scientific Corporation, Ossining, New York, 1979).
13. F. M. Mirabella, Appl. Spectrosc. Rev. **21,** 45 (1985).
14. M. P. Fuller and P. Griffiths, Anal. Chem. **50,** 1906 (1978).
15. P. R. Griffiths and M. P. Fuller, Adv. Infrared Raman Spectrosc. **9,** 63 (1982).
16. J. M. Chalmers and M. W. Mackenzie, Appl. Spectrosc. **39,** 634 (1985).
17. P. R. Young, B. A. Stein, and A. C. Chang, Nat. SAMPE Symp./Exhib. [Proc.] **28,** 824 (1983).
18. P. R. Young and A. C. Chang, Nat. SAMPE Tech. Conf. **16,** 136 (1984).
19. K. C. Cole, D. Noël, and J.-J. Hechler, Nat. SAMPE Symp./Exhib. [Proc.] **30,** 624 (1985).
20. K. C. Cole, D. Noël, and J.-J. Hechler, Proc. Soc. Photo-Opt. Instrum. Eng. **553,** 114 (1985).
21. K. C. Cole, C. Lehto, and M. Yuhasz, Proc. Soc. Photo-Opt. Instrum. Eng. **665,** 243 (1986).
22. P. J. Brimmer, P. R. Griffiths, and N. J. Harrick, Appl. Spectrosc. **40,** 258 (1986).
23. M. T. McKenzie, S. R. Culler, and J. L. Koenig, Appl. Spectrosc. **38,** 786 (1984).
24. S. R. Culler, M. T. McKenzie, L. J. Fina, H. Ishida, and J. L. Koenig, Appl. Spectrosc. **38,** 791 (1984).
25. M. T. McKenzie and J. L. Koenig, Appl. Spectrosc. **39,** 408 (1985).

Reprinted with permission from *Materials Evaluation,* Vol. 41(3), pp. 302-308 (March 1983).

Nondestructive Thickness Measurement of Phosphate Coatings by Infrared Absorption

by T. M. Mansour

Abstract

The current method for determining the zinc phosphate coating thickness on steel is a gravimetric weigh/strip/weigh technique, which measures the coating weight per unit area. Because of the destructive nature of the method, it is used on test coupons that are processed in the phosphating bath along with the actual parts. Besides being destructive, this method is tedious, subject to weighing errors, and has limited utility for real-time production process control. This report presents the development and evaluation of a nondestructive method and instrument for monitoring phosphate coating thickness. The method is based upon the spectral absorption of infrared (IR) energy by the phosphate ion. The instrument provides accurate and instantaneous coating thickness readings and is applicable to test coupons as well as actual parts. It can be used in a spot measurements mode or in a recording mode to generate a thickness profile record and a measure of the coating uniformity. Data provided by this new instrument were found to be in good agreement with x-ray fluorescence data as well as the gravimetric coating weight method.

T. M. (Tony) Mansour is a staff engineer in Ford Motor Company's Engineering and Research Staff. He is a project leader in the Product Quality and Inspection Technology group. He has been involved in the development and implementation of nondestructive inspection technologies with emphasis on ultrasonics, acoustics, and optics to solve problems related to material properties and product quality. He is currently involved in the development of a microprocessor-based signal analysis and characterization scheme for use in ultrasonic weld inspection and in acoustic emission weld process monitoring. He graduated from Eastern Michigan University in 1958, majoring in physics and mathematics, and did graduate work at Wayne State University. He has written several papers and published in the journals of ASNT, SAE, and ASTM. He is an active ASNT member and is the Past Chairman of the Detroit Chapter.

INTRODUCTION

The phosphating process has long been used by the metal finishing industry for coating a wide variety of products.[1] Parts are sprayed or immersed in an aqueous mixture of phosphoric acid, zinc phosphate, and oxidizing agents, whereupon their surfaces acquire by chemical conversion a crystalline coating of zinc phosphate in a relatively short time. This coating, which is now an integral part of each surface, provides a measure of corrosion protection and also serves as an excellent base for impregnation with paints, lacquers, and enamels. The protection provided by the phosphate coatings results from the improvement of paint adhesion, restriction of moisture penetration, and an inhibiting of the spread of underfilm corrosion. Thus, there is great interest in test methods for quality assurance of phosphate coatings, particularly in automotive applications.

The quality of the phosphate coating is generally determined by the performance of phosphated and painted panels in salt spray tests. Various studies have been performed[2,3] to delineate phosphating process variables that significantly affect the coating quality. These include cleanliness of the substrate steel surface, maintenance of a certain constancy of the phosphating bath constituents and temperature, and residence time of the workpiece in the phosphating atmosphere. Other studies have been performed[3,4] to identify the phosphate coating's measurable parameters, which correlate with its quality performance. These include thickness, uniformity, porosity, and adhesion. There are currently no widely accepted instruments for real-time measurement of any of these coating parameters on the production line.

In the automotive industry, the common practice for quality monitoring of phosphated parts (car bodies) involves suspending test coupons with the actual parts as they enter the phosphating bath, then removing these coupons for subsequent testing in the laboratory. A universally performed test on these coupons is the determination of the coating weight.[5] This test, a chemical/gravimetric (weigh-strip-weigh) technique, provides the coating weight per unit area. Among the disadvantages of this method are the following: It is destructive, indirect, time-consuming, and the results are subject to operator variances. Additionally, it provides merely a single average value for the distribution of the phosphate coating over a relatively large area with no information on uniformity of deposition. An x-ray fluorescence (XRF) method[6] was evaluated for the nondestructive thickness measurement of phosphate coatings and found feasible. However, the slowness of the method and the regulatory constraints associated with the use of radioactive sources inhibited its acceptance by the manufacturing operations as a production environment measuring tool. Another phosphate coating parameter test was recently developed and reported by Zurilla.[7] This is an electrochemical test that provides a measure of the relative porosity of the coating. The porosity data has been found to correlate with the performance of phosphated and painted panels in salt spray testing.[8] However, the test cell design and partially destructive nature of this method restricts its application on actual parts in the production operations.

This paper presents a nondestructive infrared (IR) absorption Technique for measuring phosphate coating thickness. The technique has been incorporated in a specifically developed instrument for making the measurement instananeously and without physical damage on both test panels and actual parts. A description of the instrument, its calibration, and its utility is presented.

The relatively high speed of the instrument's response permits its use in either single spot measurement mode or continuous scanning and recording mode. The latter mode facilitates obtaining a profile of the coating thickness and a measure of its uniformity. In the following sections, the theory of the technique as well as instrument design considerations are discussed. Experimental data obtained on test panels with the prototype instrument and correlation with the gravimetric method are presented.

Note: Because of the universal use of coating weight units (mg/ft²) in the coating industry, the thickness data in this paper is expressed in these units. For zinc phosphate, the coating thickness to coating weight conversion factor is 1 microinch—4 mg/ft² or 1 μm—1.7 g/m.²

During the course of this development effort, the author became aware of a similar work on an IR method and instrument for phosphate coatings by Cheever.[9] Although the basic technical approach in both cases is the same (i.e., utilizing the characteristic spectral absorption of IR energy by the phosphate ion), Cheever's data showed that the optical scheme utilized and resultant instrument are applicable only to coating thicknesses above 1075 mg/m² (100 mg/ft²). The data also showed that the readings are affected by changes in the base material reflectivity. The capabilities to cover the full thickness range (0 to 3760 mg/m² or 0 to 350 mg/ft²) as well as minimizing or eliminating the effect of substrate reflectivity are two very desirable features for production process monitoring and quality inspection. As will be shown in the following sections of this paper, these features were considered in the optical scheme selection and incorporated in the instrument design.

THEORY

The approach for the measurement technique reported here is based upon IR spectral absorption of the phosphate ion and Beer and Lambert's laws, commonly used in analytical spectroscopy. When a beam of electromagnetic radiation passes through a homogeneous, isotropic material, some of the energy is absorbed at specific wavelengths due to interaction with molecular vibrations. The vibrating molecules of the material produce an alternating dipolar electric field whose magnitude changes at a frequency equal to the vibrational frequency. The interaction occurs between this dipolar electric field and the electrical component of the electromagnetic radiation, giving rise to absorption. Thus, the absorption depends upon the nature of the material and is a function of the wavelength. In many instances, the spectral position of the absorption for a given material can be reasonably predicted by calculations.

Investigators in the field of spectroscopy have observed, verified, and tabulated the characteristic absorption spectra for a vast number of organic and inorganic materials.[10] The spectral absorption for zinc phosphate has been found to occur in the 9 to 11 μm region of the IR spectrum,[11] and this absorption activity has been attributed to the phosphate radical in its crystalline environment of zinc phosphate. The effect is a crystalline field splitting of a threefold degenerate stretching vibration resulting in three absorption maxima. It was determined experimentally that the production-type phosphate compound (containing some iron as well as zinc) displayed the three absorption peaks. (See the spectrograms in Fig. 1.)

With the characteristic spectral absorption band identified for a given material, the thickness of a moderately thin film can be determined by applying the Beer-Lambert absorption law.[12,13] If $I_{o\lambda}$ is the original intensity (at a specific wavelength λ) of an energy beam entering a given material, and I_λ is its intensity after passing through a thickness x, then:

$$I_\lambda = I_{o\lambda} e^{-\alpha\lambda x} \quad (1)$$

or

$$\ln (I_\lambda / I_{o\lambda}) = -\alpha_\lambda x, \quad (2)$$

where α_λ is the spectral absorption coefficient of that material.

Two wavelengths are to be utilized to make a difference-measurement: one at an absorption peak and the other just outside the absorption band such that its value of α_λ is zero. A source with a broad spectrum is to be used, having equal intensity at the two wavelengths. Let $I_{\lambda 1}$ be the transmitted intensity outside the absorption band and $I_{\lambda 2}$ be the trans-

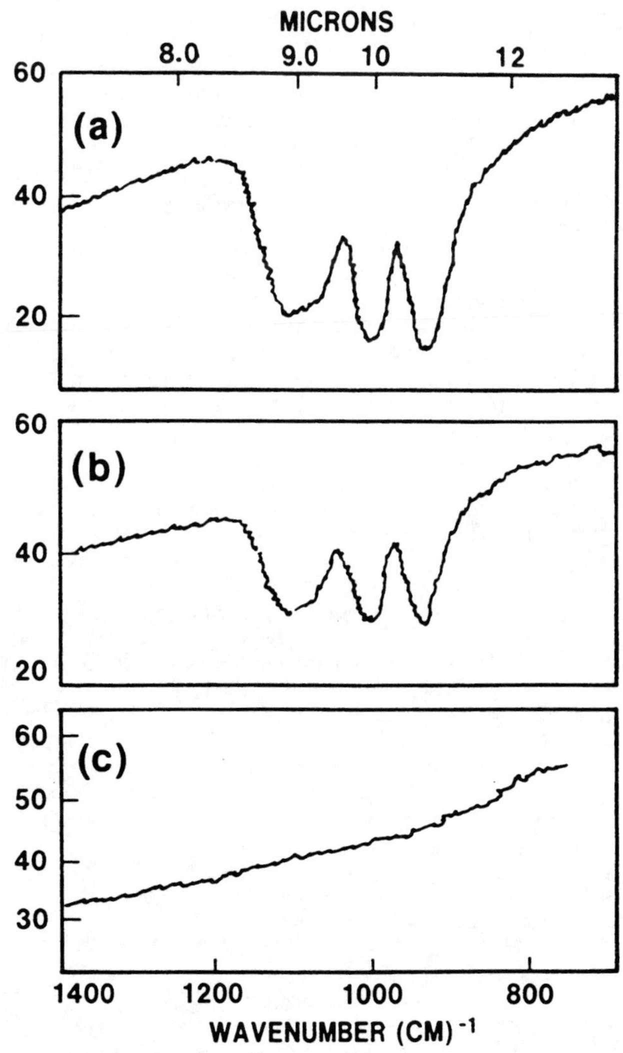

Figure 1—Spectral intensity distribution curves for two phosphated panels (a and b), and a bare panel (c) (measured; traced from strip chart).

mitted intensity at the absorption peak. Then one can use the well-known relationship

$$x = k \log R, \tag{3}$$

where

$$R = I_{\lambda 2}/I_{\lambda 1}, \tag{4}$$

$$k = 1/\alpha_{\lambda 2}, \tag{5}$$

and

$$\alpha_{\lambda 2} = \alpha_{\lambda} \text{ at the absorption peak.} \tag{6}$$

Equation 3 is the basis of IR thickness gaging. The present form is particularly useful in a practical sense because it is a linear expression for x, the phosphate thickness. A straight-line calibration "curve" or a linear instrumental calibration can be generated using Equation 3.

Electronically, components such as intensity detectors and logarithmic amplifiers can be used to build an instrument to implement the theory.

FEASIBILITY EXPERIMENTS

The initial set of test samples included five (identical) pairs of 100 by 300 mm (4 by 12 in.) zinc-phosphated steel panels, which had been specially coated to cover the coating thickness range normally encountered in the automotive body parts phosphating process. Coating weight data were obtained by the conventional (destructive) gravimetric method on a 100 by 100 mm (4 by 4 in.) section of one panel from each pair for reference. An additional set of coating thickness data was generated for these test panels by the x-ray fluorescence method. The second panel from each pair was run on a laboratory IR spectrophotometer to obtain an IR spectral transmittance/reflectance distribution curve for the phosphate coatings. Two of these distribution curves are shown in Figs. 1a and 1b. A similar spectral reflectance distribution curve was also generated for an uncoated steel test panel (see Fig. 1c). The curves for the coated panels show a strong absorption band with spectral location and features nearly identical to that tabulated for the phosphate radical discussed in Reference 11. The absorption band is between 8.5 and 11 μm, with three distinct absorption maxima at 9.1, 10.0, and 10.8 μm. The curve for the uncoated panel shows that the steel substrate reflectivity varies moderately but linearly with wavelength, with no localized absorption in the spectral region of interest.

The spectral distribution curves generated above provided the data required for computing the intensity ratios ($I_{\lambda 2}/I_{\lambda 1}$) of Equation 4 in the previous section. The initial wavelength selection was 8.35 and 9.10 μm for λ_1 and λ_2, respectively. The intensity ratios for the five panels are tabulated in Table 1, along with the gravimetric and x-ray fluorescence data previously obtained. The log of these intensity ratios (ln R) is plotted against the gravimetric coating weight data in Fig. 2 and against the x-ray fluorescence coating thickness data in Fig. 3. These plots show that, at the selected wavelengths, ln R does indeed vary linearly with coating thickness (weight), within this thickness range. Thus, it is possible to use Equation 3 to generate a usable instrument.

TABLE 1 Calculated IR Intensity Ratio of $Zn_3(PO_4)_2$ from Spectrophotometric Curves.

Panel #	Gravimetric Coating Weight (mg/m²)	Gravimetric Coating Weight (mg/ft²)	X-Ray Fluorescence Counts	IR Absorption Calculated ln R
A1	1270	118	250	0.367
B1	1732	161	343	0.425
C1	2130	198	438	0.675
D1	2776	258	499	0.835
E1	3475	323	571	1.155

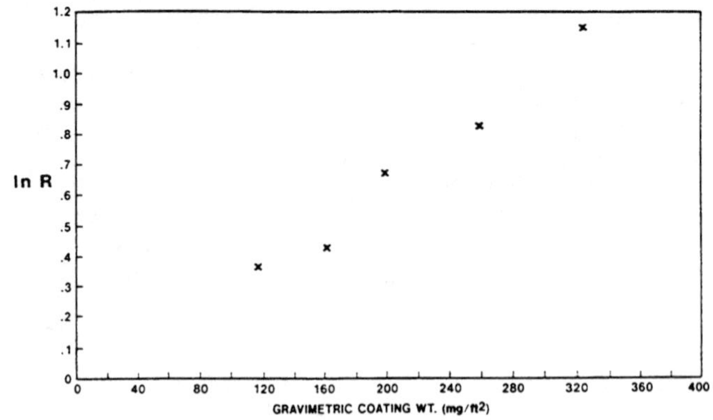

Figure 2—Calculated ln R *vs. gravimetric coating weight data (ln* R *is the material log intensity ratio).*

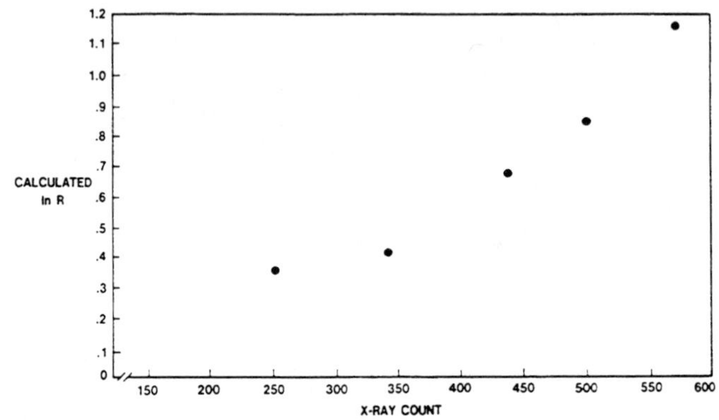

Figure 3—Calculated ln R *vs. x-ray fluorescence data (ln* R *is the material log intensity ratio).*

GENERATION OF PROTOTYPE INSTRUMENT

Design Considerations

Figure 4 shows the basic geometry and optical components for a phosphate coating thickness measuring system based upon this IR absorption technique. It includes an IR source S modulated by a chopper C. Lenses L_1 and L_2 image the source onto the test specimen. After transversing the phosphate coating, the beam is reflected by the substrate and traverses the coating a second time on its path to the receiving optics to be registered by detectors D_1 and D_2. Lens system L_3, L_4 and L_5 and beam splitter B/S image the detectors onto the test surface such that the detection image is somewhat larger than the source image. This would allow a small source to be used in such a way that some variation in sample distance could be tolerated. "OFF-BAND" filter F_1 in front of detector D_1 is a band-pass filter transmitting a narrow wavelength band with a central wavelength λ_1 just outside the phosphate spectral absorption region. "ONBAND" filter F_2, in front of detector D_2, transmits only a spectral band with a central wavelength λ_2 within the region of strong absorption by the phosphate coating. The signals from the detectors D_1 and D_2 should be proportional to $I_{\lambda 1}$ and $I_{\lambda 2}$ of Equation 4 and, when fed into a log ratio amplifier, would output a signal proportional to ln R. This layout was used for preliminary signal calculations and optical components selection.

The signal calculation for this (and any) optical layout includes the following well-known integral I:

$$I = \int T(\lambda)\, Fi(\lambda)\, \rho(\lambda)\, S(\lambda)\, D(\lambda)\, d\lambda \tag{7}$$

where

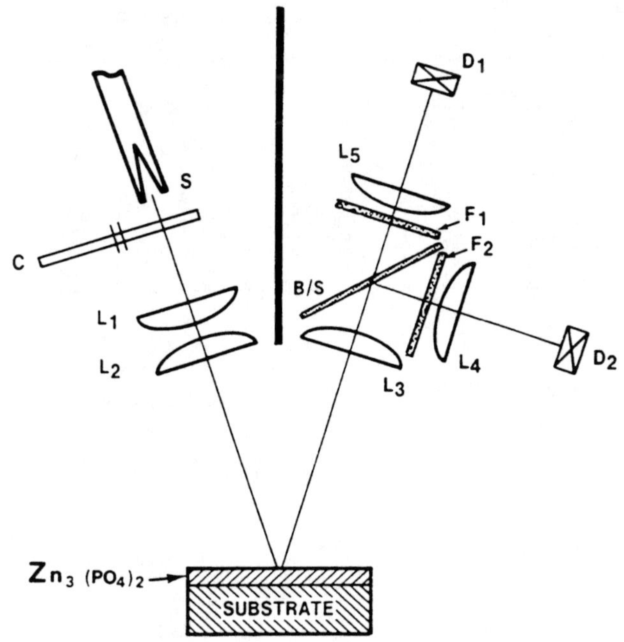

Figure 4—Basic optical layout of IR instrument.

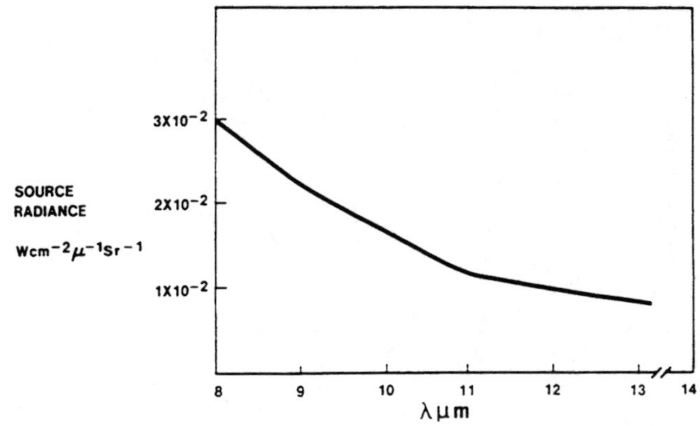

Figure 5—Spectral radiance of IR source.

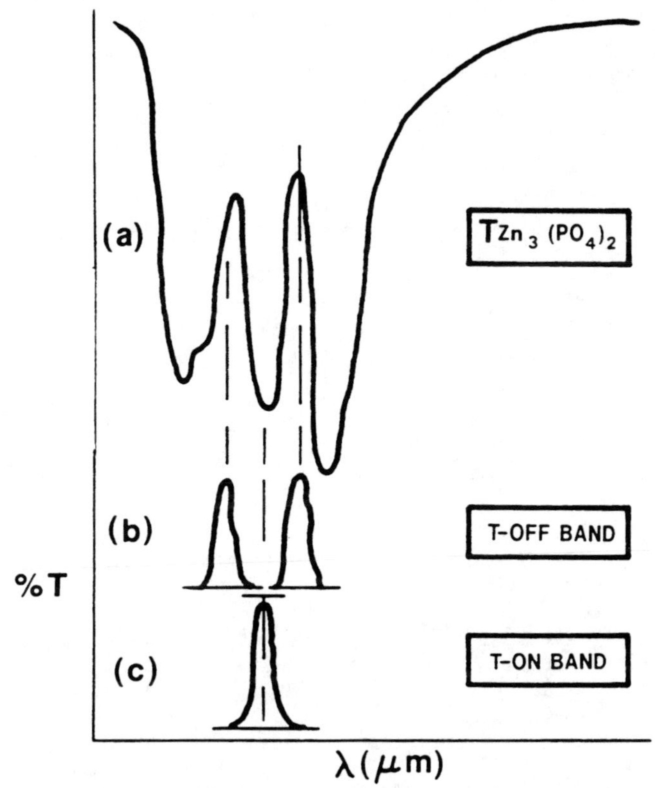

Figure 6—Spectral characteristics of "on-band" and "off-band" filters for phosphate thickness measurement.

$T(\lambda)$ is the transmission of the phosphate coating (2 passes);
$Fi(\lambda)$ is the transmission of the interference filter;
$\rho(\lambda)$ is the reflectivity of the substrate metal;
$S(\lambda)$ is the radiance of the source (W cm^{-2} μm^{-1} sr^{-1}); and
$D(\lambda)$ is the response of the detector.

Equation 7 was used to determine signal levels from components that could be procured for the design of the instrument. The results of the subsidiary experiments and calculations showed that available components were adequate to build a workable prototype instrument that would be able to make accurate measurements at a rapid rate.

A source was chosen to approximate a black body at 700°K. Its spectral radiance is shown in Fig. 5. Both the radiance of the source and the reflectance of the substrate as shown in Fig. 1c have gradual slopes in the vicinity of the phosphate absorption band. The effect of these conditions were minimized by the choice of filters. The detectors chosen have equal values of $D(\lambda)$ at the wavelengths to be used.

The effects of the slopes in source radiance and substrate reflectivity can be minimized by proper filter design and selection. This is best achieved by having the "off-band" measurement made symmetrically about the "on-band" measurement (or vice versa). In the case of the characteristic absorption band for $Zn_3(PO_4)_2$ (Fig. 6a), this can be approximated by a double-peaked "off-band" filter as shown in Fig 6b, working with an "on-band" filter as shown in Fig. 6c. While a double-peaked interference filter is difficult to make, its behavior can be simulated by a negative filtering technique. For this purpose, a broadband, flat-response filter covering the two "off-band" peaks of Fig. 6b would be used. The signal from this broadband filter is Sb. If the signal Sn from the narrow-band, "on-band" filter of Fig. 6c is subtracted from Sb, the resultant signal will be equivalent to an "off-band" signal from a double-peaked filter. The "on-band" to "off-band" signal ratio R is now given by

$$R=(k'\,Sn)/(Sb-k'Sn), \quad (8)$$

where k' is an amplification factor set so that when measurements are made on the uncoated base material, the ratio R is equal to unity. For this calculation, two filters were selected, both of which peaked at 10 μm. The band widths at half maximum were 0.2 and 0.8 μm for Sn and Sb, respectively.

For further calculations, the transmittance of a phosphate coating of 213 mg/ft^2 is shown in Fig. 7. For other thicknesses, this curve was modified by a scale factor based on the Beer-Lambert absorption law.[12,13]

Values of the integral I of Equation 7 were then calculated, with all of the assumptions and component selections given above, for coating thicknesses of 0, 1145, 2290, 3440, and 4580 mg/m.^{2} The calculated values are listed in Table 2. These intensity (I) values are all within a range easily measurable with procurable components. Thus, it was determined that the prototype instrument would be feasible.

The calculation in Table 2 also included two base spectral reflectivities, as shown in Fig 8, in addition to that previously determined for cold rolled steel. In one case, the spectral reflectivity is constant, and in the other it changes with wavelength at a greater rate than that of cold rolled steel. This was done to evaluate the effect of changes in the base reflectivities upon ln R. In Table 2, the calculated integrals are labeled I_b and I_n and correspond to the signals Sb and Sn as discussed

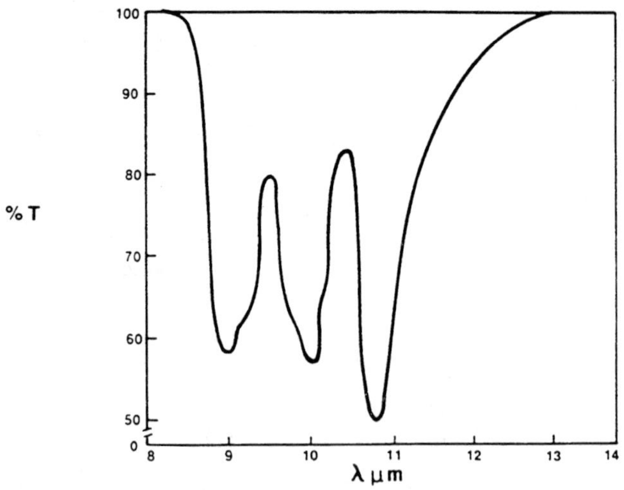

Figure 7—Calculated spectral transmission zinc phosphate coating at 213 mg/ft.[2]

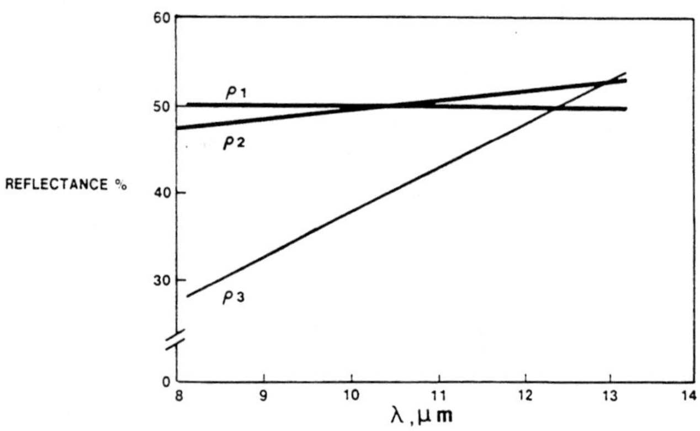

Figure 8—Spectral reflectivities of three substrates.

prior to Equation 8. The amplification factor k' was computed from Equation 8 to be 0.778 by using $R=1.0$ at zero-thickness of phosphate. Then, using this value of k', the other values of ln R were calculated from the intensities in Table 2 by means of Equation 8. The resulting values of ln R are shown in Table 3.

The values of ln R in Table 3 are expected to be essentially independent of the spectral reflectivity of the base and approximately proportional to the thickness of the $Zn_3\ (PO_4)_2$ coating. An inspection of these values shows the largest variation of ln R with base reflectivity ρ to be 5 percent for the 106.5 mg/ft^2 coating thickness case. A plot of ln R versus Zn_3 $(PO_4)_2$ thickness for the cold rolled steel substrate (ρ_2) case is shown in Fig. 9. The values of ln R fall on a straight line passing through zero. The low sensitivity to base reflectance changes shows that the instrument would be accurate for all steels to be encountered.

Building Prototype Instrument

The instrument design considerations in the preceuing section were performed cooperatively with H.S.S., Inc., Bedford, MA, an optical instruments manufacturing company. These design considerations, along with certain desirable and functional instrument features, served as the basis for developing the specifications for a laboratory prototype IR phosphate coating thickness monitor. Figure 10 shows such an instrument fabricated by H.S.S. to these specifications. H.S.S. used data and calculations provided by the author to choose the optimum optical components. The actual filter combination used in the present prototype IR instrument comprised two narrow-band filters centered on two adjacent spectral bands, in one of which there is strong absorption while in the other there is little or no absorption. In the subsequent instrument evaluation tests, this filter combination was found to provide sufficient measurement sensitivity and independence of the substrate reflectivity.

TABLE 3 In R Calculated for Five Coating Weights and Three Base Reflectivities.

$Zn_3\ (PO_4)_2$ Thickness				
mg/m^2	mg/ft^2	ρ_1	ρ_2	ρ_3
0	0	0	0	0
1145	106.6	0.1335	0.1334	0.1274
2290	213	0.2694	0.2681	0.2635
3340	319.5	0.4023	0.4015	0.3966
4580	426	0.5354	0.5347	0.5292

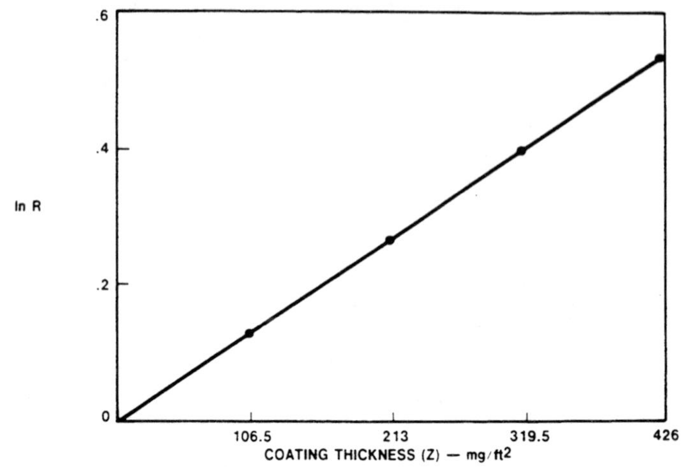

Figure 9—Calculated ln R *vs. coating thickness of substrate reflectivity* ρ_2.

The instrument comprises two modules, an optical head, and an electronic signal processing and readout unit. The optical head illuminates a 6.37 mm (¼ in.) diameter test area. The electronic signal processing unit provides a digital and an analog output proportional to ln R, with a time constant of 0.16 s. There are two calibration controls: a "zero" control for setting the reading for an uncoated substrate test panel and a "calibrate" control for adjusting the reading to a reference-coated specimen of known coating thickness. Sufficient gain adjustment was incorporated to facilitate calibrating the instrument to read directly in mg/ft^2 units.

The instrument can be calibrated to provide a direct digital readout of the coating thickness in the comon units (mg/ft^2) as provided by the gravimetric technique. The analog output

TABLE 2 Calculated Values of the Integral I (W. cm^{-2}sr^{-1}) for Three Substrate Reflectivities (ρ_1,ρ_2,ρ_3).

ZnP_2O_4 Thickness		ρ_1		ρ_2 (cold rolled steel)		ρ_3	
mg/m^2	mg/ft$_2$	I_b	I_n	I_b	I_n	I_b	I_n
0	0	3.45×10^{-3}	9.27×10^{-4}	3.41×10^{-3}	9.16×10^{-4}	2.63×10^{-3}	7.09×10^{-4}
1145	106.5	2.72×10^{-3}	6.87×10^{-4}	2.74×10^{-3}	6.88×10^{-4}	2.12×10^{-3}	5.33×10^{-4}
2290	213.0	2.23×10^{-3}	5.23×10^{-4}	2.22×10^{-3}	5.17×10^{-4}	1.71×10^{-3}	4.00×10^{-4}
3440	319.5	1.83×10^{-3}	3.94×10^{-4}	1.81×10^{-3}	3.89×10^{-4}	1.39×10^{-3}	3.02×10^{-4}
4580	426.0	1.50×10^{-3}	2.97×10^{-4}	1.48×10^{-3}	2.90×10^{-4}	1.14×10^{-3}	2.27×10^{-4}

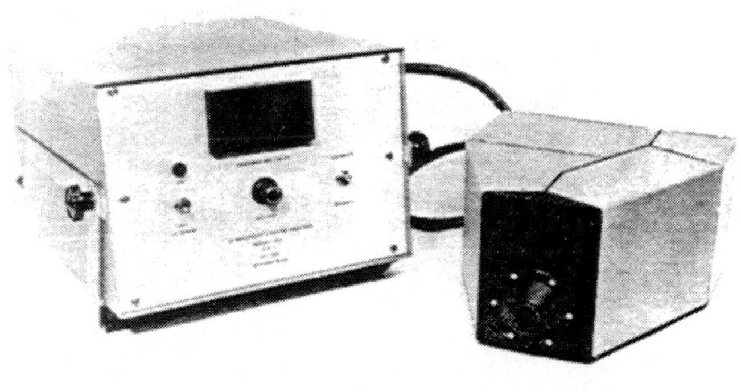

Figure 10—Prototype IR phosphate thickness monitor.

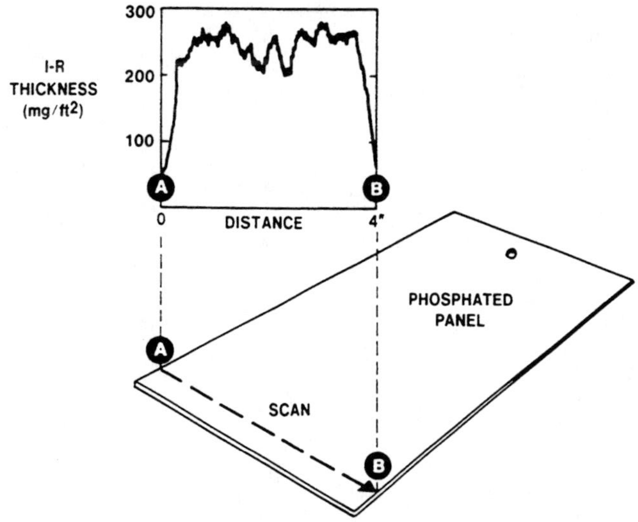

Figure 11—Typical IR scan of a phosphated test panel.

was incorporated to facilitate recording thickness profiles of the phosphate coating as the test part is scanned. The measurement range of the instrument was specified at 0 to 4300 mg/m^2 (0 to 400 mg/ft^2) to cover the full range of coating thickness normally encountered in the phosphating process.

Calibration of the Instrument

Because the instrument is not an absolute thickness measuring device, its calibration requires the use of reference-coated specimens whose coating thickness is known. In the absence of any nondestructive method for predetermining the coating thickness, the gravimetric technique is used in generating the reference calibration specimens. However, in the use of this technique for such a purpose, some precautions are required to ensure that the candidate reference panels have been uniformly coated. These precautions are necessitated by the fact that the gravimetric analysis is generally performed on a large (100 by 100 mm or 4 by 4 in.) section to provide an average coating weight per unit area, whereas the IR instrument views a very small (6.37 mm or ¼ in. diameter) area. The screening of candidate test panels for coating uniformity is best achieved by using the scanning feature of the IR instrument. This requires an external drive to move the specimen (at 2 cm/s scanning rate) while the analog output signal of the IR instrument goes to the Y-axis and a distance (or time) signal goes to the X-axis on a recorder. A typical scan is shown in Fig. 11. Several scans should be made for each candidate test panel to record the coating profiles over an area somewhat larger than the section to be used for the gravimetric analysis. The selection is then made from those panels whose coating thickness profiles show reasonably good uniformity within the scanned area. A portion of the selected panel adjacent to the section to be gravimetrically analyzed can then be used for the IR instrument calibration.

The selection procedure explained above was used to calibrate the IR instrument. A bare panel represented zero-thickness of phosphate, and a panel coated with 2800 mg/m^2 (260 mg/ft^2) of phosphate (as determined gravimetrically) provided the other calibration point for the linear calibration of ln R versus thickness.

PERFORMANCE VERIFICATION

A library of phosphated test panels was collected for correlating the coating thickness readings of the prototype IR instrument with the gravimetric coating weight data. The samples were obtained from different phosphate suppliers (and had different steel substrates). The library comprised 31 test panels that were divided into three groupings because they were received in three installments. The gravimetric data received with the test panels was thus obtained by different operators. Sample set #1 was the five panels used in the initial experimental work in this development (see the section on "Feasibility Experiments"). Set #2 included eight panels with a coating weight range of 540 to 4300 mg/m^2 (50 to 400 mg/ft^2) in approximately 540 mg/m^2 steps. Set #3 comprised nine panel pairs, where the two panels of each pair were prepared under the same conditions as a third panel, which had been used for obtaining the gravimetric data.

Preliminary spot measurements were made on a few panels from the library of samples to evaluate the coating uniformity. These initial spot measurements showed variations, which in some cases were quite large, on the same test panel. These localized variations and the fact that the gravimetric coating weight data represented an average value for a 100 by 100 mm (4 by 4 in.) coating area established the need to have the IR readings also represent an average for a large area.

IR averages over a large area were procured by making three IR scans of each panel as in Fig. 11 and generating an average IR thickness reading from these scans. The IR data generated for the three sets of test panels and the reference gravimetric coating weight data are shown in the correlation graph of Fig. 12. The correlation coefficient between the IR readings and the gravimetric data is 0.90. There appears to be no deterioration of the accuracy down to 50 mg/ft.

DISCUSSION OF RESULTS

The experimental work and resultant prototype instrument confirms Cheever's earlier work[9] by showing that phosphate coating thickness can indeed be measured nondestructively, rapidly, and reliably with an instrument based upon the IR absorption technique. The technique provides data that correlate with the destructive gravimetric method. Because of localized coating thickness variations, good correlation is obtained when a large area is measured (scanned) and averaged with the IR instrument. The need for scanning to correlate with the gravimetric data might be eliminated by using an IR aperture (test area) larger than the 6.37 mm (¼ in.) diameter aperture of the prototype instrument. However, the combination of small aperture and scanning provides better resolution of coating thickness variations within a test specimen and a measure of the coating uniformity that is unobtainable by any other technique. Such a measure of coating uniformity, which includes the lowest coating thickness reading for the test specimen, could prove very useful in predicting the performance of the coating in the salt spray performance tests. This instrument provides the opportunity to acquire such data. One notes from Fig. 12 that this instrument provides a correlation that extends down to 540 mg/m^2 (50 mg/ft^2) so that measurements can be made in that range as well as higher, which represents an improvement of a factor of 2 over the previous work.

The characteristics of the IR absorption band for the phosphate radical are such that a number of interference filter com-

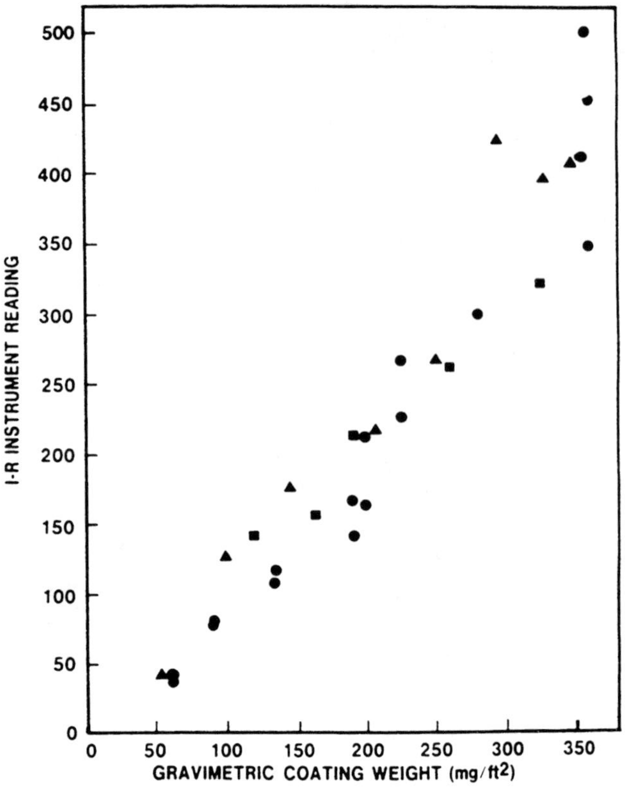

Figure 12—Prototype IR instrument readings vs. gravimetric coating weights.

bination choices are possible. The simplest choice (from an optical component availability and cost standpoint) is one where the whole spectral interval (8.5 to 11 mm) bracketing the three narrow absorption bands is used for the "on-band" filter. Such a choice, however, would result in reduced measurement sensitivity and accuracy at the low end of the thickness range. An IR instrument, developed elsewhere,[9] uses this type of filter and was found ineffective for measurements below 1975 mg/m^2 (100 mg/ft^2). Another disadvantage of this broadband filter choice for the "on-band" signal is that the measurements would be significantly affected by variations in the spectral reflectivity of the substrate. The filter combination employing the negative filtering technique as described and used for the computations in the previous section would provide the maximum thickness measurement sensitivity and minimize the effect of the base material reflectivity. However, this choice would involve the use of specially fabricated (and thus more costly) interference filters and require greater signal amplification. In the present prototype IR instrument, two narrow-band filters centered on two adjacent spectral bands were used. The spectral positions of these bands were purposely chosen such that there is strong absorption in one and little or no absorption in the other. This filter combination was found to provide sufficient measurement sensitivity and independence of the substrate reflectivity in the evaluation tests.

The present prototype IR instrument was designed and fabricated for the primary purpose of proving out the measurement approach. A production-model version of this instrument would require repackaging for compactness and portability, as well as the addition of operational features and controls appropriate to the intended application.

CONCLUSION

An IR instrument has been developed for nondestructive thickness measurement of phosphate coatings on metal substrates. The measurement approach applies Beers-Lambert absorption laws to the characteristic IR absorption of the phosphate ion. A narrow-band laboratory model instrument was designed and fabricated incorporating two operational modes: a read-and-hold mode for localized spot measurement and a continuous-reading mode, which facilitates profiling the phosphate coating thickness on test parts. Instrument readings had a good correlation with the conventional gravimetric weigh/strip/weigh technique and covered a coating weight range of 0 to 4300 mg/m^2 (0 to 400 mg/ft^2).

With appropriate packaging and electronic modification, the instrument could be configured into a compact, portable unit for manual spot checks or into an automatic, continuously recording system for production-line monitoring on car bodies. The immediate usefulness of the instrument is in timely detection of changes in the phosphating process, through scheduled spot measurements. A more rigorous application would be in real-time, 100 percent production monitoring of coating thickness level and uniformity on car body parts. The nondestructive nature of the method and the scanning capability of the instrument present a potentially valuable new tool for the ongoing laboratory studies on phosphating process and corrosion protection improvement.

Acknowledgments

The author is indebted to the following Ford activities and individuals who contributed significantly toward this development effort: Ford Body and Assembly Division Paint Operations for sample procurement and for funding the design and purchase of the prototype instrument; J. Parsons of the Engineering and Research Staff for running the spectrophotometric curves; H. L. Chesney for supplying x-ray fluorescence data; and R. Zurilla, H. Greenwood, and E. P. Papadakis for valuable discussions. The author also acknowledges the contributions made by H.S.S., Inc., in the predesign calculations and fabrication of the prototype model instrument. This work was supported by E. P. Papadakis under a Ford Research program on product quality and inspection technology.

References

1. Burns, R. M., and W. W. Bradley, *Protective Coatings for Metals*, 1959, p 545. Reinhold, New York, NY.
2. Cheever, G. D., "Formation and Growth of Zinc Phosphate Coatings," *Journal of Paint Technology*, Vol. 39, No. 504, 1967, p 1.
3. Lakeman, J. B., D. R. Gabe, and M. O. W. Richardson, "The Physical Properties of Phosphate Coatings and their Correlation with Potential/Time Measurements," *Transactions of the Institute of Metal Finishing*, Vol. 55, 1977, p 47.
4. Machu, W., "Determination of Porosity on Nonconductive Phosphate Coatings on Iron," *Korrosion Metalschutz*, Vol. 20, 1944, p 6.
5. *Ford Laboratory Test Method*, "Determination of Phosphate Coating Weights—Chronic Acid Method," Ford Motor Co., Manufacturing Standards Department, FMCC Bldg., Dearborn, MI, Document MA-P#AQ1-1, Jan. 9, 1976.
6. Chesney, H. L., E. P. Papadakis, and J. Brinkerhoff, "Nondestructive Thickness Gaging Method for Thin Layers Using X-Ray Fluorescence," *Materials Evaluation*, Vol. 39, No. 8, 1981, p 726.
7. Zurilla, R. W., and V. Hospadaruk, "Quantitative Test for Zinc Phosphate Coating Quality," paper (#780187) presented at National Society of Automotive Engineers (SAE) Meeting, Detroit, MI, March 1978.
8. ASTM Standard B-117-73, "Standard Method of Salt Spray Testing," Part 27, 1976. American Society for Testing and Materials, Philadelphia, PA.
9. Cheever, G. D., "Nondestructive Determination of Zinc Phosphate Coating Weights Using Specular Reflectance Infrared Absorbance," *Journal of Coating Technology*, Vol. 50, No. 640, May 1978, p 78.
10. Chicago Society for Paint Technology, *Infrared Spectroscopy—Its Use in the Coating Industry*, 1969. Federation of Societies for Paint Technology, Philadelphia, PA.
11. Nyquist, R. A., and R. O. Kagel, *Infrared Spectra of Inorganic Compounds*, 1971, p 173. Academic Press, New York, NY.
12. *American Institute of Physics Handbook*, 3rd edition, Section 6, p 2, 1972. McGraw-Hill Book Company, New York, NY.
13. Optical Society of America, *Handbook of Optics*, 1978, Section 8, p 12. McGraw-Hill Book Company, New York, NY.

Reprinted with permission from *Applied Optics,* Vol. 27(17), pp. 3675-3678
(September 1, 1988). ©1988 Optical Society of America.

Photoresist thickness measurement using laser-induced fluorescence

Shigeharu Kimura and Katsumi Takami

A technique is proposed to measure the thickness of a small area of photoresist film on a substrate. This technique uses the fluorescence (peak wavelength, ~600 nm) induced in the film by a focused laser beam (wavelength, 497 nm). An initial calibration line is obtained by comparing several thicknesses and their related fluorescence. The film's thickness is then determined from its fluorescence signal. The laser spot size and the accuracy of this measurement have also been estimated.

I. Introduction

Photoresists are generally used to make circuit patterns in the semiconductor device fabrication process. The photoresist thickness is an important parameter in obtaining the desired patterns. Therefore, it must be controlled and measured. The deposition of the photoresist is carried out many times in the process, so that the film of the photoresist does not always coat a flat surface; i.e., in many cases it coats circuit patterns which have asperities (roughness) of the order of microns. There is a need for measuring the thickness of the film on such asperities.

Optical profilometers have been developed in many places.[1-13] They have the merits of noncontact and nondestructive testing, unlike the mechanical stylus scanning technique,[14] and generally provide high depth sensitivity in measuring a surface profile. When a film is deposited onto a flat surface substrate, such a profilometer can measure the thickness of the film on the substrate. However, if the surface covered by the film is rough, the profilometer cannot measure the thickness.

The thickness of a photoresist film on smooth and transparent glass has been measured by Takami *et al.*[15] using absorbance of light. If this technique is used in measurement of the thickness of the film on a substrate having asperities, difficulties will be encountered due to the opaqueness of the substrate.

Kolodner *et al.*[16] have developed a method for real-time etch-rate measurement and endpoint detection during reactive-ion etching of the photoresist film. This method has monitored the intensity of the fluorescence from the film excited with a green light and could measure the thickness of the film of a large area. Another technique using fluorescence for obtaining thermal imaging[17,18] of an integrated circuit has been described, where the dependency on temperature of the fluorescence from a doped photoresist is employed.

The aim of this work is to describe one possible technique for measuring the thickness of the photoresist film on asperities. The technique employed a focused laser beam and observed the fluorescence induced from the beam spot. To determine the thickness, initial calibration using a photoresist of known thickness is necessary. In this work, the feasibility of measuring the thickness of only a very small segment on a flat substrate is shown. Since the strength of the fluorescence is considered to be proportional to the volume in the beam spot, the technique should be able to measure the thickness on asperities.

II. Experimental

The experimental setup used to measure photoresist thickness is shown in Fig. 1. The thickness of an AZ1350J photoresist (Shipley)[19] was measured. This is a positive photoresist often used in semiconductor device fabrication.

An argon-ion laser beam with a 497-nm wavelength was used to induce fluorescence. The beam was expanded, then passed through an interference filter that eliminated light at wavelengths other than 497 nm. The beam was then reflected off a beam splitter and through an objective lens with a numerical aperture of 0.9. This lens focused the beam on a photore-

---

When this work was done both authors were with Hitachi, Ltd., Central Research Laboratory, Kokubunji, Tokyo 185, Japan; K. Takami is now with Hitachi Electronics Engineering Company, Ltd., Kamisato, Kodama-gun, Saitama, Japan.

Received 30 December 1987.
0003-6935/88/173675-04$02.00/0.

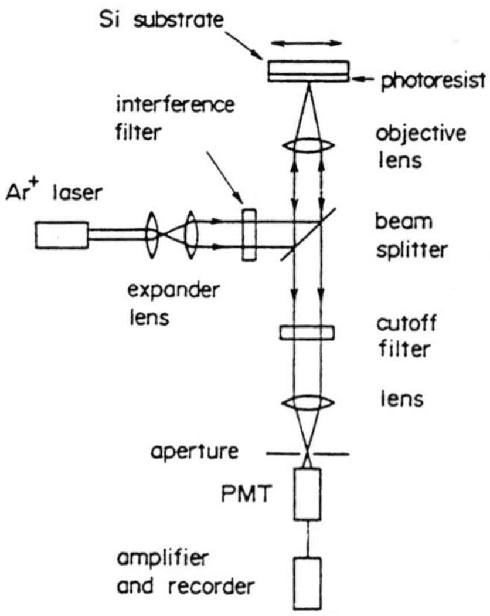

Fig. 1. Experimental setup.

sist covering a silicon substrate. The substrate was movable laterally at a speed of 40 μm/s.

The induced fluorescence from the photoresist was collected by the objective lens. It then passed through the beam splitter and a cutoff filter, which eliminated the wavelength light shorter than ~580 nm. In consequence, this filter removed the strong laser beam reflected off the silicon substrate that coincided with the fluorescence. The fluorescence then passed through a lens with a 50-mm focal length and an aperture. It was detected by a photomultiplier tube (PMT), amplified, and recorded. It should be noted that the aperture was not small enough for the optical arrangement to constitute a confocal system[11–13,20]; its role was only to obstruct background light.

III. Results and Discussion

The spectrum of the photoresist fluorescence induced by the laser beam is shown in Fig. 2. It was used to determine which cutoff filter should be used. This spectrum was obtained with a monochromator (G250, Nikon) and a PMT (R649, Hamamatsu). It was not calibrated. The small peak on the left side is the laser beam reflected off the substrate.

To obtain the best spectrum, various cutoff filters were placed in front of the monochromator. The cutoff filter that removed light with wavelength shorter than 580 nm was selected so that most of the fluorescence could pass through while making the intensity of the reflected laser beam as low as possible.

The actual procedures for gathering thickness data must be considered with this setup in mind. Procedurally, a calibration line was first obtained. It provided the relationship between the photoresist thickness and the fluorescence obtained with the experimental setup. To determine this line, the PMT output current proportional to the strength of fluorescence was measured for photoresists of various thickness on a silicon substrate. The corresponding absolute thicknesses were mechanically determined with a stylus scanning technique. The resulting calibration

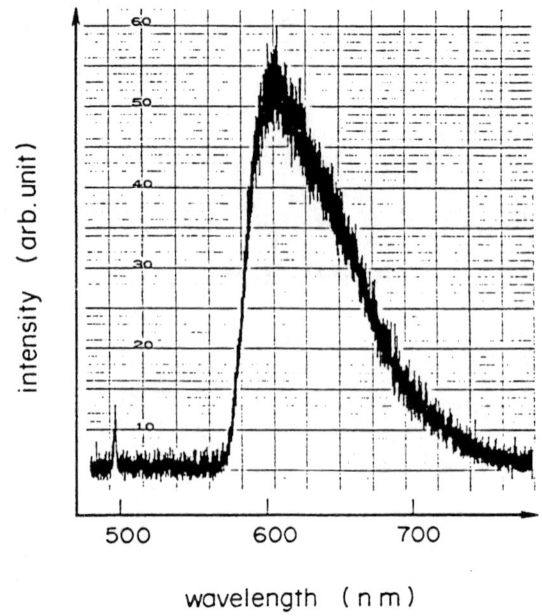

Fig. 2. Fluorescence spectrum for AZ1350J photoresist induced by a 497-nm wavelength laser beam.

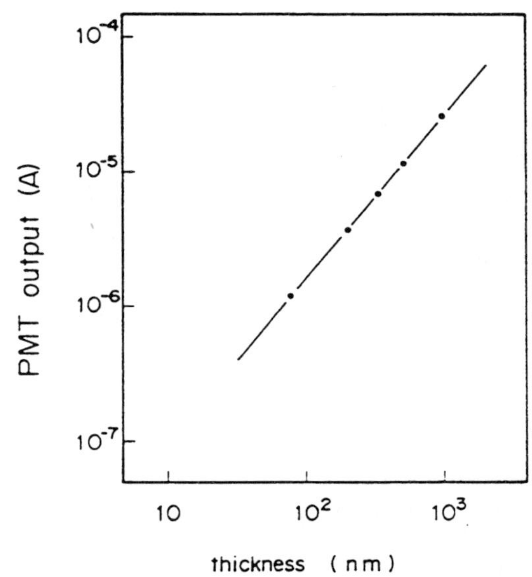

Fig. 3. Calibration line for photoresist thickness and PMT output current signal of fluorescence.

line is shown in Fig. 3. The total incident power of the laser beam on the photoresists was ~30 μW. Note that the photoresist area must be larger than that of the laser beam spot so that fluorescence can be induced at every point irradiated by the spot. If the area is smaller than the spot, a thickness measurement may be inaccurate.

There is no interference fringe signal in Fig. 3 unlike Ref. 16. The fringe is due to the interference between the fluorescence emitted directly and that reflected by the substrate and has an angular distribution.[21] Since the objective lens of a large numerical aperture was used in the experimental setup, the fluorescence up to a large angle about the optical axis was collected. Therefore, the detected fluorescence was the integral over the fringes. This might make the interference fringe signal disappear in Fig. 3.

Recorded PMT output signals on fluorescence were obtained when the focused laser beam successively irradiated two sets of tiny islands of photoresist (3×3 and $2 \times 2\ \mu m^2$) in a line. The distance between the central points of the islands was 8 μm. Results are shown in Fig. 4. Each peak corresponds to an island. The average value of the four peaks gave thicknesses of ~600 nm for the larger islands and 360 nm for the smaller ones by referring to the calibration line. The thickness of the photoresist mechanically measured using a large-area pattern was 850 nm. The experimentally obtained values were smaller than the values with the stylus scanning technique, because the islands were contained in the laser beam spot.

Using the estimated thickness of the larger islands makes it possible to determine the laser beam spot size.[22-27] Since the laser beam is operated in its fundamental TEM_{00} mode, the irradiance of the beam in an (x,y) plane perpendicular to the propagation axis is expressed as

$$I(x,y) = \frac{2P_0}{\pi\omega_0^2} \exp\left(\frac{2x^2 + 2y^2}{\omega_0^2}\right), \tag{1}$$

where P_0 is the total power and ω_0 is the spot size at the beam waist. The peak value is believed to occur when the center of the island corresponds to the axis. In this correspondence, the power incident on an island having an area of $a \times a$ is

$$P_i(\omega,a) = \int_{-\frac{a}{2}}^{\frac{a}{2}} dx \int_{-\frac{a}{2}}^{\frac{a}{2}} dy I(x,y)$$

$$= P_0 \operatorname{erf}^2\left(\frac{a}{\sqrt{2}\omega_0}\right), \tag{2}$$

with the error function

$$\operatorname{erf}(z) = \frac{2}{\sqrt{\pi}} \int_0^z \exp(-t^2) dt.$$

The PMT output current signal by fluorescence may be proportional to power $P_i(\omega_0,a)$. Since the calibration line is obtained using a photoresist of an area much larger than that of the spot size, then $P_i(\omega_0,a) = P_0$. If the real thickness of the photoresist is b and the estimated one is c, there is a relationship.

$$\frac{c}{b} = \frac{P_i(\omega_0,a)}{P_0} = \operatorname{erf}^2\left(\frac{a}{\sqrt{2}\omega_0}\right). \tag{3}$$

When $c/b = 600\ \text{nm}/850\ \text{nm} = 0.71$ and $a = 3\ \mu m$, the spot size is determined to be 2.1 μm using the result of the numerical calculation of the error function.

If $2\omega_0 = a$ is set in Eq. (3) to obtain a criterion of measurement accuracy, $c/b = 0.91$. This means that when the laser beam diameter (twice the spot size) is smaller than the length of one side of the square, thickness can be measured within an error of 9%. In this case only the spot size is taken into consideration, whereas other factors also contribute to measurement inaccuracy, for example, the relative positions of the laser illumination and the photoresist.

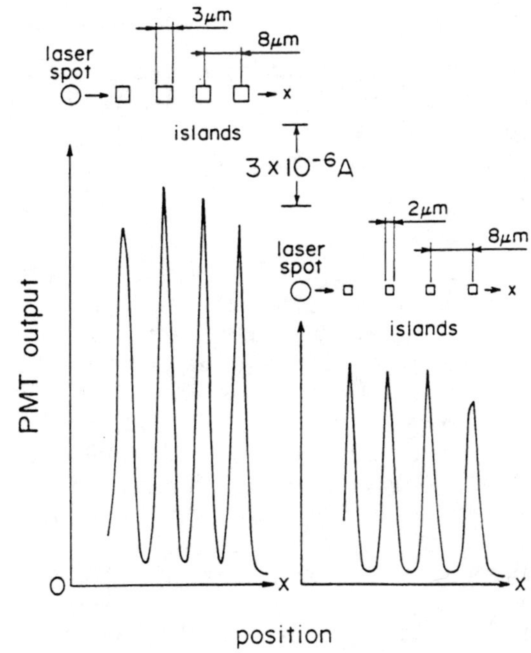

Fig. 4. PMT output signals of fluorescence from tiny islands successively irradiated by focused laser beam. The island area in the left half is $3 \times 3\ \mu m^2$; in the right half, $2 \times 2\ \mu m^2$.

A background shot noise level was measured using a silicon substrate without the photoresist film. The film thickness compatible with this level was estimated from Fig. 3 to be in the order of nanometers, which was the thinner limit of the thickness measurable by the system.

When the photoresist film becomes thicker, lateral resolution can decrease. The resolution depends on the spot size of the laser beam. As the laser beam is incident on a thick film and the beam waist agrees with the surface of the substrate, the spot size out of focus must be taken into consideration. The laser beam spot size ω is expressed as[28]

$$\omega^2 = \omega_0^2\left[1 + \left(\frac{\lambda z}{\pi\omega_0^2}\right)^2\right]. \tag{4}$$

Here λ is the wavelength of the laser beam, and the z is measured from the beam waist. Thus the spot size increases as it leaves the beam waist. If we allow the resolution to decrease by 5% and $\omega_0 = 2.1\ \mu m$, the spot size corresponding with this resolution is calculated from Eq. (4) to occur at a distance of 9.3 μm. Multiplying the distance by the refractive index of the film ($n = 1.66$), we obtain a value of 15 μm. If the thickness of the film is within this value, the lateral resolution is estimated to worsen within 5%. The absorption by the 1.0-μm thick film is a few percent at the 497-nm wavelength, since nondoped photoresist film was used. Therefore, saturation of the fluorescence is assumed to be small.

When the laser beam strikes the photoresist film at the same position, the fluorescence from the film decreases by the effect of heating or photobleaching. The fluorescence was found to decrease by 5% ~3 s after the onset of the laser beam. Here we may define a limit speed of the scanning laser beam as the diameter of the beam divided by 3 s. If the laser beam scans faster than this speed, the inaccuracy of the thickness measurement may be higher than 5%.

IV. Conclusion

One possible technique has been proposed to measure the photoresist thickness on asperities. This technique uses a visible focused laser beam with a 497-nm wavelength to illuminate the photoresist and observe the induced fluorescence. An initial calibration line is obtained by comparing several film thicknesses and their related fluorescence. The thickness of tiny islands of photoresist can be estimated.

This technique is promising for the following reasons: (1) Since the laser beam can be focused onto a submicron spot, high lateral resolution can be expected. (2) It is practical for use in the semiconductor device fabrication process. The fluorescence from a doped photoresist[16–18] does not need to be used, even though it is stronger. This is important because dopants can adversely affect the devices. (3) The laser beam used for excitation is not UV but visible. It is thus easy to arrange the experimental setup since no optical components for the UV wavelength are required.

In this experiment measurement of the thickness of the photoresist film on asperities was not performed. To do this, a focusing servo system for maintaining maximum fluorescence is required, since the asperities of the film may change the strength of the induced fluorescence depending on the position of the laser beam waist relative to the film.

The authors wish to thank Chusuke Munakata for his encouragement.

References

1. F. Laeri and T. C. Strand, "Angstrom Resolution Optical Profilometry for Microscopic Objects," Appl. Opt. **26**, 2245 (1987).
2. K. Yoshizumi, T. Murao, J. Masui, R. Imanaka, and Y. Okino, "Ultrahigh Accuracy 3-D Profilometer," Appl. Opt. **26**, 1647 (1987).
3. H. Chung, L. M. Walpita, and W. S. C. Chang, "Simplified Differential Phase Optical Microscope," Appl. Opt. **25**, 3014 (1986).
4. C. W. See, M. V. Iravani, and H. K. Wickramasinghe, "Scanning Differential Phase Contrast Optical Microscope: Application to Surface Studies," Appl. Opt. **24**, 2373 (1985).
5. G. Makosch and B. Drollinger, "Surface Profile Measurement with a Scanning Differential ac Interferometer," Appl. Opt. **23**, 4544 (1984).
6. R. W. Peterson, G. M. Robinson, R. A. Carlsen, C. D. Englund, P. J. Moran, and W. M. Wirth, "Interferometric Measurements of the Surface Profile of Moving Samples," Appl. Opt. **23**, 1464 (1984).
7. D. Y. Lou, A. Martinez, and D. Stanton, "Surface Profile Measurement with a Dual-Beam Optical System," Appl. Opt. **23**, 746 (1984).
8. D. M. Perry and G. M. Robinson, "Measurement of Surface Topography of Magnetic Recording Materials through Computer Analyzed Microscopic Interferometry," IEEE Trans. Magn. **MAG-19**, 1656 (1983).
9. Y. Fainman, E. Lenz, and J. Shamir, "Optical Profilometer: A New Method for High Sensitivity and Wide Dynamic Range," Appl. Opt. **21**, 3200 (1982).
10. G. E. Sommargren, "Optical Heterodyne Profilometry," Appl. Opt. **20**, 610 (1981).
11. H. J. Matthews, D. K. Hamilton, and C. J. R. Sheppard, "Surface Profiling by Phase-Locked Interferometry," Appl. Opt. **25**, 2372 (1986).
12. D. K. Hamilton and H. J. Matthews, "The Confocal Interference Microscope as a Surface Profilometer," Optik **71**, 31 (1985).
13. D. K. Hamilton and T. Wilson, "Surface Profile Measurement Using the Confocal Microscope," J. Appl. Phys. **53**, 5320 (1982).
14. J. M. Bennett and J. H. Dancy, "Stylus Profiling Instrument for Measuring Statistical Properties of Smooth Optical Surfaces," Appl. Opt. **20**, 1785 (1981).
15. K. Takami, C. Shinbo, A. Izumi, and Y. Tomita. "Thickness Measurement Method for Thin Photoresist Film on Transparent Material," Rev. Sci. Instrum. **54**, 690 (1983).
16. P. Kolodner, A. Katzir, and N. Hartsough, "End-Point Detection and Etch-Rate Measurement during Reactive-Ion Etching Using Fluorescent Polymer Films," J. Vac. Sci. Technol. B **1**, 501 (1983).
17. P. Kolodner and J. A. Tyson, "Microscopic Fluorescent Imaging of Surface Temperature Profiles with 0.01°C Resolution," Appl. Phys. Lett. **40**, 782 (1982).
18. P. Kolodner and J. A. Tyson, "Remote Thermal Imaging with 0.7-μm Spatial Resolution Using Temperature-Dependent Fluorescent Thin Films," Appl. Phys. Lett. **42**, 117 (1983).
19. T. K. Gupta, "Correlation of the Chemical and Electrical Properties of AZ1350J Photoresist Solution," J. Appl. Phys. **56**, 1145 (1984).
20. I. J. Cox, C. J. R. Sheppard, and T. Wilson, "Super-Resolution by Confocal Fluorescent Microscopy," Optik **60**, 391 (1982).
21. K. H. Drexhage, "Monomolecular Layers and Light," Sci. Am. **222**, 108 (1970).
22. S. Kimura and C. Munakata, "Method for Measuring the Spot Size of a Laser Beam Using a Boundary-Diffraction Wave," Opt. Lett. **12**, 552 (1987).
23. D. Letalick and I. Renhorn, "Instrument for Measuring Laser-Beam Profiles," Rev. Sci. Instrum. **58**, 765 (1987).
24. E. Stijns, "Measuring the Spot Size of a Gaussian Beam with an Oscillating Wire," IEEE J. Quantum Electron. **QE-16**, 1298 (1980).
25. A. H. Firester, M. E. Heller, and P. Sheng, "Knife-Edge Scanning Measurements of Subwavelength Focused Light Beams," Appl. Opt. **16**, 1971 (1977).
26. A. Yoshida and T. Asakura, "A Simple Technique for Quickly Measuring the Spot Size of Gaussian Laser Beams," Opt. Laser Technol. **8**, 273 (1976).
27. D. R. Skinner and R. E. Whitcher, "Measurement of the Radius of a High-Power Laser Beam near the Focus of a Lens," J. Phys. E **5**, 237 (1972).
28. H. Kogelnik and T. Li, "Laser Beams and Resonators," Appl. Opt. **5**, 1550 (1966).

Reprinted from *Optical Techniques for Industrial Inspection,* P.G. Cielo, editor, Proc. SPIE Vol. 665, pp. 258-264 (1986).

Laser Excited Fluorescence Studies of Black Liquor

J.J. Horvath and H.G. Semerjian

Center for Chemical Engineering, National Bureau of Standards, Gaithersburg, Maryland 20899

Abstract

Laser excited fluorescence of black liquor was investigated as a possible monitoring technique for pulping processes. A nitrogen pumped dye laser was used to examine the fluorescence spectrum of black liquor solutions. Various excitation wavelengths were used between 290 and 403 nm. Black liquor fluorescence spectra were found to vary with both excitation wavelength and black liquor concentration. Laser excited fluorescence was found to be a sensitive technique for measurement of black liquor with good detection limits and linear response over a large dynamic range.

Introduction

There is an increasing need for the development of measurement science and technology for on-line measurement of process variables in the pulp and paper industry. Current process techniques for measurement of delignification rates in the pulping process are based on sampling and off-line analysis systems. The resultant delay prevents the use of on-line control of the digestor, resulting in less than optimum operating conditions. The ideal process monitoring technique would be both an in-situ and real-time measurement system, so that important variables of the cooking process can be monitored and corrections applied before a cook is complete.

Due to its aromatic nature, lignin strongly absorbs ultraviolet light, which has led to the use of absorption spectrometry to follow the dissolution of lignin during a sulfate cook.[1] Unfortunately, various other compounds in the cooking liquor also absorb ultraviolet light further complicating the measurement of lignin concentration.[2] Spectra of cooking liquors have been shown to change during a cook but the lack of specificity of the absorption method has precluded its application for routine process control measurement. The fluorescence technique can provide improved selectivity, by utilizing the extra degree of freedom inherent in the technique, i.e., use of both the excitation and emission wavelengths.

The fluorescence behavior of cooking liquors has been studied by several authors. Wilson[3] determined the fluorescence spectra for diluted solutions of both sulfite and Kraft cooks and found that fluorescence could be used for quantitative determination in both processes. He also found that fluorescence was at least ten times more sensitive than the Pearl Benson Index (PBI)[4,5] for determining spent liquor in stream waters. He was also able to differentiate between Kraft and sulfite processed lignins by using selected excitation and emission wavelengths plus the effects of pH. The excellent sensitivity of fluorescence for detecting spent liquor in water was verified by Christman and Minear[6] who also found a linear relationship between fluorescence intensity and concentration for all cases studied.

The effect of excitation and emission wavelengths for detection of lignin in natural waters has been studied with several combinations. Wilander et al.[7] found the optimum wavelengths for excitation and fluorescence emission to be 285 and 405 nm respectively although other pairs such λ_{ex} = 225, λ_f = 430 nm[8] and λ_{ex} = 210, λ_f = 310 em[9] have been suggested. Bublitz and Wade[10] have studied the use of liquor fluorescence as a mill process control tool for pulping. Fluorescence of spent pulping liquors from Douglas-fir and western hemlock sulfite laboratory cooks were correlated with yields, kappa numbers and pulp lignin contents. Fluorescence of liquors from a sulfite mill had an apparent maximum intensity near the end of the cook, possibly indicating a better method for indicating the end of the cook than present procedures.

All the previous studies on lignin and liquor fluorescence were accomplished using broad band spectra excitation sources (10-20 nm bandwidth) with relatively low intensity, i.e., continuum and UV-lamps. These studies also required withdrawal of a sample from the reactor and dilution before the measurement is made. In this study, we are attempting to develop a technique which could be used as an in-situ on-line process measurement tool.

For in-situ measurements where highly concentrated solutions occur, deviations from ideal spectroscopic conditions must be considered. In the present study, we are using a pulsed laser system to allow for high power and narrow spectral excitation of black liquor solutions. We have also looked at the effects of pre- and post filter on the observed fluorescence spectrum. Due to the large concentration changes occurring during a pulp cook, the geometry of the viewing optics will play an important part in the observed results. For a real time in-situ optical probe, these effects need to be known in detail.

Experimental Apparatus

The experimental system used in the present studies is shown schematically in Figure 1. A pulsed nitrogen laser (Molectron UV 24)* was used to pump a dye laser (Molectron DL-14) to produce tunable visible light as the excitation source. Several dyes such as DPS and BBQ were used to obtain wavelengths between 380 and 410 nm. The output from the dye laser, using RB and R6G dyes, was frequency doubled using crystals of KDP to obtain wavelengths between 285 and 320 nm. Alternatively the dye head could be by-passed and the output of the nitrogen laser at 337 nm could be used as an excitation source. The laser beam is passed through a 1 cm square quartz cuvette containing liquor solutions and the fluorescence is viewed at 90° from the excitation beam. A 0.3 m scanning spectrometer (EV-700 GCA/McPherson) was used with a diffraction grating blazed at 250 nm. Entrance and exit slits were adjustable and typically set for a 1 nm bandpass. For these spectra, the volume monitored was determined by the intersection of the laser beam (3 mm diameter) with the monochromator slit width (250 μm). This collection volume, approximately 2 cubic millimeters, was located in the center of a 10 cm square fluorescence cuvette. The laser beam passed through a path length of approximately 4.5 mm before entering the collection volume where fluorescence was measured. The fluorescence emitted from the collection volume also passed through a path length of 4.5 mm before leaving the cuvette. A standard photomultiplier (R955 Hamamatsu) operated at 800V was used as the detector and its output was measured by a Boxcar Averager (Model #162, Princeton Applied Research). Output of the boxcar averager was either sent to a digital voltmeter or a x-y plotter for spectral scans. A home-built trigger interface was used to trigger the detection electronics to avoid jitter in the laser firing. Typical repetition rate of the N_2 laser was set at approximately 15 Hz with a pulse energy of 9 mJ/pulse. This resulted in a dye laser output of 800 mJ/pulse at 406 nm while the frequency doubled wavelengths (285-320 nm) had energies of approximately 25 μJ/pulse.

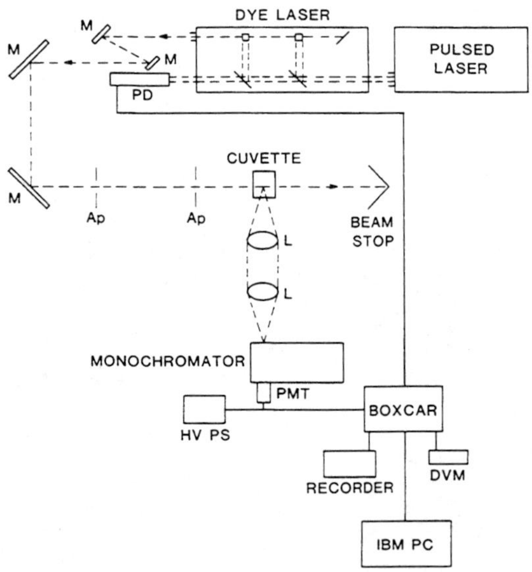

Fig. 1. Experimental apparatus.

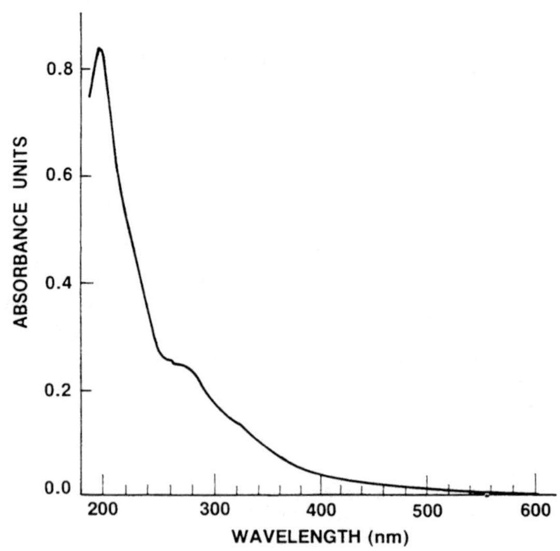

Fig. 2. Absorption spectra of 60 ppm black liquor solution.

*Certain commercial equipment, instruments, and materials are identified in this paper in order to specify adequately the experimental procedure. In no case does such identification imply recommendation or endorsement by the National Bureau of Standards, nor does it imply that the material or equipment is necessarily the best available for the purpose.

Results

Experiments with nitrogen laser excitation

Initial experiments were made using only the N_2 laser as the excitation source. Absorption spectra of typical lignins[11] isolated from black liquor of sulphate cooks show a relatively small amount of absorption at 337 nm, with the main absorption peaks occurring around 280 nm, 203 nm, and 180 nm. The absorption spectra of our black liquor sample is shown in Figure 2. Due to the small amount of absorption at the N_2 laser wavelength, a relatively concentrated solution was prepared by dissolving black liquor particles in distilled water. The black liquor particles were used as-received from the Institute of Paper Chemistry and no further drying was performed. While scanning the monochromator from 250-600 nm, with N_2 laser excitation, the only fluorescence signal was observed between 350 nm and 550 nm. This fluorescence, shown in Fig. 3, appeared as a relatively broad and intense hump with two peaks, one centered at 427 nm and the other at 450 nm. The large peak at 337 nm is elastically scattered light at the nitrogen laser wavelength.

To establish the detection limits and dynamic range of the fluorescence measurements, a series of experiments were carried out where the fluorescence intensity was measured as a function of concentration. A measured amount of black liquor particles was dissolved in distilled water and a series of standard solutions over the concentration range of 1-60,000 ppm (by weight) were prepared. Figure 4 shows the fluorescence intensity (relative units) at the 427 nm emission wavelength vs. the concentration. The curve is linear from 0.3 ppm to approximately 800 ppm, over 3 orders of magnitude, whereupon the curve levels off and then declines. An identical behavior is observed for the 450 nm peak. This decline is due to the post-filter effect where the fluorescence radiation is reabsorbed by lignin molecules in solutions not excited by the laser beam but in the detector field of view. Visual observation of the sample cell shows that, at low concentrations (1-50 ppm), the liquid is uniformly emitting along the entire beam length in the cell, but as the concentration increases, the emitting volume in the cell decreases until, at about 60,000 ppm, the region of fluorescence is confined to a very narrow region near the wall of the cuvette where the beam enters. This combination of pre- and post filter effects causes the calibration curve (fluorescence intensity vs. concentration) to level off because as the fluorescence intensity increases due to concentration, the optical collection volume is becoming smaller due to the above filtering effects. This effect could be reduced or utilized by judicious choice of fluorescence collection optics.

Absorption spectra of lignin cooking liquors have been shown to be pH dependent[9], so experiments were performed to determine if the fluorescence spectrum observed was affected by a change in pH. A 480 ppm solution of black liquor was found to be neutral by a litmus

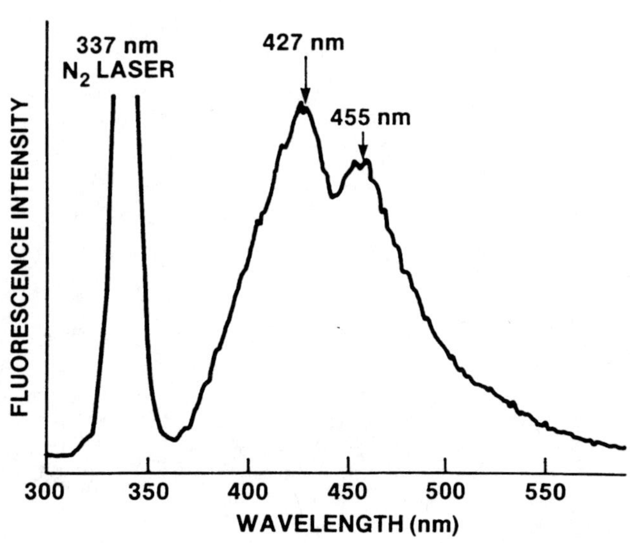

Fig. 3. Fluorescence spectrum of black liquor solution with nitrogen laser excitation.

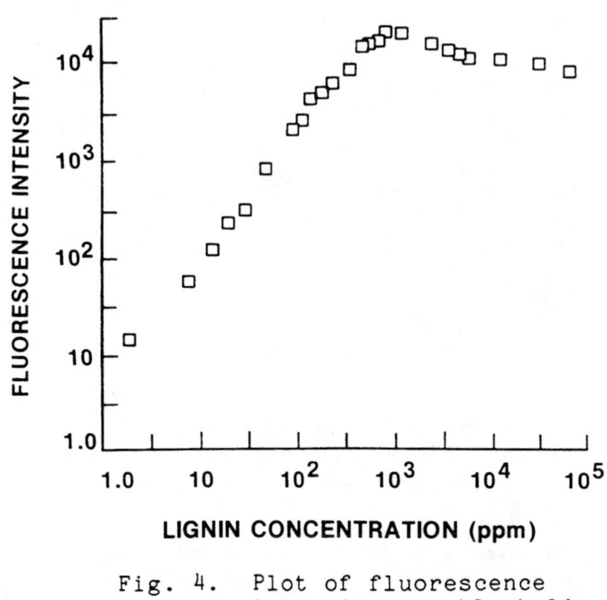

Fig. 4. Plot of fluorescence intensity vs. black liquor concentration λ_{ex} = 337 nm, λ_f = 427 nm.

paper test. To vary the pH of the sample, 1 drop of concentrated NaOH solution was added to it which made the pH of the sample >10. This was found to have no effect on the shape of the fluorescence spectrum observed.

Experiments with Dye Laser Excitation

Several studies have been carried out on the effect of the excitation wavelength on the fluorescence spectra[6-8] of lignin, and various wavelengths have been suggested as optimum for detection of lignin. A similar study was undertaken using a nitrogen pumped tunable dye laser. Excitation wavelengths both longer and shorter than the nitrogen laser excitation at 337 nm were used. The effects of excitation at longer wavelengths are shown in Figures 5 and 6 which show the fluorescence spectra obtained with excitation wavelengths of 386 nm and 403 nm, respectively. The same general broadband structure can be seen in these spectra with several significant differences. With excitation at 337 nm, two broad peaks are observed with the one centered at 427 nm greater than the one at 450 nm, but with 386 nm excitation the peak at 427 nm is decreased relative to the peak at 450 nm which now appears to have the maximum shifted to 458 nm. With excitation at 403 nm, the differences between the two peaks become even greater with the maximum intensity occurring at 470 nm with a tail extending out past 600 nm. All 3 spectra were run using concentrations less than 500 ppm to eliminate pre- and post-filter effects. The change in intensity ratios between the two peaks can be seen in Table 1.

Results with laser excitation at wavelengths shorter than 337 nm are shown in Figures 7-9 which are obtained with excitation wavelengths of 312, 297 and 290 nm, respectively. Upon excitation with 312 nm, we observe once again the two broad peaks located at approximately 427 nm and 450 nm with the 427 nm peak having the greater intensity. However, we also observe an increase in the intensity of fluorescence at wavelengths less than 400 nm whereas, with 337 nm excitation, there was a large decrease in intensity between 400 nm and 360 nm. On the other hand, with 312 nm excitation, the fluorescence is relatively constant over this range dropping to background levels by 320 nm.

Table 1. Effect of Excitation Wavelength on Peak Intensity Ratios

λ_{ex} (nm)	$I_f(427)/I_f(450)$	$I_f(427)/I_f(470)$
337	1.17	1.41
386	0.68	0.60
403	0.52	0.37

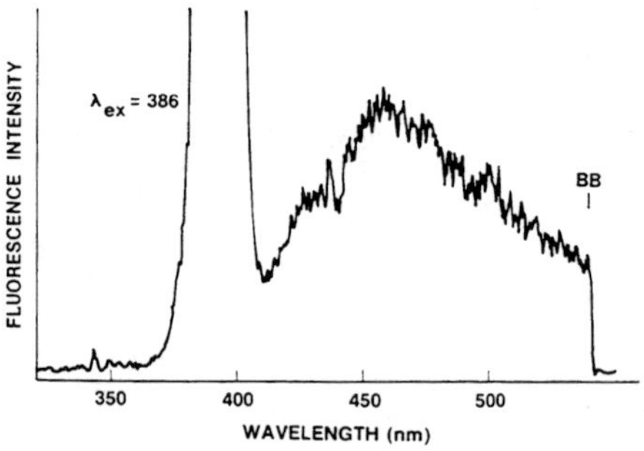

Fig. 5. Black liquor fluorescence spectrum with λ_{ex} = 386 nm. BB: laser beam blocked at 540 nm.

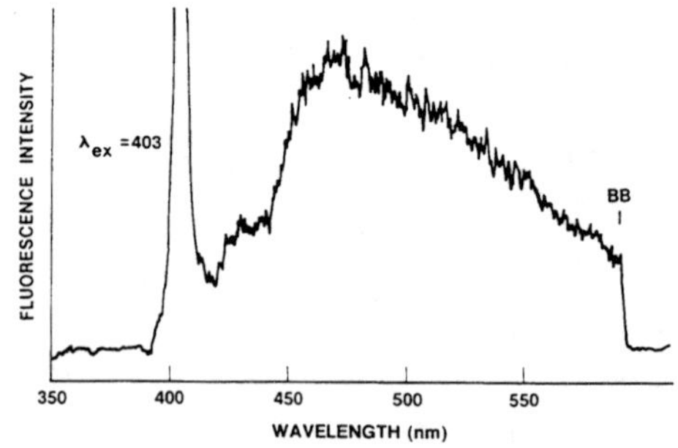

Fig. 6. Black liquor fluorescence spectrum with λ_{ex} = 403 nm. BB: laser beam blocked at 590 nm.

As the laser excitation wavelength decreases to 297 nm (Fig 8), the fluorescence intensity increases below 400 nm such that the spectrum appears to consist of three broad peaks centered at 370, 420 and 455 nm. Figure 9 shows the fluorescence spectrum with excitation at 290 nm, where the most dominant feature is the fluorescence occurring between 300 and 375 nm. Once again, we see the fluorescence peaks at 427 nm and 450 nm but with greatly reduced intensity compared to the fluorescence at the shorter wavelengths. The general trend appears to be for the maximum region of black liquor fluorescence to follow the excitation wavelength. Short wavelength excitation, below 300 nm, yields maximum fluorescence in the spectral region of 290-350 nm, while excitation wavelengths $\geq$320 nm promote maximum fluorescence intensity in the region of 375-600 nm.

The post filter effect is dramatically illustrated in Figure 10 which shows the fluorescence spectrum for two different concentrations of black liquor with laser excitation at 290 nm. At low concentrations, the peak fluorescence intensity is observed between 300 and 350 nm, but as concentration increases, the peak fluorescence shifts to longer and longer wavelengths until, finally, the peak is centered around 400 to 450 nm in high concentration solutions.

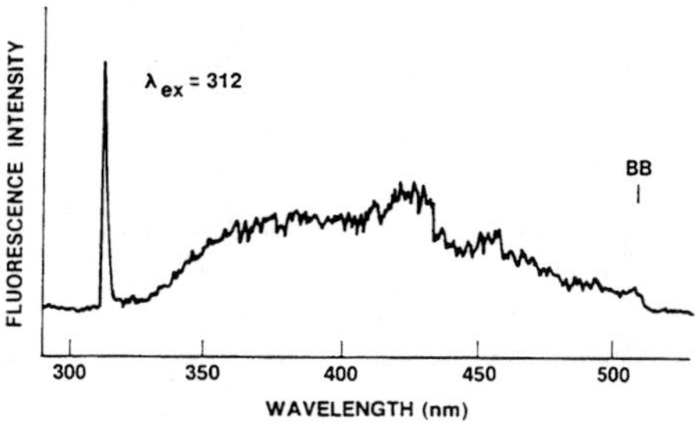

Fig. 7. Black liquor fluorescence spectrum with λ_{ex} = 312 nm. BB: laser beam blocked at 510 nm.

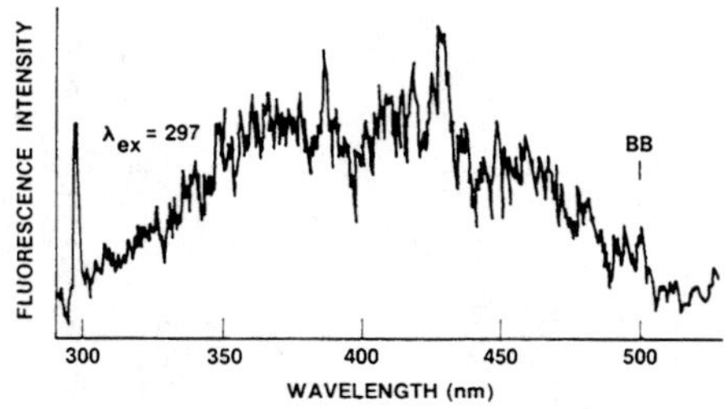

Fig. 8. Black liquor fluorescence spectrum with λ_{ex} = 297 nm. BB: laser beam blocked at 500 nm.

Discussion

The experimental results presented here indicate that the fluorescence spectrum of lignin containing solutions is a complex function of both excitation wavelength and lignin concentration. The visible fluorescence between 400 and 500 nm in this study is similar to the fluorescence observed by Christman and Minear in their studies[6] while the short wavelength fluorescence in the region of 300 nm was observed by Baumgartner et al.[9] using excitation at 210 nm. However, the present studies indicate a much greater dependence on excitation wavelength than observed by previous researchers. This may be due, in part, to the narrow spectral band of our excitation source and/or the pre- and post filter effects. Figures 8 and 9 show the large difference observed in the fluorescence spectrum when the excitation source differs by only seven nanometers. As the excitation wavelength decreases below 320 nm, the fluorescence between 300 and 350 nm dramatically increases as wavelength decreases. This indicates excitation into a different electronic level from the one responsible for fluorescence between 350 and 550 nm. This is also indicated by the absorption spectra which increases rapidly below 320 nm. This indicates the possibility of attaining a very high degree of selectivity in a fluorescence measurement, where the excitation wavelength can be tuned in order to avoid any interference from other species. An example of this is shown in Table 2. Another way of avoiding interfering species is by scanning the excitation wavelength over an interval. By monitoring the fluorescence during an excitation scan, an excitation/emission matrix can be determined which will yield a unique signature for the individual components in a pulping cook. By plotting excitation wavelength vs fluorescence emission wavelength vs fluorescence intensity, a three dimensional plot can be created which will be characteristic of that species.

As shown in Figure 10, concentration of lignin in the solution also has a significant effect on the intensity as well as the spectral disribution of the observed fluorescence emissions. The higher absorber concentrations result in a pre-filter effect, where the fluorescence emission intensity is reduced due to absorption of the excitation beam. At low concentrations, fluorescence was observed from the entire beam volume in the cuvette and total light absorbed from the beam was less than 5%. As the concentration increased, more and more of the laser beam was absorbed by the black liquor until at 60,000 ppm where the laser beam was totally absorbed in the solution. At this point, the fluorescence was confined to a narrow volume where the beam first entered the cuvette. Even though the beam was totally absorbed in the black liquor solution, we were able to measure fluoescence spectra by moving our collection volume to where the beam first entered. The reduction in fluorescence intensity (or fluorescence quenching) was observed by Wilson[3], where the fluorescence intensity decreased by an order of magnitude, as the solution concentration was increased from 300 ppm to 1,300 ppm. These results were obtained in a standard flourimeter with conventional imaging (central excitation volume). By varying the geometry of the collection optics, it is possible to obtain spectra on very concentrated solutions of black liquor. The observed shift towards longer wavelengths in the fluorescence spectra could be explained by the post-filter effect. In low concentration

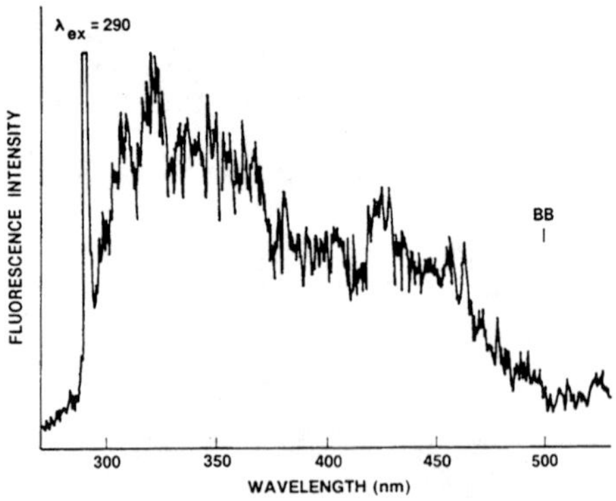

Fig. 9. Black liquor fluorescence spectrum spectrum with λ_{ex} = 290 nm. BB: laser beam blocked at 500 nm.

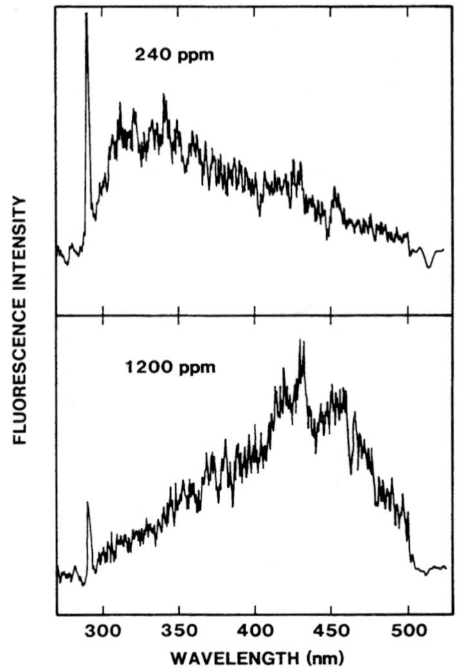

Fig. 10. Black liquor fluorescence spectra at two concentrations with λ_{ex} = 290 nm.

solutions, the emitted fluorescence leaves the solution without further significant absorption; with higher absorber concentrations, the fluorescence emissions at shorter wavelengths may be preferentially absorbed (because of much larger absorption coefficients at shorter wavelengths, see Fig. 2) and reemitted at longer wavelengths, resulting in a shift of the fluorescence peak towards longer wavelengths.

All previous studies of black liquor lignin fluorescence needed large dilutions of their samples (2,500-10,000 times) to obtain spectra. The highest concentration from which spectra were obtained in the previous studies[3,6-9] was approximately 1,300 ppm, still much below the concentrations found during actual pulping cooks. By varying the collection optics geometry, we were able to obtain spectra from black liquor solutions of 120,000 ppm with good signal strength, indicating good likelihood of obtaining usable spectra from an actual pulping cook. By using a fiber optic probe, in-situ measurements could be accomplished on-line during a cook. During a cook of wood pulp, the lignin concentration in the cooking liquor will be constantly changing which could be monitored by observing the fluorescence at two different wavelengths. By comparing intensities at

Table 2. Maximum Emission Wavelengths and Limits of Detection (LOD) for Various Excitation Wavelengths

λ_{ex} (nm)	$\lambda_{f_{max}}$ (nm)	LOD (ppm)
287	315	24
290	321	6
297	427	34
300	427	64
304	427	5
312	427	12
315	427	42
337*	427	2
386**	460	14
403**	470	15
406**	470	7

*Direct excitation with the N_2 laser.
**Excitation with the dye laser. All other wavelengths were achieved by frequency doubling the output of the dye laser.

320 and 427 nm $I_f(320)/I_f(427)$ for the two concentrations shown in Figure 10, one obtains for the ratio $I_f(320)/I_f(427)$ a value of 1.46 for 240 ppm and 0.215 for 1200 ppm. Ideally, the fluorescence from a cook could be monitored and when this ratio either reaches some preset value or becomes constant, the cook could be ended.

This work also indicates that laser excited fluorescence is a more sensitive method for determining lignin concentration than the Pearl-Benson method which is commonly used for lignin concentration measurement.[5] Fluorescence measurements could also be performed in-situ whereas the Pearl-Benson method requires withdrawal of a sample, addition of an alkali and measurement of the reaction product.

Conclusions

We have shown that laser excited fluorescence can be a useful technique for measurements in lignin solutions. Fluorescence intensities from lignin solutions are shown to be linear with concentration over several orders of magnitude and more sensitive than currently used measurement techniques. The fluorescence spectrum observed from lignin solutions has been shown to depend on both excitation wavelength and concentration of lignin. When using narrow bandwidth excitation, small changes in the excitation wavelength, less than 10 nm, result in significant changes in the emission spectra. Results indicate that the laser excited fluorescence technique could be a strong candidate for in-situ monitoring of pulping processes.

Acknowledgement

Support received from the U.S. Department of Energy, Office of Industrial Programs, under Interagency Agreement No. DE-AI01-76PR06010 is gratefully acknowledged. The project monitor was Stanley F. Sobczynski. The authors also thank Dr. David T. Clay of the Institute of Paper Chemistry for providing the black liquor samples.

References

1. Sjostrom, E. and Haglund, P., TAPPI, 47, 286 (1964).
2. Hartler, N. and Norrstrom, H., TAPPI, 52, 1712 (1969).
3. Wilson, W.C., Proc. 13th Pacific Northwest Industrial Waste Conf., Washington State University, Pullman (1967).
4. Pearl, I.A. and Benson, H.K., Paper Trade J. 111, 35 (1940).
5. Fellicetta, V.F. and McCarthy, J.L., TAPPI, 46, 337 (1963).
6. Christman, R.F. and Minear, R.A., The Trend, 19, 3 (1967).
7. Wilander, A., Kvarnas, H., and Lindell, T., Water Research, 8, 1037 (1974).
8. Bublitz, W.J. and Meng, Y.T., TAPPI, 61, 27 (1978).
9. Baumgartner, D.J., Feldman, M.H., and Gibbons, C.L., Water Research, 5, 533 (1971).
10. Bublitz, W.J., Wade, D.C., Svensk Papperstidning, 82, 535 (1979).
11. Goldschmid, O., Lignins: Occurrence, Formation, Structure and Reactions, (K.V. Sarkaknen and C.H. Ludwig, eds), J. Wiley, New York, NY (1971).
12. Norrstrom, H. and Teder, A., Svensk Papperstidning, 74, 337 (1971).

Reprinted from *Optical Techniques for Industrial Inspection,* P.G. Cielo, editor, Proc. SPIE Vol. 665, pp. 234-242 (1986). ©1986 SPIE.

Raman Spectroscopy Applied To Polymer Analysis: Low Frequency Spectra of Polyester Fibers

Bernard J. Bulkin, Frank DeBlase, and Menachem Lewin

Department of Chemistry, Polytechnic Institute of New York
Brooklyn, New York 11201

Abstract

Among the many advantages of Raman spectroscopy for polymer characterization, low frequency data have been the least exploited. Results are presented on the use of low frequency Raman data on polyester fibers for determining the relative specific heats, and the degree of amorphous orientation. Background theory is presented.

Introduction

Raman spectroscopy of polymers has long been used to characterize crystallinity. This determination is usually based upon one or more bands which split or grow in intensity as crystallinity develops. In some cases, these modes are in fact true crystallinity bands, that is, they are associated with long range order in the polymer. In other cases, crystallinity modes are more reflective of a short range order or a conformational change which precedes or accompanies crystallinity. As a result, Raman crystallinity may or may not parallel that determined by techniques such as X-ray diffraction, thermal analysis, or dilatometry.

Nonetheless, there are considerable advantages to the use of Raman spectroscopy for studying both crystallinity and the development of crystallinity in polymers. Among these are the ability of Raman spectroscopy to examine single filaments or multi-filament bundles, including fibers under stress; polarization data which may give information about orientation (1,2); the ability to acquire Raman data at high speed - microseconds to seconds where appropriate - allowing for real time kinetic studies; low frequency Raman scattering which can provide data on long range order, mechanical properties, etc.

Low frequency data on polymers have been heavily exploited in the study of longitudinal acoustic modes (LAM's) in polyethylene and related polymers (3). Relatively little has been done to understand polymer properties from low frequency Raman spectroscopy of more complex polymers. Recently, work by Viras and King (4) has indicated that it is possible to exploit the low frequency region for an understanding of polymer specific heats. This fundamental property, or even its variation between samples with different processing conditions, is not easily obtained in a routine way for fibers.

In this paper, we build on the approach of Viras and King (4) and our earlier spectroscopic study of the low frequency vibrational spectrum of poly(ethylene terephthalate) (PET) (5,6) to show that considerable information about the polymer and its specific heat can be obtained from the low frequency region. Data are presented on two series of fibers, those obtained by thermal annealing of an original partially oriented yarn which was prepared at a take-up speed of 3.3 km/min (5,6) and those obtained by processing at a series of take-up speeds from 3.0 to 5.5 km/min.

Background to the Specific Heat Calculation

The vibrational component of the specific heat, of both amorphous and crystalline solids, is calculable from the low frequency Raman spectrum through statistical thermodynamic considerations. The quasi-harmonic theory has often been applied to calculate the density of vibrational states in materials. Furthermore, the heat capacity values obtained from the Debye and Einstein theories often compare closely to calorimetric results (7). The vibrational density of states can be considered to arise from the distribution of phonons (quantized molecular vibrations) within the solid. This phonon distribution is obtained, for both amorphous and crystalline material, from the low frequency Raman spectrum. These considerations are based on the fact that the low frequency Stokes spectrum arises from direct phonon - light interactions. Because of the disorder in these materials, all the low frequency modes are Raman active. Furthermore, since the phonons within any material can be considered a system of non interacting bosons, by invoking the Bose - Einstein equation, the vibrational component of the specific heat can be calculated. The expression of the relation between the specific heat and the density of vibrational states, derived from either the quasi-harmonic theory or the application of Bose - Einstein statistics, is the same. This expression is given as:

$$C_v\ vib = k \int_{min}^{max} \frac{g(\nu)\ (h\nu c/kT)^2}{[e^{-h\nu c/kT}/e^{-h\nu c/kT} - 1]^2}\, d\nu \tag{1}$$

Where: ν = wavenumber cm^{-1} and
$g(\nu)$ = density of vibrational states.

In order to calculate the specific heat from the low frequency spectrum the exact relationship between the density of vibrational states and the Raman scattering intensity is needed. According to the theory of Shuker and Gammon (8) a reduced Stokes low frequency intensity I_r should be used in order to calculate specific heat from the Raman scattering. This reduced intensity takes into account the effect of thermal fluctuations on intensity. The observed Stokes Raman spectrum I_{obs} (ν) is related to the vibrational density of states by:

$$I_{obs}(\nu) = \sum_b O_b\ (1/\nu)(1 - e^{-h\nu c/kT})\, g_b\ (\nu) \tag{2}$$

where O_b equals the optical coupling coefficient, which is summed over all the bands (b), and g_b (ν) equals the density of vibrational states.

This gives the reduced intensity as:

$$I_r = I_{obs}\ (\nu)\quad (1 - e^{-h\nu c/kT})^{-1} \tag{3}$$

In this expression, the contribution to the first order scattering from thermodynamic fluctuations is accounted for by the Bose factor $[1 - e^{-h\nu c/kT}]$.

The expression in Equation 2 holds if the scattering is due to vibrations of short correlation ranges compared to the optical wavelength. This is a reasonable assumption for polymer systems, with imperfect crystalline and amorphous domains. This theory is also based on the scattering being first order, disorder allowed, which is true for polymers as well as inorganic glasses (4, 9, 10). Equation 2 gives a direct relationship between the Raman spectrum and the $g_b(\nu)$ in the final term. The density of states can then be calculated upon rearranging Equation 2 to:

$$g_b\ (\nu) = \left[\sum_b O_b\ (1/\nu)\right]^{-1} I_{obs}\ (\nu)\ (1 - e^{-hc\nu/kT})^{-1} \tag{4}$$

or in terms of $I_r(\nu)$ from Equation 3 as:

$$\sum_b g_b\ (\nu) = [I_r\ (\nu)]\ (\nu)\ [\Sigma O_b]^{-1} \tag{5}$$

This expression for the relationship between the vibrational density of states and frequency is similar to the theory of Debye which states the density of vibrational states is proportional to the frequency squared $(\sum_b g_b\ \alpha\ (\nu)^2)$.

From Equation 5 the expression for the vibrational component of the specific heat in Equation 1 becomes:

$$C_v\ vib = k \int_{min}^{max} \frac{N/nI_r(\nu)\ \ \nu\ O_b^{-1}\ (h\nu c/kT)^2}{[e^{-h\nu c/kT}/e^{-h\nu c/kT} - 1]^2}\, d\nu \tag{6}$$

Where N/n represents the number of vibrational units per gram, and the relationship between the Raman intensity and density of vibrational states is used.

The expressions in Equations 5 and 6 for the density of states and C_v^{vib}, hold for both amorphous and crystalline samples, however the nature of the O_b optical constant is ambiguous. O_b may vary, both between amorphous and crystalline samples, and among different vibrational bands. In spite of these complications Viras and King (4, 9, 10) have calculated the density of vibrational states and the specific heat for several polymer systems by assuming the O_b optical coupling constant is proportional to $1/\nu^2$. This gives the density of vibrational states as $g_b(\nu) = I_r/\nu^2$.

It may be argued however, that this assumption violates Debye theory and should not be considered as a general approach. An alternative is to consider the constant O_b as an unknown constant and calculate the relative values of specific heat between different samples. If some additional information is available concerning the calorimetric value of the specific heat, the O_b values may be estimated in order to fit the spectroscopic results to the calorimetric values. The values of the C_v^{vib} obtained by both the method of Viras and King and by considering O_b as a general constant will be presented.

According to the approach of Viras and King the phonon vibrational unit can be associated with one or more repeat units depending on the nature of the vibration. In a study of polycarbonate Viras and King considered the low frequency molecular vibrations of the polymer to be associated with one repeat unit. In that study, the origin of the Raman scattering was related to the pendant aromatic groups of the main polymer backbone in polycarbonate. However, one polymer repeat unit is not a strict requirement for every polymer system, and a low frequency vibrational unit with greater delocalization can exist.

Results

Figures 1 and 2 present the reduced spectra calculated according to Equation 3 at 5 cm^{-1} intervals, for thermally annealed and take-up speed samples, respectively. This representation of the low frequency spectrum removes the influence of the Rayleigh exciting line and introduces the appropriate statistical weights previously discussed to each wave-number value. Extrapolation of the shoulder of the Rayleigh line, which extends past 25 cm^{-1} was still necessary. This method was used by Viras and King (4, 9, 10) and Fraser (11) and Snyder et al. (13) and is based upon constructing an extrapolated background.

Changes in the reduced spectrum with processing condition are seen in Figures 1 and 2. The most striking change is the peak development at 73 and 129 cm^{-1} which occurs as crystallinity increases. Although the development of these features is also observed in the observed spectra I (ν), the difference in the rate of increases between the 73 and 129 cm^{-1} regions is seen only in the reduced spectra.

The reduced spectrum also demonstrates the asymmetry in the entire region as measured by the ratio of intensities at low and high frequency values of 60 and 160 cm^{-1}. These regions appear to be the least overlapped by either the Rayleigh background or the 73 cm^{-1} or 129 cm^{-1} modes.

The intensity ratio $(I_{160}/I_{60})r$ from reduced Raman spectra of these samples displays changes which relate to the degree of amorphous orientation. The lowest ratio is observed for both the most amorphous unoriented samples (bulk and low spinning speed) and most crystalline samples. The ratio increases only as amorphous orientation of fibers increases while crystallinity remains low. This will be discussed further.

The values of the fully reduced scattering according to Viras and King assuming the optical coupling constant $O_b = 1/(\nu)^2$ are plotted in Figures 3 and 4. Discussion of the calculation of C_v^{vib} by the two approaches will follow.

Discussion

Sensitivity of the reduced spectra to amorphous orientation and crystallinity

The reduced spectra lend themselves to measurement of amorphous orientation by means of the ratio of scattering intensity at 160 cm^{-1} to that at 60 cm^{-1}. This is listed along with the amorphous orientation factor of the fiber samples in Table 1. Figure 5 shows a correlation between the ratios and the amorphous orientation factor. The ratio and amorphous orientation factor increase with the degree of orientation imparted by spinning speed stress up to the point where crystallization takes place. As crystallinity occurs in the fibers either by thermal annealing or increased spinning speed stress, a corresponding decrease in amorphous orientation occurs. This is apparent from both the decrease in amorphous orientation factor f_a and the decrease in the ratio, for fibers annealed at temperatures greater than T_g and spinning speeds greater than 4.5 km/min.

This sensitivity of the reduced low frequency I_{160}/I_{60} ratio to amorphous orientation ties the DSC and density data with the Raman data. The changes in the crystallization temperature and density of the fibers, particularly where abrupt changes in amorphous orientation occur, can be predicted by the I_{160}/I_{60} ratio. In addition, the sensitivity to crystallinity by the development of the features at 73 and 129 cm^{-1} is also visible from the reduced spectra of crystalline samples.

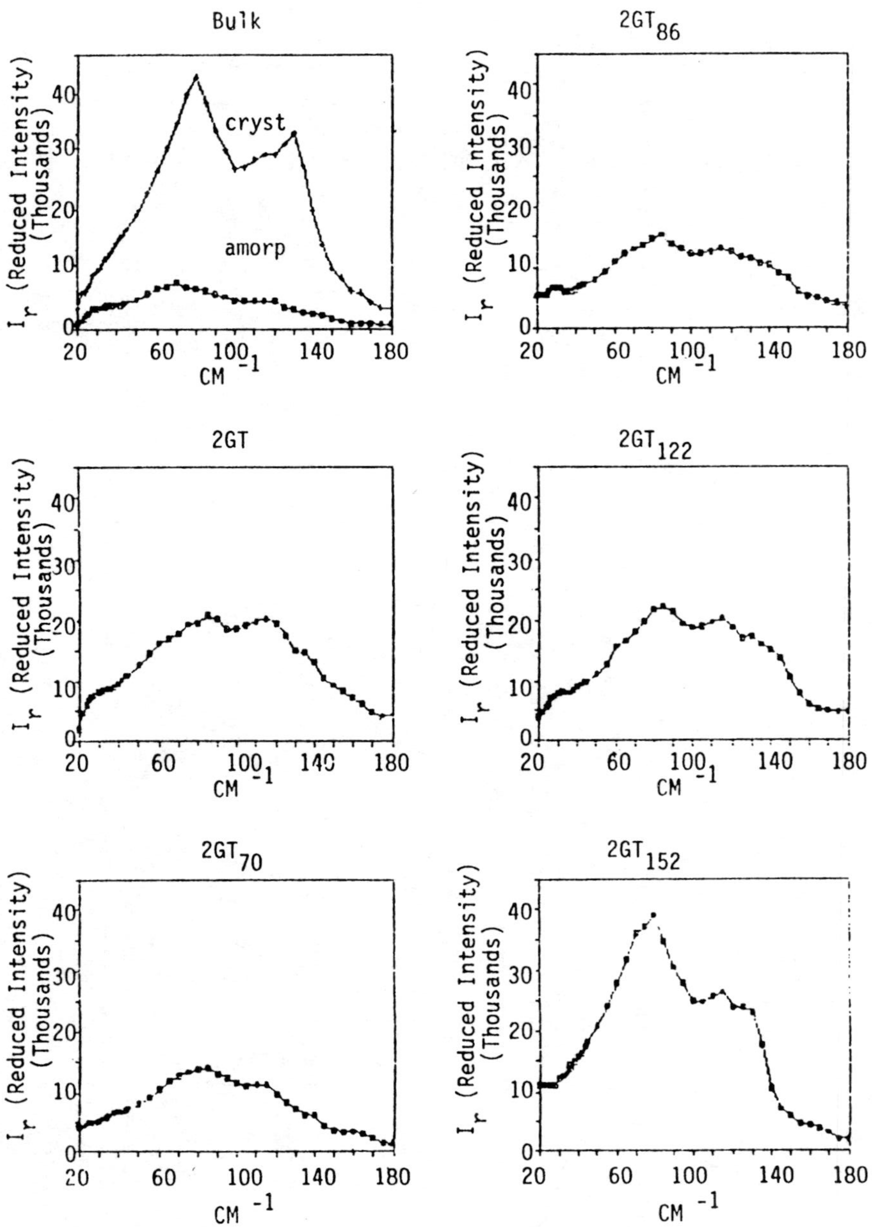

Figure 1. Reduced low frequency Raman spectra of amorphous and crystalline bulk and fiber samples: bulk samples are a quenched amorphous and a highly annealed crystalline material. 2GT is an original fiber, and the other samples are prepared from it. (See Reference 1)

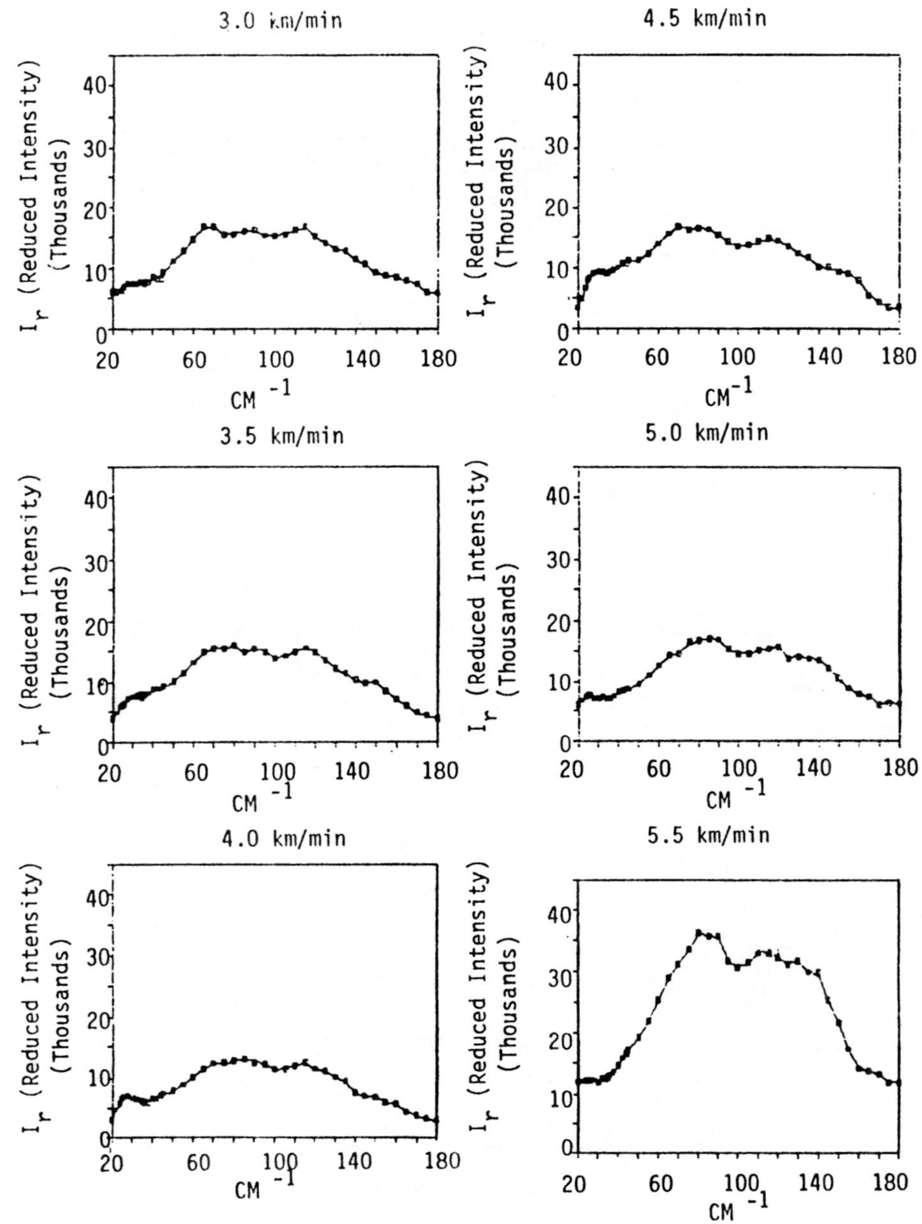

Figure 2. Reduced low frequency Raman spectra of PET fibers prepared with increasing take-up speeds, as noted.

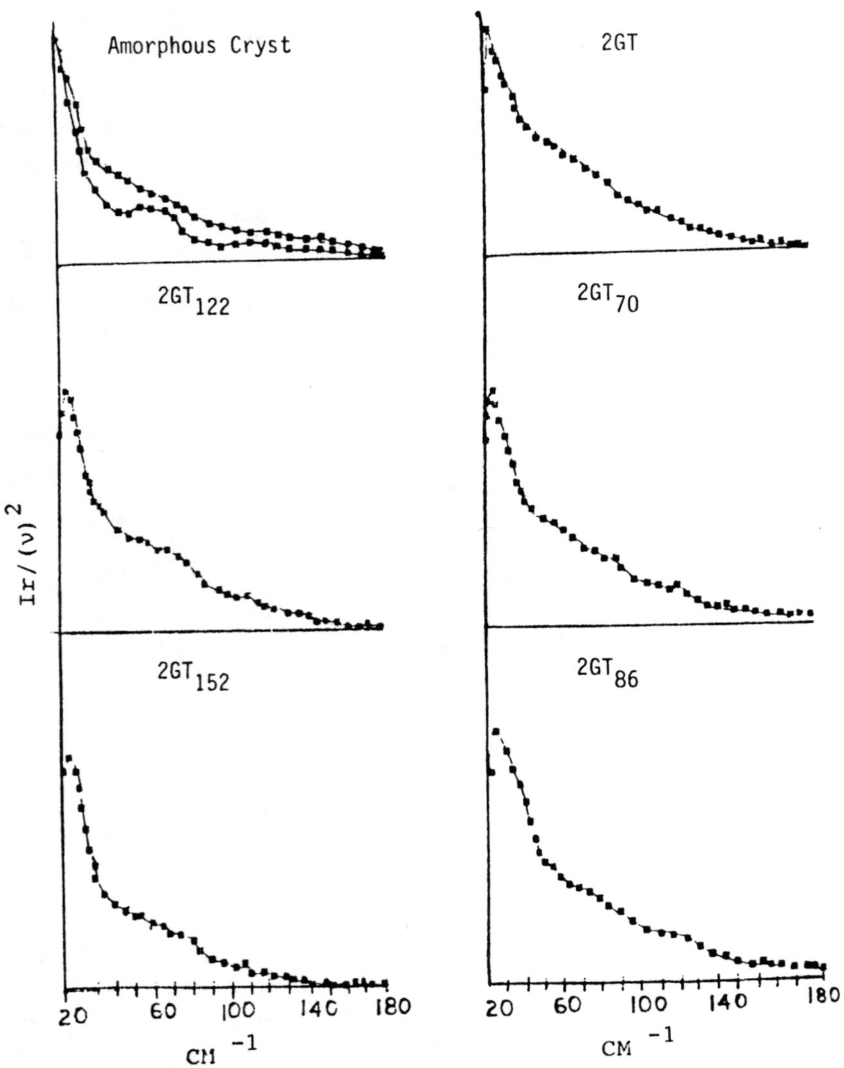

Figure 3. Plot of $Ir/(\nu)^2$ against increasing wavenumber of amorphous and crystalline bulk and fiber samples.

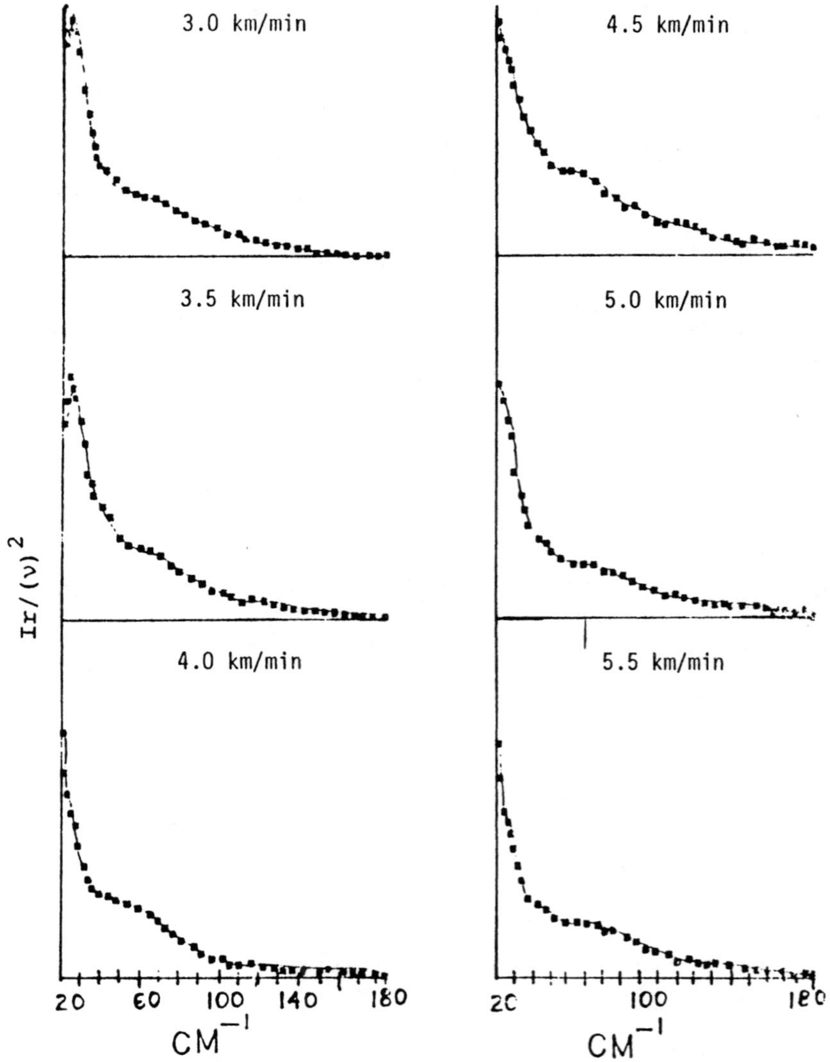

Figure 4. Plot of $Ir/(\nu)^2$ against increasing cm^{-1} for the variable take-up speed PET fiber samples.

Specific heat

According to the approach used by Viras and King O_b, the optical coupling coefficient of a band (b), is assumed equal to $1/(\nu)^2$. Then the expression of the density of states g_{N1b} of a single vibrational unit (N_1) of a band (b) in relation to the Raman intensity $I_r(\nu)$ can be expressed according to Equation 3 as $I_r/\nu^2 = g_{N1b}$. Since $g_{N1}(\nu)$ is the density of states of a single vibrational unit. $N/n\ g_{N1}(\nu) = g(\nu)$ is the total density of states in a gram of material. Here N/n is the ratio of the number of repeat units in a gram of material (N) divided by the integer (n = 1,2,3...) which determines the number of repeat units contained in the vibrational unit. Therefore N/n is the total number of vibrational units per gram of material.

From Equation 3, $N/n\ I_r\ (\nu)/\nu^2 = g(\nu)$, Equation 6 becomes:

$$C_v^{vib} = k\int_{min}^{max} \frac{N/n\ I_r/(\nu)^2\ (h\nu c/kT)^2}{[e^{-h\nu c/kT}/e^{-h\nu c/kT}-1]}\, d\nu \quad (7)$$

For the samples of PET used in this work the intrinsic viscosity, $[\eta]$, is 0.64, corresponding to a M_n of 21,059 as calculated from the equation for PET:

$$[\eta] = KM^{n}_{a} \quad \text{where } K=3 \times 10^{-4} \text{ and } a = 0.77$$

and the degree of polymerization is 109. The number of chains per gram will accordingly be 3.00×10^{19} and the number of monomeric units per gram $N=109 \times 3.00 \times 10^{19} = 3.27 \times 10^{21}$. This is the N used in Equation 7.

The C_v^{vib} values for the PET samples studied were calculated with different values of n from 1 to 3 and are listed in Table 2. The specific heat decreased with increase in crystallinity of the samples studied, and increased with large changes in orientation associated in the transition that exists between bulk amorphous and oriented amorphous fibers. Since the extent of delocalization of the low frequency vibrational unit was incorporated with the equation used to calculate C_v, a comparison of calorimetric values with the calculated C_v values allowed a prediction to the extent of delocalization. For PET the comparison indicates that the vibrational delocalization is greater than one repeat unit, and most likely extends over 3 repeat units.

The value of C_v for the amorphous but oriented fibers are higher than the calculated value for bulk amorphous material. As crystallinity increases the calculated value of C_v decreases if one adjusts the degrees of delocalization of the mode.

The specific heat of PET fibers has also been measured by calorimetry by Smith and Dole (13) for various samples, which were also subjected to thermal annealing. The C_p heat capacity data from equations found for samples with treatment similar to that performed on the fibers of this study are listed in Table 3. The calorimetric values of C_p can be used to adjust the parametric n in N/n to reduce the calculated value of C_v to a value in the range of the measured C_p. These C_v^{vib} values are given in Table 2 along with the calorimetric data of similar fibers studied by Smith and Dole.

According to the second approach O_b is left as a constant and the relative specific heat values are calculated from the reduced intensity curves according to Equation 6. As in the Viras and King method (10) the reduced intensity is normalized to unity prior to the integration to give the specific heat. The only difference is that I_r/ν^2 is replaced by the term $O_b I_r$ in the integrand and the values of the specific heat are calculated as $O_b C_v^{vib}$. These relative specific heats are also given in Table 2.

The calculated C_v^{vib} values from the low frequency Raman spectra appear to increase when amorphous bulk PET is melt spun into an oriented fiber. Also, as crystallinity increases by annealing fibers, the calculated C_v^{vib} decreases. A comparison of the Viras and King method shows the vibrational unit extends over three repeat units.

Acknowledgments

This work was supported by the Polymers Program of the National Science Foundation, Grant DMR-8304220. We are grateful to Drs. D. Selivansky and M. Deeg for helpful discussions and cooperation.

References

1. Bulkin, B. J., Lewin, M., and DeBlase, F. J., "Conformational Change, Chain Orientation, and Crystallinity in Poly(ethylene terephthalate) Yarns: Raman Spectroscopic Study", Macromolecules, 18, 2587 (1985).

2. Adar, F., and Noether, H., "Raman Microprobe Spectra of Spin-Oriented and Drawn Filaments of Polyethylene Terephthalate: Conformation, Orientation, and Crystallization", Microbeam Analysis, 20, 41 (1985).

3. Hsu, S. L., Krimm, S., "Longitudinal Acoustic Mode in Polymers", J. Appl Phys., 47, 4265 (1976)

4. Viras, F., Viras, K. G., and King, T. A., "Low-Frequency Excitations in Amorphous Poly(vinyl acetate)", J. Polym. Sci., Polym. Phys. Ed., 23, 609 (1985).

5. DeBlase, F. J., McKelvy, M. L., Lewin, M., and Bulkin, B. J., "Low-Frequency Raman Spectra of Poly(ethylene terephthalate)", J. Polym. Sci., Polym. Lett. Ed., 23, 109 (1985).

6. DeBlase, F. J., Ph.D. Thesis, Polytechnic University, 1986.

7. McQuarrie, D. A., "Statistical Thermodynamics", (1973) Harper and Row, New York.

8. Shuker, R., and Gammon, R. W., "Raman Scattering Selection-Rule Breaking and the Density of States in Amorphous Materials", Phys. Rev. Lett., 25, 222 (1970).

9. Viras, F., and King, T. A., "Low-Frequency Excitations in Amorphous Polycarbonate Studied by Raman Spectroscopy", Polymer, 25, 1411 (1984).

10. Viras, F., and King, T. A., "Low-Frequency Excitation in Amorphous Acrylic Polymers", Polymer, 25, 899 (1984).

11. Fraser, G. V., "Intensities of Longitudinal Acoustic Modes and Their Structural Implications", Polymer, 19, 857 (1978).

12. Snyder, R. G., Scherer, J. R., "Interpretation of Longitudinal-Acoustical-Mode Spectra of Polymers", J. Polym. Sci., Polym. Phys. Ed., 18, 421 (1980).

13. Smith, C. W., and Dole, M., "Specific Heat of Synthetic High Polymers, VII. Poly(ethylene terephthalate)", J. Polym. Sci., 20, 37 (1956).

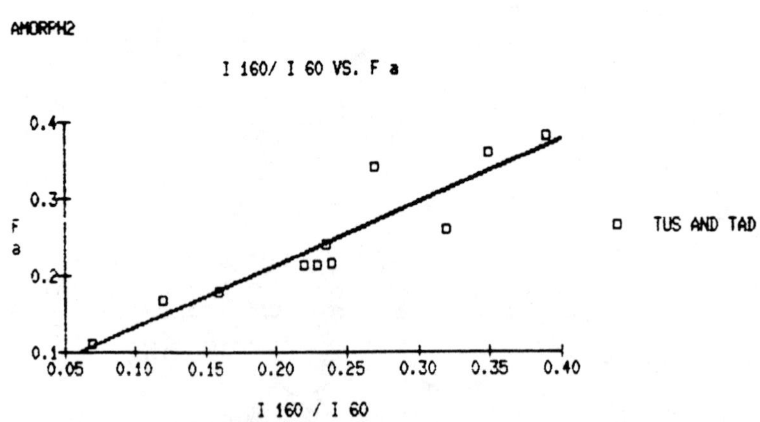

Figure 5. Plot of I_{160}/I_{60} vs. amorphous orientation factor f_a.

TABLE 1

Reduced Low-Frequency $[I_{160}/I_{60}]$ Data vs. Amorphous Orientation (f_a)

Sample	$[I_{160}/I_{60}]r^{(1)}$	f_a
2GT	0.240	0.217
$2GT_{70}$	0.230	0.215
$2GT_{86}$	0.220	0.214
$2GT_{122}$	0.120	0.167
$2GT_{152}$	0.070	0.112
3.0 km/min	0.160	0.180
3.5 km/min	0.236	0.240
4.0 km/min	0.270	0.340
4.5 km/min	0.390	0.380
5.0 km/min	0.350	0.360
5.5 km/min	0.320	0.260
Bulk Amorphous	0.060	——
Bulk Crystalline	0.080	——

(1) The $[I_{160}/I_{60}]r$ was calculated as the ratio of scattering intensity measure above the base line drawn from 20 cm^{-1} to 180 cm^{-1}.

TABLE 2

C_v^{vib} (by Viras and King Method). C_v^{vib} Calculated as $O_b - C_v^{vib}$

Sample:	C_v^{vib} (Viras and King Method) n=no. of rpts/vibrational unit			$O_b - C_v^{vib}$
Fibers	n=1	n=2	n=3	(Relative Method)
2GT	3.21	1.61	1.07	4149
$2GT_{70}$	2.94	1.50	0.98	3497
$2GT_{86}$	2.95	1.47	0.98	3020
$2GT_{122}$	2.91	1.45	0.97	1712
$2GT_{152}$	2.88	1.44	0.96	1212
3.0 km/min	2.96	1.48	0.99	2817
3.5 km/min	3.09	1.54	1.03	3034
4.0 km/min	3.16	1.58	1.05	3802
4.5 km/min	3.39	1.69	1.13	4321
5.0 km/min	3.32	1.66	1.11	2857
5.5 km/min	2.92	1.46	0.97	1383
Bulk				
Amorphous	2.86	1.43	0.95	1811
Crystalline	2.80	1.40	0.93	1035

TABLE 3

Cp Data (Literature from Smith and Dole)

	KJ/kg-K
Undrawn Yarn	1.138
Cold Drawn Yarn	1.154
Commercially Drawn Yarn	1.1419
Annealed From Melt	1.100
Annealed at Temp below T_m	1.114

Reprinted from *Optically Based Methods for Process Analysis,* D.S. Bomse et al., editors, Proc. SPIE Vol. 1681, pp. 276-290 (1992).

Applications of Raman spectroscopy to industrial processes

Stuart Farquharson
Advanced Fuel Research, Inc.
87 Church St., East Hartford, Connecticut 06108

and Stan F. Simpson
Dow Chemical USA
Building B-1400, Freeport, Texas 77541

ABSTRACT

The role of on-line chemical analyzers is vital to process monitoring and control and product quality. Although traditional optical filter methods such as UV-VIS, NIR, and IR have enjoyed considerable success when applied on-line, they often require inconvenient and complex sampling schemes. These restrictions can be largely eliminated by fiber optic probes as evidenced by their growing popularity. Fiber optic technology also allows remote location of full spectrum analyzers which in turn facilitates multicomponent analysis.

Recently, we have developed a Raman spectrograph which utilizes fiber optic probes and a CCD detector. We have been most successful with this system when it is applied to processes in a short term, investigative role. Examples of reaction intermediates and products, contaminant identification, and process optimization will be given.

2. INTRODUCTION

A variety of analyzers exist today for monitoring and controlling pharmaceutical, chemical, and petrochemical processes.[1] These range from traditional physical (mass, flow, pressure and temperature) and chemical (gas chromatography and photometers) measurement devices to more sophisticated analyzers which employ micro capillary gas chromatography, flow injection analysis, and Fourier transform infrared spectroscopy (FTIR).[2-4] The complexity of the latter techniques, however, often requires protective housing and extensive sample conditioning systems to ensure long-term instrument stability and reliability.

Raman spectroscopy, when coupled to sample probes via fiber optics, has the potential of overcoming these stringent requirements, and offers a unique set of advantages for process monitoring. First, Raman scattered radiation depends on molecular vibrations allowing chemical identification, and the signal is linearly proportional to the chemical concentration.[5-7] Second, Raman spectroscopy is performed typically in the visible region of the spectrum allowing the use of highly transmittant, inexpensive communication fiber optics.[8-14] This would allow placement of the instrument in a stable laboratory environment remote from the process.[8] Third, a fiber optic probe could be placed within a process vessel or pipe to perform reversible, in-situ analysis without requiring sample preparation, separation, disposal or addition of reagents.[10,15] Fourth, the recent advances in array detector technology,[16,17] particularly charge coupled device (CCD) detectors allow measurement of complete spectral regions simultaneously.[11,12] This allows for multicomponent analysis. Fifth, the fast response of these detectors, allows near real-time analysis on a continuous basis. Finally, partitioning the CCDs allows simultaneous, multiple sample point analysis.[13]

However, the application of Raman spectroscopy to industrial process monitoring requires overcoming several significant limitations: long term instrument instability, fluorescence interference, and sensitivity. Recent technological advances in lasers, optics, and detectors has significantly reduced the severity of these limitations. Instrument instability is typically caused by laser power fluctuations and beam to sample alignment. Today air cooled argon ion (Ar+), helium neon (HeNe), and diode lasers are capable of delivering greater than 50 mW of power without requiring excessive electrical power or water cooling.[18-20] New designs minimize power drift and extend operational lifetimes.[20] Direct coupling of fiber optic probes to the laser head and the spectrograph virtually eliminate sample alignment problems. The broad spectral quantum efficiencies (QE) of CCD detectors allow measurement at the red end of the spectrum, thereby minimizing fluorescence interference (Figure 1).[21-23] The associated fast electronics of these detectors also allow addition of numerous spectra to improve signal-to-noise ratios by as much as two orders of magnitude. Nevertheless, the detection limit (S/N = 2) of a solvent in water is approximately 1000 ppm for a typical system (Figures 2 and 3).[24] For this reason, surface enhanced Raman has been employed for measurements of organic contaminants in water,[25-27] and intracavity Raman has been used for gas phase measurements to increase sensitivity.[28,29]

Here, we report laboratory, research facility, and short term studies of process streams aimed at employing Raman spectroscopy for monitoring and eventually closed loop control of industrial chemical processes.

3. EXPERIMENTAL

Benzene and dichloromethane were reagent grade (Fisher) and used as received. All other samples were supplied by Dow process plants. The Raman spectrometer used for these studies consisted of a Lyconix model 5301 krypton ion laser, an NRC x,y,z, pitch and yaw model F-915 fiber optic alignment holder, an in-house designed and constructed fiber optic probe consisting of a 200 μ source fiber surrounded by four 600 μ collecting fibers having an overall length of 5 meters, contained in a stainless steel body, and a SPEX model 1877 triple monochrometer, fitted with a Photometrics model PM512 thermoelectrically cooled CCD. Typical conditions were as follows: 200 mW of 647.1 nm at the sample, 100 second collection time, 1 cm^{-1} resolution, and the fiber optic probe placed either into the process samples or 2 mm above the samples.

A portable compact Raman spectrograph employing similar fiber optic sampling probes and a HeNe laser was also used for short term on-line measurements for demonstration purposes.[30]

4. APPLICATIONS

Recently, extruders have been used for the continuous production of epoxy resins. Since the physical properties of epoxy resins are largely dependent upon the extent of polymerization (chain length and branching, see reaction below), the extruder design (mixing elements) and operating conditions (reactant feed rates) are critical to its performance.

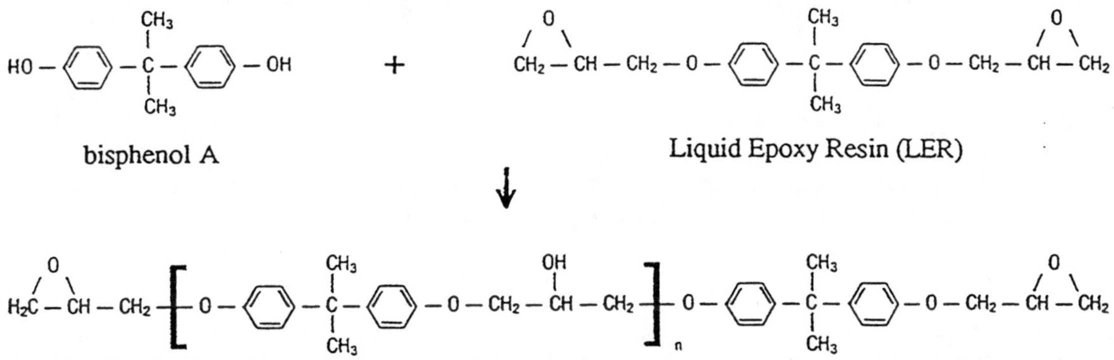

Previously, we coupled a visible spectrometer via fiber optics to a reaction extruder to optimize the engineering design,[31] however off-line chemical analysis was performed to optimize operating conditions. In an effort to eliminate the lag time between this analysis and extruder 'tuning' we investigated fiber optic Raman spectroscopy to perform the required measurements on-line and in real-time. Figure 4 shows the spectra of cured (DER) and uncured epoxy resin (LER). These preliminary results suggest that the epoxy ring mode at approximately 1250 cm^{-1} can be used to measure the extent of polymerization.

The synthesis of some polymers requires multiple steps in which intermediate products are formed, cleaned of excess reactants and solvents, followed by the addition of new reactants initiating the next step. The end point of these intermediate steps have traditionally been determined by collecting samples and performing a laboratory analysis typically once per hour. In an effort to eliminate lag times between intermediate formation and the initiation of the next reaction step, an on-line analysis was desired. Figure 5 shows the Raman spectrum of the maximum concentration of intermediate generated in a laboratory scale reactor signaling the completion of the first step. Installation of an in-line fiber optic Raman probe is being pursued.

In a particular polymerization reactor, the reaction occurs at the interface between an aqueous and organic phase. In order to generate high molecular weight product, a phase transfer catalyst (PTC) is required to increase the polarity of the non-polar oligomer fractions thereby holding them at the interface for further polymerization. Thus the molecular weight can be controlled by the concentration of PTC. This process also contains a solvent recovery recycle line. It is therefore necessary to measure the amount of PTC in the recycle solvent, before additional PTC is added to the system. Laboratory Raman measurements of neat PTC and the organic recovery solvent, allowed correct determination of PTC in the recycle stream (Figure 6).

Continuous polymerization processes require precise system control. Traditionally, this has been performed using engineering principles based on measurement of reactant feed rates, process conditions, and mass balance of the system. However, elimination of off-spec product and better quality control require more precise measurement of reactant concentrations entering and within the reactor, as well as the concentration of product within and exiting the reactor. A series of standards covering the operating range were prepared in a 316 stainless steel vessel pressured to 100 psig with the fiber optic Raman probe inserted through a Swagelok fitting. Raman spectra of typical reactant concentrations (solvent spectrally subtracted out) and generated product are shown in Figure 7.

A trace contaminant or side product can often have deleterious effects on a process. Severe corrosion was observed in a process water recycle stream, and several possible chemicals were suggested. Figure 8 shows the positive identification between a prepared standard of a suspect chemical and a process sample. To monitor the fluctuations of this chemical on-line, a compact portable Raman system equipped with a HeNe laser and a 30 meter fiber optic probe was used to measure the recycle stream (Figure 8c).

The mechanical properties of many polymers are often determined by tests, such as tear, stress and strain measured on the final product. In an effort to correlate process conditions of a polyethylene blown film extruder to product properties, on-line measurements of percent morphology were obtained using a fiber optic Raman probe (Figure 9). Spectra were recorded for the methylene twisting and bending modes (1300 and 1440 cm^{-1}, respectively, Figure 10), which have been correlated to polymer orientation and crystallinity.[32,33] In the amorphous state these bands are relatively broad, indicative of a large distribution of molecular environments. Upon crystallization, the number of environments decreases and the bands sharpen and shift to lower wavenumber. For example, the methylene twisting band shifts from 1303 to 1297 cm^{-1}, with a decrease in band width at half maximum from 26 to 16 cm^{-1}. A simplified percent morphology can be obtained by fitting this mode with two Lorentzian curves and normalizing the total area to one.[34] Kinetics can be obtained by fitting this data with a Sigmoidal curve:

$$y = A - Be^{-Cx^n}$$

where y corresponds to the percent morphology, x is the height along the bubble, and n is a rate factor (Figure 11). This equation allows determination of the maximum ordering rate (slope), at what height it occurs (second derivative), and the maximum order (A) for the first crystallization step.[33] Variations in operating conditions are directly observed as changes in coefficients A, B and C, and have been correlated to physical properties of the final product.

5. CONCLUSSIONS

A fiber optic based Raman spectrometer equipped with a CCD detector was used to perform both qualitative and quantitative chemical analysis for a variety of process streams. The use of fiber optic probes largely eliminated the complexity of sample systems associated with FTIR systems, yet maintained the spectral richness often cited for the latter technique. However, it should be noted that these measurements were largely performed in research facilities with the longest continuous data analysis lasting two days. Nevertheless, the ability to perform these measurements demonstrates the potential for Raman spectroscopy to develop and optimize process conditions, measure process kinetics, monitor process chemical concentrations, and eventually to be used for closed loop process control.

6. ACKNOWLEDGEMENTS

This research was supported by Dow, U.S.A., specifically several chemical processes of Texas Operations. The author appreciates the assistance of Anne Leugers (Michigan Division), Barbara Kirsch and Peter Arnoudse in performing these experiments. The author is also grateful to Kaelyn Koch, JoLynn Higganbotham, Dulanay Cortelek, Joe Stout and Kelly Jones (Louisiana Division) for making their research and process facilities available.

7. REFERENCES

1. K. J. Clevett, in "Process Analyzer Technology", John Wiley & Sons, New York (1986)

2. S. Farquharson, P. B. Arnoudse, M. H. Wyckoff, and P. T. Keillor III, in "Chemical, Biological, and Environmental Fiber Sensors", R. A. Lieberman and M. T. Wlodarczyk, Eds., Proc. SPIE, 1172, 164-173 (1989)

3. S. Farquharson and J. P. Chauvel, Jr., in "Environmental Sensing and Combustion Diagnostics", J. J. Santoreli, Ed., Proc. SPIE, 1434, 135-146 (1991)

4. S. Farquharson, P.D. Swaim, C. P. Christenson, M. McCloud, and H. Freizer in "Chemical, Biological, and Environmental Fiber Sensors III", R. A. Lieberman and M. T. Wlodarczyk, Eds., Proc. SPIE, 1587, 232-241 (1991)

5. G. Placzek, UCRL Translation No. 526L from Handbuch der Radiologie, 2, E. Marx, Ed. Akademische Verlags-Gesellschaft, Leipsag (1934),

6. R. S. Tobias, J. Chem. Ed., 44, 2-8 (1967)

7. A C. Albrecht and M.. C. Hutley, J. Chem. Phys.,55, 4438-4443 (1971)

8. R.L. McCreery, M. Fleischmann, and P. Hendra, Anal. Chem., 55, 146-148 (1983)

9. R. D. McLachlin, G. L. Jewett, and J. C. Evans, *U. S. Patent 4,573,761* (1986)

10. *A paper presented by S. Farquharson for* M. A. Leugers and R. D. McLachlan, "Chemical, Biological, and Environmental Applications of Fibers", R. A. Leiberman and M. T. Wlodarczyk, Eds. Proc. SPIE, 990, 88-95 (1988)

11. J. M. Williamson, R. J. Bowling, and R.L. McCreery, *Appl. Spectrosc.*, *43*, 372-375 (1989)

12. C. D. Allred and R.L. McCreery, *Appl. Spectrosc.*, *44*, 1229-1231 (1990)

13. S. M. Angel, T. M. Vess, and M. L. Myrick, in "Chemical, Biological, and Environmental Fiber Sensors III", R. A. Lieberman and M. T. Wlodarczyk, Eds., Proc. SPIE, 1587, 219-231(1991)

14. M. M. Carrabba, K. M. Spencer, and R. D. Rauh, in "Environmental Sensing and Combustion Diagnostics", J. J. Santoleri, Ed., Proc. SPIE, 1434, 127-134 (1991)

15. A. A. Garrison, E. C. Muly, M. J. Roberts, D. S. Trimble, and C. F. Moore, *ISA*, Research Triangle Park, 357-363 (1989)

16. W. H. Woodruff and S. Farquharson, in "New Applications of Lasers to Chemistry", G. H. Hieftje, Ed., American Chemical Society, Wash. D. C., chapter 12 (1978)

17. W. H. Woodruff and S. Farquharson, *Anal. Chem.*, *50*, 1389 (1978)

18. Product literature, Ion Laser Technologies (1992)

19. Product literature, Uniphase (1992)

20. Product literature, Lyconix (1990) and Spectra Diode Labs (1992)

21. Product literature, Photometrics (1990)

22. Product literature, SPEX (1991)

23. Product literature, Princeton Instruments (1992)

24. The conditions in Figure 2 (250 mW at 647.1 nm, a SPEX Triplemate with 21% throughput) are comparable to 50 mW at 514.5 nm and a single stage monochrometer (65% throughput). Similarly, we obtain a detection limit of approximately 50% for nitrogen at STP, with 100 mW at 647.1 nm, 60 second accumulation time, with 90° collection into a SPEX Triplemate, fitted with a Photometrics CCD

25. T. Vo-Dinh, D. L. Stokes, Y. S. Li, G. H. Miller, in "Chemical, Biological, and Environmental Fiber Sensors II", R. A. Lieberman and M. T. Wlodarczyk, Eds., Proc. SPIE, 1368, 203-209 (1990)

26. M. L. Myrick and S. M. Angel, *Appl. Spectrosc.*, *44*, 565-570 (1990).

27. M. M. Carrabba, K. M. Spencer, C. Rich, and R. D. Rauh, *Appl. Spectrosc.*, *44*, 1558-1561 (1990)

28. R. A. Van Wagenen, D. R. Westenskow, R. E. Benner, D. E. Gregonis, and D.L. Coleman, *J. Clin. Monit.*, *2*, 215-222 (1986)

29. D. R. Westenskow, K. W. Smith, D.L. Coleman, D. E. Gregonis, and R. A. Van Wagenen, *Anesthesiology*, *70*, 350-355 (1989)

30. Measurements employing the compact Raman spectrograph were performed by M. A. Leugers, Dow U.S.A., Midland, MI

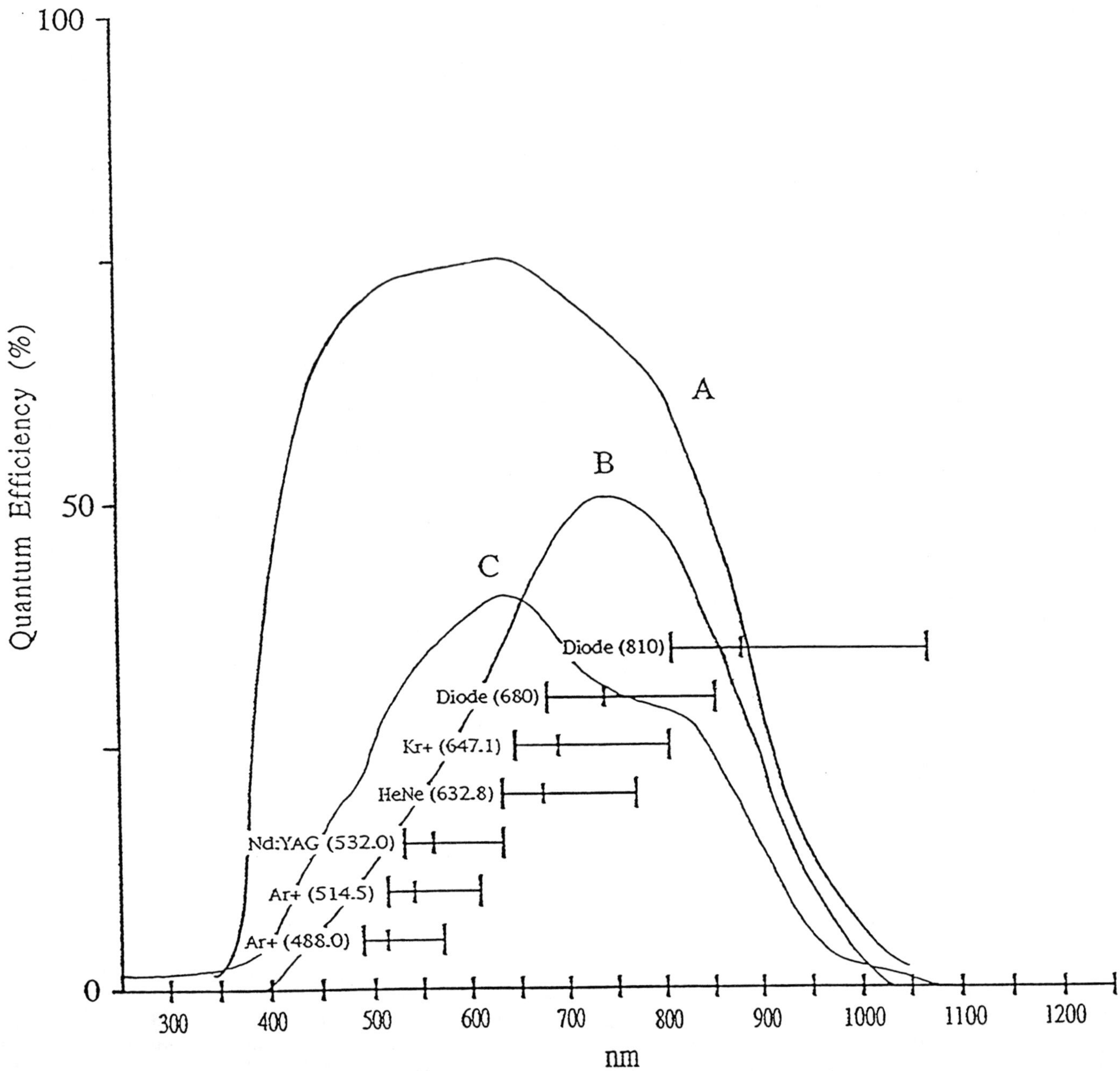

Figure 1. CCD quantum efficiencies for A) a Princeton Instruments back-illuminated model CCD-500/b (1992), B) a SPEX SPECTRUM ONE (1991), and C) a Photometrics Thompson TH7882 CDA (1990), reproduced from product literature. Laser excitation wavelengths and their resultant Raman spectral range (0, 1000, and 3000 cm^{-1}) are included. The dates indicate improvements in CCD technology, whereas the curves do not necessarily represent the "state of the art" CCD detectors for these companies.

31. S. Farquharson and P. T. Keillor III, in "Chemical, Biological, and Environmental Applications of Fibers", R. A. Lieberman and M. T. Wlodarczyk, Eds., Proc. SPIE, 990, 96-102 (1988)

32. F. J. DeBlase, M. L. McKelvy, M. Lewin, and B. J. Bulkin, *J. Polym. Sci.: Polym. Let. Ed.*, *23*, 109-115 (1985)

33. B. J. Bulkin, M. Lewin, and F. J. DeBlase, *Macromolecules*, *18*, 2587-2594 (1985)

34. Since x-ray crystallographic measurements do not yield similar percent crystallinity as these Raman spectra, we prefer to ascribe our measurements to percent morphology based on molecular ordering within a distribution of environmental states.

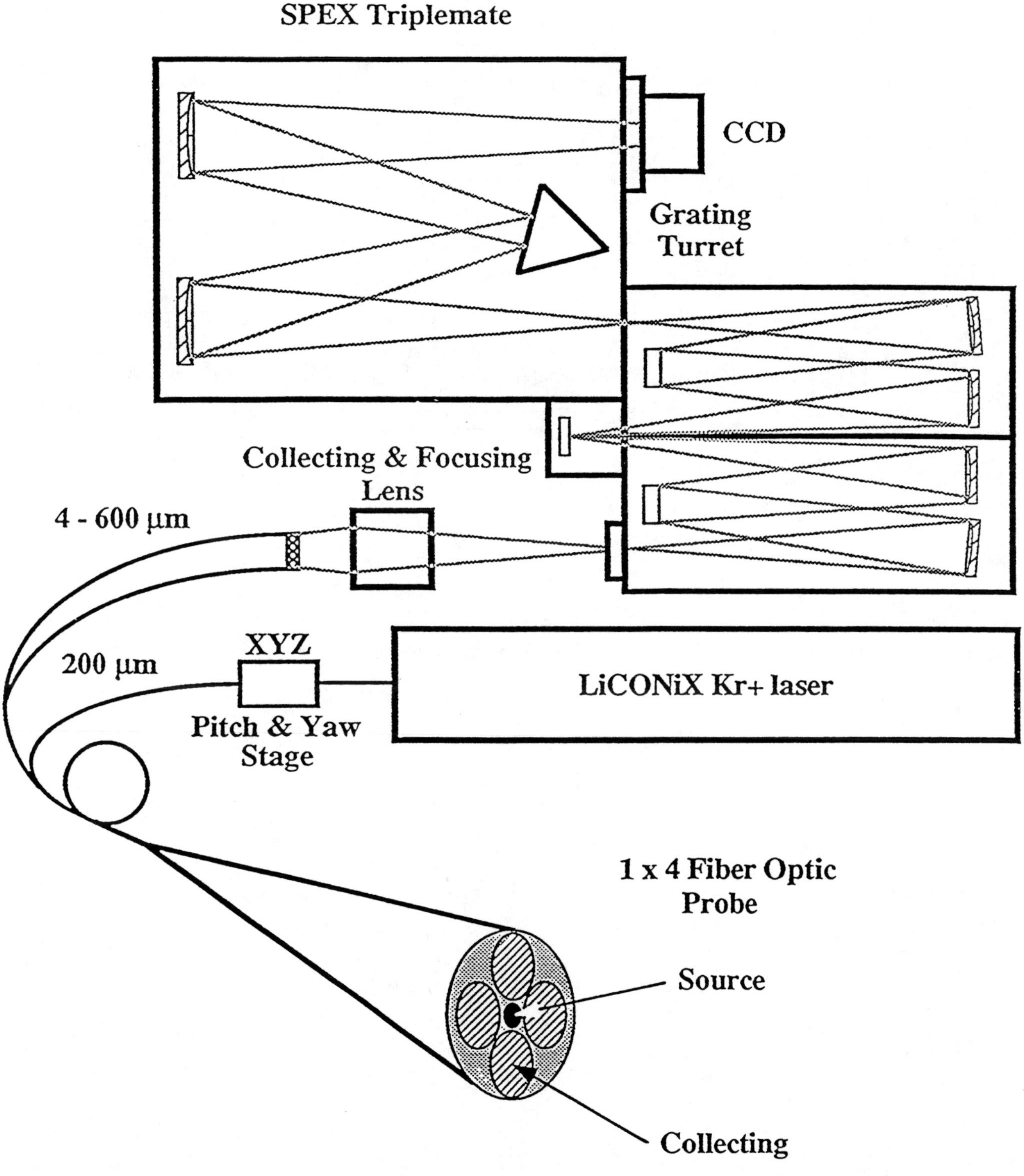

Figure 2. The laboratory Raman system used for feasibility and research facility studies. See Experimental section for component details.

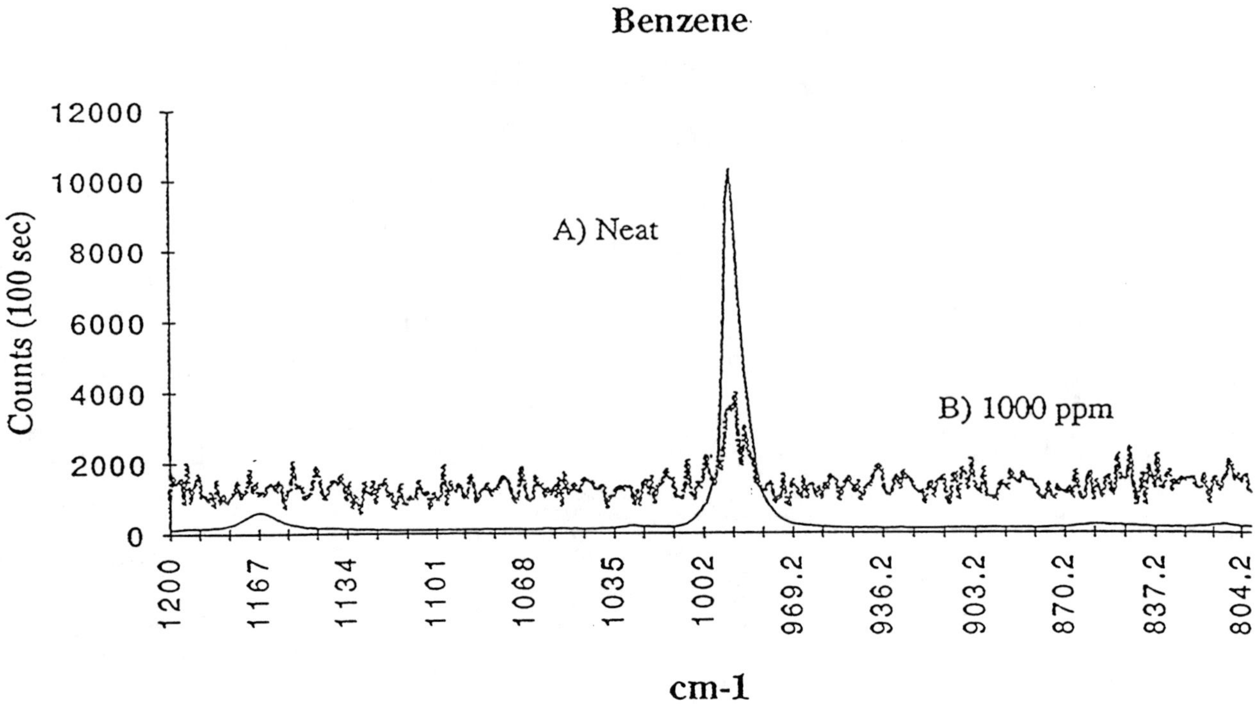

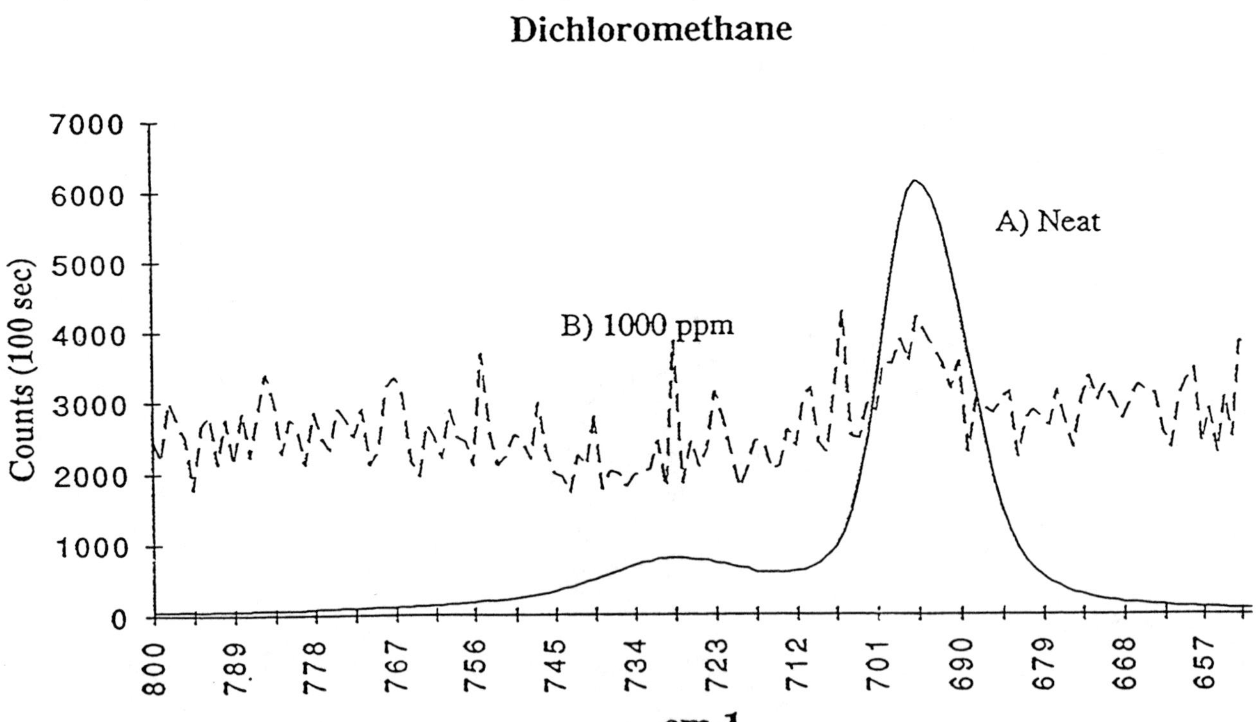

Figure 3. Raman spectra of benzene and dichloromethane in water, illustrating detection limits. The system in Figure 2 was used with the following conditions: 200 mW of 647.1 nm at the sample, 100 second collection time, 1 by 4 five meter fiber optic probe in a 100 mL beaker.

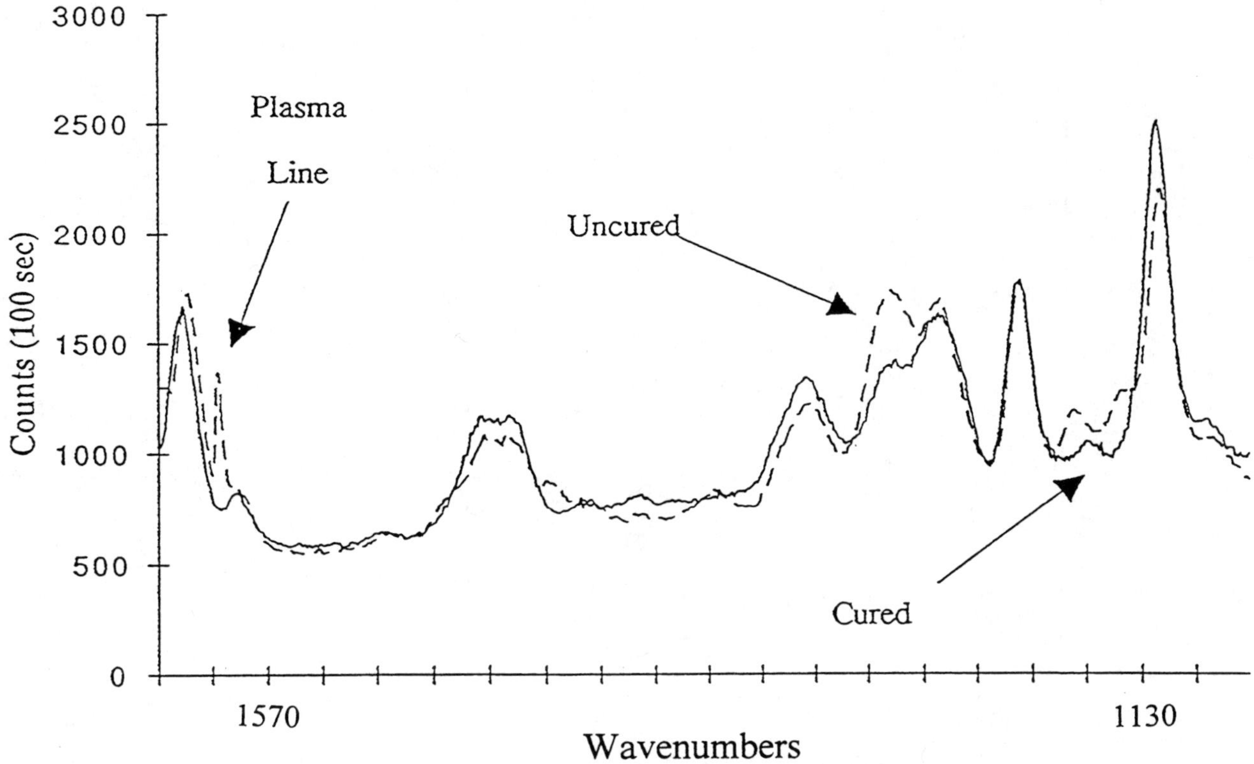

Figure 4. Raman spectra of uncured (LER) and cured (DER) epoxy resins. Conditions are the same as in Figure 3, except the fiber probe was placed above the sample, but not in contact.

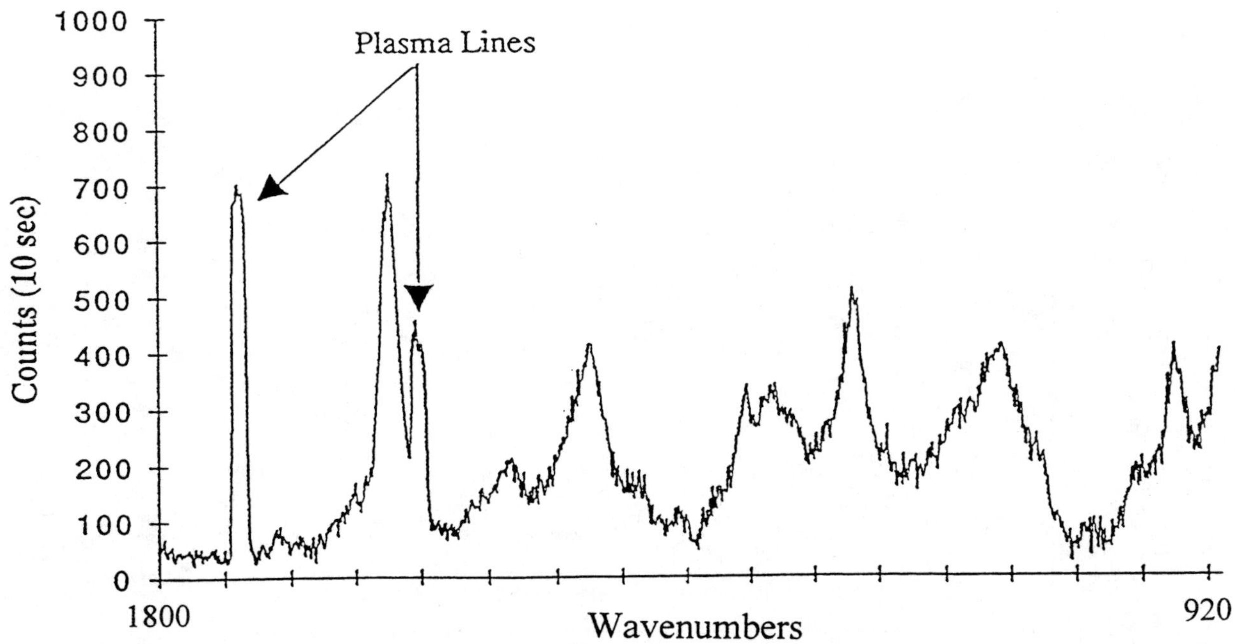

Figure 5. Raman spectrum of a polymer intermediate during kinetic measurements. Conditions are the same as Figure 3, except the collection time was 10 seconds and the sample was contained in a reaction flask.

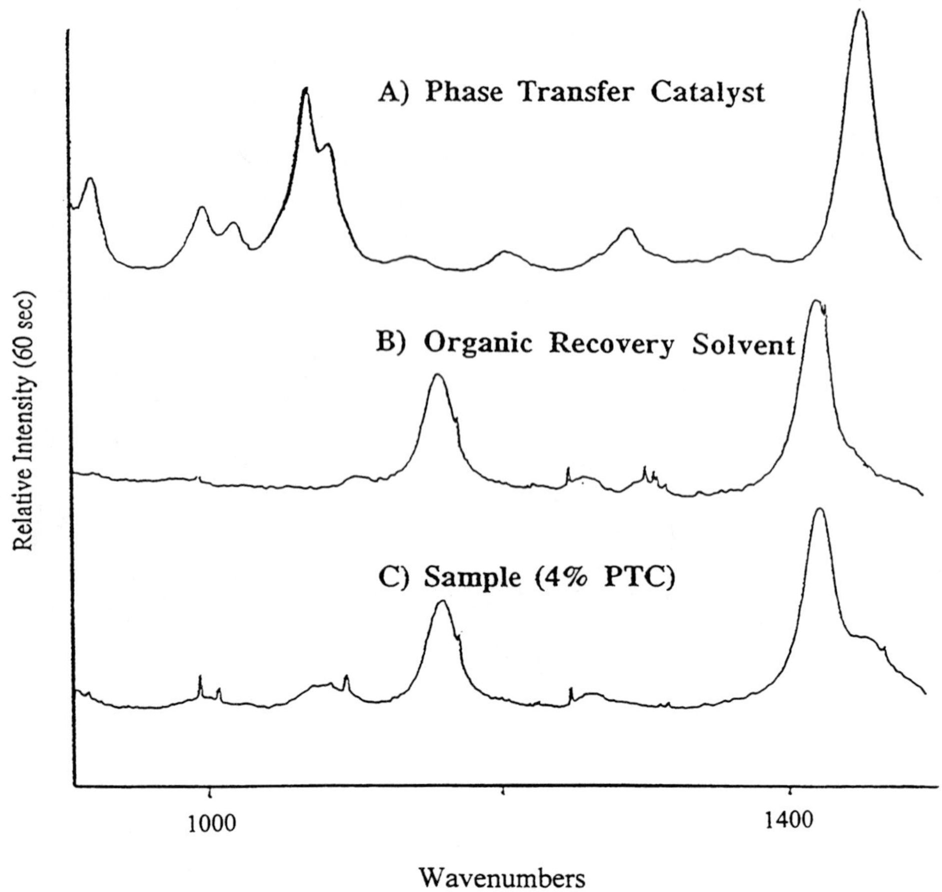

Figure 6. Raman spectra of A) pure phase transfer catalyst, B) pure organic recovery solvent, and C) a plant sample of 4% catalyst in the solvent. Conditions similar to Figure 3, except a 60 second collection time was used.

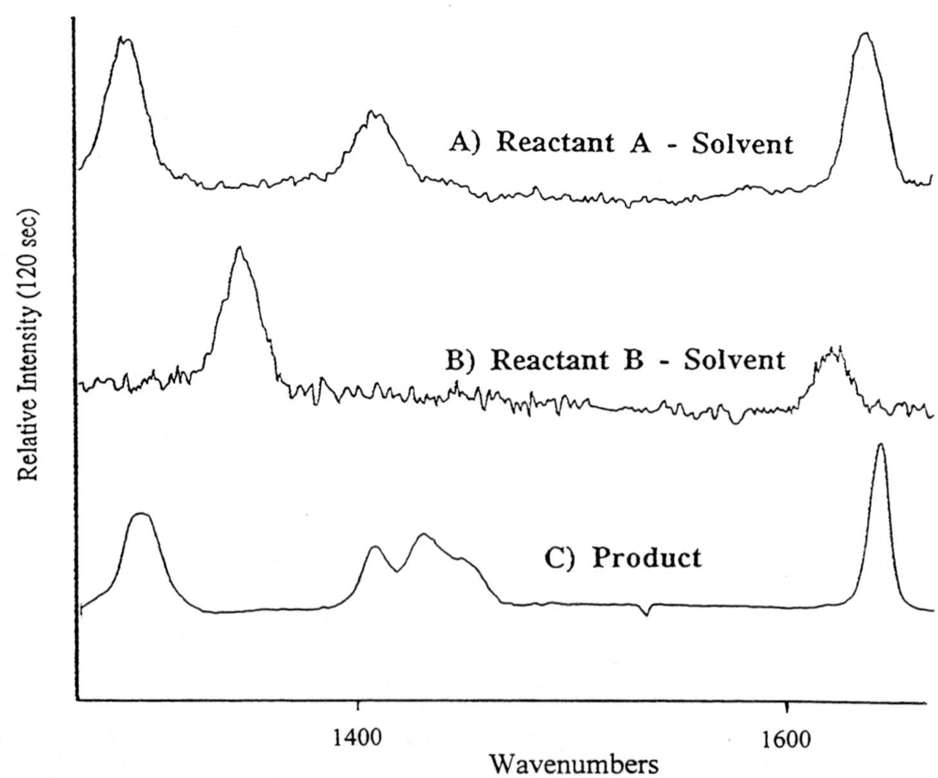

Figure 7. Raman spectra of reactants A and B with solvent bands spectrally subtracted, and C) the product. Conditions similar to Figure 3, except the probe was 'swaged' into a 316 SS reaction vessel operated at 100 psig.

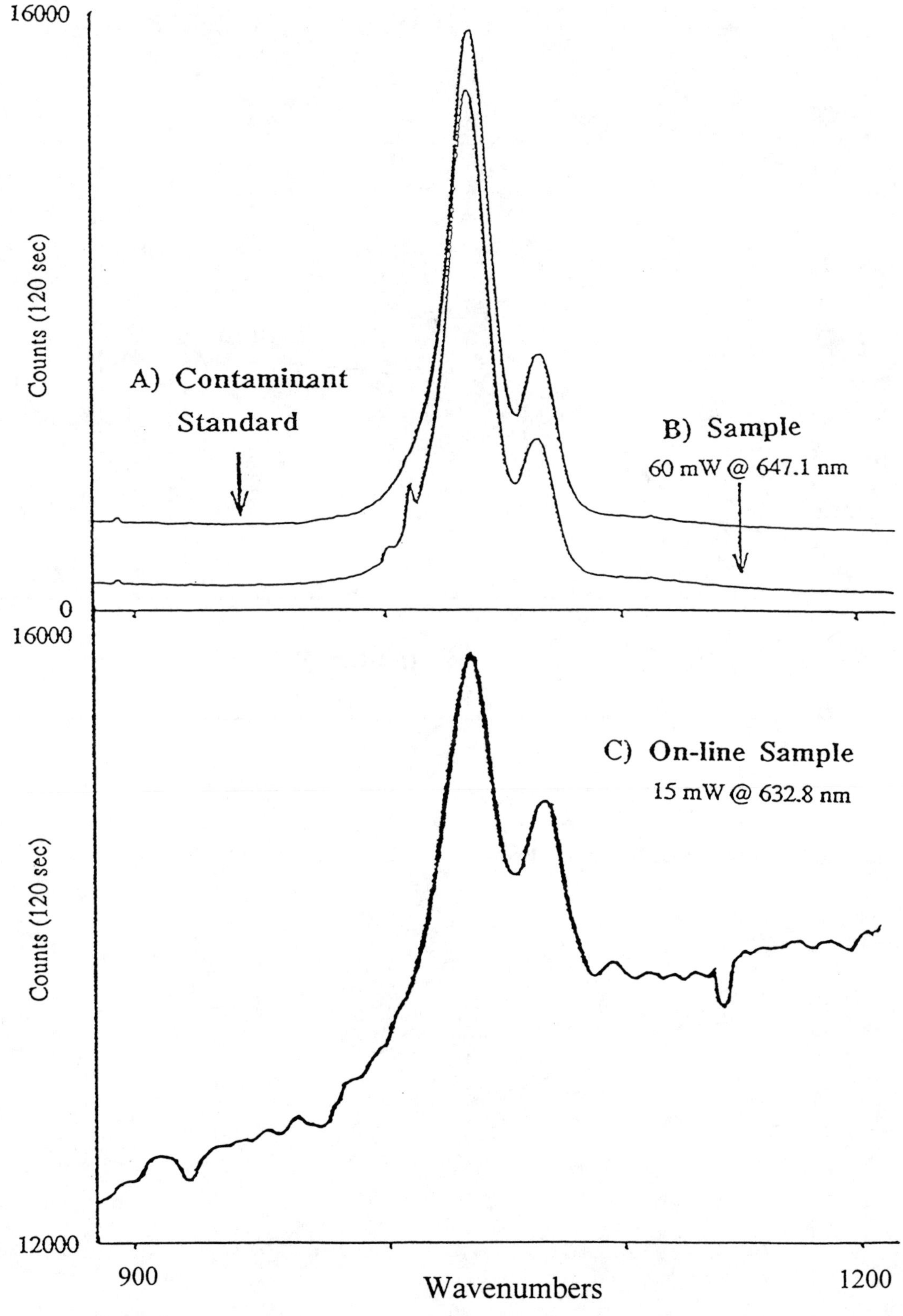

Figure 8. Raman spectra of A) a water contaminant standard used to identify and quantitate B) the process sample, and C) a process sample measured with a compact portable Raman system employing 30 meter fibers. Conditions are shown.

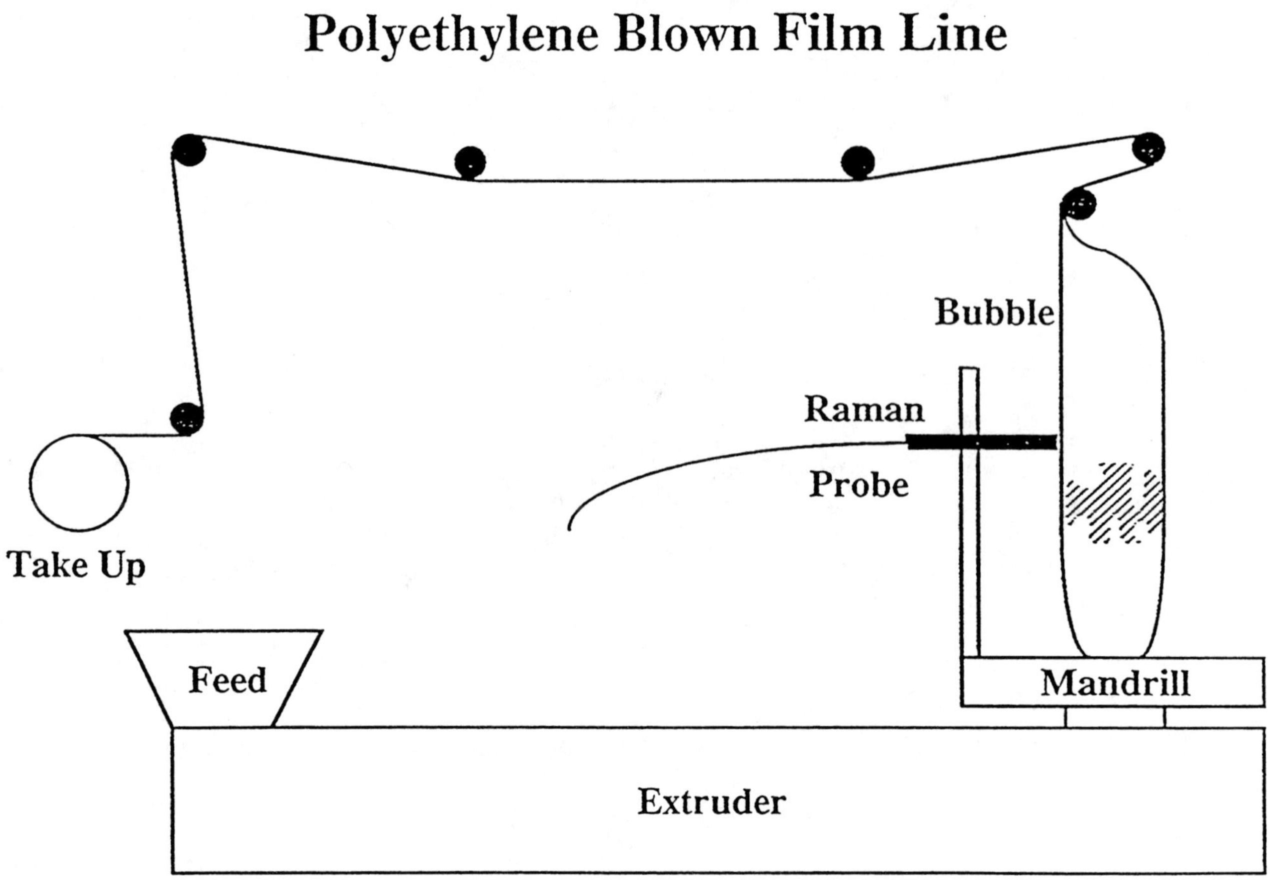

Figure 9. Schematic of a polyethylene blown film line. The position of the fiber optic Raman probe is shown.

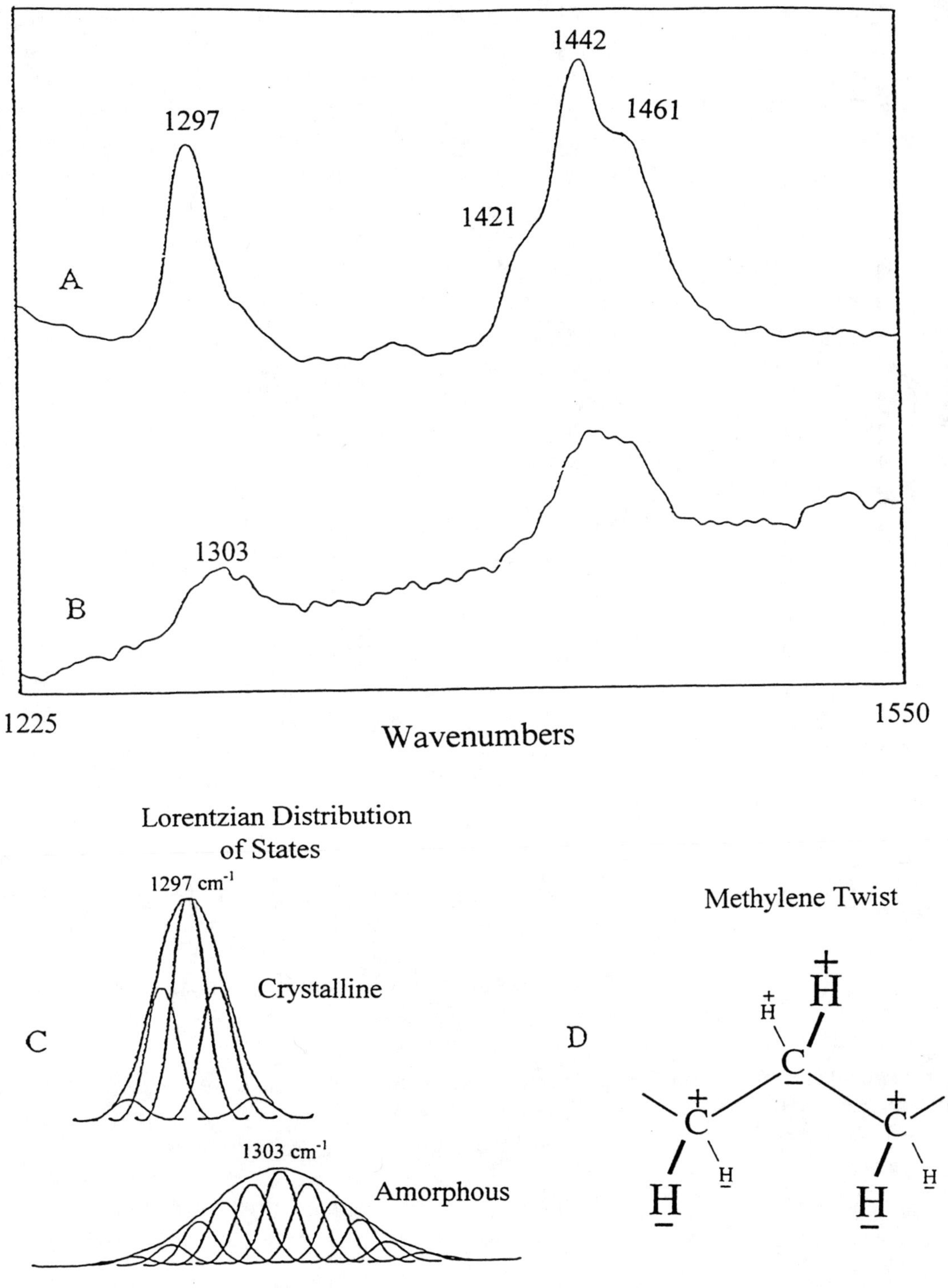

Figure 10. On-line fiber optic Raman spectra of a polyethylene blown film. A) Crystalline at the "take up role", B) amorphous, 1 inch above the mandrill, prior to the frost line, C) illustration depicting a possible distribution of environmental states for the D) methylene twisting mode. Conditions are similar to Figure 3.

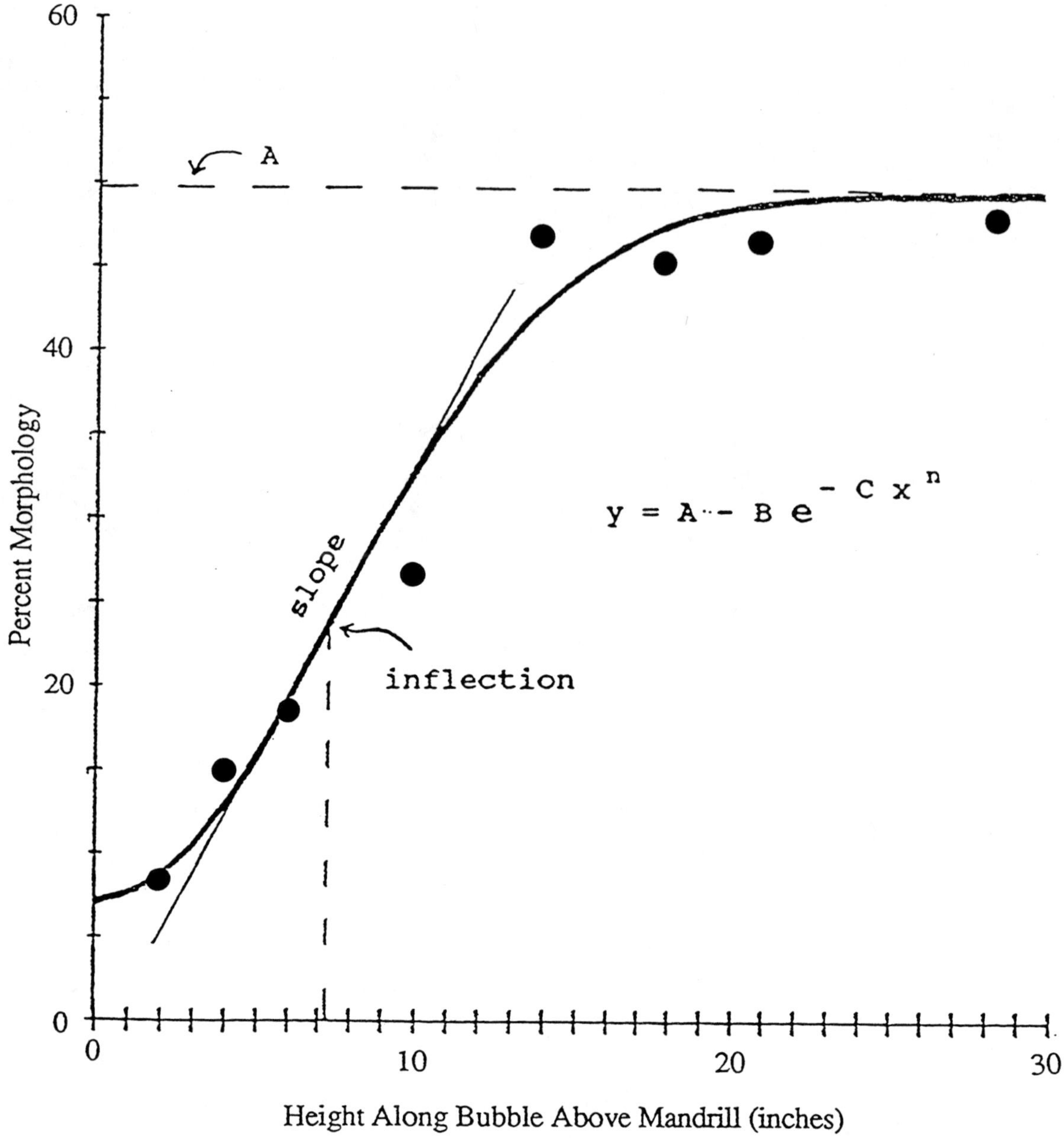

Figure 11. A Sigmoid curve fit to the calculated percent morphology of the methylene twisting mode as a function of height along the bubble. Key parameters are shown.

Reprinted with permission from *Spectrochimica Acta,* Vol. 22(9), pp. 1678-1680 (September 1966).

Spectrochemical analysis of molten metal using a pulsed laser source

(*Received* 20 *August* 1965)

Introduction

The possibility of determining the chemical composition of a material situated in a hostile environment, such as molten metal within a furnace, by a method which does not require the removal and cooling of the sample, has long intrigued both melters and analysts. The practicality of performing such an analysis by means of a laser has been conjectured by the authors since finding, in 1963, that a laser can excite useful spectra from solid metals without an auxiliary electrical discharge. If a giant-pulse laser can coincidentally vaporize and excite atomic emission spectra from solid samples, it is highly probable that it will function equally well with molten samples. However, there is the question of whether the intense radiant continuum from the surface of the molten sample will prevent detection of spectrum lines above this background radiation. The experiment to be described was designed to test the quantitative nature of pure laser excitation of major constituents in molten metal samples located within a small laboratory-size induction furnace.

Apparatus

The ruby laser is operated in a Q-spoiled, giant pulse mode. When the laser is operated in this mode, the metastable chromium atom population is prevented from radiating, except in a random fashion, until most of the atoms are excited. A rotating mirror then feeds back the random radiation down the long axis of the ruby rod within the proper angle to deplete the energy of the entire excited chromium atom population. The output beam in this condition exhibits peak power in the 3–5 MW range and durations of 15–20 nsec with energies of 50–75 mJ. Using a laser in the giant pulse mode allows improved beam collimation, often approaching the theoretical diffraction limit of the ruby rod and resulting in sharper focus than obtainable from a non-giant pulse laser. As a result, the power density near a focus is in excess of 10^{13} W/cm^2.

The samples to be analysed are melted in a 2·5 KW Leco (Laboratory Equipment Corp.,

St. Joseph Michigan) induction furnace and the spectra are produced with a large Littrow quartz spectrograph. The spectrography, plate processing and densitometry are carried out according to standard practices. Figure 1 is a schematic diagram indicating the arrangement of the system by which the laser is optically coupled with the spectrograph, and showing the relative location of the induction furnace.

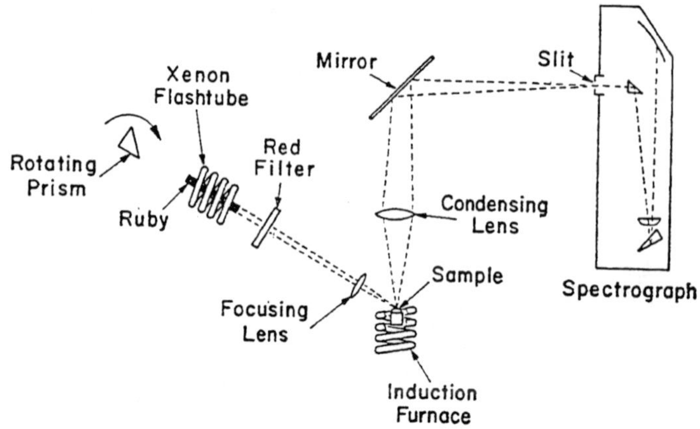

Fig. 1. Schematic diagram of the optical arrangement used in laser emission spectroscopy of molten metal.

Experimental

The laser is mounted pointing downward at an angle of 30° from the horizontal to allow the beam to clear the coil of the induction furnace and strike the surface of the sample. The laser beam is first passed through a filter (Corning No. 2–58) to remove the non-coherent xenon flash tube radiation that accompanies it. The beam is then brought to a sharp focus on the sample surface by means of a short (1·5 cm) focal length lens. A thin, optically flat, quartz plate is placed between this lens and the sample to prevent damage to the lens from spattering of the molten metal. Glass optics are used in the laser system, since the only radiation of interest is the 6943 Å line of the ruby rod. However, once excitation of the sample has been produced at the focal point of the laser beam, a quartz lens and a first surface mirror are used to pass the ultra-violet light, resulting from the sample excitation, to the spectrograph.

The optical axis of the spectrograph is in its normal horizontal position (Fig. 1 implies a vertical arrangement), and the spectrographic slit is displaced 12 cm vertically from the focal point of the laser beam on the sample. The excitation spot on the sample and the spectrographic slit lie at conjugate foci of a quartz lens used to image the sample light at the spectrograph. The first surface mirror matches the two optical axes.

The samples received only that preparation normally used in spectrochemical analysis. In this experiment cylindrical sections of stainless steel rod standards (6·4 mm long by 5·6 mm dia.) were used with one end surface ground flat. The samples were located in 6·4 mm deep by 6·4 mm dia. holes drilled in a graphite cylinder 2·5 cm thick by 2·5 cm dia. Three laser-excited exposures per sample were superimposed on the spectrographic plate (Kodak 103–0) to produce lines of adequate intensity to be read on the densitometer. The volume of sample vaporized by the laser is calculated to be of the order of 10^{-6} cm^3 per discharge.

Results

Three stainless steel standards of differing composition were used to prepare analytical curves for nickel and chromium. In order to compare the results obtained from molten material with those of solid material, spectra of each of the three standard samples were first recorded while they were at room temperature. The samples were then each melted in turn, and a second series of spectra from the laser excitation of the molten metal was recorded. The transmissions of the line pairs Ni 3414·77 Å/Fe 3765·54 Å and Cr 3963·69 Å/Fe 3765·54 Å were measured in each of the two sets of spectra with a recording microphotometer, and after background correction the log intensity ratios were plotted on semi-logarithmic paper against the known nickel and chromium percentages to produce the analytical curves in Fig. 2. The relationships are straight lines over the concentration ranges covered by the standards in both the solid and molten states.

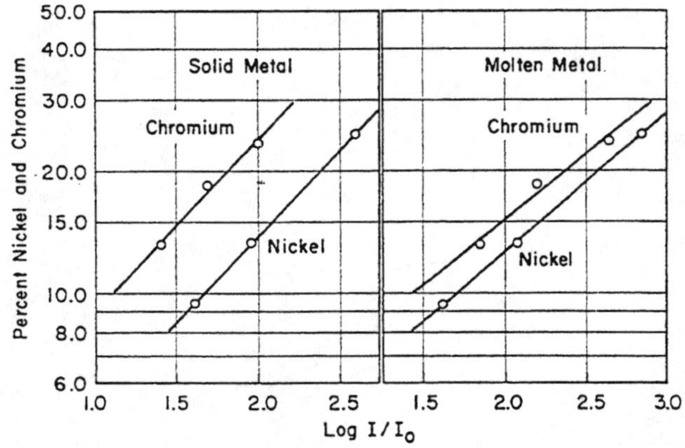

Fig. 2. Analytical curves for nickel and chromium based on laser-excited spectra of solid and molten samples.

Discussion

It is encouraging to find that background continuum from the surface of the molten metal does not prevent measurement of lines of major constituents of the melt. Analytical curves for nickel have approximately the same slope whether the material is in solid or molten state. The chromium curves differ, however, in both slope and position (possibly due to a thin oxide layer on the surface of the molten metal), indicating that it may be difficult to analyse molten

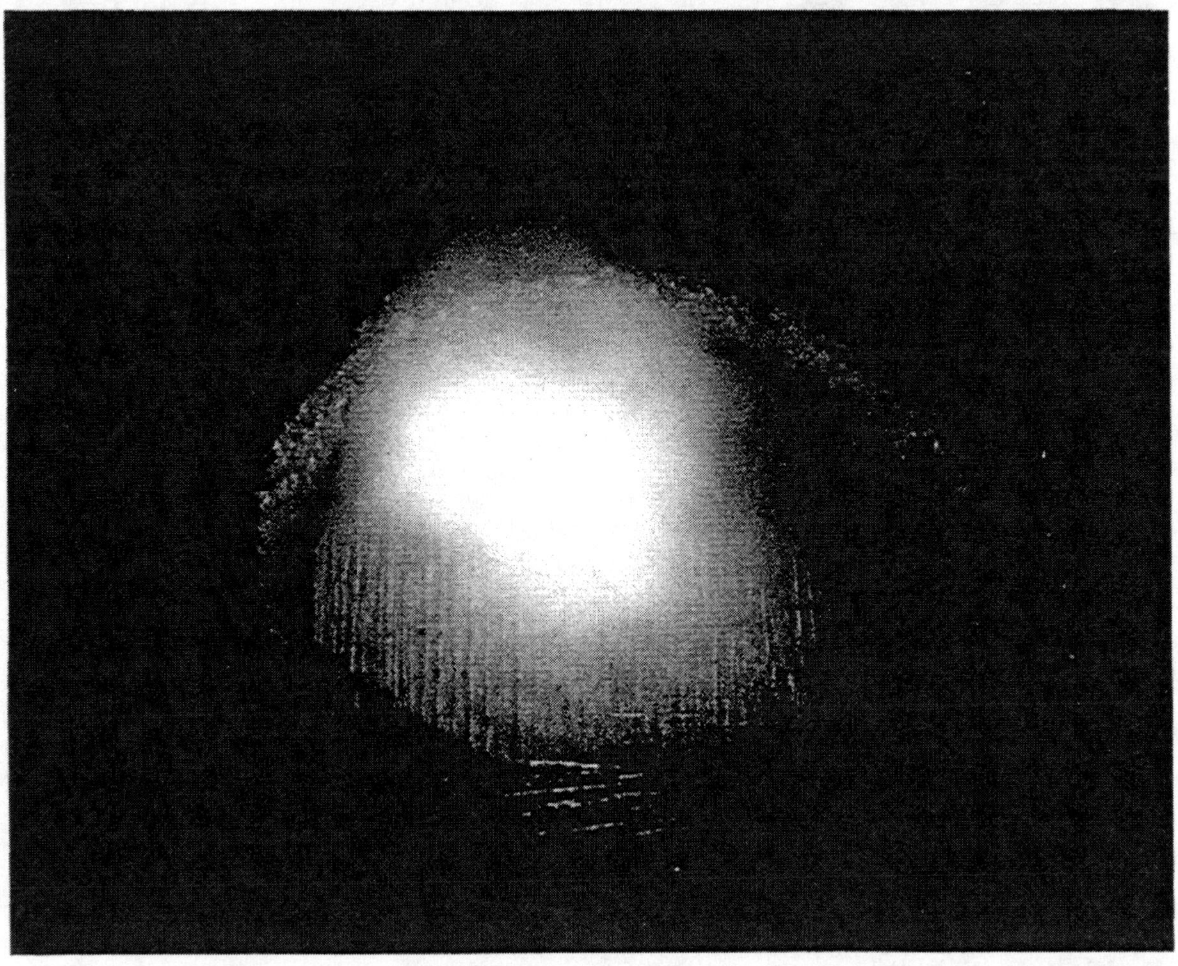

Fig. 3. Focal hot spot and plasma cloud of solid metal sample during laser-excitation.

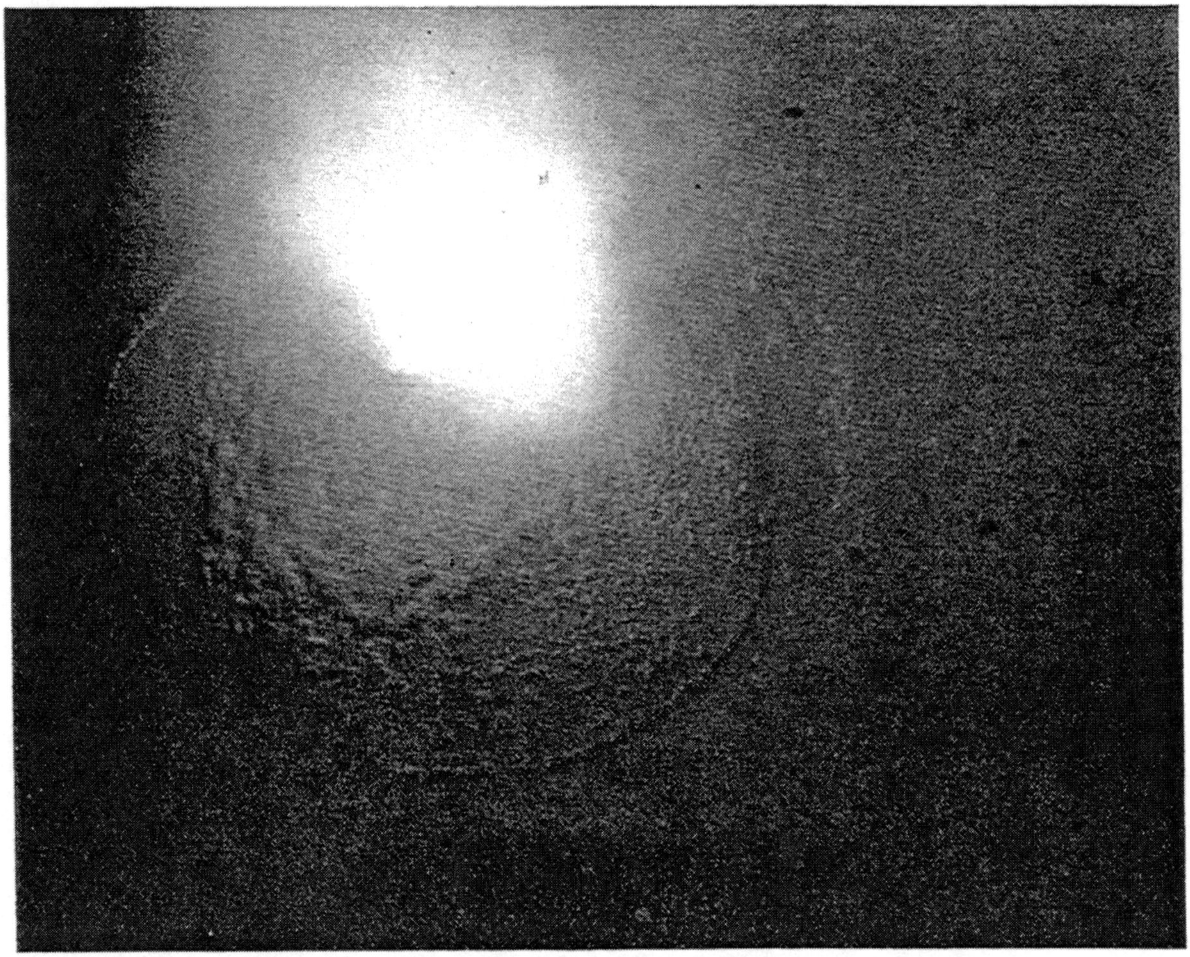

Fig. 4. Focal hot spot and plasma cloud of molten metal sample during laser-excitation.

materials by calibrating with solid standards. The fact that the entire sample is molten appears to have little effect on the spectrum. Background radiation would appear to be a problem with minor constituents, however, whether samples are solid or liquid.

Photographs of the plasma cloud formed during vaporization–excitation of both solid and molten samples (Fig. 3 and 4) show the diameter of the luminous cloud to be several times that of the crater diameter. In this case, the crater is about 0·5 mm dia, the cloud about 3 mm dia. Color photographs reveal the cloud to be blue-white in color, in contrast to the yellow-red molten metal background. The plasma is seen moving upward, perpendicular to the sample surface; not in the forward direction along the reflected axis of the primary laser beam. This large vapor cloud could account for the sometimes conspicuous number of self-reversed lines appearing in laser emission spectra.

Maintaining the simplest imaging system, using a single lens only, the experiment described needs to be scaled upward by perhaps two orders of magnitude to be of practical significance in commercial melting procedures. The laser would need to be of sufficient power to function at a distance of several feet from the molten bath, and excite vapor over a considerably larger focal depth. Laser technology is rapidly advancing, and sufficiently powerful lasers are becoming available. A major problem in analysing molten metal remotely by means of a laser would appear to be trace element detectability in the presence of background due to the intense hot spot produced by the laser discharge itself, rather than background due to the molten condition of the bulk metal.

Ford Motor Company
Engineering and Research Staff

E. F. Runge
S. Bonfiglio
F. R. Bryan

Reprinted with permission from *Journal of Analytical Atomic Spectrometry,* Vol. 7, pp. 1029-1035 (September 1992).

Applications of Laser-induced Emission Spectral Analysis for Industrial Process and Quality Control*

Claus J. Lorenzen, Christoph Carlhoff, Ulrich Hahn and Martin Jogwich
Krupp Forschungsinstitut GmbH, Postfach 10 22 52, D-4300 Essen 1, Germany

Laser-induced emission spectral analysis (LIESA, a registered trademark of instruments developed by Krupp), better known in the literature as laser microanalysis or laser-induced breakdown spectroscopy, is a suitable method for the direct in-process measurement of elemental concentrations in various solid and liquid materials. This method has been developed recently by Krupp for in-process quality assurance and process control in different industrial branches such as steel production and plant making. As a result several LIESA instruments have already been developed or are under development for marketing. In all cases on-line and in-process elemental analysis of materials at various stages of production yield information on the quality of the material and the fabrication process. The beam of a pulsed high-power laser (irradiance: 1×10^8–5×10^9 W cm^{-2}), focused onto the solid or liquid sample surface in an ambient gas atmosphere of normal pressure (focus area$\approx$ablation area, 0.1–6 mm^2), produces a hot bright plasma (early electron temperatures, 20 000–30 000 K). The emitted plasma light is observed end-on and passes by way of an optical fibre bundle to a spectrometer, where it is detected in the focal plane by means of an optical multichannel analyser with high time resolution (on the microsecond scale). A fast computer evaluates the measured spectra and calculates the element concentrations *via* calibration procedures. Relative detection limits of between 10 and 100 ppm can be achieved for most of the detectable elements in various matrices (steel, rubber, rock and glass). Procedures are available to convert relative measurements with relative standard deviations of between 1 and 2% into absolute concentration values with relative accuracies of about 3%.

Laser ablation of solid or liquid samples and optical emission spectrometry of the microplasma produced is a simple and fast technique for direct elemental analysis. Time-consuming sample preparation can be omitted. Bulk and surface analysis in addition to depth profiling are possible with the same apparatus. Tight focusing of the laser beam even allows microanalysis of sample surfaces. Recently, two textbooks[1,2] have been published briefly describing the historical background and the present status of scientific investigations and applications. In the period 1989–1992 Niemax and co-workers[3–8] published a series of papers presenting systematic investigations of the spatial and temporal evolution of laser-produced microplasmas and ablation of the related material. By using a Q-switched Nd:YAG laser operating at its fundamental wavelength (1 064 nm) Niemax and co-workers found the following experimental conditions and parameters to be best-suited to laser-induced emission spectral analysis: (*i*) use of a buffer gas at reduced pressure (most appropriate, argon at 140 hPa); (*ii*) use of reduced laser irradiance (no electrical breakdown in the surrounding atmosphere when the sample is removed); (*iii*) use of long delay times ($\geqslant$30 μs) between laser pulse and detector gate pulse; and (*iv*) use of analyte and reference spectral lines with comparable excitation energies.

A buffer gas plasma formed simultaneously with the ablation process serves as an energy reservoir for the atomization of ablated droplets and particles, and for the excitation of atoms and ions. The heating of the buffer gas plasma by inverse bremsstrahlung is most effective with argon. Since a buffer gas keeps ablated material in the observation region for a longer time and a larger fraction of this material can penetrate into the buffer gas plasma at reduced pressure, best analytical results were achieved with argon at 140 hPa. Recombination of the buffer gas plasma is completed after approximately 1 μs. At this time ablated material from the sample surface reaches the centre of the buffer gas plasma at almost supersonic velocities. Atomization and excitation processes start to develop. Complete atomization is achieved after 20–30 μs with number densities of free atoms and ions and plasma temperatures remaining high (>6 000 K). The buffer gas–sample material plasma can be described by partial local thermodynamic equilibrium verified by plasma temperature measurements using Boltzmann plots. Internal standardization can be applied and matrix independent measurements (common calibration graphs) can be performed when delay times greater than 30 μs and analyte and reference spectral lines of comparable excitation energies are used in optical emission spectrometry.

In 1986 Krupp started with basic investigations of the laser analysis method applied to solid and liquid steel samples (Carlhoff *et al.*[9]). A pilot system for on-line and in-process carbon and temperature monitoring was developed and successfully operated in a Krupp steel plant (Carlhoff and Kirchhoff[10]). This was the first successful attempt worldwide to install a laser-induced emission spectral analysis (LIESA) system in the harsh environment of a production steel plant. Meanwhile Krupp started the development of a laboratory prototype for the in-process monitoring of element distributions in polymeric materials (Lorenzen *et al.*[11]). Further activities in this field are the on-line control of geological raw materials on conveyor belts and investigations of depth profiling of multi-layers on metallic substrates.

The following sections describe the basics and some details of the different industrial LIESA applications at Krupp. Emphasis is placed mainly on instrument development rather than on fundamental scientific investigations.

Instrumentation

The general set-up of all LIESA instruments is shown in Fig. 1. In some applications where element distribution measurements are performed, a linear scanner system is

*Presented at the 1992 Winter Conference on Plasma Spectrochemistry, San Diego, CA, USA, January 6–11, 1992.

Table 1 Hardware components and ranges of operating conditions of LIESA instruments

Q-switched Nd:YAG laser—	
Wavelength/nm	1064 or 355
Mode	Multimode
Pulse energy/mJ	25–300 (on sample surface)
Pulse width/ns	6–9
Repetition rate/Hz	10–25
or	
Excimer laser—	
Wavelength/nm	248 (KrF)
Pulse energy/mJ	50–110 (on sample surface)
Pulse duration/ns	26
Repetition rate/Hz	15–50
Optical multichannel analyser—	
Detector	Gated MCP-intensified diode array (1024 pixel, UV enhanced)
Sensitive wavelength range/nm	150–900
Wavelength capture range/nm (see spectrometer)	14–42 (in the UV)
Gate widths/μs	0.2–10
Delay times (laser gate)/μs	0–30
Data transfer speed/spectra per second	Maximum 50
Spectrometer—	
Mounting	Czerny–Turner
Focal length/m	0.5 (*f*/4)
Grating/rules per mm	3600 or 2400
Reciprocal dispersion/nm mm^{-1}	0.6 or 1.7 (in the UV)
Entrance slit-width/μm	≤80
Linear scanner system—	
Scan range/m	0–1.2
Scan velocity/m s^{-1}	0–2
Scan acceleration/m s^{-2}	0–6
Load/kg	20

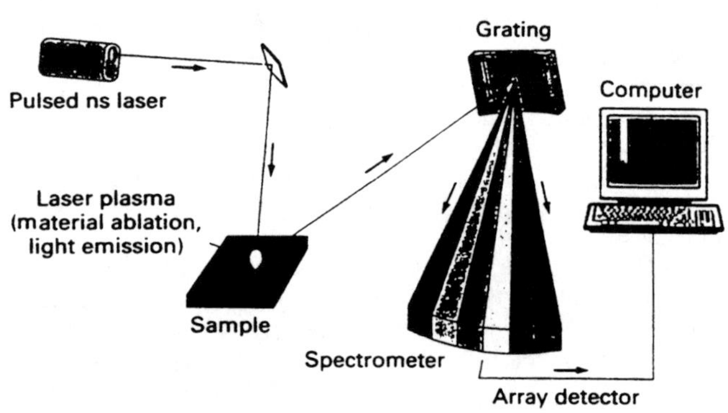

Fig. 1 General set-up of LIESA instruments

employed to scan the laser beam across the sample surface.

The specifications for the hardware components and the operating conditions are summarized in Table 1. Since the operating parameters are dependent on the application of the instrument, parameter ranges are given in most instances for reasons of simplicity.

When designing a LIESA instrument for a specific application special attention has to be paid to the selection of the appropriate laser system. Commercial Nd:YAG and excimer lasers were found to be most suitable for operation in industrial environments owing to their sturdiness and ease of operation. High pulse-to-pulse stability of both lasers [about 2% relative standard deviation (RSD)] and long lifetimes of the flashlamp of the Nd:YAG laser and of the gas fills and resonator mirrors of the excimer laser (up to

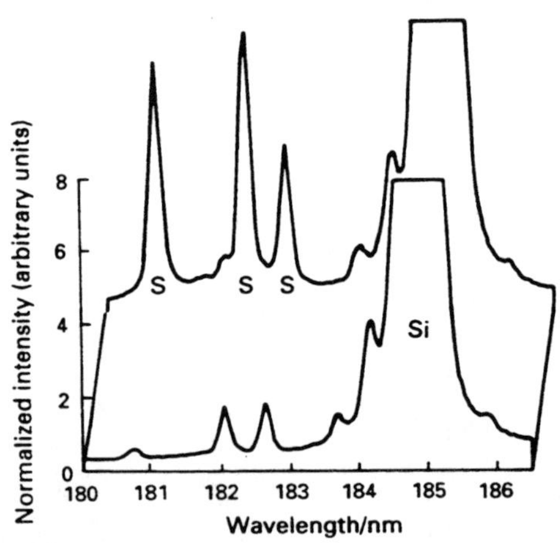

Fig. 2 Comparison of detector signals in the spectral region 180–186 nm obtained with use of conventional optics (upper trace) and 3 m fibre optics (lower trace) from a rubber sample. The sulfur concentration is 1.12%

1×10^8 laser pulses in both cases) are additional features making these lasers useful for industrial applications.

In the next step the appropriate operating laser wavelength is most important to obtain the best analytical results. It was found that the 1024 nm Nd:YAG laser produces plasmas with strong emission on metallic and glass surfaces. Ultraviolet (UV) excimer laser wavelengths must be employed on polymeric surfaces since only then are sharp and regular ablation patterns produced without any thermal side-effects. For geological samples infrared

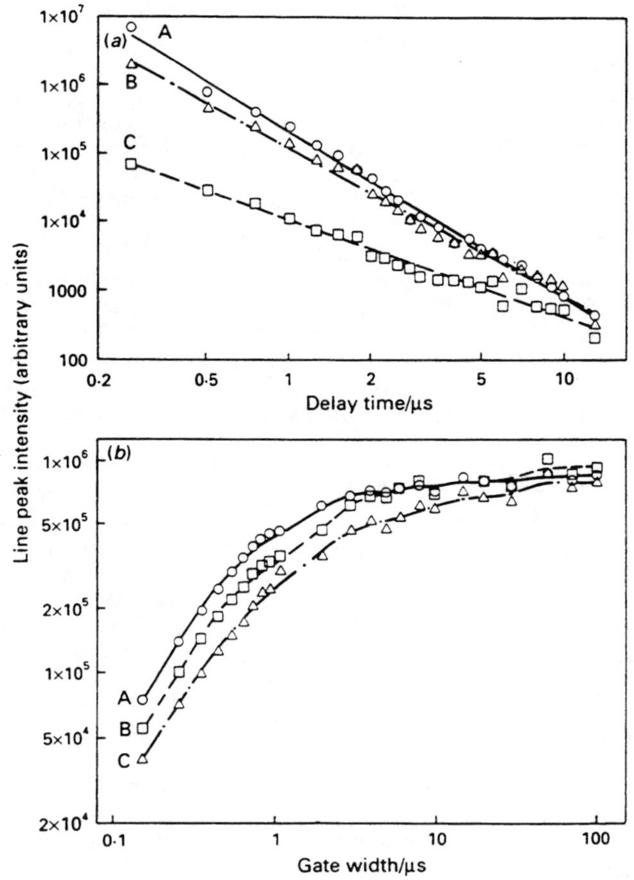

Fig. 3 (*a*) Double-logarithmic plot of peak intensities of the lines: A, Si I 288.2 nm; B, Ca II 393.4 nm; and C, Al I 396.2 nm *versus* delay time. The gate width is always 200 ns. A rubber sample was used as the target. (*b*) Double-logarithmic plot of peak intensities of the lines: A, Mg II 279.6 nm; B, Mg I 285.2 nm; and C, Si I 288.2 nm *versus* gate width. The delay time is always 200 ns. A rubber sample was used as the target

(IR) as well as UV laser light give comparable analytical results.

Another key component in the LIESA instruments are the optical fibres used for guidance of the plasma emission light to the spectrometer system, thus greatly simplifying the opto-mechanical design and adjustment of the apparatus. The fibre output is directly coupled to the entrance slit of a 0.5 m spectrometer. It can be seen from Fig. 2 that the transmission of the fibre guide near the sulfur triplet rapidly decreases with decreasing wavelength (lower trace of Fig. 2) compared with the spectrum taken with direct lens imaging (upper trace of Fig. 2). The beam paths and spectrometer are purged with Ar and N_2, respectively. In addition, the cut-off transmission wavelength at about 180 nm is also governed by limited detector window transmission and detector sensitivity in addition to limited transmission through residual oxygen in the purged beam paths.

Transmission of nanosecond laser pulses through optical fibres with high peak power densities (>0.1 GW cm^{-2}) causes problems even with core diameters of up to 2 mm.

As described below, in some instances, element distribution measurements are performed. Here the laser beam is scanned across the sample surface by means of an optical system carried by a high speed scanning machine (single axis linear positioning stage). This industrial device is designed for long life under high duty cycle and continuous operating conditions. Linear drive actuation is effected by a brushless linear d.c. servomotor. Precision position feedback is achieved by an optical incremental linear encoder (for specifications see Table 1).

Data Acquisition and Evaluation

The main component from the analytical point of view is the detection system connected to a host computer. The plasma emission light is detected with high time resolution. It is necessary to create a trigger signal coincident with the nanosecond laser pulse in order to start the delay and gate generator in the optical multichannel analyser. In the case of the Nd:YAG laser this trigger signal is directly available from the Q-switch electronics of the laser power supply, whereas in the excimer laser a fast UV sensitive photodiode mounted behind a laser beam aperture creates the trigger signal. A constant delay due to electrical propagation differences in the detector electronics always has to be taken into account when calculating the true delay time. As the laser analysis is restricted to atmospheric pressures in most instances, only delay times much shorter than 30 µs can be used. In general, the spectral line intensities decrease by at least two orders of magnitude within 3 µs after the nanosecond laser pulse [Fig. 3(*a*)]. Therefore, delay times must be kept short in order to obtain detector signals (count rates) with sufficient signal-to-noise ratio, particularly when taking spectra of single laser shots. In general, the width of the detector gate pulse should not be greater than 10 µs [see Fig. 3(*b*)].

Only two data acquisition modes are used throughout the measurements. Either the spectra are taken from each single laser plasma and transferred to the host computer or a certain number of spectra are taken from successive laser plasmas and then accumulated into one resulting spectrum. The first mode is used for measurements with a scanning laser beam while the second mode is used for measurements at a fixed sample position. When a certain loss of spatial resolution is acceptable during element distribution measurements, spectra accumulation can also be employed.

For the determination of net signals (central peak intensities of spectral lines), a mathematical procedure was developed. This model takes into account the underlying continuum radiation and the contributions of the wings of neighbouring spectral lines. Concentrations of analytes are determined *via* internal standardization. The ratio of net peak intensities of an analyte line to a reference line is found. The corresponding concentration ratio of the analyte to the internal standard is then calculated by using a calibration graph. However, owing to the limited wavelength range captured simultaneously by the detector array,

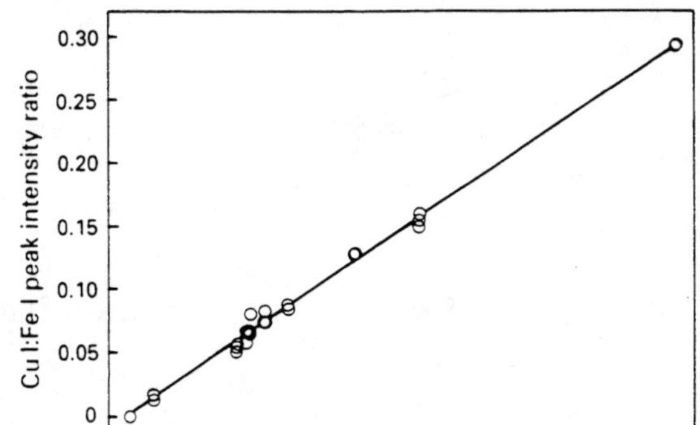

Fig. 4 Calibration graph of copper (Cu I 327.4 nm) to iron (Fe I 327.1 nm) (in solid steel) peak intensity ratio *versus* copper concentration. Copper varies between 30 and 700 ppm in the certified reference samples

Table 2 Selection of prominent analyte and reference lines used with LIESA for a variety of sample matrices

Solid/liquid steel in laboratory—	
Analyte line	Si I 288.2 nm Mn I 403.3 nm Cr I 425.4 nm Ni I 341.5 nm Cu I 327.4 nm C I 193.1 nm Mo II 281.6 nm
Reference line	Fe I 439.4 nm
Reference element concentration	>90% iron
Liquid steel in steel plant—	
Analyte line	C I 193.1 nm
Reference line	Fe II 193.2 nm Fe I 196.0 nm Fe II 201.1 nm
Reference element concentration	>90% iron
Rubber mixtures—	
Analyte line	Si I 212.4 nm Zn I 208.7 nm S I 182.6 nm Co II 228.6 nm
Reference line	C I 193.1 nm C I 247.9 nm
Reference element concentration	⩾80% carbon
Geological samples—	
Analyte line	Si I 288.2 nm Al I 308.2 nm Mg I 285.2 nm Fe I 302.0 nm Na I 589.0 nm
Reference line	Ca II 315.9 nm and lines of other elements

a presumption of internal standardization, *i.e.*, comparable excitation energies of analyte and reference lines, cannot be fulfilled in many cases.

As an example, the linear calibration graph of the Cu I 327.4 nm:Fe I 327.1 nm line intensity ratio *versus* the copper concentration in the range between 30 and 700 ppm is shown in Fig. 4. The relative detection limit is around 10 ppm.

To achieve accurate absolute concentration measurements the most abundant element with a high concentration should serve as the internal standard. The prominent analyte and reference lines and the internal standards used with the different LIESA instruments are summarized in Table 2. The analytical figures of merit as a result of many experimental investigations are described in Table 3.

In order to determine absolute analyte concentrations either the concentration of the reference element must be known, or all elements must be measured relative to the reference element with unknown concentration. The latter can only be achieved when the major constituents of the sample are detected simultaneously in one spectrum. Then the sum of all concentrations is near to 100%. High concentration of the most abundant reference element significantly improves the accuracy of the absolute analyte concentration. In such cases the relative error of the analyte to reference concentration ratio is mainly governed by the relative error of the analyte concentration; relative accuracies of 3% are achievable when measuring absolute element contents (see Table 3).

All measurements should ideally be performed under constant operating conditions. However, in reality, changes in system parameters may occur from time to time, resulting in different plasma conditions (*e.g.*, temperatures and number densities). In such cases the measured calibration graphs are no longer valid. To overcome these problems and to allow for longer time periods between system recalibrations, the plasma conditions can frequently be monitored spectroscopically. In all spectral regions of interest atomic and ionic spectral lines of one element can be simultaneously detected. The intensity ratios of such atomic/ionic line pairs are a sensitive monitor of the averaged plasma temperature. By permanent observation of these intensity ratios, changes in plasma temperatures can be easily detected and a correction factor can be derived to calculate back to the plasma conditions for which the calibration graphs were measured (Carlhoff *et al.*[12]).

Applications

Direct Analysis of Liquid Steel

Steelmakers have long been searching for ways to obtain data on the condition of molten metal throughout production. Since conventional methods such as lance sampling lead to an interruption of the process there is a strong need for on-line control. By using laser-induced emission spectrometry it is possible to determine directly the concentration of chemical elements in the melt. A LIESA instrument has been adapted to an 80-t AOD (argon–oxygen decarburization) converter at a Krupp steel plant in Germany. This system measures the carbon content (by laser analysis) and the temperature of the melt (by pyrometry). Additional laboratory measurements were carried out at the University of Madrid.

Table 3 Analytical figures of merit. In most cases ranges are given to cover all investigated applications of laser-induced emission spectral analysis

On-line capability	Detection, transfer and partial evaluation of a maximum of 50 single spectra per second (1 spectrum per laser shot for 1 024 14-bit integers)
In-process capability	Direct elemental analysis, simultaneous multi-element analysis, no sample preparation, access to targets in harsh and hazardous industrial environments
Micro/macro sampling (Nd:YAG, excimer laser)	Ablated masses of 0.1–7 μg per laser shot, ablation areas of 0.1–6 mm^2, linear scanning of maximum 2.5 $m\ s^{-1}$ scan velocity
Relative detection limit	10–100 $\mu g\ g^{-1}$
Absolute detection limit	1–100 pg
RSD value	1–2%
Relative accuracy	3%
Dynamic range	Linearity over 1–3 orders of magnitude

Fig. 5 Water-cooled box at converter wall containing laser head, optics and control electronics

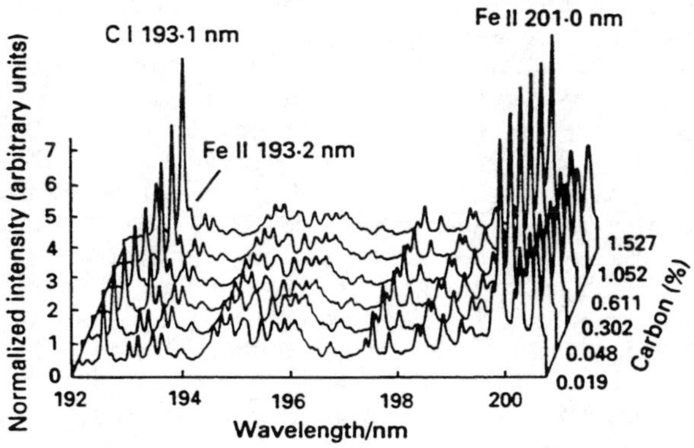

Fig. 6 Spectra of liquid steel with increasing carbon content. The signal intensities are normalized to the Fe II 193.2 nm line. The carbon analyte line at 193.1 nm and one of the iron reference lines at 201.0 nm are indicated

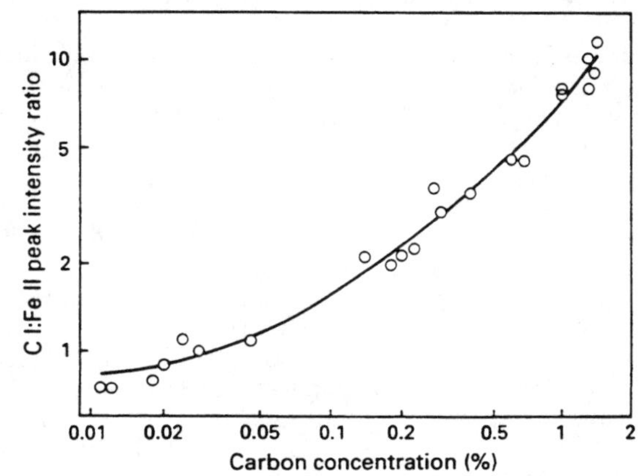

Fig. 7 Calibration graph of carbon (C I 1931 nm) to iron (Fe II 193.2 nm) (in liquid steel) peak intensity ratio *versus* carbon concentration

The system design for the converter is as follows: the beam of a high-power Nd:YAG laser (wavelength, 1 064 nm) is focused onto the surface of the melt through a narrow channel in the side wall of the converter. Pre-heated argon gas ($T > 800$ °C) is injected to prevent the steel from escaping and the melt at the end of the channel from solidifying. The laser beam produces a hot plasma. The light generated passes by way of a 7.5 m long optical fibre to the spectrometer where it is detected with an optical multichannel analyser (see above).

A computer is used to calculate the concentrations of the elements from the detected spectral lines. Melt temperatures are determined by a pyrometer. The results are transferred to the process computer for optimum production control.

To achieve correct alignment of the laser beam through the converter channel it was necessary to attach laser head and beam guiding optics directly to the converter wall. These parts of the system are installed in a closed, water-cooled box (Fig. 5) while the remaining parts (host computer, electronic control units, spectrometer, laser power supply and gas handling system) are contained in a housing on the converter platform.

During many runs, the LIESA system proved that it can sustain the extreme environmental conditions of a steel production furnace.

The system has been operated at the converter for several months. A set of spectra from liquid steel in the near vacuum UV spectral region with the 193.1 nm carbon analyte line at different carbon concentrations is shown in Fig. 6. All remaining lines are atomic and ionic iron lines. By accumulating the optical emission of 200 laser shots within 10 s, repeatability is significantly improved. Despite the very small observation angle and the long optical path of the plasma light, which has to pass several mirrors, lenses and the optical fibre, resolution and signal-to-noise ratio of the resulting spectra are excellent.

The concentration of carbon can be evaluated by calculating the net line peak intensity ratio of the carbon line to an isolated iron line as internal standard. A calibration graph for the decarburization as a result of several melts is shown in Fig. 7. After lance sampling, carbon concentrations could be measured off-line in the analytical laboratory with X-ray fluorescence. At very low carbon contents the graph flattens because of coincidence with a small iron line. The carbon detection limit from this graph is less than 200 ppm. It will be possible in the near future to avoid occasional solidification of metal in the converter channel by optimizing the argon gas pre-heating process.

The application of LIESA to the converter will permit optimum control of the melt process. Thus, tap-to-tap times can be shortened, reducing consumption of gases, fluxes and refractories. This will enhance cost efficiency in steelmaking. More accurate adjustment of the composition will improve steel quality. The LIESA system appears to be suitable not only for the converter process but also for other components in a steel production line.

Homogeneity Measurements of Tyre Rubber in the Mixing Shop

The European tyre industry is today faced with two major trends: a constant growth in raw materials being processed and a decrease in profit compared with previous years.

Since the market rigorously demands low price and high quality, there are various approaches that can be used to improve profitability. The reduction of material costs is limited owing to their close relation to product performance. However, processing techniques offer a number of possibilities for higher economy. Increased productivity and improved consistency of quality can be achieved by better preparation and performance during the compounding stages. The tolerances in compounding and mixing must be kept as tight as possible so that the rejection rate is decreased to a minimum. Additionally, automation must be increased in order to reduce the human influence on the result.

For industrial rubber mixing processes, it is desirable to evaluate on-line the dispersion of the different ingredients in the polymer matrix (*i.e.*, compound homogeneity). Therefore, Krupp, Pirelli (as a tyre manufacturer) and King's College London (as a scientific supporter) started the development of a laboratory system capable of in-process monitoring of the homogeneity of rubber slabs in the open mill and other locations in the early stage of tyre production.

By scanning the focused beam of a pulsed excimer laser across the surface of rubber slabs in a rubber mixing line, spatial element distributions can be measured and evaluated on-line. Thus, the actual element composition and homogeneity of the rubber material can be monitored frequently during the mixing and forming process. Each laser pulse produces a hot bright plasma on the rubber surface in a buffer gas flow of constant flow rate (argon at normal pressure). The optical plasma emission is detected and the spectra are evaluated as described above.

Systematic investigations of the optical plasma emission were performed in the wavelength region from 180 to 800 nm. All measured spectral features in that range could be identified and assigned to appropriate elements and molecules. Atomic and ionic spectral lines of C, H, S, Si, Zn, Co, Mg, Al, Ca, Na and K and molecular bands of C_2 and CN have been detected. The temporal behaviour of the laser-induced plasma under different experimental conditions has been studied. It has been observed with tyre rubber that spectral line peak intensities exhibit roughly the same behaviour as the lines of all the other materials analysed [see Fig. 3(*a*) and (*b*)].

The interaction of the UV laser with rubber was also investigated in detail. As expected the UV laser ablation pattern on the rubber surface exhibits sharp edges without thermal side-effects. This is not the case with the IR Nd:YAG laser beam. Excimer laser peak powers as low as 1×10^8 W cm^{-2} were sufficient to produce plasmas with strong spectral emission. Ablation depths of 2 μm per laser pulse with an ablation area of 3 mm^2 have been achieved with this irradiance employing the KrF excimer laser at 248 nm.

A typical calibration graph is shown in Fig. 8 where the line peak intensity ratio of a zinc to a carbon line is plotted *versus* the zinc oxide to carbon concentration ratio. Such calibration graphs reveal RSD values of between 0.5 and 2%. The corresponding spectra are shown in Fig. 9. Since each spectrum is normalized to the C I 193.1 nm line, variations in the sulfur and silicon (oxide) content of the rubber samples can also be seen.

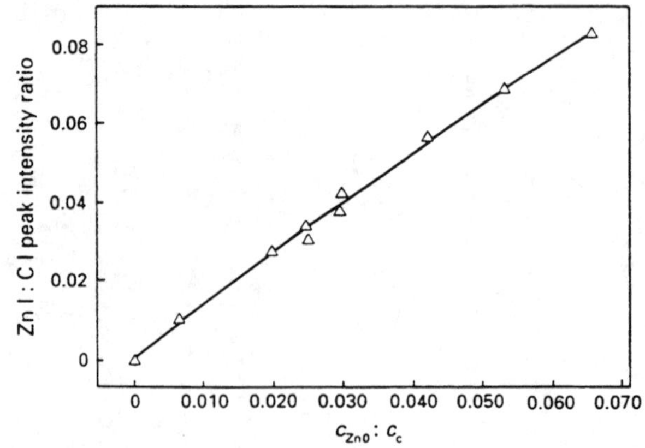

Fig. 8 Calibration graph of zinc oxide in solid tyre rubber. The peak intensity ratio of the Zn I 208.7 to the C I 193.1 nm line is plotted *versus* the zinc oxide to carbon concentration ratio

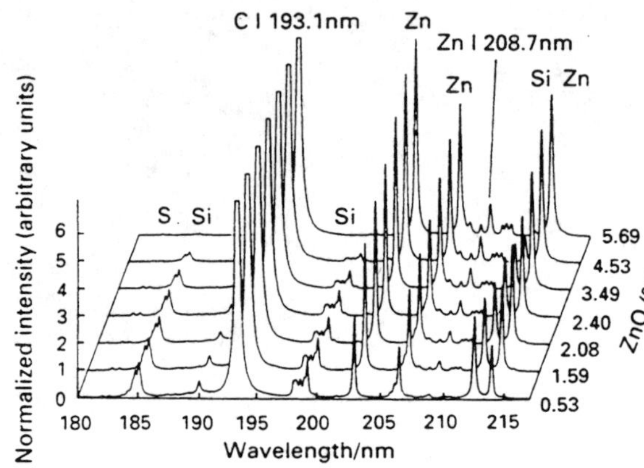

Fig. 9 Spectra of solid tyre rubber with increasing zinc oxide content. The signal intensities are normalized to the C I 193.1 nm line. The zinc analyte line at 208.7 nm, the carbon reference line at 193.1 nm and spectral lines of S and Si are indicated

Element Distribution Measurments of Geological Materials on Conveyor Belts

There is a great need in industry for the on-line control of continuous mass streams of geological raw materials. By permanent monitoring of element distributions, raw materials with narrow composition tolerances can be available ahead of further processing. In the case of bad or incorrect compositions, measures will be taken to reject or divert the materials that are outside of the range of tolerance.

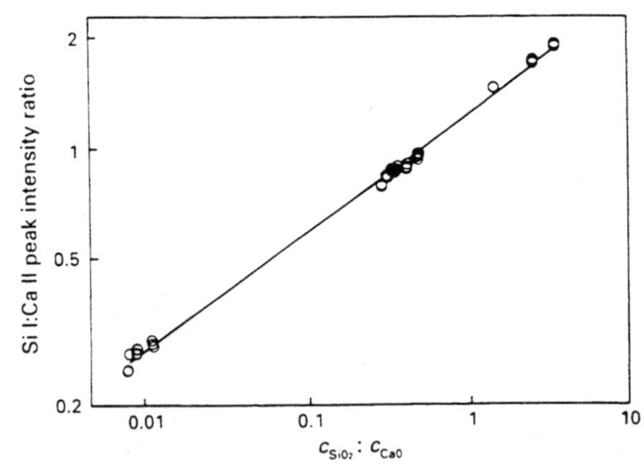

Fig. 10 Calibration graph of silicon oxide in geological samples. The peak intensity ratio of the Si I 288.2 to the Ca II 317.9 nm line is plotted *versus* the silicon oxide to calcium oxide concentration ratio

Preliminary studies of different samples led to encouraging results. A calibration graph of SiO_2 *versus* CaO employing the Si I 288.2 nm analyte and Ca II 317.9 nm reference line is shown in Fig. 10. Several data points lying close together at a given concentration ratio demonstrate the very good precision of the measurements.

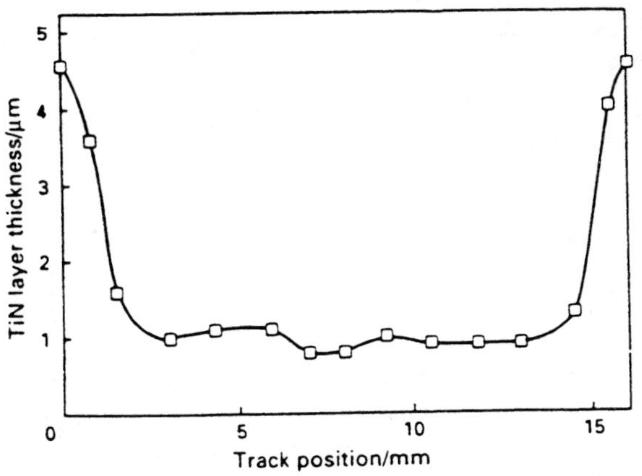

Fig. 11 Thickness profile of TiN coated on a flat metallic substrate along a linear track on the sample surface

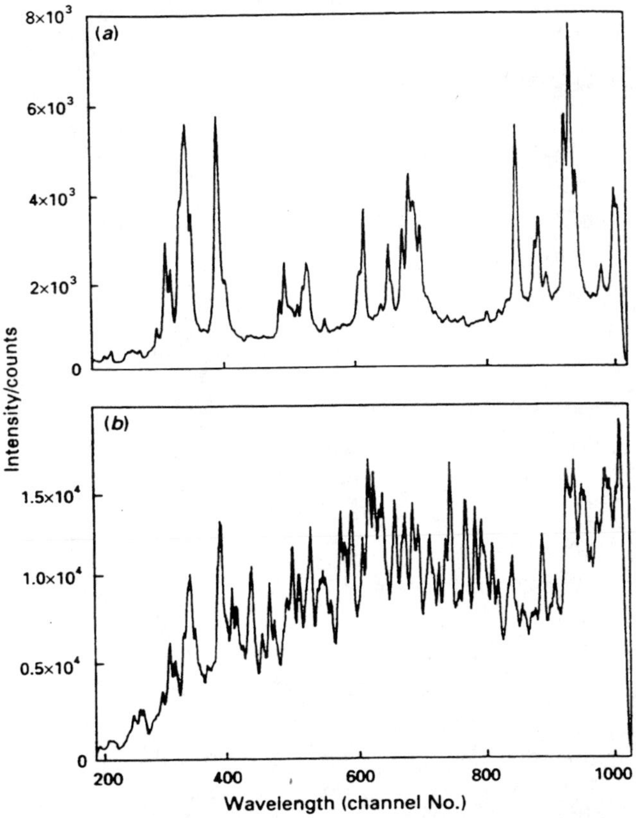

Fig. 12 (*a*) Spectrum of Ti in the UV spectral region emitted after the tenth laser pulse and (*b*) spectrum of the metallic substrate in the same spectral region as in (*a*) emitted after the hundredth laser pulse

Depth Profiling of Muilti-layers on Metallic Substrates

Intensive investigations at Krupp showed that laser analysis is not only suitable for elemental analysis but also for the depth profiling of coated substrates.

During laser–matter interaction, material is ablated in a defined way. The ablation pattern and ablation depth per laser pulse can be controlled *via* the laser parameters (spatial intensity distribution of the laser beam, pulse energy, pulse repetition rate and shape and size of laser focus). If each laser plasma is analysed a depth profile of the substrate layer(s) is created. The change in a layer is indicated by the change in the detected spectral pattern.

A thickness profile (single TiN layer on a metallic substrate) along a straight line on the flat surface of the coated sample is shown in Fig. 11. The ablation depth of the pulsed laser was 0.1 μm per pulse. Corresponding spectra are presented in Fig. 12(*a*) and (*b*) (same wavelength region in the UV). A Ti spectrum emitted after the tenth laser pulse is shown in Fig. 12(*a*) while a metal spectrum emitted after the hundredth laser pulse deep in the substrate can be seen in Fig. 12(*b*). The measured profile was in excellent agreement with results of off-line standard measurement techniques.

Conclusions

Laser-induced emission spectral analysis appears to be a very promising method for on-line analysis in those industrial fields where element distribution measurements of materials at all stages of production yield information on the quality of material and production process. At present Krupp is developing a whole family of LIESA instruments to cover the needs of different industrial branches.

Financial support for the development of the LIESA systems by the German Minister of Research and Technology (BMFT), by the European Communities for Steel and Coal (ECSC) and by the BRITE/EURAM programme of the Commission of the European Communities is gratefully acknowledged.

References

1 Moenke-Blankenburg, L., *Laser Microanalysis*, Wiley, New York, 1989.
2 *Laser-Induced Plasmas and Applications*, eds. Radziemski, L. J., and Cremers, D. A., Marcel Dekker, New York, 1989, pp. 295–346.
3 Ko, J. B., Sdorra, W., and Niemax, K., *Fresenius' Z. Anal. Chem.*, 1989, **335**, 648.
4 Leis, F., Sdorra, W., Ko, J. B., and Niemax, K., *Mikrochim. Acta*, 1989, **II**, 185.
5 Sdorra, W., and Niemax, K., *Spectrochim. Acta, Part B*, 1990, **45**, 917.
6 Niemax, K., and Sdorra, W., *Appl. Opt.*, 1990, **29**, 5000.
7 Sdorra, W., and Niemax, K., *Mikrochim. Acta*, 1992, **107**, 319.
8 Sdorra, W., Brust, J., and Niemax, K., *Mikrochim. Acta*, 1992, in the press.
9 Carlhoff, C., Lorenzen, C.-J., Nick, K.-P., and Siebeneck, H.-J., in *In-Process Optical Measurements*, ed. Spring, K. H., *Proc. SPIE*, SPIE Publications, Bellingham, vol. 1012, 1989, pp. 194–196.
10 Carlhoff, C., and Kirchhoff, S., *Laser Optoelektronik*, 1991, **23**, 50.
11 Lorenzen, C.-J., Jogwich, M., Burge, R. E., Michette, A. G., Daghooghi, R., and Nahmias, M., paper presented at the BRITE/EURAM-EUREKA Workshop on Polymer Technology and Industrial Applications, Ferrara, Italy, May 13–15, 1991, Book of Abstracts 2, pp. 122–125.
12 Carlhoff, C., Lorenzen, C.-J., and Nick, K.-P., German patent No. DE 39 11 965 C2; date of application: April 12, 1989.

Paper 2/00713D,
Received February 11, 1992
Accepted April 13, 1992

Reprinted with permission from *Journal of Polymer Science,* Vol. 27, pp. 219-229 (January 1958).

Theory of Light Scattering by Polymer Solutions*

P. J. FLORY, *Department of Chemistry, Cornell University, Ithaca, N. Y.* and A. M. BUECHE, *Research Laboratory, General Electric Company, Schenectady, N. Y.*

INTRODUCTION

The manner in which the intensity of light scattered by a dilute polymer solution depends on its concentration and on the angle of observation has been treated by Zimm[1] in the single contact approximation. He thus found the reciprocal intensity I to be represented by:

$$Kc/I = 1/MP(\theta) + 2A_2c + \ldots \tag{1}$$

where K is a constant, c is the concentration in g./cc., $P(\theta)$ is a function[2] of the scattering angle θ, M is the molecular weight and A_2 is the second virial coefficient. The second term of the series (but not higher terms, which have been omitted from equation 1) was found to be independent of the concentration in the single contact approximation. Zimm[1] surmised that higher approximations would yield a dependence of this term on angle.

An alternative basis for treating properties of dilute polymer solutions is afforded by the Gaussian smoothed density model for a polymer chain.[3,4] According to this model, the density of chain segments is considered to depend only on the distance from its center of gravity, the density being a Gaussian function of that distance. The smoothed Gaussian model has met with considerable success in treating both intramolecular interactions within a polymer chain[3] and the intermolecular interactions manifested in the second virial coefficient.[3,5,6] The model is inaccurate in detail, but these inaccuracies do not impair its application to solutions in which the interaction is large, *i.e.*, to good solvent media. The single contact approximation is necessarily restricted to very small interactions, *i.e.*, to poor solvents in the vicinity of the theta point.

In this paper we present a treatment of the angular dependence of the intermolecular contribution to the scattering intensity based on the smoothed density Gaussian model. In particular, we investigate the dependence of the second virial term on angle for higher interactions than would be consistent with the use of the single contact approximation.

* This investigation was supported in part by the National Science Foundation.

THEORY

The intensity of the radiation scattered by a polymer solution may be readily separated into intramolecular and intermolecular contributions (see Zimm[1]). Thus, the total intensity may be expressed as the square of the sum of the amplitudes of the rays scattered by all polymer segments, or, equivalently, as the sum of products of amplitudes $\sum_{i,j} A_i A_j$ for every pair (i, j) of polymer segments. This sum is immediately separable into intramolecular terms arising from segments belonging to the same molecule and intermolecular terms for segments of different molecules.[1] We shall be concerned here only with the latter terms.

Consider the contribution to the intermolecular scattering arising from segments of molecule 1 in volume element $d\tau_1$ (Fig. 1) and from segments of *other* molecules in volume element $d\tau_2$. Let initial considerations be restricted further to those segments in $d\tau_2$ which belong to molecules whose centers occur within volume element $d\omega_2$ of Figure 1. The segment densities are assumed to be Gaussian† about the molecular centers of mass,

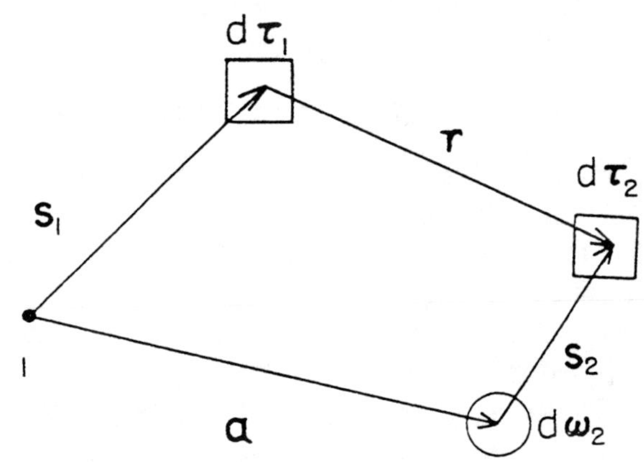

Figure 1.

situated in Figure 1 at point 1 and in volume element $d\omega_2$, respectively. Thus, the segment density at a distance s from the molecular center is assumed to be given with sufficient accuracy by:

$$\rho(\mathrm{s}) = x(\beta'/\pi^{1/2})^{3/2} \exp\{-(\beta')^2 s^2\} \tag{2}$$

where x is the number of segments in a molecule and $(\beta')^2 = 3/2\overline{s^2}$, $(\overline{s^2})^{1/2}$ being the radius of gyration of the molecule. The simultaneous expectation of segments of molecule 1 in $d\tau_1$ and of segments of other molecules in $d\tau_2$, the molecular center of the latter occurring within $d\omega_2$, is given by:

† The molecular distribution will be distorted somewhat, according to unpublished work, by the mutual interactions between the molecules when in close proximity. It is possible to show however that the distortion should not be great except in good solvents at distances of approach which occur with small probability in dilute solutions. The effect of such distortions is neglected here.

$$\rho_1(\mathbf{s}_1)d\tau_1 \cdot \rho_2(\mathbf{s}_2)d\tau_2 \cdot C_a d\omega_2$$

where C_a is the concentration of molecules with centers situated at the distance a from the center of a specified molecule (*i.e.*, $4\pi r^2 C_a$ expresses the radial distribution of molecular centers). Introduction of Gaussian distributions (2) for $\rho_1(\mathbf{s}_1)$ and for $\rho_2(\mathbf{s}_2)$, integration over $d\omega_2$ for fixed distance a, followed by integration over $d\tau_1$ at fixed distance r from $d\tau_2$, and, finally, integration over all coordinates $d\tau_2$ yields:

$$xg_{12}(a,r)dadr = 2^{3/2}\pi^{1/2}x^2\beta' C_a ar\,[\exp\{-[(\beta')^2/2](a-r)^2\} - \exp\{[-[(\beta')^2/2](a+r)^2\}]\,dadr \quad (3)$$

for the incidence of segment pairs separated by a distance r to $r + dr$, one member of the pair belonging to molecule 1 and the other to a molecule whose center is at a distance a to $a + da$ from the center of molecule 1. The function $g_{12}(a, r)$ represents the mean radial distribution function for the latterly specified segments about a segment of molecule 1.

The intensity of light scattered per unit volume of dilute solution, after subtracting the scattering of the pure solvent, is given by the well-known formula:

$$I = K'Cx \int g(r)(k\sigma r)^{-1} \sin(k\sigma r)dr \quad (4)$$

where K' is a constant for the given system, $k = 2\pi/\lambda$, $\sigma = 2 \sin(\theta/2)$, λ is the wave length of the radiation in the scattering medium, θ is the scattering angle, C is the molecular concentration per cc., and $g(r)$ is the radial distribution function for scattering elements. As already noted, we separate $g(r)$ into $g_1(r)$ for the radial distribution of segments of the same molecule, and $g_{12}(r)$ for segments belonging to other molecules. Substitution of the former into equation (4) yields I_1, the intermolecular scattering intensity. The latter leads to the contribution I_{12} due to intermolecular pairs.[1]

The portion of I_{12} due to molecules with centers separated by the distance a to $a + da$ is obtained from equation (3) by substitution of $g(r) = g_{12}(a,r)$ as given by equation (3). The result is:

$$I_{12}(a)da = K'CC_a x^2 P^*(\theta)(k\sigma a)^{-1} \sin(k\sigma a) 4\pi a^2 da \quad (5)$$

where:

$$P^*(\theta) = \exp\{-k^2\sigma^2/2(\beta')^2\} \quad (6)$$

Equation (5) has the usual form for interparticle (or interatomic) scattering, $P^*(\theta)$ appearing as the particle (or atomic) scattering factor.

According to the theory of imperfect gases,[7] the results of which are at once applicable to the dilute polymer solution:

$$C_a = C(1 + f_{12})[1 + C\int f_{13}f_{23}d\omega_3 + \ldots] \quad (7)$$

where a is the distance between molecules 1 and 2, and:

$$f_{ij} = \exp\{-\Delta F_{ij}/kT) - 1\} \qquad (8)$$

ΔF_{ij} is the interaction free energy for the pair of molecules i,j, and $d\omega_i$ specifies the coordinates of the center of molecule i. Equation (7) includes triple interactions (*i.e.*, molecules 1, 2, and 3) and accordingly is accurate to terms in C^2. Substitution of equation (7) in (5) yields:

$$\begin{aligned} I_{12} = K'C^2x^2P^*(\theta)\{&\int (k\sigma a)^{-1} \sin (k\sigma a) d\omega_2 \\ &+ \int (k\sigma a)^{-1} \sin (k\sigma a) f_{12} d\omega_2 \\ &+ C\int\int (k\sigma a)^{-1} \sin (k\sigma a)(1 + f_{12}) f_{13} f_{23} d\omega_2 d\omega_3\} \end{aligned} \qquad (9)$$

where higher terms have been disregarded and $4\pi a^2 da$ is replaced by $d\omega_2$.

The first term in equation (9) is negligible for a macroscopic sample. For representation of the second and third terms we introduce the definitions:

$$\begin{aligned} \int (k\sigma a)^{-1} \sin (k\sigma a) f_{12} d\omega_2 &= Q'(\theta) \int f_{12} d\omega_2 \qquad (10) \\ &= -2mQ'(\theta)\Gamma_2 \qquad (10') \end{aligned}$$

and:

$$\begin{aligned} &\int\int (k\sigma a)^{-1} \sin (k\sigma a)(f_{12} + 1) f_{13} f_{23} d\omega_2 d\omega_3 \\ &= R'(\theta) \int\int (f_{12} + 1) f_{13} f_{23} d\omega_2 d\omega_3 \\ &= R'(\theta)[\int\int f_{12} f_{13} f_{23} d\omega_2 d\omega_3 + \int f_{13} d\omega_1 \int f_{23} d\omega_2] \qquad (11) \\ &= -m^2 R'(\theta)[3\Gamma_3 - (2\Gamma_2)^2] \qquad (11') \end{aligned}$$

where m is the mass of the polymer molecule, and Γ_2 and Γ_3 are the modified virial coefficients in the expansion of the osmotic pressure:

$$\pi/c = (\mathbf{k}T/m)(1 + \Gamma_2 c + \Gamma_3 c^2 + \ldots)$$

where c is in g. cc.$^{-1}$. Thus:

$$I_{12} = -K\, MP^*(\theta)\{2Q'(\theta)\Gamma_2 c^2 + R'(\theta)[3\Gamma_3 - (2\Gamma_2)^2]c^3 + \ldots\} \qquad (12)$$

where $K = K'N_a/M_0^2$; N_a is Avogadro's number and M_0 is the molecular weight of a segment of the polymer chain.

We return now to consideration of the individual molecule scattering I_1. In the interests of consistency the Gaussian smoothed density model will be retained, although it is known to yield a result which becomes inaccurate at larger values of $(k\sigma/\beta')^2$, *cf. seq.* Then by calculation of the intramolecular $g(r)$ from equation (2) and substitution of the result in equation (3) there is obtained (see Hermans[8]):

$$\begin{aligned} I_1 &= K'Cx^2P^*(\theta) \\ &= KcMP^*(\theta) \end{aligned} \qquad (13)$$

(The same result is obtainable from the preceding equations by observing the identity of the intermolecular scattering I_{12} for two superimposed

Gaussian distributions with the intramolecular scattering I_1 from one alone. Thus, replacing $C_a 4\pi a^2 da$ in equation (5) by unity and setting $a = 0$, we obtain equation (13) at once.)

Combination of equations (13) and (12) for I_1 and I_{12}, respectively, yields for the total scattering:

$$I = KcMP^*(\theta)\{1 - 2P^*(\theta)Q(\theta)\Gamma_2 c - P^*(\theta)R(\theta)[3\Gamma_3 - 4\Gamma_2^2]c^2 - \ldots\} \quad (14)$$

or:

$$Kc/I = 1/MP^*(\theta) + 2Q(\theta)A_2 c + \{3R(\theta)A_3 + 4[P^*(\theta)Q^2(\theta) - R(\theta)]A_2^2 M\}c^2 + \ldots \quad (15)$$

where:

$$Q(\theta) = Q'(\theta)/P^*(\theta) \quad (16)$$

$$R(\theta) = R'(\theta)/P^*(\theta) \quad (17)$$

and $A_2 = \Gamma_2/M$ and $A_3 = \Gamma_3/M$ are the usual virial coefficients. (The present notation for the factors describing the angle dependence differs from that used by Zimm.[1] Working in the single contact approximation, Zimm found the second term of equation (15) to be independent of angle, this result corresponding to reduction of our $Q(\theta)$ to unity. Zimm assigned the symbol $Q(\theta)$ to represent the angle dependence of the third term in the virial expansion.)

Before proceeding with the reduction of $Q(\theta)$, we pause to examine certain limitations of the present model. More realistically, each pair of chain segments should be considered to be separated by distances subject to Gaussian distributions appropriate to the number of intervening segments along the chain. The *intra*molecular scattering intensity thus obtained is:

$$I_1 = KcMP(\theta) \quad (18)$$

with:

$$P(\theta) = (2/v^2)[\exp\{-v\} - (1 - v)] \quad (19)$$

where $v = (3/2)(k\sigma/\beta')^2$. Comparing the series expansions of equations (19) and (6), we have:

$$P(\theta) = 1 - v/3 + v^2/12 - v^3/60 + \ldots \quad (19')$$

and:[8]

$$P^*(\theta) = \exp\{-v/3\} = 1 - v/3 + v^2/18 - v^3/162 + \ldots \quad (6')$$

Series expansion of the ratio leads to:

$$P^*(\theta)/P(\theta) = 1 - v^2/36 + v^3/810 - \ldots \quad (20)$$

The difference between P^* and P becomes serious for large v, *i.e.*, for large angles or small wave lengths. Benoit and Goldstein[9] attributed the disparity to deviations from spherical symmetry, implicitly assumed in the model comprising a Gaussian distribution of segments about the center of gravity. This explanation is manifestly incorrect, inasmuch as the deviation of $P^*(\theta)$ from $P(\theta)$ is most serious for large v where distances which are small compared to the over-all molecular dimension $(\overline{s^2})^{1/2}$ are most important in contributing to $1 - P(\theta)$. The relevant inadequacy of the Gaussian *spatial* distribution is rather to be found in its failure to recognize the contiguity of the segments in the polymer chain. Thus, the expectancy $g_1(r)dr$ of a segment at a distance r to $r + dr$ from a given segment is assumed to depend only on the distance (s) from the center of the molecule and not otherwise on r. Actually, the presence of the given segment at its given location enhances the expectancy of finding another segment of the same molecule nearby. If the distance r is small compared to $(\overline{s^2})^{1/2}$, this expectancy may easily be several times that given by the smoothed density (Gaussian) model; if r is of the order of $(\overline{s^2})^{1/2}$, this conditionality becomes negligibly small. It is for this reason that P^* deviates from P for larger values of v (*i.e.*, for comparatively large values of $\overline{s^2} \sin^2 (\theta/2)/\lambda^2$). Equations (18) and (19) should therefore be used for the single molecule scattering.

The foregoing considerations naturally call into question the use of the smoothed Gaussian segment density model for the calculation of I_{12}. In computing the modification of the scattering by phase correlations of rays scattered by segments belonging to *different* molecules, as required for evaluation for I_{12}, the circumstances differ very significantly from those prevailing in the computation of the intramolecular scattering in that the two segments concerned are not connected, even indirectly. We require the *intermolecular* radial distribution $g_{12}(a,r)$ of segments distant r to $r + dr$ from the segment in question. Given that a second molecule is located with its center in a certain volume element distant a from the center of the first, the knowledge that a segment of the first molecule is located within $d\tau_1$ (Fig. 1) has relatively little effect on the expectancy of segments of molecule 2 in $d\tau_2$. To be sure, if r is small, the expected density ρ_2 in $d\tau_2$ may be *diminished* somewhat by the proximity of segment 1 (owing to the necessity of accommodating its adjoining segments in the same vicinity), but the correction factor will not be much less than unity for any value of r. We conclude therefore that the smoothed Gaussian density model should be reasonably satisfactory for the calculation of I_{12}, and, accordingly, that the appearance of $P^*(\theta)$ rather than $P(\theta)$ in equation (4), *et seq.*, is justified by the considerations mentioned. (The small error introduced by use of the smoothed Gaussian model in this connection is in the direction of rendering $|I_{12}|$ slightly too small.)

In this situation we may expect to achieve close approximation to reality by deliberately replacing $P^*(\theta)$ in equation (15) by $P(\theta)$ with the result:

$$Kc/I = 1/MP(\theta) + 2Q(\theta)A_2c + \{3R(\theta)A_3 + 4[P(\theta)Q^2(\theta) - R(\theta)]A_2^2M\}c^2 + \ldots \tag{15'}$$

The first term is then correctly expressed for a polymer chain comprising segments whose distances apart are Gaussian, and the higher terms in the reciprocal intensity expansion are represented in accordance with the Gaussian smoothed density model, except for replacement of $P^*(\theta)$ by $P(\theta)$ in the last term.

Alternatively, we might have combined the preferred expression for I_1, equation (18), with I_{12} according to equation (12) to obtain for the total scattering:

$$I = KcM\{P(\theta) - 2P^*(\theta)Q'(\theta)\Gamma_2c - P^*(\theta)R'(\theta)[3\Gamma_3 - 4\Gamma_2^2]c^2 - \ldots\} \tag{14'}$$

The reciprocal intensity expression (15′) follows from (14′) provided $R(\theta)$ and $Q(\theta)$ are redefined as follows:

$$Q(\theta) = Q'(\theta)P^*(\theta)/P^2(\theta) \tag{16'}$$

$$R(\theta) = R'(\theta)P^*(\theta)/P^2(\theta) \tag{17'}$$

It may be shown that the previous definitions yield $Q(\theta) = 1$ and $R(\theta) = 1$ in the limit of vanishing interactions between chain segments (*i.e.*, at the theta point where $X_1 = 0$, *cf. seq.*). The definitions (16′) and (17′) yield $Q(\theta) = R(\theta) = [P^*(\theta)/P(\theta)]^2 < 1$ in this limit. The former result brings about convergence to the results of the single contact approximation[1] in the limit of vanishing interactions. We therefore adopt equation (15′) in conjunction with equations (16) and (17). In subsequent discussion these definitions of $Q(\theta)$ and $R(\theta)$ will be retained.

According to theory previously developed[3,4] for the interaction between a pair of polymer molecules treated as smoothed Gaussian spheres, the free energy of their interaction when their centers are separated by the distance a may be written:

$$\Delta F/kT = X_1 \exp\{-y^2\} \tag{21}$$

where

$$y = \beta' a/2^{1/2} \tag{22}$$

$$X_1 = \psi_1(1 - \Theta/T)x^2 v_1(\beta')^3/2^{1/2}\pi^{3/2}N_a \tag{23}$$

ψ_1 and Θ being constants characterizing the solvent-polymer interaction, and v_1 the molar volume of the solvent.[3] Substitution of equation (21) for ΔF_{ij} in equation (8) and employing the result in equation (10), we obtain with further aid of (16):

$$Q(\theta) = [P^*(\theta)]^{-1} \frac{\int_0^\infty (by)^{-1} \sin (by)[1 - exp\{-X_1 \exp(-y^2)\}]y^2 dy}{\int_0^\infty [1 - \exp\{-X_1 \exp(-y^2)\}]y^2 dy} \quad (24)$$

where:

$$b^2 = 2(k\sigma/\beta')^2 = (4/3)r \quad (25)$$

Through the use of equations (11) and (21), $R(\theta)$ may likewise be expressed as a function of X_1 and b. The resulting integrals offer little hope of reduction to usable form, hence we shall not be concerned further with the angle dependence of the third virial term.

Series expansion of the lower exponentials occurring in both numerator and denominator of equation (24) yields:

$$Q(\theta) = [P^*(\theta)]^{-1} \sum_{i=1}^{\infty} [(-X_1)^i/i!i^{3/2}] \exp\{-(b/2)^2/i\} \Big/ \sum_{i=1}^{\infty} (-X_1)^i/i!i^{3/2} \quad (26)$$

The slow convergence of these alternating series, except for small values of X_1 (*i.e.*, for poor solvents), restricts the usefulness of this equation.

If $(by)^{-1} \sin (by)$ is expanded in series, equation (24) may be written:

$$Q(\theta) = [P^*(\theta)]^{-1} \sum_{n=1}^{\infty} (-1)^{n-1}(b/2)^{2(n-1)}\Phi_n/\Phi_1 \quad (27)$$

where:

$$\Phi_n = [2^{2(n-1)}/(2n-1)!] \int_0^\infty [1 - \exp\{-X_1 \exp(-y^2)\}]y^{2n} dy \quad (28)$$

By series expansion of the lower exponential in equation (28), we obtain:

$$\Phi_n = [\pi^{1/2}/4(n-1)!] \{X_1 - X^2/2!\,2^{(2n+1)/2} + X_1^3/3!\,3^{(2n+1)/2} - \ldots\} \quad (29)$$

The slow convergence of this alternating series renders it of little use for larger values of X_1. The recursion relationship:

$$\Phi_n = (n-1)^{-1} \int (\Phi_{n-1}/X_1) dX_1 = [(n-1)!]^{-1} \int \ldots \int \Phi_1 (d \ln X_1)^{n-1} \quad (30)$$

offers an alternative device for evaluation of the Φ_n, provided Φ_1 is known as a function of X_1. Values of Φ_1 have been determined previously[3,4] in connection with the treatment of the second virial coefficient. (Thus, $\Phi_1 \equiv (\pi^{1/2}/4)X_1\mathfrak{F}(X_1)$, where $\mathfrak{F}(X_1)$ is the function previously introduced and calculated.[3,4]) The series in equation (27) ordinarily will converge rapidly, hence terms beyond the third or fourth will rarely be needed.

We have found it expeditious to evaluate the Φ_n by Simpson's rule integration using the IBM type-650 computer (assistance of Mr. R. C. Lesser of the Cornell Computing Center is gratefully acknowledged). Results are given in Figure 2, where Φ_n/Φ_1 is plotted for $n = 2, 3, 4$, and 5, against X_1 over the range 0 to 50.

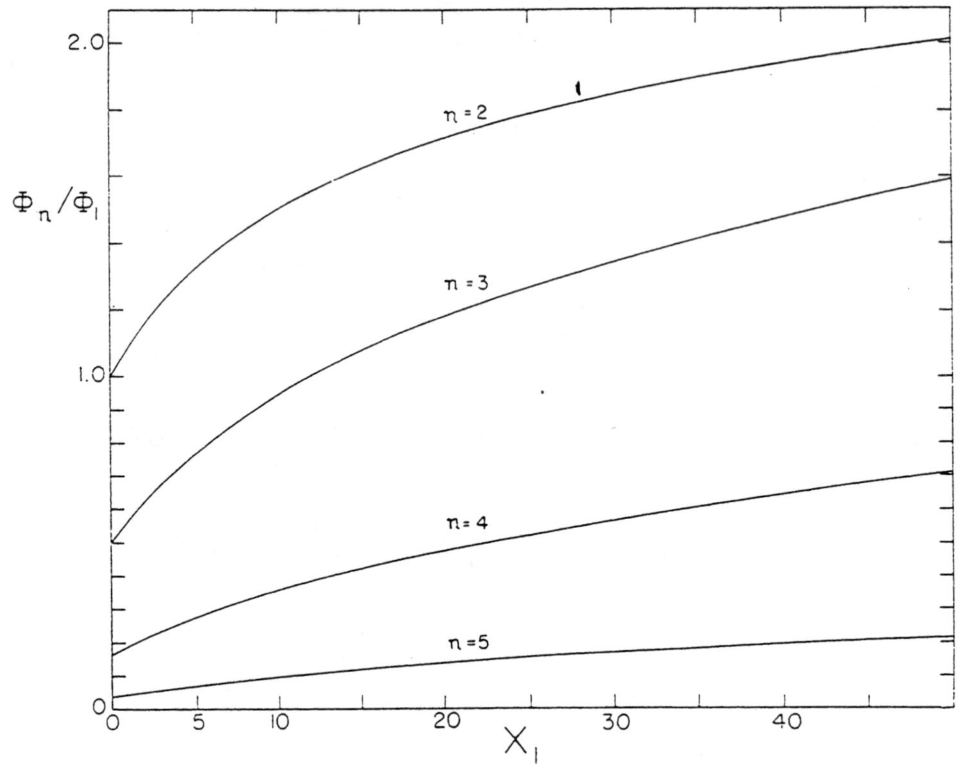

Figure 2.

For lower values of X_1 the following series obtained from equation (29) may be used:

$$\Phi_n/\Phi_1 = [1/(n-1)!]\{1 + (1 - 1/2^{n-1})2^{-5/2}X_1 - [(1 - 1/3^{n-1})2^{-1}3^{-5/2} - (1 - 1/2^{n-1})2^{-5}]X_1^2 + \dots\} \quad (31)$$

or:

$$\left.\begin{aligned} \Phi_2/\Phi_1 &= 1 + 0.0884X_1 - 0.00576X_1^2 + \dots \\ \Phi_3/\Phi_1 &= 0.5000 + 0.0663X_1 - 0.00254X_1^2 + \dots \\ \Phi_4/\Phi_1 &= 0.1667 + 0.0258X_1 - 0.00059X_1^2 + \dots \\ \Phi_5/\Phi_1 &= 0.0417 + 0.0069X_1 - 0.00010X_1^2 + \dots \end{aligned}\right\} \quad (32)$$

Neglect of terms beyond those given explicitly in equations (32) is permissible for $|X_1| \leq 5$; the error within this range in no case exceeds 2%. For larger values of X_1 these parabolic expressions rapidly become inadequate; it then becomes necessary to utilize the computational results shown in Figure 2.

Sample calculations of $Q(\theta)$ carried out according to equation (27) together with values of Φ_n/Φ_1 from equations (32) for small X_1 and from Figure 2 for $X_1 > 5$ are given in Table I. The values of $(b/2)^2$ chosen are representative of the radius of gyration–wave length ratio and scattering angle commonly used in light scattering studies on polymers. By definitions set forth above:

$$(b/2)^2 = (16\pi^2/3)(\overline{s^2}/\lambda^2) \sin^2 (\theta/2)$$

TABLE I

Sample Calculations of $Q(\theta)$

$(b/2)^2$	$(\overline{s^2})^{1/2}/\lambda$ for $\theta = 90°$	$Q(\theta)$ $X_1 = 1$	$X_1 = 3$	$X_1 = 10$
0.25	0.098	0.98	0.94	0.87
0.50	.138	.95	.88	.74
1.00	.195	.89	.74	.43

Hence, if $\theta = 90°$, $(b/2)^2 = (8\pi^2/3)\ (\overline{s^2}/\lambda^2)$, and we obtain the ratios given in the second column. (These are to be multiplied by $6^{1/2}$ to obtain the more familiar ratio $(\overline{r^2})^{1/2}/\lambda$ of the root-mean-square end-to-end distance to the wave length.) The values found for $Q(\theta)$ deviate appreciably from unity, especially for larger $(b/2)^2$ and for higher X_1 (*i.e.*, better solvents).

In the limit $X_1 = 0$, $\Phi_n/\Phi_1 = 1/(n-1)!$ and $Q(\theta)$ reduces to unity as previously stated.

On the basis of the theory and calculations here presented, assumption of parallel concentration lines in the Zimm plot of the reciprocal turbidity may lead to appreciable error in good solvents, particular when the molecular weight is large. Knowledge of $(\overline{s^2})^{1/2}$, acquired either by the dissymmetry method (with extrapolation to infinite dilution) or from the intrinsic viscosity, will enable evaluation of $(b/2)^2$. From provisional values of the second virial coefficient and molecular weight, X_1 can be estimated.[6] Corrections of the plots based on equation (27) could then be applied by a method of successive approximations.

References

1. B. H. Zimm, *J. Chem. Phys.*, **16,** 1093 (1948).
2. P. Debye, *J. Phys. & Coll. Chem.*, **51,** 18 (1947). B. H. Zimm, R. S. Stein, and P. Doty, *Polymer Bull.*, **1,** 90 (1945).
3. P. J. Flory, *Principles of Polymer Chemistry,* Cornell Univ. Press, Ithaca, N. Y., 1953, Chapters XII and XIV.
4. P. J. Flory, *J. Chem. Phys.*, **17,** 1347 (1949); P. J. Flory and W. R. Krigbaum, *ibid.*, **18,** 1086 (1950).
5. W. R. Krigbaum and P. J. Flory, *J. Am. Chem. Soc.*, **75,** 1775, 5254 (1953). W. R. Krigbaum, *ibid.*, **76,** 3758 (1954).
6. T. A. Orofino and P. J. Flory, *J. Chem. Phys.*, **26,** 1067 (1957).
7. J. O. Hirschfelder, C. F. Curtiss, and R. B. Bird, *Molecular Theory of Gases and Liquids*, Wiley, New York, 1954, p. 320.
8. J. J. Hermans, *Rec. trav. chim.*, **68,** 859 (1949).
9. H. Benoit and M. Goldstein, *J. Chem. Phys.*, **21,** 947 (1953).

Synopsis

The intermolecular contribution to light scattering by polymer solutions has been treated according to the smoothed Gaussian sphere model. Combining the results with the usual term for the individual molecule scattering we obtain for the reciprocal turbidity:

$$Kc/I = 1/MP(\theta) + 2Q(\theta)A_2c + \ldots$$

The third virial term is also formulated. The angle dependent factor $Q(\theta)$ falls appreciably below unity at higher angles for large molecules in good solvents. Calculation of $Q(\theta)$ according to the theory is illustrated. Its replacement by unity may under some circumstances invite appreciable error.

Résumé

La contribution intermoléculaire à la diffusion lumineuse de solutions polymériques a été traitée en admettant un modele sphérique gaussien. En combinant les résultats avec le terme usuai correspondant aux molécules individuals de diffusion, on obtient pour l'inverse de la turbidité

$$Kc/I = 1/MP(\theta) + 2Q(\theta)A_2c + \ldots$$

Le troisième terme viriel a également été formulé. Le facteur $Q(\theta)$ dépendant de l'angle d'observation est nettement inférieur à l'unité aux grands angles pour les molécules de poids moléculaire élevé, dissoutes dans de bons solvants. Le calcul de $Q(\theta)$ suivant la théorie a été illustré. L'attribution d'une valeur unitaire à ce facteur peut causer des erreurs appréciables dans certaines circonstances.

Zusammenfassung

Der intermolekulare Beitrag zur Lichtstreuung von Polymerlösungen wurde in Übereinstimmung mit dem geglätteten Gauss'schen Kugelmodel berechnet. Wenn man die Resultate mit dem üblichen Ausdruck für die individuelle Molekül-Streuung vereinbart, erhalten wir die reziproke Trübung

$$Kc/I = 1/MP(\theta) + 2Q(\theta)A_2c + \ldots$$

Der dritte viriale Ausdruck wird auch formuliert. Der Winkel-abhängige Faktor $Q(\theta)$ fällt bei höheren Winkeln für grosse Moleküle in guten Lösungsmitteln erheblich unterhalb eins. Die Berechnung von $Q(\theta)$ nach der Theorie wird gezeigt. Seine Ersetzung durch eins kann unter Umständen zu erheblichen Fehlern führen.

Reprinted with permission from *Journal of Polymer Science C*, Vol. 27, pp. 195-205 (1969).

Particle Size Distribution by Turbidimetry

L. D. MAXIM, A. KLEIN, M. E. MEYER, and C. H. KUIST, *Alexander Laboratories, National Starch and Chemical Corporation, Plainfield, New Jersey 07063*

Synopsis

A turbidimetric method for rapidly estimating particle size distribution parameters for latex systems has received intensive evaluation. This method originally proposed by W. Heller and co-workers [J. Chem. Phys. 34, 1796 (1961)], and based on Mie theory, involves the generation of theoretical turbidity functions versus wavelength and the comparison of these theoretical curves with ones determined experimentally. A digital computer program involving the use of a lattice search technique by minimization of residual sums of squares over a response surface was developed and used for comparing these curves. It has been found that this curve matching leads to multivalued solutions and, unless prior estimates of the answer from some other technique are available, there are no criteria for choosing between alternative solutions. The response surface is in the form of a "trench" rather than a unimodel "depression." Consideration of typical experimental errors for this technique yields quite large areas of uncertainty in the final answers further complicating the analysis of results. These findings indicate that this method, though very attractive because of the simplicity of the experimental technique, should not be used alone for particle size analysis and hence cannot be considered a routine or simple procedure.

Introduction

The various averages (moments) of particle size distribution have long been recognized as important parameters in both characterizing and studying the fundamental behavior of latex systems. With increasing industrial interest in synthetic latices, a great need exists for a relatively simple means of determining particle size averages and distributions. Among the better-known techniques; ultracentrifugation (sedimentation), fractional creaming, electron microscopy and light scattering, the latter two have shared the most attention in the literature. Electron microscopy which is viewed as the absolute method is plagued in most instances with added laborious techniques to avoid systematic errors (1) and bias (2). Light-scattering techniques, such as the polarization ratio method developed by M. Kerker, et al. (16,17) have proven to be quite useful. A turbidimetric method developed by W. Heller et al. (3,4), appeared of interest due to the experimentally simple turbidity measurement, the computerizable data analysis, and the added advantage of rapidity, enabling us to measure particle size distributions as a function of time. The method is based on the Mie theory and the assumption is made that the distribution follows the function below; which presumably is commonly found in emulsions and latices:

$$Cf_{(r)} = (r - r_0) \exp - \{(r - r_0)/S\}^{+3} \text{ for } r > r_0 \qquad (1)$$

$$= 0 \qquad \text{otherwise}$$

Where $Cf_{(r)}$ dr is the number of particles per unit volume with radii between r and r + dr, C is a normalization constant, r_0 is the radius of the smallest particles present in statistically significant numbers, and S is a parameter describing the width of the distribution.

From theoretical considerations a dimensionless quantity is derived, for constant m:

$$(\lambda\tau/\phi)_m \qquad (2)$$

where λ = wavelength; τ = turbidity; ϕ = volume fraction of polymer; m = refractive index of particle/refractive index of medium.
This is experimentally determined as well as theoretically evaluated from the following two parameters:

$$P = 2\pi r_0/\lambda$$
$$Q = 2\pi s/\lambda \qquad (3)$$

The function (2) has been tabulated as a function of p and q in ref (3). By normalizing against an arbitrarily selected wavelength the volume fraction term is deleted.

The normalized turbidity function obtained experimentally is plotted against wavelength and matched up with a theoretically obtained curve in order to derive the proper distribution. The theoretical curves are constructed by varying p_r and q_r as defined:

$$P_r = p\lambda/\lambda r$$
$$Q_r = q\lambda/\lambda r \qquad (4)$$

Where λ_r is the reference wavelength.

Estimation of the Parameters

The conventional method of obtaining estimates of the parameters of nonlinear equations is to select some objective function based on a suitable transformation of the residuals. This objective function is minimized either analytically, or more typically numerically, by a sequential search procedure. The parameter values corresponding to the minimum in the objective function are chosen as estimators of the population parameters.

The class of objective functions most generally used are of the type:

$$\psi = \sum_{i=1}^{i=n} w_i |y_i - y_p|^{\alpha} \tag{5}$$

where y_i = observed value of function; y_p = predicted value of function; w_i = weighting factor usually chosen so that

$$\sum_{i=1}^{i=n} w_i = 1.0;$$

α = exponential weighting factor. When $w_1 = w_2 = w_3 \ldots\ldots = w_n$, and $\alpha = 2.0$.

ψ is simply the sum of squares of deviations. When $\alpha = 1.0$, ψ is the sum of absolute deviations. By assigning different weights to selected regions in the experimental space, various weighted residual functions are obtained. In the computer program w_i $(i = 1, n)$ and α can be entered as variables so that any objective function of the class represented by eq. (5) could be selected.

The residual sum of squares (w_i equal) is probably the most commonly used objective function for a variety of reasons:

(a) If the data are normally distributed the estimates obtained in this manner will be maximum likelihood estimates.

(b) If, in addition to (a), the prior distribution is nearly constant over a region where the likelihood function has appreciable value, and sufficiently small over the region where the likelihood function is itself small (the so-called "locally uniform" prior distribution), then the resulting estimates correspond to Bayesian estimates.

Of the two considerations, condition (b) is almost always fulfilled in industrial experimentation and in particular in the case of measurement of particle size. If we had a highly nonuniform distribution (e.g., by knowing in advance electron microscope data for the sample), it would be questionable whether to run the experiment at all, except for correlation purposes. Condition (a) is much more difficult to fulfill in practice. For this study the sum of squares is used as the objective function.

It is normally assumed that the residual sum of squares surface within

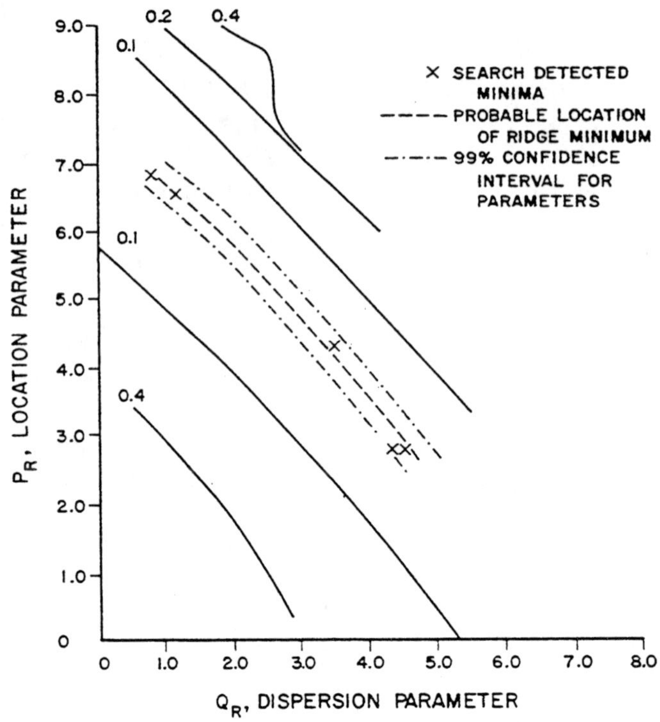

Fig. 1. Residual error map for sample A using interpolation routine.

a given experimental space is unimodal and most search techniques are predicated on this assumption

Barnard (5) has pointed out that it is always best to examine the entire response surface to ensure finding the true minimum.

If the concept of "approximate confidence regions" defined by Beale (6) or "likelihood regions" defined by Hudson (7) is used, it is possible to establish point confidence intervals for the parameters p_r and q_r in the following manner. A "likelihood region" for p_r and q_r is defined as the surface of the p_r, q_r plane enclosed by "isolikelihood" contours, or contours of equal residual sum of squares. These regions may not connect.

The probability content of these regions can be estimated by using a contour at a height of:

$$\sigma^2[\chi_n{}^2]_\alpha \tag{6}$$

above the minimum value of R. Estimating σ^2 by R_{min}/(d.f.) we obtain the following expression from eq. (4):

$$h_\alpha = R_{min}\,(1 + \chi^2_{n\alpha}/\text{d.f.}) \tag{7}$$

where h_α = contour height to enclose α probability region of q_r, p_r; R_{min} = minimum residual sum of squares; n = number of parameters to be estimated (in this case 2); d.f. = degrees of freedom. For the ex-

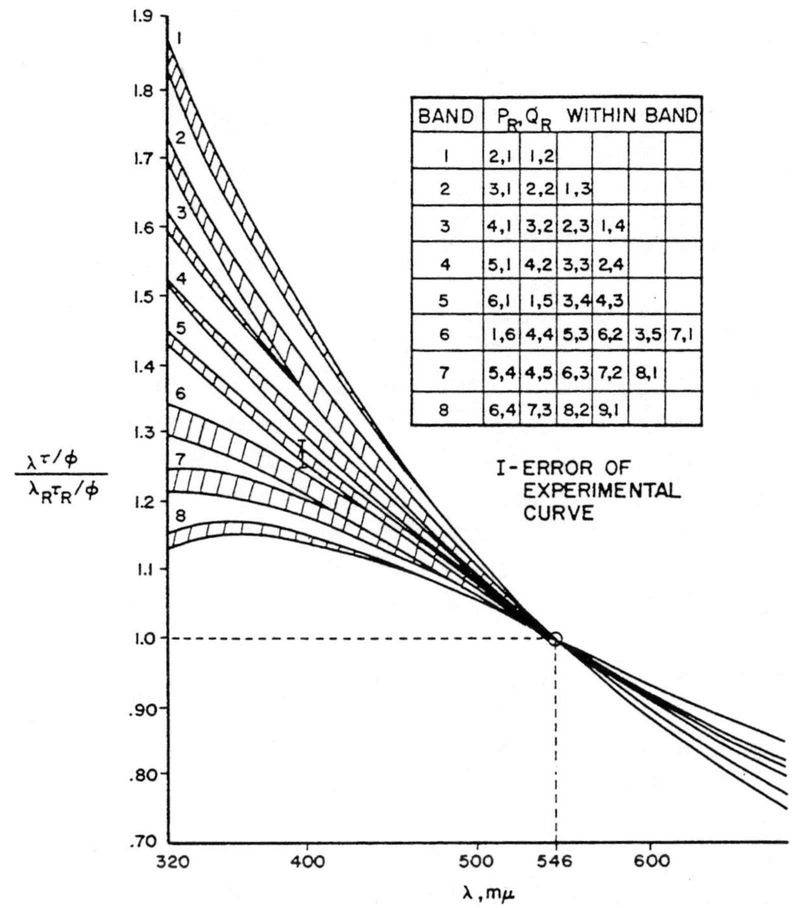

Fig. 2. Wavelength dependency of turbidity function.

ample shown in Figure 1 the region enclosed by the dotted line represents the 99% joint confidence interval for p_r, q_r. This joint confidence interval follows an RSS trough corresponding roughly to those pairs (p_r, q_r) for which the sum ($p_r + q_r$) is a constant. Thus, we might expect the actual distribution to fall anywhere within this envelope (an infinite number of alternative distributions), so that the results of this experiment do not specify a unique distribution.

Perhaps this problem in discrimination can better be visualized by reference to Figure 2. This shows various $(\lambda\tau/\phi)/(\lambda\tau/\phi)_r$ curves as $f(\lambda)$. As can be seen, many alternative distributions are quite far apart, and therefore (with a reasonable number of experimental points) effective discrimination is possible. On the other hand, curves for certain pairs (roughly those for which sum $p_r + q_r$ is constant) are quite close together, however, the actual distributions they represent are quite different.

The probable error of a typical function value $(\lambda\tau/\phi)/(\lambda\tau/\phi)_r$ can be derived as follows:

If $F = (\lambda\tau/\phi)/(\lambda\tau/\phi)_r$ is the function under consideration then the variance of F is given by (8):

$$VAR(F) = [\frac{\partial F}{\partial \lambda}]^2 var(\lambda) + [\frac{\partial F}{\partial \tau}]^2 var(\tau) + [\frac{\partial F}{\partial \lambda_r}]^2 var(\lambda_r) + [\frac{\partial F}{\partial \tau}]^2 var(\tau_r) \qquad (8)$$

Thus the proportional variance $(\sigma_F/F)^2$ is simply:

$$(\sigma_F/F)^2 = (\sigma\lambda/\lambda)^2 + (\sigma\lambda_r/\lambda_r)^2 + (\sigma_\tau/\tau)^2 + (\sigma_{\tau r}/\tau_r)^2 = \Sigma(\sigma_f/f)^2 \qquad (9)$$

For a specific case let $\lambda = 400$ mμ, $\lambda_R = 546$ mμ and the corresponding turbidity values $\tau = 0.90$ and $\tau_R = 0.51$. The proportional error (the square root of the proportional variance) was calculated from eq. (9) using values for the accuracy of λ and T% provided by the spectrophotometer manufacturer. The proportional error was found to be 3% and the error of the experimental curve at the 99% confidence level was found to be 2%. The relationship between these errors is approximately

$$(\text{Error of the experimental curve})_{99\%} = 3/\sqrt{n} \text{ proportional error} \qquad (10)$$

where n is number of observations on which the experimental curve is based. Figure 2 illustrates the fact that this 2% error brackets a number of alternative distributions.

If results of a turbidimetric determination can at least specify a family of alternative distributions, then why not combine this with some other technique for measuring a given moment of the distribution in order to specify a unique member of this family? This should lead to satisfactory results, although questions are raised on the joint reliability of two techniques. Further, resulting confidence limits may still be quite broad. Any technique for measurement of average particle size yields an estimate d_1 together with an error term, ϵ_1. Recasting this into a suitable form, we observe that moments of the distribution are of the form $\alpha P_r + \beta Q_r$. For example, the number average particle size is given by (9).

$$D_n = 0.06515\ (1.476q_r + 2p_r) \qquad (11)$$

and the weight average

$$D_w = 0.06515\ (2.214q_r + 2p_r) \qquad (12)$$

Thus a likelihood region for p_r and q_r will be another trough of p_r, q_r possibilities making an acute angle with the original joint likelihood region. To minimize the area of the joint likelihood region, we should

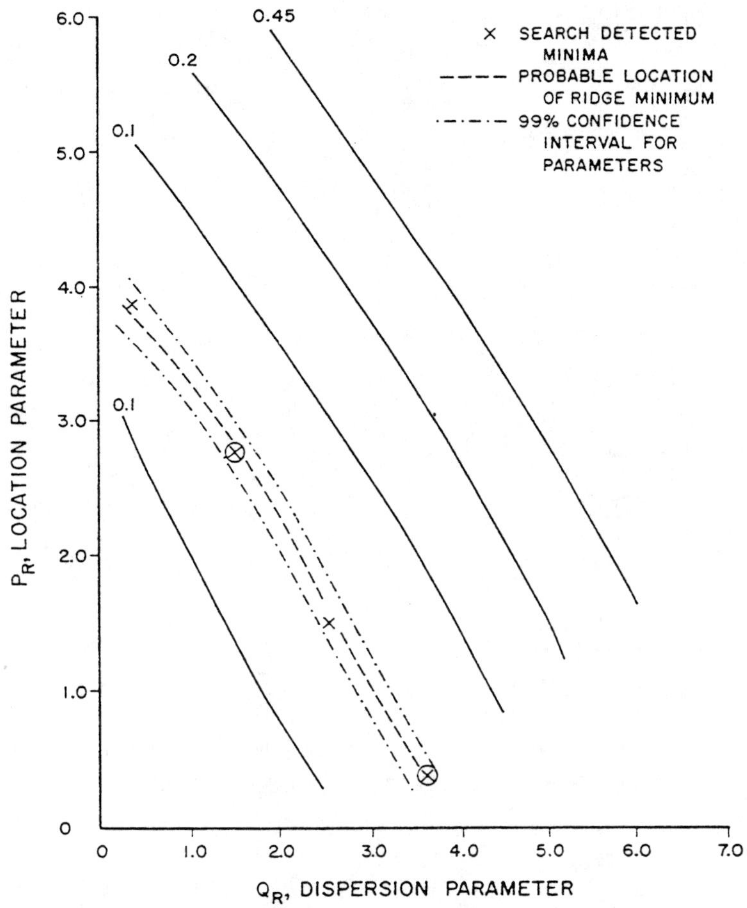

Fig. 3. Residual error map for sample B.

seek another trough which is orthogonal to the original. In other words, we seek a moment of the distribution which has the form:

$$\alpha P_r - \beta Q_r$$

This unfortunately does not correspond to any measurable quantity.

Using this method, a distribution curve of a certain shape must be assumed. The assumption used by Heller et al. (9) is a positively skewed distribution limited to a maximum dispersion (D_w/D_n) of 1.5. The assumption of positive skewness is valid for most commercial latexes. The restriction of dispersity to 1.5 is unreasonable. Typical values of dispersity for commercial poly(vinyl acetate) latexes will range from near 1 to as high as 7 or 8 or more.

Experimental

Measurements were carried out on commercial vinyl acetate latexes

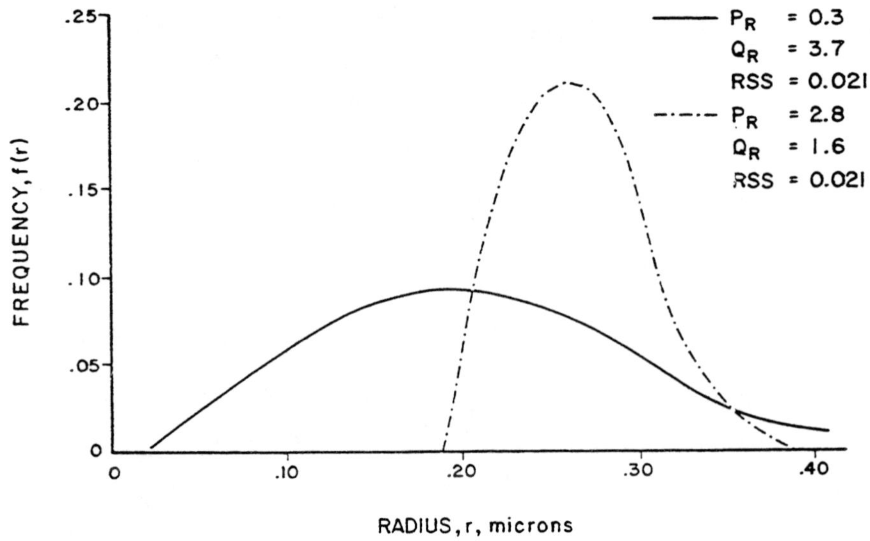

Fig. 4. Sample B distributions.

of average particle size 0.1–1.0 μ. The sample was diluted with distilled water. A Beckman DB spectrophotometer was used for turbidity measurements at 20 mμ intervals between 320 and 720 mμ. This instrument is uniquely suited for this work because of the unusually long path length (18.5 cm) between the sample cell and the photocell. This minimization of effects due to forward scattering, etc., is illustrated by the fact that the relationship between turbidity and concentration is linear. This eliminates the necessity of extrapolation to zero concentration (4). The relative refractive index m, was found to be 1.100 from measurements of the refractive index of the polymer with an Abbe refractometer.

A latex sample which, by electron microscopy, was found to have a D_w of 0.257 and a D_n of 0.188 giving a dispersity of 1.36 (within the limitations of the method) produced an error map shown in Figure 3. The X points shown on this map were search detected minima and have the following values: D_w 0.551, D_n 0.537; D_w 0.596, D_n 0.519; D_w 0.543, D_n 0.422; and D_w 0.573, D_n 0.395. The circled X values lead to the distributions shown in Figure 4.

Calculations and Computer Programs

The calculations consist of: (a) constructing an experimental curve based on the data; (b) constructing a family of theoretical curves based on different p_r and q_r values; (c) finding the theoretical curve that best matches the experimental curve; (d) computing the particle size distribution based on the p_r and q_r values underlying the theoretical curve found in step c.

A computer program entitled CRAPS (Computer Routine for Analysis

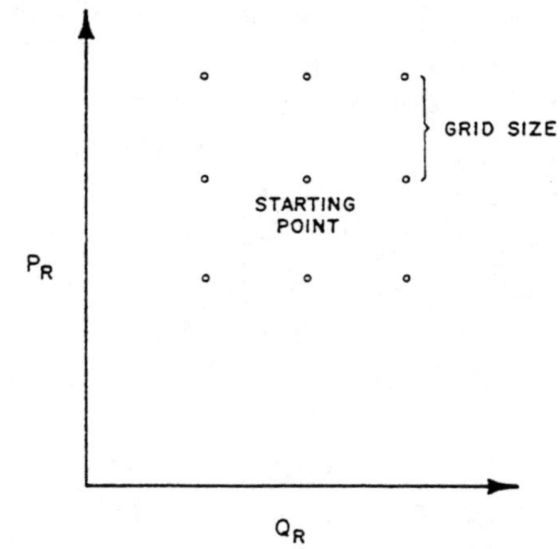

Fig. 5. Computer search technique.

of Particle Size) was written to take care of the above computations which are otherwise quite laborious.

A number of numerical search techniques exist for solving a problem of this type, e.g., one at a time (10), steepest ascent, direct search (11), gradient search (12), contour tangent (13), and lattice search (13). The advantages and disadvantages of each are discussed by Boas (14) and Wilde (15). CRAPS incorporates the lattice search technique.

The lattice search technique consists of selecting a starting point somewhere on the plane to be explored (the p_r, q_r plane) and computing the value of the objective function at that point. Next values of the function corresponding to points forming a grid about the starting point are computed as illustrated in Figure 5.

These nine results are now compared and the midpoint shifted to the point having the minimum value of the objective function. This process continues until the minimum value occurs at the center point of the grid. At this stage the grid size is reduced (to refine the estimates of the location of the minimum) and the process repeated. The p_r and q_r coordinates of this minimum point are chosen as estimates of the population parameters.

Running time of the program will vary with the number of iterations required, but averaged about 4 min on a Honeywell 200.

Conclusions

(a) Matching experimental and theoretical ($\lambda\tau/\phi$) curves to determine appropriate p_r and q_r values may result in multivalued solutions. Thus, conventional minimization algorithms will give misleading estimates.

(b) Consideration of typical errors associated with the necessary measurements yield likelihood regions which are decidedly nonspecific.

(c) Combination of this technique with another which gives only one average of the distribution may not reduce the number of alternative distributions sufficiently.

These considerations point up the fact that this turbidity method should not be used exclusively for particle size determination. This negates the value of the simplicity of experimental technique, and effectively removes this from the domain of routine measurements. Other light-scattering techniques, however, such as that of Jacobsen et al. (18), have been found to have single-valued solutions and should prove more satisfactory for particle size distribution analysis.

The authors express their appreciation to M. Q. Stills for performing the many turbidity measurements necessary for this work.

References

(1) E. B. Bradford and J. W. Vanderhoff, J. Colloid Sci., 14, 543 (1959); E. Vanzo, Ph.D. Thesis, State University College of Forestry of Syracuse University, June 1963.

(2) J. E. Gwyn, E. J. Crosby, and W. R. Marshall, Ind. Eng. Chem. Fundamentals, 4, 204 (1965).

(3) M. L. Wallach, W. Heller, and A. F. Stevenson, J. Chem. Phys., 34, 1796 (1961).

(4) M. L. Wallach and W. Heller, J. Phys. Chem., 68, 924 (1964).

(5) G. A. Barnard, papers presented at the Fifth Special Seminar in Recent Advances in Statistical Methods, University of Tennessee, Knoxville, Tenn., March 28–April 1, 1966.

(6) E. M. L. Beale, "Confidence Regions in Nonlinear Estimation," J. Roy. Statistical Soc., B22, 41–81 (1960).

(7) D. J. Hudson, Statistics Lectures II: Maximum Likelihood and Least Squares Theory, CERN 64–18, C.E.R.N., Geneva, 1964.

(8) C. A. Benett and N. L. Franklin, Statistical Analysis in Chemistry and Chemical Industry, Wiley, New York, 1963.

(9) A. F. Stevenson, W. Heller, and M. L. Wallach, J. Chem. Phys., 34, 1789 (1961).

(10) M. Friedman and L. S. Savage, Selected Techniques in Statistical Analysis, McGraw-Hill, New York, 1947.

(11) R. Hooke and T. A. Jeeves, J. Assoc. Computing Mach., 8, No. 2, 212 (1961).

(12) H. E. Zellnik, N. E. Sondak, and R. S. Davis, Chem. Eng. Prog., 58, No 8, 35 (1962).

(13) D. J. Wilde, "Gradient Free Search for an Optimum," paper presented at A.I.Ch.E. National Meeting, Los Angeles, California, Feb. 5, 1962.

(14) A. H. Boas, "Optimizing Multivariable Functions," Chem. Eng., March 4, 1963, p. 97.

(15) D. J. Wilde, Optimum Seeking Methods, Prentice Hall, New Jersey, 1964.
(16) M. Kerker, E. Matijević, W. Espenscheid, W. A. Farone, and S. Kitani, J. Colloid Sci., 19, 213 (1964).
(17) J. P. Kratohvil and C. Smart, J. Colloid Interface Sci., 20, 875 (1965).
(18) R. T. Jacobsen, M. Kerker, and E. Matijević, J. Phys. Chem., 71, 514 (1967).

Reprinted with permission from *Journal of Applied Physics,* Vol. 45(1), pp. 216-220 (January 1974). ©1974 American Institute of Physics.

Light scattering by pores in polycrystalline materials: Transmission properties of alumina

J. G. J. Peelen and R. Metselaar

Philips Research Laboratories, Eindhoven, The Netherlands

The in-line transmission of polycrystalline high-density ceramics is discussed in terms of light scattering by pores. The Mie theory for light scattering is applied to calculate scattering coefficients using generalized parameters. Results are shown for scattering by spherical pores with uniform size and with a lognormal size distribution. Specific examples are given of the calculated transmission of alumina in the wavelength region 0.4–5 μm. Experimentally determined transmission curves of normally sintered and of hot-pressed alumina are compared with calculated curves. The porosity, the position of the maximum, and the width of the lognormal distribution are treated as variables in the calculation. Good agreement with the experimental data is obtained.

I. INTRODUCTION

During recent years it has become possible to obtain transparent or nearly transparent ceramics by using improved sintering techniques.[1–3] Alumina in particular has attracted much interest since its application as a lamp envelope for the high-pressure sodium lamp. This has stimulated investigations into the dependence of the optical properties of polycrystalline alumina on parameters of the microstructure. Possible causes of transmission losses are scattering by pores or second phase, scattering by grain boundaries, absorption, birefringence, and surface roughness. Several authors[4,5] have investigated these contributions to determine their relative importance. It has been shown that scattering by residual porosity is the main cause of the attenuation of the intensity of the light on its passage through a slice of alumina. The absorption in the visible and near-infrared parts of the spectrum is negligible; grain boundaries play a role only when they contain a second phase. The effect of birefringence in alumina ($\Delta n/n = 4.5\times10^{-3}$) is not known quantitatively but its influence is of secondary importance compared with the role of the pores.[3,5,6] This optical anisotropy becomes significant only when the porosity is extremely low. The surface scattering contribution can be neglected after a careful polishing treatment.

The transmission depends not only on the total porosity but, especially in the visible part of the spectrum, the actual pore size is of very great importance.[3] It is therefore likely that a pore size distribution, which is always present in a sintered material, will also have a marked influence on the light transmission. The purpose of this work is to calculate the on-axis or in-line transmission for a given pore size distribution, and to compare the calculated transmission with the experimentally measured transmission of polycrystalline alumina. Section II deals with the preparation method of the alumina samples. In Sec. III we first describe calculations of the scattering by pores with a fixed radius, and after a discussion of the pore size distribution we describe the computation of light scattering by pores with a lognormal size distribution. In Sec. IV experimental transmission curves for two different alumina samples are compared with calculated curves.

II. EXPERIMENTAL

The alumina samples with different pore sizes were prepared by using the normal sintering technique and the technique of continuous hot pressing.[7] To sinter the alumina powder[8] to high densities it is necessary to add 0.2% of magnesia,[9] by making a suspension in isopropanol. The powder is then pressed into pellets, pressed isostatically at a pressure of 1000 kg/cm² and fired in an externally wound Al_2O_3 tube furnace. Sintering takes place in a moist hydrogen atmosphere (dew point, 20 °C) for 24 h at 1900 °C. To hot press the alumina powder, addition of magnesia as a sintering aid is not necessary. The hot pressing takes place in an oxygen atmosphere and typical process parameters are temperature 1400 °C, pressure 1200 kg/cm², and velocity 2.5 cm/h. It is noteworthy that hot pressing of the alumina powders has proved to be possible in an alumina die without any permanent sticking to the die wall.

We found the x-ray density of alumina to be $d_{th} = 3.986$ g/cm³. Densities of the sintered and hot-pressed products are determined by the immersion method. With much care it is possible to determine the density with an accuracy of 0.001 g/cm³. This means a rather large relative inaccuracy in the porosity, defined as

$$V_p = (d_{th} - d_{expt})/d_{th}.$$

After sintering, the residual porosity is about 0.2% and after hot pressing about 0.1%.

A typical microstructure of a sintered specimen shows a rather homogeneous grain size of about 20 μm and a pore radius of 0.5–1.0 μm. The pores are spherical and most of them lie inside the grains. The microstructure of a hot-pressed specimen shows a less homogeneous grain size (no magnesia present) of about 1 or 2 μm. The pores are very small and have a radius of about 0.1 μm. They are found predominantly on the grain boundaries. Photomicrographs of our samples are shown in Figs. 1 and 2. For transmission measurements, slices with a thickness of 0.5 mm were made from the sintered pellets and hot-pressed rods. The slices were given a final polishing on both sides with 0–2 μm diamond paste to obtain a good surface finish.

Transmission measurements were carried out in the visible and near-infrared parts of the spectrum, using a Beckman DK 2 spectrophotometer. The apparatus had a special transmission unit, and the detector was situated 25 cm behind the sample to minimize the measure-

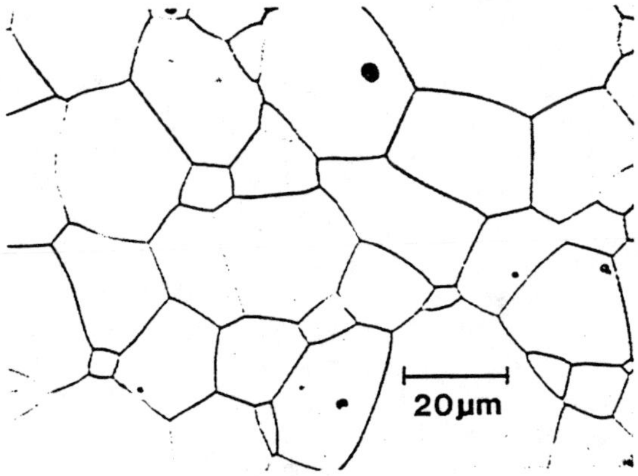

FIG. 1. Photomicrograph of normally sintered alumina.

ment of scattered light. As the light-sensitive surface of of the detector was 7.2 mm^2, only light scattered in a cone of a half-angle less than 0.01 rad could interfere with the transmission measurement.

III. SCATTERING BY PORES

A. Scattering by pores with a fixed radius

When an electromagnetic wave passes through a medium, part of the energy is absorbed within the medium, and a second part may be scattered. The total extinction cross section C_{ext} in this case can be written $C_{ext} = C_{abs} + C_{sca}$. Generally, the cross sections are functions of the wavelength λ of the incident light. In our study of the scattering of light by pores in alumina the absorption is negligible in the wavelength range 0.4–5.0 μm ($C_{ext} = C_{sca}$). If the geometrical cross section of the scattering particles is G, we can define a dimensionless efficiency factor $Q_{ext} = C_{ext}/G$. This quantity depends on the refractive index m and on the size and orientation of the scattering particles. For our purpose we can assume that m is isotropic, so if we restrict our discussion to spherical pores, Q_{sca} is independent of the particle orientation. The total geometrical cross section is related to the pore radius r and the pore density N_0 by $G = N_0 \pi r^2$.

A general treatise on the scattering by spherical particles was formulated by Mie.[10,11] The Mie expressions are written in terms of the size parameter $x = 2\pi r/\lambda$ and the relative refractive index m. In our case we call m_1 the refractive index of the gas in the pores and m_2 the refractive index of the Al_2O_3 medium, so $m = m_1/m_2$ ($m < 1$). For the wavelength λ in the medium we have $\lambda = \lambda_0/m_2$, where λ_0 is the wavelength in vacuum. The computing program will not be described here, but we refer to the literature.[12]

A generalized representation of Q_{sca} is given in Fig. 3, where the scattering efficiency is plotted against the function ρ defined by

$$\rho = 2x|m-1| = (4\pi r/\lambda_0)|m_1 - m_2|. \tag{1}$$

The physical meaning of this quantity is the phase lag experienced by a central light ray that travels through the sphere along a full diameter. The result of our calculations for scattering by spheres with a fixed radius

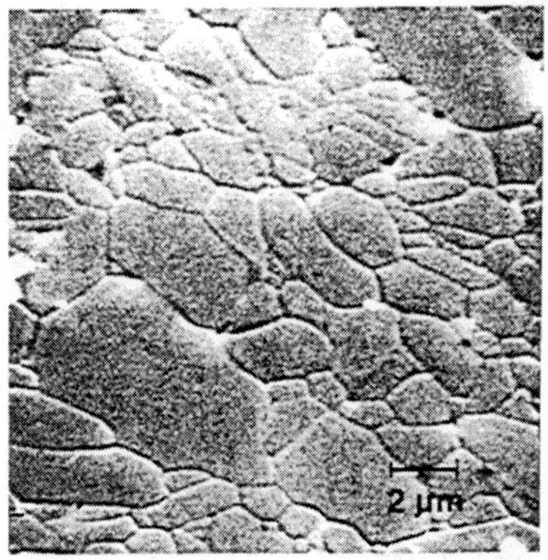

FIG. 2. Scanning electron micrograph of hot-pressed alumina.

r and relative refractive index $m = 0.565$ is indicated in Fig. 3 by the curve with parameter $\sigma = 0$. The most noteworthy feature of this curve is the oscillatory behavior of the scattering efficiency for $\rho > 4$. For comparison with the experimentally determined transmission curves, a plot of the transmission T against the wavelength λ_0 is more suited. For a sample thickness t we have

$$T = (1-R)^2 \exp(-C_{sca} t), \tag{2}$$

where R is the reflectivity. This notation implies that the scattering is due to independent scattering centers. The exponential thickness dependence of the transmission has been experimentally verified by Grimm *et al.*[5] The geometrical cross section $G = N_0 \pi r^2$ can be express-

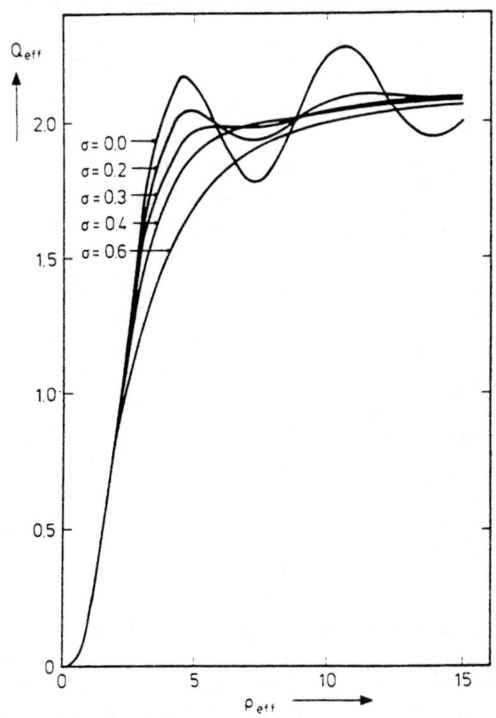

FIG. 3. The effective efficiency factor for scattering by pores in alumina as a function of $\rho_{eff} = 4\pi r_{eff}|m-1|/\lambda$ with $m = 0.565$ (cf. Sec. IIIA), for a lognormal pore size distribution. The width of the distribution increases with increasing σ values.

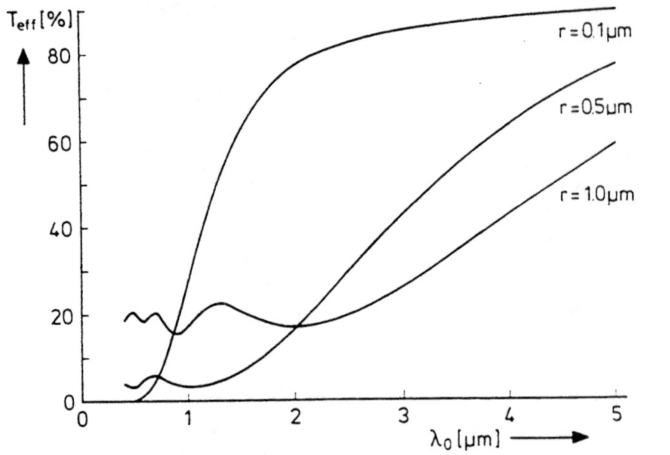

FIG. 4. Calculated transmission curves for alumina with a single pore radius r, as a function of wavelength. Porosity $V_p = 2\times10^{-3}$, sample thickness $t = 0.5$ mm.

ed in the experimentally more accessible quantity V_p, the volume porosity, by $G = 3V_p/4r$. In our calculations for alumina the refractive index m_1 of the gas in the pores was set equal to unity, while for m_2 we used the dispersion equation[13]

$$m_2 = \left(1 + \frac{1.023798\lambda_0^2}{\lambda_0^2 - 0.00377588} + \frac{1.058264\lambda_0^2}{\lambda_0^2 + 0.0122544} + \frac{5.280792\lambda_0^2}{\lambda_0^2 - 321.3616}\right)^{1/2}. \quad (3)$$

This equation was also used to calculate the reflection loss:

$$R = (m_2 - 1)^2/(m_2 + 1)^2. \quad (4)$$

In Fig. 4 results are shown for three different values of the pore radius r, with a porosity $V_p = 0.002$ and thickness $t = 0.5$ mm. It is seen that under these conditions the oscillations of the scattering coefficient should be present in the transmission spectrum above 0.4 μm for pores with radius $r \geq 0.2$ μm. The fact that we never observed such oscillations in our experimental transmission curves suggests that the assumption of uniform pore size was not realistic. In fact van de Hulst[10] has shown, for the limiting case of Rayleigh-Gans scattering ($|m-1| \ll 1$), that the amplitude of the oscillations in Q_{sca} decreases if we no longer assume the presence of pores with a fixed radius, but instead introduce a pore size distribution.

B. Pore size distribution

The assumption of uniform pore size is not realistic in actual ceramic materials: there is always a spread in pore size, just as there is always a spread in grain size. Determination of the pore size distribution is very laborious: many measurements of generally very small pores are necessary to estimate the distribution with some accuracy. It is probably for this reason that hardly any results on pore size distributions have been published. Moreover the derivation of the true size distribution from the observed distribution in a polished section is a complex problem. An approximate method for this conversion, valid for spherical pores, has been given by Tomandl.[14]

For grain size and powder particle size distributions a lognormal distribution is often assumed. We do not know of any data published on pore size distribution functions. However, figures shown by Oel[15] can be fitted reasonably well with a lognormal distribution. We have tested the assumption of a lognormal pore size distribution in a vacuum-sintered alumina specimen with large pores. Pore sizes were determined with a planimeter from photographs of a polished section. The resulting two-dimensional distribution was then converted into a three-dimensional one.[14] A lognormal distribution with a standard deviation $\sigma \approx 0.25$ proved to fit the experimental distribution with reasonable accuracy. For this purpose we used the following representation of a logarithmic distribution, called ZOLD function[11]:

$$f(r) = [(2\pi)^{1/2}\sigma r_m \exp(\tfrac{1}{2}\sigma^2)]^{-1} \exp[-(\ln r - \ln r_m)^2/2\sigma^2], \quad (5)$$

where σ is a measure of both width and skewness of the distribution. The parameter r_m denotes the value of the radius r for which the function $f(r)$ reaches its maximum value. The preexponential factor ensures normalization according to $\int_0^\infty f(r)\,dr = 1$. The mean pore radius r_{av} is found to be

$$r_{av} = \int_0^\infty r f(r)\,dr = r_m \exp(\tfrac{3}{2}\sigma^2). \quad (6)$$

C. Influence of pore size distribution on the scattering coefficient

Upon introduction of a pore size distribution $f(r)$, the expression for the scattering coefficient becomes

$$C_{sca} = (3V_p/4r_m)\left(\int_0^\infty Q_{sca}(u) f(u) u^2\,du\right)\left(\int_0^\infty f(u) u^3\,du\right)^{-1}, \quad (7)$$

where for convenience we have introduced the dimensionless parameter $u = r/r_m$.

Using the ZOLD distribution [Eq. (5)] with parameters σ and r_m, and using the Mie expressions to calculate $Q_{sca}(u)$, we can numerically integrate Eq. (7) to obtain C_{sca} as a function of wavelength. A general representa-

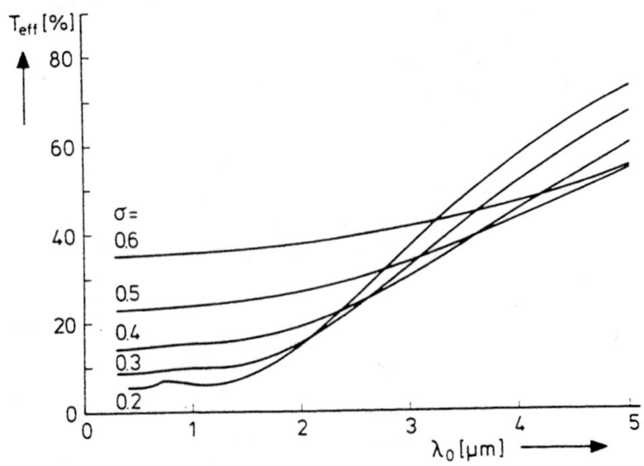

FIG. 5. Calculated transmission curves for alumina with a lognormal pore size distribution, as a function of wavelength. The width of the distribution increases with increasing σ values. The maximum of the distribution $r_m = 0.5$ μm, porosity $V_p = 2\times10^{-3}$, and sample thickness $t = 0.5$ mm.

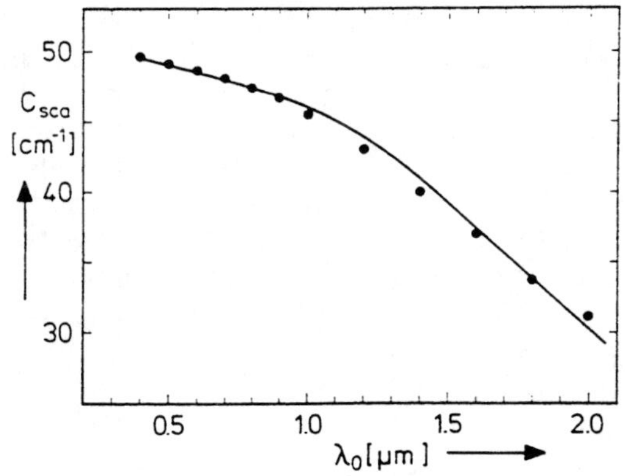

FIG. 6. ●: scattering coefficients as a function of wavelength as measured on a normally sintered sample of alumina. Solid line: calculated scattering coefficients assuming a lognormal pore size distribution with $r_m = 0.34$ μm, $\sigma = 0.4$, and a porosity $V_p = 1.9\times10^{-3}$.

tion of the results can be obtained by introducing the parameters

$$\rho_{eff} = (4\pi r_{eff}/\lambda_0)(m_2 - 1),$$

with

$$r_{eff} = (3V_p/4G) = r_m\left(\int_0^\infty f(u)u^3\,du\right)\left(\int_0^\infty f(u)u^2\,du\right)^{-1}. \tag{8}$$

Substituting Eq. (5) we have $r_{eff} = r_m \exp(\frac{7}{2}\sigma^2)$. We further define

$$Q_{eff} = \left(\int_0^\infty Q_{sca}(u) f(u)u^2\,du\right)\left(\int_0^\infty f(u)u^2\,du\right)^{-1}. \tag{9}$$

The results of the numerical calculation of Q_{eff} as a function of ρ_{eff} are shown in Fig. 3. We see that the oscillations in Q_{eff} are damped out with increasing values of the parameter σ.

Curves of Q_{eff} as a function of ρ_{eff}, with m and σ as parameters, are quite generally applicable. For instance one of us has applied these results to investigate the infrared transmission of polycrystalline yttrium-iron garnet.[16]

IV. DISCUSSION

In this section we shall compare our calculations with experimental results obtained on alumina. The influence of the spread in pore sizes on the transmission of alumina is illustrated in Fig. 5, where the calculated in-line transmission is given as a function of wavelength. The transmission T_{eff} is obtained from Eq. (2) after substitution of

$$C_{sca} = \frac{3V_pQ_{eff}}{4r_{eff}} = \frac{3V_pQ_{eff}}{4r_m\exp(3.5\sigma^2)}. \tag{10}$$

With increasing values of σ the curves are flattened out more and more, while the over-all transmission increases in the visible part of the spectrum and decreases at longer wavelengths.

C_{sca} must be corrected for the light scattered in a cone of half-angle θ in the forward direction.[17] As already mentioned θ is less than 0.01 rad in our experimental setup. This means a correction of 0.5% upon C_{sca} in the worst case of $x = 20$. Hence, we have neglected this correction in the following.

With our assumptions the on-axis or in-line transmission of polycrystalline alumina is a function of three parameters of the microstructure: the value of the radius where the pore size distribution assumes its maximum, r_m; the width of the distribution, σ; and the porosity, V_p. We must treat V_p as a parameter as well, because the inaccuracy in the porosity determination can be high for very dense materials. For each parameter set, it is possible to calculate the transmission as a function of wavelength. Conversely, we can try to determine these parameters from the measured transmission curve.

It is advantageous to plot the scattering coefficient C_{sca} instead of the transmission against the wavelength, because C_{sca} is directly proportional to the porosity V_p according to Eq. (10). From the calculated scattering curves one can derive graphs which give the scattering coefficient at a given wavelength as a function of pore radius r_m with σ and V_p as parameters. Comparison with the experimentally obtained C_{sca} at some discrete wavelengths gives generally a number of possible combinations of r_m, σ, and V_p. For each of these combinations we can calculate the scattering coefficient as a function of wavelength, and by comparison with the experimentally determined curve one can find the parameter set that gives the best fit over the whole wavelength region. As an example Fig. 6 gives a plot of the experimental and calculated C_{sca} versus λ_0 of a sintered alumina. Its measured density is 3.978 g/cm^3, which gives $V_p = (2\pm0.25)\times10^{-3}$. The best fit with the experimental transmission could be obtained with the parameters $\sigma = 0.4$, $V_p = 1.9\times10^{-3}$, and $r_m = 0.34$ μm (this gives $r_{av} = 0.43$ μm). The value found for the porosity in this way is in good agreement with the experimental V_p. The same calculations were made for a sample of hot-pressed alumina. This material is highly transparent, in contrast with normally sintered alumina which is translucent. The measured density of the transparent material amounts to 3.984 g/cm^3, which means a poros-

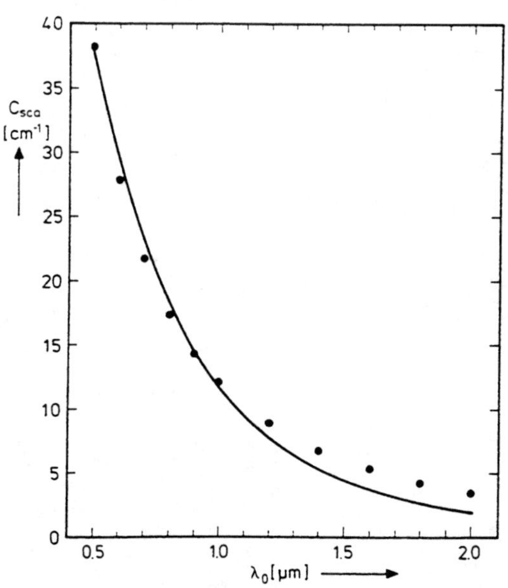

FIG. 7. ●: scattering coefficients as a function of wavelength as measured on a hot-pressed sample of alumina. Solid line: calculated scattering coefficients assuming a lognormal pore size distribution with $r_m = 0.11$ μm, $\sigma = 0.3$, and porosity $V_p = 4.9\times10^{-4}$.

ity $V_p = (5 \pm 2.5) \times 10^{-4}$. For several values of σ a combination of V_p and r_m can be found that gives a reasonable agreement between calculated and measured transmission curves. In Fig. 7 the experimental transmission is shown together with the curve belonging to the set $\sigma = 0.3$, $V_p = 4.9 \times 10^{-4}$, and $r_m = 0.11$ μm. However, the standard deviation of a least-squares fit for this curve is equal to the standard deviation for a curve with the parameter set $r_m = 0.03$ μm, $\sigma = 0.6$, and $V_p = 6.6 \times 10^{-4}$. The porosities found do not differ much and are again in good agreement with the experimentally found porosity. To be able to make a choice between the possible values of r_m and σ more information about the pore size and pore size distribution in this very dense material is required.

V. CONCLUSION

The light transmission of polycrystalline alumina is governed by the scattering of light by the residual porosity. It is possible to obtain highly transparent alumina by reducing the porosity and the pore size. The Mie theory allows a calculation of the transmission losses due to scattering by pores. The experimental data can be described by this theory if we assume a spread in pore size. We have calculated curves which give the scattering coefficient for nonabsorbing materials with a lognormal pore size distribution. We have also calculated curves of transmission versus wavelength for alumina with a given value of the porosity V_p, pore radius r_m, and width of the distribution σ.

Finally we have tried to fit experimental transmission curves for normally sintered and hot-pressed alumina, using V_p, r_m, and σ as parameters. For the normally sintered sample a unique set of parameters was established. The value of the porosity determined in this way agrees with the experimental value. This good agreement supports our assumption that the transmission loss in alumina is mainly due to scattering by pores and that the contribution of scattering by grain boundaries is negligible. In this approximation we can also neglect the influence of birefringence. For the hot-pressed sample we could not decide which combination of r_m and σ gave best agreement with the experiments, but again the calculated and experimentally determined porosities agreed well. For further quantitative calculations, however, we need more information on the actual pore size distributions in this kind of high-density ceramics.

[1]R.L. Coble, J. Appl. Phys. 32, 793 (1961).
[2]G. Winkler, IEEE Trans. Magn. MAG-7, 773 (1971).
[3]J.G.J. Peelen, Sci. Ceram. 6, XVII (1973).
[4]D.W. Lee and W.D. Kingery, J. Am. Ceram. Soc. 43, 594 (1960).
[5]N. Grimm, G.E. Scott, and J.D. Sibold, Ceram. Bull. 50, 962 (1971).
[6]D.W. Budworth, Spec. Ceram. 5, 185 (1970).
[7]G.J. Oudemans, Proc. Brit. Ceram. Soc. 12, 83 (1969).
[8]Rubis Synthétique des Alpes, Extra Pure A6.
[9]Union Chimique Belge, *pro analysi*.
[10]H.C. van de Hulst, *Light Scattering by Small Particles* (Wiley, New York, 1957).
[11]M. Kerker, *The Scattering of Light and Other Electromagnetic Radiation* (Academic, New York, 1969).
[12]J.V. Dave, IBM J. Res. Dev. 13, 302 (1969).
[13]I.H. Malitson, J. Opt. Soc. Am. 52, 1377 (1962).
[14]G. Tomandl, Ber. Dtsch. Keram. Ges. 48, 222 (1971).
[15]H.J. Oel, Ber. Dtsch. Keram. Ges. 43, 624 (1966).
[16]R. Metselaar, P.J. Rijnierse, and U. Enz, Ber. Dtsch. Keram. Ges. 47, 663 (1970).
[17]R.O. Grumprecht and C.M. Sliepcevich, J. Phys. Chem. 57, 90 (1953). According to these authors the scattering cross section C_{sca} has to be corrected with a factor $F = 1 - i\theta^2/Qx^2$. Here i is the intensity function of the scattered wave, which can be calculated from the Mie theory (Ref. 10). We find that F increases with the size parameter x. In our experiments the highest value of x is about 20 and this gives $1 - F = 5 \times 10^{-3}$. This means a correction of 0.5% upon C_{sca}.

Reprinted from *Effective Utilization of Optics in Quality Assurance,* J. Cruz, editor, Proc. SPIE Vol. 129, pp. 49-58 (1977). ©1977 SPIE.

LIGHT SCATTERING INSTRUMENTATION FOR PARTICULATE MEASUREMENTS IN PROCESSES

A. L. Wertheimer, H. N. Frock, E. C. Muly
Leeds & Northrup Company
North Wales, PA 19454

Abstract

A line of new instruments is described which use light scattering to measure various parameters of particulate distributions. Employing laser illumination of a flowing stream of particles, a Fraunhofer diffraction pattern is produced and processed with optical and electronic means. Various parameters of the particle size distribution are measured, including a thirteen segment histogram of the volume size distribution. The measurement is rapid and continuous. Technical background and instrumentation is discussed. Many applications have been made, and this paper includes three examples relating to industrial quality control involving particles.

Introduction

A surprisingly large number of materials produced by industry involve either materials, intermediates or end products in powder or particulate form. Examples range from basic minerals to a variety of foods, chemicals, metal powders, ceramics, pigments, etc. In many of these processes, the size of the particulate is important or even critical. The importance ranges from strength of cement, appearance of paints and coatings, mouthfeel of confectionaries, to electrical properties of insulators, and so on. Also, in many of these processes, considerable energy is consumed in achieving the desired particle size through milling, grinding, precipitating, crystallizing or some other size-determining process. To achieve optimum product quality and throughput for minimum energy consumption requires a rapid, precise means of on-line measurement of particle size. Until recently, the necessary measurement capability existed only in the laboratory through painstaking measurements by specially trained, skilled operators. Through new applications of long-established optical principles, however, a new instrument has been developed which has demonstrated its suitability to perform the necessary measurements, either on-line in the process or off-line in the laboratory. This instrument, the MICROTRAC Particle Size Analyzer, has been tested successfully in a variety of industrial applications and environments.

The MICROTRAC Analyzers use optical methods based on diffraction which have been defined for measuring particle size since the last century. However, the combination of laser light source, special optical masks and modern micro-electronics has produced a new generation of light scattering instruments. While previous optical instruments are used for measuring individual particles, a large number of applications exist where the concentration is too high to permit a practical measurement of the individual particles. Rather, some composite value must be determined. Recently, analysis of the diffraction pattern has permitted extraction of both statistically meaningful particle distribution parameters, and relatively narrow volumetric histogram data, even when particles of many different sizes are present in the light beam at each instant of time.

This paper discusses some of the techniques and applications of this capability. The first portion provides a summary of the technical background. Then, several instrumental configurations are discussed, and finally, three of the many field-proven examples relating to industrial quality control are presented.

Theoretical Background

To understand the theory of operation it is convenient to consider a single spherical particle, illuminated by a plane monochromatic wavefront of wavelength, λ. The region of interest is for particles that are approximately 2-200 μm in diameter, with specific attention to particles that are nominally opaque. For such a particle whose radius, a, is sufficiently large compared with the wavelength, λ, the intensity distribution of the Fraunhofer diffraction pattern is given by the familiar Airy formula[1]

$$I(w) = Ek^2a^4\left[\frac{J_1(kaw)}{kaw}\right]^2 \tag{1}$$

where E is the flux per unit area of the incident beam, $k = 2\pi/\lambda$, $w = \sin\theta$ with θ the angle relative to the direction of the incident beam, and J_1 is the first order Bessel function of the first kind.

It is well known that if the total diffracted flux is collected and measured, a signal will be generated that is proportional to the square of the particle diameter, permitting a direct measurement of the cross-sectional or surface area of the sphere.

If many particles are randomly located within the illuminated area, and the average intensity is measured, this average value is equal to the sum of the individual diffracted light intensities. Accordingly, the scattered flux is proportional to the sum of the cross section areas.

Many industrial processes, however, require measurement of the total mass or volume of particulates. Selective attenuation and collection of the diffracted flux can produce a signal proportional to the cube of the particle diameter[2], and hence a measure of the volume, or mass, if the specific gravity is known. A Fraunhofer plane mask which passes flux in proportion to the diameter cubed is shown in Fig. 1.

The a^3 response is only one of a number of radius dependent responses possible by selective attenuation of diffracted flux. Both a second power (a^2) response mentioned here, and a fourth power (a^4) response can readily be obtained. All three response characteristics have been combined into a single instrumental execution[3], shown schematically in Fig. 2. Three apertures, giving responses proportional to a^2, a^3, and a^4, can be combined in a single mask. The signals from the different apertures must then be distinguished by some means, and a convenient way is to use a fixed sector disk, and to rotate the mask behind the disk so that the apertures are exposed in succession.

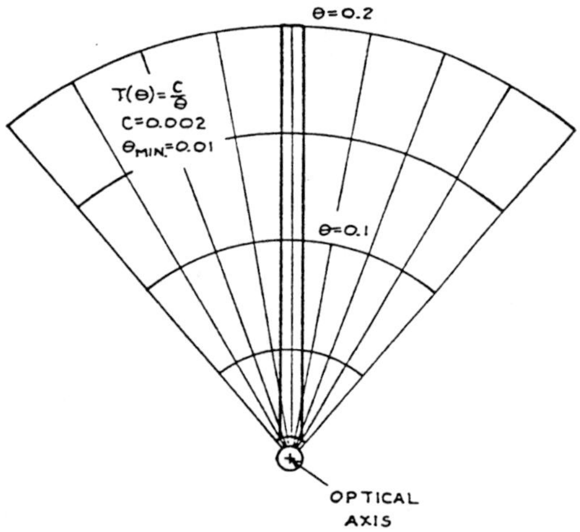

Figure 1: Polar diagram of transmission mask for a volumetric response

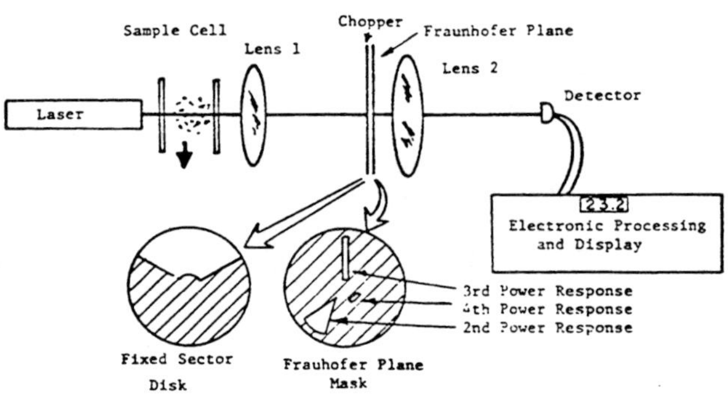

Figure 2: Optical system showing transmission mask for various power responses

This device can be used to measure directly parameters of surface area and volume for any distribution of particles within the size range of 2 to 200 microns in diameter. In addition, it is possible to measure statistical parameters of the distribution by taking ratios of flux measurements from the three masks.

The volume mean diameter, a_V, the area mean diameter, a_A, and the area standard deviation, σ_A, can be used to characterize and monitor a distribution. These three parameters are measureable in a direct and continuous manner. If S_2, S_3, and S_4 are the scattered light signals obtained through the second, third, and fourth power masks, respectively, and K_A and K_V are instrumental constants, then,

$$a_V = K_V S_4 / S_3 \tag{2}$$

$$a_A = K_A S_3 / S_2 \tag{3}$$

$$\sigma_A = \sqrt{a_A (a_V - a_A)} \tag{4}$$

These are derived in more detail in Ref. 3.

Histogram Approach

The measurements described in the last section are especially useful for situations such as process control applications where it is necessary to monitor volume loading or characteristic size parameters of a volume or area distribution. In some situations, however, one wants to determine the volume distribution as a histogram of volume components in various particle size categories. It is feasible to extract this information from the composite diffraction pattern, since all the information about the particle sizes is preserved in the diffraction pattern of the particles.

Several workers have developed histogram-type data processing schemes based on selective sampling of the diffraction pattern.[4,5,6] All these methods require mathematical manipulation of the measured data.

In developing this approach, special attention was given to the development of a dedicated microprocessor system for rapid and efficient processing of the data. The system outlined here is used for both histogram and distribution parameter measuring instrumentation.

The 3-function mask shown in Fig. 2 is replaced by a multi-opening mask, as shown in Fig. 3. Measurements of the pattern are taken sequentially, as the mask rotates through 360 degrees behind a fixed slit. Lens 2 collects the transmitted light and directs it onto a silicon photodetector. This photodetector is the interface between the analog optical data collection system and the digital electronic processor, which is shown in Fig. 4.

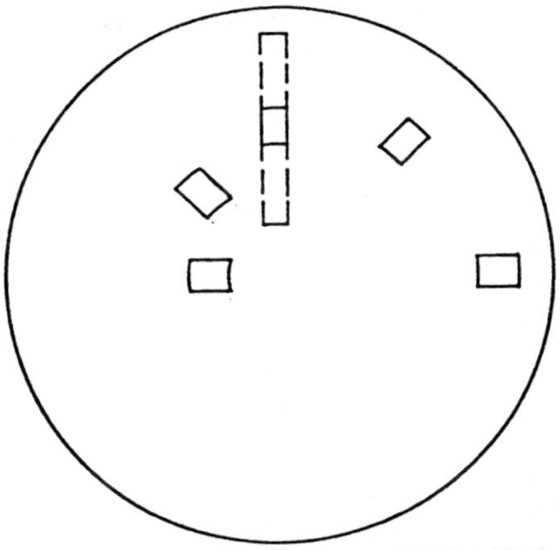

Figure 3: Multi-opening Fraunhofer plane mask for generation of a histogram of the volume distribution

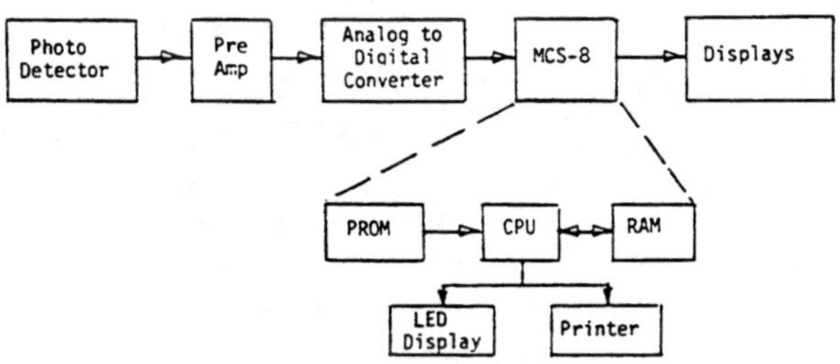

Figure 4: Block diagram of the electronic system for measurement and analysis of the scattered light signals

All the computations are carried out in digital form, after the optical signal is amplified and converted by the ADC (Analog to Digital Converter). The microcomputer itself consists of programmable read-only memories (PROM), random-access memories (RAM), input and output circuitry, and a central processing unit (CPU) which is an MCS-8. Program instructions are inserted into the PROM's. The CPU carries out these instructions, taking data, making computations, and storing them in the RAM's. Output is selected by the operator, and it is displayed on an LED display or it can be directed to a printer for hard copy.

The microprocessor provides extensive computation capacity in addition to controlling the data acquisition sequence for the entire measuring process. The microprocessor is a vital part in the generation of a histogram of the volume distribution. A thirteen segment histogram output is possible, covering a dynamic size range from 1.9 microns to 176 microns with channel centers at intervals of the square root of two. The channel centers were chosen to be consistent with sieving practice of root two size intervals.

To extract the histogram information from the diffraction pattern, raw scattered flux levels are measured at 13 radially distributed points in the diffraction pattern. The microprocessor selects groups of raw flux signals in a predetermined sequence, and computes 13 histogram channels.[7]

Figure 5 shows a data printout from a typical measurement sequence. This entire sequence can be remeasured or updated as often as three seconds; or up to as long as 800 seconds.

While it is possible to generate histogram data as easily as the less involved distribution measurements described in the last section, some advance thought should be devoted to the nature of the information required from a process so as to obtain the most meaningful data for the process. Examples are shown later using both types of data analysis.

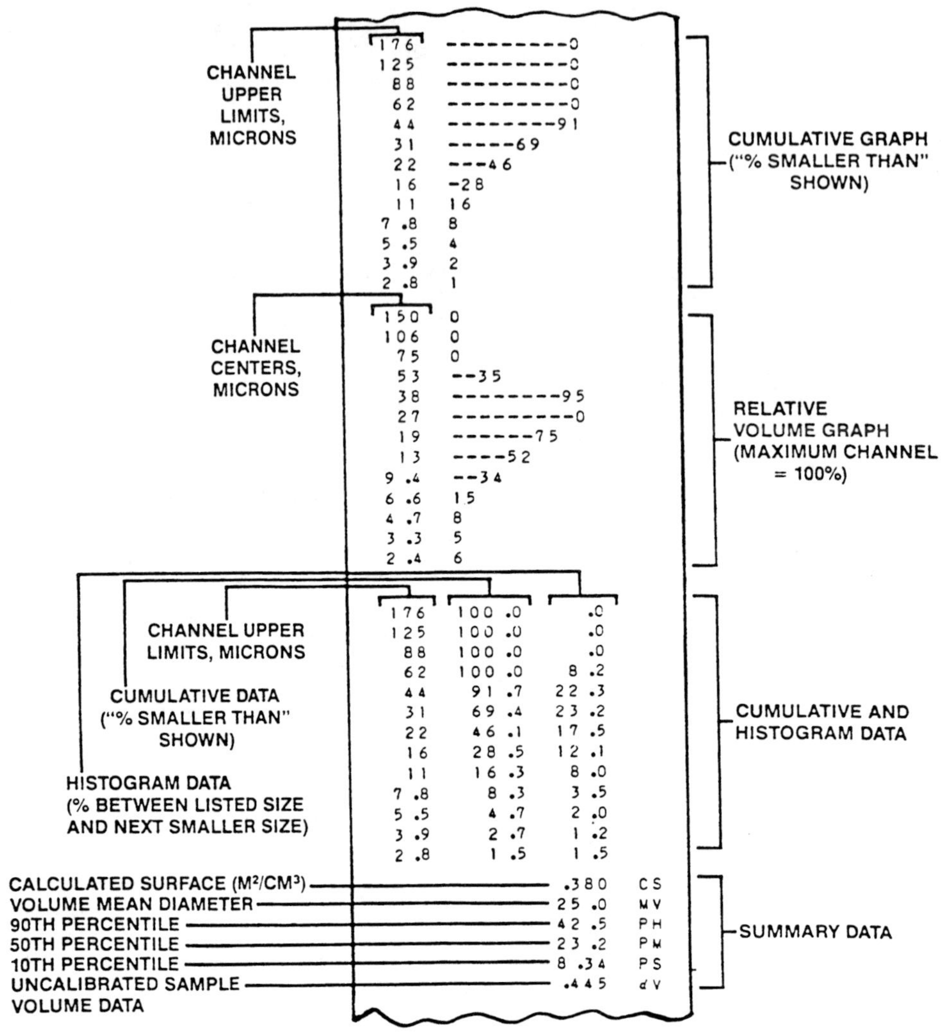

Figure 5: Data printout from MICROTRAC Particle Analyzer

Laboratory Measurements - Wet and Dry

The previous sections have concentrated on optical principles relevant to making optical scattering measurements of particulate matter. The next few sections are devoted to various sampling and measuring techniques. Many applications of this technology arise in laboratory analysis of various particulate grinding processes, such as taconite, ceramics, and minerals processing, where the grinding takes place in a water slurry. Frequently, a small sample is scooped from the bulk material and brought to a laboratory for analysis. The first commercial unit contained a basin into which this sample was added. The sample was mixed thoroughly with water and continuously circulated through a glass plate sample cell with an optical path of approximately 2 mm. Measurements were made using the optical system and electronic data processor shown in Figs. 2 and 4, respectively. At completion of the measuring sequence, typically 10 to 30 seconds, the basin could be drained, rinsed, and refilled for the next sample, if desired, or the measurement could be repeated. The basic unit is shown in the inset to Figure 6.

For many applications, however, a liquid suspension of particles is not practical. Substances such as sugar, flour, toners, and cement are best measured in powder form, and a dry powder sample conditioner can be added as shown in Fig. 6. This provides a means for dispersing dry powder samples in a representative powder/air mixture and transporting this mixture through the laser probe beam. The sample is not constrained between windows, but is drawn through the beam by a suction tube on the opposite side. The air stream is typically 1 cm in diameter at the laser beam.

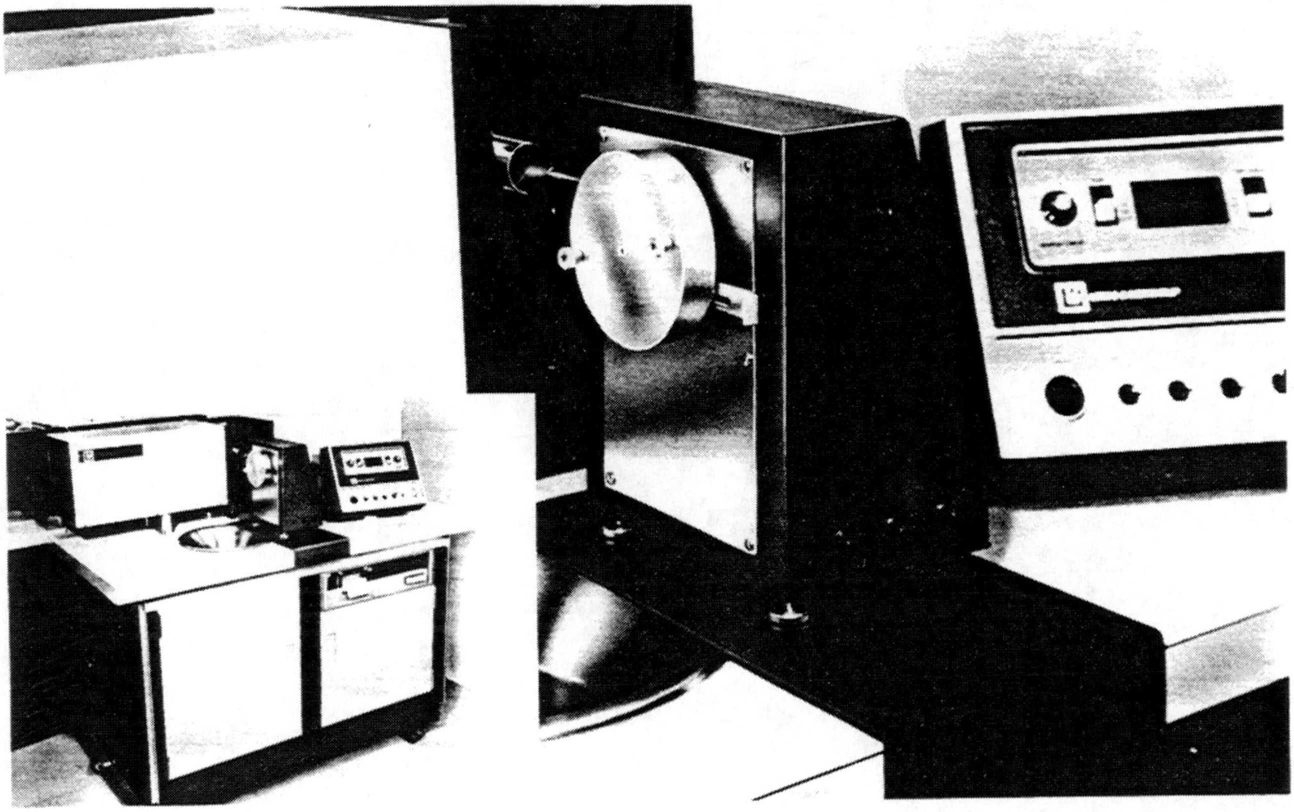

Figure 6: Dry sample conditioner and delivery system on standard MICROTRAC™ unit

Inset: MICROTRAC™ Laboratory Analyzer

On-line Instrumentation

A more sophisticated sample handling system is available for use as an on-line monitor for slurry applications.[8] The basic optical train is unchanged, but capabilities for demagnetization of magnetic ores, on-line sampling interfaces, and special control system output signals have been added. This unit is designed to provide the on-line signals necessary for feedback and control of grinding circuits. A block diagram of the system is shown in Fig. 7.

Suspended Solids Monitor

A considerably less complex system is the Suspended Solids Monitor, designed to produce volumetric concentration of particles as a primary output. Using the same optical principles as the other instrumentation, it provides a direct readout of particulate matter suspended in a liquid in parts per million by volume. With a single point calibration to measure specific gravity, the output can be scaled to produce mass loading. The interior of the unit is shown in Fig. 8. A one inch diameter pipe can be connected directly to a process stream, and the material flowing through the pipe can be continuously monitored. Additionally, it is possible to read particle size by measuring the volume mean diameter over the range 2 to 60 microns. This unit is automatically compensated for variations in transmission, and is unaffected by the color of the liquid. The instrument produces a linear concentration output which is independent of particle size over its operating range. Field tests have shown it to produce a continuous, accurate monitoring of suspended solids concentration in both waste treatment and industrial process applications.

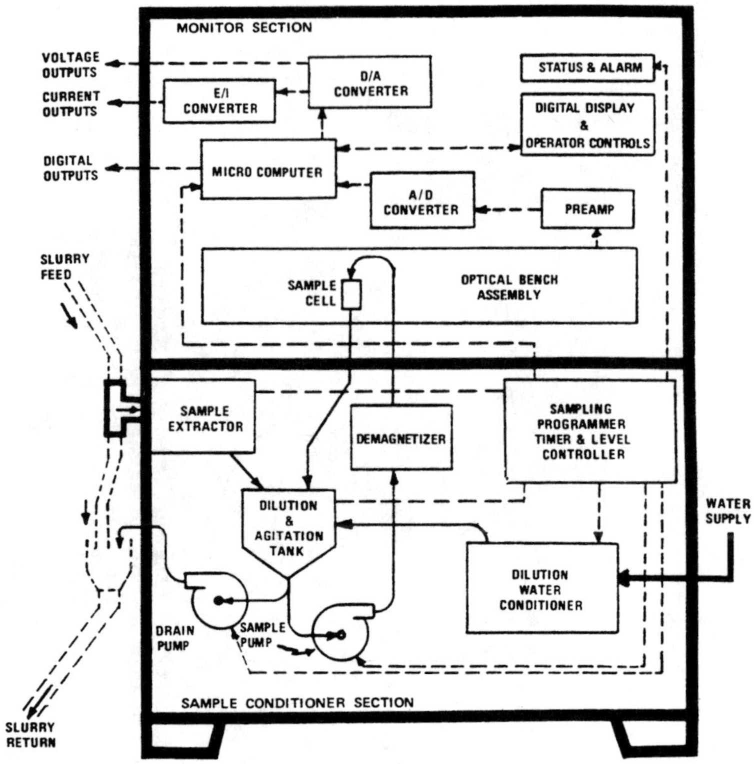

Figure 7: Block Diagram of On-line Monitor

Figure 8: Interior of Suspended Solids Monitor

Food and Drug Processing

Another application of forward angle scattering is in the detection of contaminants in liquids at levels insufficient to cause visual turbidity or cloudiness.[9,10] Such instrumentation has been built under a contract with the U. S. Food and Drug Administration to detect microbial contaminants in sealed glass containers of a blood component, human serum albumin. While this unit is dedicated to a specific packaging size and configuration, the basic principle is that pre-visible levels of contamination (in air or in liquid) can be detected using laser light scattering. This instrument is being field tested as a final inspection tool in the demanding control of quality in parenteral drug manufacturing.

Other Special Applications

Principles of light scattering are being applied in several other prototype instruments. Using a configuration very similar to the main photograph in Fig. 6, an instrument has been built under contract to ERDA for measuring particulate loading in hot gas streams from pressurized coal combustion systems. Small particle entrainment in gases derived from coal combustion are believed to contribute significantly to turbine erosion. Particles in the size range of 4 to 10 microns are especially suspect. The ERDA instrument uses a [illegible]ium-cadmium laser at .4416 microns and a narrow band interference filter to reduce the black body radiation from the duct, which may be as hot as 1500°F.

Other particulate measurement programs are underway with EPA, to measure respirable particles in the 0.2 to 10 micron size range, and with NIOSH, for discriminating and counting asbestos fibers.

Industrial Applications

There is a diverse range of processes where particle size is an important factor in determining the final product quality. Examples are measurements of yeast budding rates and chocolate fineness in food processing,[11] and monitoring of grinding loops in the extraction of iron from taconite in the mining industry.[12] Included here are three additional examples of industrial quality assurance problems.

Quality Assurance Measurements of Metal Powders

In the manufacture of electrical contacts, a variety of powdered metals can be used. One manufacturer had noticed changes in the final product with different batches of silver powder. Chemical assay indicated all four were acceptable, and microscopic analysis suggested that particle size was the key difference. Four samples were measured by the MICROTRAC™ Analyzer, and the results are shown in Fig. 9. The curves are normalized to 1.0 at the maximum, and the plots show volume per log diameter interval as a function of particle diameter. Powder B was considered "good" when the final inspection was made on the finished product, while powders A, C, and D resulted in product surface imperfections. By measuring the particle size before the product is made, a considerable cost and time savings can be realized, especially when the raw material is expensive. Powders which contain excessive coarse or fine material can be rejected at incoming inspection.

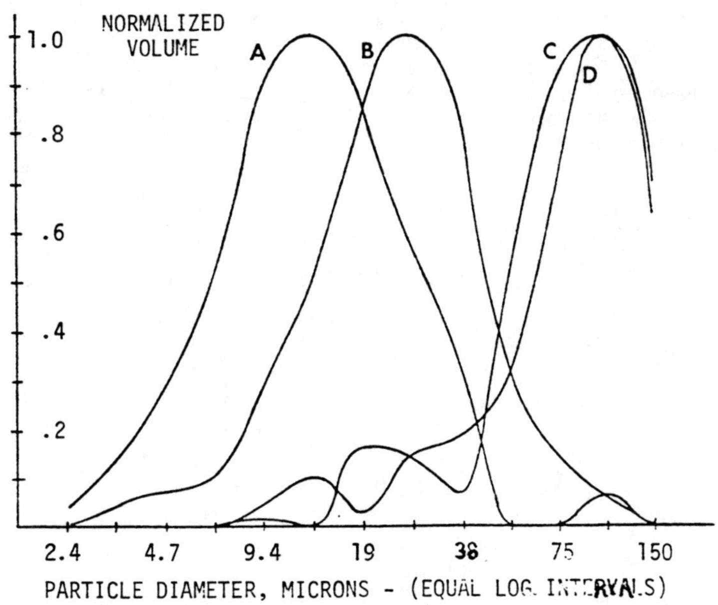

Figure 9: Volume distribution for four samples of silver powder

While it is interesting to see a continuous curve of particle size, single parameter measurements can provide the same pass or reject information. Parametric measurements of volume mean diameter a_V , and the characteristic width, σ_A, can readily discriminate between powders. This data is shown in Table 1.

Table 1

Distribution Parameters for Various Silver Powders

Measured Samples	Volume Mean Diameter a_V	Area Standard Deviation, σ_A (Characteristic Width)
Powder A (too many small particles)	16.6 microns	7.9 microns
Powder B (good final product)	28.3 microns	13.4 microns
Powder C (too coarse)	89.3 microns	42.9 microns
Powder D (too coarse)	89.0 microns	43.8 microns

Monitoring of Grinding for Ceramics Manufacturing

An alternate form for presentation of data is "percent passing". This has developed from sieving practice, where it is common to specify the percentage of total mass passing through a sieve of a certain mesh size. The accumulated percentage less than a given diameter is plotted as a function of particle diameter.

In the grinding of clay materials for ceramics it is essential to get the right fineness of material to produce desirable properties of the intermediate and final product. Particle size can influence the amount of water retained in the molding operations, the shrinkage of the material and its mechanical strength. Normally, the raw material is ground in a ball mill, and accurate analysis of particle size is required to assess whether to stop or continue the grinding process.

A series of material taken at intervals of 500 revolutions from 2000 to 6000 was analyzed and the "percent passing" data are shown in Fig. 10. For clarity, only a portion of the data is presented on an expanded scale, covering a size range from 5.5 to 31 microns. The distinction between samples is measureable, and the data is sufficiently accurate to enable proper ordering of material versus grinding time.

The ability to rapidly establish the precise size distribution is valuable insuring consistent manufacturing characteristics, while minimizing the time and energy consumed in overgrinding the material.

Quality Control Abrasives

As a final example, the standard specifications for abrasives have been established in terms of percent of mass above or below a certain particle size. Proper control of abrasive size and size distribution is crucial in processes such as the rough grinding of optical elements. Here again the MICROTRAC™ instrumentation and data format is well suited for a rapid and continuous determination of volume (or mass) distribution by particle size. Fig. 11 includes several families of curves showing, in this case, percentage of material volume greater than a specific size for four different grit sizes; 240, 320, 400, and 600. Each band represents between two and four individual samples from different batches of materials. The various samples produced curves only within the ranges shown here. The tests demonstrated the ability to distinguish between grit sizes and to tell if a product is outside of the acceptable range for that specific grit. This is particularly important since a relatively small excess of large material can result in grinding characteristics inappropriate for the fineness of the grit. Continuous monitoring in manufacture and use of abrasives can provide quality control advantages unavailable through time consuming microscopic or sedimentation analysis.

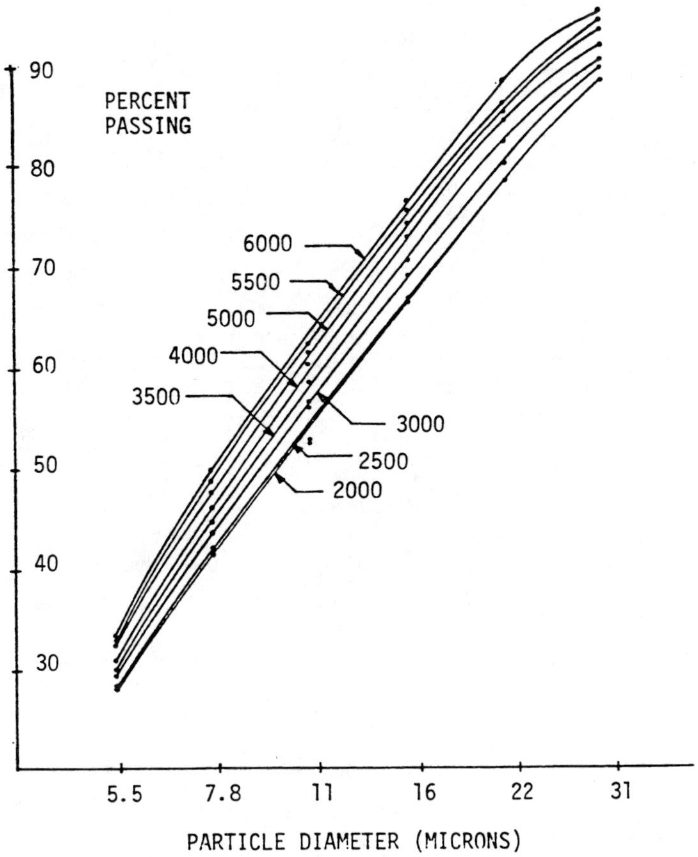

Figure 10: Cumulative volume distribution for clay material, for various numbers of revolutions of a ball mill

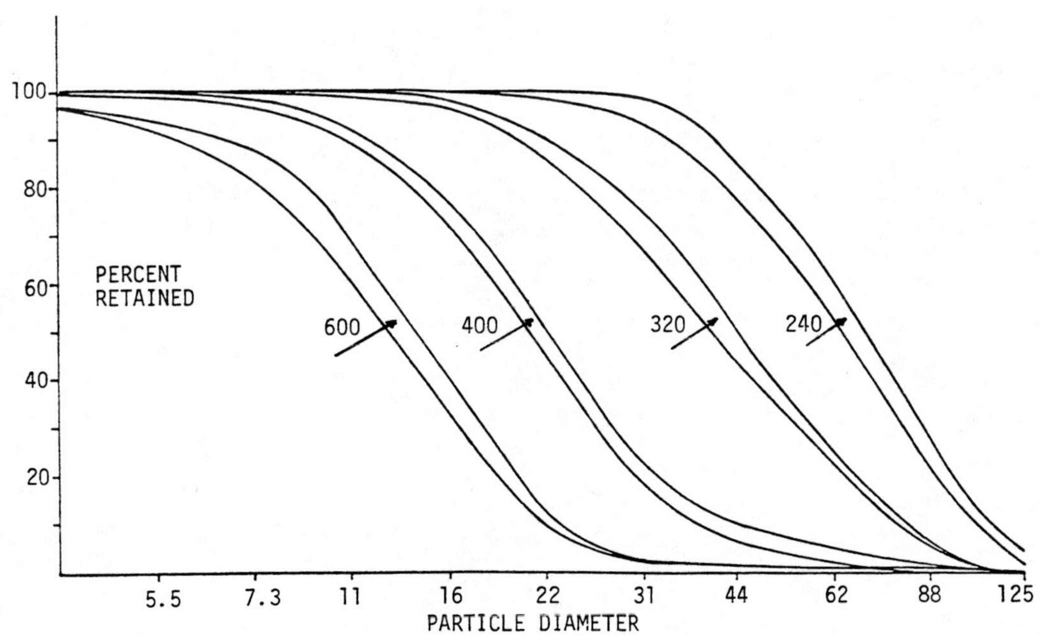

Figure 11: Cumulative volume distribution for abrasives, showing ranges of measured data for four grit sizes

Summary

Numerous industrial processes depend on rapid and accurate determination of particle size. This paper has presented a summary of concepts and instrument configurations capable of making volumetric measurements from light scattering in dynamic processes, and shown three examples of applications. In some cases, one or two parameters describing the particle distribution may be sufficient for control purposes, while other situations may require a measurement of the complete particle distribution as provided by a histogram format. The ability to measure these parameters in or near real time, without the need to evaluate and count individual particles, offers a unique capability for efficient quality control in industrial applications.

The examples given represent only a few of the wide variety of process and quality control applications in which the MICROTRAC™ product line has found rapid acceptance. Other examples (in addition to metal powders, ceramics, and abrasives) include mineral processing, powdered catalysts, flour, food powders, coal preparations, metal flake pigments, oil/water emulsions, aluminum refining, oil/sand extraction, powdered resins and brake lining materials. In each case, the instrument has been shown to provide repeatable, consistent measurements entirely independent of operator skill levels.

References

(1) M. Born and E. Wolf, Principles of Optics (Pergamon, New York, 1959), p. 395.

(2) W. L. Wilcock, U.S. Patent 3,873,206 (1975).

(3) A. L. Wertheimer and W. L. Wilcock, Applied Optics 15, 1616 (1976).

(4) J. Cornillault, Applied Optics 11, 265 (1972).

(5) J. Swithenbank, et. al., AIAA Paper No. 76-69, AIAA 14th Aerospace Sciences Meeting, Washington, DC, January 26-28, 1976.

(6) J. H. Ward, J.O.S.A. 58, 1566 (1968).

(7) E. C. Muly, H. N. Frock, E. L. Weiss, "Application of Fourier Imaging Systems to Fine Particles," submitted to the Journal of Powder and Bulk Solids Technology.

(8) T. Bay, H. J. Kortright, and E. C. Muly, "Continuous Particle Size Analysis and Grinding Control," presented at ISA-75 in Milwaukee, WI, October, 1975.

(9) A. L. Wertheimer, J.O.S.A. 66, 1124 (1976).

(10) A. L. Wertheimer, et. al., "In-Situ Contamination Detection Using Laser Light," submitted to Applied & Environmental Microbiology.

(11) J. W. Stitley, T. Morris, and E. K., Food Product Development, December, 1976.

(12) E. L. Weiss and H. N. Frock, Powder Technology, 14, 287 (1976).

Author Index

Subject Index